COLD SPRING HARBOR SYMPOSIA ON QUANTITATIVE BIOLOGY

VOLUME XLVIII

COLD SPRING HARBOR SYMPOSIA ON QUANTITATIVE BIOLOGY

VOLUME XLVIII

MOLECULAR NEUROBIOLOGY

COLD SPRING HARBOR LABORATORY

1983

COLD SPRING HARBOR SYMPOSIA ON QUANTITATIVE BIOLOGY
VOLUME XLVIII

International Standard Book Number 0-87969-047-X
International Standard Serial Number 0091-7451
Library of Congress Catalog Card Number 34-8174

Printed in the United States of America

COLD SPRING HARBOR SYMPOSIA ON QUANTITATIVE BIOLOGY

Founded in 1933 by
REGINALD G. HARRIS
Director of the Biological Laboratory 1924 to 1936

Previous Symposia Volumes

I (1933) Surface Phenomena
II (1934) Aspects of Growth
III (1935) Photochemical Reactions
IV (1936) Excitation Phenomena
V (1937) Internal Secretions
VI (1938) Protein Chemistry
VII (1939) Biological Oxidations
VIII (1940) Permeability and the Nature of Cell Membranes
IX (1941) Genes and Chromosomes: Structure and Organization
X (1942) The Relation of Hormones to Development
XI (1946) Heredity and Variation in Microorganisms
XII (1947) Nucleic Acids and Nucleoproteins
XIII (1948) Biological Applications of Tracer Elements
XIV (1949) Amino Acids and Proteins
XV (1950) Origin and Evolution of Man
XVI (1951) Genes and Mutations
XVII (1952) The Neuron
XVIII (1953) Viruses
XIX (1954) The Mammalian Fetus: Physiological Aspects of Development
XX (1955) Population Genetics: The Nature and Causes of Genetic Variability in Population
XXI (1956) Genetic Mechanisms: Structure and Function
XXII (1957) Population Studies: Animal Ecology and Demography
XXIII (1958) Exchange of Genetic Material: Mechanism and Consequences
XXIV (1959) Genetics and Twentieth Century Darwinism
XXV (1960) Biological Clocks
XXVI (1961) Cellular Regulatory Mechanisms
XXVII (1962) Basic Mechanisms in Animal Virus Biology
XXVIII (1963) Synthesis and Structure of Macromolecules
XXIX (1964) Human Genetics
XXX (1965) Sensory Receptors
XXXI (1966) The Genetic Code
XXXII (1967) Antibodies
XXXIII (1968) Replication of DNA in Microorganisms
XXXIV (1969) The Mechanism of Protein Synthesis
XXXV (1970) Transcription of Genetic Material
XXXVI (1971) Structure and Function of Proteins at the Three-dimensional Level
XXXVII (1972) The Mechanism of Muscle Contraction
XXXVIII (1973) Chromosome Structure and Function
XXXIX (1974) Tumor Viruses
XL (1975) The Synapse
XLI (1976) Origins of Lymphocyte Diversity
XLII (1977) Chromatin
XLIII (1978) DNA: Replication and Recombination
XLIV (1979) Viral Oncogenes
XLV (1980) Movable Genetic Elements
XLVI (1981) Organization of the Cytoplasm
XLVII (1982) Structures of DNA

All Cold Spring Harbor Laboratory publications are available through booksellers or may be ordered directly from Cold Spring Harbor Laboratory, Box 100, Cold Spring Harbor, New York 11724. SAN 203-6185

Symposium Participants

ABRAMS, THOMAS, Dept. of Neurobiology and Behavior, Columbia University, New York, New York
ADLER, JULIUS, Dept. of Biochemistry, University of Wisconsin, Madison
AGNEW, WILLIAM, Dept. of Physiology, Yale University School of Medicine, New Haven, Connecticut
AKERBLOM, INGRID, University of California, San Diego
ALDRICH, RICHARD, Dept. of Physiology, Yale University School of Medicine, New Haven, Connecticut
AMANO, TAKEHIKO, Mitsubishi-Kasei Institute of Life Sciences, Tokyo, Japan
ANDERSON, DAVID, Cancer Research Institute, Columbia University College of Physicians & Surgeons, New York, New York
APPLEBURY, MEREDITHE, Dept. of Biological Sciences, Purdue University, West Lafayette, Indiana
AXEL, RICHARD, Institute of Cancer Research, Columbia University College of Physicians & Surgeons, New York, New York
BALLINGER, DENNIS, Dept. of Biology, Massachusetts Institute of Technology, Cambridge
BALLIVET, MARC, Dept. of Biochemistry, University of Geneva, Switzerland
BANKER, GARY, Dept. of Anatomy, Albany Medical College, New York
BARKER, J.L., NINCDS, National Institutes of Health, Bethesda, Maryland
BARNARD, E.A., Dept. of Biochemistry, Imperial College, London, England
BARNSTABLE, COLIN J., Dept. of Neurobiology, Harvard Medical School, Boston, Massachusetts
BENTLEY, DAVID, Dept. of Zoology, University of California, Berkeley
BENZER, SEYMOUR, Division of Biology, California Institute of Technology, Pasadena
BERNIER, LISE, Dept. of Neurobiology and Behavior, Columbia University College of Physicians & Surgeons, New York, New York
BIRDSALL, NIGEL, Dept. of Molecular Pharmacology, Medical Research Council, London, England
BLUM, MARIANN, Rockefeller University, New York, New York
BOWERS, CHAUNCEY, Dept. of Physiology, University of California Medical School, San Francisco
BRACKENBURY, ROBERT, Rockefeller University, New York, New York
BROSIUS, JURGEN, Dept. of Neurobiology and Behavior, Columbia University College of Physicians & Surgeons, New York, New York
BUCK, LINDA, Institute for Cancer Research, Columbia University, New York, New York
BUCKLEY, KATHLEEN, Dept. of Biochemistry and Biophysics, University of California, San Francisco
CARLSON, STEVEN, Dept. of Biochemistry and Biophysics, University of California, San Francisco
CARONI, PICO, Dept. of Biochemistry and Biophysics, University of California, San Francisco
CASPER, DIANA, Dept. of Neuroscience, Albert Einstein College of Medicine, Bronx, New York
CASTELLUCCI, VINCENT, Dept. of Neurobiology and Behavior, Columbia University College of Physicians & Surgeons, New York, New York
CATTERALL, WILLIAM, Dept. of Pharmacology, University of Washington School of Medicine, Seattle
CHAN, SHEW Y., Dept. of Pharmacology, Harvard Medical School, Boston, Massachusetts
CHANGEUX, J.P., Institut Pasteur, Paris, France
CHIKARAISHI, DONA, Dept. of Molecular and Cell Biology, National Jewish Hospital/National Asthma Center, Denver, Colorado
CHIU, ARLENE, Dept. of Physiology and Biophysics, Washington University Medical School, St. Louis, Missouri
CHUONG, C.-M., Rockefeller University, New York, New York
CLAUDIO, TONI, Institute of Cancer Research, Columbia University, New York, New York
CLAYTON, DAVID, Rockefeller University, New York, New York
CLAYTON, DAVID F., Rockefeller University, New York, New York
COREY, DAVID P., Dept. of Physiology, Yale University School of Medicine, New Haven, Connecticut
COVAULT, JONATHAN, Dept. of Physiology and Biophysics, Washington University Medical School, St. Louis, Missouri
COWAN, MAX, Salk Institute, San Diego, California
CULL-CANDY, STUART, Dept. of Pharmacology, University College, London, England
CUNNINGHAM, BRUCE A., Bronk Laboratory, Rockefeller University, New York, New York
DAVIDSON, NORMAN, Dept. of Chemistry, California Institute of Technology, Pasadena
DAVIS, RON, Dept. of Biochemistry, Michigan State University, East Lansing
DE GENNARO, LOUIS, Dept. of Psychiatry, Max-Planck-Institute, Munich, Federal Republic of Germany
DOBNER, PAUL, Dept. of Molecular Genetics, University of Massachusetts Medical Center, Worcester
DODD, JANE, Dept. of Neurobiology, Harvard Medical School, Boston, Massachusetts
EBERWINE, JAMES, Dept. of Biochemistry, Columbia University, New York, New York
EDELMAN, GERALD, Rockefeller University, New York, New York
ELLIS, LELAND, Dept. of Anatomy and Cell Biology,

Columbia University College of Physicians & Surgeons, New York, New York
EVANS, RONALD, Salk Institute, San Diego, California
EVINGER, MARIAN, Columbia University College of Physicians & Surgeons, New York, New York
FAMBROUGH, DOUGLAS, Dept. of Embryology, Carnegie Institution of Washington, Baltimore, Maryland
FELDMAN, DANIEL, Dept. of Physiology and Biophysics, Washington University School of Medicine, St. Louis, Missouri
FISHER, J., Institute for Cancer Research, Columbia University, New York, New York
FORREST, JOHN, Dept. of Molecular Neurobiology, Salk Institute, San Diego, California
FRANK, BENNET, Baylor College of Medicine, Houston, Texas
FRITZ, LAWRENCE, Dept. of Chemistry, California Institute of Technology, Pasadena
FUKUI, HIROYUKI, NEI, National Institutes of Health, Bethesda, Maryland
GAGNER, J.-P., Dept. of Molecular Biology, Clinical Research Laboratory of Montreal, Canada
GALLAY, BRIAN J., Yonkers, New York
GASIC, GREGORY, Rockefeller University, New York, New York
GEFFEN, IRIS, Dept. of Biochemistry, Albert Einstein College of Medicine, Bronx, New York
GELLER, HERBERT, Dept. of Pharmacology, UMDNJ-Rutgers Medical School, Piscataway, New Jersey
GOELET, PHILIP, Laboratory of Molecular Biology, Medical Research Council, Cambridge, England
GOLDIN, STANLEY, Dept. of Pharmacology, Harvard Medical School, Boston, Massachusetts
GOMBOS, GIORGIO, CNRS Centre de Neurochimie, Strasbourg, France
GOODMAN, COREY S., Dept. of Biological Sciences, Stanford University, Palo Alto, California
GORIDIS, C., Centre d'Immunologie, INSERM-CNRS, Marseille, France
GOULD, ROBERT, Dept. of Neuroscience, Johns Hopkins University School of Medicine, Baltimore, Maryland
GREENBERG, SHARON, Dept. of Anatomy, Case Western Reserve University Medical School, Cleveland, Ohio
GREENGARD, PAUL, Dept. of Pharmacology, Yale University School of Medicine, New Haven, Connecticut
GRUMET, MARTIN, Rockefeller University, New York, New York
GURLING, H.M.D., Institut of Psychiatry de Crespigny Park, London, England
HAHN, WILLIAM, Dept. of Anatomy, University of Colorado Health Sciences Center, Denver
HALL, ZACH W., Dept. of Physiology, University of California, San Francisco
HARRELSON, ALLAN, Rockefeller University, New York, New York
HARRIS, DAVID A., Dept. of Neurobiology and Behavior, Columbia University College of Physicians & Surgeons, New York, New York
HAWROT, EDWARD, Dept. of Pharmacology, Yale University School of Medicine, New Haven, Connecticut
HEINEMANN, STEPHEN, Salk Institute, San Diego, California
HERBERT, EDWARD, Dept. of Chemistry, University of Oregon, Eugene
HERSHEY, DAVIS, California Institute of Technology, Pasadena
HIGGINS, GERALD A., Neurobiology Group, E.I. Du Pont de Nemours & Company, Glenolden, Pennsylvania
HOCKFIELD, SUSAN, Cold Spring Harbor Laboratory, New York
HOFFMAN, STANLEY, Rockefeller University, New York, New York
HOFSTEIN, RAFI, Dept. of Neurobiology, Harvard Medical School, Boston, Massachusetts
HOGAN, PATRICK, Dept. of Neurobiology, Harvard Medical School, Boston, Massachusetts
HOLLIDAY, JANET, New Haven, Connecticut
HOMONOFF, M., Psychiatric Center, Albert Einstein College of Medicine, Bronx, New York
HORVITZ, ROBERT, Dept. of Biology, Massachusetts Institute of Technology, Cambridge
HUGANIR, RICHARD, Dept. of Pharmacology, Yale University School of Medicine, New Haven, Connecticut
INGRAHAM, HOLLY, Dept. of Molecular Neurobiology, Salk Institute, San Diego, California
INGRAM, VERNON, Dept. of Biology, Massachusetts Institute of Technology, Cambridge
ISENBERG, KEITH, Dept. of Pharmacology, Washington University School of Medicine, St. Louis, Missouri
ISRAEL, MARK A., NCI, National Institutes of Health, Bethesda, Maryland
JAN, LILY, Dept. of Physiology, University of California, San Francisco
JAN, Y.N., Dept. of Physiology, University of California, San Francisco
JESSELL, THOMAS, Dept. of Neurobiology, Harvard Medical School, Boston, Massachusetts
JOH, TONG H., Neurobiology Laboratory, Cornell University Medical College, New York, New York
JONASSEN, JULIE, Columbia University College of Physicians & Surgeons, New York, New York
JONES, STEPHEN W., Dept. of Neurobiology and Behavior, State University of New York, Stony Brook
JULIUS, DAVID J., Dept. of Microbiology and Immunology, University of California, Berkeley
KALDANY, RASHAD-RUDOLF, Dept. of Biochemistry, Columbia University, New York, New York
KANAZIR, SELMA, Dept. of Pharmacology, Yale School of Medicine, New Haven, Connecticut
KANDEL, ERIC R., Dept. of Neurobiology and

Behavior, Columbia University College of Physicians & Surgeons, New York, New York
KARLIN, ARTHUR, Columbia University College of Physicians & Surgeons, New York, New York
KATZ, FLORA N., Center for Neurobiology and Behavior, Columbia College of Physicians & Surgeons, New York, New York
KELLY, REGIS B., Dept. of Biochemistry and Biophysics, University of California, San Francisco
KENNEDY, MARY B., Dept. of Biology, California Institute of Technology, Pasadena
KESHISHIAN, HAIG, Dept. of Neurobiology, University of Chicago, Illinois
KHORANA, H. GOBIND, Dept. of Biology and Chemistry, Massachusetts Institute of Technology, Cambridge
KISTLER, HENRY B., JR., Dept. of Neurobiology and Behavior, Columbia University College of Physicians & Surgeons, New York, New York
KONGSAMUT, SATHAPANA, Dept. of Pharmacology, University of Chicago, Illinois
KORF, BRUCE R., Dept. of Genetics, Children's Hospital, Boston, Massachusetts
KOSHLAND, DANIEL, JR., Dept. of Biochemistry, University of California, Berkeley
KRUEGER, KARL E., NHLBI, National Institutes of Health, Bethesda, Maryland
KUENZLE, CLIVE, Dept. of Biochemistry and Pharmacology, University of Zurich, Switzerland
KURADO, YOICHIRO, Dept. of Neurochemistry, Tokyo Metropolitan Institute of Neurosciences, Japan
LANCASTER, BARRIE, Dept. of Neurobiology and Behavior, State University of New York, Stony Brook
LASEK, RAYMOND, Dept. of Anatomy, Case Western Reserve University Medical School, Cleveland, Ohio
LAUGHREA, MICHAEL, Lady Davis Institute for Medical Research, Montreal, Canada
LEWIS, ROBERT, Dept. of Pharmacology, Yale University School of Medicine, New Haven, Connecticut
LINDSTROM, JON, Salk Institute, San Diego, California
LISTON, DAME, CNRS, Laboratoire de Physiologie Nerveuse, Gif-sur-Yvette, France
LITTAUER, URIEL, Memorial Sloan-Kettering Cancer Center, New York, New York
LOBEL, PETER, Dept. of Biochemistry, Columbia University, New York, New York
MADAULE, PASCAL, Institute for Cancer Research, Columbia University, New York, New York
MALLET, JACQUES, Dept. of Microbiology, University of Paris-Sud, Orsay, France
MATUS, ANDREW, Friedrich Miescher-Institute, Basel, Switzerland
MAYO, KELLY E., Dept. of Molecular Biology and Virology, Salk Institute, San Diego, California
MCKAY, RON, Cold Spring Harbor Laboratory, New York
MCMAHAN, U. JACK, Dept. of Neurobiology, Stanford University School of Medicine, Palo Alto, California
MEIRI, KARINA, Dept. of Anatomy and Neurobiology, Washington University Medical School, St. Louis, Missouri
MERLIE, JOHN P., Dept. of Pharmacology, Washington University School of Medicine, St. Louis, Missouri
MILNER, ROBERT J., Salk Institute, San Diego, California
MIXTER, KATHERINE, Dept. of Biology, California Institute of Technology, Pasadena
NELSON, JAMES C., Dept. of Biology, Yale University, New Haven, Connecticut
NIRENBERG, MARSHALL, NHLBI, National Institutes of Health, Bethesda, Maryland
NODA, MASAHARU, Dept. of Medical Chemistry, Kyoto University Faculty of Medicine, Japan
NUMA, SHOSAKU, Dept. of Medical Chemistry, Kyoto University Faculty of Medicine, Japan
O'MALLEY, K., V.A. Hospital, Stanford University School of Medicine, Palo Alto, California
O'FARRELL, PATRICK, Dept. of Biochemistry, University of California Medical School, San Francisco
OBLINGER, MONICA, Dept. of Anatomy, Case Western Reserve University, Cleveland, Ohio
PALAZZOLO, MICHAEL, Dept. of Neurobiology and Behavior, Columbia University, New York, New York
PAPAZIAN, DIANE M., Dept. of Pharmacology, Harvard Medical School, Boston, Massachusetts
PATRICK, JAMES, Dept. of Molecular Neurobiology, Salk Institute, San Diego, California
PATTERSON, PAUL, Dept. of Neurobiology, Harvard Medical School, Boston, Massachusetts
POLLOCK, JONATHAN, Dept. of Neurobiology and Behavior, Columbia University, New York, New York
POTTER, D., Dept. of Neurobiology, Harvard University, Cambridge, Massachusetts
QUINN, WILLIAM, Dept. of Biology, Princeton University, New Jersey
RAFF, MARTIN, Dept. of Zoology, University College, London, England
RAFTERY, MICHAEL, Dept. of Chemistry and Chemical Engineering, California Institute of Technology, Pasadena
RANDO, THOMAS, Anesthesia Research Laboratories, Harvard Medical School, Boston, Massachusetts
RATNER, NANCY, Dept. of Anatomy and Neurobiology, Washington University Medical School, St. Louis, Missouri
REESE, JAMES, Dept. of Pharmacology, Yale University School of Medicine, New Haven, Connecticut
REICHARDT, LOUIS, Dept. of Physiology, University of California, San Francisco
RESTIFO, LINDA L., Lidy Laboratories, University of Pennsylvania, Philadelphia
REUTER, HARALD, Dept. of Pharmacology, University of Berne, Switzerland

ROBERTS, JAMES, Dept. of Biochemistry, Columbia University College of Physicians & Surgeons, New York, New York
ROBERTSON, MIRANDA, *Nature*, London, England
ROOF, DOROTHY, Dept. of Biological Sciences, Purdue University, West Lafayette, Indiana
ROSENBERG, ROBERT, Dept. of Physiology, Yale University School of Medicine, New Haven, Connecticut
ROSSIER, JEAN, Laboratoire de Physiologie Nerveuse, CNRS, Gif-sur-Yvette, France
ROYDEN, CONSTANCE, Dept. of Physiology, University of California School of Medicine, San Francisco
RUTISHAUSER, URS, Dept. of Anatomy, Case Western Reserve School of Medicine, Cleveland, Ohio
SABOL, STEVEN, NHLBI, National Institutes of Health, Bethesda, Maryland
SAITOH, TSUNAO, Dept. of Neurobiology and Behavior, Columbia University College of Physicians & Surgeons, New York, New York
SAKMANN, BERT, Dept. of Biophysical Chemistry, Max-Planck Institute, Goettingen, Federal Republic of Germany
SALKOFF, LARRY, Dept. of Physiology, University of California, San Francisco
SANDROCK, A., Dept. of Neurobiology, Harvard Medical School, Boston, Massachusetts
SANES, JOSHUA, Dept. of Physiology, Washington University Medical Center, St. Louis, Missouri
SARTHY, P. VIJAY, Dept. of Ophthalmology, University of Washington, Seattle
SCHACHNER, MELITTA, Dept. of Neurobiology, University of Heidelberg, Federal Republic of Germany
SCHELLER, RICHARD, Dept. of Biological Sciences, Stanford University, California
SCHENKER, CHRISTINA, Dept. of Physiology, Harvard Medical School, Boston, Massachusetts
SCHRAMM, MICHAEL, Dept. of Biological Chemistry, Hebrew University of Jerusalem, Israel
SCHUBERT, DAVID, Salk Institute, San Diego, California
SCHWARTZ, JAMES, Dept. of Neurobiology and Behavior, Columbia University College of Physicians & Surgeons, New York, New York
SCHWIETZER, BARRY, Orange, Connecticut
SEBBANE, ROLAND, Dept. of Biological Sciences, University of Pittsburgh, Pennsylvania
SHAW, GERRY, Dept. of Biophysical Chemistry, Max-Planck Institute, Goettingen, Federal Republic of Germany
SHINE, J., Australian National University, Canberra
SHOOTER, ERIC M., Dept. of Neurobiology, Stanford University School of Medicine, Palo Alto, California
SIEGELBAUM, STEVEN, Dept. of Pharmacology, Columbia University, New York, New York
SIGWORTH, F.J., Dept. of Biophysical Chemistry, Max-Planck Institute, Goettingen, Federal Republic of Germany
SIMON, MELVIN, Dept. of Biology, California Institute of Technology, Pasadena
STALLCUP, WILLIAM, Dept. of Molecular Neurobiology, Salk Institute, San Diego, California
STAPLETON, ANN E., New York State Psychiatric Institute, New York
STEIN, REUBEN, Cancer Research Institute, Columbia University, New York, New York
STENGELIN, SIEGFRIED, Salk Institute, San Diego, California
STEVENS, CHARLES, Dept. of Physiology, Yale University School of Medicine, New Haven, Connecticut
STEWARD, OSWALD, Dept. of Neurological Surgery, University of Virginia School of Medicine, Charlottesville
STROUD, ROBERT, Dept. of Biochemistry, University of California, San Francisco
STRYER, LUBERT, Dept. of Structural Biology, Stanford University School of Medicine, Palo Alto, California
SULSTON, J.E., Dept. of Molecular Biology, Medical Research Council, Cambridge, England
SUTCLIFFE, J. GREGOR, Dept. of Immunology, Scripps Clinic and Research Foundation, La Jolla, California
SWANSON, RICHARD, Dept. of Biology, Purdue University, West Lafayette, Indiana
TAYLOR, SHIRLEY, Pasadena, California
THAYER, GREGORY C., Dept. of Neurobiology and Behavior, Columbia University, New York, New York
THOENEN, HANS, Dept. of Neurochemistry, Max-Planck Institute for Psychiatry, Martinsried, Federal Republic of Germany
TIMPE, LESLIE C., Dept. of Physiology, University of California Medical School, San Francisco
TRIGGLE, DAVID, Dept. of Biochemistry and Pharmacology, State University of New York, Buffalo
TSIEN, RICHARD, Dept. of Physiology, Yale University School of Medicine, New Haven, Connecticut
TURNER, DAVID L., Dept. of Neuroscience, Harvard Medical School, Boston, Massachusetts
UENO, SATASHI, Salk Institute, San Diego, California
ULLRICH, AXEL, Genentech Inc., South San Francisco, California
VENKATESH, T.R., Dept. of Biology, California Institute of Technology, Pasadena
WALLACE, WILLIAM, Dept. of Pharmacology, Yale University School of Medicine, New Haven, Connecticut
WEBER, KLAUS, Dept. of Biophysical Chemistry, Max-Planck Institute, Goettingen, Federal Republic of Germany
WHITE, J.G., Dept. of Molecular Biology, Medical Research Council, Cambridge, England
WIESEL, T., Dept. of Neurobiology, Harvard Medical School, Boston, Massachusetts
WILCOX, JOSIAH, Dept. of Reproductive Science, Columbia University College of Physicians & Surgeons, New York, New York

Willard, Mark, Dept. of Anatomy and Neurobiology, Washington University School of Medicine, St. Louis, Missouri

Winter, Janet, Dept. of Neurobiology, University of California, San Francisco

Wyman, Robert, Dept. of Biology, Yale University, New Haven, Connecticut

Yamamoto, Miyuki, Dept. of Neurobiology, Harvard Medical School, Boston, Massachusetts

Yee, Amy, Dept. of Biochemistry and Biophysics, University of California, Davis

Yellen, Gary, Dept. of Physiology, Yale University School of Medicine, New Haven, Connecticut

Young, Anthony, Dept. of Pharmacology, Stanford University, Palo Alto, California

Zipser, Birgit, Cold Spring Harbor Laboratory, New York

Zipursky, Lawrence, Dept. of Biology, California Institute of Technology, Pasadena

First row: B. Sackmann, M. Schramm; J. Patrick, M.A. Raftery
Second row: E.R. Kandel, C. Barnstable, R.D.G. McKay
Third row: D.J. Anderson, R.B. Kelly; U.J. McMahan, Y.N. Jan
Fourth row: Beach picnic

First row: T. Broker, L. Chow, N. Davidson; Z.W. Hall, R.J. Lasek
Second row: S. Numa, W. Hahn, J. Brosius; L. Stryer, J.D. Watson
Third row: D. Koshland, T. Wiesel; S. Goldin, W.A. Catterall; H. Thoenen, M. Schachner
Fourth row: Volleyball game

Foreword

Ever since the genetic code was solved in 1966, the brain and the phenomena it oversees—perception, memory, and thinking—have stood out like Himalayan peaks to the world of molecular biologists. Until very recently, however, most of us have believed that we have little unique to offer those long professionally involved in neurobiological phenomena and that the brain in particular was far beyond our capabilities. Those few molecular biologists who did move into neurobiology we have in the past regarded as frightfully brave or recklessly silly.

Apparently particularly hopeless was the approach that should be the most natural to us, that of focussing first on the structure and functioning of the molecules out of which the nerve cells are made. Although such research of itself could become fun to do, for many years the chances seemed nil that it would be through molecular biology that the intimate secrets of the brain would fall out. So an audacious few of us had moved into the fields of psychophysics, hoping that bouts of hard thinking could do the job. But here again the odds that fundamental revelations would quickly emerge seemed painfully low.

Today, however, we sense a new mood among those molecular biologists who gaze upwards toward the brain. By using recombinant DNA, monoclonal antibodies, and the facts of receptor biochemistry to the fullest, we should have a fighting chance to understand the uniqueness of nerve cells and the ways they come together to form functional networks. To mark this new mood, we decided to hold the 48th Symposium on Molecular Neurobiology.

In putting together this meeting, we received excellent advice from many colleagues. We wish to thank Eric Kandel and Charles Stevens, both long associated with the Cold Spring Harbor summer courses in neurobiology. The final result was a program of 91 speakers that started with an introduction by Max Cowen and concluded with a summary by Eric Kandel. The total attendance was 233 participants, all of whom thought that this Symposium would be long remembered as a decisive turning point in the development of neurobiology.

Our Meetings Office—Gladys Kist, Barbara Ward, Maureen Berejka, Merilyn Simkins—with their usual cheerful competence made sure that everyone was well received, well fed, and well housed. Herb Parsons expertly oversaw the audio-visual aspects of the meeting.

Financial support for this meeting was provided by the National Institutes of Health, National Science Foundation, and the Department of Energy. These much-needed funds enabled us to cover the meeting costs of our invited speakers and to provide travel expenses for many of our overseas speakers.

This year we have worked especially hard to publish these volumes in the same year as our June meeting, an interval of less than 7 months. So we have been firm in sticking to our requests that the manuscripts must be handed in during the Symposium week or immediately after. We note the great cooperation we have received from our speakers, a happy situation to which we must also credit the friendly persistence and speedy editing of Judith Cuddihy, Dorothy Brown, Karen Sundin, and Joan Ebert of our Publications Department, which continues to be ably headed by Nancy Ford.

J.D. Watson
R. McKay
November 17, 1983

Contents

Part 1

Part 2

Genetic Analysis of the Nervous System

Neuronal Surface Molecules

The Regulation of Axon Organization

COLD SPRING HARBOR SYMPOSIA ON QUANTITATIVE BIOLOGY

VOLUME XLVIII

The Arrangement and Functions of the Chains of the Acetylcholine Receptor of *Torpedo* Electric Tissue

A. KARLIN,* R. COX,* R.-R. KALDANY,* P. LOBEL,* AND E. HOLTZMAN†
*Departments of Biochemistry and Neurology and †Department of Biological Sciences, Columbia University, New York, New York 10032

This paper is concerned with the five polypeptide chains that constitute the nicotinic acetylcholine receptor (AChR). We have used electron microscopy to locate the chains relative to one another. We have also used affinity labels and time-resolved photolabeling to identify functional sites on the chains other than the ACh-binding sites. The background for these studies has been reviewed by Karlin (1980), Changeux (1981), and Conti-Tronconi and Raftery (1982). Our conclusions are (1) that the five chains of the receptor are arranged around the central axis in the order $\alpha\gamma\alpha\beta\delta$ and (2) that the α- and β-chains contribute to functionally significant local anesthetic binding sites.

Receptor Structure

The monomeric form of the receptor from *Torpedo* electric tissue has a molecular weight of about 250,000 and contains two ACh-binding sites (Reynolds and Karlin 1978). It is composed of four types of polypeptide chains of *apparent* molecular weights (determined by gel electrophoresis in SDS) 39,000 (α), 48,000 (β), 58,000 (γ), and 64,000 (δ) (Weill et al. 1974) in the stoichiometry $\alpha_2\beta\gamma\delta$ (Reynolds and Karlin 1978; Lindstrom et al. 1979; Raftery et al. 1980). The receptor from species of *Torpedo*, but not that from other genera examined, occurs in native membrane predominantly as a dimer cross-linked by a disulfide bond between δ-chains (see references in Hamilton et al. 1979).

The overall shape of the receptor has been examined in membrane and in detergent solution. The predominant feature of negatively stained, receptor-rich membrane is the presence of randomly close-packed, discoid particles of 80–90-Å diameter, each with a central stain-filled pit (references in Changeux 1981). The disk is the image of the monomeric unit of the receptor; the isolated receptor dimer appears as two contiguous disks (Cartaud et al. 1980; Wise et al. 1981a). The average distance between the centers of the disks in dimer, measured in the electron micrographs, is about 95 Å, in agreement with the distance estimated by neutron-scattering analysis (Wise et al. 1981a). In the direction normal to the plane of the membrane, the overall length of the receptor is about 110 Å. It extends about 50 Å above the bilayer on the extracellular side and about 20 Å on the intracellular side (references in Karlin 1980; Kistler et al. 1982).

A neutron-scattering analysis of the shape of the receptor in solution yielded a radius of gyration for the protein portion of the *Torpedo californica* receptor monomer-detergent complex of 46 ± 1 Å (Wise et al. 1979). This value is consistent only with an extended shape, such as a lopsided dumbbell (Fig. 1), in which the extracellular domain of the receptor and the intracellular domain are connected by a narrower, membrane-spanning domain (Wise et al. 1979; cf. Kistler et al. 1982).

Each of the receptor chains spans the membrane (Anderson and Blobel 1981). In fact, a comparison of the sequences of the chains with those of other membrane-spanning proteins indicates that each chain contains four hydrophobic segments of lengths between 19 and 27 residues that are likely to be membrane-spanning α-helices (Claudio et al. 1983; Devillers-Thiery et al. 1983; Noda et al. 1983). Thus, the portion of the receptor within the hydrophobic core of the membrane may consist of 20 α-helices.

Arrangement of Chains around the Central Axis

Since each of the chains spans the membrane and since the sequences of each chain in the regions of the inferred membrane-spanning helices are very similar (Noda et al. 1983), it seems likely that these membrane-spanning helices are arranged pseudosymmetrically around a central axis passing through the central pit. We assume further that the chains do not significantly interdigitate or overlap. Each chain then would occupy a sector of the ring around the central pit. In this case, there are 12 possible orders of the five chains around the central pit, the six shown in Figure 2 and their mirror images (Karlin et al. 1983).

If the above assumptions about the chains are valid and if we were able to tag the individual chains for identification in the electron microscope, we would be able to determine which of the possible orders is correct. For this purpose, a useful landmark is the region of closest approach of the monomers in receptor dimer. The monomers in the dimer are covalently cross-linked via a disulfide bond between δ-chains (see above), and the working hypothesis is that each chain is confined within a sector of the ring around the central pit. Thus, we take the region of closest approach of the monomers in native dimer, the site of the disulfide cross-link, to mark the δ-chains.

We also devised a method for tagging the α-chains, utilizing the high-affinity binding of curarimimetic snake toxins to the ACh-binding sites. We derivatized

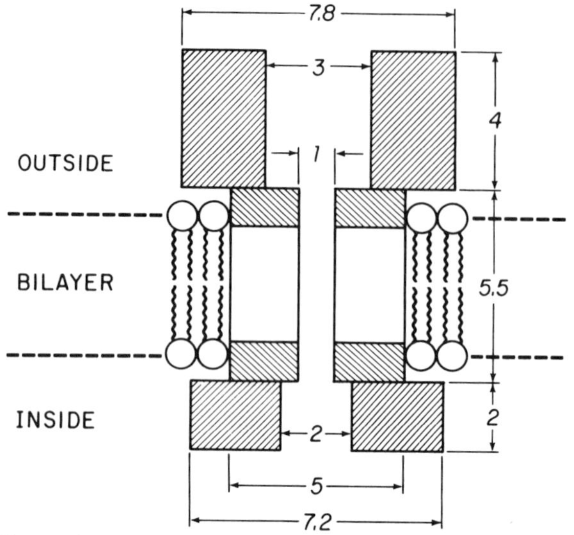

Figure 1. An axial cross-section of the receptor modeled as three cylinders. The model is similar to Fig. 7e in Wise et al. (1979), except that the sequences and the inferred folding of the chains (Claudio et al. 1983; Devillers-Thiery et al. 1983; Noda et al. 1983) were taken into account. The central cylinder is composed of the four membrane-spanning helices per chain suggested by Noda et al. (1983), which contain 87 residues per chain. These helices fit approximately into the unshaded portion of the central cylinder, spanning about 3.3 nm of the bilayer. Added to the central cylinder are 60 residues per chain (61 in the case of γ), sufficient to extend the central cylinder to 5.5 nm. The extracellular and intracellular cylinders contain the remaining residues, assigned as suggested by Noda et al. (1983). We have also included in the extracellular cylinder 60 hexoses and 20 *N*-acetylhexosamines (Lindstrom et al. 1979; Vandlen et al. 1979). The volumes of the cylinders were determined by their content of amino acid (and sugar) residues and by the individual residue volumes. The dimensions were chosen to be close to those determined by electron microscopy and X-ray analysis and to yield a radius of gyration of neutron-scattering density of 4.6 nm (see text). The calculated radius of gyration is based on the neutron-scattering lengths and volumes of the residues in the three cylinders, from which the average neutron-scattering density of each of the cylinders was calculated. Finally, the neutron-scattering density of the medium in which the radius of gyration was determined (18% D_2O) was subtracted to yield the average neutron-scattering contrast in this medium of each of the cylinders. The average contrast of each of the cylinders was used in the calculation of the radius of gyration of the model. The average contrast of the membrane-spanning cylinder, containing predominantly nonpolar residues, is 30% lower than that of the other two cylinders.

the principal curarimimetic toxin of *Naja naja siamensis* venom with biotin and formed the complex of receptor, biotinylated toxin, and avidin (Holtzman et al. 1982; Karlin et al. 1983). The toxin has five lysines and a free terminal amino group (Karlsson et al. 1971). Of the six monobiotinylated derivatives of this toxin formed by the reaction with *N*-hydroxysuccinimidyl-biotin, two bind to receptor nearly as well as native toxin and, when bound, still bind avidin (P. Lobel and A. Karlin, in prep.). Avidin (M_r 67,000) is clearly resolved in the negatively stained complex of receptor, biotinylated toxin, and avidin (Fig. 3). In our analysis of these electron micrographs, we assume that the line connecting the center of a bound avidin and the center of the monomer to which it is bound passes through the toxin-binding site on the α-chain.

We examined complexes of avidin with three forms of receptor: the monomer formed by reduction of the native dimer (not shown), native δ-δ cross-linked dimer (Fig. 3A–D; Holtzman et al. 1982), and β-β cross-linked dimer (Fig. 3E–H; Karlin et al. 1983). The angles between the lines connecting the centers of bound avidins and the center of the monomer to which they were bound and between these lines and the line connecting the centers of the monomers in dimer were measured on the electron micrographs. We found, first, that the average angle between two avidins bound to the same monomer was about 110° ± 30° in each of the three forms of receptor. Second, avidin was found at

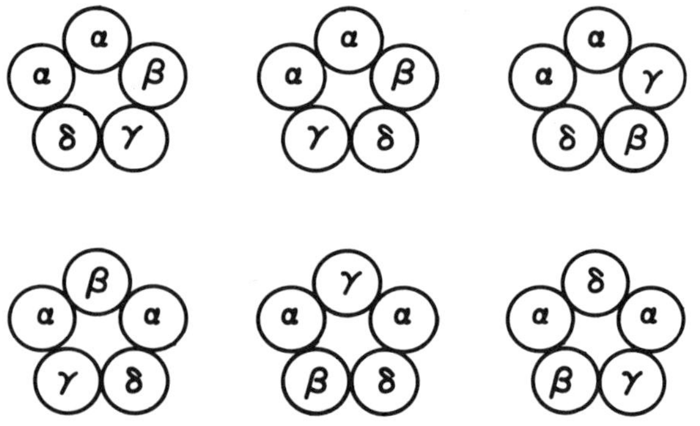

Figure 2. The possible arrangements of the receptor chains around the central pit, as projected onto the plane of the membrane. There are twelve possible arrangements, the six shown and their mirror images. (Reprinted, with permission, from Karlin et al. 1983.)

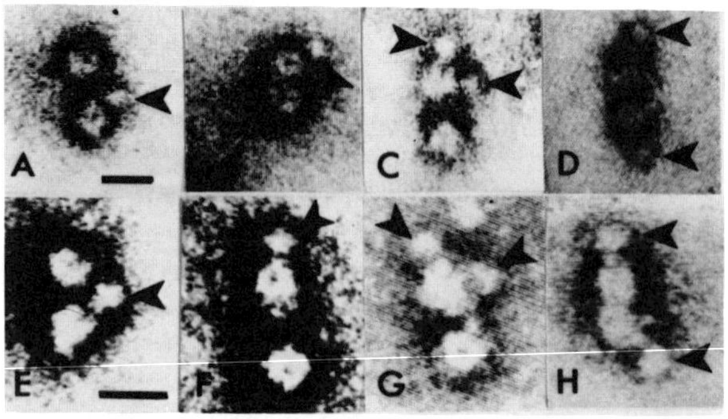

Figure 3. Representative electron micrographs of complexes of receptor dimer, biotinylated toxin, and avidin. (*A–D*) Complexes with native δ-δ dimer (Holtzman et al. 1982); (*E–H*) complexes with β-β dimer (Karlin et al. 1983). β-β dimer was generated by reducing receptor in membrane to monomer and by cross-linking the monomers with diamide (Hamilton et al. 1979). The receptor was solubilized, affinity chromatographed, and resolved into oligomeric species by sucrose density gradient centrifugation. Both types of dimer were incubated with biotinylated toxin and then with avidin. The complexes were isolated by sedimentation in a sucrose density gradient. They were applied to a carbon-coated grid and negatively stained with uranyl sulfate. The arrows point to bound avidins. The magnifications of *A–D* and of *E–H* are indicated by the bars in *A* and *E*, respectively, representing 10 nm.

two general positions relative to the cross-link in dimer: close to the cross-link (at about 60°) and close to the end opposite the cross-link (at about 170°). The same results were obtained with δ-δ and with β-β dimers.

Given our assumptions, these findings seem to eliminate all possible arrangements of the chains (Fig. 2) except one. First of all, the α-chains are not likely to be adjacent, for if they were, the 110° measured between the toxin-binding sites would imply that each α-chain subtended an angle of 110° at the central axis of the monomer. The two α-chains would then subtend 220°, leaving only 140° for β, γ, and δ, all of which are larger than α. Thus, the arrangements in which the α-chains are adjacent (Fig. 2, top row) appear to be unlikely. (An additional assumption made here is that the α-chains, and hence the toxin-binding sites, are more or less superimposable by rotation around the central axis of the receptor; i.e., that the two α-chains are oriented similarly in the radial direction from the central pit to the hydrophobic interior of the membrane.)

We must then choose among the three arrangements in which one of the other three chains is sandwiched between the two α-chains (Fig. 2, bottom row). If δ were between the α-chains, then we would expect to find avidins bound only close to the δ-δ cross-link and not to the ends of the monomers opposite to the cross-link. Avidin, however, is frequently observed bound to the ends of the dimer (Fig. 3B–D; Holtzman et al. 1982). Thus, the locations of the α-chains relative to the δ-chain were found to be consistent with the arrangements in which either β or γ (but not δ) lies between the α-chains (Wise et al. 1981b; Holtzman et al. 1982). Others, using image-reconstruction techniques, have also reached the conclusion that the α-chains are not adjacent; they have suggested that β lies between the α-chains (Kistler et al. 1982; Zingsheim et al. 1982).

The electron micrographs of the complexes of avidin and β-β dimer (Fig. 3E–H), which are quite similar to those of the complexes of avidin and δ-δ dimer (Fig. 3A–D), are inconsistent, however, with the arrangement in which β lies between the α-chains. If β lay between the α-chains, then the bound avidins would be found only at the waist of the dimer, close to the β-β cross-link. Avidins are frequently found, however, at the ends of the monomers opposite to the cross-link as well. By elimination, therefore, the arrangement in which γ lies between the α-chains (Fig. 2, middle of bottom row) seems most probable (Karlin et al. 1983). This arrangement is also consistent with the angular separation of 50–80° between β and δ (Wise et al. 1981b).

Functional Sites: ACh-binding Sites

A chain of apparent molecular weight of about 40,000 (α) is specifically and uniquely tagged by affinity labels of the ACh-binding site in receptor from *Electrophorus electricus, Torpedo* species, and mammalian muscle (Reiter et al. 1972; Karlin et al. 1976; Froehner et al. 1977; Wolosin et al. 1980; Gullick and Lindstrom 1982). Activating and inhibiting affinity labels react at the same site, which is thus the site of activation by ACh (Damle et al. 1978).

There are two ACh-binding sites and two α-chains per receptor monomer (Reynolds and Karlin 1978). Nevertheless, [^{3}H]-4-(*N*-maleimido)benzyltrimethylammonium (Damle and Karlin 1978) and [^{3}H]bromoacetylcholine (Damle et al. 1978; Wolosin et al. 1980) label a particular one of the two sites much more readily than the other. This asymmetry in the reactivity of the two ACh-binding sites may be a consequence of the structural asymmetry of the receptor, since the two α-chains in the complex $\alpha_2\beta\gamma\delta$ cannot have the same subunit interactions.

Noncompetitive Binding Sites

ACh, *d*-tubocurarine, and curarimimetic snake toxins all act primarily at the ACh-binding site. Another class of effectors, however, acts primarily elsewhere on the receptor. These are the noncompetitive inhibitors, a diverse collection of local anesthetics (i.e., inhibitors of conduction), other aromatic tertiary amines, anti-muscarinic agents, anti-arrhythmic agents, psychogenic agents, anti-viral agents, alkaloid toxins, detergents, and alcohols (for review, see Changeux 1981). Their most salient feature in common is their amphipathicity. No one of these agents is specific for the AChR. On the basis of the effects of some of these agents on the kinetics of end-plate currents and of single-channel currents, it has been suggested that they bind within and block the open channel (for review, see Steinbach 1980; Adams 1981). There is evidence that some of these agents also bind to the receptor in the closed state and decrease the rate of opening of the channel. In addition, many of these agents promote the transition of the receptor to the high-affinity (Cohen et al. 1980; Heidmann and Changeux 1981), desensitized state (Sine and Taylor 1982). Despite the complexity of their actions, the noncompetitive inhibitors provide useful probes for functionally significant sites other than the ACh-binding sites.

The binding of radioactive local anesthetics and congeners has been studied in receptor-rich membrane from *Torpedo* electric tissue. The specific binding, defined as that blocked by a large excess of nonradioactive local anesthetics, saturates and is characteristically of higher affinity in the presence of agonists and competitive antagonists than in their absence (Krodel et al. 1979). The extent and sites of photolabeling of receptor by local anesthetic analogs are also strongly affected by agonists and competitive antagonists (Oswald and Changeux 1981a,b; Muhn and Hucho 1983). For example, the photolytic incorporation of trimethisoquin azide is less into α and more into δ in the presence of either agonists or competitive antagonists than in their absence (Oswald and Changeux 1981a). These alterations in the binding of and photolabeling by local anesthetic analogs are unlikely to be consequences of receptor activation and channel opening since competitive antagonists do not open the channel, and agonists, under the conditions of these studies, put the receptor predominantly into the desensitized state.

Quinacrine mustard (QM), an alkylating derivative of the potent local anesthetic quinacrine (Grunhagen and Changeux 1976; Adams and Feltz 1980; Oxford and Hudson 1982), has been reported to block the receptor-mediated permeability increase of *Torpedo* membrane under conditions that also result in 50% block of ACh binding and labeling of all of the receptor chains (Lauffer et al. 1979). Under milder conditions, QM blocks channel activity far more readily than ACh binding (Fig. 4) (Kaldany and Karlin 1983). Furthermore, the reaction of QM that inhibits channel activity is enhanced in the presence of either the agonist carbamylcholine or the competitive antagonist *d*-tubocurarine (Fig. 5); by contrast, there is no reaction of QM with the ACh-binding sites in the presence of these ligands. The reaction of QM that inhibits channel activity is retarded in the presence of the local anesthetics quinacrine and proadifen, both in the presence and in the absence of carbamylcholine and *d*-tubocurarine.

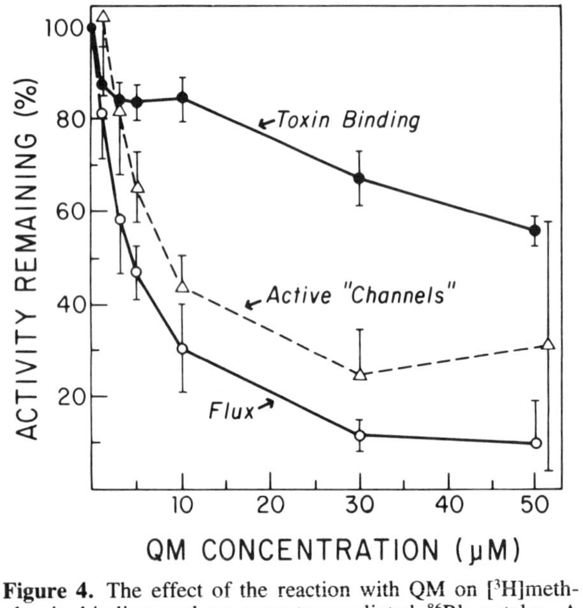

Figure 4. The effect of the reaction with QM on [^{3}H]methyltoxin binding and on receptor-mediated ^{86}Rb uptake. A membrane suspension (1.2 pmoles of toxin-binding sites per μg of protein) was reacted with 20 μM [^{14}C]*N*-ethylmaleimide for 15 min to provide a marker for protein and then with 2 mM *N*-ethylmaleimide for 15 min to block remaining accessible sulfhydryls. QM was reacted with the membrane (1 μM in toxin-binding sites) in 50 mM NaCl, 10 mM $NaPO_4$, and 1 mM EDTA (pH 7.0) at 23°C. After 5 min, 10 mM ethanolamine and 100 μM quinacrine were added to quench the reaction. The membrane was washed once with liposomes and assayed for [^{3}H]methyltoxin binding (●) or dissolved in 2% cholate-2% asolectin and reconstituted by dialysis. ^{86}Rb uptake was determined on the reconstituted membrane (○). The uptake corrected for block of ACh-binding sites (△) was calculated as described in Kaldany and Karlin (1983). (Reprinted, with permission, from Kaldany and Karlin 1983.)

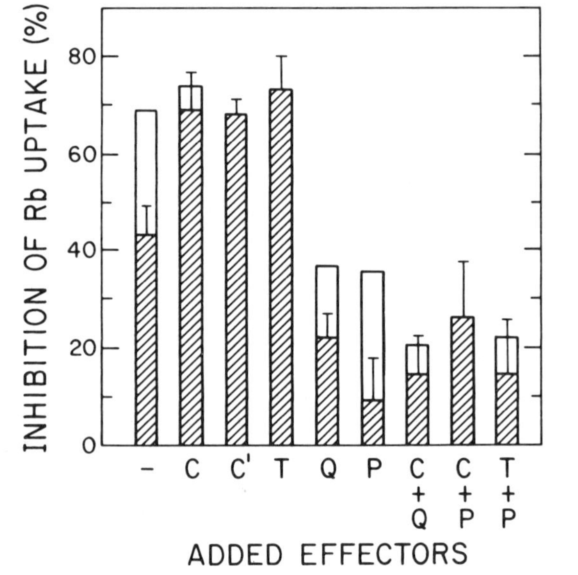

Figure 5. The influence of ligands on the inhibition by QM of receptor-mediated ^{86}Rb uptake. Membrane was reacted with 10 μM QM in the presence of the indicated ligands, washed with liposomes, and reconstituted, as in Fig. 4. The ligands were added 5 min before QM and were present during the reaction. The upper horizontal bars are the measured extents of inhibition of the uptake. The lower horizontal bars are the extents of inhibition of channel activity; i.e., ^{86}Rb uptake corrected for the extents of block of the toxin (ACh)-binding sites. The standard errors in the extents of corrected inhibition are shown. These include the errors in both the uptake and the toxin-binding determinations. C is 10 μM carbamylcholine; C′ is 100 μM carbamylcholine; T is 10 μM *d*-tubocurarine; Q is 100 μM quinacrine; and P is 100 μM proadifen. (Reprinted, with permission, from Kaldany and Karlin 1983.)

Thus, the inhibitory reaction of QM is enhanced by the occupation of the ACh-binding sites and is retarded by the occupation of the local anesthetic-binding sites.

It was possible to determine which chains of the receptor are labeled by the reaction of QM that blocks channel activity. We labeled membrane with [^{3}H]QM under the same conditions that we used in the functional studies (Kaldany and Karlin 1983). The receptor was isolated from the membrane, and the chains were separated by gel electrophoresis and visualized by staining. Finally, the gel was sliced, and the slices were counted. All the chains react with [^{3}H]QM. The presence of ACh-binding site ligands, however, causes an increase in the labeling of α and β, and the presence of local anesthetics causes a decrease in the labeling of α and β (Fig. 6). Each ligand causes either no change or a decrease in the labeling of γ and δ. Looking at the differences in the labeling of the chains in the presence of different ligands, we found that ligands that enhance the inhibitory reaction of QM are associated with an increase in the labeling of α and β and a decrease in the labeling of γ and δ, whereas ligands that retard the inhibitory reaction of QM decrease the labeling of α and β and only to a lesser extent decrease the labeling of γ and δ (Fig. 7).

These results show a correlation between the inhibi-

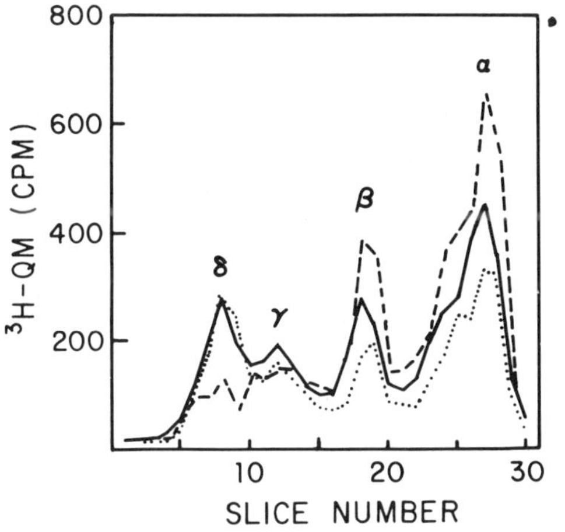

Figure 6. The effects of carbamylcholine and of proadifen on the incorporation of [^{3}H]QM into the subunits of membrane-bound receptor. Membrane was reacted with 10 μM [^{3}H]QM (see Fig. 4) in the presence of 10 μM carbamylcholine (----), 100 μM proadifen (· · · ·), or in the absence of effectors (——). The receptor was purified by affinity chromatography and analyzed by SDS-polyacrylamide gel electrophoresis. The ^{3}H and ^{14}C were determined in 1-mm slices. Shown are the ^{3}H counts multiplied by the ratio of the average ^{14}C counts in the β-chain in the three cases over the ^{14}C counts in the β-chain in each individual case. This corrects for differences in recovery. (Reprinted, with permission, from Kaldany and Karlin 1983.)

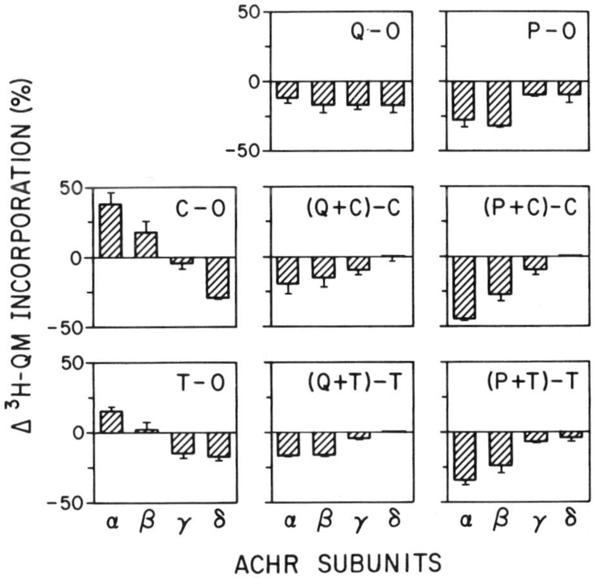

Figure 7. The effects of ligands on the incorporation of [^{3}H]QM into the subunits of the receptor expressed as differences in the relative incorporation in the presence and absence of the ligands. Membrane was labeled and washed, and the receptor was purified and its labeling analyzed by SDS-polyacrylamide gel electrophoresis, as in Fig. 6. The ^{3}H counts in each of the four subunits were divided by the ^{14}C counts in the β peak, to normalize for the quantity of receptor on the gel. These quotients were then expressed as a fraction (%) of the quotient, similarly determined, for the α-chain in the control; this serves to scale the quotients. For each condition of labeling, then, we calculate a normalized and scaled quotient for each chain, which is proportional to the incorporation of [^{3}H]QM. We then calculate the differences of these quotients for pairs of labeling conditions. These differences are plotted. The conditions of labeling are represented as follows: O represents the control in which labeling was carried out in the absence of effectors, and C, T, Q, and P represent labeling carried out in the presence of the respective ligands, as defined in Fig. 5; e.g., panel C – O represents the differences between the relative incorporation into each chain in the presence of 10 μM carbamylcholine and that in the absence of ligands, and (P + C) – C represents the difference between the relative incorporation in the presence of both 100 μM proadifen and 10 μM carbamylcholine and that in the presence of 10 μM carbamylcholine alone. (Reprinted, with permission, from Kaldany and Karlin 1983.)

tion of channel activity and the labeling by QM of both the α- and β-chains. Either there is one site of binding of QM from which the reactive aziridinium moiety (generated from a 2-chloroethylamino group of QM) has access to amino acid residues of both chains, or there are multiple sites of binding and reaction of QM. In the first case, we can imagine that both types of chains contribute to the functionally significant local anesthetic-binding site. In fact, if this hypothetical site were in the channel and if the walls of the channel were formed by all five chains, then all chains might contribute to the local anesthetic-binding site. This would be consistent with the labeling of the γ- and δ-chains, as well as the α- and β-chains, in the absence of ACh-binding site ligands (Fig. 6). Furthermore, ACh-binding site ligands might change the conformation of this site to increase the reactivity of the α- and β-chains and decrease the reactivity of the γ- and δ-chains (Fig. 7, C – O, T – O). Quinacrine or proadifen, alone, blocks labeling of all the chains (Fig. 7, Q – O, P – O), whereas in the presence of carbamylcholine or *d*-tubocurarine, the labeling of γ and δ is already suppressed, and the predominant differential effect of quinacrine and proadifen is to block the labeling of α and β (e.g., Fig. 7, [P + C] – C). These results, therefore, provide evidence for the contribution of α and β to functionally significant local anesthetic binding sites but do not exclude the possibile contribution of γ and δ to such sites.

What is the meaning of the similar influence of carbamylcholine and *d*-tubocurarine on the extent and location of reaction of local anesthetic analogs with the receptor? It is not a consequence of activation, since under the conditions of the reaction the receptor is predominantly in either the desensitized state or the resting state. It appears that the increased reactivity is a reflection of conformational changes that are elicited by both agonists and some competitive antagonists. Examples of other effects on receptor elicited by both agonists and competitive antagonists include an increase in the affinity for local anesthetics (Krodel et al. 1979), an increase in the fluorescence of bound probes, including quinacrine (Grunhagen et al. 1977) and QM (Kaldany 1982), an increase in tetraphenylphosphonium (reversible) uptake (Davis et al. 1983), and a decrease in the rate of reduction of the disulfide near the binding site (Damle and Karlin 1980). (In general, the rate of onset and amplitude of the effects are greater with agonists than with competitive antagonists.)

Time-resolved Photolabeling

Can we detect differences in susceptibility to labeling between the channel-open (active) state of the receptor and the channel-closed (resting and desensitized) states of the receptor? Such differences would arise, for example, as a consequence of altered accessibility of a site of reaction. The active state of the receptor is transitory. In the presence of agonist, the receptor initially undergoes rapid (sub-millisecond) transitions between the channel-closed and the channel-open states (Colquhoun and Sakmann 1981). Eventually, on the time scale of tenths of seconds to seconds, the receptor desensitizes, and the channel remains closed (Neubig et al. 1982; Walker et al. 1982, and see for references). We have used rapid mixing techniques and photolytic reactions to try to label the receptor in the active state.

The continuous-flow, rapid-mixing device we constructed for this purpose was based on a design of Thayer and Hinkle (1979) and consists, in series, of a pneumatic ram, two syringes, an off-set T-mixer, a delay tube, a quartz photolysis tube, and a collecting tube (R.N. Cox et al., in prep.). The light from a 1000-watt mercury (xenon) arc lamp is collimated, filtered through 10 cm of water and a broad-band filter cutting off below 290 nm, and focused on the quartz tube. Typically, a suspension of receptor-rich membrane is loaded into one syringe and effector and photolabel are loaded into the other syringe. The ram is actuated and the contents of the two syringes mix, pass through the delay tube, through the quartz tube, and finally into the collecting tube. Delays as short as 20 msec between mixing and exposure to light in the quartz tube are achievable. The mixture is irradiated for 1–3 msec, depending on the rate of flow (2–5 ml/sec) and the length of the quartz tube exposed. The collecting tube contains a final concentration of dithiothreitol (DTT) of 10 mM, which quenches long-lived reactive products of photolysis and reduces the azido group (Staros et al. 1978).

The photolabel used in these experiments was the tritiated derivative of quinacrine azide (QA) (Mueller et al. 1982), 3-azido-9-[{4-(diethylamino)-1-methyl-[1-^{3}H]butyl}amino]-7-methoxyacridine, the synthesis of which will be described elsewhere. It was used at a final concentration of 5–10 μM. Control experiments showed that [^{3}H]QA is completely photolyzed during a 1-msec exposure in the quartz tube.

[^{3}H]QA labels all of the chains of the receptor, but predominantly α, at all delay times. When ACh is in the syringe with [^{3}H]QA, the labeling of α and β increases relative to the control. When *d*-tubocurarine is in the syringe with [^{3}H]QA, the extents of labeling of α, β, and γ are decreased relative to the control. The processes giving rise to the differences in the effects

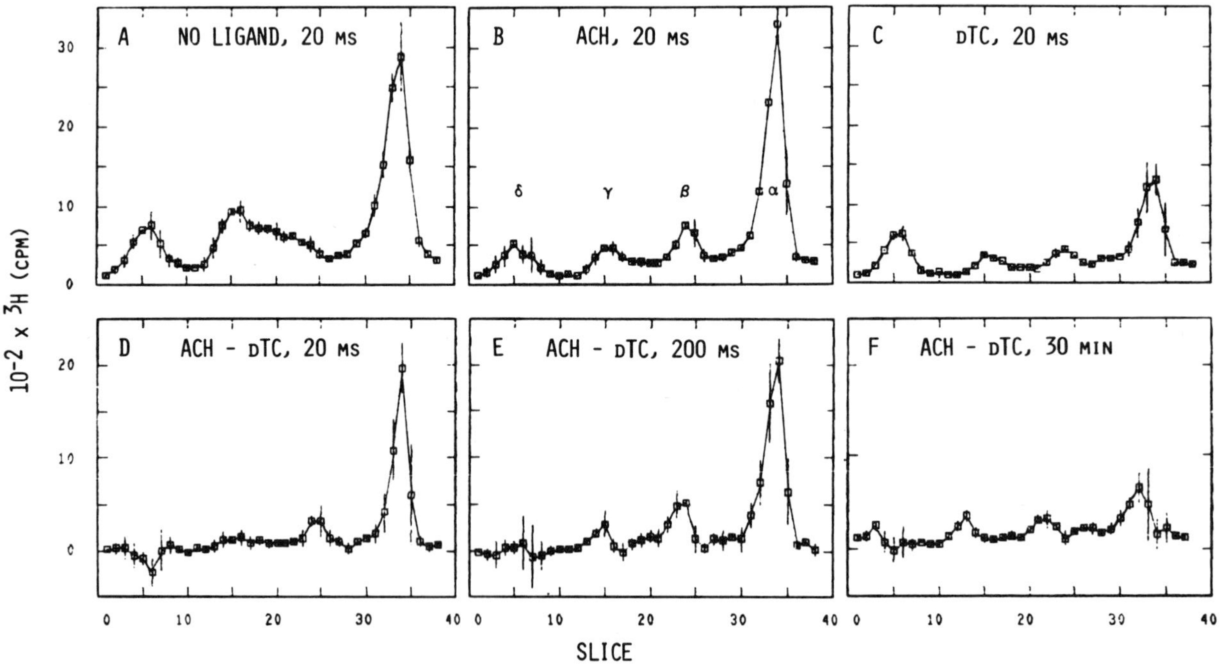

Figure 8. The influence of ligand and the time between mixing and photolysis on the photolabeling of the receptor chains by [^{3}H]quinacrine azide (QA). Receptor-rich membrane (1.8 pmoles toxin sites/μg) in one syringe of the rapid-mixing device and [^{3}H]QA ± ligand in the other syringe were mixed and, after a delay of 20–200 msec, passed through the photolysis tube and into the collection tube (see text). The distribution of label in the receptor chains was determined as in Fig. 6, except that the receptor was not affinity-purified before electrophoresis. (*A*) No ligand; (*B*) 50 μM ACh; (*C*) 50 μM *d*-tubocurarine. In *A*, *B*, and *C*, the delay was 20 msec. (*D*) The difference between the distribution in *B* and *C*. (*E*) As *D* except with 200-msec delay. (*F*) As *D* except with a 30-min preincubation with the ligands and a 100-msec delay after mixing with [^{3}H]QA. In all cases (*A–F*), the average and average errors of duplicate experiments are plotted.

of ACh and *d*-tubocurarine on the incorporation of [^{3}H]QA occur within delay times between mixing and photolysis of as little as 20 msec (Fig. 8); these differences persist to delay times of at least 200 msec. Infinite delay time points were devised as follows: Receptor was desensitized by incubation with 50 μM ACh (and 200 μM diisopropylfluorophosphate) or 50 μM carbamylcholine for 30 minutes, mixed with photolabel, and irradiated 20–200 msec later. Alternatively, receptor was incubated with 50 μM *d*-tubocurarine for 30 minutes, mixed with photolabel, and irradiated 20–200 msec later. Under these conditions, the extents of labeling of α and β in the presence of agonist minus the extents of labeling in the presence of *d*-tubocurarine are one half of those differences obtained without the 30-minute incubation (Fig. 8). In all cases, the photolabel is present only during the delay time of 20–200 msec.

Both ACh and *d*-tubocurarine protect the ACh-binding sites against photolabeling by QA. The decrease in the labeling of α in the presence of *d*-tubocurarine compared with the control is at least in part due to this effect. The increase in the labeling in the presence of ACh occurs despite the protection of the ACh-binding sites. The difference between the labeling in the presence of ACh and of *d*-tubocurarine, arising within the shortest delay time that we have tested (20 msec), may be correlated with the open-channel state of the receptor induced by ACh. That the difference is considerably less after long exposure to these ligands may be a reflection of desensitization or other long-term changes induced by both agonists and some competitive antagonists (see above). In summary, both QM and QA label α and β. In both cases, this labeling is increased by ACh and blocked by proadifen. The labeling of α and β by QM is correlated with inhibition of channel activity, and the photolabeling of α and β by QA is enhanced by a rapid change in the receptor induced by ACh and not by *d*-tubocurarine. It appears that α has at least two roles. It binds ACh and also, along with β, contributes to functionally significant local anesthetic-binding sites.

ACKNOWLEDGMENTS

This work was supported in part by research grants from the National Institutes of Health (NS-07065 and EY-03168) and the Muscular Dystrophy Association, Inc. We thank D. Wise for his helpful comments on the model, Q. Al-Awqati, H. Bayley, P. Brandt, H. Chase, A. Dwork, B. Ferren, and P. Kao for their valuable help with the photolabeling equipment, R. Hudson for his generous gift of quinacrine azide precursor, and N. Yodh for her expert technical assistance.

REFERENCES

Adams, P.R. 1981. Acetylcholine receptor kinetics. *J. Membr. Biol.* **58:** 161.

Adams, P.R. and A. Feltz. 1980. Quinacrine (mepacrine) action at frog end-plate. *J. Physiol.* **306:** 261.

Anderson, D.J. and G. Blobel. 1981. In vitro synthesis, glycosylation, and membrane insertion of the four subunits of *Torpedo* acetylcholine receptor. *Proc. Natl. Acad. Sci.* **78:** 5598.

Cartaud, J., J.L. Popot, and J.P. Changeux. 1980. Light and heavy forms of the acetylcholine receptor from *Torpedo marmorata* electric organ. *FEBS Lett.* **121:** 327.

Changeux, J.-P. 1981. The acetylcholine receptor: An 'allosteric' membrane protein. *Harvey Lect.* **75:** 85.

Claudio, T., M. Ballivet, J. Patrick, and S. Heinemann. 1983. Nucleotide and deduced amino acid sequences of *Torpedo californica* acetylcholine receptor γ subunit. *Proc. Natl. Acad. Sci.* **80:** 1111.

Cohen, J.B., N.D. Boyd, and N.S. Shera. 1980. Interactions of anesthetics with nicotinic postsynaptic membranes isolated from *Torpedo* electric tissue. In *Molecular mechanisms of anesthesia* (ed. B.R. Fink), p. 165. Raven Press, New York.

Colquhoun, D. and B. Sakmann. 1981. Fluctuations in the microsecond time range of the current through single acetylcholine receptor ion channels. *Nature* **294:** 464.

Conti-Tronconi, B.M. and M.A. Raftery. 1982. The nicotinic cholinergic receptor: Correlation of molecular structure with functional properties. *Annu. Rev. Biochem.* **51:** 491.

Damle, V. and A. Karlin. 1978. Affinity labeling of one of two α-neurotoxin binding sites in acetylcholine receptor from *Torpedo californica. Biochemistry* **17:** 2039.

———. 1980. Effects of agonists and antagonists on the reactivity of the binding site disulfide in acetylcholine receptor from *Torpedo californica. Biochemistry* **19:** 3924.

Damle, V.N., M. McLaughlin, and A. Karlin. 1978. Bromoacetylcholine as an affinity label of the acetylcholine receptor from *Torpedo californica. Biochem. Biophys. Res. Commun.* **84:** 845.

Davis, C.G., S. Hestrin, H. Landahl, A.S. Gordon, I. Diamond, and J.I. Korenbrot. 1983. Activation of acetylcholine receptors causes the partition of hydrophobic cations into postsynaptic membrane vesicles. *Nature* **302:** 525.

Devillers-Thiery, A., J. Giraudat, M. Bentaboulet, and J.-P. Changeux. 1983. Complete mRNA coding sequence of the acetylcholine binding α-subunit of *Torpedo marmorata* acetylcholine receptor: A model for the transmembrane organization of the polypeptide chain. *Proc. Natl. Acad. Sci.* **80:** 2067.

Froehner, S.C, A. Karlin, and Z.W. Hall. 1977. Affinity alkylation labels two subunits of the reduced acetylcholine receptor from mammalian muscle. *Proc. Natl. Acad. Sci.* **74:** 4685.

Grunhagen, H.H. and J.-P. Changeux. 1976. Studies on the electrogenic action of acetylcholine with *Torpedo marmorata* electric organ. IV. Quinacrine: A fluorescent probe for the conformational transitions of the cholinergic receptor protein in its membrane bound state. *J. Mol. Biol.* **106:** 497.

Grunhagen, H.H., M. Iwatsubo, and J.-P. Changeux. 1977. Fast kinetic studies on the interaction of cholinergic agonists with the membrane-bound acetylcholine receptor from *Torpedo marmorata* as revealed by quinacrine fluorescence. *Eur. J. Biochem.* **80:** 225.

Gullick, W.J. and J.M. Lindstrom 1982. Structural similarities between acetylcholine receptors from fish electric organs and mammalian muscle. *Biochemistry* **21:** 4563.

Hamilton, S.L., M. McLaughlin, and A. Karlin. 1979. Formation of disulfide-linked oligomers of acetylcholine receptor in membrane from *Torpedo* electric tissue. *Biochemistry* **18:** 155.

Heidmann, T. and J.-P. Changeux. 1981. Stabilization of the high affinity state of the membrane-bound acetylcholine receptor from *Torpedo marmorata* by noncompetitive blockers. *FEBS Lett.* **131:** 239.

HOLTZMAN, E., D. WISE, J. WALL, and A. KARLIN. 1982. Electron microscopy of complexes of isolated acetylcholine receptor, biotinyl-toxin and avidin. *Proc. Natl. Acad. Sci.* **79:** 310.

KALDANY, R.-R. 1982. "Affinity labeling by quinacrine mustard of a functionally significant local anesthetic binding site of the acetylcholine receptor." Ph.D. thesis, Columbia University, New York, New York.

KALDANY, R.-R. and A. KARLIN. 1983. Reaction of quinacrine mustard with the acetylcholine receptor from *Torpedo californica:* Functional consequences and sites of labeling. *J. Biol. Chem.* **258:** 6232.

KARLIN, A. 1980. Molecular properties of nicotinic acetylcholine receptors. In *The cell surface and neuronal function* (ed. C.W. Cotman et al.), p. 191. Elsevier/North Holland, Amsterdam.

KARLIN, A., C.L. WEILL, M.G. MCNAMEE, and R. VALDERRAMA. 1976. Facets of the structures of acetylcholine receptors from *Electrophorus* and *Torpedo. Cold Spring Harbor Symp. Quant. Biol.* **40:** 203.

KARLIN, A., E. HOLTZMAN, N. YODH, P. LOBEL, J. WALL, and J. HAINFELD. 1983. The arrangement of the subunits of the acetylcholine receptor of *Torpedo californica. J. Biol. Chem.* **258:** 6678.

KARLSSON, E., H. ARNBERG, and D. EAKER. 1971. Isolation of the principal neurotoxins of two *Naja naja* subspecies. *J. Biol. Chem.* **21:** 1.

KISTLER, J., R.M. STROUD, M.W. KLYMKOWSKY, R.A. LALANCETTE, and R.H. FAIRCLOUGH. 1982. Structure and function of an acetylcholine receptor. *Biophys. J.* **37:** 371.

KRODEL, E.K., R.A. BECKMAN, and J.B. COHEN. 1979. Identification of a local anesthetic binding site in nicotinic post-synaptic membranes isolated from *Torpedo marmorata* electric tissue. *Mol. Pharmacol.* **15:** 294.

LAUFFER, L., K.-H. WEBER, and F. HUCHO. 1979. Acetylcholine receptor. Binding properties and ion permeability response after covalent attachment of the local anesthetic quinacrine. *Biochim. Biophys. Acta* **587:** 42.

LINDSTROM, J., J. MERLIE, and G. YOGEESWARAN. 1979. Biochemical properties of acetylcholine receptor subunits from *Torpedo californica. Biochemistry* **18:** 4465.

MUELLER, D.M., R.A. HUDSON, and C. LEE. 1982. An azido analogue of quinacrine: Its synthesis and utilization as a photoaffinity probe with submitochondrial particles. *Biochemistry* **21:** 1445.

MUHN, P. and F. HUCHO. 1983. Covalent labeling of the acetylcholine receptor from *Torpedo* electric tissue with the channel blocker [^{3}H]triphenylmethylphosphonium by ultraviolet irradiation. *Biochemistry* **22:** 421.

NEUBIG, R.R., N.D. BOYD, and J.B. COHEN. 1982. Conformations of *Torpedo* acetylcholine receptor associated with ion transport and desensitization. *Biochemistry* **21:** 3460.

NODA, M., H. TAKAHASHI, T. TANABE, M. TOYOSATO, S. KIKYOTANI, Y. FURUTANI, T. HIROSE, H. TAKASHIMA, S. INAYAMA, T. MIYATA, and S. NUMA. 1983. Structural homology of *Torpedo californica* acetylcholine receptor subunits. *Nature* **302:** 528.

OSWALD, R.E. and J.-P. CHANGEUX. 1981a. Selective labeling of the δ subunit of the acetylcholine receptor by a covalent local anesthetic. *Biochemistry* **20:** 7166.

——. 1981b. Ultraviolet light-induced labeling by noncompetitive blockers of the acetylcholine receptor from *Torpedo marmorata. Proc. Natl. Acad. Sci.* **78:** 3925.

OXFORD, G.S. and R.A. HUDSON. 1982. Azidoacridines as photoaffinity probes for ionic channels in excitable membranes. *Biochem. Biophys. Res. Commun.* **104:** 1579.

RAFTERY, M.A., M.W. HUNKAPILLER, C.D. STRADER, and L.E. HOOD. 1980. Acetylcholine receptor: Complex of homologous subunits. *Science* **208:** 1454.

REITER, M.J., D.A. COWBURN, J.M. PRIVES, and A. KARLIN. 1972. Affinity labeling of the acetylcholine receptor in the electroplax: Electrophoretic separation in sodium dodecyl sulfate. *Proc. Natl. Acad. Sci.* **69:** 1168.

REYNOLDS, J. and A. KARLIN. 1978. Molecular weight in detergent solution of acetylcholine receptor from *Torpedo californica. Biochemistry* **17:** 2035.

SINE, S.M. and P. TAYLOR. 1982. Local anesthetics and histrionicotoxin are allosteric inhibitors of the acetylcholine receptor. *J. Biol. Chem.* **257:** 8106.

STAROS, J.V., H. BAYLEY, D.N. STANDRING, and J.R. KNOWLES. 1978. Reduction of aryl azides by thiols: Implications for the use of photoaffinity reagents. *Biochem. Biophys. Res. Commun.* **80:** 568.

STEINBACH, J.H. 1980. Activation of nicotinic acetylcholine receptors. In *The cell surface and neuronal function* (ed. C.W. Cotman et al.), p. 119. Elsevier/North Holland, Amsterdam.

THAYER, W.S. and P.C. HINKLE. 1979. Design of a simple rapid mixing and quenching device and its use for measurement of the kinetics of ATP synthesis by submitochondrial particles. *Methods Enzymol.* **56G:** 492.

VANDLEN, R.L., W.C.-S. WU, J.C. EISENACH, and M.A. RAFTERY. 1979. Studies of the composition of purified *Torpedo californica* acetylcholine receptor and of its subunits. *Biochemistry* **18:** 1845.

WALKER, J.W., K. TAKEYASU, and M.G. MCNAMEE. 1982. Activation and inactivation kinetics of *Torpedo californica* acetylcholine receptor in reconstituted membranes. *Biochemistry* **21:** 5384.

WEILL, C.L., M.G. MCNAMEE, and A. KARLIN. 1974. Affinity labeling of purified acetylcholine receptor from *Torpedo californica. Biochem. Biophys. Res. Commun.* **61:** 997.

WISE, D.S., A. KARLIN, and B.P. SCHOENBORN. 1979. An analysis by low angle neutron scattering of the structure of the acetylcholine receptor from *Torpedo californica* in detergent solution. *Biophys. J.* **28:** 473.

WISE, D.S., B.P. SCHOENBORN, and A. KARLIN. 1981a. Structure of acetylcholine receptor dimer determined by neutron scattering and electron microscopy. *J. Biol. Chem.* **256:** 4124.

WISE, D.S., J. WALL, and A. KARLIN. 1981b. Relative locations of the β and δ chains of the acetylcholine receptor determined by electron microscopy of isolated receptor trimer. *J. Biol. Chem.* **256:** 12624.

WOLOSIN, J.M., A. LYDDIATT, J.O. DOLLY, and E.A. BARNARD. 1980. Stoichiometry of the ligand-binding sites in the acetylcholine-receptor oligomer from muscle and from electric organ. Measurement by affinity alkylation with bromoacetylcholine. *Eur. J. Biochem.* **109:** 495.

ZINGSHEIM, H.P., F.J. BARRANTES, J. FRANK, W. HANICKE, and D.C. NEUGEBAUER. 1982. Direct structural localization of two toxin-recognition sites on an ACh receptor protein. *Nature* **299:** 81.

Subunit Organization and Structure of an Acetylcholine Receptor

R.H. FAIRCLOUGH, J. FINER-MOORE, R.A. LOVE, D. KRISTOFFERSON, P.J. DESMEULES, AND R.M. STROUD
Department of Biochemistry and Biophysics, University of California School of Medicine, San Francisco, California 94143

The nicotinic acetylcholine receptor (AChR) from the electric organ of *Torpedo* is the best-characterized macromolecule involved in synaptic communication. Both functionally and biochemically, this receptor closely resembles the AChR of the mammalian neuromuscular junction (Einarson et al. 1982). The AChR from *Torpedo californica* is a 255-kD (Reynolds and Karlin 1978) integral membrane protein composed of four glycosylated polypeptide chains with apparent molecular weights of 40,000 (α), 50,000 (β), 60,000 (γ), and 65,000 (δ) in an $\alpha_2\beta\gamma\delta$ stoichiometry (Reynolds and Karlin 1978; Lindstrom et al. 1979; Raftery et al. 1980). In *Torpedo* the receptor dimerizes via one or more disulfide bridges between δ subunits (Chang and Bock 1977; Hucho et al. 1978; Hamilton et al. 1979). The sequences of the primary transcript of the four polypeptides have recently been elucidated through sequencing of the corresponding cDNA (Noda et al. 1982, 1983a,b; Sumikawa et al. 1982; Claudio et al. 1983; Devillers-Thiery et al. 1983). A three-dimensional structure of the receptor has been obtained to a resolution of 18 Å in the membrane plane and 13 Å perpendicular to the plane (see Fig. 1) (Stroud 1981; Kistler et al. 1982). Particular features of this model include the asymmetric distribution of protein across the lipid bilayer, the transmembrane character of the protein subunits, and the rather regular distribution of protein about a central well that narrows to the 7-Å-diameter transmembrane ion channel (Kistler et al. 1982).

The physiological and biochemical properties of the AChR are reviewed by Karlin (1980), Changeux (1981), Conti-Tronconi and Raftery (1982), and Anholt et al. (1983). Current structural analysis of AChR is reviewed by Stroud (1983). In this paper, we focus on the polypeptide threading of the subunit chains relative to the bilayer, subunit secondary structure, assembly of the receptor from membrane-bound subunits, receptor quaternary structure, and the location and nature of the receptor's toxin-binding surface.

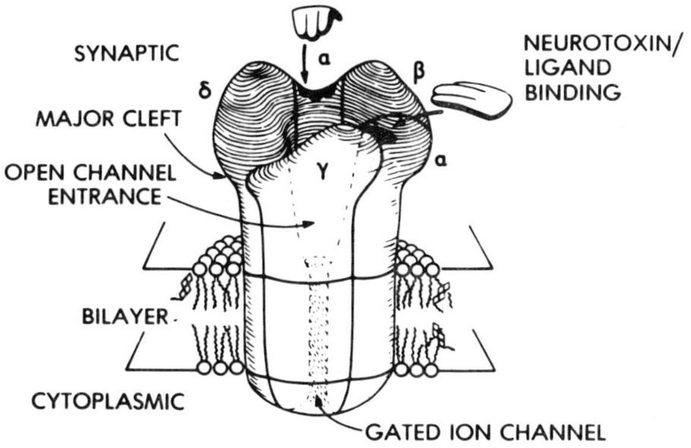

Figure 1. Three-dimensional model of AChR from *Torpedo californica*. The subunit locations around the central well are tentative assignments. (Reprinted, with permission, from Stroud 1981.)

MATERIALS AND METHODS

Sequence analysis. The sequences of the four receptor polypeptides are aligned using the hydrophobicity correlation and the comparison matrix of McLachlan (1971) (see Finer-Moore and Stroud 1983). Hydrophobicity plots are calculated using

$$H_i = 1/7 \sum_{j=i-3}^{i+3} h_j$$

in which h_j is the hydrophobicity of residue j in kcal/mole from a consensus set of values developed by Eisenberg et al. (1983b). Periodic amphipathic quality is probed by computing amphipathic spectra $I_k(\nu)$ for 25 residue stretches of sequence:

$$I_k(\nu) = \left| \sum_{j=k-12}^{k+12} (h_j - \bar{h}_k) \exp(2\pi ij\nu) \right|^2$$

in which $\bar{h}_k$ is the average hydrophobicity of $k-12$ to $k+12$. Spectra are scaled by $1/I_m$ in which I_m is the mean intensity in the power spectrum of a random arrangement of the 25 amino acids in the window (McLachlan and Stewart 1976).

Electron microscopy of Fab-labeled AChR subunits. AChR α-subunits are labeled with Fab fragments of rat monoclonal antibody no. 35 prepared in the laboratory of Dr. J. Lindstrom (Conti-Tronconi et al. 1981). The entire AChR-labeling procedure will be described elsewhere (D. Kristofferson et al., in prep.). In brief, AChR-enriched membranes containing dispersed dimers (Kistler et al. 1982) are incubated with a fivefold excess of Fab (on the basis of titer) for 2 hours at room temperature. Vesicles with bound Fabs are separated from unbound Fabs by centrifugation in a Beckman airfuge. The pellet is resuspended, applied to a glow-discharged, carbon-parlodian grid, and negatively stained with uranyl acetate. Grids are examined at 117,000× magnification in a Philips EM 400 operated at 100 kV with a 70-μm objective aperture. Micrograph negatives are digitized with a Gould/Deanza IP6400 image processor interfaced with a VAX 11/780. AChR dimers are selected for determination of Fab positions if they are clearly labeled and well separated from neighboring dimers. The center of each receptor and the Fab label are located interactively in the digitized image. The Fab angle is measured counterclockwise around the center of the rosette from the center-to-center line in a dimer to the center-to-Fab line.

To remove systematic broadening of angle distributions, we first selected dimers with two Fabs bound to a single rosette. These dimers can be oriented relative to one another by aligning them so that the smaller angle between the two Fabs is consistently oriented. We next adopted an angle-measuring convention that assigns the same angle to equivalent positions on translationally related monomers (see Results for further details).

Small-angle X-ray diffraction of oriented AChR membranes. The detailed procedures of the small-angle diffraction experiments will be reported elsewhere (R.H. Fairclough et al., in prep.). Briefly, alkali-stripped AChR membrane vesicles are oriented by ultracentrifugation. The membrane pellet is aligned in the focused 100-μm^2 beam of an Elliott GX-6 X-ray generator at the front end of a small-angle camera of our design (Ross et al. 1977). The average orientation of the membrane planes is parallel to the X-ray beam. The small-angle camera is purged with water-saturated helium, and the diffraction patterns are recorded on Kodak "No Screen" X-ray film. Films are scanned on a Syntex AD-1 microdensitometer. The intensity profile is generated by radial integration and averaged over 20° sectors along the two axes of the diffraction pattern: the meridian perpendicular to and the equator parallel to the membrane planes. Raw intensity profiles are background-subtracted, corrected for camera geometry and sample disorder, and converted to amplitudes. The meridional amplitudes are subjected to constrained iterative refinement (Stroud and Agard 1979), starting with the electron density profile across an AChR membrane reported by Ross et al. (1977).

RESULTS AND DISCUSSION

Hydrophobicity and Transmembrane Crossings

The recent determination of the nucleotide sequences of the cDNA and hence the complete amino acid sequence for each of the four receptor subunits (Noda et al. 1982, 1983a,b; Sumikawa et al. 1982; Claudio et al. 1983; Devillers-Thiery et al. 1983) prompts a comparison and evaluation of these sequences in light of

Table 1. Aligned Sequences of the Four Receptor Subunits

```
        -24                     -1
ALPHA   MILCSYWHVGLVLLLFSCCGLVLG
BETA    MENVRRMALGLVVMMALALSGVGA
GAMMA         MVLTLLLIICLALEVRS
DELTA   MGNIHFVYLLISCLYYSGCS   G

1
SEHETRLVANLL EN YNKVIRPVEHHTHFVDITVGLQLIQLISVDEVNQIVETNVRLRQQWIDVRLRWNPADYGGIKKIRLPSDDV
SVMEDTLLSVLF ET YNPKVRPAQTVGDKVTVRVGLTLTNLLILNEKIEEMTTNVFLNLAWTDYRLQWDPAAYEGIKDLRIPSSDV
ENEEGRLIEKLL GD YDKRIIPAKTLDHIIDVTLKLTLTNLISLNEKEEALTTNVWIEIQWNDYRLSWNTSEYEGIDLVRIPSELL
VNEEERLINDLLIVNKYNKHVRPVKHNNEVVNIALSLTLSNLISLKETDETLTSNVWMDHAWYDHRLTWNASEYSDISILRLPPELV

         100
WLPDLVLYNNADGDFAIVHMTKLLLDYTGKIMWTPPAIFKSYCEIIVTHFPFDQQNCTMKLGIWTYDGTKVSIS         PESDRP
WQPDIVLMNNNDGSFEITLHVNVLVQHTGAVSWQPSAIYRSSCTIKVMYFPFDWQNCTMVFKSYTYDTSEVTLQ     HALDAKGERE
WLPDVVLENNVDGQFEVAYYANVLVYNDGSMYWLPPAIYRSTCPIAVTYFPFDWQNCSLVFRSQTYNAHEVNLQLS       AEEGEA
WIPDIVLQNNNDGQYHVAYFCNVLVRPNGYVTWLPPAIFRSSCPINVLYFPFDWQNCSLKFTALNYDANEITMDLMTDTIDGK DYP

                  200
     DLSTFMESGEWVMKDYRGWKH WVYYTCCPD TPYLDITYHFIMQRIPLYFVVNVIIPCLLFSFLTGLVFYLPTDSG EKM
VKEIVINKDAFTENGQWSIEH KPSRKNW RSD DP  S YEDVTFYLIIQRKPLFYIVYTIIPCILISILAILVFYLPPDAG EKM
VEWIHIDPEDFTENGEWTIRH RPAKKNYNWQLTKDD TDFQEIIFFLIIQRKPLFYIININIIAPCVLISSLVVLVYFLPAQAGGQKC
IEWIIIDPEAFTENGEWEIIH KPAKKN IYPDKFPNGTNYQDVTFYLIIRRKPLFYVINFITPCVLISFLASLAFYLPAESG EKM

                                    300
TLSISVLLSLTVFLLVIVELIPSTSSAVPLIGKYMLFTMIFVISSIIITVVVINTHHRSPSTHTMPQWVRKIFIDTIPNV
SLSISALLAVTVFLLLLADKVPETSLSVPIIIRYLMFIMILVAFSVILSVVVLNLHHRSPNTHTMPNWIRQIFIETLPPFLWIQRPV
TLSISVLLAQTIFLFLIAQKVPETSLNVSLIGKYLIFVMFVSMLIVMNCVIVLNVSLRTPNTHSLSEKIKHLFLGFLPKYLGMQLEP
STAISVLLAQAVFLLLTSQRLPETALAVPLIGKYLMFIMSLVTGVIVNCGIVLNFHFRTPSTHVLSTRVKQIFLEKLPRILHMSRAD

                                                 400
                                    MFFSTMKRASKEKQENKIFADDIDISDISGKQVTGEVIFQTPLIKNP DV
TTPSPD         SKPTIISRANDEYFIRKPA    GDFVCPVDNARVAVQPERLFSEMKWHLNG          LTQPVTLPQDL
SEETPEKPQ PR RRSSFGIMI KA EEYILKKPRSELMFEEQKDRHGLKRVNK MTSDIDI       GTTVDLYKDLANFAP EI
ESEQPDWQNDLKLRRSSSVGYISKAQ EYFNIKSR    SELMFEKQSERHGLVP RVTPRIGFGNNNEN       IAASDQLHDEI

                                                             500
KSAIEGVKYIAEHMKSDEESSNAAEEWKYVAMVIDHILLCVFMLICIIGTVSVFAGRLIELSQEG
KEAVEAIKYIAEQLESASEFDDLKKDWQYVAMVADRLFLYVFFVICSIGTFSIFLDASHNVPPDNPFA
KSCVEACNFIAKSTKEQNDSGSENENWVLIGKVIDKACFWIALLLFSIGTLAIFLTGHFNQVPEFPFPGDPRKYVP
KSGIDSTNYIVKQIKEKNAYDEEVGNWNLVGQTIDRLSMFIITPVMVLGTIFIFVMGNFNHPPAKPFEGDPFDYSSDHPRCA
```

our three-dimensional model of the receptor presented in Figure 1. A comparison of the four amino acid sequences as aligned by Noda et al. (1983a) or by us in Table 1 shows that sequence homology extends through most of the primary structure of the subunits, with only the region between residues 342 and 425 showing little homology. Such homology suggests a similar structural motif for each subunit and a similar contribution of each subunit to the total structure.

Insight into the polypeptide threading pattern of each subunit relative to the bilayer derives from consideration of the distribution of polar and nonpolar residues in the sequences, as represented in the hydrophobicity plots of Figure 2. Here, the ordinate is the average hydrophobicity of a stretch of seven residues, and the abscissa is the residue number at the center of the stretch (Kyte and Doolittle 1982). Each subunit is characterized by four unusually hydrophobic regions (besides the signal peptide): residues 228–254, 261–279, 289–316, and 472–491. On the basis of this observation, Claudio et al. (1983), Devillers-Thiery et al. (1983), and Noda et al. (1983a) have proposed a structural scheme for the subunits of AChR in which each of the four hydrophobic regions traverses the membrane once.

The similarity of the hydrophobicity plots of the subunits and the consequent similarity in chain threading relative to the bilayer supports the transmembrane character of the subunits and further emphasizes the similar contribution each subunit makes to the total structure. The arrangement of the five subunits can be envisioned much as staves about the central well. This threading pattern with four transmembrane segments would organize 52% of the protein mass on the synaptic side, 28% on the cytoplasmic side, and 20% buried in the lipid bilayer. Our three-dimensional model organizes 61% on the synaptic side, 12% on the cytoplasmic side, and 27% within the lipid bilayer (Finer-Moore and Stroud 1983).

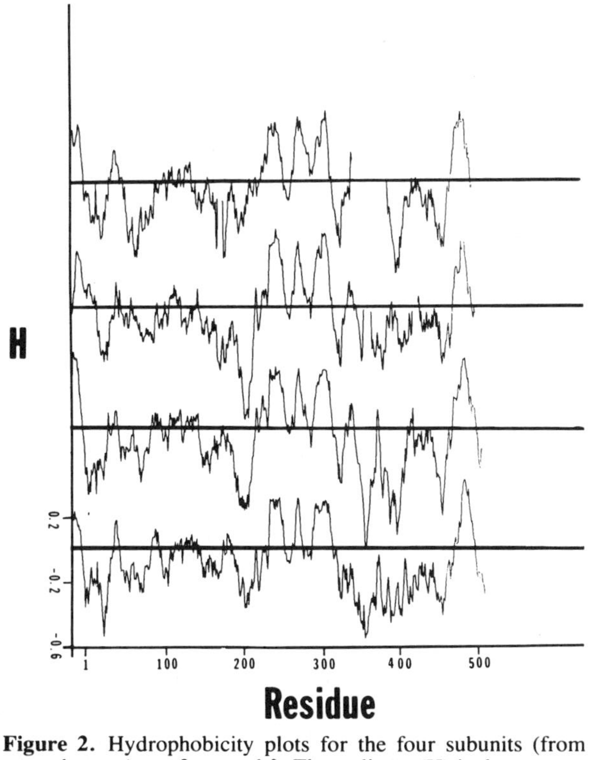

Figure 2. Hydrophobicity plots for the four subunits (from top to bottom): α, β, γ, and δ. The ordinate (H) is the average hydrophobicity in kcal/mole for "windows" of seven residues. The gaps correspond to deletions inserted when aligning the four sequences.

Sequence Organization

Despite the overall consistency of the structural conclusions from sequence analysis with the three-dimensional model, the scheme places both the amino and carboxyl termini of the subunits on the synaptic side of the membrane. In contrast, six integral membrane proteins studied to date have extracellular exposure of the amino terminus and *cytoplasmic* exposure for the carboxyl terminus (Sabatini et al. 1982). A second consequence of the scheme is the rather large fraction of protein on the cytoplasmic side of the membrane (28%) compared to the model (12%). Finally, the four-transmembrane-segment model provides no explanation for the formation of a hydrophilic ion channel that necessarily spans the bilayer.

The relative permeability of various organic cations and nonelectrolytes suggests that the AChR channel is a water-filled pore containing hydrogen-bonding groups and at least one negatively charged site (Huang et al. 1978). The 6.5-Å minimum diameter of the channel deduced from cationic selectivity (Furukawa and Furukawa 1959; Maeno et al. 1977; Dwyer et al. 1980) allows simultaneous occupation of even the narrowest part of the channel by an alkali metal cation and at least one water molecule (Horn and Stevens 1980). Also, C. Lewis and C.F. Stevens (in prep.) have concluded that the channel's selectivity region provides an environment essentially identical to free water. The AChR ion channel, then, is perhaps not formed by the very hydrophobic regions identified from hydrophobicity plots but rather by one or more segments that are hydrophobic on one face and hydrophilic on the other.

Ion Channel Formation and Subunit Assembly

To identify alternately hydrophilic/hydrophobic segments, we used Fourier analysis (McLachlan and Stewart 1976; Finer-Moore and Stroud 1983) to search for periodicities in hydrophobicity that characterize amphipathic secondary structure. We systematically examined the entire sequences of the four subunits in 25-residue stretches (an α-helix of 25 residues is 38 Å long, approximately the width of the lipid bilayer). The results are plotted in Figure 3 as two-dimensional contour maps with frequency (1/periodicity) as the abscissa and residue number (for the central residue in the stretch of 25) as the ordinate. Local periodicities of a given frequency are identified by peaks in the map at

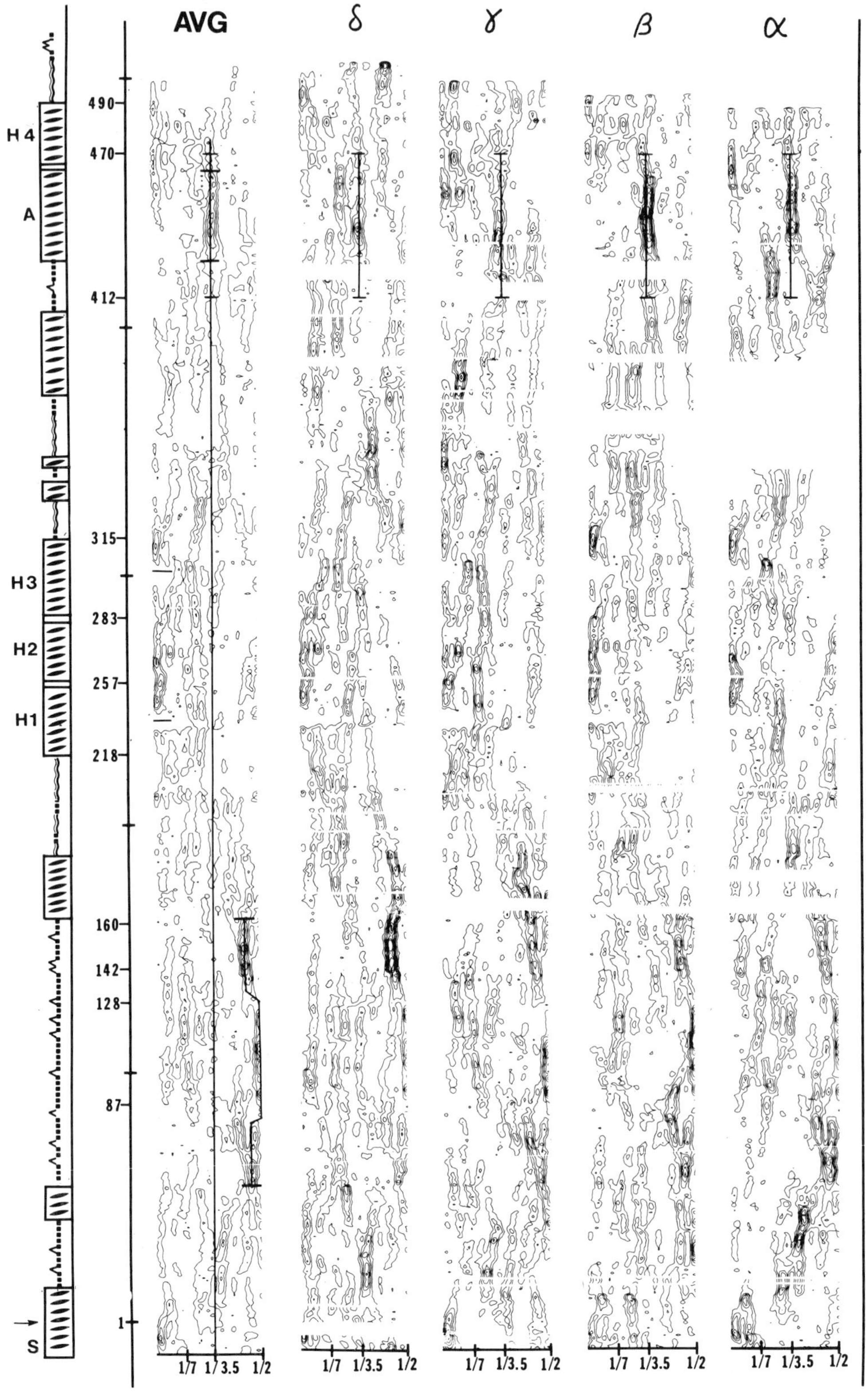

Figure 3. Fourier transform of the hydrophobicity values of individual amino acids in the α, β, γ, and δ sequences and averaged for all sequences (AVG). $I_k(\nu)$ (horizontal) is calculated for 25 residues around the sequence position k, shown on the ordinate. Sequences were aligned as in Table 1, and interruptions in the spectra represent gaps in sequences after alignment. An overall prediction (Finer-Moore and Stroud 1983) of α (\\\\ ), β (▪▪▪▪), turn (<), and coil (〰) secondary structure is shown to the left of the ordinate. (H1, H2, H3, H4) Hydrophobic regions; (A) amphipathic helix.

that frequency. The Fourier transforms for the four subunit sequences have been averaged and plotted in Figure 3, AVG, in order to highlight their common features.

The strong peak in Figure 3, AVG, centered at residue 440 with a frequency of 1/3.5 residues^{-1} indicates that the region between residues 412 and 470 has a pattern of hydrophobicities characteristic of amphipathic α-helices (3.5 residues/turn). Examination of the subunit sequences in this region shows that in each, residues 425 to 451 can form an α-helix with a continuous hydrophobic face on one side and a hydrophilic face on the other. The length of this segment is ideal for spanning the lipid bilayer. We propose that this segment is a transmembranous, channel-forming helix as displayed in the helical net diagram of Figure 4 (right). Figure 4 (left) displays the consequent five-crossing model, which places the carboxyl termini of the subunits on the cytoplasmic side of the membrane and has a distribution of protein mass among the synaptic (53%), cytoplasmic (22%), and lipid (25%) regions that is slightly more consistent with the composite three-dimensional structure of the AChR than the distribution in the four-helix model (Finer-Moore and Stroud 1983).

The amphipathic, channel-forming helices from the five subunits contain a total of 21 positive, 19 negative, and 10 uncharged hydrophilic groups. The remarkable balance of charges in this region and the clustering of the charged residues on the sides of the putative helices suggest that the charged residues could be involved in intersubunit ion pairing. Thus, as the five subunits assemble to form the receptor monomer, each would contribute an amphipathic helix to the wall of the central channel. The result is a tight complex with a central channel diameter of approximately 7 Å between five amphipathic helices, close to the diameter estimated from electrophysiological experiments for the narrowest part of the AChR ion channel (Furukawa and Furukawa 1959; Maeno et al. 1977; Huang et al. 1978; Dwyer et al. 1980). This arrangement also suggests an attractive model for multi-subunit assembly, in which the amphipathic helix and the last hydrophobic helix fold into the structure only when balanced by other subunits. This provides for insertion and assembly of a water-filled channel among the five subunits.

The Central Well Lining

The most intense features in the averaged Fourier map (Fig. 3) are peaks centered at residues 62, 110, and 146 with frequencies 1/2.3, 1/2.0, and 1/2.3 residues^{-1}, respectively. Eisenberg et al. (1983a) have correlated hydrophobicity periodicities in the range from 1/2.5 to 1/2.0 residues^{-1} with β-secondary structure in protein crystal structures. We reason that periodicities of 2.3 residues are characteristic of twisted strands of a β-barrel structure interacting with another

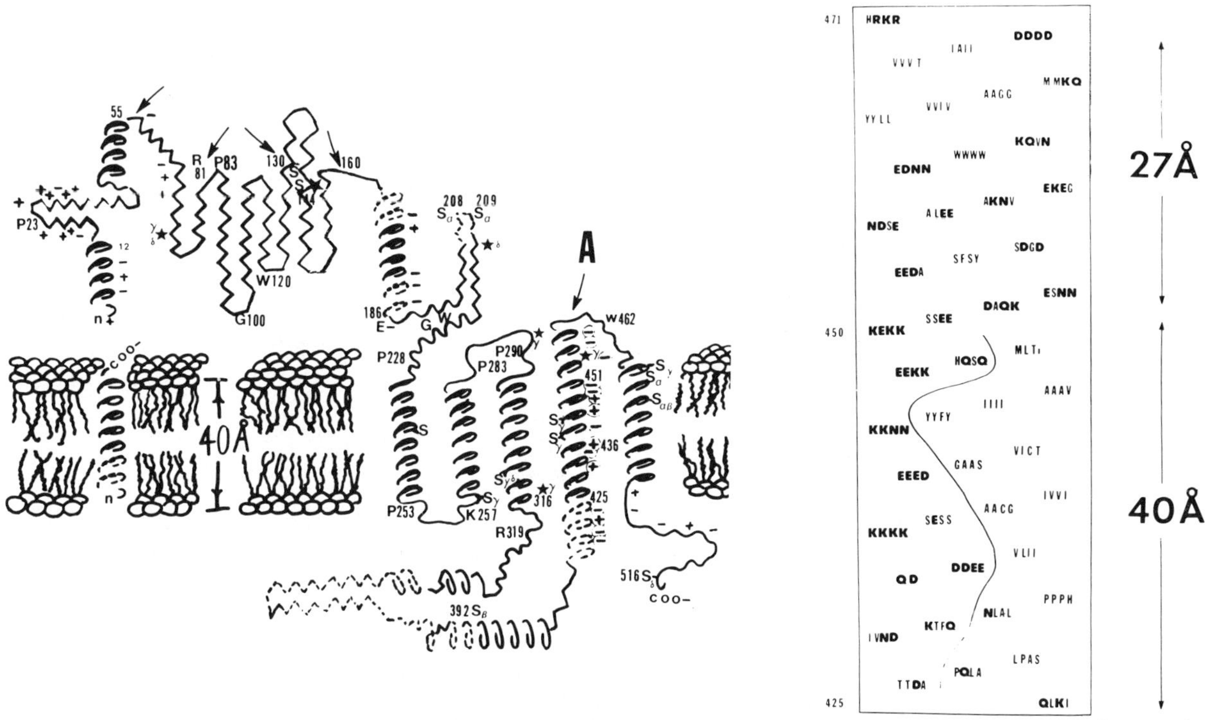

Figure 4. Schematic drawing showing predicted secondary structure for the receptor subunits and the proposed five-crossing model of polypeptide chain threading through the bilayer (Finer-Moore and Stroud 1983). (S) Sulfhydryl groups; (★) possible glycosylation sites; (A) predicted amphipathic helix; (→) boundaries of the 2.3- and 2.1-residue peaks of Fig. 3; (----) regions in the sequences that have gaps after alignment. Residues 425–470 in the four sequences are plotted on a helical net to the right of the threading scheme, using bold letters for the hydrophilic residues.

domain by hydrophobic forces, whereas periodicities of 2.0 are characteristic of strands in contact with solvent (Finer-Moore and Stroud 1983). Periodicity is usually less pronounced in β-strands than in an α-helix (Eisenberg et al. 1983a). The strong intensities of the features at residues 62, 110, and 145 are then an indication of an unusually amphipathic β-structure in the regions between residues 70 and 160. The large window size used for the Fourier analysis of the sequences may exaggerate the length of these features in the map and mask out stretches of reverse-turn secondary structure. The actual lengths of the β-strands, positions of turns in the sequence, and the arrangement of the β-strands in the tertiary structure are probably best predicted by direct inspection of the sequence and by model building.

The regions between residues 70 and 160 in the four subunit sequences are highly homologous (27 residues are totally conserved). These domains carry a total of 50 negative charges and only 36 positive charges for a net −14 charge. Both the four- and five-membrane-crossing schemes put this region on the synaptic side of the membrane. Included in this stretch is a site for *N*-glycosylation (nucleotides 143–145) and a candidate for the reducible disulfide 4-(*N*-maleimido)benzyltrimethylammonium (MBTA) iodide site (C130, C144) on the α-subunit (Noda et al. 1982). Thus, this stretch of amphipathic β-sheet common to all the subunits possibly forms a six- or eight-strand β-barrel domain that interfaces with solvent between residues 87 and 128 and with other (hydrophobic) protein surfaces for residues 51–87 and 128–160. The solvent side of the β-barrel could line the central well of the receptor. In the α-subunit a β-bend in this barrel near disulfides 130–144 could form part of the ACh-binding site as suggested by Noda et al. (1982). The excess negative charge in the vestibule of the receptor might then contribute to the cation specificity of the receptor.

Subunit Localization in AChR Monomer

Sequence homology among the four different receptor polypeptides emphasizes the similar role each subunit plays in the organization of the receptor structure. Hence, when viewed from above the membrane

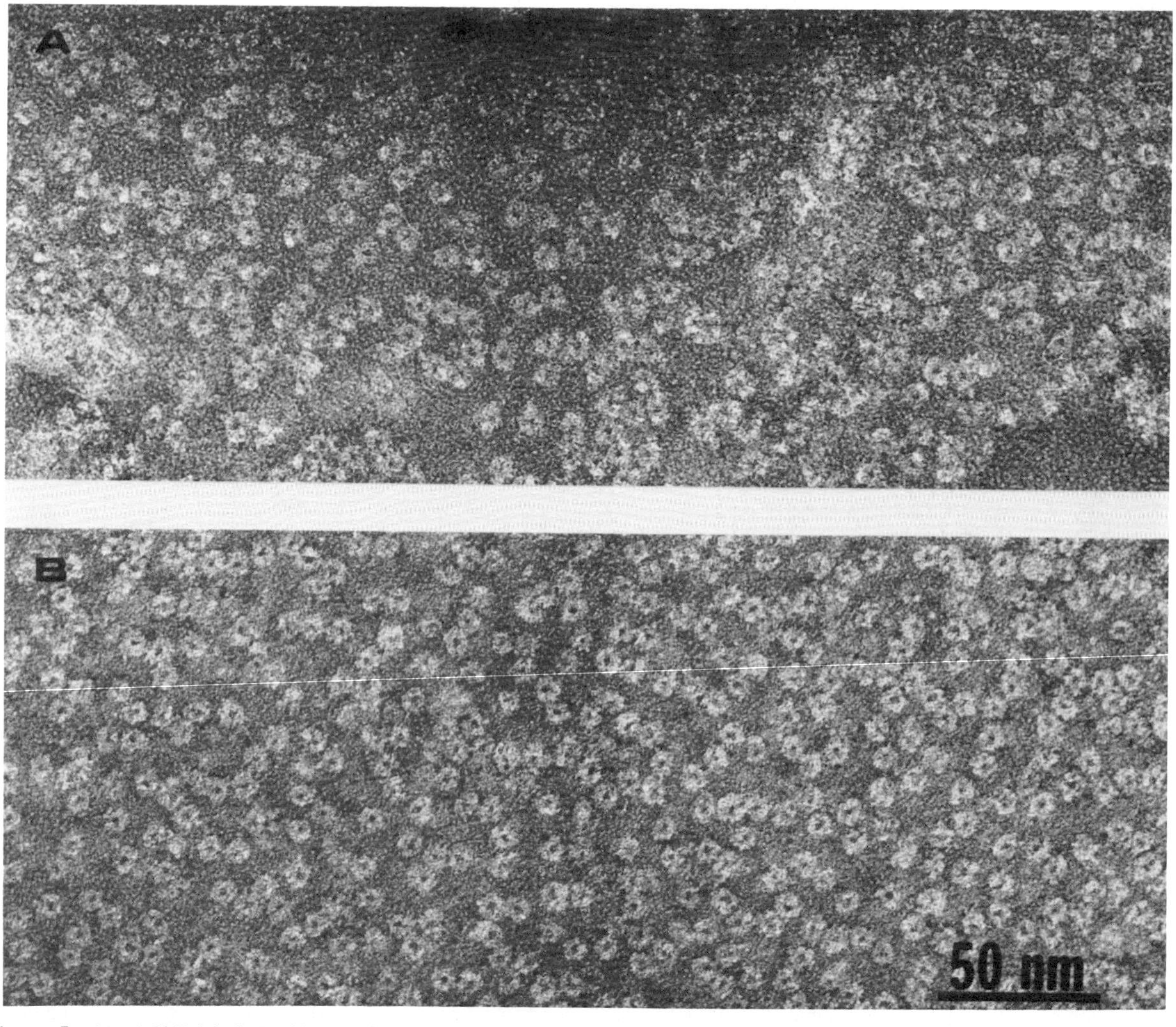

Figure 5. (*A*) AChR labeled with Fab fragments against the α-subunit; Fabs appear as white, stain-excluding regions on the receptor rosettes. (*B*) Unlabeled AChR for comparison. Magnification, 410,000×.

plane at low resolution, the five receptor subunits occupy nearly equivalent 72° sector positions around the central well. At higher resolution, the strict $\alpha_2\beta\gamma\delta$ subunit stoichiometry implies a unique environment for each of the five subunits. In collaboration with Dr. J. Lindstrom of The Salk Institute, we are using subunit-specific, monoclonal Fab fragments to visualize the relative locations of the subunits in electron micrographs. In Figure 5A, Fab fragments specific to the α-subunit appear as stain-excluding white dots on the membrane-bound AChR rosettes (D. Kristofferson et al., in prep.). For comparison, Figure 5B is a micrograph of AChR in the absence of Fab fragments.

The angles measured between two Fabs attached to the same rosette are distributed as in Figure 6C when the recorded angle is taken to be the smaller of the two possible angles. The mean of this distribution is 144° with a standard deviation $\sigma = 22°$ and an estimate of the error in the mean of 4° ($\epsilon = \sigma/\sqrt{n}$). For five equivalent 72° sector subunits, one expects nearest neighbors to be separated by 72° and next-nearest neighbors, by 144°. Our result clearly locates the α-subunits as noncontiguous, next-nearest neighbors about the receptor rosettes. Holtzman et al. (1982) came to a similar conclusion by using biotinylated cobra toxin complexed to receptors and visualizing the toxins with avidin by negative-stain electron microscopy. However, the inter-α-subunit angle they measured was 100°, in contrast to our value of 144° ± 4°.

Orientation of Monomers in AChR Dimer

Our micrographs display δ-δ covalent dimers with Fabs on each monomer, some with Fabs on the same side of the line of centers (Fig. 6A, positions 1 and 2) and some with Fabs on opposite sides (Fig. 6A, positions 1 and 3). This indicates that all α-subunits are accessible to the Fabs. We determined the distribution of α-subunits relative to the dimer line of centers by the angle-measuring convention described in Materials and Methods. The sample consisted of 86 Fabs on singly, doubly, and triply labeled dimers. The result is the bimodal distribution of angles displayed in Figure 6B, with means of the two modes being 99° and 275° ($\sigma = 30°$ in each case, and $\epsilon = 5°$). The angular difference of the means, 176° ± 7°, is substantially larger than the 144° inter-α angle obtained above. The 176° difference in the means thus reflects a structural feature of receptor dimers other than the inter-α angle of a monomer. This feature is the relative orientation of the monomers in a dimer. A C2 axis at the midpoint of the line of centers between two AChR monomers in a dimer would place equivalent α-subunits at positions 1 and 3 as well as at 2 and 4 of Figure 6A. Measuring Fab angles on such a system would result in a bimodal distribution of Fab angles with a mean difference of 144° (or 216°). These are quite different from our measured difference of 176° ± 7°. Dimers with monomers related to their partners by translation along the line of centers with little or no rotation would have

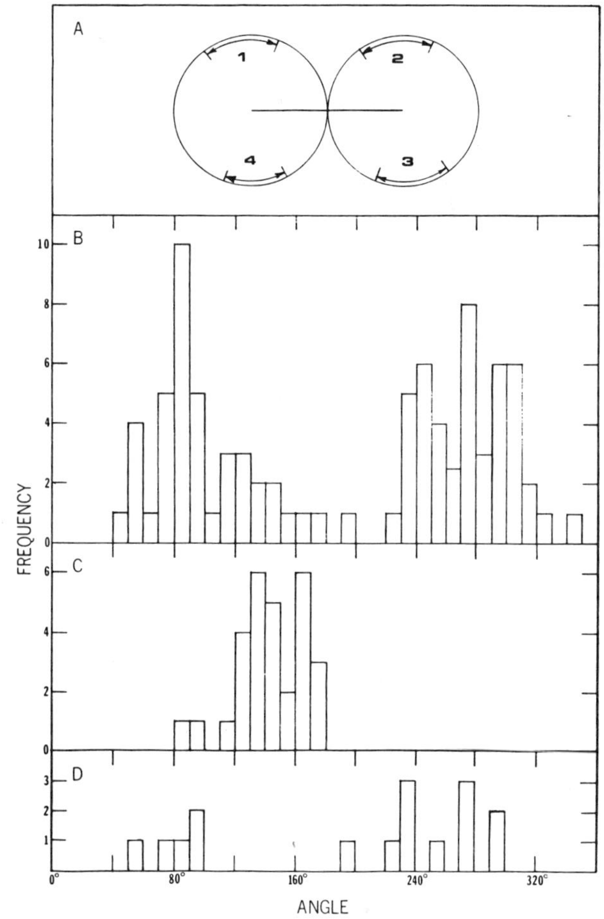

Figure 6. (*A*) Schematic of AChR dimer, illustrating the average range of Fab positions derived from the raw data in *B*. (*B*) Fab positions determined by the counterclockwise angle around a rosette from the line of centers in the dimer to a Fab. The average angle for the lower distribution is 99°, $\sigma = 32°$, $n = 40$, and $\sigma/\sqrt{n} = 5°$. The average of the higher distribution is 275°, $\sigma = 28°$, $n = 46$, and $\sigma/\sqrt{n} = 4°$. (*C*) Inter-α-subunit angle determined by the smallest angle between two Fabs on the same monomer. The average angle is 144°, $\sigma = 22°$, $n = 29$, and $\sigma/\sqrt{n} = 4°$. (*D*) Corrected position of a single α-subunit in doubly labeled monomers in dimers. If the smallest angle between the Fabs faces the other monomer in the dimer, the smallest counterclockwise Fab angle is recorded. If the inter-Fab angle faces away from the neighboring monomer, the larger counterclockwise angle is used. The average angle and σ and n for each of the distributions are 81°, 16°, 5, and 254°, 30°, and 11, respectively. See text for the discussion of dimer symmetry.

equivalent α-subunits at positions 1 and 2 and at 3 and 4 of Figure 6A. Such an arrangement also predicts a bimodal distribution of measured Fab angles, but now the mean difference would be 180°, very close to our measured value of 176° ± 7°. This difference is not significantly altered by the presence of a subpopulation of C2-related dimers in the field. Evidence for such a subpopulation (<25%) is presented in Figure 6D. Monomers with two Fabs attached were preferentially distributed such that the inter-Fab angle faced away from the other monomer twice as frequently as it faced its neighbor. This emerges from the population of 11 in the second mode of the distribution of Figure 6D compared with the population of 5 in the first mode. This statistic is most easily accounted for by a mixture of translationally related dimers with a subpopulation of C2-related dimers. Kistler and Stroud (1981), who reconstructed images of AChR arrays in 85% dimer-rich preparations, and Bon et al. (1982), who rotationally aligned digitized images of monomer partners, also concluded that the major preferred packing in δ-δ dimeric AChR is a translational relation between the two monomers: R–R as opposed to R–ꓤ.

Flexibility of the δ-δ Cross-link

A translational relation between monomer partners suggests that the δ-δ cross-link would have to be long (>90 Å) and flexible. The Fab angle distributions displayed in Figure 6B are broadened by three major contributions, one of which is the δ-δ flexibility. The two others are (1) the variability in locating the Fab itself and (2) the systematic broadening resulting from two possible dimer orientations (R–R and ꓤ–ꓤ) and the adopted angle-measuring convention. This systematic broadening can be removed by dimer orientation and a new angle convention as described in Materials and Methods. The $\sigma = 27°$ of the corrected angle distribution now depends on σ_α, the variability in Fab angle determination, and σ_δ, the δ-δ flexibility: $\sigma^2 = \sigma_\alpha^2 + \sigma_\delta^2$. The $\sigma = 22°$ from the inter-α-subunit measurements (see Subunit Localization in AChR Monomer) can be used to calculate σ_α, since these measurements are not affected by σ_δ. The value of σ_α is $22°/\sqrt{2} = 16°$, and one can then calculate the rotational flexibility of the dimer bond: $\sigma_\delta = 22°$ (D. Kristofferson et al., in prep.). Flexibility about the δ-δ cross-link alone contributes ±22° variation in Fab angle measurements.

Localization of Toxin-binding Sites

The AChR binding site for the 8000-dalton snake neurotoxic protein α-bungarotoxin was localized by immuno-electron microscopy on the synaptic side of receptor-enriched membranes; antitoxin antibodies complexed with colloidal gold were used to highlight the toxin (Klymkowsky and Stroud 1979).

Small-angle X-ray diffraction from oriented, receptor-enriched membranes in the presence and absence of α-bungarotoxin provides evidence for the location of the toxin-binding site on the receptor, perpendicular to the membrane plane. The X-ray patterns of oriented membranes with and without α-bungarotoxin are presented in Figure 7. These two patterns exhibit similar sampling contributions to diffraction on the vertical axis, unlike those patterns published by Kistler et al. (1982). Direct comparison of the patterns and further quantitative analysis are facilitated by such similar sampling.

The toxin disrupts the 1/79-$Å^{-1}$ equatorial reflection arising from packing of receptors in the plane of the membrane and alters the *d* spacing of the meridional

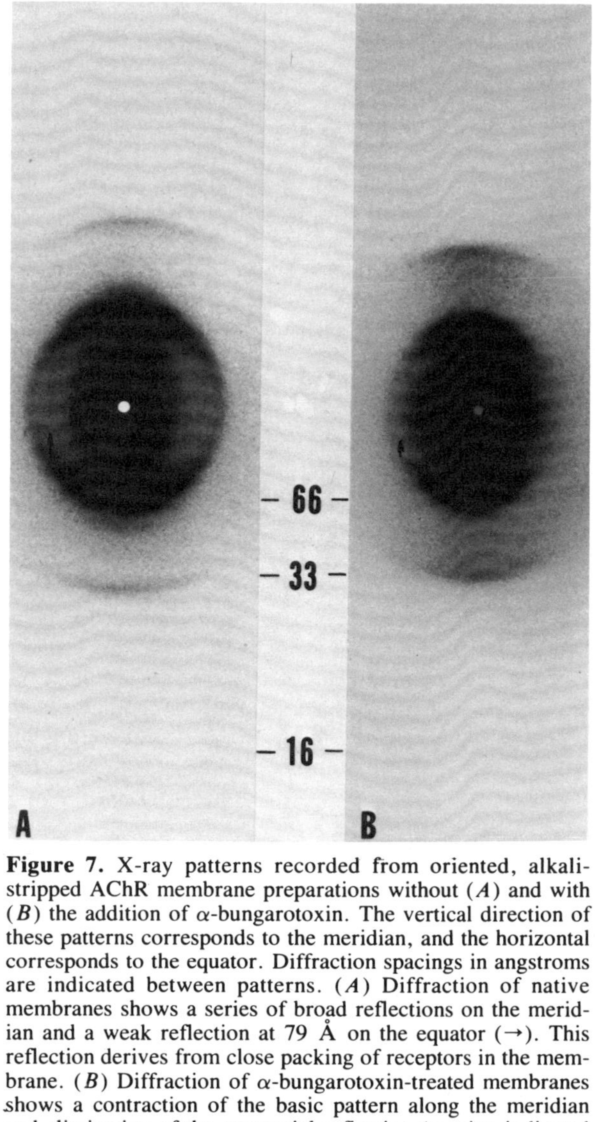

Figure 7. X-ray patterns recorded from oriented, alkali-stripped AChR membrane preparations without (*A*) and with (*B*) the addition of α-bungarotoxin. The vertical direction of these patterns corresponds to the meridian, and the horizontal corresponds to the equator. Diffraction spacings in angstroms are indicated between patterns. (*A*) Diffraction of native membranes shows a series of broad reflections on the meridian and a weak reflection at 79 Å on the equator (→). This reflection derives from close packing of receptors in the membrane. (*B*) Diffraction of α-bungarotoxin-treated membranes shows a contraction of the basic pattern along the meridian and elimination of the equatorial reflection (spacing indicated by arrow) observed in untreated membranes.

reflections. Hence, α-bungarotoxin disrupts both in-plane packing of receptors in the bilayer and the interplane stacking of membrane vesicles. This suggests a top, peripheral location of the toxin molecule on the synaptic crest of the receptor head. Constrained iterative refinement (Stroud and Agard 1979) of the meridional diffraction amplitudes presented in Figure 8A gives rise to the electron-density profiles perpendicular to the 40-Å membrane bilayer with and without α-bungarotoxin, as shown in Figure 8B (R.H. Fairclough et al., in prep.). The synaptic (S) and cytoplasmic (C) sides of the membrane are indicated in the figure. The 7–8-Å extension of electron density in α-bungarotoxin-treated membranes above the top of the synaptic head of untreated membranes most likely represents the presence of α-bungarotoxin. An alternative interpretation, that the 7–8-Å extension results from a toxin-induced structural change of the receptor,

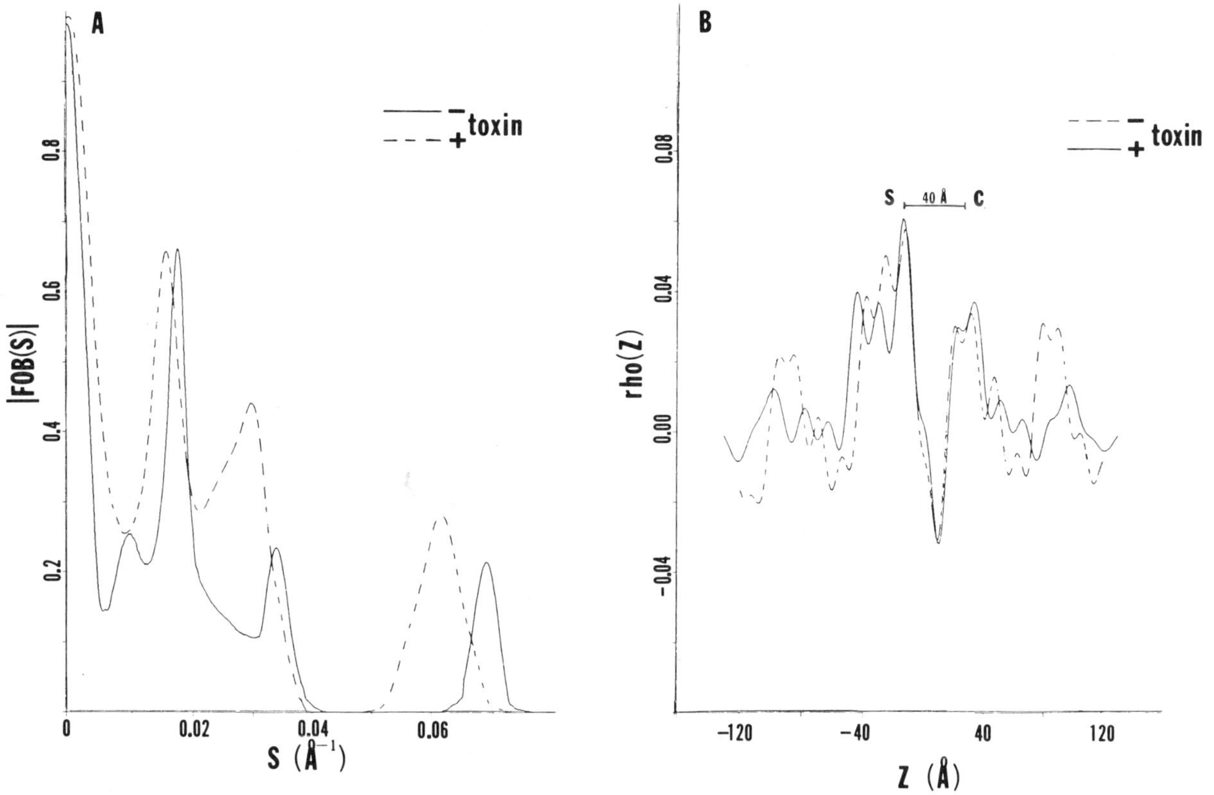

Figure 8. (*A*) Observed amplitudes of the meridional data of Fig. 7 with (– – –) and without (——) toxin. In the presence of toxin, native peaks move to smaller values of *s*, the scattering variable given by $s = 2\sin\theta/\lambda$, in which θ = Bragg angle and λ = X-ray wavelength. (*B*) The electron-density profiles relative to water in $e/Å^3$ along the direction perpendicular to the bilayer. These profiles for membranes with and without toxin are derived by constrained iterative refinement of amplitude data in *A*.

seems unlikely because *d*-tubocurarine, a 700-dalton organic AChR antagonist, causes none of the changes in the X-ray diffraction pattern of oriented membranes that α-bungarotoxin does (data not shown). In addition, Dunn and Raftery (1982) show that the α-bungarotoxin binds to AChR membrane vesicles labeled with 4-[*N*-(iodoacetoxy)ethyl-*N*-methyl]amino-7-nitrobenz-2-oxa-1,3-diazole (NBD) without altering the fluorescence of this agonist-sensitive probe. We conclude that α-bungarotoxin contributes the 7–8-Å extension of electron density on the synaptic crest of toxin-treated membranes (R.H. Fairclough et al.. n prep.). Zingsheim et al. (1982) and Changeux et al. (this volume), who localized α-bungarotoxin in negatively stained electron micrographs of AChR, also observe the toxin on the synaptic head.

The α-Bungarotoxin/AChR Interface

The binding of α-bungarotoxin to the AChR blocks the receptor's response to ACh. This binding occurs with a 2:1 stoichiometry, the same as that observed for ACh-induced channel activation (Neubig and Cohen 1980). It is also the stoichiometry of the α-subunit in the receptor itself. Competition between cholinergic agonists and neurotoxins for AChR suggests that the toxin-binding region of AChR includes the ACh-binding site. Recently, Tzartos and Changeux (1983) reported high-affinity binding of α-bungarotoxin to purified α-subunit but not β-, γ-, or δ-subunits of AChR. It is possible to map the surface domain of AChR, which must complement the extensive surface of the α-bungarotoxin structure determined at 2.5-Å resolution by X-ray crystallography (R. Love et al., in prep.). This map of the receptor surface can be used as a model of the α-subunit surface in the neighborhood of the competitive agonist site.

The α-bungarotoxin molecule is a flat, hand-shaped protein of 74 amino acids and overall dimensions of 40 Å × 30 Å × 20 Å. The toxin is composed of three polypeptide loops projecting 20 Å, 25 Å, and 10 Å from a central core of four disulfide bridges near one end. The two longest loops are held together by β-sheet hydrogen bonds. The three loops together form a curved surface. On its concave side, in the crystal structure, are side chains of functionally important residues as identified by sequence homology among snake neurotoxins or by chemical modification coupled to toxicity assays (Karlsson 1979; Low 1979; Dufton and Hider 1983). These residues map over an area 20 Å × 30 Å. Very similar structural motifs are observed in other snake neurotoxin structures: cobra toxin (Walkinshaw et al. 1980) and erabutoxin (Low et al. 1976; Tsernoglou and Petsko 1976). Direct evidence for side chains of the

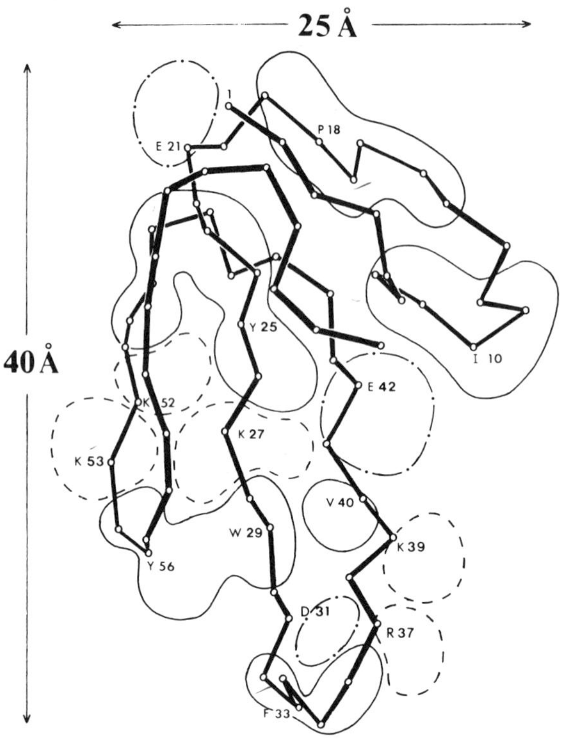

Figure 9. α-carbon skeleton of a bound α-bungarotoxin molecule, as seen from above the toxin-receptor complex. Side chains, the carboxyterminal seven residues, and disulfide bonds are not shown; the Karlsson numbering scheme is employed. Visible through the toxin is the expected complementary surface of AChR, as indicated by contours. Each contour encloses a region on AChR of all possible contacts for a certain type of toxin side chain(s). This allows for flexibility of these side chains about their crystal structure positions. (——) Contours of hydrophobic and occasional hydrogen-bonding domains on AChR that are contacted by similar type side chains of toxin; these domains are the most frequent and in the figure are often merged to show stretches of such toxin residues. (– – –) Contours of negatively charged domains that are contacted by positive toxin side chains; now each contour corresponds to a specific toxin side chain. (— · —) Contours of positively charged domains. The interface is actually curved in a convex manner, with the outer loops contacting AChR farther down than the central loop. See text for a discussion of specific toxin residues.

concave surface making contact with AChR comes from derivatizing specific residues with electron paramagnetic resonance (EPR) or fluorescent probes and monitoring changes in probe signal after the toxin binds (Tsetlin et al. 1979, 1982; Cheung et al. 1983). Also, photo-induced cross-linking of specific toxin lysines to AChR indicates that only concave-surface residues are in contact (Tsetlin et al. 1982).

The α-bungarotoxin-binding region on AChR will be complementary to the toxin's concave binding surface. The AChR surface is mapped according to all possible contacts with each toxin side chain by allowing these side chains their maximum flexibility while keeping the backbone fixed in the crystal conformation. Each part of the AChR surface is then characterized as being positive, negative, hydrophobic, or hydrogen bonding on the basis of the most attractive interaction with the respective toxin side chain. The result is presented in Figure 9 with the toxin backbone overlaying the receptor surface. Only small readjustments in the other neurotoxin structures would be needed in order to place their invariant, functionally important side chains onto this AChR surface. Qualitatively, the surface provides manifold interactions, which result in the specific and tight binding of α-bungarotoxin.

Near the end of the long central loop in α-bungarotoxin, the highly conserved side chains of Asp-31 and Arg-37 form a triangle with the side chain of Phe-33 (position 33 is Phe, His, or Trp in toxins). This arrangement of side chains bears a resemblance to the structure of cholinergic agonists (Tsernoglou et al. 1978; Kimball et al. 1979). The interatomic distance between the positive quaternary ammonium of ACh and its hydrogen-accepting carbonyl is about 5 Å (Beers and Reich 1970). In α-bungarotoxin, the charges on Asp-31 and Arg-37 are separated by 4 Å, with no steric restriction preventing separation of these charges up to 6 Å. Chemical modification of the invariant Arg-37 reduces toxicity by 75% in neurotoxins (Yang et al. 1974). Thus, an anionic site on the AChR under the Arg-37 likely forms part of the ACh-binding site.

Other important, negatively charged sites on the AChR lie under Lys-27, Lys-53, and Lys-39. Acetylation experiments have revealed that Lys-27 and Lys-53 play major electrostatic roles in toxin binding, and Lys-39 is invariant among long neurotoxins. The centers of the areas under Lys-27 and Lys-53 are separated by 9 Å, and they are 16 Å and 21 Å, respectively, from the agonist anionic site. EPR probe labeling of Lys-27 suggests that in bound toxins this side chain is near an AChR disulfide (Tsetlin et al. 1982). A strong candidate for a disulfide in the toxin-binding region of AChR is the MBTA-reactive disulfide, since toxins compete with MBTA. The MBTA-reactive disulfide is known to be about 10 Å from the anionic agonist site on AChR (Karlin 1980); 10 Å is a possible distance between two points lying in the areas mapped under Lys-27 and Arg-37.

Positively charged areas of AChR seem to play a lesser role in the binding of α-bungarotoxin. The cationic site under Asp-31 may only be a hydrogen donor toward the carbonyl of ACh. But two major cationic sites lie under Glu-21 and Glu-42, since these positions are almost always occupied by acidic side chains in neurotoxins. The Glu-42 area along with the anionic areas under Lys-27, Lys-52, and Lys-53 form a 5–7-Å-wide patch of ionic character across the AChR toxin-binding surface.

Several functionally important aromatic residues in α-bungarotoxin will map into hydrophobic areas on the AChR. Tyr-25 is invariant among all neurotoxins, and its nitration significantly reduces toxicity. Thus, the complementary surface for Tyr-25 will be generally hydrophobic with perhaps a hydrogen-bonding group. A hydrophobic region extends from Tyr-25 to include the area under residues 46–50. Another hydrophobic region lies under the toxin's short loop. These two regions together form a row 6–8 A wide of hydrophobicity across the AChR toxin-binding surface just above the ionic patch.

A second hydrophobic row is mapped under Tyr-56 across to Val-40. Between the two residues lies the invariant Trp-29, whose chemical modification leads to substantial reduction of toxicity in all neurotoxins. In the α-bungarotoxin structure, the Trp-29 indole ring is found on the back, convex side of the molecule. However, we find that with the α-bungarotoxin–AChR complex, Trp-29 is totally protected from the tryptophan-specific reagent 2-hydroxy-5-nitrobenzyl bromide, whereas with free α-bungarotoxin the Trp-29 is easily labeled (R. Love and R.M. Stroud, in prep.). Thus, bound α-bungarotoxin probably has its Trp-29 side chain on the concave side, as in the other two toxin crystal structures.

The general pattern that emerges from the map of the AChR surface is one of alternate rows of hydrophobic and ionic character every 5–8 Å, starting from the agonist anionic site.

SUMMARY AND CONCLUSIONS

We have learned the positions of the α-subunits around the AChR rosette and the location of the toxin on the synaptic crest. A charge/hydrophobic character map of the 40 Å × 30 Å receptor surface that binds α-bungarotoxin has been constructed. A β-structure domain surrounds the agonist binding site on the α-subunits, as predicted by amphipathic Fourier sequence analysis. The ion channel may be constructed from five amphipathic helices, which insert into the bilayer as the $\alpha_2\beta\gamma\delta$ subunits come together. They form a water-filled channel on one side and interface with hydrophobic helices in each subunit on the other.

ACKNOWLEDGMENTS

We thank Steve Heinemann, Jim Patrick, and Toni Claudio for communicating their sequence data to us ahead of full publication. The studies using monoclonal antibodies are in collaboration with Jon Lindstrom (Salk Institute), who provided the Fab fragments of monoclonal antibody no. 35. We thank Robert Sweet and David Eisenberg for helpful discussions. We thank Melvin O. Jones for maintenance of the Elliott GX-6 rotating anode X-ray generator, and Peter Stroud for help with the illustrations. R.H.F. thanks S. Doniach and K. Hodgson of Stanford University for advice and partial support through National Institutes of Health grant GM-25217. J.F.-M. acknowledges support through NIH training grant 5-T32-CA-09270. D.K. acknowledges support from the Damon Runyon–Walter Winchell Cancer Fund (DRG-587). P.J.D. was supported by NIH fellowship NRSA-5-F32-AM-06602. The research reviewed here was supported by grant PCM-80-21433 from the United States National Science Foundation and by grant GM-24485 from the NIH to R.M.S.

REFERENCES

Anholt, R., J. Lindstrom, and M. Montal. 1983. The molecular basis of neurotransmission: Structure and function of the nicotinic acetylcholine receptor. In *Enzymes of biological membranes*, 2nd edition (ed. A. Martonosi). Plenum Press, New York. (In press)

Beers, W.H. and E. Reich. 1970. Structure and activity of acetylcholine. *Nature* **228:** 917.

Bon, F., E. Lebrum, J. Gomel, R. Van Rapenbusch, J. Cartaud, J.L. Popot, and J.-P. Changeux. 1982. Orientation relative des deux oligomeres constituant la forme lourde du recepteur de l'acetylcholine chex *Torpille marbree. C. R. Acad. Sci.* (ser III) **295:** 199.

Chang, H.W. and E. Bock. 1977. Molecular forms of acetylcholine receptor. Effects of calcium ions and a sulfhydryl reagent on the occurrence of oligomers. *Biochemistry* **16:** 4513.

Changeux, J.-P. 1981. The acetylcholine receptor: An allosteric membrane protein. *Harvey Lect.* **75:** 85.

Cheung, A.T., D.A. Johnson, and P. Taylor. 1983. Site-specific fluorescein-labeled cobra α-toxin: Kinetics of interaction with the acetylcholine receptor. *Biophys. J.* **41:** 137a.

Claudio, T., M. Ballivet, J. Patrick, and S. Heinemann. 1983. Nucleotide and deduced amino acid sequences of *Torpedo californica* acetylcholine receptor γ-subunit. *Proc. Natl. Acad. Sci.* **80:** 1111.

Conti-Tronconi, B.M. and M.A. Raftery. 1982. The nicotinic cholinergic receptor: Correlation of molecular structure with functional properties. *Annu. Rev. Biochem.* **51:** 491.

Conti-Tronconi, B., S. Tzartos, and J. Lindstrom. 1981. Monoclonal antibodies as probes of acetylcholine receptor structure. 2. Binding to native receptor. *Biochemistry* **20:** 2181.

Devillers-Thiery, A., J. Giraudat, M. Bentaboulet, and J.-P. Changeux. 1983. Complete mRNA coding sequence of the acetylcholine binding α-subunit of *Torpedo marmorata* acetylcholine receptor: A model for the transmembrane organization of the polypeptide chain. *Proc. Natl. Acad. Sci.* **80:** 2067.

Dufton, M.J. and R.C. Hider. 1983. Conformational properties of the neurotoxins and cytotoxins isolated from Elapid snake venoms. *CRC Crit. Rev. Biochem.* **14:** 113.

Dunn, S.M.J. and M.A. Raftery. 1982. Activation and desensitization of *Torpedo* acetylcholine receptor: Evidence for separate binding sites. *Proc. Natl. Acad. Sci.* **79:** 6757.

Dwyer, T.M., D.J. Adams, and B. Hille. 1980. The permeability of the endplate channel to organic cations in frog muscle. *J. Gen. Physiol.* **75:** 469.

Einarson, B., W. Gullick, B. Conti-Tronconi, and J. Lindstrom. 1982. Subunit composition of fetal calf muscle nicotinic acetylcholine receptor. *Biochemistry* **21:** 5295.

Eisenberg, D., R.M. Weiss, and T. Terwilliger. 1983a. The hydrophobic moment detects periodicity in protein hydrophobicity. *Proc. Natl. Acad. Sci.* (in press).

Eisenberg, D., P. Weiss, T. Terwillger, and W. Wilcox. 1983b. Hydrophobic moment and protein structure. *Faraday Symp. Chem. Soc.* **17:** (in press).

Finer-Moore, J. and R.M. Stroud. 1983. Amphipathetic analysis and possible formation of the ion channel in acetylcholine receptor. *Proc. Natl. Acad. Sci.* (in press).

Furukawa, T. and A. Furukawa. 1959. Effects of methyl- and ethyl-derivatives of NH4+ on the neuromuscular junction. *Jpn. J. Physiol.* **9:** 130.

Hamilton, S.L., M. McLaughlin, and A. Karlin. 1979. Formation of disulfide-linked oligomers of acetylcholine receptor in membrane from *Torpedo* electric tissue. *Biochemistry* **18:** 155.

Holtzman, E., D. Wise, J. Wall, and A. Karlin. 1982.

Electron microscopy of complexes of isolated acetylcholine receptor, biotinyl-toxin, and avidin. *Proc. Natl. Acad. Sci.* **79:** 310.

HORN, R. and C.F. STEVENS. 1980. Relation between structure and function of ion channels. *Comments Mol. Cell. Biophys.* **1:** 57.

HUANG, L.-Y.M., W.A. CATTERALL, and G. EHRENSTEIN. 1978. Selectivity of cations and nonelectrolytes for acetylcholine-activated channels in cultured muscle cells. *J. Gen. Physiol.* **71:** 397.

HUCHO, F., G. BANDINI, and B.A. SÚAREZ-ISLA. 1978. The acetylcholine receptor as part of a protein complex in receptor-enriched membrane fragments from *Torpedo californica* electric tissue. *Eur. J. Biochem.* **83:** 335.

KARLIN, A. 1980. Molecular properties of nicotinic acetylcholine receptors. In *The cell surface and neuronal function* (ed. C.W. Cotman et al.), vol. 6, p. 191. Elsevier/North-Holland Biomedical Press, Amsterdam.

KARLSSON, E. 1979. Chemistry of protein toxins in snake venoms. *Handbook Exp. Pharmacol.* **52:** 159.

KIMBALL, M.R., A. SATO, J.S. RICHARDSON, L.S. ROSEN, and B.W. LOW. 1979. Molecular conformation of erabutoxin b: Atomic coordinates at 2.5 Å resolution. *Biochem. Biophys. Res. Commun.* **88:** 950.

KISTLER, J. and R.M. STROUD. 1981. Crystalline arrays of membrane-bound acetylcholine receptor. *Proc. Natl. Acad. Sci.* **78:** 3678.

KISTLER, J., R.M. STROUD, M.W. KLYMKOWSKY, R.A. LALANCETTE, and R.H. FAIRCLOUGH. 1982. Structure and function of an acetylcholine receptor. *Biophys. J.* **37:** 371.

KLYMKOWSKY, M.W. and R.M. STROUD. 1979. Immunospecific identification and three-dimensional structure of a membrane-bound acetylcholine receptor from *Torpedo californica. J. Mol. Biol.* **128:** 319.

KYTE, J. and R.F. DOOLITTLE. 1982. A simple method for displaying the hydropathic character of a protein. *J. Mol. Biol.* **157:** 105.

LINDSTROM, J., J. MERLIE, and G. YOGEESWARAN. 1979. Biochemical properties of acetylcholine receptor subunits from *Torpedo californica. Biochemistry* **18:** 4465.

LOW, B.W. 1979. The three-dimensional structure of postsynaptic snake neurotoxins: Consideration of structure and function. *Handbook Exp. Pharmacol.* **52:** 213.

LOW, B.W., H.S. PRESTON, A. SATO, L.S. ROSEN, J.E. SEARL, A.D. RUDKO, and J.S. RICHARDSON. 1976. Three dimensional structure of erabutoxin b neurotoxic protein: Inhibitor of acetylcholine receptor. *Proc. Natl. Acad. Sci.* **73:** 2991.

MAENO, T., C. EDWARDS, and M. ANRAKU. 1977. Permeability of the endplate membrane activated by acetylcholine to some organic cations. *J. Neurobiol.* **8:** 173.

MCLACHLAN, A.D. 1971. Tests for comparing related amino-acid sequences: Cytochrome *c* and cytochrome c_{551}. *J. Mol. Biol.* **61:** 409.

MCLACHLAN, A.D. and M. STEWART. 1976. The 14-fold periodicity in α-tropomyosin and the interaction with actin. *J. Mol. Biol.* **103:** 271.

NEUBIG, R.R. and J.B. COHEN. 1980. Permeability control by cholinergic receptors in *Torpedo* postsynaptic membranes: Agonist dose-response relations measured at second and millisecond times. *Biochemistry* **19:** 2770.

NODA, M., H. TAKAHASHI, T. TANABE, M. TOYOSATO, Y. FURUTANI, T. HIROSE, M. ASAI, S. INAYAMA, T. MIYATA, and S. NUMA. 1982. Primary structure of α-subunit precursor of *Torpedo californica* acetylcholine receptor deduced from cDNA sequence. *Nature* **299:** 793.

NODA, M., H. TAKAHASHI, T. TANABE, M. TOYOSATO, S. KIKYOTANI, Y. FURUTANI, T. HIROSE, H. TAKASHIMA, S. INAYAMA, T. MIYATA, and S. NUMA. 1983a. Structural homology of *Torpedo californica* acetylcholine receptor subunits. *Nature* **302:** 528.

NODA, M., H. TAKAHASHI, T. TANABE, M. TOYOSATO, S. KIKYOTANI, T. HIROSE, M. ASAI, H. TAKASHIMA, S. INAYAMA, T. MIYATA, and S. NUMA. 1983b. Primary strucures of β- and δ-subunit precursors of *Torpedo californica* acetylcholine receptor deduced from cDNA sequences. *Nature* **301:** 251.

RAFTERY, M.A., M.W. HUNKAPILLER, C.D. STRADER, and L.E. HOOD. 1980. Acetylcholine receptor: Complex of homologous subunits. *Science* **208:** 1454.

REYNOLDS, J. and A. KARLIN. 1978. Molecular weight in detergent solution of acetylcholine receptor from *Torpedo californica. Biochemistry* **17:** 2035.

ROSS, M.J., M.W. KLYMKOWSKY, D.A. AGARD, and R.M. STROUD. 1977. Structural studies of a membrane-bound acetylcholine receptor from *Torpedo californica. J. Mol. Biol.* **116:** 635.

SABATINI, D.D., G. KREIBICH, T. MORIMOTO, and M. ADESNIK. 1982. Mechanisms for the incorporation of proteins in membranes and organelles. *J. Cell Biol.* **92:** 1.

STROUD, R.M. 1981. Structure of an acetylcholine receptor, a hypothesis for a dynamic mechanism of its action. In *Proceedings of the Second SUNYA Conversation in the Discipline of Biomolecular Stereodynamics* (ed. R.H. Sarma), vol. 2, p. 55. Adenine Press, New York.

———. 1983. Acetylcholine receptor structure. *Neurosci. Comment.* **1:** (in press).

STROUD, R.M. and D.A. AGARD. 1979. Structure determination of asymmetric membrane profiles using an iterative Fourier method. *Biophys. J.* **25:** 495.

SUMIKAWA, K., M. HOUGHTON, J.C. SMITH, L. BELL, B.M. RICHARDS, and E.A. BARNARD. 1982. The molecular cloning and characterization of cDNA coding for the α subunit of the acetylcholine receptor. *Nucleic Acids Res.* **10:** 5809.

TSERNOGLOU, D. and G.A. PETSKO. 1976. The crystal structure of post-synaptic neurotoxin from sea snake at 2.2 Å resolution. *FEBS Lett.* **68:** 1.

TSERNOGLOU, D., G.A. PETSKO, and R.A. HUDSON. 1978. Structure and function of snake venom curarimimetic neurotoxins. *Mol. Pharmacol.* **14:** 710.

TSETLIN, V.I., E. KARLSSON, A.S. ARSENIEV, Y.N. UTKIN, A.M. SURIN, V.S. PASHKOV, K.A. PLUZHNIKOV, V.T. IVANOV, V.F. BYSTROV, and Y.A. OVCHINNIKOV. 1979. EPR and fluorescence study of interaction of *Naja naja oxiana* neurotoxin II and its derivatives with acetylcholine receptor protein from *Torpedo marmorata. FEBS Lett.* **106:** 47.

TSETLIN, V.I., E. KARLSSON, Y.N. UTKIN, K.A. PLUZHNIKOV, A.S. ARSENIEV, A.M. SURIN, V.V. KONDAKOV, V.F. BYSTROV, V.T. IVANOV, and Y.A. OVCHINNIKOV. 1982. Interacting surfaces of neurotoxins and acetylcholine receptor. *Toxicon* **20:** 83.

TZARTOS, S. and J.-P. CHANGEUX. 1983. High affinity binding of α-bungarotoxin to the purified α-subunit and to its 27,000-dalton proteolytic peptide from *Torpedo marmorata* acetylcholine receptor. Requirement for sodium dodecyl sulfate. *EMBO J.* **2:** 381.

WALKINSHAW, M.D., W. SAENGER, and A. MAELICKE. 1980. Three-dimensional structure of the "long" neurotoxin from cobra venom. *Proc. Natl. Acad. Sci.* **77:** 2400.

YANG, C.C., C.C. CHANG, and I.F. LIOU. 1974. Studies on the status of arginine residues in cobrotoxin. *Biochim. Biophys. Acta* **365:** 1.

ZINGSHEIM, H.P., F.J. BARRANTES, J. FRANK, W. HANICKE, and D.C. NEUGEBAUER. 1982. Direct structural localization of two toxin-recognition sites on an ACh receptor protein. *Nature* **299:** 81.

The Nicotinic Acetylcholine Receptor: Subunit Structure, Functional Binding Sites, and Ion Transport Properties

M.A. RAFTERY, S.M.J. DUNN, B.M. CONTI-TRONCONI, D.S. MIDDLEMAS, AND R.D. CRAWFORD
Division of Chemistry and Chemical Engineering, California Institute of Technology, Pasadena, California 91125

The acetylcholine receptor (AChR) from *Torpedo* electroplax can be isolated in a pure form either in its native membrane-bound state or by affinity chromatography after solubilization. This is due to very dense packing of the AChR molecules in the postsynaptic membrane that excludes other intrinsic membrane components. Purified membrane fragments can reseal, forming closed, right-side-out vesicles, which can be used for functional and structural studies of the AChR.

The major physicochemical properties of *Torpedo californica* AChR are summarized in Figure 1. AChRs from electroplax of other *Torpedo* species have very similar properties (see Conti-Tronconi and Raftery 1982).

Protein Components of *Torpedo* AChR

It is now established that *Torpedo* AChR is formed by four different proteins that in *T. californica* have, upon SDS gel electrophoresis, apparent molecular weights of 40,000, 50,000, 60,000, and 65,000, and are commonly referred to as α, β, γ, and δ, respectively (see Conti-Tronconi and Raftery 1982). This complex subunit structure was first described by this laboratory (Raftery et al. 1974) and confirmed shortly thereafter by another group (Weill et al. 1974).

A model of AChR formed by four different subunits of which one (α) would be present in two copies to account for the M_r of the AChR complex (Reynolds and Karlin 1978) was challenged for many years by the persistent reports of pure AChR preparations composed of only one subunit of M_r 40,000 (Sobel et al. 1977). An alternative model of the AChR as a hexamer of six identical subunits was suggested (Sobel et al. 1979), and due to its symmetry and simplicity, it held a great measure of appeal.

The dilemma of the subunit composition of AChR was resolved by the demonstration that the four peptides present in pure AChR preparations are highly homologous proteins (Raftery et al. 1980). The aminoterminal sequences (until residue 54) of the four subunits of *T. californica* AChR are reported in Figure 1. Identical residues are present in 11 positions, and in many other positions either two or three of the residues are identical or related, so that in only 4 of the 54 positions do all four peptides contain different, unrelated amino acids.

By aminoterminal microsequencing, it was possible to demonstrate that these four subunits are present in the AChR molecule with a stoichiometry of $\alpha_2\beta\gamma\delta$ (Raftery et al. 1980). This could be determined directly by simultaneous quantitative sequencing of the mixture of polypeptides obtained by SDS denaturation of purified, intact AChR (Table 1). This approach eliminates the uncertainties arising from staining artifacts, proteolysis of higher-molecular-weight subunits, or incomplete elution from the gel.

The extensive homology of the four AChR subunits argues for a shared ancestry and allows the generation of a pseudosymmetric complex similar (Raftery et al. 1980) to that observed for many complex proteins (Matthews and Bernhard 1973).

The AChR preparations reported by others, which had a simpler subunit pattern, were due to proteolytic artifacts (see Conti-Tronconi and Raftery 1982). The subunit pattern of *Torpedo* AChR is very sensitive to proteolysis, and it can be cleaved to the extent that only the α-chain retains its integrity, without significant effect on the functional and structural characteristics of AChR, including its binding capacity for α-neurotoxins, antibodies, or cholinergic ligands (Lindstrom et al. 1980; Conti-Tronconi et al. 1982a). This indicates that extensive proteolytic cleavage of peptide bonds results in minimal fragmentation until the protein is denatured in SDS and indicates a strong, intimate interaction between the AChR subunits.

In the recent past, the complete amino acid sequence of the precursors of all *Torpedo* subunits has been deduced from corresponding nucleic acid clones obtained by recombinant DNA technology. The first of such deduced sequences has been for the α-subunit of *Torpedo marmorata* AChR (Sumikawa et al. 1982). Shortly thereafter, sequences were published for precursors of all the subunits of *T. californica* AChR (Noda et al. 1982, 1983a,b; Claudio et al 1983). The calculated M_r from the cDNA sequences of all four *T. californica* AChR subunits leads to an M_r of 270,000 for the intact AChR, in excellent agreement with our experimental value of $270{,}000 \pm 30{,}000$ for a preparation consisting of essentially all 9S AChR (Martinez-Carrion et al. 1975). Eventually the sequence of the precursor of the α-subunit of *T. marmorata* AChR was confirmed by another laboratory (Devillers-Thiery et al. 1983).

Nonequivalence of the Two α-Subunits

Each α-subunit has a high-affinity binding site for cholinergic ligands (see below) but the two sites are not

STRUCTURE

SUBUNIT COMPOSITION .... 40, 50, 60, 65 x 10^3 DALTONS
SUBUNIT STOICHIOMETRY 2:1:1:1
MOLECULAR WEIGHT .... 270 ± 30 x 10^3 DALTONS (EXPERIMENTAL)
.... 255 x 10^3 DALTONS (CALCULATED)

PHYSICAL PROPERTIES

S VALUE 9S
.... 13.7S AS DIMER
STOKES RADIUS .... 72 Å
pI 4.9
SPECIFIC ACTIVITY .... ONE α-BUTX PER 110 ± 15 x 10^3 DALTONS

SEQUENCE HOMOLOGY AT N-TERMINI OF ACETYLCHOLINE RECEPTOR SUBUNITS

2 4 6 8 10 12 14 16 18 20 22 24 26 28 30 32 34 36 38 40 42 44 46 48 50 52 54 56

40K S E H E T R L V A N L L E N Y N K V I R P V E H H T H F V D I T V G L Q L I Q L I S V D E V N Q I V E T N V

50K S V M E D T L L S V L F E T Y N P K V R P A Q T V G D K V T V R V G L T L T N L L I L N E K I E E M R T N V

60K E N E E G R L I E K L L G D Y D K R I I P A K T L D H I I D V T L K L T L T N L I S L N E M E E A L T T N V

65K V N E E E R L I N D L L I V N K Y N K H V R P V K H N N E V V N I A L S L T L S N L I S L K E T D E T L T S N V

CARBOHYDRATE .... ~75 RESIDUES/MOLECULE
O-SUBSTITUTED SER .... ~22 RESIDUES/MOLECULE
O-SUBSTITUTED THR .... ~23 RESIDUES/MOLECULE
PHOSPHOSERINE .... ~7 RESIDUES/MOLECULE
} OF AcChR

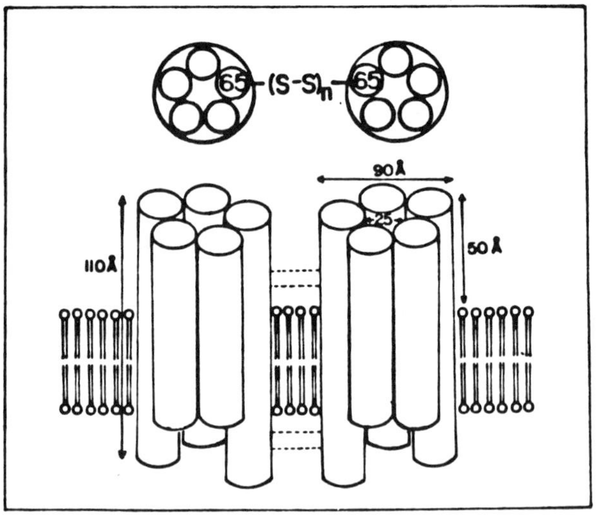

Figure 1. Properties of *Torpedo* AChR (see Raftery et al. 1980).

equivalent. Bromoacetylcholine (BrACh) can be used to label one or both α-subunits, whereas [4-(*N*-maleimido)benzyl] trimethylammonium diiodide (MBTA) labels only one (see Conti-Tronconi and Raftery 1982). The nonequivalence of these two sites can be explained by their different microenvironments, since each α-chain must be flanked by different subunits. In addition, the two α-subunits could be chemically different, in spite of their identical primary structure. We purified the α-subunits of *T. californica* by preparative SDS gel electrophoresis and submitted them to proteolysis using V8 protease. The peptide pattern obtained was identical to that described by Gullick et. al. (1981) and two prominent peptides of approximate M_r 17,000 and 19,000 (V8 17 and V8 19) were consistently present in similar amounts. It has been reported (Gullick et al. 1981) that only the V8 19 peptide can be labeled by MBTA under mild reducing conditions and that only the V8 17 peptide can be stained for carbohydrates. We have determined the aminoterminal sequence of all the peptides obtained after V8 digestion and found that V8 17 and V8 19 had the same aminoterminal sequence starting at residue 47. Since even extensive hydrolysis does not convert V8 19 to V8 17, it is likely that these two peptides differ only in their degree of glycosylation. These results indicate that the two α-chains differ in their extent of glycosylation. The binding site on the less glycosylated α-subunit is likely to be more accessible to reducing and affinity-labeling reagents. These results serve to explain the nonequivalence of high-affinity ligand binding to the α-subunits (see Conti-Tronconi and Raftery 1982).

The AChR Is a Transmembrane Protein

Since the AChR formed by the four subunits contains both the binding sites for cholinergic ligands and the cation gating unit (see below), it must span the postsynaptic membrane. A direct demonstration that the AChR molecule protrudes from both sides of the postsynaptic membrane has come from morphological studies of binding of the anti-AChR antibodies to the outside and inside surfaces of *Torpedo* postsynaptic membrane fragments. By examining replicas of intact and sheared membrane vesicles (Strader et al. 1979), we found that both surfaces were labeled by antibodies, in agreement with the results obtained by thin sectioning by Tarrab-Hazdai et al. (1978).

To determine which of the AChR subunits is a transmembrane protein, we studied the effect of proteases acting either inside or outside membrane vesicles containing all four AChR peptides (Strader and Raftery 1980; Conti-Tronconi et al. 1982a). All subunits were susceptible to tryptic degradation from either side of the membrane; but when trypsin was added outside, all four peptides disappeared at the same rate, whereas when the enzyme was acting from within the vesicles, they were degraded at a rate proportional to their molecular weight, with the α-chain being most resistant. Therefore, it may be concluded that all the AChR subunits span the membrane and that they protrude outside to the same extent and inside to an extent approximately proportional to their M_r (scheme of Fig. 1).

Investigation of the Exposure of AChR Subunits to the Lipid Bilayer

We have labeled all four AChR subunits in membrane fragments from *T. californica* using [^{3}H]adamantanediazirine (Fig. 2). This hydrophobic probe, which upon irradiation generates [^{3}H]adamantylidene, was proposed as an alternative to aryl nitrene probes (Bayley and Knowles 1978). Carbenes label both saturated and unsaturated membrane lipids more efficiently than nitrenes, suggesting they may be superior for labeling lipid-exposed regions of intrinsic membrane proteins (Bayley and Knowles 1980). As all four of AChRs' homologous subunits are in contact with the hydrocarbon core of the lipid bilayer, it is possible that all the subunits interact with the surrounding membrane in a related fashion (see Fig. 2).

Is the AChR from *Torpedo* a Good Model for Other Nicotinic AChRs?

Due to the ease with which *Torpedo* AChR can be purified, both its structure and its function have been investigated to a remarkable degree of sophistication. Study of the AChR from other sources has been hampered by difficulties in obtaining suitable amounts of intact AChR, due to the much lower AChR content and to high levels of protease activity. Similarities in the pharmacology, morphology, antigenicity, and physical properties as well as the frequent presence upon

Table 1. AChR Subunit Stoichiometry

(a) Torpedo californica *electric organ*

Subunit	Residues			Triton-solubilized AChR[a] preparation 1	Triton-solubilized AChR[a] preparation 2	Membrane-bound AChR[a] preparation 3	Membrane-bound AChR[a] preparation 4	Average
α	Ala-9	Asn-10	Asn-14	1.93 ± 0.13	1.92 ± 0.14	1.96 ± 0.04	2.05 ± 0.16	1.97 ± 0.12
β	Ser-9	Val-10	Thr-14	1.02 ± 0.08	1.07 ± 0.09	1.03 ± 0.04	1.02 ± 0.01	1.03 ± 0.06
γ	Glu-9	Lys-10	Asp-14	1.00 ± 0.10	1.02 ± 0.21	1.01 ± 0.03	1.00 ± 0.07	1.01 ± 0.10
δ	Asn-9	Asp-10	Val-14	1.04 ± 0.07	1.00 ± 0.13	1.01 ± 0.08	0.93 ± 0.08	0.99 ± 0.09

(b) Electrophorus electricus *electric organ*

Subunit	Residues	Preparation 1	Preparation 2	Average
α_I	Val-8, Gly-14	1.90 ± 0.19	1.96 ± 0.18	1.93 ± 0.10
β_I	Met-8, Ala-14	1.02 ± 0.02	0.99 ± 0.01	1.01 ± 0.02
γ_I	Ala-8, Tyr-14	1.10 ± 0.27	1.04 ± 0.12	1.07 ± 0.20
δ_I	Ile-8, Glu-14	1.02 ± 0.07	1.00 ± 0.05	1.04 ± 0.06

(c) *Mammalian muscle (fetal calf)*

Subunit	Residue (cycle 8)	Ratio
α	Val-8	2.16
β	Leu-8	0.95
x	Gln-8	0.92
y	Ile-8	0.98

[a]Values are means ± S.E.M..

SDS gel electrophoresis of complex polypeptide patterns (see Conti-Tronconi and Raftery 1982), reminiscent of the subunit pattern of *Torpedo* AChR, suggest the likelihood of close structural and functional similarities between the AChRs from different sources. Such evidence is crucial, because it would justify the use of analogy from *Torpedo* AChR to structural and functional characteristics of AChRs from other species.

Subunit Structure of the AChR from *Electrophorus electricus*

Torpedo (a marine elasmobranch) and *Electrophorus* (a freshwater teleost) are highly diverged species whose evolution arose separately from the primordial vertebrate stock (~400 million years ago), and accordingly the presence of electric organs in these two species is due to convergent evolution. Sufficient AChR can be isolated from *Electrophorus* to conduct structural analysis of its constituent subunits. In Figure 3, the SDS-polyacrylamide gel pattern and the morphology of the purified AChR are shown. It contains four main polypeptides in the same relative molecular weight range of *Torpedo* AChR, and the AChR molecules have the usual rosettelike morphology (see Conti-Tronconi and Raftery 1982). For each of the four peptides, the aminoterminal amino acid sequence was determined (Conti-Tronconi et al. 1982d) (Fig. 4[1]). The four subunits have distinct but homologous sequences, and the degree of identity between pairs of subunits ranges between 47.5% and 37.5%. Numerous conservative substitutions tend to increase further the degree of similarity among subunits. In Figure 4(2), comparison is made between the aminoterminal sequences of *Torpedo* and *Electrophorus* AChR of comparable relative molecular weight. The extent of sequence identity (up to 62.5%) is indicated in the figure. Although all subunits from both species are homologous polypeptides, the greatest level of identity was with the corresponding subunits of each species, i.e., α to α_I, etc. Simultaneous quantitative sequencing of the peptides present in preparations of intact *Electrophorus* AChR was used to determine the subunit stoichiometry (Table 1). Molar ratios of 2:1:1:1 were obtained for the subunits of M_r 41,000 (α_I), 50,000 (β_I), 55,000 (γ_I), and 62,000 (δ_I), respectively.

Electrophorus AChR is, therefore, a pseudosymmetric pentametric complex from four different polypeptides, and from their 2:1:1:1 stoichiometry, a M_r of 249,000 can be calculated that fits with experimental determinations and is consistent with the size of *Electrophorus* AChR, as determined by electron microscopy (see Conti-Tronconi and Raftery 1982).

The similarity in the primary structure of AChR subunits from *Torpedo* and *Electrophorus* explains the difficulties encountered in the past in obtaining antisera specific for the individual subunits; some degree of cross-reaction with other subunits was consistently obtained even using monoclonal antibodies (see Conti-Tronconi and Raftery 1982). The high degree of identity between corresponding AChR subunits of *Torpedo* and *Electrophorus* explains the extensive cross-reactivity of antisera and monoclonal antibodies raised against the subunits of either of these AChRs.

A genealogical tree showing the evolutionary pathway by which the four contemporary subunits of both *Electrophorus* and *Torpedo* AChRs can be generated from a single ancestral sequence via minimum nucleo-

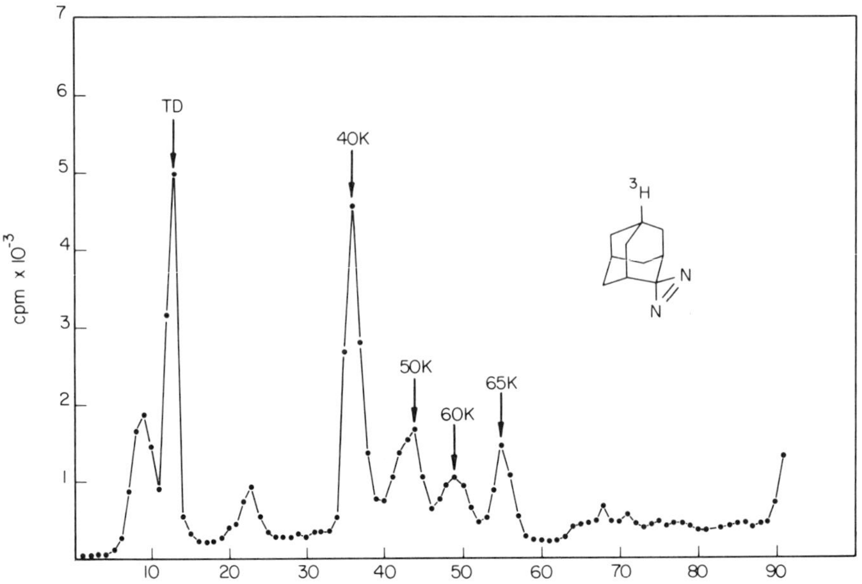

Figure 2. Polyacrylamide gel electrophoresis scan of AChR-enriched membrane fragments labeled with [^{3}H]adamantanediazirine (596 mCi/mmole). Membrane fragments were incubated in the dark with 33 μM [^{3}H]adamantanediazirine and 20 mM reduced glutathione followed by UV irradiation. After electrophoresis, the gel was sliced and counted. (TD) Tracking dye.

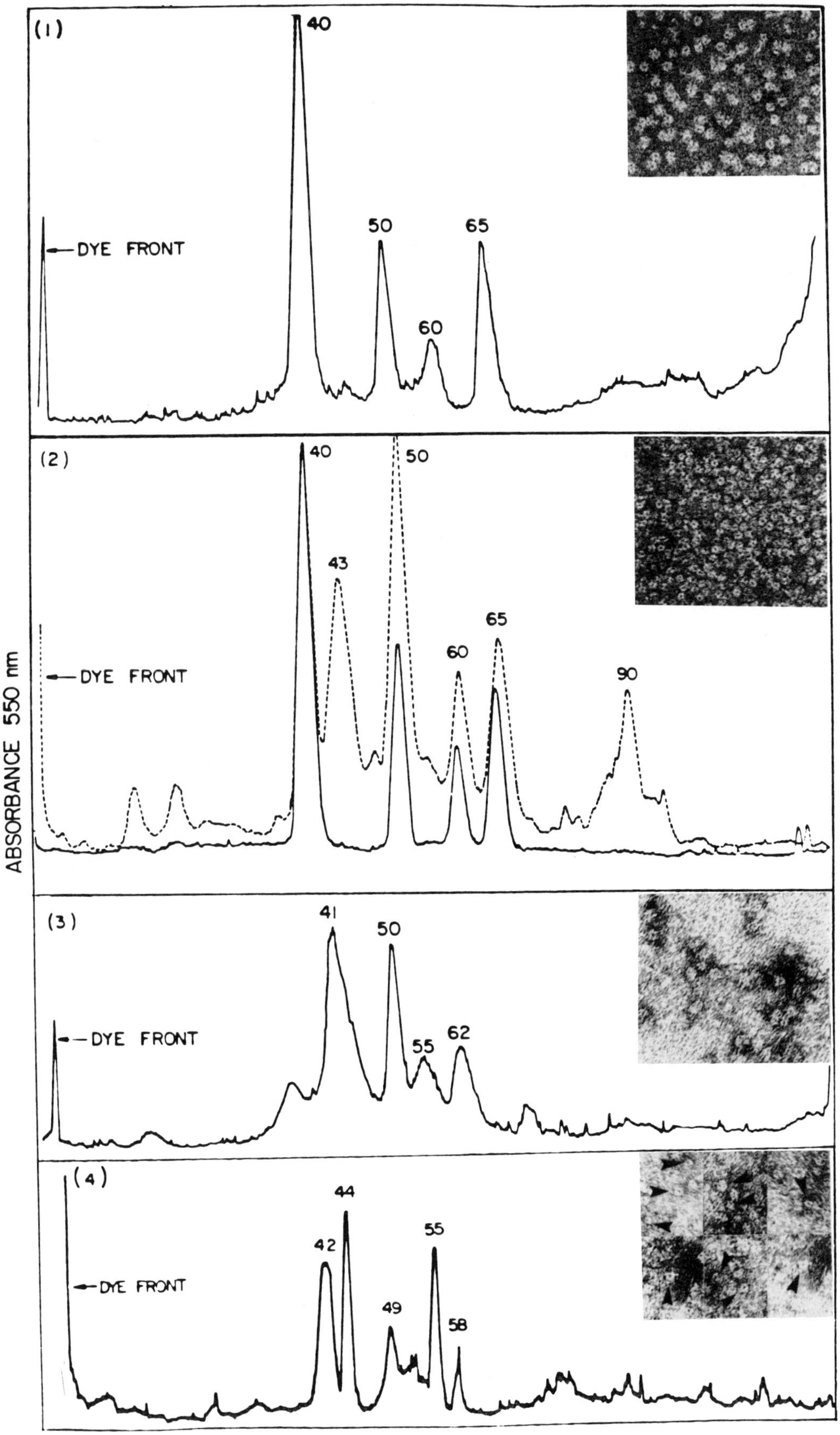

Figure 3. SDS gel electrophoresis scans of *T. californica* AChR (*1*), *Electrophorus* AChR (*3*), and fetal calf AChR (*4*). Lane *2*, SDS-gel scans of purified AChR-rich membrane fragments before (----) and after (——) alkali extraction (Elliott et al. 1980). All the extrinsic membrane proteins are removed by this treatment and the AChR subunits are the only polypeptides left. All gels were stained with Coomassie blue. Numbers shown are $M_r \times 10^{-3}$. (*Insets*) Electromicrographs of these AChRs (negative staining 1,3,4, 250,000×; 2, 125,000×).

tide substitution is shown in Figure 4(3). The tree demonstrates that the two gene duplications that produced the four subunits present in the AChR happened before the divergence of *Electrophorus* from *Torpedo* at the very beginning of vertebrate evolution, if not before then.

The Molecular Structure of Mammalian Muscle AChR

To elucidate the molecular structure of mammalian muscle AChR, we purified this receptor from fetal calf muscle. The purified solubilized AChR had physical properties similar to the *Torpedo* AChR monomers (Gotti et al. 1982). Negatively stained preparations of purified receptor show the presence of rosettelike structures, with a diameter of ~95 Å and an electron-dense central pit very similar to those found in *Torpedo* and *Electrophorus* AChR (Fig. 3).

Upon SDS gel electrophoresis, the purified AChR resolved into five major polypeptides having relative molecular weights of 42,000, 44,000, 49,000, 55,000, and 58,000 (Fig 3). The peptide of M_r 44,000 is actin because it contains 3-methyl-histidine and binds anti-actin antibodies and the peptide of M_r 42,000 is labeled by [^{3}H]BrACh (Conti-Tronconi et al. 1982c).

Each polypeptide of fetal calf AChR was submitted to aminoterminal sequence analysis (Conti-Tronconi et al. 1982c). Three subunits (M_r 42,000, 49,000, and 53,000) yielded distinct but homologous sequences. Due to the lack of identifiable sequences associated with the 55K and 58K polypeptides (most likely due to blockage of their amino termini during isolation), intact AChR preparations were analyzed. Four homologous sequences were obtained (Fig. 4). Three were identical with those independently determined and the fourth sequence could be deduced by difference. The four sequences were present in a stoichiometry of 2:1:1:1 (Table 1), which demonstrates that mammalian muscle nicotinic receptor is a pentameric complex composed of two equivalent and three pseudoequivalent subunits.

In Figure 4, comparison is made between the aminoterminal sequences of the two lighter subunits of AChR from fetal calf muscle, *Torpedo* and *Electrophorus*. Among the α-subunits, 51% of the residues were identical, and, in an additional 36% of the positions, two of the three polypeptides had the same residue. In the case of β-subunits, 38% of the residues were identical. A comparison of all six polypeptide sequences shows that 23% of all positions were identical for the first 26 residues and an additional 19% identity was observed in five out of the six polypeptides at other positions.

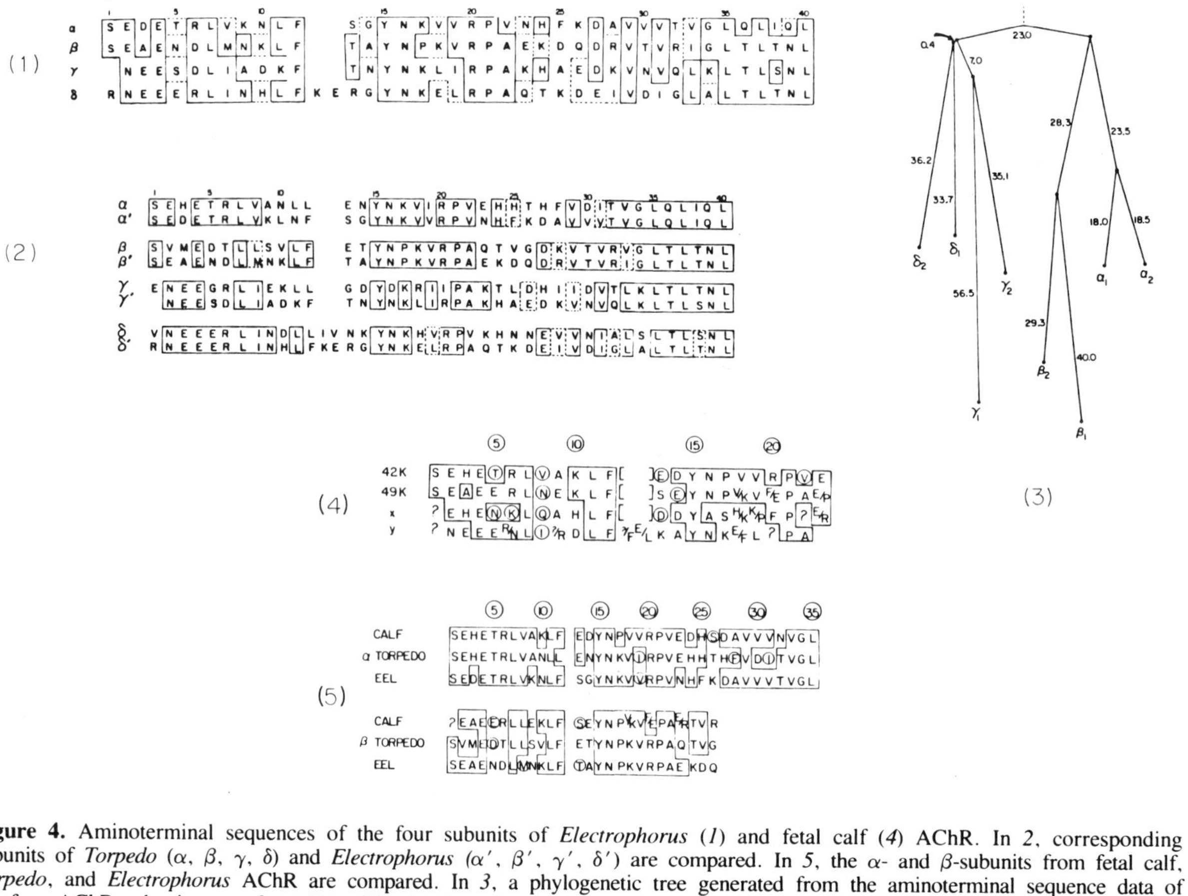

Figure 4. Aminoterminal sequences of the four subunits of *Electrophorus* (*1*) and fetal calf (*4*) AChR. In *2*, corresponding subunits of *Torpedo* (α, β, γ, δ) and *Electrophorus* (α′, β′, γ′, δ′) are compared. In *5*, the α- and β-subunits from fetal calf, *Torpedo*, and *Electrophorus* AChR are compared. In *3*, a phylogenetic tree generated from the aminoterminal sequence data of the four AChR subunit types from *T. californica* (α_1, β_1, γ_1, δ_1) and *E. electricus* (α_2, β_2, γ_2, δ_2) by using the best-fit matrix method. Each branch length represents the "accepted point mutations" (PAMs) per 100 amino acid residues that occurred in generating the contemporary subunits of both *Torpedo* and *Electrophorus* AChRs.

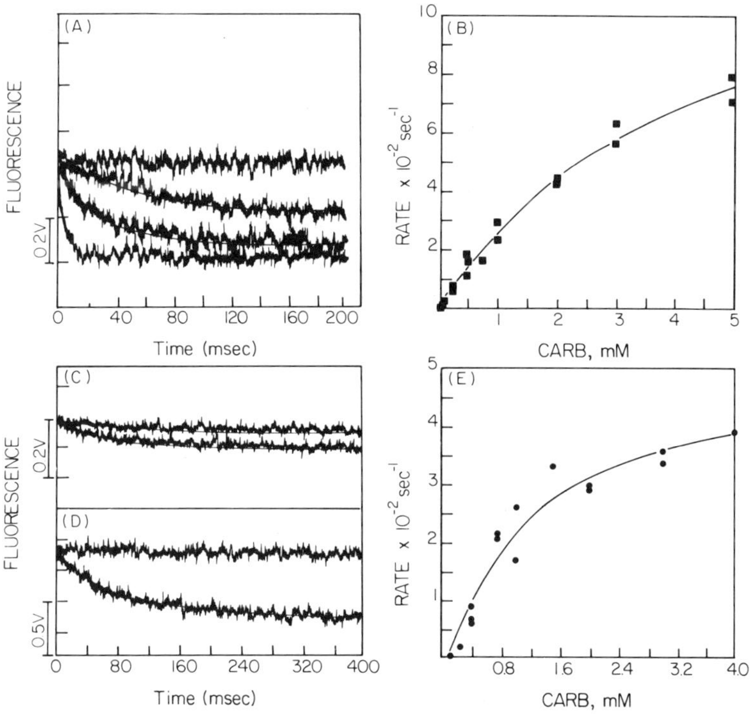

Figure 5. (*A*) Kinetics of Tl^+ influx into ANTS-loaded vesicles induced by Carb concentrations (*top to bottom trace*) of 0, 50, 100, and 500 μM Carb. (*B*) Effect of Carb concentration on flux rate constant. Solid line is best fit to two-ligand-binding model $k_{app} = k_{max}/(1 + [L]/K_0)$ using $k_{max} = 1100\ s^{-1}$ and $K_d = 1$ mM. (*C*, *D*) Comparison of Tl^+ influx observed using AChR-enriched vesicles before (*C*) and after (*D*) alkali extraction of nonreceptor proteins. Upper trace in each panel is leakage in absence of agonist and lower is in presence of 100 μM Carb. (*D*) Effect of Carb concentration on flux rate using reconstituted vesicles. Best fit to two-ligand-binding model with $k_{max} = 490\ s^{-1}$ and $K_d = 500\ \mu$M.

Cation Gating Function of AChR

Correlation of the known subunit structure of the AChR with its physiological function necessitates the development of quantitative methods to evaluate in vitro the efficiency of receptor-mediated cation transport. A rapid kinetic method was developed for this purpose to allow spectroscopic detection of monovalent cation transport (Moore and Raftery 1980). A water-soluble fluorophore (8-aminonaphthalene-1,3,6-trisulfonate [ANTS]) is trapped within AChR-enriched vesicles and the quench of its fluorescence caused by agonist-mediated inward transport of thallium(I) ion is monitored in stopped-flow experiments allowing detection on a millisecond time scale. The rate of fluorescence decay depends on the number of activated AChRs, and this can be used to determine the level of ion transport mediated by a single AChR molecule. In Figure 5A, the effect of Carb on the Tl^+-mediated fluorescence decay rate of ANTS is shown and the resultant dose-response curve is plotted in Figure 5B, demonstrating saturation of the rate ($\sim 1500\ s^{-1}$) at high Carb concentrations. This was shown (Moore and Raftery 1980) to correspond to 7×10^6 ions per second per receptor in *Torpedo* Ringer's. This value is very close to that ($\sim 10^7$) estimated for the AChR at the neuromuscular junction in vivo (see Lester 1977) and demonstrates that the receptor in isolated vesicles is fully functional. Preparations of membrane-bound AChR composed only of the four homologous polypeptides previously discussed transported Tl^+ as efficiently (Fig. 5C,D) as membranes containing other protein components such as the 43-kD protein discovered by Sobel et al. (1977). This 43-kD protein was suggested to be a separate ionophore (Sobel et al. 1978) associated with the ACh-binding protein (ACh modulator). Such notions are eliminated by the above results, which show that the pentameric complex $\alpha_2\beta\gamma\delta$ constitutes a complete physiological receptor for postsynaptic depolarization by ACh. This conclusion is confirmed by reconstitution studies using detergent-solubilized AChR preparations where the AChR polypeptides were the only protein components reassociated with phospholipid vesicles (Wu et al. 1981). The quantitative Tl^+ flux assay was also used to analyze the efficiency of cation transport by AChR monomers (9S) and dimers (13.7S) (Wu and Raftery 1981b). The specific cation transport per AChR molecule in such reconstituted systems was within a factor of two of those reported earlier for native membrane preparations (Moore and Raftery 1980). Using a different approach, Boheim et al. (1981) confirmed that only the four receptor proteins form the active complex.

It should be noted that the midpoints of the dose-response curves for Carb-mediated flux in native membranes (Fig. 5B) or in reconstituted vesicles (Fig. 5E) are in the millimolar range in agreement with physiological studies (see Adams 1981). Similarly, values

Table 2. Effect of Agonists on Fluorescence of NBD-labeled AChR before and after Solubilization and Comparison with K_d Values from Thallium Flux Data

	K_d (fluorescence) (mM)		K_d (thallium flux) (mM)	
Agonist	membrane bound	solubilized	(A)[a]	(B)[b]
ACh	0.09 ± 0.02	1.5 ± 0.2	0.16	0.057
Carb	0.96 ± 0.17	7.9 ± 1.2	5.0	1.0
Nicotine	0.60 ± 0.21	0.24 ± 0.17	0.26	0.13
PTA[c]	0.47 ± 0.09	—	0.29	0.11
Choline	35	—	—	—

[a]Data fitted to single ligand-binding model: $k_{app} = k_{max}[L]/(K_d + [L])$.
[b]Data fitted to model where two ligand-binding sites must be occupied for channel to open: $k_{app} = k_{max}([L]/(K_d + [L]))^2$.
[c]PTA, phenyltrimethylammonium chloride.

for ACh (~100 μM) and other agonists (see Table 2) are in close agreement. Such high concentrations of agonist necessary for activation of ion transport are at variance with many direct studies of ligand binding to AChR in either the resting or desensitized state.

An important aspect of the flux response is the shape of the dose-response curves. The data in Figure 5 can be adequately fit by mechanisms that assume a hyperbolic or slightly cooperative response. More recent results (Fig. 6) obtained using a different fluorophore (pyrenetetrasulfonic acid) suggested by Haugland (1981) as a substitute for ANTS clearly demonstrate that the flux response is a cooperative phenomenon with a Hill coefficient of 1.7. A major question with respect to mechanism is whether the cooperative flux response is due to multiple ligand binding and if so whether such binding is positively cooperative (see below).

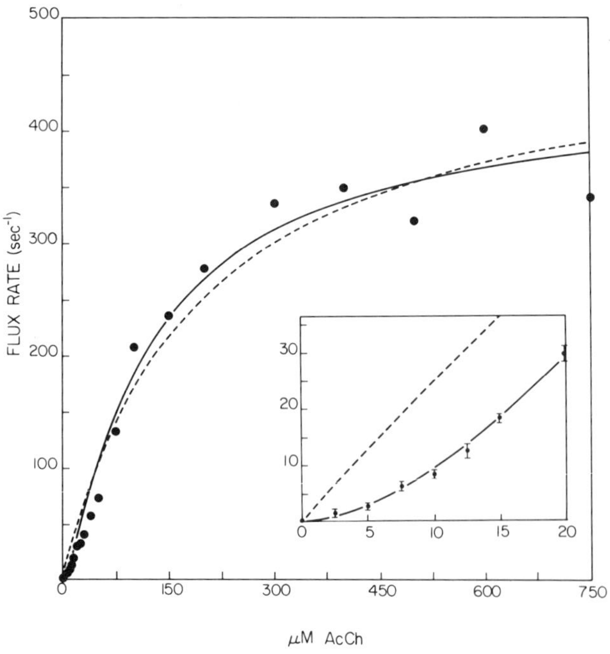

Figure 6. Effect of ACh on kinetics of thallium influx (50 mM after mixing) monitored by fluorescence changes of pyrene tetrasulfonic acid (17 mM) trapped within AChR-enriched membrane vesicles. (——) Fit to two-ligand-binding model (see legend to Fig. 5b) using $k_{max} = 440\ s^{-1}$ and $K_d = 57$ μM; (- - - -) fit to single-ligand-binding model $k_{app} = k_{max}[L]/(K_d + [L])$ with $k_{max} = 486\ s^{-1}$ and $K_d = 186$ μM. (*Inset*) Expansion of low-concentration region and lines obtained from best-fit parameters above.

Agonist Binding to AChR

Measurements of the equilibrium binding of agonists to the AChR have shown that under these conditions in which the receptor is desensitized and the ion channel is closed, the affinity for agonists is high (K_d ~10 nM for ACh). A variety of stopped-flow fluorescence techniques have been used to investigate conformational transitions of the AChR occurring under preequilibrium conditions that may be important in channel activation.

In one study of agonist-binding kinetics, the membrane-bound AChR was first covalently labeled by a fluorescent probe, 5-iodoacetamidosalicylic acid (IAS), after mild reduction. The fluorescence of this probe was enhanced on the binding of agonists (Dunn et al. 1980) and at least three kinetic processes were observed that could be described by the following mechanism (mechanism 1):

$$R + L \underset{}{\overset{14.6\ \mu M}{\rightleftharpoons}} RL \underset{0.12\ s^{-1}}{\overset{13.1\ s^{-1}}{\rightleftharpoons}} R'L \overset{17.4\ \mu M}{\rightleftharpoons} R'L_2 \qquad (1)$$
$$RL \underset{0.006\ s^{-1}}{\overset{0.2\ s^{-1}}{\rightleftharpoons}} C_1 \qquad R'L_2 \underset{0.65\ s^{-1}}{\overset{2\ s^{-1}}{\rightleftharpoons}} C_2$$

where the kinetic parameters given are for ACh binding (Blanchard et al. 1982). This mechanism is in good agreement with one previously proposed to account for the kinetics observed when ethidium was used as an extrinsic probe (Quast et al. 1979) except for the inclusion of an additional conformational change (RL ⇌ R′L). The resting state of the AChR has initially a low affinity for agonists but two sequential conformational changes result in a tightly bound C_1 complex. This C_1 complex can be correlated with the high-affinity binding site(s) observed under equilibrium conditions, since the overall dissociation constant for its formation (4 nM) is in agreement with K_d values obtained in equilib-

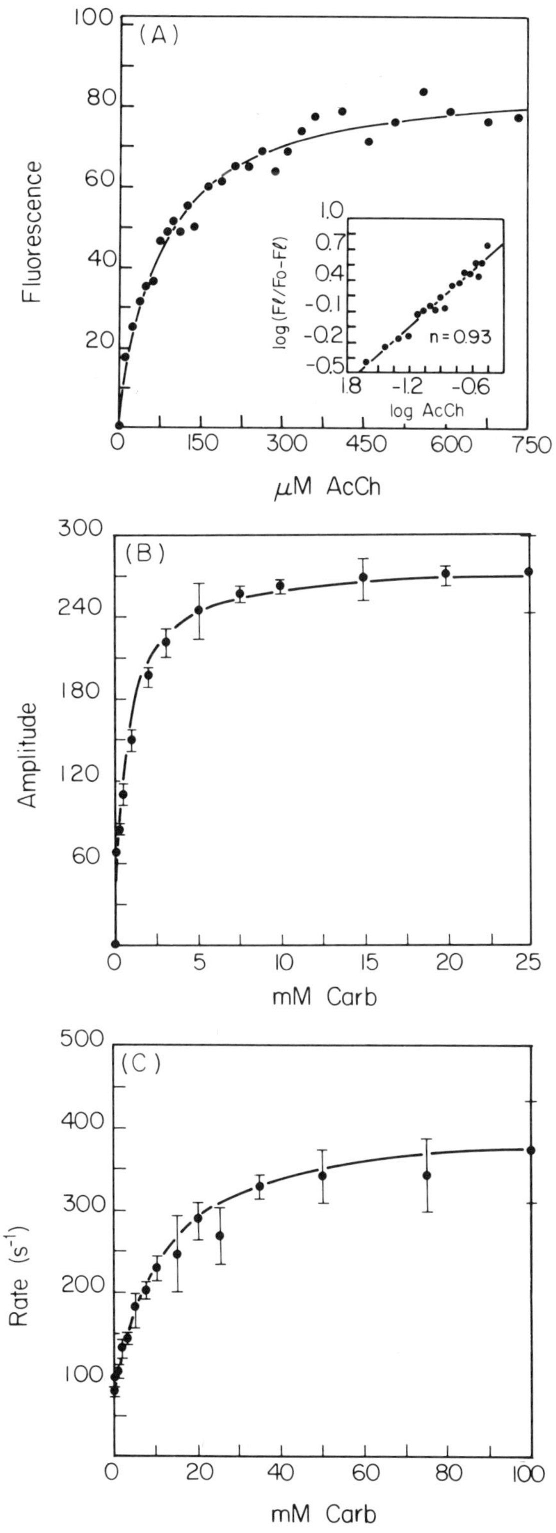

Figure 7. (*A*) Equilibrium fluorescence titration of NBD-labeled AChR by ACh. (——) Fit to $Fl = Fo[L]/(K_d + [L])$ (where $K_d = 75 \pm 2 \mu M$); Fl is observed fluorescence and Fo is maximum fluorescence change. (*Inset*) Hill plot of the same data. (*B*) Effect of Carb concentration on amplitude and (*C*) on rate of binding to NBD-labeled AChR measured in stopped-flow fluorescence experiments. Solid lines are calculated from best-fit parameters obtained from fit of mechanism 2 as described in the text.

rium experiments. Binding of the second ligand occurs only at higher ligand concentrations when the formation of C_2, which has lower affinity for agonist, becomes the predominant kinetic process.

In the above study and in other rapid kinetic experiments (see Conti-Tronconi and Raftery 1982), there are two major problems in correlating any observed receptor conformation with the open-channel state: (1) Apparent dissociation constants for agonist binding to the resting state of the AChR are lower than those obtained from the concentration dependence of the permeability response. (2) No observed conformational change is fast enough to be identified with channel opening, which must occur on a millisecond time scale. It is therefore likely that these slow processes are related to desensitization or other inactivation mechanisms, which in electrophysiological experiments have been shown to occur on similar time scales (see Adams 1981).

A high-affinity binding site for agonists has been assigned to each subunit of M_r ~40,000, since following reduction of a reactive disulfide bond near the site it can be labeled by affinity-alkylating agents such as BrACh (see Conti-Tronconi and Raftery 1982). The fluorescent probe IAS was shown to react with the same reduced disulfide bond near these sites because after the labeling procedures, covalent binding of $[^3H]$BrACh was much reduced (Dunn et al. 1980). In view of the proximity of the fluorophore, agonist binding to these 40K sites was probably measured and it is notable that all observed conformational changes were too slow to be primary events in channel activation.

The general assumption that occupancy of a single class of sites leads to both functional responses of channel opening and desensitization has resulted in proposals of complex kinetic schemes that include both a low-affinity state in which the ion channel is open and a high-affinity, desensitized state reached by sequential conformational transitions of the AChR-agonist complex (Hess et al. 1979; Neubig and Cohen 1980). Recently, evidence has been obtained for the existence of a low-affinity site(s) for agonists that is present under equilibrium conditions and is distinct from the high-affinity sites that can be labeled by BrACh (Conti-Tronconi et al. 1982b; Dunn and Raftery 1982a,b; Dunn et al. 1983).

The membrane-bound AChR was covalently labeled by the fluorescence probe 4-([(iodoacetoxy)ethyl]methylamino)-7-nitro-2,1,3-benzoxadiazole (IANBD), and it was shown that the fluorescence of this probe was enhanced in a saturable manner on the binding of agonists (Dunn and Raftery 1982a,b) with the same dose dependency as the flux response (see Table 2). A typical fluorescence titration curve for ACh is shown in Figure 7A. The enhancement has a simple hyperbolic dependence on agonist concentration with a K_d of 75 μM and a Hill coefficient of approximately 1. Kinetic experiments have shown that the fluorescence change is a monophasic process occurring on a rapid time scale. The rate and amplitude each had a hyperbolic dependence on agonist concentration (Fig. 7B) and the observed transition must therefore be a conformational change of the receptor-ligand complex.

The low-affinity site(s) is distinct from the high-affinity sites on the α-subunits since even when these latter sites were maximally labeled by BrACh, the fluorescence enhancement and therefore agonist binding to the low-affinity site was unaltered (Conti-Tronconi et al. 1982b; Dunn et al. 1983). In the presence of covalently bound BrACh, no ion flux was observed. The conformational change occurring on agonist binding to the low-affinity site is therefore independent of other transitions that inhibit channel opening.

The binding site revealed by 7-nitro-2,1,3-benzoxadiazole (NBD) fluorescence is present after solubilization of the AChR in cholate or Triton X-100. The affinity of solubilized receptor for ACh and Carb is about tenfold lower and the affinity for nicotine is unaltered (Table 2). Reconstitution of cholate-solubilized AChR with asolectin (Wu and Raftery 1981a) had little effect on the K_d for Carb binding, which was 5.5 ± 2.7 mM in the solubilized state and 3.7 ± 1.6 mM after reconstitution.

Several lines of evidence implicate the low-affinity site in channel opening and suggest that activation and desensitization are parallel pathways that are mediated by agonist binding to different sites. (1) The fluorescence enhancement is specific for agonists, is abolished by prior incubation with α-bungarotoxin, and reflects a conformational transition of the receptor-agonist complex; (2) K_d values for agonist binding correspond to those for activation (Table 2); (3) The conformational change is rapid, reaching about 400 s^{-1} for Carb and about 600 s^{-1} for ACh; (4) Q_{10} is about 2.5, in agreement with electrophysiological measurements for channel opening; (5) the binding is unaffected by desensitization, covalently bound BrACh, or prior incubation with physiologically active concentrations of curare, histrionicotoxin (HTX), or local anesthetics.

Mechanism of Agonist Binding to the Low-Affinity Site

The simplest model that predicts a hyperbolic dependence of both rate and amplitude on agonist concentration is one in which rapid binding is followed by a conformational change of the receptor-agonist complex

$$R + L \underset{}{\overset{K_1}{\rightleftharpoons}} RL \underset{k_2}{\overset{k_2}{\rightleftharpoons}} R^*L$$

This mechanism gave adequate fits to the Carb-binding data shown in Figure 7B, but there was a discrepancy between the overall K_d values obtained from the amplitude data (~ 0.73 mM) and those calculated from the best-fit rate parameters (Dunn and Raftery 1982b). An extension of this model to include a preequilibrium between R and R^* in the absence of ligand removes this discrepancy and the following model (mechanism 2) adequately describes the data.

$$\begin{array}{ccc} R + L & \overset{K_1}{\rightleftharpoons} & RL \\ k_0 \updownarrow k_{-0} & & k_2 \updownarrow k_{-2} \\ R^* + L & \overset{K_3}{\rightleftharpoons} & R^*L \end{array} \qquad (2)$$

In this model, it is assumed that the ligand-binding steps are unobservably fast and that the fluorescence change occurs in the R $\rightleftharpoons$ R^* and RL $\rightleftharpoons$ R^*L transitions. Under circumstances where $K_0(= k_{-0}/k_0) >> 1 >> K_0K_3/K_1$, i.e., that the equilibrium lies in favor of R in the absence of ligand and that agonist binds more tightly to R^* than to R (see Janin 1973), the amplitude is approximated by

$$\frac{\overline{R^*} + \overline{R^*L}}{R_0} \sim \frac{[L]}{[L] + K_0K_3} \qquad (3)$$

Equations describing the behavior of the apparent rate and procedures used for fitting this model have been described previously (Dunn and Raftery 1982b) and the best-fit parameters for the rate data for Carb binding (Fig. 7B) are given by $k_{-0} = 130$ s^{-1}, $k_0 = 18$ s^{-1}, $k_{-2} = 18.3$ s^{-1}, $k_2 = 385$ s^{-1}, $K_1 = 9.1$ mM, and $K_3 = 0.11$ mM, giving an overall equilibrium constant of approximately 0.78 mM in good agreement with the concentration dependence of the amplitude data (0.73 mM).

Conformational Coupling between Agonist Binding and Channel Opening

The observed binding of agonists to the low-affinity site revealed by NBD fluorescence showed no evidence of cooperativity, and, as illustrated in Figure 7A, Hill coefficients not significantly different from 1 were obtained. However, the concentration dependence of the flux response was clearly sigmoidal (Fig. 6) and had a Hill coefficient of 1.7, suggesting that two ligand molecules must bind for channel opening to occur. The simplest explanation for this discrepancy is that not one but two subunits of the AChR must undergo the conformational change before channel opening occurs. In terms of mechanism 2 above using the same assumptions as before, the fraction of subunits in the R^* conformation in the presence of ligand concentration [L] is approximated by

$$\frac{[L]}{[L] + K_0K_3} \qquad (4)$$

but the probability of two subunits of the AChR being simultaneously in the R^* conformation, i.e., the number of open channels, is

$$p(\text{open}) \sim \left(\frac{[L]}{[L] + K_0K_3}\right)^2 \qquad (5)$$

This model predicts that ligand binding to individual subunits follows a simple Langmuir isotherm and displays no cooperativity, but the rate of flux, since it is dependent on the number of open channels, has a sigmoidal dependency on agonist concentration.

This model is illustrated schematically in Figure 8A. The location of the low-affinity agonist-binding site(s) has not been elucidated and, although in Figure 8A both subunits are depicted as being identical, it is possible that two binding sites may exist on nonidentical but homologous subunits that have similar affinities for agonist.

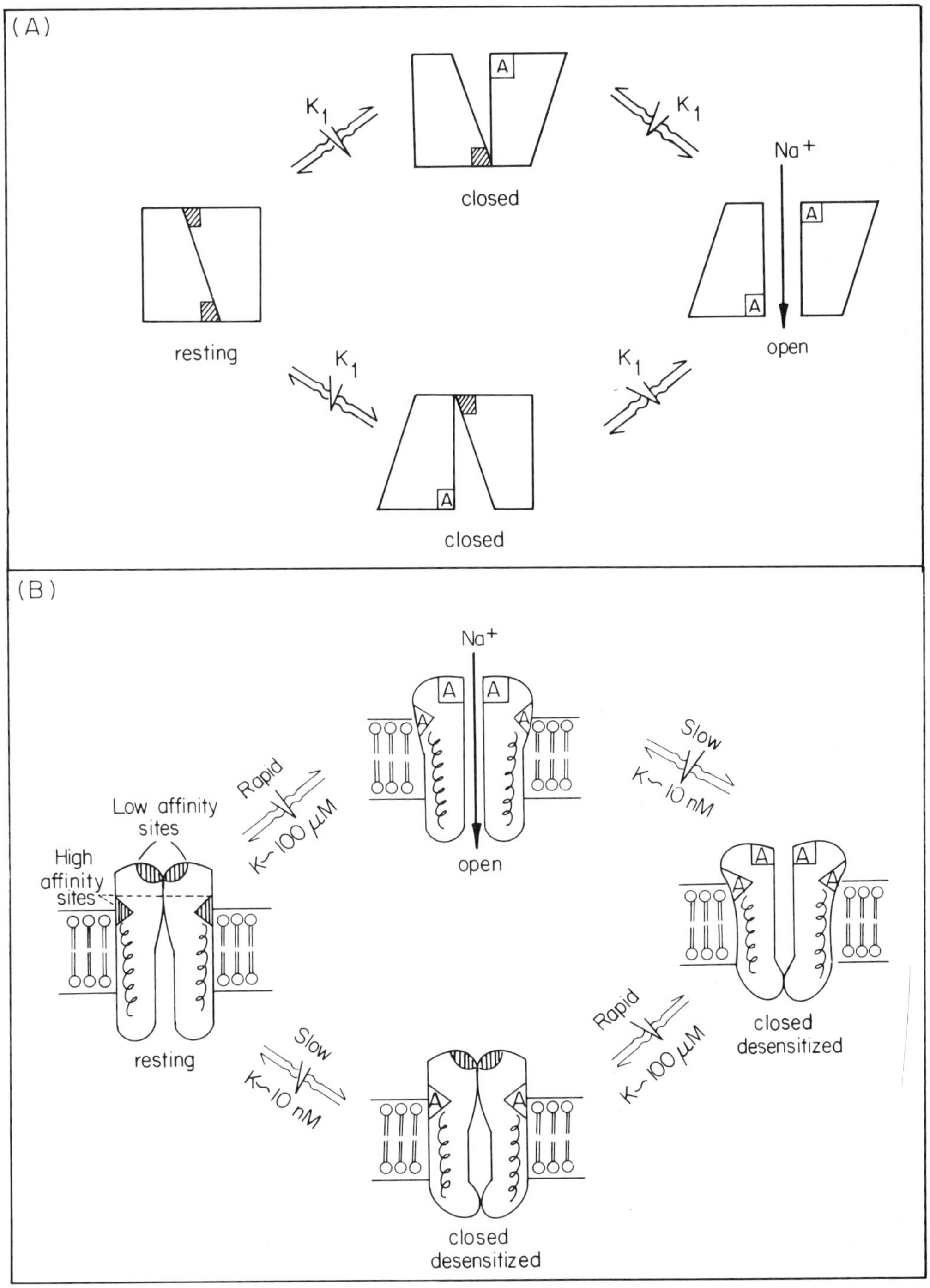

Figure 8. Schematic representation of ligand-binding mechanisms. (*A*) Model in which two independent subunits must undergo identical conformational changes before channel opens. (*B*) Model in which activation and desensitization are parallel and independent mechanisms mediated by agonist binding to different sites.

Independent Pathways of Channel Activation and Desensitization

Torpedo AChR appears to have two classes of agonist binding sites—those of high affinity on the 40K subunits, which may be labeled by BrACh, and those of low affinity, which are revealed by NBD fluorescence changes. The conformational change occurring on agonist binding to the low-affinity site is unaffected by desensitization or by covalent labeling by BrACh and the two pathways must therefore be independent. Such a model is illustrated pictorially in Figure 8B. In the resting state, the "activation gate" is closed but in the presence of high concentrations of agonists the low-affinity sites are occupied and the AChR undergoes a rapid conformational transition to an open-channel state. Over longer time scales, slow conformational transitions in another part of the molecule mediated by agonist binding to other sites (possibly those of high-equilibrium affinity on the α-subunits) cause the channel to close. Alternatively, at lower concentrations of agonist, occupancy of these latter sites causes slow conformational changes which close an "inactivation gate." Under these circumstances, the same conformational change may be induced by agonist binding to the low-affinity sites, but this transition cannot now lead to opening of the ion channel.

CONCLUSION

The structure of the nicotinic AChR has been highly conserved during animal evolution, and in all the species and tissues studied so far, including mammals, it is a pseudosymmetric, pentameric complex of related subunits with very similar physical properties. All subunits of these nicotinic receptors were derived from a common ancestral gene, probably by way of gene duplications occurring very early in animal evolution. The likely existence of such a unique ACh-binding ancestral protein raises the possibility of a shared ancestry for many or even all proteins able to bind ACh, such as the muscarinic receptor, acetylcholinesterase, etc., whose genes would have diverged very early to perform different receptor-mediated or enzymatic functions. The relatively early divergence of the genes coding for the subunits of the nicotinic AChR, as well as their highly conserved primary structure and molecular weight, supports the possibility that the AChR subunits themselves evolved separately to perform discrete functions, such as activation, inactivation, and desensitization. The demonstration of multiple binding sites for agonists in *Torpedo* AChR provides strong support for this hypothesis.

ACKNOWLEDGMENTS

This research was supported by U.S. Public Health Service grant NS-10294, ARO Contract DAMD17-82-C-2175, and by a grant from the Myasthenia Gravis Foundation, Los Angeles Chapter.

REFERENCES

ADAMS, P.R. 1981. Acetylcholine receptor kinetics. *J. Membr. Biol.* **58:** 161.

BAYLEY, H. and J.R. KNOWLES. 1978. Photogenerated reagents for membrane labeling. 2. Phenylcarbene and adamantylidene formed within the lipid bilayer. *Biochemistry* **17:** 2420.

———. 1980. Photogenerated reagents for membranes: Selective labeling of intrinsic membrane proteins in the human erythrocyte membrane. *Biochemistry* **19:** 3883.

BLANCHARD, S.G., S.M.J. DUNN, and M.A. RAFTERY. 1982. Effects of reduction and alkylation on ligand binding and cation transport by *Torpedo californica* acetylcholine receptor. *Biochemistry* **24:** 6258.

BOHEIM, G., W. HANKE, F.J. BARRANTES, H. EIBL, B. SAKMANN, G. FELLS, and A. MAELICKE. 1981. Agonist activated ionic channels in acetylcholine receptor reconstituted into planar lipid bilayers. *Proc. Natl. Acad. Sci.* **78:** 3586.

CLAUDIO, T., M. BALLIVET, J. PATRICK, and S. HEINEMANN. 1983. Nucleotide and deduced amino acid sequences of *Torpedo californica* acetylcholine receptor γ subunit. *Proc. Natl. Acad. Sci.* **80:** 1111.

CONTI-TRONCONI, B.M. and M.A. RAFTERY. 1982. The nicotinic cholinergic receptor: Correlation of molecular structure with functional properties. *Annu. Rev. Biochem.* **51:** 491.

CONTI-TRONCONI, B.M., S.M.J. DUNN, and M.A. RAFTERY. 1982a. Functional stability of *Torpedo* acetylcholine receptor: Effects of protease treatment. *Biochemistry* **21:** 893.

———. 1982b. Independent sites of low and high affinity for agonists on *Torpedo californica* acetylcholine receptor. *Biochem. Biophys. Res. Commun.* **107:** 123.

CONTI-TRONCONI, B.M., C. GOTTI, M. HUNKAPILLER, and M.A. RAFTERY. 1982c. Mammalian muscle acetylcholine receptor: A supramolecular structure formed by four related proteins. *Science* **218:** 1227.

CONTI-TRONCONI, B.M., M.W. HUNKAPILLER, J.M. LINDSTROM, and M.A. RAFTERY. 1982d. Subunit structure of the acetylcholine receptor from *Electrophorus electricus*. *Proc. Natl. Acad. Sci.* **79:** 6489.

DEVILLERS-THIERY, A., J. GIRAUDAT, M. BENTABOULET, and J.-P. CHANGEUX. 1983. Complete mRNA coding sequence of the acetylcholine binding α-subunit of *Torpedo marmorata* acetylcholine receptor: A model for the transmembrane organization of the polypeptide chain. *Proc. Natl. Acad. Sci.* **80:** 2067.

DUNN, S.M.J. and M.A. RAFTERY. 1982a. Activation and desensitization of *Torpedo* acetylcholine receptor: Evidence for separate binding sites. *Proc. Natl. Acad. Sci.* **79:** 6757.

———. 1982b. Multiple binding sites for agonists on *Torpedo californica* AcChR. *Biochemistry* **21:** 6264.

DUNN, S.M.J., S.G. BLANCHARD, and M.A. RAFTERY. 1980. Kinetics of carbamylcholine binding to membrane-bound acetylcholine receptor monitored by fluorescence changes of a covalently bound probe. *Biochemistry* **19:** 5645.

DUNN, S.M.J., B.M. CONTI-TRONCONI, and M.A. RAFTERY. 1983. Separate sites of low and high affinity for agonists on *Torpedo californica* acetylcholine receptor. *Biochemistry* **22:** 2512.

ELLIOT, J., S.G. BLANCHARD, W.C.-S. WU, J. MILLER, C. STRADER, P. HARTIG, H.-P.H. MOORE, J. RACS, and M.A. RAFTERY. 1980. Purification of *Torpedo californica* postsynaptic membranes and fractionation of their constituent proteins. *Biochem. J.* **185:** 667.

GOTTI, C., B.M. CONTI-TRONCONI, and M.A. RAFTERY. 1982. Mammalian muscle acetylcholine receptor purification and characterization. *Biochemistry* **21:** 3148.

GULLICK, W., S. TZARTOS, and J. LINDSTROM. 1981. Monoclonal antibodies as probes of acetylcholine receptor structure. 1. Peptide mapping. *Biochemistry* **20:** 2173.

HAUGLAND, R.P. 1981. *Molecular probes handbook of fluorescent probes*. p. 43. Molecular Probes, Plano, Texas.

HESS, G.P., D.G. CASH, and H. AOSHIMA. 1979. Acetylcholine receptor-controlled ion fluxes in membrane vesicles investigated by fast reaction techniques. *Nature* **282:** 329.

JANIN, J. 1973. The study of allosteric proteins. *Prog. Biophys. Mol. Biol.* **27:** 77.

LESTER, H.A. 1977. The response to acetylcholine. *Sci. Am.* **236:** 106.

LINDSTROM, J., W. GULLICK, B. CONTI-TRONCONI, and M. ELLISMAN. 1980. Proteolytic nicking of the acetylcholine receptor. *Biochemistry* **19:** 4791.

MARTINEZ-CARRION, M., V. SATOR, and M.A. RAFTERY. 1975. The molecular weight of an acetylcholine receptor isolated from *Torpedo californica*. *Biochem. Biophys. Res. Commun.* **65:** 129.

MATTHEWS, B.W. and S.A. BERNHARD, 1973. Structure and symmetry of oligomeric enzymes. *Annu. Rev. Biophys. Bioeng.* **4:** 257.

MOORE, H.-P.H. and M.A. RAFTERY. 1980. Direct spectroscopic studies of cation translocation by *Torpedo* acetylcholine receptor on a time scale of physiological relevance. *Proc. Natl. Acad. Sci.* **77:** 4509.

NEUBIG, R.R. and J.B. COHEN. 1980. Permeability control by cholinergic receptor in *Torpedo* postsynaptic membranes: Agonist dose-response relations measured at second and millisecond times. *Biochemistry* **19:** 2770.

NODA, M., H. TAKAHASHI, T. TANABE, M. TOYOSATO, Y. FURUTANI, T. HIROSE, M. ASAI, S. INAYAMA, T. MIYATA, AND S. NUMA. 1982. Primary structure of α-subunit precursor of *T. californica* acetylcholine receptor deduced from cDNA sequence. *Nature* **299:** 793.

NODA, M., H. TAKAHASHI, T. TANABE, M. TOYOSATO, S. KIKYOTANI, Y. FURUTANI, T. HIROSE, H. TAKASHIMA, S. MAYAMA, T. MIYATA, and S. NUMA. 1983a. Structural homology of *Torpedo californica* acetylcholine receptor subunits. *Nature* **302:** 528.

NODA, M., H. TAKAHASHI, T. TANABE, M. TOYOSATO, S. KIKYOTANI, T. HIROSE, M. ASAI, H. TAKASHIMA, S. INAYAME, T. MIYATA, and S. NUMA. 1983b. Primary structures of β and δ subunit precursors of *Torpedo californica* acetylcholine receptor deduced from cDNA sequences. *Nature* **301:** 251.

QUAST, U., M.I. SCHIMERLIK, and M.A. RAFTERY. 1979. Ligand-induced changes in membrane-bound acetylcholine receptor observed by ethidium fluorescence. 2. Stopped-flow studies with agonists and antagonists. *Biochemistry* **18:** 1891.

RAFTERY, M.A., M.W. HUNKAPILLER, C.D. STRADER, and L.E. HOOD. 1980. Acetylcholine receptor: Complex of homologous subunits. *Science* **208:** 1454.

RAFTERY, M.A., R. VANDLEN, D. MICHAELSON, J. BODE, T. MOODY, Y. CHAO, K. REED, J. DEUTSCH, and J. DUGUID. 1974. The biochemistry of an acetylcholine receptor. *J. Supramol. Struct.* **2:** 582.

REYNOLDS, J.A. and A. KARLIN. 1978. Molecular weight in detergent solution of acetylcholine receptor from *Torpedo californica*. *Biochemistry* **17:** 2035.

SOBEL, A., M. WEBER, and J.-P. CHANGEUX. 1977. Large-scale purification of the acetylcholine-receptor protein in its membrane-bound and detergent-extracted forms from *Torpedo marmorata* electric organ. *Eur. J. Biochem.* **80:** 215.

SOBEL, A., T. HEIDMANN, J. HOFLER, and J.-P. CHANGEUX. 1978. Distinct protein components from *Torpedo marmorata* membranes carry the acetylcholine receptor site and the binding site for local anesthetics and histrionicotoxin. *Proc. Natl. Acad. Sci.* **75:** 510.

SOBEL, A., J. HOFLER, T. HEIDMANN, and J.-P. CHANGEUX. 1979. Structural and functional properties of the acetylcholine regulator. *Adv. Cytopharmacol.* **3:** 191.

STRADER, C.D. and M.A. RAFTERY. 1980. Topographic studies of *Torpedo* acetylcholine receptor subunits as a transmembrane complex. *Proc. Natl. Acad. Sci.* **77:** 5807.

STRADER, C.D., J.-P. REVEL, and M.A. RAFTERY. 1979. Demonstration of the transmembrane nature of the acetylcholine receptor by labeling with anti-receptor antibodies. *J. Cell Biol.* **83:** 499.

SUMIKAWA, K., M. HOUGHTON, J.C. SMITH, L. BELL, B.M. RICHARDS, and E.A. BARNARD. 1982. The molecular cloning and characterization of cDNA coding for the α-subunit of the acetylcholine receptor. *Nucleic Acids Res.* **10:** 5809.

TARRAB-HAZDAI, R., B. GEIGER, S. FUCHS, and A. AMSTERDAM. 1978. Localization of acetylcholine receptor in excitable membrane from the electric organ of *Torpedo*: Evidence for exposure of receptor antigenic sites on both sides of the membrane. *Proc. Natl. Acad. Sci.* **75:** 2497.

WEILL, C.L., M.G. MCNAMEE, and A. KARLIN. 1974. Affinity-labeling of purified acetylcholine receptor from *Torpedo californica*. *Biochem. Biophys. Res. Commun.* **61:** 997.

WU, W.C.-S. and M.A. RAFTERY. 1981a. Reconsitution of acetylcholine receptor function using purified receptor protein. *Biochemistry* **20:** 694.

———. 1981b. Functional properties of acetylcholine receptor monomeric and dimeric forms in reconstituted membranes. *Biochem. Biophys. Res. Commun.* **99:** 436.

WU, W.C.-S., H.-P.H. MOORE, and M.A. RAFTERY. 1981. Quantitation of cation transport by reconstituted membrane vesicles containing purified acetylcholine receptor. *Proc. Natl. Acad. Sci.* **78:** 775.

Allosteric Properties of the Acetylcholine Receptor Protein from *Torpedo marmorata*

J.-P. CHANGEUX,* F. BON,‡ J. CARTAUD,† A. DEVILLERS-THIÉRY,*
J. GIRAUDAT,* T. HEIDMANN,* B. HOLTON,* H.-O. NGHIÊM,* J.-L. POPOT,*§
R. VAN RAPENBUSCH,‡ AND S. TZARTOS*

**Neurobiologie Moléculaire, Institut Pasteur et Laboratoire associé Interactions Cellulaires et Moléculaires du C.N.R.S., Paris, 75015 France; †Laboratoire de Microscopie Electronique, Institut Jacques Monod, Université Paris VII, Paris, 75005 France; ‡Laboratoire de Biologie Structurale, Institut Jacques Monod, Université Paris VII, Paris, 75005 France*

Since the initial characterization of the acetylcholine receptor (AChR) in crude extracts of fish electric organ (Changeux et al. 1970a,b; Miledi et al. 1971; Meunier et al. 1971a,b, 1972a,b; Raftery et al. 1971; Reiter et al. 1972), knowledge about this important regulatory protein developed in an exceptionally fast manner (for review, see Changeux 1981; Conti-Tronconi and Raftery 1982; Anholt et al. 1983). It has been established that the basic unit, or light form, of the AChR consists of four different polypeptide chains of exact molecular weight: α (50,116), β (53,681), γ (56,279), and δ (57,565) (Noda et al. 1982, 1983a,b) in a 2.1.1.1. stoichiometry (Reynolds and Karlin 1978; Lindstrom et al. 1979; Raftery et al. 1980; Saitoh et al. 1980). In extracts of adult *Torpedo* electric organ, the receptor protein primarily occurs in a species of higher molecular weight or heavy form (Raftery et al. 1972) that results from the association of two light forms by a disulfide bridge between δ-chains (Chang and Bock 1977; Suarez-Isla and Hucho 1977; Hamilton et al. 1977).

Deciphering of the covalent structure of the four chains began by the chemical identification of the sequence of the aminoterminal amino acids of the α-chain from *Torpedo marmorata* (Devillers-Thiéry et al. 1979) and, then, of the α-, β-, γ-, and δ-chains from *Torpedo californica* (Hunkapiller et al. 1979; Raftery et al. 1980). These partial sequence data already revealed striking sequence homologies between the four chains (see also Tzartos and Lindstrom 1980). Then, the isolation of cDNA clones coding for the α-chain in *T. marmorata* (Giraudat et al. 1982; Sumikawa et al. 1982) and for the four chains in *T. californica* (Ballivet et al. 1982; Noda et al. 1982, 1983a,b) led to the determination of the complete amino acid sequence of the four chains (Noda et al. 1982, 1983a,b; Claudio et al. 1983; Devillers-Thiéry et al. 1983). As a consequence, the molecular weight of the chains was reevaluated (compare Noda et al. 1983a and Karlin 1980), and significant differences in sequence were noticed between the two most commonly used species of *Torpedo* (Devillers-Thiéry et al. 1983).

In parallel, the functional properties of the receptor protein were explored in vitro by measuring the state of opening of the associated ion channel following ion transport (Kasai and Changeux 1971; Hazelbauer and Changeux 1974; Popot et al. 1976) and the interaction of radioactive or fluorescent ligands (for review, see Changeux 1975, 1981) with the receptor in its native membrane environment or in reconstituted lipid vesicles (Hazelbauer and Changeux 1974; Epstein and Racker 1978; Changeux et al. 1979; Huganir et al. 1979; Wu and Raftery 1979; Anholt et al. 1980, 1981, 1982; for review, see Anholt et al. 1983; Popot 1983).

These studies resulted in the demonstration that the $\alpha_2\beta\gamma\delta$ oligomer contains the ion-selective channel and thus constitutes the minimal structural unit responsible for the electrogenic action of acetylcholine (ACh). The application to the AChR of rapid kinetic techniques for ligand binding (Weber et al. 1975; for review, see Heidmann et al. 1983b) and ion transport (Hess et al. 1979, 1981; Neubig and Cohen 1980; Neubig et al. 1982; Heidmann et al. 1983b) then made possible an extensive analysis of the molecular mechanisms that determine the action of ACh at the synapse in the physiological millisecond time scale.

Therefore, it seems timely to reconsider the proposal (Changeux 1966, 1969; Changeux et al. 1967b, 1976; Karlin 1967; Colquhoun 1973) that the AChR is an allosteric protein (Monod et al. 1963, 1965).

Symmetry Properties of the AChR Heavy and Light Forms from *T. marmorata**

Two structural features characterize the quaternary organization of typical allosteric proteins: (1) the assembly of a fixed number of identical subunits into a finite oligomer and (2) the presence of symmetry properties in the mode of assembly of the subunits (Monod et al. 1965). The vectorial integration of the receptor protein into the subsynaptic membrane is, however, expected to constrain these symmetry properties (Changeux and Thiéry 1968; Changeux 1981). Thus, it appeared of special interest to identify the symmetry of the two forms of the AChR: the light-form ($\alpha_2\beta\gamma\delta$) oligomer and its heavy-form dimer, which results from the association of two light forms by a δ-δ disulfide

§Present address: Yale University, Department of Molecular Biophysics and Biochemistry, New Haven, Connecticut 06511.

*Bon et al. 1982 and in prep.

bridge (Chang and Bock 1977; Suarez-Isla and Hucho 1977; Hamilton et al. 1977, 1979).

Observed with the electron microscope, the detergent-purified light form of the AChR (Cartaud et al. 1973) appears as a rosette of about 9 nm in diameter with a central depression, or funnel, that retains the dye used in the negative staining (for review, see Cartaud 1980; Kistler et al. 1982). This rosette represents the *en face* view (or axial projection) of an ~11-nm-long, rodlike structure that spans the postsynaptic membrane and projects to about 5.5 nm into the synaptic cleft (Cartaud et al. 1978; Schiebler and Hucho 1978). The largest end of the rod always faces the synaptic cleft and carries the ACh-binding site. Thus, the receptor molecule possesses a transverse polarity (Changeux and Thiéry 1968).

Preparations of the receptor heavy form (from *T. marmorata* and *T. californica*) contain coplanar pairs or doublets of light-form rosettes with a center-to-center distance of 9 ± 2 nm (Cartaud et al. 1980; Wise et al. 1981; see also Lindstrom et al. 1980; Barrantes 1982; Kistler et al. 1982). These doublets are identified, on a quantitative basis, with *en face* views of the 500-kD heavy form (Cartaud et al. 1980; Wise et al. 1981). Biochemical evidence indicates that the two light forms comprising a heavy-form doublet possess the same transmembrane orientation (F. Bon et al., in prep.). Axes of symmetry coplanar with the membrane are excluded, the only accessible ones being perpendicular to the membrane. To test for the presence of such axes of symmetry, the relative rotational orientation of the two rosettes within a doublet was examined by image analysis of negatively stained preparations of purified heavy form from *T. marmorata* (Bon et al. 1982 and in prep.). Optical densities of the rosette images were digitalized with an automatic flat bed scanner using, for each rosette, the annulus delimited by two circles of radii 0.82 nm and 4.38 nm. The average density of each annulus segment of $\pi/32$ radian yielded a 64-term vector. The phasing of the two rosettes from a doublet, arbitrarily using the right one as reference, was done by variance analysis of the distribution of the differences between homologous coordinates of the two vectors. The minimal value was taken as optimal positioning. Figure 1a shows that the angle of the relative orientation of the two rosettes in a doublet did not follow the sharp unimodal distribution expected if the doublet possessed a fixed geometry with, for example, a C_2 dyad axis of symmetry perpendicular to the plane of the membrane (in that case, $\theta = \pi$). Their spread, instead, indicates a variability in the relative position of the rosettes within the doublets, with a tendency to favor the side-by-side R·R disposition. The most likely interpretation of this observation is that the attachment of the two rosettes via the δ-chains involves a flexible polypeptide tail or loop. This view is consistent with the independent inactivation by electron irradiation of each of the two 250-kD units of the heavy form (Lo et al. 1982) and with the absence of major functional difference between the heavy and light forms in reconstitution experiments (for review, see Anholt et al. 1981, 1982, 1983; Popot 1983). It is also compatible with the sequence data on the *T. californica* δ-chain (Noda et al. 1983a) and with the model of transmembrane organization proposed for the α-chain (Devillers-Thiéry et al. 1983) and the other polypeptide chains of the receptor (Noda et al. 1983b). Accordingly, the only cysteinyl residue available for cross-linking the δ-chains and consistent with the data on the proteolysis of the δ-chain (Wennogle and Changeux 1980; Wennogle et al. 1981) is found at position 500 (Noda et al. 1983a), one

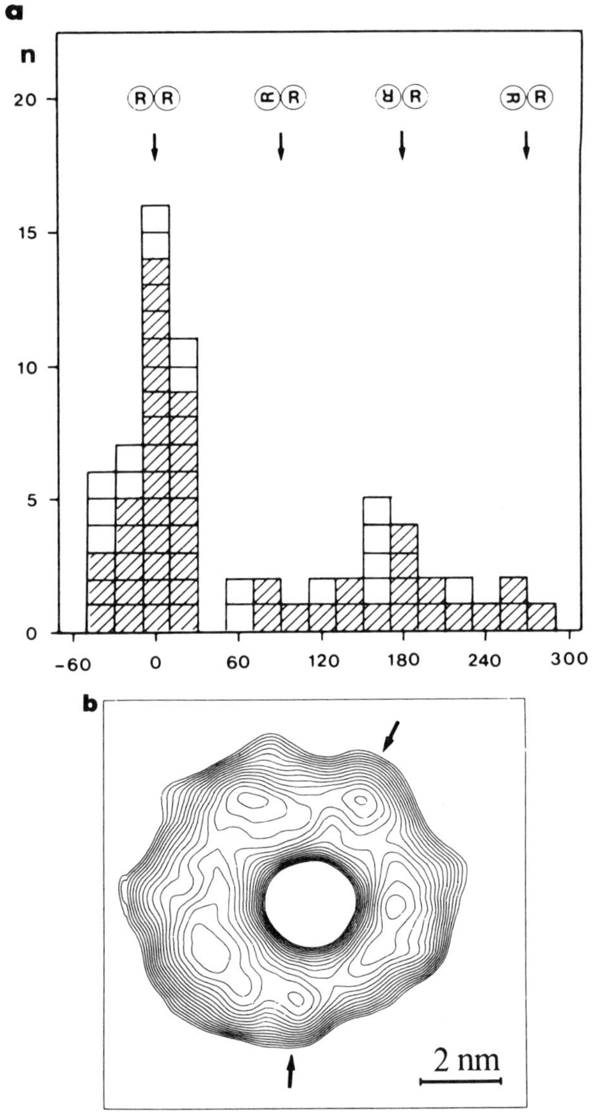

Figure 1. (*a*) Distribution of the diphasing angles between the two rosettes forming the heavy-form doublets from *T. marmorata*. In the majority of the doublets, $\theta = 0°$; only a small fraction possesses a C_2 dyad axis of symmetry ($\theta = 180°$). The link between the two rosettes is thus flexible (from Bon et al. 1982). (*b*) Averaged image of 46 light-form rosettes drawn in equidistant contour lines. Five peaks are clearly distinguished. The small peaks marked by an arrow are modified in the presence of α-toxin and thus most likely correspond to the α-chains (from Bon et al. 1982).

residue from the carboxyl terminus and 18–20 residues from the probable end of the nearest transmembrane helix. Fully extended, this extra membrane tail could contribute 6.5–7.0 nm for each δ-chain to the link connecting the two rosettes in a doublet and thus may exhibit a significant flexibility.

After reorientation, the images of the two rosettes of a doublet can be averaged, and, the "averaged doublets" obtained, reoriented by 180° rotations around two binary axes. One normal to the plane of projection defines the rotation of the doublet in the membrane; the second, identical to the center-to-center axis of the doublet, relates the image of molecules located on the opposite faces of the flattened lipidic vesicles. The averaged rosette obtained (Fig. 1b) displays the three major peaks and two major grooves previously reported by Zingsheim et al. (1980), Kistler and Stroud (1981), and Kistler et al. (1982). In addition, two minor peaks become apparent in the vicinity of the major grooves (at 13 and 18 hr in Fig. 1b at the resolution of 1/1.75 nm^{-1}). These two peaks become more manifest after preincubation of the receptor with snake venom α-bungarotoxin (F. Bon et al., in prep.; see also Zingsheim et al. 1982). In addition, their angular distance (~160°) fits with that found for the two binding sites of biotinylated α-bungarotoxin by Holtzman et al. (1982). Thus, they most likely correspond to, or at least are in the vicinity of, the two α-chains.

Thus, the morphological data are compatible with the accepted pentameric $\alpha_2\beta\gamma\delta$ organization (Reynolds and Karlin 1978) of the light-form rosette. Moreover, comparison with the crystallographic data of Kistler and Stroud (1981) and Zingsheim et al. (1982) leads to the conclusion that the image presented in Figure 1b corresponds to an extracellular view of the molecule. Accordingly, part of the differences in molecular weight noticed between the chains (Noda et al. 1983b) would become apparent on the cleft side of the molecule. On the other hand, the presence of important sequence homologies between the four chains (Raftery et al. 1980; Noda et al. 1983b) might serve as an index of a cryptic symmetry (Changeux 1981) of the molecule. Since the four chains traverse the membrane (Strader and Raftery 1980; Wennogle and Changeux 1980; Froehner 1981; Saint-John et al. 1982), the only organization of the light-form oligomer compatible with the available data is that of a heterologous (Monod et al. 1965) pentamer with a fivefold axis of cryptosymmetry perpendicular to the plane of the membrane and thus passing through the center of the rosette.

Structural Features of the ACh- and α-Toxin-binding Site*

ACh regulates cation transport through the postsynaptic membrane when it binds to a "regulatory" site referred to as the AChR site. Snake venom α-toxins (Lee and Chang 1966) have been used as highly selective labels to identify this site and to isolate, purify, and characterize the protein that carries this site (Changeux et al. 1970a,b; for review, see Changeux 1981). At the level of the receptor site, the α-toxins interact with the cholinergic ligands in a competitive manner (for review, see Changeux 1975, 1981). Moreover, the three-dimensional organization of their folded polypeptide chain discloses locally some analogies with the structure of curare and its congeners (see Low 1979). It was postulated that, despite important differences in chemical structure and molecular weight, ACh and the α-toxins bind to the same, or overlapping, areas on receptor surface. Affinity-labeling reagents derived from typical cholinergic agonists or antagonists (such as *p*-(trimethyl ammonium)benzene diazonium fluoroborate [TDF], Changeux et al. 1967a; Weiland et al. 1979; 4-(*N*-maleimido)benzyltrimethyl ammonium iodide [MBTA], Karlin and Cowburn 1973; Weill et al. 1974; 4-(*N*-maleimido)phenyltrimethyl ammonium iodide [MPTA], Sobel et al. 1977; BrACh, Damle et al. 1978; bis(3-azidopyridinium)-1,10-decane ferchlorate [DAPA], Witzemann and Raftery 1977) covalently react with the α-chain, and this chain is also labeled by photolabile derivatives of α-bungarotoxin (Hucho 1979; Witzemann et al. 1979) or in the presence of radioactive α-bungarotoxin and added cross-linking agents (Hamilton et al. 1978). Thus, the α-chain carries at least part of the ACh α-toxin-binding site.

To investigate further the role of each individual subunit, a powerful method would be to disperse these subunits and reproducibly reassemble them into a functional oligomer. An early report (Meunier et al. 1972b) mentioned that the ability to bind snake venom toxins could be preserved after dispersion of *Electrophorus electricus* receptor into lower-molecular-weight units by SDS. Haggerty and Froehner (1981) also reported that the α-subunit isolated by SDS-polyacrylamide gel electrophoresis exhibited some α-bungarotoxin binding (K_d ~0.2 μM) after elimination of SDS.

Attempts have been made to recover high-affinity α-toxin binding to the *T. marmorata* α-subunit purified by gel electrophoresis in SDS (Tzartos and Changeux 1983). Unexpectedly, the denaturing detergent SDS (but not Triton X-100 or sodium cholate) strikingly enhanced the binding of the toxin to the pure α-subunit. In the presence of 0.02% SDS, nearly 100% of the sites were recovered with K_d of ~3 nM (still 10–130 times higher than the values reported with the intact receptor). Under the same conditions, α-bungarotoxin did not significantly bind to the β-, γ-, or δ-subunits (Fig. 2). Unlabeled α-toxins from different snakes competed with labeled α-bungarotoxin at the level of the isolated α-subunit, but α-bungarotoxin was more effective than the other toxins tested. Cholinergic ligands also displaced α-bungarotoxin bound to the pure α-subunit but in a manner different from that found with the native oligomer. The bisquaternary agonist decamethonium and the bisquaternary antagonist hexamethonium efficiently inhibited α-bungarotoxin binding to the pure α-subunit, but at variance with what was found with the native receptor, the monoquaternary

*Oswald and Changeux 1982; Tzartos and Changeux 1983 and in prep.

agonist carbamylcholine had nearly the same effect as NaCl, i.e., no significant effect. One possible explanation for this difference is that in the state and conformation adopted by the isolated subunit in SDS, the anionic site where carbamylcholine, ACh, and one of the quaternary groups of deca- or hexamethonium bind does not coincide with the main area to which α-bungarotoxin attaches. In contrast, on the isolated subunit, the second anionic subsite to which the bisquaternary ligands bind would be more closely related to the α-bungarotoxin site.

The availability of a large library of monoclonal antibodies directed against conformation-dependent and -independent sites from the various subunits of the AChR (Tzartos and Lindstrom 1980; Tzartos et al. 1981, 1982) offered the opportunity to investigate the conformation of the isolated α-subunit. Most of the monoclonal antibodies derived from rats immunized against SDS-denatured α-subunit bound well to the isolated toxin–α-subunit complex. On the other hand, most conformation-dependent anti-α-subunit monoclonal antibodies (2, 6, 47, 50) bound weakly, although much stronger than those reported with the denatured and reduced ^{125}I-labeled α-subunit from *T. californica* (Tzartos and Lindstrom 1980; Gullick and Lindstrom 1982). Thus, under the present conditions of assay, the α-subunit recovered a high-affinity α-toxin-binding site but only some of the features characteristic of its native conformation (Fig. 2).

Further substantial improvement toward the renaturation of the SDS-isolated α-subunit was achieved by diluting the purified subunit into 1% lipids (asolectin) plus 0.5% sodium cholate (S. Tzartos and J.P. Changeux, in prep.). After elimination of the detergent by an anion exchange resin, the lipid-stabilized α-subunit exhibited a very high affinity for [^{125}I]α-bungarotoxin (K_d ~0.5 nM). In addition, both the monoquaternary agonist carbamylcholine and the bisquarternary antagonist hexamethonium efficiently competed with the toxin. Furthermore, the noncompetitive blockers, meproadifen, trimethisoquin phencyclidine, and, to some extent, decamethonium (known to possess both agonistic and noncompetitive blocking properties) enhanced α-bungarotoxin binding, indicating that the isolated α-subunit possesses a site for these allosteric effectors (see next section). Also, the reactivity of several monoclonal antibodies against the main immunogenic region, a segment of the chain particularly sensitive to conformation, was recovered with this lipid-stabilized α-subunit. Thus, the α-subunit carries sites for the snake α-toxin, the cholinergic agonists and antagonists, and also for several noncompetitive blockers.

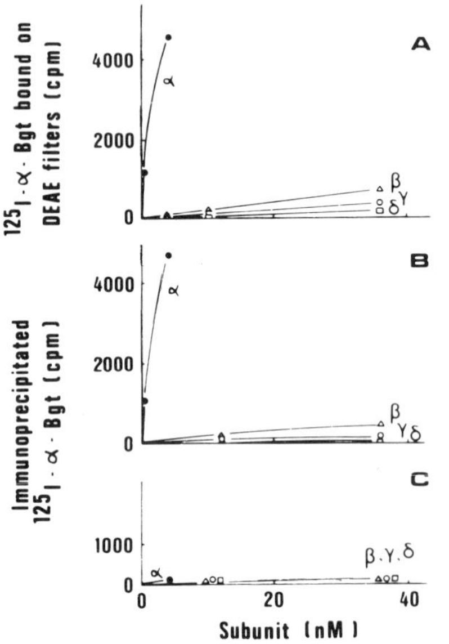

Figure 2. Binding of [^{125}I]α-bungarotoxin to the isolated subunits of *T. marmorata* AChR dispersed in SDS (0.02% SDS, 0.04% cholate, 50 mM Tris [pH 7.4]). (*A*) Specific binding of [^{125}I]α-bungarotoxin to the α-chain measured by DEAE filter assay. (*B*) Immunoprecipitation of the α-chain–α-bungarotoxin complex by anti-α-subunit monoclonal antibody 142. (*C*) Absence of precipitation of the α-chain–[^{125}I]α-bungarotoxin complex by the anti-β-, γ-, and δ-subunit monoclonal antibodies. (Reprinted, with permission, from Tzartos and Changeux 1983.)

The other subunits of the AChR oligomer, the β-, γ-, and δ-chains, may also contribute to the binding areas of the snake α-toxins. The already mentioned photolabile derivatives of α-bungarotoxin (Hucho 1979; Hucho et al. 1979; Witzemann et al. 1979) or radioactive α-bungarotoxin in the presence of cross-linking agents (Hamilton et al. 1978; Karlin et al. 1979) gave covalently linked products of the β-, γ-, and δ-chains, in addition to a prominent labeling of the α-chain. However, the interpretation of the cross-linked chains in terms of binding site topology was, in these studies, complicated by the presence of an intervening cross-linking arm. A simple and direct technique of UV-irradiation (Oswald and Changeux 1981a) was therefore used to investigate the close topology of the receptor–α-bungarotoxin complex (Oswald and Changeux 1982).

UV exposure of [^{125}I]α-bungarotoxin bound to native receptor-rich membranes from *T. marmorata* indeed resulted in the covalent labeling of several bands with apparent molecular weights on SDS gels that coincided with those of the α-, γ-, and δ-chains increased by the molecular weight of the α-toxin. Approximative labeling ratios of the chains were α:4, β:0, γ:1, δ:2. Pretreatment with unlabeled α-bungarotoxin or, after reduction, with MPTA completely abolished the labeling of the three chains. On the other hand, the noncompetitive blockers tested (phencyclidine and 5-azidotrimethisoquin; see next section) did not exert any significant protective effect. The labeling thus takes place at the level of the α-toxin–ACh-binding area. The simplest interpretation of the observed multichain labeling is that α-bungarotoxin attaches primarily to amino acid side chains from the α-subunit but also to neighboring γ- and δ-chains; accordingly, since two molecules of α-bungarotoxin are known to bind to the $\alpha_2\beta\gamma\delta$ oligomer, one of the sites might be built from one of the α-chains plus the δ-chain, the other by the second α-chain and the γ-chain. This interpretation is consistent with the observation that the ACh- or α-toxin-binding sites exhibit different binding properties (for review, see Changeux 1981; Sine and Taylor 1981). The relatively large size of the α-toxins most likely accounts for the extensive overlapping of their binding areas over several receptor subunits, yet no evidence exists for a direct contribution of subunits other than the α-chain to the site where ACh and the competitive antagonists exert their pharmacological action.

The Allosteric Sites for Noncompetitive Blockers*

Reconstitution experiments have shown that the AChR carries both the ACh-binding sites and the ion channel or ionophore (for review, see Changeux 1981; Anholt et al. 1983; Popot 1983). A class of pharmacological agents distinct from the typical competitive antagonists, and referred to as noncompetitive blockers, has been postulated to interfere, directly and/or indirectly, with the channel (for review, see Changeux 1981). Early experiments with *E. electricus* organ failed to show any correlation between the blocking of the permeability response by local anesthetics and their effect on the binding of snake venom α-toxins (Weber and Changeux 1974), indicating that the binding site(s) for these noncompetitive blockers differs from the ACh-binding site (see also Cohen et al. 1974; Krodel et al. 1979; for review, see Changeux 1981; Heidmann et al. 1983a). The sites involved have been explored with a fluorescent noncompetitive blocker quinacrine (Grünhagen and Changeux 1976), and a variety of radiolabeled derivatives of local anesthetics (Cohen 1978; Krodel et al. 1979; Sobel et al. 1980), of the frog toxin perhydrohistrionicotoxin ($H_{12}HTX$) (Eldefrawi et al. 1977, 1978, 1980a; Elliott and Raftery 1977, 1979; Aronstam et al. 1981), and of the hallucinogen phencyclidine (Eldefrawi et al. 1980b; Kloog et al. 1980; Oswald and Changeux 1981a,b). Interestingly, under equilibrium conditions, these noncompetitive blockers reversibly enhance (to different extents) the binding of cholinergic ligands to the AChR site (Cohen et al. 1974), and conversely cholinergic ligands potentiate their binding (Grünhagen and Changeux 1976; Krodel et al. 1979; for review, see Changeux 1981; Heidmann et al. 1983a). Reciprocal positive heterotropic interactions thus take place between the AChR sites and the allosteric sites for noncompetitive blockers.

Detailed equilibrium binding studies with radioactive derivatives of $H_{12}HTX$, phencyclidine, two local anesthetics (meproadifen and trimethisoquin), the neuroleptic chlorpromazine, and the detergent Triton X-100 recently led to the distinction of two main categories of allosteric sites for noncompetitive blockers on the receptor protein (Heidmann and Changeux 1981; Heidmann et al. 1983a).

*Oswald and Changeux 1981a,b; Heidmann et al. 1982, 1983a.

High-affinity sites. The noncompetitive blockers tested bind to this population of sites following a simple hyperbolic law and with an equilibrium dissociation constant of a fraction of to several micromolar. This binding is inhibited by $H_{12}HTX$ and takes place with the rank order of potency $H_{12}HTX$ (0.2 μM) > meproadifen (0.5 μM) > phencyclidine (0.8 μM) > Triton X-100 (5 μM). Careful measurements give a stoichiometry of one common high-affinity site per two α-toxin sites, i.e., per $\alpha_2\beta\gamma\delta$ light form (Fig. 3a). Carbamylcholine but not erabutoxin b enhances the affinity of the noncompetitive blockers for this site (1.5-fold for $H_{12}HTX$ up to 10-fold for Triton X-100, chlorpromazine, and

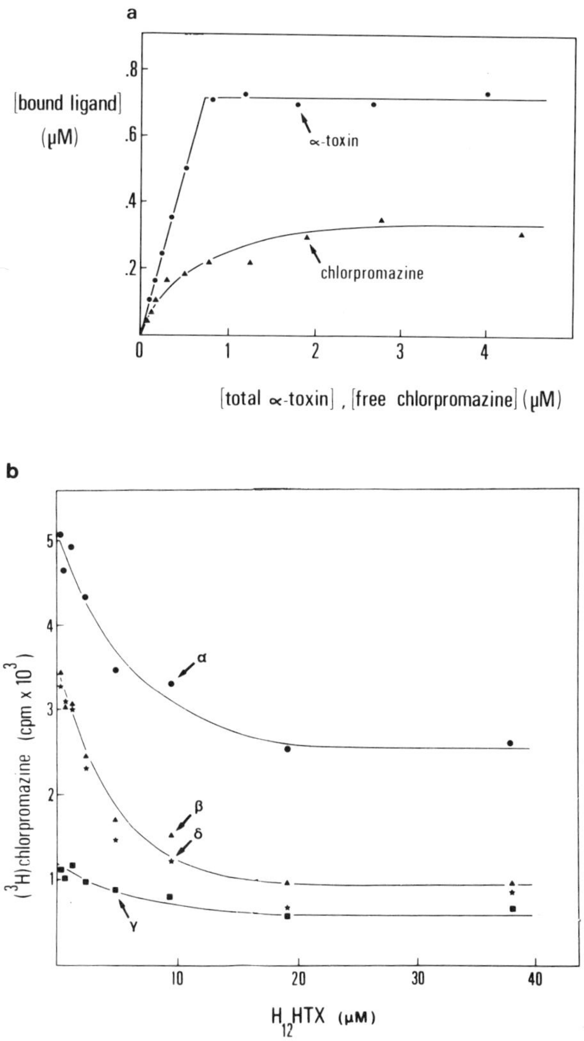

Figure 3. (*a*) Stoichiometry of the α-toxin (or ACh)-binding sites and of the $H_{12}HTX$-sensitive high-affinity site for [^{3}H]chlorpromazine. There are two α-toxin sites per [^{3}H]chlorpromazine high-affinity site. (Reprinted, with permission, from Heidmann et al. 1982, 1983a.) (*b*) Covalent labeling by [^{3}H]chlorpromazine of the four chains of the AChR from *T. marmorata* and its inhibition by perhydrohistrionicotoxin. The labeling was achieved by UV-irradiation of a suspension of membrane fragments in the presence of 1 mM carbamylcholine, 0.5 μM [^{3}H]chlorpromazine, and the concentration of $H_{12}HTX$ indicated in the abscissa. (Reprinted, with permission, from Heidmann et al. 1982.)

meproadifen and with an intermediate value of 5 for phencyclidine).

Low-affinity sites. All the noncompetitive blockers tested also interact with a population of low-affinity sites with a rank order different from that found with the high-affinity site. Chlorpromazine (~19 μM) and trimethisoquin (~39 μM) are the most potent, whereas phencyclidine (~170 μM) and mostly H_{12}HTX (>100 μM) are the least. A large number of sites are involved with native receptor-rich membranes (~10–20 times the number of α-toxin sites), and cholinergic ligands have only a limited effect on the binding of the noncompetitive blockers to these sites. Moreover, reconstitution experiments with the purified receptor and lipids (asolectin) have shown that the number of these sites depend on the lipid-to-protein ratio, suggesting that they are located at the interface of the receptor protein with membrane lipids.

Since reconstituted vesicles containing only the purified receptor and lipids exhibit almost identical binding properties for noncompetitive blockers as the native membrane (Sobel et al. 1980; Heidmann et al. 1983a), their site must be carried by the receptor protein itself and not by nonreceptor peptides (Neubig et al. 1979), such as the 43K protein (Sobel et al. 1977, 1978) present in the receptor-rich membranes.

An unambiguous demonstration that the high-affinity site for noncompetitive blockers is localized on the AChR protein resulted from labeling studies done with a radioactive photoaffinity derivative of the potent local anesthetic trimethisoquin (5-azido-trimethisoquin synthesized by Waksman et al. [1980]). With *T. marmorata* receptor, this compound selectively labeled the δ-chain; moreover, this labeling was inhibited by an excess of trimethisoquin (and H_{12}HTX) and enhanced by cholinergic agonists (and some antagonists) but not by the snake α-toxins (Oswald and Changeux 1981a,b). Similar results were obtained with [^{3}H]H_{12}HTX, [^{3}H]phencyclidine, [^{3}H]trimethisoquin, and [^{3}H]chlorpromazine. In all cases, labeling was enhanced by the agonist carbamylcholine and decreased by H_{12}HTX.

With [^{3}H]chlorpromazine, all four chains of the receptor were labeled (after UV-irradiation) in addition to the δ-chain (Oswald and Changeux 1981a; Heidmann et al. 1982, 1983a). The incorporation of [^{3}H]chlorpromazine was not identical in all chains but decreased in a concentration-dependent manner and with identical inhibition constants for H_{12}HTX or phencyclidine (Fig. 3b). With all four chains, the constants were the same as those found for the displacement of reversibly bound [^{3}H]chlorpromazine to its high-affinity H_{12}HTX-sensitive site. In addition, the amount of [^{3}H]chlorpromazine incorporated in an H_{12}HTX-sensitive manner was found close to the amount reversibly bound to the nonirradiated membranes. Conclusively, [^{3}H]chlorpromazine shares the same high-affinity site as H_{12}HTX or phencyclidine. When bound to this site, it can be attached covalently to any of the four chains of the receptor.

The exact mechanism of this covalent attachment is not yet completely understood, but (1) the excited states of chlorpromazine are short lived (<μsec, Navaratnam et al. 1978), (2) the wavelength of UV-irradiation corresponds to a major peak of free chlorpromazine absorption, and (3) the rate of dissociation of the noncompetitive blockers from this site is rather slow.

The data presently available support the conclusion that the covalent attachment takes place locally and therefore that the four polypeptide chains contribute to the unique high-affinity site for the noncompetitive blockers on the $\alpha_2\beta\gamma\delta$ oligomer.

Recently, Haring et al. (1983) have reported striking species differences in the pattern of the chains labeled by an azido derivative of phencyclidine. For instance, this ligand is primarily incorporated in the α-chain with *T. californica* receptor and, in agreement with the results of Oswald and Changeux (1981a,b), in the δ-chain with *T. marmorata* receptor. Similar observations have been made independently by Muhn and Hucho (1983) and Kaldany and Karlin (1983) with covalent derivatives of other noncompetitive blockers. Significant differences in the amino acid sequence of the α-chain (11 mutations in the mRNA coding sequence among which 6 introduce changes in the amino acid sequence) exist between these two species of *Torpedo* (Devillers-Thiéry et al. 1983). The various labeling patterns noticed between the different species of *Torpedo* might thus be explained by differences in the reactivity and/or accessibility of the amino acid side chains able to establish a covalent bond with the photosensitive noncompetitive blocker positioned at its multichain, high-affinity site.

A Model for the Transmembrane Organization of the AChR Oligomer*

This model is based on the known structural and functional properties of the AChR (for review, see Changeux 1981; Kistler et al. 1982; Anholt et al. 1983; and present paper).

The five subunits of the light-form oligomer traverse the membrane bilayer (Strader and Raftery 1980; Wennogle and Changeux 1980; Froehner 1981; Saint-John et al. 1982) and are packed as a cylindrical bundle with the long axis of fivefold cryptosymmetry perpendicular to the plane of the membrane. The five subunits are disposed around this axis with the two nonneighboring α-chains oriented at ~160° from each other (see High-affinity sites, above, and Bon et al. 1982 and in prep.).

In a side view, the molecule resembles a thick funnel with a central channel at least 5.5 nm deep, for a total length of 11 nm (Cartaud et al. 1978; Klymkovsky and Stroud 1979; Kistler et al. 1982). According to Klymkovsky and Stroud (1979), approximately 51–56% of the protein mass would extend into the synaptic cleft, 36–45% would be incorporated in the membrane, and 6–9% would face the cytoplasm.

The two α-subunits carry, on their synaptic face, the

*Changeux 1981; Bon et al. 1982 and in prep.; Devillers-Thiéry et al. 1983; Heidmann et al. 1983a).

two binding sites for cholinergic ligands and snake α-toxins. The unique high-affinity site for noncompetitive blockers (see Allosteric Sites, above, and Heidmann et al. 1983a) is located within the central depression that constitutes the unique region of the molecule where the distances to all five chains are minimum (a similar location has been described for the unique allosteric site for diphosphoglycerate in hemoglobin [see Arnone 1972]). The multiple low-affinity sites for noncompetitive blockers are located at the lipid-protein interface (see Allosteric Sites, above, and Heidmann et al. 1983a).

Straightforward inspection of the complete mRNA coding sequence of the α-subunit from *T. marmorata* (Devillers-Thiéry et al. 1983) leads to the subdivision of the chain into several characteristic domains: first, a 24-amino acid-long aminoterminal signal peptide that is cleaved off in the mature protein, two domains (residues 1–209 and 299–408) rich in hydrophilic amino acids, and two other domains (residues 210–298 and 409–437) with a highly hydrophobic character. The hydrophobic domains can be themselves subdivided into four α-helices of about 20 amino acids which traverse the lipid bilayer. The first aminoterminal hydrophilic domain is assigned to the cleft face of the subunit, since trypsin attacks the α-chain from the cytoplasmic side at its carboxyterminal end (Wennogle and Changeux 1980). The second hydrophilic domain thus would face the cytoplasm and expectedly contains 14 lysine or arginine residues that give potential cleavage sites for trypsin. The first hydrophilic domain contains four cysteine residues, of which two are plausibly located in the vicinity of the ACh-binding site. The only possible site of *N*-glycosylation (asparagine 141) is also present on the same domain (but several other residues could be *O*-glycosylated) (Fig. 4).

Helix III presents an interesting feature. Projected in two dimensions, it shows a vertical line of polar residues: tyrosine 277, threonine 281, serine 288, threonine 292, and histidine 299. A similar disposition is also found on helix III from other subunits (see Noda et al. 1983a,b). These hydrophilic faces could delimit a transmembrane hydrophilic channel with an internal diameter of ~6.6 Å, which could, of course, serve as the ionophore. Interestingly, this line begins (on all four subunits) by a tyrosine that could photoreact with chlorpromazine. However, as stressed by Devillers-Thiéry et al. (1983), the known cationic selectivity of

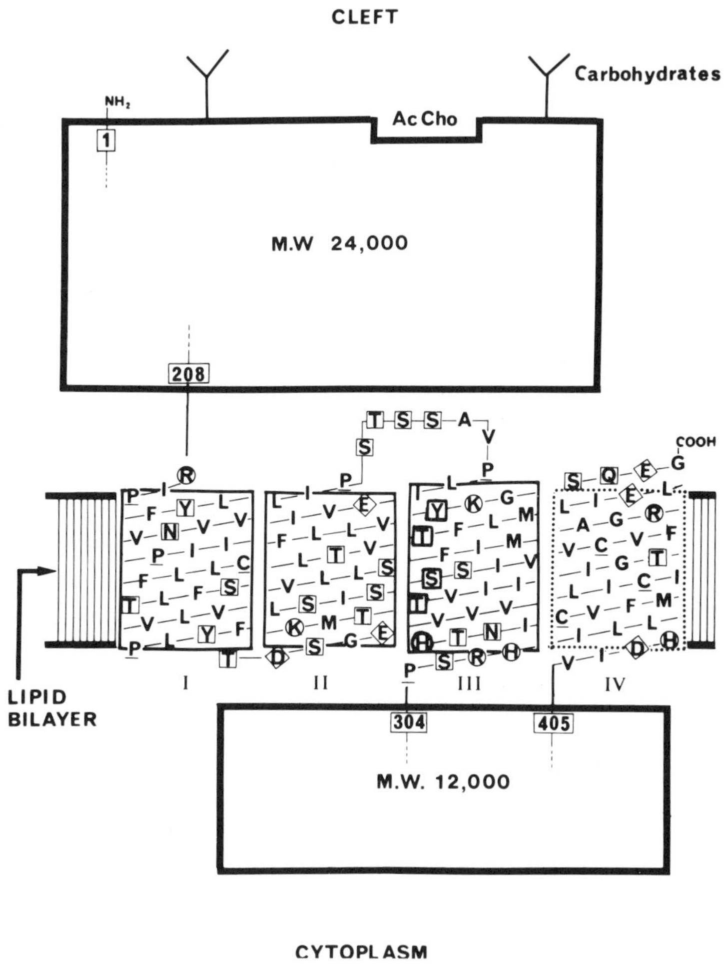

Figure 4. A model for the transmembrane organization of the α-chain from *T. marmorata*. (Reprinted, with permission, from Devillers-Thiéry et al. 1983.)

the ionophore can hardly be explained solely on the basis of the uncharged amino acids present on the hydrophilic surface of helix III, and parts of the extrinsic synaptic and cytoplasmic domains might also be involved in the selectivity process. Accordingly, Kosower (1983), R. Guy (in prep.), and Fairclough et al. (this volume) have suggested that regions of the cytoplasmic hydrophilic domain make an additional transmembrane loop that would directly contribute to the ion transport. As a consequence, the position of helix IV would be inverted and the carboxyterminal amino acid would then face the cytoplasm.

A model similar to that of Devillers-Thiéry et al. (1983) has been proposed independently by Claudio et al. (1983) for the γ-chain and by Noda et al. (1983b) for the α-, β-, γ-, and δ-chains. These latter authors favor helix I as a possible component of the ion channel, since it appears to be the most conserved helix in the four subunits and between different species. But, the most conserved sequences might also be the basis of the symmetrical assembly of the subunits and thus contribute to the areas of association between subunits.

In any case, the models discussed are consonant with the concept that the interaction between the ACh-binding site and ion channel is indirect, or allosteric, since each of these two sites would be carried by distinct segments of the polypeptide chains. The model proposed and its detail variations (concerning in particular the position of helix IV) should be easy to test by selective peptide labeling.

Allosteric Transitions of the AChR Molecule*

The model proposed in the section on allosteric sites (see above) posits that the ACh-binding sites and the ion channel belong to different domains of the receptor molecule. The regulation of the state of opening of the ion channel by ACh then has to be mediated by an indirect coupling mechanism, i.e., by one (or several) conformational transitions of the receptor molecule (see Monod et al. 1963, 1965; Changeux et al. 1976; Changeux 1981). The AChR-rich membrane fragments from *Torpedo* (Cohen et al. 1972; Duguid and Raftery 1973; Sobel et al. 1977) and the purified AChR protein reconstituted into artificial lipid bilayers (see Heidmann et al. 1980; Popot et al. 1981; for review, see Anholt et al. 1983; Popot 1983) are particularly adequate systems to analyze these transitions since, on the same preparation, one can measure, in parallel, the permeability response (e.g., by following the transport of radioactive cations [Kasai and Changeux 1971; Popot et al. 1981]) and the structural changes of the receptor protein by the relevant physicochemical technique (see Changeux 1981). Moreover, the recent introduction (Hess et al. 1979, 1981; Neubig and Cohen 1980; Neubig et al. 1982; see also Moore and Raftery 1980) of rapid mixing and quench-flow techniques allows a quantitative analysis of the rates of agonist-triggered ion transport in the millisecond to minute time scale.

*Heidmann et al. 1983a,b; Oswald et al. 1983.

Rapid kinetics of Dns C6 Cho binding (Heidmann and Changeux 1979, 1980; Heidmann et al. 1983a,b). Early work on the AChR revealed a striking variability of the binding properties of the molecule, suggesting its eventual interconversion between different affinity states (see Meunier and Changeux 1973; Changeux 1975). Following the binding of radiolabeled snake α-toxins, Weber et al. (1975) provided the first evidence that, in the resting membrane, the receptor protein is present under a state of low affinity and that cholinergic agonists trigger a slow (second-minute) transition that results in the stabilization, at equilibrium, of a state D of high affinity for agonists (K_d ~3 nM for ACh [see also Weiland et al. 1977]). Subsequently, rapid kinetic measurements with fluorescent (Dns C6 Cho) (Heidmann and Changeux 1979) or radiolabeled (Boyd and Cohen 1980; Neubig et al. 1982) agonists led to the identification of a faster transition (0.1–1 sec) toward a second conformation with lower affinity (K_d ~1 μM for ACh). A very low-affinity binding (to the R and/or A states) was characterized by following the binding kinetics of Dns C6 Cho in the millisecond time range (Heidmann et al. 1983b). Finally, a third conformation A was characterized by following the binding kinetics of Dns C6 Cho in the millisecond time range (Heidmann et al. 1983b). The equilibrium dissociation constant of Dns C6 Cho for this state was found in the 50–100 μM range, i.e., close to the concentration of ACh in the synaptic cleft during transmission.

Detailed analysis of the rate constant of the intermediate relaxation process (leading to the I state) as a function of Dns C6 Cho concentration further revealed positive cooperative effects (Hill coefficient $n_H = 1.9$) for the binding of Dns C6 Cho to the states that precede the I state. Thus, in agreement with the known stoichiometry of the ACh-binding sites, and as expected for a typical allosteric protein (see Monod et al. 1965; Changeux 1981), at least two molecules of Dns C6 Cho cooperate in a homotropic manner to this low-affinity binding.

The data presently available on the rapid kinetics of agonist binding are thus simply fitted by a minimal four-state model of the AChR (see Fig. 6) with two strongly interacting ACh-binding sites (Heidmann and Changeux 1980; Neubig and Cohen 1980).

Rapid ion flux measurements (Heidmann et al. 1983b). Regulation of ion transport by ACh is a complex, multistep process. Mixing of the closed membrane fragments with a cholinergic agonist markedly accelerates ion transport (in or out) (Kasai and Changeux 1971). On the other hand, preincubation of the membranes with an agonist causes a subsequent time-dependent decline of the amplitude of the permeability response; desensitization follows the activation of the ion channel. Both the in vitro electrophysiological recordings (of single channels [Sakmann et al. 1980] or of end-plate conductance decay [Feltz and

Trautmann 1980, 1982]) and in vitro rapid ion flux measurements (Walker et al. 1981, 1982) revealed that desensitization is a two-step process that includes a "rapid" (10–100 msec) and a "slow" (seconds) component. The fluorescent agonist Dns C6 Cho elicits the activation and the two-step desensitization processes, thus making possible a comparison, within the same time scale and with the same membrane preparation, of the rapid binding and flux data (Heidmann et al. 1983b) (Fig. 5). The results are the following: (1) The initial ion flux increases in a cooperative manner ($n_H \sim 1.8$) with Dns C6 Cho concentration and levels off in the 150–200 μM range with a half-maximum rate at ~90 μM. The curve closely follows the evolution of the apparent rate constant of the intermediate relaxation process which, in the fluorescence experiments, was assumed proportional to the fraction of receptor molecules in the R-A state. Thus, in the low-affinity R-A state, the ion channel is in its open conformation. (2) The rate constants for the rapid and slow desensitizations of the permeability response closely follow the isomerization rate constant of the receptor toward the I and D states, as measured by fluorescence experiment. Thus, in both the I and D states, the ion channel is closed. In conclusion, the minimum four-state model (Fig. 6) adequately fits the rapid binding and flux measurements.

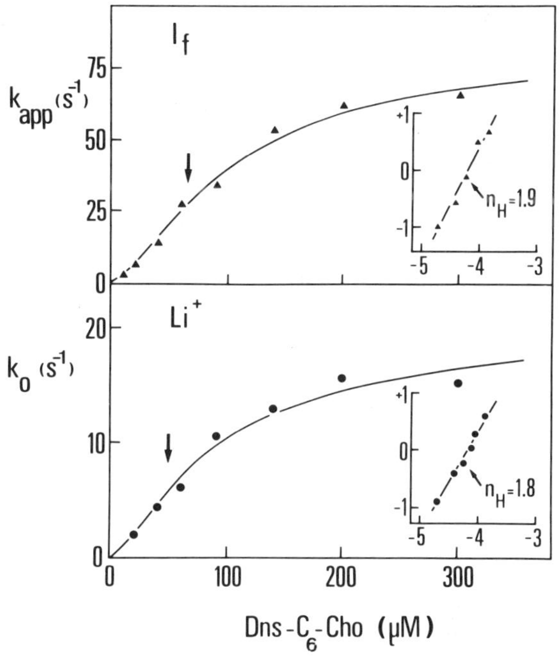

Figure 5. Comparison of the rapid kinetic data for binding of the fluorescent agonist Dns C6 Cho (I_f, *top*) and for opening of the ion channel (Li^+, *bottom*). K_{app} is the rate constant of the intermediate relaxation process directly proportional to the fraction of liganded receptor molecules, and K_0 is the initial flux rate.

Interactions with noncompetitive blockers (Heidmann et al. 1983a; Oswald et al. 1983). As mentioned in the section on allosteric sites (above), under equilibrium conditions, the noncompetitive blockers enhance the binding of cholinergic ligands to the AChR site and conversely. Reciprocal effects between the two categories of ligands are also observed for their binding kinetics:

1. The noncompetitive blockers accelerate the slow transition toward the high-affinity D state of the AChR (Grünhagen and Changeux 1976; Weiland et al. 1977; Heidmann and Changeux 1979; Sine and Taylor 1982). In the absence of the agonist, they stabilize the high-affinity D state. Thus, as expected from a typical allosteric transition (Monod et al. 1965), this transition takes place between preexisting discrete states (Heidmann and Changeux 1979; Cohen et al. 1980; Young and Sigman 1981; Sine and Taylor 1982). They also accelerate the "intermediate" transition toward the I state (Heidmann and Changeux 1979) and thus bind to this state in addition to the high-affinity D conformation.

   The noncompetitive blockers may regulate these transitions when they bind to their high-affinity site but also when they interact with the low-affinity, H_{12}HTX-insensitive ones (Heidmann et al. 1983a). In this last case, positive cooperative effects are also noticed. Interestingly, the kinetic data are fitted by the simplest model which postulates that the same D conformation is stabilized by the noncompetitive blockers when they interact with either their high- or their low-affinity sites. Thus, the conformational transitions of the receptor result from interconver-

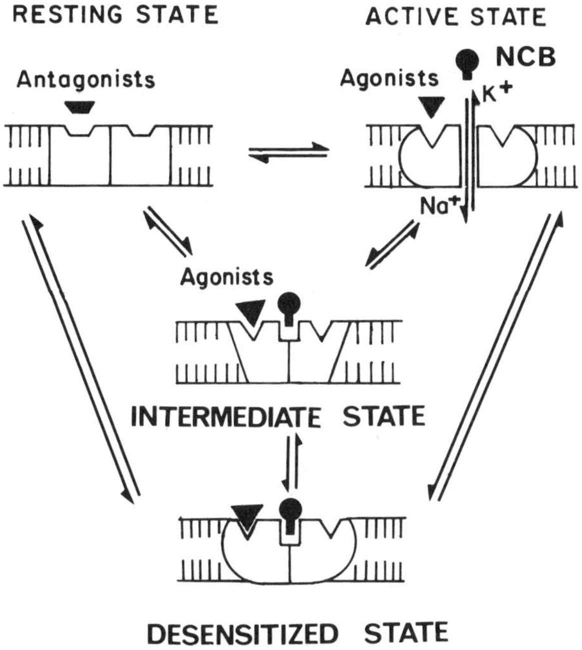

Figure 6. Minimum four-state model for the allosteric transitions of *Torpedo* AChR. The channel is open only in the active state. (NCB) Noncompetitive blocker.

sions between a small number of discrete states (Monod et al. 1965).

2. Rapid mixing with cholinergic agonists transiently enhances, in the millisecond time scale, the $H_{12}HTX$-sensitive fluorescence response of the receptor to agonists, and this effect is not observed with the competitive antagonists (Grünhagen et al. 1977). Agonists also transiently accelerate the binding of radiolabeled $H_{12}HTX$ to its high-affinity site upon rapid mixing with the AChR (Eldefrawi et al. 1980a; Aronstam et al. 1981). A similar effect has been observed with [^{3}H]phencyclidine. Its detailed quantitative analysis (Oswald et al. 1983) shows that when the receptor-rich membranes are equilibrated with the agonist carbamylcholine, the association and dissociation rate constants for [^{3}H]phencyclidine binding increase 6-fold and 1.3-fold, respectively. But when carbamylcholine is mixed simultaneously with [^{3}H]phencyclidine, its association and dissociation rate constants increase 10^3–10^4-fold and 4–10-fold, respectively. These effects are not observed with *d*-tubocurarine (Fig. 7). Moreover, the data are consistent with the already mentioned four-state model. They indicate that, in addition to the stabilization of the D state, phencyclidine binds with the fastest rate to the A state where the channel is open (and also accelerates the transition toward the I state). The most likely interpretation of this accelerated binding is that the transition to the A state releases barriers that restrict the diffusion of phencyclidine to its site. A correlation exists between the absolute values of the rates of phencyclidine binding and the state of opening of the ion channel. Thus, these observations support the concept that the unique and multichain high-affinity site for phencyclidine (and $H_{12}HTX$) is part of the ion channel itself. They are also consistent with the view that the noncompetitive blockers may interfere with the permeability response both by "plugging" the ion channel in its open conformation and by accelerating the transitions to the desensitized conformations.

A Hypothesis about the Physiological Significance of Desensitization: A Short-term Regulation of Synapse Efficacy at the Postsynaptic Level*

Even though desensitization of receptor response to a neurotransmitter is a widespread experimental phenomenon (at both peripheral and central synapses), its physiological significance remains obscure. A hypothesis inspired by its molecular mechanism (but not yet documented by functional studies) is that desensitization plays a role in the regulation of synapse efficacy at the postsynaptic level (Heidmann and Changeux 1982).

The central postulate (see previous section) is that the receptor spontaneously exists under at least, and to simplify, two states in reversible equilibrium: one (e.g., R) susceptible to activation by the agonists (i.e., R→A) and another one refractory to activation, or desensitized (e.g., D). If the amplitude of the postsynaptic response is directly proportional to the fraction of receptor in the R state, then the R/D ratio will simply determine the efficacy of the synapse and, of course, the ability of the postsynaptic cell to fire, when a threshold value of the membrane potential is reached.

The ratio of the two states might, itself, be regulated by allosteric effectors, such as the membrane potential (when a difference of dipolar moment exists between the two states) or ligands (when the affinity for the considered ligand differs in the two states). The regulatory ligands might be the neurotransmitters. They might also include internal effectors such as Ca^{++} or cyclic nucleotides. The binding site(s) for the regulatory ligands might be either the receptor site for the neurotransmitter or allosteric sites, such as those identified for the noncompetitive blockers in the case of the AChR (see section on allosteric sites, above) (Fig. 8).

*Heidmann and Changeux 1982.

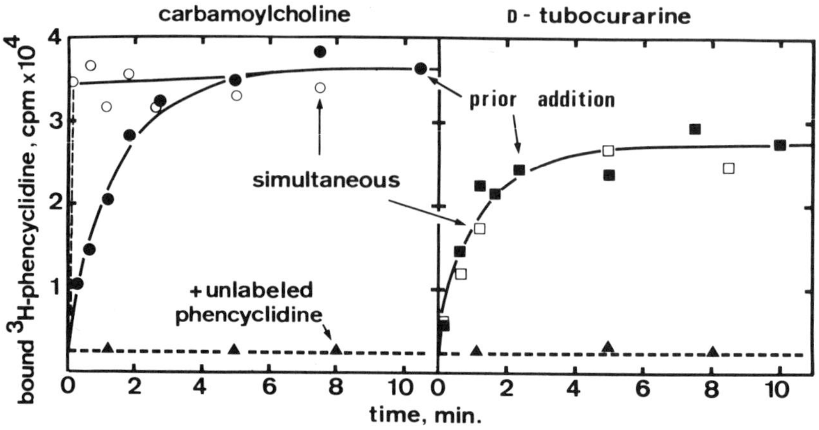

Figure 7. Compared effects of the agonist carbamylcholine and of the antagonist *d*-tubocurarine on the kinetics of [^{3}H]phencyclidine binding. Simultaneous addition of carbamylcholine and [^{3}H]phencyclidine causes a very fast (here, unresolved) binding of [^{3}H]phencyclidine that is absent with *d*-tubocurarine or when carbamylcholine is added first. (Reprinted, with permission, from Oswald et al. 1983.)

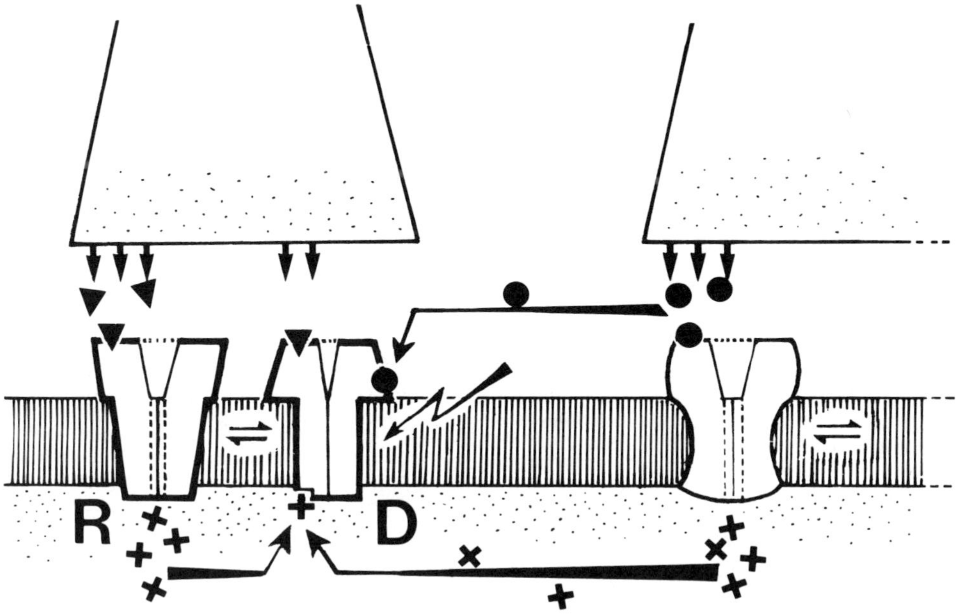

Figure 8. A model for the regulation of synapse efficacy at the postsynaptic level. The ratio of the desensitized (D) to resting (R) conformations determines the efficacy of the synapse. This ratio can be regulated by the neurotransmitter itself (▲) when it binds to the D state with a high affinity, by the neurotransmitter from a neighboring synapse (●) when it binds to an allosteric site, by an internal chemical signal (+) (e.g., Ca^{++} ions or a cyclic nucleotide), and by the membrane potential if a difference of dipole moment exists between the D and R states.

Detailed schemes have been derived on this basis for the regulation of efficacy of a given synapse by its own state of activity (homosynaptic regulation) or by that of neighboring synapses (heterosynaptic regulation) (Heidmann and Changeux 1982). The rate constants of the $R \rightleftharpoons D$ transition would then determine the duration of the change of synapse efficacy. For instance, taking the values of the constants found for *T. marmorata* AChR, effects lasting several minutes can be predicted. Finally, timing relationships between the activity of neighboring synapses may create changes of synapse efficacy (via the shift of the $R \rightleftharpoons D$ equilibrium) that are those expected in the case of the well-known experimental paradigm of classical conditioning (Heidmann and Changeux 1982). Finally, long-lasting effects might be achieved via covalent modification of the receptor protein. Such modifications (like phosphorylation-dephosphorylation reactions) are already known to play a fundamental role in the regulation of metabolic pathways (see Krebs and Beavo 1979) and might contribute to the regulation of synaptic properties via similar molecular mechanisms.

Assembly, Stabilization, and Maturation of the AChR in the Subsynaptic Membrane*

In the adult neuromuscular junction and electromotor synapse, the AChR molecules are clustered under the nerve terminal with an approximate surface density of 10–25,000 light forms per μm^2 (see Bourgeois et al. 1973, 1978 for *E. electricus*; for review, see Fertuck and Salpeter 1976, for the neuromuscular junction), whereas outside the synapse or in the embryonic muscle fiber their density is 10–1000 times lower. After denervation of the electric organ (Bourgeois et al. 1973, 1978) and of the neuromuscular junction (Frank et al. 1975), these dense accumulations of receptor persist for days (even weeks) without signs of lateral diffusion. This strong immobilization contrasts with the fast motion of the AChR in the embryonic muscle membrane (see Axelrod et al. 1976). Also, in the adult neuromuscular junction, the metabolic degradation rate of the receptor protein is longer than 10 days, whereas it is about 18 hours in embryonic myotubes (for review, see Fambrough 1979; Weinberg et al. 1981). In addition, the ionic channel associated with the receptor (with rat and frog but not with chick) has a mean open time three to five times shorter in the adult junction than in the nonsynaptic areas (for review, see Michler and Sakmann 1980). Several molecular mechanisms may contribute to these differences: (1) differential gene expression (see Giraudat et al. 1982; Ballivet et al. 1982); (2) posttranscriptional epigenetic (Changeux and Danchin 1976; Changeux 1983) events—for instance, covalent modifications such as phosphorylation (Teichberg and Changeux 1977; Saitoh and Changeux 1981; Davis et al. 1982; Huganir and Greengard 1983), glycosylation (Roisin et al. 1983), and acetylation and/or changes in membrane environment (Rousselet et al. 1979; Saitoh et al. 1979; Cartaud et al. 1981). Only two such epigenetic processes will be mentioned here: the transition of the AChR from the light to the heavy form and its association with the 43K ν_1 protein.

To identify possible structural differences between

*Cartaud et al. 1981, 1982; Rousselet et al. 1982; Nghiêm et al. 1983; B. Holton et al., in prep.

embryonic and adult receptors, a library of 38 monoclonal antibodies directed against the different chains of *T. californica* receptor by Tzartos and Lindstrom (1980) and by Tzartos et al. (1981, 1982) was tested against extracts of embryonic and adult electric organs from *T. marmorata* (B. Holton et al., in prep.). Interestingly, anti-α-subunit antibodies did not distinguish between embryonic and adult receptors, but nearly all the anti-β-, γ-, and δ-subunit antibodies precipitated much more of the adult than of the embryonic receptor. In the same extract, the ratio of the heavy to light forms of the receptor resembled this immunoprecipitation pattern. Embryonic electric organ extracts contained a higher proportion of light form (70%) than did adult electric tissue (20%). By immunoprecipitation, embryonic and adult light forms or embryonic and adult heavy forms did not differ. Hence, different proportions of heavy and light forms of the receptor caused the precipitation differences noticed between embryonic and adult receptors. The explanation for different proportions of light and heavy forms in adult and embryonic tissue is not evident. Inclusion of the alkylating agent *N*-ethylmaleimide in electric organ preparations increased the percentage of heavy forms in embryonic but not in adult extracts. When [^{125}I]α-toxin-labeled adult receptor was mixed with embryonic electric organ and present throughout the membrane preparation, the ratio of heavy forms and light forms did not change. Thus, the transition from the light form to the heavy form of the receptor via the δ-chain disulfide bridge may constitute a critical step in the embryonic developmental maturation of the receptor protein and its assembly into the subsynaptic membrane (B. Holton et al., in prep.).

Receptor-rich membrane fragments purified from *Torpedo* electric organ contain, in addition to the receptor polypeptides, a protein band with an apparent molecular weight of 43,000 (Sobel et al. 1977, 1978) distinct from actin (Sobel et al. 1978; Strader et al. 1980). On two-dimensional gels, this protein (also named ν by Karlin et al. 1979) yields one major component that focuses at alkaline pH (ν_1) and two others (ν_2 and ν_3) that focus at more acidic pH (Saitoh and Changeux 1980; Gysin et al. 1981). Of these three species, only ν_1 appears strictly membrane bound (Gysin et al. 1981). Brief exposure of the membrane fragments to pH 11 releases this protein without interfering significantly with the main functional properties of the still membrane-bound receptor (Neubig et al. 1979). Evidence that the 43K (ν_1) protein plays a structural role in the subsynaptic membrane was suggested by the observation that its removal enhanced the thermal inactivation of the membrane-bound receptor (Saitoh et al. 1979) and its digestion by proteolytic enzymes (Klymkovsky et al. 1980). Moreover, the rotational motion of the membrane proteins labeled by a spin-labeled maleimide (Rousselet et al. 1979, 1981) or by a phosphorescent derivative of α-bungarotoxin (Lo et al. 1980) increased as well. Also, electron microscopy of the alkaline-treated membrane fragments showed a reorganization of the receptor rosettes in the plane of the membrane (Rousselet et al. 1979; Barrantes et al. 1980; Cartaud et al. 1981).

Bartholdi et al. (1981), working with membrane fragments labeled by a phosphorescent α-toxin, detected apparent changes of motion of their probe only at very high temperature with pH 11-treated membranes, or after reduction of intact membranes with DTT. The rotational diffusion of the membrane-bound receptor was thus reinvestigated with a spin-labeled α-bungarotoxin that exclusively bound to the AChR site (Rousselet et al. 1982). With native membranes, a complete immobilization of the toxin and thus of the receptor was observed. Reduction of the membrane by dithiothreitol had little effect on this motion, but alkaline extraction of the 43K protein(s), as expected, enhanced the rotational diffusion with correlation times in the 0.5×10^{-4} to 1×10^{-4}-second range. The interaction of the 43K (ν_1) protein with the membrane thus strongly immobilizes the AChR.

The 43K protein is distributed in an asymmetric manner on the subsynaptic membrane. It resists proteolytic attack (Wennogle and Changeux 1980) and iodination (Saint-John et al. 1982) unless the receptor-rich membrane vesicles are opened by sonication or detergent. Since the membrane fragments close with the right side out, the 43K protein(s) must lie on the inner face of the subsynaptic membrane. This observation was, however, recently challenged by Conti-Tronconi et al. (1982). Monoclonal antibodies directed against the 43K proteins were thus developed in mice (Nghiêm et al. 1983). One of them reacted exclusively with the 43K ν_1 polypeptide but not with ν_2 and ν_3 and was used for localization studies at the electron microscope level by the immunogold and immunoperoxidase techniques. Interestingly, this antibody only labeled the innervated membrane of *T. marmorata* electroplaque and exclusively on its inner cytoplasmic face (Nghiêm et al. 1983).

Electron microscope observations of *Torpedo* electroplaque disclose after glutaraldehyde tannic acid fixation a dense "coating" on the inner cytoplasmic face of the subsynaptic membrane often referred to as "postsynaptic densities" (Cartaud 1980; Sealock and Kavookjian 1980; Cartaud et al. 1981; Sealock 1982). These densities that still adhere to the purified receptor-rich membranes disappear after the alkaline treatment (Sobel et al. 1980; Cartaud et al. 1981). Intermediate filaments often appear intermingled with these condensations (Cartaud et al. 1981). A filamentous network also becomes apparent after cross-linking of the ν_1 polypeptide by the bifunctional reagent *para*-azidophenacyl bromide (Cartaud et al. 1982); globular entities, which most likely represent receptor molecules, are found in contact with the meshwork. Thus, it appears plausible that an interaction of the receptor protein with the cytoskeleton contributes to the immobilization of the receptor protein in the subsynaptic membrane. Accordingly, the 43K ν_1 protein would play a role of "intermediate piece" between the receptor protein and some components of the cytoskeleton and, in this respect, would resemble ankyrin in the red blood cell membrane (for review, see Branton et al. 1981) (Fig. 9).

CONCLUSIONS

The AChR possesses properties characteristic of typical allosteric proteins, directly relevant to its physiological role as a chemoelectrical transducer. On the other hand, it displays structural features that one may plausibly relate to its vectorial integration into the postsynaptic membrane and to its localization at a well-defined cellular locus.

Cytoplasmic oligomeric proteins most often present a C_2 dyad axis of rotational symmetry. Such a disposition rarely occurs in the heavy-form dimer of the AChR where the two associated light forms can, most often, be deduced from each other by translation. Moreover, the heterologous, light-form pentamer cannot possess any true axis of symmetry. Nevertheless, the presently available data (in particular, the important sequence homologies between the chains) are compatible with the presence of a C_5 rotational axis of cryptosymmetry perpendicular to the plane of the membrane with a regular and fixed disposition of the five subunits.

The two regulatory sites that bind ACh (and the snake α-toxins) are carried by the two, nonneighboring α-subunits and exposed to the synaptic cleft. Cross-linking experiments also show that other subunits contribute to the two α-toxin sites. Cooperative homotropic interactions take place between the two ACh sites.

The $\alpha_2\beta\gamma\delta$ oligomer contains the ion channel, but we still ignore its precise location. The central axial "funnel" might be part of the channel. Most likely, the unique and multichain high-affinity site for noncompetitive blockers lies at its level. Yet, alternative structural models (such as, for instance, multiple ionic pores traversing the individual subunits or crossing the areas of association between neighboring subunits) cannot be excluded.

In addition to the unique high-affinity and H_{12}HTX-sensitive site for noncompetitive blockers, multiple low-affinity and H_{12}HTX-insensitive sites for these ligands are present on the receptor molecule and are most likely located at the interface with the lipid bilayer. These sites and the ACh sites interact with each other in an allosteric, heterotropic manner.

The receptor protein undergoes a cascade of discrete conformational transitions upon exposure to cholinergic ligands and noncompetitive blockers. The kinetic data on agonist binding and ion transport in the millisecond-to-minute time scale are fitted by a minimum four-state model. The fast transition from the resting to the active conformation accounts for the opening of the channel and takes place in the 100 μM concentration range of ACh, i.e., close to that of ACh in the cleft during synaptic transmission. It includes the positive homotropic interactions between at least the two ACh sites and the release of barriers that hinder the access of noncompetitive blockers to their high-affinity sites. In addition, the receptor may, more slowly, interconvert to conformations (at least two) where the channel is closed and the affinity for ACh higher (K_d from μM to nM). Similar (sometimes slightly different) states of high affinity for agonists are stabilized by noncompetitive blockers, in the absence of the agonists, suggesting that the receptor protein, as expected for a typical allosteric protein, undergoes structural transitions between major discrete states which preexist to agonist binding. These transitions account for the two-step desensitization of the permeability response. Future work should decide whether or not these slow transitions play a physiological role in a short-term regulation of synapse efficacy.

The molecular mechanisms responsible for the localization, stabilization, and maturation of the receptor protein in the postsynaptic membrane during development are not yet identified. The covalent binding of the receptor light form into the heavy form and the association with the 43K ν_1 polypeptide on the in-

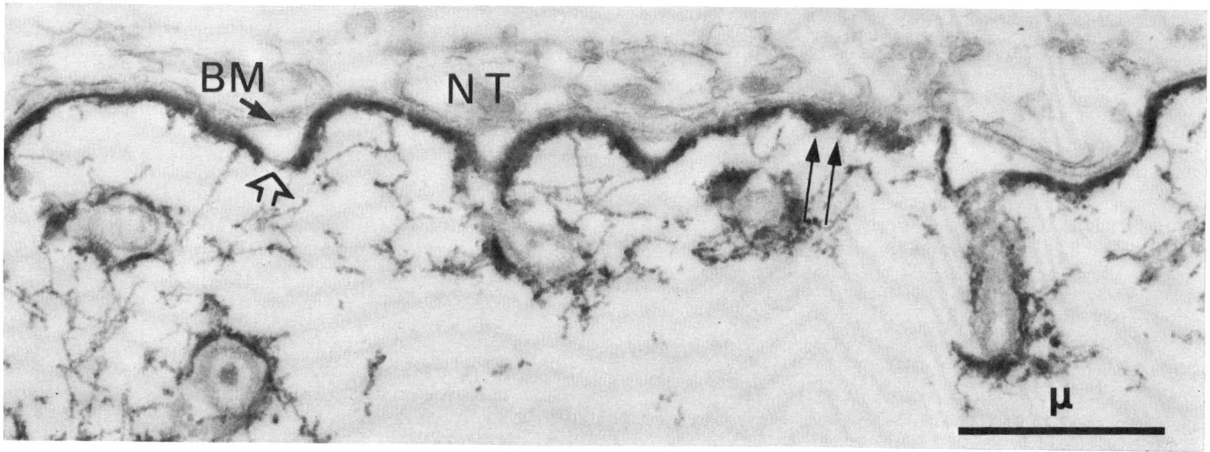

Figure 9. Labeling of a fragment of *T. marmorata* electroplaque by a monoclonal antibody directed against the 43K ν_1 polypeptide revealed by the immunoperoxydase method. (NT) Remnant of nerve terminal; (BM) basal lamina. Only the cytoplasmic face of the postsynaptic membrane is labeled with occasionally some interruption (open arrow).

ner face of the postsynaptic membrane might contribute to the assembly of this supramacromolecular structure.

REFERENCES

ANHOLT, R., J. LINDSTROM, and M. MONTAL. 1980. Functional equivalence of monomeric and dimeric forms of purified acetylcholine receptors from *Torpedo californica* in reconstituted lipid vesicles. *Eur. J. Biochem.* **109:** 481.

——. 1981. Stabilization of acetylcholine receptor channels by lipids in cholate solution. *J. Biol. Chem.* **256:** 4377.

——. 1983. The molecular basis of neurotransmission: Structure and function of the nicotinic acetylcholine receptor. In *Enzymes of biological membranes.* New York. (In press)

ANHOLT, R., D.R. FREDKIN, T. DEERINCK, M. ELLISMAN, M. MONTAL, and J. LINDSTROM. 1982. Incorporation of acetylcholine receptors into liposomes. *J. Biol. Chem.* **257:** 7122.

ARNONE, A. 1972. X-ray diffraction study of binding of 2,3-diphosphoglycerate to human deoxy haemoglobin. *Nature* **237:** 146.

ARONSTAM, R.S., A.T. ELDEFRAWI, I.N. PESSAH, J.W. DALY, E.X. ALBUQUERQUE, and M.E. ELDEFRAWI. 1981. Regulation of ^3H perhydrohistrionicotoxin binding to *Torpedo* electroplax by effectors of the acetylcholine receptor. *J. Biol. Chem.* **256:** 2843.

AXELROD, D., P. RAVDIN, D.E. KOPPEL, J. SCHLESSINGER, W.W. WEBB, E.L. ELSON, and T.R. PODLESKI. 1976. Lateralisation of fluorescently labelled acetylcholine receptors in membranes of developing muscle fibers. *Proc. Natl. Acad. Sci.* **73:** 4594.

BALLIVET, M., J. PATRICK, J. LEE, and S. HEINEMANN. 1982. Molecular cloning of cDNA coding for the γ subunit of *Torpedo* acetylcholine receptor. *Proc. Natl. Acad. Sci.* **79:** 4466.

BARRANTES, F.J. 1982. Oligomeric forms of the membrane-bound acetylcholine receptor disclosed upon extraction of the M_r 43,000 nonreceptor peptide. *J. Cell Biol.* **92:** 60.

BARRANTES, F.J., D.-C. NEUGEBAUER, and H.P. ZINGSHEIM. 1980. Peptide extraction by alkaline treatment is accompanied by rearrangement of the membrane-bound acetylcholine receptor from *Torpedo marmorata. FEBS Lett.* **112:** 73.

BARTHOLDI, M., F.J. BARRANTES, and T.M. JOVIN. 1981. Rotational molecular dynamics of the membrane-bound acetylcholine receptor revealed by phosphorescence spectroscopy. *Eur. J. Biochem.* **120:** 389.

BON, F., E. LEBRUN, J. GOMEL, R. VAN RAPPENBUSCH, J. CARTAUD, J.L. POPOT, and J.P. CHANGEUX. 1982. Orientation relative de deux oligomères constituant la forme lourde du récepteur de l'acetylcholine chez la *Torpille* marbrée. *C.R. Acad. Sci.* **295:** 199.

BOURGEOIS, J.P., J.L. POPOT, A. RYTER, and J.P. CHANGEUX. 1973. Consequences of denervation on the distribution of the cholinergic (nicotinic) receptor sites from *Electrophorus electricus* revealed by high resolution autoradiography. *Brain Res.* **62:** 557.

——. 1978. Quantitative studies on the localisation of the cholinergic receptor protein in the normal and denervated electroplaque from *Electrophorus electricus. J. Cell Biol.* **79:** 200.

BOYD, N.D. and J.B. COHEN. 1980. Kinetics of binding of ^3H-acetylcholine and ^3H-carbamylcholine to *Torpedo* postsynaptic membranes. *Biochemistry* **19:** 5344.

BRANTON, D., C.M. COHEN, and J. TYLER. 1981. Interaction of cytoskeletal proteins on the human erythrocyte membrane. *Cell* **24:** 24.

CARTAUD, J. 1980. A critical reevaluation of the structural organization of the excitable membrane in *Torpedo marmorata* electric organ. In *Ontogenesis and functional mechanisms of peripheral synapses* (ed. J. Taxi), p. 199. Elsevier/North-Holland, Amsterdam.

CARTAUD, J., J.L. POPOT, and J.P. CHANGEUX. 1980. Light and heavy forms of the acetylcholine receptor from *Torpedo marmorata* electric organ. *FEBS Lett.* **121:** 327.

CARTAUD, J., L. BENEDETTI, A. SOBEL, and J.P. CHANGEUX. 1978. A morphological study of the cholinergic receptor protein from *Torpedo marmorata* in its membrane environment and in its detergent extracted purified form. *J. Cell Sci.* **29:** 313.

CARTAUD, J., R. OSWALD, G. CLÉMENT, and J.P. CHANGEUX. 1982. Evidence for a skeleton in acetylcholine receptor rich membranes from *T. marmorata. FEBS Lett.* **145:** 250.

CARTAUD, J., L. BENEDETTI, J.B. COHEN, J.C. MEUNIER, and J.P. CHANGEUX. 1973. Presence of a lattice structure in membrane fragments rich in nicotinic receptor protein from the electric organ of *Torpedo marmorata. FEBS Lett.* **33:** 109.

CARTAUD, J., A. SOBEL, A. ROUSSELET, P.F. DEVAUX, and J.P. CHANGEUX. 1981. Consequences of alkaline treatment for the ultra structure of the acetylcholine-receptor rich membranes from *Torpedo marmorata* electric organ. *J. Cell Biol.* **90:** 418.

CHANG, H.W. and E. BOCK. 1977. Molecular forms of acetylcholine receptor. Effect of calcium ions and of a sulfhydryl reagent on the occurrence of oligomers. *Biochemistry* **16:** 4513.

CHANGEUX, J.P. 1966. Responses of acetylcholinesterase from *Torpedo marmorata* to salts and curarizing drugs. *J. Mol. Pharmacol.* **2:** 369.

——. 1969. Remarks on the symmetry and cooperative properties of biological membranes. *Nobel Symp.* **11:** 235.

——. 1975. The cholinergic receptor protein from fish electric organ. In *Handbook of psychopharmacology* (ed. L.L. Iversen), vol. 6, p. 235. Plenum Press, New York.

——. 1981. The acetylcholine receptor: An allosteric membrane protein. *Harvey Lect.* **75:** 85.

——. 1983. Remarks about the singularity of nerve cells and its ontogenesis. *Prog. Brain Res.* **58:** 465.

CHANGEUX, J.P. and A. DANCHIN. 1976. Selective stabilisation of developing synapses as a mechanism for the specification of neuronal networks. *Nature* **264:** 705.

CHANGEUX, J.P. and J. THIERY. 1968. On the excitability and cooperativity of biological membranes. *BBA Libr.* **11:** 116.

CHANGEUX, J.P., M. KASAI, and C.Y. LEE. 1970a. The use of a snake venom toxin to characterize the cholinergic receptor protein. *Proc. Natl. Acad. Sci.* **67:** 1241.

CHANGEUX, J.P., T.R. PODLESKI, and L. WOFSY. 1967a. Affinity labeling of the acetylcholine receptor. *Proc. Natl. Acad. Sci.* **58:** 2063.

CHANGEUX, J.P., T. HEIDMANN, J.L. POPOT, and A. SOBEL. 1979. Reconstitution of a functional acetylcholine regulator under defined conditions. *FEBS Lett.* **105:** 181.

CHANGEUX, J.P., M. KASAI, M. HUCHET, and J.C. MEUNIER. 1970b. Extraction à partir du tissu électrique de gymnote d'une proteine présentant plusieurs propriétés charactéristiques du récepteur physiologique de l'acétylcholine. *C.R. Acad. Sci.* **270D:** 2864.

CHANGEUX, J.P., J.P. THIERY, Y. TUNG, and C. KITTEL. 1967b. On the cooperativity of biological membranes. *Proc. Natl. Acad. Sci.* **57:** 335.

CHANGEUX, J.P., L. BENEDETTI, J.P. BOURGEOIS, A. BRISSION, J. CARTAUD, P. DEVAUX, H. GRÜNHAGEN, M. MOREAU, J.L. POPOT, A. SOBEL, and M. WEBER. 1976. Some structural properties of the cholinergic receptor protein in its membrane environment relevant to its function as a pharmacological receptor. *Cold Spring Harbor Symp. Quant. Biol.* **40:** 211.

CLAUDIO, T., M. BALLIVET, J. PATRICK, and S. HEINEMANN. 1983. Nucleotide and deduced amino acid sequences of *Torpedo californica* acetylcholine receptor γ-subunit. *Proc. Natl. Acad. Sci.* **80:** 1111.

COHEN, J.B. 1978. Ligand-binding properties of membrane-bound cholinergic receptors of *Torpedo marmorata*. In *Molecular specialization and symmetry in membrane function* (ed. A.K. Solomon and M. Karnovsky), p. 99. Harvard University Press, Cambridge, Massachusetts.

COHEN, J.B., M. WEBER, and J.P. CHANGEUX. 1974. Effet des anesthésiques locaux sur les propriétés de liaison du récepteur cholinergique de *Torpille*. *C.R. Acad. Sci.* **278D:** 1269.

COHEN, J.B., M. WEBER, M. HUCHET, and J.P. CHANGEUX. 1972. Purification from *Torpedo marmorata* electric tissue of membrane fragments particularly rich in cholinergic receptor. *FEBS Lett.* **26:** 43.

COHEN, J.B., N.D. BOYD, A.Y. YENG, D.C. MEDYNSKI, R.R. NEUBIG, P. ST. JOHN, and N.S. SHERA. 1980. Permeability, control by *Torpedo* acetylcholine receptors. In *Psychopharmacology and biochemistry of neurotransmitter receptors* (ed. H. Yamamura), p. 47. Elsevier/North-Holland, Amsterdam.

COLQUHOUN, D. 1973. The relation between classical and cooperative models for drug action. In *Drug receptors* (ed. H.P. Rang), p. 149. Macmillan, London.

CONTI-TRONCONI, B. and M.A. RAFTERY. 1982. The nicotinic receptor: Correlation of molecular structure with functional properties. *Annu. Rev. Biochem.* **51:** 491.

CONTI-TRONCONI, B., S. DUNN, and M.A. RAFTERY. 1982. Functional stability of *Torpedo* acetylcholine receptor. Effects of protease treatment. *Biochemistry* **21:** 893.

DAMLE, V., M. MCLAUGHLIN, and A. KARLIN. 1978. Bromoacetylcholine as an affinity label of the acetylcholine receptor from *Torpedo californica*. *Biochem. Biophys. Res. Commun.* **84:** 845.

DAVIS, G.C, A.S. GORDON, and I. DIAMOND. 1982. Specificity and localization of the acetylcholine receptor kinase. *Proc. Natl. Acad. Sci.* **79:** 3666.

DEVILLERS-THIERY, A., J.P. CHANGEUX, P. PAROUTAUD, and D. STROSBERG. 1979. The amino-terminal sequence of the 40K subunit of the acetylcholine receptor protein from *Torpedo marmorata*. *FEBS Lett.* **104:** 99.

DEVILLERS-THIERY, A., J. GIRAUDAT, M. BENTABOULET, and J.P. CHANGEUX. 1983. Complete mRNA coding sequence of the ACh binding α subunit from *T. marmorata* AChR. A model for the transmembrane organization of the polypeptide chain. *Proc. Natl. Acad. Sci.* **80:** 2067.

DUGUID, J.R. and M.A. RAFTERY. 1973. Fractionation and partial characterization of membrane particles from *Torpedo californica* electroplax. *Biochemistry* **12:** 3593.

ELDEFRAWI, M.E., R.S. ARONSTAM, N. BAKRY, A.T. ELDEFRAWI, and E.X. ALBUQUERQUE. 1980a. Activation, inactivation and desensitization of acetylcholine receptor channel complex detected by binding of perhydrohistrionicotoxin. *Proc. Natl. Acad. Sci.* **77:** 2309.

ELDEFRAWI, M.E., A.T. ELDEFRAWI, R.S. ARONSTAM, M.A. MALEQUE, J.E. WARNICK, and E.X. ALBUQUERQUE. 1980b. ^{3}H Phencyclidine: A probe for the ionic channel of the nicotinic receptor. *Proc. Natl. Acad. Sci.* **77:** 7458.

ELDEFRAWI, M.E., A.T. ELDEFRAWI, N.A. MANS UR, J.W. DALY, B. WITKOP, and E.X. ALBUQUERQUE. 1978. Acetylcholine receptor and ionic channel of *Torpedo* electroplax: Binding of perhydrohistrionicotoxin to membrane and solubilized preparations. *Biochemistry* **17:** 5474.

ELDEFRAWI, A.T., M.E. ELDEFRAWI, E.X. ALBUQUERQUE, A.C. OLIVEIRA, N. MANSÔUR, M. ADLER, J.W. DALY, G.B. BROWN, W.B. BURGERMEISTER, and B. WITKOP. 1977. Perhydrohistrionicotoxin: A potential ligand for the ion conductance modulator of the acetylcholine receptor. *Proc. Natl. Acad. Sci.* **74:** 2172.

ELLIOT, J. and M.A. RAFTERY. 1977. Interactions of perhydrohistrionicotoxin with postsynaptic membranes. *Biochem. Biophys. Res. Commun.* **77:** 1347.

——. 1979. Binding of perhydrohistrionicotoxin to intact and detergent solubilized membranes enriched in nicotinic acetylcholine receptor. *Biochemistry* **18:** 1868.

EPSTEIN, M. and E. RACKER. 1978. Reconstitution of carbamylcholine-dependent sodium ion flux and desensitization of the acetylcholine receptor from *Torpedo californica*. *J. Biol. Chem.* **253:** 6660.

FAMBROUGH, D. 1979. Control of acetylcholine receptors in skeletal muscle. *Physiol. Rev.* **59:** 165.

FELTZ, A. and A. TRAUTMANN. 1980. Interaction between nerve-released acetylcholine and bath applied agonists at the frog endplate. *J. Physiol.* **299:** 533.

——. 1982. Desensitization at the frog neuromuscular junction: A biphasic process. *J. Physiol.* **322:** 257.

FERTUCK, H.C. and M.M. SALPETER. 1976. Quantitation of junctional and extrajunctional acetylcholine receptors by electron microscope autoradiography after ^{125}I-α-bungarotoxin binding at mouse neuromuscular junctions. *J. Cell Biol.* **69:** 144.

FRANK, E., K. GAUTVIK, and H. SOMMERSCHILD. 1975. Cholinergic receptors at denervated mammalian endplates. *Acta Physiol. Scand.* **95:** 66.

FROEHNER, S. 1981. Identification of exposed and buried determinants of the membrane-bound acetylcholine receptor from *Torpedo californica*. *Biochemistry* **20:** 4905.

GIRAUDAT, J., A. DEVILLERS-THIERY, C. AUFFRAY, F. ROUGEON, and J.P. CHANGEUX. 1982. Identification of a cDNA clone coding for the ACh binding subunit of *Torpedo marmorata* AChR. *EMBO J.* **1:** 713.

GRÜNHAGEN, H.H. and J.P. CHANGEUX. 1976. Studies on the electrogenic action of acetylcholine with *T. marmorata* electric organ. *J. Mol. Biol.* **106:** 497.

GRÜNHAGEN, H.H., M. IWATSUBO, and J.P CHANGEUX. 1977. Fast kinetic studies on the interaction of cholinergic agonists with the membrane-bound acetylcholine receptor from *T. marmorata* as revealed by quinacrine fluorescence. *Eur. J. Biochem.* **80:** 225.

GULLICK, W. and J. LINDSTROM. 1982. Antigenic structure of the acetylcholine receptor from *Torpedo californica*. *J. Cell. Biochem.* **19:** 223.

GYSIN, R., M. WIRTH, and S. FLANAGAN. 1981. Structural heterogeneity and subcellular distribution of nicotinic synapse associated proteins. *J. Biol. Chem.* **256:** 11373.

HAGGERTY, J.G. and S.C. FROEHNER. 1981. Restoration of ^{125}I-α-bungarotoxin binding activity to the α-subunit of *Torpedo* acetylcholine receptor isolated by gel electrophoresis in sodium dodecyl sulfate. *J. Biol. Chem.* **255:** 8294.

HAMILTON, S.L., M. MCLAUGHLIN, and A. KARLIN. 1977. Disulfide bond cross-linked dimer in acetylcholine receptor from *Torpedo californica*. *Biochem. Biophys. Res. Comm.* **79:** 692.

——. 1978. Crosslinking of the acetylcholine receptor from *Torpedo* electric tissue. *Fed. Proc.* **37:** 529a.

——. 1979. Formation of disulfide-linked oligomers of acetylcholine receptor in membranes from *Torpedo* electric tissue. *Biochemistry* **18:** 155.

HARING, R., Y. KLOOG, A. KALIR, and M. SOKOLOVSKY. 1983. Species differences determine azido phencyclidine labeling pattern in desensitized nicotinic acetylcholine receptors. *Biochem. Biophys. Res. Commun.* **113:** 723.

HAZELBAUER, G. and J.P. CHANGEUX. 1974. Reconstitution of a chemically excitable membrane. *Proc. Natl. Acad. Sci.* **71:** 1479.

HEIDMANN, T. and J.P. CHANGEUX. 1979. Fast kinetic studies on the interaction of a fluorescent agonist with the membrane-bound acetylcholine receptor from *T. marmorata*. *Eur. J. Biochem.* **94:** 255.

——. 1980. Interaction of a fluorescent agonist with the membrane-bound acetylcholine receptor from *Torpedo marmorata* in the millisecond time range: Resolution of an "intermediate" conformational transition and evidence for positive cooperative effects. *Biochem. Biophys. Res. Commun.* **97:** 889.

——. 1981. Stabilization of the high affinity state of the membrane-bound acetylcholine receptor from *Torpedo marmorata* by non competitive blockers. *FEBS Lett.* **131:** 239.

——. 1982. Un modèle moléculaire de régulation d'efficacité d'une synapse chimique au niveau postsynaptique. *C.R. Acad. Sci.* **295:** 665.

HEIDMANN. T., R. OSWALD and J.P. CHANGEUX. 1982. Le site de liaison de haute affinité de la chlorpromazine est présent à un seul exemplaire par molécule de récepteur cholinergique et est commun aux 4 chaînes polypeptidiques. *C.R. Acad. Sci.* **295:** 345.

——. 1983a. Multiple sites of action for non competitive blockers on acetylcholine receptor rich membrane fragments from *Torpedo marmorata. Biochemistry* **22:** 3112.

HEIDMANN, T., J. BERNHARDT, E. NEUMANN, and J.P. CHANGEUX. 1983b. Rapid kinetics of agonist binding and permeability response analyzed in parallel on acetylcholine receptor-rich membranes from *Torpedo marmorata. Biochemistry* (in press).

HEIDMANN, T., A. SOBEL, J.L. POPOT, and J.P. CHANGEUX. 1980. Reconstitution of a functional acetylcholine receptor: Conservation of the conformational and allosteric transitions and recovery of the permeability response: Role of lipids. *Eur. J. Biochem.* **110:** 35.

HESS, G.P., D.J. CASH, and H. AOSHIMA. 1979. Acetylcholine receptor-controlled ion fluxes in membrane vesicles investigated by fast reaction techniques. *Nature* **282:** 329.

HESS, G.P., H. AOSHIMA, D.J. CASH, and B. LENCHITZ. 1981. Specific reaction rate of acetylcholine receptor-controlled ion translocation: A comparison of measurements with membrane vesicles and with muscle cells. *Proc. Natl. Acad. Sci.* **78:** 1361.

HOLTZMAN, E., D. WISE, J. WALL, and A. KARLIN. 1982. Electron microscopy of complexes of isolated acetylcholine receptor, biotinyl-toxin, and avidin. *Proc. Natl. Acad. Sci.* **79:** 310.

HUCHO, F. 1979. Photoaffinity derivatives of α-bungarotoxin and α-Naja Siamensis toxin. *FEBS Lett.* **103:** 27.

HUCHO, F., L. BERESS, and S. STENGELIN. 1979. Covalently reacting derivatives of neurotoxins and ion channel ligands. In *Function and molecular aspects of biomembrane transport* (ed. E. Quagliariello), p. 43. Elsevier/North-Holland, Amsterdam.

HUGANIR, R.L. and P. GREENGARD. 1983. cAMP-dependent protein kinase phosphorylates the nicotinic acetylcholine receptor. *Proc. Natl. Acad. Sci.* **80:** 1130.

HUGANIR, R., M. SCHELL, and E. RACKER. 1979. Reconstitution of the purified acetylcholine receptor from *Torpedo californica. FEBS Lett.* **108:** 155.

HUNKAPILLER, M.W., C.D. STRADER, L. MOODY, and M.A. RAFTERY. 1979. Amino terminal amino acid sequence of the major subunit of *Torpedo californica* acetylcholine receptor. *Biochem. Biophys. Res. Commun.* **91:** 164.

KALDANY, R.-R. and A. KARLIN. 1983. Reaction of quinacrine mustard with the acetylcholine receptor from *Torpedo californica. J. Biol. Chem.* **258:** 6232.

KARLIN, A. 1967. On the application of a plausible model of allosteric proteins to the receptor for acetylcholine. *J. Theor. Biol.* **16:** 306.

——. 1980. Molecular properties of nicotinic acetylcholine receptors. *Cell Surf. Rev.* **6:** 191.

KARLIN, A. and D. COWBURN. 1973. The affinity-labeling of partially purified acetylcholine receptor from electric tissue of *Electrophorus. Proc. Natl. Acad. Sci.* **70:** 3636.

KARLIN, A., V. DAMLE, S. HAMILTON, M. MCLAUGHLIN, R. VALDERAMMA, and D. WISE. 1979. Acetylcholine receptors in and out of membranes. *Adv. Cytopharmacol.* **3:** 183.

KASAI, M. and J.P. CHANGEUX. 1971. In vitro excitation of purified membrane fragments by cholinergic agonists. *J. Membr. Biol.* **6:** 1.

KISTLER, J. and R.M. STROUD. 1981. Crystalline arrays of membrane-bound acetylcholine receptor. *Proc. Natl. Acad. Sci.* **78:** 3678.

KISTLER, J., R.M. STROUD, M.W. KLYMKOWSKY, R. LALANCETTE, and R.H. FAIRCLOUGH. 1982. Structure and function of an acetylcholine receptor. *Biophys. J.* **37:** 371.

KLOOG, Y., A. KALIR, O. BUCHMAN, and M. SOKOLOVSKY. 1980. Specific binding of ^{3}H phencyclidine to membrane preparation. *FEBS Lett.* **109:** 125.

KLYMKOVSKY, M.W. and R.M. STROUD. 1979. Immunospecific identification and three-dimensional structure of a membrane-bound acetylcholine receptor from *Torpedo californica. J. Mol. Biol.* **128:** 319.

KLYMKOVSKY, M.W., T.E. HEUSER, and R.M. STROUD. 1980. Protease effects on the structure of acetylcholine receptor membranes from *Torpedo californica. J. Cell Biol.* **85:** 823.

KOSOWER, E. 1983. Partial tertiary structure assignments for the β, γ and δ subunits of the acetylcholine receptor on the basis of the hydrophobicity of amino acid sequences and channel location using single group rotation theory. *FEBS Lett.* **155:** 245.

KREBS, G. and J. BEAVO. 1979. Phosphorylation, dephosphorylation of enzymes. *Annu. Rev. Biochem.* **48:** 923.

KRODEL, E.K., R.A. BECKMAN, and J.B. COHEN. 1979. Identification of a local anesthetic binding site in nicotinic post-synaptic membranes isolated from *Torpedo marmorata* electric tissue. *Mol. Pharmacol.* **15:** 294.

LEE, C.Y. and C.C. CHANG. 1966. Modes of actions of purified toxins from elapid venoms on neuromuscular transmission. *Mem. Inst. Butantan* **33:** 555.

LINDSTROM, J., J. COOPER, and S. TZARTOS. 1980. Acetylcholine receptors from *Torpedo* and *Electrophorus* have similar subunit structures. *Biochemistry* **19:** 1454.

LINDSTROM J., J. MERLIE, and G. YOGEESWARAN. 1979. Biochemical properties of acetylcholine receptor subunits from *Torpedo californica. Biochemistry* **18:** 4465.

LO, M.M., E. BARNARD, and J.O. DOLLY. 1982. Size of acetylcholine receptors in the membrane. An improved version of the radiation inactivation method. *Biochemistry* **21:** 2210.

LO, M.M., P.B. GARLAND, J. LAMPRECHT, and E.A. BARNARD. 1980. Rotational mobility of the membrane-bound acetylcholine receptor of *Torpedo* electric organ measured by phosphorescence depolarisation. *FEBS Lett.* **111:** 407.

LOW, B. 1979. Three dimensional structure of erabutoxin b, prototype structure of the snake venom postsynaptic neurotoxins: Consideration of structure and function; description of the reactive site. *Adv. Cytopharmacol.* **3:** 141.

MEUNIER, J.C. and J.P. CHANGEUX. 1973. Comparison between the affinities for reversible cholinergic ligands of a purified and membrane bound state of the acetylcholine-receptor protein from *Electrophorus electricus. FEBS Lett.* **32:** 143.

MEUNIER, J.C., R.W. OLSEN, and J.P. CHANGEUX. 1972a. Studies on the cholinergic receptor protein from *Electrophorus electricus. FEBS Lett.* **24:** 63.

MEUNIER, J.C., M. HUCHET, P. BOQUET, and J.P. CHANGEUX. 1971a. Séparation de la protéine réceptrice de l'acétylcholine et de l'acétylcholinestérase. *C.R. Acad. Sci.* **272D:** 117.

MEUNIER, J.C., R.W. OLSEN, A. MENEZ, P. FROMAGEOT, P. BOQUET, and J.P. CHANGEUX. 1972b. Studies on the cholinergic receptor protein of *Electrophorus electricus. Biochemistry* **11:** 1200.

MEUNIER, J.C., R.W. OLSEN, A. MENEZ, J.L. MORGAT, P. FROMAGEOT, A.M. RONSERAY, P. BOQUET, and J.P. CHANGEUX. 1971b. Quelques propriétés physiques de la protéine réceptrice de l'acétylcholine étudiées à l'aide d'une neurotoxine radioactive. *C.R. Acad. Sci.* **273D:** 595.

MICHLER, A. and B. SAKMANN. 1980. Receptor stability and channel conversion in the subsynaptic membrane of the developing mammalian neuromuscular junction. *Dev. Biol.* **80:** 1.

MILEDI, R., P. MOLINOFF, and L.T. POTTER. 1971. Isolation

of the cholinergic receptor protein of *Torpedo* electric tissue. *Nature* **229:** 554.

MONOD, J., J.P. CHANGEUX, and F. JACOB. 1963. Allosteric proteins and cellular control systems. *J. Mol. Biol.* **6:** 306.

MONOD, J., J. WYMAN, and J.P. CHANGEUX. 1965. On the nature of allosteric transitions: A plausible model. *J. Mol. Biol.* **12:** 88.

MOORE, H. and M. RAFTERY. 1980. Direct spectroscopic studies of cation translocation by *Torpedo* acetylcholine receptor on a time scale of physiological relevance. *Proc. Natl. Acad. Sci.* **77:** 4509.

MUHN, P. and F. HUCHO. 1983. Covalent labeling of the acetylcholine receptor from *Torpedo* electric tissue with the channel blocker (^{3}H) triphenylmethylphosphonium by ultraviolet irradiation. *Biochemistry* **22:** 421.

NAVARATNAM, S., B.J. PARSONS, and G.O. PHILLIPS. 1978. Laser flash photolysis study of the photoionisation of chlorpromazine and promazine in solution. *J. Chem. Soc.* **74:** 1811.

NEUBIG, R. and J. B. COHEN. 1980. Permeability control by cholinergic receptors in *Torpedo* postsynaptic membranes. *Biochemistry* **19:** 2770.

NEUBIG, R., N.D. BOYD, and J.B. COHEN. 1982. Conformation of *Torpedo* acetylcholine receptor associated with ion transport and desensitization. *Biochemistry* **21:** 3460.

NEUBIG, R., E. KRODEL, N. BOYD, and J.B. COHEN. 1979. Acetylcholine and local anesthetic binding to *Torpedo* nicotinic postsynaptic membranes after removal of nonreceptor peptides. *Proc. Natl. Acad. Sci.* **76:** 690.

NGHIÊM, H.O., J.C. CARTAUD, C. DUBREUIL, C. KORDELI, G. BUTTIN, and J.P. CHANGEUX. 1983. Production and characterization of a monoclonal antibody directed against the 43,000 M.W. ν_1 polypeptide from *Torpedo marmorata* electric organ. *Proc. Natl. Acad. Sci.* (in press).

NODA, M., H. TAKAHASHI, T. TANABE, M. TOYOSATO, S. KIKYOTANI, T. MIYATA, and S. NUMA. 1983a. Primary structures of β and δ subunit precursors of *T. californica* acetylcholine receptor deduced from cDNA sequences. *Nature* **301:** 251.

NODA, M., H. TAKAHASHI, T. TANABE, M. TOYOSATO, Y. FURUTANI, T. HIROSE, M. ASAI, S. INAYAMA, T. MIYATA, and S. NUMA. 1982. Primary structure of α subunit precursor of *Torpedo californica* AChR deduced from cDNA sequence. *Nature* **299:** 793.

NODA, M., H. TAKAHASHI, T. TANABE, M. TOYOSATA, S. KIKYOTANI, Y. FURUTANI, T. HIROSE, H. TAKASHIMA, S. INAYAMA, T. MIYATA, and S. NUMA. 1983b. Structural homology of *Torpedo californica* AChR subunits. *Nature* **302:** 528.

OSWALD, R. and J.P. CHANGEUX. 1981a. Ultraviolet light induced labeling by non competitive blockers of the acetylcholine receptor from *Torpedo marmorata*. *Proc. Natl. Acad. Sci.* **78:** 3925.

——. 1981b. Selective labeling of the δ subunit of the acetylcholine receptor by a covalent local anesthetic. *Biochemistry* **20:** 7166.

——. 1982. Crosslinking of α-bungarotoxin to the acetylcholine receptor from *Torpedo marmorata* by ultraviolet light irradiation. *FEBS Lett.* **139:** 225.

OSWALD, R.E., T. HEIDMANN, and J.P. CHANGEUX. 1983. Multiple affinity states for non competitive blockers revealed by (^{3}H) phencyclidine binding to acetylcholine receptor rich membrane fragments from *Torpedo marmorata*. *Biochemistry* **22:** 3128.

POPOT, J.L. 1983. Structural and functional properties of the nicotinic acetylcholine receptor studied using reconstituted vesicles. In *Proceedings of the Symposium Basic Mechanisms of Neuronal Hyperexcitability* (ed. H.H. Jasper and N.M. Van Gelden), p. 137. A.R. Liss, New York.

POPOT, J.L., J. CARTAUD, and J.P. CHANGEUX. 1981. Reconstitution of a functional acetylcholine receptor. *Eur. J. Biochem.* **118:** 203.

POPOT, J.L., H. SUGIYAMA, and J.P. CHANGEUX. 1976. Pharmacological desensitization *in vitro* of the receptor rich-membrane fragments by cholinergic agonists. *J. Mol. Biol.* **106:** 485.

RAFTERY, M.A., J. SCHMIDT, and D.G. CLARK. 1972. Specificity of α-bungarotoxin binding to *Torpedo californica* electroplax. *Arch. Biochem. Biophys.* **152:** 882.

RAFTERY, M.A., M. HUNKAPILLER, C. STRADER, and L.E. HOOD. 1980. Acetylcholine receptor: Complex of homologous subunits. *Science* **208:** 1454.

RAFTERY, M.A., J. SCHMIDT, D.G. CLARK, and R.G. WOLCOTT. 1971. Demonstration of a specific α-bungarotoxin binding component in *Electrophorus electricus* electroplax membranes. *Biochem. Biophys. Res. Commun.* **45:** 1622.

REITER, M.J., D.A. COWBURN, J.M. PRIVES, and A. KARLIN. 1972. Affinity labeling of the acetylcholine receptor in the electroplax. *Proc. Natl. Acad. Sci.* **69:** 1168.

REYNOLDS, J.A. and A. KARLIN. 1978. Molecular weight in detergent solution of acetylcholine receptor from *Torpedo californica*. *Biochemistry* **17:** 2035.

ROISIN, M.P., Y. GU, and Z.W. HALL. 1983. The specificity of a myasthenic serum for developmentally different forms of the AChR. *Soc. Neurosci. Abstr.* (in press).

ROUSSELET, A., J. CARTAUD, and P.F. DEVAUX. 1979. Importance des interactions protéine-protéine dans le maintien de la structure des fragments excitables de l'organe électrique de *Torpedo marmorata*. *C.R. Acad. Sci.* **289D:** 461.

——. 1981. Effects of temperature, lipid modification and pH on the mobility of the major proteins of the receptor-rich membranes from *Torpedo marmorata*. *Biochim. Biophys. Acta* **648:** 169.

ROUSSELET, A., J. CARTAUD, P. DEVAUX, and J.P. CHANGEUX. 1982. The rotational diffusion of the acetylcholine receptor in *T. marmorata* membrane fragments studied with a spin-labelled α-toxin. Importance of the 43K protein. *EMBO J.* **1:** 439.

SAINT-JOHN, P.A., S.C. FROEHNER, A. GOODENOUGH, and J.B. COHEN. 1982. Nicotinic postsynaptic membranes from *Torpedo*. *J. Cell Biol.* **92:** 333.

SAITOH, T. and J.P. CHANGEUX. 1980. *In vitro* phosphorylation of acetyl choline receptor rich membrane from *Torpedo marmorata*: Effect on membrane solubilization by detergents. *Eur. J. Biochem.* **105:** 51.

——. 1981. Change in the state of phosphorylation of the acetylcholine receptor during maturation of the electromotor synapse in *Torpedo marmorata* electric organ. *Proc. Natl. Acad. Sci.* **78:** 4430.

SAITOH, T., L. WENNOGLE, and J.P. CHANGEUX. 1979. Factors regulating the susceptibility of the acetylcholine receptor protein to heat inactivation. *FEBS Lett.* **108:** 489.

SAITOH, T., R. OSWALD, L.P. WENNOGLE, and J.P. CHANGEUX. 1980. Conditions of the selective labeling of the 66,000 dalton chain of the acetylcholine receptor. *FEBS Lett.* **116:** 30.

SAKMANN, B., J. PATLAK, and E. NEHER. 1980. Single acetylcholine-activated channels show burst kinetics in presence of desensitizing concentrations of agonist. *Nature* **286:** 71.

SCHIEBLER, W. and F. HUCHO. 1978. Membranes rich in acetylcholine receptor: Characterization and reconstitution to excitable membranes from exogenous lipid. *Eur. J. Biochem.* **85:** 55.

SEALOCK, R. 1982. Cytoplasmic surface structure in postsynaptic membranes from electric tissue visualized by tannic acid-mediated negative contrasting. *J. Cell Biol.* **92:** 514.

SEALOCK, R. and A. KAVOOKJIAN. 1980. Postsynaptic distribution of acetylcholine receptors in electroplax of the torpedine ray, *Narcine brasiliensis*. *Brain Res.* **190:** 81.

SINE, S.M. and P. TAYLOR. 1981. Relationship between reversible antagonist occupancy and the functional capacity of the acetylcholine receptor. *J. Biol. Chem.* **256:** 6692.

——. 1982. Local anesthetics and histrionicotoxin on allosteric inhibition of the acetylcholine receptor. *J. Biol. Chem.* **257:** 8106.

SOBEL, A., A. HEIDMANN, and J.P. CHANGEUX. 1977. Purification d'une protéine liant la quinacrine et l'histrionicotoxine à partir de fragments de membranes. *C.R. Acad. Sci.* **285D:** 1255.

SOBEL, A., T. HEIDMANN, J. CARTAUD, and J.P. CHANGEUX. 1980. Reconstitution of a functional acetylcholine receptor. *Eur. J. Biochem.* **110:** 13.

SOBEL, A., T. HEIDMANN, J. HOFLER, and J.P. CHANGEUX. 1978. Distinct protein components from *T. marmorata* membranes carry the acetylcholine receptor site and the binding site for local anesthetics and histrionicotoxin. *Proc. Natl. Acad. Sci.* **75:** 510.

STRADER, C. and M. RAFTERY. 1980. Topographic studies of *Torpedo* acetylcholine receptor subunits as a transmembrane complex. *Proc. Natl. Acad. Sci.* **77:** 5807.

STRADER, C.D., E. LAZARIDES, and M.A. RAFTERY. 1980. The characterization of actin associated with postsynaptic membranes from *Torpedo californica. Biochem. Biophys. Res. Commun.* **92:** 365.

SUAREZ-ISLA, B.A. and F. HUCHO. 1977. Acetylcholine receptor: -SH group reactivity as indicator of conformational changes and functional states. *FEBS Lett.* **75:** 65.

SUMIKAWA, K., M. HOUGHTON, J.C. SMITH, L. BELL, B.M. RICHARDS, and E.A. BARNARD. 1982. The molecular cloning and characterisation of cDNA coding for the α subunit of the acetylcholine receptor. *Nucleic Acids Res.* **10:** 5809.

TEICHBERG, V. and J.P. CHANGEUX. 1977. Evidence for protein phosphorylation and dephosphorylation in membrane fragments isolated from the electric organ of *Electrophorus electricus. FEBS Lett.* **74:** 71.

TZARTOS, S.J. and J.-P. CHANGEUX. 1983. High affinity binding of α-bungarotoxin to the purified α-subunit and to its 27,000-dalton proteolytic peptide from *Torpedo marmorata* acetylcholine receptor. Requirement for sodium dodecyl sulfate. *EMBO J.* **2:** 381.

TZARTOS, S. and J. LINDSTROM. 1980. Monoclonal antibodies used to probe acetylcholine receptor structure. *Proc. Natl. Acad. Sci.* **77:** 755.

TZARTOS, S., M.E. SEYBOLD, and J. LINDSTROM. 1982. Specificities of antibodies to acetylcholine receptors in sera from myasthenia gravis patients measured by monoclonal antibodies. *Proc. Natl. Acad. Sci.* **79:** 188.

TZARTOS, S., D.E. RAND, B.L. EINARSON, and J. LINDSTROM. 1981. Mapping of surface structures of *Electrophorus* acetylcholine receptor using monoclonal antibodies. *J. Biol. Chem.* **256:** 8635.

WAKSMAN, G., R. OSWALD, J.P. CHANGEUX, and P. ROQUES. 1980. Synthesis and pharmacological activity on *Electrophorus electricus* electroplaque of photoaffinity labelling derivatives of the non-competitive blockers di- and tri-methisoquin. *FEBS Lett.* **111:** 23.

WALKER, J.W., K. TAKEYASU, and M.G. MCNAMEE. 1982. Activation and inactivation kinetics of *Torpedo californica* acetylcholine receptor in reconstituted membranes. *Biochemistry* **21:** 5384.

WALKER, J.W., M.G. MCNAMEE, E. PASQUALE, D.J. CASH, and G.P. HESS. 1981. Acetylcholine receptor inactivation in *Torpedo californica* electroplax membrane vesicles. Detection of two processes in the millisecond and second time range. *Biochem. Biophys. Res. Commun.* **100:** 86.

WEBER, M. and J.P. CHANGEUX. 1974. Binding of *Naja nigricollis* ^{3}H α-toxin to membrane fragments from *Electrophorus* and *Torpedo* electric organs. *Mol. Pharmacol.* **10:** 1.

WEBER, M., T. DAVID-PFEUTY, and J.P. CHANGEUX. 1975. Regulation of binding properties of the nicotinic receptor protein by cholinergic ligands in membrane fragments from *Torpedo. Proc. Natl. Acad. Sci.* **72:** 3443.

WEILAND, G., D. FRISMAN, and P. TAYLOR. 1979. Affinity labeling of the subunits of the membrane associated cholinergic receptor. *Mol. Pharmacol.* **15:** 213.

WEILAND, G., B. GEORGIA, S. LAPPI, G.F. CHIGNELL, and P. TAYLOR. 1977. Kinetics of agonist mediated transitions in state of the cholinergic receptor. *J. Biol. Chem.* **252:** 7648.

WEILL, C.L., M.G. MCNAMEE, and A. KARLIN. 1974. Affinity labeling of purified acetylcholine receptor from *Torpedo californica. Biochem. Biophys. Res. Commun.* **61:** 997.

WEINBERG, C., J. SANES, and Z. HALL. 1981. Formation of neuromuscular junctions in adult rats: Accumulation of acetylcholine receptors, acetylcholinesterase and components of synaptic basal lamina. *Dev. Biol.* **84:** 255.

WENNOGLE, L. and J.P. CHANGEUX. 1980. Transmembrane orientation of proteins present in acetylcholine receptor-rich membranes from *Torpedo marmorata* studied by selective proteolysis. *Eur. J. Biochem.* **106:** 381.

WENNOGLE, L.P., R. OSWALD, T. SAITOH, and J.P. CHANGEUX. 1981. Dissection of the 66,000 dalton subunit of the acetylcholine receptor. *Biochemistry* **20:** 2492.

WISE, D.S., B.P. SCHOENBORN, and A. KARLIN. 1981. Structure of acetylcholine receptor dimer determined by neutron scattering and electron miroscopy. *J. Biol. Chem.* **256:** 4124.

WITZEMANN, V. and M.A. RAFTERY. 1977. Selective photoaffinity labeling of acetylcholine receptor using a cholinergic analogue. *Biochemistry* **16:** 5862.

WITZEMANN, V., D. MUCHMORE, and M.A. RAFTERY. 1979. Affinity directed cross-linking of membrane-bound acetylcholine receptor polypeptides with photolabile α-bungarotoxin derivatives. *Biochemistry* **18:** 5511.

WU, W.C.S. and M. RAFTERY. 1979. Carbamylcholine induced rapid cation efflux from reconstituted membrane vesicles containing purified acetylcholine receptor. *Biochem. Biophys. Res. Commun.* **89:** 26.

YOUNG, A. and D. SIGMAN. 1981. Allosteric effects of volatile anesthetics on the membrane-bound acetylcholine receptor protein. I. Stabilization of the high affinity state. II. Alteration of α bungarotoxin binding kinetics. *Mol. Pharmacol.* **20:** 498.

ZINGSHEIM, M.P., D.C. NEUGEBAUER, F.J. BARRANTES, and J. FRANK. 1980. Structural details of membrane bound acetylcholine receptor from *Torpedo marmorata. Proc. Natl. Acad. Sci.* **77:** 952.

ZINGSHEIM, H.P, F. BARRANTES, J. FRANK, W. HÄNICKE, and D.C. NEUGEBAUER. 1982. Direct structural localization of two toxin recognition sites on an ACh-receptor protein. *Nature* **299:** 81.

Muscarinic Receptor Subclasses: Allosteric Interactions

N.J.M. Birdsall, E.C. Hulme, and J.M. Stockton
Division of Molecular Pharmacology, National Institute for Medical Research, Mill Hill, London NW7 1AA, England

In the simplest conceptual terms, a receptor system consists of a receptor-binding site, R, and an effector system, E. These two components, which may be on the same or different molecules, are conformationally coupled and may be considered as existing in two states, a ground state, R.E, and an activated state, R′E*. Hence, the activation of the effector, E→E*, is associated with a change in the conformation of the receptor, R→R′. Conversely, any drug that will induce the R→R′ conformation change will activate the receptor. Agonists may be considered as binding preferentially to the R′E* state; antagonists bind either equally to the two states or selectively to the ground state.

There are a number of pharmacological implications of this scheme. First, as the conformations of R and R′ are different, the structure-binding characteristics of these two states must be different. This has been discussed in some detail previously for the interaction of competitive agonists and antagonists with muscarinic acetylcholine receptors (AChR) (Burgen et al. 1979).

Second, a ligand that interacts with the effector will allosterically regulate the binding of drugs to the receptor if the interaction between receptor and effector is perturbed. If the ligand perturbs the energetics of effector activation (E→E*), then agonist binding to the receptor is primarily affected. Hence, allosteric effects on the receptor-binding properties of drugs, especially agonists, may give information on the nature of effector molecules that are directly linked to receptors.

The third implication of the scheme is that, although it is normally considered that the way in which receptor systems are activated is by the binding of agonists to the neurotransmitter (or hormone)-binding site, it may be possible to activate the receptor in other ways. In fact, any interaction that favors E* formation will activate a receptor system. This may be accomplished by the binding of a ligand either to E or to any site or sites on R that are conformationally coupled to E. Similarly, it may be possible to modulate receptor activation by agonists (or the action of antagonists) by the binding of ligands to a second binding site on the receptor (or effector). As the existence of two separate binding sites is implied, the interaction between ligands binding to these sites will not be competitive but will manifest positive or negative heterotropic cooperativity. In order that there be allosteric interactions, the conformation of the binding sites must be changed. As this involves the spatial displacement of one or more residues at the binding sites, the structure-binding relationships must be modified. In this paper, we consider some of the evidence of such allosteric interactions at muscarinic receptors that has helped to confirm the existence of muscarinic receptor subclasses.

Allosteric Modulation by Divalent Cations and Guanine Nucleotides

In common with a number of other neurotransmitter receptor systems, the binding of agonists to muscarinic receptors in some tissues is enhanced by divalent cations, notably Mg^{++} and Mn^{++} (Gurwitz and Sokolovsky 1980; Hulme et al. 1980, 1981b; Wei and Sulakhe 1980a; Ehlert et al. 1981; Harden et al. 1982), and is inhibited by guanine nucleotides (Berrie et al. 1979; Wei and Sulakhe 1979, 1980a,b; Ehlert et al. 1980a,b,c, 1981; Gurwitz and Sokolovsky 1980; Hulme et al. 1980, 1981b; Sokolovsky et al. 1980; Waelbroeck et al. 1982; Harden et al. 1982; Burgisser et al. 1982). The modulation is thought to be due to changes in the interaction of the receptor with a guanine nucleotide regulatory protein complex, which is an intermediary in the mechanism of receptor control of adenylate cyclase activity. Indeed, in the myocardium, many groups have shown that muscarinic agonists will inhibit adenylate cyclase, and this is the tissue where the largest effects of guanine nucleotides are observed. This regulation of agonist binding shows the characteristics of negative heterotropic cooperativity (Hulme et al. 1980), and under certain conditions it is also possible to demonstrate a reciprocal, positively cooperative regulation of antagonist binding (Ehlert et al. 1981; Hulme et al. 1981a; Burgisser et al. 1982; Hosey 1982). Divalent cations appear to be required and they themselves produce an opposite allosteric regulation of binding of agonists and antagonists from that exhibited by guanyl nucleotides. It is also possible to demonstrate that different guanine nucleotides have both varying potency and extents of negative cooperativity (Hulme et al. 1980). Therefore, not only has the nature of the effector linked to muscarinic receptors in the myocardium been elucidated, but the structure-binding relationships of guanine nucleotides for the binding site on the effector have been described.

In other tissues, the effects of guanine nucleotides and divalent cations are much less pronounced and are absent in, for example, the hypothalamus (E.C. Hulme et al., in prep.). Does this mean that there is a subclass of muscarinic binding site that is not regulated by M^{++}/GTP? Studies in the cerebral cortex have shown that there are two classes of muscarinic receptor site that have 10–20-fold different affinities for the selective antagonist pirenzepine (Hammer et al. 1980). The higher-affinity site can be monitored using [^{3}H]pirenzepine (Watson et al. 1982; Birdsall et al. 1983b; J.M. Stockton et al., in prep.), and somewhat surprisingly we have found that only the agonist-binding properties of the high-affinity pirenzepine site are regulated by divalent ions and nucleotides. In contrast, the myocardial muscarinic receptors have been found to have a low affinity for pirenzepine in both functional (Fuder 1981) and binding studies (Hammer et al. 1980). There are, therefore, two subclasses of receptor sites, both regulated by guanine nucleotides but distinguished by pirenzepine and one receptor site that is not regulated by guanine nucleotides and has a low affinity for pirenzepine but whose function and effector mechanism (if any) is not known at present.

A Second Binding Site on Muscarinic Receptors

Gallamine, a neuromuscular blocking drug, has been known for some time to inhibit the effects of muscarinic agonists on the force and rate of contraction of heart muscle (Riker and Wescoe 1951). This side effect occurs at doses comparable with those that produce neuromuscular blockade. However, under these conditions, no effects on smooth muscle or the salivary glands are observed (Rathbun and Hamilton 1970; Clark and Mitchelson 1976; Li and Mitchelson 1980).

We have investigated the effects of gallamine on the binding of muscarinic drugs to their receptors in different tissues and have found that gallamine interacts with a site on muscarinic receptors, distinct from the classical muscarinic ligand binding site to which ACh and atropine bind (Stockton et al. 1983). The effect of the binding of gallamine is to change the binding of agonist and antagonists allosterically. For example, the binding constant of *N*-methylscopolamine for myocardial muscarinic receptors is decreased by gallamine, a maximum factor of 15 being observed at high concentrations of gallamine. In a binding experiment in which the ability of varying concentrations of gallamine to inhibit the binding of a fixed concentration of a tritiated classical muscarinic drug is monitored, this phenomenon is manifest as an inability of gallamine to inhibit 100% of the receptor-specific binding of the tritiated ligand. The nondisplaceable component represents the tritiated ligand-receptor-gallamine ternary complex.

There are two parameters that characterize the gallamine interaction; first, the affinity of gallamine for the receptor in the absence of other drugs, K_1, and second, the affinity of gallamine for the receptor-muscarinic ligand complex, K_2. Whereas K_1 is a characteristic of the muscarinic receptors in the tissue, K_2 depends both on the receptor and the nature of the muscarinic ligand that forms a ternary complex with the ligand. The ratio K_2/K_1 is the cooperativity. We have examined the cooperativity between gallamine and 20 muscarinic drugs (Stockton et al. 1983). In the heart, the negative heterotropic cooperativity (K_1/K_2) varies from 8 for 4-diphenylacetoxy-*N*-methylpiperidine methiodide, a selective muscarinic antagonist (Barlow et al. 1976; Birdsall et al. 1980b), to 90 for cyproheptidine. Hence, gallamine has produced a conformational change at the classical binding site on muscarinic receptors and has changed the structure-binding relationships. In general, the cooperativity is somewhat lower in the cerebral cortex. Furthermore, the value of K_1 is 20-fold lower in the cerebral cortex. Hence, there are clear differences between the binding properties of gallamine in these two tissues. In fact, the binding picture in a number of peripheral tissues mirrors the pharmacological picture, with the most potent effects of gallamine being observed in the myocardium and much smaller effects in salivary glands, lacrimal glands, and smooth muscle. For example, at 10^{-6} M gallamine, the affinity of *N*-methylscopolamine in the heart is halved, whereas that in the submandibular gland is only decreased by 2%.

For ligands, such as agonists (Birdsall et al. 1978, 1980a) and pirenzepine (Hammer et al. 1980), which manifest heterogeneous binding curves in their interaction with muscarinic receptors in certain tissues, gallamine has selective effects on the subclasses of receptor-binding sites or conformational states of the receptor (Stockton et al. 1983). It therefore seems that gallamine not only has the ability to distinguish, by itself, between muscarinic receptor subtypes, but is also capable of changing structure-binding relationships of classical muscarinic drugs and amplifying differences in the receptors that are sensed by these drugs.

We have also demonstrated that a number of other drugs, including other neuromuscular blockers, appear to bind to the same allosteric site (M.J. Zigmond et al., in prep.). As yet, we have not discovered a sufficiently potent ligand to allow us to attempt to label this site directly, nor have we discovered a function for this site. Although the binding constant of gallamine is very sensitive to ionic strength, up to 1000-fold variations in K_1 being readily observed, the value of K_1 is not selectively affected by the presence of Na^+ (vs. K^+) or Ca^{++} (vs. Mg^{++}). Neither is the binding of gallamine affected by guanine nucleotides. Therefore, there is no evidence that the gallamine site is associated with a specific ion binding site or a site that might be important for receptor-guanine nucleotide regulatory protein interaction. It appears that the gallamine site is on or very closely associated with the same protein as the ACh-binding site because the gallamine effects can be observed after solubilization under conditions that yield an apparently monomeric binding species with a molecular weight of approximately 80,000 (E.C. Hulme et al., unpubl.).

We have also found that a selective agonist, McN-A-343 (Roszkowski 1961), exhibits the same behavior as gallamine in its interaction with myocardial muscarinic receptors (Birdsall et al. 1983a), the negative cooperativity with *N*-methylscopolamine being about 40. It apparently behaves as a competitive agonist in the cerebral cortex. In common with other muscarinic agonists, its binding in the heart is decreased by guanine nucleotides but still remains noncompetitive with *N*-methylscopolamine. As McN-A-343 is an agonist on myocardial muscarinic receptors (Fozard and Muscholl 1972) and hence must generate the active state of the receptor (R′E*), these findings suggest that it may be possible, as postulated in the introductory section, to activate muscarinic receptors by the binding to a site different to the one to which ACh binds. We have, however, not proved that the site to which McN-A-343 binds is the same as that to which gallamine binds, but it is possible that these two drugs could be a novel agonist-antagonist pair.

REFERENCES

Barlow, R.B., K.J. Berry, P.A.M. Glenton, N.M. Nikolaou, and K.S. Soh. 1976. A comparison of affinity constants for muscarine-sensitive acetylcholine receptors in guinea-pig atrial pacemaker cells at 29°C and in ileum at 29°C and 37°C. *Br. J. Pharmacol.* **58:** 613.

Berrie, C.P., N.J.M. Birdsall, A.S.V. Burgen, and E.C. Hulme. 1979. Guanine mucleotides modulate muscarinic receptor binding in the heart. *Biochem. Biohphys. Res. Commun.* **87:** 1000.

Birdsall, N.J.M., A.S.V. Burgen, and E.C. Hulme. 1978. The binding of agonists to brain muscarinic receptors. *Mol. Pharmacol.* **14:** 723.

Birdsall, N.J.M., E.C. Hulme, and A.S.V. Burgen. 1980a. The character of the muscarinic receptors in different regions of the rat brain. *Proc. R. Soc. Lond. B* **207:** 1.

Birdsall, N.J.M., E.C. Hulme, R. Hammer, and J.M. Stockton. 1980b. Subclasses of muscarinic receptors. In *Psychopharmacology and biochemistry of neurotransmitter receptors* (ed. H.I. Yamamura et al.), p. 97. Elsevier/North-Holland, New York.

Birdsall, N.J.M., A.S.V. Burgen, E.C. Hulme, J.M. Stockton, and M.J. Zigmond. 1983a. The effect of McN-A-343 on muscarinic receptors in the cerebral cortex and heart. *Br. J. Pharmacol.* **78:** 257.

Birdsall, N.J.M., E.C. Hulme, J.M. Stockton, A.S.V. Burgen, C.P. Berrie, R. Hammer, E.H.F. Wong, and M.J. Zigmond. 1983b. Muscarinic receptor subclasses—Evidence from binding studies. In *CNS receptors—From molecular pharmacology to behavior* (ed. F.V. DeFeudis and P. Mandel). Raven Press, New York. (In press)

Burgen A.S.V., N.J.M. Birdsall, and E.C. Hulme. 1979. Muscarinic receptors and the nature of their reaction with agonists and antagonists. In *IUPAC Medicinal Chemistry Symposium* (ed. M.A. Simkins), p. 25. Oxford University Press, Oxford.

Burgisser, E., A. de Lean, and R.J. Lefkowitz. 1982. Reciprocal modulation of agonist and antagonist binding to muscarinic cholinergic receptor by guanine nucleotides. *Proc. Natl. Acad. Sci.* **79:** 1732.

Clark, A.L. and F. Mitchelson. 1976. The inhibitory effect of gallamine on muscarinic receptors. *Br. J. Pharmacol.* **58:** 323.

Ehlert, F.J., W.R. Roeske, and H.I. Yamamura. 1980a. Regulation of muscarinic receptor binding by guanine nucleotides and *N*-ethylmaleimide. *J. Supramol. Struct.* **14:** 149.

———. 1981. Muscarinic receptor: Regulation by guanine nucleotides, ions and *N*-ethylmaleimide. *Fed. Proc.* **40:** 153.

Ehlert, F.J., W.R. Roeske, L.B. Rosenberger, and H.I. Yamamura. 1980b. The influence of guanyl-5′-yl-imidodiphosphate and sodium on muscarinic receptor bindng in the rat brain and longitudinal muscle of the rat ileum. *Life Sci.* **26:** 245.

Ehlert, F.J., H.I. Yamamura, D.J. Triggle, and W.R. Roeske. 1980c. The influence of guanyl-5′-yl-imidodiphosphate and sodium chloride on the binding of the muscarinic agonist [^{3}H] cis methyldioxolane. *Eur. J. Pharmacol.* **61:** 317.

Fozard, J.R. and E. Muscholl. 1972. Effects of several muscarinic agonists on cardiac performance and the release of noradrenaline from sympathetic nerves of the perfused rabbit heart. *Br. J. Pharmacol.* **45:** 616.

Fuder, H. 1981. The affinity of pirenzepine and other antimuscarinic compounds in pre- and postsynaptic muscarine receptors of the isolated rabbit and rat ileum. *Scand. J. Gastroenterol.* (suppl.) **72:** 79.

Gurwitz, D. and M. Sokolovsky. 1980. Agonist-specific reverse regulation of muscarinic receptors by transition metal ions and guanine nucleotides. *Biochem. Biophys. Res. Commun.* **96:** 1296.

Hammer, R., C.P. Berrie, N.J.M. Birdsall, A.S.V. Burgen, and E.C. Hulme. 1980. Pirenzepine distinguishes between subclasses of muscarinic receptors. *Nature* **283:** 90.

Harden, T.K., A.G. Scheer, and M.M. Smith. 1982. Differential modification of the interactions of cardiac muscarinic cholinergic and β-adrenergic receptors with a guanine nucleotide binding component(s). *Mol. Pharmacol.* **21:** 570.

Hosey, M.M. 1982. Regulation of antagonist binding to cardiac muscarinic receptors. *Biochem. Biophys. Res. Commun.* **107:** 314.

Hulme, E.C., C.P. Berrie, N.J.M. Birdsall, and A.S.V. Burgen. 1981a. Two populations of binding sites for muscarinic antagonists in the rat heart. *Eur. J. Pharmacol.* **73:** 137.

———. 1981b. Interactions of muscarinic receptors with guanine nucleotides and adenylate cyclase. In *Drug receptors and their effectors* (ed. N.J.M. Birdsall), p. 23. Macmillan, London.

Hulme, E.C., N.J.M. Birdsall, C.P. Berrie, and A.S.V. Burgen. 1980. Muscarinic receptor binding: Modulation by nucleotides and ions. In *Neurotransmitters and their receptors* (ed. V.Z. Littauer et al.), p. 241. Wiley, New York.

Li, C.K. and F. Mitchelson. 1980. The selective antimuscarinic action of stercuronium. *Br. J. Pharmacol.* **70:** 313.

Rathbun, F.J. and J.T. Hamilton. 1970. Effect of gallamine on cholinergic receptors. *Can. Anaesth. Soc. J.* **17:** 574.

Riker, W.F. and W.C. Wescoe. 1951. The pharmacology of flaxedil with observations on certain analogs. *Ann. N.Y. Acad. Sci.* **54:** 373.

Roszkowski, A.P. 1961. An unusual type of ganglionic stimulant. *J. Pharmacol. Exp. Ther.* **132:** 156.

Sokolovsky, M., D. Gurwitz, and R. Galron. 1980. Muscarinic receptor binding in mouse brain: Regulation by guanine nucleotides. *Biochem. Biophys. Res. Commun.* **94:** 487.

Stockton, J.M., N.J.M. Birdsall, A.S.V. Burgen, and E.C. Hulme. 1983. Modification of the binding properties of muscarinic receptors by gallamine. *Mol. Pharmacol.* **23:** 551.

Waelbroeck, M., P. Robberecht, P. Chatelain, and J. Christophe. 1982. Rat cardiac muscarinic receptors: Effects of guanine nucleotides on high and low affinity sites. *Mol. Pharmacol.* **21:** 581.

Watson, M., W.R. Roeske, and H.I. Yamamura. 1982. ^{3}H-Pirenzepine selectively identifies a high affinity population of muscarinic cholinergic receptors in the rat cerebral cortex. *Life Sci.* **31:** 2019.

Wei, J.-W. and P.V. Sulakhe. 1979. Agonist-antagonist interactions with rat atrial muscarinic cholinergic receptor sites: Differential regulation by guanine nucleotides. *Eur. J. Pharmacol.* **58:** 91.

———. 1980a. Cardiac muscarinic cholinergic receptor sites: Opposing regulation by divalent cations and guanine

nucleotides of receptor-agonist interaction. *Eur. J. Pharmacol.* **62:** 345.
——. 1980b. Requirement for sulfhydryl groups in the differentiated effects of magnesium ion and GTP on agonist-binding of muscarinic cholinergic receptor sites in rat atrial membrane fractions. *Naunyn-Schmiedebergs Arch. Pharmacol.* **314:** 51.

Molecular Structure of the Nicotinic Acetylcholine Receptor

S. NUMA, M. NODA, H. TAKAHASHI, T. TANABE, M. TOYOSATO, Y. FURUTANI, AND S. KIKYOTANI
Department of Medical Chemistry, Kyoto University Faculty of Medicine, Kyoto 606, Japan

The nicotinic acetylcholine receptor (AChR) is a postsynaptic membrane protein complex that alters the ion permeability of the membrane as a consequence of binding acetylcholine (ACh) (for review, see Karlin 1980; Changeux 1981; Conti-Tronconi and Raftery 1982). The AChR from the electroplax of the ray *Torpedo californica* is composed of five subunits present in a molar stoichiometry of $\alpha_2\beta\gamma\delta$ (Reynolds and Karlin 1978; Lindstrom et al. 1979; Raftery et al. 1980). The pentameric protein complex is responsible for the binding of agonists, cholinergic antagonists, α-toxins, histrionicotoxin, and local anesthetics as well as for cation translocation (for review, see Karlin 1980; Changeux 1981; Conti-Tronconi and Raftery 1982). A prerequisite to the understanding of the molecular basis for these various functions is to elucidate the primary structures of the constituent polypeptides of the AChR. The aminoterminal 54–56 amino acids of all four polypeptides derived from *T. californica* have been sequenced (Raftery et al. 1980).

Recently, we have applied recombinant DNA techniques to study the primary structures of the subunit polypeptides of the AChR. DNA sequences complementary to the *T. californica* mRNAs coding for these polypeptides have been cloned, and nucleotide sequence analysis of the cloned cDNAs has revealed the complete primary structures of the α-, β-, γ-, and δ-subunit precursors of the AChR (Noda et al. 1982b, 1983b,c). The amino acid sequences of the four subunits exhibit conspicuous homology. The close resemblance among the hydrophilicity profiles and predicted secondary structures of all the subunits suggests that these polypeptides are oriented in a pseudosymmetric fashion across the membrane. The transmembrane topology of the subunit molecules is proposed. Each subunit contains four putative transmembrane segments that may be involved in the ionic channel. A possible ACh-binding site and a potential candidate for the "main immunogenic region" on the α-subunit molecule are postulated.

We have also cloned cDNA for the α-subunit precursor of the calf skeletal muscle AChR and a human genomic DNA segment containing the corresponding gene. Nucleotide sequence analysis of the cloned DNAs has disclosed the complete primary structures of the calf and human AChR α-subunit precursors (Noda et al. 1983a). The α-subunits of the fish and mammalian AChR exhibit marked sequence homology. The protein-coding sequence of the human AChR α-subunit precursor gene is divided by eight introns into nine exons, which seem to correspond to different structural and functional domains of the subunit precursor molecule (Noda et al. 1983a). This paper deals with our studies on the molecular structure of the AChR.

Isolation and Sequence Analysis of cDNA Clones for *T. californica* AChR Subunit Precursors

A library of cDNA clones derived from poly(A)-containing RNA from an electric organ of *T. californica* was constructed with the use of the plasmid DNA vector elaborated by Okayama and Berg (1982). The cDNA library was screened by hybridization with a mixture of synthetic oligodeoxyribonucleotides representing all possible cDNA sequences predicted from known amino acid sequences in the aminoterminal region. Two kinds of oligodeoxyribonucleotide probes composed of 14–16 bases were used to screen the cDNA library for clones carrying cDNA for each subunit precursor (see Fig. 1). cDNA clones that hybridized with both probes were isolated. Because the initial clone obtained for the β-subunit precursor did not carry the entire protein-coding sequence, the cDNA library was further screened by hybridization with a restriction fragment derived from an upstream region of this clone to yield clones harboring cDNA extending beyond the translational initiation site. The nucleotide sequences of the cloned cDNAs were determined by the procedure of Maxam and Gilbert (1980).

Primary Structures of *T. californica* AChR Subunit Precursors Deduced from cDNA Sequences

The primary structures of the *T. californica* mRNAs coding for the α-, β-, γ-, and δ-subunit precursors of the AChR, deduced from the cDNA sequences, are shown in Figure 1, A, B, C, and D, respectively. The complete amino acid sequences of the AChR subunit precursors have been assigned by using the reading frame corresponding to the known aminoterminal 54- to 56-amino acid sequences of the mature subunits. The α-, β-, γ-, and δ-subunit precursors are composed of 461, 493, 506, and 522 amino acids, including a hydrophobic prepeptide of 24, 24, 17, and 21 amino acids, respectively. The molecular weights of the mature α-, β-, γ-, and δ-subunits are calculated to be 50,116, 53,681, 56,279, and 57,565, respectively. The prepeptides of the AChR subunit precursors resemble the signal peptide of secretory proteins (Blobel and Dobberstein 1975) in that they contain a region rich in hydrophobic amino acids having large side chains in their central portion and terminate in a residue with a small neutral side chain, i.e., glycine, alanine, or serine (Steiner et al. 1980). Thus, these prepeptides are most probably involved in the translocation of the AChR subunits into the cytoplasmic membrane.

The alignment of the amino acid sequences of the α-, β-, γ-, and δ-subunit precursors of the *T. californica* AChR shows conspicuous homology among them (Fig. 2). Of the positions aligned, 19% and 20% are occupied by four and three identical residues, respectively, and 54% of the positions with three identical residues have a chemically similar residue (Dayhoff et al. 1978) to the fourth one. These findings strongly suggest that the genes encoding the four subunit precursors descended from a single common ancestor by gene duplications. Computer analysis of the nucleotide and the amino acid sequences of the AChR subunit precursors did not detect such internal homology units as observed for the immunoglobulin polypeptides (Poljak et al. 1973).

The amino acid sequences of the four AChR subunit precursors were analyzed for local hydrophilicity (Hopp and Woods 1981) and for secondary structure (Chou and Fasman 1978). The hydrophilicity profiles along the aligned amino acid sequences (Fig. 3) as well as the predicted secondary structures are similar for all the subunits, suggesting that they are oriented in a pseudosymmetric fashion with respect to the membrane. The segments that can form secondary structures extend over a wide range of the subunit molecules. We have initially identified four strongly hydrophobic regions with predicted secondary structure in the α-subunit molecule and have assumed that they represent transmembrane segments or are involved in intersubunit interaction. Similarly, four markedly hydrophobic regions with putative secondary structure are found at equivalent positions in the β-, γ-, and δ-subunit molecules. These hydrophobic regions of the four subunits correspond to a continuous stretch of 19–27 uncharged amino acid residues including many nonpolar residues (68–93%) (Figs. 2 and 4). These segments, designated as M1, M2, M3, and M4, are bounded by charged residues. Because such a sequence is characteristic of transmembrane protein segments (Robb et al. 1978; Engelman et al. 1980; Rogers et al. 1980; Uehara et al. 1981; Rice et al. 1982; Ross et al. 1982), the segments M1–M4 most probably traverse the membrane. The α-helical structure requires about 24 residues to span the hydrocarbon area of the lipid bilayer, and the extended β-configuration requires about 11 residues. In view of the lengths of the segments M1–M4 and by analogy with the transmembrane segments of bacteriorhodopsin (Henderson and Unwin 1975) and glycophorin A (Egmond et al. 1979; Schulte and Marchesi 1979), it seems reasonable to assume that the segments M1–M4 form α-helical structures. The segment M1 contains three conserved proline residues (positions 253, 263, and 278), which introduce bends with an angle of 20° in the helix axis. (Note that the numbers indicating positions in the aligned amino acid sequences differ from the amino acid numbers given in Fig. 1.)

Studies on selective proteolysis of the *Torpedo* AChR subunits have indicated that the aminoterminal two thirds of the subunit molecules are resistant to proteolysis and are glycosylated (Strader and Raftery 1980; Wennogle and Changeux 1980; Wennogle et al. 1981). In fact, this region contains the single possible *N*-glycosidic linkage site (position 169) in the α- and β-subunits, all of the three possible glycosylation sites (positions 96, 169, and 236) in the δ-subunit, and two of the four possible glycosylation sites (positions 96, 169, 343, and 486) in the γ-subunit (Marshall 1974), as indicated in Figures 2 and 3. Thus, the aminoterminal half of the mature subunit molecules preceding the transmembrane segment M1 is most probably exposed on the extracellular surface of the membrane, forming a rigid conformation. There is also evidence indicating that the carboxyterminal region of the δ-subunit carries the disulfide bond responsible for the dimerization of the AChR (Wennogle et al. 1981) and that the segment linking the δ-subunits of adjacent receptor molecules is located outside the membrane (Conti-Tronconi et al. 1982). The cysteine residue (position 550) next to the carboxyl terminus of the δ-subunit is the only cysteine that meets these requirements. On the basis of these findings, the orientation of the transmembrane segments M1–M4 has been assigned as shown in Figure 4. Thus, the region between the segments M3 and M4, constituting about one fourth of the mature subunit molecules, is located on the cytoplasmic side of the membrane. This region may have a looser conformation because it is susceptible to proteolysis (Strader and

Figure 1. Primary structures of the *T. californica* mRNAs coding for the α- (*A*), β- (*B*), γ- (*C*), and δ- (*D*) subunit precursors of the AChR. The nucleotide sequences of the mRNAs were deduced from those of the cloned cDNAs. Nucleotide residues are numbered in the 5′→ 3′ direction, beginning with the first residue of the codon specifying the aminoterminal residue of the mature subunits, and the nucleotides on the 5′ side of residue 1 are indicated by negative numbers. The 5′-terminal sequences presented do not extend to the 5′ end of the mRNAs, whereas the 3′-terminal sequences shown apparently represent the complete sequences of this region except in *D*; the 3′-terminal sequence of the γ-subunit precursor mRNA seems to have two alternative polyadenylation sites (indicated by arrows), one of which exhibits no known polyadenylation signal upstream. The sizes of the α-, β-, γ-, and δ-subunit precursor mRNAs, estimated by blot hybridization analysis of electroplax poly(A) RNA are ~2300, ~2000, ~2300, and ~6000 nucleotides, respectively. The predicted amino acid sequences are displayed above the nucleotide sequences. Amino acid residues are numbered beginning with the aminoterminal residue of the mature subunits, and the preceding residues are indicated by negative numbers. The amino acid sequences corresponding to the synthetic oligodeoxyribonucleotide probes used for screening the cDNA library are overlined. The nucleotide and amino acid differences observed among individual cDNA clones for the same subunit precursor are shown; they are probably due to polymorphism in the respective genes, although possible errors occurring during cDNA synthesis in vitro or DNA cloning cannot be excluded. The sequence of the γ-subunit precursor has also been deduced from the cDNA sequence by Claudio et al. (1983), who have assigned a serine residue instead of the proline residue 281. (Reprinted, with permission, from Noda et al. 1982b, 1983b,c.)

A

```
5'---CAGCUAUCAGCUGUCGCUGAGACAGGUGGCAUAAGAGUGGAACAGAGAGUUGAAAAGGCAGGAAACUGGCUUAUCUCUUCACUAGAAAAGAGCUGAACACAGAAGUCCAGAAGAU
                -240                -220                -200                -180                -160

                                                                                     -20
                                                                 Met Ile Leu Cys Ser Tyr Trp His Val Gly Leu Val
CUAACAAGUUCAUCGUUUAGUUAUUAGAAGUGGCAGAUUUGCUUGAAAAGCCAAUUAUUGAAAGCUGAAGA AUG AUU CUG UGC AGU UAU UGG CAU GUA GGG UUG GUG
-140                -120                -100                -80                     -60                     -40

    -10                                 -1  1                            10
Leu Leu Leu Phe Ser Cys Cys Gly Leu Val Leu Gly Ser Glu His Glu Thr Arg Leu Val Ala Asn Leu Leu Glu Asn Tyr Asn Lys Val
CUA CUG UUA UUU UCG UGU UGU GGU CUG GUA CUA GGU UCU GAA CAU GAA ACA CGU UUG GUU GCU AAU UUA UUA GAA AAU UAU AAC AAG GUG
                    -20                 -1  1                    20                      40

20                      30                                      40
Ile Arg Pro Val Glu His His Thr His Phe Val Asp Ile Thr Val Gly Leu Gln Leu Ile Gln Leu Ile Ser Val Asp Glu Val Asn Gln
AUU CGU CCA GUG GAG CAU CAC ACC CAC UUU GUA GAU AUU ACA GUG GGG CUA CAG CUG AUA CAA CUC AUC AGU GUG GAU GAA GUA AAU CAA
 60                      80                      100                     120                     140

50                                      60                                      70
Ile Val Glu Thr Asn Val Arg Leu Arg Gln Gln Trp Ile Asp Val Arg Leu Arg Trp Asn Pro Ala Asp Tyr Gly Gly Ile Lys Lys Ile
AUU GUG GAA ACA AAU GUG CGC CUA AGG CAG CAA UGG AUU GAU GUG AGG CUU CGC UGG AAU CCA GCC GAU UAU GGU GGA AUU AAA AAG AUC
                    160                     180                     200                     220

80                                      90                                      100
Arg Leu Pro Ser Asp Asp Val Trp Leu Pro Asp Leu Val Leu Tyr Asn Asn Ala Asp Gly Asp Phe Ala Ile Val His Met Thr Lys Leu
AGA CUG CCU UCU GAU GAU GUU UGG CUG CCA GAU UUA GUU CUG UAC AAC AAU GCU GAU GGU GAU UUU GCC AUU GUU CAC AUG ACC AAA CUG
240                     260                     280                     300                     320

110                                     120                                     130
Leu Leu Asp Tyr Thr Gly Lys Ile Met Trp Thr Pro Pro Ala Ile Phe Lys Ser Tyr Cys Glu Ile Ile Val Thr His Phe Pro Phe Asp
CUU UUG GAU UAU ACG GGA AAA AUA AUG UGG ACA CCU CCA GCA AUC UUC AAA AGC UAU UGU GAA AUU AUU GUA ACA CAU UUC CCA UUU GAU
                    340                     360                     380                     400

140                                     150                                     160
Gln Gln Asn Cys Thr Met Lys Leu Gly Ile Trp Thr Tyr Asp Gly Thr Lys Val Ser Ile Ser Pro Glu Ser Asp Arg Pro Asp Leu Ser
CAA CAA AAU UGC ACU AUG AAG UUG GGA AUC UGG ACG UAC GAU GGG ACA AAA GUU UCC AUA UCC CCG GAA AGU GAC CGU CCG GAU CUG AGU
420                     440                     460                     480                     500

170                                     180                                     190
Thr Phe Met Glu Ser Gly Glu Trp Val Met Lys Asp Tyr Arg Gly Trp Lys His Trp Val Tyr Tyr Thr Cys Cys Pro Asp Thr Pro Tyr
ACA UUU AUG GAA AGU GGA GAG UGG GUA AUG AAA GAU UAU CGU GGA UGG AAG CAC UGG GUG UAU UAU ACC UGC UGU CCU GAC ACU CCU UAC
                    520                     540                     560                     580

200                                     210                                     220
Leu Asp Ile Thr Tyr His Phe Ile Met Gln Arg Ile Pro Leu Tyr Phe Val Val Asn Val Ile Ile Pro Cys Leu Leu Phe Ser Phe Leu
CUG GAU AUC ACC UAC CAU UUU AUC AUG CAG CGU AUU CCU CUU UAU UUU GUU GUG AAU GUC AUC AUU CCU UGU CUG CUU UUU UCA UUU UUA
600                     620                     640                     660                     680

230                                     240                                     250
Thr Gly Leu Val Phe Tyr Leu Pro Thr Asp Ser Gly Glu Lys Met Thr Leu Ser Ile Ser Val Leu Leu Ser Leu Thr Val Phe Leu Leu
ACU GGA UUA GUA UUU UAC UUA CCA ACU GAU UCA GGU GAG AAG AUG ACU UUG AGU AUU UCC GUU UUG CUG UCU CUG ACU GUG UUC CUU CUG
                    700                     720                     740                     760

260                                     270                                     280
Val Ile Val Glu Leu Ile Pro Ser Thr Ser Ser Ala Val Pro Leu Ile Gly Lys Tyr Met Leu Phe Thr Met Ile Phe Val Ile Ser Ser
GUU AUU GUU GAG CUG AUC CCC UCA ACU UCC AGC GCU GUG CCU UUG AUU GGC AAA UAC AUG CUU UUU ACA AUG AUU UUU GUC AUC AGU UCA
780                     800                     820                     840                     860

290                                     300                                     310
Ile Ile Ile Thr Val Val Val Ile Asn Thr His His Arg Ser Pro Ser Thr His Thr Met Pro Gln Trp Val Arg Lys Ile Phe Ile Asp
AUC AUC AUU ACU GUU GUU GUA AUU AAU ACU CAC CAU CGC UCU CCA AGU ACA CAU ACA AUG CCA CAA UGG GUA CGA AAG AUC UUU AUU GAU
                    880                     900                     920                     940

320                                     330                                     340
Thr Ile Pro Asn Val Met Phe Phe Ser Thr Met Lys Arg Ala Ser Lys Glu Lys Gln Glu Asn Lys Ile Phe Ala Asp Asp Ile Asp Ile
ACU AUA CCC AAU GUU AUG UUU UUC UCA ACA AUG AAA CGA GCU UCU AAG GAA AAG CAA GAA AAU AAG AUA UUU GCU GAU GAC AUU GAU AUC
960                     980                   1,000                   1,020                   1,040

350                                     360                                     370
Ser Asp Ile Ser Gly Lys Gln Val Thr Gly Glu Val Ile Phe Gln Thr Pro Leu Ile Lys Asn Pro Asp Val Lys Ser Ala Ile Glu Gly
UCU GAC AUU UCU GGA AAG CAA GUG ACA GGA GAA GUA AUU UUU CAA ACA CCU CUC AUU AAA AAU CCA GAU GUC AAA AGU GCU AUU GAG GGA
                  1,060                   1,080                   1,100                   1,120

380                                     390                                     400
Val Lys Tyr Ile Ala Glu His Met Lys Ser Asp Glu Ser Ser Asn Ala Ala Glu Glu Trp Lys Tyr Val Ala Met Val Ile Asp His
GUC AAA UAU AUU GCA GAG CAC AUG AAG UCU GAU GAG GAA UCA AGC AAU GCU GCA GAG GAA UGG AAA UAU GUU GCA AUG GUG AUU GAU CAC
1,140                 1,160                   1,180                   1,200                   1,220

410                                     420                                     430
Ile Leu Leu Cys Val Phe Met Leu Ile Cys Ile Ile Gly Thr Val Ser Val Phe Ala Gly Arg Leu Ile Glu Leu Ser Gln Glu Gly
AUU CUG CUG UGU GUC UUC AUG CUG AUU UGU AUA AUU GGU ACA GUU AGC GUG UUU GCU GGC CGU CUC AUU GAA CUC AGU CAA GAG GGC UAA
                  1,240                   1,260                   1,280                   1,300

AUCUUCAUUGUGAGCAAAAAAGGCAAUACUGGAAUAAGGGAUGGAUAUCACUCCACAGAAAAGAUGUGUGGGUUUAGUGUUGCAAUUGUAGUCUGUUUUAUGAGAUAUAUAGUUUGCUUU
  1,320               1,340               1,360               1,380               1,400               1,420

GUUUUACAAUGAAAUGUACUUAAGGUAUUUGAAUAUGUAAAAAAAAGUAAUGAAAUAACAGUAAGUGAAAAAUGUUAUUAUGCAAGUACCUGAAACGUGUAAUAAGUGGAACAACUUUUU
  1,440               1,460               1,480               1,500               1,520               1,540

AAUACAUUACAUAAAAGUAAGCAAAAAAUAAGUUUAACAAAUUAUGAGGGUAGUCAUUUGAAAUGUAACAGAGAAAUGAAAAUUAUUAGAAAUAUAAACAGUAUAUUAAGUUAAACAA
  1,560               1,580               1,600               1,620               1,640               1,660

AGUUAAUCCAUUCUUUUAUAUCCAAAGUUUGUAUUAUACAUUUAGAAGUGUAGUUCUAUUGUAUAAUUUAAAGUAUGUUUUACAGAUCAUUAAUAAAUAUUCAAUGCAUUACU---3'
  1,680               1,700               1,720               1,740               1,760               1,780
```

Figure 1A. *(See facing page for legend.)*

B

Asp -20 Val -10

Met Glu Asn Val Arg Arg Met Ala Leu Gly Leu Val Val Met Met Ala Leu

5'------AUCAGGCGGCGGCGCCCGCUGGUGCAGGGAGAAGUCGUGGGCG AUG GAG AAC GUG AGG AGG AUG GCG CUG GGU CUG GUG GUG AUG AUG GCG CUG

-100 -80 G -60 G -40

-1 1 10 20

Ala Leu Ser Gly Val Gly Ala Ser Val Met Glu Asp Thr Leu Leu Ser Val Leu Phe Glu Thr Tyr Asn Pro Lys Val Arg Pro Ala Gln

GCC CUC AGC GGC GUG GGG GCG UCG GUG AUG GAG GAC ACA CUC CUG AGC GUG CUG UUC GAG ACC UAC AAC CCG AAG GUG CGG CCG GCG CAG

-20 -1 1 20 40 60

30 40 50

Thr Val Gly Asp Lys Val Thr Val Arg Val Gly Leu Thr Leu Thr Asn Leu Leu Ile Leu Asn Glu Lys Ile Glu Glu Met Thr Thr Asn

ACG GUG GGC GAU AAA GUC ACG GUG AGG GUG GGG CUG ACC CUC ACA AAC CUC CUC AUC CUG AAU GAA AAA AUU GAA GAG AUG ACG ACC AAC

80 100 120 140

60 70 80

Val Phe Leu Asn Leu Ala Trp Thr Asp Tyr Arg Leu Gln Trp Asp Pro Ala Ala Tyr Glu Gly Ile Lys Asp Leu Arg Ile Pro Ser Ser

GUC UUC UUG AAC CUG GCG UGG ACA GAC UAC CGC CUG CAG UGG GAC CCU GCU GCC UAC GAA GGC AUC AAA GAU CUC CGC AUU CCC UCC UCC

160 180 200 220 240

90 100 110

Asp Val Trp Gln Pro Asp Ile Val Leu Met Asn Asn Asn Asp Gly Ser Phe Glu Ile Thr Leu His Val Asn Val Leu Val Gln His Thr

GAC GUC UGG CAG CCC GAC AUC GUC CUG AUG AAC AAU AAC GAC GGG UCC UUU GAA AUU ACG CUC CAC GUG AAC GUG UUA GUG CAG CAU ACC

260 280 300 320

120 130 140

Gly Ala Val Ser Trp Gln Pro Ser Ala Ile Tyr Arg Ser Ser Cys Thr Ile Lys Val Met Tyr Phe Pro Phe Asp Trp Gln Asn Cys Thr

GGC GCA GUC AGC UGG CAG CCC UCU GCC AUC UAC CGC AGU UCC UGC ACC AUC AAA GUG AUG UAC UUC CCC UUC GAU UGG CAA AAC UGC ACC

340 360 380 400 420

150 160 170

Met Val Phe Lys Ser Tyr Thr Tyr Asp Thr Ser Glu Val Thr Leu Gln His Ala Leu Asp Ala Lys Gly Glu Arg Glu Val Lys Glu Ile

AUG GUC UUC AAG UCC UAC ACG UAC GAC ACA UCC GAG GUG ACC CUG CAG CAC GCC CUG GAU GCC AAG GGG GAG CGG GAG GUC AAG GAG AUC

440 460 480 500

180 190 200

Val Ile Asn Lys Asp Ala Phe Thr Glu Asn Gly Gln Trp Ser Ile Glu His Lys Pro Ser Arg Lys Asn Trp Arg Ser Asp Asp Pro Ser

GUG AUC AAC AAG GAC GCC UUC ACG GAG AAC GGG CAG UGG AGC AUC GAA CAC AAG CCG UCC CGC AAG AAC UGG CGG AGC GAU GAC CCC AGC

520 540 560 580 600

210 220 230

Tyr Glu Asp Val Thr Phe Tyr Leu Ile Ile Gln Arg Lys Pro Leu Phe Tyr Ile Val Tyr Thr Ile Ile Pro Cys Ile Leu Ile Ser Ile

UAC GAG GAC GUG ACC UUC UAC CUG AUC AUC CAG CGC AAG CCC CUC UUU UAC AUC GUC UAC ACC AUC AUC CCC UGC AUC CUG AUC AGC AUC

620 640 660 680

240 250 260

Leu Ala Ile Leu Val Phe Tyr Leu Pro Pro Asp Ala Gly Glu Lys Met Ser Leu Ser Ile Ser Ala Leu Leu Ala Val Thr Val Phe Leu

CUC GCC AUC CUG GUC UUC UAC CUC CCU CCC GAC GCA GGU GAG AAG AUG AGC CUG UCC AUC UCG GCG CUC CUG GCU GUC ACC GUC UUC CUG

700 720 740 760 780

270 280 290

Leu Leu Leu Ala Asp Lys Val Pro Glu Thr Ser Leu Ser Val Pro Ile Ile Ile Arg Tyr Leu Met Phe Ile Met Ile Leu Val Ala Phe

CUG CUG UUG GCC GAC AAG GUC CCU GAG ACG UCA CUA AGC GUC CCU AUC AUC AUC CGC UAC CUG AUG UUC AUC AUG AUC CUG GUC GCC UUC

800 820 840 860

300 310 320

Ser Val Ile Leu Ser Val Val Val Leu Asn Leu His His Arg Ser Pro Asn Thr His Thr Met Pro Asn Trp Ile Arg Gln Ile Phe Ile

UCC GUC AUC CUG AGU GUG GUC GUG CUG AAC CUC CAU CAC CGC UCG CCC AAC ACC CAC ACC AUG CCC AAC UGG AUC AGA CAG AUC UUC AUC

880 900 920 940 960

330 340 350

Glu Thr Leu Pro Pro Phe Leu Trp Ile Gln Arg Pro Val Thr Thr Pro Ser Pro Asp Ser Lys Pro Thr Ile Ile Ser Arg Ala Asn Asp

GAG ACG UUA CCU CCA UUC CUG UGG AUC CAG AGG CCG GUG ACA ACA CCU UCC CCC GAC AGC AAG CCC ACG AUC AUC AGC CGA GCU AAC GAC

980 1,000 1,020 1,040

360 370 380

Glu Tyr Phe Ile Arg Lys Pro Ala Gly Asp Phe Val Cys Pro Val Asp Asn Ala Arg Val Ala Val Gln Pro Glu Arg Leu Phe Ser Glu

GAG UAC UUC AUC CGG AAA CCA GCC GGG GAC UUC GUC UGC CCC GUG GAC AAC GCC AGG GUC GCG GUU CAG CCA GAA CGG CUG UUC UCG GAG

1,060 1,080 1,100 1,120 1,140

390 400 410

Met Lys Trp His Leu Asn Gly Leu Thr Gln Pro Val Thr Leu Pro Gln Asp Leu Lys Glu Ala Val Glu Ala Ile Lys Tyr Ile Ala Glu

AUG AAG UGG CAC CUG AAC GGC CUG ACC CAG CCC GUG ACC CUG CCU CAG GAC CUC AAG GAA GCG GUG GAG GCC AUC AAG UAU AUC GCC GAG

1,160 1,180 1,200 1,220

420 430 440

Gln Leu Glu Ser Ala Ser Glu Phe Asp Asp Leu Lys Lys Asp Trp Gln Tyr Val Ala Met Val Ala Asp Arg Leu Phe Leu Tyr Val Phe

CAA CUG GAG UCG GCG UCG GAG UUC GAU GAU CUG AAG AAA GAC UGG CAG UAC GUG GCG AUG GUC GCC GAC CGC CUC UUC CUC UAC GUC UUC

1,240 1,260 1,280 1,300 1,320

450 460

Phe Val Ile Cys Ser Ile Gly Thr Phe Ser Ile Phe Leu Asp Ala Ser His Asn Val Pro Pro Asp Asn Pro Phe Ala

UUC GUC AUA UGC UCC AUC GGG ACC UUC AGC AUC UUC CUC GAC GCC AGC CAC AAC GUG CCC CCC GAC AAC CCC UUU GCC UGA UCUCCACCCCCC

1,340 1,360 1,380 1,400 1,420

AACCUGUGUGCGGCCUCCUGGGACAGACGGGCAGCGUUUAUCGGCGUGAUUUGGCCGAGCGUUGGGUGUUCGGACCUCAGCUUCAGAGCUUUUCGUUGCGGUUGUAUUUUCUGCAUCGUG

1,440 1,460 1,480 1,500 1,520 1,540

GGGAUCUUCCCUCCCGCGCCUUCAGAGGGAAACGCUCCGAGCUAAUUAAAACCUCCUGGAUCCCC------3'

1,560 1,580 1,600

Figure 1B. *(See p. 58 for legend.)*

C

```
5'----AAUUGUACAGCUGGUGUCUAGCAGCUGGGUGGUGUAAUCUGGCCCUGCCUGAUGACCAGAAACUAAAACCCCUCACCCAACGACUCACUCACAAGCUGAGCUAGCACACUCACU
          -160            -140            -120            -100            -80            -60

                                      -10                                        -1  1                                        10
        Met Val Leu Thr Leu Leu Leu Ile Ile Cys Leu Ala Leu Glu Val Arg Ser Glu Asn Glu Glu Gly Arg Leu Ile Glu Lys Leu
GACGACC AUG GUA CUG ACA CUU CUC CUG AUU AUU UGU CUG GCG UUA GAA GUG AGA UCA GAA AAU GAA GAA GGA AGA CUU AUC GAG AAA UUG
                    -40                             -20                    -1  1                      20

                                20                                      30                                      40
Leu Gly Asp Tyr Asp Lys Arg Ile Ile Pro Ala Lys Thr Leu Asp His Ile Ile Asp Val Thr Leu Lys Leu Thr Leu Thr Asn Leu Ile
UUG GGU GAC UAC GAU AAG AGG AUC AUA CCA GCA AAG ACC CUC GAU CAC AUU AUU GAC GUG ACA CUG AAA CUU ACA CUA ACU AAU CUC AUU
    40                          60                  80                      100                         120

                                50                                      60                                      70
Ser Leu Asn Glu Lys Glu Glu Ala Leu Thr Thr Asn Val Trp Ile Glu Ile Gln Trp Asn Asp Tyr Arg Leu Ser Trp Asn Thr Ser Glu
UCA UUG AAU GAA AAA GAG GAG GCA CUA ACC ACA AAU GUC UGG AUU GAA AUA CAA UGG AAU GAC UAC CGG UUA UCC UGG AAC ACG UCC GAA
                    140                     160                     180                     200

                                80                                      90                                      100
Tyr Glu Gly Ile Asp Leu Val Arg Ile Pro Ser Glu Leu Leu Trp Leu Pro Asp Val Val Leu Glu Asn Asn Val Asp Gly Gln Phe Glu
UAC GAA GGG AUA GAU CUG GUC AGA AUC CCC UCU GAA UUG CUU UGG CUU CCU GAU GUU GUC CUU GAG AAC AAC GUU GAU GGU CAG UUU GAA
        220                     240                     260                     280                         300

                                110                                     120                                     130
Val Ala Tyr Tyr Ala Asn Val Leu Val Tyr Asn Asp Gly Ser Met Tyr Trp Leu Pro Pro Ala Ile Tyr Arg Ser Thr Cys Pro Ile Ala
GUU GCU UAC UAU GCC AAU GUU CUG GUC UAC AAU GAU GGC AGC AUG UAC UGG CUG CCU CCU GCC AUA UAU CGA AGC ACC UGC CCC AUC GCU
                    320                     340                     360                     380

                                140                                     150                                     160
Val Thr Tyr Phe Pro Phe Asp Trp Gln Asn Cys Ser Leu Val Phe Arg Ser Gln Thr Tyr Asn Ala His Glu Val Asn Leu Gln Leu Ser
GUG ACA UAC UUC CCU UUC GAC UGG CAG AAC UGU UCC CUC GUC UUC AGA UCA CAG ACA UAC AAU GCG CAU GAA GUG AAU UUG CAA CUU UCA
        400                     420                     440                     460                         480

                                170                                     180                                     190
Ala Glu Glu Gly Glu Ala Val Glu Trp Ile His Ile Asp Pro Glu Asp Phe Thr Glu Asn Gly Glu Trp Thr Ile Arg His Arg Pro Ala
GCA GAA GAA GGA GAG GCG GUA GAA UGG AUC CAC AUC GAC CCU GAA GAC UUC ACU GAG AAU GGU GAG UGG ACA AUU AGA CAU AGG CCU GCG
                    500                     520                     540                     560

                                200                                     210                                     220
Lys Lys Asn Tyr Asn Trp Gln Leu Thr Lys Asp Asp Thr Asp Phe Gln Glu Ile Ile Phe Phe Leu Ile Ile Gln Arg Lys Pro Leu Phe
AAG AAG AAU UAC AAC UGG CAG UUG ACC AAG GAU GAC ACU GAU UUC CAG GAG AUC AUU UUC UUC CUG AUC AUU CAG AGG AAA CCA CUG UUC
        580                     600                     620                     640                         660

                                230                                     240                                     250
Tyr Ile Ile Asn Ile Ile Ala Pro Cys Val Leu Ile Ser Ser Leu Val Val Leu Val Tyr Phe Leu Pro Ala Gln Ala Gly Gly Gln Lys
UAC AUC AUC AAC AUC AUU GCU CCG UGU GUC CUC AUU UCA UCA CUG GUC GUC CUU GUC UAU UUC CUG CCU GCA CAA GCU GGG GGC CAA AAA
                    680                     700                     720                     740

                                260                                     270                                     280
Cys Thr Leu Ser Ile Ser Val Leu Leu Ala Gln Thr Ile Phe Leu Phe Leu Ile Ala Gln Lys Val Pro Glu Thr Ser Leu Asn Val Pro
UGU ACU CUA UCA AUC UCU GUU CUC CUG GCA CAA ACU AUC UUC CUC UUC CUG AUU GCA CAG AAA GUU CCG GAA ACA UCA CUG AAU GUU CCA
        760                     780                     800                     820                         840

                                290                                     300                                     310
Leu Ile Gly Lys Tyr Leu Ile Phe Val Met Phe Val Ser Met Leu Ile Val Met Asn Cys Val Ile Val Leu Asn Val Ser Leu Arg Thr
CUC AUC GGA AAA UAU UUA AUA UUU GUG AUG UUC GUA UCG AUG UUG AUC GUA AUG AAC UGC GUU AUU GUC UUG AAU GUC UCC UUG AGA ACG
                    860                     880                     900                     920

                                320                                     330                                     340
Pro Asn Thr His Ser Leu Ser Glu Lys Ile Lys His Leu Phe Leu Gly Phe Leu Pro Lys Tyr Leu Gly Met Gln Leu Glu Pro Ser Glu
CCC AAC ACU CAC UCC UUG UCU GAG AAG AUC AAA CAC CUG UUC CUG GGA UUC CUG CCC AAG UAU CUG GGU AUG CAG UUA GAG CCU UCU GAG
        940                     960                     980                     1,000                       1,020

                                350                                     360                                     370
Glu Thr Pro Glu Lys Pro Gln Pro Arg Arg Arg Ser Ser Phe Gly Ile Met Ile Lys Ala Glu Glu Tyr Ile Leu Lys Lys Pro Arg Ser
GAA ACA CCG GAA AAG CCU CAA CCA AGA CGG AGA AGU UCC UUU GGG AUU AUG AUC AAG GCA GAG GAA UAU AUU CUC AAA AAA CCA AGG AGU
                    1,040                   1,060                   1,080                   1,100

                                380                                     390                                     400
Glu Leu Met Phe Glu Glu Gln Lys Asp Arg His Gly Leu Lys Arg Val Asn Lys Met Thr Ser Asp Ile Asp Ile Gly Thr Thr Val Asp
GAG CUG AUG UUU GAA GAA CAG AAA GAC AGG CAU GGU CUG AAG AGA GUA AAC AAG AUG ACA UCA GAU AUU GAC AUU GGA ACG ACG GUC GAC
        1,120                   1,140                   1,160                   1,180                       1,200

                                410                                     420                                     430
Leu Tyr Lys Asp Leu Ala Asn Phe Ala Pro Glu Ile Lys Ser Cys Val Glu Ala Cys Asn Phe Ile Ala Lys Ser Thr Lys Glu Gln Asn
CUC UAC AAA GAC CUG GCU AAC UUC GCA CCA GAG AUA AAA UCC UGC GUC GAG GCU UGC AAC UUU AUU GCC AAA AGC ACU AAA GAA CAG AAC
                    1,220                   1,240                   1,260                   1,280

                                440                                     450                                     460
Asp Ser Gly Ser Glu Asn Glu Asn Trp Val Leu Ile Gly Lys Val Ile Asp Lys Ala Cys Phe Trp Ile Ala Leu Leu Leu Phe Ser Ile
GAU UCU GGC UCU GAA AAC GAA AAC UGG GUU CUG AUU GGA AAA GUC AUC GAC AAG GCU UGC UUC UGG AUU GCC CUU CUA CUC UUU AGU AUU
        1,300                   1,320                   1,340                   1,360                       1,380

                                470                                     480
Gly Thr Leu Ala Ile Phe Leu Thr Gly His Phe Asn Gln Val Pro Glu Phe Pro Phe Pro Gly Asp Pro Arg Lys Tyr Val Pro
GGC ACU CUG GCA AUC UUC CUC ACA GGG CAC UUC AAC CAA GUA CCG GAA UUC CCC UUC CCA GGA GAC CCC AGG AAA UAC GUC CCG UAA UAGA
                    1,400                   1,420                   1,440                   1,460

AUUCUUCGCCCAUCCUCUUCCUAAUCCCAGUUCAGAUCCAGAAUCAUUUCUGACCUCCUUAUGUUUAAUAUUUUCAAAUAAAUAAAUUACUGUUCAGCAGCUGUAAGCCACGGGGCUUCC
    1,480          1,500          1,520          1,540          1,560          1,580

AGGCAGGGCUUUAGAAAGUGAUUUCUAUGUAAAGUCUUGGCUUAUCGCUCCUUCAAUGUGUUUGAAUCUCAGUGUUUGGCAGCGUUUGAUGGUGGGGUUUAGGAAGGAGGGAGAAUGGGG
    1,600          1,620          1,640          1,660          1,680          1,700

                                                                                      ↓
UGAUUUGCAAACUACCCCACGGUCAACUUUAAAUUGGUUCUUCCGUGAAACAUUUUUAAUAAAGGACAAAACCUUCCAUCGUCCGUCCACAUUCAUGAAAAAAAAAAUCAAGGACUAUUA
    1,720          1,740          1,760          1,780          1,800          1,820

                                                                                            ↓
AUUUUUCAUUUCUGUGGUUGUACUGCAUUCUUUAAGGUUGUUUUAAUUAUGCUUAACGGACAAAUACAACAUAAUUGCUCUCUUGUU----3'
    1,840          1,860          1,880          1,900          1,920
```

Figure 1C. *(See p. 58 for legend.)*

D

5'----CUUUCUCACACCUUGCGGAUAUGGAACAAACUGCCAGAGAAAUUGAUAGAGGCAGGGAUAAACAGUGUUUAUAAGACAGUUGGACUGGUUCAUAGAUAGGAAAGGUUUAGAAGG
-520 -500 -480 -460 -440 -420

AUAGUUGGCCUAUGCAAUCAAGUGGGAUUAGCUCAGGAACGCACCUUGGUUAGCAUGGAUGAGUUGGGAUGAAGGGCCUGUUUCGGUGUUAUCUAACUCCAGCUUUAUGACUCAGAUUUG
-400 -380 -360 -340 -320 -300

ACUUUCACGUGGAAUGUGAAUCACAGAUUCCCCAACCGUGGUUCUAGCUUCCAGAAUCUUCACGUCAGAGACAGUUUAUUUUAUAUCCAUCUGCAUCAGCAGCUGUUUUUGUCAGUUUCC
-280 -260 -240 -220 -200 -180

-20
Met Gly Asn Ile
ACUCUGAAUCAGGUUUCAGAUUCUCAUUGCCCUUCAAAGAAUACCCAAUAAGACUGCAAAUGCUAUACUGCAACCAGUAAAUUCGGACCUGAGCUCUUUCAAA AUG GGG AAC AUU
-160 -140 -120 -100 -80 -60

-10 -1 1 10
His Phe Val Tyr Leu Leu Ile Ser Cys Leu Tyr Tyr Ser Gly Cys Ser Gly Val Asn Glu Glu Glu Arg Leu Ile Asn Asp Leu Leu Ile
CAU UUU GUU UAU CUU CUC AUU UCC UGC CUG UAU UAU UCA GGA UGU UCU GGU GUG AAU GAA GAG GAG AGG CUG AUU AAU GAC CUU UUA AUA
-40 -20 -1 1 20

20 30 40
Val Asn Lys Tyr Asn Lys His Val Arg Pro Val Lys His Asn Asn Glu Val Val Asn Ile Ala Leu Ser Leu Thr Leu Ser Asn Leu Ile
GUC AAC AAA UAC AAC AAA CAU GUA CGC CCA GUC AAG CAU AAC AAU GAA GUG GUG AAU AUU GCU CUG AGU UUA ACA CUC UCC AAU CUU AUC
40 60 80 100 120

50 60 70
Ser Leu Lys Glu Thr Asp Glu Thr Leu Thr Ser Asn Val Trp Met Asp His Ala Trp Tyr Asp His Arg Leu Thr Trp Asn Ala Ser Glu
UCA UUG AAA GAA ACU GAU GAG ACU CUU ACU UCG AAU GUA UGG AUG GAU CAU GCA UGG UAU GAU CAU CGG CUG ACC UGG AAU GCA UCA GAA
140 160 180 200

80 90 100
Tyr Ser Asp Ile Ser Ile Leu Arg Leu Pro Pro Glu Leu Val Trp Ile Pro Asp Ile Val Leu Gln Asn Asn Asn Asp Gly Gln Tyr His
UAC AGU GAC AUU UCC AUC CUU CGA CUU CCU CCU GAA UUA GUG UGG AUA CCA GAC AUU GUC CUA CAA AAC AAU AAC GAU GGG CAG UAC CAC
220 240 260 280 300

110 120 130
Val Ala Tyr Phe Cys Asn Val Leu Val Arg Pro Asn Gly Tyr Val Thr Trp Leu Pro Pro Ala Ile Phe Arg Ser Ser Cys Pro Ile Asn
GUG GCC UAU UUU UGC AAU GUG CUG GUC CGA CCA AAU GGA UAC GUU ACA UGG CUU CCU CCU GCA AUA UUU AGG AGU UCC UGU CCC AUC AAC
320 340 360 380

140 150 160
Val Leu Tyr Phe Pro Phe Asp Trp Gln Asn Cys Ser Leu Lys Phe Thr Ala Leu Asn Tyr Asp Ala Asn Glu Ile Thr Met Asp Leu Met
GUC CUC UAU UUU CCA UUU GAC UGG CAG AAC UGC UCC UUG AAA UUC ACU GCU UUG AAU UAC GAU GCA AAU GAG AUA ACU AUG GAC CUA AUG
400 420 440 460 480

170 180 190
Thr Asp Thr Ile Asp Gly Lys Asp Tyr Pro Ile Glu Trp Ile Ile Ile Asp Pro Glu Ala Phe Thr Glu Asn Gly Glu Trp Glu Ile Ile
ACA GAC ACU AUA GAU GGC AAG GAU UAU CCC AUU GAA UGG AUU AUU AUU GAU CCU GAA GCU UUC ACG GAG AAU GGU GAA UGG GAA AUC AUU
500 520 540 560

200 210 220
His Lys Pro Ala Lys Lys Asn Ile Tyr Pro Asp Lys Phe Pro Asn Gly Thr Asn Tyr Gln Asp Val Thr Phe Tyr Leu Ile Ile Arg Arg
CAC AAA CCU GCA AAA AAA AAC AUC UAU CCG GAU AAA UUC CCU AAC GGC ACC AAU UAC CAG GAU GUG ACU UUC UAC CUC AUA AUC AGG CGC
580 600 620 640 660

230 240 250
Lys Pro Leu Phe Tyr Val Ile Asn Phe Ile Thr Pro Cys Val Leu Ile Ser Phe Leu Ala Ser Leu Ala Phe Tyr Leu Pro Ala Glu Ser
AAA CCA UUG UUU UAU GUG AUC AAC UUC AUA ACU CCC UGU GUC CUC AUU UCC UUC CUG GCA UCA UUG GCU UUU UAU CUU CCU GCA GAG AGU
680 700 720 740

260 270 280
Gly Glu Lys Met Ser Thr Ala Ile Ser Val Leu Leu Ala Gln Ala Val Phe Leu Leu Leu Thr Ser Gln Arg Leu Pro Glu Thr Ala Leu
GGU GAG AAA AUG AGC ACG GCU AUU UCU GUC UUG CUU GCU CAG GCU GUU UUC CUG CUG CUA ACU UCA CAA AGA UUA CCA GAA ACG GCU CUA
760 780 800 820 840

290 300 310
Ala Val Pro Leu Ile Gly Lys Tyr Leu Met Phe Ile Met Ser Leu Val Thr Gly Val Ile Val Asn Cys Gly Ile Val Leu Asn Phe His
GCU GUU CCA UUA AUU GGC AAG UAC UUA AUG UUC AUC AUG AGU CUG GUC ACU GGU GUG AUC GUG AAU UGC GGC AUU GUG UUA AAC UUU CAC
860 880 900 920

320 330 340
Phe Arg Thr Pro Ser Thr His Val Leu Ser Thr Arg Val Lys Gln Ile Phe Leu Glu Lys Leu Pro Arg Ile Leu His Met Ser Arg Ala
UUC CGA ACU CCA AGC ACA CAU GUU CUC UCA ACA AGG GUU AAA CAG AUC UUC UUG GAA AAG CUU CCA CGG AUC CUC CAC AUG UCC CGU GCU
940 960 980 1,000 1,020

350 360 370
Asp Glu Ser Glu Gln Pro Asp Trp Gln Asn Asp Leu Lys Leu Arg Arg Ser Ser Ser Val Gly Tyr Ile Ser Lys Ala Gln Glu Tyr Phe
GAU GAG AGC GAG CAG CCG GAU UGG CAG AAC GAU UUG AAG CUG CGA CGC AGC AGU UCU GUU GGG UAC AUU UCC AAA GCA CAA GAA UAU UUU
1,040 1,060 1,080 1,100

380 390 400
Asn Ile Lys Ser Arg Ser Glu Leu Met Phe Glu Lys Gln Ser Glu Arg His Gly Leu Val Pro Arg Val Thr Pro Arg Ile Gly Phe Gly
AAC AUC AAA UCC AGA AGU GAA CUC AUG UUU GAA AAA CAA UCU GAA CGU CAU GGA UUG GUG CCG CGA GUC ACU CCA AGA AUU GGU UUU GGA
1,120 1,140 1,160 1,180 1,200

410 420 430
Asn Asn Asn Glu Asn Ile Ala Ala Ser Asp Gln Leu His Asp Glu Ile Lys Ser Gly Ile Asp Ser Thr Asn Tyr Ile Val Lys Gln Ile
AAC AAC AAU GAG AAC AUA GCU GCU UCA GAU CAA CUA CAU GAU GAG AUA AAG UCA GGU AUC GAC AGC ACA AAU UAC AUU GUG AAA CAA AUC
1,220 1,240 1,260 1,280

440 450 460
Lys Glu Lys Asn Ala Tyr Asp Glu Glu Val Gly Asn Trp Asn Leu Val Gly Gln Thr Ile Asp Arg Leu Ser Met Phe Ile Ile Thr Pro
AAA GAA AAG AAU GCC UAU GAU GAG GAA GUU GGG AAC UGG AAU CUA GUG GGC CAG ACU AUU GAC AGA CUG UCC AUG UUC AUC AUA ACU CCA
1,300 1,320 1,340 1,360 1,380

470 480 490
Val Met Val Leu Gly Thr Ile Phe Ile Phe Val Met Gly Asn Phe Asn His Pro Pro Ala Lys Pro Phe Glu Gly Asp Pro Phe Asp Tyr
GUG AUG GUC CUU GGA ACU AUC UUC AUA UUU GUG AUG GGA AAU UUC AAU CAC CCU CCA GCA AAG CCA UUU GAA GGU GAU CCC UUU GAC UAU
1,400 1,420 1,440 1,460

500
Ser Ser Asp His Pro Arg Cys Ala
UCA UCA GAU CAC CCA CGA UGU GCA UAA CUGAAGAACAUAGAACAUUACAGCACAGGCCCUUCAGCCCAUAAUGUUGUGCUGACCUAUGUAAACCCACUCCACAACAAUCUA
1,480 1,500 1,520 T 1,540 1,560 1,580

AUUCUCCCCAACCUUACACUCAUAAUCCACUGGUUUUUGUGUAUUCAUGUGCCUAUCUAAGAGUCUUUUGAAUGUCCCUAUUGUACCAACCUCCAAUGCAUUCCAGGCACCCAACACUCC
1,600 1,620 1,640 1,660 1,680 1,700

CUCUGU----3'

Figure 1D. *(See p. 58 for legend.)*

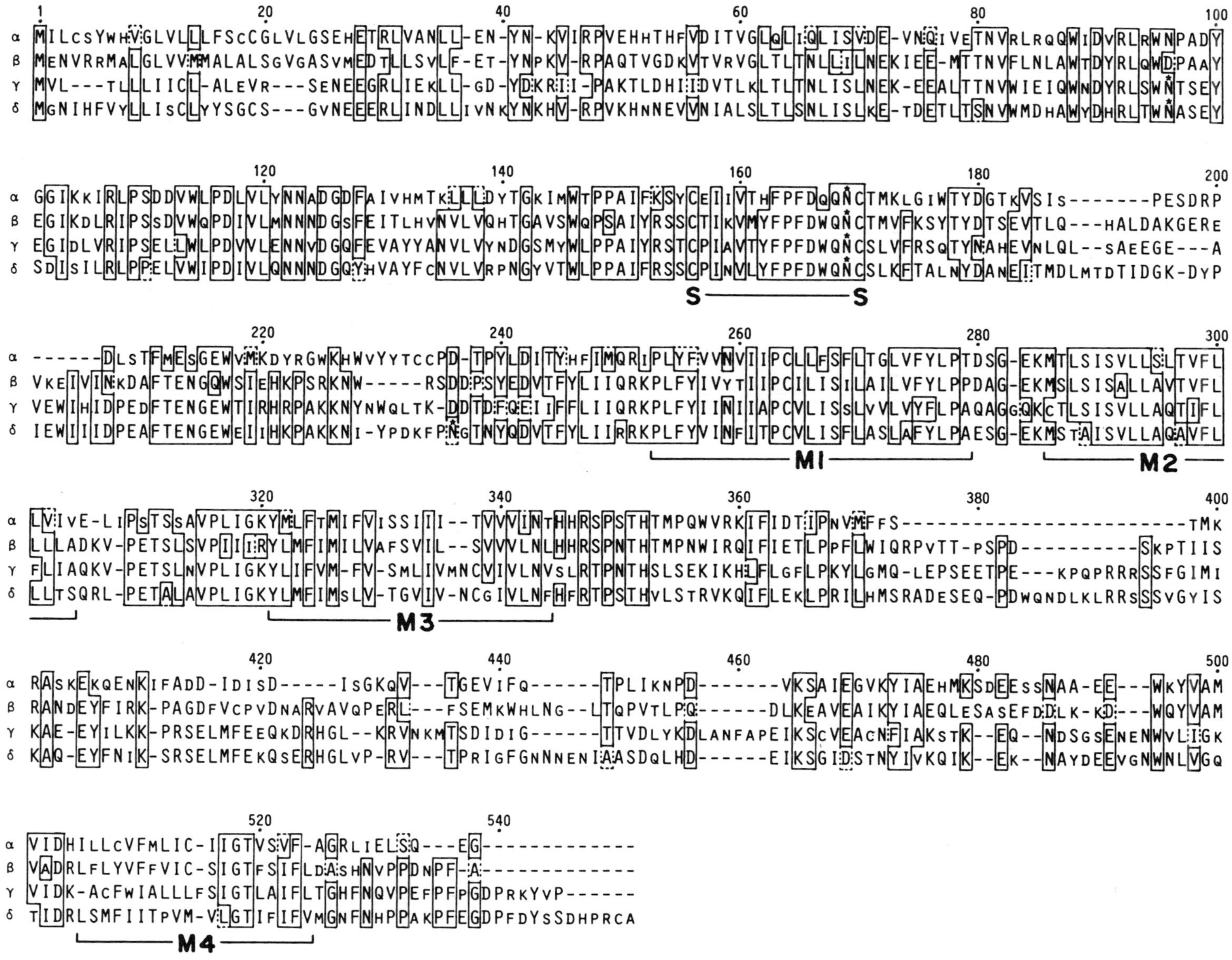

Figure 2. *(See p. 64 for legend.)*

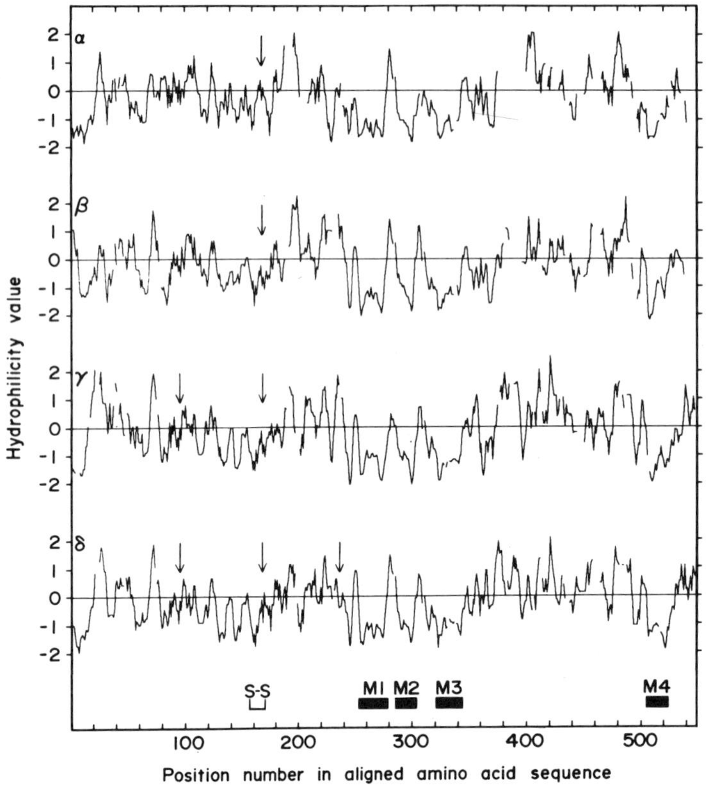

Figure 3. Hydrophilicity profiles of the α-, β-, γ-, and δ-subunit precursors (from top to bottom). The averaged hydrophilicity value of a hexapeptide composed of residues at positions $i-2$ to $i+3$ has been plotted against i, where i represents the position number in the aligned sequences (Fig. 2); no plots are given for the positions corresponding to gaps. The locations of the putative disulfide bridge near the ACh-binding site (S–S), the putative transmembrane segments (M1–M4), and the possible glycosylation sites (arrows) are shown (see Fig. 2).

Raftery 1980; Wennogle and Changeux 1980; Wennogle et al. 1981). This seems to be related to the fact that the amino acid sequence in this region is less well conserved among the four subunits, being subject to many deletions/insertions. The proposed transmembrane topology of the AChR molecule is consistent with the estimations hitherto made by physical and morphological approaches (Klymkowsky and Stroud 1979; Wise et al. 1979; Zingsheim et al. 1982).

It is tempting to assume that the putative transmembrane segments are involved in the cation channel of the AChR, the polar side of the α-helices constituting the wall of the channel. Noteworthy in this context is that the polar uncharged side chains present in the transmembrane segments M1–M4 are clustered largely on one side of an α-helix. The end-plate channel, which is permeable to monovalent and divalent cations and nonelectrolytes but discriminates against anions, is considered to be a water-filled neutral pore without high-field-strength sites inside (Adams et al. 1980). The uncharged nature of the transmembrane segments M1–M4 is consistent with this concept and may underlie the weak selectivity of the ionic channel of the AChR. In view of the diameter of the central pore being 6–25 Å (Cartaud et al. 1973; Nickel and Potter 1973; Maeno et al. 1977; Klymkowsky and Stroud 1979; Adams et al. 1980), one or two helical segments from each subunit together would suffice to construct the wall of the ionic channel, and the remaining helical segments might interact with those forming the channel. In this context, it seems more likely that the segments M1 and M2, which are highly conserved among the four subunits, are more directly involved in the ionic channel. The segment M1 contains three proline residues (see above) and one cysteine residue (position 264), which are conserved in all four subunits. This is not typical of transmembrane segments acting simply as a lipophilic membrane anchor (Rice et al. 1982), but favors the view that the segment M1 carries a specific function. The cysteine residue in the segment M1 is the only cysteine residue

Figure 2. Alignment of the amino acid sequences of the α- (*top*), β- (*second row*), γ- (*third row*), and δ- (*bottom*) subunit precursors of the *T. californica* AChR. The one-letter amino acid notation is used. Sets of identical residues and residues considered to be favored amino acid substitutions are represented by large letters, and the remaining residues by small letters. Sets of three or four identical residues at one aligned position are enclosed with solid lines, and the fourth residues regarded as favored substitutions at the same position are enclosed with dotted lines. Favored amino acid substitutions are defined as pairs of residues belonging to one of the following groups: S, T, P, A, and G; N, D, E, and Q; H, R, and K; M, I, L, and V; F, Y, and W (Dayhoff et al. 1978). Gaps (–) have been inserted to achieve maximum homology. The positions in the aligned sequences including gaps are numbered beginning with that of the initiative methionine. The positions of the putative disulfide bridge proposed as being in close proximity to the ACh-binding site on the α-subunit (S–S) and the putative transmembrane segments (M1–M4) are shown. The asparagine residues as possible sites of glycosylation are indicated by asterisks, except for the two sites in the γ-subunit assigned to the cytoplasmic side of the membrane. (Reprinted, with permission, from Noda et al. 1983b.)

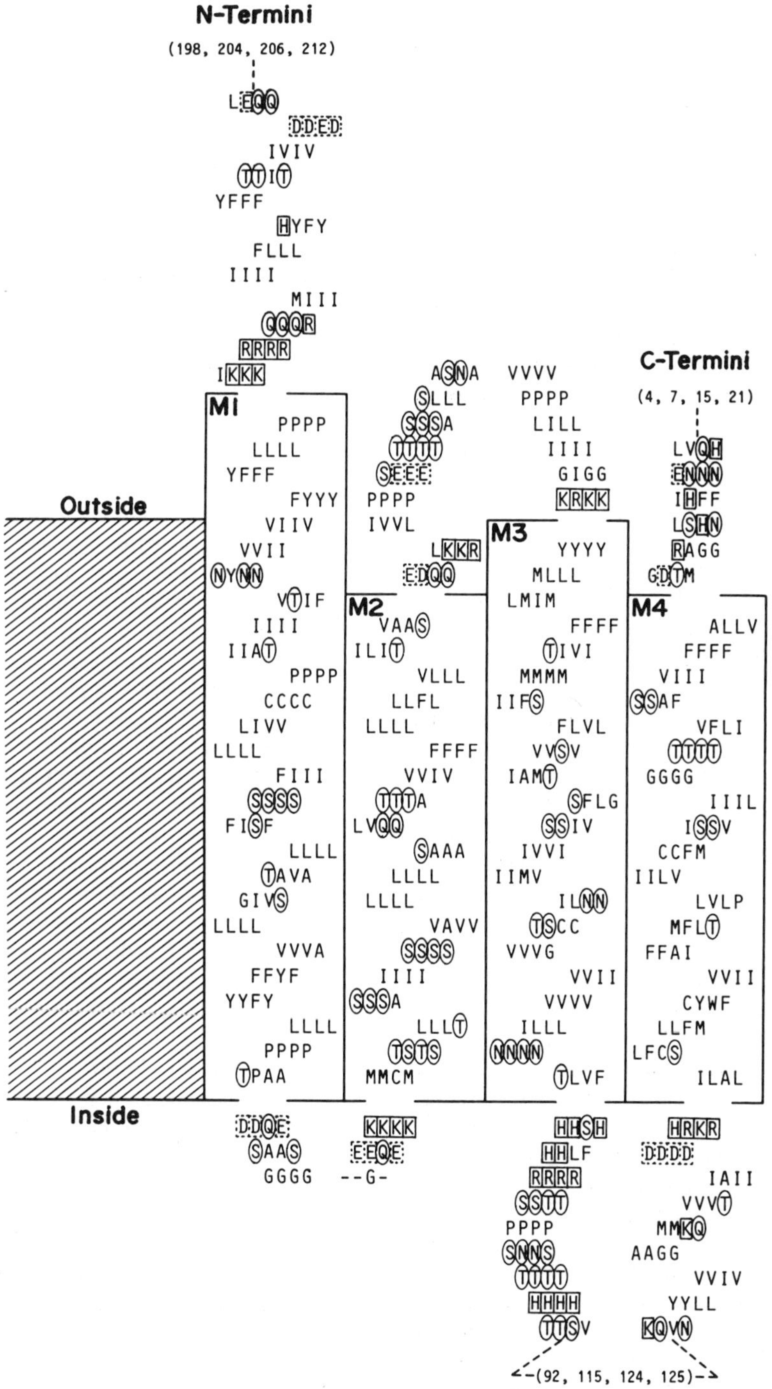

Figure 4. Proposed orientation of the *T. californica* AChR subunit polypeptides across the membrane. The amino acid sequences of the four putative transmembrane segments (M1–M4) and the adjacent regions are displayed. Sets of four letters indicate the residues in the aligned α-, β-, γ-, and δ-subunit sequences, respectively, from left to right; gaps (–) have been removed, except at position 283, to permit ready identification of the side chains extending toward one side of an α-helix. Sets of four numbers in parentheses indicate the sum of residues in the mature α-, β-, γ-, and δ-subunit sequences, respectively (from left to right), that either extend beyond the ends of the displayed sequences or are inserted between them. Uncharged polar residues (serine, threonine, asparagine, and glutamine) are enclosed with solid circles, positively charged residues with solid squares, and negatively charged residues with dotted squares. The termini of the proposed transmembrane segments have been tentatively assigned to the ends of a region composed exclusively of uncharged residues, so that the lengths of the boxed segments do not necessarily represent the real lengths of the segments embedded in the lipid hydrocarbon portion (shaded) of the membrane. (Reprinted, with permission, from Noda et al. 1983b.)

conserved in all subunits other than the two cysteine residues (positions 156 and 170) proposed as taking part in the disulfide bridge near the ACh-binding site on the α-subunit (see below), and is unlikely to be involved in this disulfide bridge because of its intramembranous location. The cysteine residue at position 264, being adjacent to one of the conserved proline residues causing a bend in the helix axis of the segment M1, may possibly be important for the function of the ionic channel.

It has been shown that the α-subunit of the fish electric organ AChR contains part or all of the ACh-binding site (Weill et al. 1974; Karlin et al. 1976; Damle and Karlin 1978). Furthermore, there is evidence indicating the presence of a disulfide bridge in the vicinity of the ACh-binding site in the α-subunit (Karlin 1969, 1980). This disulfide can be reduced and alkylated by affinity reagents. Such affinity-labeling studies suggest that, in the ACh-binding region, the distance between the negative subsite binding the quaternary ammonium group and one of the sulfhydryl groups formed by reduction is about 1 nm. In the native unreduced state, a second subsite that is involved in the binding of ACh through hydrophobic interaction or hydrogen bonding is postulated. Because three of the seven cysteine residues present in the mature α-subunit molecule (positions 264, 508, and 514) are located within the putative transmembrane segments, it seems improbable that they appear on the outer surface of the membrane or receptor molecule for ready access to affinity reagents. Two of the remaining four cysteine residues (positions 156, 170, 233, and 234), which are assigned to the extracellular side of the membrane, may take part in the disulfide bridge present in close proximity to the ACh-binding site; the formation of a disulfide bond between the adjacent cysteine residues (positions 233 and 234) is unlikely (Dayhoff 1976). Because the asparagine residue at position 169 is most probably glycosylated (see above), the region surrounding it should reside on the extracellular surface of the receptor molecule. Moreover, this region encompasses a predicted β-turn structure lying between predicted β-sheet structures, which may be tightly packed, so that its conformation would be more or less stabilized. It is known that the nucleotide-binding site of dehydrogenases occurs at an end of a pleated sheet structure on the outer surface of the molecules (Liljas and Rossmann 1974). A similar conformation is observed in the hypervariable regions of the immunoglobulin molecules (Poljak et al. 1973). Thus, it is tempting to assume that the ACh-binding site is located in the region including the cysteine residues at positions 156 and 170. Measurements on Corey-Pauling-Koltun space-filling models are consistent with the hypothesis that the aspartic acid residue at position 166 or the glutamic acid residue at position 157 as the negative subsite, together with, for example, the histidine residue at position 162 as the second subsite, may be involved in the binding of ACh. It is also possible that a hydrophobic residue in the vicinity acts as the second subsite. Interestingly, the two cysteine residues at positions 156 and 170 are conserved in all four subunits, and one of them is adjacent to the asparagine residue (position 169) that is the only glycosylation site conserved in all subunits. It is also noteworthy that only the α-subunit contains the histidine at position 162 (see above) instead of the tyrosine, which is conserved in the other three subunits. Although the possibility that one of the two remaining cysteine residues (positions 233 and 234) in the extracellular region of the α-subunit participates in the disulfide bridge near the ACh-binding site cannot be excluded, it seems most likely that the conserved cysteine residues at positions 156 and 170 form the disulfide bridge (for the third conserved cysteine residue at position 264, see above). The β-, γ-, and δ-subunits may also contain a corresponding disulfide bridge.

Knowledge of the antigenic determinants on the AChR and its subunits is important because antibodies directed to them are the primary pathological agents in myasthenia gravis and its animal model, experimental autoimmune myasthenia gravis (for review, see Lindstrom and Dau 1980). It has been shown that most of the antibodies produced by human patients as well as by experimental animals are directed to the "main immunogenic region," which is located on the extracellular surface of the α-subunit and is distinct from but may be near the ACh-binding site (Gullick et al. 1981; Tzartos et al. 1981, 1982). Hopp and Woods (1981) have reported that the point of highest local hydrophilicity is invariably located in or immediately adjacent to a protein antigenic determinant. This, in conjunction with the proposed transmembrane topology of the subunit molecules, allows us to predict possible candidates for antigenic determinants on the α-subunit molecule. The peptide segment Glu-Ser-Asp-Arg-Pro-Asp (positions 196–207 with gaps) on the α-subunit seems to be the most likely candidate for the "main immunogenic region," although other regions on the extracellular surface of the α-subunit corresponding to local upspikes of the hydrophilicity profile (Fig. 3) cannot be excluded (Hopp and Woods 1981).

Primary Structure of Mammalian Muscle AChR α-Subunit Precursor

A library of cDNA clones derived from poly(A)-containing RNA from newborn calf skeletal muscle was constructed with the Okayama-Berg vector (Okayama and Berg 1982) and was screened by hybridization with a cDNA fragment containing most of the protein-coding sequence for the *T. californica* AChR α-subunit precursor. Because the initial calf cDNA clone thus obtained did not contain the entire protein-coding sequence, additional clones carrying cDNA for a further upstream region were isolated by screening cDNA transcripts that were formed by elongation of a specific oligodeoxyribonucleotide primer and cloned into plasmid pBR322 according to the procedure described previously (Kakidani et al. 1982; Noda et al. 1982a;

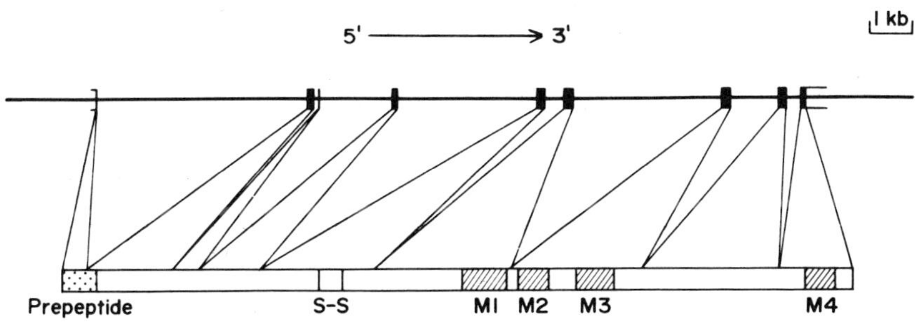

Figure 5. Arrangement of the protein-coding regions in the human AChR α-subunit precursor gene. The nine exons (■) and the regions of the α-subunit precursor molecule encoded by them are shown. The positions of the prepeptide (⊡), the putative disulfide bridge near the ACh-binding site (S–S), and the putative transmembrane segments M1–M4 (▨) are indicated; kb denotes kilobase pairs.

Furutani et al. 1983). Nucleotide sequence analysis of the cloned cDNA has revealed the complete primary structure of the α-subunit precursor of the calf muscle AChR (Noda et al. 1983a). It is composed of 457 amino acids including a prepeptide of 20 amino acids. Thus, the mature α-subunit of the calf AChR has the same number of amino acids as its *Torpedo* counterpart. The mature α-subunits of the two species exhibit marked sequence homology (81%).

A human genomic DNA library, provided by Dr. T. Maniatis (Lawn et al. 1978), was then screened for phage carrying AChR α-subunit precursor gene sequences by hybridization with a calf cDNA probe. Because the initial phage clone obtained did not contain the entire protein-coding sequence, additional clones that harbored a DNA segment extending further upstream or downstream were isolated by screening the genomic DNA library with a hybridization probe derived from the 5′- or 3′-terminal region of the DNA insert carried by the initial clone. Restriction fragments containing the gene sequences were subcloned in plasmid pBR322.

Nucleotide sequence analysis of the cloned human DNA in comparison with the calf cDNA sequence has shown that the protein-coding sequence of the human AChR α-subunit precursor gene is divided by eight introns into nine exons (Noda et al. 1983a). The protein segments encoded by distinct exons seem to correspond to different structural and functional domains of the subunit precursor molecule, as shown in Figure 5. The putative ACh-binding region, the putative transmembrane segments M1, M2 plus M3, and M4, the cytoplasmic region, and most of the prepeptide are encoded by separate exons. The gene sequence encoding the remaining aminoterminal extracellular region and a small portion of the prepeptide is divided into three exons. The protein segments corresponding to these three exons may represent as yet unknown structural or functional domains. The complete primary structure of the human AChR α-subunit precursor has been deduced from the corresponding gene sequence (Noda et al. 1983a). It also consists of 457 amino acids, including a prepeptide of 20 amino acids. The degree of sequence homology between the mature α-subunits of the human and calf AChR is 97%.

ACKNOWLEDGMENTS

We are indebted to Dr. S. Inayama, Dr. T. Hirose, Miss M. Asai, and Mr. H. Takashima for chemical synthesis of the oligodeoxyribonucleotides and to Dr. T. Miyata for computer analysis. We also thank Dr. P. Berg and Dr. H. Okayama for providing us with their high-efficiency cloning system and Dr. T. Maniatis for the human genomic DNA library. This investigation was supported in part by research grants from the Ministry of Education, Science and Culture of Japan, the Mitsubishi Foundation, and the Japanese Foundation of Metabolism and Diseases.

REFERENCES

Adams, D.J., T.M. Dwyer, and B. Hille. 1980. The permeability of endplate channels to monovalent and divalent metal cations. *J. Gen. Physiol.* **75:** 493.

Blobel, G. and B. Dobberstein. 1975. Transfer of proteins across membranes. II. Reconstitution of functional rough microsomes from heterologous components. *J. Cell Biol.* **67:** 852.

Cartaud, J., E.L. Benedetti, J.B. Cohen, J.-C. Meunier, and J.-P. Changeux. 1973. Presence of a lattice structure in membrane fragments rich in nicotinic receptor protein from the electric organ of *Torpedo marmorata. FEBS Lett.* **33:** 109.

Changeux, J.-P. 1981. The acetylcholine receptor: An "allosteric" membrane protein. *Harvey Lect.* **75:** 85.

Chou, P.Y. and G.D. Fasman. 1978. Empirical predictions of protein conformation. *Annu. Rev. Biochem.* **47:** 251.

Claudio, T., M. Ballivet, J. Patrick, and S. Heinemann. 1983. Nucleotide and deduced amino acid sequences of *Torpedo californica* acetylcholine receptor γ subunit. *Proc. Natl. Acad. Sci.* **80:** 1111.

Conti-Tronconi, B.M. and M.A. Raftery. 1982. The nicotinic cholinergic receptor: Correlation of molecular structure with functional properties. *Annu. Rev. Biochem.* **51:** 491.

Conti-Tronconi, B.M., S.M.J. Dunn, and M.A. Raftery. 1982. Functional stability of *Torpedo* acetylcholine receptor. Effects of protease treatment. *Biochemistry* **21:** 893.

Damle, V.N. and A. Karlin. 1978. Affinity labeling of one of two α-neurotoxin binding sites in acetylcholine receptor from *Torpedo californica. Biochemistry* **17:** 2039.

Dayhoff, M.O., ed. 1976. *Atlas of protein sequence and structure,* vol. 5, suppl. 2. National Biomedical Research Foundation, Silver Spring, Maryland.

Dayhoff, M.O., R.M. Schwartz, and B.C. Orcutt. 1978. A model of evolutionary change in proteins. In *Atlas of protein sequence and structure* (ed. M.O. Dayhoff), vol. 5, suppl. 3, p. 345. National Biomedical Research Foundation, Silver Spring, Maryland.

EGMOND, M.R., R.J.P. WILLIAMS, E.J. WELSH, and D.A. REES. 1979. ^{1}H-Nuclear-magnetic-resonance studies on glycophorin and its carbohydrate-containing tryptic peptides. *Eur. J. Biochem.* **97:** 73.

ENGELMAN, D.M., R. HENDERSON, A.D. MCLACHLAN, and B.A. WALLACE. 1980. Path of the polypeptide in bacteriorhodopsin. *Proc. Natl. Acad. Sci.* **77:** 2023.

FURUTANI, Y., Y. MORIMOTO, S. SHIBAHARA, M. NODA, H. TAKAHASHI, T. HIROSE, M. ASAI, S. INAYAMA, H. HAYASHIDA, T. MIYATA, and S. NUMA. 1983. Cloning and sequence analysis of cDNA for ovine corticotropin-releasing factor precursor. *Nature* **301:** 537.

GULLICK, W.J., S. TZARTOS, and J. LINDSTROM. 1981. Monoclonal antibodies as probes of acetylcholine receptor structure. 1. Peptide mapping. *Biochemistry* **20:** 2173.

HENDERSON, R. and P.N.T. UNWIN. 1975. Three-dimensional model of purple membrane obtained by electron microscopy. *Nature* **257:** 28.

HOPP, T.P. and K.R. WOODS. 1981. Prediction of protein antigenic determinants from amino acid sequences. *Proc. Natl. Acad. Sci.* **78:** 3824.

KAKIDANI, H., Y. FURUTANI, H. TAKAHASHI, M. NODA, Y. MORIMOTO, T. HIROSE, M. ASAI, S. INAYAMA, S. NAKANISHI, and S. NUMA. 1982. Cloning and sequence analysis of cDNA for porcine β-neo-endorphin/dynorphin precursor. *Nature* **298:** 245.

KARLIN, A. 1969. Chemical modification of the active site of the acetylcholine receptor. *J. Gen. Physiol.* **54:** 245s.

———. 1980. Molecular properties of nicotinic acetylcholine receptors. *Cell Surf. Rev.* **6:** 191.

KARLIN, A., C.L. WEILL, M.G. MCNAMEE, and R. VALDERRAMA. 1976. Facets of the structures of acetylcholine receptors from *Electrophorus* and *Torpedo*. *Cold Spring Harbor Symp. Quant. Biol.* **40:** 203.

KLYMKOWSKY, M.W. and R.M. STROUD. 1979. Immunospecific identification and three-dimensional structure of a membrane-bound acetylcholine receptor from *Torpedo californica*. *J. Mol. Biol.* **128:** 319.

LAWN, R.M., E.F. FRITSCH, R.C. PARKER, G. BLAKE, and T. MANIATIS. 1978. The isolation and characterization of linked δ- and β-globin genes from a cloned library of human DNA. *Cell* **15:** 1157.

LILJAS, A. and M.G. ROSSMANN. 1974. X-ray studies of protein interactions. *Annu. Rev. Biochem.* **43:** 475.

LINDSTROM, J. and P. DAU. 1980. Biology of myasthenia gravis. *Annu. Rev. Pharmacol. Toxicol.* **20:** 337.

LINDSTROM, J., J. MERLIE, and G. YOGEESWARAN. 1979. Biochemical properties of acetylcholine receptor subunits from *Torpedo californica*. *Biochemistry* **18:** 4465.

MAENO, T., C. EDWARDS, and M. ANRAKU. 1977. Permeability of the endplate membrane activated by acetylcholine to some organic cations. *J. Neurobiol.* **8:** 173.

MARSHALL, R.D. 1974. The nature and metabolism of the carbohydrate-peptide linkages of glycoproteins. *Biochem. Soc. Symp.* **40:** 17.

MAXAM, A.M. and W. GILBERT. 1980. Sequencing end-labeled DNA with base-specific chemical cleavages. *Methods Enzymol.* **65:** 499.

NICKEL, E. and L.T. POTTER. 1973. Ultrastructure of isolated membranes of *Torpedo* electric tissue. *Brain Res.* **57:** 508.

NODA, M., Y. FURUTANI, H. TAKAHASHI, M. TOYOSATO, T. HIROSE, S. INAYAMA, S. NAKANISHI, and S. NUMA. 1982a. Cloning and sequence analysis of cDNA for bovine adrenal preproenkephalin. *Nature* **295:** 202.

NODA, M., H. TAKAHASHI, T. TANABE, M. TOYOSATO, Y. FURUTANI, T. HIROSE, M. ASAI, S. INAYAMA, T. MIYATA, and S. NUMA. 1982b. Primary structure of α-subunit precursor of *Torpedo californica* acetylcholine receptor deduced from cDNA sequence. *Nature* **299:** 793.

NODA, M., Y. FURUTANI, H. TAKAHASHI, M. TOYOSATO, T. TANABE, S. SHIMIZU, S. KIKYOTANI, T. KAYANO, T. HIROSE, S. INAYAMA, and S. NUMA. 1983a. Cloning and sequence analysis of calf cDNA and human genomic DNA encoding α-subunit precursor of muscle acetylcholine receptor. *Nature* (in press).

NODA, M., H. TAKAHASHI, T. TANABE, M. TOYOSATO, S. KIKYOTANI, Y. FURUTANI, T. HIROSE, H. TAKASHIMA, S. INAYAMA, T. MIYATA, and S. NUMA. 1983b. Structural homology of *Torpedo californica* acetylcholine receptor subunits. *Nature* **302:** 528.

NODA, M., H. TAKAHASHI, T. TANABE, M. TOYOSATO, S. KIKYOTANI, T. HIROSE, M. ASAI, H. TAKASHIMA, S. INAYAMA, T. MIYATA, and S. NUMA. 1983c. Primary structures of β- and δ-subunit precursors of *Torpedo californica* acetylcholine receptor deduced from cDNA sequences. *Nature* **301:** 251.

OKAYAMA, H. and P. BERG. 1982. High-efficiency cloning of full-length cDNA. *Mol. Cell. Biol.* **2:** 161.

POLJAK, R.J., L.M. AMZEL, H.P. AVEY, B.L. CHEN, R.P. PHIZACKERLEY, and F. SAUL. 1973. Three-dimensional structure of the Fab′ fragment of a human immunoglobulin at 2.8-Å resolution. *Proc. Natl. Acad. Sci.* **70:** 3305.

RAFTERY, M.A., M.W. HUNKAPILLER, C.D. STRADER, and L.E. HOOD. 1980. Acetylcholine receptor: Complex of homologous subunits. *Science* **208:** 1454.

REYNOLDS, J.A. and A. KARLIN. 1978. Molecular weight in detergent solution of acetylcholine receptor from *Torpedo californica*. *Biochemistry* **17:** 2035.

RICE, C.M., J.R. BELL, M.W. HUNKAPILLER, E.G. STRAUSS, and J.H. STRAUSS. 1982. Isolation and characterization of the hydrophobic COOH-terminal domains of the sindbis virion glycoproteins. *J. Mol. Biol.* **154:** 355.

ROBB, R., C. TERHORST, and J.L. STROMINGER. 1978. Sequence of the COOH-terminal hydrophilic region of histocompatibility antigens HLA-A2 and HLA-B7. *J. Biol. Chem.* **253:** 5319.

ROGERS, J., P. EARLY, C. CARTER, K. CALAME, M. BOND, L. HOOD, and R. WALL. 1980. Two mRNAs with different 3′ ends encode membrane-bound and secreted forms of immunoglobulin μ chain. *Cell* **20:** 303.

ROSS, A.H., R. RADHAKRISHNAN, R.J. ROBSON, and H.G. KHORANA. 1982. The transmembrane domain of glycophorin A as studied by cross-linking using photoactivatable phospholipids. *J. Biol. Chem.* **257:** 4152.

SCHULTE, T.H. and V.T. MARCHESI. 1979. Conformation of human erythrocyte glycophorin A and its constituent peptides. *Biochemistry* **18:** 275.

STEINER, D.F., P.S. QUINN, S.J. CHAN, J. MARSH, and H.S. TAGER. 1980. Processing mechanisms in the biosynthesis of proteins. *Ann. N. Y. Acad. Sci.* **343:** 1.

STRADER, C.D. and M.A. RAFTERY. 1980. Topographic studies of *Torpedo* acetylcholine receptor subunits as a transmembrane complex. *Proc. Natl. Acad. Sci.* **77:** 5807.

TZARTOS, S.J., M.E. SEYBOLD, and J.M. LINDSTROM. 1982. Specificities of antibodies to acetylcholine receptors in sera from myasthenia gravis patients measured by monoclonal antibodies. *Proc. Natl. Acad. Sci.* **79:** 188.

TZARTOS, S.J., D.E. RAND, B.L. EINARSON, and J.M. LINDSTROM. 1981. Mapping of surface structures of *Electrophorus* acetylcholine receptor using monoclonal antibodies. *J. Biol. Chem.* **256:** 8635.

UEHARA, H., J.E. COLIGAN, and S.G. NATHENSON. 1981. Amino acid sequence of the carboxyl-terminal hydrophilic region of the H-2K^b MHC alloantigen. Completion of the entire primary structure of the H-2K^b molecule. *Biochemistry* **20:** 5940.

WEILL, C.L., M.G. MCNAMEE, and A. KARLIN. 1974. Affinity-labeling of purified acetylcholine receptor from *Torpedo californica*. *Biochem. Biophys. Res. Commun.* **61:** 997.

WENNOGLE, L.P. and J.-P. CHANGEUX. 1980. Transmembrane orientation of proteins present in acetylcholine receptor-rich membranes from *Torpedo marmorata* studied by selective proteolysis. *Eur. J. Biochem.* **106:** 381.

WENNOGLE, L.P., R. OSWALD, T. SAITOH, and J.-P. CHANGEUX. 1981. Dissection of the 66,000-dalton subunit of the acetylcholine receptor. *Biochemistry* **20:** 2492.

WISE, D.S., A. KARLIN, and B.P. SCHOENBORN. 1979. An analysis by low-angle neutron scattering of the structure of the acetylcholine receptor from *Torpedo californica* in detergent solution. *Biophys. J.* **28:** 473.

ZINGSHEIM, H.-P., D.-C. NEUGEBAUER, J. FRANK, W. HANICKE, and F.J. BARRANTES. 1982. Dimeric arrangement and structure of the membrane-bound acetylcholine receptor studied by electron microscopy. *EMBO J.* **1:** 541.

Molecular Cloning of the Acetylcholine Receptor

J. PATRICK, M. BALLIVET, L. BOAS, T. CLAUDIO, J. FORREST, H. INGRAHAM, P. MASON, S. STENGELIN, S. UENO, AND S. HEINEMANN
Molecular Neurobiology Laboratory, The Salk Institute, San Diego, California 92138

Acetylcholine (ACh) mediates synaptic transmission at the vertebrate neuromuscular junction. The synthesis and properties of the molecular components that subserve transmission at this synapse are determined by the interaction of nerve and muscle. In this respect, the neuromuscular junction may be typical of other synapses that are formed and stabilized through appropriate interaction of the pre- and postsynaptic cells. To study the mechanisms involved in synaptogenesis, we first chose to study the properties and regulation of one component of the transmission machinery, the receptor for ACh. This multimeric transmembrane glycoprotein resides in the postsynaptic membrane where it functions to convert binding of ACh into a change in the transmembrane potential.

Many properties of the ACh receptor (AChR) are determined by the presence of functional innervation (for review, see Fambrough 1979). Prior to innervation of an embryonic muscle cell by a motor neuron, AChRs are found over the entire surface of the muscle. After innervation has occurred, the receptor is localized predominately under the nerve terminal. Denervation causes receptor to appear once again diffusely over the muscle surface. Thus, in some way, the nerve controls the distribution of AChR over the muscle surface. Likewise, the nerve regulates receptor metabolism. Receptor that is localized to the end-plate region of the muscle has a metabolic half-life of about 2 weeks, whereas receptor that is distributed diffusely over the muscle surface has a half-life of about 17 hours (Berg and Hall 1974). These two classes of receptor that are synthesized and degraded at different rates also differ biochemically. The receptors that appear following denervation have an antigenic determinant that is not found on receptor localized to the adult end plate (Weinberg and Hall 1979). Likewise, these receptors focus in a pH gradient at a position about 0.15 pH units more basic than do the end-plate receptors (Brockes and Hall 1975). Lastly, the receptors that appear following denervation open an ion channel that stays open about twice as long as does the ion channel of the end-plate receptor. These well-studied instances of neural determination of AChR properties make the receptor a good tool for learning how, in general, the interaction of nerve and muscle results in the establishment of a functional synapse.

There are two general ways in which the nerve could control receptor properties. The nerve could modify, or cause the muscle to modify, extrasynaptic receptors so that they acquire properties of synaptic receptors. Alternatively, the nerve could cause the expression of a separate set of genes that code for synaptic as opposed to extrasynaptic receptors. In either case, it is very likely that innervation regulates the rate of expression of the AChR genes, since the rate of incorporation of receptor into the membrane is much slower following innervation.

In this specific example of a trophic interaction between nerve and muscle, we would like to know the basis of the difference between synaptic and extrasynaptic receptors and we would like to know how the nerve regulates the expression of the genes that code for AChR. A complete understanding of the differences between junctional and extrajunctional receptors requires knowledge of the amino acid sequence of the two types of receptor. Likewise, an understanding of how the nerve regulates expression of receptor genes requires knowledge of their structure and organization. Purification of these genes using recombinant DNA techniques would allow analysis of their structure and organization and the determination of the amino acid sequence of the receptor subunits.

Isolation of Clones Coding for *Torpedo* AChR Subunits

Extensive structural data is available for AChR isolated from the electric organ of *Torpedo* sp. Receptor from this muscle-derived tissue is composed of four different polypeptides, α, β, γ, and δ, which have apparent molecular sizes on SDS gels of 40, 50, 60, and 65 kD, respectively, and are present in the receptor as an $\alpha_2\beta\gamma\delta$ complex (for review, see Conti-Tronconi and Raftery 1982). The aminoterminal sequences of the four polypeptides have been shown to have extensive sequence homology, suggesting they are encoded by genes derived from a single ancestral sequence (Raftery et al. 1980). We began our search for nucleotide sequences coding for AChR by isolating mRNA from the electric organ of *Torpedo californica*, since we expected receptor sequences to be richly represented in this preparation.

Poly(A)$^+$ RNA from *Torpedo* electric organ was size-fractionated on a sucrose gradient and alternate fractions were tested for their ability to direct the synthesis of polypeptides recognized by anti-AChR antibody. The results in Figure 1 show the size distribution of in vitro-synthesized proteins precipitated by anti-receptor antibody in the presence and absence of competing unlabeled AChR. One fraction of the mRNA obtained from the sucrose gradient was particularly enriched in mRNA species coding for receptor, and

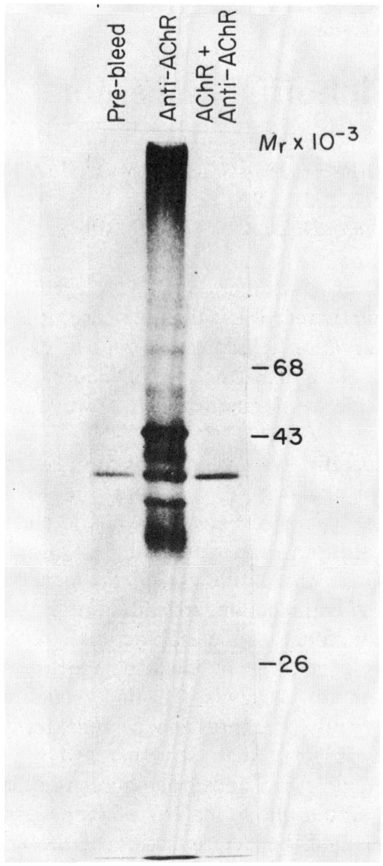

Figure 1. In vitro translation of sized electric organ mRNA. Poly(A)$^+$ RNA purified from electric organ RNA was layered in a 15–30% sucrose gradient and sedimented at 36,000 rpm in an SW-41 rotor for 17 hr. Fractions were collected and assayed for their ability to direct the synthesis of protein that was immunoprecipitable by antibody prepared against SDS-denatured *Torpedo* AChR. This figure shows translation products immunoprecipitable by preimmune serum, antireceptor antiserum, and anti-receptor antiserum plus excess purified AChR. Immune serum precipitates four major bands between 35 kD and 60 kD plus the higher-molecular-weight material seen at the top of the gel. The band of about 43 kD present in each lane represents a major band in the total in vitro translation product of this fraction and due to its abundance may contaminate each lane.

was used as template for the preparation of a cDNA library using the cloning protocols devised by Okayama and Berg (1982). In this procedure, first-strand synthesis is primed by an oligo(dT) tract that is part of the cloning vector. An oligo(dC) tract added to the newly synthesized cDNA is used to join the cDNA to a linking piece of DNA and subsequently to the free end of the cloning vector. This procedure eliminates the use of the endonuclease S1 and, in general, produces long cDNA inserts. In addition to this procedure, we prepared conventional libraries by inserting dC-tailed double-stranded DNA into pBR322 cut with *Pst*I and tailed with oligo(dG) (Bolivar et al. 1977). This library had mostly short cDNA inserts (200–500 bp) and we refer to it as the short library. These libraries were screened in two ways. The first was by selecting clones that hybridized to radioactive cDNA made from mRNA from *Torpedo* electric organ but not to cDNA made from mRNA from *Torpedo* brain. This procedure identified 30 out of 1000 clones in the library prepared using this Okayama and Berg procedure. We then screened these selected clones by the technique of hybridization-selection (Harpold et al. 1978). Clones from the short library were also screened at random by hybridization-selection. Plasmid DNA was pooled from groups of seven clones and annealed to nitrocellulose filters. The filters were then incubated with mRNA to allow hybridization of mRNA to the cDNA on the filter. Unhybridized mRNA was washed off and the specifically hybridized mRNA eluted and translated

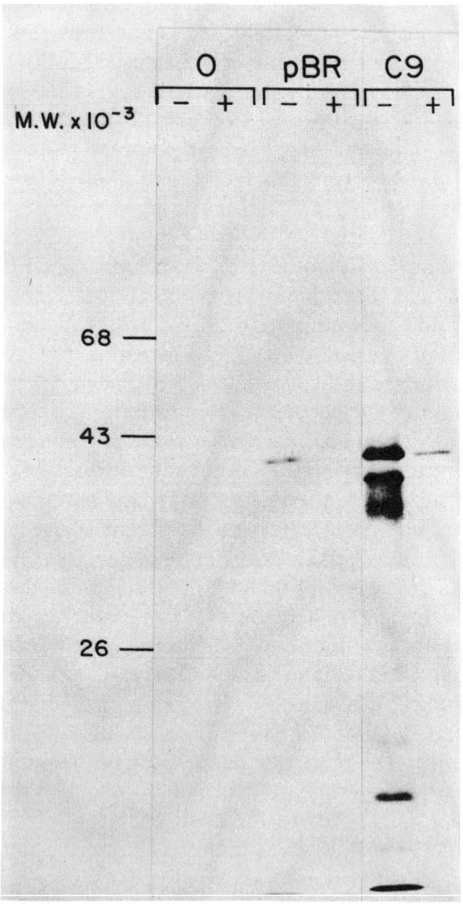

Figure 2. Hybridization-selection of RNA coding for a protein recognized by anti-AChR antibody. Pools of seven clones from a cDNA library were grown and used to prepare a cleared lysate. DNA from the pools was bound to nitrocellulose filters that were then hybridized with poly(A)$^+$ RNA from the electric organ. The hybridized RNA was eluted, translated in vitro, and the translation products tested for their ability to bind to anti-AChR antibody. Once immunoprecipitable products were detected, the individual clones in a pool were grown and tested separately. This figure shows immunoprecipitable proteins produced by mRNA which hybridized to DNA either from clone 1C9 (*right lanes*) or from pBR (*left lanes*). Lanes marked with a plus are those in which the immunoprecipitation was done in the presence of unlabeled competing purified AChR. The 43-kD band present in all lanes is the contaminating 43-kD protein seen in Fig. 1.

in vitro. Proteins made in vitro from hybridization-selected mRNA were reacted with anti-receptor antibody, precipitated, and analyzed on SDS gels. Groups of clones that hybridized to an mRNA that directed the synthesis of protein recognized by the antibody were further fractionated until a single clone was identified. The results in Figure 2 show an example of one clone, identified as 1C9, that hybridizes to mRNA and, when translated in vitro, directs synthesis of polypeptides that are recognized by anti-AChR antibody.

Application of hybridization selection to the two cDNA libraries resulted in the identification of two clones, 4D8, from the library of long inserts, and clone 1C9, from the library of short inserts. The inserted cDNA sequences were isolated from each clone and used to rescreen the library of long inserts to find related clones. Screening with the insert from clone 4D8 revealed clones that differed from 4D8 only in the length of the cDNA insert. Screening with the short insert from clone 1C9, however, revealed two classes of clones that could be distinguished by two criteria: (1) digestion with restrictive endonucleases generated different patterns of restriction fragments and (2) clones within one class failed to hybridize to clones of the other class although clones of both classes hybridize with clone 1C9. Identification of these clones, which are shown schematically in Figure 3, and evaluation of their relationships were achieved through analysis of their nucleotide sequences.

Clone 4D8. Clone 4D8 was identified as coding for the γ-subunit of *Torpedo* AChR by comparison of the sequence at the 5′ end of the clone with the aminoterminal amino acid sequence of the mature protein (Ballivet et al. 1982). A portion of this sequence codes for a methionine followed by a 16-amino acid polypeptide that is contiguous to the aminoterminal amino acid of the native protein and that has the characteristics of a leader peptide. The following 162 bases code for a protein with a sequence that differs at only one position from the sequence of the γ-subunit determined by sequencing the protein isolated from *Torpedo* electric organ (Raftery et al. 1980). Analysis of the remainder of the sequence showed that clone 4D8 contained 2010 bases of which 461 were 3′-untranslated sequences, 31 were 5′-untranslated sequences, and the remaining 1518 were coding sequences. The sequence between bases 82 and 1550 codes for a protein of 489 amino acids with a molecular mass of 56.6 kD (Claudio et al. 1983).

Analysis of the amino acid sequence derived from the nucleotide sequence of clone 4D8 made it possible to predict the distribution of the protein across the lipid bilayer. Application of the analysis devised by Kyte and Doolittle (1982) revealed four extended stretches of hydrophobicity in the primary structure (Fig. 4). Each of these stretches is sufficiently long and sufficiently hydrophobic to span the bilayer. If these sequences are within the bilayer and if the aminoterminal sequences are extracellular (Wennogle and Changeux 1980; Anderson et al. 1982), then the distribution of the remainder of the sequences must be as indicated in Figure 5. This figure shows about 51% of the protein is extracellular, 31% is intracellular, and 17% is buried within the membrane. This model is therefore consistent with results obtained from X-ray diffraction (Ross et al. 1977), immunoelectron microscopy (Klymkowsky and Stroud 1979), and neutron scattering (Wise et al. 1981).

This model is consistent with two other observations. The γ-subunit is glycosylated in at least one and possibly three positions (Anderson and Blobel 1981). The sequence of the γ-subunit deduced from clone 4D8 has five canonical glycosylation sites (Struck and Lennarz 1980), three of which are found in the extracellular sequence seen in the model in Figure 5. In addition, Noda et al. (1982) have identified a putative ACh-binding site in the sequence of the α-subunit. This sequence is highly conserved between the α- and γ-subunit and is found in the extracellular portion of the sequence in our model based upon the γ-subunit sequence. The model also specifically predicts that both the amino- and carboxyterminal sequences are extracellular. This distribution is unique among membrane proteins studied so far (Sabatini et al. 1982). However, the assignment of membrane-spanning regions may be incorrect if regions

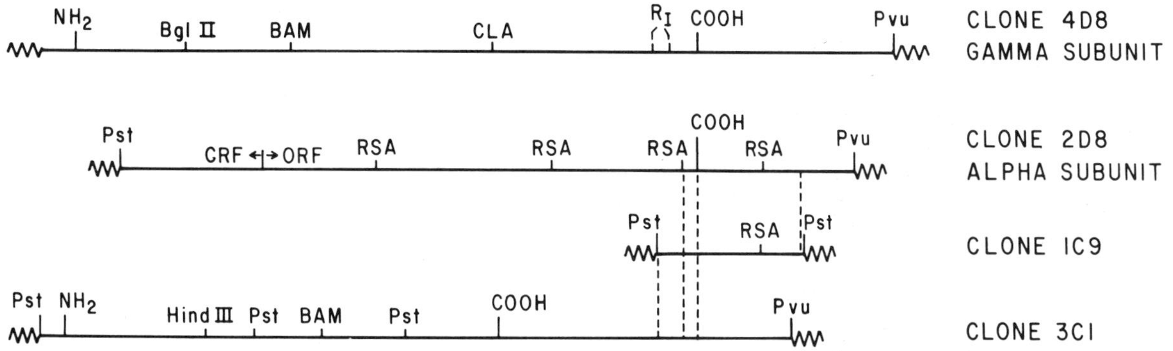

Figure 3. This figure shows the general structures of four clones isolated from a cDNA library prepared using *Torpedo* electric organ mRNA. Clone 4D8 was identified by hybridization to a clone previously identified by hybridization selection. Clone 1C9 was identified by hybridization selection and clones 2D8 and 3C1 were identified by hybridization with clone 1C9. The dotted lines indicate the portions of the 1C9 sequence homologous with 3C1 and 2D8.

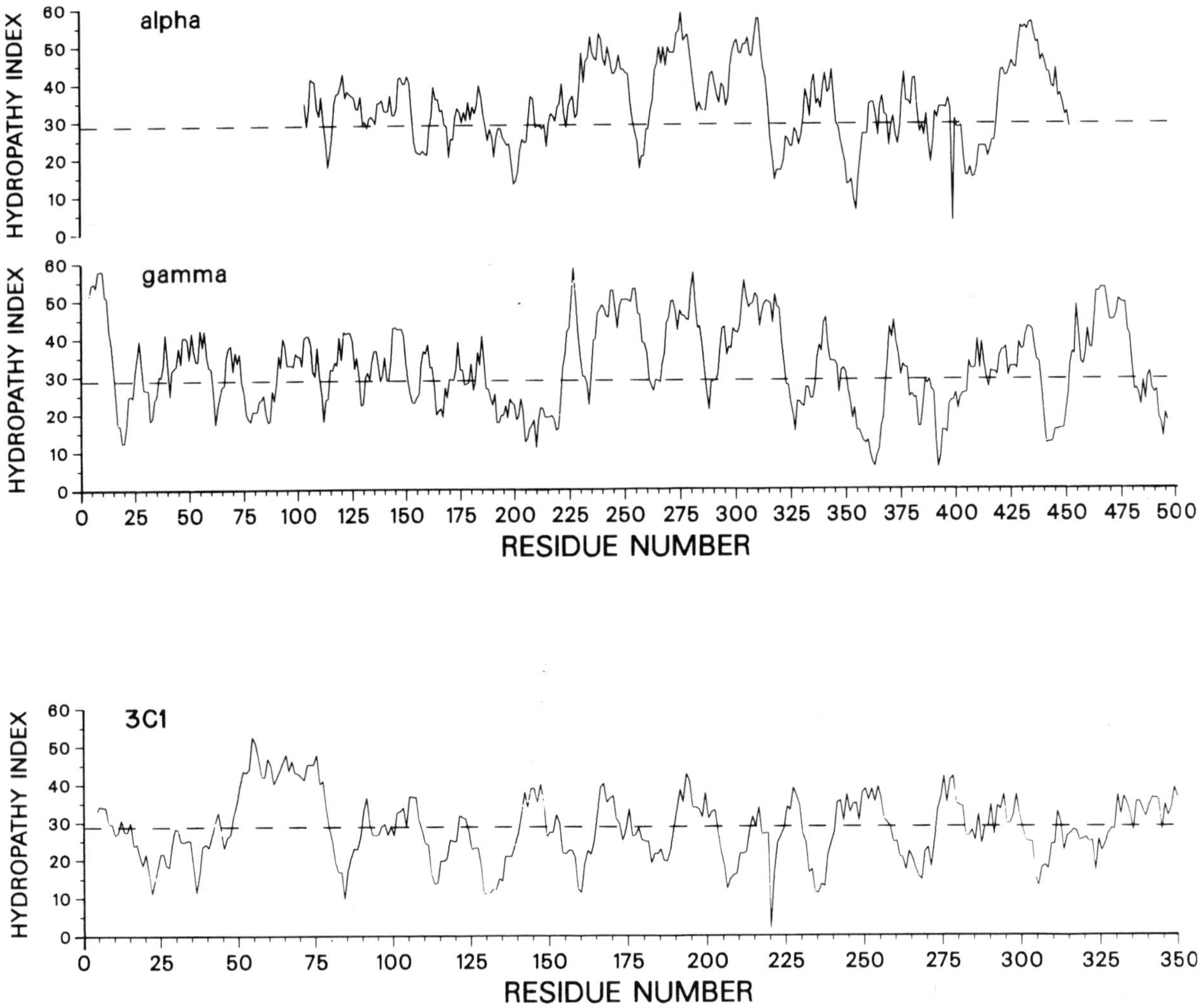

Figure 4. This figure shows the relative hydropathy of the proteins encoded by clones 4D8, 2D8, and 3C1 as a function of their sequence number. The data were generated by the SOAP program of Kyte and Doolittle (1982). Hydrophobic regions of the protein lie above the midline and hydrophilic regions lie below the midline. The midline corresponds to the average hydropathy of all the proteins in the Dayhoff data base.

of lower hydrophobicity are also found within the lipid bilayer. The addition (or deletion) of one membrane-spanning region would move the carboxyterminal sequences from the extracellular to the intracellular side of the membrane.

Clone 2D8. Clone 2D8 was more difficult to identify because the sequence of the 5′ end did not contain an open reading frame that translated into the published aminoterminal sequence of any of the receptor subunits. However, sequences near the middle of the clone contained an open reading frame that coded for a protein with a sequence that was the same as that found at the carboxyl terminus of the α-subunit (M.A. Raftery et al., unpubl.). Based upon sequences supplied by Dr. M. Raftery, we determined that clone 2D8 coded for all but 84 of the aminoterminal amino acids of the mature α-subunit. In place of the aminoterminal coding sequences in clone 2D8, we found a closed reading frame of 318 bases. Comparison of this nucleotide sequence with the 5′-untranslated sequences of the α-subunit clone isolated by Noda and his collaborators (1982) failed to reveal any significant sequence homology. Thus, our α-subunit clone, 2D8, does not contain the entire coding region for the α-subunit, not because the clone is insufficiently long, but because the 5′ sequences in this clone are not appropriate for the 5′ end of the mature mRNA coding for the α-subunit. A number of possibilities may explain this result: (1) 2D8 may represent an unspliced mRNA retaining intron sequences, (2) 2D8 may represent a pseudogene product, or (3) 2D8 may have resulted from a cloning artifact.

Clone 2D8 hybridizes to two different species of RNA. The results in Figure 6 show hybridization of clones 4D8, 2D8, and 3C1 to total and poly(A)$^+$ RNA isolated from *Torpedo* electric organ. Clones 4D8 and 3C1 hybridize to an RNA species that is more abundant in the poly(A)$^+$ RNA than in the total RNA. Clone 2D8 on the other hand hybridizes to two RNA species, one smaller species that is present in total RNA and less

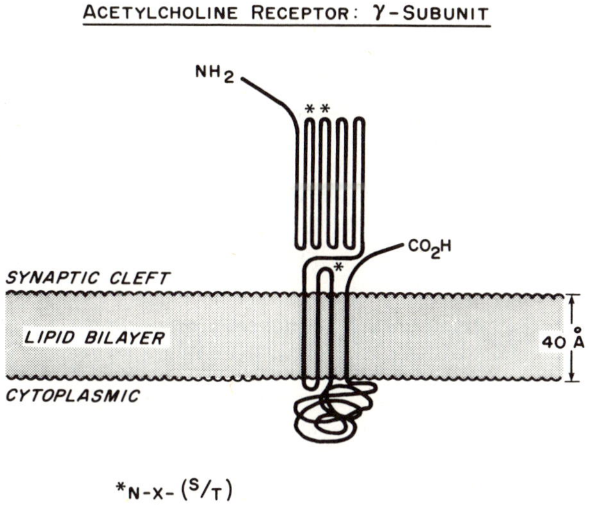

Figure 5. This figure shows a partitioning of the γ-subunit polypeptide chain across the lipid bilayer based upon the identification of membrane-spanning sequences seen in Fig. 4. The asterisks indicate the location of canonical glycosylation sites located in the extracellular side of the membrane.

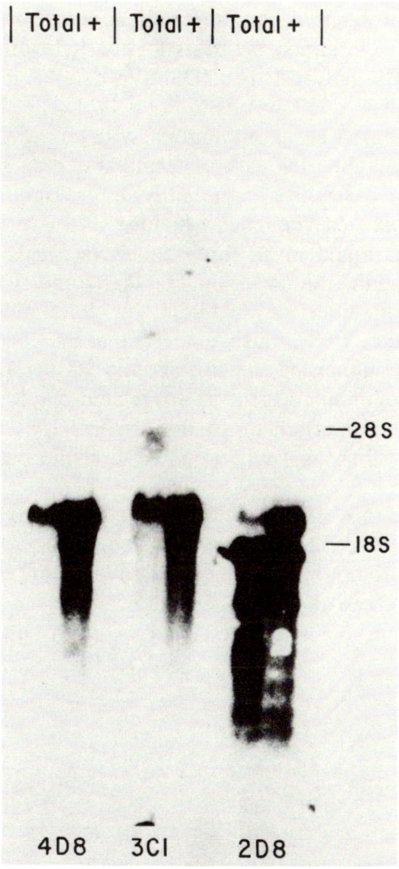

Figure 6. Hybridization of clones 4D8, 3C1, and 2D8 to total and poly(A)$^+$ RNA isolated from *Torpedo* electric organ. The RNA was fractionated in denaturing agarose gels and transferred to nitrocellulose. The transfered RNA was hybridized to probes prepared from either clone 4D8, 3C1, or 2D8. The lower band hybridizing with clone 2D8 in the total RNA lane is probably not mRNA coding for receptor since it does not appear when the probe is derived from regions of clone 2D8 that code for the α-subunit.

abundant in the poly(A)$^+$ RNA, and a slightly larger species that is more abundant in the poly(A)$^+$ RNA preparation. It seems likely that only the larger of these species codes for receptor, since hybridization with probes prepared from the protein-coding region of clone 2D8 only hybridize to the larger RNA that is also enriched in the poly(A)$^+$ RNA preparation. However, a probe prepared from the noncoding sequences at the 5′ end of the clone recognizes the smaller RNA species present in the total RNA preparation. This result suggests that the sequences at the 5′ end of clone 2D8 correspond to a sequence that appears as a nonpolyadenylated RNA species in the electric organ.

Clone 3C1. As noted above, clone 3C1 was identified by hybridization with the insert isolated from clone 1C9, which was itself identified because it hybridizes to mRNA coding for the α-subunit. Clone 3C1 contains a single open reading frame of 1011 bases coding for a 38-kD protein that has no obvious sequence homology with any of the *Torpedo* AChR subunits. The initiation codon for the encoded protein is identified by a consensus start sequence (Kozak 1981) and by the fact that sequences 5′ to the putative start codon contain a termination codon in the same reading frame as the protein-coding sequences. These observations made it seem likely that the true amino terminus of the protein has been identified and that the 58 bases 5′ of the start sequence are 5′-untranslated sequences. The amino acid residues following the methionine coded by the start codon in position 52 have none of the properties characteristic of a leader sequence, suggesting that the aminoterminal sequences are not removed by a processing event and the mature protein has an aminoterminal methionine.

Although the protein sequence derived from clone 3C1 does not contain an aminoterminal leader or signal peptide, it does contain a 28-residue sequence of hydrophobicity beginning 35 residues from the aminoterminal end of the molecule. This sequence has an average hydrophobicity characteristic of membrane-spanning regions of known integral membrane proteins and is substantially longer than is necessary to span a membrane bilayer. The length and location of this sequence is similar to that seen in influenza virus neuraminidase (Fields et al. 1981), which is thought to be an integral membrane protein. If the protein encoded by clone 3C1 spans the membrane, it might have either its short aminoterminal sequence extracellular or its longer carboxyterminal sequence on the extracellular surface. It is interesting to note that all the canonical glycosylation sites are in the sequence on the carboxyterminal side of the membrane-spanning region.

Identification of clone 3C1 by hybridization with clone 1C9 could reflect a chance homology in either coding or untranslated sequences in the two clones, or it might reflect some functional relationship between the two proteins. As a first step in addressing this question, we determined the basis for the hybridization between 1C9 and 3C1 and 2D8. We determined the sequence of the 357 bases in clone 1C9 and discovered that 88 bases are homologous with clone 2D8 in a region including the carboxyterminal coding sequences and 269 bases are completely homologous with a 3′-untranslated sequence of clone 3C1. A 10-bp sequence is found in all three clones (see Fig. 3). This observation accounts for the fact that clone 1C9 hybridizes to both clone 3C1 and 2D8, yet clone 3C1 does not hybridize to 2D8. It also suggests that the genes coding for the α-subunit of receptor and the 3C1 protein either have a common ancestor or are related one to the other through the mechanism of their expression. This suggests that there is a functional relationship between the two proteins.

The mRNA coding for the 3C1 protein is found in both *Torpedo* electric organ and muscle, but not *Torpedo* liver. The results in Figure 7 show hybridization of 3C1 to an mRNA species of about the same size in both muscle and electric organ with substantially more hybridization in electric organ. This difference in hybridization between the two tissues is similar to that seen when looking at mRNA coding for receptor subunits. This clearly suggests that the 3C1 protein is under the same regulation in both electric organ and muscle and is consistent with the idea that the protein is involved in the neuromuscular junction in some way. Definitive resolution of this issue must come from direct studies of the 3C1 protein.

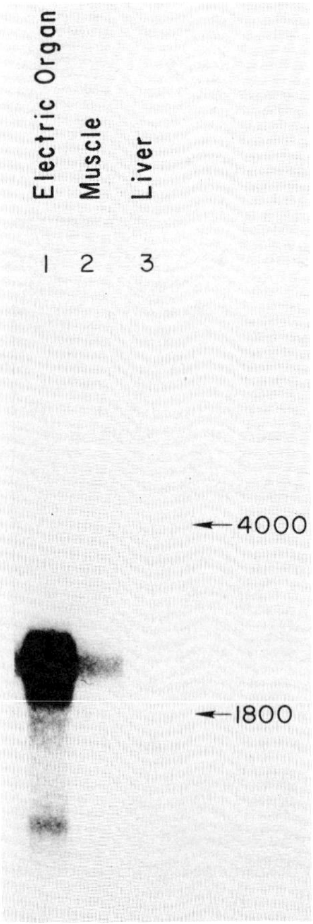

Figure 7. Hybridization of clone 3C1 to size-fractionated poly(A)$^+$ RNA isolated from *Torpedo* electric organ, muscle, and liver. Two micrograms of poly(A)$^+$ RNA from each source were fractionated in denaturing agarose gels, transferred to nicrocellulose, and hybridized with a probe made by nick-translation of the entire 3C1 clone.

Isolation of Sequences Coding for Mammalian Receptor

To address the questions raised in the introduction, it is necessary to have clones that code for mammalian muscle nicotinic AChR. We began studies with *Torpedo* AChR because the mRNA is abundant in electric organ, the sequences of *Torpedo* receptor would help identify clones, and the wealth of structural data on *Torpedo* receptor would facilitate analysis of the amino acid sequence. However, once clones for *Torpedo* receptor were identified, we began looking for clones coding for mammalian AChR. There are two general ways one could approach the problem of finding receptor sequences in mammals. One could repeat the hybridization-selection techniques we used in *Torpedo* using a cDNA library prepared from mammalian muscle. Alternatively, one could use the *Torpedo* receptor nucleotide sequences to identify mouse clones by hybridization. This latter approach depends upon the hypothesis that the sequences of AChR subunits are conserved in evolution. We already know that there is considerable homology between *Torpedo* receptor subunits. This suggests that these sequences are constrained, probably due to the requirement that the receptor subunits interact with each other as well as ligand in a precise way to form a ligand-gated ion channel. In general, it seems likely that proteins that interact, such as protein hormones and their receptors, are highly conserved in evolution. Clearly, the general structure and size of the receptor subunits are conserved between *Torpedo* and mammals (for review, see Conti-Tronconi and Raftery 1982) and the amino acid sequences of the amino termini are quite similar (Conti-Tronconi et al. 1982). This suggests that regions conserved between subunits will also be highly conserved in evolution.

If the amino acid sequence is conserved during evolution, then to some extent the nucleotide sequence coding for the protein will also be conserved. We decided to test this idea by asking whether or not the cDNA clones coding for the α- and γ-subunits would hybridize to mammalian genomic DNA. We found that when the hybridization conditions were set at low stringency, allowing a 30% mismatch, it was possible to detect hybridization of the *Torpedo* receptor cDNA clones to mouse genomic DNA. This finding encouraged us to screen a mouse genomic library made in λ Charon 28 to isolate the genes coding for mouse AChR. To increase the chance of isolating receptor genes, we made subclones containing only protein-coding sequences from the clone coding for the α-subunit. Three subclones were constructed containing (1) a 174-bp sequence including the coding region for the putative ACh-binding site, (2) a 408-bp sequence that includes the three putative central membrane-spanning regions, (3) a 328-bp sequence that includes the sequences coding for the cytoplasmic and fourth membrane-spanning portions of the molecule (see Fig. 4). When sequences coding for the ACh-binding site were used to screen 1.5×10^6 Charon 28 plaques containing inserted embryonic mouse genomic sequences, we found eight hybridizing plaques, at least two of which were also found to be positive with the subclone containing the central membrane-spanning sequences. We are currently sequencing portions of these genomic clones to determine whether there are sequences that code for a protein with amino acid sequence homology to *Torpedo* AChR.

DISCUSSION

We began cloning AChR genes to provide access to the regulation of receptor synthesis and its properties during synaptogenesis. Our first experiments used mRNA prepared from *Torpedo* electric organ because the mRNA is abundant in this tissue and, more importantly, because the aminoterminal sequences had been determined by sequencing receptor purified from the electric organ which gave us an unambiguous means of identification of cDNA clones. From these initial studies, we obtained clones coding for the γ- and α-subunits of *Torpedo* receptor (Ballivet et al. 1982). Analysis of the sequence of the clone coding for the γ-subunit allowed us to deduce the amino acid sequence of the encoded protein. From that amino acid sequence, we proposed a testable model for the distribution of the γ-subunit polypeptide across the lipid bilayer (Claudio et al. 1983).

The sequences for all four receptor polypeptides are now available (Sumikawa et al. 1982; Noda et al. 1982, 1983; Claudio et al. 1983; Devillers-Thiéry et al. 1983). These sequences point out some interesting facts about the receptor molecule. Most important is the considerable homology that exists between the various polypeptide chains that make up the receptor oligomer. The sequences can be aligned such that in 92 positions the same amino acid is found in all four subunits. The general distribution of hydrophobic sequences is the same in all four subunits, making it very likely that the model we proposed for the γ-subunit also applies to the other subunits. Some regions of the sequence are more conserved than others. The first two aminoterminal membrane-spanning regions are highly conserved with, for example, more than 50% of the amino acid residues in common between the α- and γ-subunits. Another conserved sequence appears in the proposed ACh-binding site identified by Noda et al. (1982). Of the 15 amino acids in this sequence, 10 are found in all four subunits. Another sequence of 13 residues located between the amino terminus and the proposed ACh-binding site is highly conserved, with 70% of the residues present in each of the four subunits. The conserved hydrophobic sequences have a proposed function as membrane-spanning regions, but currently there is not a good explanation for the conservation of the proposed ACh-binding site in all four polypeptides, nor is there yet a function associated with the additional conserved region between the proposed ACh-binding site and the amino terminus. It seems likely that the identification of these sequences will focus attention on them and make it possible to assign them a role in receptor function. For example, the model of receptor distribution across the membrane can be tested by synthesizing peptides corresponding to known sequences in the protein. Antibodies against these peptides can be used to locate the sequences in the cytoplasmic or extracellular face of the membrane. It is possible that antibodies against some peptides will interfere with specific receptor functions and allow one to assign a tentative role to specific receptor sequences.

An important consequence of the homology between receptor subunits is the possibility that these homologies are essential to receptor function, and consequently conserved during evolution. If this is true, we might expect we could identify mammalian receptor amino acid sequences by comparison with *Torpedo* receptor amino acid sequences. This is clearly the case at the amino terminus, where amino acid sequences that are not highly conserved between *Torpedo* subunits are conserved between *Torpedo* and fetal calf receptor (Conti-Tronconi et al. 1982). The possibility led us to begin looking for nucleotide sequences in the mouse genome that would hybridize with portions of the cDNA clone coding for the *Torpedo* α-subunit. Specifically, we chose the nucleotide sequences coding for the proposed ACh-binding site and a nucleotide sequence coding for the three most aminoterminal membrane-spanning regions. Although we have isolated genomic clones that hybridize to probes prepared from both these nucleotide sequences, we do not yet know whether they contain true receptor-coding sequences. If this approach works, it may provide access to the whole family of AChRs.

This approach to the neuromuscular junction may prove to be valuable, not only for its power in answering specific questions but also in its ability to reveal new synaptic molecules. The protein encoded by clone 3C1 may represent just such a case. In this case, the relationship between clone 3C1 and the AChR subunits is so far only apparent in the expression of the genes coding for the two proteins. This is not something that could have easily come from the traditional biochemical study of the neuromuscular junction. However, knowing the sequence of the protein encoded in clone 3C1, we have synthesized peptides corresponding to the amino- and carboxyterminal sequences. Antibodies against these peptides will make it possible to assay the

3C1 protein and determine its distribution within the cell.

ACKNOWLEDGMENTS

This work was supported by grants from the Muscular Dystrophy Association, the National Institute of Health (NS 13546 and NS 11549), the Noble Foundation, and the Whitehall Foundation to J.P. and S.H. and by a USAMRC grant to J.P.

REFERENCES

ANDERSON, D.J. and G. BLOBEL 1981. *In vitro* synthesis, glycosylation, and membrane insertion of the four subunits of *Torpedo* acetylcholine receptor. *Proc. Natl. Acad. Sci.* **78:** 5598.

ANDERSON, D.J., P. WALTER, and G. BLOBEL. 1982. Signal recognition protein is required for the integration of acetylcholine receptor delta subunit, a transmembrane glycoprotein, into the endoplasmic reticulum membrane. *J. Cell Biol.* **93:** 501.

BALLIVET, M., J. PATRICK, J. LEE, and S. HEINEMANN. 1982. Molecular cloning of cDNA coding for the gamma subunit of *Torpedo* acetylcholine receptor. *Proc. Natl. Acad. Sci.* **79:** 4466.

BERG, D.K. and Z. HALL. 1974. Fate of alpha-bungarotoxin bound to acetylcholine receptors of normal and denervated muscle. *Science* **184:** 473.

BOLIVAR, F., R.L. RODRIGUEZ, P.J. GREENE, M.C. BETLACH, H.L. HAYNECKER, H.W. BOYER, J.H. CROSO, and S. FALKOW. 1977. Construction and characterization of new cloning vehicles. II. A multipurpose cloning system. *Gene* **2:** 95.

BROCKES, J.P. and Z.W. HALL. 1975. Acetylcholine receptors in normal and denervated rat diaphragm muscle. II. Comparison of junctional and extrajunctional receptors. *Biochemistry* **14:** 2100.

CLAUDIO, T., M. BALLIVET, J. PATRICK, and S. HEINEMANN. 1983. Nucleotide and deduced amino acid sequences of *Torpedo californica* acetylcholine receptor gamma-subunit. *Proc. Natl. Acad. Sci.* **80:** 1111.

CONTI-TRONCONI, B.M. and M.A. RAFTERY. 1982. The nicotinic cholinergic receptor: Correlation of molecular structure with functional properties. *Annu. Rev. Biochem.* **51:** 491.

CONTI-TRONCONI, B.M., M. HUNKAPILLER, J. LINDSTROM, and M.A. RAFTERY. 1982. Subunit structure of the acetylcholine receptor from *Electrophorus electricus*. *Proc. Natl. Acad. Sci.* **79:** 6489.

DEVILLERS-THIÉRY, A., J. GIRAUDAT, M. BENTABOULET, and J.P. CHANGEUX. 1983. Complete mRNA coding sequence of the acetylcholine binding alpha-subunit from *Torpedo marmorata* acetylcholine receptor: A model for the transmembrane organization of the polypeptide chain. *Proc. Natl. Acad. Sci.* **80:** 2067.

FAMBROUGH, D.M. 1979. Control of acetylcholine receptors in skeletal muscle. *Physiol. Rev.* **9:** 165.

FIELDS, S., G. WINTER, and G.G. BROWNLEE. 1981. Structure of the neuraminidase gene in human influenza virus A/PR/8/34. *Nature* **290:** 213.

HARPOLD, M.M., P.R. DOBNER, R.M. EVANS, and F.C. BANCROFT. 1978. Construction and identification by positive hybridization-translation of a bacterial plasmid containing a rat growth hormone structural gene sequence. *Nucleic Acids Res.* **5:** 2039.

KLYMKOWSKY, M.W. and R.M. STROUD. 1979. Immunospecific identification and three-dimensional structure of a membrane-bound acetylcholine receptor from *Torpedo californica*. *J. Mol. Biol.* **128:** 319.

KOZAK, M. 1981. Possible role of flanking nucleotides in recognition of the AUG initiator codon by eukaryotic ribosomes. *Nucleic Acids Res.* **9:** 5233.

KYTE, J. and R.F. DOOLITTLE. 1982. A simple method for displaying the hydropathic character of a protein. *J. Mol. Biol.* **157:** 105.

NODA, M., H. TAKAHASHI, T. TANABE, M. TOYOSATO, Y. FURUTANI, T. HIROSE, M. ASAI, S. INAYAMA, T. MIYATA, and S. NUMA. 1982. Primary structure of alpha-subunit precursor of *Torpedo californica* acetylcholine receptor deduced from cDNA sequence. *Nature* **299:** 793.

NODA, M., H. TAKAHASHI, T. TANABE, M. TOYOSATO, S. KIKYOTANI, T. HIROSE, M. ASAI, H. TAKASHIMA, S. INAYAMA, T. MIYATA, and S. NUMA. 1983. Primary structures of beta- and gamma-subunit precursors of *Torpedo californica* acetylcholine receptor deduced from cDNA sequences. *Nature* **301:** 251.

OKAYAMA, H. and P. BERG. 1982. High efficiency cloning of full length cDNA. *Mol. Cell. Biol.* **2:** 161.

RAFTERY, M.A., M.W. HUNKAPILLER, C.D. STRADER, and L.E. HOOD. 1980. Acetylcholine receptor: Complex of homologous subunits. *Science* **208:** 1454.

ROSS, M.J., M.W. KLYNKOWSKY, D.A. ASARD, and R.M. STROUD. 1977. Structural studies of a membrane-bound acetylcholine receptor from *Torpedo californica*. *J. Mol. Biol.* **116:** 635.

SABATINI, D.D., G. KREIBICH, T. MORIMOTO, and M. ADESNIK. 1982. Mechanisms for the incorporation of proteins in membranes and organelles. *J. Cell Biol.* **92:** 1.

STRUCK, D.K. and W.J. LENNARZ. 1980. In *The biochemistry of glycoproteins and proteoglycans* (ed. W.J. Lennarz), p. 35. Plenum Press, New York.

SUMIKAWA, K., M. HOUGHTON, J.C. SMITH, L. BELL, B.M. RICHARDS, and E.A. BARNARD. 1982. The molecular cloning and characterisation of cDNA coding for the alpha subunit of the acetylcholine receptor. *Nucleic Acids Res.* **10:** 5809.

WEINBERG, C.B. and Z.W. HALL. 1979. Antibodies from patients with myasthenia gravis recognize determinants unique to extrajunctional acetylcholine receptors. *Proc. Natl. Acad. Sci.* **76:** 504.

WENNOGLE, L.P. and J.P. CHANGEUX. 1980. Dissection of the 66000-dalton subunit of the acetylcholine receptor. *Eur. J. Biochem.* **106:** 381.

WISE, D.S., B.P. SCHOENBORN, and A. KARLIN. 1981. Structure of acetylcholine receptor dimer determined by neutron scattering and electron microscopy. *J. Biol. Chem.* **256:** 4124.

Structure and Expression of Genomic Clones Coding for the δ-Subunit of the *Torpedo* Acetylcholine Receptor

N.D. HERSHEY, D.J. NOONAN, K.S. MIXTER, T. CLAUDIO,* AND N. DAVIDSON

Department of Chemistry and Division of Biology, California Institute of Technology, Pasadena, California 91125

Many important problems concerning the structure and the function of nicotinic acetylcholine receptors (AChR) can be advantageously approached by molecular cloning. From the nucleic acid sequences of the cDNA clones, one can conveniently determine the amino acid sequences of the polypeptide subunits of this integral membrane complex. This goal has now been accomplished in several laboratories for the *Torpedo* AChR (Noda et al. 1982, 1983; Sumikawa et al. 1982; Claudio et al. 1983; Devillers-Thiery et al. 1983). Important conclusions about the external, transmembrane, and cytoplasmic domains of each subunit and about their evolutionary relatedness can be deduced from these data.
data.

In addition, one can address the problem of the number of nonallelic genes for each subunit by the study of genomic clones. An important question to be answered is whether any particular subunit is itself a member of a multigene family, with different members being expressed in different muscle, nerve, or other tissues. Furthermore, the determination of the structures of the genomic clones is an essential initial step for the study of the developmental regulation of expression of these genes.

It will be of interest to determine whether the intron-exon structures of the natural genes correlate with domain structures deduced from cDNA sequences. These results will provide information on the evolution of these related genes and on the general question of the relationship of exons to protein domains in evolution.

MATERIALS AND METHODS

Isolation of electric organ RNA (Chirgwin et al. 1979) and its subsequent in vitro translation and immunoprecipitation (Mendez et al. 1980) were performed essentially as described previously.

Cloning of electric organ poly(A) RNA into the *Pst*I site of pBR322 (Efstratiadis and Villa-Komaroff 1979), screening of the resulting cDNA library (Hanahan and Meselson 1980), and hybridization selection analysis (Ricciardi et al. 1979) of electric-organ-specific cDNA clones were performed as described previously.

High-molecular-weight *Torpedo californica* DNA was obtained by grinding of the frozen electric organ, isolation of nuclei, lysis with Sarkosyl and proteinase K, banding in cesium chloride, RNase-A digestion, and deproteinization.

The resulting high-molecular-weight DNA (avg. length >100 kb) was partially digested with restriction endonuclease *Mbo*I. Aliquots from several different degrees of digestion were analyzed by gel electrophoresis, and appropriate samples were mixed to help obtain a representative sample. The pooled fragments were subjected to sucrose gradient velocity sedimentation. Fractions that were approximately 20 kb long were isolated, purified, and used for library construction.

Cohesive ends of bacteriophage λL47 DNA (Loenen and Brammar 1980) were annealed, the DNA was digested with *Bam*HI, and the ends were separated from the stuffer fragment by sucrose gradient velocity sedimentation. Standard methods (Maniatis et al. 1982) were used for ligation of the vector to *Torpedo* DNA. In vitro packaging was carried out as described by Mullins et al. (1981). Typical efficiencies were in the range of 5×10^5 pfu/μg of *Torpedo* DNA. Libraries were screened by standard Benton-Davis screening (Maniatis et al. 1982).

RESULTS

Isolation of cDNA Clones

cDNA libraries were prepared as described from *Torpedo* electric organ poly(A) RNA. The library was screened for clones that reacted strongly with cDNA made from electric organ poly(A) RNA and weakly with cDNA from liver or brain poly(A) RNA. Several hundred clones were isolated and tested by hybrid selection of RNA from the electric organ followed by in vitro translation in a rabbit reticulocyte system. The resulting polypeptides were immunoprecipitated with antisera specific to the SDS-denatured subunits or total receptor (Claudio and Raftery 1977) and analyzed by gel electrophoresis. Clones that code for parts of the β- and δ-receptor subunits were positively identified by this procedure. They are denoted pT50 and pT65, respectively.

By comparison with the DNA sequence determined by Noda et al. (1983) from the full-length cDNA clones, it has been determined that the clone denoted pT65 extends from nucleotide 353 to nucleotide 1715 in the coordinate system given by Noda et al. and thus includes the codons extending from amino acid residue 118 to the carboxyl terminus of the mature protein. Clone pT50 extends from coordinate 130 to about 830

*Present address: Department of Biochemistry, Institute for Cancer Research, Columbia University College of Physicians and Surgeons, New York, New York 10032.

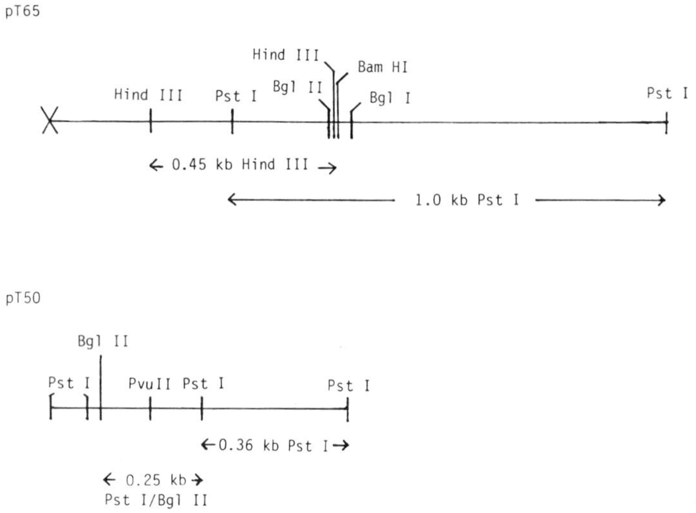

Figure 1. Restriction endonuclease maps of the cDNA clones for the δ(pT65) and the β(pT50) subunits. The *Pst*I sites at the extreme ends of the clones were introduced by the cloning procedure of Efstratiadis and Villa-Komaroff (1979). The site at the left end of the insert in pT65 was not reconstructed during cloning. In both cases, the left-to-right orientation of the clones is 5′ to 3′.

nucleotides in the coordinate system of the β-subunit (Noda et al. 1983), corresponding to codons 44 to 277 of the amino acids in the mature β-subunit. Maps of these clones are shown in Figure 1.

RNA Blots

Standard RNA blots to determine the lengths of the electric organ poly(A) RNAs that are complementary to the β and δ cDNA sequences were carried out. We observe that the β-subunit is represented by a single RNA transcript of about 1600 nucleotides in length (Fig. 2). This length is only slightly greater than that necessary to code for a polypeptide with 460 amino acids (1370 nucleotides of protein-coding sequence).

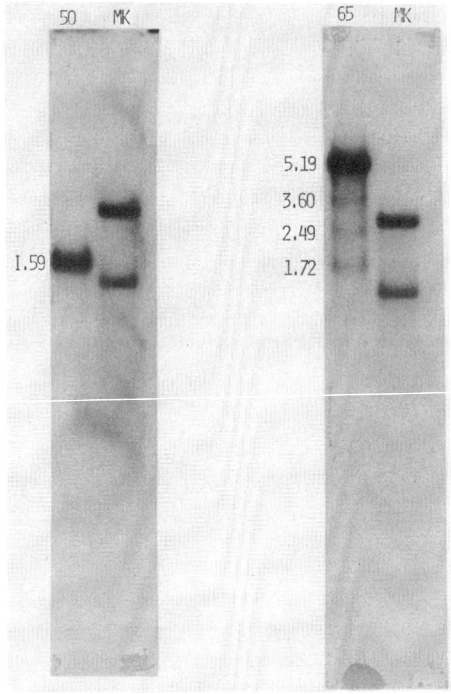

Figure 2. Autoradiographs of RNA gel blots of electric organ poly(A) RNA probed with the pT50 (*left*) and pT65 (*right*) cDNA clones. The markers (MK) are *E. coli* 16S and 23S rRNAs.

In contrast, four transcripts with lengths of 5.2, 3.6, 2.4, and 1.7 kb are observed when the δ cDNA clone is used as a probe (Fig. 2). Similar results are reported by Noda et al. (1983). At the level of sensitivity used, these RNA species are specific to electric organ and are not detected in RNA extracted from brain, liver, or muscle. The relative intensities of the weaker bands vary from preparation to preparation; the 5.2-kb transcript was always predominant, the 1.7-kb transcript was the next in intensity but much weaker, and the two transcripts of intermediate length were fainter still.

Genome Blots

Southern blots of gel-fractionated, restriction-endonuclease-digested *Torpedo* DNA have been probed with β and δ cDNA clones to obtain a preliminary indication of the number of chromosomal genes coding for each subunit and to determine their restriction endonuclease pattern. To avoid problems due to nonspecific hybridization and stickiness associated with the G·C homopolymer sequences at the junctions of the cDNA inserts with vector DNA, the central 450-bp *Hin*dIII fragment of pT65 and the 250-bp *Pst*I-*Bgl*II fragment of pT50 were used as probes. The δ probe hybridizes to only one *Hin*dIII fragment, which has a length of 5.9 kb, and to *Eco*RI fragments with lengths of 4.4 kb and 4.0 kb. The β probe hybridizes to a single 17.7-kb *Hin*dIII fragment, to *Pst*I fragments with lengths of 4.2, 3.1, 1.9, and 1.5 kb, and to *Eco*RI fragments with lengths of 1.7 kb and 3.8 kb (Fig: 3).

These results show that the δ-probe sequence used is present as a single copy in the *Torpedo* genome. The resolution of the 1% agarose gel in the 18-kb region is not high. Therefore, it is probable that the observation that there is only a single band 17.7 kb long for the

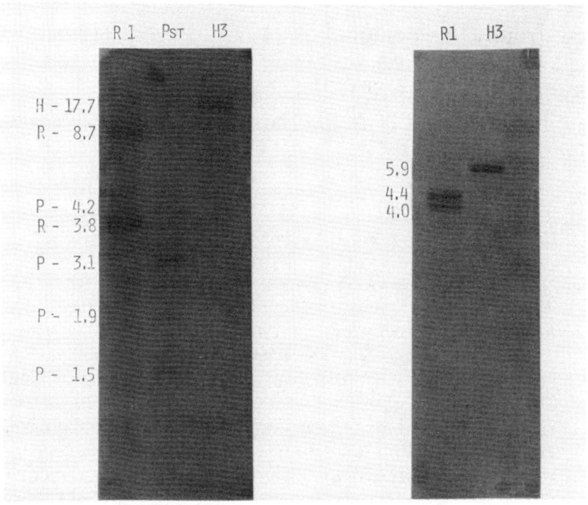

Figure 3. Gel blots of *Torpedo* genomic DNA digested with the indicated enzymes and probed with cDNA clones for the β-subunit (*left*) and the δ-subunit (*right*). These genome blots were carried out with the central 450-bp *Hin*dIII fragment of pT65 and the central 250-bp *Pst*I-*Bal*II fragment of pT50 as probes.

*Hin*dIII digest that hybridizes with the β-subunit probe indicates that this sequence is also only single copy in the genome; but the possibility remains that there are two or three unresolved bands in the genome that correspond to several nonallelic copies of this sequence. We therefore conclude that the δ gene is single copy and that the β gene is certainly not a member of a large multigene family and may be a single-copy gene.

δ Genomic Clones

The *Torpedo* haploid genome comprises 6.5×10^6 kb of DNA (Hinegardner 1976). Thus, for an average insert size of 12 kb as observed for the present λL47 library, 5.5×10^5 phage are equivalent to one haploid genome. A phage library consisting of about 1.0×10^6 plaques was screened by the standard Benton-Davis procedure with a probe prepared by nick translation of the 0.45-kb *Hin*dIII fragment of the δ-subunit cDNA clone. Four clones hybridizing to this fragment have been isolated and characterized.

Restriction maps of these clones were determined by standard methods. These maps (Fig. 4) show that the four clones are overlapping, that they define a total of 20 kb of contiguous genomic *Torpedo* DNA, and that they contain the 4.4-kb and 4.0-kb *Eco*RI fragments and the 5.9-kb *Hin*dIII fragment, which were observed to hybridize to the pT65 probe in the genome blots.

By blot hybridization of the genomic clones with the several *Hin*dIII fragments of pT65, whose orientation was determined by DNA sequence data, the 5′ to 3′ orientation of the gene was determined to be from left to right in Figure 4.

*Eco*RI fragments from the left and right ends of the cloned region were gel-isolated and hybridized to electric organ RNA gel blots. (These fragments are depicted as 14 and 10 in Fig. 4.) Both of these probes hybridized to the same four RNA bands shown in Figure 2, with the relative intensities noted above. The intensity of hybridization by probe 14 was substantially less than that for probe 10, suggesting that the exon-to-intron ratio within this particular segment is low. Thus, all four RNAs contain sequences from the far left and far right of the cloned region.

The map in Figure 4 also shows those regions of the genomic clones that do and do not hybridize with the entire pT65 probe. It should be noted particularly that, reading from left to right, there is a 1.2-kb *Eco*RI fragment at 3–4 kb from the left end, a 1.65-kb *Hin*dIII fragment at 5–7 kb from the left, and a 0.54-kb *Eco*RI-

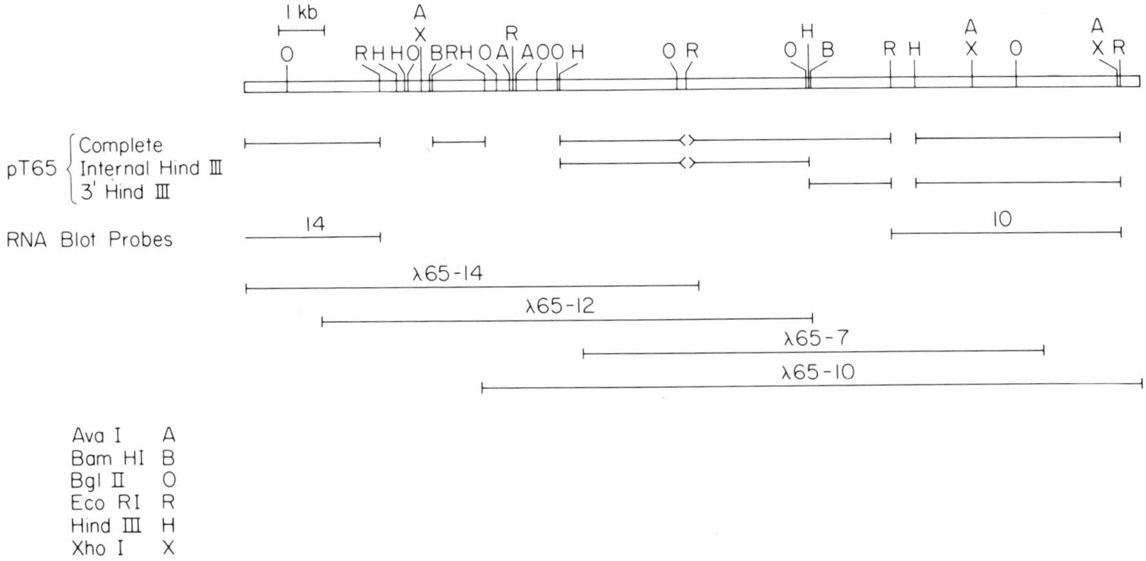

Figure 4. Restriction endonuclease map and summary of blotting experiments with the cloned inserts for the chromosomal gene for the δ subunit.

*Hin*dIII fragment at about 14 kb, all of which do not hybridize with the cDNA probe. Therefore, these fragments must lie entirely within introns. In addition, there is an *Eco*RI site in the genomic clone at 9.6 kb on the map of Figure 4 that is not present in the cDNA clone. Therefore, there must be an additional intron in this region. This observation is confirmed by the fact that the 5.9-kb genomic *Hin*dIII fragment that contains this *Eco*RI site is the only genomic segment that hybridizes with the 0.45-kb cDNA *Hin*dIII segment. Therefore, there must be about 5.45 kb of intron sequences within this genomic fragment. Thus, there are at least four introns within the 20-kb region that has been cloned.

DISCUSSION

In this preliminary report, we describe the isolation of 20 kb of overlapping genomic clones coding for part of the δ-subunit of the AChR. This region probably does not include either the 5′ end or the 3′ end of the complete gene. It contains at least four introns and at most 1 kb of exon sequences. Thus, the ratio of intron-to-exon sequences in this particular example of a structural gene in the *Torpedo* genome is approximately 20:1.

One of the more surprising results is the observation that the most abundant δ RNA transcript has a length of about 5 kb. The intensity of this band on the RNA blots is comparable to that of the 1.7-kb β-band. Since the amounts of β- and of δ-peptide synthesized by in vitro translation of total electric organ RNA in a rabbit reticulocyte system are comparable, it is highly probable that the 5-kb transcript is the mRNA for the δ-subunit in spite of its seemingly excessive length. Whether the fainter, shorter transcripts have biological significance or whether they are the results of degradation at nuclease-sensitive points during extraction remains to be determined.

The preliminary evidence presented here indicates that there is a single copy of the δ-subunit in the *Torpedo* haploid genome. Nicotinic AChRs occur in the electric organ for *Torpedo* and at skeletal muscle synapses and in the CNS for vertebrates in general. Presumably there are nicotinic AChRs in the skeletal muscles and CNS of *Torpedo.* There is evidence for pharmacological differences between such receptors in the chick (Carbonetto et al. 1978). If there is diversity in the δ-subunit of *Torpedo,* there would have to be alternate processing of mRNA transcripts derived from a single gene, as observed for the myosin heavy chain of *Drosophila* (C. Rozek and N. Davidson, pers. comm.). Alternatively, any diversity in receptors may reside in the other subunits. These important questions remain to be studied.

ACKNOWLEDGMENTS

This research has been supported by research grants from the National Institutes of Health and the Muscular Dystrophy Association to N.D., by a postdoctoral fellowship from the American Cancer Society and the National Institutes of Health for D.N., and by a graduate training grant from the National Institutes of Health for K.S.M.

REFERENCES

Carbonetto, S.T., D.M. Fambrough, and K.J. Muller. 1978. Non-equivalence of α-bungarotoxin receptors and acetylcholine receptors in chick sympathetic neurons. *Proc. Natl. Acad. Sci.* **75:** 1016.

Chirgwin, J.M., A.E. Przybyla, R.J. MacDonald, and W.J. Rutter. 1979. Isolation of biologically active ribonucleic acid from sources enriched in ribonuclease. *Biochemistry* **18:** 5294.

Claudio, T. and M.A. Raftery. 1977. Immunological comparison of acetylcholine receptors and their subunits from species of electric ray. *Arch. Biochem. Biophys.* **181:** 484.

Claudio, T., M. Ballivet, J. Patrick, and S. Heinemann. 1983. Nucleotide and deduced amino acids sequences of *Torpedo californica* acetylcholine receptor γ-subunit. *Proc. Natl. Acad. Sci.* **80:** 1111.

Devillers-Thiery, A., J. Giraudat, M. Bentaboulet, and J.P. Changeaux. 1983. Complete mRNA coding sequence of the acetylcholine binding α-subunit of *Torpedo marmorata* acetylcholine receptor: A model for the transmembrane organization of the polypeptide chain. *Proc. Natl. Acad. Sci.* **80:** 2067.

Efstratiadis, A. and L. Villa-Komaroff. 1979. Cloning of double-stranded cDNA. In *Genetic engineering, principles and methods* (ed. J.K. Setlow and A. Hollaender), vol. 1, p. 15. Plenum Press, New York.

Hanahan, D. and M. Meselson. 1980. Plasmid screening at high colony density. *Gene* **10:** 63.

Hinegardner, R. 1976. The cellular DNA content of sharks, rays, and some other fishes. *Comp. Biochem. Physiol.* **55B:** 367.

Loenen, W.A.M. and W.J. Brammar. 1980. A bacteriophage lambda vector for cloning large DNA fragments made with several restriction enzymes. *Gene* **20:** 249.

Maniatis, T., E.F. Fritsch, and J. Sambrook. 1982. *Molecular cloning, a laboratory manual.* Cold Spring Harbor Laboratory, Cold Spring Harbor, New York.

Mendez, B., P. Valenzuela, J.A. Martial, and J.D. Baxter. 1980. Cell-free synthesis of acetylcholine receptor polypeptides. *Science* **209:** 695.

Mullins, J.I., J.W. Casey, M.O. Nicolson, K.B. Burke, and N. Davidson. 1981. Sequence arrangement and biological activity of cloned feline leukemia proviruses from a virus-productive human cell line. *J. Virol.* **38:** 688.

Noda, M., H. Takahashi, T. Tanabe, M. Toyosato, Y. Furutani, T. Hirose, M. Asai, S. Inayama, T. Miyata, and S. Numa. 1982. Primary structure of α-subunit precursor of *Torpedo californica* acetylcholine receptor deduced from cDNA sequence. *Nature* **299:** 793.

Noda, M., H. Takahashi, T. Tanabe, M. Toyosato, S. Kikyotani, T. Hirose, M. Asai, H. Takashima, S. Inayama, T. Miyata, and S. Numa. 1983. Primary structure of β- and δ-subunit precursors of *Torpedo californica* acetylcholine receptor deduced from DNA sequences. *Nature* **301:** 251.

Ricciardi, R.P., J.S. Miller, and B.E. Roberts. 1979. Purification and mapping of specific mRNAs by hybridization-selection and cell-free translation. *Proc. Natl. Acad. Sci.* **76:** 4927.

Sumikawa, K., M. Houghton, J.C. Smith, L. Bell, B.M. Richards, and E.A. Barnard. 1982. The molecular cloning and characterization of cDNA coding for the α-subunit of the acetylcholine receptor. *Nucleic Acids Res.* **10:** 5809.

Genomic Sequences Encoding the α-Subunit of Acetylcholine Receptor Are Conserved in Evolution

M. Ballivet, P. Nef, R. Stalder, and B. Fulpius
Département de Biochimie, Université de Genève, 1211 Genève 4, Switzerland

Several laboratories have recently succeeded in cloning and sequencing some (Ballivet et al. 1982; Sumikawa et al. 1982; Claudio et al. 1983; Devillers-Thiéry et al. 1983) or all (Noda et al. 1982, 1983a,b) of the mRNAs encoding the four subunits of the nicotinic acetylcholine receptor (AChR). In all cases, the source of mRNA was the electric organ of the marine ray *Torpedo*, which an initial report (Mendez et al. 1980) had shown to contain appreciable amounts of receptor messenger.

Examination of the deduced protein sequences for the α-, β-, γ-, and δ-subunits of *Torpedo* AChR has yielded a wealth of new information on receptor function, structure, and phylogeny. An exciting report (Raftery et al. 1980) that the aminoterminal ends of the four subunits were related and, thus, had probably evolved from a common ancestral sequence was confirmed and extended to the whole lengths of the four proteins. All subunits were shown to have four highly hydrophobic sequences, three of them closely grouped, of sufficient length to span the membrane. This suggested a model (Claudio et al. 1983; Devillers-Thiéry et al. 1983; Noda et al. 1983a) for the transmembrane insertion of the individual subunits and allowed one to think about the ways the transmembrane sequences could interact to form the ion channel. Moreover two regions of the extracellular aspect of the α-subunit were singled out as potential ACh-binding sites (Noda et al. 1982) by virtue of the fact that they contain negatively charged residues closely apposed to cysteines, one of which is known to be alkylated by affinity reagents (Karlin 1980).

How conserved is the AChR and how general might be the conclusions and models derived from the study of *Torpedo* receptor? Two recent reports (Conti-Tronconi et al. 1982a,b) have established that the amino ends of the four subunits purified from *Electrophorus* electric organ are highly homologous to the corresponding *Torpedo* subunits as are at least two of the amino ends of AChR subunits purified from calf muscle. Since the genetic code is degenerate, it was by no means obvious, however, that cDNA clones isolated from *Torpedo* would hybridize detectably to genomic sequences from distant species and allow the isolation and characterization of the corresponding receptor genes.

In this preliminary report, we show that a probe encompassing the main transmembrane region and one of the two putative ACh-binding sites of the *Torpedo* α-subunit easily detected the corresponding sequences in a chicken genomic library. In addition, we show that introns separate the various structural domains of the chicken AChR α-subunit and that a *Torpedo* genomic clone contains at least one intron at precisely the same location as in the corresponding chicken clone. These findings establish that existing *Torpedo* cDNA probes permit the cloning and study of AChR genes in species amenable to developmental and genetic analysis.

METHODS

Construction of genomic libraries. We used the L47 vector system (Loenen and Brammar 1980) because it allows cloning of partial *Mbo*I digests of the target DNA between *Bam*HI sites in the vector and because recombinant phage can be selected on phage P2 lysogens by virtue of their *spi* phenotype (Lindahl et al. 1970). High-molecular-weight (>50 kb) DNA was prepared by a standard method (Blin and Stafford 1976) from chicken red blood cells and *Torpedo* electric organ pulverized in liquid nitrogen. Target DNA was partially digested by *Mbo*I to various degrees and the digests were pooled to ensure randomization of DNA in the range 15–25 kb. The pooled digests were centrifuged on high-salt sucrose gradients and DNA in the range 15–25 kb was recovered and used in the cloning procedure. Annealed and *Bam*HI-treated vector DNA was mixed with target DNA in the molar ratio 1:2. After ligation, the DNA mixture was packaged essentially according to Sternberg et al. (1977) in carefully optimized conditions yielding 10^8 to 5×10^8 phage per μg of native λ DNA. We obtained 5×10^4 to 2×10^5 recombinant phage per μg of ligated DNA. The ratio of *spi* to wild-type phage was 25–30%. In control experiments, *Bam*HI-digested and ligated L47 DNA was packaged and shown to yield less than 1% *spi* phage. No phage could be detected when packaging was performed in the absence of added DNA or when the added DNA had not been ligated. The average insert size in recombinants was estimated to be larger than 15 kb. Ten to twenty genome equivalents of recombinant phage were amplified on plate stocks using a P2 lysogen as host. Amplification was about $10^5\times$ and wild-type phage was less than 1% in the amplified stocks. Aliquots of the amplified libraries were stored in 7% dimethylsulfoxide (DMSO) at −70°C.

Nick-translated probes. Whole plasmids or appropriate restriction fragments electrophoresed from acrylamide gels were treated with DNase and nick-translated (Rigby et al. 1977) to a specific activity of 2×10^8 to

4×10^8 dpm/μg. Unreacted label was separated from the probe by gel filtration on P60 columns. Incorporation of label was routinely 70% and neosynthesized DNA was 20% of the added nucleic acid.

Benton-Davis screening of libraries. The original procedure (Benton and Davis 1977) has been adapted as suggested by W.J. Brammar (pers. comm.) to permit screening of up to 5×10^4 phage/10-cm plate. Since our libraries have average inserts of 15 kb, we can plate and screen several genome-equivalents on 30 plates or less. Surprisingly, although the phage plaques are practically confluent, excellent signals can be detected after 24 hours exposure with a fast screen. After transfer of the phage to nitrocellulose filters and processing, the filters are hybridized in 5× SSPE, 5× Denhardt's, 0.1% SDS containing 100 μg/ml sonicated and denatured salmon sperm DNA and 0.5×10^6 to 1.0×10^6 dml/ml denatured probe at 65°C for 18 hours. We estimate those conditions to be about 30° below the melting temperature of a long, perfect hybrid and to allow detection of imperfect hybrids with no more than 20–30% mismatch. After hybridization, the filters are washed extensively in several changes of 2× SSC, 0.01% SDS at room temperature, and at 65°C. After autoradiography, phage plugs (3 mm in diameter) corresponding to positive signals are resuspended in buffer and plated at about 500 phage per plate. The phage lawns are lifted, processed, and hybridized as above. Usually two such cycles suffice to obtain pure stocks of the positive recombinant phage.

Blotting and hybridization of recombinants and subclones. Purified DNA from positive phage recombinants or plasmid subclones were digested by appropriate combinations of restriction enzymes and the digests were electrophoresed on flat-bed agarose gels for 14 hours at 2 volt/cm in Tris-EDTA-Borate buffer. After staining with ethidium bromide and photography, the DNA fragments were transferred to GeneScreen membrane in the conditions recommended by the manufacturer. The baked and prehybridized blots were hybridized in 4× SSPE, 1× Denhardt's, 0.1% SDS, 50% deionized formamide containing 100 μg/ml denatured salmon sperm DNA, and 0.5×10^6 dpm/ml denatured probe at 42°C for 18–36 hours. We used 50 μl hybridization mix per cm^2 of blot. We washed the blots extensively in several changes of 2× SSC, 0.01% SDS at 42°C on a rocking platform. Good signals are usually obtained after 18-hour exposure with a fast screen. A given blot can be hybridized successively with several nick-translated cDNA probes of increasing lengths to determine which restriction fragments of a given genomic clone encode a particular coding sequence.

DNA sequencing. We used the chemical method of Maxam and Gilbert (1977) as modified by Smith and Calvo (1980). DNA fragments were end-labeled either by the kinase procedure or by filling in with the appropriate [α-^{32}P]dNTP (Maniatis et al. 1982).

RESULTS

Construction of a cDNA Probe Encoding the Main Transmembrane Region of the *Torpedo* α-Subunit—Its Use in Screening Genomic Libraries

We had previously isolated a cDNA clone encoding most of the *Torpedo* α-subunit (M. Ballivet et al., unpubl.). The clone was sequenced by T. Claudio. A convenient restriction fragment 448 bp in length and bounded by *Msp*I and *Bgl*II sites was subcloned between the *Cla*I and *Bam*HI sites of pBR322. This fragment was chosen because it encompasses the main transmembrane region of α, a portion of the molecule likely to be highly conserved since it must constitute part of the ion channel (Fig. 1a). In addition, the fragment encodes one of the two putative ACh-binding sites of the α-subunit (Noda et al. 1982). The probe was nick-translated to high specific activities and used to screen chicken and *Torpedo* genomic libraries by the Benton-Davis procedure in hybridization and washing conditions of moderate stringency. Positive recombinant phage were detected in both libraries at a frequency corresponding to less than five copies per genome. Of 18 chicken genome recombinants, 7 were shown by restriction mapping to have resulted from separate cloning events (since the library had been amplified, the same recombinant can be isolated repeatedly). DNA from these seven distinct recombinants contained a fragment of 2.8 kb bounded by *Eco*RI and *Bgl*II sites that hybridized to the probe. Upon further digestions and blottings, this fragment was seen to contain the same additional restriction sites and yield the same hy-

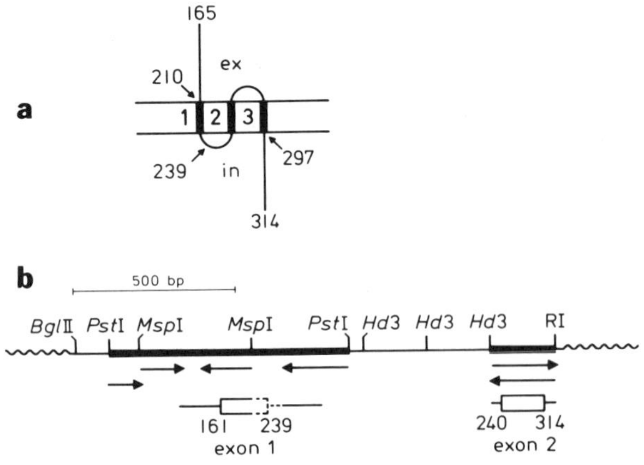

Figure 1. (*a*) The *Torpedo* cDNA probe used in these experiments has a length of 448 bp. It encodes a portion of the mature α-subunit extending between residues 165 and 314, encompassing the main transmembrane region (residues 210–297). This region is thought to span the membrane three times. (ex, in) The extracellular and cytoplasmic compartments, respectively. (*b*) Restriction map of the segment of chicken genome found to contain two exons encoding transmembrane domains of the α-subunit. Exons 1 and 2, respectively, encode residues 161–239 and 240–314. Thicker lines represent restriction fragments that hybridize to the probe; arrows describe the sequencing strategy.

bridization pattern in all recombinants (blotting data not shown). Since we were satisfied that our seven isolates contained the same positive sequence, we subcloned it from one recombinant chosen at random and submitted it to sequence analysis.

We followed an identical procedure to detect, isolate, and subclone *Torpedo* genomic sequences hybridizing to the transmembrane α probe.

Structure and Sequence of a Chicken Gene Encoding the α-Subunit

Figure 1b shows a partial restriction map of the chicken genomic subclone described above. The restriction fragments hybridizing to the screening probe are indicated. Since they are not contiguous, it was evident that at least one intron interrupted the coding sequence. Exon 2 was sequenced first to yield the data in Figure 2A. A sequence of 224 nucleotides was found to encode a polypeptide strikingly similar to the portion of the *Torpedo* α-subunit extending from residue 241 to residue 314 (Noda et al. 1982). Upstream and downstream from this region, the chicken genomic and *Torpedo* cDNA sequences diverged sharply; moreover, the points of divergence correspond closely to the canonical consensus splice sites ...A/C AG[GTAAGT ...C/TC/TTNCAG]G... (Lerner et al. 1980). In addition, a *Torpedo* genomic clone was sequenced in part and shown to code for an exon extending from residue 161 to residue 239 of the α-subunit. Its 3′ end is shown in Figure 2B. Comparison of the two sequences reveals that in both species the GGT triplet encoding the glycine residue at position 240 is split by an intron between its first and second base.

Exon 1 (see Fig. 1b) has not yet been sequenced in full. However, as can be seen in Figure 3, it also resembles closely the corresponding region of the *Torpedo* α-subunit extending from the glutamic acid at position 161 to the glutamine at position 208. Another intron might conceivably interrupt the sequence extending from position 208 to the known splice site at position 240, but this is unlikely in view of our mapping and hybridization results. Therefore, we tentatively conclude that the chicken gene encoding the AChR α-subunit consists of at least four exons. Two of these encompass the main transmembrane region of the protein: Exon 1 extends from residue 161 to residue 239 and encodes the first transmembrane segment of the protein, whereas exon 2 encodes the second and third transmembrane segments. The splice point between exons 1 and 2 is at the midpoint of the sequence encoding a short peptide connecting the first and second membrane-spanning regions. This splice point is exactly conserved in the corresponding *Torpedo* gene. There is 80% nucleotide homology in the 368 bp of *Torpedo* and chicken coding sequences that have been compared. This corresponds to 123 residues or 28% of the α-subunit. Of these 123 residues, only 7 differ in chicken and *Torpedo*. Most differences in the nucleotide sequences are at the wobble position and have no effect on the protein sequence. Of the seven differences in the protein sequences, four are conservative, replacing an aliphatic or polar side group by a side group of like nature. None of the changes affect charged side groups.

```
A
             ↓ 241                                                     255
               GLU LYS MET THR LEU SER ILE SER VAL LEU LEU SER LEU THR VAL
...GCATTGCAGGT GAG AAA ATG ACC CTC AGC ATC TCT GTC TTG CTG TCA CTG ACT GTG
                     G       T T G   T   T   C   T           T

                                                                            275
PHE LEU LEU VAL ILE VAL GLU LEU ILE PRO SER THR SER SER ALA VAL PRO LEU ILE GLY
TTC CTG CTG GTC ATC GTG GAG CTG ATT CCC TCT ACC TCC AGC GCA GTG CCT CTG ATA GGC
      T       T   T   T           C       A   T           T           T   T

                                                                            295
LYS TYR MET LEU PHE THR MET VAL PHE VAL ILE ALA SER ILE ILE ILE THR VAL ILE VAL
AAA TAC ATG TTG TTT ACC ATG GTG TTT GTC ATC GCT TCC ATC ATC ATC ACC GTC ATC GTC
            C T       A     A T             AG    A           T   T   T G T   A
                            ILE             SER                         VAL

                                                                        314 ↓
ILE ASN THR HIS HIS ARG SER PRO SER THR HIS THR MET PRO PRO TRP VAL ARG LYS
ATC AAC ACC CAC CAC CGC TCC CCC AGC ACT CAC ACC ATG CCA CCT TGG GTC AGG AAG GTG
  T   T   T       T       T   A   T   A   T   A         AA        A C A
                                                        GLN

AGCTCCAGGG...


B
                                    239 ↓
    LEU VAL PHE TYR LEU PRO THR ASP SER
... TTA GTA TTT TAC TTA CCA ACT GAT TCA GGTATGAAAA...
```

Figure 2. (*A*) Nucleotide sequence of exon 2. The predicted transmembrane regions (boxed) extend between residues 242–261 and 278–297. The corresponding *Torpedo* nucleotide and protein sequences are indicated where they differ from the chicken. Vertical arrows mark predicted splice sites. (*B*) Nucleotide sequence of the *Torpedo* genome at the splice site located between the first and second base of the triplet encoding glycine 240.

```
                                            181                             189
                                            GLU SER ASP ARG PRO ASP LEU SER ASN
                                 ...CTGTCAG GAG AGC GAT CGC CCC GAC CTG AGT AAC
                                              A           T   G   T          CA
                                                                            THR

                                                                            189
PHE MET GLU SER GLY GLU TRP VAL MET LYS ASP TYR ARG GLY TRP LYS HIS TRP VAL TYR
TTC ATG GAG AGC GGT GAG TGG GTG ATG AAG GAC TAC CGT GGC TGG AAG CAC TGG GTT TAC
  T       A   T   A           A       A   T   T       A                   G   T

                                                                        208
TYR ALA CYS CYS PRO ASP THR PRO TYR LEU ASP ILE THR TYR HIS PHE LEU MET GLN
TAC GCC TGC TGC CCT GAC ACC CCC TAC TTG GAC ATC ACG TAC CAC TTC CTC ATG CAG C..
  T A         T           T         C     T       C       T   T A
    THR                                                         ILE
```

Figure 3. Partial nucleotide sequence of exon 1 from the chicken genome. The part of the α-subunit encoded by exon 1 forms an extracellular domain anchored into the membrane by the first transmembrane region (see text). A splice site and the *Torpedo* sequence corresponding to exon 1 are indicated as in Fig. 2A.

DISCUSSION

A probe coding for the main transmembrane region of the *Torpedo* AChR α-subunit was used to screen a chicken genomic library. It detected two exons separated by about 1.5 kb and encoding residues 161–314 of the chicken AChR α-subunit. The two exons correspond to the predicted structural domains of the α-subunit. Exon 1 encodes sequences connecting the bulk of the extracellular aspect of the protein to the cytoplasm through a first transmembrane segment. Exon 2 codes for another connecting sequence that winds twice through the membrane and ends in the cytoplasmic compartment. In view of the close analogy of the four AChR subunits, it is likely that introns will also be found to separate sequences encoding the several structural domains of the β-, γ-, and δ-subunits. Exons 1 and 2 encode almost one-third of the chicken α-subunit, and the chicken and *Torpedo* proteins are almost identical in this region, differing only at 7 of 123 positions. Reports that sera raised against *Torpedo* receptor react only poorly with avian or mammalian AChR are therefore surprising unless the region we have analyzed contains very few accessible antigenic determinants.

In this preliminary report, we base our claim to have cloned and sequenced parts of the chicken AChR α-subunit gene solely on the close analogy existing between these sequences and the corresponding sequences in *Torpedo* electric organ mRNA. We plan to look for transcription of these sequences in chicken muscle and CNS during development and in various experimental conditions. In view of the astonishing conservation reported here of at least some of the sequences encoding the AChR α-subunit, it seems certain that isogenic probes will shortly be available to study expression of the AChR genes in species ideally suited for genetic and developmental analysis.

ACKNOWLEDGMENTS

This work was supported by a grant (3.685.0.80) of the Swiss National Fund for Scientific Research.

REFERENCES

Ballivet, M., J. Patrick, J. Lee, and S. Heinemann. 1982. Molecular cloning of cDNA coding for the α subunit of *Torpedo* acetylcholine receptor. *Proc. Natl. Acad. Sci.* **79:** 4466.

Benton, W.D. and R.W. Davis. 1977. Screening λgt recombinant clones by hybridization to single plaques in situ. *Science* **196:** 180.

Blin, N. and D.W. Stafford. 1976. A general method for isolation of high molecular weight DNA from eukaryotes. *Nucleic Acids Res.* **3:** 2303.

Claudio, T., M. Ballivet, J. Patrick, and S. Heinemann. 1983. Nucleotide and deduced amino acid sequences of *Torpedo californica* acetylcholine receptor α subunit. *Proc. Natl. Acad. Sci.* **80:** 1111.

Conti-Tronconi, B.M., C.M. Gotti, M.W. Hunkapiller, and M.A. Raftery. 1982a. Mammalian muscle acetylcholine receptor: A supramolecular structure formed by four related proteins. *Science* **218:** 1227.

Conti-Tronconi, B.M., M.W. Hunkapiller, J.M. Lindstrom, and M.A. Raftery. 1982b. Subunit structure of the acetylcholine receptor from *Electrophorus electricus*. *Proc. Natl. Acad. Sci.* **79:** 6489.

Devillers-Thiéry, A., J. Giraudat, M. Bentaboulet, and J.-P. Changeux. 1983. Complete mRNA coding sequence of the acetylcholine binding α-subunit of *Torpedo marmorata* acetylcholine receptor: A model for the transmembrane organization of the polypeptide chain. *Proc. Natl. Acad. Sci.* **80:** 2067.

Karlin, A. 1980. Molecular properties of nicotinic acetylcholine receptors. *Cell Surf. Rev.* **6:** 191.

Lerner, M.A., J.A. Boyle, S.M. Mount, S.L. Wolin, and J.A. Steitz. 1980. Are sn RNPs involved in splicing? *Nature* **283:** 220.

Lindahl, G., G. Sironi, H. Bialy, and R. Calendar. 1970. Bacteriophage lambda; abortive infection of bacteria lysogenic for phage P2. *Proc. Natl. Acad. Sci.* **66:** 587.

Loenen, W.A.M. and W.J. Brammar. 1980. A bacteriophage lambda vector for cloning large DNA fragments made with several restriction enzymes. *Gene* **20:** 249.

Maniatis, T., E.F. Fritsch, and J. Sambrook. 1982. *Molecular cloning. A laboratory manual*. Cold Spring Harbor Laboratory, Cold Spring Harbor, New York.

Maxam, A.M. and W. Gilbert. 1977. A new method for sequencing DNA. *Proc. Natl. Acad. Sci.* **74:** 560.

Mendez, B., P. Valenzuela, J.A. Martial, and J.D. Baxter. 1980. Cell-free synthesis of acetylcholine receptor polypeptides. *Science* **209:** 695.

Noda, M., H. Takahashi, T. Tanabe, M. Toyosato, Y. Furutani, T. Hirose, M. Asai, S. Inayama, T.

MIYATA, and S. NUMA. 1982. Primary structure of α-subunit precursor of *Torpedo californica* acetylcholine receptor deduced from cDNA sequence. *Nature* **299:** 793.

NODA, M., H. TAKAHASHI, T. TANABE, M. TOYOSATO, S. KIKYOTANI, Y. FURUTANI, T. HIROSE, H. TAKASHIMA, S. INAYAMA, T. MIYATA, and S. NUMA. 1983a. Structural homology of *Torpedo californica* acetylcholine receptor subunits. *Nature* **302:** 528.

NODA, M., H. TAKAHASHI, T. TANABE, M. TOYOSATO, S. KIKYOTANI, T. HIROSE, M. ASAI, H. TAKASHIMA, S. INAYAMA, T. MIYATA, and S. NUMA. 1983b. Primary structures of β- and δ-subunit precursors of *Torpedo californica* acetylcholine receptor deduced from cDNA sequences. *Nature* **301:** 251.

RAFTERY, M.A., M.W. HUNKAPILLER, C.D. STRADER, and L.E. HOOD. 1980. Acetylcholine receptor: Complex of homologous subunits. *Science* **208:** 1454.

RIGBY, P.W.J., M. DIECKMANN, C. RHODES, and P. BERG. 1977. Labeling DNA to high specific activity in vitro by nick translation with DNA polymerase 1. *J. Mol. Biol.* **113:** 237.

SMITH, D.R. and J. CALVO. 1980. Nucleotide sequence of the *E. coli* gene coding for dihydrofolate reductase. *Nucleic Acids Res.* **8:** 2255.

STERNBERG, N., D. TIEMEIER, and L. ENQUIST. 1977. In vitro packaging of λ D*am* vector containing *Eco*RI DNA fragments of *Escherichia coli* and phage P1. *Gene* **1:** 255.

SUMIKAWA, K., M. HOUGHTON, J.C. SMITH, L. BELL, B.M. RICHARDS, and E.A. BARNARD. 1982. The molecular cloning and characterization of cDNA coding for the α subunit of the acetylcholine receptor. *Nucleic Acids Res.* **10:** 5809.

Use of Monoclonal Antibodies to Study Acetylcholine Receptors from Electric Organs, Muscle, and Brain and the Autoimmune Response to Receptor in Myasthenia Gravis

J. LINDSTROM,* S. TZARTOS,* W. GULLICK,* S. HOCHSCHWENDER,* L. SWANSON,* P. SARGENT,† M. JACOB,‡ AND M. MONTAL‡

*The Salk Institute, San Diego, California 92138; †Stanford University School of Medicine, Stanford, California 94305; ‡University of California, San Diego, California 92093

Monoclonal antibodies have many applications in the study of acetylcholine receptors (AChRs) and the autoimmune response to AChRs. They are useful because they can provide a wide range of highly specific probes for the receptor molecule that are amenable to many types of experimental manipulation, and because they can be used as model autoantibodies. The basic approaches and techniques that we describe here should prove valuable in the study of many other receptors and molecules of interest to neurobiologists.

We have generated a library of more than 160 monoclonal antibodies to native AChRs purified from the electric organs of *Torpedo californica* (Tzartos and Lindstrom 1980) and *Electrophorus electricus* (Tzartos et al. 1981) and the muscles of cattle (Tzartos and Lindstrom 1981; S. Tzartos et al., unpubl.) and humans (Tzartos et al. 1983) and to the denatured subunits of receptors from both electric organs (Tzartos and Lindstrom 1980; S. Tzartos et al., unpubl.) and muscle (S. Hochschwender et al., unpubl.). To obtain all of the monoclonal antibodies we require, we are continuing to generate and characterize other carefully selected anti-receptor monoclonal antibodies. This intensive approach of generating and carefully characterizing a large library of monoclonal antibodies to a single type of receptor is very laborious, and quite different from the approach frequently used in neurobiology of immunizing with crude tissue and using any interesting cell-specific monoclonal antibodies that arise. However, this strategy permits many types of experiments that would otherwise not be possible or where the results would not be so unambiguous.

There are at least five classes of applications in which we are using monoclonal antibodies for studies of AChRs from electric organ tissue and muscle: (1) to identify receptor subunits and localize substructures within receptor subunits, (2) to test the role of the substructures to which they bind in receptor function, (3) to compare the structures of receptors from different tissues and species, (4) to localize subunits within the quaternary structure of the receptor molecule, and (5) to study the synthesis, conformational maturation, and assembly of receptor subunits. This paper reviews some of the progress and current approaches involved in the first three applications listed that use monoclonal antibodies to study the structure and function of receptors from electric organ and muscle. Some data are presented on the transmembrane orientation of antigenic determinants and the effects of bound monoclonal antibodies on receptor function. Subunit-specific monoclonal antibodies to receptor from *T. californica* electric organ are being employed in the elegant high-resolution electron microscopy studies of receptor structure described by Stroud et al. (this volume). Anderson et al. (1983 and this volume) describe the use of monoclonal antibodies to determine the transmembrane orientation of antigenic determinants in early biosynthetic forms of subunits of receptor from *Torpedo* synthesized in vitro. Merlie et al. (1982, 1983, and this volume; Sebbane et al. 1983; J.P. Merlie and J. Lindstrom, in prep. 1983) describe the use of monoclonal antibodies to study the synthesis, conformational maturation, and assembly of subunits of receptor in muscle cells grown in tissue culture. Subunit-specific monoclonal antibodies are the critical tool that makes it possible to recognize newly synthesized receptor subunits, and they provide methods for detecting their conformational maturation and assembly.

There are three main applications of monoclonal antibodies in the study of the autoimmune response to AChRs in myasthenia gravis and its animal model: (1) to determine the antigenic structure of the receptor, (2) to determine the specificities of autoantibodies to receptor, and (3) as model autoantibodies to determine the pathological effects of various autoantibody specificities. The results of these studies are briefly reviewed below.

There are four applications to studies of nicotinic AChRs in neurons in which we are exploring the use of monoclonal antibodies raised against AChRs from electric organ tissue and skeletal muscle: (1) to identify putative nicotinic receptors in neurons, (2) to localize these neurons, (3) to purify these putative receptors, and (4) to characterize these putative receptors. These studies are reviewed and some results are presented below.

Mapping the Binding Sites of Monoclonal Antibodies on AChRs

The binding properties of a library of monoclonal antibodies must be determined for the library to be useful. We have used a number of techniques in mapping the specificities of anti-receptor monoclonal antibodies and are also using other techniques and the availability of amino acid sequence data on all four receptor subunits (Noda et al. 1983) to map the binding sites of especially interesting monoclonal antibodies with increasing precision. These mapping techniques have recently been reviewed (Lindstrom 1983). In summary, we have determined subunit specificity by immunoprecipitation using ^{125}I-labeled subunits (Tzartos and Lindstrom 1980; Gullick and Lindstrom 1983a), by immunoblotting (Gullick and Lindstrom 1982), and by ELISA assays using purified subunits fixed to microwells (S. Hochschwender et al., unpubl.). Specificities within a subunit have been determined by both immunoprecipitation of ^{125}I-labeled fragments of purified subunits (Gullick et al. 1981) and by immunoblotting of Cleveland-type peptide maps (Gullick and Lindstrom 1983a). Precise mapping will require generation of very small peptide fragments purified by affinity chromatography on monoclonal antibody columns and other techniques. Antisera and monoclonal antibodies raised to synthetic peptides based on subunit amino acid sequences deduced from cDNAs (Noda et al. 1983) will no doubt prove useful both in their own right and in mapping other monoclonal antibodies. Monoclonal antibodies against conformationally dependent determinants have been mapped by competitive binding with other monoclonal antibodies (Tzartos and Lindstrom 1980; Tzartos et al. 1981). These studies have revealed the presence of a "main immunogenic region" (MIR) on receptor toward which many of the monoclonal antibodies raised against intact receptor are directed. To avoid monoclonal antibodies of this specificity and to obtain monoclonal antibodies that can recognize nascent or SDS-denatured subunits, many of our monoclonal antibodies have been raised against denatured receptors or receptor subunits. The number of binding sites for a monoclonal antibody on the receptor and the ability of monoclonal antibodies to cross-link receptors have been evaluated by centrifugation of complexes of monoclonal antibodies with [^{125}I]α-bungarotoxin-labeled receptors on sucrose gradients at varying antigen:antibody ratios (Conti-Tronconi et al. 1981; Lindstrom et al. 1981). The transmembrane orientation of binding sites has been determined by studying the binding of monoclonal antibodies to receptor-rich membrane vesicles in the presence or absence of saponin to permeabilize them (J. Lindstrom and D. Schneider, unpubl.), by the binding of monoclonal antibodies to frog muscle in the presence or absence of saponin (P. Sargent et al., in prep.), and by binding to radiolabeled peptide fragments of early biosynthetic forms of *Torpedo* receptor subunits synthesized in vitro (Anderson et al. 1983).

The MIR is so named because many of the monoclonal antibodies to native receptor are mutually competitive for binding to receptor and any one of them can inhibit the binding of more than half of the antibodies in an anti-receptor serum (Tzartos and Lindstrom 1980; Tzartos et al. 1981, 1982). Of 70 monoclonal antibodies in our library directed at native receptors, 45 are directed at the MIR, whereas of 90 monoclonal antibodies in our library to SDS-denatured receptor, only 6 are directed at this region. The MIR is located on the extracellular surface of α-subunits, but is neither the ACh-binding site nor the attached carbohydrate (Gullick et al. 1981). Because receptor monomers are composed of two α-subunits and one each of β-, γ-, and δ-subunits (Lindstrom et al. 1979; Raftery et al. 1980), there are two MIRs on each receptor. These are oriented so that they cannot be cross-linked by a single monoclonal antibody, thus monoclonal antibodies to this region, unlike monoclonal antibodies specific for β-, γ-, and δ-subunits, can cross-link receptors into large aggregates. Because a MIR has been detected on receptors from all species tested, we presume that it is conserved through evolution because of its vital function, but we do not know what this function is. We have speculated (Tzartos et al. 1981) and have a bit of evidence (E. Godfrey et al., unpubl.) that this region may be involved in specific interactions with extracellular matrix. It is interesting to recall that this structure and perhaps many other functionally important structures on the receptor would go unrecognized in the absence of specific monoclonal antibody probes.

Monoclonal antibodies specific for each subunit can be made. Of 154 monoclonal antibodies in the library to receptors from various species, 74 bound detectably to SDS-denatured ^{125}I-labeled subunits from receptor of *T. californica* (Gullick and Lindstrom 1983a). There are substantial sequence homologies between the receptor subunits, no doubt as a result of their evolution by duplication and reduplication of a primordial subunit (Raftery et al. 1980; Conti-Tronconi et al. 1982). Despite this, 94% of the monoclonal antibodies tested were at least 95% specific for a single subunit (Gullick and Lindstrom 1983a). However, some monoclonal antibodies recognize other subunits with lower affinity, and two monoclonal antibodies bind to some extent to all four receptor subunits. By the different patterns of peptide fragments of each subunit bound by various monoclonal antibodies, it was deduced that in the library there are monoclonal antibodies recognizing a total of between five and nine antigenic determinants on each subunit.

Binding of monoclonal antibodies to peptide fragments has been most extensively studied with α-subunits. There is an ACh-binding site on each of the two α-subunits in a receptor monomer; one of these sites is more reactive with the affinity label 4-(*N*-maleimide)benzyltrimethylammonium iodide (MBTA) than the other (Karlin 1980), whereas one of these sites has higher affinity for curare than the other (Neubig and Cohen 1979). We have observed that two peptide fragments of α are characteristically bound by monoclonal antibodies directed at the MIR (Gullick et al. 1981). One fragment of 17,000 apparent molecular weight contains all the carbohydrate found on α, whereas the other fragment of 19,000 apparent molecular weight is readily labeled by MBTA. Recently, we discovered that both peptides begin at amino acid 42 and have virtually identical peptide maps (B. Gullick et al., unpubl.).

These results suggest that these peptides are identical, except that one is glycosylated, and that only one of the two α-subunits in a receptor is glycosylated. Further, these results suggest that glycosylation causes the difference in binding properties of the two α-subunits, and that the glycosylated α-subunit contains the ACh-binding site that reacts poorly with MBTA, whereas the unglycosylated α-subunit contains the site that has high affinity for curare. This explains the observations that an antibody against carbohydrate blocks toxin binding to the high-affinity curare site (Hall et al., this volume) and that concanavalin A blocks one of the toxin-binding sites on receptor (Boulter and Patrick 1979).

Although the majority of monoclonal antibodies made to native receptor recognize the MIR on its extracellular surface, by contrast, the majority of monoclonal antibodies to denatured receptor appear to be directed at its cytoplasmic surface. Froehner (1981) noticed that antisera to SDS-denatured receptor did not bind well to receptor-rich membrane vesicles of *Torpedo* unless saponin was added to permeabilize the vesicles. Similarly, we (J. Lindstrom and D. Schneider, unpubl.) observed that, of 81 monoclonal antibodies to determinants other than the main immunogenic region tested, only 7 bound well to vesicles in the absence of saponin. Further, Anderson et al. (1983) studied *Torpedo* receptor subunits synthesized in dog pancreas microsomes in vitro and then proteolytically cleaved these into cytoplasmic, membrane-associated, and extracellular domains and found that all δ-specific monoclonal antibodies raised against denatured receptor bound to the cytoplasmic domain. Monoclonal antibodies to both the cytoplasmic and extracellular domains of β were observed. In intact frog muscle, all 11 monoclonal antibodies directed at the MIR tested bound to the extracellular surface, whereas all 20 monoclonal antibodies specific for other sites on α-, β-, γ-, and δ-subunits tested bound only in the presence of saponin (P. Sargent et al., in prep.). Further, electron microscopy showed that all of the saponin-requiring monoclonal antibodies tested bound on the cytoplasmic surface of the postsynaptic membrane. An example of such an experiment is shown in Figure 1 which uses colloidal gold-labeled anti-antibody to localize a monoclonal antibody directed at the MIR on the extracellular surface and a monoclonal antibody raised against SDS-denatured β-subunits on the cytoplasmic surface of the postsynaptic membrane. These experiments (P. Sargent et al., in prep.) not only show that receptor in muscle contains antigenic determinants corresponding to the α-, β-, γ-, and δ-subunits of receptor from *Torpedo* but also show that these determinants are similarly oriented across the postsynaptic membrane.

Comparing AChRs between Tissues and Species

Antigenic determinants corresponding to all four receptor subunits from *Torpedo* have been detected in receptors purified from bovine and rat muscle (Einarson et al. 1982), *Electrophorus* electric organ (Lindstrom et al. 1980) and muscle (Lindstrom et al. 1983), and in intact frog muscle (P. Sargent et al., in prep.). In all cases where the receptor was purified, these determinants were on subunits similar, but distinct, in apparent molecular weight from those of *Torpedo*. These results are expected because aminoterminal amino acid sequence homologies in corresponding receptor subunits from *Torpedo* and *Electrophorus* suggest that the subunit structure of receptor was established before the evolution of this cartilaginous ray and this bony fish diverged at the level of a primordial vertebrate some 400 million years ago (Conti-Tronconi et al. 1982).

The recently discovered (Lindstrom et al. 1983) high concentration of receptor in the muscle of *Electrophorus* provides a unique opportunity to compare the structure of homologous receptors in two different tissues. Receptor purified from *Electrophorus* muscle (Lindstrom et al. 1983) has subunits of about the same apparent molecular weights as receptor purified from its electric organ (Lindstrom et al. 1980). Further, all the subunits react with the same anti-subunit sera and monoclonal antibodies. The peptide maps of the corresponding subunits of receptor from each tissue observed by an immunoblotting technique using anti-subunit sera or by affinity labeling of site-specific peptides are essentially identical (Gullick and Lindstrom 1983b). This is quite a sensitive test for sequence identity throughout the sequence of each subunit, because such peptide maps of homologous receptor subunits between species are not identical (Gullick and Lindstrom 1982). Further, the aminoterminal 30-amino-acid sequence-corresponding subunits from electric organ and muscle are identical (B. Conti-Tronconi et al., unpubl.). Slight differences in apparent molecular weight of an α-subunit peptide fragment were observed between receptor from electric organ and muscle (Gullick and Lindstrom 1983b). Because this peptide contained all of the carbohydrate for the subunit, it was concluded that the receptors in the two distinct but closely related tissues are identical gene products that may differ slightly in glycosylation.

This conclusion is consistent with the observations from several laboratories that indicate the presence of single genomic copies of receptor subunits (Merlie et al. 1983; Davidson et al.; Ballivet et al.; Numa et al.; all this volume). Moreover, the evidence presented by Hall et al. (this volume), which indicates that junctional and extrajunctional forms of receptor muscle differ by glycosylation near an ACh-binding site, is strikingly consistent with our observations on the difference between AChR forms in muscle and electric organ.

Effects of Monoclonal Antibodies on Receptor Function

The observation that labeling 74% of the receptors on muscle cells in culture with antibody reduced the mean ACh-induced conductance only 15% and the mean channel open time by only 23% (Heinemann et al.

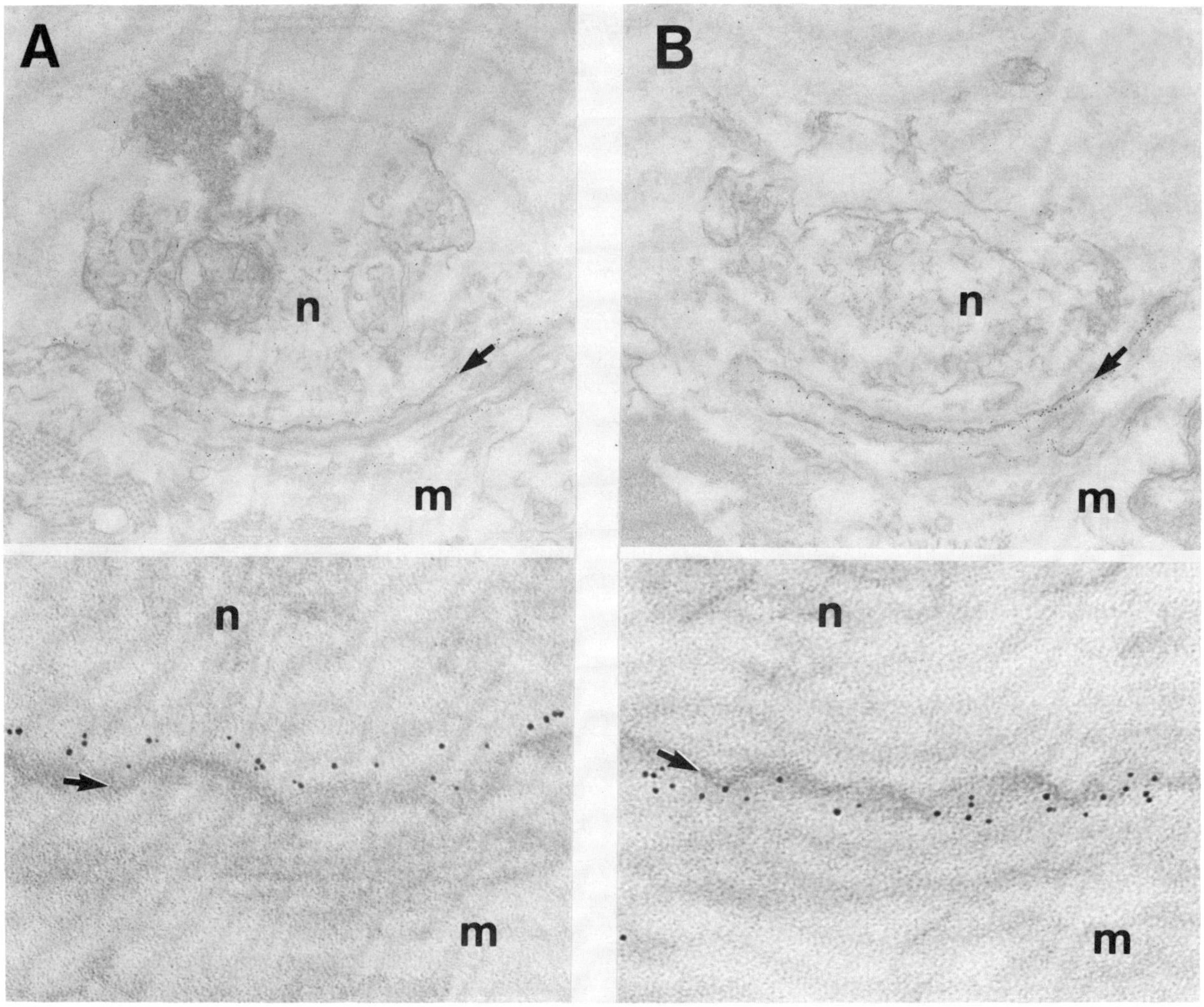

Figure 1. Binding of monoclonal antibodies to the extracellular and cytoplasmic surfaces of the postsynaptic membrane of muscle. *A* and *B* show saponin-permeabilized frog neuromuscular junctions at a magnification of 35,000× in the top panels and 185,000× in the bottom panels. The muscle shown in *A* was labeled with mAb 42, which is directed at the MIR. The muscle shown in *B* was labeled with mAb 111, which is directed at β-subunits. Bound monoclonal antibodies in both cases were visualized with rabbit anti-rat IgG followed by colloidal gold-labeled Protein A. Clearly, the colloidal gold particles bound to mAb 42 are located on the extracellular surface of the postsynaptic membrane, whereas the gold particles bound to mAb 111 are on the cytoplasmic surface of the membrane. Intact cutaneous pectoris muscles from *Rana pipiens* were fixed with 0.025% formaldehyde in 115 mM Na phosphate buffer (pH 7.2). Then the muscles were incubated for 1 hr with titers of 1×10^{-7} M mAb 42 or 1×10^{-6} M mAb 111 in frog Ringer's solution supplemented with 1% BSA and 0.025% saponin. With intervening 30-min washes, the muscles were next incubated with 10^{-7} M rabbit anti-rat IgG and finally with colloidal gold-labeled Protein A in the same buffer. The Protein A labeled with 5-nm gold particles was a gift from Dr. R. Brands. After the final wash, the muscles were fixed with 1% glutaraldehyde and 1% osmium tetroxide before embedding. The sections were examined without grid staining. An extensive study of the reaction of this library of monoclonal antibodies with muscle from *Rana* and *Xenopus* and of the transmembrane orientation of the antigenic determinants is reported in P. Sargent et al. (in prep.).

1977) is consistent with other experiments in suggesting that the most frequent specificities of antibodies to receptor do not block function. Correspondingly, it was found that antibodies to the MIR do not block carbamylcholine-induced $^{22}Na^{+}$ flux into reconstituted vesicles containing receptor from *Torpedo* (Lindstrom et al. 1981). In a much more sensitive test, as shown in Figure 2 and Tables 1 and 2, it can be seen that a monoclonal antibody bound to the MIR of receptor from *Torpedo* reconstituted into planar lipid bilayers (Nelson et al. 1980; Montal et al. 1983; P. Labarca et al., in prep.) alters neither the conductance nor the duration of single receptor channels (M. Montal et al., unpubl.). The recent discovery that single-channel recordings can be made from reconstituted bilayers formed in the tip of patch pipettes directly from reconstituted vesicles containing purified receptor (Suarez-Isla et al. 1983) should permit studies of monoclonal antibodies on receptor function using defined components under conditions that are also amenable to biochemical studies.

Some monoclonal antibodies do block receptor function. For example, mAb 10 noncompetitively blocks

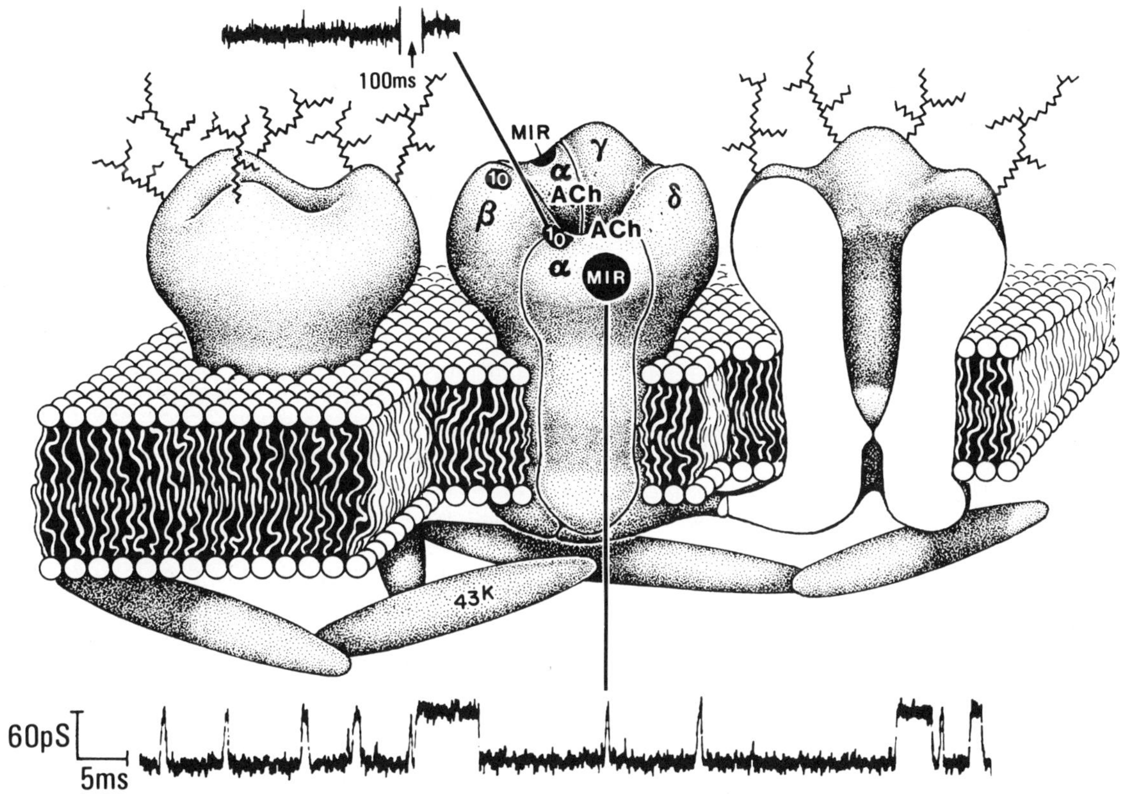

Figure 2. Single-channel records of reconstituted AChRs in the presence of monoclonal antibodies superimposed upon a cartoon of the receptor molecule which indicates the binding sites of the antibodies. Records are from planar lipid bilayers formed with receptors from *Torpedo* as indicated in the legend to Table 1. Upward deflections indicate channel openings in the presence of 10 μM carbamylcholine. The receptor structure shown is based on the model of Kistler et al. (1982). The MIR is on the extracellular surface of α-subunits, and, when mAb 6 (Tzartos and Lindstrom 1980), which is directed at the MIR, is bound, channel amplitudes and durations are normal. The binding sites for mAb 10 are on homologous parts of the extracellular surface of α- and β-subunits (Tzartos and Lindstrom 1980) and, when mAb 10 is bound, channel opening is abolished (M. Montal et al., unpubl.).

carbamylcholine-induced $^{22}Na^+$ flux into reconstituted vesicles (Lindstrom et al. 1981). Similarly, as shown in Figure 2, it completely blocks single AChR channels (M. Montal et al., unpubl.). mAb 10 was raised against β-subunits but also reacts with α-subunits (Tzartos and Lindstrom 1980) and probably cross-links homologous parts on the extracellular surface of each subunit, because at any ratio of monoclonal antibody to receptor tested, only complexes of the size of one monoclonal antibody and one receptor were formed (J. Lindstrom, unpubl.). The ideal effect of a monoclonal antibody on function would be to alter but not obliterate the amplitude or duration of receptor channels. Then precisely mapping the binding site of the rare monoclonal antibody that alters function should give insight into the roles of receptor substructures in opening and closing channels. We are beginning to screen many monoclonal antibodies in our library for their effect on receptor

Table 1. Effect of mAb 6 on the Conductance of Receptor Channels

	Channel conductance (pS)	
Agonist	control	plus mAb 6
Carbamylcholine	60 ± 2	62 ± 3
Suberyldicholine	60 ± 2	63 ± 5

Planar lipid bilayers were formed and studied as previously described (Nelson et al. 1980; Montal et al. 1983; P. Labarca et al., in prep.). Membranes were formed in 1.0 M NaCl, 5 mM $CaCl_2$, and 2.5 mM Tricine (pH 7.4). Prior to bilayer formation, mAb 6 was incubated for 20 min at 4°C with the reconstituted vesicles containing purified receptor (Anholt et al. 1982) in fivefold molar excess. Carbamylcholine was used at 1 μM to 1 mM, and single-channel currents were recorded at 80 mV. Suberyldicholine was used at 0.1 μM to 1 μM and recordings were obtained at 100 mV.

Table 2. Effect of mAb 6 on the Open Times of Receptor Channels

	Control				Plus mAb 6			
Agonist	A_s	τ_s	A_l	τ_l	A_s	τ_s	A_l	τ_l
Carbamylcholine (100 μM)	157	2.55 ± 0.15	36	22.7 ± 1	120	2.85 ± 0.2	87	27 ± 0.8
Suberyldicholine (1 μM)	654	4.7 ± 0.16	437	27 ± 0.16	448	4.0 ± 0.1	36	20 ± 2.5

Conditions were as described for Table 1. A is the number of channels measured and τ is the open time in msec. The subscripts s and l refer to the more frequent short (s) and the less frequent long (l) channel open times, respectively. The time constants were determined by fitting two computer-generated exponentials to the open time distribution.

function (K. Wan et al., unpubl.) by allowing them to bind to purified receptors in cholate/lipid solution before they are reconstituted into native vesicles (Anholt et al. 1982) so that monoclonal antibodies directed at the cytoplasmic surface can bind. Also, several subunit-specific monoclonal antibodies that appear to block receptor function partially have been detected by microinjection of monoclonal antibodies into frog muscle cells in culture (M. Tamaki et al., unpubl.). Clearly, immunological studies of the structural basis of receptor function are at an early, but promising, stage.

Using Monoclonal Antibodies to Study Myasthenia Gravis

Myasthenia gravis (MG) is caused by an antibody-mediated autoimmune response to AChRs in skeletal muscle and can be modeled by immunizing animals with purified receptor. (See Lindstrom and Engel (1981) for a recent detailed review.) There is evidence that the primary cause of the characteristic muscular weakness and fatigue of MG is loss of AChRs and that blockage of receptor function by bound antibodies and simplification of the folded structure of the postsynaptic membrane are less important in impairing neuromuscular transmission. Receptor loss occurs both by antigenic modulation, and by complement-mediated focal lysis of the postsynaptic membrane. Antigenic modulation is an increase in the rate of receptor destruction caused by antibody cross-linking of receptors which facilitates their endocytosis. What initiates the autoimmune response to receptor is not known. Despite overwhelming evidence that MG is mediated by anti-receptor antibodies, the absolute concentration of anti-receptor antibody in a patient's serum does not accurately predict the severity of his muscular weakness. Thus far, no specific immunosuppressive therapy for MG has been developed.

Subunit-specific monoclonal antibodies have been used to determine the spectrum of autoantibody specificities to receptor in MG patients' sera (Tzartos et al. 1982). The fraction of the patients' antibodies that could be inhibited from binding to receptors from human muscle saturated with various monoclonal antibodies was determined. These experiments showed that the autoantibody response recognized many parts of the receptor molecule, but that the MIR was preferentially immunogenic. Thus, MG is not due to loss of immunoregulatory control over a single antibody specificity or to cross-reaction with a single microbial epitope, but to some mechanism which provokes immunization against receptor. A similar average spectrum of anti-receptor specificities dominated by the MIR was found in all degrees of clinical severity.

Antibodies to the MIR are pathologically significant. These antibodies do not directly block receptor function (e.g., Fig. 2, Tables 1 and 2). However, all monoclonal antibodies to the MIR tested could passively transfer the complement-mediated acute form of experimental autoimmune MG (Tzartos and Lindstrom 1980; S. Tzartos et al., unpubl.). In addition, all of these monoclonal antibodies tested also caused antigenic modulation of AChRs in muscle cells in culture (Conti-Tronconi et al. 1981; S. Tzartos et al., unpubl.). This is because, unlike monoclonal antibodies to β-, γ-, and δ-subunits, these bivalent anti-α monoclonal antibodies can effectively link the two α-subunits of receptor monomers into large arrays which are readily endocytosed (Conti-Tronconi et al. 1981). Thus, the most frequent autoantibody specificity is capable of the most prominent pathological mechanisms in MG. The variations in clinical response in populations of MG patients with similar concentrations of anti-receptor antibodies may be accounted for by the variety of complexly regulated mechanisms by which antibody can impair transmission.

A predominant autoantibody specificity in experimental autoimmune myasthenia gravis (EAMG) that is pathologically significant might suggest that the presence of high concentrations of anti-idiotype antibody against a monoclonal antibody with this specificity would effectively suppress disease. This is not the case (Lennon et al. 1981; J. Killen and J. Lindstrom, unpubl.). This is not surprising, because all antibodies binding to the same region of the receptor need not have identical binding sites. Protein antigenic determinants are complex and comprised of six to eight amino acids (Atassi 1978), and the MIR may well be larger. Certainly, the area occluded by the end of a bound antibody is much larger. Even a simple hapten like azobenzenearsonate provokes more than one idiotype of antibody in inbred mice (Nelles et al. 1981). mAb 35 bound to receptor can prevent the binding of more than half of the antibodies in an anti-receptor serum, yet only 1–2% of serum antibodies to receptor from *Elec-*

trophorus cross-react with antibodies specific for the binding site of mAb 35 (S. Hochschwender et al., unpubl.). Similarly, there is little or no cross-reaction between a number of monoclonal antibodies to the MIR and anti-idiotype sera specific for other monoclonal antibodies to the MIR (J. Killen et al.; M. DeBaets et al.; both unpubl.). Thus, therapy using anti-idiotype antibodies to monoclonal antibodies, and probably anti-idiotype therapy in the general case, does not seem practical in MG.

Neuronal AChRs

α-Bungarotoxin binds very specifically to the ACh-binding sites of nicotinic AChRs in muscle and electric organ tissue, but it has limitations not shared by some

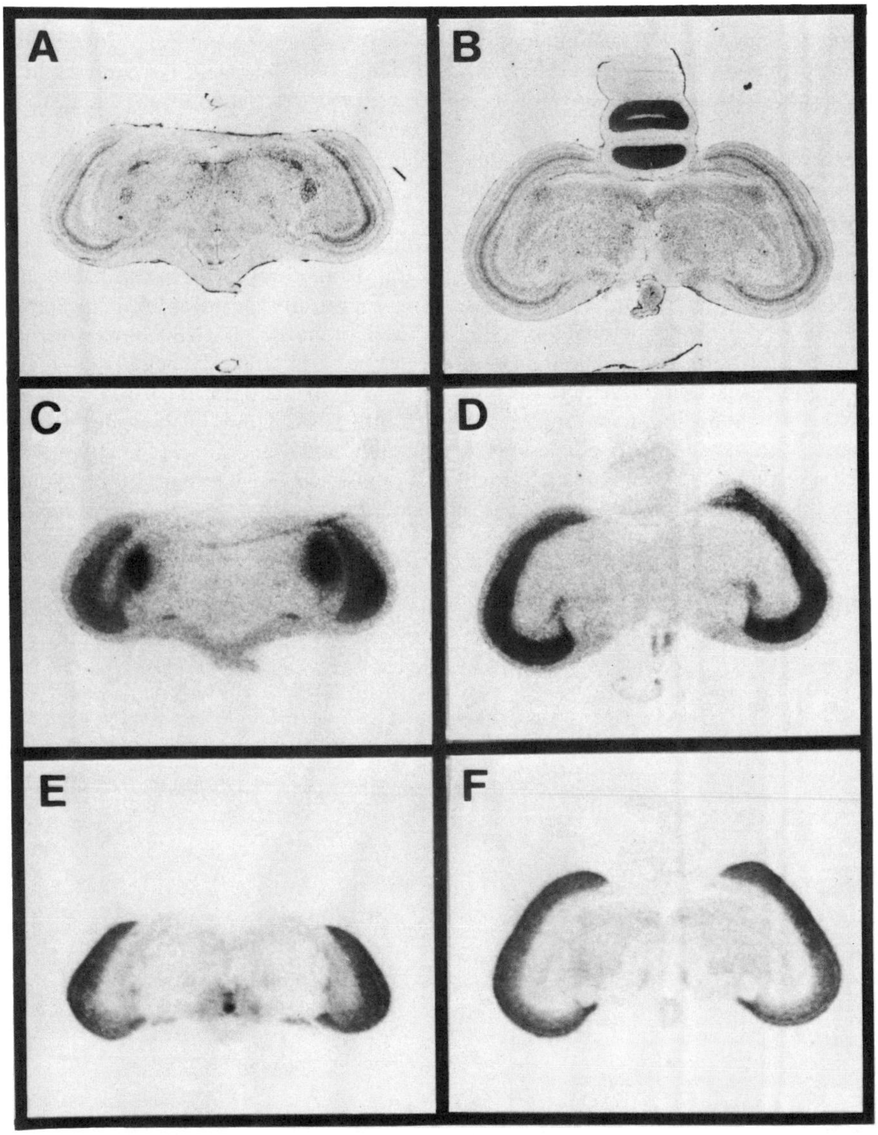

Figure 3. Labeling of chicken midbrain by ^{125}I-labeled mAb 35 and by [^{125}I]α-bungarotoxin. *A*, *C*, and *E* are 18-μm frozen sections from the brain of a young chicken taken relatively rostrally through chicken midbrain whereas *B*, *D*, and *F* are sections taken more caudally. The actual width of the brain is about 1.5 cm. *A* and *B* are Nissl stained to reveal anatomic features. *C* and *D* were incubated overnight with 1×10^{-8} M ^{125}I-labeled F(mAb 35) (4.8×10^{17} cpm/mole) in 10% normal rat serum in 100 mM NaCl and 10 mM Na phosphate (pH 7.5) and then washed with six changes of buffer over 30 min. Labeling could be specifically inhibited by other monoclonal antibodies to the MIR. *E* and *F* were similarly incubated with 1×10^{-8} M [^{125}I]α-bungarotoxin (3.6×10^{17} cpm/mole). Labeling could be specifically inhibited with curare or carbamylcholine. Contact autoradiograms were made on preflashed Kodak XAR-5 film for 48 hr. Note labeling of both the spiriform nucleus and its tectal projection in *C*, whereas section *D* occurs behind the nucleus and shows only the tectal labeling with this ^{125}I-labeled monoclonal antibody to the MIR. Note also that the regions in sections *E* and *F* labeled with [^{125}I]α-bungarotoxin are clearly distinct from those labeled with ^{125}I-labeled mAb 35 in sections *C* and *D*.

anti-receptor monoclonal antibodies. α-Bungarotoxin does not recognize early synthetic conformations of receptor α-subunits and, of course, does not recognize other subunits at all, but monoclonal antibodies can be used for these purposes (J.P. Merlie and J. Lindstrom 1983). Further, AChRs in snake skeletal muscle do not bind α-bungarotoxin, but are labeled by anti-receptor sera (Ishikawa and Shimada 1981). Worse, both chick ganglia (Carbonetto et al. 1978) and the mouse neuronal cell line PC12 (Patrick and Stallcup 1977) bind α-bungarotoxin to sites other than their AChRs. Receptor function on PC12 cells is blocked by anti-receptor antibodies (Patrick and Stallcup 1977). Thus, although α-bungarotoxin binds to many sites in brain (Hunt and Schmidt 1978), it is uncertain which, if any, of these sites are physiologically significant receptors.

Recently we observed that anti-receptor monoclonal antibodies bound to neurons in chicken brain (Swanson et al. 1983) and ganglia (M. Jacob et al., unpubl.) at sites that do not bind α-bungarotoxin. A selection of 57 monoclonal antibodies was first screened for its ability to bind to receptors in frozen sections of chick muscle (Swanson et al. 1983). Of the 17 monoclonal antibodies that cross-reacted, 14 also labeled neurons in the lateral spiriform nucleus and its projection to the optic tectum, whereas none of the 40 monoclonal antibodies that failed to cross-react with receptors in chick muscle reacted in the spiriform nucleus. The number of monoclonal antibodies in this library strongly argues against spurious cross-reaction. Further, monoclonal antibodies specific for the MIR and determinants on the β-, γ-, and δ-subunits of receptor cross-reacted. This suggests that a receptor containing analogs of all four subunits observed in receptors from electric organ and muscle exists on neurons. Figure 3 shows that the areas of the chick midbrain labeled by [^{125}I]α-bungarotoxin and by an ^{125}I-labeled monoclonal antibody to the MIR are quite distinct. Recently Jacob and Berg (1983) showed that in chick ciliary ganglia α-bungarotoxin is not bound at synapses (where receptors are presumably located) but, instead, along the surface membrane of short processes emerging from the cell bodies in the vicinity of synapses. By contrast, it has recently been observed (M. Jacob et al., unpubl.) that a monoclonal antibody to the MIR does bind to the synaptic regions, as shown in Figure 4.

^{125}I-labeled Fab fragments of a monoclonal antibody to the MIR can be used like [^{125}I]α-bungarotoxin to label this putative receptor specifically after solubilization from chicken brains with Triton X-100, as shown in Figure 5. The solubilized receptor is slightly larger than monomers of receptor from *Torpedo* (276,000 mw) and is similarly negatively charged. The concentration of receptor (1×10^{-13} to 2×10^{-13} moles/g brain) is very low, of the order of receptor in normally innervated rat muscle (3×10^{-13} moles/g muscle) (Lindstrom and Lambert 1978). Nonetheless, this putative receptor has recently been very substantially puri-

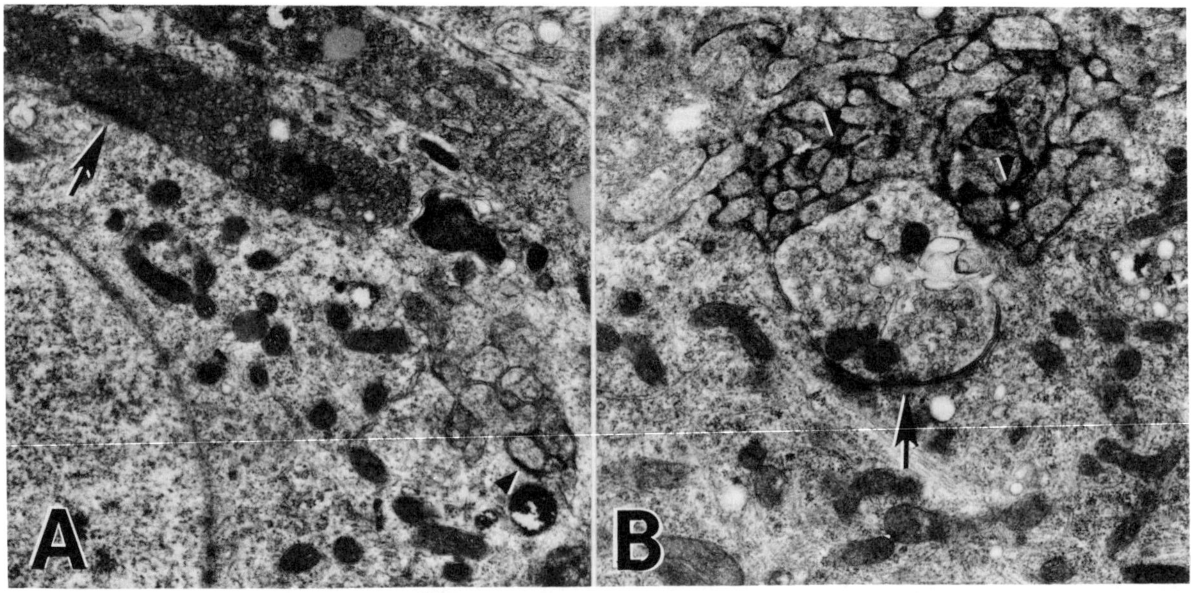

Figure 4. Localization of horseradish peroxidase-labeled mAb 35 and α-bungarotoxin on neurons in ciliary ganglia from a 16-day chicken embryo. (*A*) Dense accumulation of peroxidase reaction product at a synapse (→) in a ganglion that was incubated with 10^{-7} M peroxidase-labeled mAb 35, a monoclonal antibody directed at the MIR. This labeling was specifically blocked by coincubation with excesses of either of two other monoclonal antibodies directed at the MIR. There is also some fainter labeling of the numerous short processes (▲) that project from the postsynaptic cell. (*B*) Dense accumulation of peroxidase reaction product over most of the short processes that project from the postsynaptic cell in a ganglion which was incubated with 10^{-7} M peroxidase-labeled α-bungarotoxin. There is no labeling at the synapse by α-bungarotoxin. (*B* is reprinted, with permission, from Jacob and Berg [1983] and the techniques for the peroxidase reaction, etc., described in this source were used in *A* as well.)

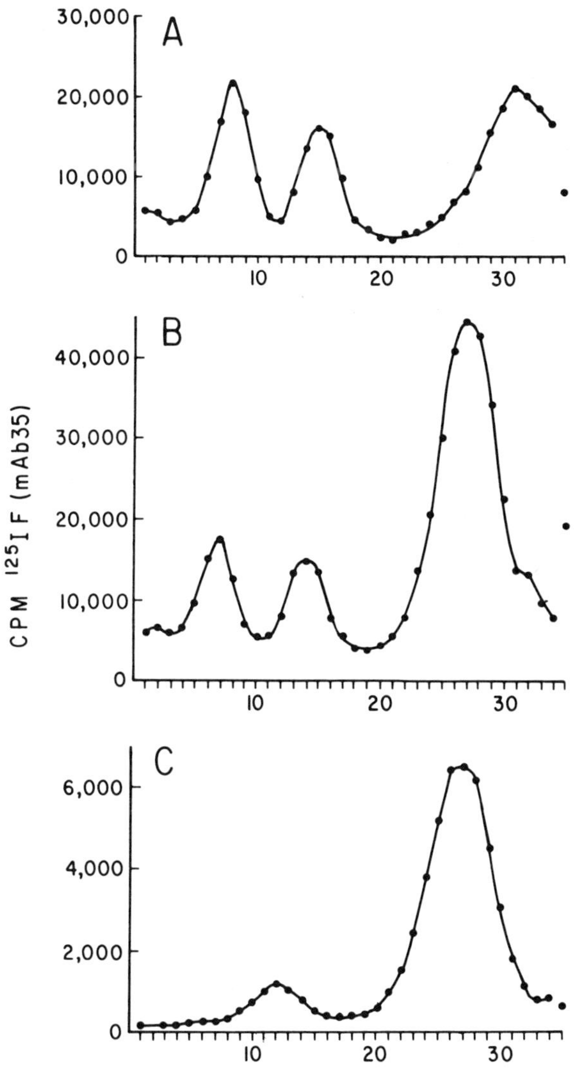

Figure 5. Sucrose gradient sedimentation of receptors from *T. californica* and chick brain. (*A*) Affinity-purified AChRs from *Torpedo* labeled with [^{125}I]α-bungarotoxin and sedimented on a sucrose gradient. Receptor dimers (mw = 540,000) peak in tube 8 and receptor monomers (mw = 270,000) peak in tube 14, whereas free [^{125}I]α-bungarotoxin (mw = 8,000) peaks in tube 31. (*B*) Receptors from *Torpedo* labeled with ^{125}I-labeled Fab fragments of mAb 35, a monoclonal antibody to the MIR. Sedimentation of receptor monomers and dimers labeled with ^{125}I-labeled F(mAb 35) (mw ~50,000) is only slightly more rapid than is sedimentation of [^{125}I]α-bungarotoxin-labeled receptors, apparently because only one Fab is bound per monomer (observed stoichiometry = 0.65 bound Fab/bound α-bungarotoxin). (*C*) Labeling by ^{125}I-labeled F(mAb 35) of a putative receptor extracted from chicken brains in Triton X-100 and partially purified by elution from DEAE. This putative receptor appears to be slightly larger than monomers of receptor from *Torpedo*. In *A* and *B*, receptor at 1×10^{-8} M in toxin-binding sites was incubated with either 2×10^{-8} M [^{125}I]α-bungarotoxin (3.6×10^{17} cpm/mole) or 2×10^{-8} M ^{125}I-labeled F(mAb 35) (3.3×10^{17} cpm/mole). (*C*) Chick extract was incubated with 2×10^{-9} M ^{125}I-labeled F(mAb 35). Then 100-μl aliquots were layered on 4.9 ml 5–20% sucrose gradients and centrifuged for 70 min at 65,000 rpm at 4°C in a Beckman VTi65 rotor before 14 drop fractions were collected from the bottoms of the tubes.

fied by affinity chromatography on a monoclonal antibody affinity column (J. Lindstrom et al., unpubl.).

These studies suggest that monoclonal antibodies can be used to identify, localize, purify, and characterize a previously unrecognized nicotinic receptor on neurons.The evidence that this component is a receptor is that it appears to be localized at synapses, appears to contain structures corresponding to all four subunits of nicotinic receptors from electric organs and muscle, and has a size and charge similar to that of authentic receptor monomers. It does not bind α-bungarotoxin. Its actual subunit structure has not been determined. Most important, it has not been proven that this component is a physiologically significant nicotinic receptor. Therefore, on this problem and a number of others alluded to in the preceding sections, a number of interesting experiments await.

ACKNOWLEDGMENTS

We thank Dr. Rick Rotondo for helpful advice on extracting membrane proteins from chicken brains and Myrta Montal for studies of monoclonal antibodies on planar bilayers. The laboratory of J.L. is supported by grants from the National Institutes of Health (NS11323), the Muscular Dystrophy Association (MDA), the McKnight and Onassis Foundation, and the California and Los Angeles Chapters of the MG Foundation. S.T., W.G., and S.H. are or were recipients of MDA postdoctoral fellowships. The laboratory of L.S. is supported by the NIH (NS16686 and EY-03653) and the Clayton Foundation. The laboratory of M.M. is supported by the NIH (EY-02084), the Department of the Army Medical Research, and the Guggenheim Foundation. M.J. is a recipient of an MDA postdoctoral fellowship in the laboratory of Dr. Darwin Berg. The laboratory of P.S. is supported by the National Science Foundation, the March of Dimes, and the American Heart Asociation, California Affiliate.

REFERENCES

Anderson, D., G. Blobel, S. Tzartos, W. Gullick, and J. Lindstrom. 1983. Transmembrane orientation of an early biosynthetic form of acetylcholine receptor δ subunit determined by proteolytic dissection in conjunction with monoclonal antibodies. *J. Neurosci.***3:** 1773.

Anholt, R., D.R. Fredkin, T. Deerinck, M. Ellisman, M. Montal, and J. Lindstrom. 1982. Incorporation of acetylcholine receptors into liposomes: Vesicle structure and acetylcholine receptor function. *J. Biol. Chem.* **257:** 7122.

Atassi, M.Z. 1978. Precise determination of the entire antigenic structure of lysozyme. *Immunochemistry* **15:** 909.

Boulter, J. and J. Patrick. 1979. Concanavalin A inhibition of α bungarotoxin binding to a nonfusing muscle cell line. *J. Biol. Chem.* **254:** 5652.

Carbonetto, S.T., D.M. Fambrough, and K.F. Muller. 1978. Nonequivalence of α bungarotoxin receptors and acetylcholine receptors in chick sympathetic neurons. *Proc. Natl. Acad. Sci.* **75:** 1016.

Conti-Tronconi, B., S. Tzartos, and J. Lindstrom. 1981.

Monoclonal antibodies as probes of acetylcholine receptor structure. II. Binding to native receptor. *Biochemistry* **20:** 2181.

CONTI-TRONCONI, B., M. HUNKAPILLER, J. LINDSTROM, and M. RAFTERY. 1982. Subunit structure of the acetylcholine receptor from *Electrophorus electricus*. *Proc. Natl. Acad. Sci.* **79:** 6489.

EINARSON, B., W. GULLICK, B. CONTI-TRONCONI, and J. LINDSTROM. 1982. Subunit composition of fetal calf muscle nicotinic acetylcholine receptor. *Biochemistry* **21:** 5295.

FROEHNER, S. 1981. Identification of exposed and buried determinants of the membrane bound acetylcholine receptor from *Torpedo californica*. *Biochemistry* **20:** 4905.

GULLICK, W.J. and J.M. LINDSTSROM. 1982. Structural similarities between acetylcholine receptors from fish electric organs and mammalian muscle. *Biochemistry* **21:** 4563.

———. 1983a. Mapping the binding of monoclonal antibodies to the acetylcholine receptor from *Torpedo californica*. *Biochemistry* **22:** 3312.

———. 1983b. Comparison of the subunit structure of acetylcholine receptors from muscle and electric organ of *Electrophorus electricus*. *Biochemistry* **22:** 3801.

GULLICK, W.J., S. TZARTOS, and J. LINDSTROM. 1981. Monoclonal antibodies as probes of acetylcholine receptor structure. I. Peptide mapping. *Biochemistry* **20:** 2173.

HEINEMANN, S., S. BEVAN, R. KULLBERG, J. LINDSTROM, and J. RICE. 1977. Modulation of the acetylcholine receptor by anti-receptor antibody. *Proc. Natl. Acad. Sci.* **74:** 3090.

HUNT, S. and J. SCHMIDT. 1978. Some observations on the binding patterns of α bungarotoxin in the central nervous system of the rat. *Brain Res.* **157:** 213.

ISHIKAWA, Y. and Y. SHIMADA. 1981. Fluorescent staining of neuromuscular junctions by using the antibody against acetylcholine receptors of *Narke japonica* and double staining with the antibody and Erubatoxin b. *Brain Res.* **224:** 45.

JACOB, M. and D. BERG. 1983. The ultrastructural localization of α bungarotoxin binding sites in relation to synapses on chick ciliary ganglion neurons. *J. Neurosci.* **3:** 260.

KARLIN, A. 1980. Molecular properties of nicotinic acetylcholine receptors. In *The cell surface and neuronal function* (ed. G. Poste et al.), p. 191. Elsevier/North-Holland, New York.

KISTLER, J., R. STROUD, M. KLYMKOWSKY, R. LALANCETTE, and R. FAIRCLOUGH. 1982. Structure and function of an acetylcholine receptor. *Biophys. J.* **37:** 371.

LENNON, V. and E. LAMBERT. 1981. Monoclonal autoantibodies to acetylcholine receptors: Evidence for a dominant idiotype and requirement of complement for pathogenicity. *Ann. N.Y. Acad. Sci.* **377:** 77.

LINDSTROM, J. 1983. Use of monoclonal antibodies in the study of myasthenia gravis. In *Monoclonal antibodies: Probes for the study of autoimmunity and immunodeficiency* (ed. G.S. Eisenbarth and B.F. Haynes). (In press)

LINDSTROM, J. and A. ENGEL. 1981. Myasthenia gravis and the nicotinic cholinergic receptor. In *Receptors and recognition* (ed. P. Cuatrecasas and M. Greaves), series B, vol. 13, p. 161. Chapman and Hall, London.

LINDSTROM, J.M. and E.H. LAMBERT. 1978. Content of acetylcholine receptor and antibodies bound to receptor in myasthenia gravis, experimental autoimmune myasthenia gravis and in Eaton-Lambert syndrome. *Neurology* **28:** 130.

LINDSTROM, J., J. COOPER, and J. SWANSON. 1983. Purification of acetylcholine receptors from the muscles of *Electrophorus electricus*. *Biochemistry* **22:** 3796.

LINDSTROM, J., J. COOPER, and S. TZARTOS. 1980. Acetylcholine receptors from *Torpedo* and *Electrophorus* have similar subunit structures. *Biochemistry* **19:** 1454.

LINDSTROM, J., J. MERLIE, and G. YOGEESWARAN. 1979. Biochemical properties of acetylcholine receptor subunits from *Torpedo californica*. *Biochemistry* **18:** 4465.

LINDSTROM, J., S. TZARTOS, and B. GULLICK. 1982. Structure and function of acetylcholine receptors studied using monoclonal antibodies. *Ann. N.Y. Acad. Sci.* **377:** 1.

MERLIE, J. and J. LINDSTROM. 1983. Assembly *in vivo* of mouse muscle acetylcholine receptor: Identification of an α subunit species which may be an assembly intermediate. *Cell* (in press).

MERLIE, J., R. SEBBANE, S. GARDNER, and J. LINDSTROM. 1983. Regulation of acetylcholine receptor gene expression: Molecular cloning of a cDNA specific for α subunit of the receptor from the mouse muscle cell line BC3H-1. *Proc. Natl. Acad. Sci.* (in press).

MERLIE, J.P., R. SEBBANE, S. TZARTOS, and J. LINDSTROM. 1982. Inhibition of glycosylation with tunicamycin blocks assembly of newly synthesized acetylcholine receptor subunits in muscle cells. *J. Biol. Chem.* **257:** 2694.

MONTAL, M., P. LA BARCA, D. FREDKIN, B. SUAREZ-ISLA, and J. LINDSTROM. 1983. Channel properties of the purified acetylcholine receptor from *Torpedo californica* reconstituted in planar lipid bilayer membranes. *Biophys. J.* (in press).

NELLES, M., L. GILL-PAZARIS, and A. NISONOFF. 1981. Monoclonal anti-idiotypic antibodies reactive with a highly conserved determinant on A/J serum anti-*p*-azophenylarsonate antibodies. *J. Exp. Med.* **154:** 1752.

NELSON, N., R. ANHOLT, J. LINDSTROM, and M. MONTAL. 1980. Reconstitution of purified acetylcholine receptors with functional ion channels in planar lipid bilayers. *Proc. Natl. Acad. Sci.* **77:** 3057.

NEUBIG, R. and J.B. COHEN. 1979. Equilibrium binding of ^{3}H-tubocurarine and ^{3}H acetylcholine by *Torpedo* postsynaptic membranes: Stoichiometry and ligand interactions. *Biochemistry* **18:** 5464.

NODA, M., H. TAKAHASHI, T. TANABE, M. TOYOSATO, S. KIKYOTANI, Y. FURUTANI, T. HIROSE, H. TAKASHIMA, S. INAYAMA, T. MIYATA, and S. NUMA. 1983. Structural homology of *Torpedo californica* acetylcholine receptor subunits. *Nature* **302:** 528.

PATRICK, J. and W. STALLCUP. 1977. Immunological distinction between acetylcholine receptor and the α bungarotoxin binding component on sympathetic neurons. *Proc. Natl. Acad. Sci.* **74:** 4689.

RAFTERY, M., M. HUNKAPILLER, C. STRADER, AND L. HOOD. 1980. Acetylcholine receptor: Complex of homologous subunits. *Science* **208:** 1454.

SEBBANE, R., G. CLOKEY, J.P. MERLIE, S. TZARTOS, and J. LINDSTROM. 1983. Characterization of the mRNA for mouse muscle acetylcholine receptor α subunit by quantitative translation in vitro. *J. Biol. Chem.* **258:** 3294.

SUAREZ-ISLA, B., K. WAN, J. LINDSTROM, and M. MONTAL. 1983. Single channel recordings from purified acetylcholine receptors reconstituted in bilayers formed at the tip of patch pipettes. *Biochemistry* **22:** 2319.

SWANSON, L., J. LINDSTROM, S. TZARTOS, L. SCHMUED, D. O'LEARY, and W. COWAN. 1983. Immunohistochemical localization of monoclonal antibodies to the nicotinic acetylcholine receptor in the midbrain of the chick. *Proc. Natl. Acad. Sci.* **80:** 4532.

TZARTOS, S. and J.M. LINDSTROM. 1980. Monoclonal antibodies used to probe acetylcholine receptor structure: Localization of the main immunogenic region and detection of similarities between subunits. *Proc. Natl. Acad. Sci.* **77:** 755.

———. 1981. Production and characterization of monoclonal antibodies for use as probes of acetylcholine receptors. In *Monoclonal antibodies in endocrine research* (ed. R. Fellows and G. Eisenbarth), p. 69. Raven Press, New York.

TZARTOS, S.J., M. SEYBOLD, and J. LINDSTROM. 1981a.

Specificity of antibodies to acetylcholine receptors in sera from myasthenia gravis patients measured by monoclonal antibodies. *Proc. Natl. Acad. Sci.* **79:** 188.

TZARTOS, S., L. LANGEBERG, S. HOCHSCHWENDER, and J. LINDSTROM. 1983. Demonstration of a main immunogenic region on acetylcholine receptors from human muscle using monoclonal antibodies to human receptor. *FEBS Lett.* **158:** 116.

TZARTOS, S.J., D.E. RAND, B.E EINARSON, and J.M. LINDSTROM. 1981b. Mapping of surface structures of *Electrophorus* acetylcholine receptor using monoclonal antibodies. *J. Biol. Chem.* **256:** 8635.

A Developmental Change in the Immunological Properties of Acetylcholine Receptors at the Rat Neuromuscular Junction

Z.W. HALL, M.-P. ROISIN,* Y. GU, AND P.D. GORIN
Division of Neurobiology, Department of Physiology, University of California, San Francisco, California 94143

The model of a fluid, mosaic membrane, originally proposed by Singer and Nicholson (1972), emphasized the mobility of membrane components and their ability to exchange with each other. More recently, there has been increasing recognition that surface membranes often contain specialized domains with distinctive properties. These domains are stable and can have lifetimes that are longer than those of their constitutents. Such domains are not limited to nerve cells (see, e.g., Fambrough, this volume), but neuronal function depends to an unusual degree on the ability of neurons to construct and maintain specialized regions on the cell surface. The axon hillock, the nodes of Ranvier, and the membrane of the presynaptic terminal are all examples of such specialized domains.

The specialized region of excitable cells that has been most thoroughly studied is the postsynaptic membrane of the neuromuscular junction and of synapses related to it. The major protein constituent of this membrane is the acetylcholine receptor (AChR), which is packed in the membrane in paracrystalline array. The density of AChRs in the postsynaptic membrane exceeds by several orders of magnitude their density elsewhere on the muscle cell (Fambrough 1979). The AChR-rich regions of the postsynaptic membrane are not continuous but are divided by AChR-poor regions. In many species, the postsynaptic membrane is highly enfolded: The AChR-rich regions form the crests of the folds, and the AChR-poor regions form their depths (Fertuck and Salpeter 1976). These two regions can be morphologically distinguished both by the presence of AChRs, which can be visualized after appropriate staining, and by a postsynaptic dense material that is coextensive with the AChRs (Sealock 1980, 1982). Recent immunocytochemical and ultrastructural evidence indicates that not only the postsynaptic membrane, but also the cytoskeleton and the extracellular matrix with which it is associated are specialized (Sanes and Hall 1979; Froehner et al. 1981; Hall et al. 1981; Anderson and Fambrough 1982; Hirokawa and Heuser 1982; Bloch and Hall 1983).

Since the AChR can occur both as part of the specialized postsynaptic membrane and as a component of the entire surface membrane, study of the biochemical and functional properties of this protein in the two different locations may further our understanding of how the postsynaptic membrane is constructed and maintained.

*Permanent address: Institut de Biologie Physico-Chimique, 75005 Paris, France.

Developmental Changes in the Functional Properties of the AChR

In the rat, the formation of a mature neuromuscular junction takes approximately 3 weeks; during this time the AChR undergoes several sequential changes in its properties. Nerve-muscle contact is first detected at embryonic day 14, a time when no postsynaptic specializations are observed (Dennis et al. 1981). Within 48 hours, however, AChR clusters appear underneath nerve terminals (Bevan and Steinbach 1977; Braithwaite and Harris 1979). These clustered receptors arise in part from the redistribution of AChRs in the extrajunctional muscle membrane (Anderson and Cohen 1977; L. Ziskind-Conhaim et al., in prep.). Shortly afterward, focal accumulation of acetylcholinesterase activity is detected at the new end plates (Bevan and Steinbach 1977; Harris 1981). The AChRs at the new end plates initially undergo metabolic turnover at the same rate as AChRs in the extrajunctional membrane (Burden 1977; Reiness and Weinberg 1981), but in the rat, within several days of clustering, the turnover time is dramatically slowed (Reiness and Weinberg 1981). Thus, at approximately embryonic day 20, the turnover time of the AChRs at end plates increases from a half-time of about 1 day to a half-time of about 10 days. At roughly the same time, the cluster itself appears to become stabilized and becomes more resistant to dispersal by high potassium concentrations or energy-blocking agents (Bloch and Steinbach 1981). A final developmental stage occurs postnatally. The mean channel-open time of the AChR changes from 3–4 msec to about 1 msec. This change occurs gradually during the second and third postnatal weeks (Sakmann and Brenner 1978; Fischbach and Schuetze 1980; Michler and Sakmann 1980). During this same period, the subsynaptic folds are formed (Teravainen 1968; Kelly and Zacks 1969). In several different physiological situations, these two properties, the duration of the mean channel-open time, and the presence or absence of synaptic folds, appear to change simultaneously (Schuetze 1980; Brenner and Sakmann 1983).

Changes in Molecular Properties of the AChR

The change in functional properties of the AChR during development could reflect changes in its molecular structure or in its association with other components of the membrane, the cytoskeleton, or the extracellular matrix. Comparison of AChRs purified from adult end plates and from extrajunctional membrane has shown them to be very similar (Nathanson and Hall 1979) but has also revealed two differences between them that suggest that structural changes in the AChR may occur during development: a difference in pI and a difference in immunological properties. Isoelectric focusing of toxin-receptor complexes has shown that the junctional receptor is slightly more acidic than the extrajunctional receptor (Brockes and Hall 1975; Nathanson and Hall 1979). Although end-plate AChRs have a more acidic pI as early as birth (Brockes and Hall 1975), the time of the first appearance of AChRs with an acidic pI is not known.

The difference in immunological properties between AChRs in the two locations is revealed by sera from patients with myasthenia gravis. The sera of most myasthenic patients contain antibodies that recognize a unique determinant on the extrajunctional receptor (Almon and Appel 1975; Weinberg and Hall 1979; Dwyer et al. 1981). The determinant recognized by these antibodies is present on AChRs in extrajunctional regions of both denervated and embryonic muscles.

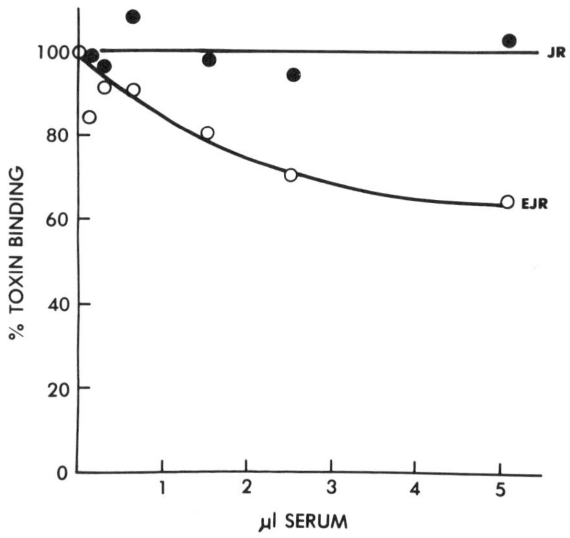

Figure 1. A myasthenia gravis serum inhibits binding of α-bungarotoxin to AChR purified from denervated rat muscle but not from normal rat muscle. AChRs were partially purified from normal (JR) or denervated (EJR) rat leg muscle, using the protocol as described in Froehner et al. (1977), except that the concanavalin-A–Sepharose step was omitted and Triton X-100 was used as the detergent throughout the procedure. Approximately 200 fmoles AChR was incubated with the indicated amount of serum overnight at 4°C. ^{125}I-labeled α-bungarotoxin was then added to a final concentration of 10 nM, and the reaction mixture (30 μl) was incubated for 60 min at 37°C. Toxin-receptor complexes were then assayed by DEAE-cellulose filtration (Brockes and Hall 1975).

RESULTS

A Developmental Change in Immunological Properties Occurs Postnatally

To determine when during development the immunological properties of AChRs at end plates change and whether this change coincides with functional changes, we have sought an antibody preparation that was highly specific for the extrajunctional AChR. In one approach, we have attempted to make monoclonal antibodies with such specificity; these efforts have been so far unsuccessful. In a second approach, we have screened the sera of a large number of patients with myasthenia gravis for their ability to distinguish AChRs purified from normal and denervated adult rat muscle. Following an earlier observation by Dwyer et al. (1981), we specifically looked for antibodies that blocked the binding of α-bungarotoxin to the AChR. Because the neurotoxin site is exposed on the external surface, antibodies that recognize this site should also bind the AChR in intact cells and thus be suitable for immunocytochemistry. One of the sera that we screened appears to be highly specific for the AChR of denervated rat muscle, and we have used this serum in all subsequent studies. Because this patient had undergone plasmapheresis, large amounts of the serum were available. (We are grateful to Dr. P. Dau, who was then at Children's Hospital, San Francisco, for providing us with the plasmapheresis serum.)

The properties of the serum are shown in Figure 1. When incubated with AChR purified from denervated muscle, a progressive inhibition of toxin binding was seen with increasing amounts of antibody; the antibody had no effect, however, on toxin binding by AChR purified from normal muscle. Inhibition of toxin binding to the AChR from denervated muscle never exceeded 50%; we have interpreted this to mean that the antibody blocks binding to one of the two toxin-binding sites that are present on each AChR (see below). When antibody binding to toxin-receptor complexes formed by AChRs from the two sources was examined, using a second antibody and IgG-Sorb (Enzyme Center) to precipitate the complex, no binding to either receptor type was observed (data not shown). Thus, the only antibodies in the serum that bind the AChR are those directed against the toxin-binding site. The effect of the serum was also examined on AChRs purified from the myotubes of the C2 mouse muscle cell line (Yaffe and Saxel 1977; Inestrosa et al. 1983) and on AChRs purified from adult mouse muscle. Inhibition of toxin binding to C2 AChRs was similar to that seen with AChRs from denervated rat muscle; the serum had no effect on binding to AChRs from adult mouse muscle (data not shown).

We then used this serum to stain end plates in frozen sections of developing rat muscle (Fig. 2). Binding of antibody was detected by a fluorescein-labeled second antibody; because the myasthenic serum only partially blocks the binding of α-bungarotoxin, rhodamine-α-bungarotoxin added after the myasthenic antibody could be used to identify end plates. AChRs at end plates in neonatal and 6-day-old rats bound antibody. Binding was diminished at some end plates by day 10, was decreased at most end plates by day 14, and was absent by day 18 (Fig. 2). Thus, AChRs at end plates cease to be recognized by the myasthenic antibodies during the second and third postnatal weeks of life. To

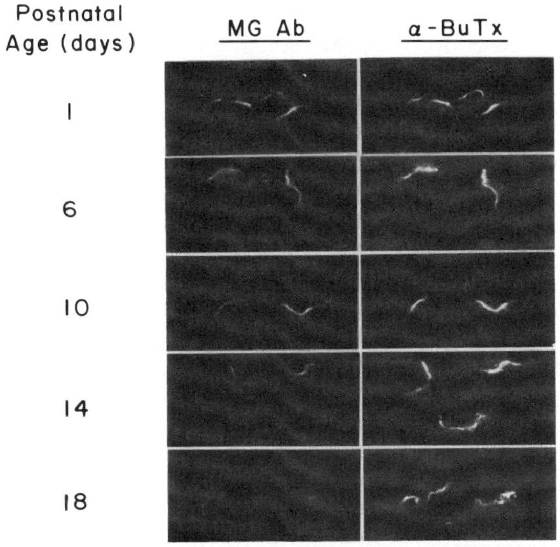

Figure 2. Binding of antibody to AChRs at end plates in developing rat muscle declines during the second and third postnatal weeks. Frozen sections (4 μm) of rat diaphragm muscle from postnatal rats of varying ages were incubated with an immunoglobulin fraction from the myasthenic serum (2 mg/ml protein), followed by a mixture of fluorescein-conjugated goat anti-human IgG (Fab fragment) (Cappel Laboratories) and rhodamine-labeled α-bungarotoxin. Sections were then viewed under rhodamine or fluorescein optics. Further details of the general procedures are given in Sanes and Hall (1979) and Hall et al. (1981).

determine if the bound antibodies observed by immunocytochemistry had the same specificity as that seen when using purified AChR, sections of 10-day-old muscles were incubated with rhodamine-α-bungarotoxin before adding antibody. After this procedure, end plates were seen that were stained with rhodamine-α-bungarotoxin, but no fluorescein labeling was observed. The toxin thus blocked all antibody binding. These results confirm those obtained with the solubilized AChR in demonstrating that the only antibodies that recognize the AChR are those directed against a determinant that is near, or part of, the toxin-binding site.

The change in immunological properties that the receptor undergoes during the second and third postnatal weeks is clearly not associated with AChR clustering or with the change in metabolic turnover time of the AChR, because these events occur prenatally (Bevan and Steinbach 1977; Reiness and Weinberg 1981). The developmental change in immunological properties also probably represents a difference of AChR structure that is distinct from that detected by isoelectric focusing, since AChRs at end plates in neonatal muscle already have an acidic pI (Brockes and Hall 1975). The immunological properties of the AChRs change at a time during development that corresponds most closely to the time during which the mean channel-open time of the AChR changes and during which the subsynaptic folds appear, suggesting that the change in immunological reactivity may be related to one or both of these properties. Consistent with this hypothesis, AChRs at end plates in denervated rat muscle, which have folds and whose AChRs have a short channel-open time, do not bind the antibody, whereas AChRs at adult chicken end plates, which are without folds and whose AChRs have a slow channel-open time, do bind the antibody (Table 1).

The Antibodies Recognize One of the Two Toxin-binding Sites on the AChR

We have attempted to define the specificity of the antibodies for the AChR in this serum in two ways. First, we have used myotubes of C2 cells to investigate the partial inhibition by the antibodies of α-bungarotoxin binding to the AChR. Because each AChR monomer contains two toxin-binding sites, corresponding to the ACh sites associated with each of the two α-chains of the AChR, we were interested in the possibility that the antibodies specifically recognize one of the two sites. Recent experiments by Sine and Taylor (1979, 1981) on the AChRs of the BC3H-1 cell line have shown that these two sites can be distinguished pharmacologically

Table 1. Properties of the Postsynaptic Membrane

Muscle	AChR metabolic turnover time[a]	Prominent folds[b]	AChR channel-open time[c]	Immunological reactivity
Neonatal rat diaphragm	slow	–	long	+ +
Adult rat diaphragm	slow	+	short	–
Denervated adult rat diaphragm	fast	+	short	+ / –[d]
Adult chicken ALD	slow	–	long	+ +[d]

[a]Berg and Hall 1975; Chang and Huang 1975; Burden 1977; Loring and Salpeter 1980; Brett et al. 1982; see Fambrough 1979 for additional references.

[b]Nickel and Waser 1968; Teravainen 1968; Kelly and Zacks 1969; Miledi and Slater 1969; Padykula and Gauthier 1970; Korneliussen and Jansen 1976; Atsumi 1977.

[c]Sakmann and Brenner 1978; Fischbach and Schuetze 1980; Michler and Sakmann 1980; Schuetze 1980; Brenner and Sakmann 1983.

[d]Z. Hall (unpubl.). Immunohistochemical examination of sections of adult chicken anterior latissimus dorsi (ALD) muscle and of denervated (21-day) rat diaphragm muscle was carried out as described in the legend to Fig. 3.

in muscle cells. They found that although the association rates of α-neurotoxin binding for the two sites are indistinguishable, their affinities for several agonists and antagonists are different. We have carried out an analysis of the AChRs on myotubes of C2 cells with similar results. ^{125}I-labeled α-bungarotoxin binds to C2 myotubes with a single association rate constant of 2.4×10^4 M^{-1}s^{-1}; thus, both sites are identical with respect to the on-rate for toxin binding. The two sites differ, however, in their affinities for *d*-tubocurarine. *d*-Tubocurarine inhibits the initial rate of binding of ^{125}I-labeled α-bungarotoxin over a broad concentration range (Fig. 3). The inhibition data are well fit by assuming that there are two sites whose concentrations are equal and whose dissociation constants are 5.6×10^{-8} M and 1.6×10^{-6} M.

The effect of the myasthenic serum on the binding of ^{125}I-labeled α-bungarotoxin to C2 cells is similar to that seen with solubilized AChR; toxin binding is inhibited to a maximum of approximately 50% (Fig. 4). Kinetic analysis shows that the rate constant of binding to the remaining sites is 1.6×10^4 M^{-1}s^{-1}, a value only slightly reduced from normal. To determine whether the antibodies bind specifically to one of the two sites or bind to both sites with equal probability, we analyzed the inhibition of toxin binding by *d*-tubocurarine after treatment of the AChR with antibody. In the experiment shown in Figure 5, addition of antibody resulted in 40% inhibition of the total number of toxin molecules bound. This inhibition could result from the selective loss of sites with a low affinity for *d*-tubocurarine, from the selective loss of sites with high affinity for *d*-tubocurarine, or from the loss of equal numbers of both sites. Theoretical curves for each of these possibilities, assuming that the antibodies cause no change in the value of the dissociation constants, are shown in Figure 5. As can be seen, the data are very well fit by the curve that is based on the selective loss of high-affinity sites. We therefore conclude that the antibodies recognize specifically a determinant near one of the two toxin-binding sites and that this site corresponds to the site with high-affinity for *d*-tubocurarine.

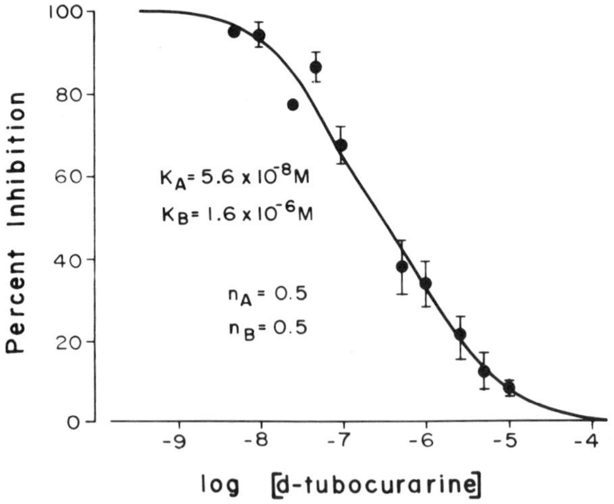

Figure 3. *d*-Tubocurarine inhibits the initial rate of binding of α-bungarotoxin to the AChRs on C2 myotubes. This can be expressed by a model in which *d*-tubocurarine binds to two independent sites of equal concentrations and different affinities. The general procedure used was similar to that described by Sine and Taylor (1981). C2 myotubes were grown in 24-well tissue-culture plates (Falcon) as described by Inestrosa et al. (1983). After equilibrating the cells for 20 min with 0.2 ml of assay buffer (140 mM NaCl, 5.4 mM KCl, 1.8 mM $CaCl_2$, 1.7 mM $MgSO_4$, 1.0 mM $NaHPO_4$, 5.5 mM glucose, 25 mM HEPES, 60 μg/ml bovine serum albumin, pH 7.4), cells were incubated at room temperature for 20 min with the desired concentration of *d*-tubocurarine in assay buffer, followed by the same solution containing 10 nM ^{125}I-labeled α-bungarotoxin. After 10 min, the cells were washed three times, taken up in 2×0.5 ml 0.1 N NaOH, and counted in a gamma counter. The ratio of the initial rate of toxin binding with and without *d*-tubocurarine was then plotted as a function of *d*-tubocurarine concentration. Percent inhibition $= 100[n_A(K_A/K_A + L) + n_B(K_B/K_B + L)]$, where n_A and n_B are the fractional numbers of high-affinity and low-affinity sites, respectively, K_A and K_B are their dissociation constants, and L is the ligand concentration. All individual values were duplicates and each point represents the mean ± S.E. of data taken from five experiments. The solid line is a theoretical curve expressing inhibition by two independent sites of equal concentrations, whose dissociation constants are 5.6×10^{-8} M and 1.56×10^{-6} M. The curve was derived by fitting the data at each extreme of the curve to a least squares linear algorithm.

The Antibody Apparently Recognizes a Carbohydrate Side Chain of the AChR

To investigate further the biochemical specificity of the antibody in this myasthenic serum, we have used AChRs purified from *Torpedo* electric organ. As with

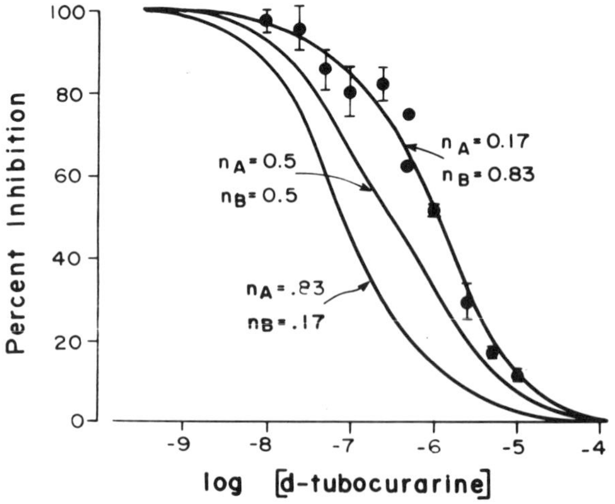

Figure 4. *d*-Tubocurarine inhibition of antibody-blocked AChRs. The myasthenic serum inhibits toxin binding to AChRs in situ in C2 myotubes. C2 myotubes were grown as described in Inestrosa et al. (1983) in 24-well tissue culture plates (Falcon). The cells were incubated for 2 hr at 37°C in 0.2 ml of myasthenic serum at the indicated protein concentrations. ^{125}I-labeled α-bungarotoxin was then added to a final concentration of 3 nM and the incubation was continued for 60 min. The cells were then washed three times, removed with 0.2 ml of 0.1 N NaOH, and counted in a gamma counter. Determinations were made in duplicate. In different experiments, each well contained 50–100 fmoles AChR.

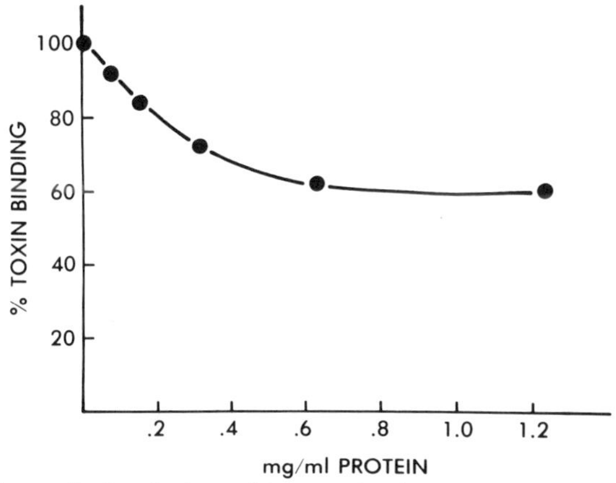

Figure 5. Incubation with myasthenic serum specifically blocks toxin binding to the site with high affinity for *d*-tubocurarine. Experiments were carried out as described in the legend to Fig. 3, except that cells were incubated with 0.2 ml of myasthenic serum (4 mg/ml) for 2 hr at 37°C prior to assay. Each point represents the mean ± S.E. of four determinations. Incubations with the antibody reduced total toxin binding, assayed under saturating conditions, to 60% of that seen without antibody. The solid curves were calculated by using the values for the dissociation constants obtained in Fig. 3 and by assuming that the 40% of blocked sites were all low-affinity sites ($n_A = 0.83$; $n_B = 0.17$), all high-affinity sites ($n_A = 0.17$; $n_B = 0.83$), or equally shared between the two sites ($n_A = 0.5$; $n_B = 0.5$).

the muscle AChR, the serum inhibits toxin binding to the *Torpedo* AChR by a maximum of 50% (Fig. 6A). The *Torpedo* AChR, which is derived from adult synapses, thus resembles the AChRs at neonatal rat or adult chicken end plates in its affinity for the antibodies.

Upon examination of the amino acid sequence of the α-subunit of the *Torpedo* AChR that has been derived from the gene sequence (Noda et al. 1982), we were intrigued by the presence of a potential site of glycosylation near the presumptive active site. We therefore tested the ability of each of the monosaccharides that are associated with the purified AChR from *Torpedo* (Vandlen et al. 1979) to inhibit antibody binding. One of these, *N*-acetylglucosamine, was effective (Fig. 6A); the others, *N*-acetylgalactosamine, mannose, galactose, and sialic acid, were ineffective. For the *Torpedo* AChR, antibody binding was inhibited 50% at approximately 8 mM *N*-acetylglucosamine and completely inhibited at 50 mM. The sugar also inhibited antibody binding to the AChR from denervated muscle but was less effective; 100 mM *N*-acetylglucosamine gave only partial inhibition (Fig. 6B). These results suggest that the antibody recognizes a glycosyl side chain on the AChR.

DISCUSSION

We have found a myasthenic serum that displays remarkable specificity. The serum does not affect toxin binding to AChRs purified from adult rat or mouse muscle but blocks approximately 50% of the toxin bound by AChRs purified from denervated rat muscle, from C2 myotubes, or from *Torpedo* electric organ. Toxin binding to AChRs in situ on C2 myotubes is also inhibited by 50%. Both experiments on the solubilized AChR and immunocytochemical experiments on AChRs in situ show that antibody binding is completely blocked by α-bungarotoxin. Thus, virtually the only antibodies in this serum that recognize AChR from rat, mouse, or *Torpedo* are those that bind at or near a toxin site.

AChR monomers have two toxin-binding sites. These are thought to be close to, or part of, the ACh-binding site associated with each of the two α-subunits that are constituents of the AChR monomer (for review, see Karlin 1980; Conti-Tronconi and Raftery 1982). The two α-subunits have identical polypeptide sequences, as far as is known (Raftery et al. 1980), but because of the oligomeric arrangement of the subunits, each has different neighbors and thus a different environment (Karlin et al.; Fairclough et al.; both this volume). The two sites bind toxin at equal rates but have different binding constants for several affinity alkylating reagents (Damle and Karlin 1978; Wolosin et al. 1980) and for a variety of agonists and antagonists (Neubig and Cohen 1979; Sine and Taylor 1979, 1980, 1981). We have used the pharmacological difference between the two sites to show that the antibody specifically recognizes one of them. Incubation of C2 myotubes with the serum apparently blocks toxin binding to the AChR site with high affinity for *d*-tubocurarine but does not affect toxin binding to the low-affinity *d*-tubocurarine site.

Experiments with muscle AChR and with the *Torpedo* AChR suggest that the antibody recognizes a portion of a glycosyl side chain on the AChR. The antibody inhibition of toxin binding to the *Torpedo* AChR is completely blocked by a simple sugar, *N*-acetylglucosamine. None of the other sugars that have been reported to be associated with the AChR from *Torpedo* (Vandlen et al. 1979) or from fetal calf muscle (Gotti et al. 1982) are effective. Although the concentration of sugar required to block antibody binding is high, the antibodies almost certainly have a more complex specificity (Kabat 1966), and the concentration of sugar is comparable to those of other simple sugars that block antibodies (Watkins and Morgan 1952; Ginsberg et al. 1971; McKenzie et al. 1977; Cheng and Bennett 1980). These results, combined with those using C2 myotubes, suggest that the two toxin-binding sites of the AChR may differ with respect to the oligosaccharides in their vicinity. One interesting possibility is that the two α-chains may differ in their glycosyl side chains. The gene sequence data for the *Torpedo* AChR suggest that each α-chain has a single asparagine-linked oligosaccharide that is near the presumptive binding site for ACh (Noda et al. 1982; Devillers-Thiery et al. 1983). Experiments on AChR synthesis in BC3H-1 cells are also consistent with the hypothesis that each α-chain has a single *N*-glycosyl side chain (Merlie et al. 1982). Since we have yet to determine which polypeptide

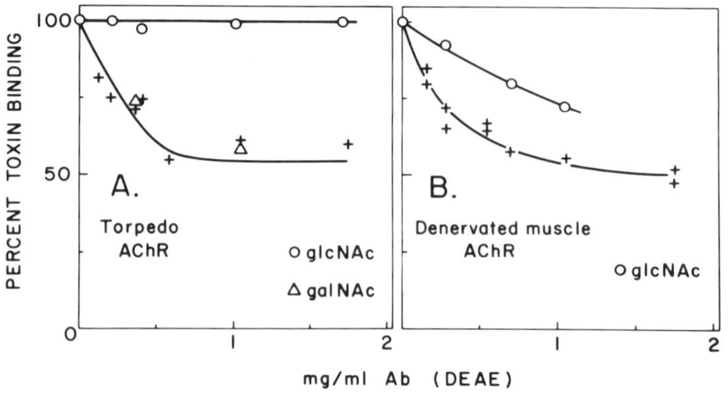

Figure 6. Binding of antibody to AChR from *Torpedo* and muscle AChR is blocked by *N*-acetylglucosamine. (*A*) AChR (200 fmoles) purified from *Torpedo* electric organ as described in Nathanson and Hall (1979) was incubated for 2 hr at 37°C with the indicated concentrations of an IgG fraction of the myasthenic serum, containing either 0.1 M *N*-acetylglucosamine, 0.1 M *N*-acetylgalactosamine, or without sugar. Toxin binding was then assayed as described in the legend to Fig. 1. (*B*) AChR (87 fmoles) from denervated rat muscle purified as described in the legend to Fig. 1 was incubated for 2 hr at 37°C with an IgG fraction of the myasthenic serum at the indicated concentrations, either with (○) or without (+) 0.1 M *N*-acetylglucosamine. Following the incubation, toxin binding was assayed as described in the legend to Fig. 1.

chain the antibody recognizes, it is also possible that the antibody binds a subunit that is a neighbor for one of the α-chains but not the other, thus selectively blocking binding to a single site. Our results suggest that the glycosyl side chain on the AChR that is recognized by the antibody has an exposed *N*-acetylglucosamine. *N*-acetylglucosamine residues occur in *N*-linked oligosaccharides both as terminal sugars and as part of the disaccharide that links the remainder of the chain to asparagine. The determinant could also be part of an *O*-linked sugar, although *N*-acetylglucosamine is not a common component of *O*-linked oligosaccharides (Kornfeld and Kornfeld 1980).

Whatever the exact nature of the determinant, our results clearly show that it changes during development of the neuromuscular junction. Thus, in the rat, the AChRs at end plates in neonatal animals bind the antibody, but AChR at end plates in 18-day-old animals do not. This change in immunological reactivity most likely reflects a change in the glycosylation of the AChR. Such a change could occur, for instance, by a change in the oligosaccharide that is associated with the AChR such that a terminal *N*-acetylglucosamine is removed or is blocked. It is possible that a conformational change in the protein could block access to the sugar, but only if the change were a stable one that survived solubilization of the AChR.

The change in immunological properties of the AChR during development in the rat is clearly not related to the clustering of AChRs that occurs at the end plate. Clustered AChRs in C2 myotubes (Silberstein et al. 1983) and at the end plates in neonatal rat muscle and in adult chicken muscle bind the antibodies. The difference in immunological properties is also unrelated to metabolic turnover time. Thus, AChRs that are bound by the antibodies in situ have both fast (C2 myotubes) and slow (neonatal rat muscle and adult chicken muscle) turnover times, and AChRs that do not bind the antibodies can also have slow and fast turnover times (AChRs at end plates in normal and denervated adult rat muscle, respectively) (Table 1).

The developmental changes that are best correlated with the change in immunological reactivity to this serum are the change in channel properties of the AChR and the appearance of folds in the subsynaptic membrane. Both of these occur during approximately the second and third postnatal weeks in the rat. These changes do not occur in the anterior latissimus dorsi muscle of the chicken (Atsumi 1977; Schuetze 1980), and, correspondingly, AChRs at end plates in this muscle bind the antibodies. In denervated rat muscle, the metabolic turnover time of the AChR is increased (Brett et al. 1982), but the folds and the mean channel open time are unchanged, and, likewise, the AChRs remain unreactive to the antibody. Thus, the circumstantial evidence relating these three phenomena—the folds, the mean channel-open time, and the immunological properties of the AChR—is strong. An additional, late developmental change that could also be related is the conversion of *d*-tubocurarine from an agonist to an antagonist (Ziskind-Conhaim and Dennis 1978). Using a different myasthenic serum to analyze crude extracts of muscle, we have previously reported that an immunological change can be detected as early as 2 days after birth, well before the channel properties of the AChR change (Reiness and Hall 1981). Whether the discrepancy between our previous and present results arises from different specificities of the two sera that we have used or represents different sensitivities of the two methods remains to be determined.

Finally, the change in immunological properties of the AChR in developing muscles clearly occurs later than the change in isoelectric point (Brockes and Hall 1975). This suggests that there may be a third, hybrid form of the AChR that has the immunological properties of the embryonic, extrajunctional AChR and the pI of the adult end plate form of the AChR. Thus, there may be a group of AChR forms that appear sequentially during muscle development, such as has recently been found for the heavy and light chains of myosin (Obinata et al. 1980; Whalen et al. 1981; Gauthier et al. 1982; Crow et al. 1983). These forms could result from the differential expression of related genes, from posttranslational modification, as in the present case, or from a combination of the two.

ACKNOWLEDGMENTS

We thank Charis Bennett for expert technical assistance, Liz Neville and Kem Wefer for help with the manuscript, and Drs. Laura Silberstein, A.J. Hudspeth, Jon Lindstrom, and Marsha Rosner for helpful discussions and advice. This work was supported by grants from the National Institutes of Health and the Muscular Dystrophy Association.

REFERENCES

ALMON, R.R. and S.H. APPEL. 1975. Interaction of myasthenic serum globulin with the acetylcholine receptor. *Biochim. Biophys. Acta* **393:** 66.

ANDERSON, M.J. and M.W. COHEN. 1977. Nerve-induced and spontaneous redistribution of acetylcholine receptors on cultured muscle cells. *J. Physiol.* **268:** 757.

ANDERSON, M.J. and D.M. FAMBROUGH. 1982. Nerve-induced deposition of basal lamina proteoheparan sulphate during development of the neuromuscular junction. *J. Cell Biol.* **95:** 120a.

ATSUMI, S. 1977. Development of neuromuscular junctions of fast and slow muscles in the chick embryo: A light and electron microscopic study. *J. Neurocytol.* **6:** 691.

BERG, D.K. and Z.W. HALL. 1975. Loss of α-bungarotoxin from junctional and extrajunctional acetylcholine receptors in rat diaphragm muscle *in vivo* and in organ culture. *J. Physiol.* **252:** 771.

BEVAN, S. and J.H. STEINBACH. 1977. The distribution of α-bungarotoxin binding sites on mammalian skeletal muscle developing *in vivo*. *J. Physiol.* **267:** 195.

BLOCH, R.J. and Z.W. HALL. 1983. Cytoskeletal components of the vertebrate neuromuscular junction: Vinculin, alpha-actinin, filamin and myosin. *J. Cell Biol.* **97:** 217.

BLOCH, R.J. and J.H. STEINBACH. 1981. Reversible loss of acetylcholine receptor clusters at the developing rat neuromuscular junction. *Dev. Biol.* **81:** 386.

BRAITHWAITE, A.W. and A.J. HARRIS. 1979. Neural influence on acetylcholine receptor clusters in embryonic development of skeletal muscles. *Nature* **279:** 549.

BRENNER, H.R. and B. SAKMANN. 1983. Neurotrophic control of channel properties at neuromuscular synapses of rat muscle. *J. Physiol.* **337:** 159.

BRETT, R.S., S.G. YOUNKIN, M. KONIECZKOWSKI, and R.M. SLUGG. 1982. Accelerated degradation of junctional acetylcholine receptor–α-bungarotoxin complexes in denervated rat diaphragm. *Brain Res.* **233:** 133.

BROCKES, J. and Z.W. HALL. 1975. Acetylcholine receptors in normal and denervated rat diaphragm muscle. II. Comparison of junctional and extrajunctional receptors. *Biochemistry* **14:** 2100.

BURDEN, S. 1977. Acetylcholine receptors at the neuromuscular junction: Developmental change in receptor turnover. *Dev. Biol.* **61:** 79.

CHANG, C.C. and M.C. HUANG. 1975. Turnover of junctional and extrajunctional acetylcholine receptors of the rat diaphragm. *Nature* **253:** 643.

CHENG, C.C. and D. BENNETT. 1980. Nature of the antigenic determinants of T locus antigens. *Cell* **19:** 537.

CONTI-TRONCONI, B.M. and M.S. RAFTERY. 1982. The nicotinic acetylcholine receptor: Correlation of molecular structure with functional properties. *Annu. Rev. Biochem.* **51:** 491.

CROW, M.T., P.S. OLSON, and F.E. STOCKDALE. 1983. Myosin light chain expression during avian muscle development. *J. Cell Biol.* **96:** 736.

DAMLE, V.N. and A. KARLIN. 1978. Affinity labeling of one or two α-neurotoxin binding sites in acetylcholine receptor from *Torpedo californica*. *Biochemistry* **17:** 2039.

DENNIS, M.J., L. ZISKIND-CONHAIM, and A.J. HARRIS. 1981. Development of neuromuscular junctions in rat embryo. *Dev. Biol.* **81:** 266.

DEVILLERS-THIERY, A., J. GIRAUDAT, M. BENTABOULET, and J.P. CHANGEUX. 1983. Complete mRNA coding sequence of the acetylcholine binding α-subunit of *Torpedo marmorata* acetylcholine receptor: A model for the transmembrane organization of the polypeptide chain. *Proc. Natl. Acad. Sci.* **80:** 2067.

DWYER, D.S., R.L. BRADLEY, L. FURNER, and G.E. KEMP. 1981. Immunochemical properties of junctional and extrajunctional acetylcholine receptor. *Brain Res.* **217:** 23.

FAMBROUGH, D.M. 1979. Control of acetylcholine receptors in skeletal muscle. *Physiol. Rev.* **59:** 165.

FERTUCK, H.C. and M.M. SALPETER. 1976. Quantitation of junctional and extrajunctional acetylcholine receptors by electron microscope autoradiography after ^{125}I-α-bungarotoxin binding at mouse neuromuscular junctions. *J. Cell Biol.* **69:** 144.

FISCHBACH, G.D. and S.M. SCHUETZE. 1980. A postnatal decrease in acetylcholine channel open time at rat endplates. *J. Physiol.* **303:** 125.

FROEHNER, S.C., C.G. REINESS, and Z.W. HALL. 1977. Subunit structure of the acetylcholine receptor from denervated rat skeletal muscle. *J. Biol. Chem.* **252:** 8589.

FROEHNER, S.C., V. GULBRANDSEN, C. HYMAN, A.Y. JENE, R.R. NEUBIG, and J.B. COHEN. 1981. Immunofluorescence localization at the mammalian neuromuscular junction of the M_r 43000 protein of *Torpedo* postsynaptic membrane. *Proc. Natl. Acad. Sci.* **78:** 5230.

GAUTHIER, G.F., S. LOWEY, D.A. BENFIELD, and A.W. HOBBS. 1982. Distribution and properties of myosin isozymes in developing avian and mammalian skeletal muscle fibers. *J. Cell Biol.* **92:** 471.

GINSBERG, V., A. KABATA, C. HICKEY, and T. SAWECKA. 1971. Biochemical basis for blood types in man. In *Glycoproteins of blood cells and plasma* (ed. G.A. Jamieson and T.L. Greenwalt), p. 114. J.P. Lippincott, Philadelphia.

GOTTI, C., B.M. CONTI-TRONCONI, and M.A. RAFTERY. 1982. Mammalian muscle acetylcholine receptor purification and characterization. *Biochemistry* **21:** 3148.

HALL, Z.W., B.W. LUBIT, and J.H. SCHWARTZ. 1981. Cytoplasmic actin in postsynaptic structures at the neuromuscular junction. *J. Cell Biol.* **90:** 789.

HARRIS, A.J. 1981. Embryonic growth and innervation of rat skeletal muscles. II. Neural regulation of muscle cholinesterase. *Philos. Trans. R. Soc. Lond. B* **293:** 279.

HIROKAWA, N. and J.E. HEUSER. 1982. Internal and external differentiation of the postsynaptic membrane at the neuromuscular junction. *J. Neurocytol.* **11:** 487.

INESTROSA, N.C., J.B. MILLER, L. SILBERSTEIN, L. ZISKIND-CONHAIM, and Z.W. HALL. 1983. Developmental regulation of 16S acetylcholinesterase and acetylcholine receptors in a mouse cell line. *Exp. Cell Res.* **147:** 393.

KABAT, E.A. 1966. The nature of an antigenic determinant. *J. Immunol.* **97:** 1.

KARLIN, A. 1980. Molecular properties of nicotinic acetylcholine receptors. In *The cell surface and neuronal function* (ed. C.W. Cotman et al.), p. 191. Elsevier/North-Holland, New York.

KELLY, A.M. and S.I. ZACKS. 1969. The fine structure of motor endplate morphogenesis. *J. Cell Biol.* **42:** 154.

KORNELIUSSEN, H. and J.K.S. JANSEN. 1976. Morphological aspects of the elimination of polyneuronal innervation of skeletal muscle of newborn rats. *J. Neurocytol.* **5:** 591.

KORNFELD, R. and S. KORNFELD. 1980. Structure of glycoproteins and their oligosaccharide units. In *The biochemistry of glycoproteins and proteoglycans* (ed. W.J. Lennarz), p. 1. Plenum Press, New York.

LORING, R.H. and M.M. SALPETER. 1980. Denervation increases turnover rate of junctional acetylcholine receptors. *Proc. Natl. Acad. Sci.* **77:** 2293.

MCKENZIE, I.F.C., A. CLARKE, and C.R. PARISH. 1977. Ia antigenic specificities are oligosaccharide in nature: Haptene-inhibition studies. *J. Exp. Med.* **145:** 1039.

MERLIE, J.P., R. SEBBANE, S. TZARTOS, and J. LINDSTROM. 1982. Inhibition of glycosylation with tunicamycin blocks assembly of newly synthesized acetylcholine receptor subunits in muscle cells. *J. Biol. Chem.* **257:** 2694.

MICHLER, A. and B. SAKMANN. 1980. Receptor stability and channel conversion in the subsynaptic membrane of the developing mammalian neuromuscular junction. *Dev. Biol.* **80:** 1.

MILEDI, R. and C.R. SLATER. 1969. Electron-microscopic structure of denervated skeletal muscle. *Proc. R. Soc. Lond. B* **174:** 253.

NATHANSON, N.M. and Z.W. HALL. 1979. Subunit structure and peptide mapping of junctional and extrajunctional acetylcholine receptors from rat muscle. *Biochemistry* **15:** 3392.

NEUBIG, R.R. and J.B. COHEN. 1979. Equilibrium binding of ^{3}H-tubocurarine and ^{3}H-acetylcholine by *Torpedo* postsynaptic membranes: Stoichiometry and ligand interactions. *Biochemistry* **18:** 5465.

NICKEL, E. and P.G. WASER. 1968. Electronenmikroskopische Untersuchungen am diaphragma der Maus nach einseitiger Phrenikotomie. I. Die degenerierende motorische Endplatte. *Z. Zellforsch. Mikrosk. Anat.* **88:** 278.

NODA, M., H. TAKAHASHI, T. TANABE, M. TOYOSATO, Y. FURUTANI, T. HIROSE, M. ASAI, S. INAYAMA, T. MIYATA, and S. NUMA. 1982. Primary structure of alpha-subunit precursor of *Torpedo californica* acetylcholine receptor deduced from cDNA sequence. *Nature* **299:** 793.

OBINATA, T., T. MASAKI, and H. TAKANO. 1980. Types of myosin light chains present during the development of fast skeletal muscle in chick embryo. *J. Biochem.* **87:** 81.

PADYKULA, H.A. and G.F. GAUTHIER. 1970. The ultrastructure of the neuromuscular junctions of mammalian red, white and intermediate skeletal muscle fibers. *J. Cell Biol.* **46:** 27.

RAFTERY, M.A., M.W. HUNKAPILLER, C.D. STRADER, and L.E. HOOD. 1980. Acetylcholine receptor: Complex of homologous subunits. *Science* **208:** 1454.

REINESS, C.G. and Z.W. HALL. 1981. The developmental change in immunological properties of the acetylcholine receptor in rat muscle. *Dev. Biol.* **81:** 324.

REINESS, C.G. and C.B. WEINBERG. 1981. Metabolic stabilization of acetylcholine receptors at newly formed neuromuscular junctions in rat. *Dev. Biol.* **84:** 247.

SAKMANN, B. and H.R. BRENNER. 1978. Changes in synaptic channel gating during neuromuscular development. *Nature* **274:** 68.

SANES, J.B. and Z.W. HALL. 1979. Antibodies that bind specifically to synaptic sites on muscle fiber basal lamina. *J. Cell Biol.* **83:** 357.

SCHUETZE, S.M. 1980. The acetylcholine channel open time in chick muscle is not decreased following innervation. *J. Physiol.* **303:** 111.

SEALOCK, R. 1980. Identification of regions of high acetylcholine receptor density in tannic acid–fixed postsynaptic membranes from electric tissue. *Brain Res.* **199:** 267.

——. 1982. Visualization at the mouse neuromuscular junction of a submembrane structure in common with *Torpedo* postsynaptic membranes. *J. Neurosci.* **2:** 918.

SILBERSTEIN, L., P.D. GORIN, and Z.W. HALL. 1983. Differentiation of the acetylcholine receptor at developing neuromuscular junctions. In *5th International Congress on Neuromuscular Diseases,* Marseille (ed. G. Serratrice et al.). (In press)

SINE, S.M. and P. TAYLOR. 1979. Functional consequences of agonist-mediated state transitions in the cholinergic receptor. Studies in cultured muscle cells. *J. Biol. Chem.* **254:** 3315.

——. 1980. The relationship between agonist occupation and the permeability response of the cholinergic receptor revealed by bound cobra α-toxin. *J. Biol. Chem.* **255:** 10144.

——. 1981. Relationship between reversible antagonist occupancy and the functional capacity of the acetylcholine receptor. *J. Biol. Chem.* **256:** 6692.

SINGER, S.J. and G.L. NICHOLSON. 1972. The fluid mosaic model of the structure of cell membranes. *Science* **175:** 720.

TERAVAINEN, H. 1968. Development of the myoneural junction in the rat. *Z. Zellforsch. Mikrosk. Anat.* **87:** 249.

VANDLEN, R.L., W.C.-S. WU, J. EISENACH, and M.A. RAFTERY. 1979. Studies of the composition of purified *Torpedo californica* acetylcholine receptor and its subunits. *Biochemistry* **18:** 1845.

WATKINS, W.M. and W.T.J. MORGAN. 1952. Neutralization of the anti-H-glutin in eel serum by simple sugars. *Nature* **169:** 825.

WEINBERG, C.B. and Z.W. HALL. 1979. Antibodies from patients with myasthenia gravis recognize determinants unique to extrajunctional acetylcholine receptors. *Proc. Natl. Acad. Sci.* **76:** 504.

WHALEN, R.G., S.M. SELL, G.S. BUTLER-BROWNE, K. SCHWARTZ, P. BOUVERET, and I. PINSET-HARSTROM. 1981. Three myosin heavy chain isozymes appear sequentially in rat muscle development. *Nature* **292:** 805.

WOLOSIN, J.M., A. LYDDIATT, J.O. DOLLY, and E.A. BARNARD. 1980. Stoichiometry of the ligand-binding sites in the acetylcholine-receptor oligomer from muscle and from electric organ. *Eur. J. Biochem.* **109:** 495.

YAFFE, D. and O. SAXEL. 1977. Serial passaging and differentiation of myogenic cells isolated from dystrophic mouse muscle. *Nature* **270:** 725.

ZISKIND-CONHAIM, L. and M.J. DENNIS. 1978. Depolarizing effects of curare on embryonic rat muscles. *Nature* **276:** 622.

Acetylcholine and GABA Receptors: Subunits of Central and Peripheral Receptors and Their Encoding Nucleic Acids

E.A. Barnard,* D. Beeson,*† G. Bilbe,*† D.A. Brown,‡ A. Constanti,‡ B.M. Conti-Tronconi,§ J.O. Dolly,* S.M.J. Dunn,*§ F. Mehraban,* B.M. Richards,† and T.G. Smart‡

**Department of Biochemistry, Imperial College of Science and Technology, London SW7 2AZ, England; †Searle Research and Development, High Wycombe, Buckinghamshire HP12 4HL, England; ‡Department of Pharmacology, The School of Pharmacy, University of London, London WC1N 1AX, England; §Church Laboratory of Chemical Biology, California Institute of Technology, Pasadena, California 91125*

Insights at the molecular level into the complex proteins that constitute the receptors for neurotransmitters and hormones—which should be securely based on an intimate knowledge of their architecture—have hitherto been severely limited, although they have been attained in the cases of many enzymes. An understanding in depth of receptor function has been elusive because of a number of special and major difficulties inherent in the nature and role of these systems:

1. These receptors in nearly every case are present in extremely minute amounts, which may reach down to the level of only a few molecules per active cell.
2. The receptors under consideration here are embedded in the cell membrane, and hence can only be extracted and studied in isolation by replacing the natural lipid environment with a detergent envelope; this frustrates many types of investigations.
3. In many cases, the last-named factor has impeded a purification of the receptor, if the detergent interferes with specific ligand binding or with affinity chromatography.
4. X-ray diffraction analysis, the method that has deciphered the secondary, tertiary, and quaternary structures of other proteins, is not applicable at sufficiently high resolution to membrane receptors, at least in the present state of the art, since (due to factor 2) they have resisted crystallization.
5. The components responsible for the transduction or other transmembrane signaling by the receptor are, in most cases, another set of proteins working in tandem. Thus, many receptors employ cyclic nucleotide second messengers. Present models indicate that these may open channels by interacting with at least four other proteins: adenyl cyclase and guanine nucleotide regulatory protein (Rodbell 1980; Limbird 1981) and further, it is presumed, a specific protein kinase and a protein substrate of phosphorylation (Greengard 1980; Schwartz et al., this volume). Hence, the link between ligand binding and membrane channel opening is not direct but rather complex and is dependent on the steric relationships and mobilities of these components in the membrane. There are other classes of receptors where not even an identification of the components involved has been possible by conventional techniques. Obviously, extra orders of difficulty are introduced into the analysis of a receptor in all of these cases.
6. After solubilization of a receptor, its overall activity (as distinct from ligand binding alone) can, by necessity, only be monitored in some fully reconstituted system in a lipid bilayer. This has been accomplished only in a very few cases, and in most attempts, with only limited fidelity.

For all of these reasons, most progress has been made with a limited class of neurotransmitter receptors that carry a gated ion channel in their own structure. The case par excellence here is, of course, the nicotinic acetylcholine receptor (AChR), which, from a great deal of painstaking evidence accumulated in a number of laboratories, is known to contain an integral small-cation-selective gated channel. The limitation of quantity (factor 1) is far less onerous when fish electric organ is used as a source. Furthermore, the functions of this receptor are exceptionally resistant to nondenaturing detergents (Conti-Tronconi and Raftery 1982). However, some degree of complexity is introduced by the presence of four different types of subunits (of stoichiometry $\alpha_2\beta\gamma\delta$) combined in its "monomer" structure, a multiplicity that is being encountered now with other receptors. For this and related reasons, chemical amino acid sequencing has proven difficult, and published results have not gone beyond 54 aminoterminal residues of any of the chains from *Torpedo* as source (Raftery et al. 1980) or 24–35 residues from other chains (Conti-Tronconi et al. 1982), despite the exploitation of current gas-phase microsequencing technology. Obviously, therefore, for all of the other enormously less abundant receptors noted above, determination of the entire structure by chemical sequencing is not a realistic goal.

It cannot be assumed that the principles deduced from the wealth of studies of the AChR of the electric organ, a highly unusual tissue, apply to all receptors of the direct channel-gating class. Nor can this be assumed even for other nicotinic AChRs, especially those in the brain. The latter, for example, are not blocked by α-bungarotoxin or, in certain cases, α-bungarotoxin binds and blocks but far more reversibly than it does at the electric organ receptor (Oswald and Freeman 1981; Norman et al. 1982). We shall review here, therefore, our recent studies on three other receptors that likewise have the advantage of being in the direct channel-gating class but are, biologically, progressively more distant from the electric organ receptor. These are the nicotinic AChRs of vertebrate skeletal muscle, the equivalent receptor of chick brain, and the receptor for γ-aminobutyric acid (GABA) in mammalian brain (which carries a gated chloride channel). In particular, this paper focuses on initial applications of molecular genetics in analyzing those systems.

The immensely powerful methods of current molecular genetics can, indeed, provide an escape from many of the problems enumerated above. If the cDNAs encoding a receptor are sought and can be obtained, its entire protein sequence is rapidly provided. This can be expected, due to the considerations outlined above, to become the method of choice for structural analyses of receptor proteins. In general, it may be possible to operate at the nucleic acid level for many (though not all) types of study, and hence avoid the major difficulties for receptors in operating at the level of the purified protein. A variety of specific questions on structure-function relations can be framed and answered in terms of translations of the mRNAs for the receptor subunits, as will be illustrated below. Illuminating homologies can be sought, for example, to ask whether structural similarities occur between receptor types having a given type of ion channel in common, or an agonist in common (e.g., nicotinic vs. muscarinic receptors). Furthermore, studies on receptor DNA and mRNA can offer new types of information on receptors, at the level of their genetic control.

Torpedo Receptor mRNA and cDNA

Translation of receptor mRNA in a foreign environment. When we first attempted the cell-free translation of mRNA from *Torpedo marmorata* electric organ, using immunoprecipitation by polyclonal anti-receptor antibodies of the products, followed by SDS-polyacrylamide gel electrophoresis, no bands could ever be obtained at the positions of the four receptor subunits, and there was no α-bungarotoxin-binding activity. The evidence indicated that synthesis of the polypeptides in the reticulocyte lysate (or wheat germ) translation system is insufficient for their assembly into the active receptor oligomer; we pursued instead, therefore, synthesis in a cellular system, the *Xenopus* oocyte. Only inactive smaller subunits, immuno-cross-reactive with the native subunits, were produced in the cell-free system (Sumikawa et al. 1981), in agreement with the findings of Mendez et al. (1980) and Anderson and Blobel (1981) for cell-free translation of this mRNA.

In the *Xenopus* oocyte (with [^{35}S]methionine in its medium), microinjection of the *Torpedo* mRNA led to the synthesis of four mature subunits (^{35}S-labeled and visualized in autoradiographs of SDS gels) which now comigrated with the α-, β-, γ-, and δ-subunits from purified *Torpedo* receptor protein (Sumikawa et al. 1981). Binding of [^{125}I]α-bungarotoxin occurred normally to the product, nicotinic ligands competed specifically with this binding, and the product was in the form of the 9S receptor monomer (Fig. 1). The induced receptor is incorporated in the oocyte membranes and can be demonstrated on the external surface of that cell (Barnard et al. 1982). The synthesis of these subunits on the *Xenopus* polysomes is truly directed by the *Torpedo* mRNA, and not by activation of the *Xenopus* genome, as shown by the characteristic peptide map of the *Torpedo* α-subunit, which is, in fact, given by the corresponding subunit produced in the oocyte by the mRNA injection (Fig. 2A). The peptide map of a given subunit of this receptor is highly species-specific, since the amino acid sequence varies considerably with the

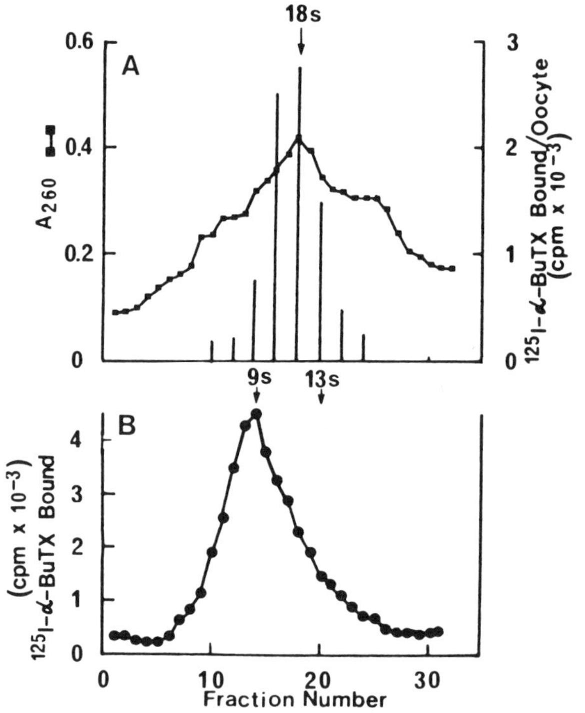

Figure 1. (*A*) RNA fractions containing the mRNAs coding for the AChR. Poly(A)$^+$ mRNA, prepared from *T. marmorata* electric organ, is shown after separation on a 10–31% isokinetic sucrose gradient containing also 10 mM HEPES (pH 7.5), 1 mM EDTA, and 0.1% Li dodecyl sulfate at 2°C. The nucleic acid content is shown by A_{260} nm (■), with human fibroblast rRNA used (in a parallel gradient) as the 18S marker. The vertical bars show the activity of the fractions in directing, in the oocyte translation assay, the synthesis of the product that binds [^{125}I]α-bungarotoxin. (*B*) The oligomeric (9S) form of the AChR is produced in the oocyte translation. After injection of the peak active fraction of mRNA from *A* into 30 oocytes, followed by receptor extraction, the extract was sedimented on a 5–20% sucrose density gradient. The [^{125}I]α-bungarotoxin-binding activity was assayed in every fraction. The arrows indicate the positions to which the 9S (monomer) and the 13S (dimer) forms of the native AChR purified from *Torpedo* sediment in a parallel sucrose gradient (from Sumikawa et al. 1982d). The same profile was obtained when the oocytes used had been injected, instead, with the unfractionated poly(A)$^+$ mRNA from the electric organ. Note that the 9S form assembled in the oocyte corresponds to the pentameric receptor ($\alpha_2\beta\gamma\delta$), and that the dimer of that pentamer (13S form) is not produced.

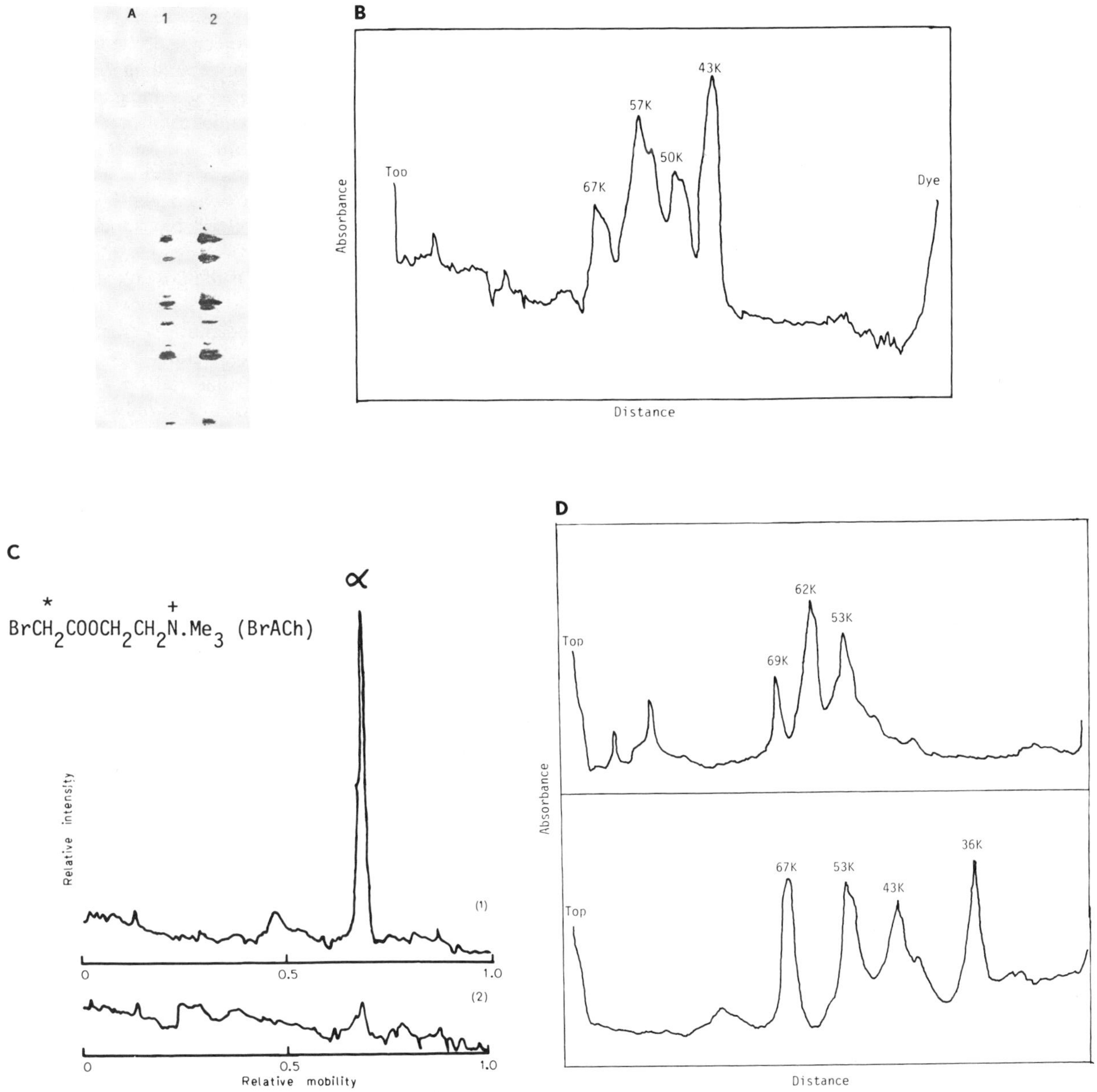

Figure 2. (*A*) Comparison by peptide mapping of the α-subunit produced in *Xenopus* oocytes after injection with *Torpedo* electric organ poly(A)$^+$ mRNA (track *2*) with the α-subunit of the purified AChR from *Torpedo* (track *1*) (Sumikawa et al. 1982c). The former preparation was labeled biosynthetically with [^{35}S]methionine and the latter by ^{125}I iodination on the purified receptor protein. In each case, the four subunits were separated by SDS gel electrophoresis and isolated; limited proteolysis by *S. aureus* V8 protease of the α-subunit produced the peptide maps, using the methods of Sumikawa et al. (1982a). Peptide pattern *1* is highly characteristic of the *Torpedo* α-subunit, and its identity with pattern *2* is evidence for the identity of the *Torpedo* subunits produced by translation in the foreign cell. (*B*) Subunits of the AChR from chicken muscle. The purified (Sumikawa et al. 1982b) receptor from chicken muscle was subjected to electrophoresis in an SDS-polyacrylamide (1-mm-thick) slab gel using the procedures of Sumikawa et al. (1982a), except that 8.75% acrylamide was used. Staining was with Coomassie Brilliant Blue: A scan at 550 nm is shown. M_r values were obtained from a linear plot given by four protein standards run in a parallel track (as shown in *D*, which also shows typical background). The four labeled bands represent (*right to left*) the α-, β-, γ-, and δ-subunits of this receptor. The 43,000 M_r (43K or α) polypeptide, when eluted and analyzed by microsequencing, gave the sequence shown in Table 2. The 50K and 57K polypeptides have been tested similarly: They have given aminoterminal sequences homologous with that of the 43K, confirming that all are receptor subunits. The sequencing has shown that there are four homologous polypeptides in the receptor, in all. (*C*) Affinity labeling of the muscle receptor α-subunit. The affinity reagent, bromoacetylcholine (BrACh), and its position of ^{3}H labeling (*) are shown. Trace *1* shows purified chicken muscle receptor, prelabeled with [^{3}H]BrACh after dithiothreitol reduction, analyzed on a gel as in *B*, but showing (by densitometry of a fluorogram) the radioactivity present. Trace *2* shows a control in which an equal quantity of the same receptor was labeled likewise but in the presence of an excess of α-cobra toxin. The result shows that of the four subunits, only α is labeled in the specific binding reaction. For details, see Sumikawa et al. (1982a). (*D*) Subunits of the chick optic lobe AChR. (*Upper trace*) The receptor, purified by affinity chromatography on α-bungarotoxin-Sepharose and elution (method modified from Norman et al. 1982), was analyzed as in *B*. (*Lower trace*) Four protein standards used for the calibration.

species. The maps obtained after the oocyte translation are fingerprints characteristic of the *Torpedo* subunits.

Further studies have been directed to the question of whether the specific ion channel of the peripheral nicotinic AChR is encoded by the same mRNA. An interesting finding emerged here. When Professor R. Miledi was asked to test the oocytes injected with *Torpedo* electric organ poly(A)$^+$ mRNA for a possible response to ACh, it was discovered that, indeed, the AChR channel had become inserted in the oocyte membrane. Microiontophoretic application of ACh to the voltage-clamped surface of the preinjected oocytes elicited a conductance change characteristic of nicotinic AChRs (Barnard et al. 1982). This response has since been obtained reproducibly. It is (1) specific for nicotinic agonists; (2) absent when the injection is of the other gradient fractions of the *Torpedo* mRNA that do not yield receptor subunits or of fibroblast mRNA; (3) absent when ACh or other agonists are applied, instead, to the inside of the oocyte membrane; and (4) yields an ACh dose-response curve similar to that obtained for receptors on a muscle surface. The response is blocked by *d*-tubocurarine and by α-bungarotoxin (4 min at 1 μg/ml) (Sumikawa et al. 1982c). The effects of modifying the bathing ions, and the reversal potential of -8 mV found for the response (Barnard et al. 1982), indicate that the channel has the original Na^+, K^+ selectivity. The uninjected oocyte has no nicotinic ACh channels of its own, and we must conclude that when this mRNA fraction produces the subunits and the 9S α-bungarotoxin-binding oligomer, the reception ion channel spontaneously forms in the membrane.

The binding of lectins (Sumikawa et al. 1981) showed that the receptor produced in the oocyte is glycosylated as is (Vandlen et al. 1979; Lindstrom et al. 1979) the native *Torpedo* receptor on all of its subunits. Moreover, when tunicamycin (2 μg/ml) was present in the medium during the normal incubation of injected oocytes, production of assembled receptor was totally blocked (K. Sumikawa and E.A. Barnard, unpubl.). The translational capacity to form the AChR resides in a restricted fraction of the poly(A)$^+$ RNA, centered on the 18S position when the latter is separated on a density gradient (Sumikawa et al. 1982d). From all of the above-mentioned results, we conclude that the four mRNAs (see below) coding for the α-, β-, γ-, and δ-subunits in the electric organ can be faithfully translated by the oocyte, processed, transported to the cell membrane, assembled to the full receptor structure with all of its functions present, and inserted with correct orientation in that foreign cell membrane. We presume that all of the information necessary for these steps is contained in the mRNA and subunit sequences. However, a final proof of that will come from experiments in which the pure mRNAs are obtained from their cloned cDNAs and are microinjected into oocytes with no contaminating *Torpedo* messengers present.

Cloning of receptor subunit cDNA. An approach to gene cloning that has facilitated the exploitation of other low-abundance mRNAs is the use of even a short stretch of known amino acid sequence of the gene product for construction of the corresponding DNA fragment as a probe for hybridization screening. We therefore used the aminoterminal sequence of the α-subunit, known from the work of Raftery et al. (1980) and Devillers-Thiery et al. (1979) and identical in *Torpedo californica* and *T. marmorata* (which we used) for the first 20 residues. ^{32}P-labeled oligodeoxyribonucleotides were synthesized corresponding to codons 22–25 of the mature α-subunit (Fig. 3A). Of the possible 12-nucleotide probes, 6 were tested for the ability to act as a primer with the fractionated receptor-specific poly(A)$^+$ mRNA. The products of reverse transcription primed with those components were analyzed on DNA gels, and a positive result was obtained with only one probe, α2, giving a single band corresponding to a segment of 275 nucleotides. This was specific, also, to the mRNA fractions that were active in receptor biosynthesis (Fig. 1B). The sequence of this DNA was obtained, and it showed an encoded amino acid sequence that agreed with the published aminoterminal sequence of the α-subunit, together with a further pre-aminoterminal sequence (Fig. 3B) corresponding to a signal peptide (Sumikawa et al. 1982d). The 12-nucleotide primer itself was insufficiently selective as a hybridization screening probe, but the 275-base fragment of genomic sequence thus recognized was used to construct an authentic 19-nucleotide probe (Fig. 3B). This acted as a very efficient screening probe. A *Torpedo* recombinant DNA library was cloned in the disabled *E. coli* strain DH1, using the plasmid pAT153 (Sumikawa et al. 1982d). Using the 19-base probe, a series of clones were identified that proved, on DNA sequencing, to contain partial and full-length inserts of the α-subunit cDNA. Using those to cover both DNA strands, the α-subunit cDNA sequence was determined. This revealed a single open reading frame, which starts with the codons for a typical signal peptide (of the type [Blobel 1980] that determines membrane insertion) of 24 mainly hydrophobic residues (Fig. 3B). This is followed immediately by codons for the known aminoterminal sequence of the α-subunit. This confirmed the identity of the cDNA cloned.

The entire cDNA sequence to the termination codon codes for a mature subunit of 437 amino acid residues. This sequence differs at five positions from the otherwise identical sequence of the α-subunit of *T. californica* published by Noda et al. (1982) shortly after our cloning was reported (Sumikawa et al. 1982d) (Table 1). These five substitutions are all due to 1-base changes and are regarded as true species differences. This is confirmed at one of them (our Asn-42) where Ser in *T. californica* is present both in the cDNA-cloning sequence and in the chemical sequencing of Raftery et al. (1980) for that species.

The full sequences for the three other subunits determined by cDNA cloning have since been reported, and there is considerable homology among all four, as discussed by others in this volume. (The sequences for the

A

Primers		
	α1	3′ CAT·CTT·GTG·GTG 5′
	α2	3′ CAT·CTC·GTG·GTG 5′
	α3	3′ CAG·CTT·GTG·GTG 5′
	α4	3′ CAG·CTC·GTG·GTG 5′
	α5	3′ CAA·CTT·GTG·GTG 5′
	α6	3′ CAA·CTC·GTG·GTG 5′

B

					−20										−10					
	Met	Ile	Leu	Cys	Ser	Tyr	Trp	His	Val	Gly	Leu	Val	Leu	Leu	Leu	Phe	Ser	Cys	Cys
GAAGA	AUG	AUU	CUG	UGC	AGU	UAU	UGG	CAU	GUC	GGG	UUG	GUG	CUA	CUG	UUA	UUU	UCG	UGU	UGU
	1			10			20			30				40			50		

				−1	1									10					
Gly	Leu	Val	Leu	Gly	Ser	Glu	His	Glu	Thr	Arg	Leu	Val	Ala	Asn	Leu	Leu	Glu	Asn	Tyr
GGU	CUG	GUA	CUA	GGU	UCU	GAA	CAU	GAA	ACA	CGU	UUG	GUU	GCU	AAU	UUA	UUA	GAA	AAU	UAU
							←	─	─	─	─	─	─	→					
60				70			80			90				100			110		

				20										30					
						─	─	─	─										
Asn	Lys	Val	Ile	Arg	Pro	Val	Glu	His	His	Thr	His	Phe	Val	Asp	Ile	Thr	Val	Gly	Leu
AAC	AAG	GUG	AUU	CGU	CCA	GUG	GAG	CAU	CAC	ACC	CAC	UUU	GUA	GAU	AUU	ACA	GUG	GGG	CUA
120				130			140			150				160			170		

				40										50					
Gln	Leu	Ile	Gln	Leu	Ile	Asn	Val	Asp	Glu	Val	Asn	Gln	Ile	Val	Glu	Thr	Asn	Val	Arg
CAG	CUG	AUA	CAA	CUC	AUC	AAU	GUG	GAU	GAA	GUA	AAU	CAA	AUU	GUG	GAA	ACA	AAU	GUG	CGC
180				190			200			210				220			230		

Figure 3. Construction of a probe for screening the clones, and identification of the cloned cDNA for the α-subunit. (*A*) Residues 22–25 of the known aminoterminal sequence of the mature α-subunit of *Torpedo* receptor (shown in *B*), Val Glu His His, were chosen and the corresponding 12-base deoxyribonucleotide primers shown were synthesized in ^{32}P-labeled form. Primer α2 was the only one of these to give a positive result in the priming of the *Torpedo* 18S poly(A)$^+$ mRNA. (*B*) Pre-aminoterminal and aminoterminal regions of the α-subunit and the corresponding mRNA sequence. The numbered amino acid residues (1–54) are those known (Raftery et al. 1980) to be at the amino terminus of *Torpedo californica* receptor mature α-subunit. The bar shows the tetrapeptide sequence used in *A* for the priming. The cDNA obtained from the α2 priming as the sole electrophoretic product (275 nucleotides) was identified by sequence correspondence in the appropriate region of the α-subunit. This led to the choice of the 19-base sequence marked by arrows, for synthesis of a ^{32}P-labeled DNA fragment (CTTTGTGCAAACCAACGAT, 3′-5′) and its use as a hybridization screening probe. The cloned cDNA thus obtained included in its sequence the entire region shown (in mRNA form). The pre-aminoterminal peptide (codons −24 to −1) is a typical signal sequence, in that 11 of its 24 residues are Leu, Ile, or Val, it contains Cys, and it ends (at −1) in a very small residue. (The codon found at position 42 codes for Asn, not the Ser in the known *T. californica* α-subunit aminoterminal segment.) (Data from Sumikawa et al. 1982d.)

Table 1. *Torpedo* α-Subunit Protein Sequence Differences

	T. californica[a]	*T. marmorata*[b]
Number of residues	437	437
Molecular weight	50,116	50,186
Coding sequence differences	1. Ser-42	Asn-42
	AGU	AAU
	2. Gly-230	Val-230
	GGA	GUA
	3. Ile-291	Val-291
	AUU	GUU
	4. Asp-318	Asn-318
	GAU	AAU
	5. Ser-424	Cys-424
	AGC	UGC
Untranslated sequence differences	three 1-base changes in first 85 bases at 5′ end of mRNA	

[a]Data from Noda et al. (1982).

[b]Data from Sumikawa et al. (1982d) and unpublished results, and confirmed by Devillers-Thiery et al. (1983). In the latter study, there is one coding difference from ours, at codon 323, given as Leu (CUU) in contrast to our Val (GUU); the latter, however, is found both by us and (in *T. californica*) by Noda et al. (1982).

α-subunit obtained in our laboratories [Sumikawa et al. 1982d] and by Noda et al. [1982] have recently been further confirmed by Devillers-Thiery et al. [1983].) Hence, the entire primary structure (other than the disulfide bridge locations) of this receptor has been determined with remarkable speed, once the cloning technology was found applicable to it.

Antisera raised against the four *Torpedo* subunits, when renatured from SDS, show considerable cross-reactivity among all four chains (Mehraban et al. 1982). This is to be expected from the degree of homology now seen between the four subunits.

Folded structure and membrane insertion. Where a polypeptide is destined to remain in the membrane and not to be secreted through it, it is assumed that there is additional information to direct this in the form of "stop-transfer" and "insertion" sequences in the mature polypeptide chain (Blobel 1980). These are likely to be one or more hydrophobic transmembrane helices flanked by polar residues. It is interesting to note that the amphibian oocyte recognizes whatever the insertion signal is in this *Torpedo* sequence, since we found that the AChR in the microinjected oocytes was not secreted but accumulated as a firmly bound membrane protein.

The aminoterminal half of the sequence contains the only site (Asn-141) in the subunit for N-glycosylation, a modification that is known (Vandlen et al. 1979) to occur on that subunit. The amino terminus formed by signal peptide cleavage is expected to be on the extracellular side of the membrane (Blobel 1980). An ACh-binding site is present on each of the two α-subunits, from the evidence on the BrACh affinity labeling (Wolosin et al. 1980). This binding site is adjacent to a disulfide (Karlin et al. 1976) and must be located in an extracellular projection on the aminoterminal half.

Attempts to use the amino acid sequence in the secondary structure prediction method of Chou and Fasman (1978) or in assigning transmembrane helices by inspection of hydrophobic regions (methods since used by others [Noda et al. 1982, 1983a; Devillers-Thiery et al. 1983]) gave ambiguous results. The initiation points of the α-helices are uncertain, as are their precise lengths, and, depending on exactly how they are chosen in the hydrophobicity or secondary structure plots, different helices can be equally well predicted. These would present different distributions of the polar residues present, over a given face. In fact, it has been shown (Green and Flanagan 1976; Geisow and Roberts 1980) that the Chou-Fasman procedure, which uses a globular protein data-base, considerably underpredicts the occurrence of α-helices in α-helix-rich membrane proteins. Also, different hydrophobicity indices in use give different profiles here. The aminoterminal half of the subunit sequence shows much greater polarity than the carboxyterminal half, as assessed by all those indices, and very probably is nonhelical. We conclude that it is not safe to predict the topology of such a membrane subunit, nor the alignment of the channel, by the above-mentioned methods.

Subunit sizes. It is interesting to note that the true molecular weight of the *T. marmorata* α-subunit is 50,186, as calculated from the sequence. The apparent molecular weight, deduced from mobilities in SDS-polyacrylamide gels, has long been found to be 39,000–40,000 (Karlin et al. 1976; Raftery et al. 1980). That value is for the fully glycosylated subunit, but cell-free synthesis without glycosylation (but with the signal peptide attached) gives an apparent size that is even smaller (38,000), not greater (Anderson and Blobel 1981). Hence, the anomaly seems to come from its particular sequence. Likewise, the δ-subunit of *T. californica* has a sequence molecular weight (Noda et al. 1983b) of 57,565, about 7000 below that usually found in the gels. These effects in opposite directions demonstrate that errors of up to at least 20% in M_r are liable to be made on either side of the true value, in gel analysis of receptor subunits.

Furthermore, the molecular weights from the four sequences of *T. californica* (Noda et al. 1982, 1983a,b; Claudio et al. 1983) give a sum of 267,751. This compares, for example, with a value of 250,000 ± 7,000 (± S.E.) obtained in solution by a rigorous sedimentation equilibrium method using D_2O balance (Reynolds and Karlin 1978). Values in the range 240,000–270,000 were obtained by several other techniques in which the interference of the detergent is eliminated (for review, see Barnard 1983).

These comparisons, and also model building based on the sequences, would be in error if the primary translation product of the mRNA is cleaved at sites other than at the signal peptide. The occurrence of paired basic residues present toward the carboxyl terminus has suggested (Noda et al. 1982) the possibility that the α-subunit might be further processed to a smaller size, closer to that estimated from gel analysis. This argument can be discarded, however, since some peptides that have been obtained and identified in chemical sequencing studies of the *T. californica* α-subunit (B.M. Conti-Tronconi and M.A. Raftery, pers. comm.) fit our amino acid sequence up to Glu-397.

Muscle AChRs

The receptor from innervated muscle is present at a very low level compared with that in *Torpedo* electric organ—often down to one-thousandth, as in chicken muscle (Sumikawa et al. 1982b). Obviously, cDNA technology can be of real advantage here. Chicken muscle is suitable for such use, since the chick embryo is a good source of muscle-specific mRNAs. Even so, the receptor mRNA is a very minor component of these. AChR from chick muscle (adult innervated, denervated, or embryonic) has been purified to homogeneity and several criteria of purity met (Fig. 4). The AChR species purified from those three types of chicken muscle could not be distinguished physically (Fig. 4B) nor by subunit analysis (Sumikawa et al. 1982a).

In SDS gel electrophoresis, three subunits were clearly seen (after taking full precautions against proteolysis during the isolation) (Sumikawa et al. 1982a);

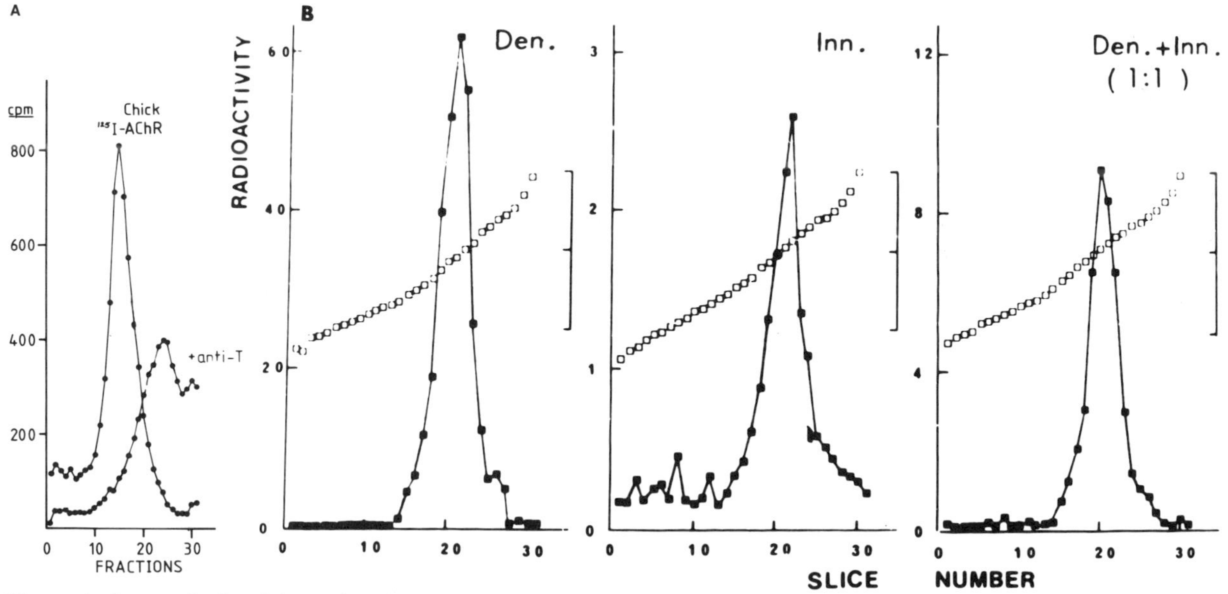

Figure 4. Some criteria of the purity of the isolated AChR from chicken muscle, synaptic and extrasynaptic. (*A*) The free receptor, ^{125}I-labeled ([^{125}I]AChR), sediments on a sucrose density gradient as a single peak of sedimentation constant 9S. Further, on addition of an equivalent amount of a cross-reacting antibody directed against the pure AChR from *T. marmorata* electric organ to the peak fractions from the gradient in a parallel sedimentation, all the ^{125}I present is seen to be moved to a zone lower in the gradient, corresponding to receptor-antibody complexes. Hence, all of the protein in the purified preparation is immunochemically characterized as AChR. (*B*) The complex of the pure, unlabeled receptor with [^{125}I]α-bungarotoxin was isolated and subjected to gel isoelectric focusing (narrow-range, pH 4–6 [□] on right-hand scales). The radioactivity in each slice (■) was measured (in cpm $\times 10^{-2}$, on left-hand scales). (*Left panel*) Extrasynaptic receptor, from denervated (adult) chicken muscle; (*center panel*) synaptic receptor, from innervated (adult) chicken muscle; (*right panel*) a 1:1 mixture of these two forms. Note that a single species (by charge) is present, indistinguishable in the two forms. (Reprinted, with permission, from Sumikawa et al. 1982b.)

with recent improvements in the procedure, however, four are obtained (Fig. 2B), the heaviest band now becoming clear. Similar methods applied to a muscle from the guinea pig yield a receptor with protein purity and four subunits, similar to those shown here for the chicken. The band at 40,000–43,000 M_r is the α-subunit, since this is labeled specifically with [^{3}H]BrACh after disulfide reduction (Fig. 2C).

Amino acid sequence of muscle receptor subunits. The pure chicken muscle receptor was separated in preparative SDS gel electrophoresis, and the individual subunits electroeluted. Automated analysis by gas-phase sequencer carried out at the California Institute of Technology yielded the aminoterminal sequence shown in Table 2. It can be seen that the α-subunit is homologous to the *Torpedo* α-subunit. Other subunits of the chick muscle AChR, for which shorter stretches of sequence have recently been determined, also show homologies with the corresponding *Torpedo* subunit. Also marked in Table 2 are positions in this region that are completely conserved in all the subunits in the AChRs from all of the sources where sequencing has been performed. It can be seen that a degree of homology exists between all of the chains.

mRNA studies of the chick receptor α-subunit. mRNA from embryonic chick muscle was extracted (as for *Torpedo* mRNA) and microinjected into the *Xenopus* oocyte. AChR was again produced, this being shown by (1) the binding of [^{125}I]α-bungarotoxin to the product, and the blockade of that binding by cholinergic ligands; (2) total precipitation of the product complexed with [^{125}I]α-bungarotoxin by anti-AChR antibody; (3) a dependence of the receptor production on mRNA added, which was the same as that (Sumikawa et al. 1981) found for *Torpedo* mRNA. The production in the oocyte of AChR of muscle is illustrated in Figure 5A, where the activity is seen to be restricted to a fraction of the total poly(A)$^+$ mRNA. Similar results have been obtained with the mRNA from denervated cat muscle.

The aminoterminal sequences of the subunits of AChR from chick muscle have been used for an approach to cloning similar to that described above. For example, the α-subunit sequence in Table 2 was used to select a 14-base sequence as marked there, and 16 suitable probes were synthesized. One batch of four of these gave positive priming with gradient-purified mRNA from chick muscle (Fig. 5B). As before, a 275-nucleotide segment was recognized specifically, corresponding to a pre-aminoterminal and aminoterminal stretch of this subunit. The 14-base probe, which was positive for this, could itself act highly selectively in hybridization screening. By cloning as before, but using chick DNA, positive clones were obtained. Determination of the full sequence of subunits of the chick muscle receptor is thus in progress.

Table 2. Aminoterminal Sequences of Muscle and *Torpedo* AChR α-Subunits

```
1                             10                                 20                                 30                                 40
   *  *  *  *  *
S  E  H  E  T  R  L  V  D  D  L  F  R  D  Y  S  K  V  V  R  P  V  E  N  H  R  D  A  V  K  V (N) V  G  L  A (L) I  ?  L  I  N  V  D  E^a
S  E  H  E  T  R  L  V  A  N  L  L  E  N  Y  N  K  V  I  R  P  V  E  H  H  T  H  F  V  D  I  T  V  G  L  Q  L  I  Q  L  I  N  V  D  E^b
```

[a]Chick muscle AChR determined by chemical microsequencing on the purified α-subunit. ? = residue uncertain. ( ) = major residue detected, but other(s) significant. The asterisks mark the set of residues used for construction of oligonucleotide probes based on possible codons (Fig. 5B).

[b]*T. marmorata* (from Raftery et al. 1980). Single underline = residue identical with chick α-subunit residue (29 out of 45). Double underline = residue identical in all of the ACh receptor α-subunit sequences so far known (8 out of 45 positions).

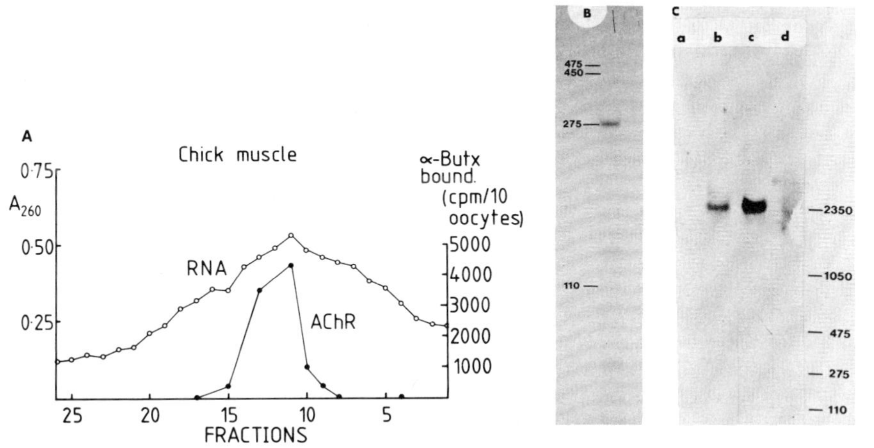

Figure 5. Chicken muscle receptor α-subunit nucleic acids. (*A*) Identification of the mRNA for this subunit. The poly(A)$^+$ mRNA isolated from the thigh muscles of 12-day chick embryos was fractionated (○) on a sucrose gradient (as in Fig. 1A). In the oocyte translation assay, the activity (●) of the fractions in directing the synthesis of the receptor was determined by the specific binding of [^{125}I]α-bungarotoxin that appeared, in batches of 10 oocytes (right-hand scale). The peak position corresponds to 18S. The fractions here are numbered from the bottom of the gradient. (*B*) cDNA segment coding for chicken muscle receptor α-subunit. The codons marked in the aminoterminal region of this subunit in Table 2 were used to construct a set of 14-base deoxyribonucleotides. One of these, (CTTCGTCTTTGGGC, 3′–5′) only, was positive in the priming of the reverse transcription of the poly(A)$^+$ mRNA from embryonic chick muscle (peak fraction 11 from *A*). This yielded a 275-base product, as shown in the DNA gel electrophoresis, with positions of markers indicated. In an adjacent track (not visible because totally negative), the products were run from similar priming of mRNA fraction 6 of the gradient in *A*, corresponding to much larger mRNA. Other procedures were as noted for Fig. 3. (*C*) RNA hybridization with 14-base DNA probes specific for chicken muscle α-subunit and for *Torpedo* α-subunit. The ^{32}P-labeled 14-base probe corresponding to codons 2–6 in each case was hybridized to glyoxal-denatured poly(A)$^+$ mRNA size-fractionated on a 1.4% agarose gel, by standard Northern blotting technique. The chicken α-subunit probe was that used in *B*, and the *Torpedo* probe contained the corresponding nucleotides as shown in the sequence in Fig. 3B, i.e., they differed in three bases. (Track *a*) *T. marmorata* electric organ poly(A)$^+$ mRNA was run (negative with the chick probe); (*b*) 12-day embryonic chicken muscle poly(A)$^+$ mRNA (positive with the chick probe); (*c*) the *Torpedo* sample again (positive with the *Torpedo* probe); (*d*) the chicken sample again (negative with the *Torpedo* probe). The positions marked were calibrated using denatured *Hin*dIII-restricted phage PM2 DNA; RNA sizes are slightly smaller than these. It is seen that these probes are specific for their corresponding receptor mRNAs and that these mRNAs are of equal size in the two sources.

Brain Nicotinic AChRs

As noted above, mammalian brain contains nicotinic AChRs that fail to show physiological blockade by α-bungarotoxin or related toxins; in the optic lobe of lower vertebrates, however, they do show such blockade (Oswald and Freeman 1981). We have therefore purified the receptor from chick optic lobe, by methods adapted (to the very low concentrations involved and the reversibility of α-bungarotoxin binding there) from those used with muscle (Norman et al. 1982). The same receptor from the rest of the chick brain (where there is no electrophysiological evidence that α-bungarotoxin blocks) was purified similarly. The products satisfy criteria of purity, behaving as a single component (after ^{125}I iodination) in narrow-range isoelectric focusing or density-gradient centrifugation (Norman et al. 1982). The pharmacological properties of the product, and the immunological cross-reactivity observed with the muscle AChR (discussed below), provided evidence that a brain nicotinic receptor has been isolated from the optic lobe.

In the AChR of chick optic lobe, three subunits can be demonstrated in SDS gels, of apparent M_r 53,000, 60,000, and about 70,000 (Fig. 2D). The 53,000 M_r subunit can be labeled specifically with [^{3}H]BrACh after disulfide reduction (Norman et al. 1982), indicating it is the equivalent of the muscle α-subunit. The higher subunit weights and sedimentation constant (in Triton X-100 medium) suggest a somewhat different structure for this receptor. It may be that a fourth subunit is present as in muscle and *Torpedo* AChRs, but is poorly resolved or poorly stained or labeled in the gels. Separation of the subunits and sequencing thereof is in progress. Clearly, one can proceed eventually to the structure determination via cloning as before. The chick embryonic optic lobe is a suitable source of mRNA for this task, despite the fact that levels of the brain receptor are lower than in muscle. Comparison of the coding sequences and flanking genomic sequences for the subunits of peripheral and brain AChRs should be of interest with respect to the neurobiology of brain receptors in general.

Relatedness of brain and muscle receptors and the use of monoclonal antibodies. The question of possible cross-hybridization of muscle and brain DNA sequences in the AChR genes, as a short-cut in coding the latter, also merits investigation. The choice of the same species, chick, for both is aimed to facilitate this. Relatedness of the two receptor proteins is evident from their antibody cross-reactivity. Polyclonal antibodies to muscle AChR generally show little cross-reactivity to brain receptors. However, one out of ten antibodies to muscle receptor (in this case cat) showed definite cross-reactivity with chick optic lobe receptor (Norman et al. 1982).

Hybridomas secreting monoclonal antibodies provide more opportunity to find such a probe and to prepare it in adequate quantities. Apart from the information yielded directly, such an antibody could then be used also in translation assays. More recently, therefore, we raised monoclonal antibodies to the pure AChR of chick muscle (Mehraban et al. 1983) and have been able to apply them as probes of this kind. Thus: (1) Purified AChR from chick optic lobe was specifically precipitated by two of the monoclonal antibodies so far characterized (Table 3; Fig. 6A). The same extent of reactivity was seen with a given monoclonal antibody toward both the optic lobe receptor and the similar component from chick brain cortex. (2) In purified preparations from CNS, the total protein present (as shown by radiolabeling with ^{125}I) was immunoprecipitated in parallel with the immunoprecipitation of the [^{125}I]α-bungarotoxin–receptor complex (Fig. 6A). (3) Ultracentrifugation after complexing of these CNS proteins with a reactive monoclonal antibody showed that they could be shifted to a lower position on the gradient (Fig. 6B), in the same way that the muscle AChR (Fig. 4A) can be. The size of the immune complexes with antibody 7B2 corresponds to a 1:1 attachment of this monoclonal antibody to the AChR (see Fig. 6B), for both the brain and the muscle AChRs. (4) The subunits ($M_r = 53,000$, 60,000, and 70,000, discussed above) detectable in the purified AChR from chick optic lobe could be specifically and simultaneously precipitated in the reaction of monoclonal antibody 7B2 with the native AChR. In the immunoprecipitate, these subunits were visualized by means of SDS gel electrophoresis and autoradiography.

These observations confirm that these two types of brain "AChR" are related structurally to the peripheral AChR. Observations 2, 3, and 4 confirm that all of the protein isolated in the purification from optic lobe is AChR. Observation 4 shows that the subunits identified therein are all part of the intact receptor molecule.

Other properties of these monoclonal antibodies are themselves of interest here. Very varied cross-reactivities are displayed (Table 3) toward the AChRs from diverse sources by the different monoclonal antibodies. The one exception is for the optic lobe receptor, compared with the "receptor" from the rest of the chick brain. These two components always behaved identically in antibody reactions. Another feature of interest is that all of the mice inoculated with the cloned cells producing such an anti-muscle-receptor antibody developed experimental autoimmune myasthenia gravis (Mehraban et al. 1983). This indicates that the mouse AChR (which is the origin of this response) possesses an antigenic determinant in common with the AChR from chick muscle, which is exposed extracellularly.

We conclude that the optic lobe receptor and the component from the rest of the brain which is purified similarly on immobilized α-bungarotoxin are so far indistinguishable. There is no difference discernible in their protein properties, subunit structures, binding parameters for a series of ligands (Barnard and Dolly 1982; Norman et al. 1982), and immunological cross-reactivities with a series of antibodies. They are both labeled in the same way, and on the 54,000 M_r subunit, with BrACh (Norman et al. 1982). They may be truly identical; any apparent difference in blockade could then arise from different cellular locations with respect to electrophysiological testing. Alternatively, they may be very similar proteins but with a subtle difference in their toxin-binding sites.

Table 3. Cross-reactivities with Various AChRs' Proteins of Monoclonal Antibodies Raised against the AChR from Chicken Muscle

	Titer (%)[b]			
Receptor antigen[a]	2A3	5B2	7B2	7B3
Chicken muscle				
denervated	100	100	100	100
innervated	100	126	100	200
embryonic	92	100	85	100
Chicken				
optic lobe[c]	0	0	11	15
brain cortex[d]	0	0	11	12
Cat muscle (denervated)[c]	0	4	25	12
T. marmorata electric organ	0	5	5	34

[a]The AChR in a crude extract was used, except where noted, as the source, prepared in the presence of a set of protease inhibitors as in Sumikawa et al. (1982b). In all cases the [^{125}I]α-bungarotoxin receptor complex was prepared and used as the antigen.

[b]Expressed as a percentage of the apparent titer against the homologous antigen, for each of the four monoclonal antibodies listed, all raised against the AChR purified from chicken denervated muscle. The values quoted are the mean of three or more experiments with duplicate samples; % errors were approximately 5%. Apparent titers were measured as described by Mehraban et al. (1983).

[c]The pure receptor was used as the antigen here.

[d]α-Bungarotoxin-binding component (pure) was used as the antigen here.

Brain GABA Receptors

The GABA$_A$ receptor of vertebrate brain is known from electrophysiological analysis (for review, see Study and Barker 1981; Haefely and Polc 1983) to open a chloride channel at a major class of GABA-mediated inhibitory synapses of the brain. The GABA$_A$ receptor is distinct from the GABA$_B$ receptor: The latter class has been defined by Hill and Bowery (1981) as another form of receptor, binding GABA but (unlike GABA$_A$) requiring Ca^{++} for this, sensitive to baclofen, and insensitive to bicuculline. The GABA$_B$ receptors have a different mechanism of action and a different distribution in the CNS. The GABA$_A$ receptors themselves probably comprise more than one subtype (Olsen 1981; Haefely and Polc 1983); we shall be concerned here with the major subtype of the GABA$_A$ receptor.

At these GABA$_A$ receptors in the neuronal membrane, the response to stimulation is extremely fast, typical rapidly gated channels are opened under the in-

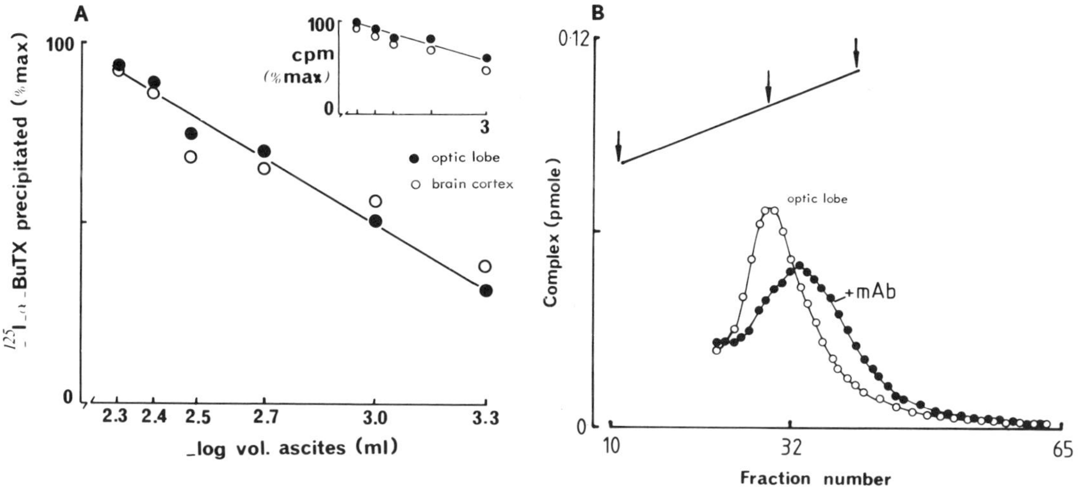

Figure 6. Reaction of chick brain receptors with monoclonal antibody directed against muscle AChR. (*A*) The receptor purified from 1-day chicks, in the form of its complex with [^{125}I]α-bungarotoxin, was incubated (8 nM) with dilutions of the ascites fluid containing monoclonal antibody 7B2 (see Table 3) overnight at 4°C. The product was immunoprecipitated in a procedure using Protein A. The percentage of the amount precipitable at the maximum, on the linear ordinate, is plotted against a logarithmic scale representing decreasing antibody additions. This titration was performed with the optic lobe AChR (○) or the α-bungarotoxin-binding "receptor" from the rest of the chick brain (●) as the antigen. (*Insert*) A similar plot was constructed, but using another preparation of each antigen, in the form of the free receptors, the protein being ^{125}I-labeled. In both cases, it is seen that the behaviors of the optic lobe receptor and of the brain "receptor" are identical. They give the same high affinity in the binding to this monoclonal antibody, which was raised against the authentic AChR from chick muscle. (*B*) AChR purified from the chick optic lobe (1 pmole) in Triton X-100 solution was complexed with [^{125}I]α-bungarotoxin. An aliquot was treated (4 hr at 23°C) with a 600-fold excess of monoclonal antibody 7B2 (see Table 3), raised against chick muscle AChR. Sedimentation on a 5–20% sucrose gradient is shown. It is seen that all of the complex is moved to a position toward the heavier (high fraction numbers) side which corresponds roughly to a 1:1 receptor-IgG complex. Unlike the other reactive antibodies, excess of this antibody did not lead to aggregates appearing near the bottom of the gradient (fraction 65); hence, it is concluded that this particular monoclonal antibody binds to the AChR in a 1:1 mode, i.e., its two binding sites recognize two copies of a determinant on one receptor molecule (i.e., on the two α-subunits or in two extremely homologous regions on two different subunits). This same behavior was shown by this monoclonal antibody, also, with the "receptor" from the rest of the chick brain, and likewise with its homologous muscle receptor antigen. Hence, the double determinant pattern is common to the chick muscle, chick optic lobe, and chick brain AChR oligomers. This reinforces the conclusion that the muscle and brain AChRs are related proteins. Arrows show the linear calibration with three marker proteins. (The fractions before 20 contained some free toxin only, which is not shown.)

fluence of GABA, and no evidence for involvement of any second messenger system is obtainable (Olsen 1981). All of the properties of this response (for review, see Krogsgaard-Laarsen et al. 1979; Haefely and Polc 1983) indicate that it involves the direct opening of a chloride channel by the binding of agonist to the receptor. Furthermore, the benzodiazepine (BZ) drugs (tranquilizers) potentiate the electrophysiological response to GABA (Brown and Constanti 1978; Study and Barker 1981; Haefely and Polc 1983) and are believed to exert therapeutic effects by means of their binding to the GABA$_A$ receptor (Olsen 1981). Binding of those drugs at nanomolar levels has been demonstrated to occur at a specific set of sites in brain membranes, and—highly significantly—this BZ binding interacts with GABA agonist binding, with the affinity for the BZ being increased severalfold when the GABA sites are all occupied (for review, see Olsen 1981). From these and a variety of other findings that testify to an interaction of the two sites, it is generally considered now that the GABA$_A$ receptor contains in its total structure a BZ-binding unit (for references, see Olsen 1981; Chang and Barnard 1982).

Other sites therein can also be defined pharmacologically. Barbiturates (Barb) bind to a distinct site associated with the GABA$_A$ receptor; this binding enhances both BZ binding and GABA agonist binding (Leeb-Lundberg et al. 1980; Olsen 1981). Barbiturates also enhance the electrophysiological response to GABA application (Brown and Constanti 1978; Study and Baker 1981; Haefely and Polc 1983). Specific blockers of the chloride channel are also known: Picrotoxinin binds selectively, does not affect any of the aforementioned ligand sites, but blocks channel opening (Olsen 1981; Haefely and Polc 1983). The GABA$_A$ receptor therefore contains at least three interacting sites for different ligand classes, plus a chloride channel structure that a picrotoxinin-binding site can regulate. We can represent it as a complex (G–BZ–Barb)Ch$_{pic}$, where G = GABA, indicating an unknown relationship of those three ligand sites in the structure, and the simultaneous presence within the complex of a chloride channel with its regulatory picrotoxinin site (Ch$_{pic}$).

Molecular size. When the GABA$_A$ receptor is in the membrane, the complex has a molecular weight of about 220,000, as determined by radiation inactivation target size analysis; moreover, the binding of BZ and the binding of a GABA$_A$ agonist were lost at precisely the same dosage rate (Chang and Barnard 1982). This supports the concept of a single protein structure in the membrane containing both sites. When solubilized in Triton X-100, the properties were again consistent with those of a unit with an approximate molecular weight of 220,000. Furthermore, the binding capacity for [^{3}H] muscimol, a $GABA_A$ agonist, migrated either in sucrose gradient centrifugation or on a gel filtration column as a single peak coinciding exactly with the BZ-binding capacities (Chang and Barnard 1982). All of these observations support the existence of the single complex noted above.

Purification of the $GABA_A$ receptor complex. Complete purification of this receptor has recently been obtained for the first time by E. Sigel and A. Stephenson at our Imperial College laboratory (Sigel et al. 1982, 1983). Bovine brain cortex was the source, the receptor being extracted from it by sodium deoxycholate. An immobilized BZ ligand was used for affinity chromatography, with specific displacement by another BZ compound, followed by ion-exchange chromatography. The product, 1800-fold enriched in binding sites, is a protein carrying both the BZ site and the $GABA_A$ agonist site, both retaining their high affinity. In SDS gels, the protein yields two major bands of apparent M_r 53,000 (α) and 57,000 (β; Fig. 7). The α-subunit becomes strongly labeled by photoaffinity reaction with the photosensitive BZ ligand [^{3}H]flunitrazepam, both in the brain membranes and in the purified product, suggesting that the BZ site resides, at least in part, on this α-subunit (Sigel et al. 1983). This subunit has been isolated from the pure preparation run in SDS gels and its aminoterminal sequence determined by microsequencing (B.M. Conti-Tronconi et al., unpubl.). The application of sequence data for probe construction and gene cloning of the α-subunit of the $GABA_A$ receptor is in progress.

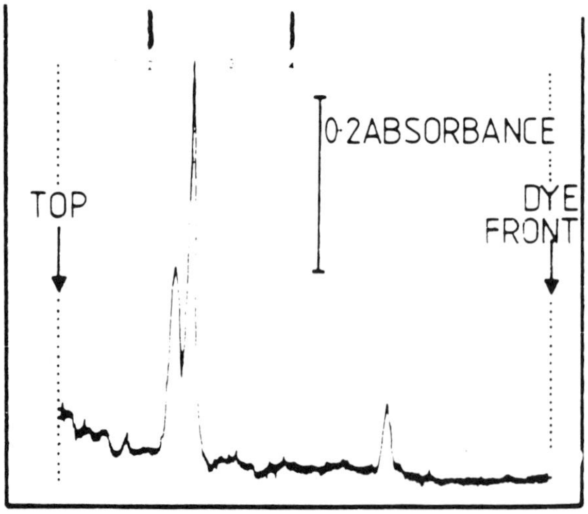

Figure 7. Detectable subunits of the GABA receptor. The purified GABA/BZ receptor was analyzed similarly. Two receptor subunits, at apparent M_r 53,000 and 57,000, are seen. (The small peak at 22,000 M_r is from a trace of soybean trypsin inhibitor, added earlier as an anti-protease.) The upper (parallel) track shows the positions of the five standards used, with two of them (serum albumin and carbonic anhydrase) seen. (Reprinted, with permission, from Sigel et al. 1983.)

The purification of the GABA-binding protein on a BZ affinity column, along with all of the other observations reviewed here, confirms that the GABA-binding and BZ-binding components are together in a single complex. Specific ligand binding of the picrotoxinin-binding structure (Ch_{pic}) has been demonstrated to be retained in this receptor isolated in detergent solution (E. Sigel and E.A. Barnard, unpubl.), and reconstitution experiments in lipid bilayers have likewise resulted in a product in which the pure receptor retains all of the interacting sites described above.

To screen for a cDNA that encodes the α-subunit, it is not necessary that the receptor preparation used as the starting point have all of the true receptor subunits present and identified in it. It could be, for example, that additional subunits exist but are not seen in the gel analysis of the pure receptor protein, for one reason or another. However, so long as at least one subunit is isolated, this can be used either to design an oligonucleotide probe or to raise antibodies for applications, as discussed below, in the cDNA cloning procedures.

Channel formation induced by mRNA. Identification of mRNA coding for the proteins of the $GABA_A$ receptor complex was again achieved in the oocyte cellular translation system. Initially, the chick optic lobe has been used as the source, because this receptor complex accumulates in vivo faster in the developing chick than in the developing rat. This difference is associated with the known development in the rodent of many brain functions and components only over the first 2 weeks of postnatal life, whereas the newly hatched chick is more active and precocious. In fact, the linked GABA- and BZ-binding components were found to develop particularly rapidly in the optic lobe of the chick just before hatching (Smart et al. 1983).

Poly(A)$^+$ mRNA was prepared from chick optic lobe and was microinjected into *Xenopus* oocytes (40 ng per cell). After incubation for a day, the oocyte surface was shown to have developed functional GABA receptors (which do not exist in these oocytes prior to this treatment). Application of GABA produced an increase in Cl^- conductance (Fig. 8A): This response is dose-dependent (compare control responses to GABA at 5 μM and 40 μM in Fig. 8A). A smooth dose-conductance curve has been obtained with the half-maximal response at 15 μM GABA (Smart et al. 1983), in accord with the sensitivity to GABA of brain neurons in situ. The response can be blocked by physiologically active concentrations of the $GABA_A$ antagonist bicuculline or of the GABA receptor chloride channel blocker, picrotoxinin (Fig. 8A). Moreover, the response to GABA that is induced in the oocytes by the mRNA shows the specific phenomena of potentiation by BZ drugs and by barbiturates (Fig. 8B). These effects are all reversible, as the responses after wash-out (Fig. 8A,B) show. The pharmacology of the GABA receptor induced in the

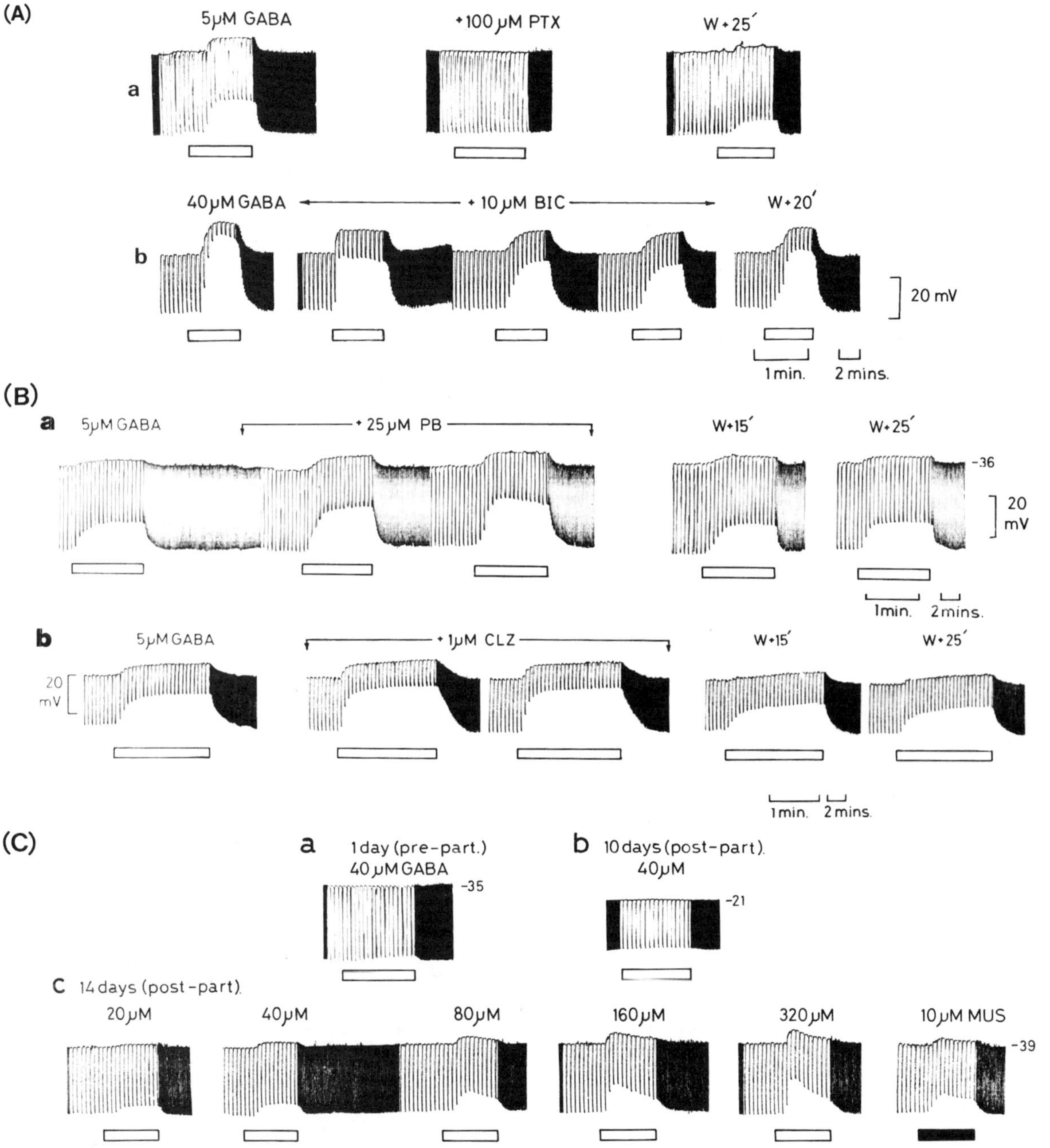

Figure 8. Responses of GABA receptors induced in the *Xenopus* oocyte cell membrane by the injection of brain mRNA. (*A*) Antagonism by picrotoxinin (PTX) or bicuculline (BIC) of the GABA-dependent membrane conductance change in oocytes 1 day after the injection of poly(A)$^+$ mRNA from the optic lobe of 19-day chick embryos. Downward deflections are hyperpolarizing electrotonic potentials evoked by injecting 35-nA current pulses (1 sec width; 0.2 Hz). Upward movements in the baseline represent membrane depolarization. Chart speed was reduced for drug offsets. (*a*) Control response to 5 µM GABA (applied in the superfused amphibian Ringer at room temperature) after 15 min in 100 µM PTX; further, after washing for 25 min in normal Ringer. (*b*) (Different cell) Control response to 40 µM GABA; three successive responses after 10 min in 10 µM BIC; after a 20-min wash. Open bars indicate periods of GABA application. Note the partial recovery obtained by washing, after either antagonist. (*B*) Enhancement of the GABA-induced conductance change by pentobarbitone (PB) or the benzodiazepine drug chlorazepate (CLZ) in oocytes injected (as in *A*) with chick optic lobe mRNA. (*a*) Response to 5 µM GABA in normal Ringer; two successive responses to 25 µM PB after 10 min (note the onset of desensitization in the later conductance change); two recovery responses after washing for 15 min and 25 min, respectively. (*b*) (Different cell) Response to GABA (5 µM) in control solution; two successive responses in 1 µM CLZ after 15 min (note a small enhancement); two recovery responses at 15 min and 25 min. The recovery from the enhancement due to CLZ or PB is always complete. (*C*) The GABA-dependent conductance change in oocytes injected with mRNA extracted from the whole rat brain during development. (*a*) From rats 1 day preparturition; (*b*) 10 days postparturition; (*C*) 14 days postparturition, with dose-dependent responses to the GABA concentrations shown and to muscimol (MUS). Note that only the oocytes injected with the 14-day brain mRNA exhibit a significant response to GABA. Note also that muscimol is more potent than GABA, as in the rat brain, and that the conductance change desensitizes at 80 µM GABA concentration and at 10 µM muscimol. (Scales as in *B,a*.)

oocyte has recently been found (Smart et al. 1983) to be as for the neuronal $GABA_A$ receptor. Baclofen (1 mM, in the presence of Ca^{++}) has no effect on it. All of the observations cited confirm that a functional receptor of the $GABA_A$ class is formed and becomes inserted in the oocyte membrane when the exogenous mRNA is translated. The entire complex (G–BZ–Barb)Ch_{pic}, is obviously assembled correctly in the membrane following the translation of the mRNA fraction used.

These observations have recently been extended to the $GABA_A$ receptor of mammalian brain (the source used for the subunit analyses). Rat whole brain was used as the source of poly(A)$^+$ mRNA, at different developmental stages. Only by 14 days after birth was the receptor mRNA activity fully present, and no activity was ever detectable in the embryonic brain (Fig. 8C). This is in accord with the ontogeny of the BZ-linked $GABA_A$ receptor complex in the rat brain (Olsen 1981). The formation of the $GABA_A$ receptor complex in the oocyte membrane showed the same properties when rat brain mRNA was used as when chick optic lobe mRNA was used. The increasing responses, up to saturation, to increasing doses of GABA are illustrated in Figure 8C. As also shown there, the $GABA_A$ agonist muscimol is about four times more potent than GABA, in agreement with the behavior seen at GABA-mediated neurons of the mammalian brain (Brown and Constanti 1978).

Hence, in two very different and complex receptor structures—the nicotinic AChR and the $GABA_A$ receptor (in each case from diverse sources)—the mRNAs are able to specify in the oocyte the synthesis of all of the component parts and the correct assembly in the membrane of the complex with its ion channel.

CONCLUSIONS

Requirements for Cloning of Receptor Genes

The major difficulties in the analysis of receptors, enumerated at the beginning of this paper, should to a considerable degree be overcome by exploiting the DNAs coding for receptor subunits and also by exploiting their mRNAs. The approaches at those two levels that have been adopted in our own initial studies are designed to be applicable to receptors generally, i.e., all of the other receptors of nervous systems that do not enjoy the unique tissue selectivity and concentration of the nicotinic AChR in the electroplax of *Torpedo*—a receptor source of which we can observe sadly, with Hamlet, "We shall not look upon his like again."

In this approach, there are three initial requirements:

1. Purification of a small quantity of the receptor protein and separation of at least some of its subunits.
2. The amino acid sequence of a short segment of one or more of those subunits, determined at present by aminoterminal microsequencing, applied to the subunit eluted after SDS-polyacrylamide electrophoretic separation.
3. An assay system for mRNA that monitors the receptor functions. Preferably this should detect binding functions *and* overall activity (e.g., channel opening). This is the value of the oocyte system.

Point 1 is needed for point 2. It is also needed for obtaining antibodies that would be used to aid the assays, perhaps also for immunoadsorptive purification of more receptor—especially employing anti-subunit antibodies—and perhaps for use in gene expression screening. The raising of monoclonal antibodies against the receptor in impure form will in many (but not all) cases be a rapid route to achieving requirement 1. Another good reason for pursuing requirement 1 as far as possible is that the information obtained, on the size of the oligomer and the number and types of subunits and their approximate sizes, is invaluable in searching for the correct gene products and in interpreting what is found.

Point 2, amino acid sequence information, is needed for identifying the clones unequivocally. It is also used, of course, for constructing primers or probes for screening. Examples of the power of the latter approach to receptor gene cloning are seen in the work reviewed above and in the remarkable progress of Dr. Numa and his colleagues (Noda et al. 1982, 1983a,b) on various subunits of nicotinic AChRs.

The above requirements need not apply where a given receptor gene, once obtained, is sought in other species. At least if the nicotinic AChR is a general precedent, a given subunit of a receptor is likely to be strongly conserved in structure. The muscle AChRs of different vertebrates generally strongly cross-react immunologically (see an example in Table 2; also Lennon and Lambert 1980; Souroujon et al. 1982). Likewise, their α-subunits are all in the molecular-weight range (apparent value, in SDS gels) of 40,000–44,000 and strong homology is expected. Cross-hybridization will be the method of choice here for obtaining the cDNAs for the corresponding subunits of the receptor from other species, once one muscle receptor cDNA is obtained. More generally, this possibility constitutes one of the important advantages of the cDNA approach over conventional protein isolation and analysis. The isolation of the receptor protein from the first source is a dead end so far as other species go; one must start the isolation anew, and the problems noted earlier are encountered with equal force for every new source. Not so at the DNA level. One of the laws of molecular biology that neurobiologists must constantly recall is that like nucleic acids hybridize with each other, but like proteins do not. Therefore, we should remain at the nucleic acid level as far as possible in comparing sequences from different species—one cDNA being used to isolate the next, by cross-hybridization. At least for each subunit type, say of the AChR, we should proceed by steps of cross-hybridization between species, and conceivably between ACh-binding proteins of all types.

The Oocyte Translation and Assembly System

If requirement 3 in the list above is met, this can greatly facilitate the molecular genetic analyses for receptors at two different levels. First, it can identify an mRNA population active in producing receptor subunits, for use in DNA cloning. We have seen that translation to form the subunits of a receptor protein in a *cell-free* assay system is unlikely to produce the active receptor. In the case of relatively unknown receptors, therefore, the production of the receptor chains in conventional cell-free translation may be difficult to recognize unequivocally.

Second, the oocyte system can verify in such a case that the mRNAs produced lead not only to the structure in the receptor that binds a ligand, but also to the full structure that maintains the channel gating or other effector function. We can test here for the coding for all of the components that contribute to an assembly of functions, as in the GABA receptor example. These binding activities for one ligand might arise by translation to give only a part of the receptor structure.

Related to this second level of mRNA analysis, the oocyte system is one in which a range of different types of expression experiments can readily be conducted. Amenable to analysis, there will be the product of an individual subunit mRNA, or of mRNAs for selected incomplete sets of subunits, or of specifically altered mRNAs derived through DNA mutagenesis experiments or by hybridization to synthetic probes. The evidence so far is encouraging for the prospect for such defined manipulations based on the sequence and on structure predictions. That evidence suggests that they may be pursued in the oocyte system, at the levels of binding functions, oligomer assembly, plasma membrane insertion of the receptor, channel function, and regulation of channel function.

ACKNOWLEDGMENT

The work at Imperial College and at the School of Pharmacy (London) was supported by the Medical Research Council.

REFERENCES

Anderson, D.J. and G. Blobel. 1981. *In vitro* glycosylation and membrane insertion of the four subunits of *Torpedo* acetylcholine receptor. *Proc. Natl. Acad. Sci.* **78:** 5598.

Barnard, E.A. 1983. Molecular weight of receptors in solution and in the membrane. In *Cell membrane receptors* (ed. E.H. Reid). Pitman, Bath, England. (In press)

Barnard, E.A. and J.O. Dolly. 1982. Peripheral and central nicotinic ACh receptors—How similar are they? *Trends Neurosci.* **5:** 325.

Barnard, E.A., R. Miledi, and K. Sumikawa. 1982. Translation of exogenous messenger RNA coding for nicotinic acetylcholine receptors produces functional receptors in *Xenopus* oocytes. *Proc. R. Soc. Lond. B* **215:** 241.

Blobel, G. 1980. Intracellular protein topogenesis. *Proc. Natl. Acad. Sci.* **77:** 1496.

Brown, D.A. and A. Constanti. 1978. Interaction of pentobarbitone and γ-aminobutyric acid in mammalian sympathetic ganglion cells. *Brit. J. Pharmacol.* **63:** 217.

Chang, L.-R. and E.A. Barnard. 1982. The benzodiazepine/GABA receptors complex: Molecular weight in brain synaptic membranes and in solution. *J. Neurochem.* **39:** 1507.

Chou, P.Y. and G.D. Fasman. 1978. Prediction of the secondary structure of proteins from their amino acid sequence. *Adv. Enzymol.* **47:** 45.

Claudio, T., M. Ballivet, J. Patrick, and S. Heinemann. 1983. Nucleotide and deduced amino acid sequences of *Torpedo californica* acetylcholine receptor γ subunit. *Proc. Natl. Acad. Sci.* **80:** 1111.

Conti-Tronconi, B.M. and M.A. Raftery. 1982. The nicotinic cholinergic receptor. *Annu. Rev. Biochem.* **51:** 491.

Conti-Tronconi, B.M., C.M. Gotti, M.W. Hunkapiller, and M.A. Raftery. 1982. Mammalian muscle acetylcholine receptor: A supramolecular structure formed by four related proteins. *Science* **218:** 1227.

Devillers-Thiery, A., J.-P. Changeux, P. Parout, and D. Strosberg. 1979. The amino-terminal sequence of the 40,000 molecular weight subunit of the acetylcholine receptor protein from *Torpedo marmorata. FEBS Lett.* **104:** 99.

Devillers-Thiery, A., J. Giraudat, M. Bentaboulet, and J.-P. Changeux. 1983. Complete mRNA coding sequence of the acetylcholine receptor binding α-subunit of *Torpedo marmorata* acetylcholine receptor: A model for the transmembrane organization of the polypeptide chain. *Proc. Natl. Acad. Sci.* **80:** 2067.

Geisow, M.J. and R.D.B. Roberts. 1980. Amino acid preferences for secondary structure vary with protein class. *Int. J. Biol. Macromol.* **2:** 388.

Green, N.M. and M.T. Flanagan. 1976. Prediction and conformation of membrane proteins from the sequence of amino acids. *Biochem. J.* **153:** 729.

Greengard, P. 1980. Intracellular signals in the brain. *Harvey Lect.* **75:** 277.

Haefely, W. and P. Polc. 1983. Electrophysiological studies on the interaction of anxiolytic drugs with GABAergic mechanisms. In *Anxiolytics: Neurochemical, behavioural and clinical perspectives* (ed. J.B. Malick et al.), p. 113. Raven Press, New York.

Hill, D.R. and N.G. Bowery. 1981. ^{3}H-Baclofen and ^{3}H-GABA bind to bicuculline-insensitive GABA sites in rat brain. *Nature* **290:** 149.

Karlin, A., C.L. Weill, M.G. McNamee, and R. Valderrama. 1976. Facets of the structures of acetylcholine receptors from *Electrophorus* and *Torpedo. Cold Spring Harbor Symp. Quant. Biol.* **40:** 203.

Krogsgaard-Laarsen, P., J. Scheel-Kruger, and H. Kofod, eds. 1979. *GABA neurotransmitters.* Munksgaard, Copenhagen.

Leeb-Lundberg, F., A. Snowman, and R.W. Olsen. 1980. Barbiturate receptors are coupled to benzodiazepine receptors. *Proc. Natl. Acad. Sci.* **77:** 7468.

Lennon, V.A. and E.H. Lambert. 1980. Myasthenia gravis induced by monoclonal antibodies to acetylcholine receptors. *Nature* **285:** 238.

Limbird, L.E. 1981. Activation and attenuation of adenylate cyclase. *Biochem. J.* **145:** 1.

Lindstrom, J., J. Merlie, and G. Yogeeswaran. 1979. Biochemical properties of acetylcholine receptor subunits from *Torpedo californica. Biochemistry* **18:** 4465.

Mehraban, F., J.O. Dolly, and E.A. Barnard. 1982. Antigen similarities between the subunits of acetylcholine receptors from *Torpedo marmorata. FEBS Lett.* **141:** 1.

Mehraban, F., J.T. Kemshead, and J.O. Dolly. 1983. Properties of monoclonal antibodies to nicotinic acetylcholine receptors from chick muscle. *Eur. J. Biochem.* (in press).

Mendez, B., P. Valenzuela, J.A. Martial, and J.D. Baxter. 1980. Cell-free synthesis of acetylcholine receptor polypeptides. *Science* **209:** 695.

Noda, M., H. Takahashi, T. Tanabe, M. Toyosato, Y. Furutani, T. Hirose, M. Asai, S. Inayama, T. Miyata, and S. Numa. 1982. Primary structure of α-subunit precursor of *Torpedo californica* acetylcholine receptor deduced from cDNA sequence. *Nature* **299:** 793.

NODA, M., H. TAKAHASHI, T. TANABE, M. TOYOSATO, S. KIKYOTANI, Y. FURUTANI, T. HIROSE, H. TAKASHIMA, S. INAYAMA, T. MIYATA, and S. NUMA. 1983a. Structural homology of *Torpedo californica* acetylcholine receptor subunits. *Nature* **302:** 528.

NODA, M., H. TAKAHASHI, T. TANABE, M. TOYOSATO, S. KIKYOTANI, T. HIROSE, M. ASAI, H. TAKASHIMA, S. INAYAMA, T. MIYATA, and S. NUMA. 1983b. Primary structures of β- and δ-subunit precursors of *Torpedo californica* deduced from cDNA sequences. *Nature* **301:** 251.

NORMAN, R.I., F. MEHRABAN, E.A. BARNARD, and J.O. DOLLY. 1982. Nicotinic acetylcholine receptor from chick optic lobe. *Proc. Natl. Acad. Sci.* **79:** 1321.

OLSEN, R.W. 1981. GABA-benzodiazepine-barbiturate receptor interactions. *J. Neurochem.* **37:** 1.

OSWALD, R.E. and J.A. FREEMAN. 1981. Alpha-bungarotoxin binding and central nervous system nicotinic acetylcholine receptors. *Neuroscience* **6:** 1.

RAFTERY, M.A., M.W. HUNKAPILLER, C.D. STRADER, and L.E. HOOD. 1980. Acetylcholine receptor: Complex of homologous subunits. *Science* **208:** 1454.

REYNOLDS, J.A. and A. KARLIN. 1978. Molecular weight in detergent solution of acetylcholine receptor from *Torpedo californica*. *Biochemistry* **17:** 2035.

RODBELL, M. 1980. The role of hormone receptors and GTP-regulatory proteins in membrane transduction. *Nature* **284:** 17.

SIGEL, E., C. MAMALAKI, and E.A. BARNARD. 1982. Isolation of a GABA receptor from bovine brain using a benzodiazepine affinity column. *FEBS Lett.* **147:** 45.

SIGEL, E., F.A. STEPHENSON, C. MAMALAKI, and E.A. BARNARD. 1983. A γ-aminobutyric acid/benzodiazepine receptor complex of bovine cerebral cortex. Purification and partial characterization. *J. Biol. Chem.* **258:** 6965.

SMART, T.G., A. CONSTANTI, G. BILBE, D.A. BROWN, and E.A. BARNARD. 1983. Synthesis of the functional chick brain benzodiazepine–barbiturate–GABA receptor complex in mRNA-injected *Xenopus* oocytes. *Neurosci. Lett.* **40:** 55.

SOUROUJON, M.C., D. MOCHLY-ROSEN, D. BARTFELD, and S. FUCHS. 1982. Experimental autoimmune receptor. In *Muscle development: Molecular and cellular control* (ed. M.L. Pearson and H.F. Epstein), p. 527. Cold Spring Harbor Laboratory, Cold Spring Harbor, New York.

STUDY, R.E. and J.L. BARKER. 1981. Diazepam and (−)-pentobarbital: Fluctuation analysis reveals different mechanisms for potentiation of γ-aminobutyric acid responses in cultured central neurons. *Proc. Natl. Acad. Sci.* **78:** 7180.

SUMIKAWA, K., E.A. BARNARD, and J.O. DOLLY. 1982a. Similarity of acetylcholine receptors of denervated, innervated and embryonic chicken muscles. 2. Subunit compositions. *Eur. J. Biochem.* **126:** 473.

SUMIKAWA, K., F. MEHRABAN, J.O. DOLLY, and E.A. BARNARD. 1982b. Similarity of acetylcholine receptors of denervated, innervated and embryonic chicken muscles. 1. Molecular species and their purification. *Eur. J. Biochem.* **126:** 465.

SUMIKAWA, K., R. MILEDI, M. HOUGHTON, and E.A. BARNARD. 1982c. A study of the mRNA and genes coding for the nicotinic acetylcholine receptor. In *Cell surface receptors* (ed. P. Strange). MacMillan, London.

SUMIKAWA, K., M. HOUGHTON, J.S. EMTAGE, B.M. RICHARDS, and E.A. BARNARD. 1981. Active multi-subunit ACh receptor assembled by translation of heterologous mRNA in *Xenopus* oocytes. *Nature* **292:** 862.

SUMIKAWA, K., M. HOUGHTON, J.C. SMITH, L. BELL, B.M. RICHARDS, and E.A. BARNARD. 1982d. The molecular cloning and characterisation of cDNA coding for the α subunit of the acetylcholine receptor. *Nucleic Acids Res.* **10:** 5809.

VANDLEN, R.L., W.C.-S. WU, J.C. EISENACH, and M.A. RAFTERY. 1979. Studies of the composition of purified *Torpedo californica* acetylcholine receptor and its subunits. *Biochemistry* **18:** 1845.

WOLOSIN, J.M., A. LYDDIATT, J.O. DOLLY, and E.A. BARNARD. 1980. Stoichiometry of the ligand-binding sites in the acetylcholine-receptor oligomer from muscle and from electric organ. *Eur. J. Biochem.* **109:** 495.

Molecular Events in the Synthesis and Assembly of a Nicotinic Acetylcholine Receptor

D.J. ANDERSON* AND G. BLOBEL†

*Institute of Cancer Research, Columbia University College of Physicians and Surgeons, New York, New York 10032; †Laboratory of Cell Biology, The Rockefeller University, New York, New York 10021

The acetylcholine receptor (AChR) has proven the most accessible model system for structure-function studies of a transmitter-activated ion channel (Karlin 1980; Conti-Tronconi and Raftery 1982). However, it also presents a good opportunity to study molecular events in the assembly of a multi-subunit integral membrane protein. Moreover, the synthesis of AChR is regulated during development, both by trophic factors (Jessel et al. 1979) and by activity (Brockes and Hall 1975). To understand the mechanistic bases of these regulatory influences, it would be helpful to know all the molecular events and intermediates in AChR biogenesis. We can presume that these events encompass gene transcription, mRNA processing and transport out of the nucleus, translation, subunit assembly, and intracellular transport of the subunits. The following discussion will concern what we have learned about the stages including and following mRNA translation.

To approach this problem we have used cell-free systems originally developed to study the transfer of secretory proteins across the rough endoplasmic reticulum (RER) membrane (Blobel and Dobberstein 1975). These systems consist of a standard mRNA in vitro translation system (from wheat germ or reticulocyte lysate), supplemented with canine pancreas rough microsomal membranes. These membranes are endowed with a set of specific proteins that can effect transfer of nascent secretory proteins across the lipid bilayer. First in the case of a viral spike glycoprotein, this system was also shown to reproduce accurately the synthesis, glycosylation, and asymmetric transmembrane insertion of a membrane protein. It has subsequently been successfully extended to eukaryotic membrane proteins, such as the H-2 antigen (Dobberstein et al. 1979), opsin (Goldman and Blobel 1981), and erythrocyte band III (Braell and Lodish 1982). Thus, this system is firmly established as a legitimate alternative to pulse-chase experiments in living cells as a means for studying membrane protein biogenesis. Moreover, it has the advantage that, because it constitutes a biosynthetic "dead end" (in the RER), certain intermediates can be trapped and studied in stable form in vitro, which would in living cells be transient, low-abundance species inaccessible for biochemical analysis.

To apply this approach, one needs a source of mRNA coding for the AChR subunits and antibodies with which to purify the translation products from the reaction mixtures by indirect immunoprecipitation (Anderson and Blobel 1983b). In the case of the AChR, the electric organs of rays such as *Torpedo* have long been known to contain copious quantities of AChR protein and were therefore apparently the tissue of choice for extracting AChR mRNA. Unexpectedly, this tissue in fact contains very small amounts of mRNA (Anderson and Blobel 1983c). Apparently, the high steady-state levels of AChR in electric organ result from slow turnover of the protein. Although the tissue yields only small amounts of mRNA, this mRNA is enriched in AChR messages, each subunit's mRNA comprising about 0.5% of the total (Mendez et al. 1980).

Subunit-specific antibodies were obtained by isolating each subunit of the AChR from preparative SDS-polyacrylamide gels and injecting them into rats (Claudio and Raftery 1977; Lindstrom et al. 1978). As these antibodies had been raised against SDS-denatured material, we were able to demonstrate their subunit specificity using AChR in which the chains had been separated by SDS dissociation (Anderson and Blobel 1981). Furthermore, we could analyze the in vitro translation products by immunoprecipitation following SDS denaturation, a procedure that has proven helpful in analyzing the primary translation products of integral membrane proteins (Anderson and Blobel 1983b).

Translation of AChR mRNA and Initial Membrane Insertion

Analysis of the AChR subunits' primary translation products indicated that each polypeptide chain is translated from a separate mRNA. Thus, antibodies to each subunit immunoprecipitated a distinct translation product (Figs. 1 and 2). These products could each be labeled by formylated initiator methionyl-tRNA (Anderson and Blobel 1981), indicating that each subunit's mRNA has a distinct 5′ AUG initiation codon. More recently, cloned cDNA probes have been used in Northern blot experiments to identify directly the separate mRNAs coding for each subunit (Noda et al. 1982, 1983; Sumikawa et al. 1982; Claudio et al. 1983; Devillers-Thiery et al. 1983). The synthesis of the AChR subunits can be contrasted with that of another multi-subunit receptor, the insulin receptor. In that case, the two subunits are synthesized as a polyprotein that is later cleaved into two distinct chains (Jacobs et al. 1983). These two alternative mechanisms for the synthesis of multi-subunit integral membrane proteins

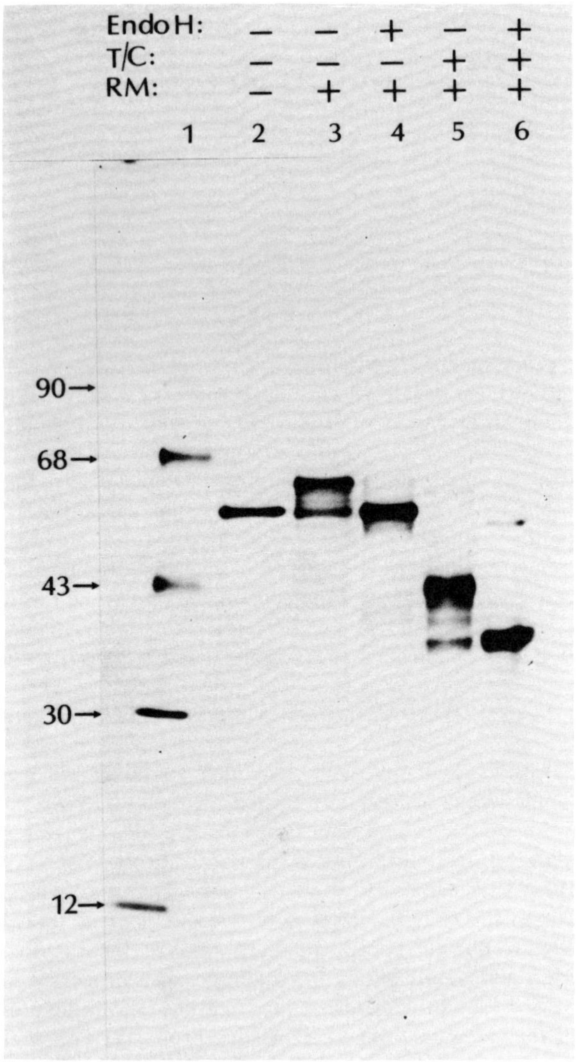

Figure 1. In vitro synthesis, glycosylation, and membrane insertion of AChR-δ subunit. *Torpedo californica* total cellular RNA was translated in a wheat germ cell-free system; AChR-δ was purified by immunoprecipitation with anti-δ antibodies, and displayed by SDS-polyacrylamide gel electrophoresis and autoradiography. (Endo H, +) Immunoprecipitates were digested with Endo H before electrophoresis (Anderson and Blobel 1983c); (Endo H, −) samples were incubated in parallel without enzyme; (T/C) posttranslational incubation with trypsin and chymotrypsin; (RM) dog pancreas rough microsomes were included in the translation at 2 A_{280} units/ml final concentration. (Lane *1*) ^{14}C-labeled molecular weight markers. Note that the smallest protease-resistant fragment of AChR-δ in lane *5* comigrates with deglycosylated material in lane *6* and is therefore probably derived from a nonglycosylated, membrane-integrated form of AChR-δ.

have important consequences when considering the problem of quaternary assembly, as we shall discuss shortly.

The insertion of each subunit into the RER membrane occurs cotranslationally; that is, it is coupled to nascent polypeptide chain elongation. Insertion results in an asymmetric transmembrane orientation for each AChR subunit, an orientation in which part of each subunit is exposed to the cytoplasmic milieu and the remainder is either embedded in the membrane or exposed to the extracellular space. This has been demonstrated by so-called "protease-protection" experiments. Protease digestion of AChR subunits synthesized in the absence of microsomal membranes results in complete degradation of the chains, even if the chains are incubated with membranes after translation is complete (Anderson and Blobel 1981). In contrast, when microsomal membranes are included from the beginning of the translation incubation, posttranslational proteolysis results in only partial digestion of each AChR subunit (see, for example, Fig. 1, lane 5). This partial digestion is due to protection of most of the chain by the permeability barrier of the microsomal membrane, as evidenced by the fact that digestion is complete when the membranes are disrupted by detergent at the time of proteolysis. Thus, the heterologous cell-free system is able to reproduce the initial integration of the AChR subunits into the lipid bilayer of the RER. That this integration is correct is shown by the agreement between our data and those obtained for mature AChR in *Torpedo* plasma membranes (Strader and Raftery 1980; Wennogle and Changeux 1980; Froehner 1981; Wennogle et al. 1981).

Glycosylation of the AChR Subunits

Concomitant with their integration into the membrane, the four AChR subunits become core-glycosylated (Figs. 1 and 2). This glycosylation represents the first stage in a multistep process and consists of the attachment of preformed "core" oligosaccharide groups (Robbins et al. 1977) en bloc to Asn residues of the growing polypeptide chain (Rothman and Lodish 1977). Thus, core-glycosylation is a quantal event, and the AChR subunits differ in the number of these quanta that they receive. The δ-subunit, for example, receives three core groups (Fig. 1, lane 3) that are susceptible to enzymatic removal by β-*N*-acetylglucosaminidase H (Endo H) (Fig. 1, lane 4), whereas the α- and β-subunits receive only one core group each (Fig. 2, lanes RM+). The number of these core oligosaccharide groups appears to be proportional to the increase in apparent molecular weight exhibited by the subunits when translated in the presence of microsomal membranes (compare Figs. 1 and 2, lanes RM,−, and RM,+, Endo H,−). These data are in good agreement with the cDNA-deduced amino acid sequences for these subunits, which reveal three Asn-X-Ser(Thr) sites in the δ-subunit and one each in the α- and β-subunits (Noda et al. 1982, 1983). (The Asn-X-Ser[Thr] sequence constitutes the recognition site for the enzymes catalyzing core glycosylation.) The case of the γ-subunit is less clear-cut. The increase in the apparent M_r of the γ-subunit due to glycosylation appears commensurate with the attachment of three or four core oligosaccharides, by comparison to the δ-, α-, and β-subunits (Anderson and Blobel 1981). Indeed, the cDNA sequence of the γ-subunit shows four possible glycosylation sites (Noda

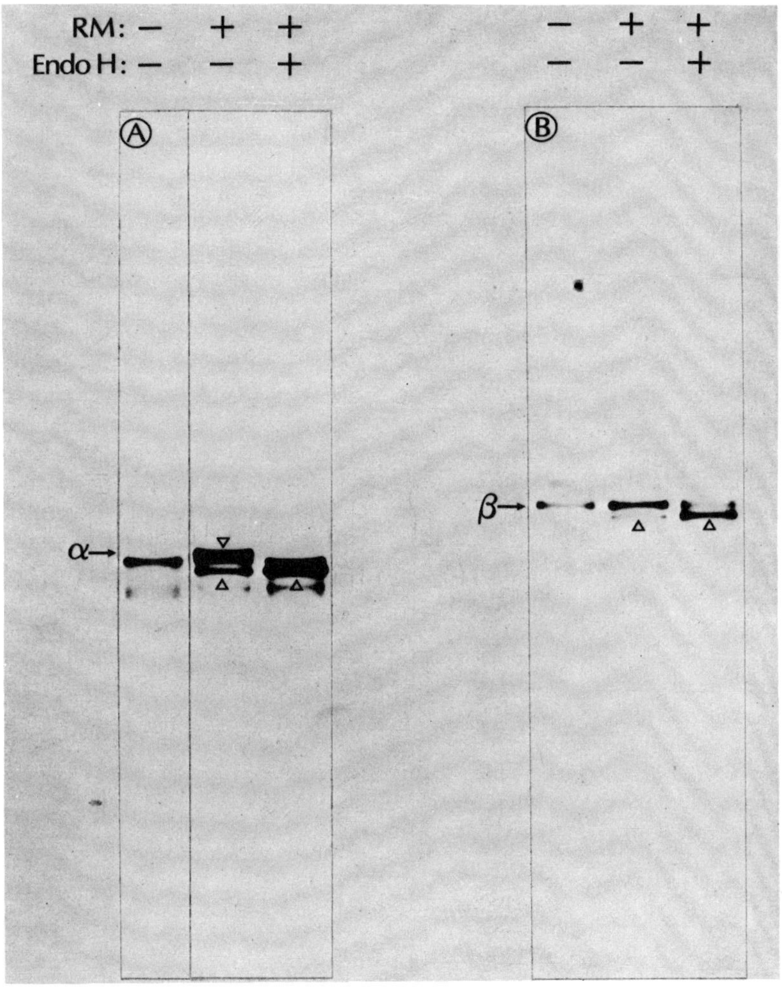

Figure 2. Glycosylation and signal peptidase cleavage of the α- and β-subunits. Immunoprecipitations were performed using either anti-α (*A*) or anti-β (*B*) antisera. (RM) Rough microsomes were either included (+) or omitted (−) from the translation. Endo H incubations (+ or −) were as per Fig. 1. The position of the glycosylated form of each subunit is marked by a Greek character and horizontal arrow. Upward arrowheads indicate the signal peptidase-cleaved, but nonglycosylated form of each subunit. Note that the intensity of these bands increases after Endo H digestion.

et al. 1983). However, some secondary structure predictions for the γ-subunit place only two of these four sites extracellularly, where they could be accessible to the glycosyltransferases (Noda et al. 1983). It remains to be determined whether the error lies in the secondary structure model or in the inferred number of core sugar groups in the γ-subunit.

As expected (from the intracellular location of the core glycosylating enzymes), the carbohydrate residues of the subunits are located on the protease-resistant domain of each chain. Thus, the protease-resistant 44-kD fragment of the δ-subunit (Fig. 1, lane 5) is Endo-H-sensitive (Fig. 1, lane 6). Interestingly, all four subunits' protease-resistant domains are converted to bands of about the same size (ca. 37 kD) by enzymatic removal of the carbohydrate (D.J. Anderson, unpubl.). As the intact subunits differ in their apparent M_r, this observation of similar-sized membrane-protected domains suggests that the mass differences between the subunits can be accounted for by the relative sizes of their protease-susceptible cytoplasmic domains, as suggested by others (Conti-Tronconi and Raftery 1982). This finding is consistent with secondary structure predictions for the four subunits based on cDNA sequence analysis (Noda et al. 1983). It is further supported by the direct identification of different-sized cytoplasmic domains, released from some of the subunits by proteolysis and recovered by monoclonal antibody precipitation (Anderson et al. 1983).

Role of the Signal Peptide in Membrane Insertion of the AChR Subunits

Many of the single-subunit membrane proteins studied previously are synthesized with short, amino-terminal peptide extensions termed signal sequences (Lingappa et al. 1978). These extensions are removed by a specific endoprotease shortly after insertion of the protein into the membrane (Blobel and Dobberstein 1975). They are shared by all proteins that undergo cotranslational transfer across the RER membrane, including secretory, lysosomal, and prokaryotic secretory proteins (Blobel 1980). Without exception, those membrane proteins whose signal peptide is proteolytically removed have their amino termini exposed on the outer surface of the membrane. There is also a class of membrane proteins that do not undergo proteolytic removal of a signal peptide (Mostov et al. 1981; Braell and Lodish 1982). Several proteins in this class have their amino termini exposed on the cytoplasmic side of the membrane (Braell and Lodish 1982).

To which class of proteins do the AChR subunits belong? We determined by direct radiosequence analysis that AChR-δ contained a 21-amino-acid-long signal peptide that was proteolytically removed following membrane insertion (Anderson et al. 1982). Consistent with this finding, the amino terminus of the δ-subunit was found to be protected by the microsomal membrane from proteolysis, and thus is most likely extracellularly exposed (Wennogle et al. 1981; Anderson et al. 1982). Our partial sequence for the δ signal peptide was subsequently confirmed by the cDNA sequence (Noda et al. 1983), and cDNA clones revealed signal sequences for the other three subunits as well (Sumikawa et al. 1982; Noda et al. 1982; Claudio et al. 1983; Devillers-Thiery et al. 1983). The secondary structure models based on these cDNA sequences also predict an extracellular disposition for the subunits' amino termini, in agreement with our data for the δ-subunit.

What role does the signal peptide play in the insertion of the AChR subunits into the RER membrane? Early on (Milstein et al. 1972; Blobel and Dobberstein 1975), it was proposed that signal sequences contain the information that "targets" nascent proteins specifically to the RER. It was also suggested that the signal peptide (perhaps by virtue of its hydrophobicity) initiated transfer of the growing polypeptide across the membrane, once a functional ribosome-membrane junction had formed (Blobel 1980). Recently, an 11S ribonucleoprotein complex has been purified that appears to decode the information contained in secretory protein signal peptides (Walter and Blobel 1980, 1982). Termed SRP (signal recognition particle), this complex consists of six protein subunits together with one molecule of 7S RNA, and forms a high-affinity "bridge" between polysomes whose nascent proteins contain signal sequences, and an integral membrane protein receptor in the RER (Gilmore et al. 1982; Meyer et al. 1982). This arrangement somehow permits the initiation of translocation of the nascent chain across the lipid bilayer.

SRP is likewise required for the insertion of the AChR subunits into the RER membrane. This was expected based on the structural similarities between signal sequences from secretory and membrane proteins and from functional similarities suggested by competition experiments (Lingappa et al. 1978). Thus, rough microsomes (RM) from which SRP has been extracted by a 0.5 M KOAc wash (K-RM) (Warren and Dobberstein 1978; Walter and Blobel 1980) are incapable of integrating or glycosylating AChR-δ, for example (Fig. 3, lane 3). When purified SRP is added back to the system at the beginning of translation, the integration and glycosylation of AChR-δ are restored (Fig. 3, lanes 4 and 5). Interestingly, in the absence of K-RM, SRP inhibited the synthesis of AChR-δ (Fig. 3, lane 2). In the case of a secretory protein, prolactin, this inhibition was shown to be due to a site-specific elongation arrest that occurs as soon as the signal sequence emerges from the ribosome (Walter and Blobel 1981), but which is released by the SRP receptor (Gilmore et al. 1982). We may assume that a similar arrest accounts for the observed inhibition of AChR-δ synthesis. Thus, the insertion of the AChR subunits into the membrane utilizes the same protein machinery that initiates the complete transfer of secretory proteins across the membrane.

In the case of membrane proteins, however, transfer of the nascent polypeptide across the membrane is interrupted, presumably due to the emergence of appro-

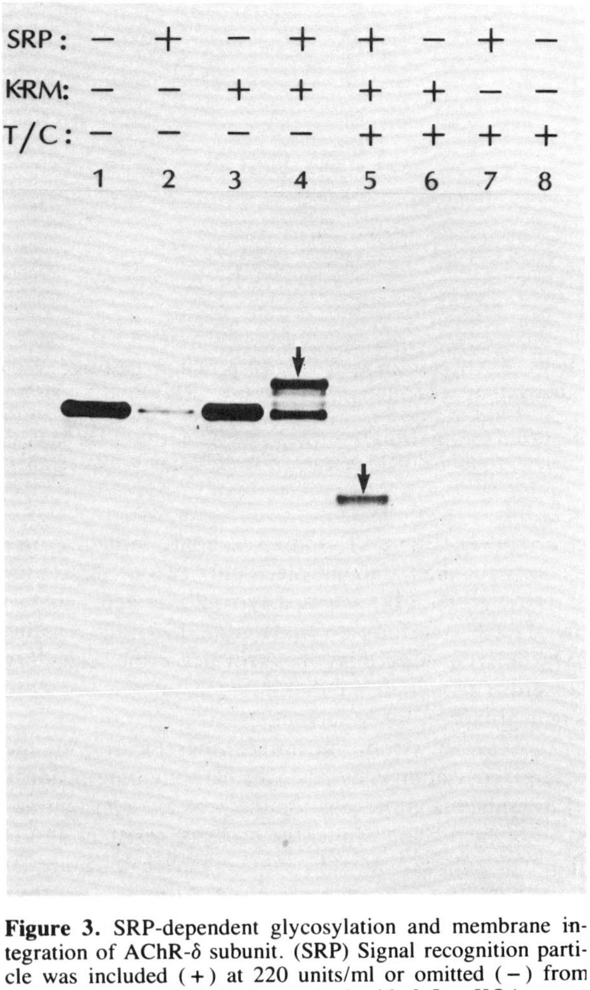

Figure 3. SRP-dependent glycosylation and membrane integration of AChR-δ subunit. (SRP) Signal recognition particle was included (+) at 220 units/ml or omitted (−) from the reaction; (K-RM) RM extracted with 0.5 M KOAc were included (+) or omitted (−); (T/C) trypsin/chymotrypsin digestion as per Fig. 1. Arrows in lanes *4* and *5* indicate the fully glycosylated form of AChR-δ and its corresponding membrane-protected fragment, respectively.

priate hydrophobic "membrane anchor" sequences or "stop-transfer" sequences from the ribosome (Blobel 1980; Boeke and Model 1982). The case of the AChR subunits is even more complex, in that secondary structure models predict four transmembrane segments for each subunit (Claudio et al. 1983; Devillers-Thiery et al. 1983; Noda et al. 1983). It is generally agreed upon by theorists that such multiple transmembrane segments are likely to be inserted into the membrane in pairs, as hairpin loops (Blobel 1980; Engleman and Steitz 1981). However, it is not yet known whether the insertion of such loops requires an internal signal sequence that is acted upon by the SRP-SRP receptor machinery. If the exclusive function of the signal sequence is that of targeting the protein to the correct intracellular membrane, then one per protein should be sufficient. The integration of subsequent helices into the membrane would then occur by insertion sequences (Blobel 1980), such as those that anchor the protein cytochrome b_5 in the membrane (Rogers and Strittmatter 1975) and that apparently partition spontaneously into the lipid bilayer by virtue of their hydrophobicity.

Posttranslational Assembly of AChR

The AChR subunits are not assembled with one another into a functional complex immediately upon insertion into the RER membrane. Rather, multimeric assembly of AChR appears to be a lengthy, posttranslational process. Thus, the in-vitro-synthesized and membrane-integrated α-subunit does not exhibit high-affinity binding to α-bungarotoxin (Anderson and Blobel 1981). Nor is any inter-subunit interaction observed in this system (see below). As it is currently not feasible to reconstruct the movement of proteins from the RER to the Golgi apparatus in vitro, we turned to a cell-culture system to investigate the formation of the toxin-binding site. Initially, we identified the 40-kD (α) subunit of the AChR in cultured chick embryonic myotubes by a two-step affinity purification procedure involving first cobra toxin-Sepharose and then concanavalin A-Sepharose (Fig. 4, lanes 1 and 2). The 40-kD polypeptide enriched by this procedure was eliminated from the preparation by including competing cholinergic ligands in the toxin-Sepharose step (Fig. 4, lane 3), and was specifically immunoprecipitated by antisera to the *Torpedo* AChR (Fig. 4, lane 4).

We used this two-step procedure to purify AChR from cultures that were pulsed for 5 minutes with [^{35}S]methionine and then chased for various periods of time in the presence of cold methionine and puromycin (to inhibit further protein synthesis). In agreement with our in vitro data, no toxin-binding 40-kD material could be recovered from cells immediately following the pulse (Fig. 5B, lane 1). With increasing times of chase, however, increasing amounts of the 40-kD subunit bound to the toxin-Sepharose column (Fig. 5B, lanes 2–4), although no further protein synthesis occurred during this period (Fig. 5A, lanes 1–4). Thus, we may conclude from these data that the formation of

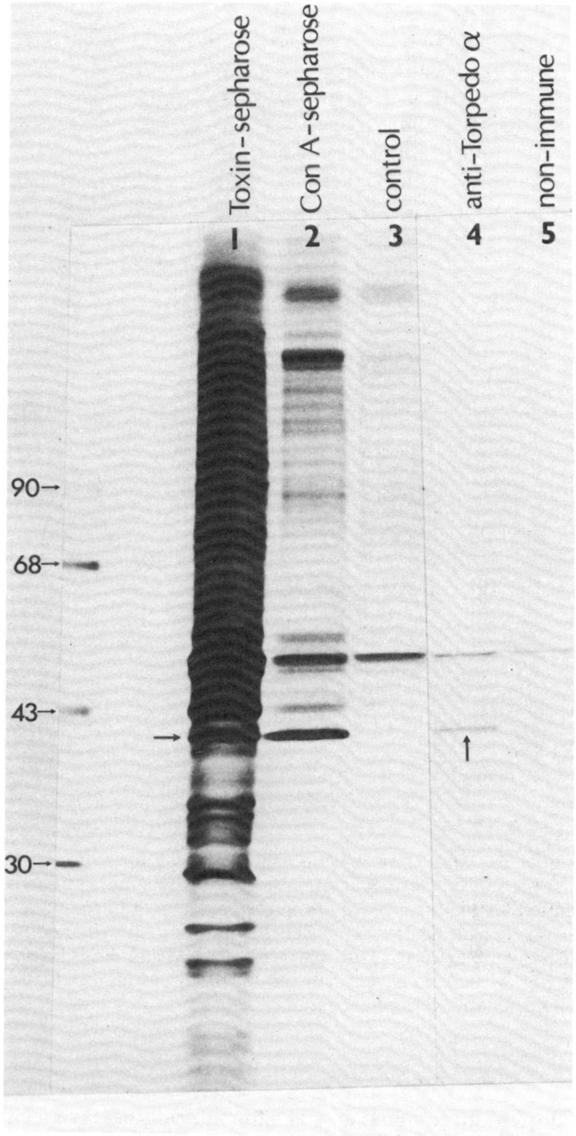

Figure 4. Identification and partial purification of an M_r 40,000 subunit of embryonic chick muscle AChR. (Lane *1*) Material retained by the cobra toxin-Sepharose affinity resin. (Lane *2*) An aliquot of the material in lane *1* was adsorbed to Con A-Sepharose, and bound material was eluted directly into SDS gel sample buffer. (Lane *3*) As in lane *2* except that 125 nM free α-bungarotoxin was present during the initial incubation with cobra toxin-Sepharose. (Lanes *4* and *5*) Aliquots of the material in lane *2* were immunoprecipitated with anti-*Torpedo* α-subunit antiserum (lane *4*) or nonimmune serum (lane *5*).

the high-affinity toxin-binding site is a posttranslational event that occurs in the absence of ongoing protein synthesis, over a period of about 30 minutes. A similar observation has been made by Merlie and Sebbane (1981) using a tumor cell line BC3H-1.

We attempted to relate temporally the formation of the toxin binding-site to some carbohydrate modifications that occur in the Golgi apparatus. Specifically, the core oligosaccharides of some membrane proteins have been shown to become resistant to Endo H cleavage as

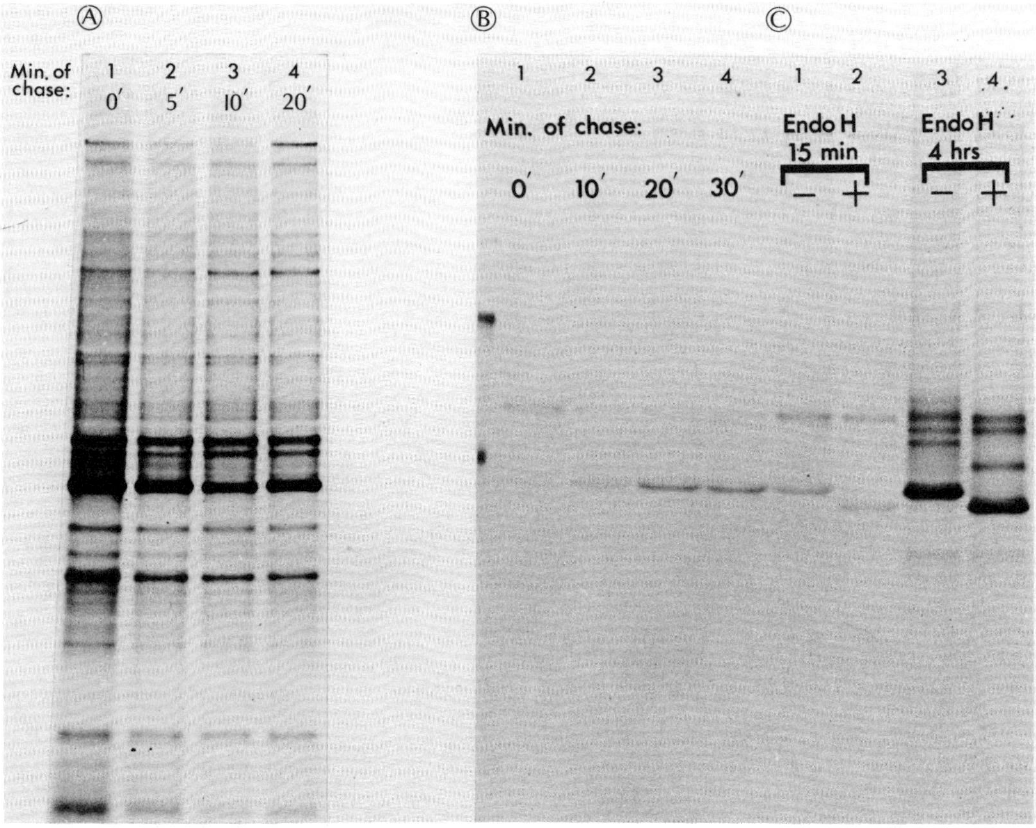

Figure 5. Pulse-chase kinetic analysis of the formation of the toxin-binding site in the 40-kD polypeptide. (*A*) Demonstration that protein synthesis is instantaneously stopped by addition of puromycin. Cells were pulsed for 5 min with [^{35}S]methionine and chased for the indicated times in the presence of puromycin. Labeled proteins solubilized by 1% Triton X-100 were analyzed by SDS gel electrophoresis and fluorography for each time point. (*B*) Cells were pulsed and chased as in *A*. For each time point, the dishes were extracted with 1% Triton X-100 and the extracts subjected to the toxin-Sepharose/Con A-Sepharose isolation scheme of Fig. 4. (*C*) Cells were labeled for the times indicated above the brackets, AChR isolated, and subjected to Endo H digestion (+) prior to electrophoresis.

the result of terminal glycosylation during passage from the *cis* to *trans* compartments of the Golgi apparatus (Dunphy et al. 1981). Unfortunately, AChR-α showed no such changes in Endo H sensitivity during its passage through the cell (Fig. 5C, lanes 2 and 4). Thus, it is not possible to say whether formation of the toxin-binding site in the α-subunit precedes transport to the Golgi apparatus. As toxin-binding sites have been seen by electron microscope autoradiography over the Golgi region but not the RER or transitional elements (Fambrough and Devreotes 1978), it is possible that the toxin-binding site forms after transport to the Golgi.

Formation of the toxin-binding site does not require association of the α-subunit with other subunits (Merlie et al., this volume). Indeed, α-subunit purified by preparative SDS-polyacrylamide gel electrophoresis of the mature AChR complex can be reconstituted, after detergent removal, into a form that displays some (but not all) characteristics of high-affinity α-bungarotoxin binding (Tzartos and Changeux 1983). This is to be contrasted with the α-subunit in the RER, which apparently displays no such activity. The difference be-

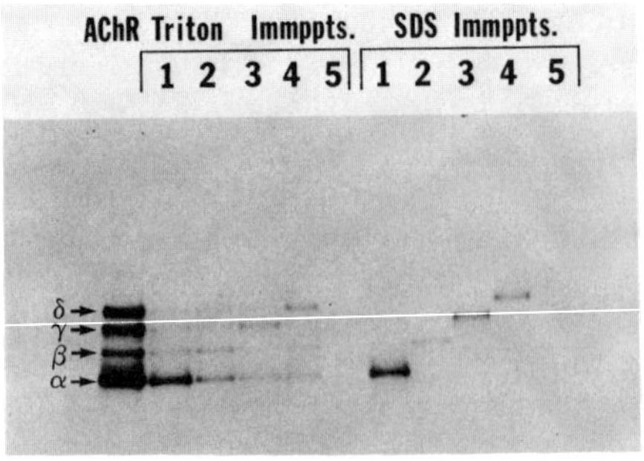

Figure 6. Characterization of anti-AChR subunit antisera under denaturing versus nondenaturing conditions. Triton X-100 immunoprecipitates: Immunoprecipitation was performed after dilution of 9S, ^{125}I-labeled nAChR into immunoprecipitation buffer containing 1% Triton X-100 and 0.2% SDS. SDS immunoprecipitates: Immunoprecipitation was performed after prior denaturation of the ^{125}I-labeled AChR in 1% SDS. Antisera used in lanes *1–5* were anti-α, -β, -γ, -δ, and nonimmune, respectively. The nonstoichiometric recovery of the four subunits in each of the Triton X-100 immunoprecipitates is most likely due to partial dissociation of the AChR complex arising from the iodination procedure (Anderson and Blobel 1981).

tween the two is likely to be due to some as yet undetected posttranslational, covalent (i.e., SDS-resistant) modification that occurs after transport out of the RER. A possible candidate for this is *O*-glycosylation of serine and threonine hydroxyl groups by hexoses and hexosamines, which is catalyzed by glycosyltransferases located in the Golgi apparatus (Hunt and Summers 1976).

AChR Subunits Appear to Self-associate in the RER

Quaternary assembly of the AChR subunits may take at least as long as acquisition of the toxin-binding site by the α-subunit, if not longer (Merlie et al., this volume). When we immunoprecipitated in-vitro-synthesized and membrane-inserted AChR subunits after solubilization of the membrane with Triton X-100 (which does not dissociate the mature AChR complex), only one subunit was brought down by each subunit-specific antibody (not shown), even though all four subunits of mature, 9S AChR were coprecipitated by this procedure as expected (Fig. 6, Triton Immppts.). Surprisingly, however, when we analyzed these Triton-solubilized in-vitro-synthesized AChR subunits by sucrose gradient centrifugation, we found that they migrated as greater than monomeric species (shown for the δ-subunit in Fig. 7). Immunoprecipitation and electrophoretic analysis showed that these complexes were formed by each subunit, independent of the others, even though all four subunits were synthesized in the same reaction mixture (Anderson and Blobel 1981). These complexes, or homo-oligomers, were seen under a variety of conditions of salt, pH, divalent cation, and type of detergent (Anderson and Blobel 1983a). Most important, they were not observed for another unrelated membrane protein, opsin, studied in exactly the same way (Fig. 8). As opsin shares the cotranslational integration process in common with the AChR subunits (Goldman and Blobel 1981), the high S values seen for the AChR subunits are unlikely to be due to an association with nonradioactive components of the RER translocation machinery (such as SRP or SRP receptor), as opsin would in that case exhibit a similar sedimentation profile. However, we cannot rigorously rule out the possibility that the AChR subunits all aggregate nonspecifically with some unlabeled RER proteins, which do not interact with opsin.

Although it is not yet clear that the homo-oligomeric forms of the AChR subunits seen in vitro exist in vivo (cf. Merlie et al., this volume), these structures can be rationalized as potential metastable intermediates in quaternary assembly. Structural studies of AChR (Kistler et al. 1982; Karlin et al. 1983) suggest that the subunits surround a central ion channel like barrel staves. If each subunit formed a boundary of the ionophore, it would be expected to contain a surface rich in charged or polar amino acid side chains within the membrane (cf. Fairclough et al., this volume). As the subunits are each inserted separately into the membrane, assembly of the ion channel cannot occur co-

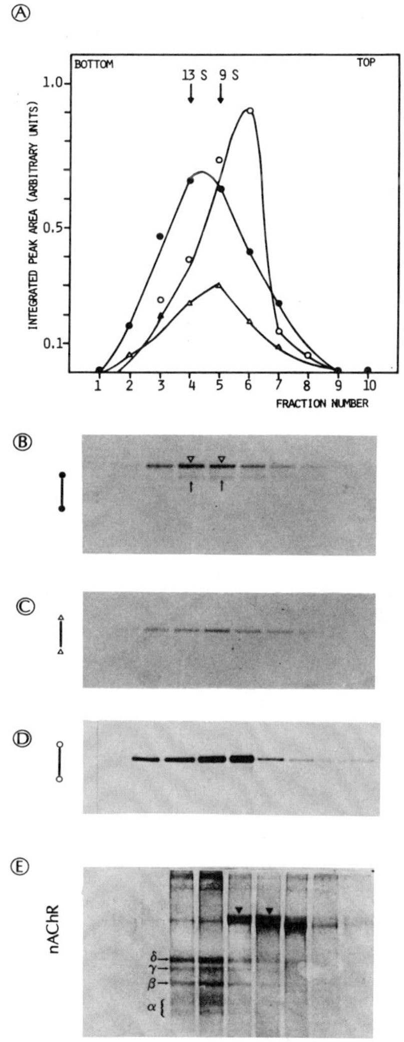

Figure 7. Sucrose gradient analysis of in-vitro-synthesized AChR-δ. (*A*) Quantitative representation of the gel profiles shown in *B–D*, obtained by a densitometric scan of the autoradiographic films. The band quantitated by this procedure is indicated (in all ensuing figures as well) by an open arrowhead, wherever multiple forms appear on the gel. (*A* [●] and *B*) AChR-δ obtained by analysis of total reaction mixture. Upward arrows in *B* indicate the nonglycosylated and partially glycosylated forms characterized previously; open arrowheads indicate the fully glycosylated form (Fig. 1). (*A* [△] and *C*) AChR-δ, solubilized from reisolated microsomes at a fourfold greater dilution than that used in *B*. (*A* [○] and *D*) Pre-AChR-δ, synthesized in the absence of microsomal membranes, analyzed on gradients not containing detergent. (*E*) AChR-enriched *Torpedo* membranes solubilized in 1% Triton X-100/0.3 M KOAc and analyzed in parallel with the samples in *B–D*. However, the gradient fractions were precipitated with trichloracetic acid for analysis of the total protein components; shown is the Coomassie blue-stained gel. The four AChR subunits are indicated by horizontal arrows; the α-chain runs as a smeared doublet due to heating and trichloroacetic acid preparation. Arrowheads indicate the putative large subunit of the Na^+-K^+ ATPase. Arrows in *A* indicate the peak fractions (4 and 5) containing nAChR (see *E*). Fraction 4 is calculated to be 12S, whereas fraction 5 is 9S. Although we have assumed (arrows) that fraction 4 contains some 13S AChR, the resolution of our fractionation is not sufficient for us to determine the actual amount of dimeric receptor. Each fraction spans ca. 3S units.

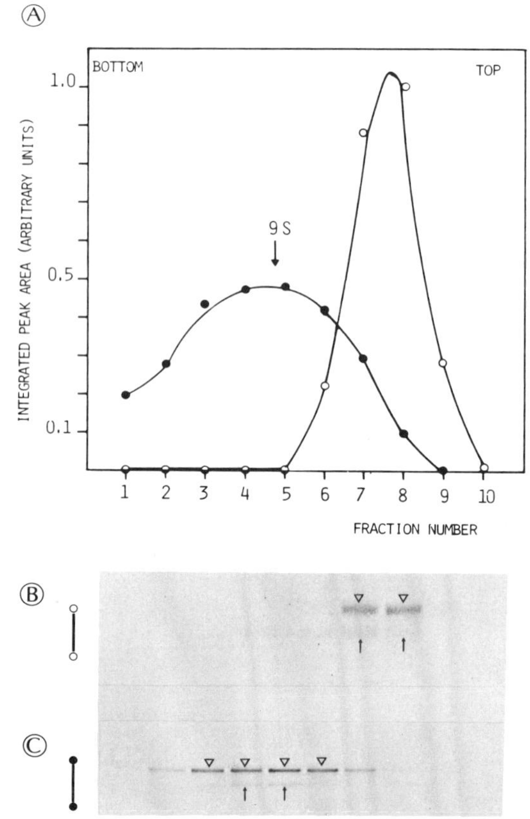

Figure 8. Comparative sedimentation analysis of AChR-δ and opsin. (*A*) Sedimentation profiles reconstructed from densitometric scans of the gels shown in *B* and *C*. (*A* [○] and *B*) Opsin. Arrowheads indicate the core-glycosylated, membrane-integrated form of opsin (Goldman and Blobel 1981); upward arrows indicate the residual unintegrated precursor (or integrated, nonglycosylated material), which smears slightly into the gradient. (*A* [●] and *C*) AChR-δ analyzed in parallel. Arrowheads indicate the fully glycosylated form; upward arrows indicate the residual precursor. AChR-δ in this experiment assumed a slightly broader distribution than on other occasions.

translationally and these ionophore boundary surfaces could in the interim be in direct contact with the interior of the lipid bilayer. However, it might be energetically preferred for the subunits to adopt a homo-oligomeric configuration, in which the hydrophilic ionophore surfaces were sequestered in the interior of aggregates formed by multiple copies of the same subunit. This configuration would maximize contact between the polar ionophore surfaces and minimize their interaction with the hydrophobic interior of the bilayer. These homologous associations could form either during or immediately after translation, when the local concentration of subunit in the membrane would be extremely high (due to the multiple copies of each chain being synthesized from the same mRNA). Irrespective of the thermodynamic rationale, homo-oligomers could provide a useful way of "packaging" the subunits for transport out of the RER to the Golgi apparatus. Finally, the homo-oligomer configuration is attractive for evolutionary reasons: As the four AChR subunits seem to have evolved by a process of gene duplication and divergence (Raftery et al. 1980; Noda et al. 1983), it is likely that the ancestral AChR may have itself been a homo-oligomer, as is the contemporary gap junction channel (Hertzberg and Gilula 1982). The assembly of the contemporary AChR could in this light be considered (tongue-in-cheek) as an example of "ontogeny recapitulates phylogeny" at the molecular level.

If the homo-oligomer precursor model is correct, subunit-subunit exchange must eventually occur in AChR assembly. These exchanges must result in heterologous subunit associations that are of higher affinity than the homologous ones, to be stable. This would occur if the homologous associations were particularly weak, or if the heterologous chain interactions were somehow rendered irreversible once they occurred. The broad sedimentation profile of the homo-oligomers (Figs. 7 and 8) is consistent with their being weakly associated. On the other hand, the heterologous associations could be rendered irreversible by a covalent modification. If they indeed occurred in the Golgi apparatus, for example, heterologous chain interactions might cause conformational changes exposing Ser or Thr hydroxyl groups to glycosyltransferases; glycosylation of these side chains could render this conformational change irreversible. The strong prediction of this model is that inhibition of *O*-glycosylation would block assembly of the AChR subunits. It has been shown that inhibition of *N*-glycosylation by tunicamycin blocks maturation of the α-bungarotoxin-binding site in the α-subunit (Merlie et al. 1981); however, it is possible that the direct effect in this case was simply to prevent transport of AChR out of the RER to the Golgi apparatus. This result points up the current lack of a good cell culture system in which to study the actual assembly of all the AChR subunits. Expression of the *Torpedo* AChR genes in animal cells may alleviate this problem. Intrinsic in this approach is the assumption that the genes coding for the four different subunits are sufficient for the assembly of the receptor-ionophore complex. There are currently no data to support this assumption, although it is a parsimonious one.

ACKNOWLEDGMENTS

This work was supported by a National Science Foundation predoctoral fellowship (to D.J.A.) and a grant from the Muscular Dystrophy Foundation. We thank Drs. Arthur Karlin and Reid Gilmore for helpful discussions during the course of this work.

REFERENCES

Anderson, D.J. and G. Blobel. 1981. In vitro synthesis, glycosylation and membrane insertion of the four subunits of *Torpedo* acetylcholine receptor. *Proc. Natl. Acad. Sci.* **78:** 5598.

——. 1983a. Identification of homo-oligomers as potential intermediates in acetylcholine receptor subunit assembly. *Proc. Natl. Acad. Sci.* **80:** 4359.

——. 1983b. Immunoprecipitation of proteins from cell-free translations. *Methods Enzymol.* **96:** 111.

——. 1983c. Biosynthesis of acetylcholine receptor *in vitro*. *Methods Enzymol.* **96:** 367.

ANDERSON, D.J., P. WALTER, and G. BLOBEL. 1982. Signal recognition protein is required for the integration of acetylcholine receptor δ subunit, a transmembrane glycoprotein, into the endoplasmic reticulum membrane. *J. Cell Biol.* **93:** 501.

ANDERSON, D.J., G. BLOBEL, S. TZARTOS, W. GULLICK, and J. LINDSTROM. 1983. Transmembrane topology of an early biosynthetic form of acetylcholine receptor δ subunit studied by limited proteolysis and monoclonal antibodies. *J. Neurosci.* **3:** 1773.

BLOBEL, G. 1980. Intracellular protein topogenesis. *Proc. Natl. Acad. Sci.* **77:** 1496.

BLOBEL, G. and B. DOBBERSTEIN. 1975. Transfer of proteins across membranes. II. Reconstitution of functional rough microsomes from heterologous components. *J. Cell Biol.* **67:** 852.

BOEKE, J.D. and P. MODEL. 1982. Identification of prokaryotic membrane anchor sequence. *Proc. Natl. Acad. Sci.* **79:** 5200.

BRAELL, W. and H.F. LODISH. 1982. Co-translational insertion of erythrocyte band III into microsomal membranes. *Cell* **28:** 23.

BROCKES, J.P. and Z.W. HALL. 1975. Synthesis of acetylcholine receptor by denervated rat diaphragm muscle. *Proc. Natl. Acad. Sci.* **72:** 1368.

CLAUDIO, T. and M. RAFTERY. 1977. Immunological comparison of acetylcholine receptors and their subunits from species of electric ray. *Arch. Biochem. Biophys.* **181:** 484.

CLAUDIO, T., M. BALLIVET, J. PATRICK, and S. HEINEMANN. 1983. Nucleotide and deduced amino acid sequences of *Torpedo californica* acetylcholine receptor γ subunit. *Proc. Natl. Acad. Sci.* **80:** 1111.

CONTI-TRONCONI, B.M. and M. RAFTERY. 1982. The nicotinic cholinergic receptor: Correlation of molecular structure with functional properties. *Annu. Rev. Biochem.* **51:** 491.

DEVILLERS-THIERY, A., J. GIRAUDAT, M. BENTABOULET, and J.-P. CHANGEUX. 1983. Complete mRNA coding sequence of the acetylcholine binding α-subunit of *Torpedo marmorata* acetylcholine receptor: A model for the transmembrane organization of the polypeptide chain. *Proc. Natl. Acad. Sci.* **80:** 2067.

DOBBERSTEIN, B., H. GAROFF, G. WARREN, and P.J. ROBINSON. 1979. Cell-free synthesis and membrane insertion of the mouse H-2d^k major histocompatibility antigen. *Cell* **17:** 759.

DUNPHY, W.G., E. FRIES, L.J. URBANI, and J.E. ROTHMAN. 1981. Early and late functions associated with the Golgi apparatus reside in distinct compartments. *Proc. Natl. Acad. Sci.* **78:** 7453.

ENGLEMAN, D.M. and T.A. STEITZ. 1981. The spontaneous insertion of proteins into and across membranes: The helical hairpin hypothesis. *Cel* **23:** 411.

FAMBROUGH, D.M. and P.N. DEVREOTES. 1978. Newly synthesized acetylcholine receptors are located in the Golgi apparatus. *J. Cell Biol.* **76:** 237.

FROEHNER, S.C. 1981. Identification of exposed and buried determinants of the membrane-bound acetylcholine receptor from *Torpedo californica*. *Biochemistry* **20:** 4905.

GILMORE, R., P. WALTER, and G. BLOBEL. 1982. Protein translocation across the endoplasmic reticulum. II. Isolation and characterization of the signal recognition particle receptor. *J. Cell Biol.* **95:** 470.

GOLDMAN, B.M. and G. BLOBEL. 1981. In vitro biosynthesis, core glycosylation and membrane integration of opsin. *J. Cell Biol.* **90:** 236.

HERTZBERG, E.L. and N.B. GILULA. 1982. Liver gap junctions and lens fiber junctions: Comparative analysis and calmodulin interaction. *Cold Spring Harbor Symp. Quant. Biol.* **46:** 639.

HUNT, L.A. and D.F. SUMMERS. 1976. Glycosylation of vesicular stomatitis virus glycoprotein in virus-infected HeLa cells. *J. Virol.* **20:** 646.

JACOBS, S., F.C. KULL, and P. CUATRECASAS. 1983. Monensin blocks the maturation of receptors for insulin and somatomedin C: Identification of receptor precursors. *Proc. Natl. Acad. Sci.* **80:** 1228.

JESSEL, T.M., R.E. SIEGEL, and G.D. FISHBACH. 1979. Induction of acetylcholine receptors on cultured skeletal muscle by a factor extracted from brain and spinal cord. *Proc. Natl. Acad. Sci.* **76:** 5397.

KARLIN, A. 1980. Molecular properties of nicotinic acetylcholine receptors. In *The cell surface and neuronal function* (ed. C.W. Cotman et al.), p. 191. Elsevier/ North-Holland, Amsterdam.

KARLIN, A., E. HOLTZMAN, N. YODH, P. LOBEL, J. WALL, and J. HAINFELD. 1983. The arrangement of the subunits of the acetylcholine receptor of *Torpedo californica*. *J. Biol. Chem.* **258:** 6678.

KISTLER, J., R.M. STROUD, M.W. KLYMKOWSKI, R.A. LALANCETTE, and R.M. FAIRCLOUGH. 1982. Structure and function of an acetylcholine receptor. *Biophys. J.* **37:** 371.

LINDSTROM, J., B. EINARSON, and J.P. MERLIE. 1978. Immunization of rats with polypeptide chains from *Torpedo* acetylcholine receptors causes an immune response to receptors in rat muscle. *Proc. Natl. Acad. Sci.* **74:** 3278.

LINGAPPA, V.R., F.N. KATZ, H.F. LODISH, and G. BLOBEL. 1978. A signal sequence for the insertion of a transmembrane glycoprotein. *J. Biol. Chem.* **253:** 8667.

MENDEZ, B., P. VALENZUELA, J.A. MARTIAL, and J.D. BAXTER. 1980. Cell-free synthesis of acetylcholine receptor polypeptides. *Science* **209:** 695.

MERLIE, J.P. and R. SEBBANE. 1981. Acetylcholine receptor subunits transmit a precursor pool before acquiring α-bungarotoxin binding. *J. Biol. Chem.* **256:** 3605.

MERLIE, J.P., J. HOFFLER, and R. SEBBANE. 1981. Inhibition of glycosylation by tunicamycin blocks the assembly of acetylcholine receptor subunits. *J. Biol. Chem.* **256:** 6995.

MEYER, D.I., E. KRAUSE, and B. DOBBERSTEIN. 1982. Secretory protein translocation across membranes—The role of the 'docking protein.' *Nature* **297:** 647.

MILSTEIN, C., G.G. BROWNLEE, T.M. HARRISON, and M.B. MATHEWS. 1972. A possible precursor of immunoglobulin light chain. *Nat. New Biol.* **239:** 117.

MOSTOV, K.E., P. DEFOOR, S. FLEISCHER, and G. BLOBEL. 1981. Co-translation integration of calcium pump protein without signal sequence cleavage. *Nature* **292:** 87.

NODA, M., H. TAKAHASHI, T. TANAKA, M. TOYOSATO, Y. FURUTANI, T. MIROSE, M. ASAI, S. INAYAMA, T. MIYATA, and S. NUMA. 1982. Primary structure of the α-subunit precursor of *Torpedo californica* acetylcholine receptor deduced from cDNA sequence. *Nature* **299:** 793.

NODA, M., M. TAKAHASHI, T. TANABE, M. TOYOSATO, S. KIKYOTANI, Y. FURUTANI, T. HIROSE, H. TAKASHIMA, S. INAYAMA, T. MIYATA, and S. NUMA. 1983. Structural homology of *Torpedo californica* acetylcholine receptor subunits. *Nature* **302:** 528.

RAFTERY, M.A., M.W. HUNKAPILLER, C.B.D. STRADER, and L.E. HOOD. 1980. Acetylcholine receptor: Complex of homologous subunits. *Science* **208:** 1454.

ROBBINS, P.W., S.C. HUBBARD, J. TURCO, and D.F. WIRTH. 1977. Proposal for a common oligosaccharide intermediate in the synthesis of membrane glycoproteins. *Cell* **12:** 893.

ROGERS, M.J. and P. STRITTMATTER. 1975. The insertion of cytochrome b_5 into natural and artificial lipid bilayer membranes. *J. Biol. Chem.* **250:** 5713.

ROTHMAN, J.E. and H.F. LODISH. 1977. Synchronised transmembrane insertion and glycosylation of a nascent membrane protein. *Nature* **269:** 775.

STRADER, C.D. and M.A. RAFTERY. 1980. Topographic studies of *Torpedo* acetylcholine receptor subunits as a transmembrane complex. *Proc. Natl. Acad. Sci.* **77:** 5807.

SUMIKAWA, K., M. HOUGHTON, J.C. SMITH, L. BELL, B.M. RICHARDS, and E.A. BARNARD. 1982. The molecular cloning and characterisation of cDNA coding for the α subunit of the acetylcholine receptor. *Nucleic Acids Res.* **10:** 5809.

TZARTOS, S.J. and J.-P. CHANGEUX. 1983. High affinity binding of α-bungarotoxin to the purified α subunit and to its 27,000-dalton proteolytic peptide from *Torpedo marmorata* acetylcholine receptor. Requirement for sodium dodecyl sulphate. *EMBO J.* **2:** 381.

WALTER, P. and G. BLOBEL. 1980. Purification of a membrane-associated protein complex required for protein translocation across the endoplasmic reticulum. *Proc. Natl. Acad. Sci.* **77:** 7112.

——. 1981. Translocation of proteins across the endoplasmic reticulum. III. Signal recognition protein (SRP) causes signal sequence-dependent and site-specific arrest of chain elongation that is released by microsomal membranes. *J. Cell Biol.* **91:** 557.

——. 1982. Signal recognition particle contains a 7S RNA essential for protein translocation across the endoplasmic reticulum membrane. *Nature* **299:** 691.

WARREN, G. and B. DOBBERSTEIN. 1978. Protein transfer across microsomal membranes reassembled from separated membrane components. *Nature* **273:** 569.

WENNOGLE, L.P. and J.-P. CHANGEUX. 1980. Transmembrane orientation of proteins present in acetylcholine receptor-rich membranes from *Torpedo marmorata* studied by selective proteolysis. *Eur. J. Biochem.* **106:** 381.

WENNOGLE, L.P., R. OSWALD, T. SAITOH, and J.-P. CHANGEUX. 1981. Dissection of the 66,000 dalton subunit of the acetylcholine receptor. *Biochemistry* **20:** 2492.

The Regulation of Acetylcholine Receptor Expression in Mammalian Muscle

J.P. MERLIE,* R. SEBBANE,† S. GARDNER,† E. OLSON,‡ AND J. LINDSTROM§

*Departments of *Pharmacology and ‡Biochemistry, Washington University School of Medicine, St. Louis, Missouri 63110; †Department of Biological Sciences, University of Pittsburgh, Pittsburgh, Pennsylvania 15260; §The Salk Institute, San Diego, California 92138*

The expression of functional acetylcholine receptors (AChRs) involves a long pathway, beginning in the nucleus with the synthesis and processing of a specific set of receptor subunit mRNA precursors. As yet, virtually nothing is known about the transcription and processing of receptor mRNA sequences except by analogy to other eukaryotic mRNAs. Studies of translation and posttranslation processing of receptor subunits have contributed recently to an appreciation of the special problems involved in the synthesis of a multisubunit plasma-membrane-specific protein (see Fig. 9). The AChR is composed of four different but homologous subunits in the stoichiometry $\alpha_2\beta\gamma\delta$ (see Karlin et al.; Raftery et al.; both this volume). Translation of receptor subunit mRNAs in vitro has demonstrated that each subunit is initiated separately (Anderson and Blobel 1981) and that the nascent subunits are cotranslationally inserted and modified in the rough endoplasmic reticulum (Anderson and Blobel 1981; Merlie et al. 1981; Anderson et al. 1982; Sebbane et al. 1983). In vivo pulse-chase-labeling experiments have shown that newly synthesized α-subunits acquire the ability to bind toxin slowly ($t_{1/2}$ = 15 min) and with relatively poor efficiency. An important issue raised by these studies is, How is the assembly of a multisubunit membrane complex accomplished? To understand the process of assembly of the AChR, we have attempted to dissect the steps involved by parallel in vivo and in vitro approaches. We have found that assembly involves several discrete steps and that each of these may be regulated in ways that are physiologically relevant. The results of these studies are reviewed. Finally, using recombinant DNA technology, we have begun to prepare specific reagents for the characterization of the molecular events involved in regulation of receptor synthesis. Results of preliminary experiments indicating that AChR expression may be regulated at translational or pretranslational steps are also presented.

Identification of Four Subunits in the AChR of Mouse Muscle Cell Line BC3H-1

The subunit polypeptides of BC3H-1 were identified by immunoprecipitation with α-bungarotoxin and affinity-purified rabbit anti-α-bungarotoxin antibody, toxin-anti-toxin (Merlie and Sebbane 1981). AChRs immunoprecipitated from cells that had been labeled for 2 hours with [^{35}S]methionine exhibited four bands with apparent molecular weights of 42,000, 46,000, 48,000, and 60,000, which were precipitated by specific binding of α-bungarotoxin (Fig. 1A) (Merlie and Sebanne 1981; Merlie et al. 1982). We have used the Greek letters α, β, γ, and δ to name the mouse subunits according to increasing order of apparent molecular weight using the convention employed for *Torpedo* receptor subunits. The subunits with apparent molecular weights of 42,000 and 46,000 have been shown to be immunologically cross-reactive with α- and β-subunits, respectively, of electric organ receptors. We have not yet determined the exact homology between BC3H-1 γ- and δ-subunits and those of electric organ; however, Lindstrom et al. (1979) showed that mammalian AChRs have antigenic determinants that cross-react with antisera specific for each of the *Torpedo* receptor subunits.

α-Subunits Are Cotranslationally Modified in the Rough Endoplasmic Reticulum

The oligosaccharide components of α- and β-subunits of AChRs labeled in vivo with [^{35}S]methionine and immunoprecipitated with toxin-anti-toxin were found to be endoglycosidase-H-sensitive, indicating the presence of at least one "high-mannose" asparagine-linked chain in each of the α- and β-subunits of the mature receptor (Fig. 1A,B) (Merlie et al. 1982). When nascent membrane polysomes from BC3H-1 cells were allowed to "run off" in vitro in the absence of membranes, the major α-subunit species was found to be the glycosylated form, indicating that core glycosylation had occurred on most of the nascent chains in vivo before extraction of the polysomes. We interpreted this to mean that the single N-linked oligosaccharide chain of the α-subunit was relatively close to the aminoterminal end (Merlie et al. 1981). When purified poly(A)$^+$ BC3H-1 RNA was translated in a nuclease-treated reticulocyte lysate, a single α-subunit polypeptide with an apparent molecular weight of 41,000 was immunoprecipitated with mAb61, an α-subunit-specific monoclonal antibody raised using SDS-denatured *Electrophorus electricus* receptor as immunogen (Fig. 2) (Tzartos et al. 1981). Comparison by peptide mapping of the reticulocyte lysate translation product with the α-subunit labeled in vivo indicated that the in vitro product was not glycosylated and retained a prepeptide or signal sequence. Cotranslational incubation with dog pancreas microsomes resulted in modifications of the in vitro

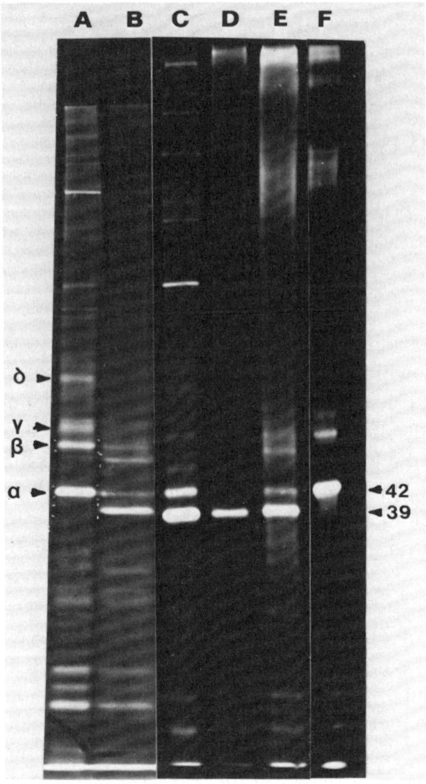

Figure 1. SDS-polyacrylamide gel electrophoresis of glycosylated and nonglycosylated forms of the AChR. (Lanes *A* and *B*) Normal control cells (one 10-cm dish) were labeled for 2 hr with 1.5 ml of [^{35}S]methionine medium containing 250 μCi at 8.3 Ci/mmole. A Triton X-100 extract was prepared, and AChRs were immunoprecipitated by the toxin-anti-toxin method. Radioactive protein was eluted with SDS. Half was treated with endoglycosidase H before electrophoresis (lane *B*), whereas the remaining half was applied directly to the 10% SDS-polyacrylamide gel (lane *A*). The dried gel was autoradiographed for 1 week. Numerical values along the vertical edges indicate apparent molecular-weight estimates ($\times 10^{-3}$ determined from comparison with standard proteins). Greek letters indicate the subunit designations we have assigned to each of the BC3H-1 subunits. (Lanes *C–F*) In a separate experiment, one 10-cm dish of normal control cells was labeled as described above for 2 hr, and two other dishes, pretreated for 6 hr with tunicamycin (Tm) were labeled for 15 min. AChRs were precipitated from control-cell extracts with toxin-anti-toxin as described and from Tm-treated cell extracts with anti-α-subunit mAb61. Radioactive proteins were eluted from the immunoprecipitates with SDS and incubated with and without endoglycosidase H before electrophoresis. The gel was exposed by fluorography for 6 weeks. (Lane *C*) AChRs from Tm-treated cells; (lane *D*) endoglycosidase-H digestion of AChRs from Tm-treated cells; (lane *E*) AChRs from control cells digested with endoglycosidase H; (lane *F*) AChR from control cells. (Reprinted, with permission, from Merlie et al. 1982.)

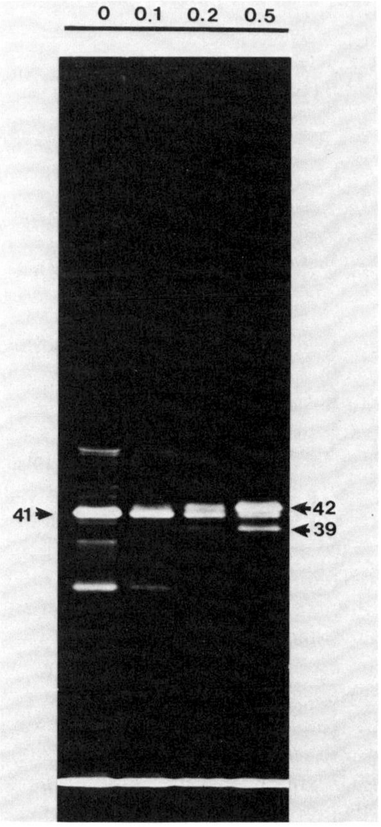

Figure 2. Effect of cotranslational incubation with dog pancreas microsomal vesicles on molecular weight of the α-subunit. Total membrane polysome RNA (4 μg) was translated in 30-μl reactions. Each of the incubations contained the volume in microliters, indicated above the lane, of a membrane suspension with an A_{260} nm = 200, b = 1 cm. Translation products were immunoprecipitated with mAb61 and analyzed by SDS-polyacrylamide gel electrophoresis. The 41,000-molecular-weight product synthesized in the absence of membranes is converted to products with molecular weights of 42,000 and 39,000 upon co-incubation with increasing amounts of membranes. The new species comigrate, respectively, with the glycosylated and nonglycosylated α-subunits labeled in vivo. (Reprinted, with permission, from Sebbane et al. 1983.)

product, producing forms with apparent molecular weights of (1) 42,000, identical with that of the mature α-subunit immunoprecipitated from BC3H-1 cells as part of the functional AChR complex, and (2) 39,000, identical with the α-subunit species generated by endoglycosidase-H digestion of mature AChR (Fig. 2) (Sebbane et al. 1983). As has been suggested by Anderson and Blobel (1981) for the *Torpedo* subunit species synthesized in a microsome-supplemented cell-free system, the polypeptide with an apparent molecular weight of 39,000 probably represents the α-subunit from which the signal peptide had been cleaved but that had failed to be glycosylated.

In Vitro Products Lack α-Bungarotoxin-binding Activity

Anderson and Blobel (1981) reported that *Torpedo* subunits synthesized in vitro were inactive for high-affinity binding of α-bungarotoxin. We have obtained similar negative results with BC3H-1 receptor subunits synthesized in vitro in the presence or absence of dog pancreas microsomes. Furthermore, we have been unable to detect synthesis in vitro of α- or β-subunits with the conformations required for binding monoclonal antibodies specific for native structure (see Tzartos et al. 1981). The dog pancreas microsome-supplemented system is capable of synthesis and cotranslational processing of receptor subunits, but it is clear that these in vitro products can be detected only with antibodies that react with denatured subunits. It will be interesting to discover the molecular bases for the acquisition of biological functions such as toxin (and other ligands)-binding and ion channel activity. We will discuss this problem in more detail below.

Newly Synthesized α-Subunits Acquire α-Bungarotoxin-binding Activity In Vivo

In vivo, α-subunits do acquire high-affinity toxin binding, as they must when they are components of functional AChRs. To our surprise, we discovered that acquisition of high-affinity toxin binding by newly synthesized α-subunits was a relatively slow and inefficient process. After a brief pulse label, the α-subunit polypeptides were detected only by mAb61. A chase period of 15–30 minutes was required before the α-subunits acquired high-affinity toxin binding, and only 30% of the total α-subunits synthesized were converted to a form with high-affinity toxin binding (Figs. 3 and 4) (Merlie and Sebbane 1981; Merlie et al. 1982). Furthermore, the 70% of the newly synthesized α-subunits that were not converted to a toxin-binding form were rapidly degraded ($t_{1/2} \cong 0.5$ hr). These findings have raised several interesting questions, many of which remain to be answered. What is the molecular basis of acquisition of toxin binding? Why is this process inefficient? Is this a rate-determining step for AChR formation and is it regulated? Why is the α-subunit that does not acquire toxin binding degraded rapidly, whereas the α-subunit in the AChR is degraded 10–20 times more slowly (Patrick et al. 1977; Merlie et al. 1978)?

α-Subunit Acquires Toxin-binding Activity Before Association with the β-Subunit

Three separate lines of evidence indicate that an α-subunit species with toxin-binding activity is formed as an intermediate in the assembly of the AChR complex. These data have been submitted for publication (J.P.

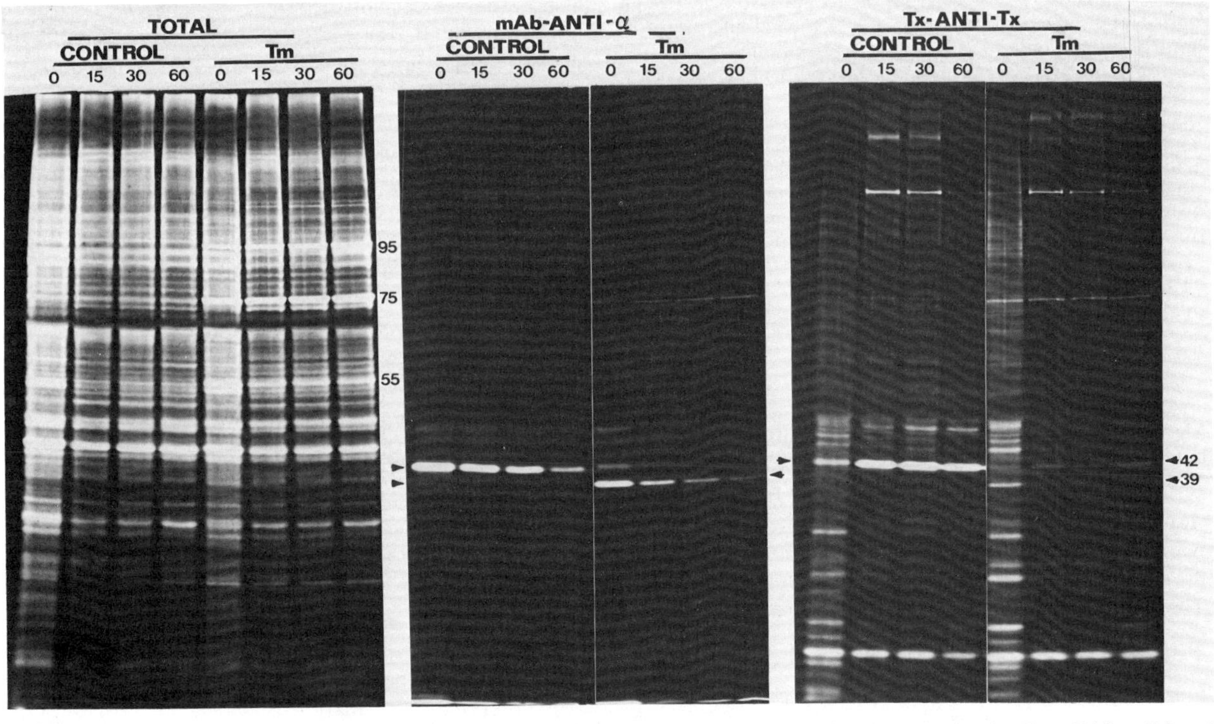

Figure 3. Pulse-chase labeling of AChR subunits and subunit precursors in cells treated with tunicamycin (Tm). Dishes (10 cm) of 7-day-old BC3H-1 cells were pretreated for 6 hr with 0.5 μg/ml Tm (or left as untreated controls) in complete growth medium. Each dish was pulse-labeled for 5 min with 1.5 ml of labeling medium containing 250 μCi of [^{35}S]methionine (8.3 Ci/mmole), the medium was replaced with chase medium consisting of conditioned, complete growth medium supplemented with 0.5 mM L-methionine. After the minutes of chase indicated over each lane, the cells were washed with phosphate-buffered saline (PBS) at 4°C, scraped, pelleted at 2000*g* for 5 min, and extracted for 5 min at 4% in PBS containing 1% Triton X-100, 1 mM each of the protease inhibitors phenylmethylsulfonyl fluoride, sodium ethylenediaminetetraacetic acid, *N*-ethylmaleimide, and sodium tetrathionate. The extract was centrifuged at 15,600*g* for 30 sec, divided in half, and immunoprecipitated with either toxin-anti-toxin (Tx-ANTI-Tx) or the anti-α-subunit antibody mAb61. The immunoprecipitates were solubilized, and the entire sample was loaded onto a 10% SDS-polyacrylamide gel. Total proteins were analyzed from 5 μl of 0.66 ml of each total extract by dilution into SDS sample buffer and electrophoresed directly. The gels were fixed and fluorographed for the following times: total proteins, 4 days; mAb61 immunoprecipitates, 4 days; toxin-anti-toxin, 26 days. Only the α-subunit is indicated in these gels. The β-, γ-, and δ-subunits are, in general, not visible because of the short labeling period and the relatively short exposure times. A trace of the β-subunit is visible in the 15-, 30-, and 60-min lanes of the toxin-anti-toxin gel of control cells. Most of the non-α-subunit bands seen in other lanes (e.g., 0 min of the toxin-anti-toxin gel of Tm-treated cells) are background contaminants of immunoprecipitation. Three polypeptides with apparent molecular weights of 55,000, 75,000, and 95,000, which accumulated in Tm-treated cells, are indicated at the left (55, 75, and 95). (Reprinted, with permission, from Merlie et al. 1982.)

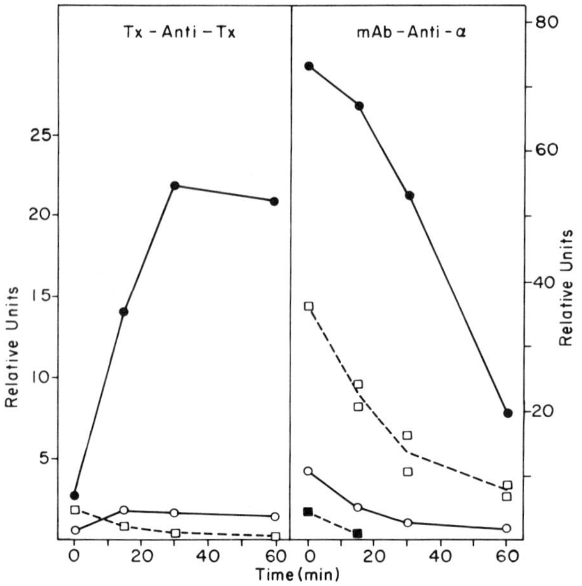

Figure 4. Pulse-chase labeling of AChR subunits and subunit precursors in cells treated with tunicamycin (Tm)—A quantitative summary. X-ray films of the gels prepared in Fig. 3 were exposed for periods of time at which band intensity was linearly proportional to radioactivity applied and time of exposure. Films were scanned with a Joyce-Loebl densitometer and the relative peak areas were determined by weighing. The values for peak area were normalized to time of exposure and the total incorporation of [^{35}S]methionine in the corresponding cell extract. The data are plotted as relative units. The duplicate series of open squares at the right were derived from the same gel exposed for 1 or 4 days to give an idea of the error introduced by scanning and normalization. (●) 42-kD band in control cells; (■) 39-kD band in control cells; (○) 42-kD band in Tm-treated cells; (□) 39-kD band in Tm-treated cells; (Tx-ANTI-Tx) toxin-anti-toxin. (Reprinted, with permission, from Merlie et al. 1982.)

Merlie and J. Lindstrom, in prep.) and are summarized below.

First, pulse-chase-labeling experiments of the type shown in Figures 3 and 4 were repeated, but 60-day as well as 4-day exposures of the fluorograms were prepared in order to quantitate both β- and α-subunits. These experiments showed that, whereas the newly synthesized α-subunits chased into a form immunoprecipitated by toxin-anti-toxin with a $t_{1/2}$ of 15–30 minutes, the time course of chase for the β-subunits was much slower ($t_{1/2} = 60$–90 min).

In a second set of experiments, we determined the kinetics of association of α- and β-subunits by measuring α-subunit coimmunoprecipitation with the β-subunit-specific monoclonal mAb73, which was raised using the fetal calf AChR as an immunogen (Tzartos et al. 1982). The kinetics of subunit association could thus be compared directly with the kinetics of acquisition of toxin binding by the α-subunit. As in the previous series of experiments, the pulse-labeled α-subunit acquired maximum toxin binding by 30 minutes, but α-β-subunit association was not complete until after 90 minutes. In this and previous experiments, we were able to show that α- and β-subunits appeared at the cell surface over a 3-hour time course, consistent with the data of Devreotes et al. (1977). These experiments indicate that newly synthesized α-subunits acquire toxin-binding activity before association with β-subunits and that both of these processes occur in intracellular compartments before the receptors are inserted into the plasma membrane.

Finally, we reasoned that if α-subunits acquired toxin-binding activity before association with β-subunits, then such α-subunits must be in a smaller complex. This prediction has been proven correct by analysis of newly synthesized subunits by velocity sedimentation. After pulse-chase labeling of cells, Triton X-100 extracts were prepared and fractionated by centrifugation on glycerol gradients. The fractions were analyzed for subunit content by immunoprecipitation with (1) toxin-anti-toxin; (2) mAb61, an anti-SDS-denatured α-subunit antibody; (3) mAb73, an anti-native β-subunit antibody (Tzartos et al. 1982); and (4) mAb35, an anti-native α-subunit antibody (Tzartos et al. 1981). Immunoprecipitates were analyzed by SDS-polyacrylamide gel electrophoresis in order to identify subunits. These analyses established several facts. At time zero after pulse labeling, only α-subunits reactive with mAb61 sedimenting as a broad peak of approximately 5S could be detected. A 5S peak of α-subunits with toxin-binding activity began to appear shortly after the beginning of the chase. This 5S α-subunit species increased until about 30 minutes and disappeared by 90 minutes. After a 30-minute lag, the 9S AChR—which was reactive with toxin-anti-toxin and all three monoclonal antibodies and which exhibited both α- and β-subunits (γ- and δ-subunits could not be detected for technical reasons; see J.P. Merlie and J. Lindstrom, in prep.)—was observed to accumulate until about 90 minutes. mAb35 gave results similar to toxin-anti-toxin. This indicated that a conformation(s) necessary for binding α-bungarotoxin and mAb35 was acquired by a 5S α-subunit before association with the β-subunit. As in previous experiments, only about 30% of the newly synthesized α-subunits detected by mAb61 were converted to a 9S AChR complex.

The species of 5S α-subunit with toxin-binding activity appears to fit many of the criteria of an intermediate in the pathway for assembly of the AChR: It is formed from the newly synthesized α-species that binds only mAb61; its formation precedes the formation of the AChR, and its level decreases as the AChRs accumulate to a maximum; finally, from its sedimentation coefficient, it appears to be a monomer or at most a dimer of the α-subunits.

Inhibition of Glycosylation with Tunicamycin Blocks the Acquisition of Toxin Binding by the 5S α-Subunit

Inhibition of glycosylation in BC3H-1 cells resulted in a tenfold reduction in the formation of the α-subunit with toxin-binding activity (Merlie et al. 1982). In contrast, the amount of nonglycosylated α-subunit detected by mAb61 was reduced only twofold (Figs. 3 and 4). In light of what we have learned recently about the assembly pathway, we conclude that inhibition of glycosylation results in a block in the conversion of newly synthesized α-subunits to the 5S toxin-binding intermediate. Unassembled, nonglycosylated α-subunits were degraded at a rate similar to that of unassembled, glycosylated subunits ($t_{1/2} \cong 0.5$ hr) and therefore were not accumulated. These findings suggest that the structural changes that accompany acquisition of toxin binding and/or assembly of mature AChRs result in metabolic stabilization of the component subunits.

Reinitiation of Growth Results in Inhibition of Subunit Assembly and Transport of Receptors to the Plasma Membrane

Muscle cells in tissue culture synthesize and accumulate AChRs upon differentiation, normally after cell division has ceased (Merlie et al. 1978). BC3H-1 cells can be induced to cease cell division and differentiate by transfer to medium containing low serum (Munson et al. 1982). Cultures of BC3H-1 induced to differentiate in low serum can be restimulated to divide by readdition of high concentrations of serum (Munson et al. 1982; Olson et al. 1983). Upon reinitiation of growth, BC3H-1 appears to lose its differentiated phenotype (Olson et al. 1983). This provides a novel way by which to examine the regulation of genes specific to differentiation. When serum-depleted BC3H-1 cells were allowed to express AChRs for 2 days, followed by reinitiation of growth with 20% fetal calf serum, the level of AChRs fell dramatically (Fig. 5A) (Olson et al. 1983). We have investigated the basis of the rapid and dramatic decline in surface receptors after stimulation of growth. A decrease in the steady-state level of surface AChRs could be due to an increase in the rate of degradation and/or to a decrease *in* the rate(s) of synthesis, maturation, assembly, or transport of new receptors to the plasma membrane. An effect on the rate of receptor degradation was eliminated by direct comparison of the rates of surface receptor degradation. By pulse-chase labeling and immunoprecipitation of α-subunits with toxin-anti-toxin, mAb61, and mAb73, we determined that total protein synthesis was stimulated by 20% serum (Fig. 5B), and α-subunit synthesis detected by mAb61 was stimulated in proportion to total protein synthesis (see Fig. 5C and Table 1). However, when α-subunits reactive with toxin-anti-toxin or mAb73 were measured, protein synthesis was found to be dramatically reduced (see Fig. 5D,E and Table 1). In several experiments it was found that subunit association measured by α-subunit coimmunoprecipitation with mAb73 was more severely inhibited than was the formation of α-subunits immunoprecipitated by toxin-anti-toxin (Table 1). We interpret this to mean that serum stimulation results in a reduction in subunit assembly, but that the formation of the 5S α-subunit intermediate with high-affinity toxin binding is relatively unaffected.

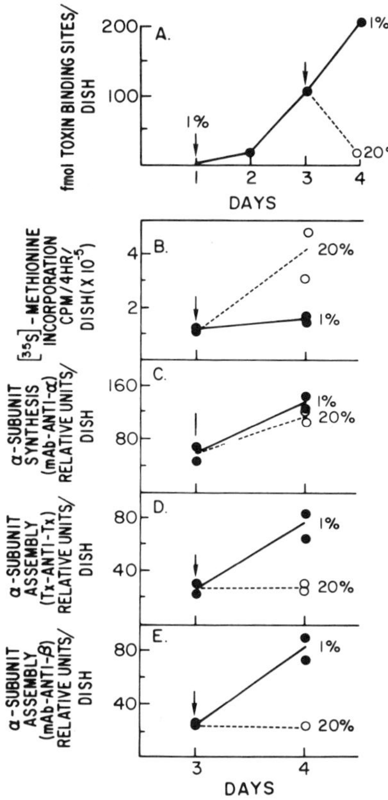

Figure 5. Rate of α-subunit synthesis and assembly. Cells in low-density culture were induced to differentiate in 1% fetal calf serum, and 2 days later, growth was reinitiated by addition of 20% fetal calf serum as described by Olson et al. (1983). (*A*) Total surface AChRs were determined by [^{125}I]α-bungarotoxin binding. To determine rates of α-subunit synthesis, cultures were labeled for 4 hr with 100 μCi/ml (5 Ci/mmole) [^{35}S]methionine. (*B*) Total [^{35}S]methionine incorporation. At the end of labeling, Triton X-100 extracts were prepared and immunoprecipitated with the indicated antibodies as described (Merlie et al. 1982; J.P. Merlie and J. Lindstrom, in prep.). Immunoprecipitates were solubilized, and radioactive polypeptides were analyzed by 10% SDS-polyacrylamide gel electrophoresis. Gels were treated for fluorography as described and exposed to Kodak X-Omat AR film. Fluorograms were scanned with a Joyce-Loebl densitometer, and the relative peak heights of α-subunit bands were quantitated (values are expressed as relative units of α-subunit/culture). (*C*) Relative rate of α-subunit synthesis per culture, determined by immunoprecipitation with mAb61 (anti-α-subunit antibody). (*D*) Relative rate of appearance of α-subunit with toxin-binding activity, determined by immunoprecipitation with toxin-anti-toxin. (*E*) Relative rate of assembly of α-subunit per culture into a complex that contains the β-subunit, determined by immunoprecipitation of α-subunit with mAb73 (anti-β-subunit antibody). (●) Cultures in 1% fetal calf serum; (○) 20% fetal calf serum. Note that 1 day after addition of 20% serum to differentiated cultures (day 4), the rate of α-subunit synthesis decreases slightly (*C*), whereas the fraction of α-subunit present in the form of a multi-subunit complex is reduced approximately 50% (*E*).

The degree of inhibition of subunit assembly (about threefold reduction relative to controls) was not sufficient to account quantitatively for the dramatic decrease in surface receptors observed in 20% of serum-stimulated cells. It was found that after serum stimulation, AChRs that were assembled intracellularly were not

Table 1. Relative Rates of α-Subunit Synthesis and Assembly

Culture condition	α-Subunit synthesis (mAb61)	α-Subunit assembly into AChR complex (mAb73)	α-Subunit with toxin-binding activity (toxin-anti-toxin)	Efficiency of α-subunit assembly (%) (mAb73/mAb61)
Experiment 1				
1% serum, day 3	69	40	60	58
1% serum, day 4	155	85	135	55
20% serum, day 4	59	19	35	32
Experiment 2				
1% serum, day 3	109	67	102	61
20% serum, day 4	170	35	112	20

Rates of α-subunit synthesis and assembly were determined in experiments similar to those described in Fig. 5 and represent relative peak heights of α-subunit bands as determined by densitometry. Efficiency of α-subunit assembly is equivalent to α-subunit assembly, as measured with mAb73 (anti-β-subunit), divided by α-subunit synthesis, as measured with mAb61 (anti-α-subunit), and represents the fraction of total α-subunits that become incorporated into a complex with β-subunit during the 4-hr labeling period. Note that 1 day following addition of 20% serum to differentiating cultures (compare 1% serum, day 3, with 20% serum, day 4), the efficiency of α-subunit assembly is reduced to approximately one half of that on day 3 in 1% serum. In contrast, the rate of α-subunit synthesis decreases only slightly and in some experiments (e.g., Experiment 2 above) actually increases.

transported to the cell surface (Olson et al. 1983). Thus, reinitiation of cell growth resulted in reduced subunit assembly as well as an inhibition of intracellular transport of assembled receptors to the cell surface.

A cDNA Clone for the α-Subunit of BC3H-1 AChR

We have undertaken the molecular cloning of BC3H-1 AChR sequences in order to prepare recombinant DNAs with which to study the regulation of receptor expression. BC3H-1 cells had been shown to be a relatively rich source of α-subunit RNA; approximately 0.2% of the total cellular protein synthesis was α-subunit (Merlie et al. 1982). Fractionation of membrane polysomes resulted in a threefold to tenfold enrichment of the α-subunit mRNA as detected by cell-free translation (Merlie et al. 1981; Sebbane et al. 1983). In addition, translation in vitro of membrane polysome mRNA resulted in a single, unique polypeptide that was immunoprecipitated by mAb61 (Sebbane et al. 1983). It seemed a realistic goal, therefore, to screen a cDNA library by hybrid-selected translation (Parnes et al. 1981) for a unique species that was present at a frequency of approximately 1–2%. The protocols used for construction of the cDNA library and for screening by hybrid-selected translation are detailed in Merlie et al. (1983).

The results of screening 80 independent cDNA clones by hybrid-selected translation are shown in Figure 6. Panel A shows results obtained from SDS-polyacrylamide gel electrophoresis analysis of the total translation products of mRNAs hybridized to pools of cDNAs bound to nitrocellulose filters. In many cases only protein products of RNAs endogenous to the reticulocyte lysate were detected. Less frequently, discrete bands specific for hybrid-selected RNAs were observed (Fig. 6, lanes 7–9). The bands with approximate molecular weights of 41,000 observed in Figure 6, lanes 8 and 9, were immunoprecipitated specifically by mAb61 (Fig. 6B). The 16 individual clones represented in these two pools of eight DNAs were grown and rescreened. Only one, pA59, was determined to be an authentic positive by repeated hybrid-selected translation. The identity of pA59 as an α-subunit cDNA clone was established by peptide map comparison of the hybrid-selected translation product (Fig. 7). Recently, this has been confirmed by determining the sequence of pA59 and comparing the translated amino acid sequence with those of α-, β-, γ-, and δ-subunits of *Torpedo* (K. Isenberg and J.P. Merlie, unpubl.). In the region of amino acid residues 450–480 (format of Noda et al. 1983), we find 75% sequence homology between pA59 and the *Torpedo* α-subunit. Since this region is one where homology among different subunits is low, the identity of pA59 as an α-specific clone is certain.

Detection of α-Subunit Sequence in the Mouse Genome

Hybridization of ^{32}P-labeled pA59 DNA to restriction enzyme fragments of mouse genomic DNA indicates the existence of a single α-subunit gene complementary to pA59. DNA prepared from BALB/c, C57BL, and C3H mouse liver cells as well as undifferentiated and differentiated BC3H-1 cells was digested with *Bam*HI, *Hin*dIII, *Pvu*II, or *Eco*RI (these enzymes do not cut the pA59 insert). The digestion products were fractionated by agarose gel electrophoresis and transferred to membrane filters by the method of Southern (1975). In all cases a single band that hybridized to pA59 under moderately stringent hybridization conditions was detected. No differences due to the strain or cellular origin of the DNA were observed. The sizes of the DNA fragments generated by digestion were as follows: *Bam*HI, 8 kb; *Hin*dIII, 4 kb; *Pvu*II, 4 kb; and *Eco*RI, 15 kb (Merlie et al. 1983). These data indicate that the cDNA insert in pA59 is complementary to only a single gene in the mouse genome (see also Numa et al.; Hershey et al.; Ballivet et al.; all this volume). However, the nucleotide sequence of pA59 identifies it as a partial-length copy of the α-subunit mRNA, representing a region of low intersubunit ho-

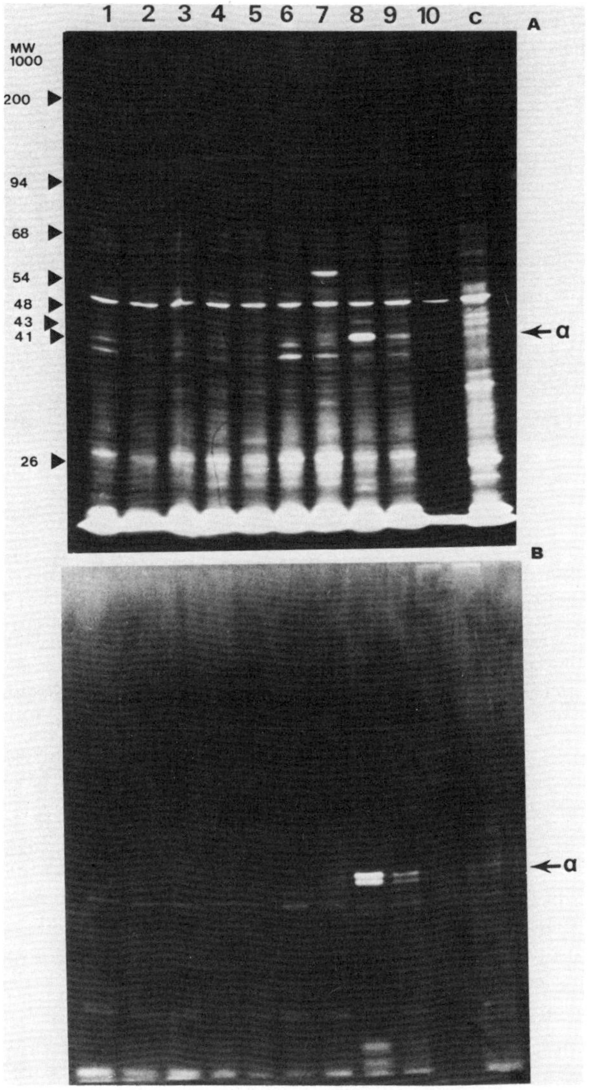

Figure 6. Translation in vitro of hybrid-selected mRNA. Nitrocellulose filters (0.5 cm) were prepared as described (Parnes et al. 1981). Each filter contained plasmid DNAs from a pool of eight independent recombinant clones; 20 filters, two each of ten different pools of eight were allowed to hybridize for 8 hr to 0.5 mg of total membrane-bound-polysome RNA (2 mg/ml). Hybridization, washing, and elution of RNA from filters were carried out as described by Parnes et al. (1981). All RNA eluted from duplicate filters was translated in a single 30-μl rabbit reticulocyte reaction mixture containing mAb61, as described by Sebbane et al. (1983). (*A*) An aliquot (10 μl) of each completed incubation was mixed with SDS_4 sample buffer and analyzed directly by electrophoresis. (*B*) The remaining 20 μl was immunoprecipitated in the presence of protease inhibitors before electrophoresis. A doublet of α-subunit was caused by *N*-ethylmaleimide modification (Anderson and Blobel 1981). The migration positions of known molecular-weight standards are shown on the left, and the nonglycosylated pre-α-subunit is indicated on the right. The last lane (*c*) contained 5 μg of unhybridized RNA from the hybridization reaction as a control for mRNA degradation. Fluorograms were exposed for 30 days. (Reprinted, with permission, from Merlie et al. 1982.)

mology. It is possible that sequences within the α-subunit mRNA not represented in pA59 may be homologous to more than one gene in the mouse genome. It will be essential to extend these experiments to the entire coding sequence of all four subunit mRNAs so that the occurrence of an AChR gene family can be tested definitively.

Detection of mRNA Coding for the α-Subunit by Hybridization with Cloned pA59 DNA

We have used ^{32}P-labeled pA59 DNA in preliminary experiments to identify α-subunit mRNAs. Analysis of BC3H-1 RNA denatured with formamide/formaldehyde and fractionated on agarose gels containing formaldehyde (Lehrach et al. 1977) indicated that a major mRNA species of 2000 bases was complementary to the pA59 DNA insert. This species was found in the poly(A)$^+$ fraction exclusively and was enriched tenfold in the membrane polysome fraction. These characteristics were in agreement with results obtained by a cell-free translation assay for α-subunit mRNA (Sebbane et al. 1983). A second, minor species of 4–5 kb was found to hybridize to pA59 DNA. It was also enriched in the membrane polysome poly(A)$^+$ fraction, but its relationship to α-subunit mRNA has not been determined (Merlie et al. 1983).

We have attempted to use pA59 to identify α-subunit mRNA from mouse muscle. The results of analyses by the gel-transfer or dot-blot procedures indicate that RNA prepared from embryonic, newborn, or adult mouse muscle tissue contains 100–1000-fold less pA59-hybridizing sequences than does BC3H-1. Since we have shown that mouse contains genomic sequences that are complementary to pA59, we must conclude that α-subunit mRNAs are expressed at a very low level in the muscle tissues we have examined (i.e., 16-day gestation to adult). We have not characterized the muscle tissue RNAs further because they are present in all cases in quantities that are at the lower limit of detection.

Finally, we have used A59 DNA to examine the question of whether terminal differentiation of muscle cells in tissue culture results in changes in the levels of mRNA for the AChR. Surprisingly, we found no difference in the relative levels of RNA hybridizing to ^{32}P-labeled pA59 DNA in dot blots prepared from undifferentiated BC3H-1 cells that were in log-phase growth versus fully differentiated cells (Fig. 8). These results pose an interesting problem. It is known that BC3H-1 cells growing in log phase do not accumulate significant levels of AChR as detected by ^{125}I-labeled α-bungarotoxin binding (Patrick et al. 1977; Munson et al. 1982). Furthermore, we have found previously that the relative level of translatable mRNA for the α-subunit increased greater than tenfold in undifferentiated, log-phase-growing cells compared with differentiated cells (Table 2) (Sebbane et al. 1983). Therefore, dividing BC3H-1 cells appear to have α-subunit mRNA that is not used for the synthesis of AChRs in vivo and that is not functional in vitro in a reticulocyte lysate translation system.

Similar results were obtained with another cell line,

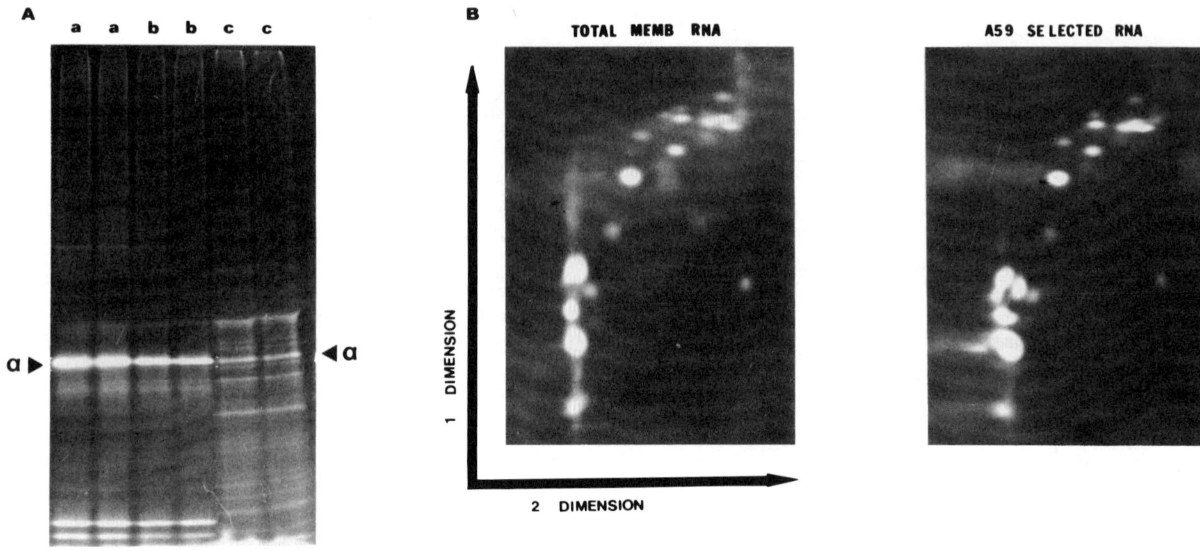

Figure 7. Peptide maps of α-subunits immunoprecipitated from total membrane polysome RNA and pA59-selected RNA. (*A*) Preparative translations were done in 0.3-ml reaction mixtures containing 0.5 mCi (1 Ci = 3.7×10^{10} Bq) [^{35}S]methionine and 10 μl of mAb61. (Lanes *a*) 50 μg of total membrane (memb) polysome RNA; (lanes *b*) RNA selected from 0.5 mg/0.25 ml of total membrane polysome RNA by two 0.5-cm filters saturated with CsCl-gradient-purified pA59 DNA; (lane *c*) RNA prepared as for lanes *b* with two 0.5-cm filters containing pA59 as one of a pool of eight plasmids (as for Fig. 1). The autoradiogram was exposed for 3 days. (*B*) The α-subunit bands marked by the arrows in *A* (for lanes *a* and *b* only) were subjected to trypsin digestion and peptide mapping as described (Palva et al. 1981). The fluorograms were exposed 60 days. (Reprinted, with permission, from Merlie et al. 1982.)

C-2, also derived from a C3H mouse (Fig. 8) (Yaffe and Saxel 1977). We do not know the solution to this apparent discrepancy yet, but the experiment in the lower part of Figure 8 suggests that an alteration in RNA processing may be involved. Undifferentiated 2-day-old BC3H-1 cells and differentiated 7-day-old BC3H-1 cells were prepared. Cytoplasmic- and membrane-bound polysomes were fractionated as described (Sebbane et al. 1983). In addition to the polysome fractions, the nuclear fractions were extracted, and the RNA was purified by the guanidinium thiocyanate procedure (Chirgwin et al. 1979) and centrifuged through a cushion of cesium chloride (Glisin et al. 1974). When RNA from these fractions was hybridized to ^{32}P-labeled pA59 DNA, we found that a significant fraction of the pA59-hybridizing sequences were retained in the nuclear RNA fraction of 2-day-old but not 7-day-old BC3H-1 cells. It will be necessary to determine if undifferentiated muscle cells regulate AChR expression at an RNA-processing step.

SUMMARY AND PERSPECTIVES

The synthesis of functional AChRs can be described as a pathway leading from the translation of subunit mRNAs to the plasma membrane forms of extrajunctional and junctional receptors (Fig. 9). We have not included in this scheme pretranslational steps for the synthesis and processing of RNA coding for receptor subunits because very little is known about such processes. Several aspects of Figure 9 are worthy of note: It is now well established that polypeptide synthesis is initiated on free cytoplasmic polysomes and that once sufficient nascent subunits bearing signal peptides at the amino terminus is formed, polysomes assemble with the membranes of the rough endoplasmic reticulum via a mechanism that employs the signal recognition particle (Anderson et al. 1982). Nascent subunits undergo cotranslational insertion through the rough endoplasmic reticulum membrane, signal peptide removal, and core glycosylation (Anderson and Blobel 1981; Merlie et al. 1981; Anderson et al. 1982; Sebbane et al. 1983). Anderson and Blobel (this volume) have demonstrated that subunits synthesized in vitro and inserted into membrane vesicles do not undergo heterologous subunit-subunit associations. We have shown that α- and β-subunits newly synthesized in vivo are not associated with each other. Our data indicate that the α-subunit is initially present in vivo in a conformation that is radically different from its native conformation in the mature receptor complex. We assume that β-, γ-, and δ-subunits also are synthesized as conformationally "immature" forms, but verification of this point must await the availability of new monoclonal antibody specificities. Our data indicate that only a fraction of the newly synthesized α-subunit undergoes conformational maturation to the 5S species which binds both α-bungarotoxin and anti-main immunogenic region monoclonal antibodies. α-Subunits synthesized during a 5-minute pulse labeling require 30 minutes for completion of this process. α-Subunits that do not undergo conformational maturation are degraded rapidly ($t_{1/2} = 0.5$ hr) (Merlie

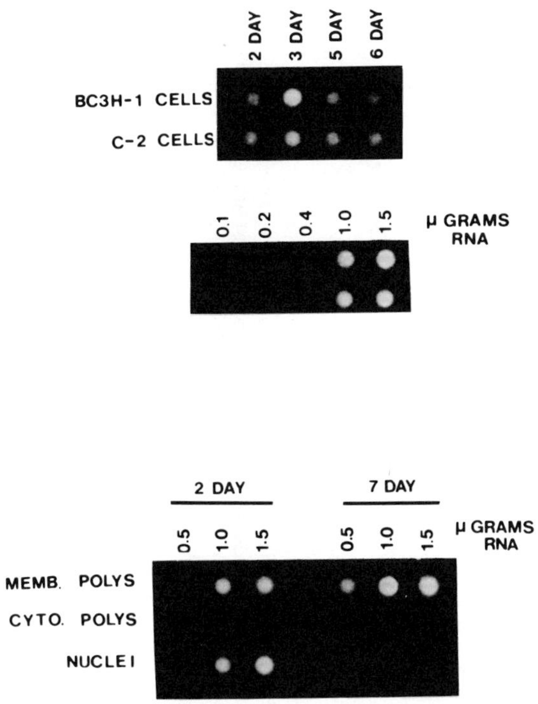

Figure 8. Quantitation of RNA complementary to pA59 DNA in undifferentiated and differentiated muscle cells in tissue culture. RNA was prepared as total cellular RNA or from fractionated membrane and cytoplasmic polysomes by the quanidinium thiocyanate procedure (Chirgwin et al. 1979) as described by Sebbane et al. (1983). In addition, the nuclear fraction from polysome fractionation was also extracted. All RNAs were purified by a centrifugation through a cushion of CsCl (Glisin et al. 1974). RNAs were fixed to a nitrocellulose filter (Thomas 1980). The filter was hybridized to ^{32}P-labeled pA59 DNA and subjected to autoradiography as described by Merlie et al. (1983). (*Top* panel) 1 μg of total RNA extracted from BC3H-1 and C-2 cells at 2, 3, 5, and 6 days after plating. At days 2 and 3, both cell types are in log-phase growth in the culture conditions used. On days 5 and 6, both cell types are confluent and display significant signs of differentiation, and C-2 cells have fused and formed large myotubes. (*Middle* panel) Hybridization signal obtained with quantities of total 7-day-old BC3H-1 RNA ranging from 0.1 μg to 5 μg. (*Bottom* panel) Results of a fractionation experiment designed to determine the intracellular distribution of α-subunit RNA sequences in BC3H-1 cells.

et al. 1982). Assembly of α- and β-subunits synthesized during a 5-minute pulse labeling lags for approximately 30 minutes and is not complete until 90 minutes. Finally, assembled receptors are transported to the surface and appear in the plasma membrane. These processes occur during expression of AChRs in differentiated myoblasts. We do not know how undifferentiated myogenic cells, in vivo or in tissue culture, differ with regard to any of these steps. At present it is possible that regulation of any (or all) of the steps described above could be responsible for the low level of expression of AChRs in undifferentiated muscle cells (see Smilowitz and Fischbach 1978). In experiments with differentiated BC3H-1 cells in which growth has been reinitiated with high serum, we have been able to observe the inhibition of receptor assembly and transport of receptors to the surface (Olson et al. 1983).

Several crucial issues remain to be resolved concerning the synthesis and assembly of AChRs. Is the low efficiency of subunit maturation observed for α-subunits in BC3H-1 cells true for all subunits and in other cell types? Alternatively, is assembly of AChRs regulated by the synthesis of a limiting β-, γ-, and/or δ-subunit? If all subunits are overproduced, and if subunit maturation is a universally inefficient process, what kind of mechanism could be responsible for the high degree of inefficiency? The observation that subunit maturation with respect to acquisition of toxin and anti-main immunogenic region binding does not occur in a cell-free system supplemented with dog pancreas microsomes suggests that the required conformation is not a property entirely intrinsic to the individual subunits. Conformational maturation may require heterologous subunit-subunit interactions or some covalent modification not readily performed by the translation systems and microsomal membranes used. Our results suggest, but do not prove, that subunit-subunit interactions are not important until after conformational maturation (but see Anderson and Blobel, this volume). We do not know if covalent modifications to oligosaccharide side chains, addition of O-linked sugars, or as yet unidentified modifications could be required for conformational

Table 2. α-Subunit mRNA Activity Is Developmentally Regulated

Culture age (days)	RNA content (μg/plate)	Efficiency of translation (10^{-5} × cpm/μg RNA)	Efficiency of α-subunit synthesis (relative units/μg RNA)
1	2	1.86	4
2	8	2.03	6
3	14	2.03	10
4	38	2.02	36
5	62	1.52	40
7	57	1.85	42
8	58	1.67	47
9	58	1.63	31

Total cellular RNA was prepared from cells of the ages indicated. To assay mRNA activity, 4-μg samples of each RNA was translated in a 30-μl reaction containing 2 μl of mAb61 (Sebbane et al. 1983). α-Subunit was immunoprecipitated and analyzed by SDS-polyacrylamide gel electrophoresis. These data were quantitated by densitometry and are presented here as (1) the efficiency of translation and (2) the efficiency of α-subunit synthesis. All translations listed contained 4 μg of RNA and 2 μl of mAb61.

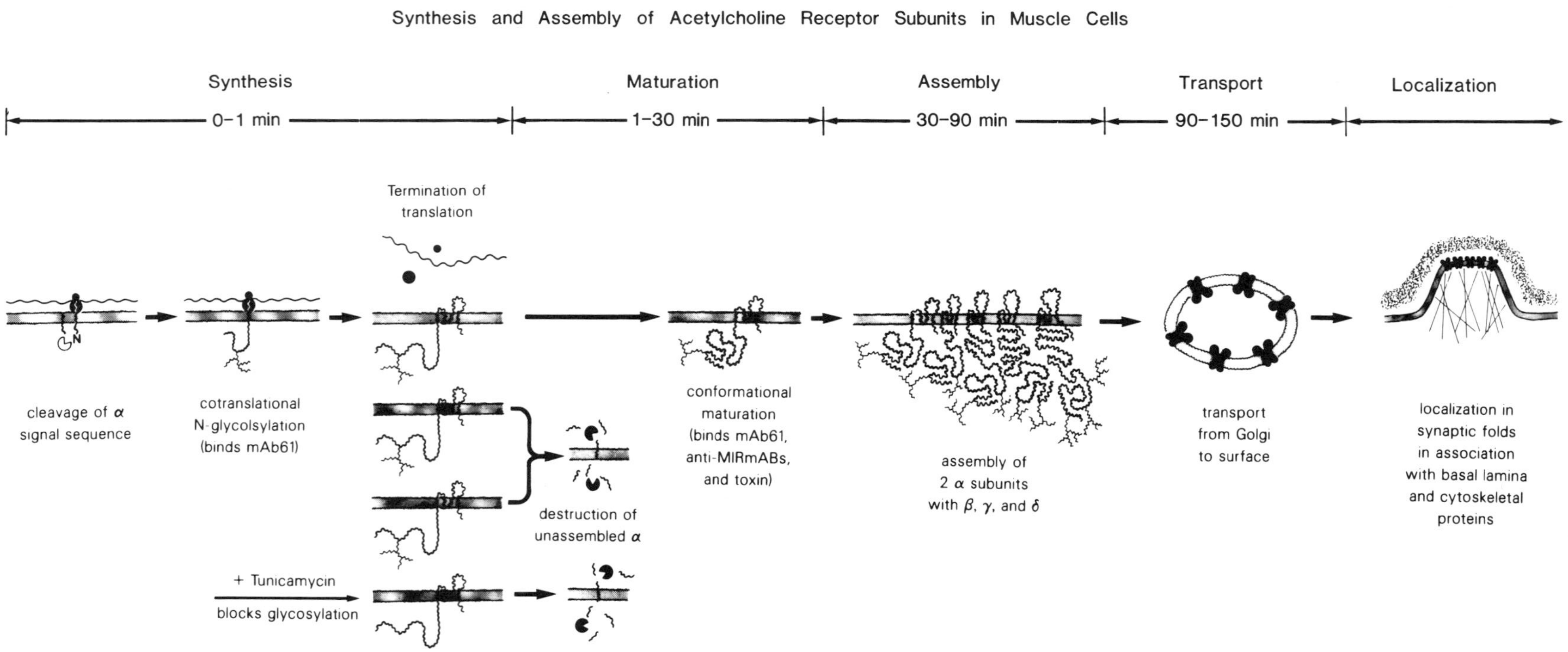

Figure 9. Synthesis and assembly of AChR subunits in muscle cells. In this summary of the events involved in receptor expression, we have attempted to fit the steps for which there is experimental support to a time line. There is, in fact, overlap between steps in several cases, and the times indicated are meant to indicate the period during which most of the subunits newly synthesized during an in vivo, 5-min [^{35}S]methionine pulse label undergo maturation, assembly, or transport. Approximately 1 min is required for polymerization of a single polypeptide of 500 amino acids in length. Clearly, maturation, assembly, and transport are much more time consuming.

maturation. One covalent modification known to be essential for functions of native AChR is disulfide bond formation (see Noda et al. 1983; Karlin et al., this volume). A recent report (Scheele and Jocoby 1982) indicated that cell-free translation systems as normally prepared are dependent on added oxidized glutathione for correct formation of disulfide bonds. We think this is an interesting line of investigation to be pursued in identifying the mechanism of subunit conformational maturation.

Receptor assembly occurs in vivo with slow kinetics. We have proposed previously that subunit assembly may occur in the Golgi apparatus. If assembly does occur in a Golgi subcompartment, it will be important to determine whether this is due to covalent modifications (or some other active processes) that are Golgi-specific or whether passage through this organelle is acting as a concentrating mechanism to increase the frequency of subunit collisions. Finally, it will be important to understand the mechanisms of membrane transport involved in subunit movement to and through the Golgi and movement of mature receptors to the plasma membrane.

At what level is AChR expression regulated during development and synapse formation in a physiologically relevant tissue? Obviously, this question cannot be answered yet. From the pathway for subunit synthesis presented, there are some specific alternatives that can be tested in developing embryonic muscle tissue and in denervated adult diaphragms.

Our preliminary results using recombinant DNA hybridization probes as a means of studying RNA metabolism suggest that RNA processing may be important in the regulation of AChR expression. Recombinant DNA techniques can be expected to play a critical role in the analysis of all aspects of the regulation of AChR expression.

ACKNOWLEDGMENTS

This work was supported by grants from the National Institutes of Health and the Muscular Dystrophy Association to J.P.M. and J.L. and by grants from the NIH to the laboratory of L. Glaser. E.O. was supported by a postdoctoral fellowship from the NIH. We thank L. Hammond and V. Farrue for preparation of the manuscript and Dr. L. Glaser for guidance and insight during the course of this work.

REFERENCES

ANDERSON, D. and G. BLOBEL. 1981. In vitro synthesis, glycosylation and membrane insertion of the four subunits of *Torpedo* acetylcholine receptor. *Proc. Natl. Acad. Sci.* **78:** 5598.

ANDERSON, D., P. WALTER, and G. BLOBEL. 1982. Signal recognition protein is required for the integration of acetylcholine σ subunit, a transmembrane glycoprotein, into the endoplasmic reticulum membrane. *J. Cell Biol.* **93:** 501.

CHIRGWIN, J.M., A.E. PRZYBYLA, R.J. MACDONALD, and W.J. RUTTER. 1979. Isolation of biologically active ribonucleic acid from sources enriched in ribonuclease. *Biochemistry* **18:** 5294.

DEVREOTES, P.N., J.M. GARDNER, and D.M. FAMBROUGH. 1977. Kinetics of biosythesis of acetylcholine receptor and subsequent incorporation into plasma membrane of cultured chick skeletal muscle. *Cell* **10:** 365.

GLISIN, V., R. CRKVENJAKOV, and C. BYUS. 1974. Ribonucleic acid isolation by cesium chloride centrifugation. *Biochemistry* **13:** 2633.

LEHRACH, H., D. DIAMOND, J.M. WOZNEY, and H. BOEDTKER. 1977. RNA molecular weight determinations by gel electrophoresis under denaturing conditions, a critical reexamination. *Biochemistry* **16:** 4743.

LINDSTROM, J., B. WALTER, and B. EINARSON. 1979. Immunochemical similarities between subunits of acetylcholine receptors from *Torpedo, Electrophorus* and mammalian muscle. *Biochemistry* **18:** 4470.

MERLIE, J.P. and R. SEBBANE. 1981. Acetylcholine receptor subunits transit a precursor pool before acquiring α-bungarotoxin binding activity. *J. Biol. Chem.* **256:** 3605.

MERLIE, J.P., J.P. CHANGEUX, and F. GROS. 1978. Skeletal muscle acetylcholine receptor: Purification, characterization and turnover in muscle cell cultures. *J. Biol. Chem.* **253:** 2882.

MERLIE, J.P., J.G. HOFLER, and R. SEBBANE. 1981. Acetylcholine receptor synthesis from membrane polysomes. *J. Biol. Chem.* **256:** 6995.

MERLIE, J.P., R. SEBBANE, S. GARDNER, and J. LINDSTROM. 1983. cDNA clone for the α-subunit of the acetylcholine receptor from the mouse muscle cell line BC3H-1. *Proc. Natl. Acad. Sci.* **80:** 3845.

MERLIE, J.P., R. SEBBANE, S. TZARTOS, and J. LINDSTROM. 1982. Inhibition of glycosylation with tunicamycin blocks assembly of newly synthesized acetylcholine receptor subunits in muscle cells. *J. Biol. Chem.* **257:** 2694.

MUNSON, R., JR., K.L. CALDWELL, and L. GLASER. 1982. Multiple controls for synthesis of muscle-specific proteins in BC3H-1. *J. Cell Biol.* **92:** 350.

NODA, M., H. TAKAHASHI, T. TANABE, M. TOYOSATO, S. KIKYOTANI, Y. FURUTANI, T. HIROSE, H. TAKASHIMA, S. INAYAMA, T. MIYATA, and S. NUMA. 1983. Structural homology of *Torpedo californica* acetylcholine receptor subunits. *Nature* **302:** 528.

OLSON, E.N., K.L. CALDWELL, J.I. GORDON, and L. GLASER. 1983. Regulation of creatin phosphokinase expression during differentiation of BC3H-1 cells. *J. Biol. Chem.* **258:** 2644.

OLSON, E.N., L. GLASER, J.P. MERLIE, R. SEBBANE, and J. LINDSTROM. 1983. Regulation of surface expression of acetylcholine receptors in response to serum and cell growth in the BC3H-1 muscle cell line. *J. Biol. Chem.* (in press).

PALVA, E.T., T.R. HIRST, S.J.S. HARDY, J. HOLMGREN, and L. RANDALL. 1981. Synthesis of a precursor to the β subunit of heat-labile enterotoxin in *Escherichia coli. J. Bacteriol.* **146:** 325.

PARNES, J.R., B. VELAN, A. FELSENFELD, L. RAMANATHAN, U. FERRINI, E. APELLA, and J.G. SIDMAN. 1981. Mouse B_2-microglobulin cDNA clones: A screening procedure for cDNA clones corresponding to rare mRNAs. *Proc. Natl. Acad. Sci.* **78:** 2253.

PATRICK, J., J. MCMILLAN, H. WOLFSON, and J.C. O'BRIEN. 1977. Acetylcholine receptor metabolism in a nonfusing muscle cell line. *J. Biol. Chem.* **252:** 2143.

SEBBANE, R., G. CLOKEY, J.P. MERLIE, S. TZARTOS, and J. LINDSTROM. 1983. Characterization of the mRNA for mouse muscle acetylcholine receptor α-subunit by quantitative translation *in vitro. J. Biol. Chem.* **258:** 3294.

SCHEELE, G. and R. JACOBY. 1982. Conformational changes associated with proteolytic processing of presecretory proteins allow glutathione-catalyzed formation of native disulfide bonds. *J. Biol. Chem.* **257:** 12277.

SMILOWITZ, H. and G.D. FISCHBACH. 1978. Acetylcholine

receptors on chick mononucleated muscle precursor cells. *Dev. Biol.* **66:** 539.

SOUTHERN, E.M. 1975. Detection of specific sequences among DNA fragments separated by gel electrophoresis. *J. Mol. Biol.* **98:** 503.

THOMAS, P.S. 1980. Hybridization of denatured RNA and small DNA fragments transferred to nitrocellulose. *Proc. Natl. Acad. Sci.* **77:** 5201.

TZARTOS, S.J., M.E. SEYBOLD, and J.M. LINDSTROM. 1982. Specificities of antibodies to acetylcholine receptors in sera from myasthenia gravis patients measured by monoclonal antibodies. *Proc. Natl. Acad. Sci.* **79:** 188.

TZARTOS, S., D.E. RAND, B.L. EINARSON, and J.M. LINDSTROM. 1981. Mapping of surface structures of electrophorus acetylcholine receptor using monoclonal antibodies. *J. Biol. Chem.* **256:** 8635.

YAFFE, D. and O. SAXEL. 1977. Serial passaging and differentiation of myogenic cells isolated from dystrophic mouse muscle. *Nature* **270:** 725.

Inactivation of Open and Closed Sodium Channels Determined Separately

R.W. ALDRICH AND C.F. STEVENS
Section of Molecular Neurobiology, Yale University School of Medicine, New Haven, Connecticut 06510

The goal of work in our laboratory is to characterize the function of a particular integral membrane protein, the voltage-dependent Na^+ channel, which underlies the nerve impulse. To achieve this goal, we must identify all of the major functionally relevant conformational states of the protein, determine which transitions are permitted between these states, measure the rate constants for each permitted transition and their voltage dependence, and find the conductance for ions associated with each state (see Stevens 1978). We describe here results of single-channel recording experiments that are part of this program to characterize Na^+ channel function.

Three classes of states have long been recognized for the Na^+ channel: resting, open, and inactivated (Hodgkin and Huxley 1952a,b). At negative values of membrane potential, all Na^+ channels are found in a resting state that does not conduct ions. On depolarization, the channels can progress to the open state, which is permeable to Na^+. Depolarization also causes channels to enter the nonconducting inactivated state, a state from which escape is impossible until the membrane potential has been returned to a sufficiently negative value for an appropriate length of time.

Kinetic evidence indicates that two or more distinct resting states exist; the open and inactivated states have often generally been considered to contain only a single state each, although little direct evidence is available in the literature and a number of alternate proposals have been made.

Hodgkin and Huxley (1952a,b) originally proposed that transitions could occur from any other state into the inactivated state and that the rate of this transition is independent of the state from which the transition occurs. Their assumption has been challenged several times, and it has been claimed that channels must open before they can inactivate, or that the rates of inactivation depend upon the particular state occupied before the transition is made (for reviews, see Goldman 1976; Meves 1978; Armstrong 1981; Khodorov 1981).

We have used the single-channel recording method, first described by Sigworth and Neher (1980), to study inactivation in Na^+ channels. We shall conclude that the inactivated state is accessible from both resting and open states (channels need not open to inactivate), but that the rates at which inactivation occurs can be very different from these two classes of states (open channels inactivate more rapidly than resting).

METHODS

Most of our investigations have used neuroblastoma N1E115, but we have occasionally studied cells of a pituitary line (GH_4C_1) and primary cultured rat myotubes. Na^+ channels seem not to differ between these preparations. The single-channel methods used are essentially like those described by Hamill et al. (1981); most experiments have been performed on cell-attached patches, but occasionally we have studied channels in inside-out or outside-out patches. Temperature was maintained near 15°C for most experiments.

We have usually limited our detailed analysis to patches containing no more than about four active Na^+ channels. This number has been determined by counting the maximum number of channels seen to be open simultaneously in a large number of records obtained with appropriate holding potentials and depolarizations sufficient to open channels with a relatively high probability. Typically, currents were sampled once per 100 μsec, but longer or shorter sample intervals were used when required. The usual depolarization epoch lasted 15 msec. For each record, we measured every open duration as well as the number of channel openings. Groups of records, usually 64, were averaged together to generate an ensemble "macroscopic" current of the sort that might be measured with the traditional voltage clamp.

In experiments designed to investigate transitions from resting to inactivated states, we used a two- or three-pulse protocol: Voltage was stepped from the holding level (a value negative enough so that no channels were inactivated) to a prepulse level sufficient to send some channels to the inactivated state, and finally to a test level (a voltage that would reveal the number of channels still available to open). For the three-pulse protocol, a brief reset level (a value sufficiently negative to restore noninactivated channels to their normal resting state, and brief enough not to permit any inactivated channels to escape from the inactivated state) was interposed between the prepulse voltage and the test level. Test pulses were usually 5 msec or longer, and reset pulses were 1–4 msec in duration. The duration of the prepulse ranged from 0 msec to 300 msec, and was changed systematically at each voltage level to map out the time course of inactivation. The fraction of channels that entered the inactivated state during the prepulse was estimated with the test pulse by: (1) the peak probability of opening, (2) the number opening, or (3) the integral of the current. Details of these methods are presented as required with the results.

RESULTS

Figure 1 presents 63 successive records of Na^+ channel behavior in a patch that contained two channels. One of the records (at the position of the blank space) has been omitted because electrical interference caused a large transient during that trace. It is apparent from this figure that opening of channels is a random process and that the dwell time in the open state is also random. The variable length of time in the open state presumably reveals how long is required for the protein to receive enough energy by collisions with water and other surrounding molecules to surmount the energy barrier separating adjacent conformational states.

The main results are presented in two sections: The first section describes experiments designed to measure the inactivation rate from open channels, and the second deals with closed channels.

Inactivation of Open Channels

The approach and main conclusions presented in this section have already appeared (Aldrich et al. 1983); the experimental observations, however, are new, and the precise method used here is somewhat different from that published.

Figure 2 illustrates the situation considered: We have three classes of states, R (resting), O (open), and I (inactivated), and we are concerned with the rate b of transitions from O to I. Transitions from O back to the R class are assumed to occur at the rate a. The mean dwell time in the open state (assuming a single member of that class) is $1/(a+b)$. We estimate mean open time directly from our records by forming a distribution of open times, as illustrated in Figure 3. Here the distribution function has been fitted by a single exponen-

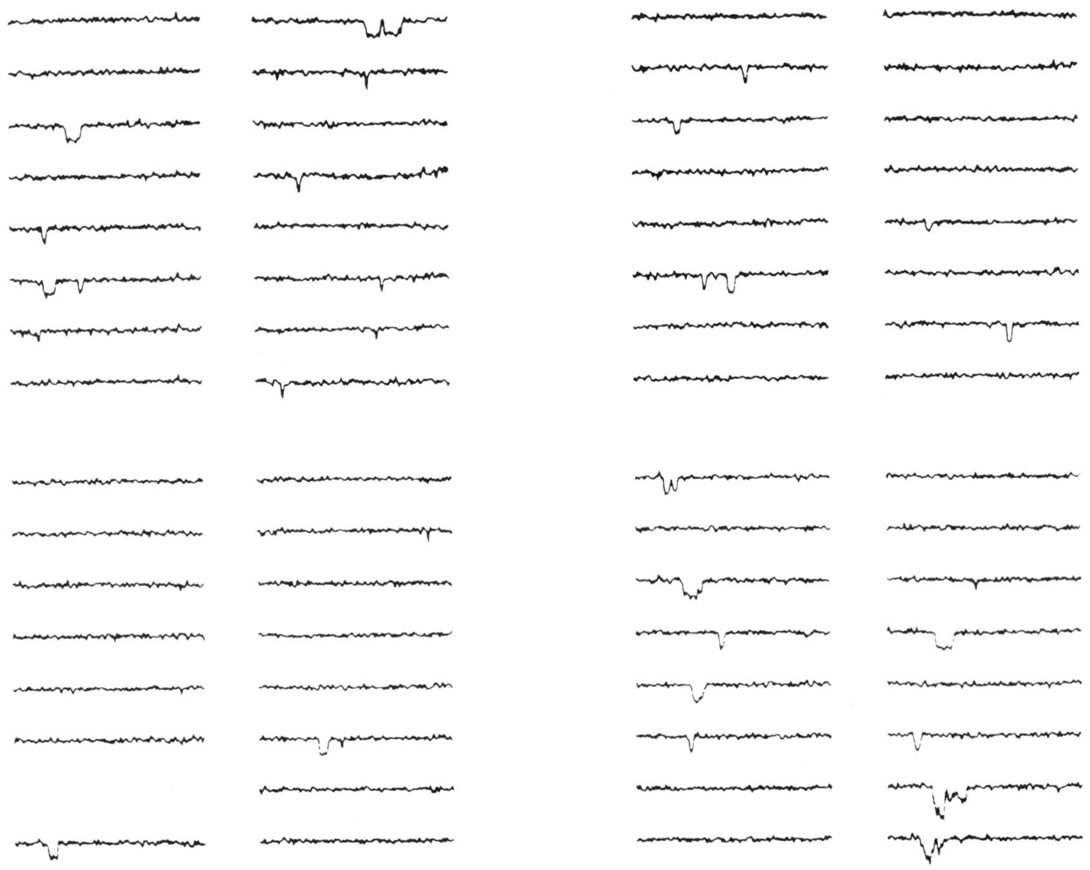

Figure 1. Single Na^+ channel currents. Repetitive 50-mV depolarizations were applied to a cell-attached patch that contained two channels. The holding voltage was 60 mV negative to the cell's resting potential. Groups of 64 identical pulses were applied. One of these was deleted due to electrical artifacts. Each trace lasts 19 msec. The pulses occurred 2.2 msec after the beginning of the record and lasted 15 msec. Leakage and uncompensated capacity current were subtracted digitally. Channel openings are seen as downward current steps of approximately 1.5 picoamps. The data was filtered at 2 kHz and digitized at a rate of once per 100 μsec. The temperature was 20°C.

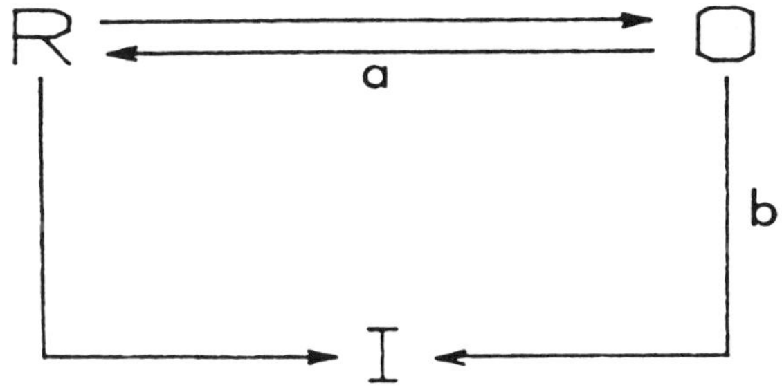

Figure 2. Possible transitions between the three classes of states. R is a group of resting states; O is a single open state; I is a group of inactivated states from which returns are unlikely at depolarized potentials; I is treated as an absorbing state. The lower-case letters represent rate constants. Inactivation of open channels occurs at rate *b* and return of open channels to resting states occurs at rate *a*.

tial after correction for the limited frequency response of our current recording apparatus. We generally find that a single exponential adequately describes our duration distributions, but a careful examination reveals that slightly too many long durations are present for an exponential to be precisely accurate.

From the duration distributions, then, we can find the mean open time and thus the sum $(a+b)$. To obtain an estimate of the inactivation rate *b*, we must discover some way to estimate *a* and *b* separately. We now describe how that can be acomplished.

If an open channel is observed for a sufficiently long time, it will eventually make a transition either to an I or R state. Dwell times in the open state are exponentially distributed, so that 98% of the open channels will make a transition in a time equal to 4*x* (mean dwell time). Because the mean dwell time is about 0.5 msec

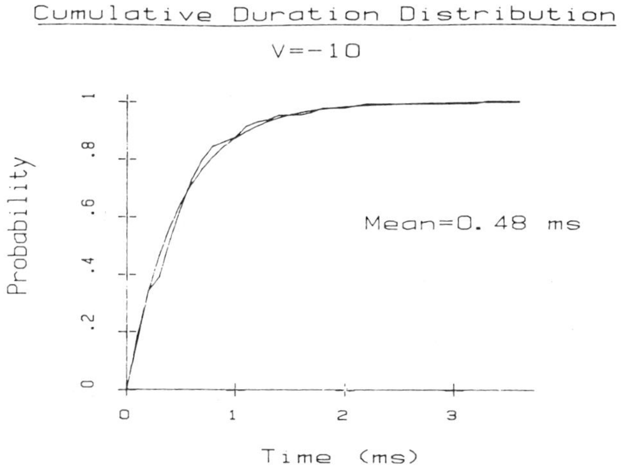

Figure 3. Distribution of open durations. The duration of each channel opening event is measured at half amplitude. The distribution function represents the probability that a given duration is less than time *t*. An exponential function that has been corrected for the limited frequency response of our current measuring apparatus is superimposed on the experimental curve. The mean duration derived from the fit was 0.48 msec. Temperature was 17°C.

under the conditions of our experiments (at 15°C), virtually all channels will make a transition within a few milliseconds. When a transition occurs, what is the probability that an open channel will enter the inactivated state? This probability, *B*, is given by the ratio of the rate into the I state to the total rate of making transitions: $B = b/(a+b)$. If *B* can be estimated, then we have two equations and two unknowns (*a* and *b*) and can determine both rates separately.

The fraction *B* of channels that move from O to I, when a transition occurs, can be estimated from three quantities: the fraction, *F*, of records in which no channels open; the number, *N*, of channels in the patch; and the average number, *K*, of channel openings per depolarization epoch. The fraction, *A*, of channels that pass from the resting into the inactivated state is calculated by taking the *N*th root of *F*. The quantity, *B*, is then found from the equation

$$K = N(1-A)/[1-(1-A)(1-B)]$$

This equation is derived by a straightforward identification of the states O, I, and R as representing a Markov chain with state R occupied initially and state I absorbing. *K* is then the average number of entrances into O before absorption.

Table 1 presents data from an inside-out patch that contained only a single Na^+ channel ($N=1$). Entries in this table show, for three separate runs at different voltages, the various quantities needed to calculate *B*. The table also presents values for the mean channel open time and estimates of the individual transition rates to the R and I states. At the two voltages used in this experiment, the channel almost always closes by entering the inactivated state. This same conclusion holds for voltages that span the normal activation range of Na^+ channels and has been confirmed in patches from about a dozen different cells (Aldrich et al. 1983). Thus, the rate of inactivation of open channels is about $2/\text{ms}^{-1}$ (at 15°C) and is not detectably dependent on membrane potential over the range of voltages that spans most of the activation curve.

The equation used to obtain this result depends upon a variety of assumptions. For example, it is assumed that (1) no channels are inactivated at the negative voltages at which the membranes were held, (2) no channels returned from the inactivated state during the depolarization epoch, and (3) the resting channels could, for purposes of the analysis, all be treated equivalently. These assumptions all give errors in the same direction: Our estimate of *B* is a lower limit. Thus, the actual values of *B* are at least as large as our estimates, and we conclude channels mostly close by entering the inactivated state rather than returning to one of the resting states.

Inactivation of Closed Channels

The results of a typical experiment using a three-pulse protocol are illustrated in Figure 4, where aver-

Table 1. Determination of Open Channel Inactivation Rates

V_h	V_c	A	K	B	t	b	a
−130	−20	0.65	0.37	0.85	0.83	1.02	0.18
−140	−20	0.63	0.38	0.93	0.93	1.0	0.15
−140	−40	0.77	0.23	0.99	0.80	1.24	0.21

V_h is the holding potential (mV), V_c is the command potential (mV), A is the probability of a transition R to I, K is the mean number of channel openings/depolarization epoch, B is the probability of an O to I transition, t is the mean open time (msec), a is the rate for O to R transitions (transitions/msec), and b is the rate of O to I transitions (transitions/msec). Note that the temperature for this experiment was approximately 5°C.

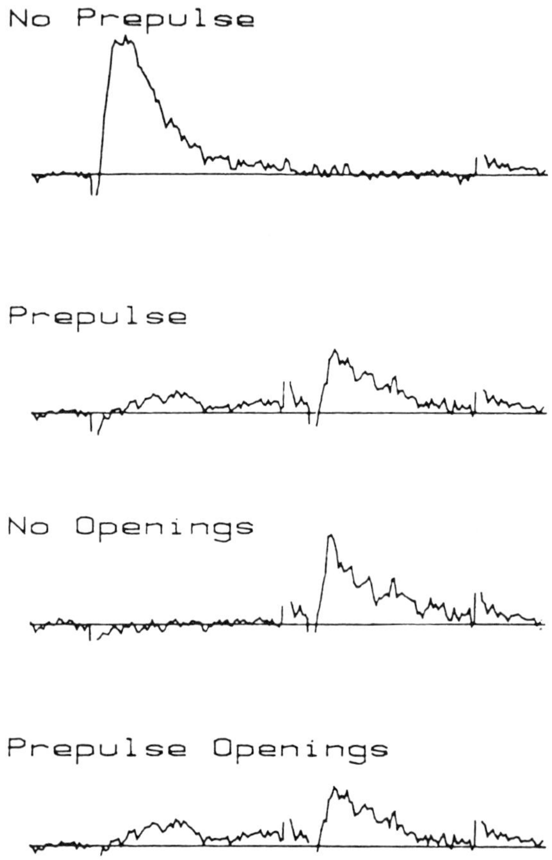

Figure 4. Inactivation of channels by a preceding voltage pulse. (*Top trace*, No Prepulse) Average current from an ensemble of 64 records similar to those shown in Fig. 1 for a 70-mV step from a holding potential of −70 mV (relative to rest). (*Second trace*, Prepulse) shows a similar average for an identical pulse that was preceded by a 7.5-msec prepulse to −20 mV (relative) and a 1-msec reset pulse. (*Lower two traces*) Conditional averages constructed by sorting the records that contributed to the prepulse average into two categories: (1) those in which no openings occurred during the prepulse and (2) those in which at least one opening occurred. The patch contained at least six channels. Opening activity is plotted upwards. Capacity currents have been blanked.

ages of individual records are displayed. The top trace shows the average current for a 70-mV test voltage step from a holding potential 70 mV more negative than the cell's resting potential. The second trace shows a similar average obtained when the test pulse is preceded by a 50-mV prepulse lasting 7.5 msec. The average current during the test pulse is depressed due to inactivation of channels during the prepulse. This inactivation can occur in two ways: Channels that have not opened can directly enter the inactivated state (R→I) or resting channels can open and then inactivate (R→O→I). We can examine these possibilities by constructing average currents during the test pulse after sorting the individual records into those that did and did not have openings during the prepulse. These conditional averages can be seen in the bottom two traces of the figure. Since the average current is depressed in both cases, we conclude that both pathways to the inactivated state are important and that inactivation of resting channels (R→I) is substantial.

The voltage dependence of resting channel inactivation can be obtained by doing the same experiment with different prepulse voltages. The results of such an experiment are seen in the left-hand side of Figure 5. This curve plots the ratio of integrated average test pulse currents after a prepulse to the integral with the prepulse omitted. Only those cases in which no opening occurred during the prepulse were used for the averages. This inactivation function (usually denoted h) shows that the R→I pathway has a substantial voltage dependence.

The patch that gave the data for Figure 4 contained six or more channels. At the prepulse voltage used, very few channels opened. Averages with and without openings during the prepulse are not very different for this patch, although the average is slightly smaller than when there were prepulse openings. In a patch containing fewer channels, even a single opening represents a substantial fraction of the channels in the patch. In this case, the amount of inactivation that occurs when channels open during the prepulse will be considerably increased compared with inactivation in records in which no channel opens during the prepulse, and fewer channels will open during the test pulse. This phenomenon is illustrated in Figure 5 (right) where the inactivation function, h, is plotted separately for groups of records in which channels did and did not open during the prepulse. Only two channels were present in this patch and, as can be seen from the figure, inactivation was much increased when channels opened. This observation implies that open channels inactivate more rapidly than do closed channels, at least on average. Other-

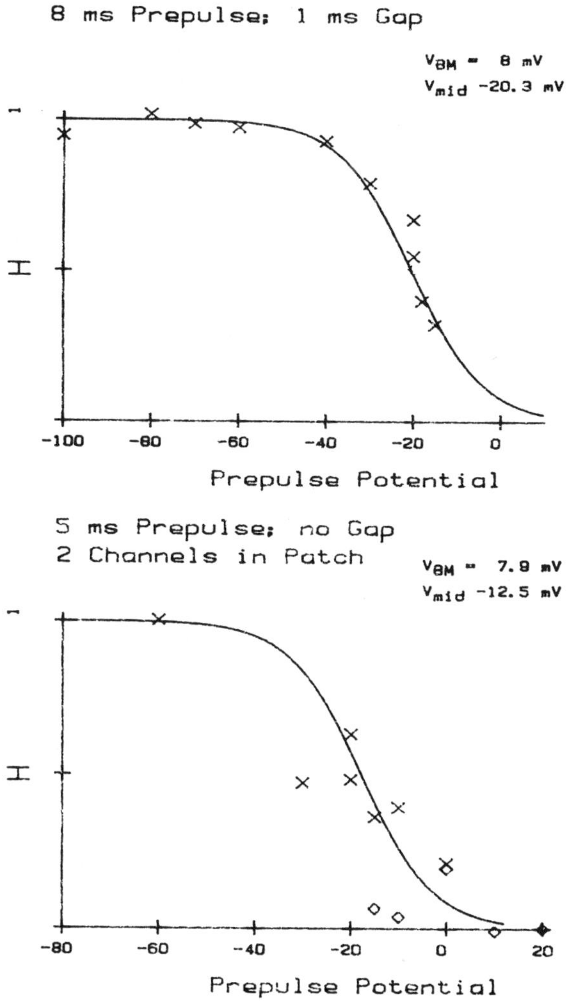

Figure 5. Voltage dependence of inactivation of resting channels. Average currents such as those shown in Fig. 4 are integrated and plotted versus the prepulse potential. These integrals are normalized by the integral of the average current obtained in the absence of a prepulse. Smaller values of this normalized integral signify greater inactivation during the first pulse. (x) Averages in which there were no openings during the prepulse; (◇) averages in which openings did occur during the prepulse. The two curves were obtained from separate patches. Both curves have been fitted by eye with Boltzmann distributions with the parameters indicated.

wise, the fact of channel opening during the prepulse would not increase the degree of inactivation at the time of the test pulse.

Using a modification of the experiment illustrated in Figure 4, we have determined the time constant of inactivation of closed channels. Inactivation is estimated from the size of the test-pulse response in averages of those records for which no openings occurred during the prepulse, and prepulse durations ranging from 0 msec to 300 msec were used to map out the progress of the inactivation process. Figure 6 shows the result of such an experiment carried out at two different prepulse voltage levels: The time constants for the development of inactivation ranged from 65.9 msec to 1.7 msec as the prepulse voltage was changed by 40 mV (from − 50 mV to − 10 mV, relative to the cell's resting potential). At − 10 mV, the time constant for inactivation from the open state was about 0.5 msec. Too few openings occurred during the first pulse at − 50 mV to provide an estimate for the inactivation rate of open channels at this voltage. The absence of voltage dependence for open-channel inactivation at higher voltages leads us to expect a time constant of about 0.5 msec at this voltage as well. Thus, inactivation from the class of closed states is much slower than from the open state, at least over the voltage range that causes few channels to open but produces essentially complete inactivation.

This same general result has been obtained in 10 experiments on 9 different patches. We have used the single-channel recording method, then, to demonstrate directly that channels can inactivate without opening—this agrees with conclusions reached from less direct

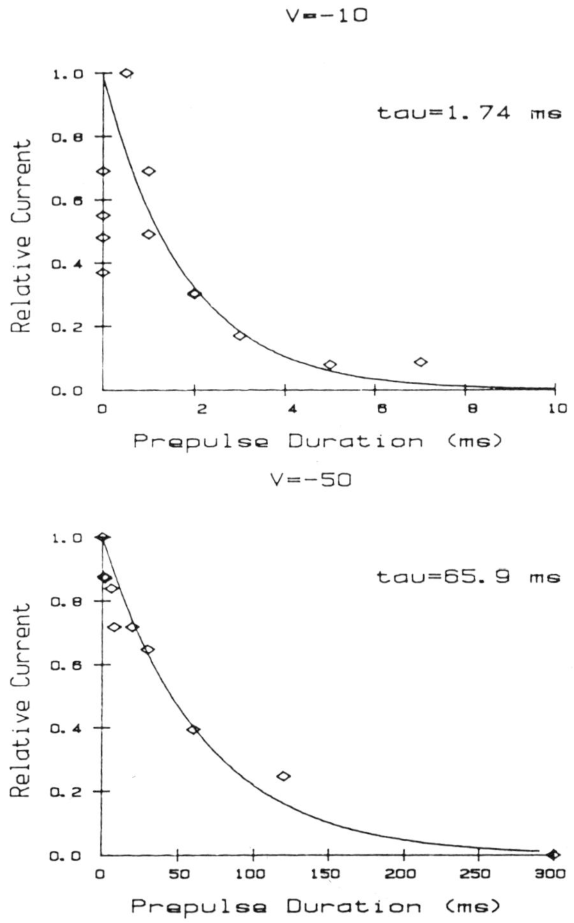

Figure 6. Time course of inactivation of resting channels. Each curve shows the normalized current integral plotted against the duration of the prepulse. Only cases in which no openings during the prepulse occurred were used to construct the averages. The top curve was obtained with a prepulse to − 10 mV (relative to rest) whereas the bottom one was with a prepulse to − 50 mV. In both cases, the test pulse potential was 0 mV. Single exponential fits are superimposed on each set of data points. The time constants of these fits are indicated.

macroscopic measurements—and have shown that the rates of inactivation can be very different, by as much as two orders of magnitude, between open and closed states.

DISCUSSION

An important distinction between the two methods of measuring inactivation rates should be stressed. Because the distribution of dwell times in the open state is exponential, we believe that only one major open state exists. The more complex distribution of first passage times from resting states to open, however, implies that the class of resting states consists of more than a single major conformation (see Horn et al. 1981; Aldrich et al. 1983). Furthermore, return from the inactivated state to open occurs at undetectably low rates. Inactivation rates estimated from open channels, then, represent an individual rate constant for transitions from O to I.

Inactivation rates estimated for closed channels, on the other hand, are more complex and cannot be identified with a single rate constant from a unique state. For example, it may be that a particular closed conformation has a very high transition rate to the I states, but that this closed conformation is visited infrequently and that the dwell time is very short. An additional complication is that, at some of the voltages at which we have studied inactivation rates, inactivation is not complete; that is, we know that channels sometimes made the transition from I back to R. In these instances, the measured inactivation rate would be a lower limit for the true individual rate constant from R to I. Although we have not measured the rate constants from individual members of R into I, we can state that the average rate is much lower than that from O to I. This average could represent, for example, very slow transitions from some closed states, and perhaps more rapid ones from others.

An important difference in the properties of inactivation from open and resting channels relates to the voltage dependence. The rate at which resting channels inactivate is strikingly voltage dependent, whereas open channels inactivate at about the same rate over a wide voltage range. We cannot, from the data presented, decide if inactivation from closed channels is actually voltage dependent or if the apparent voltage dependence arises through voltage effects on transitions between members of the R class. The precise nature of inactivation voltage dependence is an important problem that must be approached through more complicated methods of analysis. The presence or absence of voltage dependence of particular rate constants is of interest because membrane potential is coupled to conformational transitions through dipole moment changes of the protein. If we can determine the voltage sensitivity of a rate constant, we can infer the magnitude of the dipole moment change that occurs between the two states involved in the transition. Such information is necessary ultimately for relating the function of molecules to their structures.

The exact time course with which inactivation develops from closed states also needs more detailed investigation. Inactivation appears to develop exponentially, but we have the impression that the actual time course deviates from an exponential in at least one important regard: At some voltages, especially the most negative ones, inactivation of resting channels seems to occur only after a delay. If this is so, it might mean that transitions between the normal resting state and I are very slow or impossible, and that a transition between members of the R class of conformations is required before inactivation can occur. We have also sometimes noticed that, at short times after the start of the first pulse, channels seem to be less inactivated than at rest; this effect is possibly an artifact of the measuring method, but needs to be more carefully investigated.

Although a number of questions about inactivation remain, we feel that the methods we have used have provided clear answers to several questions. Inactivation definitely does occur from closed states, but the average rate is much slower than from open states. This is in agreement with the conclusions of Bean (1981) from traditional voltage-clamp experiments and is compatible with the general features of models proposed by Armstrong and Gilly (1979) and Nonner (1980). Furthermore, rates of inactivation from open states are larger than anticipated, and we have provided direct evidence that this transition is not strongly voltage dependent, as was first suggested by Bezanilla and Armstrong (1977). We anticipate that further experiments using the single-channel recording method will be able to provide a more complete picture not only of the inactivation process, but of other aspects of Na^+ channel gating.

ACKNOWLEDGMENTS

This research was supported by National Institutes of Health grants NS-12961 and NS-07102.

REFERENCES

Aldrich, R.W., D.P. Corey, and C.F. Stevens. 1983. A reinterpretation of mammalian sodium channel gating based on single channel recording. *Nature* (in press).

Armstrong, C.M. 1981. Sodium channels and gating currents. *Physiol. Rev.* **61:** 3.

Armstrong, C.M. and F. Bezanilla. 1977. Inactivation of the sodium channel. II. Gating current experiments. *J. Gen. Physiol.* **70:** 567.

Armstrong, C.M. and W.F. Gilly. 1979. Fast and slow steps in the activation of sodium channels. *J. Gen. Physiol.* **74:** 691.

Bean, B.P. 1981. Sodium channel inactivation in the crayfish giant axon. *Biophys. J.* **35:** 595.

Bezanilla, F. and C.M. Armstrong. 1977. Inactivation of the sodium channel. I. Sodium current experiments. *J. Gen. Physiol.* **70:** 549.

Goldman, L. 1976. Kinetics of channel gating in excitable membranes. *Q. Rev. Biophys.* **9:** 491.

HAMILL, O.P., A. MARTY, E. NEHER, B. SAKMANN, and F.J. SIGWORTH. 1981. Improved patch-clamp techniques for high-resolution current recording from cells and cell-free membrane patches. *Pfluegers Arch.* **391:** 85.

HODGKIN, A.L. and A.F. HUXLEY. 1952a. The dual effect of membrane potential on sodium conductance in the giant axon of *Loligo. J. Physiol.* **116:** 497.

——. 1952b. A quantitative description of membrane current and its application to conduction and excitation in nerve. *J. Physiol.* **117:** 500.

HORN, R., J. PATLAK, and C.F. STEVENS. 1981. Sodium channels need not open before they inactivate. *Nature* **291:** 426.

KHODOROV, B.I. 1981. Sodium inactivation and drug-induced immobilization of the gating charge in nerve membrane. *Prog. Biophys. Mol. Biol.* **37:** 49.

MEVES, H. 1978. Inactivation of the sodium permeability in squid giant nerve fibres. *Prog. Biophys. Mol. Biol.* **33:** 207.

NONNER, W. 1980. Relations between the inactivation of sodium channels and the immobilization of gating charge in frog myelinated nerve. *J. Physiol.* **299:** 573.

SIGWORTH, F. and E. NEHER. 1980. Single Na^+ channel currents observed in cultured rat muscle cells. *Nature* **287:** 447.

STEVENS, C.F. 1978. Interactions between intrinsic membrane protein and electric field. *Biophys. J.* **22:** 295.

Structure and Functional Reconstitution of the Voltage-sensitive Sodium Channel from Rat Brain

J.A. TALVENHEIMO, M.M. TAMKUN, R.P. HARTSHORNE, D.J. MESSNER, R.G. SHARKEY, M.R.C. COSTA, AND W.A. CATTERALL

Department of Pharmacology, School of Medicine, University of Washington, Seattle, Washington 98195

Electrophysiological experiments using the voltage-clamp technique have shown that the initial rapid depolarization during an action potential in nerve, skeletal muscle, and heart muscle results from rapid voltage-dependent increases in membrane permeability to Na^+. Many different lines of evidence indicate that a selective transmembrane Na^+ channel is responsible for the rapid Na^+ permeability increase during the action potential. Selective ion permeation is mediated by a hydrophilic pore containing a Na^+-selective ion coordination site, designated the ion selectivity filter (Hille 1971, 1972). Ion conductance through the Na^+ channel is regulated, or gated, by two separate processes: activation, which controls the rate and voltage-dependence of the Na^+ permeability increase following depolarization, and inactivation, which controls the rate and voltage-dependence of the subsequent return of Na^+ permeability to the resting level during a maintained depolarization (Hodgkin and Huxley 1952). Membrane charge movements associated with voltage-dependent activation of Na^+ channels have been detected as gating currents and are considered to represent movement of charged components of the Na^+ channel protein during the conformational changes leading to channel activation (Armstrong and Benzanilla 1974). Na^+ channel inactivation is thought to be a sequential, relatively voltage-insensitive process that is initiated by the voltage-dependent conformational changes that lead to channel activation (Armstrong and Bezanilla 1977; Aldrich and Stevens, this volume).

Analysis of Na^+ channel properties by voltage-clamp methods has provided a detailed description of the three essential functional properties of Na^+ channels: voltage-dependent activation, voltage-dependent inactivation, and selective ion transport. However, an understanding of the molecular basis of electrical excitability requires identification of the membrane macromolecules that comprise the ionic channels, solubilization and purification of these channel components, and correlation of their structural features with the known functional properties of Na^+ channels. This task has required the development of new approaches.

Neurotoxins as Molecular Probes of Na^+ Channels

Neurotoxins that bind with high affinity and specificity to voltage-sensitive Na^+ channels and modify their properties have provided the essential tools for identification and purification of Na^+ channels. There are three well-characterized neurotoxin receptor sites on the Na^+ channel (Table 1; for review, see Catterall 1980). Neurotoxin receptor site 1 binds the water-soluble, heterocyclic guanidines, tetrodotoxin (TTX) and saxitoxin (STX). These toxins inhibit Na^+ channel ion transport by binding to a common receptor site that is thought to be located near the extracellular opening of the ion-conducting pore of the Na^+ channel (for review, see Narahashi 1974; Ritchie and Rogart 1977; Catterall 1980).

Neurotoxin receptor site 2 binds several lipid-soluble toxins including grayanotoxin and the alkaloids veratridine, aconitine, and batrachotoxin (reviewed in Albuquerque and Daly 1976; Catterall 1980; Lazdunski and Renaud 1982). The competitive interactions among these four toxins at neurotoxin receptor site 2 have been confirmed by direct measurements of specific binding of [^{3}H]batrachotoxinin A 20-α-benzoate to Na^+ channels (Brown et al. 1981; Catterall et al. 1981). These toxins cause persistent activation of Na^+ channels at the resting membrane potential by blocking Na^+ channel inactivation and shifting the voltage dependence of Na^+ channel activation to more negative membrane potentials (for review, see Catterall 1980). Therefore, neurotoxin receptor site 2 is likely to be localized on a region of the Na^+ channel involved in voltage-dependent activation and inactivation.

Neurotoxin receptor site 3 binds polypeptide toxins purified from North African scorpion venoms or sea anemone nematocysts. These toxins slow or block Na^+ channel inactivation. They also markedly enhance persistent activation of Na^+ channels by the lipid-soluble toxins acting at neurotoxin receptor site 2 (for review, see Catterall 1980; Lazdunski and Renaud 1982). The affinity for binding of ^{125}I-labeled derivatives of the polypeptide toxins to neurotoxin receptor site 3 is reduced by depolarization. The voltage dependence of

Table 1. Subunit Composition of the Sodium Channel Purified from Rat Brain

	Molecular weight	Stain intensity (%)	Predicted stain intensity for 1:1:1 stoichiometry (%)
Intact Na^+ channel	316,000		
α-Subunit	260,000	79	77
β1-Subunit	39,000	11	12
β2-Subunit	37,000	10	11

scorpion toxin binding is closely correlated with the voltage dependence of Na^+ channel activation (Catterall 1979; Mozhayeva et al. 1980). These experiments indicate that neurotoxin receptor site 3 is located on part of the Na^+ channel that undergoes a conformational change during voltage-dependent channel activation, leading to markedly reduced affinity for scorpion toxin. Therefore, scorpion toxin and sea anemone toxin bind to voltage-sensing or gating structures of Na^+ channels.

In addition to these neurotoxins acting at receptor sites 1–3, new neurotoxins have been described that act at one or more additional receptor sites. Of these, a new class of scorpion toxins has already proven valuable in studies of Na^+ channels. Cahalan (1975) showed that the venom of the American scorpion *Centruroides sculpturatus* modifies Na^+ channel activation rather than inactivation. Pure toxins from several American scorpions have a similar action (Couraud et al. 1982; Meves et al. 1982; Wang and Strichartz 1982). These toxins bind to a new receptor site on the Na^+ channel (Jover et al. 1980; Barhanin et al. 1982) and have therefore been designated β-scorpion toxins.

These several neurotoxins provide specific, high-affinity probes for distinct regions of the Na^+ channel structure. Our strategy has been to identify and purify the protein components of Na^+ channels that bind these toxins and then to analyze their structural and functional properties.

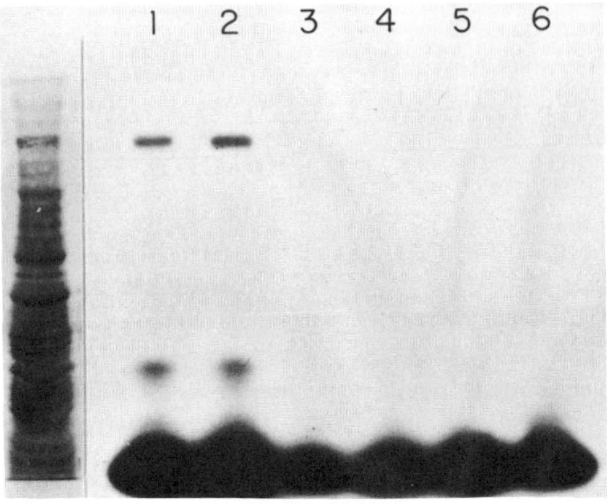

Figure 1. Covalent labeling of the α- and β1-subunits of the Na^+ channel with azidonitrobenzoyl mono-[^{125}I]iodo scorpion toxin. Intact synaptosomes were covalently labeled with 2 nM azidonitrolenzoyl mono-[^{125}I]iodo scorpion toxin, as described by Beneski and Catterall (1980), under the following experimental conditions: (lane *1*) 1 μM TTX; (lane *2*) 1 μM TTX and 1 μM batrachotoxin; (lane *3*) 1 μM TTX and 200 nM unlabeled scorpion toxin; (lane *4*) 1 μM TTX, 1 μM batrachotoxin, and 200 nM unlabeled scorpion toxin; (lane *5*) 135 mM KCl, 1 μM TTX, and 1 μM batrachotoxin; and (lane *6*) 135 mM KCl, 1 μM TTX, 1 μM batrachotoxin, and 200 nM unlabeled scorpion toxin. Samples were analyzed by SDS gel electrophoresis and covalently bound ^{125}I-scorpion toxin was detected by autoradiography. A Coomassie blue-stained gel lane is shown on the left.

Identification of Protein Components of Na^+ Channels in Excitable Membranes

Direct chemical identification of Na^+ channel components in situ was first achieved by specific covalent labeling of neurotoxin receptor site 3 with a photoreactive azidonitrobenzoyl derivative of the scorpion toxin from the North African scorpion *Leiurus quinquestriatus* (Beneski and Catterall 1980). Irradiation with UV light causes covalent attachment of the bound toxin derivative. Analysis of covalently labeled synaptosomes by polyacrylamide gel electrophoresis under denaturing conditions in SDS reveals specific covalent labeling of two polypeptides that have subsequently been designated the α- and β1-subunits of the Na^+ channel (Fig. 1, lanes 1 and 2). These proteins, as assessed by SDS-polyacrylamide gel electrophoresis, are 270 kD and 39 kD, respectively (Beneski and Catterall 1980; Hartshorne and Catterall 1981; Hartshorne et al. 1982).

The covalent labeling of these two polypeptides in synaptosomes was shown to be specific by inhibition with unlabeled scorpion toxin or by block of voltage-dependent binding of scorpion toxin by membrane depolarization (Fig. 1, lanes 3–6). The α-subunit of the Na^+ channel could also be covalently labeled with azidonitrobenzoyl scorpion toxin in electrically excitable neuroblastoma cells. In contrast, in mutant neuroblastoma cells that were selected for neurotoxin resistance and lack functional voltage-sensitive Na^+ channels, the 270-kD polypeptide corresponding to the α-subunit was not detectably labeled (Beneski and Catterall 1980). These data provide additional evidence for the specificity of photoaffinity labeling. More recently, methods have been developed to identify the protein spot corresponding to the α-subunit in two-dimensional isoelectric focusing-SDS gel electrophoresis analyses of [^{35}S]methionine-labeled proteins of normal and mutant neuroblastoma cells. These experiments show that the α-subunit is not synthesized in mutant neuroblastoma clones that lack functional Na^+ channels, providing further evidence that this 270-kD polypeptide is an essential component of the functional Na^+ channel (Costa and Catterall 1982).

The β-scorpion toxins derived from American scorpion venoms have been used to label a new receptor site on the Na^+ channel (Barhanin et al. 1983). Toxin α from *Tityus serrulatus* was covalently attached to its receptor site by covalent cross-linking with disuccinimidyl suberate. A single polypeptide of 270 kD was labeled in rat brain synaptosomes. Thus, the receptor site for the β-scorpion toxins is located on or near the α-subunit of the Na^+ channel, as previously found for the α-scorpion toxins acting at neurotoxin receptor site 3.

The Molecular Size of the Na^+ Channel

The first indications of the molecular size of the Na^+ channel in situ were derived from radiation inactivation studies (Levinson and Ellory 1973). In these experiments, membrane preparations from pig brain or eel electroplax were irradiated with X rays and the decrease in the number of functional TTX-binding sites was measured as a function of radiation dose. Applying target theory, Levinson and Ellory concluded that a structure of 230 kD was required for toxin binding. These experiments have recently been repeated by Barhanin et al. (1983), who compared the target size of the Na$^+$ channel assessed by either TTX binding or *T. serrulatus* toxin binding. In each case, the target size was approximately 280 kD, in reasonable agreement with the earlier work. This size estimate might correspond to the molecular weight of the entire Na$^+$ channel or to the molecular weight of a subunit that is essential for binding these neurotoxins.

The molecular size of the intact Na$^+$ channel protein has been measured by hydrodynamic studies of the detergent-solubilized channel. The STX- and TTX-binding component of Na$^+$ channels in garfish olfactory nerve was first solubilized by nonionic detergents with retention of high affinity and specificity of toxin binding (Henderson and Wang 1972; Benzer and Raftery 1973). Subsequent work has extended these findings to Na$^+$ channels in eel electroplax (Agnew et al. 1978), mammalian brain (Catterall et al. 1979; Krueger et al. 1979), and mammalian skeletal muscle (Barchi et al. 1980). In contrast to the ease of solubilization of neurotoxin receptor site 1 with retention of STX- and TTX-binding activity, both neurotoxin receptor site 2 (W.A. Catterall, unpubl.) and neurotoxin receptor site 3 (Catterall et al. 1979) lose high-affinity neurotoxin-binding activity on solubilization.

The size of the solubilized STX receptor has been estimated by hydrodynamic studies. The solubilized STX receptor-Triton X-100 complex from rat brain is 601 kD (Hartshorne et al. 1980). Since the detergent-receptor complex contains 0.9 g Triton X-100 and phosphatidyl choline per gram of protein, the STX receptor protein solubilized from rat brain is 316 ± 63 kD (Hartshorne et al. 1980). A similar analysis of the STX receptor from skeletal muscle solubilized in Lubrol PX arrived at an estimate of 314 kD (Barchi and Murphy 1981). Thus, the saxitoxin receptors solubilized from rat brain and skeletal muscle have similar sizes.

Purification of the Na$^+$ Channel

Although the STX receptor of the Na$^+$ channel was successfully solubilized several years ago, the marked instability of the toxin-binding activity prevented progress on the purification of the Na$^+$ channel until the discovery that addition of phospholipid markedly stabilized the STX receptor from electroplax (Agnew et al. 1978). These investigators were able to achieve substantial purification of the STX receptor by ion exchange chromatography and gel filtration (Agnew et al. 1980). The only protein component of this partially purified preparation that was clearly correlated with toxin-binding activity was a 260-kD polypeptide. This protein is likely to be analogous to the α-subunit of the Na$^+$ channel from mammalian brain that was identified in intact membranes by covalent labeling with a photoreactive scorpion toxin derivative (Beneski and Catterall 1980).

Application of similar procedures to Na$^+$ channels solubilized from mammalian skeletal muscle (Barchi et al. 1980) and brain (Hartshorne and Catterall 1980, 1981) also gives substantial purification. In addition, affinity chromatography on wheat germ lectin-Sepharose columns provides a further enrichment of the STX-binding activity from these tissues. The specific adsorption and elution of the STX receptor from this lectin affinity column indicates that it is a glycoprotein containing *N*-acetylglucosamine or sialic acid residues, or both. STX receptor preparations obtained from either rat skeletal muscle or rat brain by these procedures are 30–50% pure, as assessed by their specific binding activity for STX.

Recent experiments in our laboratory have focused on obtaining an essentially homogeneous preparation of Na$^+$ channels from mammalian brain that retain their functional properties. We have now developed a purification procedure involving solubilization in Triton X-100, ion-exchange chromatography on DEAE-Sephadex, adsorption chromatography on hydroxylapatite, affinity chromatography on wheat germ agglutinin-Sepharose, and finally sedimentation through sucrose gradients (Hartshorne et al. 1982; Hartshorne and Catterall 1983). After this series of purification steps, the purified preparation binds 2910 pmoles STX/mg protein or 0.9 mole STX per mole of Na$^+$ channels of 316 kD. This specific binding capacity is 90% of that expected for a pure Na$^+$ channel protein of 316 kD that binds 1 mole of STX. By this functional criterion, the Na$^+$ channel is 90% pure.

Analysis of these highly purified preparations by SDS-polyacrylamide gel electrophoresis reveals two protein bands of 260 kD and 38 kD (Fig. 2A; Hartshorne et al. 1982; Hartshorne and Catterall 1983). These bands correspond in molecular weight to the α- and β-subunits identified initially by scorpion toxin photoaffinity labeling (Beneski and Catterall 1980) and subsequently as the major polypeptide components of partially purified Na$^+$ channel preparations from rat brain (Hartshorne and Catterall 1980, 1981). Antibodies raised against the purified Na$^+$ channel in rabbits specifically precipitate the scorpion toxin-labeled Na$^+$ channel subunits from detergent extracts of labeled synaptosomes, confirming the identity of these proteins (Sharkey et al. 1983). These two protein bands comprise 71% and 19% of the stain intensity. No other band comprises as much as 1% of the total stain intensity. Evidently, the Na$^+$ channel is 90% pure by chemical analysis as well as by functional test. Further analysis has shown that the β-subunit band (Fig. 2A) contains two nonidentical subunits, $\beta 1$ and $\beta 2$, of similar molecular weight that are poorly resolved by SDS gel electrophoresis but are distinguished by the linkage

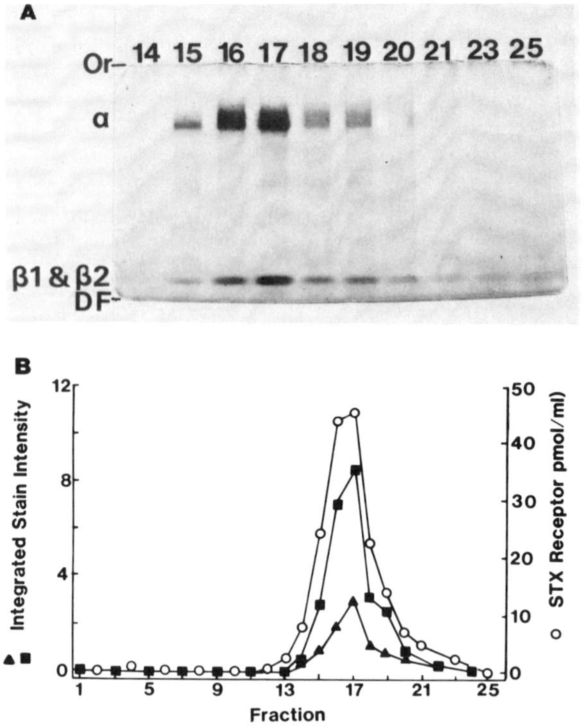

Figure 2. Analysis of the purified Na^+ channel by SDS gel electrophoresis. (*A*) Purified Na^+ channels were sedimented through a 5–20% sucrose gradient and samples were analyzed by SDS gel electrophoresis and visualized by silver stain, as described by Hartshorne and Catterall (1981, 1983). (*B*) The stain intensity of the α- and β1- + β2-bands was determined by densitometric scanning. STX-binding activity (○), stain intensity of α (■), and stain intensity of β1 + β2 (▲) are plotted versus fraction number from the bottom of the gradient.

of β2 to α by disulfide bonds (Hartshorne et al. 1982; and see below).

If these subunits are present as a stoichiometric complex with the α-subunit, they should migrate in constant ratio during separation procedures. Figure 2 illustrates analysis of several different fractions from a sucrose gradient velocity sedimentation of purified Na^+ channels. The peak of STX-binding activity was centered on fraction 17. The α- and β-subunit bands are present in constant ratio through the peak beginning at fraction 15, increasing to maximum concentration in fraction 17, and declining together through fraction 20. The close correspondence of subunit concentration and STX-binding activity is illustrated quantitatively in Figure 2B where the stain intensity of the α- and β-subunit bands is plotted together with STX-binding activity in each fraction from this sucrose gradient. These data show clearly that the α- and β-subunits migrate as a stoichiometric complex coincident with the STX-binding activity of the Na^+ channel.

Subunit Composition of the Na^+ Channel

The availability of an essentially homogenous preparation of the Na^+ channel allows a more detailed analysis of subunit composition and properties than possible previously. Figure 3 illustrates the effect of reduction of disulfide bonds with β-mercaptoethanol (βME) on the migration of the α- and β-subunit bands. As the Na^+ channel is progressively reduced by βME (Fig. 3, lanes 1–5), the high-molecular-weight band containing α migrates more rapidly consistent with a reduction in apparent molecular weight of 22,000. Concomitantly, the stain intensity of this band is reduced. These results suggest that reduction with βME causes cleavage of disulfide bonds between α and either β1 or β2.

In the absence of βME (Fig. 3, lane 1), the β-band is well resolved and is only slightly more diffuse than soluble protein standards of similar molecular weight (Fig. 3, lane 6). As reduction with βME decreases the apparent molecular weight of the α-band (Fig. 3, lanes 3–5), the β-band becomes more diffuse and stains more intensely. There is a quantitative correlation between the increase in the mobility of α and the increase in stain intensity of β as a function of βME concentration (Hartshorne et al. 1982). The stain intensity in the β-band is increased 1.9-fold. Since the stain intensity in the α-band is decreased by βME treatment, βME does not cause a general enhancement of the stain intensity of all protein bands. Therefore, these results indicate that one of the two β-subunits is covalently linked to the α-subunit by disulfide bonds. Upon reduction with βME, this β-subunit is separated from α, causing a decrease of the apparent molecular weight and stain intensity of the high-molecular-weight band and a concomitant broadening and increase in stain intensity in the β-band.

To provide further evidence in support of this conclusion, the noncovalently linked β-subunit was separated from α by gel filtration in SDS. The purified Na^+ channel was dissociated by incubation at 100°C in 2% SDS in the absence of βME. The dissociated and denatured subunits were separated by gel filtration in SDS on Sepharose 4B-CL and each column fraction

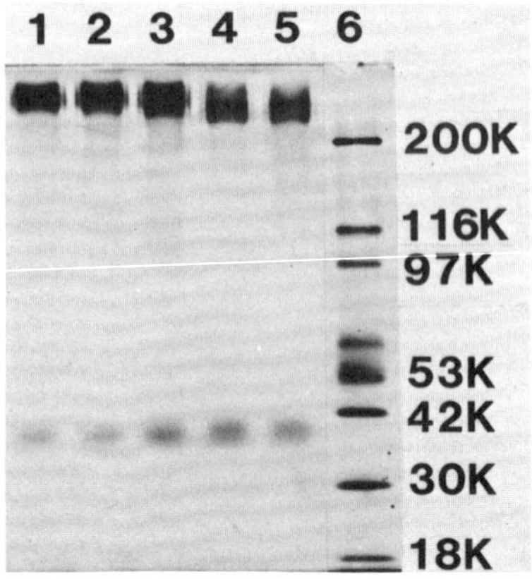

Figure 3. Effect of βME on migration of the α- and β1 + β2-subunit bands. Purified Na^+ channels were depolymerized by incubation for 5 min at 100°C with 2% SDS and 0 mM βME (lanes *1* and *2*), 0.1 mM βME (lane *3*), 1.1 mM βME (lane *4*), or 6.6 mM βME (lane *5*), separated by SDS gel electrophoresis, and visualized with silver stain as described by Hartshorne et al. (1982). Lane *6* contains molecular weight standards.

was analyzed by gel electrophoresis in SDS with and without prior reduction with βME. As expected from the data described above, one β-subunit was separated from the α-subunit by this procedure (Fig. 4). Lane 1 shows the polypeptide components of the high-molecular-weight fraction analyzed without prior reduction. Only a single band of M_r ~300,000 is observed. Reduction of the same fraction prior to electrophoresis splits this high-molecular-weight band into two smaller bands: α with M_r ~270,000 and β2 with M_r ~37,000 (Fig. 4, lane 2). The low-molecular-weight gel filtration fraction containing the β1-subunit was analyzed with and without reduction. After reduction with βME (lane 3), the β1-subunit (M_r ~39,000) migrates slightly slower than β2. Without reduction with βME (lane 4), the β1-subunit migrates more rapidly, suggesting that intrachain disulfide bonds do not allow complete unfolding without reduction. Comparison of the relative migration of the β1- and β2-subunits separated in this way (lanes 2 and 3) illustrates the difficulty of resolving them by SDS gel electrophoresis when they are present in the same sample.

These experiments provide strong evidence that the α- and β2-subunits of the purified Na^+ channels are linked by disulfide bonds and that the β1-subunit is attached to the complex of the α- and β2-subunits by noncovalent forces. The disulfide bond between α and β2 is unlikely to be an artifact produced during the purification because iodoacetamide (1 mM), which is present throughout the purification, should rapidly react with any free sulfhydryl groups preventing formation of additional disulfide bonds in vitro.

Further evidence that the β1- and β2-subunits are

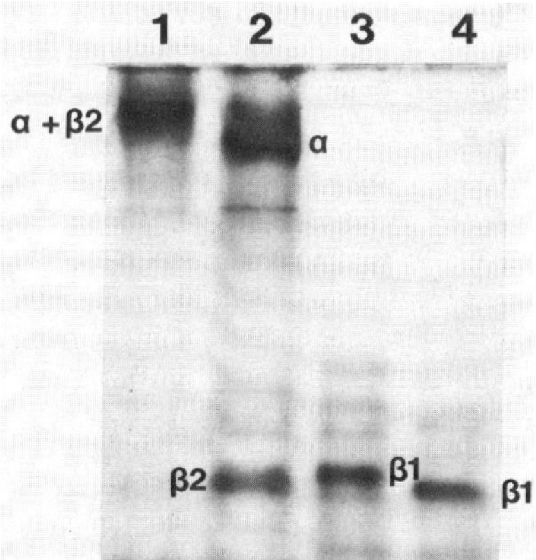

Figure 4. Separation of β1 from the complex of α and β2 by gel filtration in SDS. Purified Na^+ channels were depolymerized by incubation at 100°C in 2% SDS in the absence of βME and separated into a high-molecular-weight fraction and a low-molecular-weight fraction by gel filtration on Sepharose 4B-CL. These two fractions were analyzed by SDS gel electrophoresis, and visualized by silver stain as described by Hartshorne et al. (1982). (Lane *1*) high-molecular-weight fraction without reduction; (lane *2*) high-molecular-weight fraction with reduction; (lane *3*) low-molecular-weight fraction with reduction; (lane *4*) low-molecular-weight fraction without reduction.

components of the Na^+ channel in situ is derived from studies of the effect of reduction with βME on Na^+ channels in synaptosomes covalently labeled with azidonitrobenzoyl mono-[^{125}I]iodo scorpion toxin (Hartshorne et al. 1982). These experiments show the β1-subunit of the channel, which is noncovalently linked to α, is labeled by scorpion toxin whereas the β2-subunit, which is attached by disulfide bonds, is not labeled. In addition, reduction with βME reduces the apparent molecular weight of the α-subunit band as expected if β2 is also disulfide-linked to α in situ in synaptosomes. These experiments provide evidence for a Na^+ channel complex of α, β1, and β2 in intact synaptosomes where Na^+ channels are functional.

Table 1 summarizes our current view of the structure of the Na^+ channel from mammalian brain. The complex of 316 kD can contain only one α-subunit of 260 kD. The results of Figure 3 show that reduction with βME decreases the apparent molecular weight of the α-band by 22,000, consistent with the conclusion that each α-subunit is linked by a disulfide bond to one and only one β2-subunit. The ratio of stain intensity in the different subunit bands is also very close to that expected for a 1:1:1 stoichiometry of the α-, β1-, and β2-subunits (Table 1). Therefore, we propose that the subunit composition of the Na^+ channels, both purified and in situ in synaptosomes, is $\alpha(\beta1)_1(\beta2)_1$.

Recent work has provided further definition of the subunit composition of the Na^+ channel preparations from electroplax and skeletal muscle. Monoclonal antibody prepared against the 260-kD component of the TTX-binding protein recognizes only a single protein in electroplax membranes, precipitates STX-binding activity, and can be used in an immunospecific purification method to obtain highly purified complexes of the 260-kD subunit and monoclonal antibody (Moore et al. 1982; Nakayama et al. 1982). Miller et al. (1983) have also obtained highly purified preparations of the TTX-binding protein from electroplax that consist primarily of a 260-kD polypeptide. Although some smaller protein components remain in these preparations, these authors conclude that only the 260-kD polypeptide is required for TTX- and STX-binding activity. Since the 260-kD polypeptide purified from electroplax binds these two toxins, it is likely that the analogous α-subunit of the Na^+ channel from rat brain contains neurotoxin receptor site 1, the TTX-STX receptor site.

The initial Na^+ channel preparations from rat skeletal muscle were not reported to contain protein components analogous to the α- and β-subunits of the Na^+ channel from brain (Barchi et al. 1980). However, more recent work (Barchi 1983) indicates the presence of a large polypeptide with a molecular weight in the range of 150–250 kD, a polypeptide of 45 kD, and a doublet of polypeptides of 38 kD. It seems probable that the large polypeptide of uncertain molecular weight corresponds to the α-subunit and the doublet of polypeptides of 38 kD to the β1- and β2-subunits of the rat brain Na^+ channel. There is no counterpart of the 45-kD polypeptide in the Na^+ channel from rat brain.

Functional Reconstitution of the Na^+ Channel from Purified Components

An important goal of purification of the protein components of Na^+ channels is to restore Na^+ channel function in an experimental system containing only purified protein and lipid. Reconstitution of Na^+ channel function from purified components is the only rigorous proof that the proteins identified and purified on the basis of their neurotoxin-binding activity are indeed sufficient to form a functional voltage-sensitive ion channel. In addition, successful reconstitution will provide a valuable experimental preparation for biochemical analysis of the structure and function of Na^+ channels.

The earliest reports on the reconstitution of Na^+ channel function described the recovery of veratridine-stimulated $^{22}Na^+$ transport from lobster nerve membrane fragments (Villegas et al. 1977) or cholate extracts of brain membrane (Malysheva et al. 1980) incorporated into liposomes. Goldin et al. (1980), using a density-shift technique, demonstrated veratridine-stimulated Cs^+ transport into phosphatidyl choline vesicles containing cholate-solubilized brain Na^+ channels. Our laboratory (Tamkun and Catterall 1981a) reconstituted cholate-solubilized Na^+ channels from rat brain synaptosomes into phosphatidyl choline vesicles. Complete recovery of neurotoxin binding and action at all three toxin receptor sites, as well as Na^+ transport by the reconstituted channel was obtained. These studies demonstrated that Na^+ channel function can be recovered from soluble membrane extracts following the removal of detergent and incorporation of the crude extracts into phospholipid vesicles.

More recently, Na^+ channel ion transport has been successfully reconstituted from Na^+ channels substantially purified from rat muscle (Weigele and Barchi 1982) and from rat brain membranes (Talvenheimo et al. 1982). We have now applied these reconstitution methods to essentially homogeneous preparations of Na^+ channels from rat brain. Purified Na^+ channels in reconstitution medium (67.5 mM Na_2SO_4, 25 mM HEPES [adjusted to pH 7.4 with Tris base], 0.5 mM $MgSO_4$, and 150 mM sucrose) containing 0.1% Triton X-100 and 0.025% phosphatidyl choline are supplemented with egg phosphatidyl choline dispersed in Triton X-100. Phosphatidyl choline vesicles containing Na^+ channels are then formed by adsorption of the Triton X-100 to polystyrene beads (Bio Beads, Bio Rad Co.) during an overnight incubation. Vesicles with a mean diameter of 1800 Å containing an average of 0.75–2 Na^+ channels each are formed (Talvenheimo et al. 1983). The recovery of functional activities of the Na^+ channel that are lost upon detergent solubilization is then assessed in ion flux and neurotoxin-binding experiments.

Whereas Na^+ channels solubilized and purified in Triton X-100 retain high affinity for STX, their binding activity becomes labile to incubation at 36°C (Fig. 5, ○). Essentially all activity is lost in 5 minutes at 36°C.

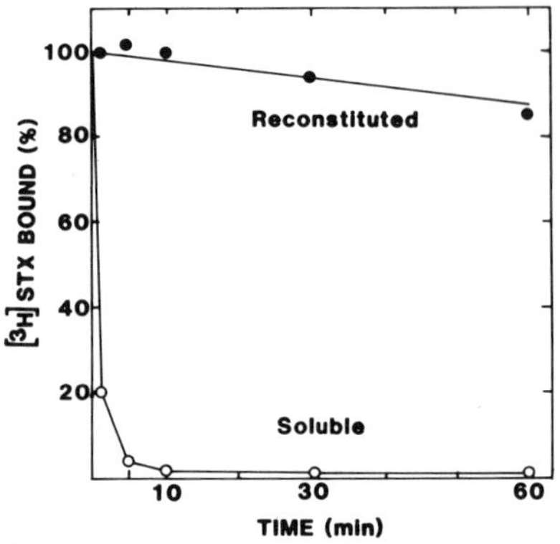

Figure 5. Thermal stability of STX-binding activity of purified and reconstituted Na^+ channels. The purified Na^+ channel in soluble form in Triton X-100 solution (○) or after reconstitution in phosphatidyl choline vesicles (●) was incubated at 36°C for the indicated times and then specific binding of [^{3}H]STX was measured as described by Talvenheimo et al. (1983).

In contrast, all of the STX-binding activity of Na^+ channels reconstituted into phosphatidyl choline vesicles is stable to incubation at 36°C for 60 minutes (Fig. 5, ●) as is the STX-binding activity of the native Na^+ channel. Thus, reconstitution into phosphatidyl choline vesicles successfully restores the thermal stability of all of the purified Na^+ channels. These results indicate that the channels are quantitatively incorporated into the vesicle bilayer in a state resembling that in native membranes.

Ion flux mediated by the purified and reconstituted Na^+ channels is illustrated in Figure 6. In these experiments, the vesicle preparation was incubated for 2 minutes with veratridine to activate Na^+ channels and then diluted into medium containing $^{22}Na^+$ to initiate influx into vesicles. After time intervals of 10 seconds to 15 minutes, the samples were applied to columns of Dowex 50W-X8 and washed through with isotonic sucrose. The vesicles and intravesicular ^{22}Na pass through the column, whereas extravesicular ^{22}Na is quantitatively bound to the column. $^{22}Na^+$ influx is then measured by gamma counting. Influx into vesicles under control conditions is slow (Fig. 6A). Incubation with veratridine increases the initial rate of influx 10–15-fold. When TTX is present in both the intravesicular and extravesicular phases, the veratridine-dependent increase in initial rate of $^{22}Na^+$ influx is nearly completely blocked (Fig. 6A). Half-maximal activation is observed with 28 μM veratridine and half-maximal inhibition with 18 nM TTX in close agreement with $K_{0.5}$ values for the action of these toxins on native Na^+ channels. These results show that the purified Na^+ channel regains the ability to mediate neurotoxin-stimulated ion flux after incorporation into phosphatidyl choline vesicles. Evidently, the purified channel retains neurotoxin receptor site 2 and the ion-conducting pore of the Na^+ channel in functional form.

Ion transport by neurotoxin-activated Na^+ channels in neural membranes is selective, although the rate of

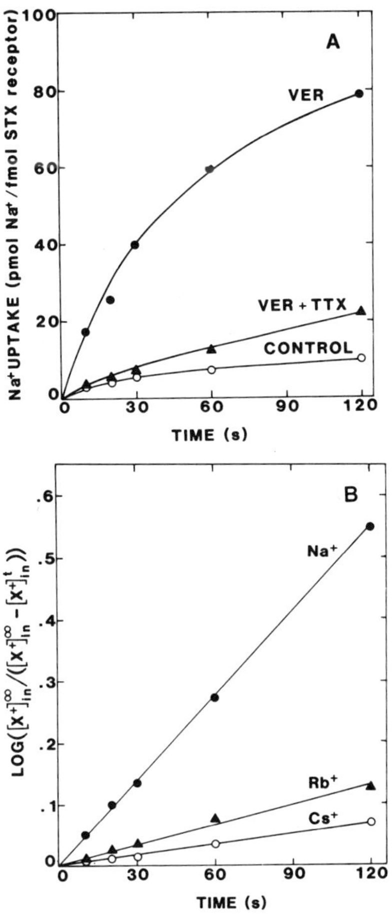

Figure 6. Rate and ion selectivity of veratridine-activated ion flux by the purified and reconstituted Na^+ channel. (*A*) Phosphatidyl choline vesicles containing purified Na^+ channels were incubated for 2 min at 36°C with 100 μM veratridine and the initial rate of $^{22}Na^+$ influx was measured as described by Talvenheimo et al. (1983). An identical sample of vesicles was prepared with 1 μM TTX inside, incubated for 2 min at 36°C with 100 μM veratridine and 1 μM TTX, and the initial rate of $^{22}Na^+$ influx was measured. (*B*) Veratridine-activated ion flux was measured as in *A* and the control rate of influx was subtracted for ^{22}Na, ^{86}Rb, and ^{137}Cs. The data are presented as a linear first-order rate plot.

transport of large cations such as Rb^+ and Cs^+ relative to the rate of transport of Na^+ is significantly greater than when channels are activated by membrane depolarization (Khodorov 1978; Huang et al. 1979; Frelin et al. 1981). Figure 6B illustrates the initial rate of influx of $^{22}Na^+$, $^{86}Rb^+$, and $^{137}Cs^+$ through veratridine-activated Na^+ channels in reconstituted vesicles. The purified and reconstituted Na^+ channel retains ion selectivity with permeability ratios of 0.25 for Rb^+ and 0.12 for Cs^+ relative to Na^+. These permeability ratios compare favorably with those of native Na^+ channels activated by neurotoxins (Talvenheimo et al. 1983).

The results of Figure 6 show that at least a fraction of the Na^+ channels in our most highly purified preparations can mediate selective neurotoxin-activated ion transport after incorporation into phospholipid vesicles. We have taken two approaches to estimating how many of the reconstituted Na^+ channels contribute to our ion flux measurements. First, we have compared the ion-transport rates measured in our purified and reconstituted Na^+ channel preparations to those measured for veratridine-activated Na^+ channels in neuroblastoma cells and synaptosomes (Table 2). This comparison shows that the transport rate measured in ions/min• STX receptor site is 33–70% of that in native membranes, suggesting that at least 33–70% of the reconstituted Na^+ channels are active. In a second approach, we have compared the fraction of vesicles that contain Na^+ channels to the fraction of vesicles whose internal volume is accessible to veratridine-activated Na^+ channels (Talvenheimo et al. 1983). Assuming that Na^+ channels are distributed among vesicles according to a Poisson distribution, this comparison leads to a range of 30–70% for the fraction of active channels, depending upon whether active vesicles containing more than one channel are assumed to have one active channel or all active channels. Both of these estimates indicate that a minimum of 30% of the reconstituted Na^+ channels are active. Since the Na^+ channel preparation is 90% pure and no single contaminant comprises as much as 1% of the protein, we conclude that the purified complex of α, $\beta1$, and $\beta2$ is sufficient to mediate selective neurotoxin-activated ion flux.

Although Na^+ channels reconstituted into phosphatidyl choline vesicles can transport Na^+, these channels do not bind scorpion toxin (Fig. 7A). In contrast, if purified Na^+ channels are reconstituted into a mixture of phosphatidyl choline and brain lipids, scorpion

Table 2. Ion Flux Rates of Veratridine-activated Na^+ Channels In Situ and in Reconstituted Vesicles

Preparation	Purity (%)	Channel density[a] (μm^{-2})	Flux rate[b] (ions/min•site)
Neuroblastoma cells	—	67.5	2.9×10^6
Synaptosomes	—	129.5	1.4×10^6
Reconstituted, partially purified Na^+ channels	15–20	—	0.6×10^6
Reconstituted, pure Na^+ channels	>90	6.2	1.0×10^6

[a]The number of STX binding sites per μm^2 of membranes surface.
[b]The number of ions transported per STX-binding site extrapolated to a physiological Na^+ concentration of 150 mM.

toxin binding is recovered. In this experiment with partially purified Na^+ channels, 0.5 mole of scorpion toxin was bound per mole of reconstituted Na^+ channels with a K_d of approximately 50 nM. With our most highly purified channel preparation, a similar level of scorpion toxin binding is observed after reconstitution in a mixture of phosphatidyl choline and brain lipid (Fig. 8). Scatchard analysis of this experiment indicates a K_d of 60 nM and a binding capacity of 0.86 mole scorpion toxin per mole STX receptor. In several similar experiments, a mean of 0.76 ± 0.08 mole scorpion toxin was bound per mole of purified Na^+ channel (Tamkun et al. 1983). The brain lipid fraction was prepared by $CHCl_3/CH_3OH$ extraction followed by silicic acid chromatography and was found to be protein free by gel electrophoresis and sensitive silver staining.

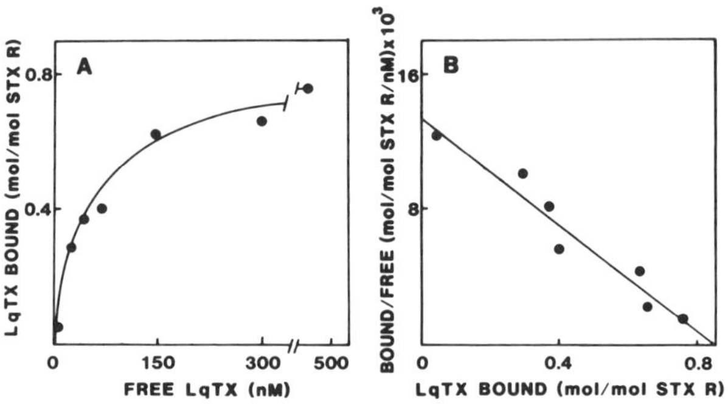

Figure 8. Scorpion toxin binding to purified and reconstituted Na^+ channels. (*A*) Purified Na^+ channels were reconstituted in vesicles composed of phosphatidyl choline and brain lipid and specific scorpion toxin binding was measured at increasing toxin concentrations. (*B*) The same data are presented as a Scatchard plot.

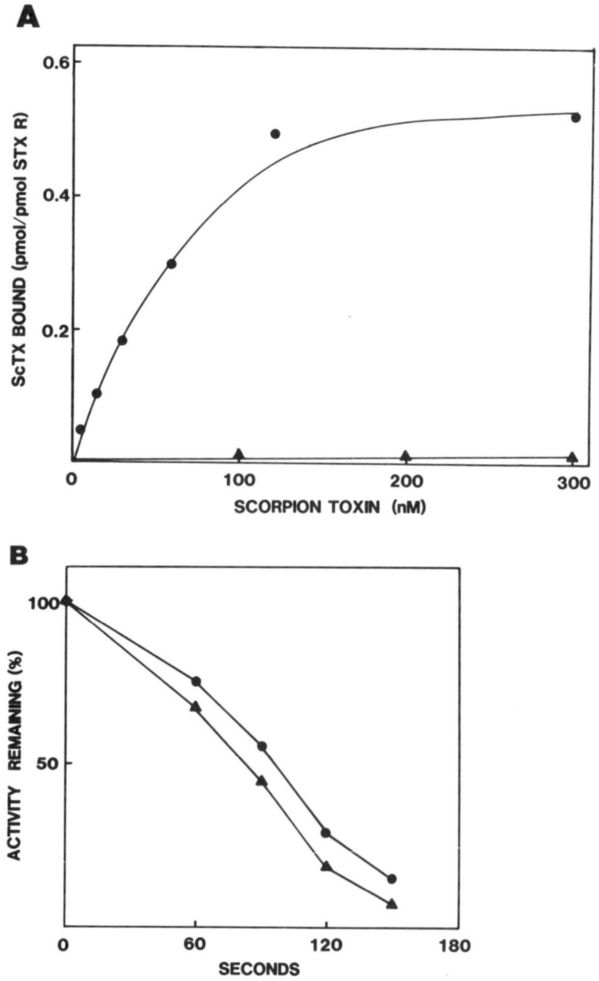

Figure 7. Reconstitution of scorpion toxin binding requires brain lipid. (*A*) Specific scorpion toxin binding to partially purified Na^+ channels reconstituted in phosphatidyl choline (▲) or in phosphatidyl choline plus a protein-free rat brain lipid extract (●) was measured using a rapid filtration assay as described by Tamkun et al. (1983). (*B*) The fully purified Na^+ channel in Triton X-100 solution was incubated at 36°C for the indicated times, reconstituted into vesicles composed of phosphatidyl choline and brain lipid, and specific binding of [^{3}H]STX (●) and [^{125}I]scorpion toxin (▲) was measured as described by Tamkun et al. (1983).

Therefore, we conclude that brain lipids are essential to restore the scorpion toxin receptor site to the same functional state as in native membranes. Since the affinity for scorpion toxin binding to synaptosomal Na^+ channels is dependent upon the functional state of the Na^+ channel, as reflected in the voltage-dependence of toxin binding and its allosteric modulation by alkaloid toxins such as veratridine (Ray et al. 1978; Catterall 1979; Tamkun and Catterall 1981b), brain lipid may also be required for other functional activities of the channel.

One possible interpretation of these results is that scorpion toxin actually binds to the brain lipid rather than the Na^+ channel. The high affinity and low binding capacity of the reconstituted vesicles for scorpion toxin are most consistent with binding to Na^+ channels. We have obtained direct evidence for that by studying the thermal denaturation of STX- and scorpion toxin-binding activities (Fig. 7B). Incubation of the purified Na^+ channel preparation at 36°C prior to reconstitution causes a rapid reduction in the STX-binding activity recovered in vesicles following reconstitution (Fig. 7B, ▲). The ability to recover scorpion toxin-binding activity following reconstitution into a mixture of phosphatidyl choline and brain lipids is also lost during incubation of soluble channel preparation at 36°C (Fig. 7B, ●). The time course of loss of the two activities is similar. These results show directly that the active Na^+ channels are required for reconstitution of scorpion toxin-binding activity. The brain lipid must be required to provide an adequate membrane environment for normal channel function.

CONCLUSION

Our results show that the Na^+ channel from rat brain consists of a stoichiometric complex of the α-, $\beta 1$-, and $\beta 2$-subunits both in purified preparations and in situ in synaptosomes. Essentially homogeneous preparations of this complex are sufficient to mediate most of the functions of the Na^+ channel that can be measured using biochemical methods, including neurotoxin binding and action at three neurotoxin receptor sites and neurotoxin-activated ion flux when incorporated into appropriate phospholipid vesicles. Although most of the functional properties of the purified and reconstituted Na^+ channels that we have measured are retained, we have discovered some potentially important functional deficits. Although scorpion toxin binding to Na^+ channels in native membranes is voltage dependent (Catterall 1979; Catterall et al. 1979), there is insignificant voltage sensitivity of the binding to purified and reconstituted channels (Tamkun et al. 1983), even in the presence of brain lipid. In addition, the allosteric interaction between neurotoxin receptor sites 2 and 3 observed in native Na^+ channels is not detectable with purified and reconstituted Na^+ channels. We believe these remaining functional deficits reflect an inappropriate lipid environment in the reconstituted vesicles and/or the presence of residual Triton X-100. Work is in progress to optimize the lipid composition of the reconstituted vesicles to restore these remaining aspects of channel function.

In excitable membranes, Na^+ channels are activated and inactivated by changes in membrane potential. Purified and reconstituted Na^+ channels have not yet been tested for their ability to activate and inactivate on depolarization. Assessment of this important aspect of channel function awaits application of methods to record electrically from purified channels in large reconstituted vesicles or planar bilayers. These experiments will provide the final test for the functional integrity of the purified and reconstituted channels.

ACKNOWLEDGMENTS

The work described was supported by Research Grant RO1-NS15751 from the National Institutes of Health and Research Grant BNS80-21619 from the National Science Foundation to W.A. Catterall. J.A. Talvenheimo was supported by National Research Service Award F32-NS06816, M.M. Tamkun and D.J. Messner by Predoctoral Training Grant T32-GM07270, and R.G. Sharkey by National Research Service Award F32-HL06328 from the National Institutes of Health. R.P. Hartshorne was supported by Predoctoral Training Grant T32-GM07750 from the National Institutes of Health and by a Research Fellowship from the Merck Foundation.

REFERENCES

AGNEW, W.S., S.R. LEVINSON, J.S. BRABSON, and M.A. RAFTERY. 1978. Purification of the tetrodotoxin binding component associated with the voltage-sensitive sodium channel from *Electrophorus electricus* electroplax membranes. *Proc. Natl. Acad. Sci.* **75:** 2606.

AGNEW, W.S., A.C. MOORE, S.R. LEVINSON, and M.A. RAFTERY. 1980. Identification of a large molecular weight peptide associated with a tetrodotoxin binding protein from the electroplax of *Electrophorus electricus*. *Biochem. Biophys. Res. Commun.* **92:** 860.

ALBUQUERQUE, E.X. and J.W. DALY. 1976. Batrachotoxin, a selective probe for channels modulating sodium conductance in electrogenic membranes. In *Receptors and recognition* (ed. P. Cuatrecasas), ser. B, vol. 1, p. 299. Chapman and Hall, London.

ARMSTRONG, C.M. and F. BEZANILLA. 1974. Charge movement associated with the opening and closing of the activation gates of the Na channels. *J. Gen. Physiol.* **63:** 533.

——. 1977. Inactivation of the sodium channel. II. Gating current experiments. *J. Gen. Physiol.* **70:** 567.

BARCHI, R.L. 1983. Protein components of the purified sodium channel from rat skeletal muscle sarcolemma. *J. Neurochem.* **40:** 1377.

BARCHI, R.L. and L.E. MURPHY. 1981. Estimate of the molecular weight of the sarcolemmal sodium channel using H_2O-D_2O centrifugation. *J. Neurochem.* **36:** 2097.

BARCHI, R.L., S.A. COHEN, and L.E. MURPHY. 1980. Purification from rat sarcolemma of the saxitoxin binding component of the excitable membrane sodium channel. *Proc. Natl. Acad. Sci.* **77:** 1306.

BARHANIN, J., A. SCHMID, A. LOMBET, K.P. WHEELER, and M. LAZDUNSKI. 1983. Molecular size of different neurotoxin receptors of the sodium channel. *J. Biol. Chem.* **258:** 700.

BARHANIN, J., J.R. GIGLIO, P. LEOPOLD, A. SCHMID, S.V. SAMPAIO, and M. LAZDUNSKI. 1982. *Tityus serrulatus* venom contains two classes of toxins. *J. Biol. Chem.* **257:** 12553.

BENESKI, D. and W.A. CATTERALL. 1980. Covalent labelling of protein components of the sodium channel with a photoactivable derivative of scorpion toxin. *Proc. Natl. Acad. Sci.* **77:** 639.

BENZER, T.I. and M.A. RAFTERY. 1973. Solubilization and partial characterization of the tetrodotoxin binding component from nerve axons. *Biochem. Biophys. Res. Commun.* **51:** 939.

BROWN, G.B., S.C. TIESZEN, J.W. DALY, J.E. WARNICK, and E.X. ALBUQUERQUE. 1981. Batrachotoxinin A 20-α-benzoate: A new radioactive ligand for voltage-sensitive sodium channels. *Cell. Mol. Neurobiol.* **1:** 19.

CAHALAN, M.D. 1975. Modification of sodium channel gating in frog myelinated nerve fibers by *Centruroides sculpturatus* scorpion venom. *J. Physiol.* **244:** 511.

CATTERALL, W.A. 1979. Binding of scorpion toxin to receptor sites associated with sodium channels in frog muscle. Correlation of voltage dependent binding with activation. *J. Gen. Physiol.* **74:** 375.

——. 1980. Neurotoxins that act on voltage-sensitive sodium channels. *Annu. Rev. Pharmacol. Toxicol.* **20:** 15.

CATTERALL, W.A., C.S. MORROW, and R.P. HARTSHORNE. 1979. Neurotoxin binding to receptor sites associated with sodium channels in intact, lysed, and detergent solubilized brain membranes. *J. Biol. Chem.* **254:** 11379.

CATTERALL, W.A., C.S. MORROW, J.W. DALY, and G.B. BROWN. 1981. Binding of batrachotoxinin A 20-α-benzoate to a receptor site associated with sodium channels in synaptic nerve ending particles. *J. Biol. Chem.* **256:** 8922.

COSTA, M.R.C. and W.A. CATTERALL. 1982. Characterization of variant neuroblastoma clones with missing or altered sodium channels. *Mol. Pharmacol.* **22:** 196.

COURAUD, F., E. JOVER, J.M. DUBOIS, and H. ROCHAT. 1982. Two types of scorpion toxin receptor sites, one related to activation, the other to inactivation of the action potential sodium channel. *Toxicon* **20:** 9.

FRELIN, C., P. VIGNE, and M. LAZDUNSKI. 1981. The specificity of the sodium channel for monovalent cations. *Eur. J. Biochem.* **119:** 437.

GOLDIN, S.M., V. RHODEN, and E.J. HESS. 1980. Molecular characterization, reconstitution, and "transport-specific fractionation" of the saxitoxin binding protein/Na^+ gate of mammalian brain. *Proc. Natl. Acad. Sci.* **77:** 6884.

HARTSHORNE, R.P. and W.A. CATTERALL. 1980. Purification of the sodium channel from rat brain. *Soc. Neurosci. Abstr.* **6:** 174.

——. 1981. Purification of the saxitoxin receptor of the sodium channel from mammalian brain. *Proc. Natl. Acad. Sci.* **78:** 4620.

——. 1983. The sodium channel from rat brain. Purification and subunit composition. *J. Biol. Chem.* (in press).

HARTSHORNE, R.P., J. COPPERSMITH, and W.A. CATTERALL. 1980. Size characteristics of the solubilized saxitoxin receptor of the voltage sensitive sodium channel from rat brain. *J. Biol. Chem.* **255:** 10572.

HARTSHORNE, R.P., D.J. MESSNER, J.C. COPPERSMITH, and W.A. CATTERALL. 1982. The saxitoxin receptor of the sodium channel from rat brain. Evidence for two nonidentical β subunits. *J. Biol. Chem.* **257:** 13888.

HENDERSON, R. and J.H. WANG. 1972. Solubilization of a specific tetrodotoxin binding component from garfish olfactory nerve membranes. *Biochemistry* **11:** 4565.

HILLE, B. 1971. The permeability of the sodium channel to organic cations in myelinated nerve. *J. Gen. Physiol.* **58:** 599.

——. 1972. The permeability of the sodium channel to metal cations in myelinated nerve. *J. Gen. Physiol.* **59:** 637.

HODGKIN, A.L. and A.F. HUXLEY. 1952. A quantitative description of membrane current and its application to conduction and excitation in nerve. *J. Physiol.* **117:** 500.

HUANG, L.M., W.A. CATTERALL, and G. EHRENSTEIN. 1979. Comparison of ionic selectivity of batrachotoxin-activated sodium channels with different tetrodotoxin dissociation constants. *J. Gen. Physiol.* **73:** 839.

JOVER, E., F. COURAUD, and H. ROCHAT. 1980. Two types of scorpion neurotoxins characterized by their binding to two separate receptor sites on rat brain synaptosomes. *Biochem. Biophys. Res. Commun.* **95:** 1607.

KHODOROV, B.I. 1978. Chemicals as tools to study nerve fiber sodium channels: Effects of batrachotoxin and some local anesthetics. In *Membrane transport processes* (ed. D.C. Tosteson et al.), vol. 2, p. 153. Raven Press, New York.

KRUEGER, B.K., R.W. RATZLAFF, and G.R. STRICHARTZ. 1979. Saxitoxin binding to synaptosomes, membranes, and solubilized binding sites from rat brain. *J. Membr. Biol.* **50:** 287.

LAZDUNSKI, M. and J.F. RENAUD. 1982. The action of cardiotoxins on cardiac plasma membranes. *Annu. Rev. Physiol.* **44:** 463.

LEVINSON, S.R. and J.C. ELLORY. 1973. Molecular size of the tetrodotoxin binding site estimated by irradiation inactivation. *Nature* **245:** 111.

MALYSHEVA, M.K., U.K. LISHKO, and A.M. CHAGOVETZ. 1980. The association of tetrodotoxin-sensitive, sodium-selective ionophore of brain membranes with liposomes. *Biochim. Biophys. Acta* **602:** 70.

MEVES, H., N. RUBLY, and D.D. WATT. 1982. Effect of toxins isolated from the venom of the scorpion *Centruroides sculpturatus* on the Na currents of the node of Ranvier. *Pfluegers Arch.* **393:** 56.

MILLER, J.A., W.S. AGNEW, and S.R. LEVINSON. 1983. Principal glycopeptide of the tetrodotoxin/saxitoxin binding protein from *Electrophorus electricus*: Isolation and partial chemical and physical characterization. *Biochemistry* **22:** 462.

MOORE, H.P.H., L.C. FRITZ, M.A. RAFTERY, and J.P. BROCKES. 1982. Isolation and characterization of a monoclonal antibody against the saxitoxin-binding component from the electric organ of the eel *Electrophorus electricus*. *Proc. Natl. Acad. Sci.* **79:** 1673.

MOZHAYEVA, G.N., A.P. NAUMOV, E.D. NOSYREVA, and E.V. GRISHIN. 1980. Potential dependent interaction of toxin from the venom of the scorpion *Buthus eupeus* with sodium channels in myelinated fiber. Voltage clamp experiments. *Biochim. Biophys. Acta* **597:** 587.

NAKAYAMA, H., R.M. WITHY, and M.A. RAFTERY. 1982. Use of a monoclonal antibody to purify the tetrodotoxin binding component from the electroplax of *Electrophorus electricus*. *Proc. Natl. Acad. Sci.* **79:** 7575.

NARAHASHI, T. 1974. Chemicals as tools in the study of excitable membranes. *Physiol. Rev.* **54:** 813.

RAY, R., C.S. MORROW, and W.A. CATTERALL. 1978. Binding of scorpion toxin to receptor sites associated with voltage-sensitive sodium channels in synaptic nerve ending particles. *J. Biol. Chem.* **253:** 7301.

RITCHIE, J.M. and R.B. ROGART. 1977. The binding of saxitoxin and tetrodotoxin to excitable tissue. *Rev. Physiol. Biochem. Pharmacol.* **79:** 1.

SHARKEY, R.G., D.A. BENESKI, and W.A. CATTERALL. 1983. Specific labeling of the α and $\beta 1$ subunits of the sodium channel by photoreactive derivatives of scorpion toxin. *J. Biol. Chem.* (in press).

TALVENHEIMO, J.A., M.M. TAMKUN, and W.A. CATTERALL. 1982. Reconstitution of neurotoxin-stimulated sodium transport by the voltage-sensitive sodium channel purified from rat brain. *J. Biol. Chem.* **257:** 11868.

——. 1983. The sodium channel from rat brain. Reconstitution of neurotoxin-stimulated ion flux from purified components. *J. Biol. Chem.* (in press).

TAMKUN, M.M. and W.A. CATTERALL. 1981a. Reconstitution of the voltage-sensitive sodium channel of rat brain from solubilized components. *J. Biol. Chem.* **256:** 11457.

——. 1981b. Ion flux studies of voltage-sensitive sodium channels in synaptic nerve ending particles. *Mol. Pharmacol.* **19:** 78.

TAMKUN, M.M., J.A. TALVENHEIMO, and W.A. CATTERALL. 1983. The sodium channel from rat brain. Reconstitution of scorpion toxin binding. *J. Biol. Chem.* (in press).

VILLEGAS, R., G.M. VILLEGAS, F.V. BARNOLA, and E. RACKER. 1977. Incorporation of the sodium channel of lobster nerve into artificial liposomes. *Biochem. Biophys. Res. Commun.* **79:** 210.

WANG, G.K. and G. STRICHARTZ. 1982. Simultaneous modifications of sodium channel gating by two scorpion toxins. *Biophys. J.* **40:** 175.

WEIGELE, J.B. and R.L. BARCHI. 1982. Functional reconstitution of the purified sodium channel protein from rat sarcolemma. *Proc. Natl. Acad. Sci.* **79:** 3651.

Errata Sheet
Cold Spring Harbor Symposia on Quantitative Biology
Volume XLVIII, p. 165

The Voltage-regulated Sodium Channel from the Electroplax of *Electrophorus electricus*

W.S. AGNEW,† J.A. MILLER,† M.H. ELLISMAN,‡ R.L. ROSENBERG,† S.A. TOMIKO,† AND S.R. LEVINSON*

*†Department of Physiology, Yale University School of Medicine, New Haven, Connecticut 06510; ‡Department of Neuroscience, University of California, San Diego, La Jolla, California 92093; *Department of Physiology, University of Colorado Medical Center, Denver, Colorado 80262*

The basis of the electrical excitability exhibited by cells from nerve, muscle, and related tissues resides in the properties of complex membrane proteins that act as ion-selective conductance channels. An understanding of the events in the membrane governing the electrical behavior will require knowledge of the molecular structures and mechanisms of these ion channel proteins. The studies discussed here are focused on the voltage-regulated Na^+ channel that mediates the early Na^+ currents of the propagated action potential (Hodgkin and Huxley 1952; Almers 1978; Armstrong 1981; Cahalan 1981).

The approach being pursued involves isolation of the protein in chemically pure form. These preparations permit compositional and structural analysis and are suitable for studies of functional reconstitution. In addition to enabling these mechanistic studies, purified preparations can be used to elicit antibodies that will have biochemical and immunocytochemical applications. Recently, rapid progress has been made in the study of Na^+ channel proteins from the electric organ of electric eel (Agnew et al. 1978; Miller et al. 1983), mammalian brain (Hartshorne and Catterall 1981; Talvenheimo et al. 1982), and mammalian skeletal muscle (Barchi et al. 1980; Weigele and Barchi 1982; Barchi 1983). This paper summarizes results from studies of the protein from the electric eel, with some reference to findings from other laboratories working with the other tissues. No attempt has been made, however, to provide a comprehensive review of progress in the field (for a brief review, see Agnew 1984).

The Na^+ channel is a mechanistically complex intrinsic membrane protein with four broad characteristics essential to its role in impulse propagation. These are high single-channel conductances ($\gamma \cong 8$ pS, representing $\sim 10^7$ ions/sec at physiological membrane potentials and ionic conditions), ion permeation selectivity ($Li \geq Na^+ > K^+ > Rb^+ > Cs^+$), and time and voltage-dependent activation and inactivation gating mechanisms. When activated, the channel forms an ion-selective pore through which Na^+ ions preferentially diffuse in passive response to electrochemical driving forces. At resting membrane potentials, the channel is closed. Upon rapid depolarization, however, the channel is rapidly activated by a voltage-dependent mechanism that responds with apparently second- or third-order kinetics. A secondary process, probably not completely independent of the activation mechanism and which effectively responds more slowly than activation, serves to inactivate the channel, leaving it in a closed or refractory state. The reader is referred to two recent reviews for a comprehensive discussion of these precise and sophisticated regulatory mechanisms (Almers 1978; Armstrong 1981) and to Aldrich and Stevens (this volume).

MATERIALS AND METHODS

Biochemical Markers

Tritiated derivatives of the neurotoxins tetrodotoxin (TTX) and saxitoxin (STX) have been used in binding assays to follow the solubilization, stabilization, and isolation of the Na^+ channel protein. These selective toxins bind to a single class of sites on the outside of the channel in or near the mouth of the ion pathway (Hille 1975; Ritchie and Rogart 1977; Kao and Walker 1982). Binding of TTX and STX is reversible, mutually competitive, and has K_d values of 1–10 nM. Binding stoichiometry appears to be one toxin per channel, and neither toxin seems to interact with channel gating mechanisms. Aside from antibodies raised to the pure protein, TTX and STX derivatives are the only ligands presently identified that associate equally well with both the membrane-bound and detergent-solubilized preparations of the molecule.

Tissue

The main electric organ of the freshwater eel, *Electrophorus electricus*, serves as a nearly homogeneous tissue source for preparative scale isolation. The organ (200–500 g) is formed by stacks of large electroplax cells that are derived during development from neonatal muscle. The rostral face of each cell (see, for example, Fig. 8A,B) is noninnervated, convoluted, and is enriched in the Na^+-K^+ ATPase that generates the ionic gradients across the membrane (Keynes and Martins-Ferreira 1953; Nakamura et al. 1965). The caudal surface is flattened, but with multiple caveolar projections toward the cell interior. This face is innervated by cholinergic electromotor neurons, with 1–2% of the surface being subsynaptic. The tissue is rich in nicotinic acetylcholine receptors (AChR) and was one of the first

The Voltage-regulated Sodium Channel from the Electroplax of *Electrophorus electricus*

W.S. AGNEW,† J.A. MILLER,† M.H. ELLISMAN,‡ R.L. ROSENBERG,† S.A. TOMIKO,† AND S.R. LEVINSON*

*†Department of Physiology, Yale University School of Medicine, New Haven, Connecticut 06510; ‡Department of Neuroscience, University of California, San Diego, La Jolla, California 92093; *Department of Physiology, University of Colorado Medical Center, Denver, Colorado 80262*

The basis of the electrical excitability exhibited by cells from nerve, muscle, and related tissues resides in the properties of complex membrane proteins that act as ion-selective conductance channels. An understanding of the events in the membrane governing the electrical behavior will require knowledge of the molecular structures and mechanisms of these ion channel proteins. The studies discussed here are focused on the voltage-regulated Na^+ channel that mediates the early Na^+ currents of the propagated action potential (Hodgkin and Huxley 1952; Almers 1978; Armstrong 1981; Cahalan 1981).

The approach being pursued involves isolation of the protein in chemically pure form. These preparations permit compositional and structural analysis and are suitable for studies of functional reconstitution. In addition to enabling these mechanistic studies, purified preparations can be used to elicit antibodies that will have biochemical and immunocytochemical applications. Recently, rapid progress has been made in the study of Na^+ channel proteins from the electric organ of electric eel (Agnew et al. 1978; Miller et al. 1983), mammalian brain (Hartshorne and Catterall 1981; Talvenheimo et al. 1982), and mammalian skeletal muscle (Barchi et al. 1980; Weigele and Barchi 1982; Barchi 1983). This paper summarizes results from studies of the protein from the electric eel, with some reference to findings from other laboratories working with the other tissues. No attempt has been made, however, to provide a comprehensive review of progress in the field (for a brief review, see Agnew 1984).

The Na^+ channel is a mechanistically complex intrinsic membrane protein with four broad characteristics essential to its role in impulse propagation. These are high single-channel conductances ($\gamma \cong 8$ pS, representing $\sim 10^7$ ions/sec at physiological membrane potentials and ionic conditions), ion permeation selectivity ($Li \geq Na^+ > K^+ > Rb^+ > Cs^+$), and time and voltage-dependent activation and inactivation gating mechanisms. When activated, the channel forms an ion-selective pore through which Na^+ ions preferentially diffuse in passive response to electrochemical driving forces. At resting membrane potentials, the channel is closed. Upon rapid depolarization, however, the channel is rapidly activated by a voltage-dependent mechanism that responds with apparently second- or third-order kinetics. A secondary process, probably not completely independent of the activation mechanism cholate column is illustrated in Figure 1B. The resolution of TTX-binding protein was far better than with Lubrol due to elution of contaminants as smaller particles than in Lubrol and due to a sharpening of the elution profile for each protein. The specific activities through the peak of toxin binding were between 1800 and 2000 pmoles/mg protein (i.e., ~0.6 that of the

MATERIALS AND METHODS

Biochemical Markers

Tritiated derivatives of the neurotoxins tetrodotoxin (TTX) and saxitoxin (STX) have been used in binding assays to follow the solubilization, stabilization, and isolation of the Na^+ channel protein. These selective toxins bind to a single class of sites on the outside of the channel in or near the mouth of the ion pathway (Hille 1975; Ritchie and Rogart 1977; Kao and Walker 1982). Binding of TTX and STX is reversible, mutually competitive, and has K_d values of 1–10 nM. Binding stoichiometry appears to be one toxin per channel, and neither toxin seems to interact with channel gating mechanisms. Aside from antibodies raised to the pure protein, TTX and STX derivatives are the only ligands presently identified that associate equally well with both the membrane-bound and detergent-solubilized preparations of the molecule.

Tissue

The main electric organ of the freshwater eel, *Electrophorus electricus*, serves as a nearly homogeneous tissue source for preparative scale isolation. The organ (200–500 g) is formed by stacks of large electroplax cells that are derived during development from neonatal muscle. The rostral face of each cell (see, for example, Fig. 8A,B) is noninnervated, convoluted, and is enriched in the Na^+-K^+ ATPase that generates the ionic gradients across the membrane (Keynes and Martins-Ferreira 1953; Nakamura et al. 1965). The caudal surface is flattened, but with multiple caveolar projections toward the cell interior. This face is innervated by cholinergic electromotor neurons, with 1–2% of the surface being subsynaptic. The tissue is rich in nicotinic acetylcholine receptors (AChR) and was one of the first sources from which the receptor was isolated (Olsen et al. 1972). The extrajunctional regions of the innervated face are electrically excitable and contain TTX- and STX-sensitive Na^+ channels that are pharmacologically and biophysically similar to those from nerve and muscle cells (Nakamura et al. 1965). The abundance of TTX- and STX-binding sites is very high ($\geq 500\ \mu M^{-2}$, 50–100 pmoles of sites per gram of tissue). This is about two to three times the abundance of the AChR, making electroplax a rich source for large-scale purification.

Purification Methods

The tissue is minced and homogenized and crude membranes are isolated by differential sedimentation. The TTX-binding sites can be solubilized by a variety of nonionic detergents, including Triton X-100, NP-40, Brij 56, and Lubrol-PX. The latter consistently solubilizes a larger fraction of the sites (>90%), also yielding the most stable extracts. The bile salts cholate and deoxycholate solubilize the protein with about half the efficiency of Lubrol, and stability is significantly lower in these detergents.

The solubilized material, even in Lubrol, is less stable than the membrane-bound form. This is at least in part due to disruption of the membrane microenvironment (Agnew and Raftery 1979). Stability is a quantitative function of the lipid: detergent ratio in the mixed micelles used for solubilization. A 1:7 molar ratio of phosphatidyl choline or phosphatidyl ethanolamine to Lubrol-PX leaves the protein relatively stable, yet in a soluble form permitting fractionation. Additional stabilization results from saturating the toxin-binding sites with TTX or STX (Agnew et al. 1978).

The two criteria used to assess the degree of purification of the protein are the specific activity (pmoles [^{3}H]TTX-binding sites/mg protein) and the peptide composition of the purified material. Molecular-weight estimates based on irradiation inactivation of TTX-binding sites (electroplax and synaptosomes) range from 240,000 to 266,000 (Levinson and Ellory 1972; Barhanin et al. 1983). Preliminary hydrodynamic estimates from synaptosomal and sarcolemmal STX-binding proteins have yielded estimates of $315{,}000 \pm 60{,}000$ (Hartshorne et al. 1980; Barchi and Murphy 1981). The sedimentation coefficient of the electroplax protein is approximately 8S, and the Stokes' radius is 95 Å, in reasonable agreement with those of the brain and muscle proteins. On calibrated Sepharose 4B gel filtration columns, the synaptosomal and electroplax proteins elute as almost precisely the same sized particles (M. Emerick and W. Agnew, unpubl.). Thus, assuming one toxin binding site per 315-kD protein, the specific activity of pure, fully active protein would be 3170 pmoles/mg protein.

Isolation of native protein in nonionic detergent. The purification of the TTX-binding protein solubilized in Lubrol-PX can be achieved with conventional purification techniques. The procedure that we have most frequently employed involves, first, fractionation by DEAE-Sephadex ion-exchange chromatography followed by two steps of gel filtration over Sepharose 6B. Initial 1% (w/v) detergent extracts from the membranes range from 15 to 30 pmoles [^{3}H]TTX-binding sites per mg protein, indicating that a 1–200-fold purification is required to achieve homogeneity.

To begin the isolation, an initial extract is made 0.2 M in NaCl and mixed with an equal volume of packed DEAE-Sephadex A-25. Normally, less than 2% of the protein, but all of the TTX-binding protein, is adsorbed. The gel is subsequently washed free of unbound protein and packed in a short column, and the toxin-binding protein is eluted stepwise with high salt. Recoveries from this step are from 50% to 70%, with increases in specific activity of 15–40-fold (see Table 1). The eluted material is concentrated by pressure dialysis and at this stage preparations have specific activities of 300 pmoles/mg protein to occasionally as high as 1000 pmoles/mg protein. The concentrate is fractionated on a column of Sepharose 6B, equilibrated with 0.1% Lubrol: phosphatidyl choline in phosphate buffer, with 20 nM [^{3}H]TTX. Recoveries from this step are generally 70–90% of the applied binding activity. Peak specific activities are often over 2000 pmoles TTX-binding sites/mg protein (Table 1), with values over 3000 pmoles/mg protein seen only on rare occasions. A profile of such a Sepharose column is illustrated in Figure 1, and SDS-polyacrylamide gels of the indicated fractions are shown below. In the illustrated experiment, the leading fractions from the peak of toxin binding had specific activities of over 2000 pmoles/mg protein. The peptide composition of this pooled material consisted of one large peptide together with lesser amounts of smaller peptides (Fig. 2A). Inspection of the gels from across the column profile illustrates that only the large peptide (apparent $M_r \sim 260{,}000$) correlated in distribution with the TTX-binding activity (see also Agnew et al. 1980). When the purest fractions from a column such as that shown in Figure 1A and with peptide composition seen in Figure 2A were subjected to a second step of gel filtration on a smaller Sepharose 6B column, the resulting protein exhibited a specific activity of 2200 pmoles/mg protein, which would be equivalent to 0.69 TTX-binding sites per protein of $M_r \sim 315{,}000$. The SDS-polyacrylamide gels of this material purified in this way revealed only a single type of large peptide, with no evidence for lower-molecular-weight species (Fig. 2B).

It seems that these preparations are homogeneous in that they contain one type of polypeptide and bind nearly stoichiometric amounts of TTX. The latter determination is subject to error in that the precise protein molecular weight has not been determined, and the accurate quantitation of protein by conventional assays can vary for pure protein species. In addition, a certain fraction of the purified material may be inactive.

The principal difficulty in the purification of the native protein has been incomplete resolution of the toxin-binding protein from contaminants on Sepharose 6B, so that yields of extremely pure native material are

Table 1. Purification of Native TTX- and STX-binding Protein

	[^{3}H]TTX bound (pmoles)	Protein (mg)	Specific activity (pmoles/mg of protein)	Percent yield
Lubrol-PX				
extract	5885	352	16.8	100
DEAE	3519	—	—	60
amicon concentrate	3004	6.1	495	51
Sepharose 6B column fraction				
19	90	0.68	1331	
20	254	0.119	2130	
21	428	0.208	2061	
22	543	0.289	1884	
23	496	0.374	1329	
24	333	0.481	690	
column pool	2448			42
Na$^+$ cholate				
extract	6248	260	24	100
DEAE				
adsorbed	4878			
recovered	1985	3.5	567	(41)[a]
amicon concentrate	1741	2.5	688	(36)[a]
Sepharose 6B column fraction				
35	—	—	1941	
36	—	—	2093	
37	—	—	1863	
38	—	—	1781	
39	—	—	1788	
40	—	—	1820	
column pool 1	931	0.531	1847	(19)[a]

Comparative data are presented for purifications of the protein in Lubrol PX, as described in Miller et al. (1983), and by the procedure employing Na$^+$ cholate, as summarized in the text.
[a]Corrected yields adjusted for initial adsorption to DEAE.

often low (~0.1 mg/purification). Two alternative procedures allow preparation of larger quantities of the pure protein for subsequent characterization. The first of these permits isolation of native material in sodium cholate suspensions and the second allows the preparation of substantial quantities (~0.5 mg) of the large peptide in SDS-denatured form.

Isolation of native protein in sodium cholate. The use of sodium cholate as a detergent offers two important advantages. First, because the detergent forms far smaller micelles than the nonionic detergents, lower amounts of detergent are expected to bind to the proteins. Thus, cholate is less likely to obscure size differences between the TTX-binding protein and smaller contaminant species during gel filtration. This expected enhanced resolution has been observed. In addition, cholate is a readily dialyzed detergent, a property often exploited for reconstitution protocols with membrane proteins. The principal disadvantage is that the binding activity is far less stable in cholate suspensions. It was found, however, that by performing all separation steps at low temperatures (0°C), recoveries and specific activities were not significantly lower than with Lubrol-PX.

The cholate purification illustrated in Figure 1B and Table 1 was carried out as follows. The TTX-binding protein was extracted in Lubrol-PX as before, because of the greater solubilization efficiency of this detergent. Again, after making the extract 0.2 M in NaCl, the protein was bound to DEAE-Sephadex. All subsequent steps were carried out as described previously, except that 1% cholate:phosphatidyl choline (7:1 molar ratio) was used rather than 0.1% Lubrol-PX. After desorption in 0.4 M NaCl and concentration, one step of purification over Sepharose 6B was performed. The column, however, was jacketed and kept at 0°C. Fractions were collected in a refrigerated aluminum fraction collector (0°C). A typical elution profile for such a cholate column is illustrated in Figure 1B. The resolution of TTX-binding protein was far better than with Lubrol due to elution of contaminants as smaller particles than in Lubrol and due to a sharpening of the elution profile for each protein. The specific activities through the peak of toxin binding were between 1800 and 2000 pmoles/mg protein (i.e., ~0.6 that of the theoretical maximum) with little variation through the peak (see Table 1). When the bulk of the binding activity was pooled and analyzed for peptide composition by SDS gel electrophoresis (>80% of the peak), the gels indicated that nearly all of the protein was in the large peptide, with low amounts of smaller peptides present (Fig. 2C). The densitometer scan of a silver-stained gel from a peak fraction from this column (Fig. 2D) showed no significant evidence of lower-molecular-weight peptides.

Isolation of the large peptide in SDS-denatured form. Because preparations of the molecule isolated either in Lubrol-PX or in cholate bind [^{3}H]TTX nearly

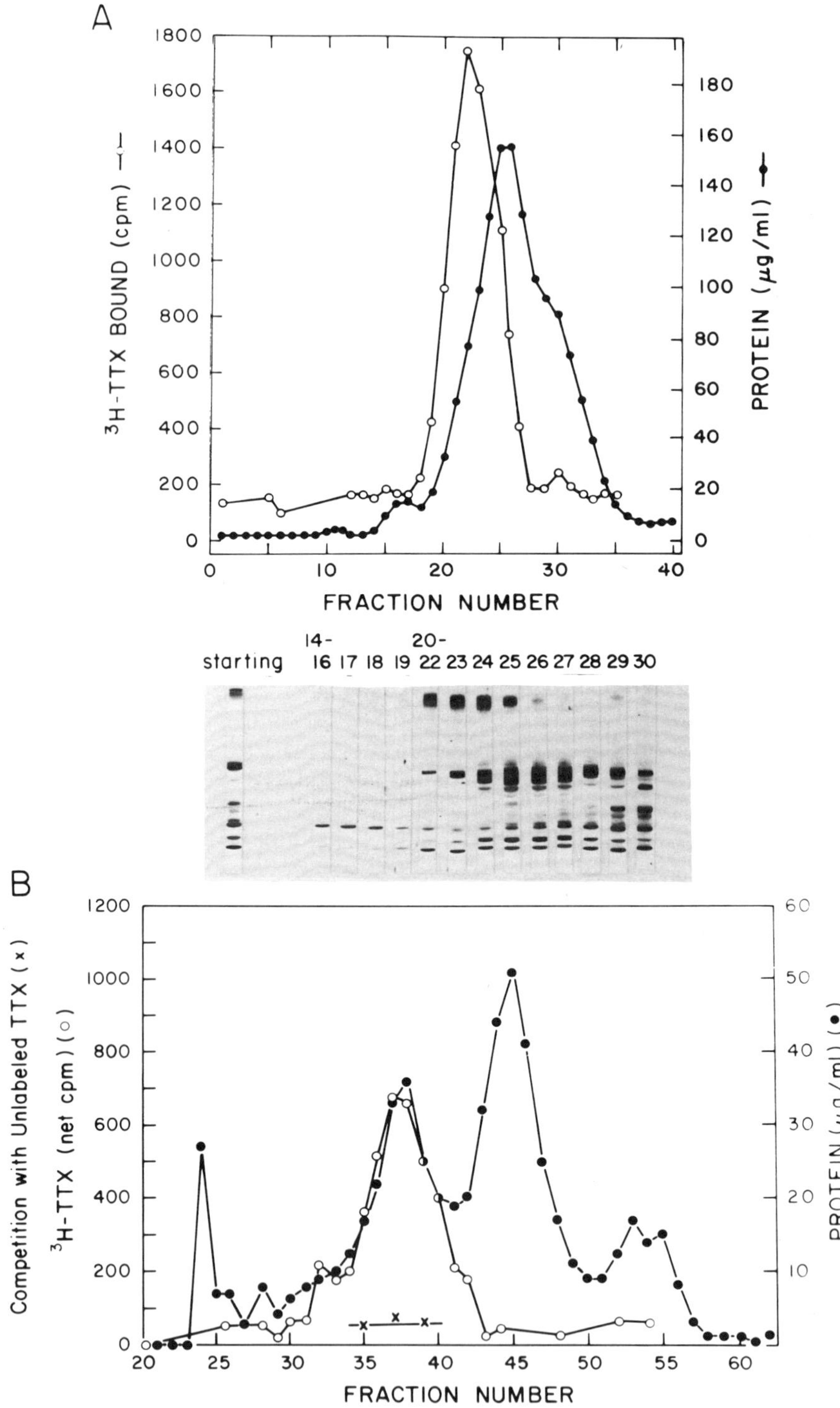

Figure 1. Sepharose 6B column chromatography of DEAE-fractionated material. (*A*) Column profile of sample run in 0.1% (w/v) Lubrol-PX:phosphatidyl choline (7:1 molar ratio). (○) [^{3}H]TTX binding; (●) protein as described in text. Below the figure are shown SDS-polyacrylamide gels to illustrate the polypeptide composition of the column fractions. Equal volumes of each fraction were applied to each gel, except for fractions 20–22 which were combined before gel analysis to conserve material for later analysis. (*B*) Column profile run in 1% (w/v) sodium cholate : phosphatidyl choline (12:1 molar ratio) as described in text. (○) [^{3}H]TTX-binding activity; (×) [^{3}H]TTX-binding activity in presence of ~1 μM cold TTX; (●) protein. Note the enhanced resolution in comparison to the Lubrol-PX column in *A*.

stoichiometrically, and contain only the high-molecular-weight species of peptide, it is possible to conclude that this peptide forms the TTX-binding site. Having identified this large peptide as belonging to the Na$^+$ channel, methods were set up for isolating it in large quantities, in SDS-denatured form. When fractions

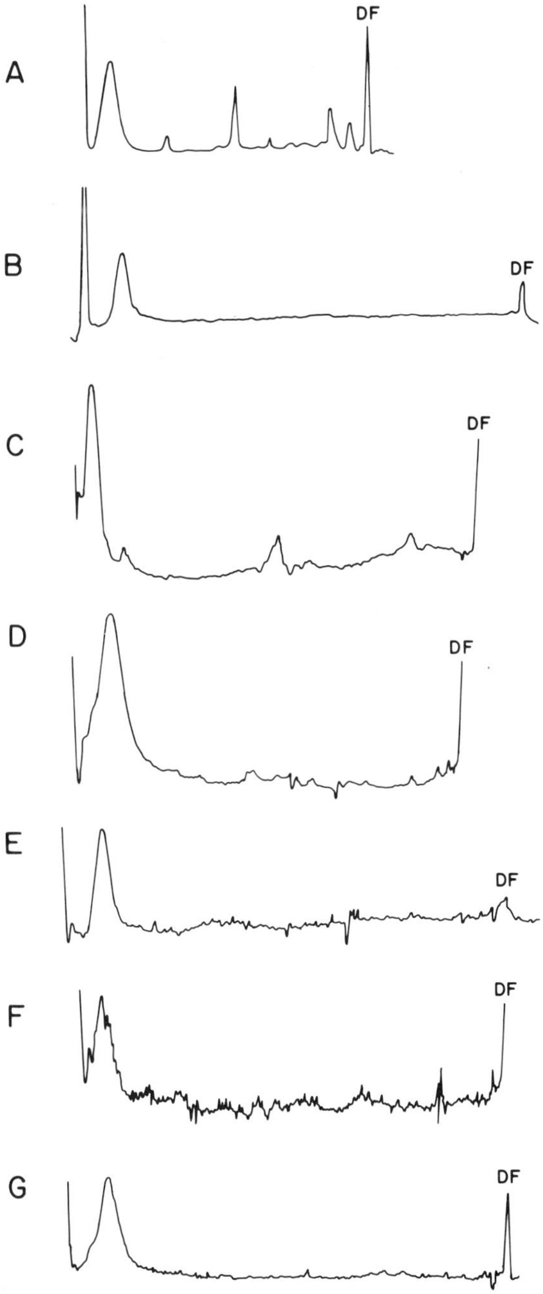

Figure 2. Densitometer scans of SDS-polyacrylamide gels. Absorbance readings were made at 650 nM, except for *F*, run at 560 nM. (*A*) Material purified by one step of ion-exchange chromatography and one step of gel filtration. This is a scan of the pooled, purest fractions (20–22 in Fig. 1A) from a Sepharose 6B chromatography column run in 0.1% Lubrol-PX : phosphatidyl choline. This represents the purest ~30% of total binding activity from the column. (*B*) Scan of material in *A*, above, after rechromatography on a second, smaller column of Sepharose 6B. Specific activity, 2200 pmoles [^{3}H]TTX-binding sites/mg protein. Note the absence of low-molecular-weight peptides. (*C*) Material purified by one step of DEAE ion-exchange chromatography and one step of gel filtration. This is a scan of pooled fractions, representing about 80% of the [^{3}H]TTX-binding activity, from a Sepharose 6B column run in 1% (w/v) cholate : phosphatidyl choline. (*D*) A scan of one of the purest fractions from a Sepharose 6B column run in cholate : phosphatidyl choline. Specific activity ~2100 pmoles [^{3}H]TTX-binding sites/mg protein. Note the absence of low-molecular-weight peptides. (*E*) A scan of material (pool) fractionated by SDS-Sepharose 4B column chromatography (cf. Fig. 3) as described in the text. (*F*) A densitometer scan of the SDS-Sepharose 4B-purified peptide stained by the periodic acid-Schiff procedure to test for the presence of carbohydrate. (*G*) Densitometer scan of material purified by DEAE-ion exchange, followed by cholate : phosphatidyl choline Sepharose 6B chromatography, followed by denaturation without reduction, and fractionation by SDS-Sepharose 4B chromatography. The gel was run with 2-mercaptoethanol to release any disulfide-linked low-molecular-weight peptides.

containing the entire peak of TTX-binding activity from a Sepharose 6B column such as that in Figure 1A are pooled, concentrated, and denatured in SDS, they can be fractionated quite effectively by gel filtration over Sepharose 4B. The large peptide is well resolved from lower-molecular-weight species. The profile of such a secondary fractionation is shown in Figure 3. The SDS gel of the individual peaks of protein and the pooled fraction I indicate that it is possible to obtain approximately 80% of the total peptide completely free of smaller species. This material is homogeneous and is suitable for a variety of compositional and physical analyses. In addition, the peptide isolated in this way is immunogenic, eliciting antibodies that will precipitate the native TTX-binding protein (M. Emerick and W. Agnew, unpubl.).

RESULTS

Peptide Composition of the TTX-binding Protein

Recently, in a series of careful studies (Beneski and Catterall 1980; Hartshorne and Catterall 1981; Hartshorne et al. 1982), Catterall and co-workers have identified three peptides that appear to be constituents of the STX-binding protein from rat brain. One of these is a large peptide of ~270 kD (α-peptide subunit). This is noncovalently associated with a second peptide of

37 kD (β1-subunit) and covalently, via disulfide bonds, with a third peptide of 39 kD (β2-subunit). The α- and β1-peptides are both photo-cross-linked by arylazido derivatives of the peptide toxin from venom of the scorpion *Leiurus* (Beneski and Catterall 1980; Hartshorne et al. 1982). When purified preparations containing only these three peptides were reconstituted into membrane vesicles, channel-mediated ion transport could be demonstrated by activating with veratridine and measuring radiotracer $^{22}Na^+$ flux into the vesicles (Talvenheimo et al. 1982; Tamkun et al. 1983). This flux was blocked by physiological STX and TTX concentrations.

Working with the protein isolated from rat skeletal muscle, Barchi and co-workers (Barchi et al. 1980; Barchi 1983) have identified a peptide of 130–230 kD and smaller species of 45 kD, 38 kD, and 37 kD in their most purified material. When reconstituted into artificial liposomes, those preparations also exhibited neurotoxin-activated and neurotoxin-blocked $^{22}Na^+$ flux (Weigele and Barchi 1982).

In the studies carried out to this date, we have not detected low-molecular-weight peptides associated with the TTX- and STX-binding proteins from electroplax. For example, in the gels accompanying Figure 1A, no peptides other than the large peptide at the top of the gels are observed to correlate with toxin binding. Further, when such preparations were repurified, material of high specific activity (2200 pmoles [^{3}H]TTX bound/mg protein) was isolated, devoid of lower-molecular-weight species. Potentially, this could have resulted from dissociation or proteolysis. When material was isolated in cholate, however, in the presence of six protease inhibitors (phenylmethylsulfonyl fluoride, pepstatin A, orthophenanthroline, iodoacetamide, soybean trypsin inhibitor, and EDTA), the cleanest fractions showed no evidence for smaller peptides (Fig. 2D). Further, to be certain there were no β2-like peptides linked by disulfides to the large peptide, material was first purified rapidly in cholate, and then the entire pool of toxin-binding protein was denatured in SDS without reduction and purified by SDS-Sepharose 4B column chromatography. This was done on a large scale (~0.4 mg). Material so isolated was then reduced and run on SDS gels. No smaller peptides were observed even when gels were stained with sensitive silver stains (Fig. 2G; Fig. 4B, lane 0).

Thus, at present, smaller peptide subunits have not been detected with the electroplax protein. The discrepancy between this and other preparations may stem from authentic molecular distinctions in the proteins, or may yet be found to result from the methods of isolations. Preliminary data cited below, however, suggest that the purified electroplax protein is functional in vesicle flux experiments such as those successfully applied to the other protein systems.

Electrophoretic Behavior of the Large Peptide

Examination of the large peptide by SDS gel electrophoresis revealed several interesting properties. The peptide ran as a diffuse band of apparent molecular

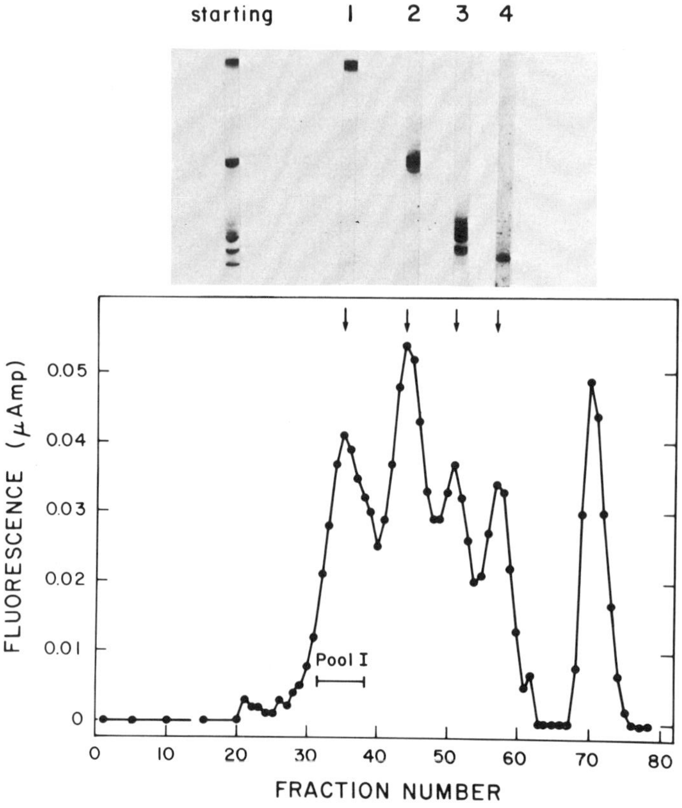

Figure 3. Purification of the major polypeptide of the TTX-binding protein by SDS gel chromatography on Sepharose 4B as described in the text. (*Bottom*) Protein profile; (*top*) SDS-polyacrylamide gels run on fractions indicated by arrows. The large peak at fraction 70 is due to Tris-HCl buffer in the applied sample, which gives a large positive reaction with fluorescamine.

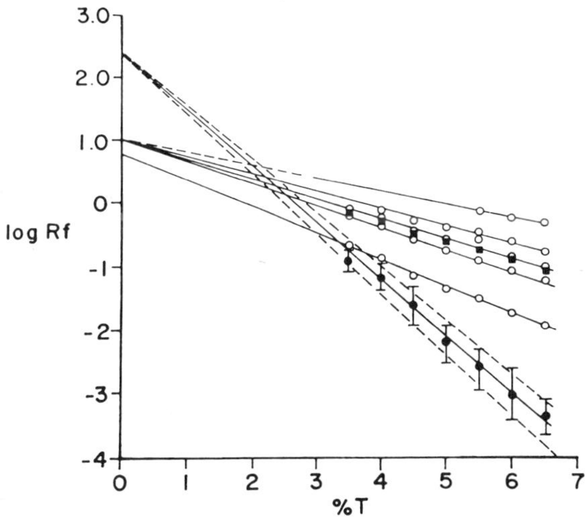

Figure 4. Ferguson analysis of the electrophoretic behavior of the large polypeptide compared with that of standards. Plot of the natural logarithm of the relative electrophoretic mobility (R_f) versus the acrylamide concentration used to cast the gels (T). Standards (○) from top to bottom: ovalbumin, bovine serum albumin, phosphorylase b, β-galactosidase, and myosin; (●) the migration of the large peptide; (■) migration of a 95-kD contaminant peptide from electroplax membranes. The bars for the large polypeptide represent the migration of the leading and trailing edges of the band (positions at half height), illustrating that all of the material comprising the broad band has the same unusual free electrophoretic mobility.

weight ~260,000 (cf. Fig. 2E). When reduced with 2-mercaptoethanol and carboxyamidated with iodoacetamide, or when run in urea-SDS gels (Swank and Munkries 1971), the peptide's behavior was not altered. Thus, diffuseness did not seem to result from failure to dissociate or from disulfide cross-linking of smaller peptides.

The peptide was analyzed on gels of different percent acrylamide by the method of Ferguson and Wallace (1961) (Fig. 4). It was found that, whereas standard peptides extrapolated to a similar free mobility at zero acrylamide, the peptide from the TTX-binding protein had a free mobility ~3.6 times that of the standards. This was suggested to result from binding of higher levels of SDS than normal (Miller et al. 1983). This has been confirmed by recent direct-binding studies showing the binding of ~3.3 g [^{14}C]SDS/g protein (J.A. Miller and W. Agnew, in prep.), more than the 1.4 g/g protein seen for conventional proteins.

The steep slope of the curve for the large peptide in this plot suggested a molecule far larger than the largest standard, myosin (210 kD). Both the leading and trailing edges of the large peptide band extrapolated to the same high free mobility, indicating that the band width was not due to contamination with unrelated peptides. These results also illustrate that it is not possible to measure accurately the peptide molecular weight by SDS gel electrophoresis, because different values are observed depending on the gel conditions used. However, analysis of these data suggests the peptide is likely to be larger than the 260 kD first estimated (Agnew et al. 1980; Miller et al. 1983).

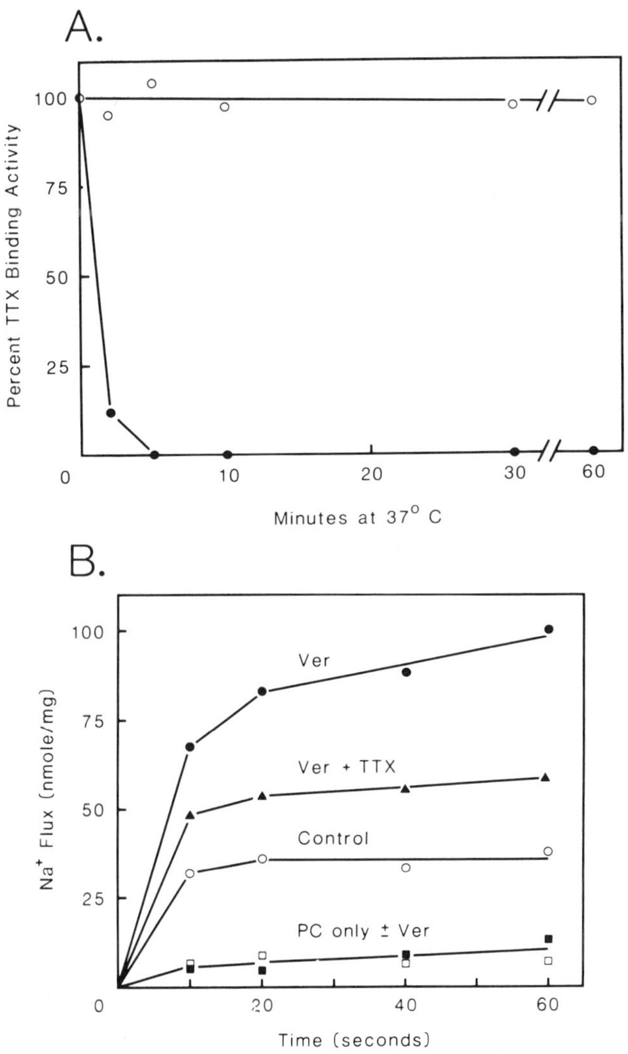

Figure 5. Reconstitution of the purified Na$^+$ channel protein into phosphatidyl choline vesicles. (*A*) Thermal stability of the TTX-binding activity before (●) and after (○) reconstitution. Samples were incubated at 37°C for the times indicated and assayed for TTX-binding activity by the gel filtration method (Agnew et al. 1978). Before assay, the reconstituted samples were solubilized under conditions that give maximal signal (0.25% Lubrol, 0°C). (*B*) $^{22}Na^+$ tracer influx into reconstituted vesicles. Vesicles containing the purified protein were incubated with 10^{-6} M TTX (▲) or no addition (●, ○) for 15 min at 0°C. Vesicles were then incubated in 10^{-4} M veratridine (●, ▲) or the veratridine carried medium (○) for 5 min at 30°C. Vesicles reconstituted in the absence of protein were also incubated at 37°C in the presence (■) and absence (□) of 10^{-4} M veratridine. At zero time, vesicles were diluted 10-fold into an osmotically balanced solution of 150 mM sucrose and 130 mM Tris-SO_4 containing 10 μCi/ml $^{22}Na^+$ and the appropriate neurotoxins. At the times indicated, 200-μl aliquots of the diluted vesicles were applied to columns of prepared Dowex 50 × 8-100 cation exchange resin and were eluted with 2 ml of osmotically balanced sucrose solution. Total eluate was counted for ^{22}Na.

It was possible to test whether the diffuse appearance

of the peptide band was due to a stable electrophoretic microheterogeneity. A sample of protein (partially purified) was run on two-directional SDS gels, under identical conditions in each dimension. The contaminant peptides behaved normally, each yielding a single symmetrical spot as expected for a nonheterogeneous species (Fig. 6A, #2). However, the large peptide yielded a sharp, long diagonal band. Thus, material running most rapidly in the first dimension also ran fast in the second dimension, and material running slowly in the first dimension ran slowly in the second. This stable microheterogeneity could result from variation in carbohydrate substitution, as the TTX-binding protein from eel does bind to lectin-affinity resins (Moore et al. 1982). When stained for carbohydrate by the periodic acid Schiff procedure, a large positive reaction was indeed observed (Fig. 2F). The peptide was then directly analyzed for amino acid and carbohydrate compositions (Miller et al. 1983). The amino acid composition was quite similar to that for other intrinsic membrane proteins, such as the nicotinic AChR (Table 2). Carbohydrate analysis, however, revealed a surprisingly large amount of carbohydrate, about 29% by weight (Table 2) (Miller et al. 1983). This carbohydrate was poor in neutral sugars such as mannose, galactose, and fucose, but rich in *N*-acetylglucosamine and *N*-acetylgalactosamine, as well as sialic acids. The large amount of sialic acid could explain the apparent acidity reported for the native protein at neutral pH (Agnew et al. 1980). In addition, variations in bulky carbohydrate moieties could underlie the electrophoretic microheterogeneity. This was confirmed in experiments in which the carbohydrate chains were cleaved from the peptide with endo-β-*N*-acetylglucosaminidase F (Endo F) (Elder and Alexander 1982).

In Figure 6B are shown gels of the large peptides before enzymatic deglycosylation with Endo F and during a time course of treatment up to 24 hours. At first, only the apparent average size of the peptide decreased, resulting in appearance of lower-molecular-weight material. With further incubation, however, a sharp band appeared at $M_r \cong 200,000$, and in parallel with this a distinct, but somewhat broad, band appeared at the very top of the gel. This latter did not stain with Coomassie blue, but did stain with silver reagents and may represent the cleaved carbohydrate chains that have not entered far into the gel. Three conclusions could be drawn from these experiments. First, more than one chain of carbohydrate must exist on the peptides, as the endoglycosidase cleaves the carbohydrate at the amino acid linkage, but nevertheless has yielded material of intermediate sizes. Second, the results indicate that carbohydrate is responsible for the microheterogeneity, because the band at ~200,000 is sharp rather than diffuse. Third, the carbohydrate composition estimated by direct analysis (29% by weight) is consistent with the decrement in apparent size on deglycosylation from 285,000 to 200,000. (These are, of course, apparent molecular weights.)

Functional Reconstitution

It is especially important to determine whether the purified protein is complete and whether it will function as an ion channel when reconstituted into artificial membranes. The assignment of the TTX-binding site to the large peptide suggests it may form the ion pathway itself. Reconstitution of the peptide allows a test of this question as well as a determination of whether other neurotoxin-binding sites that affect channel gating are present. As a first step, channel function can be assessed by measuring $^{22}Na^+$ ion flux into the vesicles when the channel is pharmacologically activated by neurotoxins such as veratridine, or activated but blocked by TTX or STX (Tamkun and Catterall 1981; Talvenheimo et al. 1982; Weigele and Barchi 1982; see Catterall 1980, for discussion of alkaloid neurotoxins). We recently have begun to have success with this approach.

The protein isolated in Lubrol-PX has been reconstituted according to protocols similar to those used for the sarcolemmal protein purified in NP-40 (Weigele and Barchi 1982) and for the synaptosomal protein in Triton X-100 (Talvenheimo et al. 1982). Briefly, purified preparations were supplemented with a phosphatidyl choline:Lubrol-PX solution followed by removal of detergent by adsorption to polystyrene beads. This results in spontaneous reassociation of membrane lipids and proteins into liposomes.

One test for reinsertion of the TTX-binding protein into membranes is the recovery of thermal stability of the toxin-binding site. TTX-binding sites in native electroplax membranes are stable at 37°C for several hours. Upon solubilization, however, the binding sites are very unstable at these elevated temperatures (Agnew and Raftery 1979). Thus, the reinsertion of the protein into the membranes should be reflected in recovery of thermal stability. As shown in Figure 5A, the purified protein was completely inactivated by less than 5 minutes at 37°C. After reconstitution into phosphatidyl choline liposomes, however, more than 97% of the binding activity survived 1 hour at this temperature.

To test for channel function of the reconstituted protein, vesicles were incubated either with no additions, with 10^{-4} M veratridine, or with veratridine plus 10^{-6} M TTX (Fig. 5B). Flux was initiated by dilution into an osmotically balanced solution containing $^{22}Na^+$, and, at the indicated times, external $^{22}Na^+$ was separated from tracer trapped within the vesicles by rapid Dowex ion-exchange chromatography according to the method of Epstein and Racker (1978). In the absence of both TTX and veratridine, vesicles reconstituted in the presence of protein showed a higher rate of influx than vesicles formed from lipid and detergent alone. Nevertheless, veratridine stimulated this flux more than twofold. The majority of the signal was blocked by 10^{-6} M TTX. Incomplete blockade of flux was consistent with parallel TTX-binding experiments which indicated that

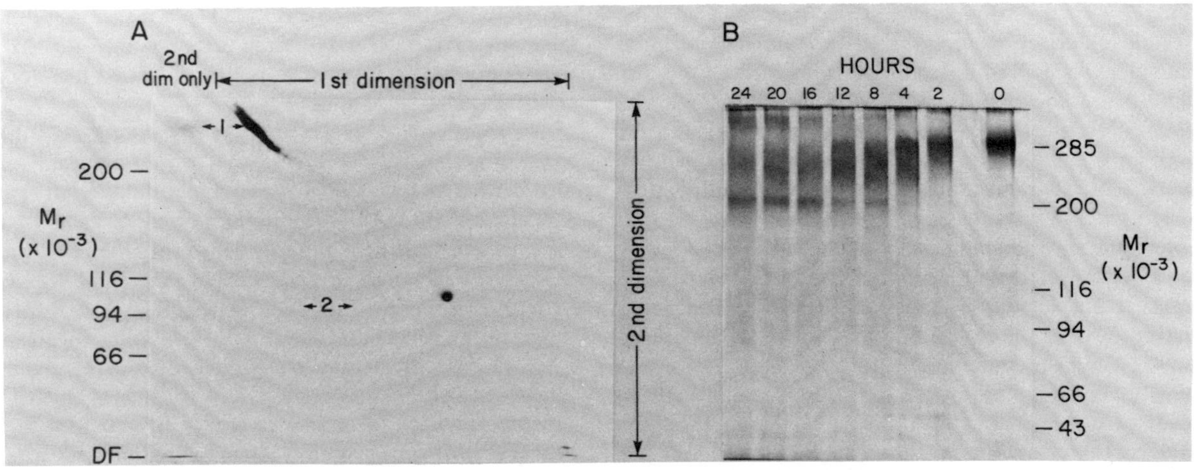

Figure 6. Electrophoretic microheterogeneity of the large peptide. (*A*) SDS × SDS two-directional gel (5% × 4.25% acrylamide) which demonstrates that the large peptide has a stable electrophoretic microheterogeneity. Material electrophoresed was purposely not homogeneous. The spot labeled 2 is a 95-kD contaminant peptide. This peptide behaves normally, yielding a single symmetrical spot after electrophoresis in the two directions. The large peptide of the Na^+ channel (labeled 1), however, banded as a sharp diagonal, indicating a stable microheterogeneity. (*B*) Treatment of the large peptide with endoglycosidase F. Pure peptide (lane 0), isolated by SDS gel chromatography (without reduction) of native cholate-purified material. (The gel was run with 2-mercaptoethanol, illustrating the absence of disulfide-linked low-molecular-weight peptides.) Samples incubated from 2 hr to 24 hr with endoglycosidase F (Endo F) were run in the lanes indicated. Trace low-molecular-weight bands at M_r ~35,000 and 50,000 were derived from the Endo F addition. Gel was stained with a silver reagent, as described in text.

Table 2. Composition of the Na^+ Channel Glycopeptide

(a) *Amino acid composition*[a]

	TTX-binding protein	Eel AChR
Asx	10.1	11.4
Thr	5.0	5.6
Ser	7.1	6.2
Glx	10.8	10.2
Pro	4.7	5.7
Gly	6.5	5.9
Ala	6.0	5.8
Cys	2.0	2.0
Val	5.3	8.6
Met	3.4	2.0
Ile	5.5	6.4
Leu	9.3	10.5
Tyr	2.8	4.0
Phe	5.9	5.7
His	2.8	2.2
Trp	3.5	ND
Lys	5.0	4.6
Arg	4.3	6.2
$H\phi_{ave}$	1073	1177
Z	0.29	0.40

(b) *Carbohydrate composition*[b]

	Percent total weight	Percent of carbohydrate
Fucose	0.5 ± 0.2	1.5 ± 0.5
Mannose	2.4 ± 0.5	8.3 ± 1.1
Galactose	1.5 ± 0.3	5.2 ± 0.6
N-acetylhexosamines	13.3 ± 0.4	45.3 ± 2.6
Sialic acid	11.8 ± 1.2	39.7 ± 1.6
Total	29.5 ± 2.4	

[a]Amino acid composition of the large glycopeptide compared with the total amino acid composition of eel nicotinic AChR. $H\phi_{ave}$ and Z are indices of hydrophobicity, as discussed in Miller et al. (1983) and references therein.

[b]Carbohydrate content of the large glycopeptide, determined as described in detail by Miller et al. (1983).

only 70% of these toxin sites were oriented toward the vesicle exterior. Whereas the lipophilic veratridine should activate all of the channels, only 70% of the flux should be blockable by the hydrophilic TTX. (*Note added*: More recent experiments have shown that TTX equilibrated with both internal and external compartments blocked transport completely.) This result is consistent with nearly identical observations with the sarcolemmal and synaptosomal proteins (Talvenheimo et al. 1982; Weigele and Barchi 1982).

These and similar experiments provide encouraging indications that the isolated protein may be functionally intact. The material so far examined has not been completely homogeneous, but nevertheless has had a lower level of contaminant peptides, especially in the low-molecular-weight class, than the preparation seen in the gel scan in Figure 2A. It now should be possible to determine whether the TTX-binding site, the ion channel, and the alkaloid neurotoxin-binding site all exist on the large peptide. The ultimate measure of channel function, however, will require application of techniques for biophysical analysis of channel activity, perhaps in reconstituted bilayers or in patch-clamped liposome preparations.

Appearance under the Electron Microscope

It was hoped that direct examination of the purified molecule might provide important information about the shape and dimensions of the protein, as well as other features that might relate to channel function or orientation in the membrane.

Homogeneous samples of the Lubrol-purified protein were examined by negative staining with uranyl acetate, employing conditions routinely used with the AChR (Ellisman et al. 1982). Under these conditions, the molecules appeared as rod-shaped structures (Fig. 7A). These rods showed a tendency to precipitate in highly ordered arrays of up to several thousand particles. They were most often observed to align side-to-side, in register, to form ribbons that often twisted on the carbonized surface of the grid. Stereoscopic imaging indicated that the particles were most likely cylinders of approximately 40 Å in diameter and 170 Å in length. No other regular structures were found in any of the preparations analyzed. The dimensions of the particles lead to a roughly estimated mass of ~257 kD, indicating that each particle might be formed from one copy of the large peptide.

The shape and dimensions of the particles did not immediately suggest a structure that would form an ion channel when inserted in a membrane bilayer. It was speculated that the molecule might orient with its long axis normal to the membrane. However, the length of the particle (170 Å) is much greater than would be required to span the membrane. More likely, however, it might be expected that the protein may alter its shape when released from the constraints of the membrane bilayer.

More recently (M. Ellisman et al., in prep.), the appearance of the protein has been examined after reconstitution into membrane vesicles from either cholate (Fig. 7) or Lubrol-PX (Fig. 7D). Reconstitution from cholate was performed by dialysis of samples of the protein and phosphatidyl choline solubilized in this detergent. Reconstitution from Lubrol was by the Biobead procedure (Talvenheimo et al. 1982; Weigele and Barchi 1982). These membranes have been studied both by freeze-fracture and by deep-etching followed by rotary platinum shadowing. Freeze-fracture revealed intramembrane particles of ~100 Å in diameter, roughly spherical in shape. No rod-shaped particles were observed lying in the bilayer plane. When deep etching was done to reveal the true surface of the vesicle followed by rotary shadowing, spherical particles of ~100 Å in diameter were observed protruding slightly (30–40 Å) beyond the membrane surface (Fig. 7F). Particles in cryoprotected and unprotected vesicles were very similar in appearance and dimensions.

The appearance of these particles was difficult to reconcile with the structures evident in the detergent-solubilized preparations. To resolve the discrepancy, pure samples of both the Lubrol-solubilized and cholate-solubilized material were examined with a battery of negative staining techniques at various pHs. It was found that when the Lubrol samples were examined at pH ~4, as before, the rod particles were again evident, almost always in the multimolecular aggregates (Fig. 7A). When cholate suspensions were examined under these conditions, particles reminiscent of the

Figure 7. Electron microscopic visualization of solubilized and reconstituted TTX-binding protein. (*A*) Field of pH 4.5 uranylacetate (UA) negatively stained TTX-binding protein in Lubrol-PX : phosphatidyl choline detergent. Gallery of individual rod-shaped particles is shown to right of field. Bar, 1000 Å in the field and 250 Å in the gallery. (*B*) Cholate-solubilized TTX-binding protein, stained with UA at pH 5, and gallery of cholate doughnuts. Bar, 1000 Å in the field and 250 Å in the gallery. (*C*) Cholate-solubilized TTX-binding protein negatively stained in UA at pH 4.5, but allowed to sit 10–30 sec in the low-pH stain before drying. This is a field of the somewhat irregular, segmented, cholate rodlike particles, and a gallery of cholate rods. Bar, 1000 Å in the field and 250 Å in the gallery. (*D*) Membrane vesicles reconstituted from Lubrol-purified samples of the TTX-binding protein were cryoprotected and then freeze-fractured and platinum-shadowed (unidirectional). The replicas show numerous ~100-Å-diameter intramembranous particles. Bar, 2000 Å; magnification, 75,000×. (*E*) Membrane vesicles were reconstituted from cholate-purified samples of the TTX-binding protein. These were frozen without cryoprotection or glutaraldehyde fixation, freeze-fractured and platinum-shadowed as above. Again, numerous ~100-Å-diameter, symmetrical particles were observed, similar to those in Lubrol-reconstituted vesicles, with and without fixation and cryoprotection. Bar, 2000 Å; magnification, 75,000×. (*F*) The true surface of a vesicle reconstituted from a Lubrol-PX-purified sample is revealed by freeze-etching, followed by rotary platinum shadowing. The surface exposed is decorated by numerous particulate projections, approximately 100 Å in diameter and protruding slightly (30–40 Å) from the true surface of the vesicle. Bar, 2000 Å; magnification, 75,000×.

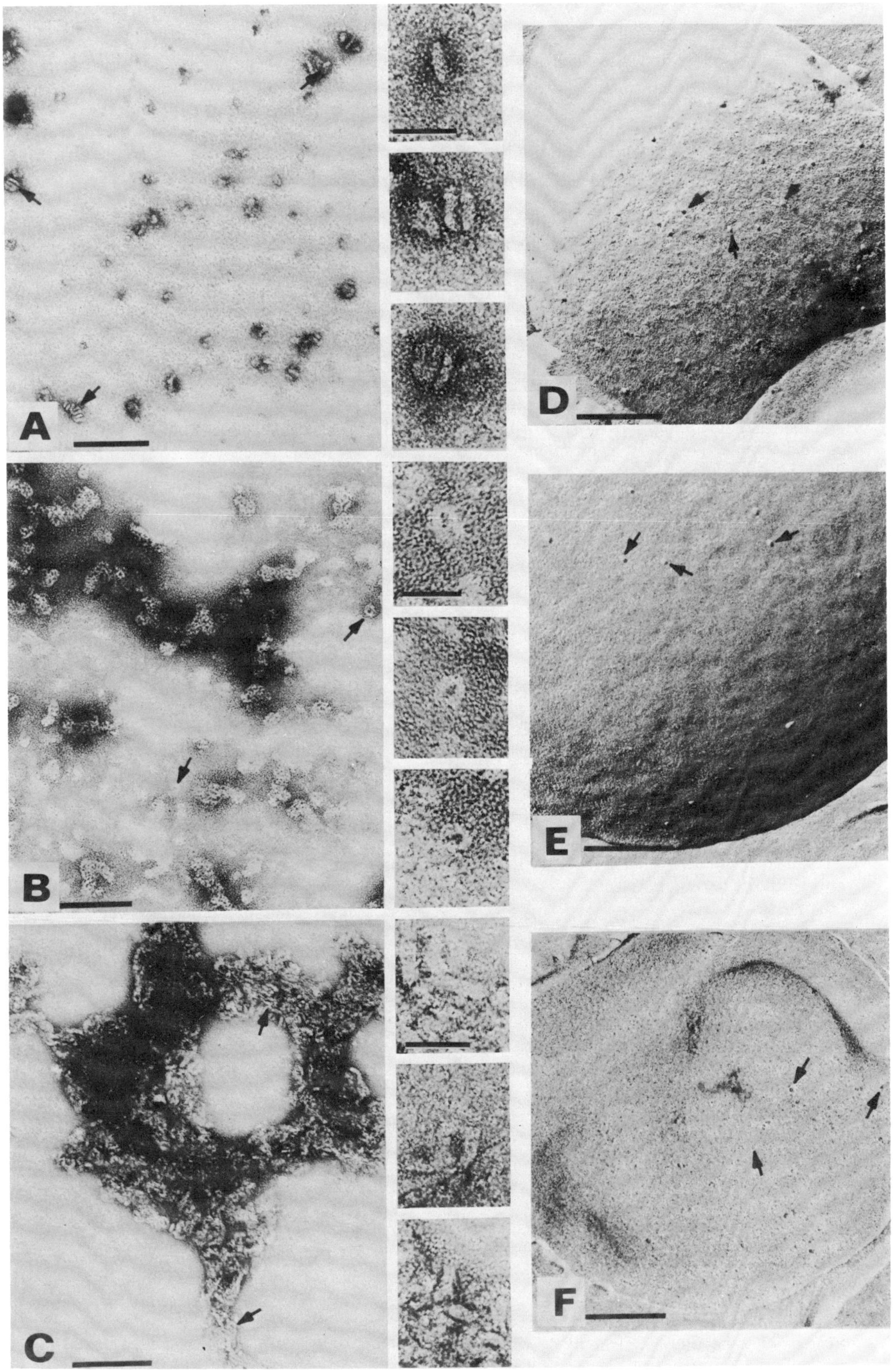

Figure 7. (*See facing page for legend.*)

Lubrol rods were also found (Fig. 7C). In cholate, however, the rods were less regular in shape, often appearing somewhat segmented in appearance, and never forming the highly ordered arrays always seen in Lubrol. When examined at more neutral pH (5.0) few if any rods could be demonstrated in Lubrol or cholate. Rather, the material appeared to aggregate in grapelike clusters. Individual particles were not easily discerned in Lubrol at these higher pHs. In cholate suspensions stained at pH 5.0, however, the particles formed the grapelike aggregates, but as was the case at lower pH, the clustering was somewhat less than in Lubrol. It was observed that the clusters were formed not from elongated particles, but appeared to be made up of doughnut-shaped particles, roughly spherical in appearance, with irregular, but pronounced, dark staining central "pits" (Fig. 7B). Numerous examples of individual particles, or groups of two or three together, could be found (Fig. 7B). The outside diameters of these particles were ~90–100 Å, consistent with the dimensions of the particles encountered in the membrane vesicles. When pure, Triton-X-100-solubilized AChR was stained at the various pHs used with the TTX-binding proteins, no pH-dependent alteration in particle morphology was observed. Thus, the TTX-binding protein shows a certain degree of morphological variability, being especially sensitive to the pH of staining conditions.

It is appropriate to be cautious about interpreting the appearance of the solubilized molecule under the electron microscope, especially insofar as conditions of pH can alter its appearance. It is clear that the large peptide can adopt an elaborate conformation that is suggestive of an ion channel. Nevertheless, even though the doughnut-shaped particles are consistent with the 100-Å-diameter particles seen in reconstituted vesicles, further studies will be required to ascertain the normal molecular appearance.

Immunological Tools

We have begun to explore the possible application of antibodies to the electroplax protein as immunocytochemical reagents with which to demarcate the conductile regions of excitable cells (Ellisman and Levinson 1982; M. Ellisman et al., unpubl.). When electroplax were stained with antibodies raised to the native protein, only the innervated face of the cell was stained, consistent with physiological evidence for channel distribution (Fig. 8A,B,C). Antibodies to the native protein have also been used to test for specific staining of myelinated nerve from spinal cord of the electric eel (Fig. 8D,E,F). When treated with antibody preblocked with the pure antigen, no specific localization of the antibody on the nerve was observed. When treated with unblocked antibody, staining of the nodal membrane was intense. When physically demyelinated nerve was treated with the unblocked antibody, strong labeling was noted in the nodal region. Thc paranodal axon was not stained, consistent with physiological and toxin-binding studies of channel distributions in myelinated nerves. This type of staining has also been demonstrated with the myelinated lateral line nerve from a related electric fish, *Sternarchus* (M. Ellisman et al., unpubl.). These preliminary observations suggest that antibodies to the electroplax protein are capable of recognizing the neuronal protein, at least in related species. The antibodies raised to the SDS-denatured peptide stain the eel CNS neurons in the same manner as antibodies to the native protein (Ellisman and Levinson 1982; M. Ellisman et al., unpubl.). However, detailed immunochemical studies will be required to determine the quantitative level of cross-reactivity between the proteins from the different tissue sources.

DISCUSSION

Our concept of the molecular structure of the voltage-regulated Na^+ channel has only recently begun to take shape based on studies such as those summarized here with the protein from eel electroplax and those carried out by other laboratories working with the proteins from mammalian brain and skeletal muscle.

When solubilized with mild nonionic detergents, the protein retains its ability to bind [^{3}H]TTX and [^{3}H]STX, permitting characterization and fractionation. The soluble eel protein exhibits a sedimentation coefficient of 8S and Stokes' radius by gel filtration of ~95 Å (Agnew et al. 1978). Best estimates of molecular weight of the protein are M_r ~300,000, although this number is not precisely determined. The protein from all sources examined is a glycoprotein, binding to lectin-affinity resins and being selectively displaced by the appropriate glycosides (Barchi et al. 1980; Hartshorne and Catterall 1981; Moore et al. 1982).

After solubilization, the protein is less stable than in the membrane-bound state, especially at elevated temperatures (Agnew and Raftery 1979). Even at ice bath temperatures, it becomes very unstable in the absence of minimal amounts of membrane phospholipids. For this reason, the protein must be purified while solubilized with mixed lipid-detergent micelles, with phosphatidyl choline or phosphatidyl ethanolamine being the most effective and convenient lipids to use.

When purified, the protein from electroplax was found to consist of a large glycopeptide of M_r (apparent) ≥260,000. This peptide may be analogous to the α-peptide subunit described for the synaptosomal protein, or the 130–230-kD band 1 of the sarcolemmal protein (Barchi 1983). No smaller peptides have been observed with the electroplax protein which might be analogous to the β1- or β2-subunits of the synaptosomal protein (37 kD and 39 kD, respectively) nor to bands 2,3A and 3B of the sarcolemmal protein (45, 38, and 37 kD). It may be that such peptides will yet be discovered in the electroplax protein, or that differences between the various preparations result from the isolation procedures employed. Alternatively, they may reflect authentic distinctions between channels from the three tissues.

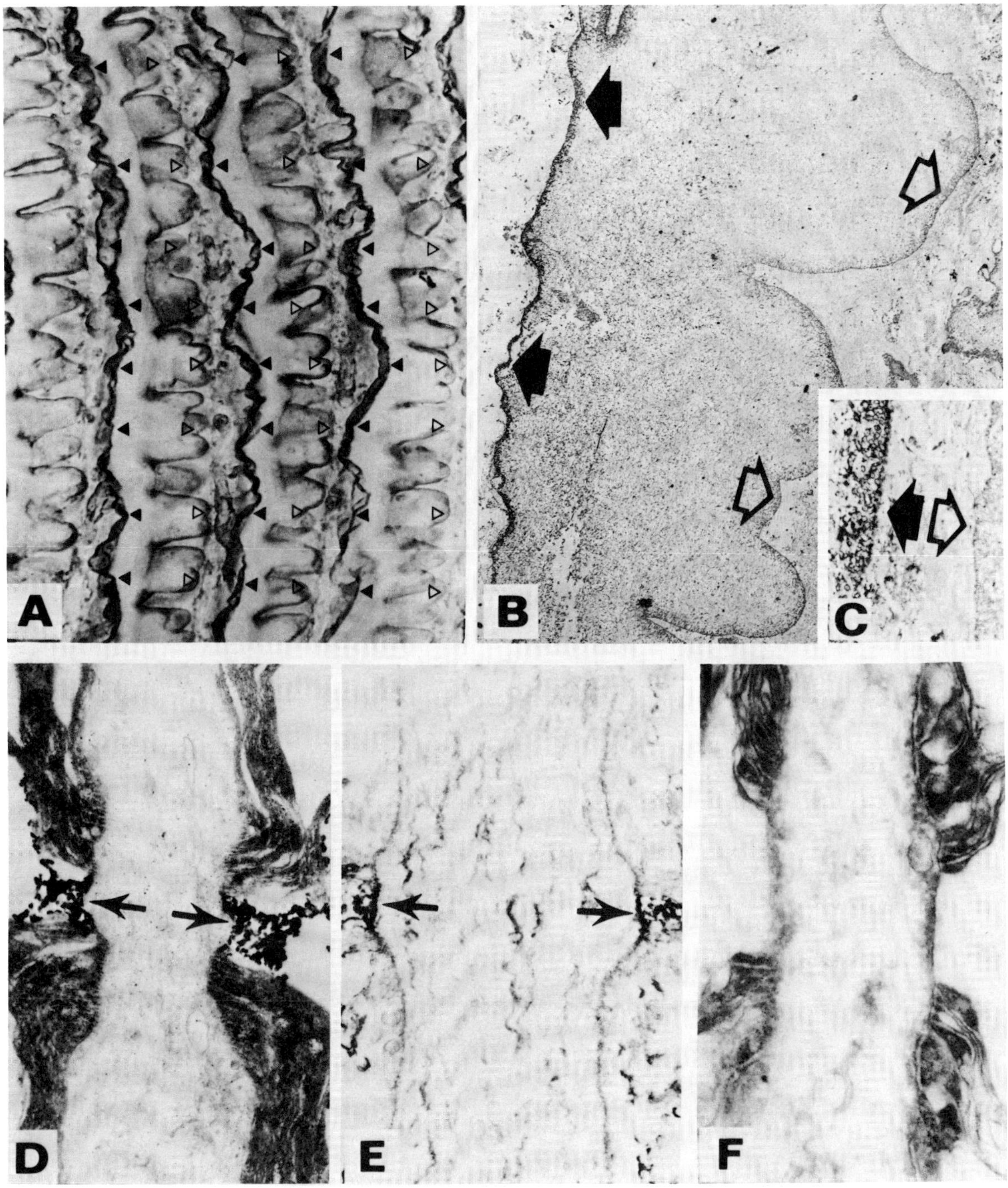

Figure 8. Immunohistochemical labeling of excitable cells in *Electrophorus electricus*. Tissue sections of eel electroplax or nerve were incubated with anti-TTX-binding protein antibodies, washed to remove membrane antibodies, followed by incubation with goat anti-rabbit IgG and a second series of washes. The sections were then incubated with peroxidase and antiperoxidase IgG, washed again, and incubated with diaminobenzidine and H_2O_2. Sections were washed and examined by light microscopy, or postfixed with osmium tetroxide, washed, dehydrated, and embedded in Epon/Araldite, for thin-section electron microscopy. (*A*) Electroplax, viewed at light microscopic level, showing heavy labeling of the innervated face (solid arrows) but not the noninnervated face (open arrows). (*B*) Electron microscopic visualization of stained thin sections. Only the innervated face is labeled (solid arrows). (*C*) Higher magnification view of abutting innervated (solid arrows) and noninnervated (open arrows) surfaces of two neighboring cells. This shows the labeling in the caveolar invaginations and represents a control to show that limited access of the reagents does not underlie the differential labeling. (*D*) Labeling of myelinated nerve from eel spinal cord. Antibodies have localized in the node of Ranvier, resulting in perfuse evolution of enzymatic reaction product from the nodal surface. (*E*) Myelinated nerve was physically stripped of myeline sheath by microtoming prior to staining. Antibodies have labeled exclusively the region of the axon that formerly was node of Ranvier, consistent with expected Na^+ channel distribution. (*F*) Myelinated nerve stained with antibodies preadsorbed to the pure TTX-binding protein, as a control for antibody entrapment. No binding of those antibodies to the nodal region is observed.

It is clear that the large peptide of the electroplax protein contains the TTX- and STX-binding sites, suggesting it may form the ion pore itself, and possibly the entire channel assembly. This conclusion is supported, but not yet proven, by the preliminary ion-flux experiments described here with the purified protein reconstituted into artificial vesicles. To establish unambiguously whether the peptide forms the channel as well as binding sites for the alkaloid neurotoxins that alter gating mechanisms, preparations that are completely homogeneous will have to be reconstituted into functional vesicles. More refined biophysical methods will undoubtedly be required to determine whether, in all senses, the channel behaves normally in response to transmembrane voltage and displays normal selectivity properties.

From the standpoint of amino acid composition, the large peptide is not unusual, being similar, for instance, to the nicotinic AChR. By contrast, the carbohydrate composition was unexpectedly high, the peptide having associated with it large amounts of the *N*-acetylaminohexoses as well as negatively charged sialic acids. The sialic acid constituents may account for the considerable negative charge carried by the protein at neutral pH (Agnew et al. 1980). The molecule has been calculated to have roughly six times the net negative charge of the AChR. The presence of a large amount of negatively charged carbohydrate at the surface of the protein, near the mouth of the ion channel, may markedly enhance the conductance of a highly selective ion channel (Miller et al. 1983). Direct carbohydrate analysis, as well as endoglycosidase experiments similar to those described here, would reveal whether the same unexpected carbohydrate composition is characteristic of the α-peptide subunit from the synaptosomal protein, or band 1 of the sarcolemmal protein.

Clearly, much remains to be resolved with respect to the molecular structure of the channel and its mechanisms for ion-selective transport and voltage-dependent gating. The conclusion that the Na^+ channel may be formed from a single large glycopeptide, or perhaps by such a central peptide in concert with two or three smaller peptides, clearly distinguishes this protein from the more thoroughly characterized nicotinic AChR, which is formed from five smaller peptide subunits. These compositional differences are perhaps not surprising in view of the fundamentally different regulatory mechanisms. These observations, however, serve to underscore our general lack of information concerning the molecular structures of the ion channels that mediate electrical excitation.

ACKNOWLEDGMENTS

Mr. Mark Emerick is thanked for expert technical assistance, Ms. Leslie Sprague for assistance in drawing figures, and Ms. Tina DeGennaro for typing the manuscript. W.S.A. was supported by a grant from the National Multiple Sclerosis Society (NMSS), by National Institutes of Health grant NS-17928, and a Merck Foundation Faculty Development Award. M.H.E. was supported by NIH grant NS-14718 and a grant from NMSS. J.A.M. is a recipient of a Muscular Dystrophy Association postdoctoral fellowship. S.R.L. was supported by NIH grant NS-15879.

REFERENCES

AGNEW, W.S. 1984. Voltage-regulated sodium channel molecules. *Annu. Rev. Physiol.* **46:** (in press).

AGNEW, W.S. and M.A. RAFTERY. 1979. Solubilized tetrodotoxin binding component from the electroplax of *Electrophorus electricus;* stability as a function of mixed lipid-detergent micelle composition. *Biochemistry* **10:** 1912.

AGNEW, W.S., S.R. LEVINSON, J.S. BRABSON, and M.A. RAFTERY. 1978. Purification of the tetrodotoxin-binding component associated with the voltage-sensitive channel from *Electrophorus electricus* electroplax membranes. *Proc. Natl. Acad. Sci.* **75:** 2606.

AGNEW, W.S., A.C. MOORE, S.R. LEVINSON, and M.A. RAFTERY. 1980. Identification of a large peptide associated with the tetrodotoxin binding protein from *Electrophorus electricus. Biochem. Biophys. Res. Commun.* **92:** 860.

ALMERS, W. 1978. Gating currents and charge movements in excitable membranes. *Rev. Physiol. Biochem. Pharmacol.* **82:** 96.

ARMSTRONG, C.M. 1981. Sodium channels and gating currents. *Physiol. Rev.* **61:** 644.

BARCHI, R.L. 1983. Protein components of the purified sodium channel from rat skeletal muscle sarcolemma. *J. Neurochem.* **40:** 1377.

BARCHI, R.L. and L.E. MURPHY. 1981. Estimate of the molecular weight of the sarcolemmal sodium channel using H_2O-D_2O centrifugation. *J. Neurochem.* **36:** 2097.

BARCHI, R.L., S.A. COHEN, and L.E. MURPHY. 1980. Purification from rat sarcolemma of the saxitoxin binding component of the excitable membrane sodium channel. *Proc. Natl. Acad. Sci.* **77:** 1306.

BARHANIN, J., A. SCHMID, A. LOMBET, K.P. WHEELER, M. LAZDUNSKI, and J.C. ELLORY. 1983. Molecular size of different neurotoxin receptors on the voltage-sensitive Na^+ channel. *J. Biol. Chem.* **258:** 700.

BENESKI, D.A. and W.A. CATTERALL. 1980. Covalent labeling of protein components of the sodium channel with a photoactivatable derivative of scorpion toxin. *Proc. Natl. Acad. Sci.* **77:** 639.

CAHALAN, M. 1981. Molecular properties of sodium channels in excitable membranes. In *The cell surface in neuronal functions* (ed. C.W. Cotman et al.), p. 1. Elsevier/North-Holland, Amsterdam.

CATTERALL, W.A. 1980. Neurotoxins that act on voltage-sensitive sodium channels in excitable membranes. *Annu. Rev. Pharmacol. Toxicol.* **20:** 15.

ELDER, J.H. and S. ALEXANDER. 1982. Endo-β-*N*-acetylglucosaminidase F: Endoglycosidase from *Flavobacterium meningosepticum* that cleaves both high mannose and complex glycoproteins. *Proc. Natl. Acad. Sci.* **79:** 4540.

ELLISMAN, M.H. and S.R. LEVINSON. 1982. Immunocytochemical localization of sodium channel (TTX binding protein) distribution in excitable membranes of *Electrophorus electricus. Proc. Natl. Acad. Sci.* **79:** 6707.

ELLISMAN, M.H., W.S. AGNEW, J.A. MILLER, and S.R. LEVINSON. 1982. Electron microscopic visualization of the tetrodotoxin binding protein from *Electrophorus electricus. Proc. Natl. Acad. Sci.* **79:** 4461.

EPSTEIN, M. and E. RACKER. 1978. Reconstituting carbamylcholine-dependent sodium ion flux and desensitization of the acetylcholine receptor from *Torpedo californica. J. Biol. Chem.* **253:** 6660.

FERGUSON, K.A. and A.L.C. WALLACE. 1961. Starch-gel

electrophoresis of anterior pituitary hormones. *Nature* **190:** 629.

Hartshorne, R. and W.A. Catterall. 1981. Purification of the saxitoxin receptor from rat brain. *Proc. Natl. Acad. Sci.* **78:** 4620.

Hartshorne, R.P., J. Coppersmith, and W.A. Catterall. 1980. Size characteristics of the solubilized receptor of the voltage-sensitive sodium channel from rat brain. *J. Biol. Chem.* **255:** 10572.

Hartshorne, R.P., D.J. Messner, J.C. Coppersmith, and W.A. Catterall. 1982. The saxitoxin receptor of the sodium channel from rat brain; evidence for two nonidentical β-subunits. *J. Biol. Chem.* **257:** 13888.

Hille, B. 1975. The receptor for tetrodotoxin and saxitoxin: A structural hypothesis. *Biophys. J.* **15:** 615.

Hodgkin, A.C. and A.F. Huxley. 1952. Quantitative description of membrane current and its application to conduction and to excitation in nerve. *J. Physiol.* **117:** 500.

Kao, C.Y. and S.E. Walker. 1982. Active groups of saxitoxin and tetrodotoxin as deduced from actions of saxitoxin analogues on frog muscle and squid axon. *J. Physiol.* **323:** 619.

Keynes, R.D. and M. Martins-Ferreira. 1953. Membrane potentials in the electroplates of the electric eels. *J. Physiol.* **119:** 315.

Levinson, S.R. and J.C. Ellory. 1972. Molecular size of the tetrodotoxin binding site estimated by irradiation inactivation. *Nat. New Biol.* **245:** 122.

Miller, J.A., W.S. Agnew, and S.R. Levinson. 1983. Principal glycopeptide of the tetrodotoxin/saxitoxin binding protein from *Electrophorus electricus*: Isolation and partial chemical and physical characterization. *Biochemistry* **22:** 462.

Moore, A.C., W.S. Agnew, and M.A. Raftery. 1982. Biochemical characterization of the tetrodotoxin binding protein from *Electrophorus electricus. Biochemistry* **24:** 6212.

Nakamura, Y., S. Nakajima, and H. Grundfest. 1965. Analysis of spike electrogenesis and depolarizing K inactivation in *Electrophorus electricus. J. Gen. Physiol.* **49:** 321.

Olsen, R.W., J.C. Meunier, and J.P. Changeux. 1972. Progress in the purification of the cholinergic receptor protein from *Electrophorus electricus* by affinity chromatography. *FEBS Lett.* **28:** 96.

Ritchie, J.M. and R.B. Rogart. 1977. The binding of saxitoxin and tetrodotoxin to excitable tissue. *Rev. Physiol. Biochem. Pharmacol.* **79:** 1.

Swank, R.T. and K.D. Munkries. 1971. Molecular weight analysis of oligopeptides by electrophoresis in polyacrylamide gel with sodium dodecyl sulfate. *Anal. Biochem.* **39:** 462.

Talvenheimo, J., M.M. Tamkun, and W.A. Catterall. 1982. Reconstitution of neurotoxin stimulated sodium transport by the voltage-sensitive sodium channel purified from rat brain. *J. Biol. Chem.* **257:** 11868.

Tamkun, M.M. and W.A. Catterall. 1981. Reconstitution of the voltage-sensitive sodium channel of rat brain from solubilized components. *J. Biol. Chem.* **256:** 11457.

Tamkun, M.M., J.A. Talvenheimo, and W.A. Catterall. 1983. Functional reconstitution of the sodium channel purified from rat brain. *Biophys. J.* **41:** 142a.

Weigele, J. and R. Barchi. 1982. Functional reconstitution of the purified sodium channel protein from rat sarcolemma. *Proc. Natl. Acad. Sci.* **79:** 3651.

Immunochemical Studies of the Voltage-sensitive Sodium Channel from the Electroplax of the Eel *Electrophorus electricus*

L.C. FRITZ,*† H.-P.H. MOORE,‡§ M.A. RAFTERY,‡ AND J.P. BROCKES*¶

*Divisions of *Biology and ‡Chemistry and Chemical Engineering, California Institute of Technology, Pasadena 91125*

Voltage-sensitive Na^+ channels are membrane components responsible for the rapid upstroke of the action potential in many excitable cells. Originally characterized by their physiological (Hodgkin and Huxley 1952) and pharmacological properties (Narahashi et al. 1964), these channel proteins have more recently been the subject of biochemical studies. Investigators in several laboratories have attempted to identify them from a variety of sources by using specific Na^+ channel neurotoxins as probes (Catterall 1980). These studies include analyses of the proteins that copurify with the peak of binding of the guanidinium compounds tetrodotoxin (TTX) and saxitoxin (STX) (Agnew et al. 1980; Barchi et al. 1980; Hartshorne and Catterall 1981) as well as analyses of proteins labeled when intact cells are treated with photoactivatable derivatives of scorpion toxin (Beneski and Catterall 1980).

Although the toxins have proven to be very useful in studying Na^+ channels, their binding properties are not ideal. For example, they have been of limited use as cytochemical reagents (Catterall 1981), and furthermore, they have not been effective affinity reagents for channel purification. In an effort to produce specific channel probes that circumvent such problems, we have isolated and characterized monoclonal and polyclonal antibodies that recognize the Na^+ channel from the electric organ of the eel *Electrophorus electricus*. In the present paper, we review this work and show how immunochemical reagents have been used (1) to identify independently a 250-kD protein as being a part of the eel Na^+ channel, (2) to study the immunological cross-reaction of Na^+ channels from a variety of sources, and (3) to investigate the distribution of Na^+ channels in the electroplax by immunocytochemistry.

Isolation and Characterization of Anti-Na^+ Channel Monoclonal Antibodies

BALB/c mice were immunized with a partially purified native preparation of Na^+ channel derived from detergent-solubilized membranes of the eel electroplax. This material, prepared by the methods of Agnew et al. (1978), was judged to be approximately 25–50% channel, based on the specific activity of TTX binding in the preparation. (This material will be referred to as the "Sepharose 6B fraction" since the final step in its purification was two successive chromatographic separations over a Sepharose 6B column.) Following priming and boosting injections with antigen, hybridomas were prepared by standard procedures and were selected in HAT medium (Moore et al. 1982).

Two hybridoma cell lines (5D10 and 5F3) secreted antibodies that were shown to bind specifically to sites on the Na^+ channel. This was demonstrated by their ability to immunoprecipitate the STX-binding component from a crude preparation of solubilized electroplax membranes. In this assay, the STX-binding sites in the solubilized membranes were saturated with ^{3}H-labeled STX and incubated with increasing amounts of monoclonal antibody. The antibody was then precipitated with an equivalent amount of rabbit anti-mouse IgG antiserum, and the immunoprecipitate was separated by centrifugation through a 30% glycerol cushion. Figure 1 shows the amount of radioactivity precipitated versus the amount of antibody added when this assay was performed with monoclonal antibody (mAb) 5D10 (Moore et al. 1982). A dose-dependent precipitation of ^{3}H-labeled STX was observed. Several controls indicate that this reflects specific binding of mAb 5D10 to the Na^+ channel. The precipitated ^{3}H-labeled STX was bound to specific toxin-binding sites, because immunoprecipitation carried out in the presence of an excess of unlabeled TTX gave a low background that was invariant with the amount of antibody added. Similarly, no dose-dependent precipitation was observed if the assay was performed with a monoclonal antibody of identical subclass (IgG1) but that recognized an unrelated antigen. Furthermore, the results were not due to the precipitation of unbound ^{3}H-labeled STX since the signal was abolished if the solubilized electroplax membranes were eliminated from the assay mixture. These results indicate that mAb 5D10 reacts with the component that specifically binds STX and therefore must bind to some portion of the eel electroplax Na^+ channel. Essentially identical precipitation results were obtained with mAb 5F3 (Fritz and Brockes 1983), indicating that it too reacts with the electroplax Na^+ channel.

Identification of the STX-binding Component by Western Blotting

We were interested in identifying the polypeptide(s) recognized by mAb 5D10 and mAb 5F3 since the im-

Present addresses: †Division of Chemistry and Chemical Engineering, California Institute of Technology, Pasadena, California 91125; §Department of Biochemistry and Biophysics, University of California, San Francisco, California 94143; ¶Medical Research Council Cell Biophysics Unit, London WC2B 5RL, England.

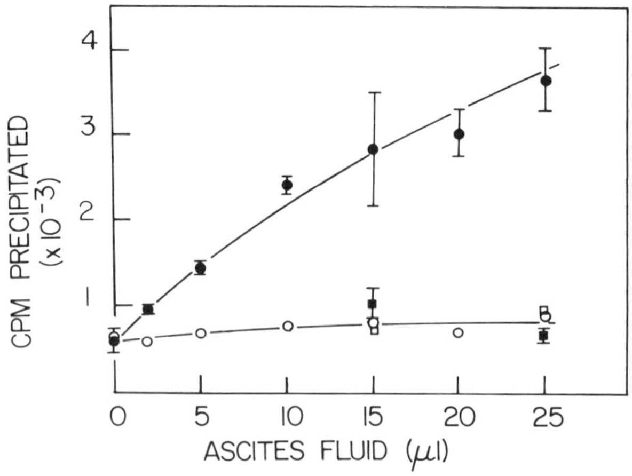

Figure 1. Immunoprecipitation of the ^{3}H-labeled STX-binding component by mAb 5D10. A crude detergent extract of electroplax membranes was saturated with 450 nM ^{3}H-labeled STX and incubated with varying concentrations of 5D10 ascites fluid. The total amount of antibody in each tube was kept constant by supplementing with purified normal mouse immunoglobulin. Immune complexes were precipitated, and the radioactivity associated with the pellet was plotted as a function of the volume of ascites fluid added. (●) 5D10 in the absence of unlabeled TTX; (○) 5D10 in the presence of 50 μM unlabeled TTX; (■) control antibody in the absence of TTX; (□) same control antibody in the presence of 50 μM TTX. Error bars show S.D. about the mean of triplicate points. (Reprinted, with permission, from Moore et al. 1982.)

munoprecipitation data (Fig. 1) indicated that those polypeptides must be components of the Na$^+$ channel. To do this, electroplax proteins from each of the four stages of channel purification (membranes, detergent extract, DEAE fraction, and Sepharose 6B fraction [Agnew et al. 1978]) were separated on SDS-polyacrylamide gels and electrophoretically transferred to nitrocellulose paper (Towbin et al. 1979). The paper was then probed with monoclonal antibody followed by rabbit anti-mouse IgG and ^{125}I-labeled protein A. An autoradiograph of such a Western blot probed with mAb 5F3 is shown in Figure 2 (Fritz and Brockes 1983). At each stage in the channel purification procedure, mAb 5F3 reacted with a diffuse band of M_r ~250,000 (the p250 protein) and with none of the many other polypeptides present in the various fractions. A control monoclonal antibody of identical subclass (IgG1) showed no binding (Fig. 2C). These results provide direct evidence that p250 is a component of the eel Na$^+$ channel, as suggested earlier by Agnew et al. (1980) on the basis of its copurification with the peak of TTX binding. The observation that mAb 5F3 binds only to p250 in crude membranes as well as in more purified fractions suggests that p250 is a bona fide component of the eel channel and does not simply result from aggregation or degradation during purification. Similar experiments indicate that mAb 5D10 also binds exclusively to p250 (Moore et al. 1982).

These results do not, however, exclude the possibility that the eel channel contains other subunits not recognized by either mAb 5F3 or mAb 5D10. This point was examined in greater detail by Nakayama et al. (1982), who used mAb 5D10 as an immunoaffinity reagent to purify the eel electroplax Na$^+$ channel. Although their results provided no clear evidence for channel subunits other than p250, the existence of other lower-molecular-weight components could not be rigorously excluded.

Cross-reaction Studies

Quantitative adsorption assays were used to assess the cross-reaction of mAb 5D10 and mAb 5F3 with Na$^+$ channels from other sources. In these assays, a limiting dilution of antibody was incubated with varying amounts of membranes prepared from different tissues and different species. The membranes were pelleted by centrifugation, and the supernatant was assayed for the level of antibody remaining. If the membranes displayed the antigenic determinant recognized by the monoclonal antibody, then the antibody level in the supernatant should have been reduced by the adsorption. Antibody concentration was assessed in a solid-phase binding assay in which the Sepharose 6B

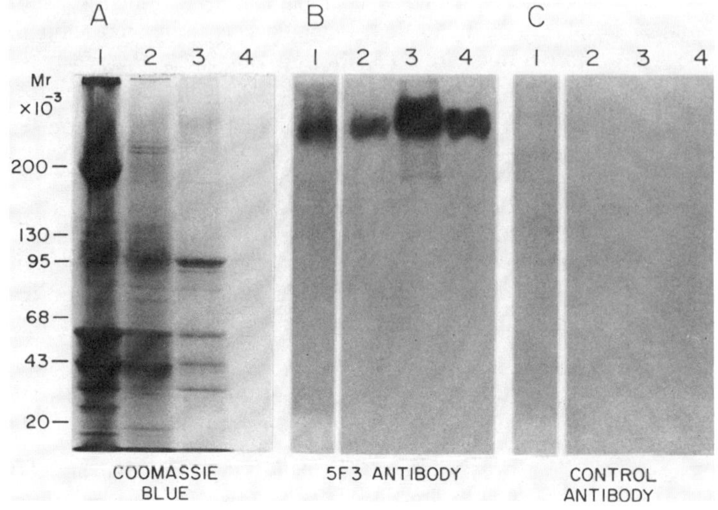

Figure 2. Western blot of electroplax proteins probed with mAb 5F3. Material from each of the four steps in the Na$^+$ channel purification procedure was fractionated in 4–12% gradient polyacrylamide gels under nonreducing conditions. Gels were either stained with Coomassie blue or electrophoretically transferred to nitrocellulose paper. The paper was then reacted with hybridoma supernatant followed by rabbit anti-mouse IgG and ^{125}I-labeled protein A. Lanes *1*, *2*, *3*, and *4* contain membranes, Lubrol extract, DEAE fraction, and Sepharose 6B fraction, respectively. (*A*) Coomassie-blue-stained gel; (*B*) autoradiograph of a nitrocellulose transfer from a gel similar to the one in *A*, stained with mAb 5F3; (*C*) Autoradiograph of an identical nitrocellulose transfer stained with a control IgG1 monoclonal antibody. Lanes *B1* and *C1* were from one experiment; lanes *B2–B4* and *C2–C4* were from an independent experiment. (Reprinted, with permission, from Fritz and Brockes 1983.)

fraction was immobilized on plastic (the Sepharose 6B RIA [Moore et al. 1982]). Results of such an experiment in which mAb 5D10 was adsorbed with eel electroplax membranes or with crude synaptosomal membranes from rat brain or frog brain are shown in Figure 3. The data are plotted as counts per minute (cpm) bound in the Sepharose 6B RIA versus the number of TTX-binding sites present in the adsorbing membranes. These data demonstrate that although electroplax membranes were effective in adsorbing out the antibody, membranes from rat brain or frog brain were not. Adsorption with chick brain membranes gave a similar negative result. The horizontal line in Figure 3 shows the result obtained when adsorption was performed with rat kidney membranes. Identical results were obtained in adsorption assays using mAb 5F3; the antibody was adsorbed only by the electroplax membranes. Similarly, Western blots of synaptosomal proteins from rat, frog, or chick brain showed no specific reaction when probed with either of these monoclonal antibodies. We conclude, therefore, that mAb 5D10 and mAb 5F3 have no detectable cross-reaction with rat, frog, or chick brain Na^+ channels. We also tested the cross-reaction of a rabbit antiserum produced against SDS-gel-purified p250 protein. In our most sensitive solid-phase assays, only a very limited reaction was observed between this antiserum and rat, frog, or chick brain synaptosomal membranes ($\leq 0.1\%$ cross-reaction) (Fritz and Brockes 1983).

The cross-reaction results demonstrate that the eel electroplax Na^+ channel is immunologically distinct from brain Na^+ channels in several species. It was

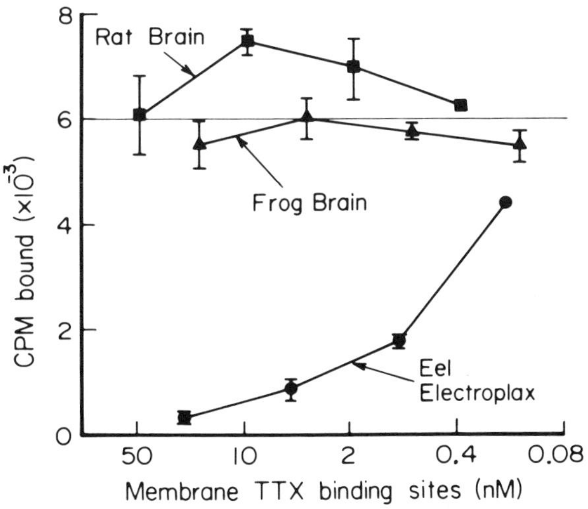

Figure 3. Cross-reaction of mAb 5D10 measured by quantitative adsorption assay. A limiting dilution of mAb 5D10 was incubated with varying amounts of membranes from the indicated sources. After pelleting the membranes, 30 μl of the supernatants was assayed for residual antibody in a Sepharose 6B RIA. Data are plotted as cpm bound in the Sepharose 6B RIA versus the concentration of TTX-binding sites in the adsorption mixture. Adsorptions were performed with eel electroplax membranes (●), rat brain membranes (■), and frog brain membranes (▲). Error bars denote the S.D. about the mean of triplicate points. (Reprinted, with permission, from Fritz and Brockes 1983.)

conceivable, however, that greater cross-reaction might be seen with muscle membranes since the eel electroplax is embryologically derived from muscle (Keynes 1957). Technical considerations limited the sensitivity of adsorption assays with rat muscle membranes (although these assays provided no evidence of cross-reaction). Antibody binding to several muscle preparations was also assessed by immunocytochemistry with negative results (see below).

Immunocytochemistry

Monoclonal antibody 5F3 proved to be effective as an immunocytochemical reagent for the visualization of the Na^+ channel distribution in the eel electroplax. Frozen sections were made from the caudal aspect of the electric organ (organ of Sachs), parallel to the long axis of the stacks of electrocytes. (See Luft [1957] for a description of the anatomy of the electroplax.) Sections were reacted sequentially with hybridoma supernatant, rabbit anti-mouse IgG, and goat anti-rabbit IgG coupled to rhodamine. Sections were then viewed under phase-contrast or rhodamine optics.

Figure 4 shows the results obtained when a section was reacted with mAb 5F3. Intense staining is observed, but the fluorescence is limited exclusively to the caudal face of the electrocyte and to projections off of the caudal face. Reaction of electrocyte sections with control IgG1 monoclonal antibodies shows no staining whatsoever, demonstrating the specificity of the immunofluorescence observed with mAb 5F3. The observation that Na^+ channels are located only in the caudal face of each electrocyte is in accord with physiological experiments in which only the caudal face was found to be electrically excitable (Keynes and Martins-Ferreira 1953). Similar immunocytochemical results have also been reported by Ellisman and Levinson (1982), who used a polyclonal antiserum that reacts with the eel Na^+ channel.

Although most of the caudal face membrane of each electrocyte was stained intensely by mAb 5F3, there were also regions of that face that were devoid of immunoreactivity. One such region is shown in Figure 5. The two caudal projections are clearly 5F3-positive; however, no reaction is seen with the membrane between the projections. Areas of caudal membrane that are 5F3-negative are generally delimited by abrupt transitions between the nonfluorescent and fluorescent regions.

The distribution of Na^+ channels in the electrocyte is clearly inhomogeneous; they are present only in one face, and even along that face, there are areas lacking channels. The mechanisms that produce this inhomogeneous distribution are of particular interest because in many cells, nonuniform channel distributions have important physiological consequences. For example, rapid saltatory conduction in myelinated axons is dependent on the clustering of Na^+ channels at nodes of Ranvier. We do not know what causes the striking Na^+ channel distribution observed in the electrocyte, but one mor-

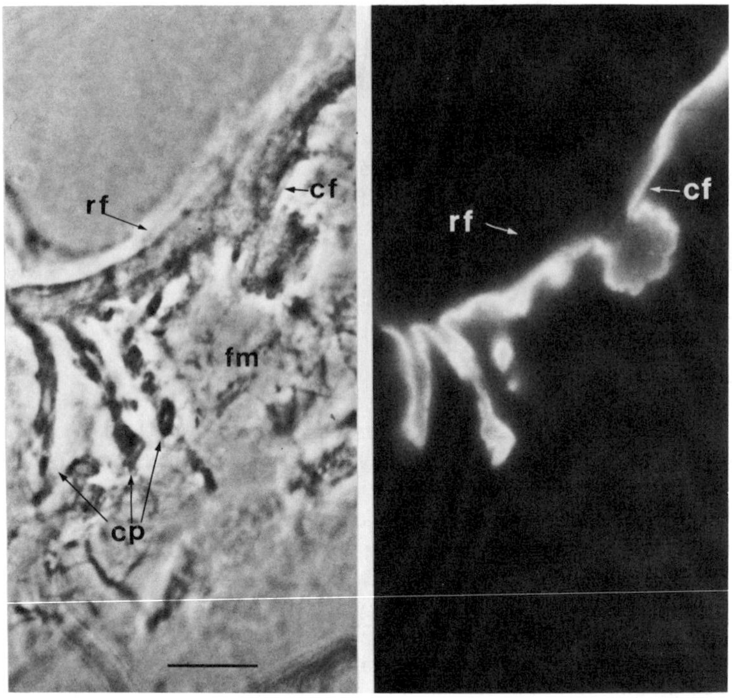

Figure 4. Eel electrocyte stained with mAb 5F3. A frozen section from the organ of Sachs was treated with 5F3 supernatant, affinity-purified rabbit anti-mouse IgG, and rhodamine-conjugated goat anti-rabbit IgG. (*Left*) Phase optics; (*right*) rhodamine optics. (cf) Caudal face; (rf) rostral face; (cp) projection off of the caudal face; (fm) filamentous material. Immunoreactivity is limited to the caudal face. Calibration, 10 μm.

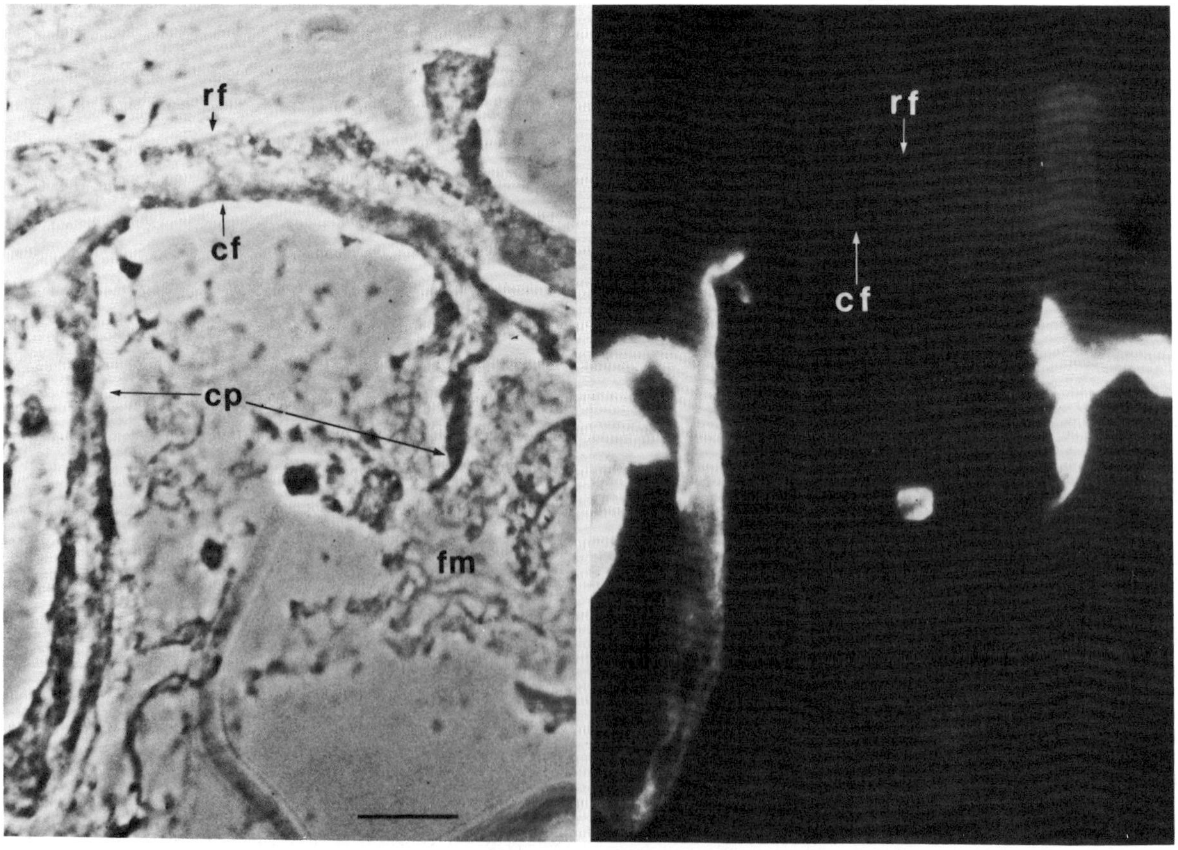

Figure 5. Frozen section of an eel electrocyte stained with mAb 5F3, affinity-purified rabbit anti-mouse IgG, and rhodamine-conjugated goat anti-rabbit IgG. (*Left*) Phase optics; (*right*) rhodamine optics. Labels are as in Fig. 4. Immunoreactivity is limited to the caudal projections and absent from the membrane between them. Calibration, 10 μm. (Reprinted, with permission, from Fritz and Brockes 1983.)

phological feature that correlates with mAb 5F3 immunoreactivity is the presence of filamentous material that is extracellular to the electrocyte. This material can be seen along the caudal face in Figure 4 and in association with the caudal projections in Figure 5. An interesting possibility is that this extracellular material interacts with the electrocyte in a process that leads to channel localization. Interactions between extracellular components and muscle membranes have been implicated in the clustering of acetylcholine receptors at neuromuscular junctions (Burden et al. 1979; Rubin and McMahan 1982).

Monoclonal antibody 5F3 and the rabbit anti-p250 antiserum were tested on a number of other excitable tissues for immunocytochemical staining. No reaction was observed using either reagent on sections of rat skeletal muscle, myelinated nerve from rat or frog, or rat hippocampus. Negative results were also obtained with cultured myotubes or neurons from rat or chick. Perhaps suprisingly, no immunocytochemical reaction was observed when mAb 5F3 was applied to sections of eel muscle or spinal cord. The rabbit anti-p250 antiserum was also negative on eel spinal cord sections and gave only the faintest staining of eel muscle (but see Ellisman and Levinson 1982). These results raise the possibility that there might be significant differences among Na^+ channels from different species and among channels from different tissues within one species.

ACKNOWLEDGMENTS

This work was supported by a postdoctoral fellowship from the Muscular Dystrophy Association of America (to L.C.F.) and grants from the National Institutes of Health (NS-12018, NS-14403, 4-RR-0700-3A), the Pew Charitable Trust, the McKnight Foundation, and the Pfeiffer Foundation.

REFERENCES

Agnew, W.S., S.R. Levinson, J.S. Brabson, and M.A. Raftery. 1978. Purification of the tetrodotoxin-binding component associated with the voltage-sensitive sodium channel from *Electrophorus electricus* electroplax membranes. *Proc. Natl. Acad. Sci.* **75:** 2606.

Agnew, W.S., A.C. Moore, S.R. Levinson, and M.A. Raftery. 1980. Identification of a large molecular weight peptide associated with a tetrodotoxin-binding protein from the electroplax of *Electrophorus electricus*. *Biochem. Biophys. Res. Commun.* **92:** 860.

Barchi, R.L., S.A. Cohen, and L.E. Murphy. 1980. Purification from rat sarcolemma of the saxitoxin-binding component of the excitable membrane sodium channel. *Proc. Natl. Acad. Sci.* **77:** 1306.

Beneski, D.A. and W.A. Catterall. 1980. Covalent labeling of protein components of the sodium channel with a photoactivatable derivative of scorpion toxin. *Proc. Natl. Acad. Sci.* **77:** 639.

Burden, S.J., P.B. Sargent, and U.J. McMahan. 1979. Acetylcholine receptors in regenerating muscle accumulate at original synaptic sites in the absence of the nerve. *J. Cell Biol.* **82:** 412.

Catterall, W.A. 1980. Neurotoxins that act on voltage-sensitive sodium channels in excitable membranes. *Annu. Rev. Pharmacol. Toxicol.* **20:** 15.

———. 1981. Localization of sodium channels in cultured neural cells. *J. Neurosci.* **1:** 777.

Ellisman, M.H. and S.R. Levinson. 1982. Immunocytochemical localization of sodium channel distributions in the excitable membranes of *Electrophorus electricus*. *Proc. Natl. Acad. Sci.* **79:** 6707.

Fritz, L.C. and J.P. Brockes. 1983. Immunochemical properties and cytochemical localization of the voltage-sensitive sodium channel from the electoplax of the eel (*Electrophorus electricus*). *J. Neurosci.* (in press).

Hartshorne, R.P. and W.A. Catterall. 1981. Purification of the saxitoxin receptor of the sodium channel from rat brain. *Proc. Natl. Acad. Sci.* **78:** 4620.

Hodgkin, A.L. and A.F. Huxley. 1952. A quantitative description of membrane current and its application to conduction and excitation in peripheral nerve. *J. Physiol.* **117:** 500.

Keynes, R.D. 1957. Electric organs. In *The physiology of fishes* (ed. M.E. Brown), vol. 2, p. 323. Academic Press, New York.

Keynes, R.D. and H. Martins-Ferreira. 1953. Membrane potentials in the electroplates of the electric eel. *J. Physiol.* **119:** 315.

Luft, J.H. 1957. The histology and cytology of the electric organ of the electric eel (*Electrophorous electricus* L.). *J. Morphol.* **100:** 113.

Moore, H.-P.H., L.C. Fritz, M.A. Raftery, and J.P. Brockes. 1982. Isolation and characterization of a monoclonal antibody against the saxitoxin-binding component from the electric organ of the eel, *Electrophorus electricus*. *Proc. Natl. Acad. Sci.* **79:** 1673.

Nakayama, H., R.M. Withy, and M.A. Raftery. 1982. Use of a monoclonal antibody to purify the tetrodotoxin binding component from the electroplax of *Electrophorus electricus*. *Proc. Natl. Acad. Sci.* **79:** 7575.

Narahashi, T., J.W. Moore, and W.R. Scott. 1964. Tetrodotoxin blockage of sodium conductance increase in lobster giant axons. *J. Gen. Physiol.* **47:** 965.

Rubin, L.L. and U.J. McMahan. 1982. Regeneration of the neuromuscular junction: Steps toward defining the molecular basis of the interaction between nerve and muscle. In *Disorders of the motor unit* (ed. D.L. Schotland), p. 187. Wiley, New York.

Towbin, H., T. Staehelin, and J. Gordon. 1979. A procedure for the electrophoretic transfer of proteins from polyacrylamide gels to nitrocellulose sheets and some applications. *Proc. Natl. Acad. Sci.* **76:** 4350.

The Molecular Mechanism of Action of the β-Adrenergic Receptor

M. SCHRAMM, M. KORNER, G. NEUFELD, AND E. NEDIVI
Department of Biological Chemistry, The Hebrew University of Jerusalem, Jerusalem 91904, Israel

Interest in the mode of action of receptors that activate adenylate cyclase stems from several considerations. The system regulates many important biological functions in various tissues (Dumont et al. 1981) and has been shown to be prominent in peripheral and central neurons (Greengard 1978). Recently, the function of a serotonin-activated adenylate cyclase has been strongly implicated in a simple learning process (Kandel and Schwartz 1982). Opiates and enkephalins inhibit adenylate cyclase (Collier 1980), and this action might be part of their biological effects.

There are additional reasons for our interest. The receptors coupled to adenylate cyclase transmit a signal through the cell membrane. It would be of considerable importance to understand the molecular mechanism of this signaling process in its own right, and also because it might shed light on the mode of action of other receptors that apparently act by transmembrane signaling, such as the muscarinic, α-adrenergic, and opiate receptors. Understanding the molecular mechanism might also help to detect abnormalities in receptor function in cases of disease or genetic defect.

Several years ago, we showed that the different receptors that activate adenylate cyclase, each specific for a certain neurotransmitter or hormone, are all interchangeable—a receptor can be transferred to a foreign adenylate cyclase and can activate it when the specific agonist is added (Schramm et al. 1977; Schramm 1979). Therefore, it seems likely that the essentials of the mechanism of action, when elucidated for one such receptor, will hold true for all, irrespective of the chemical structure of the neurotransmitter or hormone, the tissue, or the species from which the receptor is derived. The neurotransmitter or hormone (H)-activated adenylate cyclase consists of three components: the receptor, R; the guanyl nucleotide-binding protein, G (Sternweis et al. 1981); and the catalytic unit, C. A scheme describing operation of the system is shown in Figure 1 and is explained in its legend. The β-adrenergic receptor-coupled adenylate cyclase in the turkey erythrocyte membrane was chosen as a paradigm because it has several important advantages. Numerous agonists and antagonists of simple chemical structure are available, and measurement of receptor sites with ^{125}I- or ^{3}H-labeled antagonists can be readily performed. The most important consideration, however, was that activation of G by GTP in the turkey erythrocyte is almost absolutely dependent on HR action (Fig. 1), whereas in other systems, the G component undergoes considerable activation spontaneously (basal activity in absence of hormone). When the turkey erythrocyte membranes are properly prepared, a high rate of adenylate cyclase activity is produced by the synthetic hormone isoproterenol plus GTP. These features are essential for studying the mechanism of R action.

The scheme in Figure 1 suggests that the action of R takes place in the HRG complex and is limited to the activation of G to produce G_{GTP} (Fig. 1, upper right section). Thus, R action would not require C since the latter is activated indirectly, via G. Direct support for this concept was obtained after Gpp(NH)p and GTPγS, the hydrolysis-resistant analogs of GTP, were introduced (Londos et al. 1974; Pfeuffer and Helmreich 1975). The HRG complex in the membrane functioned even after C had been inactivated by *N*-ethylmaleimide (NEM) and produced activated G (G*), which accumulated (Neufeld et al. 1980):

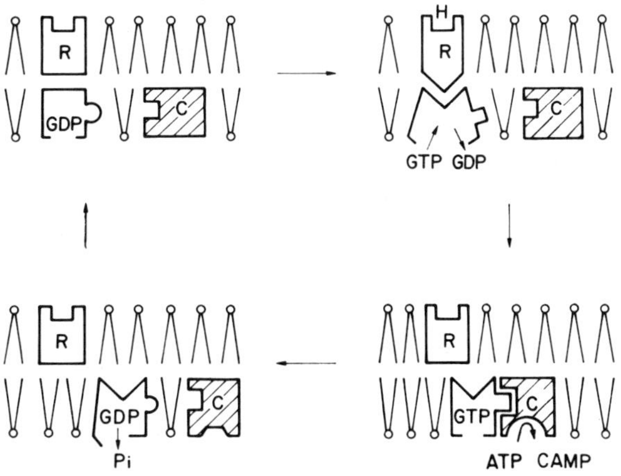

Figure 1. Scheme of the partial reactions involved in the sequence of activation-deactivation of the adenylate cyclase system. (*Upper left*) A section of the cell membrane with three components of the adenylate cyclase system: R, the receptor with the neurotransmitter-binding site facing the outside; G, the guanyl nucleotide-binding protein containing GDP; and C, the catalytic unit of the adenylate cyclase. Phospholipid is symbolized by the polar heads with twin fatty acid tails and is shown interspersed among the three components. When the hormone (H) binds to R (*upper right*), HR associates with G and thus facilitates the binding of GTP in exchange for the release of GDP. The G thus activated by GTP (Rodbell et al. 1971) associates with C (*lower right*) to form the activated enzyme GC. Hydrolysis of the GTP at the G site to release P_i results in dissociation of G from C and cessation of enzyme activity (Cassel et al. 1977) (*lower and upper left*).

$$HRG + Gpp(NH)p \rightarrow HRG_{Gpp(NH)p} \rightarrow HR + G_{Gpp(NH)p}(G^*) \quad (1)$$

The reaction was stopped readily at any time by addition of a specific receptor blocker, such as propranolol. The amount of G* produced was measured in an independent system, by implantation in another membrane that contained an excess of C (Ross et al. 1978) to produce the activated enzyme spontaneously: $G^* + C \rightarrow G^*C$ (Fig. 1, lower right). The amount of G*C was measured as the rate of cAMP production. Thus, the amount of G* is expressed in relative terms as pmoles cAMP produced per minute. By these means, it was possible for the first time to separate completely in time and space the activation of G by HR and the resulting activation of C by G*. Partial reactions in the hormone activation of adenylate cyclase could now be studied.

Support for formation of the HRG complex and its dissociation to HR + G* also came from different types of experiments (Limbird et al. 1980; Stadel et al. 1980). At this stage, attempts to reconstitute a functional HRG system seemed timely. Solubilized, separate preparations of R and of G were combined in presence of phospholipids. After removal of detergent and addition of hormone, the reconstituted system readily performed reaction 1, above (Citri and Schramm 1980). The hormone-dependent *deactivation* of $G_{Gpp(NH)p}$ via the HRG complex was also performed by the reconstituted system (Citri and Schramm 1982; Keenan et al. 1982).

$$HRG_{Gpp(NH)p} + GTP \rightarrow HRG_{GTP} + Gpp(NH)p \quad (2)$$

$$G_{GTP} \rightarrow G_{GDP} + P_i \quad (3)$$

The deactivation in the reconstituted system had characteristics similar to those described earlier for this reaction in the native membrane (Cassel and Selinger 1977; Sevilla and Levitzki 1977).

Having recognized the importance of the HRG intermediate in receptor action, we wanted to learn about the characteristics of this activated state. It seemed of particular interest to study the change that occurs in R when it binds H, leading to the HRG complex, in contrast to binding of an antagonist that does not lead to an activated HRG. Some experiments elucidating these aspects are described. The work was done with receptor in the native membranes and with crude solubilized preparations. Although a purified β-adrenergic-binding protein has been described (Caron et al. 1979), its ability to function as a receptor remains to be demonstrated.

EXPERIMENTAL RESULTS

Incubation of β-adrenergic systems with isoproterenol plus NEM leads to apparent inactivation of antagonist binding sites on R. GTP protects against such inactivation (Vauquelin et al. 1980). Table 1 shows that only 28% of the original β-receptor binding sites remained after treatment with agonist plus NEM, whereas treatment with the R blocker propranolol plus NEM left all of the receptor binding sites intact. Removal of the blocker and addition of an agonist and fresh NEM produced inactivation of R as above (not shown). This latter finding indicates that the specific –SH involved in the disappearance of receptor binding sites is only exposed in presence of agonist. It is further shown that prolonged treatment with a high concentration of GTP

Table 1. Disappearance of β-Adrenergic-binding Sites after Treatment with Isoproterenol Plus NEM and Their Reactivation by Guanyl Nucleotides

Treatment	Binding sites after treatment %	Δ%
Inactivation Experiments		
1. Isoproterenol + NEM	28	
2. Propranolol + NEM	100	
3. Treatment 2 followed by 1	30	
4. Isoproterenol + GTPγS + NEM	100	
Reactivation Experiments		
0 Time	28	(0)
5 Min, no additions	31	3
5 Min, 0.1 mM GTP	38	10
5 Min, 0.5 mM GTP	48	20
5 Min, 0.5 mM GTPγS	43	15
5 Min, 0.5 mM GMP or ATP or ITP	31	3
20 Min, no additions	42	14
20 Min, 0.5 mM GTP	82	54

Inactivation experiments. Turkey erythrocyte membranes were incubated with agonist or antagonist plus NEM at 30°C, washed, and assayed for β-adrenergic-binding sites with the blocker [^{125}I]iodocyanopindolol ([^{125}I]ICYP). In experiment 3, the membranes were first treated with propranolol plus NEM to inactivate all the exposed –SH groups, washed, and then incubated with isoproterenol plus NEM.

Reactivation experiments. Membranes were treated as in experiment 1 above, washed, and reincubated at 37°C with or without various additons for 5 min or 20 min as shown. The Δ% column shows the increase in binding sites over zero time during reactivation.

Table 2. Locking of the Agonist in R by Alkylation of a Specific -SH in G

System	Membranes (cpm/mg)	Specifically locked (pmoles/mg)
[^{3}H]Isoproterenol plus NEM	2700	0.4–0.6
[^{3}H]Isoproterenol only	900	

Turkey erythrocyte membranes (1 mg) were incubated at 30°C with 0.25 μM [^{3}H]isoproterenol in presence and absence of 1 mM NEM. After washing the membranes several times, the labeled agonist was extracted with 0.1 M formic acid and counted. Essentially the same results were obtained when the membranes were pretreated with the β-blocker propranolol plus NEM, washed, and then incubated with [^{3}H]isoproterenol plus NEM.

(0.5 mM) or GTPγS reactivates the binding sites, whereas GMP, ATP, and ITP are ineffective (Korner et al. 1982). As a further check, the turkey erythrocyte membrane containing the reactivated R was implanted by fusion in Friend cell adenylate cyclase and was shown to be fully functional, whereas the same material prior to reactivation was 50% inactive (not shown). The slow reappearance of the binding sites, enhanced by rather high concentrations of GTP, indicated that the irreversible alkylating agent had attacked G in the HRG complex and not R. The fact that an agonist must occupy the binding sites to produce their disappearance in presence of NEM raised the possibility that the agonist becomes locked in R because of the alkylation of G. Table 2 shows that this is indeed the case. As will be discussed later, locking is not a form of desensitization but rather a trapping of the activated HRG complex.

The locking of H reminded us of an earlier observation: Solubilization of a functional R in deoxycholate specifically required preincubation with an agonist (Eimerl et al. 1980). Therefore it was suspected that perhaps the deoxycholate also locks the prebound agonist in R during solubilization. The experiment presented in Table 3 confirms this notion. Dissociation of [^{3}H]isoproterenol locked in the solubilized R was extremely slow, about 50% dissociated in 9 hours. In contrast, the labeled antagonist [^{3}H]alprenolol was not locked in R to any detectable extent. It was at first assumed, in analogy to the locking by isoproterenol plus NEM, that the G component is essential and that deoxycholate somehow causes locking of the hormone in the activated HRG complex. Surprisingly, it was found, however, that deoxycholate locks prebound isoproterenol, even after the G in the membrane had been inactivated >95% by treatment at pH 11.5. Also, preactivation of G in the native membrane by GTPγS, which totally prevents locking of R with isoproterenol plus NEM, still permits specific locking of the agonist in R during solubilization in deoxycholate.

Additional studies on the turkey erythrocyte membrane revealed that low concentrations of deoxycholate that do not solubilize any R suffice to lock prebound agonist in R. Less than 10% of the membrane protein is solubilized during the procedure (Fig. 2). The process is completely reversible; when the detergent is removed, R "opens up" to release the agonist and locks again when the agonist, followed by deoxycholate, is readded. As with locking during solubilization, the G component is apparently not required and GTPγS or GTP have no effect. It should be emphasized that locking by detergent is performed in the cold, that it is agonist-specific, that antagonists are not locked, and that maintenance of the locked conformation requires Mg^{++}. Addition of deoxycholate before isoproterenol does not permit proper locking. This finding indicates that a prior interaction between H and R is further promoted by deoxycholate.

DISCUSSION

Probably the most fascinating feature of receptors is the fact that the mere binding of a neurotransmitter or hormone sets in motion a rapid sequence of reactions leading to a dramatic biological phenomenon. It is made even more intriguing by the antagonists that demonstrate that specific binding can occur also without leading to a response. Obviously, the binding of the agonist must lead to some change in the receptor

Table 3. Locking of [^{3}H]Isoproterenol in R during Solubilization of the Membranes by Deoxycholate

[^{3}H]Isoproterenol locked			[^{125}I]ICYP bound after unlocking the sites by removal of deoxycholate (pmole/mg protein)
specific (dpm)	nonspecific (dpm)	protein (pmole/mg)	
9400	600	1.3	1.0

Membranes were incubated with 1 μM [^{3}H]isoproterenol in absence and presence of 0.1 μM cyanopindolol. The latter system, containing the specific β-blocker, served to measure nonspecific binding. The preparations were solubilized in deoxycholate as described (Citri and Schramm 1982). Free [^{3}H]isoproterenol was removed by passing 100 μl of the solubilized material (about 10^6 dpm) through a Sephadex G-50 superfine column that had been "spun dry" by centrifugation (Penefsky 1977). The dpm in the table are given for the 0.1-ml sample.

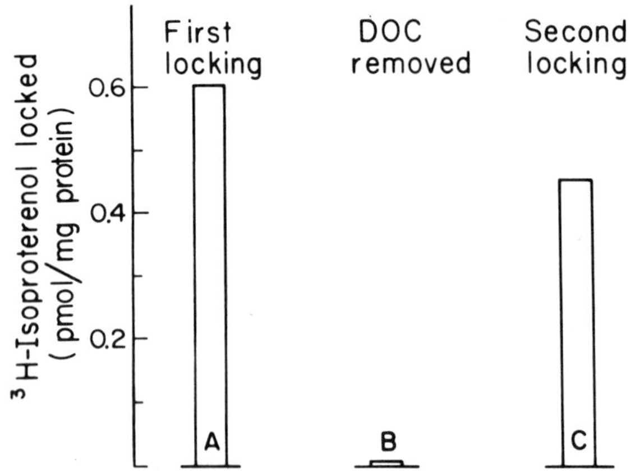

Figure 2. Reversible locking of agonist in the membrane-bound R by deoxycholate. Turkey erythrocyte membranes (1 mg) were incubated in the cold with 0.6 μM [^{3}H]isoproterenol in the absence and presence of 0.1 μM cyanopindolol. The latter system, containing the β blocker, served to measure nonspecific binding. After dilution and washing the membranes once with buffer containing 1 mg/ml deoxycholate, the membranes were suspended in the buffer containing the detergent and incubated at 4°C for 30 min in presence of 0.1 mM isoproterenol. Following centrifugation, samples of the membranes served to measure the amount of [^{3}H]isoproterenol locked (*A*). Additional samples were freed of detergents by washing in aqueous buffer, to determine residual labeled agonist retained in the receptor (*B*). Aliquots of membranes treated as in *B* were also subjected to relocking by renewed incubation with [^{3}H]isoproterenol followed by deoxycholate, as described above (*C*).

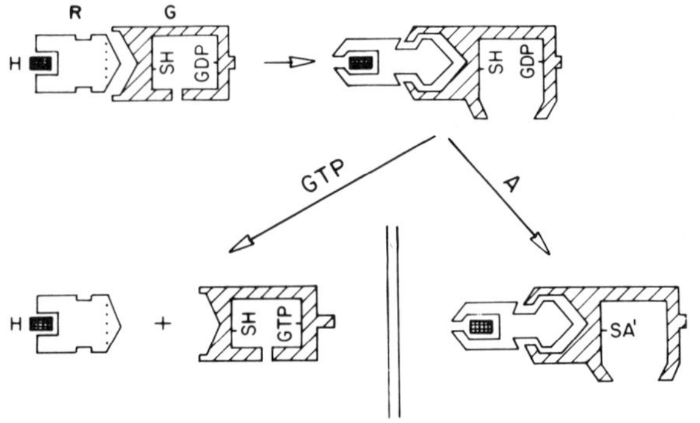

Figure 3. Locking of H in R as a transient activated state of the HRG complex. (*Upper left*) The HR complex with a pointed recognition site for G. (*Upper right*) HR and G have undergone a mutual induced-fit interaction resulting in the locking of H in R and exposure of a specific -SH as well as the guanyl nucleotide-binding site in G. (*Lower right*) Alkylating agent A (NEM or other) has reacted to form SA′ and thus traps HRG in the conformation as above. (*Lower left*) The normal course of events: GTP binds to its specific site in G causing relaxation of the structure, resulting in dissociation of the HRG* complex, and G* shuttles to combine with C (not shown).

molecule that cannot be produced by an antagonist. We are presently trying to learn what this change might be in the case of a receptor that activates an adenylate cyclase system. Earlier experiments, including reconstitution of a hormone-responsive system, pointed to the HRG complex as the activated state of the system. Figure 3 is a tentative scheme of the formation and dissociation of the HRG complex in light of the present findings (Korner et al. 1982). The binding of H, with relatively low affinity, causes a change in R that leads to a mutual, induced-fit interaction with the G component. As a result, H becomes transiently locked in R, and G exposes a unique SH group as well as a guanyl nucleotide-binding site, both of which were hitherto masked (Fig. 3, upper right).

The following observations support this concept of HRG. It is well documented that H has a higher affinity for R in absence of guanyl nucleotides, when the HRG complex predominates, than in presence of guanyl nucleotides, when HR predominates (cf. Stadel et al. 1980) (Fig. 3, lower left). Since this information comes from steady-state binding experiments, it gives the *average* affinity for all molecules dependent on the free energy change in the system. In contrast, NEM traps a specific conformation by interacting with the -SH when it becomes exposed. It seems unlikely that NEM plays any causal role in the appearance of that conformation. Although it is not possible to prove that the transient locking of H in the HRG complex is the activated intermediate, the alternative explanation—that it represents a desensitized state—appears rather improbable. Desensitization of hormone-activated adenylate cyclase systems was found to involve transition of R to a *lower*, rather than a higher, affinity for H (Pittman and Molinoff 1980; Su et al. 1980). Furthermore, the turkey erythrocyte receptor could not, so far, be induced to undergo desensitization under any conditions.

What is the evidence that NEM actually interacted with a unique -SH in G? Fully functional R could be obtained by reactivation after treatment with isoproterenol plus NEM. It therefore seems unlikely that the irreversible alkylating agent had attacked R itself. However, since R was in a special configuration, containing locked H, it must be assumed that it was trapped in a tight conformational interaction with the component that was modified by NEM. At the present time, the only known component that interacts with R is G. There are additional arguments suggesting G as the component that exposes the unique -SH with which NEM interacts upon formation of the HRG complex. Locking is dependent on the presence of a functional, nonactivated G component; destruction of G at pH 11.5, after formation of the locked complex, reactivates R. Even more significant is the finding that addition of a high concentration of GTP causes reactivation of R, in what seems to mimic the normal dissociation of the HRG complex (Fig. 3, lower right), but in slow motion. It is not surprising that the G, disabled by the attack of NEM on a crucial -SH group, would interact with GTP with a much reduced affinity and that the conformational change that normally leads to dissociation of the activated G_{GTP} in 1 second or less would now require 20 minutes.

Although G was essential for locking by isoproterenol plus NEM, it is obvious from the experiments that the structure of the R molecule is engineered to undergo quite extreme conformational modifications so as to open or close around the agonist. In that sense, it is perhaps not surprising that deoxycholate was found to cause locking of R on prebound agonist. From the findings, one would suggest that R is held in the open conformation by some hydrophobic interactions. When these are disrupted by the detergent, maximal association between groups in the agonist and in the R binding site can come into play. It seems as if deoxycholate is performing nonspecifically the function that is normally performed by the G component. Both agents caused locking that was highly specific for the agonist and it therefore seems quite likely that both induced the same conformation of the binding site. Further studies should tell to what extent locking of the agonist in the binding site reflects the response that propagates along the receptor molecule in its interaction with the next component along the pathway of signal transmission. The present work already suggests that any model of the structure of R should demonstrate the flexibility of the binding site to open and close by specific interactions with the agonist.

ACKNOWLEDGMENTS

This work was supported by grants from the National Institutes of Health, NIAMDDK AM-10451 and the United States-Israel Binational Science Foundation.

REFERENCES

CARON, M.G., Y. SRINIVASAN, J. PITHA, K. KOCIOLEK, and R.J. LEFKOWITZ. 1979. Affinity chromatography of the β-adrenergic receptor. *J. Biol. Chem.* **254:** 2923.

CASSEL, D. and Z. SELINGER. 1977. Catecholamine induced release of [^{3}H]-Gpp(NH)p from turkey erythrocyte adenylate cyclase. *J. Cyclic Nucleotide Res.* **3:** 11.

CASSEL, D., H. LEVKOVITZ, and Z. SELINGER. 1977. The regulatory GTPase cycle of turkey erythrocyte adenylate cyclase. *J. Cyclic Nucleotide Res.* **3:** 393.

CITRI, Y. and M. SCHRAMM. 1980. Resolution, reconstitution and kinetics of the primary action of a hormone receptor. *Nature* **287:** 297.

——. 1982. Probing of the coupling site of the β-adrenergic receptor. Competition between different forms of the guanyl nucleotide binding protein for interaction with the receptor. *J. Biol. Chem.* **257:** 13257.

COLLIER, H.O.J. 1980. Cellular site of opiate dependence. *Nature* **283:** 625.

DUMONT, J.E., P. GREENGARD, and G.A. ROBISON, eds. 1981. *Adv. Cyclic Nucleotide Res.* **14.**

EIMERL, S., G. NEUFELD, M. KORNER, and M. SCHRAMM. 1980. Functional implantation of a solubilized β-adrenergic receptor in the membrane of a cell. *Proc. Natl. Acad. Sci.* **77:** 760.

GREENGARD, P. 1978. *Cyclic nucleotides, phosphorylated proteins and neuronal function.* Raven Press, New York.

KANDEL, E.R. and J.H. SCHWARTZ. 1982. Molecular biology of learning: Modulation of transmitter release. *Science* **218:** 433.

KEENAN, A.K., A. GAL, and A. LEVITZKI. 1982. Reconstitution of the turkey erythrocyte adenylate cyclase sensitivity to l-epinephrine upon re-insertion of the lubrol solubilized components into phospholipid vesicles. *Biochem. Biophys. Res. Commun.* **105:** 615.

KORNER, M., C. GILON, and M. SCHRAMM. 1982. Locking of hormone in the β-adrenergic receptor by attack on a sulfhydryl in an associated component. *J. Biol. Chem.* **257:** 3389.

LIMBIRD, L.E., D.M. GILL, and R.J. LEFKOWITZ. 1980. Agonist-promoted coupling of the β-adrenergic receptor with the guanine nucleotide regulatory protein of the adenylate cyclase system. *Proc. Natl. Acad. Sci.* **77:** 775.

LONDOS, C., Y. SALOMON, M. LIN, J.P. HARWOOD, M. SCHRAMM, J. WOLFF, and M. RODBELL. 1974. 5′-Guanylylimidodiphosphate, a potent activator of adenylate cyclase system in eucaryotic cells. *Proc. Natl. Acad. Sci.* **71:** 3087.

NEUFELD, G., M. SCHRAMM, and N. WEINBERG. 1980. Hybridization of adenylate cyclase components by membrane fusion and the effect of selective digestion by trypsin. *J. Biol. Chem.* **255:** 9268.

PENEFSKY, H.S. 1977. Reversible binding of P_i by beef heart mitochondrial adenosine triphosphatase. *J. Biol. Chem.* **252:** 2891.

PITTMAN, R.N. and P.B. MOLINOFF. 1980. Interactions of agonists and antagonists with β-adrenergic receptors on intact L6 muscle cells. *J. Cyclic Nucleotide Res.* **6:** 421.

PFEUFFER, T. and E.J.M. HELMREICH. 1975. Activation of pigeon erythrocyte membrane adenylate cyclase by guanylnucleotide analogues and separation of a nucleotide binding protein. *J. Biol. Chem.* **250:** 867.

RODBELL, M., L. BIRNBAUMER, S.L. POHL, and H.M.J. KRANS. 1971. The glucagon-sensitive adenyl cyclase system in plasma membranes of rat liver: An obligatory role of guanyl nucleotides in glucagon action. *J. Biol. Chem.* **246:** 1877.

ROSS, E.M., A.C. HOWLETT, K.M. FERGUSON, and A.G. GILMAN. 1978. Reconstitution of hormone-sensitive adenylate cyclase activity with resolved components of the enzyme. *J. Biol. Chem.* **253:** 6401.

SCHRAMM, M. 1979. Transfer of glucagon receptor from liver membranes to a foreign adenylate cyclase by a membrane fusion procedure. *Proc. Natl. Acad. Sci.* **76:** 1174.

SCHRAMM, M., J. ORLY, S. EIMERL, and M. KORNER. 1977. Coupling of hormone receptors to adenylate cyclase of different cells by cell fusion. *Nature* **268:** 310.

SEVILLA, N. and A. LEVITZKI. 1977. The activation of adenylate cyclase by l-epinephrine and guanylylimidodiphosphate and its reversal by l-epinephrine and GTP. *FEBS Lett.* **76:** 129.

STADEL, J.M., A. DELEAN, and R.J. LEFKOWITZ. 1980. A high affinity agonist β-adrenergic receptor complex is an intermediate for catecholamine stimulation of adenylate cyclase in turkey and frog erythrocyte membranes. *J. Biol. Chem.* **255:** 1436.

STERNWEIS, P.C., J.K. NORTHUP, M.D. SMIGEL, and A.G. GILMAN. 1981. The regulatory component of adenylate cyclase. Purification and properties. *J. Biol. Chem.* **256:** 11517.

SU, Y.-F., T.K. HARDEN, and J.P. PERKINS. 1980. Catecholamine specific desensitization of adenylate cyclase. *J. Biol. Chem.* **255:** 7410.

VAUQUELIN, G., S. BOTTARI, C. ANDRE, B. JACOBSSON, and A.D. STROSBERG. 1980. Interaction between β-adrenergic receptors and guanine nucleotide sites in turkey erythrocyte membranes. *Proc. Natl. Acad. Sci.* **77:** 3801.

Modulation of Calcium Channels in Cultured Cardiac Cells by Isoproterenol and 8-Bromo-cAMP

H. REUTER, A.B. CACHELIN, J.E. DE PEYER, AND S. KOKUBUN
Department of Pharmacology, University of Berne, 3010 Berne, Switzerland

The central role played by Ca^{++} in the regulation of many cellular functions is widely appreciated, and the field centering around this problem is continuously growing. An important aspect of cell regulation by Ca^{++} is the fact that the free intracellular Ca^{++} concentration ($[Ca^{++}]_i$) can fluctuate over several orders of magnitude. In many cells, particularly in muscle and neurons, an increase in $[Ca^{++}]_i$ is triggered by depolarization of the cell membrane. It has long been recognized that the Ca^{++} conductance (g_{Ca}) of these cells is greatly increased during excitation (for reviews, see Reuter 1973; Hagiwara and Byerly 1981; Tsien 1983) and that this contributes to the increase in $[Ca^{++}]_i$. The rise in g_{Ca} during an action potential has commonly been described by the formalism suggested by Hodgkin and Huxley (1952) for the change in Na^+ conductance (g_{Na}) during nerve excitation. However, recent detailed analyses of g_{Ca}, and particularly of the properties of individual Ca^{++} channels, have indicated that such an analogy is inadequate. Not only are the voltage-dependent activation kinetics of Ca^{++} channels inconsistent with the Hodgkin-Huxley formalism (Fenwick et al. 1982), but also inactivation. There are at least two different mechanisms of inactivation: a voltage-dependent one (Fox 1981; Reuter et al. 1982; Cavalié et al. 1983) and one that depends on the flow of Ca^{++} current (I_{Ca}) and presumably on the resulting rise in $[Ca^{++}]_i$ (Eckert et al. 1981; Standen and Stanfield 1982).

Another feature of Ca^{++} channels, which is not shared by Na^+ channels, is their modulation by neurotransmitters and drugs. This has first been shown for cardiac cells, but it also seems to be an important regulatory mechanism in nerve cells (for review, see Reuter 1983). Detailed electrophysiological analyses of neurotransmitter and drug actions at the molecular level have become possible with the development of the patch-clamp method (Hamill et al. 1981), which permits recordings from single ion channels in cell membranes. In the present paper, we describe experiments on single Ca^{++} channels in cultured cardiac cells. We have analyzed the effects of the β-adrenoceptor agonist isoproterenol and of 8-bromo-cAMP on electrophysiological properties of these channels. We shall interpret these effects in terms of a simple kinetic model that has recently been suggested as being applicable to Ca^{++} channels (Fenwick et al. 1982; Tsien 1983).

METHODS

Cultured cardiac cells. Primary cardiac cell cultures were prepared from hearts of neonatal rats. The culture techniques used are a combination of various established methods (Mark and Strasser 1966; Blondel et al. 1971; Porzig et al. 1982). Hearts were removed under sterile conditions and the ventricles were cut into small pieces. The tissue pieces were incubated at 0°C for 20 hours in Ca^{++}-free Hank's solution containing 0.1% trypsin (Boehringer Mannheim Corp.). After warming up to 37°C for 30 minutes, the cell suspension was sedimented at 300*g* and resuspended in culture medium (Dulbecco's modified minimum essential medium containing 20% fetal calf serum). The cell suspension was filtered through cotton gauze, and myocardial cells were separated from fibroblasts by a differential attachment technique (Blondel et al. 1971). Myocardial cells were seeded at a density of about $1-3 \times 10^5$ cells/ml on glass coverslips, which were put into medium-containing petri dishes. The low density of seeding prevented the cell cultures from rapidly growing to confluency. The cultures were kept in an incubator (Forma) at 37°C in an H_2O-saturated, 5% CO_2-95% air atmosphere. Individual cells firmly adhered to the coverslip 1–3 days after seeding and many of them were fully differentiated, showing spindlelike shapes and cross-striation patterns. These cells were used for single-channel recordings.

Recording conditions. Cells grown on coverslips were put into a water-jacketed Plexiglas chamber placed on the stage of an inverted microscope (Diavert, Leitz). The cells were superfused with saline of the following composition: NaCl, 137 mM/liter; KCl, 5.4 mM/liter; $MgCl_2$, 2.0 mM/liter; $CaCl_2$, 0.02 mM/liter; HEPES buffer, 10 mM/liter; glucose, 10 mM/liter; pH 7.4; temperature 24–26°C. Isoproterenol (0.1 or 1 μM/liter) and 8-bromo-cAMP (5 or 10 μM/liter) were added to the bathing solution. Although the cells often fired spontaneous action potentials, corresponding contractions were small due to the low extracellular Ca^{++} concentration ($[Ca^{++}]_e$).

Electrophysiological recordings were made with the patch-clamp technique as extensively described by Hamill et al. (1981). The recording pipettes were filled with 96 mM/liter $BaCl_2$ solution containing 20 μM/liter tetrodotoxin, buffered at pH 7.4 with 10 mM/liter HEPES. Single-channel currents, with Ba^{++} as charge carriers, were recorded from cell-attached membrane patches with seal resistances between 10 GΩ and 100 GΩ (Hamill et al. 1981; Reuter et al. 1982).

Data acquisition and analysis. All patch currents and pipette voltages were recorded on analog tape

(Racal 4DS tape recorder). Current traces were low-pass filtered at 1.0 kHz and digitized records (sampling interval 0.2 msec) were further analyzed on a PDP 11/04 minicomputer. In a first step, records without channel openings (nulls) were averaged for each clamp potential and the averaged trace was subtracted from every single record at the same potential. After the subtraction of capacity and leakage current, the program calculated a baseline for each record (heavy lines in current traces in Figs. 1 and 4) and the mean open channel currents, i, were fitted to each record by eye (dotted lines in Figs. 1 and 4). The steady-state probability, p, that a channel enters an open state was calculated from $I = N \cdot p \cdot i$, where I is the mean current found by time-averaging the current flowing through open channels over the duration of the clamp pulse (100 msec) and N is the number of independent channels functioning in the patch during the pulse.

Quantitative information on the kinetics of Ca^{++} channels, i.e., their openings and closures under control conditions and in the presence of drugs, was obtained by plotting open-time and closed-time histograms. The program detected transitions between conducting and nonconducting states of the channels as crossings of a midline between those states. The minimum resolvable open and closed times were 0.3 msec. The calculated transition times were displayed by the computer and plotted as histograms. The histograms were fitted by nonlinear regression analysis (MODFIT; McIntosh and McIntosh 1980). Open-time histograms could be fitted by single exponentials, whereas the fit of closed-time histograms required two exponentials.

Ensemble averages of single-current traces showed drug effects consistent with those observed on macroscopic I_{Ca} measured with the single sucrose gap voltage-clamp technique (Reuter and Scholz 1977a).

RESULTS

Properties of Ca^{++} Channels

Figure 1 shows elementary currents flowing through single Ca^{++} channels in a patch of cardiac cell membrane. The charge carriers were Ba^{++} ions, and the channels switched randomly between closed and open states. With increasing depolarization (ΔV) of the membrane patch from a holding potential (V_h) 50 mV negative to the resting potential ($V_R \sim -70$ mV), single-channel currents (i) became smaller as the driving force for current flow was reduced. Ca^{++} channel openings often occur in bursts separated by longer intervals of closures. However, the channels do not open during every depolarizing clamp step, but there are periods during which no openings can be detected at all (not shown; see Reuter et al. 1982, Fig. 1). The probability, p, of the channels to open increases upon stronger depolarizations and superposition of two openings can be seen, indicating the presence of more than one functional channel in the patch.

The mean values of i (upper part) and of the average number of open channels ($N \cdot p$; lower part) at various

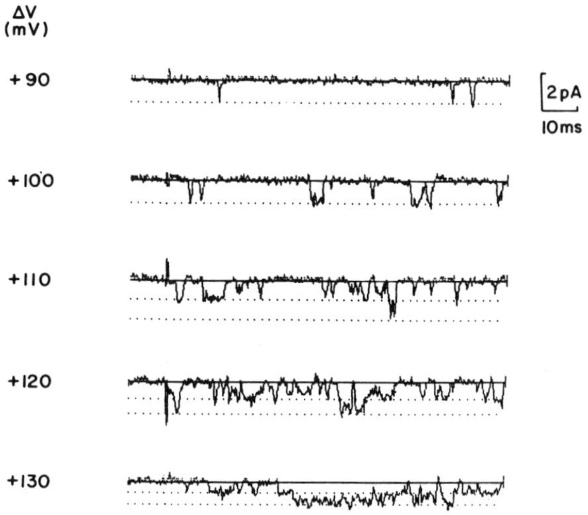

Figure 1. Elementary Ca^{++} channel currents recorded from a cell-attached membrane patch in a cultured cardiac cell (day 2). The pipette was filled with 96 mM/liter $BaCl_2$ (see Methods); thus, the current carrier was Ba^{++}. Depolarizing clamp steps (ΔV, left-hand side) were applied from a holding potential 50 mV negative to the resting potential. (.......) Mean single-channel currents at each potential. There were at least two active Ca^{++} channels in the patch.

pulse potentials have been plotted in Figure 2. As described for cardiac cells before (Reuter et al. 1982; Cavalié et al. 1983), the current-voltage (i/V) relationship is linear. However, since the voltage range, over which i can be reliably measured is limited, it cannot be decided whether or not the i/V relation has a curvature at more positive potentials (Fenwick et al. 1982). With 96 mM/liter $BaCl_2$ in the pipette, the slope conductance was 18 pS in Figure 2 (mean ± S.D. of 7 experiments: 17.7 ± 3.6 pS). The average number of open channels at each potential is a sigmoid function of voltage (Fig. 2). In experiments with only one channel in the patch, we have calculated p directly from $I = N \cdot p \cdot i$, where I is the time-averaged current flowing through a single Ca^{++} channel ($N = 1$) over 100 msec (Reuter et al. 1982). We found that p does not seem to reach a value of unity, but the p/V relationship tended to level off at values smaller than unity (~ 0.5). Similar results have been obtained by Cavalié et al. (1983). However, some uncertainty remains, since unitary currents are too small to be measured at saturating levels of p. In rat chromaffin cells (Fenwick et al. 1982) and clonal pituitary (GH_3) cells (Hagiwara and Ohmori 1983), p seems to reach a value near one.

Open times of Ca^{++} channels are exponentially distributed (see Fig. 6; Fenwick et al. 1982; Hagiwara and Ohmori 1983) with time constants, τ_0, that are little voltage dependent (Lux and Nagy 1981; Fenwick et al. 1982; Reuter et al. 1982; Hagiwara and Ohmori 1983). Closed-time histograms of Ca^{++} channels can be fitted by double exponential curves (Fig. 6). The short time constant, τ_{C1}, corresponds to the brief closures during the bursts of openings, whereas the longer time con-

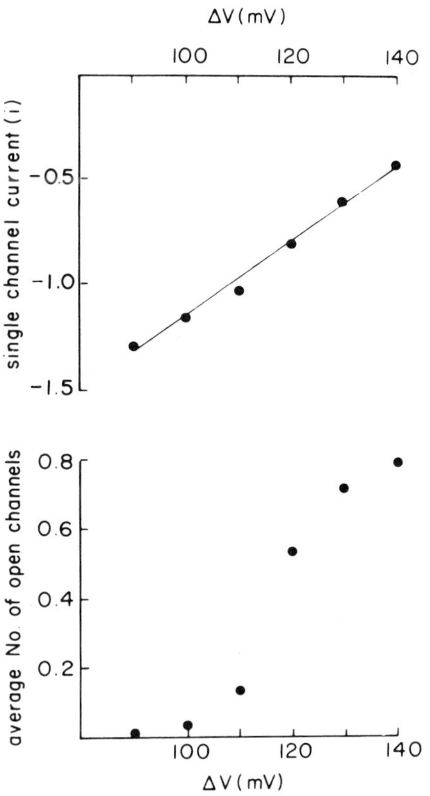

Figure 2. Voltage dependencies of open channel current (*upper part*) and of open channel probability ($N \cdot p$; $N = 2$; *lower part*). Same experiment as in Fig. 1. Each point is the average of at least 35 single-channel current traces at each potential. The straight line of the i/V relationship indicates a slope conductance of 18 pS.

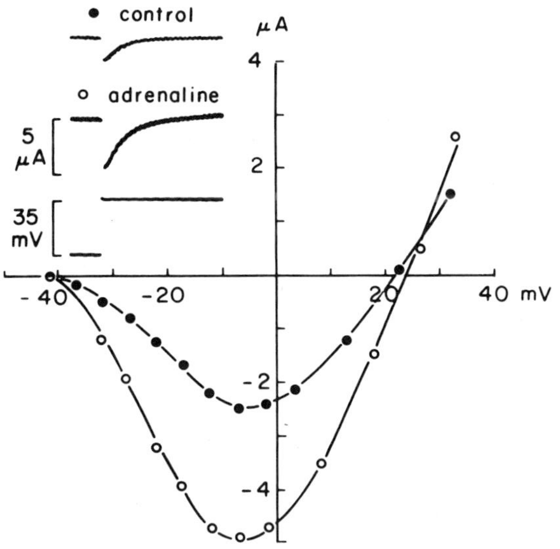

Figure 3. Current-voltage relationships of I_{Ca} measured under voltage-clamp conditions (single sucrose gap method; Reuter and Scholz 1977a) from a cow ventricular trabecula in the absence (●) and presence (○) of adrenaline (0.5 μM/liter; Tyrode solution with 1.8 mM/liter $CaCl_2$). Plotted are peak inward or outward currents (*abscissa*) against absolute membrane potential (*ordinate*). Original current traces are shown as inset and correspond to a membrane potential of −7 mV; clamp duration 250 msec. The holding potential was set at −42 mV, in order to inactivate I_{Na}.

stant, τ_{C2}, describes the distribution of closed time intervals between the bursts. Although τ_{C1} does not seem to be voltage dependent, τ_{C2} decreases with increasing depolarization (Hagiwara and Ohmori 1983).

With Ba^{++} as charge carriers, Ca^{++} channels exhibit voltage-dependent inactivation (Fig. 7; Reuter et al. 1982; Cavalié et al. 1983). There is evidence for an additional inactivation process if Ca^{++} ions are the current carriers (Tsien and Marban 1982), but this has not been analyzed in single-channel recordings.

A comprehensive review of Ca^{++} channels in excitable cells has recently been published by Tsien (1983).

Catecholamine Effects on I_{Ca}

β-Adrenoceptor agonists greatly increase the macroscopic I_{Ca} measured by voltage-clamp experiments in multicellular (for reviews, see Tisen 1977; Reuter 1979, 1983) and single-cell (Isenberg and Klöckner 1982) cardiac preparations. An example of this effect, in an experiment in a cow ventricular trabecula, is depicted in Figure 3. Original voltage-clamp and membrane current traces are shown in the inset of Figure 3. Adrenaline (0.5 μM/liter) approximately doubled the inward current elicted by a depolarizing voltage-clamp step of 35 mV amplitude (inset). The time course of the current trace, however, was not affected by the drug. An I/V plot of peak inward current against absolute membrane potential shows a large increase of I_{Ca} by adrenaline over the whole voltage range.

Kinetic analyses of I_{Ca} in multicellular (Reuter and Scholz 1977b; Noma et al. 1980; Kass and Wiegers 1982) and single-cell (Isenberg and Klöckner 1982) cardiac preparations in the presence of β-adrenoceptor agonists provided evidence that neither the voltage-dependent kinetics of I_{Ca} nor its reversal potential, V_0, are affected by these drugs, but that the limiting conductance, $\bar{g}_{Ca}$, is increased. β-Adrenoceptor agonists, however, are known to increase cAMP in cardiac cells. Therefore, it is of great interest that exogenous application of cAMP derivatives (Tsien et al. 1972; Reuter 1974), intracellular perfusion of single cardiac cells with cAMP (Irisawa and Kokubun 1983), or intracellular injection of the catalytic subunit of cAMP-dependent protein kinase (Osterrieder et al. 1982) all mimic the effect of β-adrenoceptor agonists on I_{Ca}. We have further analyzed the effect of isoproterenol and 8-bromo-cAMP on the properties of elementary Ca^{++} channel currents.

Modulation of Ca^{++} Channels by Isoproterenol and 8-Bromo-cAMP

Figure 4A shows an experiment with only one functional Ca^{++} channel in a cell-attached membrane patch. The holding potential was set 30 mV negative to the resting potential and depolarizing clamp pulses of

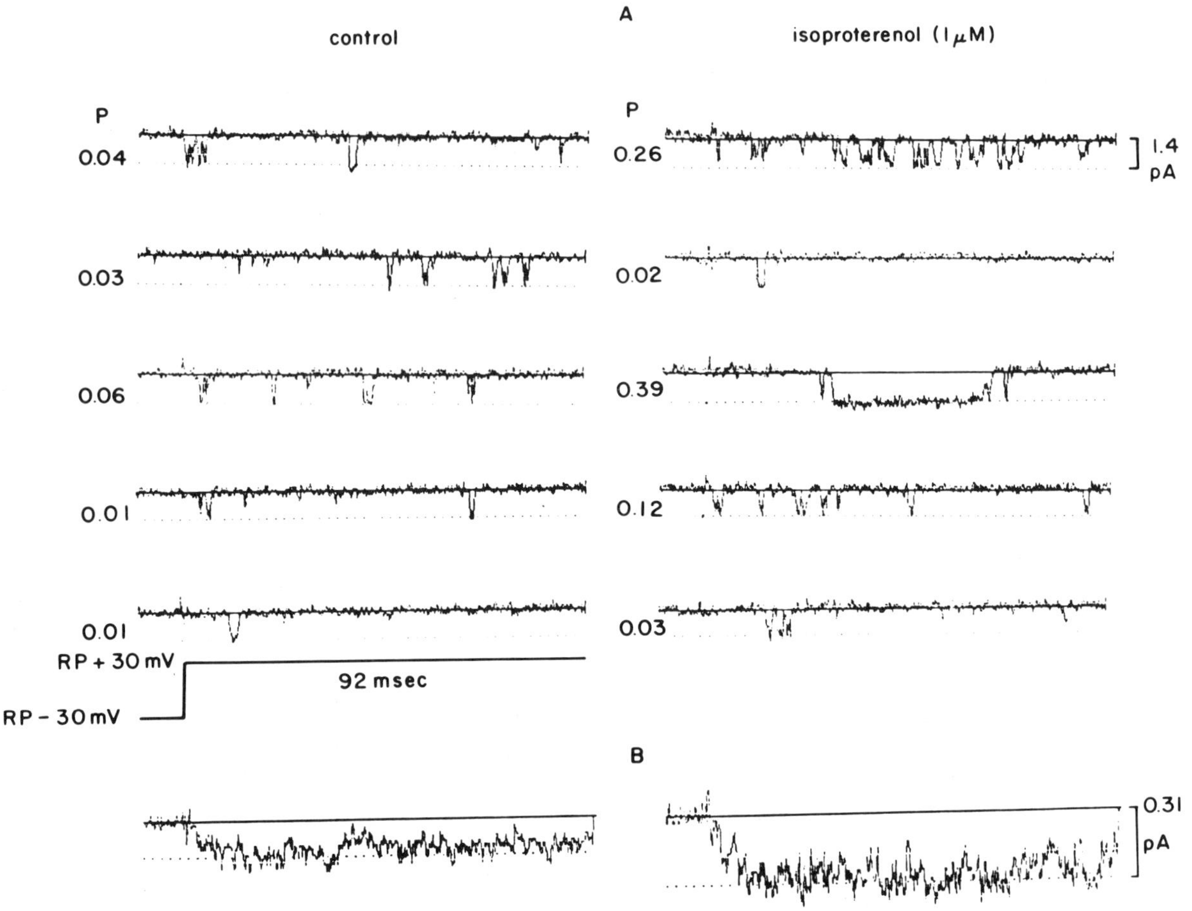

Figure 4. (*A*) Effect of isoproterenol on single Ca^{++} channel current (Ba^{++} as charge carrier) measured in a cell-attached membrane patch (cultured rat heart cell). The holding potential was 30 mV negative to the resting potential (RP), and 60 mV depolarizing voltage-clamp steps were applied under control conditions (*left*) and in the presence of isoproterenol (1 μM/liter; *right*). The probability of channel opening is indicated on the left-hand side of each record. (*B*) Ensemble average currents of 64 current traces as in *A*, but from a different experiment with at least three functional Ca^{++} channels in the patch.

60 mV amplitude were applied. The left-hand side of Figure 4 illustrates a sequence of five records with Ca^{++} channel openings under control conditions. The opening probability, p, of the channel was low at this pulse potential. Moreover, there were 21 failures of the channel to open (nulls) during 64 identical clamp steps. After application of 1 μM/liter isoproterenol, p increased significantly and there were only 5 nulls per 64 identical depolarizations. The single-channel current, however, remained unchanged. This is illustrated in Figure 5A, which shows the average i/V relations extending over a potential range of 40 mV in the upper part and the increase in the average p by isoproterenol (open circles) over the same voltage range in the lower part.

Since the macroscopic Ca^{++} current can be described by $I = N{\cdot}p{\cdot}i$, the increase in p should lead to an increase in I, even if N and i remain constant. This is illustrated in Figure 4B. The average current, I, obtained from another experiment is greatly increased by isoproterenol (0.1 μM/liter). In this experiment, there were apparently three functional channels both in the control condition as well as after application of the drug. Holding potential and pulse potential were the same as in Figure 4A. Although i remained unaffected by the drug, the average number of open channels ($N{\cdot}p$) increased. This is again due to an increase in p, since none of the single-channel traces showed more than three superimposed opening transitions ($N = 3$). Therefore, we conclude that β-adrenoceptor agonists increase I_{Ca} in cultured rat cardiac cells primarily by enhancing the opening probability of Ca^{++} channels and not by increasing the single-channel current. An increase in the number of functional channels, N, as a primary cause for the larger I_{Ca} in the presence of β-adrenergic catecholamines has previously been suggested (Reuter and Scholz 1977b). Such an effect has been observed in two experiments with isoproterenol. Therefore, an increase in N seems to be an additional factor in the enhancement of I_{Ca} by these drugs.

It is also worth mentioning that in several experiments, particularly with high concentrations of isoproterenol ($>10^{-6}$ M/liter), the channel activity decreased or the cells did not respond to the drug.

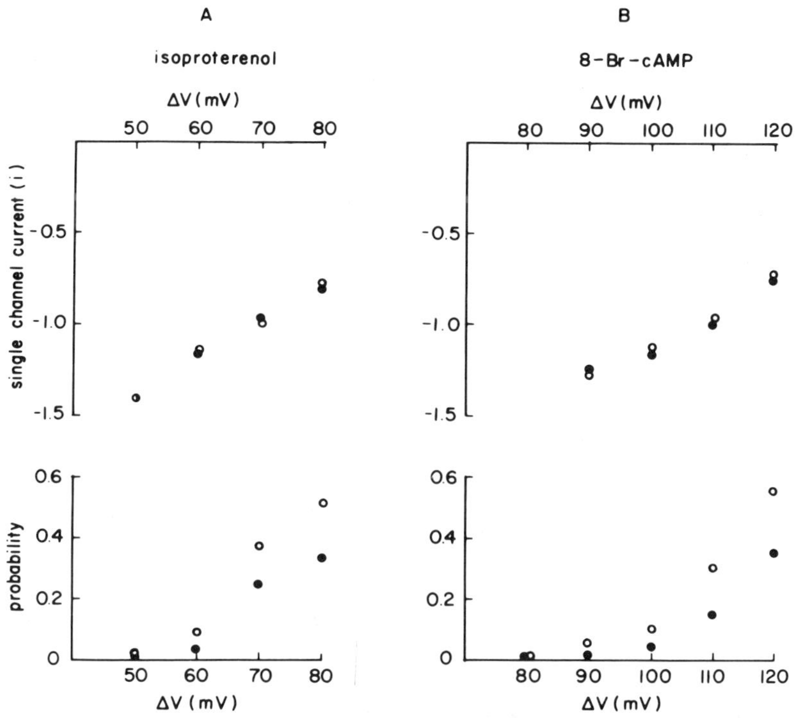

Figure 5. Voltage dependencies of open Ca^{++} channel current (*upper parts*) and of open channel probability ($N \cdot p$; $N = 1$; *lower parts*) under control conditions (●) and in the presence of isoproterenol (*A* [○]; same experiment as in Fig. 4A), or of 8-bromo-cAMP (*B* [○]). Each data point is the average of at least 64 single-channel traces. The holding potential was resting potential −30 mV in *A* and resting potential −50 mV in *B*.

Whether this is due to desensitization or experimental difficulties (e.g., rundown of the cells) is not clear. Consistent results, however, similar to those observed with isoproterenol, were obtained with 8-bromo-cAMP. This is important since cAMP is thought to be the "second messenger" after β-adrenoceptor stimulation.

In Figure 5, the effects of isoproterenol and 8-bromo-cAMP can be compared directly. In five experiments, 8-bromo-cAMP (10 μM/liter; open circles in Fig. 5B) had no effect on single-channel current, *i*, but significantly increased the opening probability, *p*. The maximum number of conductance levels seemed to be unchanged. As with isoproterenol (Fig. 4B), we found a marked increase in the average current, *I*, 10–20 minutes after application of 8-bromo-cAMP. This effect was completely reversible after washout of the nucleotide.

To obtain further information concerning the mechanism of the increase in *p* by 8-bromo-cAMP, we have analyzed the opening and closing kinetics of single Ca^{++} channels. Histogram plots of open times and closed times of a single Ca^{++} channel are shown in Figure 6 (upper part). For a statistically reliable analysis, we had to sample several hundred current traces at the same pulse potential ($\Delta V = 60$ mV) under each experimental condition. In this experiment, *p* increased from 0.046 (controls) to 0.089 (8-bromo-cAMP). The open-time histograms could be fitted by single exponentials with time constants, τ_O, of 0.6 msec (control) and 0.7 msec (8-bromo-cAMP). The closed-time histograms (Fig. 6, lower part) were fitted by the sum of two exponentials. The time constants, τ_{C1}, of the fast exponential functions were 1.0 msec (controls) and 0.9 msec (8-bromo-cAMP), respectively. The most pronounced effect of 8-bromo-cAMP, however, concerned the time constant, τ_{C2}, of the slow exponential which was reduced from 10.4 msec to 5.6 msec. This result shows that the increase in *p* by 8-bromo-cAMP is primarily due to a shortening of the longer time intervals between openings and only to a smaller extent to a lengthening of the mean open times of the channel. By contrast, on the basis of a preliminary analysis of the isoproterenol effect, Reuter et al. (1982) have suggested that the lengthening of the mean open times could be the predominant effect of the drug on Ca^{++} channels.

The voltage-dependent inactivation of Ca^{++} channels is shown in Figure 7. Application of 8-bromo-cAMP (open circles) does not affect the steady-state inactivation curve calculated from ensemble average currents. The data points have been normalized, but in this experiment 8-bromo-cAMP had increased the maximal average currents at the test potential by about 40%. The lack of an effect of 8-bromo-cAMP on steady-state inactivation of Ca^{++} channels agrees with comparable results obtained with catecholamines and cyclic nucleotide derivatives on inactivation of global Ca^{++} currents measured under voltage-clamp conditions (Reuter 1974; Noma et al. 1980).

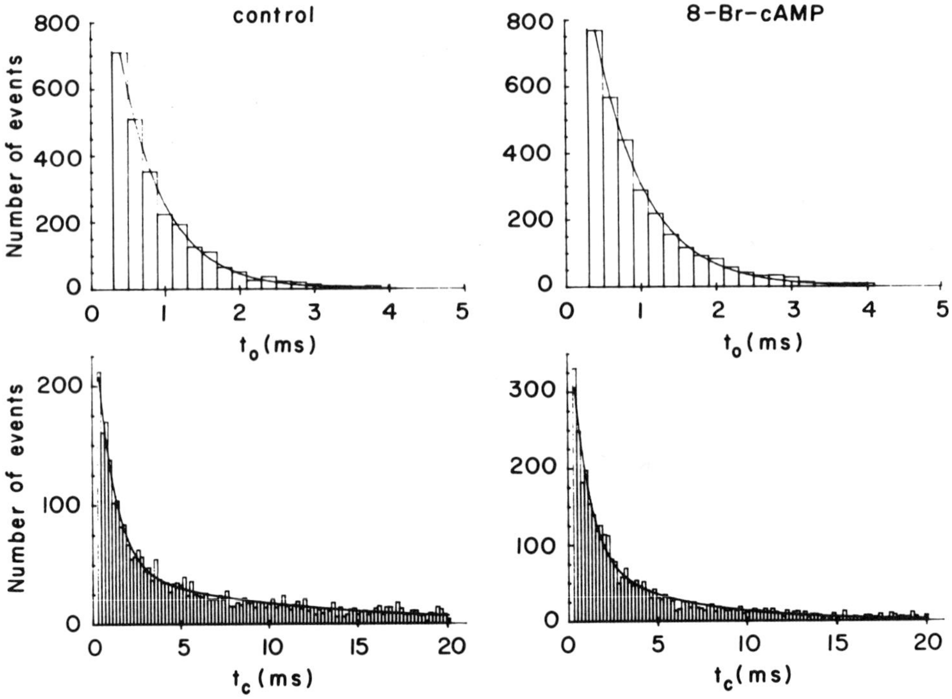

Figure 6. Open-time histograms (*upper part*) and closed-time histograms (*lower part*) of a Ca^{++} channel under control conditions (*left-hand side*) and in the presence of 8-bromo-cAMP (10 μM/liter; *right-hand side*). The histograms were fitted by nonlinear regression analysis (computer program MODFIT; McIntosh and McIntosh 1980). The open-time histograms were best fitted by single exponential curves with time constants, τ_0, of 0.6 msec (controls) and 0.7 msec (8-bromo-cAMP). Histograms of closed-time distributions were fitted by double exponential curves $[A_1 \exp(-t/\tau_{C1}) + A_2 \exp(-t/\tau_{C2})]$. The time constants, τ_{C1} and τ_{C2}, of the fast and slow components are 1.0 msec and 10.4 msec in the controls, and 0.9 msec and 5.6 msec with 8-bromo-cAMP.

DISCUSSION

The macroscopic Ca^{++} current, I_{Ca}, in excitable cells can simply be described as $I_{Ca} = N \cdot p \cdot i$ (for review, see Tsien 1983). With this formulation in mind, one can ask which of the factors is primarily responsible for the increase in I_{Ca} by β-adrenergic catecholamines and 8-bromo-cAMP. Our experiments show that in cultured rat cardiac cells, the single-channel current, i, is unaffected by these agents (see also Reuter et al. 1982). The predominant effect that we have observed is an increase in the opening probability, p, of Ca^{++} channels. Under control conditions, p does not seem to reach unity in these cells, even during strong depolarizations. A further increase in p by catecholamines or 8-bromo-cAMP therefore provides a plausible explanation for the enhancement of I_{Ca}. However, the larger opening probability alone can only account for an increase in I_{Ca} by at most a factor of 2, if under control conditions p reaches a saturating level near 0.5 (this paper; Cavalié et al. 1983). This roughly agrees with the increase in the average currents that we have observed with both agents (for isoproterenol, see Fig. 4B). On the other hand, much larger effects on I_{Ca} of β-adrenergic catecholamines have been reported in the literature (Reuter and Scholz 1977b; Kass and Wiegers 1982). Therefore, one must assume that the number, N, of functional channels can also be regulated. We have observed an increase in N in only two experiments with isoproterenol. This, however, does not exclude the possibility of a more predominant regulation of N in other cell types where the increase in I_{Ca} by catecholamines is much larger (see Tsien et al., this volume). Such a mechanism of Ca^{++} channel regulation has previously been suggested by Reuter and Scholz (1977b).

How is the opening probability, p, affected by catecholamines? Previous analyses have shown that at least two processes are involved in Ca^{++} channel gating. Openings often occur in bursts separated by longer periods of closure (Lux and Nagy 1981; Fenwick et al. 1982; Reuter et al. 1982; Cavalié et al. 1983; Hagiwara and Ohmori 1983). Probability density histograms of open times of Ca^{++} channels can be described by single exponentials, whereas closed-time histograms can be fitted by the sum of two exponentials (Fig. 6; Fenwick et al. 1982; Reuter et al. 1982; Cavalié et al. 1983; Hagiwara and Ohmori 1983). The short closed times correspond to the brief closing events when a Ca^{++} channel flickers between closed and open states during a burst, and the longer times correspond to the intervals between the bursts. Fenwick et al. (1982) suggested a kinetic scheme for Ca^{++} channel gating that may be useful for the further analysis of our results. They described activation of the channel as a sequence of transitions between two closed states, C_1 and C_2, and open state O:

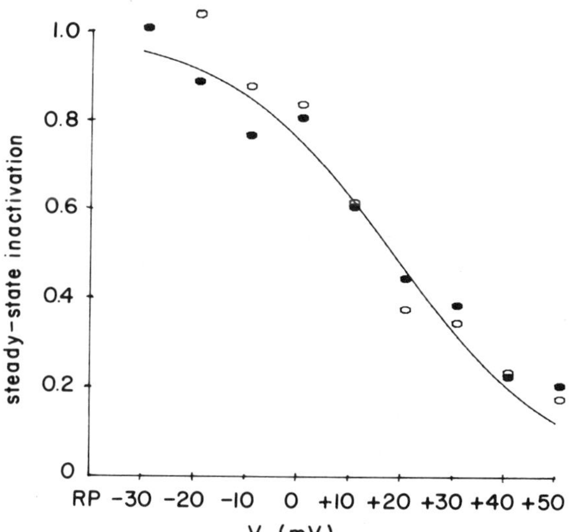

Figure 7. Voltage-dependent inactivation of averaged Ca^{++} currents (see Fig. 4B) in the absence (●) and presence (○) of 8-bromo-cAMP (10 μM/liter). The time-averaged current in the first 50 msec of a test pulse, ΔV, (resting potential [RP] + 80 mV) has been normalized and is plotted as a function of the holding potential V_h. The solid line is a Boltzmann distribution function, $f_\infty = (1 + \exp[(V_m - V_{0.5})/k])^{-1}$, fitted to the data points by nonlinear regression analysis (MODFIT). There was no statistically significant difference between controls and 8-bromo-cAMP treatment.

$$C_1 \underset{k_{-1}}{\overset{k_1}{\rightleftharpoons}} C_2 \underset{k_{-2}}{\overset{k_2}{\rightleftharpoons}} O \qquad (1)$$

where the four rate constants are voltage dependent. This model neglects inactivation, which, under our conditions, is very slow compared with activation. For such a kinetic scheme, it can be shown (Colquhoun and Hawkes 1981; Fenwick et al. 1982) that the time constant of open states

$$\tau_0 = \frac{1}{k_{-2}} \qquad (2)$$

and the fast time constant of closing events approaches

$$\tau_{C1} = \frac{1}{k_2 + k_{-1}} \qquad (3)$$

if $k_1 < k_{-1}$, k_2. The average number of brief interruptions per burst, n, is equal to the ratio of the integrals of the fast and slow components in the closed-time histograms (in Fig. 6, $n = 0.51$, controls; $n = 0.71$, 8-bromo-cAMP), and n is also related to k_2 and k_{-1} by

$$n = \frac{k_2}{k_{-1}} \qquad (4)$$

Since

$$\frac{p}{1-p} = \frac{k_1}{k_{-1}}\,\frac{k_2}{k_{-2}} \qquad (5)$$

we can derive the respective rate constants of the gating model (1). For the experiment in Figure 6, we calculated the following numerical values for the rate constants in the absence and presence of 8-bromo-cAMP, respectively: controls—k_1, 159 s^{-1}; k_2, 338 s^{-1}; k_{-1}, 662 s^{-1}; k_{-2}, 1667 s^{-1}; p, 0.046; 8-bromo-cAMP—k_1, 198 s^{-1}; k_2, 373 s^{-1}; k_{-1}, 526 s^{-1}; k_{-2}, 1429 s^{-1}; p, 0.089. This result shows that 8-bromo-cAMP increases the open state probability of a Ca^{++} channel in cultured rat cardiac cells by accelerating the forward rate constants, k_1 and k_2, and by reducing the backward rate constants, k_{-1} and k_{-2}.

Such an effect could be due to a shift of activation of Ca^{++} channels by cAMP along the voltage axis towards more negative potentials. This, however, is not a very likely explanation, since we have never seen channel openings in the presence of isoproterenol or 8-bromo-cAMP at pulse potentials more negative than in the controls. Moreover, voltage-dependent inactivation is not shifted significantly along the voltage axis (Fig. 7). Depolarization of the membrane potential by these agents is not a very likely explanation either, because i remains unchanged. Therefore, we suggest that isoproterenol and 8-bromo-cAMP have a genuine effect on the kinetics of Ca^{++} channel gating, thus shifting the reaction in model 1 towards the state O.

What is the common biochemical mechanism of Ca^{++} channel modulation by the β-adrenoceptor agonist isoproterenol and 8-bromo-cAMP? Both agents are known to increase cAMP in the cell, which in turn activates cAMP-dependent protein kinase by liberating the catalytic subunit of the enzyme (Krebs and Beavo 1979). It is therefore of great interest that injection of the catalytic subunit of cAMP-dependent protein kinase into single cardiac cells causes an increase in I_{Ca} identical to that observed with catecholamines (Osterrieder et al. 1982). An intrinsic cAMP-dependent protein kinase has recently been identified in cardiac sarcolemmal vesicles (Manalan and Jones 1982). Unfortunately, there still is considerable confusion concerning the identity of a membrane-associated phosphoprotein (M_r ~23,000) as substrate for this enzyme (Rinaldi et al. 1981; Hosey 1982; Manalan and Jones 1982). It will be interesting to see whether or not such a protein is involved in the modulation of Ca^{++} channels by cAMP (Greengard 1976; Reuter and Scholz 1977b; Tsien 1977; Reuter 1979, 1983). Phosphorylation of the channel, or of a phosphoprotein closely associated with the channel, could cause a channel conformation allowing it to open upon membrane depolarization with a higher probability than in the dephosphorylated state. We have attempted to apply the catalytic subunit of cAMP-dependent protein kinase (kindly supplied to us by Dr. E.G. Krebs, University of Washington, Seattle) directly to inside-out patches of membrane, unfortunately so far with inconclusive results. Recent evidence from Kostyuk's group (Fedulova et al. 1981; Doroshenko et al. 1982) suggests that in neuronal membranes cAMP-dependent phosphorylation of Ca^{++} channels could be an important factor for keeping them in a functional state.

ACKNOWLEDGMENTS

The support by the Swiss National Science Foundation (grant 3.154-0.81) is gratefully acknowledged. We also wish to thank the Hoffman-La Roche Animal Farm (Füllinsdorf, Switzerland) for the gift of pregnant rats.

REFERENCES

BLONDEL, B., I. ROIJEU, and J.P. CHENEVAL. 1971. Heart cells in culture; a simple method for increasing the proportion of myoblasts. *Experientia* **27:** 356.

CAVALIÉ, A., R. OCHI, D. PELZER, and W. TRAUTWEIN. 1983. Elementary currents through Ca^{2+} channels in guinea pig myocytes. *Pfluegers Arch. Eur. J. Physiol.* **398:** 284.

COLQUHOUN, D. and A.G. HAWKES. 1981. On the stochastic properties of single ion channels. *Proc. R. Soc. Lond. B.* **211:** 205.

DOROSHENKO, P.A., P.G. KOSTYUK, and A.E. MARTYNYUK. 1982. Intracellular metabolism of adenosine 3′,5′-cyclic monophosphate and calcium inward current in perfused neurones of *Helix pomatia*. *Neuroscience* **7:** 2125.

ECKERT, R., D.L. TILLOTSON, and P. BREHM. 1981. Calcium-mediated control of Ca and K currents. *Fed. Proc.* **40:** 2226.

FEDULOVA, S.A., P.G. KOSTYUK, and N.S. VESELOVSKY. 1981. Calcium channels in the somatic membrane of the rat dorsal root ganglion neurons. Effect of cAMP. *Brain Res.* **214:** 210.

FENWICK, E.M., A. MARTY, and E. NEHER. 1982. Sodium and calcium channels in bovine chromaffin cells. *J. Physiol* **331:** 599.

FOX, A.P. 1981. Voltage-dependent inactivation of a calcium channel. *Proc. Natl. Acad. Sci.* **78:** 953.

GREENGARD, P. 1976. Possible role of cyclic nucleotides and phosphorylated membrane proteins in postsynaptic actions of neurotransmitters. *Nature* **260:** 101.

HAGIWARA, S. and L. BYERLY. 1981. Calcium channel. *Annu. Rev. Neurosci.* **4:** 69.

HAGIWARA, S. and H. OHMORI. 1983. Studies of single calcium channel currents in rat clonal pituitary cells. *J. Physiol.* **336:** 649.

HAMILL, O.P., A. MARTY, E. NEHER, B. SAKMANN, and F.J. SIGWORTH. 1981. Improved patch-clamp techniques for high-resolution current recording from cells and cell free membrane patches. *Pfluegers Arch. Eur. J. Physiol.* **391:** 85.

HODGKIN, A.L. and A.F. HUXLEY. 1952. A quantitative description of membrane current and its application to conduction and excitation in nerve. *J. Physiol.* **117:** 500.

HOSEY, M.M. 1982. Chick heart plasma membranes. Isolation and analysis of autophosphorylation. *Biochim. Biophys. Acta* **690:** 106.

IRISAWA, H. and S. KOKUBUN. 1983. Modulation by intracellular ATP and cyclic AMP of the slow inward current in isolated single ventricular cells of the guinea pig. *J. Physiol.* **338:** 321.

ISENBERG, G., and U. KLÖCKNER. 1982. Calcium currents of isolated ventricular myocytes are fast and of large amplitude. *Pfluegers Arch. Eur. J. Physiol.* **395:** 30.

KASS, R.A. and S.E. WIEGERS. 1982. The ionic basis of concentration-related effects of noradrenaline on the action potential of calf cardiac Purkinje fibres. *J. Physiol.* **322:** 541.

KREBS, E.G. and J.A. BEAVO. 1979. Phosphorylation-dephosphorylation of enzymes. *Annu. Rev. Biochem.* **48:** 923.

LUX, H.D. and K. NAGY. 1981. Single channel Ca^{2+} currents in *Helix pomatia* neurons. *Pfluegers Arch. Eur. J. Physiol.* **391:** 252.

MANALAN, A.S. and L.R. JONES. 1982. Characterization of the intrinsic cAMP-dependent protein kinase activity and endogenous substrates in highly purified cardiac sarcolemmal vesicles. *J. Biol. Chem.* **257:** 10052.

MARK, G.E. and F.F. STRASSER. 1966. Pacemaker activity and mitosis in cultures of newborn rat heart ventricle cells. *Exp. Cell Res.* **44:** 217.

MCINTOSH, J.E.A. and R.P. MCINTOSH. 1980. *Mathematical modelling and computers in endocrinology*. Springer-Verlag, Berlin.

NOMA, A., H. KOTAKE, and H. IRISAWA. 1980. Slow inward current and its role mediating the chronotropic effect of epinephrine in the rabbit sinoatrial node. *Pfluegers Arch. Eur. J. Physiol.* **338:** 1.

OSTERRIEDER, W., G. BRUM, J. HESCHELER, W. TRAUTWEIN, V. FLOCKERZI, and F. HOFMANN. 1982. Injection of subunits of cyclic AMP-dependent protein kinase into cardiac myocytes modulates Ca^{2+} current. *Nature* **298:** 576.

PORZIG, H., C. BECKER, and H. REUTER. 1982. Competitive and non-competitive interactions between specific ligands and beta-adrenoceptors in living cardiac cells. *Naunyn-Schmiedeberg's Arch. Pharmacol.* **321:** 89.

REUTER, H. 1973. Divalent cations as charge carriers in excitable membranes. *Prog. Biophys. Mol. Biol.* **26:** 1.

——. 1974. Localization of *beta* adrenergic receptors and effects of noradrenaline and cyclic nucleotides on action potentials, ionic currents and tension in mammalian cardiac muscle. *J. Physiol.* **242:** 429.

——. 1979. Properties of two inward membrane currents in the heart. *Annu. Rev. Physiol.* **41:** 413.

——. 1983. Calcium channel modulation by neurotransmitters, enzymes and drugs. *Nature* **301:** 569.

REUTER, H. and H. SCHOLZ. 1977a. A study of the ion selectivity and the kinetic properties of the calcium-dependent slow inward current in mammalian cardiac muscle. *J. Physiol.* **264:** 17.

——. 1977b. The regulation of the calcium conductance of cardiac muscle by adrenaline. *J. Physiol.* **264:** 49.

REUTER, H., C.F. STEVENS, R.W. TSIEN, and G. YELLEN. 1982. Properties of single calcium channels in cardiac cell culture. *Nature* **297:** 501.

RINALDI, M.L., C.J. LE PEUCH, and J.G. DEMAILLE. 1981. The epinephrine-induced activation of the cardiac slow Ca^{2+} channel is mediated by the cAMP-dependent phosphorylation of calciductin, a 23000 Mr sarcolemmal protein. *FEBS Lett.* **129:** 277.

STANDEN, N.B. and P.R. STANFIELD. 1982. A binding model for calcium channel inactivation that depends on calcium entry. *Proc. R. Soc. Lond. B* **217:** 101.

TSIEN, R.W. 1977. Cyclic AMP and contractile activity in heart. *Adv. Cyclic Nucleotide Res.* **8:** 363.

——. 1983. Calcium channels in excitable cell membranes. *Annu. Rev. Physiol.* **45:** 341.

TSIEN, R.W. and E. MARBAN. 1982. Digitalis and slow inward current in heart muscle: Evidence for regulatory effects of intracellular calcium on calcium channels. In *Advances in pharmacology therapy* (ed. Yoshida et al.), vol. 3, p. 217. Pergamon Press, Oxford.

TSIEN, R.W., W. GILES, and P. GREENGARD. 1972. Cyclic AMP mediates the effects of adrenaline on cardiac Purkinje fibres. *Nat. New Biol.* **240:** 181.

Calcium Channels: Mechanisms of β-Adrenergic Modulation and Ion Permeation

R.W. TSIEN, B.P. BEAN,* P. HESS, AND M. NOWYCKY†

Departments of Physiology and †Neuroanatomy, Yale University School of Medicine, New Haven, Connecticut 06510

Ca^{++} channels hold a special fascination for neurobiologists, and for good reason. They are found in strategic places in the nervous system—synaptic terminals, growth cones, dendrites, cell bodies, and even axons (e.g., Katz 1969; Horn 1978; Adams et al. 1980; Llinas and Sugimori 1980; Anglister et al. 1981; Llinas et al. 1981; Caputo et al. 1983). Like many other channels, they respond to membrane depolarization by opening in a steeply voltage-dependent manner. But the entry of Ca^{++} means more than just the transfer of depolarizing positive charge, important as this may be: Ca^{++} also serves as a chemical message. By directly controlling Ca^{++} entry, Ca^{++} channels act as transducers to convert electrical information to a chemical signal. When this signal is decoded by Ca^{++}-binding proteins inside the cell (e.g., calmodulin or troponin), processes such as excitation-secretion coupling or excitation-contraction coupling are the result. We know of no other process that links membrane potential changes to variations in levels of a messenger substance. So far, this role of Ca^{++} channels seems to be unique.

Despite their importance, Ca^{++} channels have been less well understood than the Na^+, K^+, or acetylcholine receptor (AChR) channels discussed so extensively in this Symposium. For years, biophysical or biochemical analysis of Ca^{++} channels has been hampered by experimental problems (Reuter 1979; Hagiwara and Byerly 1981). No preparation has been as suitable and no drug as specific for Ca^{++} channels as the squid axon and tetrodotoxin are for Na^+ channels (Tsien 1983). But this situation has improved sharply with the development of new methods. Suction pipettes of different shapes and sizes (Kostyuk and Krishtal 1977; Lee et al. 1980; Hamill et al. 1981; Hume and Giles 1981) have been used to record Ca^{++} channel activity at the level of single cells or even single channel molecules. It is possible now to study Ca^{++} channels in isolation from other ionic pathways, with good control of membrane potential and external and internal ion composition (for review, see Hagiwara and Byerly 1983; Tsien 1983). A new class of drugs, the dihydropyridines, seems to interact with Ca^{++} channels in a potent and specific manner (see papers by Gengo et al.; Gould et al.; both this volume). Nitrendipine, an exemplar of this class, can properly be classified as a Ca^{++} channel blocker or a Ca^{++} antagonist, although questions still remain about the >100-fold disparity that can exist between half-blocking concentrations in living cells and half-binding concentrations to isolated membranes (Lee and Tsien 1983).

Recent work in our laboratory has been directed toward answering a number of general questions about Ca^{++} channels: (1) How are Ca^{++} channels modulated? (2) How do Ca^{++} channels manage to transfer ions as rapidly and selectively as they do? (3) How do Ca^{++} channels in various tissues differ in their basic properties?

METHODS

This paper is concerned mainly with properties of Ca^{++} channels in heart cells, although some preliminary work on Ca^{++} channels in sensory neurons from chick dorsal root ganglion is briefly mentioned. When studied with suitable suction pipette methods, heart cells offer several features desirable for biophysical analysis of Ca^{++} channels: (1) strong Ca^{++} current, (2) uniform and rapid voltage clamp, (3) experimental control of internal as well as external solution, (4) easy separation of Ca^{++} channel current from currents carried by other channels, and (5) a clearly defined reversal potential. These requirements, first listed by Hagiwara and Byerly (1981), seem to have been satisfied more completely in cardiac preparations than in neurons (Lee and Tsien 1982; Hagiwara and Byerly 1983; Tsien 1983; see Discussion).

Our experiments were carried out with three different heart cell preparations: neonatal rat ventricular cells kept in primary culture, adult frog ventricular cells, and adult guinea pig ventricular cells. Current signals were recorded and analyzed with Digital Equipment 11/23 or 11/03 laboratory computers. All experiments were done at room temperature (21°C).

Cultured rat heart cells. Ventricles from newborn rat hearts were enzymatically dispersed and kept in primary culture using a modification of existing methods (see, e.g., Harary et al. 1974). The approach is very similar to that described by H. Reuter et al. (this volume). Ca^{++} channel activity was recorded with cell-attached patch recording or whole-cell recording (Hamill et al. 1981).

Frog ventricular cells. Single ventricular cells were isolated from bullfrog (*Rana catesbiana*) hearts following the procedure described by Hume and Giles (1981) for atrial cells, with the following modifications. After 2–3 hours in the final 0.05% collagenase solution, the

*Present address: Department of Physiology and Biophysics, University of Iowa, Iowa City, Iowa 52242.

pieces of ventricle were triturated using a pasteur pipette. After another 2–3 hours of digestion, the cell-containing collagenase solution was diluted 1:4 into a solution containing 1 mM CaCl, 160 mM NaCl, 5 mM KCl, 1 mM $MgCl_2$, 10 mM glucose, and 5 mM HEPES (pH 7.6). Cells were used from 2–24 hours after being isolated. For recording whole-cell voltage-clamp currents, cells were placed in a solution containing 10 mM $CaCl_2$, 130 mM NaCl, 5 mM KCl, 1 mM $MgCl_2$, 10 mM glucose, and 5 mM HEPES (pH 7.6). Electrodes (1–3 MΩ) contained 120 mM CsCl, 10 mM Cs-EGTA, 5 mM $MgCl_2$, and 10 mM HEPES (pH 7.5). After formation of a gigaseal and rupture of the cell membrane, the external solution was changed to 10 mM Ba^{++}, 135 mM TEA-Cl, 1 mM $MgCl_2$, 10 mM glucose, and 10 mM HEPES (pH 7.65). Ba^{++} was chosen as the current carrier instead of Ca^{++} to avoid possible contamination from Ca^{++}-activated cation channels, to block currents through isoproterenol-sensitive background K^+ channels (Cranefield and Gadsby 1981), and to improve resolution of the fluctuation analysis by increasing the single-channel current. Series-resistance compensation was used; series-resistance (measured as the time constant of capacity transient divided by the cell capacity) was 2–8 MΩ, of which 60–80% was compensated. For patch-clamp recordings of single-channel current, cells were placed in solution containing 130 mM NaCl, 5 mM KCl, 11 mM $MgCl_2$, 2 mM Na-EGTA, 5 mM HEPES, and 10 mM glucose (pH 7.6). Electrodes (2–10 MΩ) contained 110 mM $BaCl_2$ and 2 mM HEPES (pH 7.7).

Guinea pig ventricular cells. Single ventricular cells were obtained from adult guinea pigs by enzymatic dissociation (Lee et al. 1979), and a suction pipette (Lee et al. 1980) was used for voltage-clamp and internal dialysis. To minimize current flow through channels other than the Ca^{++} channel, internal and external solutions were kept Na^+- and K^+-free. The internal solution contained 75.5 mM Cs-aspartate, 75.5 mM Cs-phosphate, 10 mM Cs-EGTA, and 6 mM HEPES (pH 7.4). The external solution had the following composition: 85 mM Cs-aspartate, 146 mM sucrose, 2.8 mM $MgCl_2$, 5 mM glucose, 5 mM HEPES (pH 7.4), and $CaCl_2$ or $BaCl_2$ as noted in the figures. Series resistance compensation was used to offset the ohmic drop resulting from current flow through the tip of the suction pipette.

RESULTS

Figure 1 shows a single Ca^{++} channel recording from a neonatal rat heart cell. The method is the same as that used by Fenwick et al. (1982) and Reuter et al. (1982). Current is carried by Ba^{++} ions moving across a small patch of membrane, defined by the opening of a pipette sealed against the heart cell surface. The rectangular pulses express the all-or-nothing nature of the opening and closing of a single Ca^{++} channel. This type of recording has already provided a rich harvest of information about Ca^{++} channels (for review, see Reuter 1983; Tsien 1983). For example, the amplitude of the current pulses tells us that the flux through a single open channel can reach transfer rates of millions of ions per second under appropriate ionic conditions. The clustered pattern of the current pulses indicates that gating, the molecular rearrangement that controls the opening and closing of the channel, is more complicated than a simple open-closed reaction: Opening seems to require at least two sequential steps.

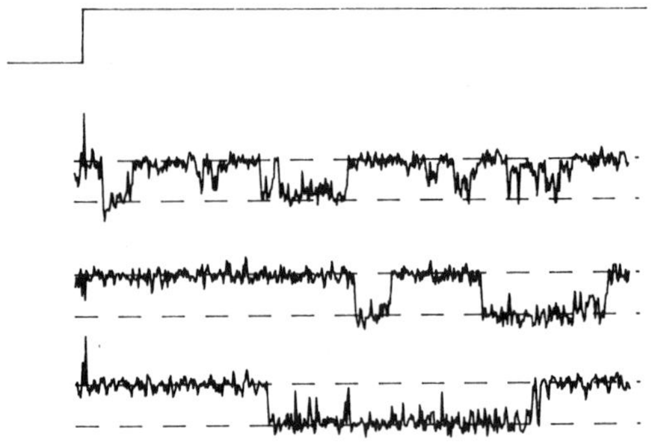

Figure 1. Single Ca^{++} channel activity. Cell-attached patch recording from a cultured neonatal rat heart cell (day 2). Pipette contained isotonic $BaCl_2$ (110 mM) plus 2 mM HEPES. Upper trace indicates voltage step from a holding potential 60 mV negative to the resting potential, up to a test potential 10 mV positive to the resting potential. Horizontal lines on current traces mark an estimated unitary current of 1.1 pA. Current traces last 150 msec. Cell M17A.

For some purposes, however, recording from single channels is a little like reading a newspaper through a high-power magnifying glass. To gain a bigger picture of Ca^{++} channel function, it is useful to supplement the cell-attached patch method with the technique of whole-cell recording (Hamill et al. 1981). Figure 2 shows such records and illustrates how information about single channels can be extracted from them. Figure 2 (b–d) shows three current signals that were evoked by repeated step depolarizations of the cell membrane. The ionic current is carried by Ba^{++}, as in Figure 1, but in this case the Ba^{++} flux expresses the activity of Ca^{++} channels throughout the cell. Closer examination reveals fluctuations in current riding on top of the same smooth time course of inward current onset. The fluctuations can be seen more clearly in Figure 2e–g, which show the difference currents that result when the overall average time course is eliminated by subtracting one current record from the next. The fluctuations reflect the statistical nature of the population of Ca^{++} channels as they open and close. It is as if thousands of simultaneous patch recordings had been added. I, the total cell current, can be described as the product of three factors: $I = N_f \cdot p \cdot i$, where N_f is the number of functional channels, p is the probability of an individual channel being open, and i is the current through a single open channel.

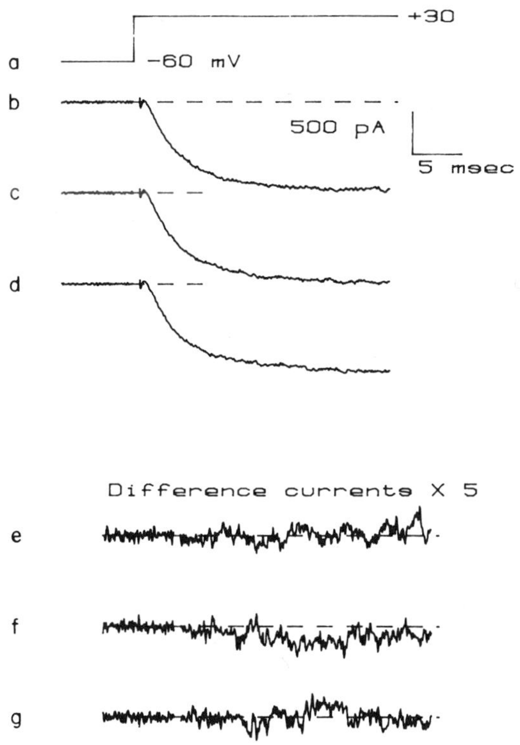

Figure 2. Fluctuation analysis of whole-cell recording of Ba^{++} current through Ca^{++} channels. Cultured neonatal rat heart cell (day 2). (*a*) Voltage protocol; (*b–d*) current traces accompanying three consecutive voltage steps; (*e–g*) difference signals, obtained by subtracting each trace from the trace immediately following after a small (<1%) correction for drift (see text). Cell M59G.

Figure 3 presents analysis of the fluctuations, using a method developed by Sigworth (1977, 1980; see also Lamb and Simon 1977). Individual difference records are squared, divided by two, and averaged. The result is the variance, marked "var" in Figure 3. Starting from a low level, the variance increases following the depolarizing step (whose timing is marked by the break in the trace). It reaches a more or less steady level after about 5 msec. The time course of the variance can be compared with the mean current, which also rises steeply at first but continues to creep upward after the variance has leveled off. The relation between mean inward current and variance is plotted at the bottom of Figure 3. The data points are well fitted by a parabolic relationship variance $= iI - I^2/N_f$. This is the behavior expected theoretically for a homogeneous population of channels that respond independently to membrane potential and show one level of nonzero conductance (Sigworth 1980).

The parabola can be given an intuitive explanation in terms of its intersections with the horizontal axis. The parabola must start at the origin: Just after the depolarization, when all channels remain closed, p is zero, and the inward current is zero; the variance is also zero, because it is certain that there is no current. At the other extreme, the parabola must curve back to cross the voltage axis. At this right-hand intercept, the lack of variance corresponds once again to complete certainty, but in this case, certainty of opening. Since $p = 1$, the total inward current I at the intercept is simply $N_f \cdot i$. Between the intercepts, the variance undergoes a maximum for $p = 1/2$, when closed and open states are equally likely. Variance piles up quickly when $p = 1/2$ because individual channels are either closed or open, but never half open.

The parabolic relationship fits the data well enough to give some confidence in the assumptions of the theory. In the exemplar experiment illustrated here, the analysis indicates that N_f is ~6350 and i is ~0.30 pA. The probability of opening reached a maximum value ($p_{max} = I_{max}/N_f \cdot i$) of about 0.55.

These values and collected results from our ensemble fluctuation analysis agree reasonably well with information from unitary Ca^{++} channel recordings by us and by others. The methods give fairly similar values for the slope conductance of open Ca^{++} channels of ~15 pS. Various estimates of the number of functional channels are also in good agreement. Fluctuation analysis in 28 neonatal rat heart cells gives an average value of N_f of ~8000. The membrane area of these cells averaged ~2000 μm^{-2} (given that the mean cell capacitance was ~20 pF, and assuming 1 $\mu F/cm^2$).

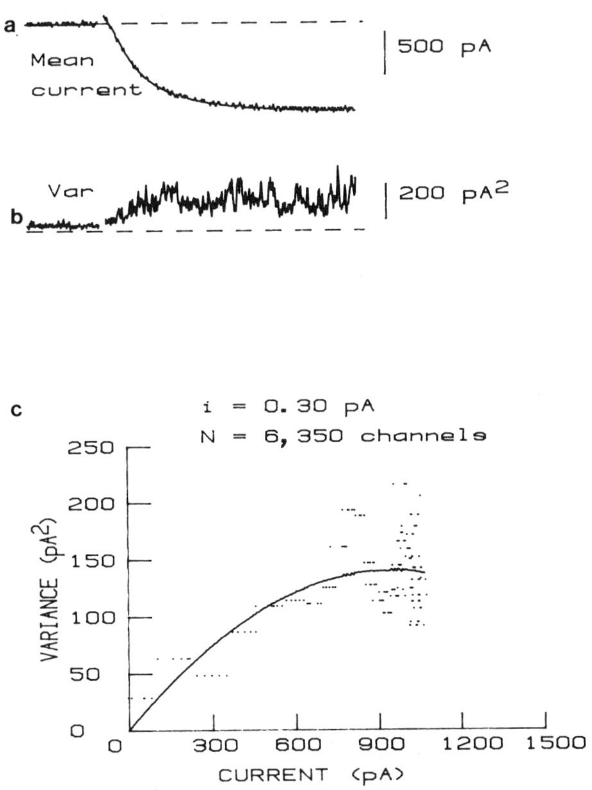

Figure 3. Ensemble fluctuation analysis of mean current and variance from the experiment illustrated in Fig. 2. (*a,b*) Time dependence of mean and variance. Dashed lines are zero baselines. (*c*) Plot of variance against mean. Background variance was subtracted. Horizontal streaks were obtained by averaging the variance over successive time bins. Smooth parabolic curve with parameters indicated was obtained by least-squares fit.

Thus, the density of functional Ca^{++} channels was ~4 μm^{-2}. A rather similar value of ~2 μm^{-2} was obtained without using fluctuation analysis by measuring total current in whole-cell recordings and dividing by $(i \cdot p)$ as determined from unitary current recordings.

Modulation of Ca^{++} Channels by β-Adrenergic Agents

Ca^{++} channels are a prime target for modulation by neurotransmitters and drugs (for review, see Reuter 1983; Siegelbaum and Tsien 1983; Tsien 1983). In neurons, Ca^{++} currents or Ca^{++}-dependent action potentials respond to a wide variety of agents, but so far all the effects have been inhibitory (Dunlap and Fischbach 1978; Horn and McAfee 1980; Shapiro et al. 1980; Galvan and Adams 1982) with one possible exception (Pellmar and Carpenter 1980). In heart cells, Ca^{++} currents can be strongly modulated in either direction. Interaction of ACh with muscarinic cholinergic receptors decreases Ca^{++} entry (Giles and Tsien 1975; Ikemoto and Goto 1975; Giles and Noble 1976). On the other hand, Ca^{++} influx is enhanced by β-adrenergic agonists (Reuter 1967; Vassort et al. 1969), angiotensin II (Kass and Blair 1981), and digitalis (Weingart et al. 1978; Marban and Tsien 1982). In this paper, we briefly review our recent experiments on the mechanism of β-adrenergic modulation. This work is described in more detail elsewhere (B.P. Bean et al., in prep.).

β-Adrenergic modulation of Ca^{++} channels plays a large part in controlling the rhythm and strength of the heartbeat. The increased Ca^{++} influx is mediated by elevation of intracellular cAMP (for reviews, see Tsien 1977; Watanabe et al. 1981; Reuter 1983) and activation of protein kinase (Osterrieder et al. 1982). The molecular mechanism of the augmented membrane Ca^{++} conductance has attracted considerable interest. An increase in the density of functional channnels has often been proposed (see, e.g., Reuter and Scholz 1977b; see also Sperelakis and Schneider 1976; Niedergerke and Page 1977). Up to now there has been no direct evidence. What single-channel recordings from cultured rat heart cells have shown is that isoproterenol or 8-bromo-cAMP increase the proportion of

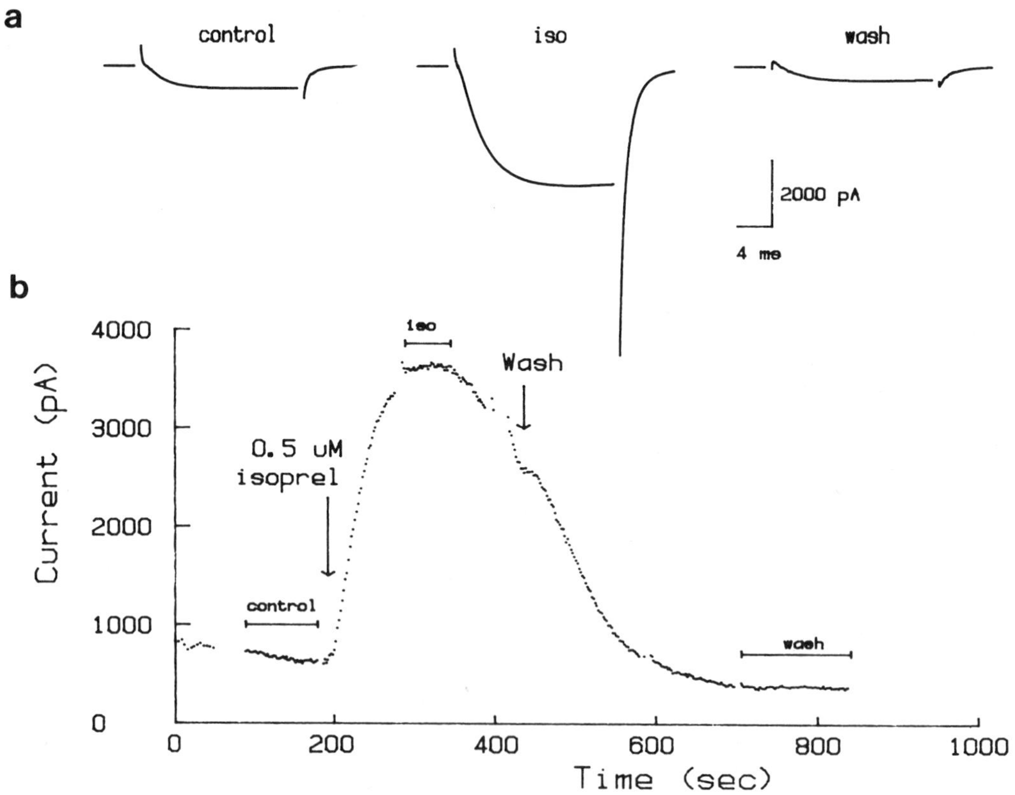

Figure 4. Isoproterenol effect on Ba^{++} currents through Ca^{++} channels in a frog ventricular cell. (*a*) Currents elicited by a voltage step to +10 mV from a holding potential of −60 mV. Tail currents evoked by a repolarization to −40 mV. Leak and capacity currents (obtained from mirror-image hyperpolarizing pulses, 10 times smaller) have been subtracted, and a 750-μsec period is blanked due to amplifier saturation during capacity transients. Each current shown is the mean current of the series of pulses indicated in *b*. (*b*) Time course of isoproterenol response. Current measured at the end of the 20-msec test pulse. Isoproterenol stock solution (200 μl) was squirted into the 1.5-ml chamber to give a final concentration of 0.5 μM. During the wash, the chamber volume was replaced 20 times with agonist-free solution within the first minute. Note that the current amplitude began to fall before the drug was washed out; in other experiments, current fell back to a stable value near its original level in 5–10 min even though the drug was not washed out, suggesting desensitization or depletion of some intracellular constituent. Cell F04B. Total capacity 75 pF. Series-resistance compensation for 2.2 MΩ of measured series resistance of 3.5–3.8 MΩ. (Reprinted, with permission, from B.P. Bean et al., in prep.)

time that individual channels spend open by prolonging openings and shortening the closed periods between openings (Reuter et al. 1982; Cachelin et al. 1983; Reuter 1983). To look for possible changes in the density of functional channels, distinct from changes in the probability of individual channels being open, we applied ensemble fluctuation analysis to whole-cell recordings of Ca^{++} channel activity from frog ventricular heart cells. We found direct evidence that β-adrenergic stimulation increases N_f, the number of functional Ca^{++} channels per cell.

β-Adrenergic Response of Frog Ventricular Cells

The β-adrenergic agonist isoproterenol produced a dramatic enhancement of the Ca^{++} channel current carried by Ba^{++} ions (Fig. 4). Panels a–c show whole-cell recordings from a frog ventricular cell during depolarizing voltage-clamp pulses from −60 mV to +20 mV. Ca^{++} channel current increased from about 0.7 nA in the absence of drug (a) to about 3.5 nA in the presence of isoproterenol (b); the effect was fully reversed following washout of the drug (c). The roughly fivefold increase of Ca^{++} channel current was fairly typical; in 25 cells exposed to 0.5 μM isoproterenol, the whole-cell current increased by a factor of 6.20 ± 0.59 (mean ± S.E.M.). Figure 4 illustrates the time course of the response and the stability of the current amplitude during long series of pulses before, during, and after the drug exposure. The largeness of the drug response and the relatively slow rundown of the Ca^{++} channel current made it easier to analyze the underlying mechanisms.

Changes in the Number of Functional Channels

Does β-adrenergic stimulation increase the number of functional channels (N_f), the probability that an individual channel is open (p), or the current carried by an open channel (i)? Figure 5 shows the answer provided by ensemble fluctuation analysis using the method illustrated in Figures 2 and 3. The mean and

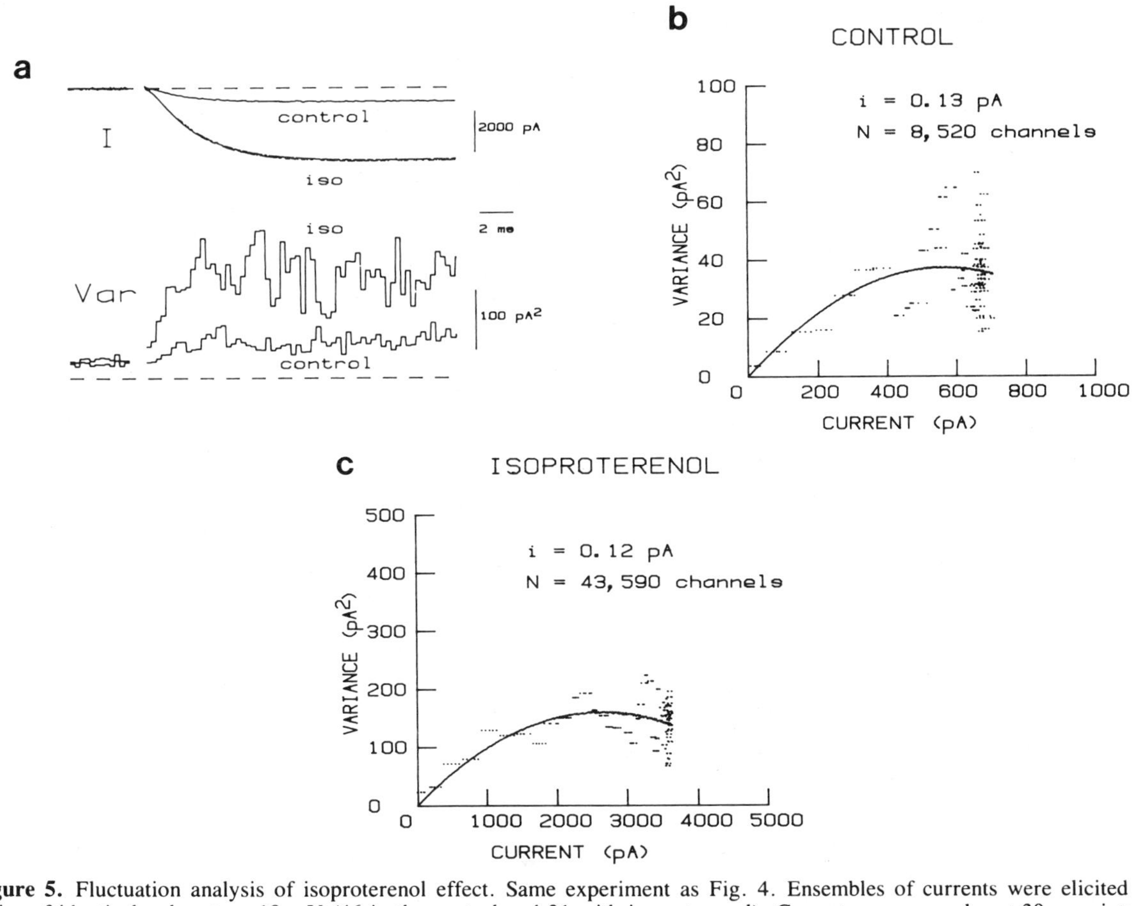

Figure 5. Fluctuation analysis of isoproterenol effect. Same experiment as Fig. 4. Ensembles of currents were elicited by a series of identical pulses to +10 mV (46 in the control and 31 with isoproterenol). Currents were samples at 30-μsec intervals after being filtered at 3 kHz (four-pole Bessel). Mean current and variance were computed for each time point. To minimize errors from slow drifts in current size, variance was calculated from successive pairs of records and averaged over the pairs; also, the first record of each pair was first scaled so that the integrated current was equal to that of the second record (Fenwick et al. 1982). The first millisecond of the test pulse was omitted from the analysis since the current showed an outward bump during this time; the variance was approximately at its background level. (*a*) Mean and variance as a function of time. Mean currents corrected for capacity and leak current. The variance was averaged over 10 successive time points. Dotted lines indicate zero levels. (*b,c*) Variance plotted vs. mean current. The background variance (computed from the current at the holding potential) was first subtracted from the variance during the test current. (Reprinted, with permission, from B.P. Bean et al., in prep.)

variance records are from control and isoproterenol runs illustrated in Figure 4. Both the mean current and the variance rise progressively following the depolarizing step. Like the mean current, the variance increased about fivefold after exposure to isoproterenol. To estimate N_f, p, and i for each run, plots of the variance versus mean current (Fig. 5c,d) were fitted by parabolic curves with a minimum least-squares curve-fitting procedure. The maximal probability, p_{max}, was then calculated as $p_{max} = I_{max}/(N_f \cdot i)$.

In the experiment shown in Figure 5, N_f increased from 8,520 to 43,590 with isoproterenol and returned to 7,470 after the drug was washed off. p_{max} increased from 0.62 to 0.70 and returned to 0.55. On the other hand, i hardly changed at all. These findings were representative of collected results from a total of eight frog ventricular cells.

This is the first direct evidence that Ca^{++} channel density is enhanced by β-adrenergic stimulation. At the membrane potential studied here (+10 mV), the overall enhancement of Ca^{++} channel current is almost completely accounted for by the increase in N_f. The absence of any great change in i under our conditions (frog, 10 mM $[Ba^{++}]_e$ +10 mV) is consistent with earlier unitary current recordings (rat, 96 mM $[Ba^{++}]_e$, about −10 to about +20 mV: Reuter et al. 1982; Cachelin et al. 1983; Reuter 1983).

Single-channel Recordings

The present results are at least somewhat surprising since increases in N_f have not shown up in previous published studies of unitary Ca^{++} channel activity in rat heart cells (Reuter et al. 1982 and this volume; Cachelin et al. 1983; Reuter 1983). Large increases in N_f would be expected to increase the maximal number of conductance levels in patch-clamp recordings, but such an increase has not been reported. This discrepancy prompted us to record unitary Ca^{++} channel activity from frog ventricular cells (Fig. 6). Channel openings were evoked by voltage-clamp depolarization of a cell-attached patch, with isotonic $BaCl_2$ (110 mM) outside to increase the unitary current through single channels, just as in Figure 2. In this recording configuration, isoproterenol was applied to the cell as a whole, not to the patch itself. The β-adrenergic agonist strongly increased the Ca^{++} channel activity in the patch, just as it did in whole-cell recordings. In Figure 6b, for example, averages of many sweeps before and after isoproterenol show an approximate sevenfold enhancement. The other traces in Figure 6 show some of the individual recordings that contributed to the averages in panel b. Horizontal lines give integer multiples of the unitary current amplitude. The control run included sweeps with no channel openings (Fig. 6d,e), as well as sweeps with one (Fig. 6c) or more unitary current levels (Fig. 6f). After β-adrenergic stimulation, sweeps with no openings are very rare; during the depolarization, at least one channel is open most of the time, and the current trace reaches a larger multiple of i. When all the sweeps from this experiment are analyzed with amplitude histograms (Fig. 6k), the isoproterenol-induced bias toward multiple simultaneous openings is evident. The percentage of time that no channel was open during depolarization drops from 88% in the control run to 25% after β-adrenergic stimulation. Within the resolution of the current recording system, four was the largest number of channels simultaneously open in control (Fig. 6f), whereas it was possible to resolve at least seven channels open at the same time after exposure to isoproterenol (Fig. 6i). In another patch, the number increased from three to eight. The increase in the number of simultaneously open channels is at least consistent with an increase in N_f.

Current through Ca^{++} Channels Is Not Simply Scaled Up by Isoproternol

According to the simplest hypothesis, changes in N_f should occur without any change in the properties of individual channels. However, when we compared whole-cell currents from control and isoproterenol-stimulated cells, we found evidence that after β-adrenergic stimulation, most or all of the channels show significantly different properties than normally functional channels. This work is described in detail elsewhere (B.P. Bean et al., in prep.), so the basic findings can be briefly summarized as follows:

1. The onset of Ca^{++} channel current (10 mM $[Ba^{++}]_e$) was consistently slowed after β-adrenergic stimulation. The slowing is not accounted for by series resistance error or by a species difference between frog and rat (B.P. Bean et al., in prep.).
2. The overall voltage-dependence of the whole-cell Ca^{++} channel current is also different after isoproterenol. The drug-induced enhancement, expressed as the ratio I(iso)/I(control), is a function of membrane potential. For weak depolarizations (e.g., −40 mV), the enhancement can be almost twice as big as it is at −10 mV, the test potential used in the fluctuation analysis. This extra enhancement at relatively negative voltages could be consistent with an increase in p (cf. Cachelin et al. 1983; Reuter et al., this volume).
3. The isoproterenol-induced enhancement also shows considerable voltage dependence at strongly positive potentials that approach or exceed the reversal potential for the Ca^{++} channel current. Here, isoproterenol increases the peak Ca^{++} channel current very little. It seems that after β-adrenergic stimulation, most or all of the channels are much less conductive at strongly positive potentials than normally functional channels.

Studies on the Mechanism of Ion Permeation through Ca^{++} Channels

Ca^{++} channels not only carry out special functions, but also face special challenges. Outside the channel, Ca^{++} ions are vastly outnumbered by the monovalent cations Na^+ and K^+ and are hardly greater in concen-

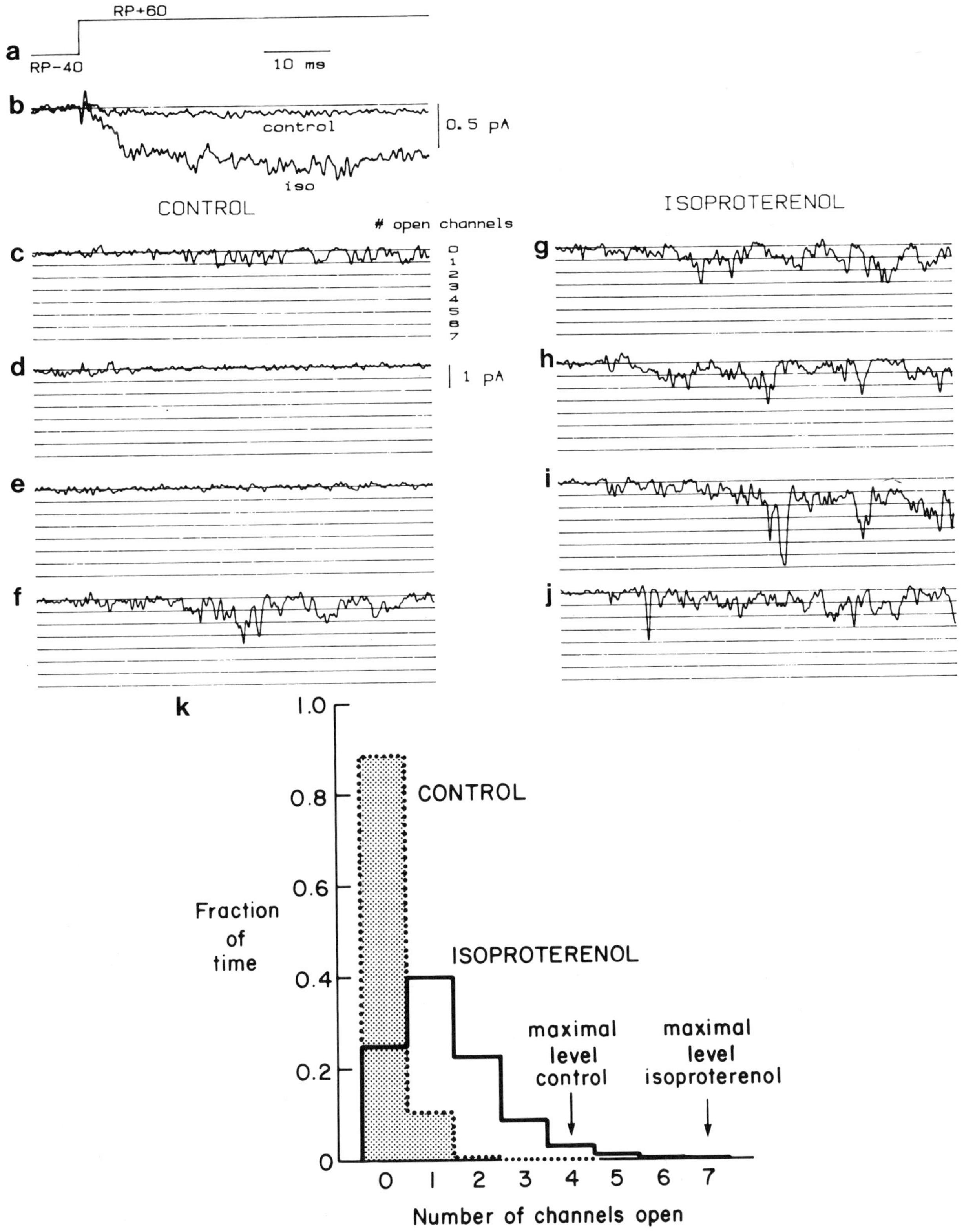

Figure 6. Effect of isoproterenol on single-channel currents in a frog ventricular cell. (*a*) Voltage protocol. Test pulses were delivered at 1 Hz. (*b*) Mean channel current before and after isoproterenol. Mean current computed from 62 records in control, 42 records with isoproterenol; isoproterenol series began 1.5 min after addition of isoproterenol. Corrected for leak and capacity current by subtracting averaged currents from nulls (traces with no channel activity). (*c–f*) Successive records from the control run. Leak and capacity current subtracted. Horizontal lines correspond to a single channel current of 0.56 pA. (*g–j*) Successive records 2 min after the addition of isoproterenol. (*k*) Histogram showing the percentage of time spent at various conductance levels. Cell F08A. (Reprinted, with permission, from B.P. Bean et al., in prep.)

tration than Mg^{++}. To allow Ca^{++} influx rather than a general cation influx, the Ca^{++} channel must be extremely selective. In fact, this seems to be the case: The selectivity of Ca^{++} channels for Ca^{++} is higher than the selectivity of Na^{+} channels for Na^{+} by at least an order of magnitude (Reuter and Scholz 1977a; Lee and Tsien 1982 and in prep.). Such performance seems particularly striking when one considers that open Ca^{++} channels are capable of transferring divalent cations at rates as rapid as 6×10^6 s^{-1} (for review, see Tsien 1983). The conductance of Ca^{++} channels in isotonic Ba^{++} is roughly equal to the conductance of Na^{+} channels in isotonic Na^{+} (see, e.g., Fenwick et al. 1982; Reuter et al. 1982 and this volume).

Our experiments have been aimed at understanding how Ca^{++} channels manage to transfer ions so rapidly yet so selectively. The results are presented in detail elsewhere (P. Hess et al., in prep.), so they can be briefly summarized here as follows:

1. Ca^{++} channel selectivity was studied in single guinea pig ventricular cells with suction pipettes for voltage clamp and internal dialysis (see Methods). Measurements of the reversal potential of the D-600-sensitive curent (E_{rev}) with 10 mM Ba^{++} outside and 150 mM monovalent cation inside gave the following selectivity sequence for the monovalents: $Na^{+} > K^{+} > Cs^{+}$.
2. Similarly, with 150 mM Cs^{+} inside, and various divalent cations outside, measurements of reversal potentials gave the following order of permeabilities: $Ca^{++} > Sr^{++} > Ba^{++} >> Mg^{++}$.
3. We tested the hypothesis that the Ca^{++} channel contains a single binding site where permeant ions and blocking ions compete (Vereecke and Carmeliet 1971; Hagiwara et al. 1974; Akaike et al. 1978; Kostyuk 1981; Edwards 1981). External Ca^{++} $[Ca^{++}]_e$ strongly inhibits Ba^{++} currents through the Ca^{++} channel: Half block of the current in 10 mM $[Ba^{++}]_e$ was obtained with as little as 0.4 mM $[Ca^{++}]_e$.
4. On the other hand, Ca^{++} currents increase significantly over a much broader range of $[Ca^{++}]_e$ from 1 mM up to 30 mM and beyond. This contradicts the single binding site hypothesis: If submillimolar $[Ca^{++}]_e$ is sufficient to block Ba^{++} current, one would expect saturation of the site (and also the Ca^{++} current) at similarly low concentrations.
5. A somewhat related approach to the same question was to look at whole cell or single Ca^{++} channel currents in mixtures where $[Ca^{++}]_e + [Ba^{++}]_e = 10$ mM. The single-site model predicts a monotonic relationship between I or i and the mole fraction of either of the divalent cations. In fact, we found an "anomalous mole-fraction dependence" (Eisenman et al. 1967; Andersen 1975; Neher 1975). In solutions with 1–3 mM $[Ca^{++}]_e$ (9–7 mM $[Ba^{++}]_e$), the current reaches a minimum that is considerably smaller than that in either 10 mM $[Ca^{++}]_e$ or 10 mM $[Ba^{++}]_e$. This is a property of open channels since it holds for unitary Ca^{++} channel current as well as for whole-cell current (P. Hess et al., in prep.).
6. These results suggest that Ca^{++} channels might be described as single-file pores containing multiple ion-binding sites, as previously suggested for gramicidin (Urban et al. 1978) and K^{+} channels (Hille and Schwarz 1978). To explore this idea, we formulated an energy barrier model for the Ca^{++} channel with two internal wells and repulsive Coulombic interactions between ions in doubly occupied channels.
7. The energy barriers and wells for each ion were adjusted to fit (1) measured reversal potentials with various combinations of internal and external ions, (2) measured values for elementary currents with only $[Ca^{++}]_e$ or $[Ba^{++}]_e$, and (3) behavior in mixtures of $[Ca^{++}]_e$ and $[Ba^{++}]_e$. Figure 8 shows putative energy profiles for the Ca^{++} channel, together with some earlier models for Na^{+} and AChR channels (Lewis and Stevens 1979; Begenisich and Cahalan 1980).
8. A satisfactory fit to all the available data was obtained with the following energy profiles: equal entry and exit barriers for Ca^{++}, Sr^{++}, and Ba^{++} but energy wells at the two "binding" sites that increase in depth from Ba^{++} to Sr^{++} to Ca^{++}; equal and shallow wells for all monovalents (consistent with nonspecific binding of these ions to the two sites); and barriers increasing in height from Na^{+} to K^{+} to Cs^{+}. The model accounted for the shape of the current-voltage relation in the various solutions and the voltage-dependence of blocking effects in mixtures of $[Ca^{++}]_e$ and $[Ba^{++}]_e$.
9. The model gives the following explanation of ion permeation through Ca^{++} channels. The channel selects for monovalent ions by its entry and exit barriers. The lower those barriers, the higher the permeability and the elementary current. The situation for the divalent ions is quite different: Selectivity occurs at the energy wells. The deeper the well, the higher the selectivity (as judged from reversal potential measurements) but the smaller the elementary current. Throughput rates are high, even for Ca^{++}, because of strong repulsion between ions.
10. The well depth for Ca^{++} (ΔG) can be expressed as an equivalent equilibrium binding constant ($K = \exp[-\Delta G/RT]$) of 3×10^6 M^{-1}. Interestingly enough, this value is quite comparable to that of the high-affinity binding sites on Ca^{++}-sensitive proteins such as calmodulin or troponin.
11. Because of the high affinity of the binding sites for Ca^{++}, the channel is occupied almost all of the time by one or more Ca^{++} ions, which help ensure that the channel is selective. At the outer site, a Ca^{++} ion simply prevents the entry of competing ions; at the inner site, a Ca^{++} slows the

entry and hastens the exit of a foreign ion from its shallower outer well. Entry of a second Ca^{++} into the channel becomes more and more likely as $[Ca^{++}]_e$ increases, and accounts for a gradual increase in I or i over a broad range of $[Ca^{++}]_e$.

12. Based on energy profiles for various ions, the model predicts that under physiological ionic conditions, more than 90% of the current is carried by Ca^{++} despite the presence of Na$^+$ and K$^+$ at much higher concentrations.

DISCUSSION

How Similar Are Ca^{++} Channels in Different Excitable Cells?

In the last 2 years there has been a remarkable convergence between studies of Ca^{++} channels in nerve cells and information on "honorary neurons" such as heart cells, adrenal chromaffin cells, and GH pituitary tumor cells (e.g., Lux and Nagy 1981; Fenwick et al. 1982; Lee and Tsien 1982; Reuter et al. 1982 and this volume; Hagiwara and Ohmori 1982; Brown et al. 1982). Some common characteristics seem to be emerging:

1. Single Ca^{++} channel slope conductance is roughly 10–20 pS in isotonic Ba^{++}.
2. Monovalent cations can carry outward Ca^{++} channel current at strongly positive voltages beyond a reversal potential.
3. Current-voltage relationship appears nonlinear when examined over a broad voltage range; conductance increases with driving force on either side of the reversal potential.
4. Channel openings appear grouped in bursts; activation kinetics are sigmoid.
5. Probability of channel opening increases steeply with depolarization and seems to level off at ~0.5–0.9.

Ca^{++} channels seem to differ from tissue to tissue in at least two respects:

1. Decay of inward Ca^{++} current during a maintained depolarization varies widely in speed and mechanism (for review, see Tsien 1983). The decay may reflect Ca^{++} depletion or accumulation in restricted spaces or a genuine decay of Ca^{++} permeability, termed "inactivation." In various preparations, inactivation may depend on membrane potential, intracellular Ca^{++}, or both factors.
2. Sensitivity of Ca^{++} channels to blocking agents also varies widely. Although 1,4-dihydropyridines, such as nifedipine or nitrendipine, bind strongly to all preparations examined so far, their blocking potency varies by several orders of magnitude, as does the potency of verapamil or D-600 (for references, see Lee and Tsien 1983 or R.J. Miller and S.M. Freedman, in prep.). On the other hand, all Ca^{++} channels seem sensitive to block by multivalent cations such as Cd^{++}, Co^{++}, or La^{3+} (Hagiwara and Byerly 1981).

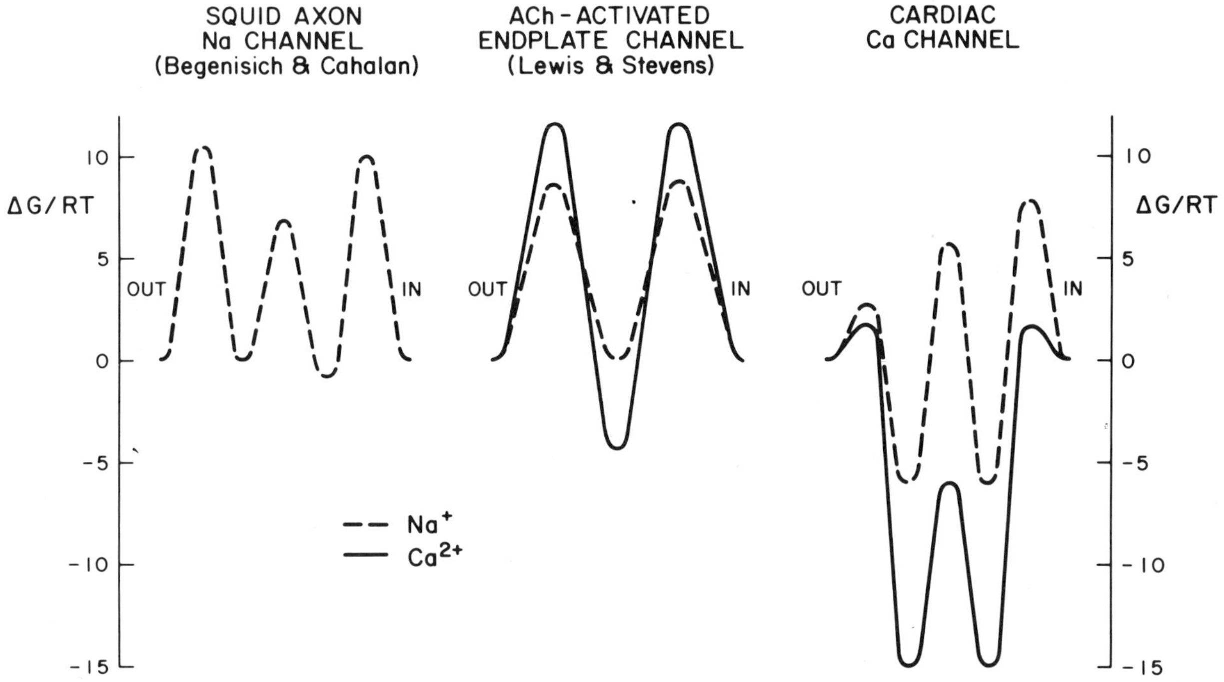

Figure 7. Energy barrier profiles for three types of ion channel: the Na$^+$ channel (Begenisich and Cahalan 1980), the AChR channel (Lewis and Stevens 1979), and the Ca^{++} channel (P. Hess et al., in prep.).

Although the work in our laboratory has focused primarily on cardiac membranes, we have already explored some of the properties of Ca^{++} channels in neurons from chick dorsal root ganglion (Nowycky et al. 1983), following the lead of Dunlap and Fischbach (1978). Using whole-cell recording, and a "kitchen-sink approach" to blocking K^+ channel currents ($[TEA]_e$, $[Cs^+]_i$, $[EGTA]_i$), we find that the neuronal Ca^{++} channels seem roughly similar in many respects to cardiac Ca^{++} channels. For example, single-channel Ba^{++} currents are blocked by $[Ca^{++}]_e$ at submillimolar concentrations in the neurons, very much as in heart cells. The most obvious difference is seen in the rate and possibly the mechanism of Ca^{++} channel inactivation. In 10 mM $[Ca^{++}]_e$, neuronal inward currents did not decay to half during 500-msec pulses from a holding potential of -60 mV. The slow decline was unchanged when $[Ca^{++}]_e$ was replaced by $[Ba^{++}]_e$. In heart cells, on the other hand, the half-time for Ca^{++} current decay is less than 50 msec in 10 mM $[Ca^{++}]_e$, but inactivation is greatly slowed when $[Ca^{++}]_e$ is replaced by $[Ba^{++}]_e$. It seems that the neurons and heart cells differ in the way that they handle Ca^{++} or in the way that Ca^{++} promotes inactivation.

Mechanism of Neurochemical Modulation of Ca^{++} Channels

Although neurochemical modulation of Ca^{++} current or Ca^{++}-dependent action potentials has been found in several types of neurons, as already mentioned, there is no information yet about the nature of the modulation at the level of single channels. It seems likely that heart cells will continue to provide an excellent system for answering general questions about molecular mechanisms. The β-adrenergic effect in frog ventricular cells seems particularly well-suited for further analysis because the Ca^{++} channel conductance increases in such a dramatic and reproducible way.

The roughly sixfold increase in Ca^{++} channel current is too large to be explained without a substantial recruitment of newly functional Ca^{++} channels, given that p exceeds 0.5 in the absence of drug and that i remains essentially the same at the $+10$ mV test potential. Changes in p may also play some role, particularly at more negative potentials (e.g., -40 mV), where the increase in inward current is even larger than at $+10$ mV. The extra enhancement at negative potentials is probably relatively unimportant for excitation-contraction coupling during a ventricular action potential, but it might be quite significant as a mechanism for adrenergic modulation of cardiac rhythm, since sinus node pacemaker activity depends on small inward currents near the foot of the Ca^{++} channel activation curve.

We have found it fruitful to use single-channel and whole-cell recordings in combination. The same techniques should help in the study of the opposing effects of adrenergic and cholinergic agents at the level of individual Ca^{++} channels. Such work in heart might lead to interesting comparisons with analysis of neuronal Ca^{++} channels and the mechanism of down-modulation by norepinephrine, somatostatin, and enkephalin (see, e.g., Dunlap and Fischbach 1978; Galvan and Adams 1982; Werz and Macdonald 1982).

Mechanism of Selectivity and Rapid Ion Transfer

Our experiments suggest that Ca^{++} channels can be occupied simultaneously by more than one ion and that interactions between ions play an important part in selectivity. Although these ideas are new for Ca^{++} channels, they owe a great deal to earlier work on gramicidin channels (e.g., Andersen 1975; Neher 1975; Urban et al. 1978) and inward rectifier K^+ channels (e.g., Hagiwara et al. 1977; Hille and Schwarz 1978). What is exciting is to see how this mechanism of ion permeation enables Ca^{++} channels to carry out their special biological functions within particularly stringent constraints. In the future, biophysical analysis of functional Ca^{++} channels will need to be fused with biochemical efforts at isolation, purification, and cloning. One obvious question will concern what structural similarities, if any, the Ca^{++} channel pore bears to Ca^{++}-binding proteins. Meanwhile, much work remains to be done with the tools at hand. It will be worthwhile, for example, to look at the response of Ca^{++} channels to pH changes and calmodulin inhibitors. It will also be interesting to characterize ion permeation through extra channels recruited by β-adrenergic stimulation to see if they differ from normally functional channels as a result of cAMP-dependent phosphorylation.

ACKNOWLEDGMENTS

We thank the U.S. Public Health Service, the American Heart Association, and the Miles Institute for support, and Barbara Frank for skillful word processing.

REFERENCES

ADAMS, D.J., S.J. SMITH, and S.H. THOMPSON. 1980. Ionic currents in molluscan soma. *Annu. Rev. Neurosci.* **3:** 141.

AKAIKE, N., K.S. LEE, and A.M. BROWN. 1978. The calcium current of *Helix* neuron. *J. Gen. Physiol.* **71:** 509.

ANDERSEN, O.S. 1975. Ion-specificity of gramicidin channels. In *Abstracts of the Fifth International Biophysics Congress* (ed. U. Lassen and J.O. Wieth), p. 112. Villadsen and Christensen, Copenhagen.

ANGLISTER, L., A. SHAHAR, I.C. FARBER, and A. GRINVALD. 1981. Growth cones expand in vitro by depolarization under conditions which permit Ca^{2+} entry. *Soc. Neurosci. Abstr.* **7:** 548.

BEGENISICH, T.B. and M.D. CAHALAN. 1980. Sodium channel permeation in squid axons. I. Reversal potential experiments. *J. Physiol.* **307:** 217.

BROWN, A.M., H. CAMERER, D.L. KUNZE, and H.D. LUX. 1982. Similarity of unitary Ca^{2+} currents in three different species. *Nature* **299:** 156.

CACHELIN, A.B., J.E. DEPEYER, S. KOKUBUN, and H. REUTER. 1983. Calcium channel modulation by 8-bromocyclic AMP in cultured heart cells. *Nature* **304:** 462.

CAPUTO, C., R. DIPOLO, and F. BEZANILLA. 1983. Electrical recording of a voltage dependent Ca conductance in the squid axon. *Biophys. J.* **41:** 294a.

CRANEFIELD, P.F. and D.C. GADSBY. 1981. Isoprenaline increases the potassium permeability of the resting membrane of canine cardiac Purkinje fibres. *J. Physiol.* **318:** 34.

DUNLAP, K. and G.D. FISCHBACH. 1978. Neurotransmitters decrease the calcium component of sensory neurone action potentials. *Nature* **276:** 837.

EDWARDS, C. 1981. The selectivity of ion channels in nerve and muscle. *Neuroscience* **7:** 1335.

EISENMAN, G., J.P. SANDBLOM, and J.L. WALKER, JR. 1967. Membrane structure and ion permeation. *Science* **155:** 965.

FENWICK, E.M., A. MARTY, and E. NEHER. 1982. Sodium and calcium channels in bovine chromaffin cells. *J. Physiol.* **331:** 599.

GALVAN, M. and P. ADAMS. 1982. Control of calcium current in rat sympathetic neurons by norepinephrine. *Brain Res.* **244:** 135.

GILES, W. and S.J. NOBLE. 1976. Changes in membrane current in bullfrog atrium produced by acetylcholine. *J. Physiol.* **261:** 103.

GILES, W. and R.W. TSIEN. 1975. Effects of acetylcholine on membrane currents in frog atrial muscle. *J. Physiol.* **246:** 64.

HAGIWARA, S. and L. BYERLY. 1981. Calcium channel. *Annu. Rev. Neurosci.* **4:** 69.

———. 1983. The calcium channel. *Trends Neurosci.* **6:** 189.

HAGIWARA, S. and H. OHMORI. 1982. Studies of Ca channels in rat clonal pituitary cells with patch electrode voltage clamp. *J. Physiol.* **331:** 231.

HAGIWARA, S., J. FUKUDA, and D.C. EATON. 1974. Membrane currents carried by Ca, Sr and Ba in barnacle muscle fiber during voltage clamp. *J. Gen. Physiol.* **63:** 564.

HAGIWARA, S., S. MIYAZAKI, S. KRASNE, and S. CIANI. 1977. Anomolous permeabilities of the egg cell membrane of a starfish in K^+-Tl^+-mixtures. *J. Gen. Physiol.* **70:** 269.

HAMILL, O.P., A. MARTY, E. NEHER, B. SAKMANN, and F.J. SIGWORTH. 1981. Improved patch-clamp techniques for high-resolution current recording from cells and cell-free membrane patches. *Pfluegers Arch. Eur. J. Physiol.* **391:** 85.

HARARY, I., F. HOOVER, and B. FARLEY. 1974. The isolation and cultivation of rat heart cells. *Methods Enzymol.* **32:** 745.

HILLE, B. and W. SCHWARZ. 1978. Potassium channels as multi-ion single-file pores. *J. Gen. Physiol.* **72:** 409.

HORN, R. 1978. Propagating calcium spikes in an axon of *Aplysia. J. Physiol.* **281:** 513.

HORN, J.P. and D.A. MCAFEE. 1980. Alpha-adrenergic inhibition of calcium-dependent potentials in rat sympathetic neurones. *J. Physiol.* **301:** 191.

HUME, J.R. and W. GILES. 1981. Active and passive electrical properties of single bullfrog atrial cells. *J. Gen. Physiol.* **78:** 19.

IKEMOTO, Y. and M. GOTO. 1975. Nature of the negative inotropic effect of acetylcholine on the myocardium. *Proc. Jpn. Acad.* **51:** 501.

KASS, R.S. and M.L. BLAIR. 1981. Effects of angiotensin II on membrane current in cardiac Purkinje fibers. *J. Mol. Cell Cardiol.* **13:** 797.

KATZ, B. 1969. *The release of neural transmitter substances.* Liverpool University Press, Liverpool.

KOSTYUK, P.G. 1981. Calcium channels in the neuronal membrane. *Biochim. Biophys. Acta* **650:** 128.

KOSTYUK, P.G. and O.A. KRISHTAL. 1977. Separation of sodium and calcium currents in the somatic membrane of mollusc neurones. *J. Physiol.* **270:** 545.

LAMB, T.D. and E.J. SIMON. 1977. Analysis of electrical noise in turtle cones. *J. Physiol.* **272:** 435.

LEE, K.S. and R.W. TSIEN. 1982. Reversal of current through calcium channels in dialysed single heart cells. *Nature* **297:** 498.

———. 1983. Mechanism of calcium channel blockade by verapamil, D600, diftiazem and nitrendipine in single dialyzed heart cells. *Nature* **302:** 790.

LEE, K.S., N. AKAIKE, and A.M. BROWN. 1980. The suction pipette method for internal perfusion and voltage clamp of small excitable cells. *J. Neurosci. Methods* **2:** 51.

LEE, K.S., T.A. WEEKS, R.L. KAO, N. AKAIKE, and A.M. BROWN. 1979. Sodium current in single heart muscle cells. *Nature* **278:** 269.

LEWIS, C.A. and C.F. STEVENS. 1979. Mechanism of ion permeation through channels in a postsynaptic membrane. In *Membrane transport processes* (ed. C.F. Stevens and R.W. Tsien), vol. 3, p. 133. Raven Press, New York.

LLINAS, R. and M. SUGIMORI. 1980. Electrophysiological properties of *in vitro* Purkinje cell dendrites in mammalian cerebellar slices. *J. Physiol.* **305:** 197.

LLINAS, R., I.Z. STEINBERG, and K. WALTON. 1981. Presynaptic calcium currents in squid giant synapse. *Biophys. J.* **33:** 289.

LUX, H.D. and K. NAGY. 1981. Single channel Ca^{2+} currents in *Helix pomatia* neurons. *Pfluegers Arch. Eur. J. Physiol.* **391:** 252.

MARBAN, E. and R.W. TSIEN. 1982. Enhancement of calcium current during digitalis inotropy in mammalian heart: Positive feedback regulation by intracellular calcium. *J. Physiol.* **329:** 589.

NEHER, E. 1975. Ionic specificity of the gramicidin channel and the thallous ion. *Biochim. Biophys. Acta* **401:** 540.

NIEDERGERKE, R. and S. PAGE. 1977. Analysis of catecholamine effects in single atrial trabeculae of the frog heart. *Proc. R. Soc. Lond. B.* **197:** 333.

NOWYCKY, M.C., B.P. BEAN, and R.W. TSIEN. 1983. Properties of calcium channels in dorsal root ganglion cells. *Soc. Neurosci. Abstr.* (in press).

OSTERRIEDER, W., G. BRUM, J. HESCHELER, W. TRAUTWEIN, V. FLOCKERZI, and F. HOFMANN. 1982. Injection of subunits of cyclic AMP-dependent protein kinase into cardiac myocytes modulates Ca^{2+} current. *Nature* **298:** 576.

PELLMAR, T.C. and D.O. CARPENTER. 1980. Serotonin induces a voltage-sensitive calcium current in neurons of *Aplysia californica. J. Neurophysiol.* **44:** 423.

REUTER, H. 1967. The dependence of the slow inward current on external calcium concentration in Purkinje fibres. *J. Physiol.* **192:** 479.

———. 1979. Properties of two inward membrane currents in the heart. *Annu. Rev. Physiol.* **41:** 413.

———. 1983. Calcium channel modulation by neurotransmitters, enzymes and drugs. *Nature* **301:** 569.

REUTER, H. and H. SCHOLZ. 1977a. A study of the ion selectivity and the kinetic properties of the calcium dependent slow inward current in mammalian cardiac muscle. *J. Physiol.* **264:** 17.

———. 1977b. The regulation of the Ca conductance of cardiac muscle by adrenaline. *J. Physiol.* **264:** 49.

REUTER, H., C.F. STEVENS, R.W. TSIEN, and G. YELLEN. 1982. Properties of single calcium channels in cultured cardiac cells. *Nature* **297:** 501.

SHAPIRO, E., V.F. CASTELLUCCI, and E.R. KANDEL. 1980. Presynaptic inhibition in *Aplysia* involves a decrease in the Ca^{2+} current of the presynaptic neuron. *Proc. Natl. Acad. Sci.* **77:** 1185.

SIEGELBAUM, S.A. and R.W. TSIEN. 1983. Modulation of gated ion channels as a mode of transmitter action. *Trends Neurosci.* **6:** 307.

SIGWORTH, F.J. 1980. The variance of sodium current fluctuations at the node of Ranvier. *J. Physiol.* **307:** 97.

———. 1977. Sodium channels in nerve apparently have two conductance states. *Nature* **270:** 265.

SPERELAKIS, N. and J.A. SCHNEIDER. 1976. A metabolic control mechanism for calcium ion influx that may protect the ventricular myocardial cell. *Am. J. Cardiol.* **37:** 1079.

TSIEN, R.W. 1977. Cyclic AMP and contractile activity in heart. *Adv. Cyclic Nucleotide Res.* **8:** 363.

——. 1983. Calcium channels in excitable cell membranes. *Annu. Rev. Physiol.* **45:** 341.

URBAN, B.W., S.B. HLADKY, and D.A. HAYDON. 1978. The kinetics of ion movements in the gramicidin channel. *Fed. Proc.* **37:** 2628.

VASSORT, G., O. ROUGIER, D. GARNIER, M.P. SAUVIAT, E. CORABOEUF, and Y.M. GARGOUIL. 1969. Effects of adrenaline on membrane inward currents during the cardiac action potential. *Pflugers Arch. Eur. J. Physiol.* **309:** 70.

VEREECKE, J. and E. CARMELIET. 1971. Sr action potentials in cardiac Purkinje fibers. II. Dependence of the Sr conductance on the external Sr concentration and Sr-Ca antagonism. *Pfluegers Arch. Eur. J. Physiol.* **322:** 73.

WATANABE, A.M., J.P. LINDEMANN, L.R. JONES, H.R. BESCH, JR., and J.C. BAILEY. 1981. Biochemical mechanisms mediating neural control of the heart. In *Disturbances in neurogenic control of the circulation* (ed. F.M. Abboud et al.), p. 189. American Physiological Society, Bethesda.

WEINGART, R., R.S. KASS, and R.W. TSIEN. 1978. Is digitalis inotropy associated with enhanced slow inward calcium current? *Nature* **273:** 389.

WERZ, M.A. and R.L. MACDONALD. 1982. Heterogeneous sensitivity of cultured dorsal root ganglion neurones to opioid peptides selective for μ- and δ-opiate receptors. *Nature* **299:** 730.

Modulation of a Specific Potassium Channel in Sensory Neurons of *Aplysia* by Serotonin and cAMP-dependent Protein Phosphorylation

J.S. CAMARDO, M.J. SHUSTER, S.A. SIEGELBAUM, AND E.R. KANDEL
Center for Neurobiology and Behavior, Columbia University, College of Physicians and Surgeons and The New York State Psychiatric Institute, New York, New York 10032

One of the important developments in the study of synaptic transmission has been the discovery that neurotransmitters produce a wide range of electrical signals by interacting with different types of ionic channels. Since 1950, it has been well established that chemical transmitters elicit short-lived postsynaptic potentials, such as the end-plate potential produced by acetylcholine (ACh) in skeletal muscle (Katz 1966; Sakmann et al., this volume). Here, ACh activates a channel that opens only in the presence of the transmitter and therefore does not normally contribute to the resting or active electrical properties of the cell (Neher and Sakmann 1976). Recently, however, transmitters have been found to exert long-lasting effects on some neurons by modulating channels that contribute to the resting potential, the action potential, or the passive membrane properties (for review, see Kehoe and Marty 1980; Siegelbaum and Tsien 1983). Modulation of these channels can dramatically change the electrophysiological behavior of the neuron. For example, such modulation can alter the integrative properties of the neuron by changing the space and time constants and the threshold for initiating the action potential. Furthermore, modulation of these channels will affect aspects of cell function that are coupled to the action potential, such as the influx of Ca^{++}, and can thereby affect transmitter release, muscle contraction, and other intracellular processes (Klein et al. 1980; Reuter 1979).

In the sensory neurons of the gill withdrawal reflex of the marine mollusc, *Aplysia californica*, a long-lasting synaptic action produced by serotonin (5-hydroxytryptamine [5-HT]) has proven to be particularly interesting because it increases the effectiveness of synaptic transmission and thereby underlies a simple form of learning: sensitization (Kandel and Schwartz 1982). This slow synaptic action involves a depression in the K^+ conductance of the sensory neurons, resulting in a small depolarization in the cell resting potential and a delay in the repolarization of the action potential (Klein and Kandel 1980). The slowing of repolarization leads to an increase in the duration of Ca^{++} influx, and therefore to more transmitter release and a greater response in the gill withdrawal reflex. This slow synaptic action and the consequent enhancement of transmitter release is simulated by serotonin. Serotonin produces its effects by acting on a specific type of K^+ channel with characteristics that differentiate it from four previously described K^+ channels: the early K^+ channel, the delayed K^+ channel, the Ca^{++}-dependent K^+ channel, and the M channel (Siegelbaum et al. 1982; Klein et al. 1982). A particularly interesting feature of the action of serotonin is that its effect on the K^+ channels in the sensory neurons is not direct but is mediated through an adenylcyclase-cAMP cascade and involves phosphorylation of a substrate protein by the cAMP-dependent protein kinase (see Kandel and Schwartz 1982).

The development of single-channel recording techniques (Neher and Sakmann 1976; Hamill et al. 1981) has allowed us to investigate at the molecular level the action of serotonin on single K^+ channels (Siegelbaum et al. 1982). In this paper, we address three questions. First, what are the properties of the K^+ current and of the K^+ channel that is modulated by serotonin? Second, on which aspect of channel function does serotonin exert its modulatory action? Third, what is the molecular mechanism whereby the channel is modulated? Is the K^+ channel itself phosphorylated by the protein kinase or does the kinase phosphorylate some intermediary regulatory protein in the cytoplasm which then influences the function of the channel?

RESULTS

Serotonin Decreases a Specific K^+ Current in *Aplysia* Sensory Neurons

The initial insight into the nature of the serotonin-sensitive current in the sensory neurons of *Aplysia* came from voltage-clamp experiments (Klein and Kandel 1980). These revealed that application of serotonin or stimulation of the pathway that produces presynaptic facilitation also decreases outward K^+ current in the sensory cells (Fig. 1). This conclusion was based on three findings. First, serotonin or stimulation of the pathway that mediates sensitization depressed an outward current that is sensitive to the extracellular concentration of K^+. Second, the action of serotonin was eliminated by pharmacologic blockade of K^+ channels. Third, the action of serotonin was blocked by replacing intracellular K^+ with the nonpermeant cation, Cs^+. Klein et al. (1982) extended this analysis and identified a specific K^+ current decreased by serotonin. They found that the membrane of the sensory cells of *Aplysia*

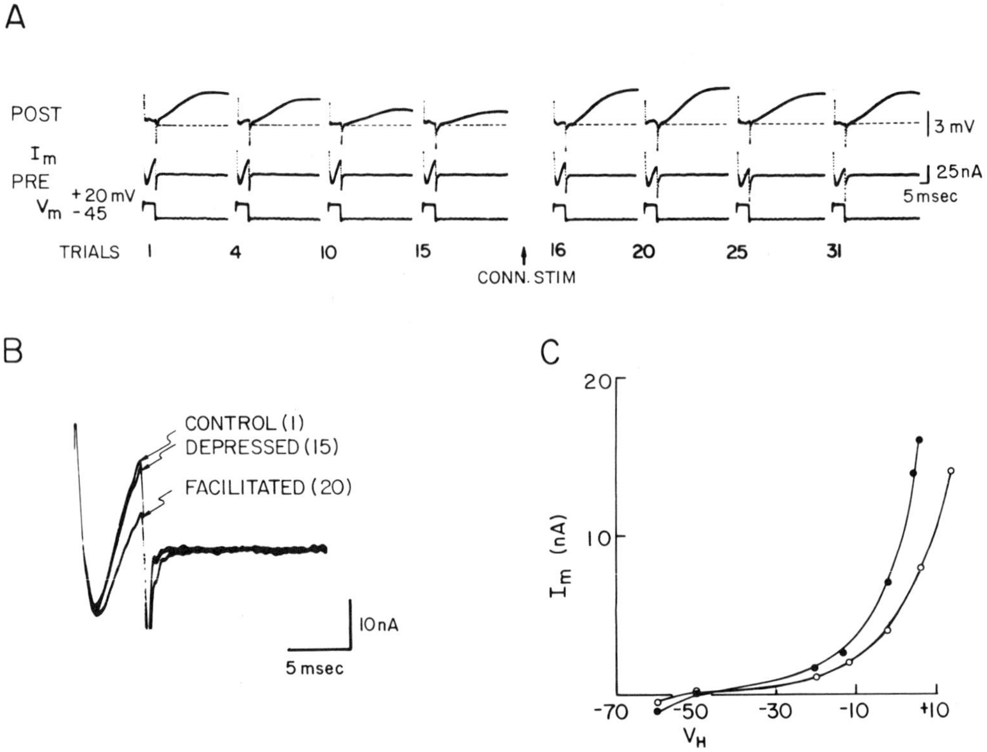

Figure 1. Presynaptic facilitation is associated with a decrease in outward current in sensory neurons. (*A*) A sensory neuron is voltage-clamped at a holding potential of −45 mV, and stepped to +20 mV, for 10 msec (*middle and lower traces*). This mimics the action potential and elicits a postsynaptic potential (PSP) in the follower cell, a motor neuron (*top trace*). The PSP decrements with repeated stimulation (trials 1–15), and is facilitated by stimulation of the connective (trials 16–31). The facilitation is always accompanied by a decrease in outward current in the sensory cell. The decrease in outward current is shown at higher resolution in *B*. (*C*) The current-voltage relation for a sensory neuron is altered by injection of cAMP. (●) Control current elicited by steps from −50 mV; (○) the current following intracellular injection of cAMP. The outward current decreases at all potentials. The results illustrated in Fig. 1B,C are identical to the results evoked by application of serotonin. Additional experiments described in the text show that both serotonin and cAMP exert their effects by decreasing a specific K^+ current in the sensory neurons.

contains the three K^+ currents described in other molluscan neurons: (1) the early K^+ current; (2) the delayed K^+ current; and (3) the Ca^{++}-activated K^+ current (Adams et al. 1980). Each of these currents can be characterized by its voltage dependence, its susceptibility to pharmacologic blocking agents, and, in the case of Ca^{++}-activated current, by its dependence on intracellular Ca^{++}. Klein et al. found that serotonin did not act on these three K^+ currents, but on a novel K^+ current, specifically sensitive to serotonin. They called the new current S current, by analogy with M current, a muscarine-sensitive current discovered in the sympathetic neurons of the frog (Brown and Adams 1980). The S current is also decreased by intracellular cAMP (Fig. 1C).

Voltage-clamp studies have provided some basic information about the serotonin-sensitive current. The current is activated at the resting potential, contributes to repolarization of the action potential, remains activated with steady-state depolarization, and is not dependent on intracellular Ca^{++}. Although the S current can be identified on the basis of these characteristics, its salient feature is that it is decreased by serotonin and cAMP.

Serotonin Modulates a Specific K^+ Channel

Single-channel recording (Neher et al. 1978; Hamill et al. 1981) allows one to measure the current that flows through a single, open ion channel in its normal membrane environment and to observe directly changes in ion flux through the channel as the channel undergoes rapid conformational changes from conducting (open) to nonconducting (closed) states. The single-channel recording offers another advantage. A high-resistance (giga-ohm) seal is formed between a fire-polished extracellular glass pipette (with a tip diameter of 0.2–0.5 μm) and the cell membrane. The giga-ohm seal is mechanically stable so that the patch of membrane under the pipette adheres to the glass when the pipette is pulled away from the cell. This allows one to record channel currents in an isolated cell-free patch of membrane where the inner surface of cell membrane (cytoplasmic side) is now exposed to the bath solution and is accessible to experimental manipulation (Hamill et al. 1981). Thus, one can investigate the action of various intracellular components including inorganic ions, second messengers, and enzymes on the function of the ion channel.

Single-channel recordings from sensory cell membranes reveal a predominant channel that has many properties expected for the S channel based on the voltage-clamp studies of the S current. By holding the patch potential at 0 mV, which inactivates the early and delayed K^+ current and fails to activate the Ca^{++}-activated K^+ current, we have been able to study the S channel in relative isolation. An illustration of the experimental design is shown in Figure 2A, and a single-channel current record is shown in Figure 2B. The channel carries outward current over a wide range of membrane potentials. At a fixed membrane potential, the unit current steps are all of a similar amplitude, indicating that a single type of channel predominates (Fig. 2C). At a membrane potential of 0 mV, the single-channel current amplitude is 3 pA. The channels display two types of closures: brief closures lasting only a few milliseconds or less (apparent as brief downward transients), and much longer closures lasting tens or hundreds of milliseconds (apparent as step changes in current level).

The S channel is open at the resting potential, and remains open with steady-state depolarization, and is therefore noninactivating (Fig. 2B). Its probability of opening is virtually independent of voltage over a range of membrane potentials from +80 mV to −80 mV, so it is aptly described as a background channel (Fig. 3A). The channel has a slope conductance of 55 pS at 0 mV in the cell-attached patch, and rectifies in the outward direction, because the K^+ concentrations across the

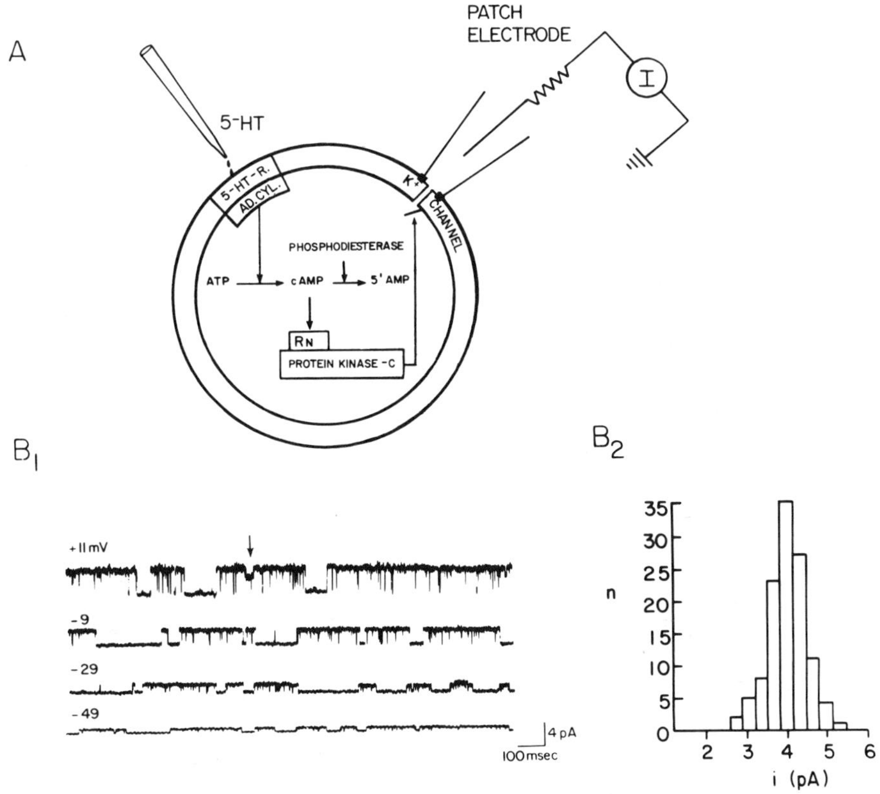

Figure 2. Patch-clamp recording of single S channel currents. (*A*) Schematic illustration of the experimental recording protocol and the cAMP-dependent model for the action of 5-HT (serotonin). A high-resistance giga-ohm seal is obtained between the extracellular patch electrode and the cell membrane to record single-channel currents in the small patch of membrane under the electrode. This is the cell-attached recording configuration. If the patch electrode is withdrawn from the cell, the seal remains intact and single-channel currents can be recorded from the cell-free patch of membrane where the intracellular surface of the membrane faces the bath solution, termed the inside-out patch configuration. Serotonin is applied to the cell outside the membrane patch but may still alter channel activity in the patch via the cAMP cascade, which is illustrated diagramatically. The substrate protein for the kinase may or may not be the S channel itself. (B_1) Serotonin-sensitive K^+ channel. Single-channel current records from mechanoreceptor sensory neurons (LE cluster) in the abdominal ganglion of *A. californica*. Both the bath and recording pipette contain artificial seawater. The patch membrane potential (left of each trace) was altered by changing the potential inside the recording pipette while the intracellular resting potential (−39 mV) was constant. Channel openings appear as step increases in outward current (outward current plotted in upward direction). The current fluctuates between two levels corresponding to the fully closed and fully open channel, and channel size increases with depolarization. At all potentials, the channel shows both brief closures (downward flickers) and longer closures but is in the open configuration for most of the time. The arrow indicates a rare event representing a closure of the channel to an intermediate conductance level. Addition of 10 μM serotonin in the bath caused this channel to close (record not shown) (from SCK82). (B_2) A histogram of channel current amplitudes at a potential of +11 mV obtained from the experiment in B_1. The channel current sizes cluster around 4 pA with a single peak in the histogram, which indicates that there is a single type of channel.

membrane patch on the cell are asymmetric (Hodgkin and Katz 1949). In experiments using the cell-free patch, this rectification disappears when the K^+ concentrations on both sides of the patch are equalized (Fig. 3B).

The S channel is highly selective for K^+ ions. Figure 3B shows the effect on the reversal potential of a change in K^+ concentration, from 360 mM on each side (where the reversal potential is zero) to 360 mM on the outside of the membrane and 40 mM on the inside of the membrane. The Nernst equation for a pure K^+ current predicts a change in reversal potential of 55 mV for this ninefold change in K^+ concentration. Our measured change was 58 mV, which, within the range of experimental error, is identical to the predicted change. An intermediate concentration of K^+ also gave the predicted reversal potential.

Channel opening is independent of intracellular Ca^{++} (Fig. 3C). The probability of the channel being open is unchanged when the Ca^{++} concentration on the inner surface of the membrane is changed from 1 nM to 1 μM. In contrast, the Ca^{++}-activated K^+ channel of the sensory neurons, qualitatively similar to that described in other cells (Lux et al. 1981; Marty 1981; Pallotta et al. 1981), is profoundly affected by intracellular Ca^{++}. It is closed at Ca^{++} concentrations below 10^{-6} M, where the S channel is open. This Ca^{++}-dependent K^+ channel gives rise to the current elicited by injection of Ca^{++} into *Aplysia* neurons (Meech 1972), and our measurements indicate that it has a 30% smaller conductance than the S channel (M.S. Shuster et al., unpubl.).

Thus, the properties of the S channel are in close agreement with the results of voltage-clamp experiments

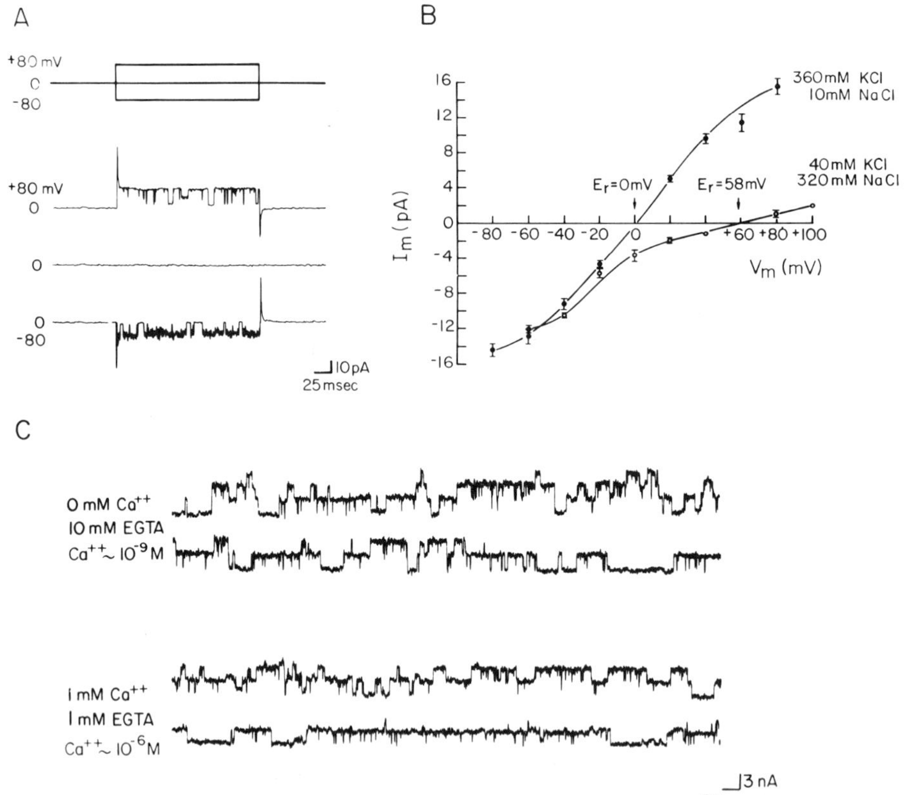

Figure 3. S channel is voltage-independent, selective for K^+ ions, and independent of intracellular Ca^{++}. (*A*) The S channel remains open over a wide range of membrane potential. With a cell-free, inside-out patch in symmetric (360 mM) KCl, the patch is held at zero potential and stepped to different membrane potentials (*top panel*). The current is recorded in the lower panel. Here, with symmetric K^+, the reversal potential for current flow is at 0 mV (*middle trace*). With depolarization, K^+ current flows outward (*top*), and with hyperpolarization, K^+ current flows inward (*bottom*). The channel is open for most of the time over a range of voltage of 160 mV, from −80 to +80; it shows no voltage threshold within this range, and almost no increase in opening with depolarization. (*B*) Reversal potential for S channel current follows the Nernst equation for K^+. (●) The current-voltage curve obtained in symmetric 360 mM KCl, where the reversal potential is at 0 mV. The inner surface of the patch was then exposed to 40 mM KCl. For a pure K^+ current, the reversal potential, from the Nernst equation, is +55 mV. (○) The reversal potential in this asymmetric solution is about +58 mV. Within the range of experimental error, this indicates that the channel is highly selective for K^+ ions. (*C*) Opening of the S channel is independent of intracellular Ca^{++}. In this experiment with an inside-out patch, the K^+ concentrations were asymmetric: 10 mM KCl in the pipette (external surface), 360 mM KCl in the bath (internal surface), and K^+ current flows outward. The Ca^{++} concentration in the bath was changed using a Ca^{++}-EGTA buffer. The current records show that the channel remains open in approximately 1 nM Ca^{++} (*top*), and its opening is not increased when the Ca^{++} is raised to approximately 1 μM (*bottom*).

for the S current. These intrinsic properties allow us to identify the S channel with some assurance, even in the absence of serotonin.

In Principle, Single-channel Currents Can Be Decreased in a Number of Different Ways

Figure 4 presents several possible ways by which single channels could be modulated to produce a decrease in K^+ current. We assume for this discussion that the channel obeys the following simple scheme:

$$\text{closed} \underset{\alpha}{\overset{\beta}{\rightleftharpoons}} \text{open}$$

where α and β represent first-order rate constants for closing and opening of the channel, respectively. The average current, through either the patch membrane or the whole-cell membrane, is the mean total current through all the channels and is given by the equation:

$$I = n \cdot p \cdot i \quad (1)$$

where n is the number of functional channels, p is the probability of finding a channel in the open state, and i is the current through a single open channel. The probability p is in turn a function of the opening and closing rates and is given by $p = \beta/(\alpha + \beta)$. It is evident from Equation 1 that a change in any of the three right-hand terms could change the total average current (Fig. 4). Thus, the transmitter could change the current through a single channel by decreasing the single-channel conductance. Alternatively, the transmitter could decrease p in two ways. First, it could shorten the mean open time of the channel, as a result of an increase in the closing rate constant (α), so that the channel closes more rapidly after opening. Second, it could decrease the probability of channel opening by decreasing the opening rate constant (β), so that channels open less frequently, but once open would stay open for their usual time. Finally, a transmitter could prevent a particular channel from contributing current at all, and therefore decrease the number of functional channels, n, in the membrane. These alternatives are not mutually exclusive. In some cases, the differences are unambiguous: One could easily observe smaller channel currents, as well as a change in the closing rate. However, certain distinctions depend on the time scale of the observations. For example, if the opening rate decreases from one channel opening per second to one channel opening per hour, so that p is drastically reduced, one might, on the time scale of a normal experiment (say 20–30 min), perceive this as a change in n.

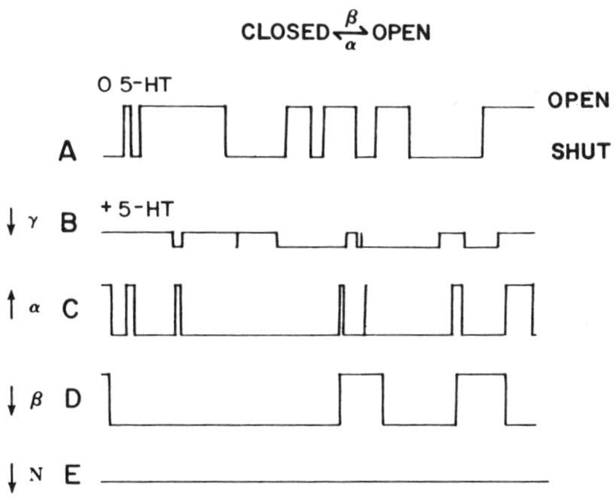

Figure 4. Simulation of channel modulation by neurotransmitter. Transmitter is assumed to decrease macroscopic ion conductance. Channel openings are displayed as upward going step changes in current. (*A*) Channel current in the absence of drug. (*B*, *C*, *D*, and *E*) Hypothetical channel current in presence of modulatory transmitter. (*B*) Expected single-channel current if transmitter causes a reduction in single-channel conductance (γ), reducing current amplitude. (*C*) Channel current with increased closing rate α. (*D*) Channel current with decreased opening rate β. (*E*) Baseline current trace with no channel openings, corresponding to a decrease in the number of active channels.

Serotonin Reduces the Number of Open Channels Present in the Membrane

To distinguish among these possibilities, single-channel currents were recorded from an identified sensory cell and serotonin was then added to the bathing solution (see Fig. 2A). Since the giga-ohm seal prevents transmitter from entering the patch pipette directly from the bath, serotonin can alter channel activity only through some intracellular second messenger. As is illustrated in Figure 5, before addition of serotonin, four channels open and close randomly, maintaining a certain average current. After addition of serotonin, and accompanying a decrease in whole-cell conductance (not shown), the average single-channel current decreased in a step-wise fashion, due to the closure of three of the four active channels. Whereas in the control period all four channels are active and contribute outward current, after the application of serotonin, only one channel remains active. Serotonin causes no change in the channel current amplitude (i remains constant) and no observable effect on channel gating (opening and closing rates are, on average, unchanged). Therefore, the best explanation for the stepwise decrease in current, in the presence of serotonin, is that some channels no longer open. The same effect is observed following the injection of cAMP (Fig. 6). Thus, serotonin and cAMP appear to decrease n, the number of active channels in the membrane. These experiments indicate that the serotonin-cAMP cascade culminates in a decrease of K^+ flux through a specific channel, as indicated by the whole-cell voltage-clamp data and that, at the molecular level, this is accomplished by prolonged closure of the channel.

The final common pathway by which cAMP is thought to mediate its effects is by means of protein phosphorylation (Greengard 1981). An important question that now remains is, How does protein phosphorylation lead to closure of the K^+ channel? Is the site of phosphorylation directly on the channel itself or does

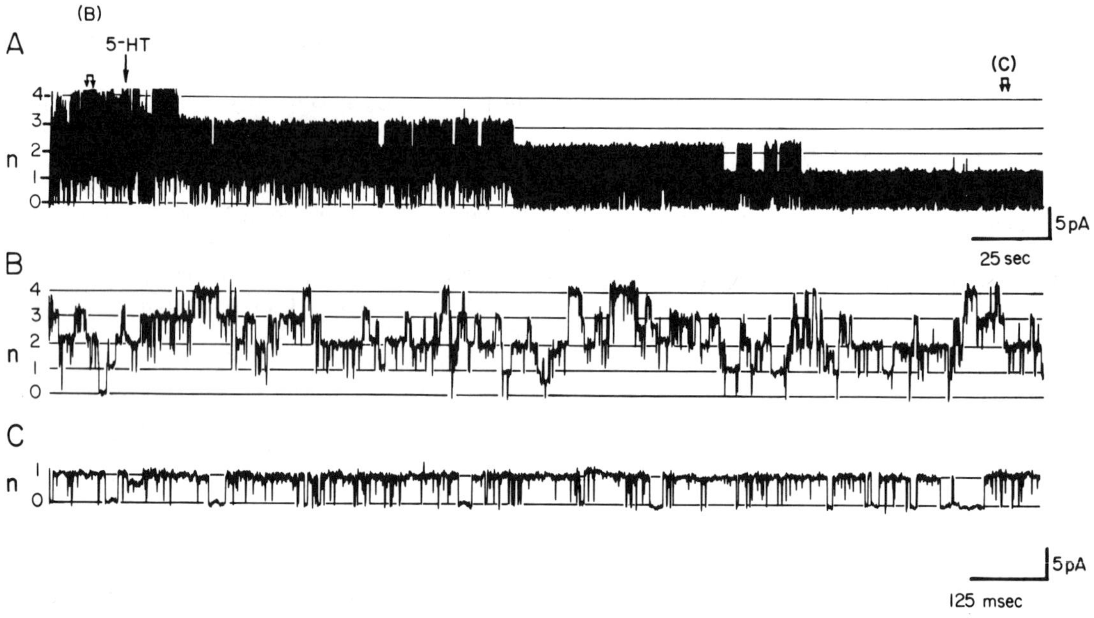

Figure 5. Effect of serotonin on single-channel current. Single-channel recordings from a sensory cell membrane patch containing four channels. (*A*) Channel currents shown on a slow time base. At the arrow, a concentrated drop of 5-HT (serotonin) was added to the bath to achieve a 5-HT concentration of 100 μM. The numbers on the left-hand ordinate indicate the number of open channels. Before addition of 5-HT, there were a maximum of four open channels in the patch. The small arrows labeled (B) and (C) indicate regions of trace *A* from which expanded records shown in traces *B* and *C* were obtained. Cell resting potential was −41 mV before 5-HT; in presence of 5-HT, the cell depolarized to −37 mV and there was a 33% increase in membrane resistance. Patch membrane potential was 0 mV before 5-HT and +4 mV after 5-HT.

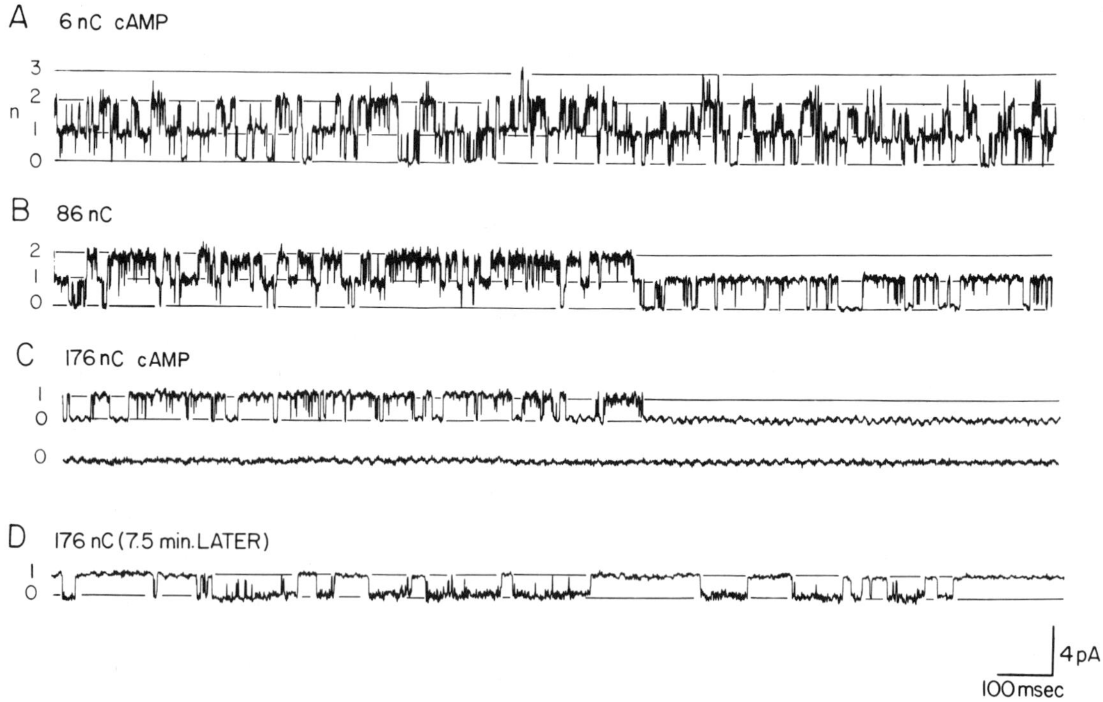

Figure 6. Effect of intracellular injection of cAMP on single-channel current. Following establishment of seal between patch electrode and membrane, the cell was impaled with a microelectrode filled with 1 M cAMP (Sigma), and cAMP was injected into the cell by hyperpolarizing current pulses (0.5–2.5 nA). Total charge injected into cell (nanocoulombs) is indicated on top of each trace. (*A*) Current record obtained soon after impalement of cell with cAMP electrode. (*B,C*) Current records after subsequent injection periods. During the records shown in *C*, the remaining active channel closed. The two records in *C* are a continuous recording. The cAMP electrode was withdrawn from the cell, and no channel openings were observed for 5 min. Channel activity then partially recovered, as shown in trace *D*. Cell membrane potential hyperpolarized somewhat during the course of the experiment. Patch membrane potential was approximately +14 mV in *D*, +11 mV in *C*, and +2 mV in *B*. The time bar in *D* equals 125 msec.

phosphorylation occur at a separate site that regulates the channel indirectly, such as an ion pump that controls the ionic concentration near the channel, or a cytoplasmic- or membrane-bound regulatory protein (Fig. 7A)?

To distinguish among these possibilities, we have recently begun a series of experiments using cell-free, inside-out patches to determine whether purified catalytic subunit of the cAMP-dependent protein kinase (Builder et al. 1981) can modulate S channel activity in vitro. The scheme for these experiments is illustrated in Figure 7B.

DISCUSSION

We have examined a slow synaptic potential mediated by serotonin in *Aplysia* neurons and have demonstrated that the action of this transmitter is limited to a specific membrane K^+ current, which participates in both the resting potential and action potential of the cell. We have identified the channel molecule that carries this K^+ current (the S channel) and have shown that it is closed for long periods of time by serotonin and by cAMP.

The ultimate goal of these studies of the S channel is to analyze the molecular basis for channel modulation by transmitters and second messengers. We still have only a limited understanding of the channel molecule itself, based mostly on descriptive measurements of its biophysical characteristics. We do not know the number of subunits that make up the channel, the amino acid sequence, how they are assembled and inserted into the membrane, and how they are regulated so that the three-dimensional structure of the protein serves its function. Thus, we do not have the biochemical understanding that is becoming available for the ACh receptor (this volume). This is largely due to the fact that the channel molecules are relatively sparse and that there is no specific ligand for any K^+ channel.

An understanding of the modulation of the S channel by the cAMP-dependent protein kinase in the sensory neurons takes on added significance because this modulation underlies short-term behavioral sensitization, a simple form of learning. While sensitization persists, an action potential that enters the sensory neuron terminal will find the serotonin-sensitive K^+ channels closed, and the repolarization of the action potential will be delayed. As a result, more Ca^{++} will flow in, more transmitter will be released, and the sensory stimulus will therefore elicit a more powerful reflex response. Thus, a simple molecular modification can have profound effects on synaptic transmission and ultimately on the behavior subserved by these synapses.

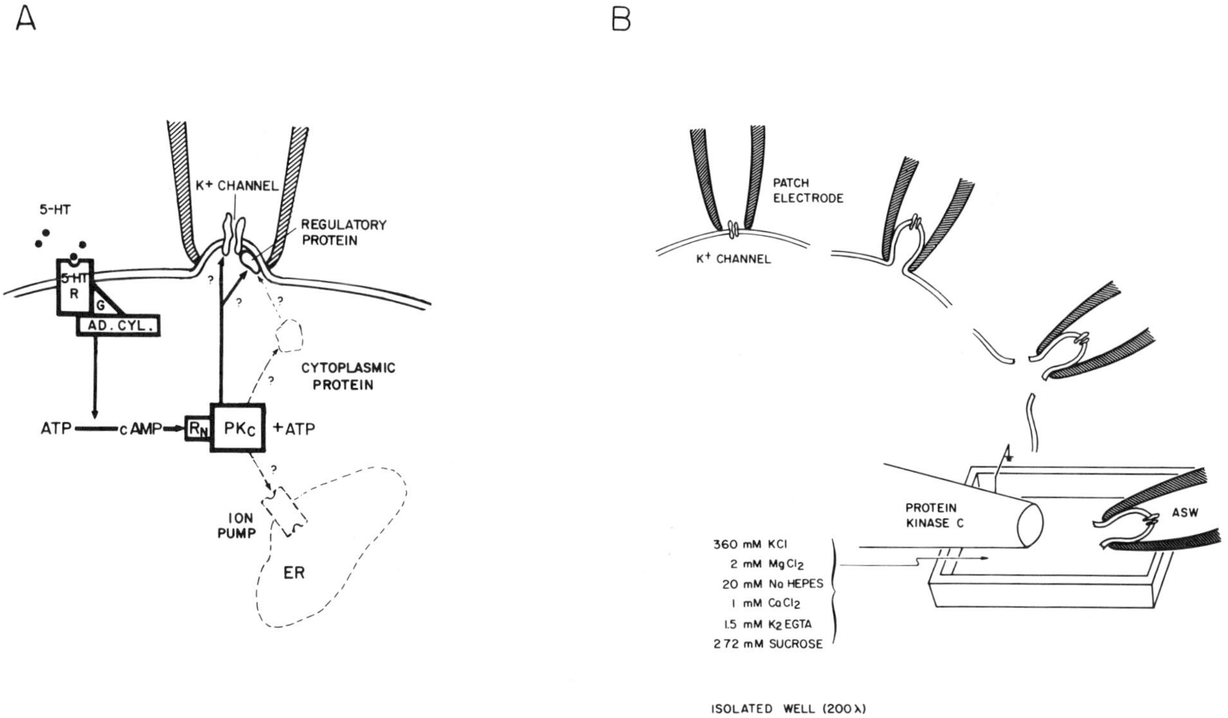

Figure 7. Cell-free patches can be used to study the interaction of protein kinase with S channels. (*A*) The serotonin-cAMP cascade and several possible substrates (indicated with question marks) for protein phosphorylation. Phosphorylation could alter the activity of an ion pump in the endoplasmic reticulum, which could modulate the channel indirectly. Other possible substrates include a cytoplasmic regulatory protein, a membrane-associated regulatory protein, or the channel itself. Cell-free patch experiments, in which the cytoplasm is washed away, and the ionic composition of the medium can be controlled, are helpful in distinguishing between these possibilities. (*B*) The design of the cell-free experiment. A patch is obtained on the cell, the pipette is withdrawn, and the patch is isolated in a 200-μl well containing a high K^+, low Ca^{++} solution. Protein kinase and ATP are injected directly onto the patch.

ACKNOWLEDGMENTS

This research was supported by National Institute of Mental Health grant MH26212 to E.R.K. and a Sloan Foundation fellowship to S.A.S. We wish to thank E. Krebs, D. Walsh, C. Rubin, and C.M. Eppler for providing catalytic subunit of protein kinase and C.M. Eppler for assaying samples of kinase. We wish to thank Harriet Ayers and Charlotte Alexander for preparing the manuscript and Kathrin Hilten and Lousie Katz for preparing the figures.

REFERENCES

Adams, D.J., S.J. Smith, and S.H. Thompson 1980. Ionic currents in molluscan soma. *Annu. Rev. Neurosci.* **3:** 141.

Brown, D.A. and P.R. Adams. 1980. Muscarinic suppression of a novel voltage-sensitive K^+ current in a vertebrate neurone. *Nature* **283:** 672.

Builder, S.E., J.A. Beavo, and E.C. Krebs. 1981. Mechanism of activation and inactivation of the cAMP-dependent protein kinases. In *Protein phosphorylation* (ed. A. Book et al.), p. 33. Cold Spring Harbor Laboratory, Cold Spring Harbor, New York.

Greengard, P. 1981. Intracellular signals in the brain. *Harvey Lect.* **75:** 277.

Hamill, O., A. Marty, E. Neher, B. Sakmann, and F.J. Sigworth. 1981. Improved patch-clamp techniques for high resolution current recording from cells and cell-free membrane patches. *Pfluegers Arch. Eur. J. Physiol.* **391:** 85.

Hodgkin, A.L. and B. Katz. 1949. The effects of sodium ions on the electrical activity of the giant axon of the squid. *J. Physiol.* **108:** 37.

Kandel, E.R. and J.H. Schwartz. 1982. Molecular biology of an elementary form of learning: Modulation of transmitter release by cyclic AMP. *Science* **218:** 433.

Katz, B. 1966. *Nerve muscle and synapse.* McGraw-Hill, New York.

Kehoe, J. and A. Marty. 1980. Certain slow postsynaptic responses: Their properties and possible underlying mechanisms. *Annu. Rev. Biophys. Bioeng.* **9:** 437.

Klein, M. and E.R. Kandel. 1980. Mechanism of calcium current modulation underlying presynaptic facilitation and behavioral sensitization in *Aplysia. Proc. Natl. Acad. Sci.* **77:** 6912.

Klein, M., J.S. Camardo, and E.R. Kandel. 1982. Serotonin modulates a specific potassium current in the sensory neurons that show presynaptic facilitation in *Aplysia. Proc. Natl. Acad. Sci.* **79:** 5713.

Klein, M., E. Shapiro, and E.R. Kandel. 1980. Synaptic plasticity and the modulation of the Ca^{2+} current. *J. Exp. Biol.* **89:** 117.

Lux, H.D., E. Neher, and A. Marty. 1981. Single channel activity associated with the calcium dependent outward current in *Helix pomatia. Pfluegers Arch. Eur. J. Physiol.* **389:** 293.

Marty, A. 1981. Ca-dependent K channels with large unitary conductance in chromaffin cell membranes. *Nature* **291:** 497.

Meech, R.W. 1972. Intracellular calcium injection causes increased potassium conductance in *Aplysia* nerve cells. *Comp. Biochem. Physiol.* **42A:** 493.

Neher, E. and B. Sakmann. 1976. Single channel currents recorded from membrane of frog muscle fibers. *Nature* **260:** 799.

Neher, E., B. Sakmann, and J.H. Steinbach. 1978. The extracellular patch clamp: A method for resolving currents through individual open channels in biological membranes. *Pfluegers Arch. Eur. J. Physiol.* **375:** 219.

Pallotta, B.S., K.L. Magleby, and J.N. Barrett. 1981. Single channel recordings of Ca^{2+}-activated K^+ currents in rat muscle cell culture. *Nature* **293:** 471.

Reuter, H. 1979. Properties of two inward membrane currents in the heart. *Annu. Rev. Physiol.* **41:** 413.

Siegelbaum, S.A. and R.W. Tsien. 1983. Modulation of gated ion channels as a mode of transmitter action. *Trends Neurosci.* (in press).

Siegelbaum, S.A., J.S. Camardo, and E.R. Kandel. 1982. Serotonin and cAMP close single K channels in *Aplysia* sensory neurones. *Nature* **299:** 413.

Genetic and Voltage-clamp Analysis of a *Drosophila* Potassium Channel

L. SALKOFF
Department of Physiology, University of California, San Francisco, California 94143

The technique of voltage clamping combined with genetic analysis presents a unique opportunity for studying ion channels. The biophysical analysis of mutationally altered ion currents can provide clues about ion channel function and structure. In addition, a better understanding of the genetic aspects of channels resulting from such studies might pave the way for a direct analysis of molecular structure via gene cloning (cf. Jan et al., this volume). With these goals in mind, the combined techniques of genetics and voltage clamping were used to analyze mutations that map to the X-linked *Shaker* locus of *Drosophila melanogaster.* As will be shown, this study has produced evidence that the *Shaker* locus codes for at least one molecular component of the K^+ channel that carries a fast transient current.

Mutations at the *Shaker* locus show a striking behavioral phenotype of violent shaking under ether anesthesia. However, even when not anesthetized, *Shaker* mutants have poor motor control. The first indication that these mutants could affect an ion channel came from experiments on neuromuscular transmission in normal and mutant (*Shaker*) *Drosophila* larvae. By characterizing normal and mutant transmission (Jan and Jan 1967a,b; Jan et al. 1977) and by mimicking the mutant defect (the abnormally large evoked release of neurotransmitter) with agents that block K^+ conductance, a hypothesis was developed that some type of K^+ conductance was altered in the mutant. Since that time, intracellular recording from identified neurons in *Shaker* mutants (Tanouye et al. 1981) showed an abnormally broadened spike, supporting that hypothesis.

Finally, the means of directly observing the development and biophysical properties of voltage-sensitive ion channels in *Drosophila* became available by the successful application of voltage-clamp technique to *Drosophila* flight muscles (Salkoff and Wyman 1980, 1981a,b, 1983; Salkoff 1983). Direct observations of the K^+ currents under voltage clamp in *Shaker* mutants showed that the mutations affected the fast, transient K^+ current (I_A) (Salkoff and Wyman 1981b; Salkoff 1983; Wu et al. 1983). Mutations of this gene specifically affect I_A. Similar effects on I_A are also seen in larval muscles (Wu and Ganetzky 1983; Wu et al. 1983). Since none of the numerous other channel types in the membrane are affected, it is quite unlikely that the mutations cause a general membrane defect. Furthermore, as will be shown, the range and types of mutant effects on I_A are indicative of the presence of a structural gene for a channel protein. Finally, data will be presented that offers hints about the molecular nature of these channels.

METHODS

The voltage-clamp technique and equipment used were described previously (Salkoff and Wyman 1983). Two intracellular electrodes were used for voltage-clamp experiments. Both voltage-recording and current-passing electrodes were filled with 3 M KCl and had resistances of 3–10 MΩ. Experiments were performed at 4°C. Both adult and pupal preparations were wax-mounted on epoxy-coated slotted metal disks. Adult animals were sealed in the slot by sticky wax and their ventral portions were exposed to an airstream to facilitate respiration. To expose the dorsal longitudinal muscles (DLM) to saline, a small portion of the dorsal thorax was opened with microdissection needles. The standard saline used contained 128 mM NaCl, 4.8 mM KCl, 1.8 mM $CaCl_2$, and 0.2 mM KH_2PO_4/K_2HPO_4 (pH 6.8). K^+ currents were observed under voltage-clamp in the following solution: 128 mM choline chloride, 4.8 mM KCl, 1.8 mM EGTA, 10 mM $MgCl_2$, and 2 mM Tris-HCl (pH 7.0). The Ca^{++} current was voltage-clamped in adult animals in the following solution: 128 mM tetraethylammonium chloride (TEA) (Kodak), 4.8 mM KCl, 1.8 mM $CaCl_2$, and 2 mM Tris-HCl (pH 7.0).

Fly stocks were maintained on a standard laboratory medium consisting primarily of cornmeal, agar, molasses, and yeast. Wild-type stocks were of the Canton-S strain. Ethylmethane sulfonate (EMS)-induced mutations of the *Shaker* locus, Sh^{KS133}, Sh^{102}, Sh^{KO120}, were obtained from the California Institute of Technology stock collection. Stocks having X-Y translocations with breakpoints in the X chromosome 16F (*Shaker*) region were obtained from Dr. John Merriam.

RESULTS AND DISCUSSION

Development of DLM Ion Channels

One very fortunate aspect of studying ion currents in *Drosophila* flight muscles is the fact that the DLMs are accessible to voltage clamp during much of their development. A study in which the DLMs were voltage-clamped at various stages of development showed that there was a staggered developmental appearance of the currents (Salkoff and Wyman 1981a). This proved to be the key to isolating and analyzing the different current-carrying systems in the membrane. In the development of the DLM, myoblasts begin to proliferate about 6–8 hours after the formation of the pupal case by the larvae (pupariation) prior to metamorphosis. At about 55 hours postpupariation, the DLMs are large and easily voltage-clamped, but they do not yet contain any of the voltage-activated channels (Fig. 1). I_A begins to develop soon afterwards and by 72 hours it is mature. This fast K^+ current is similar to the A current first characterized in molluscan neurons (Hagiwara et al. 1961; Connor and Stevens 1971; Neher 1971). The channel is striking for the rapidity of its activation and for the fact that during a maintained voltage-step pulse it also inactivates rather rapidly. This current is especially easy to study in this system because, as shown here, its early developmental onset provides a means of isolating it from the other currents. Also, because it can be studied free of contamination by the other currents, it is subject to a wide variety of experimental manipulations (Salkoff and Wyman 1983).

After 72 hours, the delayed K^+ current (I_K) begins to develop. By 96 hours, this second K^+ current also matures and the membrane current response is a composite outward response of these two currents, the early transient I_A, and the maintained I_K. It will be shown that I_K can be isolated at this stage by using a mutant that deletes I_A from the membrane. I_K is similar to the well-known delayed outward current first identified in squid giant axons and present in most animal systems (Hodgkin and Huxley 1952; Aldrich et al. 1979; Hagiwara et al. 1981; Adams 1982). This K^+ current turns on much more slowly than I_A does and, in contrast to I_A, shows no sign of inactivating during voltage-clamp step pulses for as long as 1 second. Note at 96 hours, the large, slow current tails at the cessation of the step pulse (arrowhead, Fig. 1, 96 hr), which shows the rate the I_K channels close at the holding potential of −80 mV. These tails are another obvious characteristic that distinguishes I_K from I_A. (At 96 hr a small Ca^{++} current [I_{Ca}] may also be present when Ca^{++} is included in the saline.)

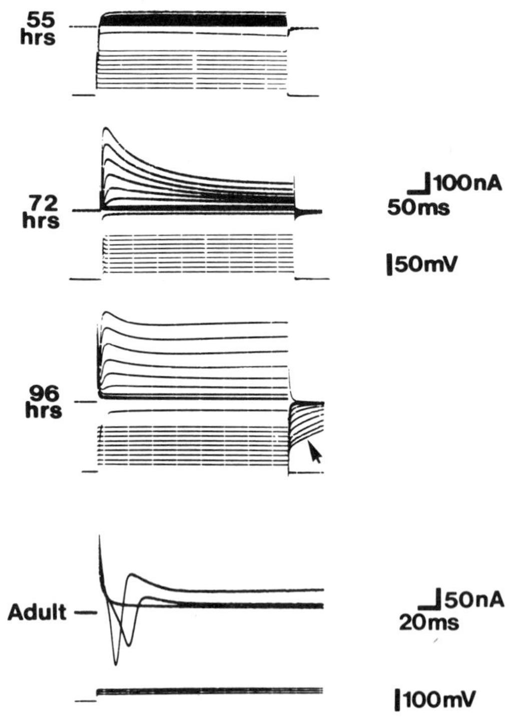

Figure 1. Development of DLM ion channels. DLM membrane ion currents (*upper traces*) are shown for different voltage-step pulses (*lower traces*). Hours are timed from puparium formation. The holding potential in this and all figures where voltage-clamp data are shown is −80 mV. Voltage traces for hyperpolarizing step pulses (to −180 mV) are not shown in these figures. At 72 hr, a large I_A is present. At 96 hr, both I_A and I_K are present. In adults, four large currents are present: I_A, I_K, I_{Ca}, and I_{Acd} (see text). The arrow at 96 hr indicates the large, slow tail currents that signal the presence of I_K.

Upon eclosion of the adult from the pupal case, the I_{Ca} matures. I_{Ca} is also a common current type found in most animals (Brehm and Eckert 1978; Adams 1982; Tsien 1983). As in other systems, the kinetic properties of this current vary markedly under different experimental conditions (Salkoff and Wyman 1983). The presence of I_{Ca}, as well as I_A and I_K, is reflected in the complex current response of the DLM membrane in adult flies. In Figure 2, the contributions of each of these currents to the adult composite current response are indicated. In the adult membrane, however, the fast, transient outward current bump (arrow I_A in B) is not entirely due to I_A. In addition to I_A, which is a voltage-activated current, there is also a fast, transient Ca^{++}-dependent K^+ current (I_{Acd}), which contributes to this early outward bump. The time course of activation of I_{Acd} almost exactly coincides with the activation of I_A. This Ca^{++}-dependent current is denoted I_{Acd} because of its similarity to I_A. (The development of I_{Acd} will be shown in Fig. 4.)

It is emphasized that none of the ion channels present in the *Drosophila* flight muscles are esoteric. Indeed, strikingly similar channels to all of these present in the *Drosophila* DLM have also been found in vertebrate neurons (Adams 1982; MacDermott and Weight 1982). Thus, *Drosophila* is a convenient and versatile system in which to study channels with almost universal properties.

Mutations at the *Shaker* Locus: Effects on I_A

At least three categories of mutant phenotypes, as assayed under voltage clamp, might be expected to result from alterations of a gene coding for a molecular component of an ion channel: (1) The current carried by the channel might be eliminated completely; (2) the current might be present but its biophysical properties changed; (3) the amount of current might be reduced but the current present has normal biophysical properties. Genetic alterations of the *Shaker* locus have been observed to produce all three types.

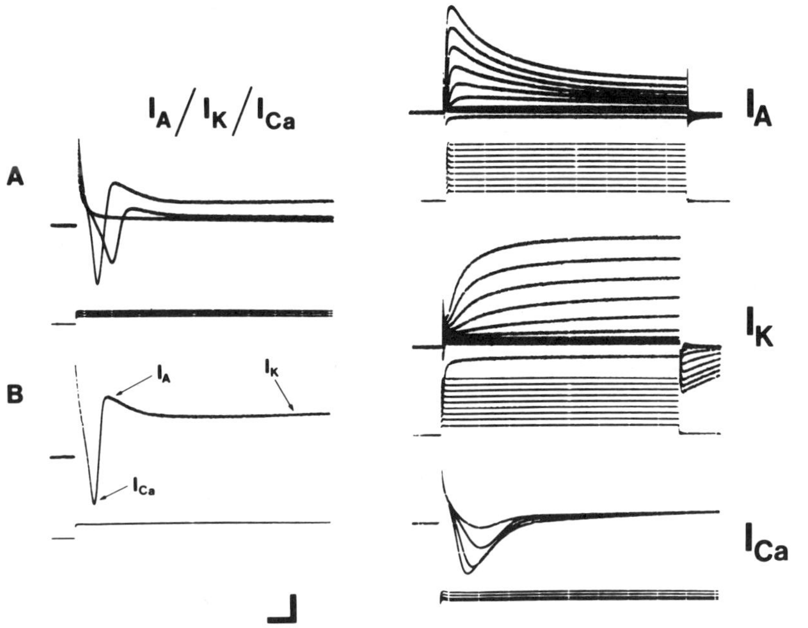

Figure 2. Voltage-activated ion currents in the DLM. The adult composite responses (*left*) are composed of three separate voltage-activated currents (*right*) and a fast transient Ca^{++}-activated current (I_{Acd}) that contributes to the early transient bump of outward current (arrow I_A in *B*). On the right, I_A is from a 72-hr pupa, I_K is from a 96-hr *Shaker* mutant that lacks I_A (see Fig. 3), and I_{Ca} was isolated in wild-type by blocking the K^+ currents with TEA^+. On the left, *A* shows the composite responses near threshold. Note that the transient inward and outward currents turn on at a more negative voltage than the maintained outward current, I_K. In *B* the prominent contributions of each current to the composite response are indicated by arrows. Note that the early outward bump arrowed from I_A is really due to both I_A and I_{Acd}. Current, voltage, and time mark: I_A and I_K = 100 nA, 50 mV, 50 msec; I_{Ca} = 50 nA, 20 mV, 20 msec; composite = 50 nA, 100 mV, 20 msec.

Mutations That Eliminate I_A

The most blatant phenotype, the total absence of current, would result from a mutation that simply wrecked the structural gene. Any number of mutagens that cause nonsense mutations or deletions of DNA could produce this result. For some channel types, this phenotype would never be seen because mutations of this kind would surely be lethal in the hemi- or homozygous condition. Fortunately, for the genetic analysis of ion channels in *Drosophila*, I_A appears to be an exception to this. The reason, as will be discussed in a later section, seems to be the existence of I_{Acd}, the similar but separate current-carrying system that may be redundant in purpose with I_A. As noted previously, this second current system, unlike I_A, which is voltage-activated, is activated by the large and rapid influx of Ca^{++} from outside the cell.

Figure 3 shows the phenotype under voltage clamp of one *Shaker* mutant that completely lacks I_A (numerous *Shaker* alleles have the same phenotype including Sh^{KS133}, Sh^{102}, and Sh^{KO120}). This phenotype is best observed during the pupal stage of development when, at 72 hours in wild-type flies, I_A is the only prominent channel in the membrane (Fig. 3A). In the homozygous mutant, however (Fig. 3E), I_A is totally absent and the membrane current response is passive. At 96 hours of development in wild type (Fig. 3B), both the early outward bump of current (I_A) as well as the maintained outward current (I_K) (with its characteristic current tails) are present. In the mutant at this developmental stage (Fig. 3F), the early outward current bump signaling the presence of I_A is totally absent.

The heterozygote animal carrying one wild-type and one mutant chromosome is perfectly intermediate between wild type and mutant at both developmental stages; half of I_A is present (Fig. 3C,D). The reduced current of the heterozygote was found to have normal kinetics and voltage-sensitive properties. Thus, in the heterozygote animal, the wild-type chromosome is apparently producing the normal *Shaker*+ product in an appropriate amount, whereas the mutant chromosome produces none of the normal product. Note that the continued presence of a normal amount of I_K is a convenient internal standard for assaying the amount of I_A at 96 hours of development. The mutant *Shaker* phenotype shown in Figure 3 illustrates the developmental, functional, and genetic independence of the two K^+ current systems, I_A and I_K. Regardless of whether I_A is present or absent, I_K develops on its normal schedule, has completely normal properties, and is thus unaffected by mutations of a gene that radically affect I_A.

The two current systems also have different sensitivities to pharmacological agents. Aminopyridine drugs completely block I_A but not I_K (Salkoff and Wyman 1981b). On the other hand, I_K is more readily blocked by TEA (Salkoff and Wyman 1983).

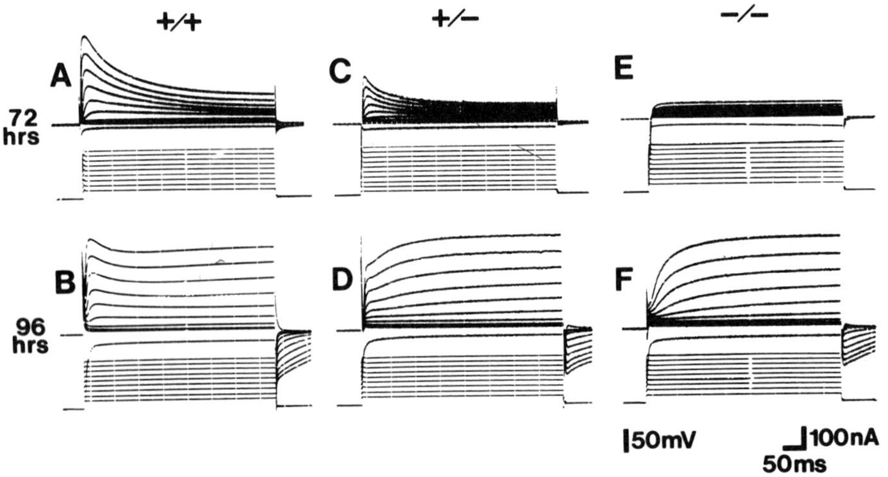

Figure 3. Comparison of DLM membrane properties during pupal development in wild-type (+/+), heterozygous wild-type/*Shaker* (+/−), and homozygous mutant *Shaker/Shaker* (−/−) pupae. The *Shaker* alleles Sh^{KS133}, Sh^{102}, and one stock of Sh^{KO120}, as shown here, all have this phenotype. (*A*) In wild-type pupae, I_A is mature at 72 hr of pupal development; (*B*) I_K is mature and is seen added to the I_A response at 96 hr of pupal development, just before the eclosion of the adult. (*C,D*) Membrane current response of heterozygous pupae. At 72 hr and 96 hr, only half the I_A response is present relative to wild type. The slow K^+ current is unaffected by the presence of the mutant gene. At 96 hr, the full current response of the slow K^+ current is present in all animals. (*E,F*) Membrane current response of homozygous mutant pupae. I_A is not present at either 72 hr or 96 hr of pupal development. Comparison of the rising phase of the current traces in *B, D,* and *F* shows, respectively, the presence of all, half, or none of I_A at 96 hr of pupal development. In *F* at 96 hr, the full current response of the slow K^+ current is present. In *A–F*, the leak current elicited by a hyperpolarizing pulse to − 180 mV is shown (below baseline trace) but the voltage record is not shown. (Adapted from Fig. 3 of Salkoff and Wyman 1981b.)

Development and Properties of I_{Acd}

The survival of mutants like Sh^{KS133}, which totally lack a major membrane ion current, appeared puzzling until further developmental changes in membrane properties were examined. Figure 4 shows the membrane response of a mutant Sh^{KS133} animal at the time of and immediately following eclosion from the pupal case. As expected, at the time of eclosion a fast transient outward current is absent; only the fast transient inward current, I_{Ca}, and the slow outward current, I_K, are present. In slightly older mutant adults, however, a small bump of early outward current is evident. Finally, several hours later in mature mutant adults, a large prominent early outward bump is present. Thus, in both wild-type and mutant mature adult animals, the composite current response is similar in that three prominent features are present: a fast, transient, inward response; a fast, transient, outward bump; and a maintained outward current. This similarity is shown in Figure 5 (top traces) for adult wild-type and Sh^{KS133} animals.

Could it be that the fast, transient, outward current of mutant adults is identical to that of wild type, but that somehow the *Shaker* mutations delay the developmental onset of the current? This was shown to be false by experiments like that in Figure 5 (middle traces) where Co^{++} was substituted for Ca^{++}. This, or any other treatment that eliminates Ca^{++} influx into the fiber, completely abolishes the transient outward current in mutant fibers, whereas the transient outward current is reduced but still present in wild type. Thus, mutant and

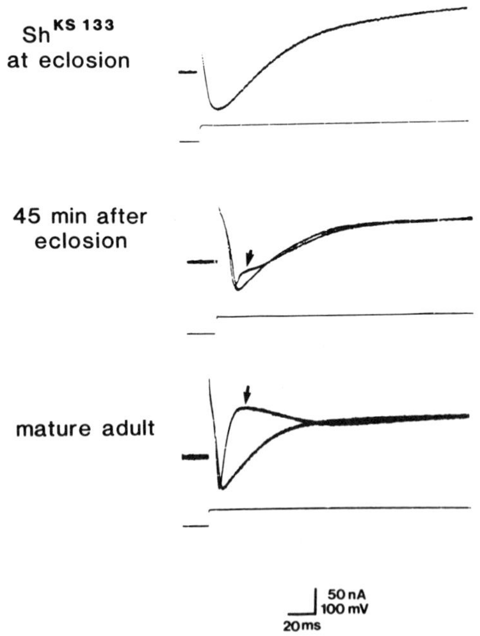

Figure 4. Development of I_{Acd} in a Sh^{KS133} fly. (*Top traces*) At the time of eclosion, only I_{Ca} and I_K were present. (*Middle traces*) At approximately 45 min after eclosion (different fly), a small I_{Acd} (arrow) was present. Two equal 200-msec clamp step pulses were applied to this preparation with an interpulse interval of 1.2 sec. Because of its slow recovery kinetics, I_{Acd} does not recover during this interval, although I_{Ca} and I_K do (Salkoff 1983). Thus, the difference between the two superimposed current traces reveals the amount of I_{Acd} in the fiber. (*Lower traces*) In a mature adult fly, a large I_{Acd} is present. Two clamp step pulses were also applied as in the traces immediately above.

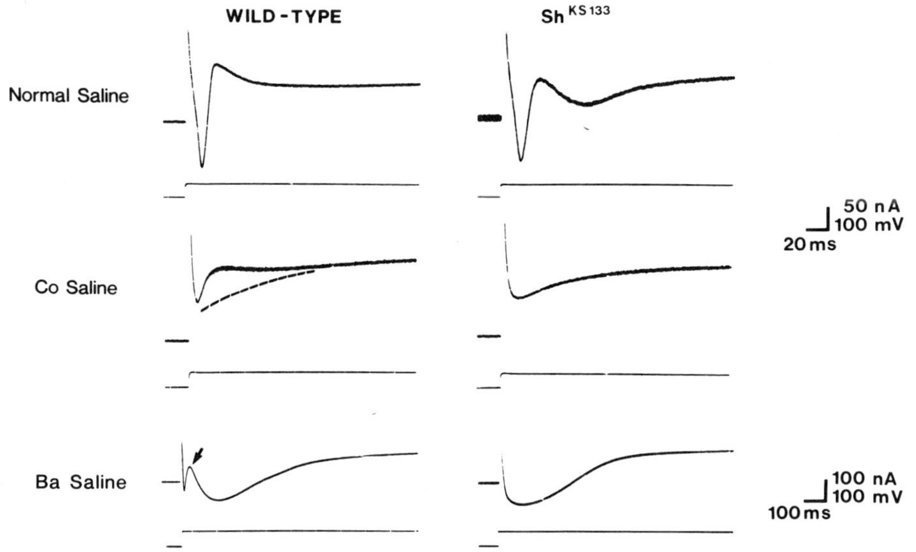

Figure 5. Transient outward currents in wild-type and Sh^{KS133} adult DLM fibers. (*Upper traces*) In normal saline, the composite current response of wild type appears similar to that of Sh^{KS133}. However, both I_A and I_{Acd} are present in wild type, but only I_{Acd} is present in Sh^{KS133}. (*Middle traces*) When Co^{++} is substituted for Ca^{++} in the saline, both I_{Acd} and I_{Ca} are eliminated, leaving I_A in wild type but no transient current whatsoever in Sh^{KS133}. Note that the traces are shifted upwards because of increased leak. The dotted line in wild type is idealized I_K in the absence of I_A. (*Lower traces*) When Ba^{++} (20 mM) is substituted for Ca^{++}, the inward current is prolonged in both wild type and mutant. I_{Acd} is abolished in both wild type and mutant. I_A, which is not abolished by Ba^{++}, is present in wild type but not in Sh^{KS133}. This is evidenced in wild type by the early interruption of the inward current by the rapid turn-on (arrow) and turn-off of transient outward current.

wild type currents differ markedly in their Ca^{++} dependence. Other major differences were also discovered. One of these, differential sensitivity to aminopyridine (AP) drugs, proved to be the key to understanding that there are actually two separate currents with similar kinetics in wild-type adults (Salkoff 1983). One of these, the voltage-sensitive I_A, develops as shown in Figure 1 during the pupal period. I_A is completely insensitive to the removal of Ca^{++} but is completely blocked by 3- or 4-AP (Salkoff and Wyman 1981b). The second is I_{Acd}, which becomes active after I_{Ca} matures in very early adults. I_{Acd} is unaffected by 3- or 4-AP. Thus, after treatment of a wild-type adult fiber with AP, the remaining transient, outward current in wild-type adults is equivalent to that of mutants. (Of course, I_{Acd} in mutants is also unaffected by AP.)

One final difference between I_A and I_{Acd} is shown in Figure 5 (lower traces), which shows the result when Ba^{++} is substituted for Ca^{++} in the saline. It is known that Ba^{++} can carry current through I_{Ca} channels (Brehm and Eckert 1978; Ashcroft and Stanfield 1981), but it does not activate Ca^{++}-dependent K^+ currents (Gorman and Hermann 1979). It was, likewise, found that in *Drosophila*, Ba^{++} would not activate I_{Acd}. Fortunately, Ba^{++} did not substantially block I_A in either wild-type pupae or adults. It, thus, provided another means of distinguishing I_A from I_{Acd}.

Ba^{++} carries a prolonged inward current in both wild-type and mutant fibers (Fig. 5, lower traces). However, with Ba^{++} carrying the inward current, a fast, transient, outward current is present only in wild type; an abrupt turn-on and turn-off of the current is seen early in the pulse in wild-type fibers (arrow) but is absent in mutant fibers. This transient current in wild type, which persists in Ba^{++}, has all the properties of I_A. For example, the current is completely abolished by the addition of AP. Thus, mutants such as Sh^{KS133}, which competely lack I_A, were useful tools for revealing the presence of a second current-carrying system, I_{Acd}. (Note that a current like I_{Acd} is also probably present in bullfrog sympathetic neurons; MacDermott and Weight 1982.) The two channels, I_A and I_{Acd}, although having similar activation kinetics, seem to be completely independent systems as evidenced by their different development times, different requirements for Ca^{++}, different sensitivities to AP and Ba^{++}, and different kinetics of recovery from activation (Salkoff 1983). Also, the fact that *Shaker* mutations such as Sh^{KS133} that completely eliminate I_A have no effect on I_{Acd} supports this conclusion.

A Mutant That Changes the Kinetics of I_A

The second type of mutant phenotype that could result from the genetic alteration of an ion channel molecule is a change in the kinetic properties of the current carried by the channel. The Sh^5 mutation, which also maps genetically to the *Shaker* locus of the X chromosome, is this type of mutation. Sh^5 could well affect the same gene as Sh^{KS133} (which eliminates I_A altogether); the two mutations could not be separated from each other by genetic recombination (Salkoff 1983). Alternatively, the two mutations could be in closely linked genes. However, unlike Sh^{KS133}, which completely deletes I_A from the pupal membrane, I_A is present in Sh^5 animals. Its kinetic properties are changed though; Sh^5 abnormally speeds up the rate of inactivation of the current. This is best observed in the DLMs at 72 hours of pupal development where I_A is isolated from the other currents. Figure 6 (upper traces) shows wild-type and mutant I_A at this development stage. The decay of the current during maintained voltage-clamp step pulses in Sh^5 animals is clearly more rapid than in wild-type flies. This obvious difference between mutant and wild type, however, is obscured at later developmental stages by the maturation of I_K (Fig. 6, lower traces). Notice that at 96 hours of development, the difference between wild type and mutant is only a more pronounced dip (arrow) between the early peak of outward current and the onset of I_K.

In addition to the accelerated rate of inactivation, the Sh^5 mutation also confers a second defect on I_A; the rate of recovery from inactivation is accelerated. For both of these abnormal kinetic properties, the relative differences between mutant and wild type vary with voltage; the differences are greatest at approximately −55 mV and become smaller at higher and lower voltages. Figure 7a plots the rates of inactivation and recovery in wild type (upper curve) and mutant (lower curve) over the voltage range of −80 mV to −20 mV. The rate of inactivation (filled symbols) is measured in a two-pulse experiment as shown in inset A. Recovery from inactivation (open symbols) is measured in a two-pulse experiment as shown in inset B. For normal wild-type channels, the T_h curve, which describes the relation between the rates of these two properties and voltage, is a steep bell-shaped relationship. In contrast, the shape of the curve for mutant Sh^5 channels is profoundly flattened, showing that, for these properties, the Sh^5 channels are far more insensitive to voltage. One striking facet of this result is that the Sh^5 mutation confers properties to I_A that are characteristic of this current in a different species. In some animal species, as in *Helix* (Neher 1971) or *Renilla* (Hagiwara et al. 1981), the T_h curve for I_A is also steeply bell-shaped as in wild-type *Drosophila.* In others, however, such as *Anisodoris* (Connor and Stevens 1971; C.F. Stevens, pers. comm.), the curve is flat as in the Sh^5 mutant. Thus, an alteration of a single gene in a single species can account for major differences in kinetic properties of I_A similar to those differences observed between species. Results such as these bear directly on the questions of genetic and evolutionary relationships among ion channel types having different kinetic properties.

Although the kinetics of inactivation and recovery are profoundly affected by the Sh^5 mutation, other properties have no obvious alteration. Wu and Ganetzky (1983) suggest that in larval muscles the Sh^5 I_A activation curve is altered from wild type. Figure 7, b and c, shows two of these properties measured in 72-hour pupal muscles. Figure 7b shows the similarity in voltage sensitivity of steady-state inactivation for wild type and mutant. Figure 7c shows the similarity in the voltage sensitivity of activation. Although Figure 7c suggests that the amount of current in mutant DLM pupal muscles does not differ from wild type, it was found that there was somewhat less I_A relative to wild type in the older adult DLM and in the body wall muscles of larvae (L. Salkoff, unpubl.; C.-F. Wu, pers. comm.). Nevertheless, the specificity of the mutant effect is suf-

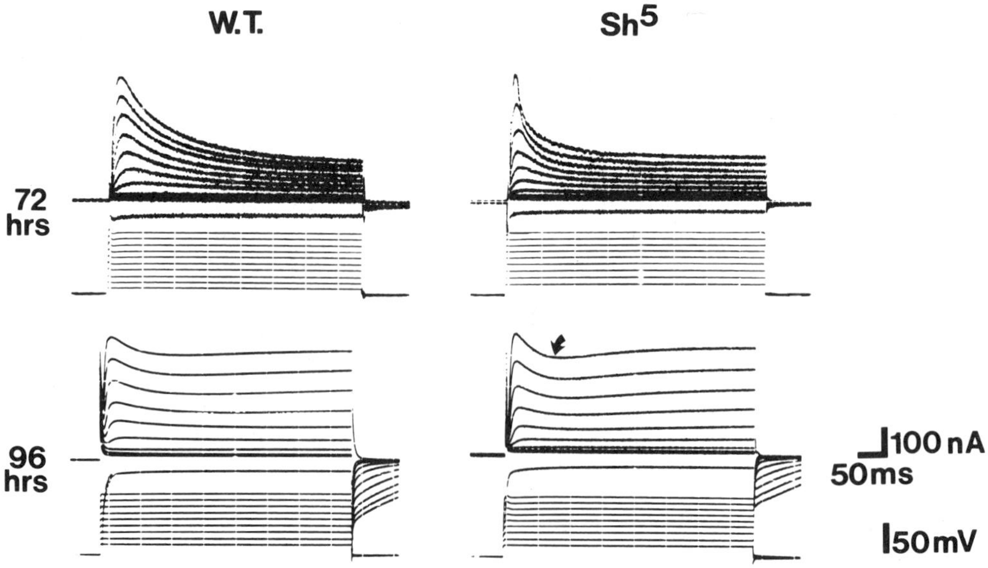

Figure 6. Comparison of I_A responses to voltage-clamp step pulses in normal (W.T.) and mutant Sh^5 animals. Each of the current traces shows the membrane responses at a different voltage level. (The voltage records for hyperpolarizing pulses to −180 mV are not shown.) At all voltages, I_A inactivates more rapidly in Sh^5 animals than in normal animals. This is obvious at the 72-hr stage of development but becomes obscured by the appearance of I_K at the 96-hr stage of development. However, at 96 hr in Sh^5 flies, there is a more pronounced dip (arrow) between the peak of I_A and the onset of I_K. Although the amount of I_A in Sh^5 DLM fibers is normal during pupal development, it is somewhat lower than normal in older adult fibers and in body wall muscle fibers of third instar larvae. (Adapted from Fig. 1 of Salkoff and Wyman 1981b.)

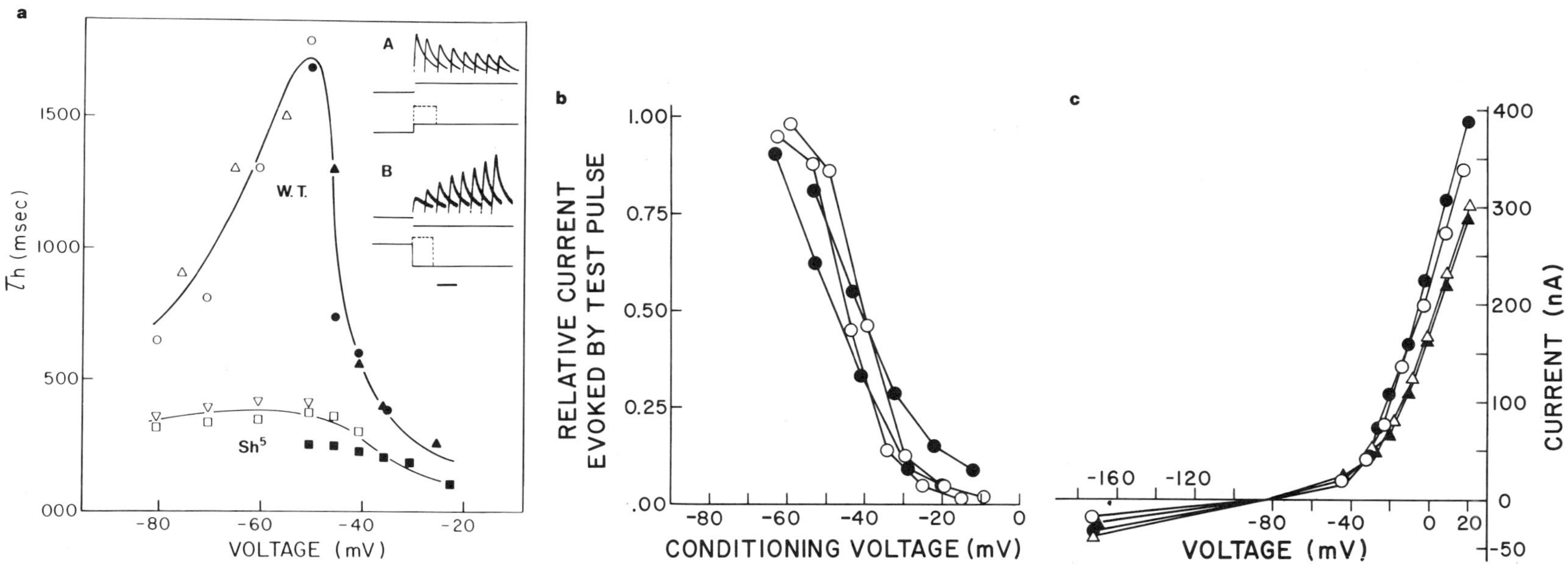

Figure 7. Biophysical properties of I_A: Wild-type and mutant (Sh^5) channels. (*a*) The time course of development and removal of inactivation (T_h), plotted on the ordinate, versus conditioning membrane voltage, plotted on the abscissa. The upper curve is from wild type. The lower curve is from Sh^5. Each symbol shape is for a different animal. (▲, ●, ■) Time constants of exponential inactivation onset, as measured by a two-pulse experiment as in *A*. (△, ○, □) Time constants of recovery from inactivation, as measured in *B*. *A* and *B* show superimposed current traces. For *A* and *B*, voltage traces are drawn below the actual currents for clarity. Only one test pulse (----) is shown for each. The test pulse was moved to the right in each subsequent trace. Test pulses were to +20 mV. Time mark, 200 msec. (*b*) Steady-state inactivation curves for I_A measured at the 72-hr stage of pupal development. (○) Wild type; (●) Sh^5. The cells were stepped from a holding potential of −80 mV to the conditioning voltage (abscissa) for 1.8 sec. The currents elicited by the conditioning voltage closely approached a steady state of inactivation within this time. The amount of current still evokable was then measured by a further depolarization (test pulse) to +20 mV. Measured currents were normalized to the current elicited by a pulse from −80 mV to +20 mV. Peak current is measured on the ordinate and conditioning voltage on the abscissa. (*c*) Current-voltage relations for wild type (○, △) and Sh^5 (●, ▲); 72-hr pupae were used. Cells were stepped from the −80 mV holding voltage to the test voltage plotted on the abscissa. I_A was measured at its peak. The leak has not been subtracted from data. (The triangles are data from smaller animals.)

ficiently great to support a hypothesis that the Sh^5 mutation primarily affects a discrete part of a channel molecule involved in the inactivation process.

One of the possible ways of getting more information about the molecular nature of the *Shaker* product is to cross mutant and wild-type animals and, thus, create a heterozygote that produces both mutant and wild-type gene products. If the resulting current, as measured in the heterozygote, behaves as if it were the algebraic sum of two separate current systems, then it might be concluded that the heterozygote membrane has either normal or mutant channels (but not hybrids) and, hence, that there is only one *Shaker* product per I_A channel. The results of experiments of this type do indeed suggest that I_A measured in heterozygote animals may behave as if it were the algebraic sum of two independent I_A systems, one with fast inactivation kinetics and one with slower (wild-type) inactivation kinetics. This can best be seen in an experiment where the mutant Sh^5 of *D. melanogaster* (*dm*) was mated to a wild-type *Drosophila simulans (ds)*, a sibling species to *dm* (Fig. 8a). Wild-type *ds* was chosen over wild-type *dm* because it was found to have even slower inactivation kinetics. Thus, the heterozygote Sh^5 *dm*/wild-type *ds* has a phenotype that is obviously intermediate and easier to analyze; the rate of decay of the current is faster than in wild-type *ds* but slower than in Sh^5 *dm*. When records such as those in Figure 8a are expanded and analyzed (Fig. 8b), it is seen that the current in Sh^5 *dm* decays as a single fast exponential and the current in wild-type *ds* decays as a single slow exponential. The heterozygote current, however, may decay as the sum of the two exponentials, the fast and the slow one. Thus, there may be two distinct types of I_A channels in the heterozygote membrane, the mutant *dm* channels and the wild-type *ds* channels, which implies that there is only one *Shaker* component per channel. These preliminary results, however, will have to be followed

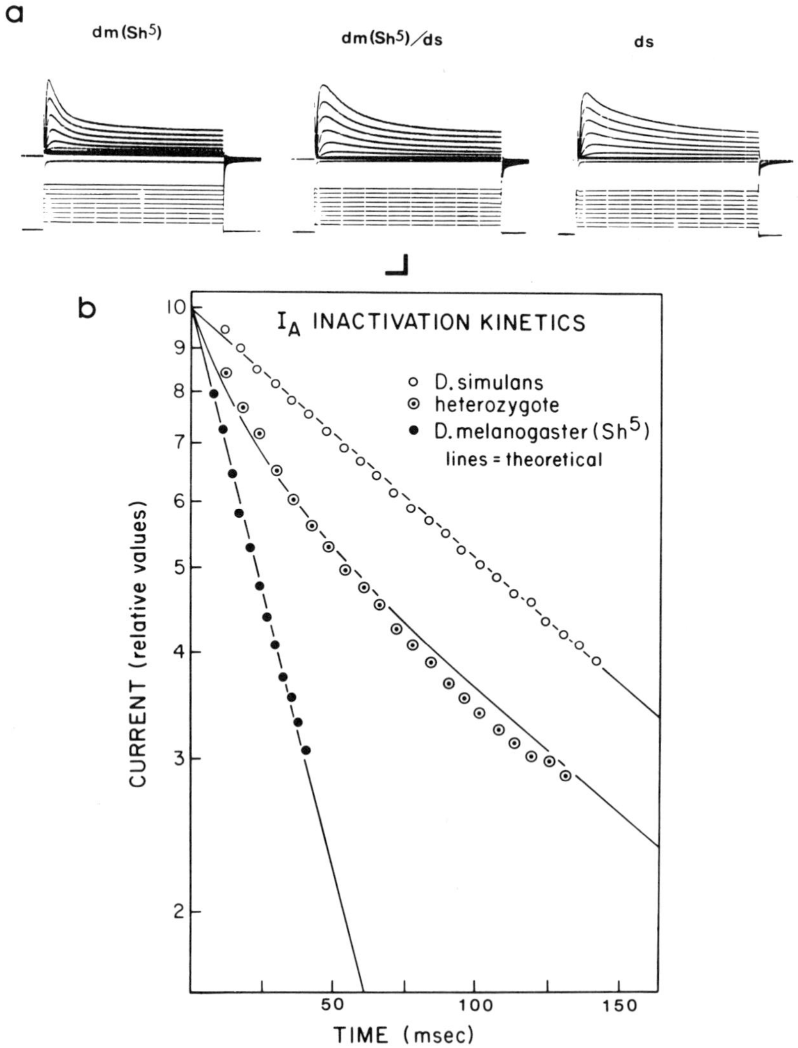

Figure 8. I_A inactivation rates for the Sh^5 mutant of *D. melanogaster* (dm[Sh^5]), wild-type *D. simulans* (ds), and the heterozygote (dm[Sh^5]/ds). (*a*) I_A currents evoked by various step pulses from a holding potential of −80 mV. (*b*) Semilog plot of current decay for each current as indicated by the legend on the graph. The rates of I_A decay during a pulse step to +20 mV at 4°C were measured for each genotype.

up with further experimentation such as patch-clamp experiments.

A Mutant Effect Reducing the Amount of I_A

The third type of mutant effect that would give evidence for the existence of a structural gene at the *Shaker* locus is a limitation in the amount of normal gene product produced. Such a phenotype is commonly produced as a result of breaks or rearrangements of chromosomal material in the vicinity of a structural gene (Baker 1968). This "position effect" need not involve the alteration of the gene itself, but only its chromosomal environment. In the case of a gene coding for a component of I_A, a position effect would be detectable as a reduced amount of an otherwise normal I_A.

To search for this phenotype, X-Y translocation chromosomes were obtained that had translocation breakpoints in the *Shaker* region of the X chromosome. These chromosomes were originally synthesized for a purpose outside of neurobiology (Stewart and Merriam 1973). Three of these chromosomes are represented in Figure 9 with the approximate sites of their breakpoints shown on the salivary chromosome map. Although these chromosomes are X-Y translocations, the Y chromosome fragments attached to the X-chromosome breakpoints are not represented. Male animals (which have only a single X chromosome) survive when they carry both parts of the broken X chromosome. The results of voltage-clamp experiments on three animals carrying different breakpoints are shown below each chromosome. The leftmost break, designated B55, reduces the amount of I_A in the membrane by approximately 50%; the middle break (W32) drastically reduces I_A; and the rightmost break (V7) has no effect on I_A in the flight muscle. Significantly, the reduced current produced by B55 has normal kinetics and voltage-sensitive properties. Thus, this reduced current appears to be caused by a position effect where the lesion is outside of the coding region of the gene. The break W32 could be either within the coding region of the gene or close enough to cause a very severe position effect.

A Deficiency at the *Shaker* Locus

A possible location of the coding region of the gene relative to the breakpoints was determined by examining the phenotypes of flies having the reciprocal parts of the broken chromosomes B55 and W32. It was found that the proximal part of the B55 chromosome (the short part that is attached to the centromere represented by a circle) conferred the B55 phenotype (a reduced I_A). Hence, flies carrying this proximal B55 segment had approximately 50% of I_A in the muscle fiber regardless of whether the distal part of W32 or B55 was present. Conversely, the W32 proximal segment paired with either its own distal segment or the B55 distal segment did not produce any I_A. Thus, the coding part

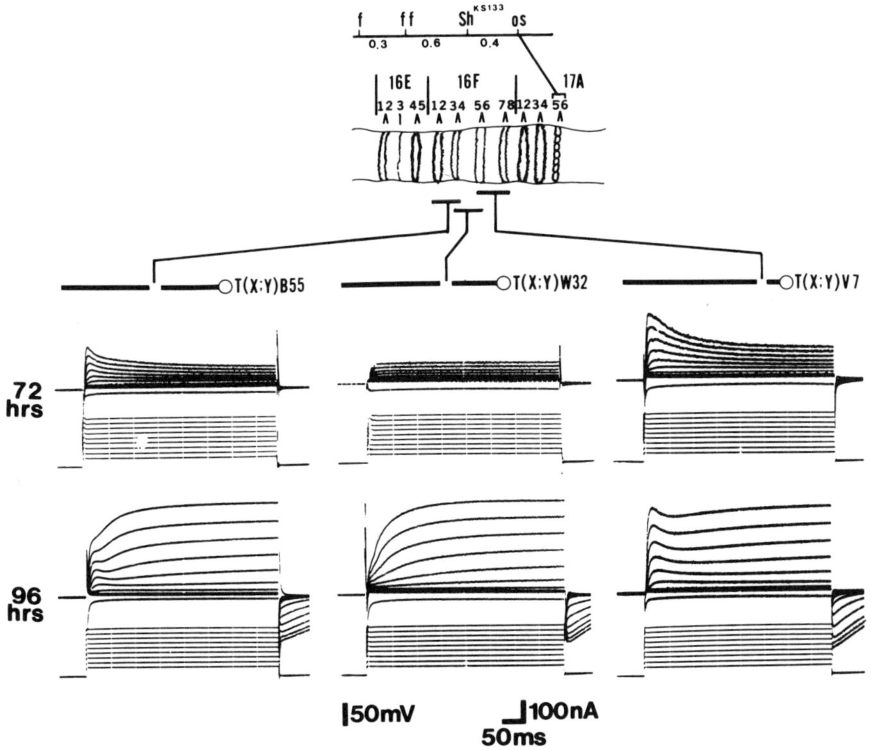

Figure 9. I_A phenotype conferred by X-chromosome breakpoints in the *Shaker* region. The approximate location of each breakpoint is indicated on the salivary chromosome map. See text for details. The positions of the mutations indicated at the top were established by Tanouye et al. (1981).

of the gene may be to the right of the B55 breakpoint on the B55 proximal segment. To create a fly stock carrying a small deletion of the *Shaker* locus, the appropriate flies were crossed to obtain males carrying the W32 proximal and B55 distal segments. These male flies, which are deficient for the region between the B55 and W32 breakpoints, although viable, have no I_A whatsoever, and shake more violently than either the B55 or W32 stocks on their own. Thus, it is likely that this deficiency deletes a part or all of the coding region of the gene.

It is, indeed, a rare and fortunate advantage to have a hemizygous viable deficiency of a gene; it will be most useful in the molecular characterization of the *Shaker* locus. In molecular studies, the DNA of this deficiency stock will be useful for localizing DNA sequences that are thought to be part of the *Shaker* region; a DNA probe from a genomic library can be tested to see if it falls in the region of the deficiency by a test hybridization to a Southern blot of DNA made from the deficiency stock. A probe that falls entirely into the deficiency region will fail to hybridize to that blot.

Duplication of the *Shaker* Locus

Although it is reasonable to expect that mutations of an ion channel gene would diminish or abolish the ion current carried by that channel, the phenotype resulting from the introduction of an extra functional gene copy is less predictable. To investigate this question, male animals carrying a small duplicated segment of the X chromosome bearing the *Shaker* locus were constructed by the appropriate crosses. Note that in the normal condition, a male fly (which has only one X chromosome) has just as much I_A as a female fly (which has two X chromosomes). This is the result of an equalization of gene expression for X-linked genes in males and females that comes about as a result of "dosage compensation" (Maroni and Lucchesi 1980). The result of such compensation is that each X chromosome in a female produces half of the total gene product produced by the single X chromosome in a male. However, it is possible to "fool" dosage compensation by including a short duplicated segment of the X chromosome in the genome. In male flies, the genes within the region of the small duplicated segment will be expressed twice as efficiently as the nonduplicated X-linked genes. Using a small chromosome segment that includes the *Shaker* locus derived from B^s (Lindsley and Grell 1968), male flies were constructed that carried a normal wild-type chromosome together with the duplicated segment. These animals were then voltage-clamped. In the limit, these flies might be expected to have twice the amount of I_A that normal males have. However, it was found that these animals had only the normal wild-type amount of I_A (Fig. 10). It appeared from this result that there was an upper limit to the number of I_A channels that the membrane could accommodate; even a single Sh^+ gene in a male could saturate the membrane. To be certain that the duplicated B^s segment was producing 100% of the Sh^+ product, male flies were constructed having the B55/W32 deletion of *Shaker* together with the duplicated B^s segment. These flies also had 100% of I_A, showing that the duplicated B^s segment produced the *Shaker*$^+$ product in a full amount.

What is the explanation for this apparent saturation phenomenon? One intriguing possibility is that the I_A channel is a heteromultimer consisting of two or more gene products from widely scattered genes. If this were the case, the duplication of any single gene would not result in the duplication of the total amount of I_A. (Note, however, that the diminution or abolition of the current by a mutation of a single gene is consistent with this theory.) Of course, it is also possible that the amount of I_A in the membrane is regulated in some very different and, as yet, unknown way.

One final note on duplication experiments: The addition of the Sh^+ duplication to male animals carrying various mutations of the *Shaker* locus appears to be a way of assaying whether or not a mutant gene is producing a defective product that competes with the Sh^+ product. The duplication paired with some mutations (e.g., Sh^{102}) shows 100% of I_A, indicating no competition from a defective product. However, Sh^{KS133} paired with the duplication produces only 50% of I_A, indicating equal competition between mutant and wild-type products. It is tempting to speculate on the site of such competition. Could it be at the level of available membrane sites? (This apparent 1:1 competition between products could be a further indication that there is a single *Sh* product per channel.)

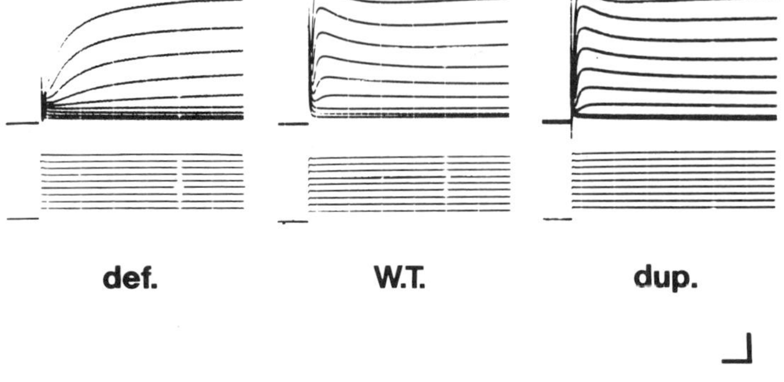

Figure 10. A duplication of the *Shaker*$^+$ gene does not increase the amount of I_A present in a fiber. (def.) Male fly carrying a deficiency at the *Shaker* locus (see text); (W.T.) normal wild-type male; (dup.) male fly carrying a wild-type chromosome and an additional chromosomal segment containing the *Shaker*$^+$ gene (see text). Current, voltage, time bar, 100 nA, 50 mV, 50 msec.

SUMMARY

These genetic and voltage-clamp studies have produced evidence that strongly supports the hypothesis that the *Shaker* locus codes for at least one molecular component of the I_A channel. Mutations at the *Shaker* locus have effects that are specific to I_A, and thus a general membrane component is not produced by this locus. Regarding the function of the *Shaker* product itself, there is evidence that it is at least involved in the inactivation and recovery process of the channel. The evidence so far is also consistent with there being a single *Shaker* product per channel. Regarding its genetic and developmental properties, I_A is apparently completely independent of the other major voltage-activated K^+ channel in the membrane, I_K.

However, only after molecular information becomes available will it become clear whether or not these experiments have revealed clues about the molecular nature of these channels. Ultimately, voltage-clamp experiments, even combined with resourceful genetic tools, can only reveal circumstantial information about the molecular structure of channels. It is hoped, however, that these studies have stimulated interest in the *Shaker* locus and that the way has been well prepared for molecular studies of the I_A channel. The actual molecular characterization of the *Shaker* locus is already under way (cf. Jan et al., this volume; M. Tanouye, pers. comm.).

ACKNOWLEDGMENTS

I express my heartfelt gratitude and appreciation to Bob Wyman for his extensive support and encouragement and to Chuck Stevens for lending me voltage-clamp equipment that he knew had a charmed life. I also thank Rick Aldrich for many helpful suggestions. This work was supported in part by a postdoctoral fellowship award to L.S. from the Muscular Dystrophy Association of America.

REFERENCES

Adams, P. 1982. Voltage-dependent conductances of vertebrate neurones. *Trends Neurosci.* **5:** 116.

Aldrich, R.W., Jr., P.A. Getting, and S.H. Thompson. 1979. Inactivation of delayed outward current in molluscan neurone somata. *J. Physiol.* **291:** 507.

Ashcroft, F.M. and P.R. Stanfield. 1981. Calcium dependence of the inactivation of calcium currents in skeletal muscle fibers of an insect. *Science* **213:** 224.

Baker, W.K. 1968. Position-effect variegation. *Adv. Genet.* **14:** 133.

Brehm, P. and R. Eckert. 1978. Calcium entry leads to inactivation of calcium channel in *Paramecium. Science* **202:** 1203.

Connor, J.A. and C.F. Stevens. 1971. Voltage clamp studies of a transient outward current in gastropod neural somata. *J. Physiol.* **213:** 21.

Gorman, A.L.F. and A. Hermann. 1979. Internal effects of divalent cations on potassium permeability in molluscan neurones. *J. Physiol.* **296:** 393.

Hagiwara, S., K. Kusano, and N. Saito. 1961. Membrane changes of *Onchidium* nerve cell in potassium-rich media. *J. Physiol.* **155:** 470.

Hagiwara, S., S. Yoshida, and M. Yoshii. 1981. Transient and delayed potassium currents in the egg cell membrane of the coelenterate, *Renilla koellikeri. J. Physiol.* **318:** 123.

Hodgkin, A.L. and A.F. Huxley. 1952. A quantitative description of membrane current and its application to conduction and excitation in nerve. *J. Physiol.* **117:** 500.

Jan, L.Y. and Y.N. Jan. 1976a. Properties of the larval neuromuscular junction in *Drosophila melanogaster. J. Physiol.* **262:** 189.

——. 1976b. L-glutamate as an excitatory transmitter at the *Drosophila* larval neuromuscular junction. *J. Physiol.* **262:** 215.

Jan, Y.N., L.Y. Jan, and M.J. Dennis. 1977. Two mutations of synaptic transmission in *Drosophila. Proc. R. Soc. Lond. B.* **198:** 87.

Lindsley, D.L. and E.H. Grell. 1968. Genetic variations of *Drosophila melanogaster. Carnegie Inst. Wash. Publ.* **627:** 382.

MacDermott, A.B. and F.F. Weight. 1982. Action potential repolarization may involve a transient, Ca^{++}-sensitive outward current in a vertebrate neurone. *Nature* **300:** 185.

Maroni, G. and J.C. Lucchesi. 1980. X-chromosome transcription in *Drosophila. Chromosoma* **77:** 253.

Neher, E. 1971. Two fast transient current components during voltage clamp on snail neurons. *J. Gen. Physiol.* **58:** 36.

Salkoff, L. 1983. *Drosophila* mutants reveal two components of fast outward current. *Nature* **302:** 249.

Salkoff, L. and R. Wyman. 1980. Facilitation of membrane electrical excitability in *Drosophila. Proc. Natl. Acad. Sci.* **77:** 6216.

——. 1981a. Outward currents in developing *Drosophila* flight muscle. *Science* **212:** 461.

——. 1981b. Genetic modifications of potassium channels in *Drosophila Shaker* mutants. *Nature* **293:** 228.

——. 1983. Ion currents in *Drosophila* flight muscle. *J. Physiol.* **337:** 687.

Stewart, B. and J.R. Merriam. 1973. Segmental aneuploidy of the X-chromosome. *Drosophila Inform. Serv.* **50:** 167.

Tanouye, M.A., A. Ferrus, and S.C. Fujita. 1981. Abnormal action potentials associated with the *Shaker* locus of *Drosophila. Proc. Natl. Acad. Sci.* **78:** 6548.

Tsien, R.W. 1983. Calcium channels in excitable cell membranes. *Annu. Rev. Physiol.* **45:** 341.

Wu, C.-F. and B. Ganetzky. 1983. Properties of potassium channels altered by mutations of two genes in *Drosophila. Biophys. J.* (in press).

Wu, C.-F., B. Ganetzky, F.N. Haugland, and A.-X. Liu. 1983. Potassium currents in *Drosophila*: Different components affected by mutations of two genes. *Science* **220:** 1076.

Mutating a Gene for a Potassium Channel by Hybrid Dysgenesis: An Approach to the Cloning of the *Shaker* Locus in *Drosophila*

L.Y. JAN, S. BARBEL, L. TIMPE, C. LAFFER, L. SALKOFF, P. O'FARRELL, AND Y.N. JAN
Departments of Physiology and Biochemistry, University of California, San Francisco, California 94143

Several different types of voltage-sensitive K^+ channels have been characterized in recent biophysical studies (Adams et al. 1980). Different neurons have been shown to have a rather different makeup of these K^+ channels, which to a large extent determines the firing pattern and shape of action potentials in neurons (Byrne 1980). Since the amount of transmitter released from a neuron can be drastically altered by slight alterations of the shape or duration of the action potential (Klein and Kandel 1980; Llinas et al. 1982), conceivably, K^+ channels may play an important role in the control of synaptic efficacy. In fact, some K^+ channels are known to be influenced not only by membrane potential, but also by transmitter molecules, such as acetylcholine or serotonin, and intracellular messengers, such as Ca^{++} or cAMP (see, e.g., Adams et al. 1982; DePeyer et al. 1982). For instance, the mechanism underlying sensitization in *Aplysia* appears to involve the inactivation of a particular K^+ channel by serotonin (Klein et al. 1982; Siegelbaum et al. 1982). To understand better how the expression of different K^+ channels is regulated in neurons and how the activity of individual K^+ channels may be controlled by various chemical messengers, one needs to characterize K^+ channels in molecular terms. Unfortunately, little information is available concerning the structure of K^+ channels or their genes, because there are no antibodies against K^+ channels and, with the exception of Ca^{++}-activated K^+ channels (Hugues et al. 1982), no toxins with high affinity have been found for K^+ channels.

One possible approach to a molecular study of K^+ channels is to clone the genes for K^+ channels. This involves the traditional genetic approach for identifying genes that code for protein molecules and can be done with animal systems with well-developed genetic tools, such as *Drosophila*. The first step in this approach is the induction of a mutation that gives a clear indication of affecting a K^+ channel. The mutant gene shown to alter normal K^+ conductance by physiological tests can be mapped genetically to a particular locus. Once the locus is known, the gene can then be cloned by a variety of techniques. In *Drosophila*, cloning may be done in a brute-force manner by "walking" along the chromosome. Alternatively, transposable elements may be used to "tag" the gene and facilitate cloning (Bingham et al. 1981; Searles et al. 1982; Modolell et al. 1983). Transposable elements are DNA fragments that apparently can generate and insert copies of themselves elsewhere in the genome, often causing mutations at the sites of insertion (*Cold Spring Harbor Symp. Quant. Biol.*, vol. 45 [1981]). The frequency of transposition of a particular transposable element, the p factor, is drastically increased by mating appropriate strains of fruit flies (Engels 1981; Rubin et al. 1982) (see Background below). This "hybrid dysgenesis" phenomenon may then be used systematically for the cloning of genes in *Drosophila*, as outlined in Figure 1.

In this paper we first review evidence for the *Shaker* locus in *Drosophila* being the site for the structural gene of a K^+ channel, and we discuss the strategies to be used for the molecular cloning of the *Shaker* locus. We then describe the recently isolated hybrid-dsygenesis-induced *Shaker* mutants (Sh^{SBn}, $n = 1, 2, \ldots, 13$; isolated by S.B.), which may be useful in the initial cloning and subsequent analysis of DNA from the *Shaker* locus.

METHODS

Physiological Experiments

Larval nerve-muscle preparation. The physiology of the normal *Drosophila* larval neuromuscular junction was described elsewhere (Jan and Jan 1976a,b). For physiological experiments, late third instar larvae were pinned out in saline and cut along the dorsal midline. The internal organs were removed, and the nerves innervating the body-wall muscles were cut near the ventral ganglion and taken up in suction electrodes for stimulation. The saline solution contained 128 mM NaCl, 2 mM KCl, 4 mM $MgCl_2$, 35.5 mM sucrose, and $CaCl_2$ as specified in each experiment. Nomarski optics were used during recording. All experiments were done at room temperature (23°–24°C).

Pupal flight-muscle preparation. The techniques for voltage clamping the *Drosophila* dorsal longitudinal muscles are described by Salkoff (this volume). The saline used contained 128 mM NaCl, 4.8 mM KCl, 1.8 mM $CaCl_2$, and 0.2 mM KH_2PO_4/K_2HPO_4 (pH 6.8). All voltage-clamp experiments were done at 4°C.

Hybrid-dysgenesis-induced *Shaker* Mutants

Isolation. The scheme as outlined in Figure 2 was used in the generation of hybrid-dysgenesis-induced mutations. The first two crosses are hybrid-dysgenic, as males carrying the p factors are mated with females of M cytotype (cf. Background below). The P-strain flies used are derived from the π_2 strain provided by

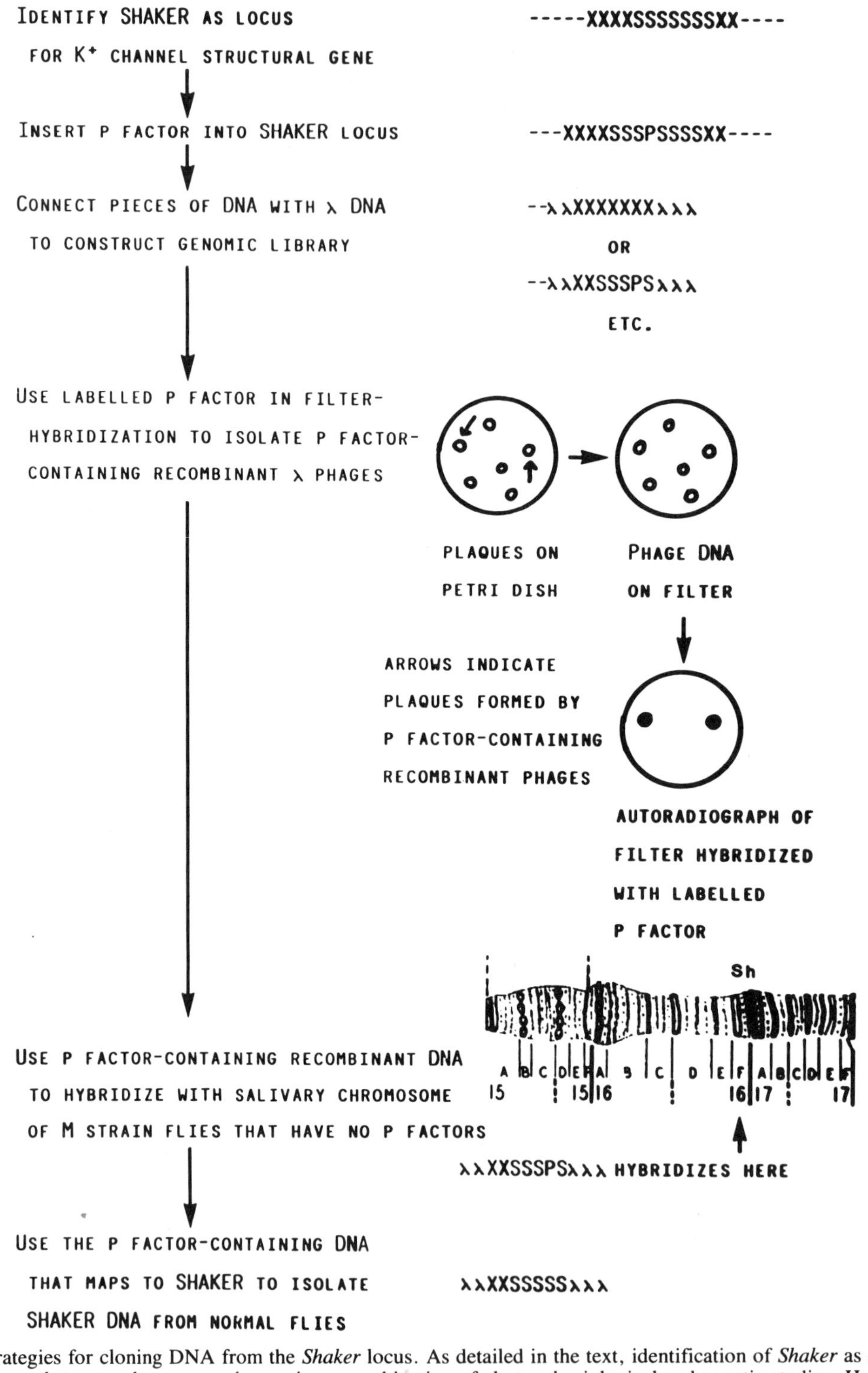

Figure 1. Strategies for cloning DNA from the *Shaker* locus. As detailed in the text, identification of *Shaker* as a possible locus for the K^+-channel structural gene was done using a combination of electrophysiological and genetic studies. Hybrid dsygenesis was used to induce the insertion of p factors into the *Shaker* locus. Techniques have been well worked out for construction of genomic libraries (Maniatis et al. 1978), filter hybridization (Benton and Davis 1977), and in situ hybridization with salivary chromosomes (Pardue and Gall 1975). P represents a transposable p factor; S represents sequences within the *Shaker* locus, X represents flanking sequences on the X chromosome, and λ represents DNA of the bacteriophage λ.

W. Engels (University of Wisconsin). In the screening for the *Shaker* mutants, the F_3 male and female progeny were anesthetized with ether and scored for their shaking behavior. Individual shaking females were mated to FM7 males of π_2 strain, so that half of the male progeny would carry the *forked*-marked X chromosome bearing the induced *Shaker* mutation (Sh^{SBn}). The shaking F_3 progeny selected as males were mated to *yf*XX virgin females of the π_2 strain, and thus all of their male progeny were of the same mutant genotype: fSh^{SBn}. Each fSh^{SBn} male of the progeny was then pair-mated to $yf\widehat{XX}$ (P) virgin females. This was repeated for a few generations until the Sh^{SBn} mutation became stabilized in the P cytotype of the π_2 strain.

Genetic mapping by recombination. The location of the Sh^{SBn} mutations on the X chromosome was de-

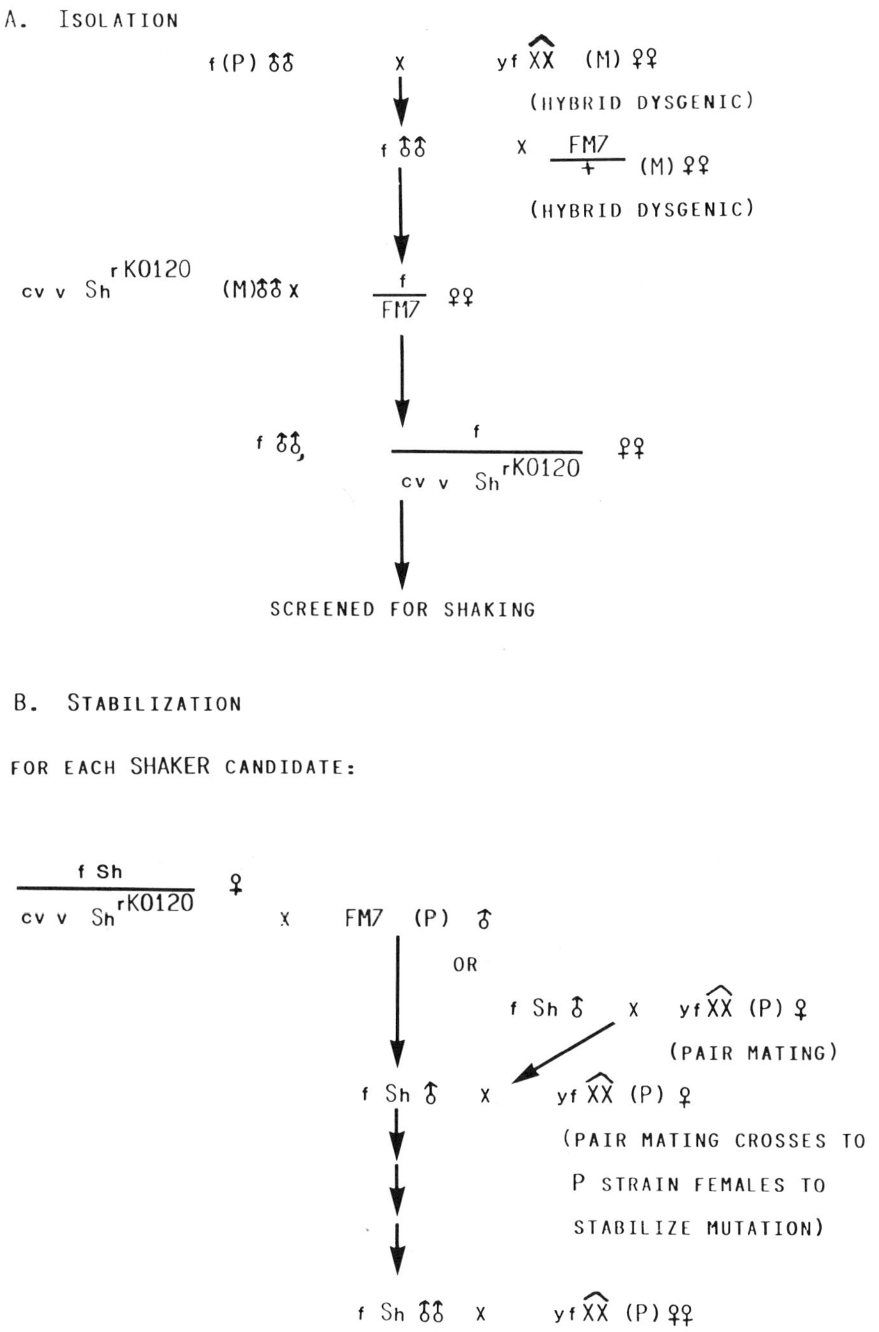

Figure 2. Scheme used to generate hybrid-dsygenesis-induced *Shaker* mutations. After two dysgenic crosses, the X chromosome marked with *f* was tested for the possible existence of a newly induced *Shaker* mutation; males carrying the *f*-marked X chromosome and females carrying that *f*-marked X chromosome and a recessive *Shaker* mutation on the other X chromosome were screened for shaking. Once a dysgenesis-induced *Shaker* candidate was isolated, it was stabilized by several generations of pair mating to P-strain females, *(f) Forked,* an X-linked bristle marker; *(cv) cross-veinless,* X-linked wing marker; *(v) vermillion,* X-linked eye color marker; *(y) yellow,* X-linked body color marker. $\widehat{XX}$ represents the attached X chromosome. FM7 is an X-chromosome balancer carrying dominant markers and multiple inversions that suppress its recombination with other X chromosomes.

Table 1. Hybrid-dysgenesis-induced *Shaker* Mutants

	Abnormally prolonged transmitter release	A current in muscles	Map units from *forked* (+1 map unit)	Between B^S and V7	P element at 16F
Sh^{SB1}	–	+		yes	yes
Sh^{SB2}		–		right of B^S	yes
Sh^{SB3}	–	+	1.2	yes	yes
Sh^{SB4}	–	+	2.3		yes
Sh^{SB5}	–	+	1.4	yes	yes
Sh^{SB6}	–	+		yes	yes
Sh^{SB7}	–	+	1.7		yes
Sh^{SB11}	–	+			
Sh^{SB12}	–	+	1.2	yes	
Sh^{SB13}		–		right of B^S	

95% confidence interval is 1 map unit. B^S: T(X;4)B^S and V7: T(X;Y)V7.

termined by recombination mapping with respect to visible markers, *yellow* (*y*, map position 0.0), *chocolate* (*cho*, 5.4), *cross-veinless* (*cv*, 13.7), *vermillion* (*v*, 33.0), and *forked* (*f*, 56.7). All flies used in these mapping experiments were P-strain flies, so that the crosses were not dysgenic. About 2000 flies were scored in each experiment. The results are presented in Table 1.

Genetic mapping by complementation tests with respect to duplicated chromosomal segments. Translocations between the X and Y chromosomes, T(X;Y) B55, T(X;Y) W32, T(X;Y) V7 (Tanouye et al. 1981), and one translocation between the X and fourth chromosomes, T(X;4) B^S (Lindsley and Grell 1968), were used. Figure 3 indicates the approximate location of breakpoints on the X chromosome. Flies carrying a normal X chromosome and the short, proximal fragment of such a translocation (from the centromere to the breakpoint on the X chromosome) are male flies. Approximately twice the normal amount of gene products are produced in these animals by the genes located on the duplicated region of the X chromosome (Maroni and Lucchesi 1980). To test whether a Sh^{SBn} mutation is located within a duplicated region like the proximal fragment of T(X;Y) V7 (V7^P), we compared the phenotype of Sh^{Sbn}/V7^P males with that of Sh^{SBn} males. (We also analyzed Sh^{SBn}/Sh^+ females.) The wild-type laboratory strain (Sh^+) used is Oregon R.

In situ hybridization. *Drosophila* polytene chromosome spreads were prepared as described by Pardue and Gall (1975) from salivary glands derived from male

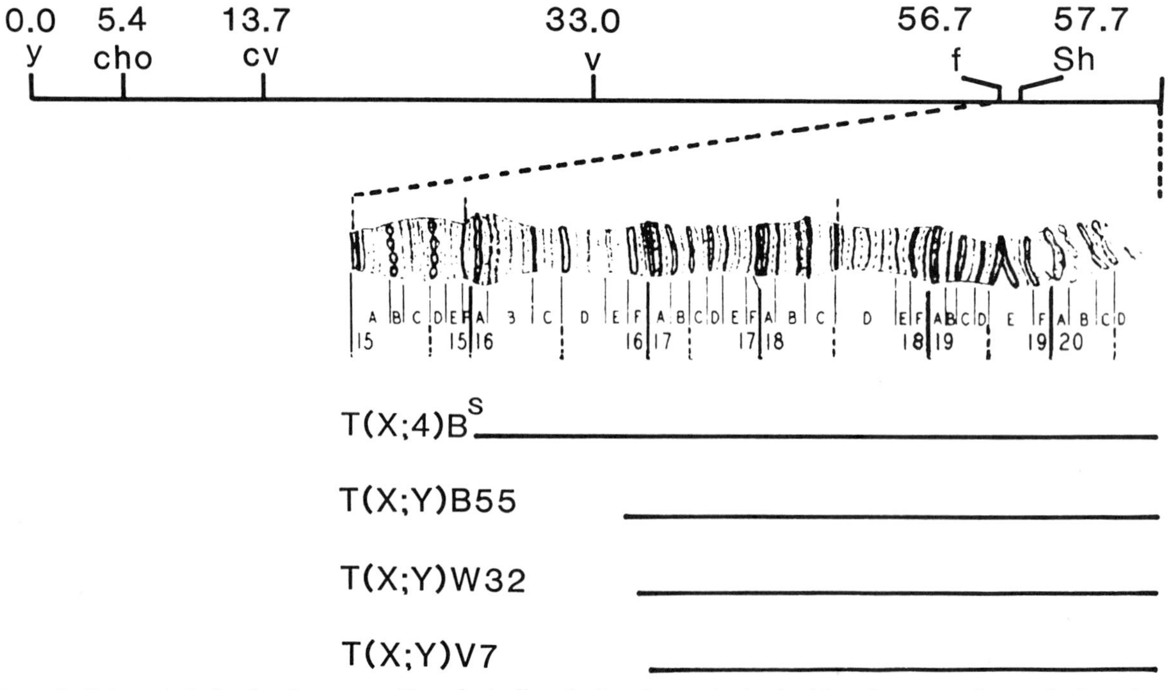

Figure 3. Schematic indicating the map position of *y* (*yellow*, body color marker), *cho* (*chocolate*, eye color marker), *cv* (*cross-veinless*, wing marker), *v* (*vermillion*, eye color marker), *f* (*forked*, bristle marker), and *Sh* (*Shaker*) on the X chromosome. The approximate locations of the breakpoints of the four translocations used are also indicated with respect to the polytene salivary chromosome.

third instar larvae of Sh^{SBn} mutants. The acetylation step described by Hayashi et al. (1978) to decrease nonspecific binding of nucleic acids was included before the chromosomes were denatured in alkali.

Hybridization probes were prepared by labeling a p-element-containing plasmid, pπ25.1, with ^{125}I-labeled dCTP (>900 Ci/mole). This plasmid contained *Drosophila* DNA retrieved from the 17C region of the X chromosome of π_2 strain flies (O'Hare and Rubin 1983). Thus, it has not only a complete p element, but also flanking sequences from the 17C region. Labeling of this plasmid was done according to the method of O'Farrell, using the T4 DNA polymerase (O'Farrell et al. 1980). The circular plasmid was cut once in the middle of the p element by the restriction enzyme *Kpn*I. The 3′ end on either side of this linear plasmid DNA was degraded by the T4 DNA polymerase, which functioned as a 3′ exonuclease in the absence of deoxyribonucleotide triphosphates. This reaction was allowed to proceed until DNA fragments with recessed 3′ termini of about 1.5 kb were formed. Deoxyribonucleotide triphosphates were then added to the reaction mixture so that the enzyme would function as a DNA polymerase and incorporate iodinated dCTP into the previously excised 3′ regions. Finally, the labeled DNA fragments were briefly exposed to λ exonuclease to excise nucleotides from the unlabeled 5′-terminal regions, resulting in DNA fragments where both ends are 1.5-kb single-stranded, labeled 3′ extensions.

Hybridization was carried out with the probe concentrations of 0.1 μg/ml·kb of probe sequence complexity. The hybridization solution contained 0.3 M NaCl, 34 mM sodium citrate (2× SSC), 40% formamide, 70 μg/ml yeast tRNA, 285 μg/ml salmon DNA, 28.5 mM sodium phosphate (pH 6.5), and 0.6 μg/ml of iodinated probe. This solution was denatured by placing in boiling water for 3 minutes, followed by quenching in an ice-water bath. The solution (20 μl) was applied to each slide of chromosome squash, covered with a coverslip, and incubated for 24 hours at 42°C. After hybridization, the squashes were washed in three changes of 2× SSC at 60°C for 30 minutes, dehydrated, air-dried, autoradiographed, and stained with Giemsa according to the method of Pardue and Gall (1975). Exposure times were between 3 and 8 days.

BACKGROUND

The *Shaker* Locus Probably Contains the Structural Gene for a K⁺ Channel: A Review

Identification of* Shaker *as a locus important for K⁺-channel function. Mutations affecting nervous function most likely would cause behavioral abnormality, paralysis, or lethality. With this expectation, over 100 behavioral mutants were screened electrophysiologically for possible defects in larval neuromuscular transmission. From that search, mutations of the *Shaker* locus were found to cause abnormally large and prolonged transmitter release, especially when the extracellular Ca^{++} concentration was low (Fig. 4) (Jan et al. 1977). Apparently, the nerve action potentials in *Shaker* mutants were prolonged and carried recurrent spikes (Jan and Jan 1980). This caused prolonged elevation of Ca^{++} conductance at the nerve terminal and prolonged transmitter release (Jan et al. 1977). The cause of the *Shaker* phenotype was attributed to a defect in K^+-channel function, because the behavioral and physiological phenotype of *Shaker* can be mimicked by treating wild-type larvae with 4-aminopyridine (4-AP), a K^+-channel blocker, whereas blocking Na^+ or Ca^{++} channels did not correct for the abnormality (Jan et al. 1977). Recent voltage-clamp studies of *Drosophila* pupal muscles showed that 4-AP specifically blocked one type of K^+ channel, the A channel (Salkoff, this volume). Moreover, various *Shaker* mutations were found to affect the A-channel function specifically (Salkoff, this volume), as discussed in the next section.

Lowering external Ca^{++} concentration exaggerated both the *Shaker* phenotype and the 4-AP-induced phenocopy (Jan et al. 1977). This may be due to a reduced Ca^{++}-activated K^+ conductance at low Ca^{++} concentration. A transient, Ca^{++}-dependent K^+ current has been found in pupal muscle (Salkoff 1983) and could play an important role in repolarizing the action potential (MacDermott and Weight 1982). If Ca^{++}-activated K^+ channels are present in larval nerve membranes, they would be functional at elevated Ca^{++} concentrations and could partially compensate for the defect in *Shaker* mutants.

Evidence for the* Shaker *locus to contain the structural gene for a K⁺ channel. In addition to abnormal synaptic transmission, *Shaker* mutants showed prolonged action potential duration in the cervical giant fibers of adult flies (Tanouye et al. 1981), as well as altered A current in pupal muscles (Salkoff and Wyman 1981). These studies have raised the possibility that the *Shaker* locus contains the structural gene for the A channel. The evidence from voltage-clamp studies of various *Shaker* mutants is presented by Salkoff (this volume). A summary of existing evidence is given as follows:

The null phenotype caused by deletion in* Shaker *locus. Deleting a small fragment within the *Shaker* locus (between the breakpoints of T[X;Y] B55 and T[X;Y] W32) eliminates the A current in pupal muscles (Salkoff, this volume) and results in prolonged transmitter release from larval nerve terminals, similar to what was found in the Sh^{KS133} mutant (cf. Fig. 5). This suggests that genes in the *Shaker* locus are necessary for the expression of functional A channels.

Gene-dosage studies. Although the Sh^{KS133} mutant exhibits the same phenotype as the deletion ($B55^D/W32^P$) (the null phenotype), the mutation apparently causes the production of an abnormal gene product, because the *Shaker* phenotype is not totally suppressed even in flies carrying enough copies of the Sh^+ gene to make close to 100% of the normal gene

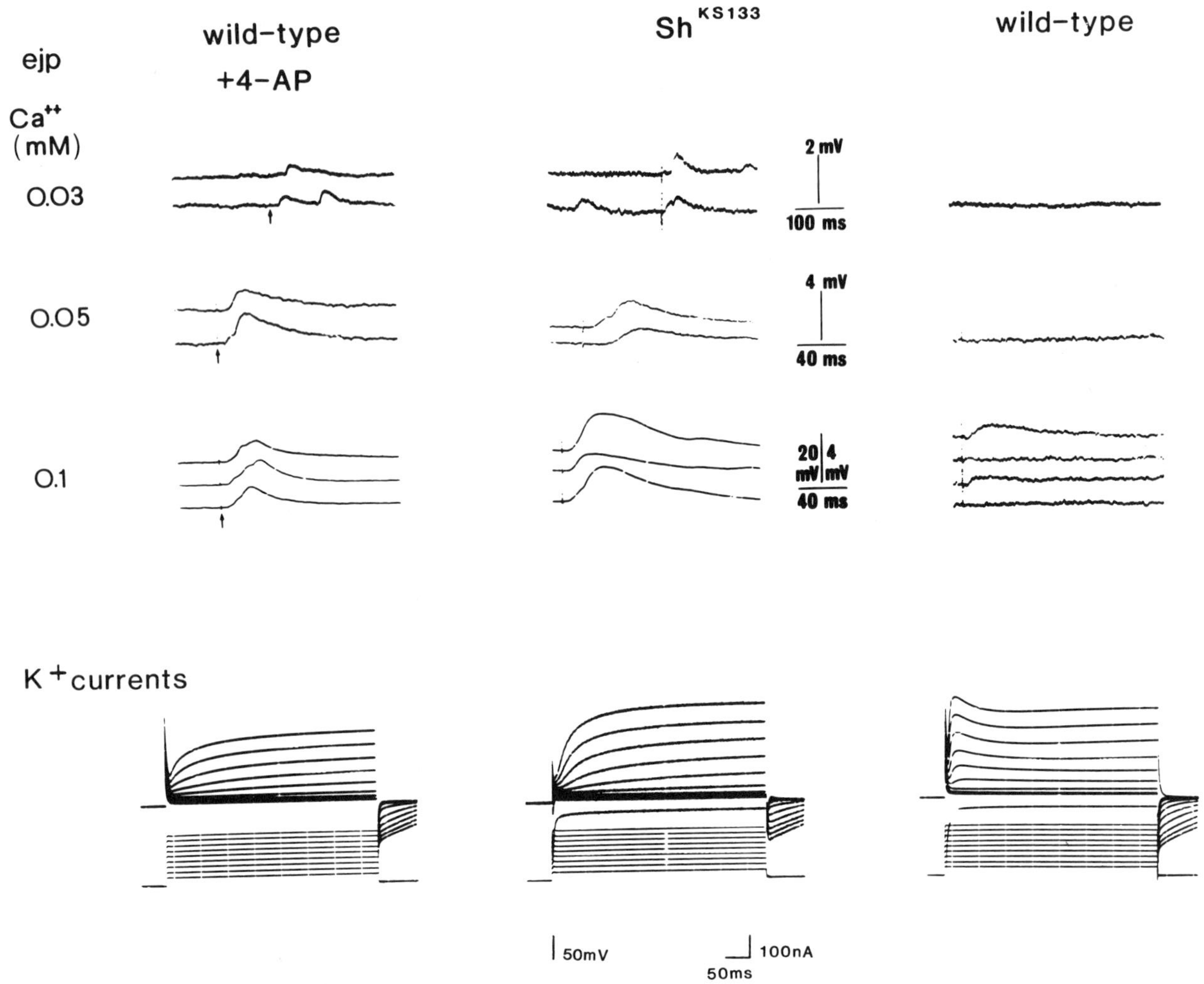

Figure 4. (*Upper three rows*) Intracellular recordings from ventral lateral longitudinal muscle fibers of third instar larvae. Cut motor nerves were stimulated with a suction electrode; arrows point to stimulus artifacts. In 0.03 mM and 0.05 mM Ca^{++}, motor-nerve stimulation failed to elicit a synaptic potential (excitatory junctional potential) in the wild-type fly, whereas stimulation succeeded in the Sh^{KS133} mutant and in the wild-type larva that was treated with 6 mM 4-AP. In 0.1 mM Ca^{++}, motor-nerve stimulation evoked a much larger and prolonged response in Sh^{KS133} and in the 4-AP-treated wild-type larva than in wild-type larvae without 4-AP. (*Bottom row*) Currents from dorsal-flight muscle of pupae, measured with a two-electrode voltage clamp. The voltage-jump protocol, the same for each experiment, is displayed beneath the current traces. The transient outward current, or A current, was present in the wild-type pupa, missing in the Sh^{KS133} pupa, and blocked by 4-AP in the wild-type pupa. The delayed rectifying K^+ current was not affected by the Sh^{KS133} mutation or by low concentrations of 4-AP.

product (Tanouye et al. 1981; Salkoff, this volume). This indicates that the gene affected by the Sh^{KS133} mutation is a structural gene.

Altered voltage dependence of A-channel inactivation in the* Sh⁵ *mutant. Study of a less-severe *Shaker* mutant, Sh^5, has provided additional evidence for this hypothesis (Salkoff, this volume). In pupal muscle of Sh^5 flies, the A current has normal amplitude and rise time, but it inactivates more abruptly than in wild type. Whereas normally the time constant for inactivation shows strong voltage dependence, in Sh^5 it is small and independent of membrane potential. Although the phenotype of Sh^5 differs from that of Sh^{KS133}, the two mutations are probably allelic; no recombination was observed between these two mutations among 8000 flies scored (Salkoff, this volume). The fact that one mutation of the *Shaker* locus alters the voltage dependence of A-channel inactivation, while another mutation eliminates the A current, suggests that the *Shaker* locus contains the structural gene of the A channel.

Use of Transposable Elements in the Cloning of Genes in *Drosophila*

The cloning of genes whose product is either not defined or not available because of difficulties in isolation can represent a formidable problem. In *Drosophila,* two general approaches have been developed to isolate the DNA sequences of genes that have been defined only by genetic criteria.

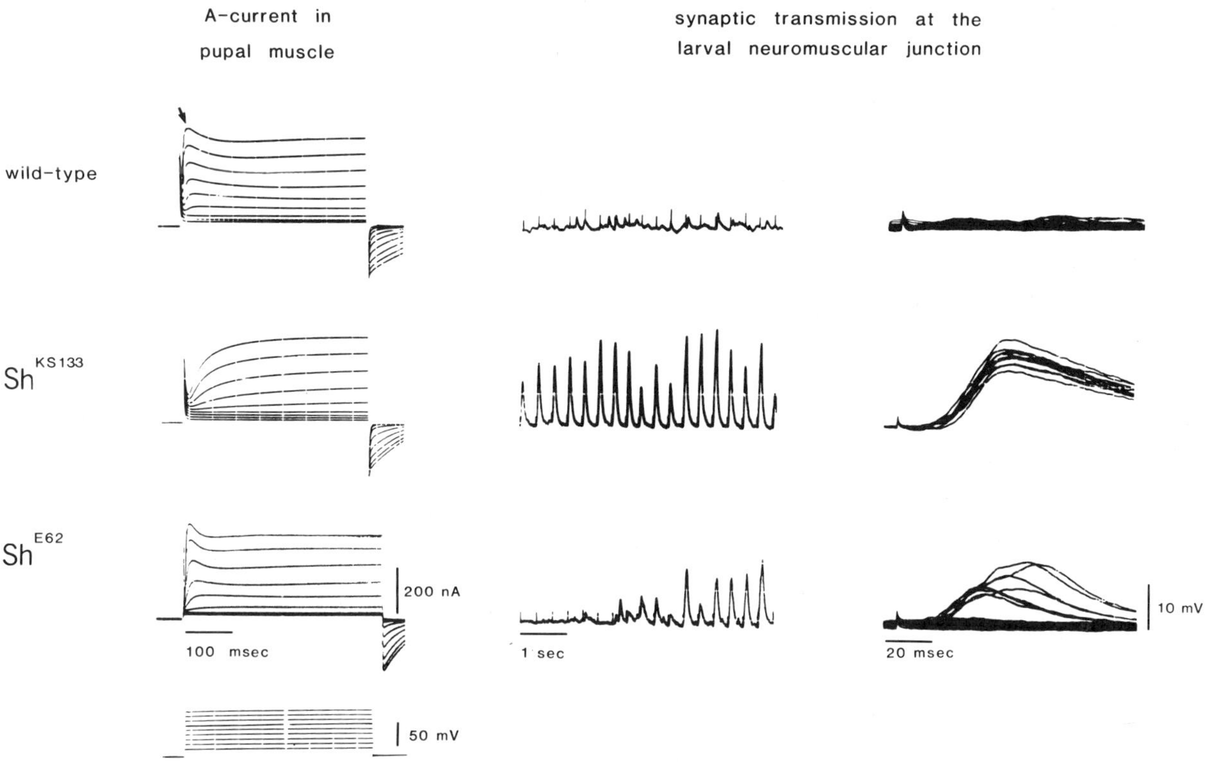

Figure 5. Mutant phenotypes in pupal muscle and at the larval neuromuscular junction. (*Left column*) Membrane currents of pupal flight muscle as recorded at 4°C with a two-electrode voltage clamp. The membrane potential was stepped from a holding potential of −80 mV to more depolarized levels, as shown by the voltage indicator traces at the bottom. In wild-type and Sh^{E62} larvae, both the A current (arrow in top trace) and the delayed rectifying K^+ current were present, whereas Sh^{KS133} showed the delayed rectifier only. (*Center and right columns*) Intracellular recording from ventral lateral longitudinal muscle fibers of third instar larvae in 0.05 mM Ca^{++}. Cut motor nerves were stimulated at 4 Hz via a suction electrode. All stimuli failed to evoke a response in the muscle of the wild-type larva, although stimulus artifacts were visible. For Sh^{KS133}, the first stimulus and then each subsequent stimulus elicited a prolonged excitatory junctional potential. In the Sh^{E62} larva, after failing for several seconds, stimuli eventually succeeded in evoking prolonged excitatory junctional potentials.

Most frequently, these genes have been cloned by first genetically defining the position of the locus of interest on the polytene chromosome map and then obtaining a nucleic acid sequence (usually a piece of cloned DNA or an mRNA) that hybridizes to the polytene chromosomes at a cytological position that is near the locus of interest. Using a process referred to as "chromosome walking," it is possible to isolate adjacent cloned sequences overlapping the start clone and by a reiterative process "walk down the chromosome" to the region of interest. This process is often exceedingly laborious, and because of the inaccuracies of cytological localization, the length of a walk cannot be reliably predicted.

An alternative proposed by Bingham et al. (1981) has been called "transposon tagging." Should transposition lead to a damaging insertion within a gene, the mutant gene sequences will be "tagged" by the presence of the transposable element. If the transposable element DNA has been isolated, clones of "tagged" sequences can be identified by hybridization between the linked tag and labeled transposable element DNA. This approach has been used to clone the genes for *white, scute,* and *RNA polymerase* (Bingham et al. 1981; Searles et al. 1982; Modolell et al. 1983).

The p element of *Drosophila* is a transposon that is particularly useful for transposon tagging because its frequency of transposition can be induced to levels that result in useful frequencies of transposon-induced mutations. Transposition is drastically increased by crosses between males from a strain that carries p elements (P strain) and females from a strain lacking p elements (M strain) (Engels 1981; Rubin et al. 1982). The progeny of these "dysgenic crosses" can be screened for any particular mutant phenotype, and these will usually be due to a damaging insertion of p-element sequences into the gene of interest. If a bank of clones is produced from the new mutant, those clones carrying tagged sequences can be identified by hybridization to labeled p-element DNA (see Fig. 1).

RESULTS

Isolation of Hybrid-dysgenesis-induced *Shaker* Mutants

Knowing that the *Shaker* locus is likely to contain the structural gene for the A channel (see Background above), we undertook the isolation of DNA from the *Shaker* locus for molecular studies. The strategy used was first to isolate *Shaker* mutations induced by hybrid dysgenesis, presumably caused by the insertion of p factors into the *Shaker* locus (see Background above). With such mutants we use labeled p factor to isolate DNA that is contiguous with the p factor within the *Shaker* locus.

The scheme of hybrid dysgenic crosses used to generate *Shaker* mutations is outlined in Figure 2. Flies (24,000) were screened, and ten candidates for *Shaker* mutations were isolated. Two were isolated as shaking *forked* males. Both showed no physiological abnormalities characteristic of *Shaker* mutants and have not been analyzed further. Eight were isolated as shaking heterozygous females carrying the recessive Sh^{rKO120} mutation on one X chromosome and a newly induced *Shaker* mutation (Sh^{SBn}, $n = 1, 3, 4, 5, 6, 7, 11, 12$) on the other dysgenic X chromosome. Each of these eight new mutants showed physiological abnormalities characteristic of existing *Shaker* mutants (see below). These mutations were further characterized as to where they mapped on the X chromosome, and whether a p factor was present near the *Shaker* locus (see below).

As expected in dysgenic crosses, instability of these dysgenesis-induced Sh^{SBn} mutations has been observed: When they were first isolated, heterogeneity in shaking phenotype was commonly observed among male progeny of single pair matings. Occasionally, a stock of mutant flies derived from a single Sh^{SBn} male ($n = 4, 5, 7$) switched from a weak physiological phenotype to a much stronger phenotype. Conversely, reversion to wild type also seemed to take place. In the case of Sh^{SB1}, a single Sh^{SB1} male gave rise to progeny showing both the *Shaker* phenotype and normal physiology, suggesting a possible reversion event. Furthermore, some stocks of Sh^{Sbn} mutants later showed the phenotype of *outstretched*, a locus closely linked to *Shaker*. This was another indication that the region near *Shaker* was affected by the dysgenic cross.

In addition to these mutants isolated from dysgenic crosses, other candidates appeared fortuitously. Two spontaneous *Shaker* mutants (Sh^{SB2} and Sh^{SB13}) were isolated from separate stocks of P-strain flies. These mutants may have arisen from spontaneous instability of p elements within the stocks. Both showed severe physiological abnormalities characteristic of the null phenotype of *Shaker*, as described below.

The Physiological Abnormalities of Sh^{SBn} Are Similar to Those of Existing *Shaker* Mutants

Having found dysgenesis-induced mutants that shake under ether anesthesia, we first wanted to know whether they showed physiological abnormalities characteristic of *Shaker* mutants: namely, an altered A current and/or prolonged transmitter release caused by prolonged nerve action potential. As described below, some Sh^{SBn} mutants showed the extreme, null phenotype of *Shaker*, whereas other dysgenesis-induced Sh^{SBn} mutants showed a phenotype typical of weak *Shaker* mutations like Sh^{E62}. The physiological analysis was then carried further to test whether some of the Sh^{SBn} mutations map to the same region and affect the same gene as existing *Shaker* mutants, as described below.

The Null Phenotype

The most extreme (null) phenotype was caused by deleting a segment from the *Shaker* locus (between T[X;Y] $B55^D$ and T[X;Y] $W32^P$). The same severe phenotype was also exhibited by the Sh^{KS133} mutant. In both cases, the A current was totally eliminated in muscles of larvae and pupae, and prolonged transmitter release at the larval neuromuscular junction was evident after each single nerve stimulus (Fig. 5). Since all these abnormalities could be mimicked by treating a normal animal with 4-AP, a blocker of A channels (Fig. 4) (see Background above), they most likely arose from the same defect in the A channels.

The spontaneous *Shaker* mutants Sh^{SB2} and Sh^{SB13} showed the extreme physiological abnormalities typical of the null *Shaker* phenotype. The A current was totally eliminated in pupal muscles, and in 0.05 mM Ca^{++}, a single stimulus to the larval nerve caused prolonged transmitter release (Figs. 6 and 7). When synaptic transmission was tested in female larvae heterozygous for Sh^{SB2} *or* Sh^{SB13} and the wild-type allele, transmission was found to occur after only a few stimuli (Fig. 6). The Sh^{SB2} and Sh^{SB13} mutations are therefore codominant. As with flies with a *Shaker* locus deletion (i.e., $B55^D/W32^P$), the *Shaker* phenotype of Sh^{SB2} and Sh^{SB13} was also suppressed in male flies carrying an extra copy of the *Shaker* locus (Fig. 7). Therefore, these mutations appeared to reduce the level of normal gene product, rather than causing the production of an abnormal gene product.

The Moderate Phenotype

A more moderate phenotype is typified by the Sh^{E62} mutant. In this mutant the A current was still present in larval and pupal muscles, and a single stimulus of the motor nerve was unable to cause transmitter release in 0.05 mM Ca^{++}. When a train of stimuli was delivered at 4 Hz, however, after several seconds, end-plate potentials appeared (Fig. 5). Extracellular recording from the larval nerve showed that by this time the nerve action potential was prolonged and carried recurrent spikes, similar to what was found in Sh^{KS133} (Jan and Jan 1980). This weaker phenotype may also be explained by a possible defect in A current of larval nerve: By doing voltage-clamp experiments, Salkoff and Wyman (1980) had shown that repetitive depolarization had cumulative effects on the inactivation of A current, causing facilitation of excitability. If the weaker *Shaker* mutation had a mild effect on A current, either reducing the amount or altering the kinetics, the defect may become more obvious after repetitive nerve stimulation had inactivated some of the A channels.

The Sh^{E62} mutation was mapped to the same region of the *Shaker* locus as Sh^{KS133} (Tanouye et al. 1981).

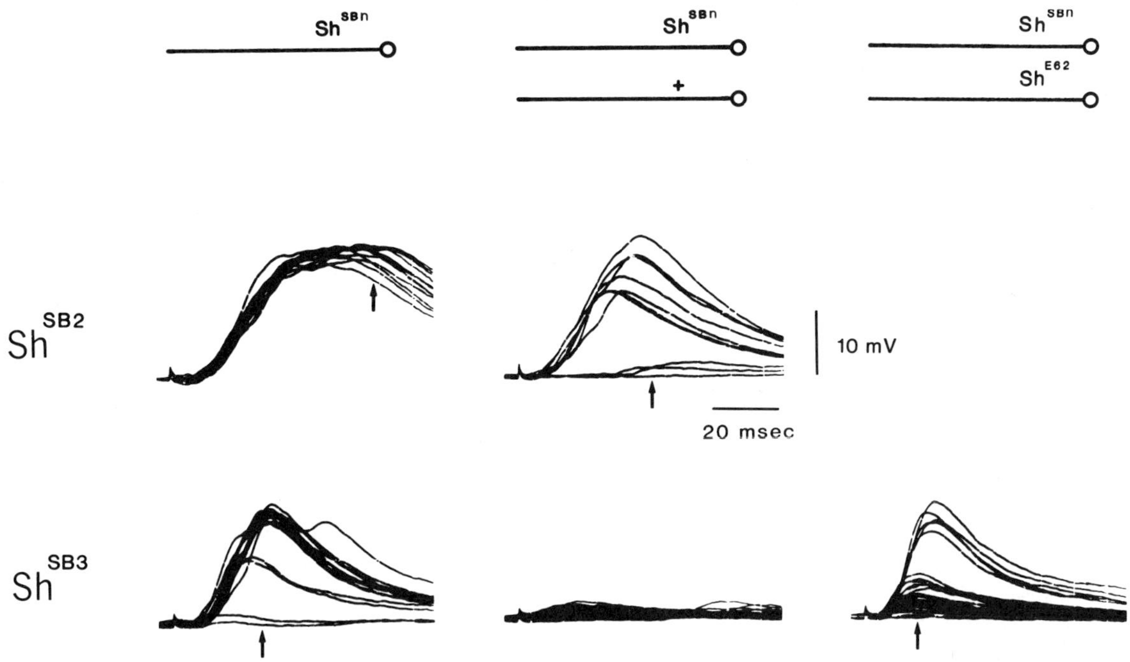

Figure 6. Neuromuscular transmission in Sh^{SB2} and Sh^{SB3}. Intracellular recordings from larval muscle fibers in 0.05 mM Ca^{++} as in Fig. 5. (*Left column*) Mutant phenotypes as they appear in hemizygous males. Arrows point to the result of the first stimulus of the motor nerve. Sh^{SB2} is the stronger mutant, with the first stimulus evoking a prolonged excitatory junctional potential. (*Center column*) Dominance test, with females heterozygous for Sh^{SB} and Sh^{+}. For Sh^{SB2}, the phenotype is still that of a mutant, whereas for Sh^{SB3} the phenotype is the same as in a wild-type larva (see Fig. 5). (*Right column*) Lack of complementation between Sh^{SB3} and Sh^{E62}.

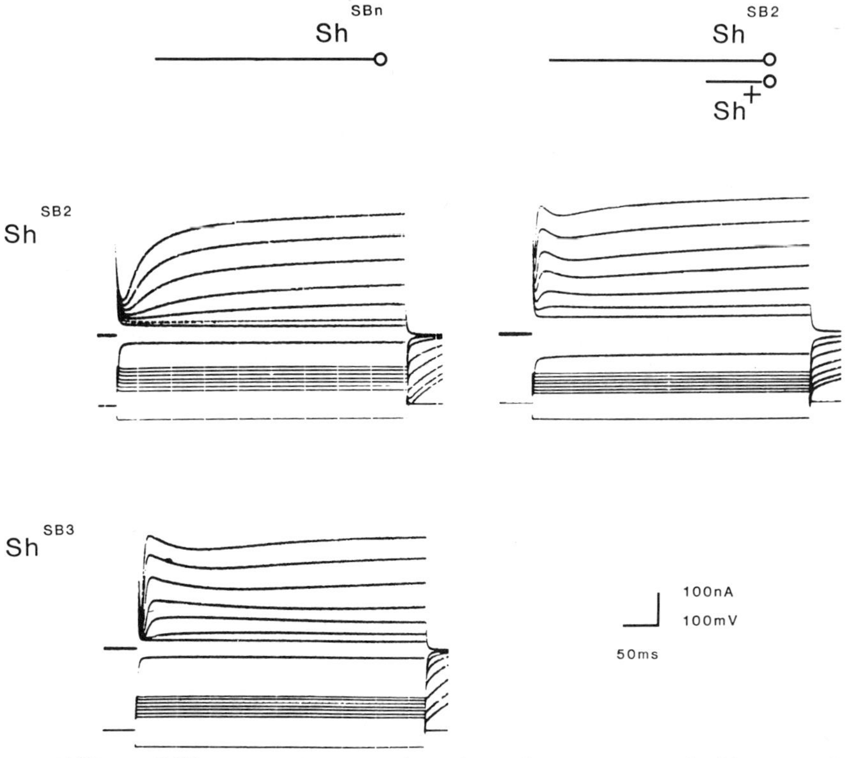

Figure 7. A current in Sh^{SB2} and Sh^{SB3} mutants. Currents of pupal muscle were measured with a two-microelectrode voltage clamp. Beneath each set of current traces is the voltage-jump protocol. (*Upper row*) The Sh^{SB2} male lacks A current (left). When the wild-type allele of *Shaker* is present on a short, proximal fragment of X chromosome, which extends to the *Bar* locus, the A current is present. The Sh^{SB2} mutation therefore affects a gene to the right of *Bar*. (*Lower trace*) Sh^{SB3}, typical of the dysgenesis-induced mutants, has A current in pupal muscles.

Chromosomal breakpoints within the *Shaker* locus, such as that of T(X;Y) B55, also produced a moderate phenotype similar to that of Sh^{E62} in larval neuromuscular transmission, as well as a partial reduction of the A current in muscles (Salkoff, this volume). Moreover, treating wild-type larvae with small amounts of 4-AP, a specific blocker of A channels, again produced a Sh^{E62}-like phenotype in neuromuscular transmission. These observations indicate that the Sh^{E62} phenotype was caused by a less-severe mutant effect at the *Shaker* locus.

The hybrid-dysgenesis-induced mutants Sh^{SBn} ($n = 1$, 3, 4, 5, 6, 7, 11, 12) had a *Shaker* phenotype similar to that of Sh^{E62} and weaker than that of Sh^{KS133}, Sh^{SB2}, or Sh^{SB13}. The A current was present in pupal muscle (Fig. 7). Although the amount of A current in pupal muscles overlapped with that of wild-type animals, a slight reduction could not be ruled out. At the larval neuromuscular junction, repetitive stimuli were required to cause prolonged transmitter release (Fig. 6). The dysgenesis-induced mutants also resembled Sh^{E62} in that, with the exception of Sh^{SB4} and Sh^{SB7}, these mutations were recessive (Fig. 6). Since both Sh^{E62} and Sh^{SB3} were recessive, a complementation test was possible to determine whether they were mutations of the same gene. Females carrying Sh^{SB3} on one chromosome and Sh^{E62} on the other X chromosome had the mutant phenotype (Fig. 6); the mutations do not complement, and therefore they probably affect the same gene.

Sh^{SBn} Mutations Appear to Affect the *Shaker* Locus

Knowing that the Sh^{SBn} mutations caused physiological abnormalities similar to those of *Shaker* mutants, we then asked whether these various Sh^{SBn} mutations mapped to the same region on the X chromosome as the *Shaker* locus. Since the *Shaker* locus is only 1 map unit to the right of *forked,* we measured the frequencies of recombination and estimated the map distance between Sh^{SBn} and *forked.* For several mutants, like Sh^{SB4}, Sh^{SB5}, Sh^{SB7}, and Sh^{SB12}, over 2000 flies have now been scored for each mutant for crossing-over events between the X chromosome carrying fSh^{SBn} and the X chromosome carrying *y cho cv v*. The distance between each of these Sh^{SBn} mutants and *forked* was around 1 to 2 map units (Table 1). In the case of Sh^{SB5} and Sh^{SB12}, clearly the *Shaker* mutation was mapped to the right of *forked.* From the existing data, we could not be sure as to whether Sh^{SB4} and Sh^{SB7} were also to the right of *forked.* Also, for these two lines an obvious asymmetry was noted among crossing-over events between *forked* and *Shaker*: There were many more f^+Sh flies than fSh^+ flies (for instance, the numbers were 50 and 1 for Sh^{SB4}, as compared with 18 and 15 for Sh^{SB5}). Whether this might arise from gene conversion is not clear. Further mapping experiments using markers flanking the *Shaker* locus will be necessary to resolve these points.

Besides genetic recombination, a complementation test was done to localize the Sh^{SBn} mutations. In this test an extra proximal fragment of the X chromosome was introduced into a male fly carrying the Sh^{SBn} mutation. Suppression of the Sh^{SBn} phenotype by the extra fragment would imply that the fragment contained the wild-type allele of the gene affected by Sh^{SBn}. Two chromosomal translocations were used. The T(X;4) B^S translocation has the X chromosome broken at the *Bar* locus and its proximal and distal fragments each connected to a fragment of the fourth chromosome (Lindsley and Grell 1968). Since the *Bar* locus is between *forked* (map position 56.7) and *Shaker* (map position 57.7), the proximal fragment of T(X;4) B^S has a complete, wild-type *Shaker* locus. The T(X;Y) V7 translocation, on the other hand, had a breakpoint within the 16F band of the X chromosome (Fig. 3). The proximal fragment of T(X;Y) V7 does not contain the wild-type allele of the existing *Shaker* mutants Sh^{KS133} and Sh^{E62}. Thus, breakpoints of these two translocations appear to be clearly outside the *Shaker* locus and the breakpoints themselves do not cause any shaking or physiological abnormalities. This helped us to localize the Sh^{SBn} mutants with respect to these breakpoints by complementation tests with the different fragments.

For Sh^{SBn}, $n = 1$, 3, 5, 6, 12, the mutation was recessive; heterozygous females carrying the *Shaker* mutation on one X chromosome and the wild-type allele on the other X chromosome did not shake. In complementation tests, male flies carrying Sh^{SBn} and the proximal fragment of T(X;4) B^S did not shake, whereas males carrying Sh^{SBn} and the proximal fragment of T(X;Y) V7 did. This result implies that the gene(s) affected by those Sh^{SBn} mutations are localized between the breakpoints of T(X;4) B^S and T(X;Y) V7 as is the *Shaker* locus (Table 1).

For Sh^{SBn}, $n = 2$, 4, 7, 13, the mutation was not recessive; the heterozygous *Shaker* females also shook. This made the interpretation of complementation tests based on behavioral assays less clear cut. However, voltage-clamp experiments showed that the defect of muscle A current in Sh^{SB2} or Sh^{SB13} was compensated in males carrying the proximal fragment of T(X;4) B^S (Fig. 7). Therefore, the genes affected by Sh^{SB2} and Sh^{SB13} are also localized to the right of *Bar* and *forked,* as is the *Shaker* locus.

In summary, all eight dysgenesis-induced Sh^{SBn} mutations and two spontaneous Sh^{SBn} mutations roughly map to the *Shaker* locus. These results of complementation tests using chromosomal breakpoints flanking the *Shaker* locus are in agreement with the results obtained from recombinant tests using multiple markers.

A p Factor Is Present in or near the *Shaker* Locus of Most Sh^{SBn} Mutants

The purpose of generating *Shaker* mutations via hybrid dysgenesis was to insert a p factor into the *Shaker* locus so that the *Shaker* DNA sequences tagged with the p factor could be identified and isolated by hybridization with the labeled p element (see Background above). We first isolated and mapped dysgenesis-induced mutations that showed the typical *Shaker* phenotype and then did in situ hybridization experiments to test whether the radioactive p-element probe would hybridize to the 16F region of the salivary chromosome, which contained the *Shaker* locus (Tanouye et al. 1981).

The radioactive p-element probe was made from a recombinant plasmid, pπ25.1, provided by G.M. Rubin (University of California, Berkeley). Because this p-element-containing DNA was retrieved from the genomic library of a π_2 strain, which had a p element inserted in the 17C region (O'Hare and G.M. Rubin, pers. comm.), it contained a complete p element as well as flanking sequences from the 17C region of the X chromosome. Besides 17C, many (around 30) other bands were labeled in Sh^{SBn} mutants. This is expected of P-strain flies that contain multiple copies of the transposable p element. For some of these bands, hybridization with the labeled p element was observed in certain Sh^{SBn} mutants but not in others. Radioactive label was found around the 16F region in each of the Sh^{SBn} mutants examined ($n = 1, 2, 3, 4, 5, 6, 7$), indicating that a p factor is inserted into or near the *Shaker* locus in these *Shaker* mutants. Figure 8 shows an in situ micrograph from Sh^{SB2} and Sh^{SB3}.

To summarize, we have isolated *Shaker* mutants through dysgenic crosses that promoted the transposition of p elements. These mutations showed both behavioral and physiological phenotypes of *Shaker* mutations and mapped to roughly the region of the *Shaker* locus. Finally, a p factor was found in or near the *Shaker* locus by in situ hybridization with the salivary chromosomes. These results suggest that the Sh^{SBn} mutations were induced by the insertion of a p factor. If we have indeed tagged the *Shaker* locus with a p element, these Sh^{SBn} mutations may prove very useful in the cloning and molecular analysis of the *Shaker* locus.

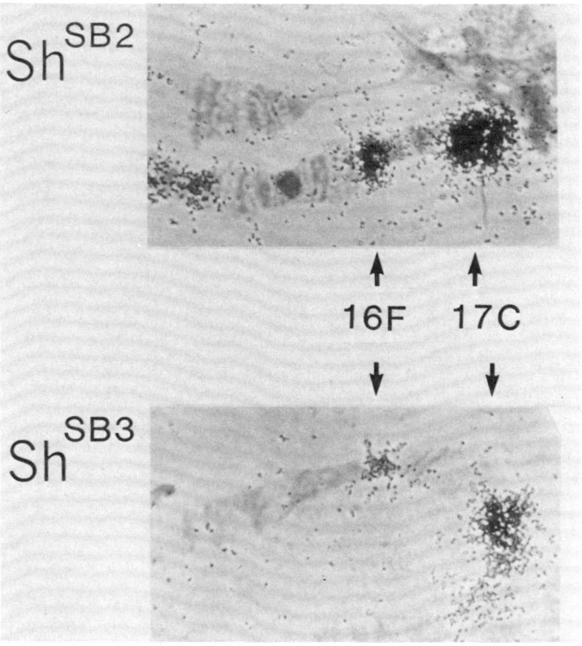

Figure 8. Hybridization of a ^{125}I-labeled p-element probe to the X chromosome of Sh^{SB2} (*top*) and Sh^{SB3} (*bottom*) mutants. The 17C region was labeled because the labeled probe contained not only a complete p element, but also flanking sequences from the 17C region of the X chromosome. In these *Shaker* mutants, radioactive label was consistently found around the 16F region, where the *Shaker* locus maps. Exposure time was 8 days. Micrographs were taken with a Zeiss 100× phase objective.

DISCUSSION

Molecular studies of K^+ channels are important because these channels are likely to play important roles in the control of neuronal activity and synaptic efficacy. The genetics of *Drosophila* and mutations of the *Shaker* locus offer a unique approach for cloning K^+ channels, because the lack of high-affinity toxins or antibodies to K^+ channels precluded the purification of these channels for biochemical analysis. Genetic analyses using electrophysiological assays, including voltage clamping, have provided strong evidence that the *Shaker* locus contains the structural gene for a K^+ channel, the A channel (Salkoff, this volume). At the same time, cloning strategies using transposable elements (Bingham et al. 1981; Searles et al. 1982; Modolell et al. 1983) combined with the recent elucidation of the molecular mechanism for the P-M hybrid dysgenesis (Rubin et al. 1982) suggest an obvious and attractive approach to the cloning of *Shaker* gene(s): If a transposable p element can be inserted into the *Shaker* locus, thereby causing a mutation of the *Shaker* gene(s), the previously cloned p element can be used to pull out DNA sequences adjacent to the p element insert in the *Shaker* locus.

We have isolated *Shaker* mutations generated by dysgenic crosses, as well as spontaneous *Shaker* mutations that arose in P-strain flies carrying multiple copies of p elements. These Sh^{SBn} mutations are behaviorally and physiologically *Shaker* in phenotype. Our initial genetic mapping places them in roughly the same region of the X chromosome as the *Shaker* locus. Moreover, in a complementation test, Sh^{SB3} is not complemented by an existing recessive *Shaker* mutation (Sh^{E62}), suggesting that this dysgenesis-induced *Shaker* mutation affects the same gene as Sh^{E62}. (Note that all previously known *Shaker* mutants are within 0.02 map units of Sh^{E62}.) Finally, a p element was found in the 16F region of the salivary chromosome in several Sh^{SBn} mutants tested. Since these Sh^{SBn} muations were induced in two different X chromosomes (Sh^{SB1} from the original unmarked X chromosomes of the π_2 strain and Sh^{SBn}, $n = 3, 4, 5, 6, 7$, from an X chromosome marked with *forked*, which maps to the 15F region), the fact that a p element was found in the 16F region of each Sh^{SBn} mutant implies that its presence is correlated with the *Shaker* mutation. The causal relationship between the insertion of the p element and the mutation of the *Shaker* locus must be further tested. This may be done by putting a Sh^{SBn} mutant through dysgenic crosses to increase the frequency of reversion, and seeing whether a reversion of the *Shaker* phenotype is accompanied by a concomitant loss of that p element in the 16F region.

In addition to providing a starting point for cloning, dysgenesis-induced *Shaker* mutants will also supply abundant new mutations that would be useful in later molecular analysis. Identification of *Shaker* gene(s) within the expected 100 kb of DNA sequences cloned from the *Shaker* locus will rely heavily on molecular analysis of mutations and rearrangments in the *Shaker* locus. By alignment of the genetic map with the physical map of cloned DNA, we will be able to define the sequences encoding particular functions. Since several Sh^{SBn} mutants appear to have a p element inserted in the region of the *Shaker* locus, the location of insertion can be mapped by comparison of the restriction fragments of DNA from wild-type and from *Shaker* mutants. This will allow us to correlate specific phenotypes with specific DNA sequences: The different Sh^{SBn} mutations have been found to show a wide spectrum of *Shaker* phenotypes. Some cause the extreme null phenotype in both nerves and muscles and others affect nerves much more than muscles. Should there be multiple structural genes (defined in a molecular analysis as sequences that could encode proteins) within the cloned region, molecular analyses of *Shaker* mutations will be crucial to the assignment of function to different coding sequences. Similarly, should there be different DNA sequences that are not expressed but play a regulatory role, controlling, for instance, the tissue specificity of expression of nearby sequences, they may be identified in these analyses as well.

There are two indications that the *Shaker* locus may be genetically complex. Rearrangement breakpoints covering a substantial region around 16F give some evidence of a shaking phenotype (Tanouye et al. 1981). Additionally, there is more than one phenotypic consequence of mutations at the *Shaker* locus. Are these different alleles of a complicated gene or are there multiple structural genes with related phenotypes encoded in the region? Because a number of *Shaker* mutations are not completely recessive, complementation studies are unable to resolve the issue. Since genetic recombination experiments thus far have not revealed any widely separated *Shaker* mutations, we believe it will be possible to clone the entire *Shaker* region and define its detailed genetic makeup by correlating coding regions with the molecularly mapped positions of numerous *Shaker* mutations of known phenotype.

How does one demonstrate that the *Shaker* locus contains the structural genes for the A channel? If the A channel were composed of a single subunit, one might expect an expression system such as the *Xenopus* oocytes to be useful, as has been the case for acetylcholine receptors (Barnard et al. 1982). In this case, one would need to use cloned *Shaker* DNA to isolate the appropriate mRNA so that the latter may be injected into cells like *Xenopus* oocytes. If the A channel were composed of more than one subunit and not all subunits were encoded by DNA from the *Shaker* locus, one might first identify the DNA sequences in the *Shaker* locus that code for one of the subunits. This might be done by identifying sequences that could hybridize to and remove an mRNA that is necessary for the formation of functional A channels. DNA transformation of *Drosophila* provides another powerful method that could be used to show that a particular DNA sequence includes the information damaged by *Shaker* mutations. In this approach, *Drosophila* embryos whose *Shaker* locus had been deleted (e.g., $B55^D/W32^P$) would be injected with different DNA segments from the *Shaker* locus that were ligated to specifically constructed fragments of the transposable p element designed to increase the frequency of integration and stable maintenance of the injected sequences (Rubin and Spradling 1982). The DNA segment that restores the A channel in the transformed fly is likely to contain the structural gene and may be used to generate large amounts of its gene products. By using specialized cloning vectors that will allow expression of the *Shaker* DNA that encodes a subunit of the A channel, one may then generate antibodies to the protein product and use them in the purification and molecular characterization of A-channel polypeptides.

We hope that this strategy will allow us to clone and characterize the gene for A channels and eventually lead to examination of channel structure, function, regulation, and evolution. Deduction of the primary sequence of the K^+ channel, together with analyses being done with other channel sequences (Ballivet et al. 1982; Noda et al. 1982), should permit us to predict the tertiary structure of the channel. Analysis of the molecular defects in certain mutants, such as Sh^5, that produce altered channels should help define functional domains in the channels. We would also like to extend these analyses to determine whether different K^+ channels in *Drosophila* share structural homology. The channel genes isolated in *Drosophila* could be used to study their evolutionary conservation and perhaps to identify other channel genes in other organisms. Finally, it would be interesting to learn whether and how the expression of these genes is regulated in different cells and whether in a given neuron the expression may vary with experience.

ACKNOWLEDGMENTS

We are grateful to Dr. W. Engels for providing the P-strain *Drosophila* stocks, Dr. G.M. Rubin for providing the p-element-containing plasmids, Ms. Louise Evans for doing chromosomal squashes and the art work, and Ms. Bonnie Iwamasa for preparing the manuscript. This study was supported by National Institutes of Health grant RO1 NS15963 to L.Y.J. and National Institutes of Health postdoctoral fellowships to C.L., L.S., and L.T.

REFERENCES

ADAMS, D.J., S.J. SMITH, and S.H. THOMPSON. 1980. Ionic currents in molluscan soma. *Annu. Rev. Neurosci.* **3:** 141.

ADAMS, P.R., D.A. BROWN, and A. CONSTANTI. 1982. Pharmacological inhibition of the M-current. *J. Physiol.* **332:** 223.

BALLIVET, M., J. PATRICK, J. LEE, and S. HEINEMANN. 1982. Molecular cloning for the γ subunit of *Torpedo* acetylcholine receptor. *Proc. Natl. Acad. Sci.* **79:** 4466.

BARNARD, E.A., R. MILEDI, and K. SUMIKAWA. 1982. Translation of exogenous messenger RNA encoding for nicotinic acetylcholine receptors produces functional receptors in *Xenopus* oocytes. *Proc. R. Soc. Lond. B.* **215:** 241.

BENTON, W.D., and R.W. DAVIS. 1977. Screening γgt recombinant clones by hybridization to single plaques in situ. *Science* **196:** 180.

BINGHAM, P.M., R. LEVIS, and G.M. RUBIN. 1981. Cloning of DNA sequences from the *white* locus of *D. melanogaster* by a novel and general method. *Cell* **25:** 693.

BYRNE, J.H. 1980. Analysis of ionic conductance mechanisms in motor cells mediating inking behavior in *Aplysia californica*. *J. Neurophysiol.* **43:** 630.

DEPEYER, J.E., A.B. CACHELIN, I.B. LEVITAN, and H. REUTER. 1982. Ca^{2+}-activated K^+ conductance in internally perfused snail neurons is enhanced by protein phosphorylation. *Proc. Natl. Acad. Sci.* **79:** 4207.

ENGELS, W.R. 1981. Hybrid dysgenesis in *Drosophila* and the stochastic loss hypothesis. *Cold Spring Harbor Symp. Quant. Biol.* **45:** 561.

HAYASHI, S., I.C. GILHAM, A.D. DELANEY, and G.M. TENER. 1978. Acetylation of chromosome squashes of *Drosophila melanogaster* decreases the background in autoradiographs from hybridization with [^{125}I]-labelled RNA. *J. Histochem.* **26:** 677.

HUGUES, M., D. DUVAL, P. KITABGI, M. LAZKUNSKI, and J.P. VINCENT. 1982. Preparation of a pure monoiodo derivative of the bee venom neurotoxin apamin and its binding properties to rat brain synaptosomes. *J. Biol. Chem.* **257:** 2762.

JAN, L.Y. and Y.N. JAN. 1976a. Properties of the larval neuromuscular junction in *Drosophila melanogaster*. *J. Physiol.* **262:** 189.

———. 1976b. L-glutamate as an excitatory transmitter at the *Drosophila* larval neuromuscular junction. *J. Physiol.* **262:** 215.

JAN, Y.N. and L.Y. JAN. 1980. Genetic dissection of synaptic transmission in *Drosophila melanogaster*. In *Insect neurobiology and pesticide action* (ed. F.E. Rickett), p. 161. The Society of Chemical Industry, London.

JAN, Y.N., L.Y. JAN, and M.J. DENNIS. 1977. Two mutations of synaptic transmission in *Drosophila*. *Proc. R. Soc. Lond. B.* **198:** 987.

KLEIN, M. and E.R. KANDEL. 1980. Mechanism of calcium current modulation underlying presynaptic facilitation and behavioral sensitization in *Aplysia*. *Proc. Natl. Acad. Sci.* **77:** 6912.

KLEIN, M., J. CAMARDO, and E.R. KANDEL. 1982. Serotonin modulates a specific potassium current in the sensory neurons that show presynaptic facilitation in *Aplysia*. *Proc. Natl. Acad. Sci.* **79:** 5713.

LINDSLEY, D.L. and E.H. GRELL. 1968. Genetic variations of *Drosophila melanogaster*. *Carnegie Inst. Wash. Publ.* **627:** 382.

LLINAS, R., M. SUGIMORI, and S.M. SIMON. 1982. Transmission by presynaptic spike-like depolarization in the squid giant synapse. *Proc. Natl. Acad. Sci.* **79:** 2415.

MACDERMOTT, A.B. and F.F. WEIGHT. 1982. Action potential repolarization may involve a transient, Ca^{++}-sensitive outward current in a vertebrate neurone. *Nature* **300:** 185.

MANIATIS, T., R.C. HARDISON, E. LACY, J. LAUER, C. O'CONNELL, D. QUON, D.K. SIM, and A. EFSTRATIADIS. 1978. The isolation of structural genes from libraries of eucaryotic DNA. *Cell* **15:** 687.

MARONI, G. and J.C. LUCCHESI. 1980. X-chromosome transcription in *Drosophila*. *Chromosoma* **77:** 253.

MODOLELL, J., W. BENDER, and M. MESELSON. 1983. *Drosophila melanogaster* mutations suppressible by the suppressor of *Hairy-wing* are insertions of a 7.3-kilobase mobile element. *Proc. Natl. Acad. Sci.* **80:** 1678.

NODA, N., H. TAKAHASHI, T. TANABE, M. TOYOSATO, Y. FURUTANI, T. HIROSE, M. ASAI, S. INAYAMA, T. MIYATA, and S. NUMA. 1982. Primary structure of α-subunit precursor of *Torpedo californica* acetylcholine receptor deduced from cDNA sequence. *Nature* **299:** 793.

O'FARRELL, P.H., E. KUTTER, and M. NAKANISHI. 1980. A restriction map of the bacteriophage T4 genome. *Mol. Gen. Genet.* **179:** 421.

O'HARE, K. and G.M. RUBIN. 1983. Structures of p transposable elements and their sites of insertion and excision in *Drosophila melanogaster* genome. *Cell* (in press).

PARDUE, M.L. and J.G. GALL. 1975. Nucleic acid hybridization to the DNA of cytological preparations. *Methods Cell Biol.* **10:** 1.

RUBIN, G.M. and A.C. SPRALDING. 1982. Genetic transformation of *Drosophila* with transposable elements vectors. *Science* **218:** 348.

RUBIN, G.M., M.G. KIDWELL, and P.M. BINGHAM. 1982. The molecular basis of P-M hybrid dysgenesis: The nature of induced mutations. *Cell* **29:** 987.

SALKOFF, L. 1983. *Drosophila* mutants reveal two components of fast outward current. *Nature* **302:** 249.

SALKOFF, L. and R. WYMAN. 1980. Facilitation of membrane electrical excitability in *Drosophila*. *Proc. Natl. Acad. Sci.* **77:** 6216.

———. 1981. Genetic modification of potassium channels in *Drosophila Shaker* mutants. *Nature* **293:** 228.

SEARLES, L.L., R.S. JOKERST, P.M. BINGHAM, R.A. VOELKER, and A.L. GREENLEAF. 1982. Molecular cloning of sequences from *Drosophila* RNA polymerase II locus by p element transposon tagging. *Cell* **31:** 585.

SIEGELBAUM, S.A., J.S. CAMARDO, and E.R. KANDEL. 1982. Serotonin and cyclic AMP close single K^+ channels in *Aplysia* sensory neurons. *Nature* **299:** 413.

TANOUYE, M.A., A. FERRUS, and S.C. FUJITA. 1981. Abnormal action potentials associated with the *Shaker* locus of *Drosophila*. *Proc. Natl. Acad. Sci.* **78:** 6548.

Ion Transport by Single Receptor Channels

B. SAKMANN, J. BORMANN, AND O.P. HAMILL
Max-Planck-Institut für biophysikalische Chemie, 3400 Göttingen, Am Fassberg, Federal Republic of Germany

Many functions of living cells depend on supramolecular structures that appear in the electron microscope as cell membranes separating the cell interior from the extracellular space. The backbone of the cell membrane is a bimolecular lipid lamella that, because of its insulation properties, provides the basis for selective ion transport processes across the membrane. Ion transport is mediated by so-called transport proteins. These are integral membrane proteins spanning the lipid bilayer.

Excitable cells like nerve cells, muscle fibers, and secretory cells are characterized by a class of transport proteins designated as channel or pore proteins. They have as a common feature that in response to a change in their environment, such as the increase in concentration of a chemical or a change in membrane voltage, they alter their structure. Small cations or anions can then flow down their electrochemical gradient across the membrane via the pores formed by these proteins. An equally plausible mechanism of ion transport would be a carrier molecule where ions are transported across the membrane one by one. The first evidence for ion transport via pores in a biological membrane came from observations made on the postsynaptic membrane of the neuromuscular junction, a highly specialized area of the skeletal muscle. Here, the neurotransmitter acetylcholine (ACh) increases the membrane conductance by several orders of magnitude. The resulting membrane current is accompanied by increased fluctuations of the current around its average value. One can account for their time and amplitude characteristics by assuming that the total current is composed of many elementary current contributions (Katz and Miledi 1972). The transport rate calculated for an elementary current is on the order of 10^7 ions/sec. This is far in excess of what can be expected for carrier-mediated transport. Hence, it was concluded that ions cross the membrane via pores. Similar experiments were subsequently made on other membrane currents gated either by voltage changes or by transmitters. The conclusion was that membrane-conductance changes in most cases are a consequence of the opening of discrete ionic pores (Neher and Stevens 1977).

The first direct measurement of the current flow through individual membrane pores was also obtained on ACh-gated channels in skeletal muscle using the patch-clamp technique (Neher and Sakmann 1976). Figure 1a shows the wave shape of the current flow through a small patch of postsynaptic membrane (5–10 μm^2 in size) in which only a few channels are present. The current switches at random intervals between two levels, marked zero and i. When the current is at zero level, all channels in the membrane patch are shut. The occurrence of a current pulse indicates the activation by ACh of a single pore. The downstroke of each current pulse marks the formation of a channel; the upstroke indicates that the channel is switched off. The average size of the current step i is 2.69 pA (Fig. 1b), representing a net transport rate of $\cong 2 \times 10^7$ positive charges/sec through the open pore.

The measurement of current through an individual channel provides the opportunity to observe the transitions of a single macromolecule between different structural states. Two classes of questions can be examined by analyzing current wave shapes like those shown in Figure 1a. One is related to the size of the current step. It reports the selectivity, the electrical profile, and the occupancy of the open channel when studied as a function of ion species and concentration and of membrane voltage (Hille 1975). The other questions relate to the duration of the current pulses and of the intervals between them. They reflect the stochastic properties of the reactions in the membrane that cause a channel to open and to close. The shape of the distribution of open and closed times depends on the number of different states a channel can adopt and the reaction rates by which these states are coupled. By analyzing such distributions, plausible reaction pathways describing the turning on and turning off of pores can be deduced (Colquhoun and Hawkes 1982). In this paper we describe experiments on the ion transport via single receptor channels. These are extracellular signaling molecules that are secreted from presynaptic nerve terminals. They play a key role in the communication between cells. The action of a transmitter on the target cell can be either excitatory or inhibitory. We focus on the end-plate channel gated by ACh, which excites the muscle fiber, and on a channel in nerve cell membrane that is gated by γ-aminobutyric acid (GABA), an inhibitory transmitter in the CNS (Peck 1980).

METHODS

The basis for measuring current signals in the range of a few picoamperes, the estimated size of single-channel currents (Anderson and Stevens 1973), is the restriction of the current measurement to a small patch of membrane of 5–10 μm^2 in size. This is achieved by electrically isolating the membrane patch via the tip of a glass pipette brought into close contact with the cell membrane. The experimental details are described in Hamill et al. (1981) and in Sakmann and Neher (1983).

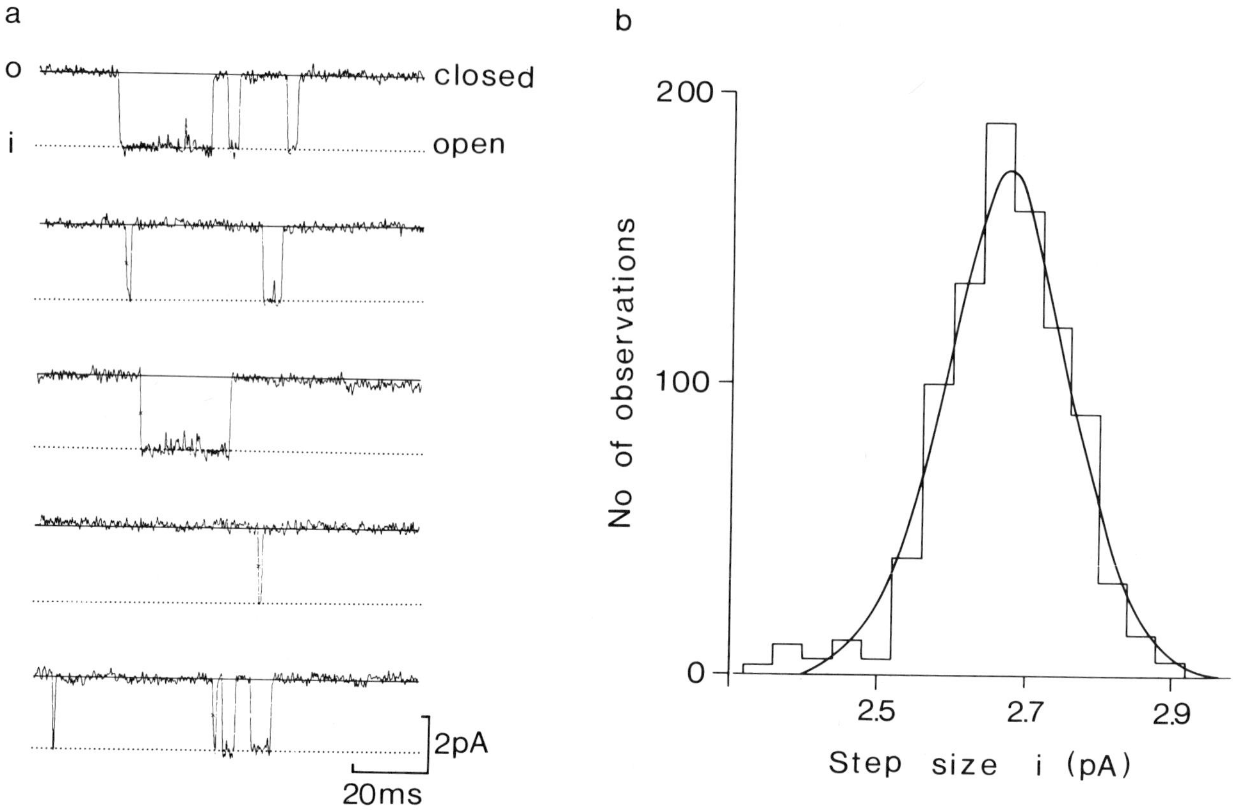

Figure 1. (*a*) Patch current recording from the postsynaptic membrane of frog (*Rana temporaria*) skeletal muscle fiber at −92 mV membrane potential (10°C); "cell-attached" patch (see Fig. 2). ACh (100 nM) activates single end-plate channels. The current jumps between two levels marked 0 and *i*. Downward deflection indicates inward current, i.e., net transport of positive charges from the extracellular solution to the cell interior. (*b*) Distribution of current pulse amplitudes. Same experiments as shown in *a*. The distribution of current-step amplitudes is singly peaked with a mean of 2.69 ± 0.1 pA.

To activate transmitter-gated channels, the transmitter is added to the pipette solution if one wishes to record "cell-attached" patches (Fig. 2a, upper part). In this configuration the physiological ion concentrations across the membrane are least disturbed. Alternatively, a membrane patch can be isolated physically from the cell membrane by simple manipulations as illustrated in Figure 2a. In the "outside-out patch" configuration, the extracellular face of a membrane patch that spans the tip opening of the pipette is exposed to the bath solution (Fig. 2a,b). Transmitter-gated channels are switched on by adding the transmitter to the bath solution, as illustrated in Figure 2c, when ACh is added to an outside-out patch isolated from a muscle cell.

RESULTS

Postsynaptic Excitation and Inhibition

Transmitter-gated membrane channels have a specific binding site (receptor) for transmitter molecules. When this site is occupied, the channel can open. Whether a transmitter is excitatory or inhibitory depends on the direction of the current flow when the channel has opened. Since ion transport is passive, the direction is determined by the ion selectivity of the channel and the electrochemical potential for the transported ions, i.e., the ion concentration and voltage gradient across the membrane. The permeant ions are Na^+, K^+, and Cl^-. They are distributed such that at the resting membrane potential (−70 mV to −90 mV), Na^+ is far from its electrochemical equilibrium potential (~ +30 mV), whereas K^+ and Cl^- are close to it. As a consequence, a transmitter is excitatory when it operates cation-selective channels and inhibitory when it gates Cl^--selective or K^+-selective channels. For example, ACh is an excitatory transmitter at the neuromuscular synapse where it depolarizes the muscle fiber by switching on cation-selective channels; it is inhibitory in the heart where it switches K^+-selective channels. GABA and glycine are inhibitory transmitters at the soma membrane of neurons in the CNS by operating Cl^--selective channels (Coombs et al. 1955). Examples of the current flow through single channels gated by ACh in a muscle cell and by GABA in a hippocampal neuron are shown in Figure 3 at different membrane voltages (V_m).

The current through ACh-activated channels is inward at the cell's resting potential (Fig. 3a). This causes depolarization by unloading of the membrane capacitance. The current decreases with decreasing membrane potential and reverses directon to outward at 0 mV membrane potential. This voltage dependence of the current size reflects the cation selectivity of the

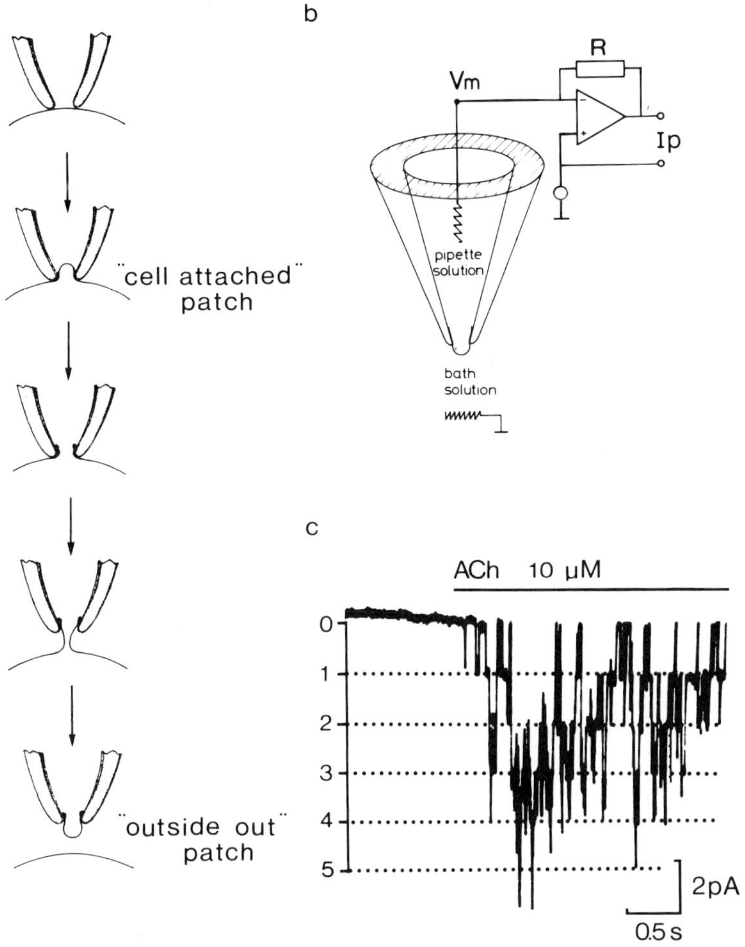

Figure 2. Technique of current recording from small membrane patches. (*a*) Schematic diagram of various patch configurations. In the "cell-attached" configuration (upper two diagrams), a tight contact is established between the tip of a glass pipette (shaded) and the cell membrane by slight suction. In the outside-out configuration (lowermost diagram), an isolated membrane patch spans the tip of the pipette with its extracellular side in contact with the bath solution. (*b*) Schematic representation of the current-measuring circuit for isolated membrane patches. The current I_p that flows between bath and pipette solution across the membrane patch is measured across the feedback resistor R. Membrane potential V_m is controlled by applying a voltage to the noninverting input of the amplifier (Sigworth and Neher 1980). (*c*) Current through an outside-out patch isolated from rat embryonic muscle fiber. The patch current increases in stepwise fashion following addition of 10 µM ACh to the bath solution. Membrane potential was set at −70 mV (cytoplasmic side of the membrane patch was negative). Numbers at the left indicate number of open channels.

channel and the unequal distribution of Na^+ and K^+ across the membrane. At negative membrane potentials, the channel transports predominantly Na^+ into the cell, at 0 mV the Na^+ inward transport is balanced by K^+ outward transport, and at positive potentials the channel transports predominantly K^+ out of the cell.

The current through GABA-activated channels shows a completely different voltage dependence (Fig. 3b). Close to the neuron's resting membrane potential, the current is not detectable. At membrane potentials more positive one measures outward currents. The outward current charges the membrane capacitance and polarizes the cell membrane. The voltage dependence reflects the Cl^- selectivity of the GABA-receptor (GABAR) channel and the distribution of Cl^- across the neurons' soma membrane: At the resting potential, Cl^- is close to electrochemical equilibrium and the net Cl^- transport through an open channel is only small. As membrane voltage approaches 0 mV, the channel transports more Cl^- into the cell because of the lower intracellular Cl^- concentration.

The open-channel current-voltage relations (*i-V* relations) shown in Figure 3c summarize the voltage-dependent transport of Na^+, K^+, and of Cl^- through ACh-receptor (AChR) and GABAR channels in their normal ionic environment and in the physiological range of membrane voltage. The current through the AChR channel is largest at the cell's resting potential. The direction of current flow is insensitive to small changes in resting membrane potential. The direction of the current through the GABAR channel is critically dependent on small alterations of the resting membrane voltage. The current is largest when the membrane potential is close to zero or positive, for example, as a result of electrical activity of summated excitatory postsynaptic potentials.

Ion Transport Through AChR and GABAR Channels

The mechanism of ion permeation through membrane pores is best understood for pore-forming oligopeptides like gramicidin A in artificial bilayers (Urban et al. 1978). One reason for this is that in bilayer experiments the species and concentration of ions on both faces of the bilayer is under control of the experimenter. Simple manipulations with the patch pipette (Fig. 2a) allow physical isolation of small patches of membrane from the surfaces of many cells and thus also permit the measurement of ion transport by receptor channels in simplified ionic gradients.

Figure 4 shows the voltage dependence of current

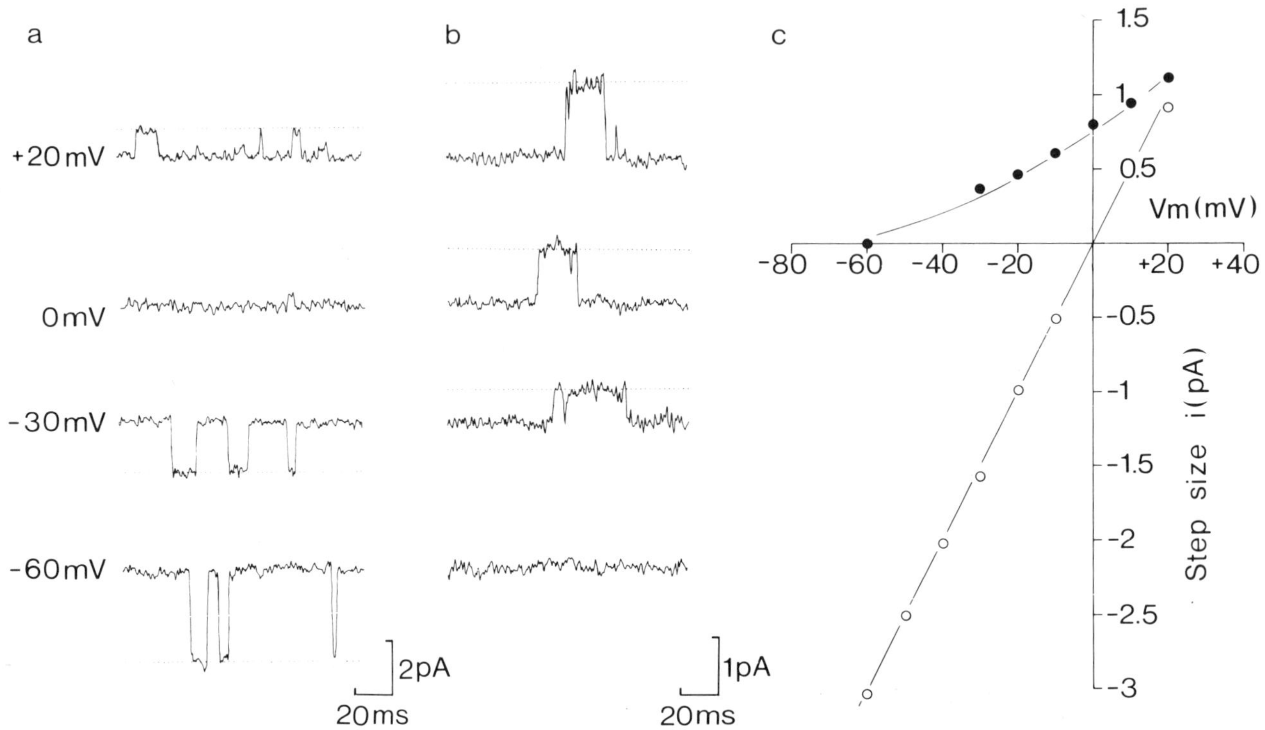

Figure 3. Current-voltage relations of ACh- and GABA-gated channels in "cell-attached" membrane patches. Both experiments were carried out at 23°C, using mammalian Ringer's solution. (*a*) ACh-activated current steps in embryonic rat muscle. Downward deflection marks inward current. Step size decreases as the membrane voltage, indicated at the left of each trace, decreases. Current inverts direction at 0 mV. (*b*) GABA-activated current steps in somal membrane of a rat hippocampal neuron. Upward deflection marks opening of a channel. It represents outward current and is caused by the inward flow of Cl^-. Membrane potentials are the same as in *a*. Step size increases as the membrane potential becomes more positive. (*c*) Open channel *i*-*V* relation of AChR (○) and GABAR channel (●). Step size *i* is plotted as a function of membrane potential, V_m. The continuous line connecting closed symbols (GABAR channel) represents the constant field *i*-*V* relation with P_{Cl} of 6×10^{-14} cm³/sec and intracellular Cl^- concentration of 13 mM.

flow through AChR and GABAR channels with equal ion concentrations on both membrane faces. The membrane patch containing AChRs is exposed to 150 mM NaCl on both faces (Fig. 4a). The patch containing GABARs is bathed in 145 mM Tris-Cl (Fig. 4b). The direction of current flow is inward at negative, zero with no membrane potential, and outward at positive potentials. The open channel *i*-*V* relations are linear. The single-channel conductance $\gamma = i/V_m$, which is a measure of the channel's capacity to transport ions, is 47 pS and 31 pS under these conditions.

The two *i*-*V* relations look very similar; however, the current steps reflect transport of Na^+ and Cl^-, respectively. Inward currents represent Na^+ transport from the extracellular to the cytoplasmic membrane side when the AChR channel is open (Fig. 4a) and Cl^- transport from the cytoplasmic side to the extracellular side when the GABAR channel is open (Fig. 4b). This is shown by decreasing the salt concentration on the cytoplasmic membrane face to 50 mM. The zero current potential is then shifted from 0 mV to +25 mV membrane potential in the case of the AChR channel and to −25 mV membrane potential in the case of the GABAR channel. These shifts are close to the expected diffusion potentials for a channel selective for cations or anions, respectively. In the case of the AChR channel, replacement of Na^+ on the cytoplasmic membrane face by K^+ mimics the physiological cation distribution. The *i*-*V* relation under this biionic condition is almost identical with the one shown in Figure 4a. The zero current potential is close (±3 mV) to 0 mV. This demonstrates in a direct way that the AChR channel does not select much between Na^+ and K^+.

The open-channel *i*-*V* relation with equal (150 mM) Na^+ or Cl^- concentrations on both sides of the membrane is linear between −70 mV and +70 mV, but shows rectification when ion concentrations are unequal. This is consistent with the physical picture of an electrically "homogeneous" channel in which the current flow is treated as a simple case of nonsaturating electrodiffusional transport (Hille 1975). For such a simple channel, the open-channel *i*-*V* relation would be described by

$$i = z^2 C \; P \; F \quad (V_m \; F/R \; T) \tag{1}$$

where z is the valency of the transported ion and F, R, and T have their usual meaning. The ion transport properties of the channel are described by the voltage and concentration independent permeability constant P. The channel conductance $\gamma = i/V_m$ should increase linearly with the concentration C of the permeant ion.

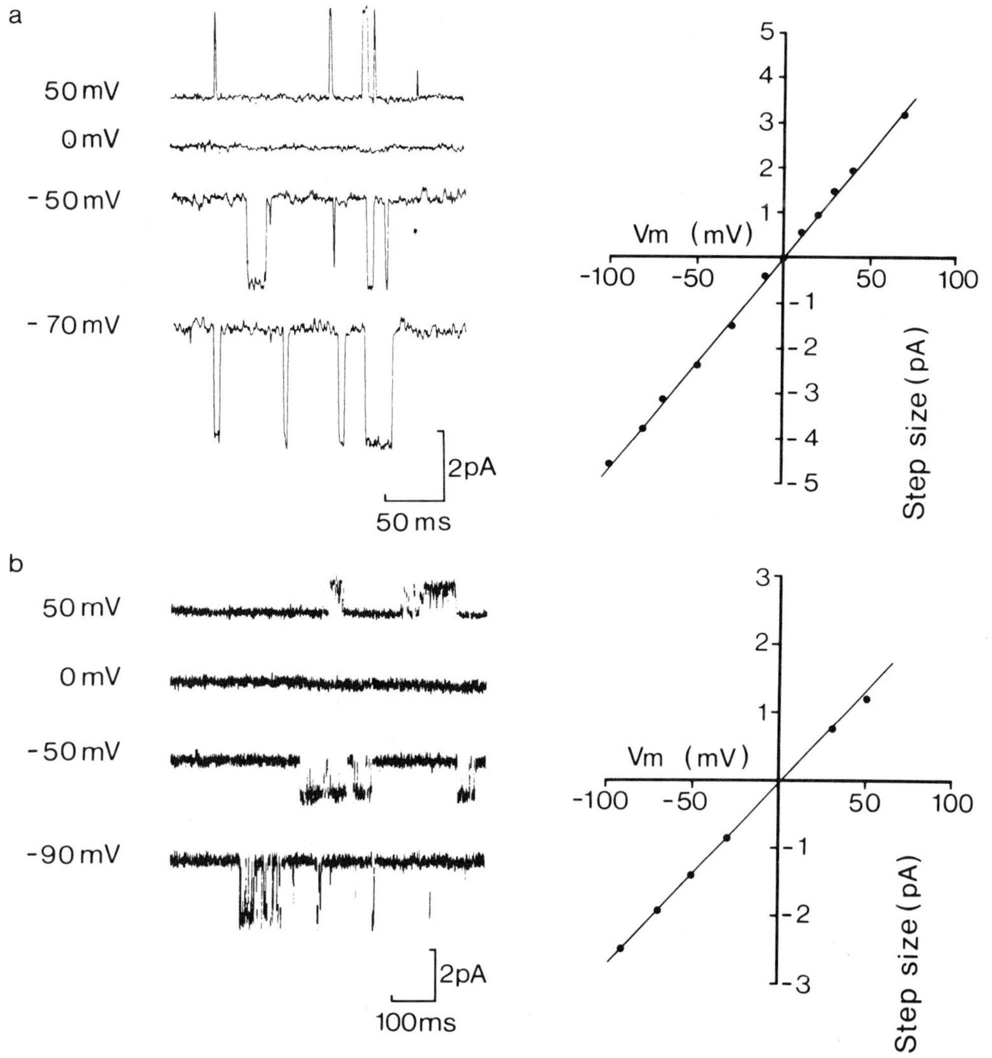

Figure 4. Current-voltage relations of ACh- and GABA-gated channels in outside-out membrane patches with symmetrical ion concentrations. (*a, left*) Currents through AChR channel. Membrane patch was isolated from embryonic rat muscle fiber. Numbers at the left side of each trace indicate membrane potential. Temperature was 23°C. Bath and pipette solution contained 150 mM NaCl. ACh (2 μM) was added to bath solution. (*a, right*) The *i-V* relation, average step size plotted as a function of membrane potential V_m, is linear. The open-channel conductance of the AChR channel is 47 pS. (*b, left*) Currents through GABAR. Membrane patch was isolated from the soma of mouse spinal neuron (Ransom et al. 1977). Temperature was 23°C. Bath and pipette solution contained 145 mM Tris-Cl. GABA (10 μM) was added to the bath solution. (*b, right*) Voltage dependence of GABA-activated currents under symmetrical transmembrane ion conditions. The *i-V* relation is linear, and the channel conductance is 31 pS.

However, when *i-V* relations of the AChR and the GABAR channels are measured with different Na^+ or Cl^- concentrations, respectively, one observes that the concentration-conductance relation deviates from linearity, since the channel conductance approaches a saturating value γ_{max} as the concentration of the permeant ion increases. Figure 5 shows the conductance of AChR channels measured in outside-out patches isolated from rat embryonic muscle fibers in symmetrical Na^+ concentrations ranging from 20 mM to 300 mM. At all concentrations, the *i-V* relations passed through the origin. They were linear and symmetrical in the potential range of ±70 mV at all Na^+ concentrations but showed a sublinear behavior at higher potentials. In Figure 5, therefore, the conductance measured at −70 mV is plotted as a function of Na^+ activity. The saturation behavior is fitted well by the empirical Michaelis-Menten relation $\gamma = \gamma_{max}\, a/(a + K_m)$, where γ_{max} is the maximal channel conductance, a the activity of Na^+, and K_m, an empirical constant specifying the ion concentration at which the channel's conductance is half-maximal. The values of γ_{max} and K_m for the predominant, nonsynaptic class of AChR channel (see below) are 77 pS and 87 mM Na^+ at 23°C (Fig. 5, closed circles) (see also Horn and Patlak 1980). A comparable saturation behavior of the Cl^- conductance of the GABAR channel of spinal cord and dorsal root ganglia cells is observed. The maximal conductance and the half-saturating Cl^- concentration for this channel are 73 pS and 178 mM Cl^-.

One possible physical picture of a channel with saturating conductance is that of a water-filled pore that

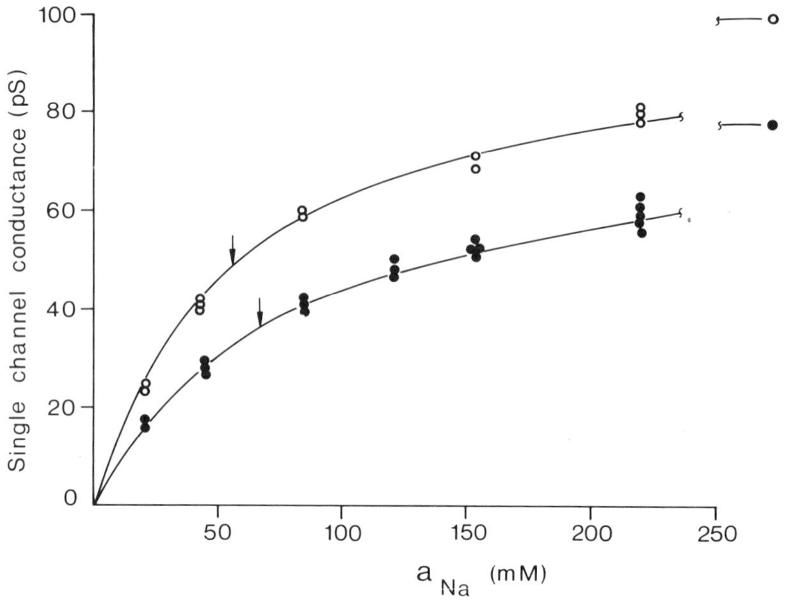

Figure 5. Ion concentration dependence of current through AChR channels. All experiments were carried out on outside-out patches isolated from embryonic rat muscle at 23°C. The concentration of NaCl was equal on both membrane faces (pH 7.3). With NaCl concentrations below 150 mM osmolarity was adjusted by glucose. Single-channel currents were activated by 1–2 μM ACh added to the bath solution. At each NaCl concentration, the *i-V* relation was measured between −100 mV and +70 mV membrane potential. Open-channel conductance was measured at −70 mV and is plotted as a function of Na^+ activity (a_{Na}). Each symbol represents one patch. (●) Nonjunctional class of AChRs; (○) junctional class of AChRs. The continuous lines represent a fit to the data by $\gamma = \gamma_{max} \cdot a_{Na}/(a_{Na} + K_m)$. To obtain γ_{max} and K_m, data were plotted according to Eadie Hofstee and fitted by linear regression. For each class of channels, K_m is indicated by arrows and γ_{max} by the horizontal line on the right.

contains only a single ion at a time (Läuger 1973). If there is only a single binding site *X* in the channel, ion permeation can be described by the two-step scheme of Hille (1975), where a permeating ion *I* crosses two energy barriers when passing through the channel. In analogy with enzyme-substrate interactions, this is represented by

$$I + X \underset{k_{-1}}{\overset{k_1}{\rightleftharpoons}} IX \underset{k_{-2}}{\overset{k_2}{\rightleftharpoons}} X + I \qquad (2)$$

The rates in this kinetic scheme are voltage-dependent and it has been used to describe the voltage dependence of the apparent AChR channel open time (Marchais and Marty 1979) and the sublinear current-voltage relation of the ACh-dependent whole-cell current (Horn and Brodwick 1980). It can, at least qualitatively, also account for the observed conductance saturations of single AChR and GABAR channels described here.

Heterogeneity of Membrane Channels

Transmitter-operated channels are distributed unevenly in the cell membrane. Their density is highest in the postsynaptic membrane in close proximity to the transmitter-secreting nerve terminal. Receptor channels are, however, already present in embryonic cells prior to innervation, albeit at much lower density than in the synaptic membrane. In noninnervated rat embryonic muscle fibers (myotubes), AChR channel currents are heterogeneous (Fig. 6a). With respect to their conductance, AChR channels fall into three discrete classes (two of them are shown in Fig. 6a). In the innervated postsynaptic membrane of the adult rat muscle fiber, on the other hand, only a single class of channels is present (Fig. 6b). Its conductance is very similar to that of the largest class observed in the embryonic fiber. This indicates that the type of AChR channel present in the muscle is controlled somehow by the nerve. In terms of the Na^+ concentration dependence of the channel conductance, the effective binding constant K_m is 71 mM Na^+. This is similar to the nonjunctional class. However, the maximal channel conductance of the junctional class of channel is 99 pS, i.e., considerably larger than that of the other, nonjunctional, class (Fig. 5, open symbols). Thus, one difference between junctional and nonjunctional AChR channels is in the rate-limiting step in ion permeation, i.e., the rate of dissociation of the permeating ion from the channel site *X* in Equation 2 is larger for junctional channels. Thus far, it is not known where in the channel this site is located and what structure is responsible for the rate-limiting step.

The heterogeneity in the transport properties of different AChR channels could reflect different states of aggregation of the receptor subunits, e.g., a different number or a different orientation of the α-helical domains that presumably form the AChR pore (Noda et al. 1983). The exact pattern of aggregation could, in part, depend on the membrane environment of the channel.

Coupling of Receptor and Channel

In the previous section, we described the transport of ions through the open channel. Here, we consider the mechanisms underlying the switching on and off of the channel. The picture of the AChR and GABAR channels that emerges from single-channel-current recordings is that they exist predominantly in either of two discrete states, closed or open. The transition between the two states is largely controlled by the occupancy of the binding site with a ligand. In both the AChR and the GABAR channels, the binding site and the ion channel

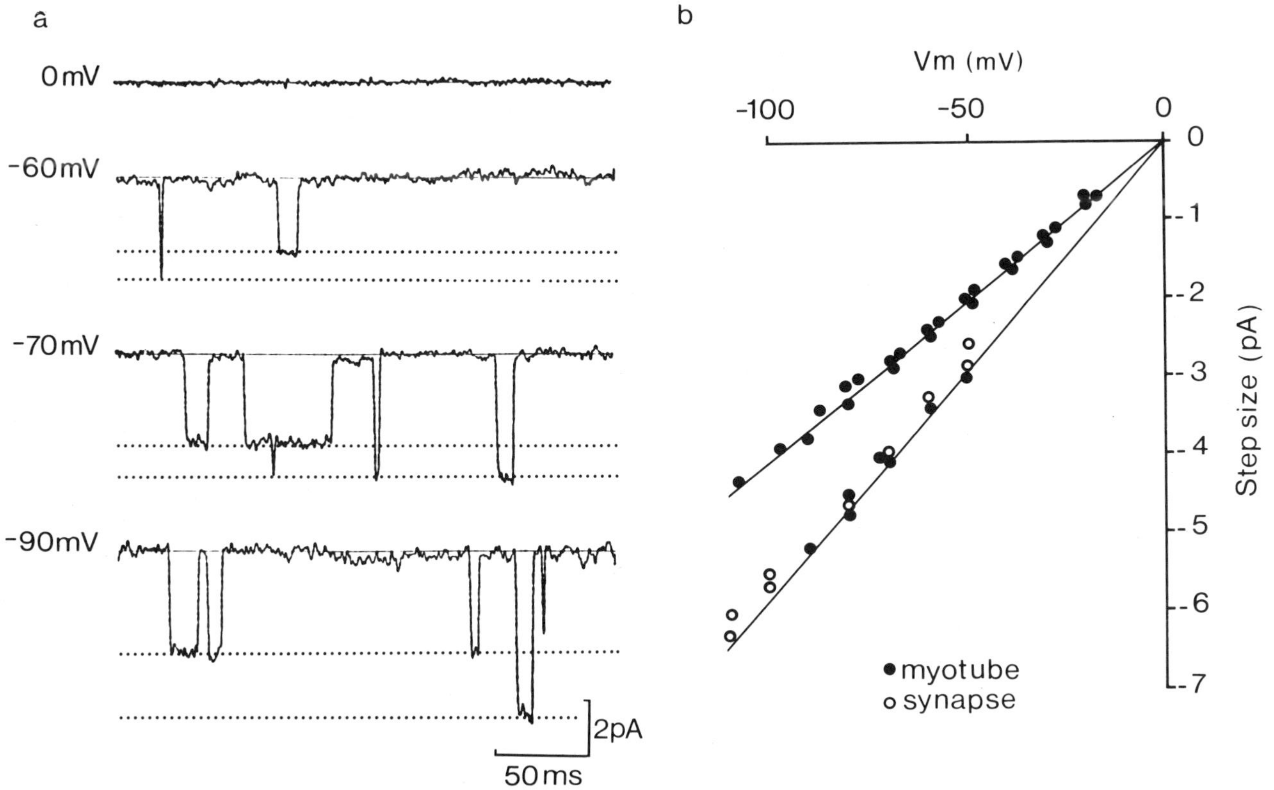

Figure 6. Heterogeneity of AChR channels in rat muscle. (*a*) Single-channel currents recorded from cell-attached patch of embryonic rat muscle (18°C, mammalian Ringer's solution). Two classes of current steps with different amplitudes are apparent. Numbers at the left indicate membrane potential. (*b*) *i-V* relations of AChR channels in embryonic (●) and adult (○) rat muscles. Data were from two patches in each case (cell-attached patches; 18°C, mammalian Ringer's solution). ACh at 2 μM (embryonic fiber) and 100 mM (adult end plate) was added to the pipette solution. The straight lines indicate single-channel conductances of 41 pS and 57 pS for the two classes under this condition.

reside in the same protein macromolecule. The simplest reaction scheme to describe coupling of binding site and channel is that suggested by del Castillo and Katz (1957), which assumes

$$nA + R \underset{k_{-1}}{\overset{k_{+1}}{\rightleftharpoons}} A_n T \underset{\alpha}{\overset{\beta}{\rightleftharpoons}} A_n T^* \qquad (3)$$

where n transmitter molecules bind to the receptor to form the liganded closed state A_nT, which isomerizes to the open A_nT^* state. This equation neglects a number of other reaction pathways occurring on a slower time scale that can be regarded as kinetically decoupled from the channel opening step. The coupling between binding site and channel is described by the opening rate β and the closing rate α. In the following sections, estimates of both rates will be derived for the AChR and the GABAR channels.

Reopening of the Channel during a Single Occupancy of the Binding Site

When the time course of the current through the ACh- or GABA-gated channel is examined at high time resolution, it becomes obvious that the flow of current is not completely smooth but is interrupted briefly at irregular intervals, as illustrated in the insets of Figure 7. These interruptions are so brief that their true time course in many cases is damped by the limited frequency response of the recording apparatus. It is unlikely that they arise from the unresolved superpositions of independent channel openings and closings, since they are observed when the average frequency of channel openings is very low. One possible explanation for these interruptions given by Colquhoun and Sakmann (1981) is that they represent brief closures and reopenings of the channel during a single occupancy of the binding site by the ligand(s). According to this suggestion, an elementary current, following ligand binding, would consist of a burst of several current pulses and not a single pulse. In terms of the reaction scheme of del Castillo and Katz, these bursts represent flickering of the occupied channel between the nonconducting A_nT and the conducting A_nT^* state. In the records shown in the inset of Figure 7a, the AChR would go from the resting state R to the A_nT state just before the current begins to flow; as long as the current is flowing, the AChR is in the A_nT^* state. It briefly returns to the A_nT state during the current interruption, reopens again to A_nT^*, closes via A_nT, and finally returns to the resting state R. In the current records through the

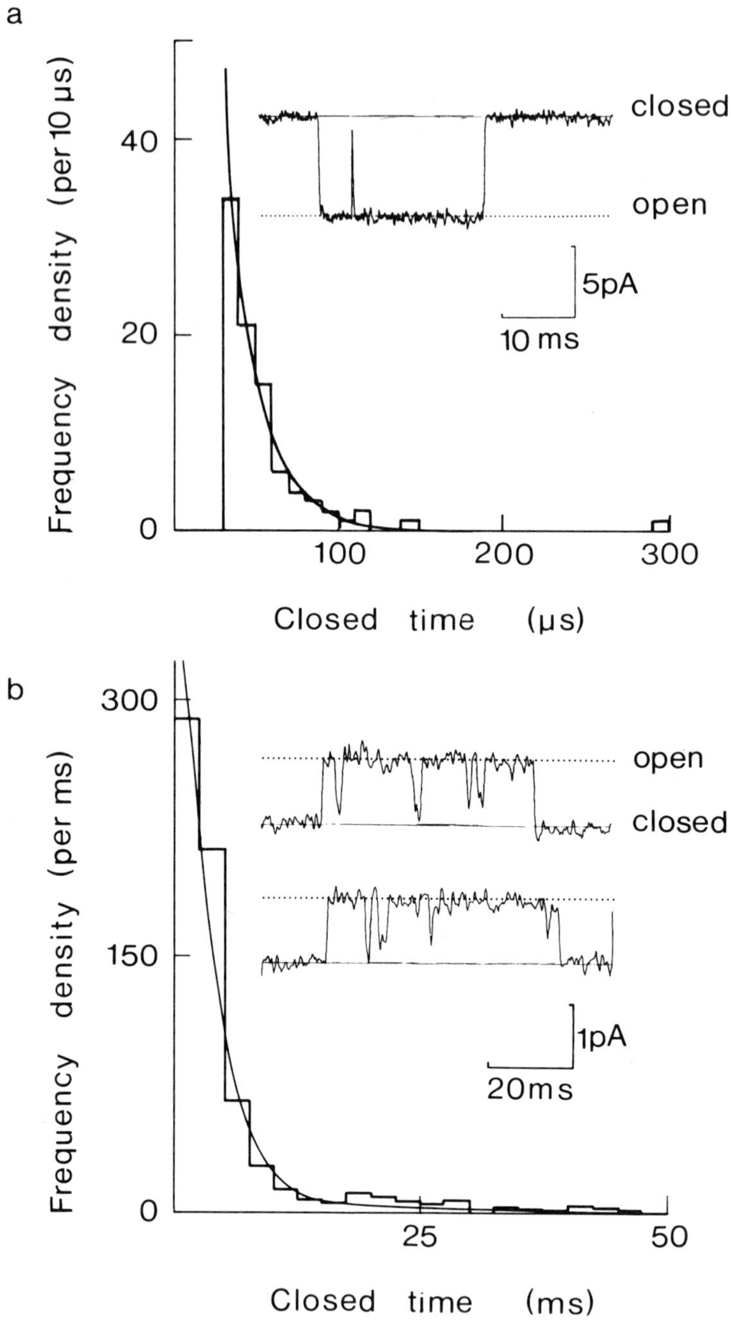

Figure 7. Bursts of current pulses in transmitter-gated channels. (*a, inset*) Current through AChR channel at high time resolution. Current is briefly interrupted once. Histogram shows the distribution of the duration of short interruption (fit by single exponential curve). Average duration is 28 μsec. Average number of interruptions is 2.3 per current burst (from Colquhoun and Sakmann 1983). (*b, inset*) Sample records of current through GABAR channel. Current is interrupted briefly four times. Histogram shows distribution of short interruptions. Average duration is 1.8 msec, average number is 4.2 per burst.

GABAR channel shown in Figure 7b, such state transitions would occur several times.

It has been suggested that the AChR channel is formed in a barrel-stave-like fashion by the α-helical domains of the receptor subunits (Noda et al. 1983). A possible interpretation of the brief bursts of the current through the AChR could be that they reflect addition and loss of one or more staves to and from the barrel. This would lead to the channel being conducting or non-conducting, respectively. Alternatively, the preferred orientation of the α-helices could be changing after occupation of the binding sites.

Colquhoun and Hawkes (1977) showed that the reaction rates in Equation 3 can be derived from the average number and duration of the closing gaps, i.e., $n_g = \beta/k_{-1}$ and $t_g = (\beta + k_{-1})^{-1}$. These are obtained from the distribution of the duration of these interruptions as shown in Figure 7 for the AChR and the GABAR. The closing rate α is estimated from the average burst duration. The result of such an analysis shows that the switching rates for the AChR are $\beta = 35{,}000$/sec and $\alpha = 900$/sec (Colquhoun and Sakmann 1983) and $\beta = 330$/sec, $\alpha = 50$/sec for the GABAR. The agonist dissociation rates k_{-1} are

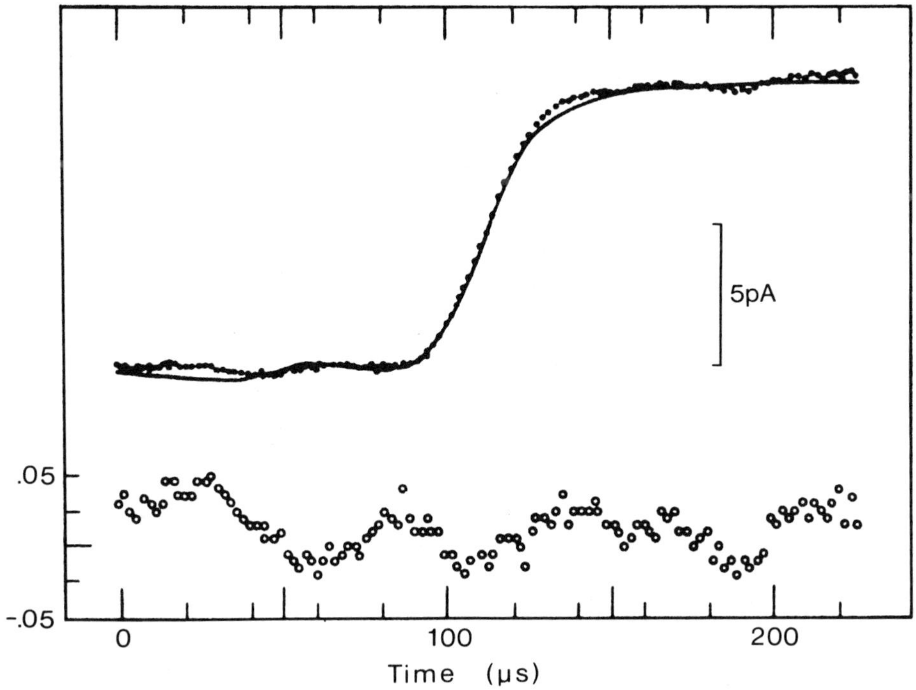

Figure 8. Opening time course of AChR channel. The filled points represent averaged time course of the current through frog end-plate channels activated by suberyldicholine, an analog of ACh. The increase in current, representing the opening of a channel, is shown as upward deflection. The superimposed continuous line is the averaged output of the recording apparatus to an ideal rectangular current step. The open points below the two traces show on a log scale the difference between a single-channel current recording and a recording of the response to rectangular current step. (Reprinted, with permission, from Hamill et al. 1981.)

8600/sec and 50/sec, respectively, assuming that the channel can open only when at least two transmitter molecules are bound.

A comparison of the various rates shows that for both receptor channels, the reaction scheme can be considered as the intermediate steady-state case common in enzyme kinetics (Bernasconi 1976). Here, two reaction steps, ligand binding and channel opening, are kinetically coupled. The average duration of an elementary current resulting from the occupancy of the binding site is determined by the reaction rates of both binding and isomerization. The difference in the isomerization rates between the GABAR and the AChR channels is that the free energy necessary to jump from the activated closed state A_nT to the open state A_nT^* and back is larger in the case of the GABAR channel.

Time Course of the Closed-open Transition

By measuring the time course of current flow through a single channel, one could, in principle, observe the time course of the structural change causing the channel to open. Figure 8 shows the wave shape of current through the AChR at high time resolution. The time course of the leading edge of an agonist-activated current pulse is compared with the time course of the response of the recording instrument to an ideal rectangular current step of the same amplitude. Both traces superimpose, indicating that the time course of the single-channel current is limited by the frequency response of the instrument. The difference between the two curves is plotted below. There is no significant difference between the two curves during the transition. Calculations of the step-response to current pulses with transition times larger than 10 μsec show that this would lead to a significant difference. Thus, the duration of the structural change causing the channel to open is shorter than 10 μsec.

Other States of the Channel

In the preceding sections we described the probability of transitions of the channel between one particular closed state, A_nT, and the open state, A_nT^*. For the physiological action of the transmitter, this is probably the most significant reaction step. The occupied receptor-channel can, however, adopt a number of other, kinetically different, states. First, there are closed states that are designated desensitized states. They are significantly populated at higher transmitter concentrations, and the transition rates to the desensitized states are much smaller (< 100/sec). This causes grouping of elementary currents in bursts and clusters (Sakmann et al. 1980). Second, the distribution of burst durations shows an excess of "short bursts" (Colquhoun and Sakmann 1981; Sakmann et al. 1983), indicating that the channel can adopt two, kinetically different, open states. Third, receptor channels adopt substates of conductance. The channel is only partially open (Hamill and Sakmann 1981; Auerbach and Sachs

1983). The significance of these states for the normal function of the channel is as yet unclear. Their analysis might, however, shed some additional light on the mechanism of how the structure of the channel is controlled by the binding site.

DISCUSSION

The measurement of the current through single-receptor pores allows one to examine ion permeation through the channel and gating of the channel independently from each other. The aim of analyzing single-channel current records is to find out what mechanism determines the size of the current through an open channel, the average duration of the current, and the frequency of occurrence of current pulses. A first step for this is to derive plausible reaction schemes for ion permeation and receptor-channel coupling. We have compared such plausible mechanisms for two receptor channels that transport different ions and that subserve different functions, i.e., postsynaptic excitation and inhibition.

The open-channel *i-V* relation of both receptor channels is, in a first approximation, consistent with an electrically homogeneous channel. Their nonlinear ion concentration–conductance relation indicates, however, that in AChR and GABAR channels, permeating ions bind to a saturatable site in the channel. A simple model of the electrical profile of the channel to account for saturation is that of a single binding site between two barriers. With this assumption, the data would indicate that the rate-limiting step for ion transport in the AChR and the GABAR channels is in the same order of magnitude. The difference between various AChR channels (i.e., junctional and nonjunctional) lies in the magnitude of the rate-limiting step. Under physiological ion conditions, the AChR channel is nearly saturated, whereas the GABAR channel is not.

The transition of receptor channels between closed and open is in its simplest form described by an isomerization of the liganded channel (i.e., when its binding sites are occupied). The time required for a transition is less than 10 μsec. Under the assumption that the brief interruptions of the current flow represent closing and reopening of the occupied channel, one can estimate the rate constants for channel isomerization, i.e., opening and closing (but see Auerbach and Sachs 1983). The estimated opening rate exceeds severalfold the closing rate. This is a very effective way to switch the channel on. The current resulting from the occupancy of the binding sites is a burst of several current pulses whose duration is determined by both the isomerization and the dissociation rates. An estimate of the microscopic equilibrium dissociation constant of the binding site can also be obtained. Assuming a conventional association rate of $10^8\ \text{M}^{-1}\cdot\text{s}^{-1}$, one arrives at estimates of 100 μM and 0.5 μM for ACh and GABA binding. These values, measured by a functional assay, are more than an order of magnitude larger than the apparent equilibrium dissociation constants derived from binding measurements (for review, see Karlin 1980; Olsen 1982). This discrepancy indicates that in the continued presence of ligands, AChR and GABAR undergo a transition into a high-affinity, probably "desensitized," state during the time required to measure binding constants by biochemical methods.

The values of the reaction rates coupling receptor and channel are much larger for AChR channels, which mediate excitation, than for inhibition-mediating GABAR channels. The difference in equilibrium and kinetic constants could explain the largely different durations of postsynaptic conductance changes in excitatory synapses and inhibitory synapses.

The description of ion permeation through the open channel and the coupling of ligand binding and channel opening in terms of plausible reaction schemes is one initial step toward the understanding of ion transport via receptor channels. To find a mechanistic basis for the observed kinetic behavior, the transitions between different functional states of a receptor pore must be related to equivalent changes in the channel's structure. One way to approach this goal is to find out how reaction pathways are altered by defined chemical modifications of different parts of the receptor-channel molecule. Another way to do this is to determine whether the difference in ion selectivity, conductance, and switching behavior between the two receptor channels described here is reflected in their molecular structure.

REFERENCES

Anderson, C.R. and C.F. Stevens. 1973. Voltage clamp analysis of acetylcholine produced end-plate current fluctuations at frog neuromuscular junction. *J. Physiol.* **235:** 655.

Auerbach, A. and F. Sachs. 1983. Flickering of a nicotinic ion channel to a subconductance state. *Biophys. J.* **42:** 1.

Bernasconi, C.F. 1976. *Relaxation kinetics.* Academic Press, New York.

Colquhoun, D. and A.G. Hawkes. 1977. Relaxation and fluctuations of membrane currents that flow through drug-operated ion channels. *Proc. R. Soc. Lond. B* **199:** 231.

———. 1982. On the stochastic properties of bursts of single ion channel openings and of clusters of bursts. *Philos. Trans. R. Soc. Lond. B* **300:** 1.

Colquhoun, D. and B. Sakmann. 1981. Fluctuations in the microsecond time range of the current through single acetylcholine receptor channels. *Nature* **294:** 464.

———. 1983. Bursts of openings in transmitter-activated channels. In *Single channel recording* (ed. B. Sakmann and E. Neher), p. 345. Plenum Press, New York.

Coombs, J.S., J.C. Eccles, and P. Fatt. 1955. The specific ionic conductances and ionic movements across the motoneuronal membrane that produce the inhibitory postsynaptic potential. *J. Physiol.* **130:** 326.

del Castillo, J. and B. Katz. 1957. Interaction at end-plate receptors between different choline derivatives. *Proc. R. Soc. Lond. B* **146:** 369.

Hamill, O.P. and B. Sakmann. 1981. Multiple conductance states of single acetylcholine receptor channels in embryonic muscle cells. *Nature* **294:** 962.

Hamill, O.P., A. Marty, E. Neher, B. Sakmann, and F.J. Sigworth. 1981. Improved patch-clamp technique for high-resolution current recording from cells and cell-free membrane patches. *Pfluegers Arch. Eur. J. Physiol.* **391:** 85.

HILLE, B. 1975. Ionic selectivity of Na and K channels of nerve membranes. In *Membranes* (ed. G. Eisenmann), vol. 3, p. 255. Marcel Dekker, New York.

HORN, R. and M.S. BRODWICK. 1980. Acetylcholine-induced current in perfused rat myoballs. *J. Gen. Physiol.* **75:** 297.

HORN, R. and J. PATLAK. 1980. Single channel current from excised patches of muscle membrane. *Proc. Natl. Acad. Sci.* **77:** 6930.

KATZ, B. and R. MILEDI. 1972. The statistical nature of the acetylcholine potential and its molecular components. *J. Physiol.* **224:** 665.

KARLIN, A. 1980. Molecular properties of nicotinic acetylcholine receptors. In *The cell surface and neuronal function* (ed. C.W. Cotman et al.), p. 191. Elsevier/North-Holland, New York.

LÄUGER, P. 1973. Ion transport through pores: A rate theory analysis. *Biochim. Biophys. Acta* **311:** 423.

MARCHAIS, D. and A. MARTY. 1979. Interaction of permeant ions with channels activated by acetylcholine in *Aplysia* neurones. *J. Physiol.* **297:** 9.

NEHER, E. and B. SAKMANN. 1976. Single-channel currents recorded from membrane of denervated frog muscle fibres. *Nature* **260:** 799.

NEHER, E. and C.F. STEVENS. 1977. Conductance fluctuations and ionic pores in membranes. *Annu. Rev. Biophys. Bioeng.* **6:** 345.

NODA, M., H. TAKAHASHI, T. TANABE, M. TOYOSATO, S. KIKYOTANI, Y. FURUTANI, T. HIVOSE, H. TAKASHIMA, S. INAYAMA, T. MIYATA, and S. NURA. 1983. Structural homology of *Torpedo californica* acetylcholine receptor subunits. *Nature* **302:** 528.

PECK, E.J. 1980. Receptors for amino acids. *Annu. Rev. Physiol.* **42:** 615.

OLSEN, R.W. 1982. Drug interactions at the GABA receptor ionophore complex. *Annu. Rev. Pharmacol. Toxicol.* **22:** 245.

RANSOM, B.R., E. NEALE, M. HENKART, P.N. BULLOCK, and P.G. NELSON. 1977. Mouse spinal cord in cell culture. I. Morphology and intrinsic neuronal electrophysiologic properties. *J. Neurophysiol.* **40:** 1132.

SAKMANN, B. and E. NEHER. 1983. Geometric parameters of pipettes and membrane patches. In *Single channel recording* (ed. B. Sakmann and E. Neher), p. 37. Plenum Press, New York.

SAKMANN, B., O.P. HAMILL, and J. BORMANN. 1983. Patch-clamp measurements of elementary chloride currents activated by the putative inhibitory transmitters GABA and glycine in mammalian spinal neurons. *J. Neural Trans.* (suppl.) **18:** 83.

SAKMANN, B., J. PATLAK, and E. NEHER. 1980. Single acetylcholine-activated channels show burst-kinetics in presence of desensitizing concentrations of agonist. *Nature* **286:** 71.

SIGWORTH, F.J. and E. NEHER. 1980. Single Na^+-channel currents observed in cultured rat muscle cells. *Nature* **286:** 447.

URBAN, B.W., S.B. HLADKY, and D.A. HAYDON. 1978. The kinetics of ion movements in the gramicidin channel. *Fed. Proc.* **37:** 2628.

Excitable Membrane Properties of Cultured Central Nervous System Neurons and Clonal Pituitary Cells

J.L. BARKER,* B. DUFY,† D.G. OWEN,* AND M. SEGAL‡
*Laboratory of Neurophysiology, National Institute of Neurological and Communicative Disorders and Stroke, National Institutes of Health, Bethesda, Maryland 20205; †Laboratoire de Neurophysiologie, Universite de Bordeaux II, 33076 Bordeaux, France; ‡Isotope Department, Weizmann Institute, Rehovot, Israel

Excitability and the communication of changes in the excitability of neurons are considered integral to the functioning of all nervous systems studied thus far. Cellular excitability involves the development and maintenance both of ionic gradients across membranes and of mechanisms to discharge and restore these gradients. The mechanisms underlying the excitability of neurons in the vertebrate CNS have been difficult to study in vivo owing to the relative complexity and inaccessibility of these cells to the appropriate electrophysiological recording techniques. Over the past 15 years, methods have been developed for isolating cellular elements from the embryonic CNS and maintaining them in culture. Since the cellular elements organize into a monolayer, they become readily accessible to a variety of electrophysiological assays at a relatively quantitative level. However, the experimental accessibility gained in monolayer culture is compromised to a degree by the absence of developmental signals in vitro. Thus, monolayer cultures of embryonic CNS neurons represent a definite, but unknown, departure from normally developed tissue.

Thus far, the results of electrophysiological experiments performed on cultured central neurons have shown that both electrically and chemically excitable membrane properties are resident in virtually all of the cells studied, and further that many neurons spontaneously exhibit different types of synaptic signals (for recent discussion, see Nelson et al. 1981). Although these properties superficially resemble those recorded in vivo, it is not yet possible to identify the types of cells being recorded in culture, and it is therefore difficult to compare, in a quantitative manner, excitable properties found in vitro with those observed in vivo.

We have begun to examine the excitable membrane properties resident in cells cultured from spinal and hippocampal regions of the embryonic mammalian CNS using electrophysiological recording techniques that allow semiquantitative and quantitative levels of analysis. We have focused on two properties that are practically ubiquitous in cultured spinal and hippocampal neurons and are thus more readily available for detailed study. Both properties function to regulate, in indirect although different ways, the triggering of regenerative action potentials. Since one of these properties is also expressed in clonal pituitary cells, we have begun to study its properties in these nonneuronal elements as well.

METHODS

Preparation of cells for maintenance in dissociated cell culture. Most of the cells we have studied have been dissected from spinal and hippocampal regions of mice and rats relatively late in embryogenesis. The cells are best maintained in culture using media supplemented initially with fetal calf serum and horse serum and later with horse serum alone. The media in the cultures are routinely changed every 2 or 3 days. With this protocol, cells can usually be maintained in culture for weeks to months without difficulty.

It should be clear that this rather primitive protocol represents a radical departure from the physiological environment of the developing CNS in vivo. The identity of the surviving cell population and the fidelity of their properties relative to those normally expressed in vivo are uncertain. Development of a strategy for maintaining specific populations of identified central neurons in culture is prerequisite to advancing the utility of monolayers as a preparation to study the cellular properties of CNS neurons.

Electrophysiological recordings of cultured neurons. For the most part we have applied single- and two-electrode recording techniques using phase-contrast optics at 250× magnification to visualize the cells. All of the cells have been recorded with somal impalements, with the experiments being carried out at room temperature. Most CNS neurons studied thus far have had cell-body diameters ranging in size from 15 μm to 30 μm. Many of these cells have extensive and complex processes, only the largest of which can be visualized with phase-contrast optics. In contrast, pituitary cells of the GH3/6 clone are usually circular, with 15-μm-diameter cell bodies and little or no process formation. Many of our two-electrode studies have utilized a multistage voltage-clamp circuit designed by Dr. T.G. Smith, Jr. (Smith et al. 1980), although some have employed a commercially available single-electrode voltage-clamp method incorporating a high-frequency switching circuit. With care, stable recordings with 50–100 MΩ microelectrodes lasting several hours can usually be obtained. However, the fidelity of the electrophysiological assay depends, in an uncompromising way, on the quality of the intracellular penetration(s). The application of voltage-clamp techniques in conjunction with relatively precise control over the complexion of the recording medium and the delivery of transmitter substances and drugs permit semiquantitative and quantitative analysis of excitable membrane properties at the level of the whole cell. Some attempts have also been made using the extracellular patch-clamp technique (Jackson et al. 1982), which restricts the area of the excitable membrane under study to about a micrometer-sized patch. This technique has thus far proved difficult to use with great consistency on neurons cultured for a month or more.

RESULTS

Cultured spinal and hippocampal neurons exhibit a complex spectrum of excitable membrane properties in addition to ubiquitous action potential activity. Two of these are expressed in the majority of cells recorded. One is an example of a chemically excitable membrane function, involving regulation of ion conductance by endogenous transmitter substances and exogenous drugs, whereas the other reflects a type of excitability that is strongly dependent on the voltage across the membrane and is therefore categorized primarily as an electrically excitable property. However, it too can be regulated by transmitters and drugs.

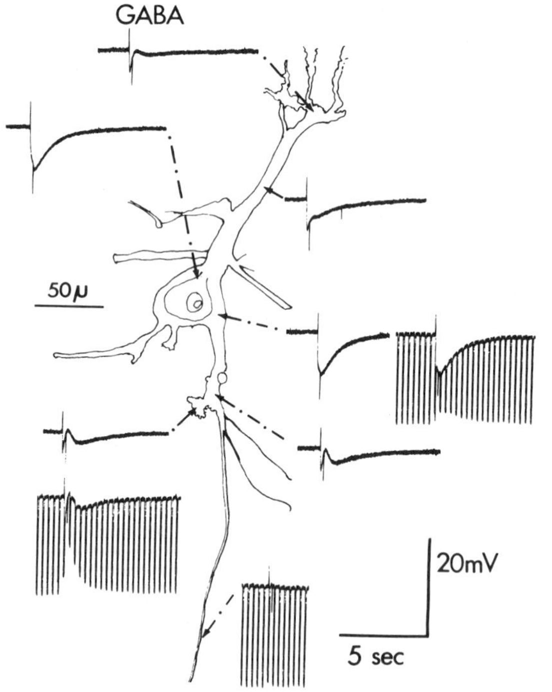

Figure 1. GABA evokes simple and complex responses in a cultured SC neuron. Intracellular recording was accomplished with a microelectrode filled with KAc, and GABA was applied by brief iontophoretic pulses of current at various sites on the cell surface. When applied at the level of the cell body, GABA evokes a monophasic, hyperpolarizing response associated with an increase in conductance, as evidenced by the decrease in voltage responses to 50-msec constant-current hyperpolarizing stimuli. Complex, biphasic responses are elicited when GABA is applied at processes, and these also are associated with an increase in membrane conductance. (Modified from Barker and Ransom 1978.)

Chemically Activated Cl^- Ion Conductances Are Common in Cultured CNS Neurons

Applications of the neutral amino acids γ-aminobutyric acid (GABA), β-alanine, and glycine evoke changes in membrane potential that are associated with an increase in conductance in most SC cells studied (Fig. 1) (Barker and Ransom 1978). The conductance and potential responses elicited at the level of the cell body are strongly dependent on the concentration of, and transmembrane gradient for, Cl^- ions (Barker and Ransom 1978). In most cells recorded with microelectrodes filled with conducting solutions containing impermeant anions, the resting-membrane potential is hyperpolarized in response to the amino acids, as illustrated in Figure 1. By hyperpolarizing the cell, the amino acids act to decrease the probability of action potential generation, an effect that has thus come to be considered as "inhibitory" to cellular excitability.

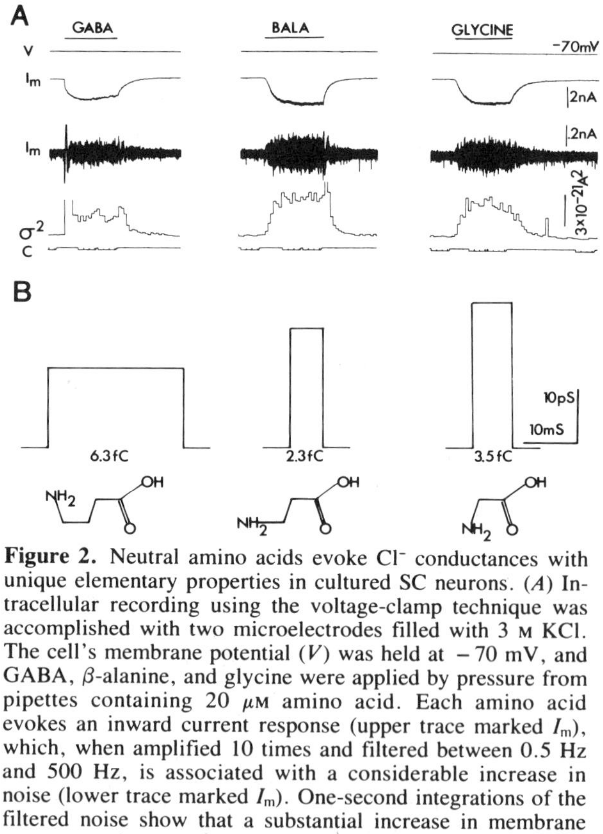

Figure 2. Neutral amino acids evoke Cl^- conductances with unique elementary properties in cultured SC neurons. (*A*) Intracellular recording using the voltage-clamp technique was accomplished with two microelectrodes filled with 3 M KCl. The cell's membrane potential (*V*) was held at −70 mV, and GABA, β-alanine, and glycine were applied by pressure from pipettes containing 20 μM amino acid. Each amino acid evokes an inward current response (upper trace marked I_m), which, when amplified 10 times and filtered between 0.5 Hz and 500 Hz, is associated with a considerable increase in noise (lower trace marked I_m). One-second integrations of the filtered noise show that a substantial increase in membrane current variance (σ^2) occurs coincident with the noise. Samples of the filtered current (downward deflections in trace marked *C*) before, during, and after the amino acid applications have been taken for computer analysis. (*B*) The average lifetime and elementary conductance of ion channels associated with the current responses evoked by the amino acids have been estimated using fluctuation analysis. The properties are schematized for each of the substances (structures shown at the bottom). The average amount of charge (in femtoCoulombs), calculated as the product of the elementary current and the channel duration, shows that GABA passes considerably more charge than either glycine or β-alanine. (Modified from Barker et al. 1982.)

Table 1. Comparison of Cl^- Ion Conductances Activated in Neurons Cultured from Spinal and Hippocampal Regions of the Embryonic CNS

	Property	DRG[a]	Spinal cord[a]	Hippocampus[b]
Pharmacologically activated GABA	γ	14 ± 1 pS	16 ± 2 pS	17 ± 3 pS
	τ	23 ± 3 msec	24 ± 4 msec	20 ± 2 msec
β-alanine	γ	no effect	24 ± 3 pS	no effect
	τ		6 ± 1 msec	
Glycine	γ	no effect	26 ± 4 pS	19 ± 3 pS
	τ		8 ± 2 msec	59 ± 6 msec
Neurally evoked	τ	none	few > 10 msec[c]	most ≅ 20 msec
	G	detected	5 ± 1 nS[c]	29 ± 4 nS
	τ		many < 10 msec	
	G		not studied	

All data are derived from two-electrode voltage-clamp recordings of cells held at −60 mV (temperature 22–24°C). Channel properties have been estimated from fluctuation analysis of membrane current responses evoked by topical application of the amino acids at the level of the cell body. DRG, Dorsal root ganglion; γ, average channel conductance; τ, average channel duration or time constant of decay in synaptic current; G, average spontaneous or evoked synaptic conductance. No effect at 200 μM.

[a]Data from Barker et al. (1982).

[b]Data from Segal and Barker (1983).

[c]Data from Barker and McBurney (1979b).

When the amino acids are applied discretely at various sites on the cell surface, it is clear that (1) cells are not uniformly sensitive to the amino acids but exhibit loci of high sensitivity and (2) more complex, biphasic responses can frequently be evoked on processes. Whether any of these responsive regions correspond to subsynaptic membranes has not yet been determined. In SC cells there is a significant correlation between the magnitudes of responses elicited by different amino acids at various test sites on the same cell. It is also evident that one amino acid can occlude the conductance response to another (Barker and McBurney 1979a), suggesting that these amino acids may utilize the same Cl^- ion conductance mechanism.

Cl^- Ion Conductances Have Different Elementary Properties

We have examined the mechanisms underlying the Cl^- ion conductances elicited by the amino acids using the strategy of fluctuation analysis first employed by Katz and Miledi (1972) and subsequently utilized by many investigators to study transmitter-evoked responses in other excitable membranes (for review, see Mathers and Barker 1982). Katz and Miledi (1972) analyzed the macroscopic "noise" they recorded during acetylcholine-evoked responses on muscle cells in a statistical manner, as if the noise were actually a complex biological signal, reflecting fluctuations in the ac-

Table 2. Clinically Important Drugs Modulate Cl^- Ion Conductance Responses Evoked by GABA in Cultured CNS Neurons

Drug	Response change	Channel mechanism	Reference
Bicuculline	↓	↓frequency	J. Barker et al. (unpubl.)
Picrotoxin	↓	↓frequency	J. Barker et al. (unpubl.)
Metrazol	↓	↓frequency (↓lifetime)	J. Barker et al. (unpubl.)
Methyl-β-carboline	↓	↓frequency	Owen et al. (1982)
Dimethoxy-β-carboline	↓	↓frequency	Owen et al. (1982)
Butyl-β-carboline	↑	↑frequency (↑lifetime)	Owen et al. (1982)
Lorazepam	↑	↑frequency	Owen et al. (1982)
Irazepine	↑	↑frequency	Owen et al. (1982)
Diazepam	↑	↑frequency (↑lifetime)	Study and Barker (1981)
(−)Pentobarbital	↑	↑lifetime (↓frequency)	Study and Barker (1981)

All data are derived from two-electrode voltage-clamp recordings of cells held at −70 mV (temperature 22–24°C). The "channel mechanism" data are derived from fluctuation analysis of membrane current responses evoked pharmacologically under control conditions and in the presence of drug.

tivity of microscopic ion conductance events of similar magnitude but exponentially distributed duration. The Cl^- ion conductance responses evoked by the amino acids in SC cells are consistently associated with an increase in membrane-current variance, or noise (Fig. 2A) (Barker et al. 1982). Analysis of the variance induced by the different amino acids in SC cells indicates that it often appears as if microscopic ion conductance changes of uniform amplitude and exponentially distributed duration comprise the macroscopic conductance response. Direct observations of unitary Cl^- ion channel currents evoked by GABA in patches of SC neurons have largely confirmed estimates made from fluctuation analysis (Jackson et al. 1982), although it also appears that in some patches, there is more than one kinetic process involved in ion channel activation. The estimated properties of the unitary conductance changes evoked by the three amino acids are significantly different from each other (Fig. 2B; Table 1).

Comparison of the estimated properties for Cl^- ion channels activated by the same amino acid (GABA) in neurons cultured from the dorsal root ganglion (DRG), SC, and hippocampus suggests that they are quite similar (Table 1). Since DRG cells in our cultures do not receive any detectable synaptic input, although SC and hippocampal cells do, the properties of GABA-activated Cl^- ion channels appear to be relatively constant and independent of synaptic innervation. However, the estimated properties of Cl^- ion channels activated by glycine in SC cells are significantly different from those activated in hippocampal neurons, which require about tenfold higher glycine concentrations to produce detectable responses (Table 1).

If these pharmacological results are relevant to synaptically operated Cl^- ion conductances mediated by the three amino acids, then we might expect that each amino acid will mediate a significantly different conductance change at synapses. Spontaneously occurring synaptic activity that involves Cl^- ion conductance mechanisms has been recorded in neurons cultured from both embryonic mouse (Ransom et al. 1977; Barker and McBurney 1979b; Nelson et al. 1981) and chick spinal cord (Choi et al. 1981). The results indicate that at least two populations of inhibitory postsynaptic potentials (IPSPs) can be distinguished on the basis of their time course. Many last less than 10 msec, but some last about 20 msec or so. The faster IPSPs are affected by drugs that alter pharmacological responses to glycine, whereas the slower IPSPs are sensitive to drugs that modulate pharmacological responses to GABA (Choi et al. 1981). From these results we tentatively conclude that glycine mediates short-lasting IPSPs and GABA mediates long-lasting IPSPs in cultured SC neurons.

Drugs Modulate the Kinetics of GABA-activated Channels

We have examined the effects of several classes of clinically important drugs on GABA-activated conductance responses evoked in SC neurons, again using fluctuation analysis to estimate the changes in the properties of Cl^- ion channels. The results show that (1) drugs with opposing pharmacological actions on CNS excitability in vivo have opposite effects on GABA-evoked responses and (2) the drug actions involve changes in the estimated kinetics, and not the conductance of Cl^- ion channels activated by GABA (Table 2) (Barker and McBurney 1979b; Study and Barker 1981; Barker et al. 1983).

We have found similar drug effects on GABA-evoked responses in hippocampal neurons and have begun to examine their effects on Cl^--dependent inhibitory postsynaptic currents (IPSCs) in these cells (Segal and Barker 1983). Diazepam increases the amplitude and the duration of these IPSCs, whereas pentobarbital increases their duration with little change in their amplitude (Fig. 3). In contrast, picrotoxin decreases IPSC amplitude without altering time course. The similarity between the average duration of GABA-activated channels and the time constant of IPSC decay

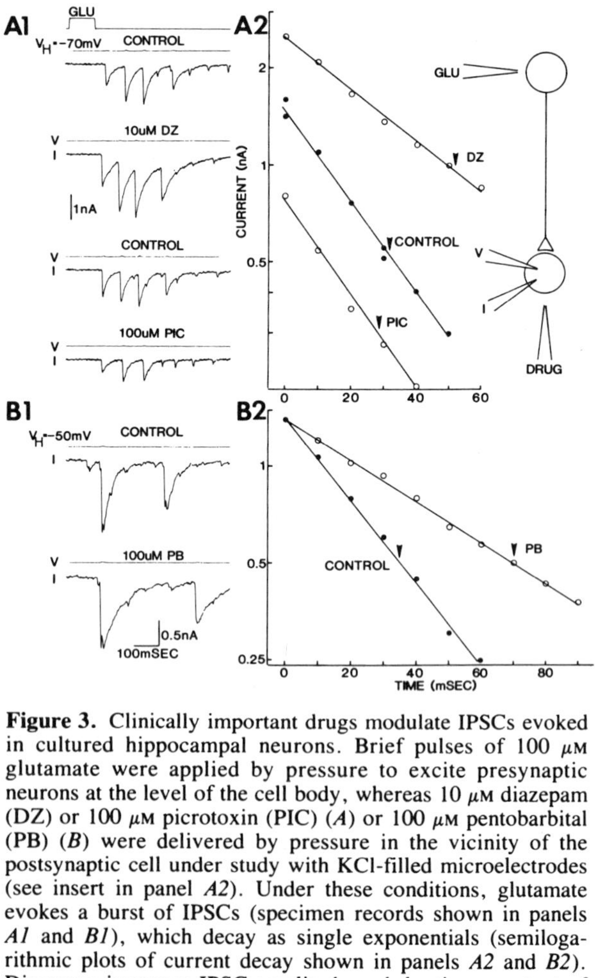

Figure 3. Clinically important drugs modulate IPSCs evoked in cultured hippocampal neurons. Brief pulses of 100 μM glutamate were applied by pressure to excite presynaptic neurons at the level of the cell body, whereas 10 μM diazepam (DZ) or 100 μM picrotoxin (PIC) (*A*) or 100 μM pentobarbital (PB) (*B*) were delivered by pressure in the vicinity of the postsynaptic cell under study with KCl-filled microelectrodes (see insert in panel *A2*). Under these conditions, glutamate evokes a burst of IPSCs (specimen records shown in panels *A1* and *B1*), which decay as single exponentials (semilogarithmic plots of current decay shown in panels *A2* and *B2*). Diazepam increases IPSC amplitude and the time constant of decay (arrowheads), whereas picrotoxin decreases IPSC amplitude and does not change decay. Pentobarbital prolongs IPSC decay evoked in another cell without changing IPSC amplitude (M. Segal and J.L. Barker, unpubl.).

and the parallelism in the drug actions on the kinetics of the pharmacologically and physiologically activated conductances suggest that the IPSCs are mediated by GABA. If this is true, then we estimate that between 1500 and 2000 ion channels are activated by GABA to generate the evoked IPSC in these cells. In a previous study, conducted on spontaneously occurring IPSCs in SC cells using some of the same drugs, we concluded that 300 channels are activated on average to generate the IPSC (Barker and McBurney 1979b). Presumably, the conductance of IPSCs evoked in hippocampal neurons is considerably greater than that calculated for spontaneous IPSCs in SC cells (Table 1) because evoked IPSCs involve more transmitter release. From the results we conclude that the rate at which the channels close in both SC and hippocampal neurons appears to determine, in large part, the time course of IPSC decay, that the properties of Cl^- ion channels activated by pharmacological applications of GABA are similar to those activated synaptically in both cell types, and that some clinically important drugs act by modulating the kinetics of channels synaptically activated by GABA.

Some Transmitters Have Multiple Actions on Membrane Excitability

In the course of our studies we have found that a variety of other putative transmitter substances including opioid peptides can activate Cl^- ion conductances in cultured SC neurons (Fig. 4) (Barker et al. 1980). However, although virtually every neuron responded to the neutral amino acids, few responded to the opioid peptides. The Cl^- conductance response evoked by Leu-enkephalin is blocked by naloxone, which, at the same concentration, does not affect GABA-activated responses. Presumably, the peptide actions are not mediated through GABA receptors, but reflect engagement of different receptors.

Leu-enkephalin has other effects on the excitable membrane properties of cultured SC neurons, all of which have been recorded in the same cell (Fig. 4B). These effects occur at concentrations lower than those required to activate Cl^- ion conductances. One action involves depression of glutamate-induced depolarization and inward current response in an apparently noncompetitive manner (Barker et al. 1980). The other effect involves either elevation (Fig. 4, panel 3) or depression of the threshold for single- and repetitive-action potential generation with little, if any, change in the steady-state current voltage (*I-V*) relations (Barker et al. 1978, 1980). All three effects act in a complementary manner to decrease the probability of action potential generation. Although we have not studied these actions in detail, we have begun to examine for the presence of electrically excitable membrane properties that might function as possible sites for peptide regulation of action potential threshold and discharge rate.

Transient Rectifications Are Present in Most Cultured Neurons

In the presence of tetrodotoxin (TTX) and tetraethylammonium (TEA) sufficent to block omnipresent Na^+ and "delayed-rectifier"-type K^+ conductance

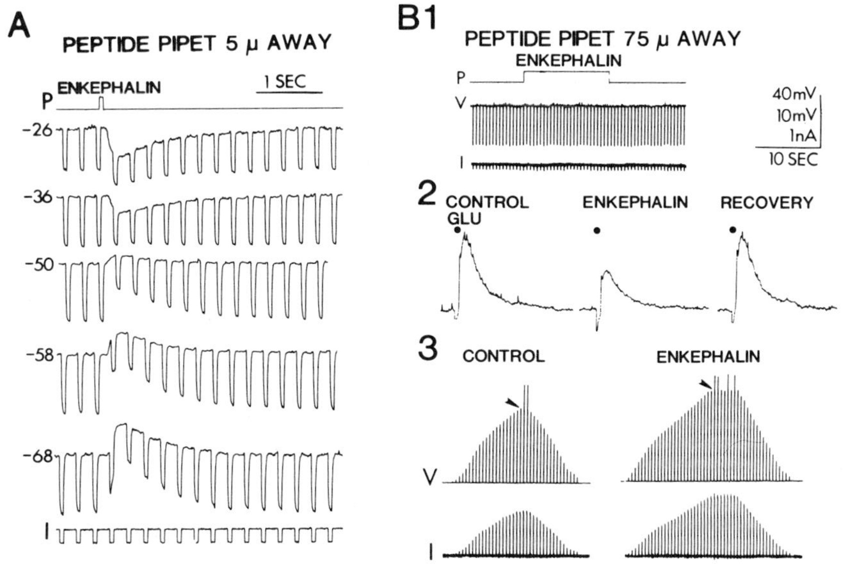

Figure 4. Leu-enkephalin has multiple actions on different excitable membrane properties in the same neuron. The peptide was applied by pressure from a pipette placed near (5 μm away), or remote from (75 μm away), the cell so as to change the concentration effective at the cell surface. (*A*) When applied briefly at the cell surface, the peptide changes membrane potential by increasing membrane conductance to Cl^- ions. (*B*) When applied from a pipette remote from the cell, there is no detectable change in membrane conductance (*B1*), but a considerable depression in the amplitude of the depolarizing response to a pulse of glutamate (*B2*) and a substantial elevation in the threshold for a single action potential (arrowhead) (*B3*). (Modified from Barker et al. 1980.)

mechanisms, we have found that the majority of neurons cultured from the spinal cord and hippocampus exhibit rapid although transient rectification in response to depolarizing current stimuli over the potential range subthreshold to action potential generation (Segal and Barker 1982; Barker and Rogawski 1982; Barker and Segal 1983). Although there is some variability in the time course of these rectifications, most occur within milliseconds following an abrupt depolarizing step and then fade rapidly (Fig. 5A, panel 1). *I-V* curves constructed in many cells often reveal marked changes in the *I-V* relationship at early (about 10 msec), but not late, times (about 50 msec). In the data illustrated in Figure 5B, changes in the *I-V* relationship occur at both times, indicating that in some cells, transient and delayed rectifications are present over the same range of membrane potential, even in 20 mM TEA.

We have found that application of 4-aminopyridine (4-AP) eliminates many, but not all, transient-type rectifications in a dose-dependent, reversible manner (Barker and Rogawski 1982; Segal and Barker 1982; Barker and Segal 1983). This can be seen in panel B of Figure 5, where the marked nonlinearity in the *I-V* relationship at 10 msec is reduced by 4-AP without change in the *I-V* curve over the hyperpolarized range of membrane potential. In contrast, 4-AP does not suppress the delayed rectification illustrated in Figure 5B.

We have applied voltage-jump protocols using the two-electrode voltage-clamp technique to study the ion conductance mechanism underlying the transient rectifications. In the presence of TTX and TEA we have found that a rapidly rising transient outward current (TOC) response is activated over the same potential range as the transient rectification (Fig. 6, upper panel). The TOC activates within a few milliseconds and then decays in tens of milliseconds, frequently as a single exponential (Fig. 6, lower panel). The decay kinetics of the TOC vary considerably from cell to cell and sometimes can be fitted with two exponentials (Barker and Segal 1983). Most, if not all, of the TOC responses can be rapidly inactivated by conditioning the cell at or near the potential required for activation (Fig. 6, lower panel), with full recovery from complete inactivation usually requiring several hundred milliseconds. The ionic conductance associated with the TOCs in

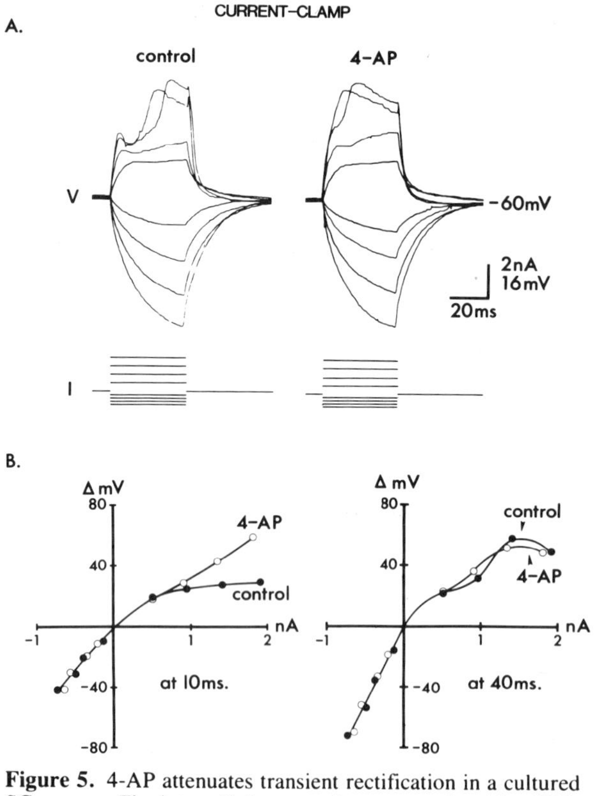

Figure 5. 4-AP attenuates transient rectification in a cultured SC neuron. The intracellular recording was made with a KCl-filled microelectrode in a medium containing 1 μM TTX and 20 μM TEA. (*A*) The cell was current-clamped at −60 mV, and a series of 40-msec current stimuli (*I*) was injected to evoke voltage responses (*V*) over a wide range of membrane potential under control conditions and in the presence of 5 mM 4-AP. (*B*) Voltage responses at 10 msec and at 40 msec are plotted as a function of injected current. 4-AP attenuates the marked rectification evident at 10 msec but not that occurring at 40 msec and has no effect on the *I-V* relations at hyperpolarized levels.

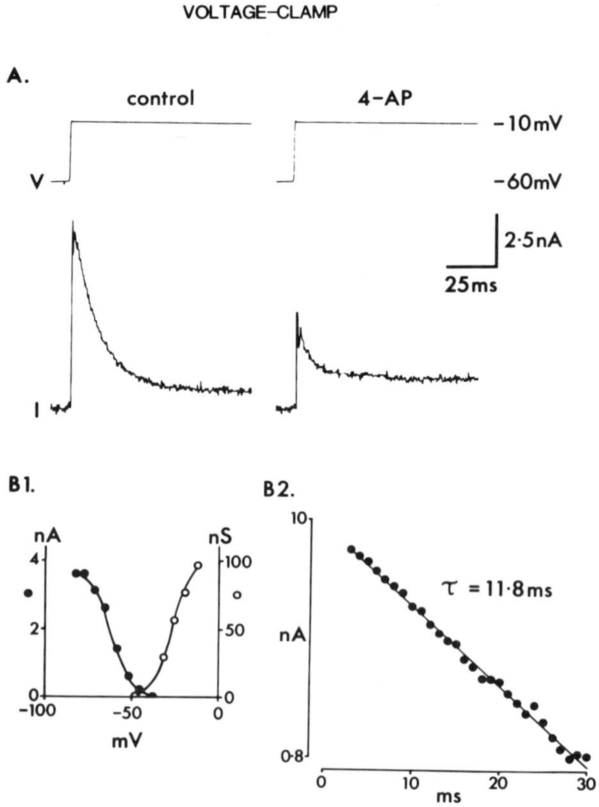

Figure 6. 4-AP attenuates TOC in a cultured SC neuron. The voltage-clamp recording was made from the same cell as illustrated in Fig. 5. (*A*) 4-AP reduces the amplitude of the TOC response evoked on stepping from −60 mV to −10 mV. A slight increase in the steady-state current is also evident. (*B1*) Activation (○) and inactivation (●) curves, constructed from voltage-jump protocols, overlap at about −50 mV. (*B2*) The TOC evoked at −10 mV decays as a single exponential with a time constant of 11.8 msec.

Table 3. Transmitters and Clinically Important Drugs Modulate Spike Frequency in Cultured CNS Neurons Independent of Other Effects on Excitability

	Spike frequency	Conductance mechanism	References
Transmitter			
Leu-enkephalin	↑	?	Barker et al. (1978)
Leu-enkephalin	↓	?	Barker et al. (1980)
cholecystokinin	↑	?	Rogawski (1982)
GABA	↑	?	Barker et al. (1980)
Drug			
4-AP	↑	↓TOC	Barker and Rogawski (1982); Segal and Barker (1983)
bicuculline	↑	?	Barker and MacDonald (1980)
picrotoxin	↑	?	Barker and MacDonald (1980)
strychnine	↑	?	Barker and MacDonald (1980)
penicillin	↑	?↓TOC	Barker and MacDonald (1980); M. Segal and J.L. Barker (unpubl.)
metrazol	↑	?	Barker and MacDonald (1980)
phenobarbital	↓	?	Barker et al. (1980)
flurazepam	↓	?	MacDonald and Barker (1982)
diazepam	↓	?	MacDonald and Barker (1982)
phenytoin	↓	?	Barker et al. (1981)

Spike frequency was measured in response to depolarizing current stimuli in cells bathed in physiological saline containing elevated [Mg^{++}].

these cells primarily involves K^+ ions, since the reversal potential of the tail current examined in a double-step protocol shifts in a depolarizing direction when $[K^+]_e$ is elevated.

4-AP depresses most but not all TOC responses in a dose-dependent, reversible manner (Fig. 6B). The drug does not usually affect the kinetics of the TOC, although the presence of another transient or delayed conductance mechanism sometimes complicates the effects of the drug on decay kinetics. This is true for the data in Figure 6, where 4-AP depresses the TOC and slightly enhances the delayed outward current. The drug causes a parallel shift in the activation curve of the TOC in the depolarizing direction. From these experiments we conclude that TOCs underlie transient rectifications recorded in many cultured SC and hippocampal neurons, that not all TOCs recorded in these cells have identical kinetics, and that there may be several distinctly different types of TOCs that are blocked by 4-AP.

We have examined the effects of 4-AP on the excitability of SC and hippocampal neurons in a physiological saline without TTX and TEA. 4-AP enhances excitability as evidenced by a decrease in the rheobase for single- and repetitive-action potential discharge (Barker and Rogawski 1982; Segal and Barker 1982). The threshold for triggering an action potential is lowered so that in the presence of 4-AP, an action potential can be evoked at a more hyperpolarized potential, nearer the level of the resting membrane potential. In most cells where 4-AP enhances action potential discharge rate, there appears to be little change in either the peak or the duration of an action potential, although an increase in action potential duration has been recorded on occasion. Functionally then, 4-AP-sensitive K^+ conductances that are activated over the range of membrane potential subthreshold to triggering an action potential regulate the rate of action potential discharge.

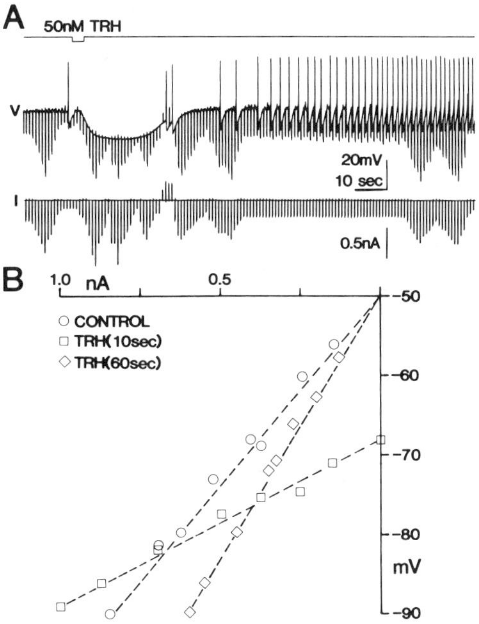

Figure 7. TRH induces spontaneous action potential discharge in a GH3/6 cell. (*A*) The intracellular recording (*V*) was made with a KCl-filled microelectrode. Brief pulses of current (*I*) of varying amplitude were injected to observe the *I-V* relations before and after peptide application. A single action potential is triggered just after positioning the pipette containing 50 nM TRH. Application of the peptide causes a transient hyperpolarization followed by the induction of spontaneous action potential discharge. (*B*) The hyperpolarization is associated with a decrease in slope resistance, and this is followed by a slight increase in slope resistance 60 sec later.

We have found that a wide variety of transmitter substances and clinically important drugs alter the rate of action potential discharge in a manner superficially similar to that of 4-AP (Table 3). As yet, we have not determined the ionic mechanism associated with these modulations. The modulatory effects of GABA and Leu-enkephalin on electrical excitability occur at concentrations considerably lower than those required for activation of Cl^- ion conductances by the same substances in the same cells.

TRH Induces Action Potentials and Attenuates TOCs in Pituitary Cells

Cells derived from the prolactin- and growth-hormone-secreting "GH" pituitary clone exhibit both electrical and chemical forms of membrane excitability (Kidokoro 1975; Dufy et al. 1979). We have begun to examine the excitable membrane properties of the GH3/6 clonal cells and the effects of hormonal transmitter substances on these properties. The majority of cells studied discharge regenerative action potentials either in response to a depolarizing current stimulus or spontaneously. These action potentials appear to be predominately Ca^{++}-dependent (Ozawa and Kimura 1979). Under voltage clamp the cells exhibit a nonlinear steady-state *I-V* relationship that is linearized by TEA and by Co^{++} or Cd^{++} (Dufy and Barker 1982). In addition, there is virtually no "resting" membrane potential if Ca^{++} ions are omitted from the bathing medium or if excess Ba^{++} ions are included. The "resting" membrane potential thus appears to be actively set by a K^+ conductance that is regulated both by membrane voltage and by Ca^{++} ions. Like central neurons these cells also exhibit 4-AP-sensitive TOCs in the potential range subthreshold to that required for triggering regenerative action potentials (Dufy and Barker 1983). Thus, GH3/6 cells possess a complex spectrum of electrically excitable membrane conductances.

A rather remarkable variety of hormonal transmitters alter the excitability of these cells and change their rate of prolactin secretion. Perhaps the best studied is thyrotropin-releasing hormone (TRH), which stimulates prolactin secretion (Dannies and Tashjian 1973) and mobilizes Ca^{++} from intracellular pools (Ronning et al. 1982). Brief application of TRH to GH3/6 cells causes a transient hyperpolarization and increase in membrane conductance, which is then followed by the induction of spontaneous action potential activity at the level of the resting potential in silent cells (Fig. 7) or an increase in the rate of discharge in spontaneously active

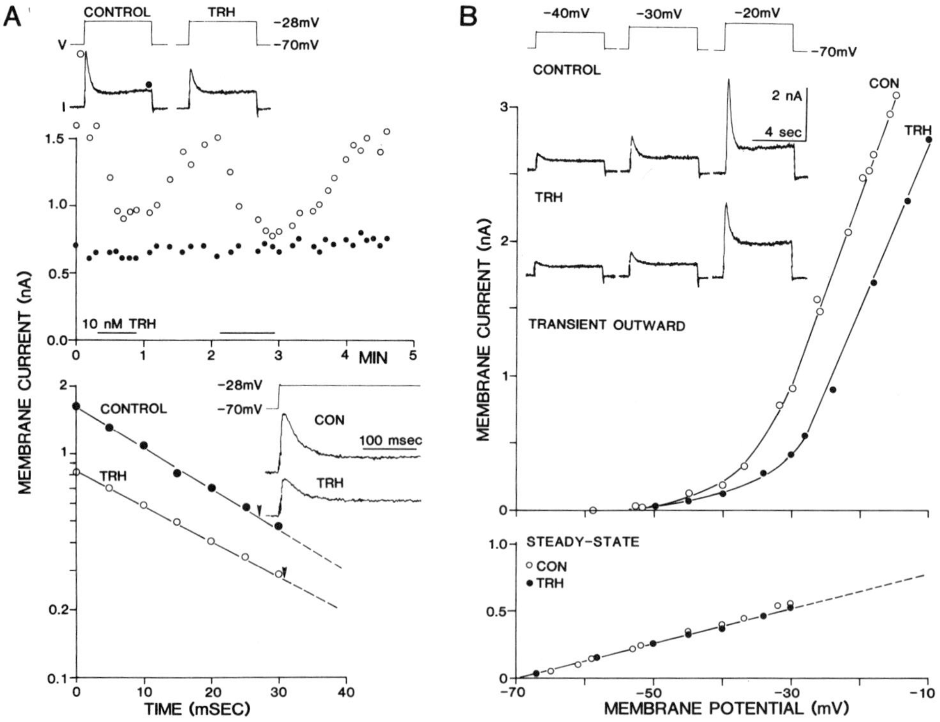

Figure 8. TRH attenuates TOC in a GH3/6 cell. The voltage-clamp recording was made with two microelectrodes filled with KCl. TRH was applied by pressure from a closely positioned pipette containing 10 nM peptide. (*A*) TRH reduces the amplitude of the TOC evoked upon stepping to −28 mV with little, if any, effect on the steady-state current flowing at the end of the 5-sec command. (*Upper panel*) Plot of TOC amplitude (○) and steady-state current before, during, and after successive applications of TRH. (*Lower panel*) Plot of the TOC decay evoked at −28 mV (see *inset*) before and during TRH application. The time constant (arrowhead) is not appreciably changed. (*B*) TOCs (*upper panel*) and steady-state currents (*lower panel*) are plotted for a series of voltage commands under control and in the presence of TRH. (*Inset*) Specimen records. TRH shifts the activation curve for the TOC in a depolarizing direction while having little effect on the steady-state current.

cells (Dufy et al. 1979). The action potentials induced are similar in amplitude and duration to those evoked prior to peptide application in some (Dufy et al. 1979; Ozawa and Kimura 1979), but not all, studies (Sand et al. 1980; Kaczorowski et al. 1983). The variability in TRH effects on action potential properties suggests that not all cells in this pituitary clone line have identical properties. The initial hyperpolarization induced by TRH is thought to be due to an increase in K^+ conductance (Ozawa and Kimura 1979; Ozawa 1981), which coincides temporally with mobilization of intracellular Ca^{++} (Ronning et al. 1982). The delayed excitability-enhancing effects can take place in the absence of any detectable changes in the steady-state *I-V* curve (see Fig. 8B), although decreases in the slope conductance sometimes occur (Fig. 7B).

The effects of TRH have recently been studied using voltage-clamp techniques. Several distinctly different actions have been observed. These include (1) an initial but evanescent outward current at the level of the resting membrane potential that is associated with increased membrane conductance, (2) a delayed and sustained decrease in the amplitude of TOC (Fig. 8), and in some cells, (3) a delayed increase in membrane current variance and a decrease in steady-state membrane conductance (B. Dufy and J.L. Barker, unpubl.). The initial outward current response likely corresponds to the reported increase in K^+ conductance frequently observed immediately after TRH application (see Fig. 7). The decrease in TOC amplitude is slow to develop, and, using 10 nM peptide, does not desensitize (Fig. 8A, top). Neither the rate of rise nor the time constant of TOC decay is altered significantly by TRH (Fig. 8A, bottom). A series of depolarizing voltage steps from a hyperpolarized holding potential reveals that the peptide shifts the activation curve of the TOC in a depolarizing direction while having little effect on the steady-state properties over the −70 mV to −30 mV range (Fig. 8B).

Thus, TRH, at concentrations required for stimulation of prolactin release, induces or enhances action potential discharge at the level of the resting potential and, under voltage clamp, suppresses activation of TOC. Presumably, suppression of the TOC blocks the transient-type rectification responsible for determining the rate of action potential discharge in these cells.

CONCLUSION

The results briefly reviewed in this paper show that monolayer culture is a suitable preparation for detailed study of ionic conductance mechanisms in cells that would otherwise be inaccessible. We have focused on two mechanisms common to many cultured cells, both of which function to regulate the pattern of action potential discharge. Neither is unique to vertebrate CNS neurons or endocrine cells, since both have been reported in many other types of excitable cells (for reviews, see Adams et al. 1980; Mathers and Barker 1982; see also Adams et al. 1982). These properties can be regulated by a variety of transmitters and drugs, and therefore are probably important both in intercellular communication and in the cellular actions of clinically important drugs in vivo. The accessibility of monolayer cultures to a variety of techniques current in cellular neurobiology suggests that we should be able to utilize a multidisciplinary strategy in vitro to study these membrane properties in more detail, as well as how they regulate cytoplasmic and secretory activities in specific cell types.

REFERENCES

Adams, D.J., S.J. Smith, and S.H. Thompson. 1980. Ionic currents in molluscan soma. *Annu. Rev. Neurosci.* **3:** 141.

Adams, P.R., D.A. Brown, and A. Constanti. 1982. M-currents and other potassium currents in bullfrog sympathetic neurones. *J. Physiol.* **330:** 537.

Barker, J.L. and J.F. MacDonald. 1980. Picrotoxin convulsions involve synaptic and non-synaptic mechanisms on cultured mouse spinal neurons. *Science* **208:** 1054.

Barker, J.L. and R.N. McBurney. 1979a. GABA and glycine may share the same conductance channel in cultured mammalian neurones. *Nature* **277:** 234.

———. 1979b. Phenobarbitone modulation of postsynaptic GABA receptor function on cultured mammalian neurones. *Proc. R. Soc. Lond. B* **206:** 319.

Barker, J.L. and B.R. Ransom. 1978. Amino acid pharmacology of mammalian central neurones grown in tissue culture. *J. Physiol.* **280:** 331.

Barker, J.L. and M.A. Rogawski. 1982. Identification and pharmacological sensitivity of a transient outward current in the somata of cultured mammalian central neurons. *Soc. Neurosci. Abstr.* **8:** 123.

Barker, J.L. and M. Segal. 1983. Three types of transient outward currents are present in cultured central mammalian neurons. *Soc. Neurosci. Abstr.* **9:** (in press).

Barker, J.L., R.N. McBurney, and J.F. MacDonald. 1982. Fluctuation analysis of neutral amino acid responses in cultured mouse spinal neurons. *J. Physiol.* **322:** 365.

Barker, J.L., T.G. Smith, and J.H. Neale. 1978. Multiple membrane mechanisms of enkephalin revealed using cultured spinal neurons. *Brain Res.* **154:** 153.

Barker, J.L., D.L. Gruol, L.M. Huang, J.F. MacDonald, and T.G. Smith. 1980. Peptides: Three forms of chemical excitability on cultured mouse spinal neurons. *Neuropeptides* **1:** 63.

Barker, J.L., J.F. MacDonald, D.A. Mathers, R.N. McBurney, and R.E. Study. 1981. Convulsant and anticonvulsant pharmacology of cultured mouse spinal neurons. In *Neurotransmitters, seizures and epilepsy* (ed. K.G. Lloyd et al.), p. 49. Raven Press, New York.

Choi, D., D. Farb, and G.D. Fischbach. 1981. GABA-mediated synaptic potentials in chick spinal cord and sensory neurons. *J. Neurophysiol.* **45:** 632.

Dannies, P.S. and A.H. Tashjian. 1973. Effects of thyrotropin-releasing hormone and hydrocortisone on synthesis and degradation of prolactin in a rat pituitary cell strain. *J. Biol. Chem.* **248:** 6174.

Dufy, B. and J.L. Barker. 1982. Calcium-activated and voltage-dependent potassium conductances in clonal pituitary cells. *Life Sci.* **30:** 1933.

———. 1983. Transient outward currents in clonal pituitary cells are regulated by peptide hormones. *Soc. Neurosci. Abstr.* **9:** (in press).

Dufy, B., J.D. Vincent, H. Fleury, P. DuPasquier, D. Gourdji, and A. Tixier-Vidal. 1979. Membrane effects of thyrotropin-releasing hormone and estrogen shown by intracellular recording from pituitary cells. *Science* **204:** 509.

JACKSON, M.B., H. LECAR, D.A. MATHERS, and J.L. BARKER. 1982. Single channel currents activated GABA, muscimol and (−)pentobarbital in cultured mouse spinal neurons. *J. Neurosci.* **2:** 889.

KACZOROWSKI, G.J., R.L. VANDLEN, G.M. KATZ, and J.P. REUBEN. 1983. Regulation of excitation-secretion coupling by thyrotropin releasing hormone (TRH): Evidence for TRH receptor-ion channel coupling in cultured pituitary cells. *J. Membr. Biol.* **71:** 109.

KATZ, B. and R. MILEDI. 1972. The statistical nature of the acetylcholine potential and its molecular components. *J. Physiol.* **224:** 665.

KIDOKORO, Y. 1975. Spontaneous calcium action potentials in a clonal pituitary cell line and their relationship to prolactin secretion. *Nature* **258:** 741.

MACDONALD, J.F. and J.L. BARKER. 1982. Multiple actions of flurazepam on the excitability of cultured mouse spinal neurons. *Brain Res.* **246:** 257.

MATHERS, D.A. and J.L. BARKER. 1982. Chemically induced ion channels in nerve cell membranes. *Int. Rev. Neurobiol.* **23:** 1.

NELSON, P.G., E.A. NEALE, and R.L. MACDONALD. 1981. Electrophysiological and structural studies of neurons in dissociated cell cultures of the central nervous system. In *Excitable cells in tissue culture* (ed. P.G. Nelson and M. Lieberman), p. 39. Plenum Press, New York.

OWEN, D., R. STUDY, E. GRATZ, and J.L. BARKER. 1982. Pharmacological modulation of GABA responses in cultured mouse spinal neurons. *Soc. Neurosci. Abstr.* **8:** 833.

OZAWA, S. 1981. Biphasic effect of thyrotropin-releasing hormone on membrane K^+ permeability in rat clonal pituitary cells. *Brain Res.* **209:** 240.

OZAWA, S. and N. KIMURA. 1979. Membrane potential changes caused by thyrotropin-releasing hormone in the clonal GH3 cell and their relationship to secretion of pituitary hormone. *Proc. Natl. Acad. Sci.* **76:** 6017.

RANSOM, B.R., C.N. CHRISTIAN, P.N. BULLOCK, and P.G. NELSON. 1977. Mouse spinal cord in cell culture. II. Synaptic activity and circuit behavior. *J. Neurophysiol.* **40:** 1151.

RONNING, S.A., G.A. HEATLEY, and T.G.J. MARTIN. 1982. Thyrotropin-releasing hormone mobilizes Ca^{2+} from endoplasmic reticulum and mitochondria of GH3 pituitary cells: Characterization of cellular Ca^{2+} pools by a method based on digitonin permeabilization. *Proc. Natl. Acad. Sci.* **79:** 6294.

ROGAWSKI, M.A. 1982. Cholecystokinin octapeptide: Effects on the excitability of cultured spinal neurons. *Peptides* **3:** 545.

SAND, O., E. HAUG, and K.M. GAUTVIK. 1980. Effects of thyroliberin and 4-aminopyridine on action potentials and prolactin release and synthesis in rat pituitary cells in culture. *Acta Physiol. Scand.* **108:** 247.

SEGAL, M. and J.L. BARKER. 1982. Voltage clamp analysis of cultured rat hippocampal neurons. *Soc. Neurosci. Abstr.* **8:** 800.

———. 1983. Voltage clamp analysis of an inhibitory synapse in cultured rat hippocampal neurons. *Soc. Neurosci. Abstr.* **9:** (in press).

SMITH, T.G., J.L. BARKER, B.M. SMITH, and T.E. COLBURN. 1980. Voltage clamping with microelectrodes. *J. Neurosci. Methods* **3:** 105.

STUDY, R.E. and J.L. BARKER. 1981. Diazepam and (−)pentobarbital: Fluctuation analysis reveals different mechanisms for potentiation of GABA responses in cultured central neurons. *Proc. Natl. Acad. Sci.* **78:** 7180.

Glutamate- and GABA-receptor Channels at the Locust Nerve–Muscle Junction: Noise Analysis and Single-channel Recording

S.G. CULL-CANDY
Department of Pharmacology, University College London, London WC1E 6BT

Various methods can be used to give information about the characteristics of synaptic channels in nerve and muscle membranes. These include the analysis of "membrane noise" associated with statistical fluctuations in the number of open channels during the presence of a steady level of chemical excitant (Katz and Miledi 1972; Anderson and Stevens 1973); analysis of the relaxation that occurs following the brief "concentration jump" caused by the release of transmitter from nerve terminals (Kordas 1969; Magleby and Stevens 1972); and the direct observation of single-channel currents by means of patch-clamp recording (Neher and Sakmann 1976). These approaches have allowed examination of amino-acid-activated synaptic channels at the locust nerve–muscle junction, where the excitatory transmitter may be glutamate or a closely related substance and the inhibitory transmitter is thought to be GABA (for review, see Usherwood and Cull-Candy 1975). The occurrence of glutamate and GABA receptors in the same muscle fibers has enabled a direct comparison of the properties of excitatory and inhibitory synaptic channels in these cells.

EXCITATORY SYNAPTIC CHANNELS

Membrane Channels Opened by Glutamate

Prolonged application of glutamate to excitatory nerve-muscle junctions in voltage-clamped locust muscle fibers produces a change in the membrane current to a new steady level, with an accompanying increase in the membrane current noise (Anderson et al. 1976). Under certain assumptions about the operation of synaptic channels (Katz and Miledi 1972; Anderson and Stevens 1973), analysis of synaptic noise indicates that the glutamate-activated channel has a mean lifetime of 2.5 msec and a conductance of 122 pS at -80 mV clamp potential ($T = 21°C$). The glutamate noise spectrum is constant at low frequencies and declines as 1/frequency^2 at higher frequencies (Fig. 1A,B). The spectral distribution of the noise variance, σ_I^2, is therefore fitted by a Lorenzian curve (continuous lines in Fig. 1) and described by the equation: $S(f) = S(0)/1 + (f/f_c)^2$ (where f is frequency, $S(f) = d\sigma_I^2/df$, and $S(0)$ is the zero frequency asymptote). The cutoff (half-power) frequency, f_c (at which spectral density is reduced to half-maximum value), gives the mean channel lifetime, τ_{noise}, according to $\tau_{noise} = 1/(2\pi f_c)$ (Katz and Miledi 1972; Stevens 1972). A change in mean channel lifetime is seen as a shift in the spectrum, along the frequency axis.

By analyzing glutamate noise, it has been possible to look, in detail, at certain of the external factors that can influence channel kinetics (see Anderson et al. 1978). Figure 1A shows glutamate-noise spectra obtained from a single junctional site at 21°C and 3.2°C. There is a marked shift in the cutoff frequency (indicated by arrow) to lower frequencies, signifying a prolongation in channel lifetime from 2.3 msec at 21°C to 11.4 msec at 3.2°C. In addition, the single-channel conductance is decreased. Channel properties are also influenced by the type of agonist that opens the channel. Figure 1B shows noise spectra obtained at a single junctional site in response to glutamate and the agonist quisqualate. The quisqualate spectrum is displaced toward lower frequencies, since the lifetime of the channel opened by quisqualate is more than twice as long as that opened by glutamate. However, the conductance of the channel is identical for the two agonists. Noise analysis has revealed a further interesting and apparently basic feature of the glutamate-activated channel in locust muscle fibers which is that the mean lifetime of the channel is exponentially dependent on membrane potential, being *decreased* by membrane hyperpolarization (see below), in contrast with the vertebrate endplate channel, where hyperpolarization increases channel lifetime (Kordas 1969; Magleby and Stevens 1972).

Excitatory Transmitter-activated Channels

Provided the transmitter concentration during a miniature excitatory junctional current (MEJC) declines rapidly, compared with the mean life span of the channel, the time constant of decay of the MEJC (τ_{MEJC}) should be equivalent to the channel lifetime. This therefore allows a simple method for examining the kinetics of transmitter-activated channels (see Katz and Miledi 1973a; Anderson and Stevens 1973). At synapses where the identity of the natural transmitter remains uncertain, this approach takes on added significance. Since it is known that channel lifetime depends on the identity of the agonist operating the channel, both at cholinergic and amino acid synapses (Katz and Miledi 1973b; Anderson et al. 1976; Crawford and McBurney 1976), similarity between

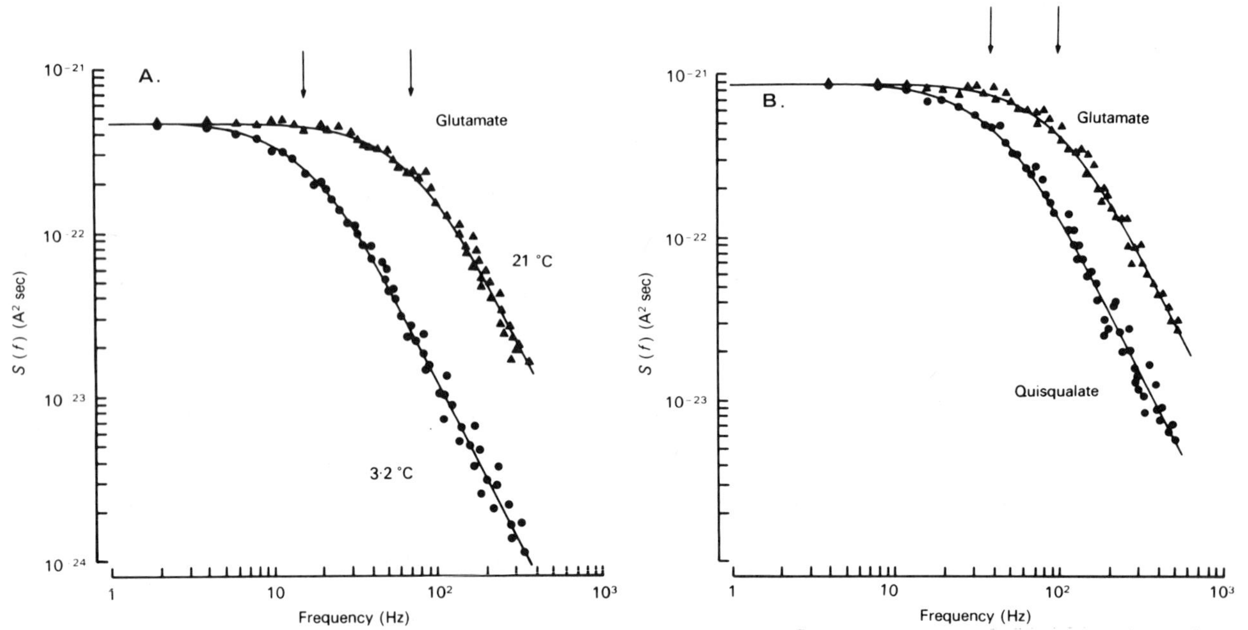

Figure 1. (*A*) Effect of temperature on current-noise spectrum produced by glutamate application to a fiber clamped at −100 mV. Spectra were obtained at the same site at 21°C (▲) and 3.2°C (●). Half-power (cutoff) frequencies (arrows) are $f_c = 70$ Hz at 21°C, hence channel lifetime = 2.27 msec; $f_c = 14$ Hz at 3.2°C, hence channel lifetime (τ_{noise}) = 11.36 msec. Single-channel conductances are $\gamma = 105$ pS at 21°C and 21 pS at 3.2°C. Vertical axis applies to 21°C curve. Multiplication of ordinate values by 0.83 gives values for 3.2°C spectrum. (*B*) Comparison of noise spectra produced by application of glutamate (▲) and quisqualate (●) to a single junctional site, in a fiber clamped at −110 mV. Half-power frequency for glutamate spectrum is $f_c = 92$ Hz, hence $\tau_{noise} = 1.73$ msec and single-channel conductance is $\gamma = 110$ pS. For quisqualate spectrum, $f_c = 40$ Hz, hence $\tau_{noise} = 3.98$ msec and $\gamma = 112$ pS. Vertical axis applies to glutamate spectrum. Multiplication of ordinate values by 0.33 gives values for quisqualate spectrum. (Reprinted, with permission, from Anderson et al. 1978.)

τ_{noise}, for a given agonist, and τ_{MEJC} supports the idea that the agonist is a suitable transmitter candidate.

MEJCs and EJCs have been analyzed at the locust nerve–muscle junction to give information about synaptic channels and to allow comparison with the channels opened by externally applied glutamate (Cull-Candy and Miledi 1982). MEJCs have been examined when the inward current is carried mainly by Na^+ (i.e., under normal conditions) or by Ca^{++}. Figure 2 shows examples of MEJCs recorded from a single muscle fiber both in normal bathing medium and after soaking the muscle for 3 hours in isotonic Ca^{++}. Under normal conditions, MEJCs rise rapidly and decay exponentially, as described at other synapses. However, a certain percentage of currents (as many as 10% at some junctions) do not fit this pattern and rise or decay more

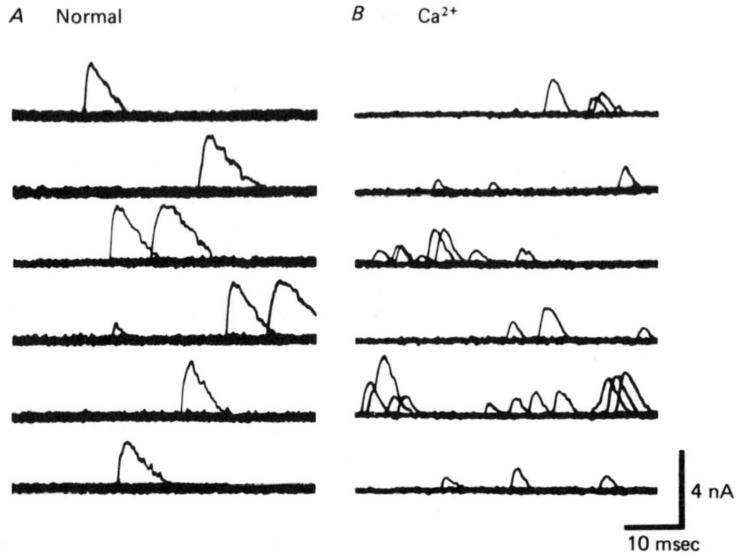

Figure 2. MEJCs from a muscle fiber in normal medium (*A*) and after soaking for several hours in isotonic calcium (*B*) at −110 mV clamp potential. Each trace shows several superimposed sweeps. In isotonic calcium (*B*) MEJCs are reduced in size and their decay time is briefer. (Reprinted, with permission, from Cull-Candy and Miledi 1982.)

slowly. Furthermore, a population of sub-MEJCs is sometimes present, as described at vertebrate end plates (Kriebel and Gross 1974).

Figure 3A shows the distribution of τ_{MEJC} values obtained in a single muscle fiber clamped at − 80 mV and − 110 mV. The histograms often show an excess of prolonged events. Nerve-impulse-evoked excitatory junctional currents are more homogeneous in their time course of decay and usually appear slightly briefer than MEJCs obtained in the same fiber. This is indicated by the arrows above the histograms (Fig. 3A,B). When all of the MEJCs visually identified as being "abnormal" in time course (i.e., prolonged, composite, or sub-MEJCs) were excluded from the analysis, the mean value of τ_{MEJC} (− 80 mV) was 2.62 ± 0.11 msec ($n = 10$ fibers at 22°C) (Cull-Candy and Miledi 1982). This value closely approaches estimates of the lifetime of the synaptic channels activated by glutamate (i.e., from glutamate noise), which is consistent with the idea that glutamate may be the transmitter.

Membrane Potential Dependence of τ_{MEJC}

The time constant of decay of MEJCs decreases with hyperpolarization (Anderson et al. 1976; Cull-Candy and Miledi 1982) as described for impulse-evoked EJCs at the crayfish nerve–muscle junction (Dudel 1974; Onodera and Takeuchi 1978). Figure 3C illustrates the dependence of τ_{MEJC} on clamp potential in two muscle fibers. τ_{MEJC} appears exponentially dependent on V_m, typically with an e-fold change in τ_{MEJC} being produced by a 180-mV change in potential.

We have also examined the decay time of MEJCs in the presence of isotonic Ca^{++}. Under these conditions, the amplitude of the MEJCs is reduced to about one-third normal size (Fig. 2B), and in addition, the decay time of the current becomes briefer, indicating a reduction in the mean lifetime of transmitter-operated channels; τ_{MEJC} (− 80 mV) = 1.01 ± 0.08 msec ($n = 14$ fibers at 22°C). A rather complex picture has emerged from experiments in which we have examined the

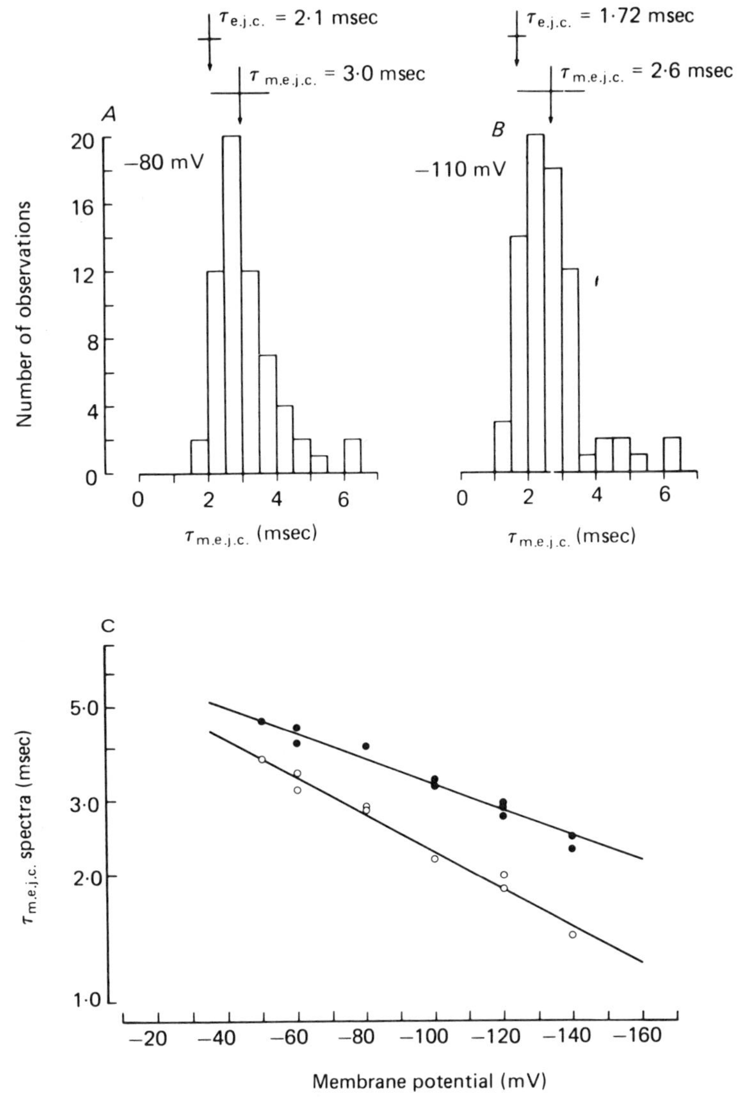

Figure 3. (*A,B*) Distribution of τ_{MEJC} values in a single muscle fiber under voltage clamp at − 80 mV (*A*) and − 110 mV (*B*). There is a small proportion of events of prolonged time course. Arrows above histograms indicate mean (and S.D.) of τ_{MEJC}, and τ_{EJC} obtained from measurements of nerve-impulse-evoked EJCs. (*C*) Dependence of τ_{MEJC} (obtained from MEJC spectra) on clamped membrane potential. Values are from two muscle fibers in which an e-fold change in τ_{MEJC} was produced by a potential change of 98.4 mV (○) and 142.8 mV (●). (Reprinted, with permission, from Cull-Candy and Miledi 1982.)

dependence of τ_{MEJC} on potential, when the inward current is carried by Ca^{++}. Membrane hyperpolarization greatly increases transmitter release from the nerve terminal, producing an increase in the number of prolonged and composite MEJCs (see Cull-Candy and Miledi 1982). The voltage dependence of the fast time-course events (which are thought to reflect more closely channel lifetime) is abolished or even reversed. The properties of the receptor channel are therefore apparently dependent on the species of permeating cation that passes through the synaptic channel.

Voltage-dependent Channel Block at Glutamate Synapses

At cholinergic synapses, classic antagonists such as tubocurarine and gallamine act not only on the postsynaptic receptor, but also on the ionic channel (Marty et al. 1976; Manalis 1977; Katz and Miledi 1978; Colquhoun et al. 1979). Although these substances have little noticeable postsynaptic action on EJCs or MEJCs at locust junctions at the resting membrane potential, there is an increasingly marked effect as the cell is hyperpolarized. In the presence of tubocurarine (5–100 μM) or gallamine (1–5 μM), fibers held at hyperpolarized levels show EJCs with an initial "fast component" followed by a "slow tail," as illustrated in Figure 4a (Cull-Candy and Miledi 1983). These resemble end-plate currents in vertebrate muscle fibers treated with local anesthetics (Katz and Miledi 1975; Ruff 1977). On hyperpolarization, the decay-time constant of the fast component decreased, whereas that of the slow component increased (Fig. 4a). In many fibers examined, the peak synaptic conductance decreased markedly on hyperpolarization beyond −50 mV. For example, Figure 4b shows the influence of tubocurarine on the relationship between MEJC amplitude and clamp potential, in which the peak MEJC amplitude decreased when the cell was hyperpolarized beyond −90 mV.

Such observations, including the voltage sensitivity of the two components comprising the EJC, are most readily explained in terms of a channel-blocking model (Adams 1976; Marty et al. 1976), and interestingly, cholinergic drugs also act in a voltage-dependent manner at other glutamate-activated synapses (Yamamoto and Washio 1979; Lingle et al. 1981).

INHIBITORY SYNAPTIC CHANNELS

Membrane Channels Opened by GABA

Steady ionophoretic application of GABA to inhibitory junctional receptors in voltage-clamped fibers produces an inward or outward current change (depending on the relative levels of the equilibrium potential and clamp potential). The membrane current change is accompanied by an increased membrane noise level, although GABA-noise variance is "smaller" than glutamate-noise variance for the same membrane current change (when the driving force $[V_m\text{-}V_{eq}]$ is similar). Analysis of GABA noise indicates that the inhibitory channel has a conductance that is considerably less than that of the excitatory channel. The single-channel conductance is approximately 22 pS, and the mean lifetime is 4–6 msec at −80 mV ($T = 21°C$).

Figure 5A shows examples of GABA-noise spectra at −74 mV and +10 mV. The spectra are similar in form to glutamate-noise spectra (although in some fibers, there may be a small high-frequency component). Membrane hyperpolarization causes the spectrum to

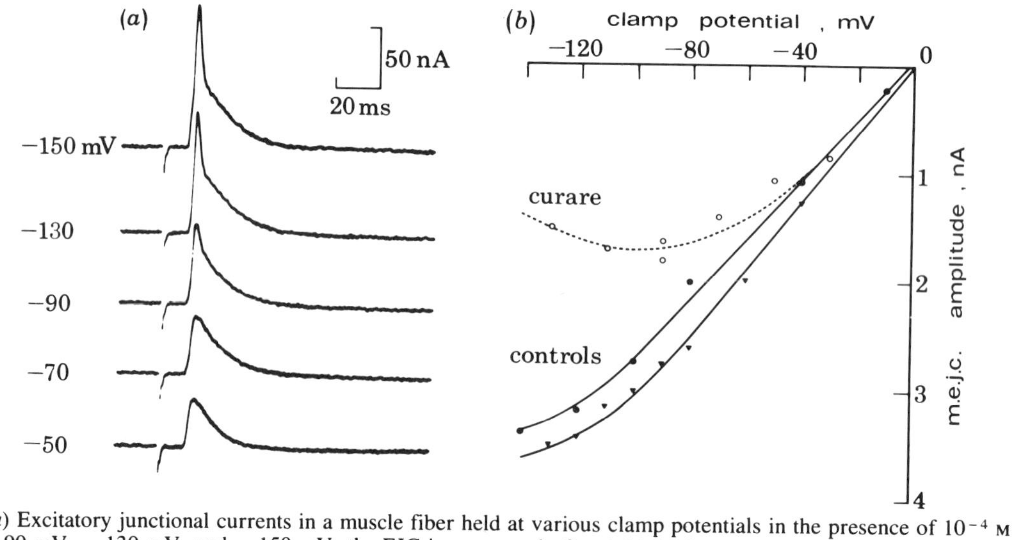

Figure 4. (*a*) Excitatory junctional currents in a muscle fiber held at various clamp potentials in the presence of 10^{-4} M tubocurarine. At −90 mV, −130 mV, and −150 mV, the EJC is composed of an initial "fast component" followed by a late "slow tail." The initial component is particularly noticeable at strongly hyperpolarized levels. (*b*) Relation between MEJC amplitude and clamp potential in two control fibers in normal medium (●, ▲) and in a third fiber in the presence of 10^{-4} M tubocurarine (○). The relationship shows slight nonlinearity at hyperpolarized potentials under control conditions and marked nonlinearity during exposure to tubocurarine. (Reprinted, with permission, from Cull-Candy and Miledi 1983.)

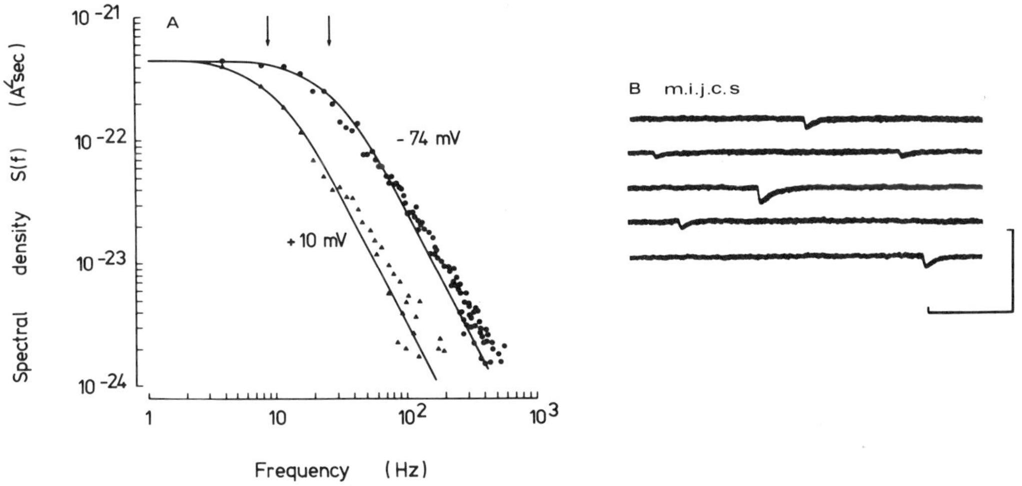

Figure. 5. (*A*) Effect of membrane potential on current-noise spectrum produced by application of GABA to a fiber clamped at −74 mV (●) and +10 mV (▲). Half-power frequency for −74 mV spectrum is $f_c = 25$ Hz, hence $\tau_{noise} = 6.37$ msec, and single-channel conductance, $\gamma = 17.1$ pS. Half-power frequency for +10 mV spectrum is $f_c = 9$ Hz, hence $\tau_{noise} = 17.7$ msec, and single-channel conductance, $\gamma = 21.1$ pS. Vertical axis applies to −74 mV curve, and multiplication by 0.27 gives spectral densities for +10-mV curve. (Reprinted, with permission, from Cull-Candy and Miledi 1981.) (*B*) MIJCs recorded in a fiber clamped at 0 mV, after soaking for several hours in 10^{-3} M glutamate. MIJCs are outward currents. Traces are not continuous, but selected for the presence of MIJCs. Calibration, 10 nA and 200 msec.

shift along the frequency axis, indicating that the duration of the channel open-time decreases with hyperpolarization.

Inhibitory Transmitter-activated Channels

The recording of miniature inhibitory junctional currents (MIJCs) and impulse-evoked inhibitory junctional currents (IJCs) should give useful information about the kinetics of inhibitory channels. However, recording MIJCs has proved difficult at many inhibitory synapses (see Finger and Stettmeier 1981), perhaps suggesting that the quantal event is rather small. To overcome this problem in locust muscle fibers, glutamate was first applied to completely desensitize the excitatory synapses. The resultant glutamate noise, as well as the MEJCs, gradually desensitized to enable stable recording. Nerve stimulation then produced IJCs in the absence of EJCs.

When membrane potential is clamped sufficiently far from resting potential, MIJCs can be recorded directly in some fibers (Cull-Candy 1983b). Figure 5B shows examples of MIJCs from a fiber clamped at 0 mV (i.e., the equilibrium potential for the excitatory transmitter). It is therefore highly unlikely that they resulted from transmitter packets released from excitatory nerve terminals. In addition, they reversed close to the Cl equilibrium potential and were abolished in Cl-free medium or picrotoxin (10^{-5} M), as previously demonstrated for impulse-evoked inhibitory junctional potentials in these fibers (Usherwood and Grundfest 1965). They were unaffected by tetrodotoxin (10^{-6} M), so it is unlikely they are multiquantal events resulting from nerve-terminal action potentials.

The time constant of decay of MIJCs is τ_{MIJC} (−80 mV) = 7.0 ± 0.2 msec ($n = 5$ fibers), which is similar to estimates of τ_{IJC} obtained under the same conditions (Cull-Candy 1983b). This value is slightly larger than the mean lifetime of the synaptic channel derived from GABA noise, although in some fibers reasonable agreement was found.

Membrane Potential Dependence of τ_{IJC}

The time course of IJCs and MIJCs was similar. However, IJCs can be more conveniently measured and were therefore used for examining the potential sensitivity of the transmitter-activated channel. As shown in Figure 6, τ_{IJC} decreased with hyperpolarization, indicating a reduction in channel lifetime, as described for IJCs at crayfish inhibitory junctions (Onodera and Takeuchi 1976; Dudel 1977). Over the range of potentials shown (−22 mV to −62 mV), the direction of the current is also reversed. The decrease in decay time was accompanied by a consistent slowing in the rise time of the synaptic current; the reason for this is less clear-cut. It can be seen from the semilogarithmic plot of the decay phase (Fig. 6B) that the inhibitory current decays exponentially from about 80% to 10% of its peak amplitude. τ_{IJC} is exponentially dependent on V_m with an *e*-fold change being produced by a 100 mV change in potential.

Although the EJC amplitude is linearly dependent on membrane potential over a wide range of potentials at crayfish and locust junctions (Onodera and Takeuchi 1978; Cull-Candy and Miledi 1982), the IJCs show a markedly nonlinear relationship (Onodera and Takeuchi 1976; Cull-Candy 1983b), and it is clear that at hyperpolarized levels, the increase in current amplitude is considerably less than expected for a given voltage change.

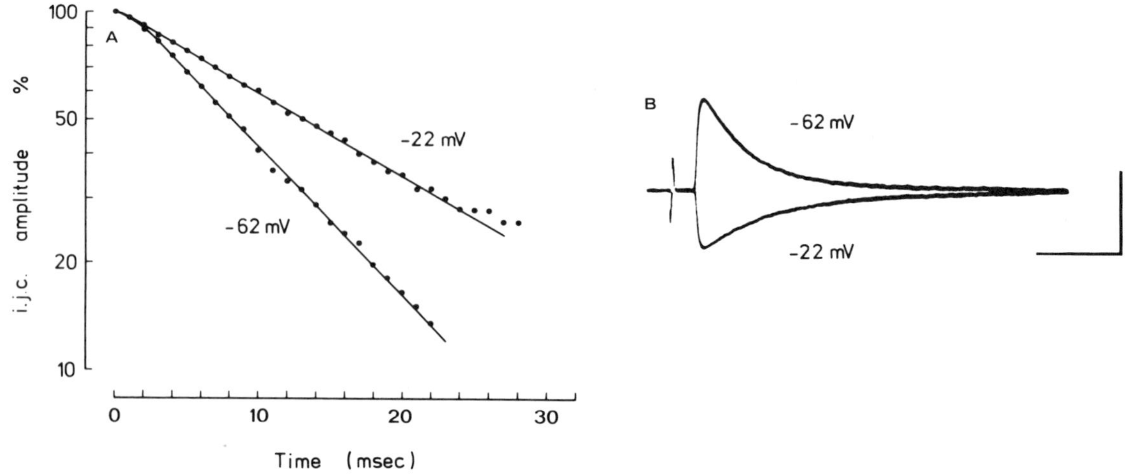

Figure 6. Dependence of τ_{IJC} on clamp potential. (*A*) Decay phases of IJCs plotted semilogarithmically. Decay time-constant τ_{IJC} (− 62 mV) = 10.3 msec and τ_{IJC} (− 22 mV) = 18.5 msec. (*B*) Nerve-impulse-evoked IJCs at − 62 mV and − 22 mV. Current is inward at − 62 mV and outward at − 22 mV. Calibration, 20 nA and 20 msec.

Number of Channels Opened by the Excitatory and Inhibitory Transmitter Packets

At the excitatory synapse, the single-channel current is roughly 10–12 pA (Anderson et al. 1978; Cull-Candy et al. 1981) and the mean amplitude of the MEJC is 2.3 nA (at − 80 mV) (Cull-Candy and Miledi 1982). Hence, a single transmitter packet opens only about 190–230 ion channels at the glutamate synapse. At the inhibitory synapse, single-channel curents are roughly 0.9 pA (Cull-Candy and Miledi 1981) and the mean MIJC amplitude is 0.6 nA (at − 80 mV when V_{eq} = − 40 mV), so a single packet of GABA opens 600–1000 ionic channels (Cull-Candy 1983a,b). This raises interesting questions as to the origin of the marked difference in the number of channels opened by the inhibitory and excitatory transmitter. It seems unlikely that the size of the transmitter packet would vary by as much as five times between the two sites. However, there is good evidence from the experiments with perfused patch-clamp electrodes (see below) that glutamate receptors have a relatively low affinity compared, for example, with acetylcholine receptors (AChR) at the vertebrate end plate (see Sakmann et al. 1980; Cull-Candy et al. 1981), and it is feasible that differences in receptor affinity may be sufficient to account for the variation in the number of channels activated. A further complication is that IJCs (and iono-phoretic GABA pulses) show a marked decrease in peak conductance with hyperpolarization. It could be that, as described in crayfish muscle fibers, the inhibitory channel rectifies (Dudel et al. 1980), although, so far, there is no evidence for this in locust muscle fibers (Cull-Candy and Miledi 1981). If, however, the channel behaves ohmically, a single quantum of GABA, which opens roughly 600–1000 ion channels at the resting potential (− 60 mV), opens less than 100 channels when the same fiber is held at − 110 mV. This clearly deserves further examination.

Patch-clamp Recording of Extrajunctional Glutamate Receptors

The patch-clamp approach allows a direct estimate of parameters of drug-operated channels (Neher and Sakmann 1976; Hamill et al. 1981), and we have used this technique to examine the properties of extrajunctional glutamate receptors (Cull-Candy et al. 1981). So far it has not been possible to patch-clamp junctional glutamate receptors (because of the intervening nerve terminal). However, locust muscle fibers possess a low density of two types of extrajunctional glutamate receptors, one of which (the D receptor) resembles the synaptic receptor in ionic selectivity (see Cull-Candy 1975, 1976). Furthermore, the relative ease with which D receptors desensitize is considerably reduced after treatment with concanavalin A (Mathers and Usherwood 1977), thereby facilitating patch-clamp recording.

Figure 7, A and B, shows examples of single-channel currents recorded with a patch-clamp electrode applied to two muscle fibers voltage-clamped at − 110 mV. High-resistance seals between the patch pipette and the cell membrane (Hamill et al. 1981) are not readily obtained in locust muscle. However, because of the large size of the single glutamate current, the ratio of signal to noise is reasonably good. The conductance of the extrajunctional channel, obtained from single-channel recording, is about 150 pS and their mean lifetime is 2.3 msec (Cull-Candy et al. 1981), which is in reasonable agreement with previous estimates from both noise analysis (Anderson et al. 1976, 1978) and patch-clamp recording (Patlak et al. 1979).

"Dose-response" Relationship for the Single Receptor Channel

It was of interest to examine the response of an individual receptor-channel complex to various glutamate

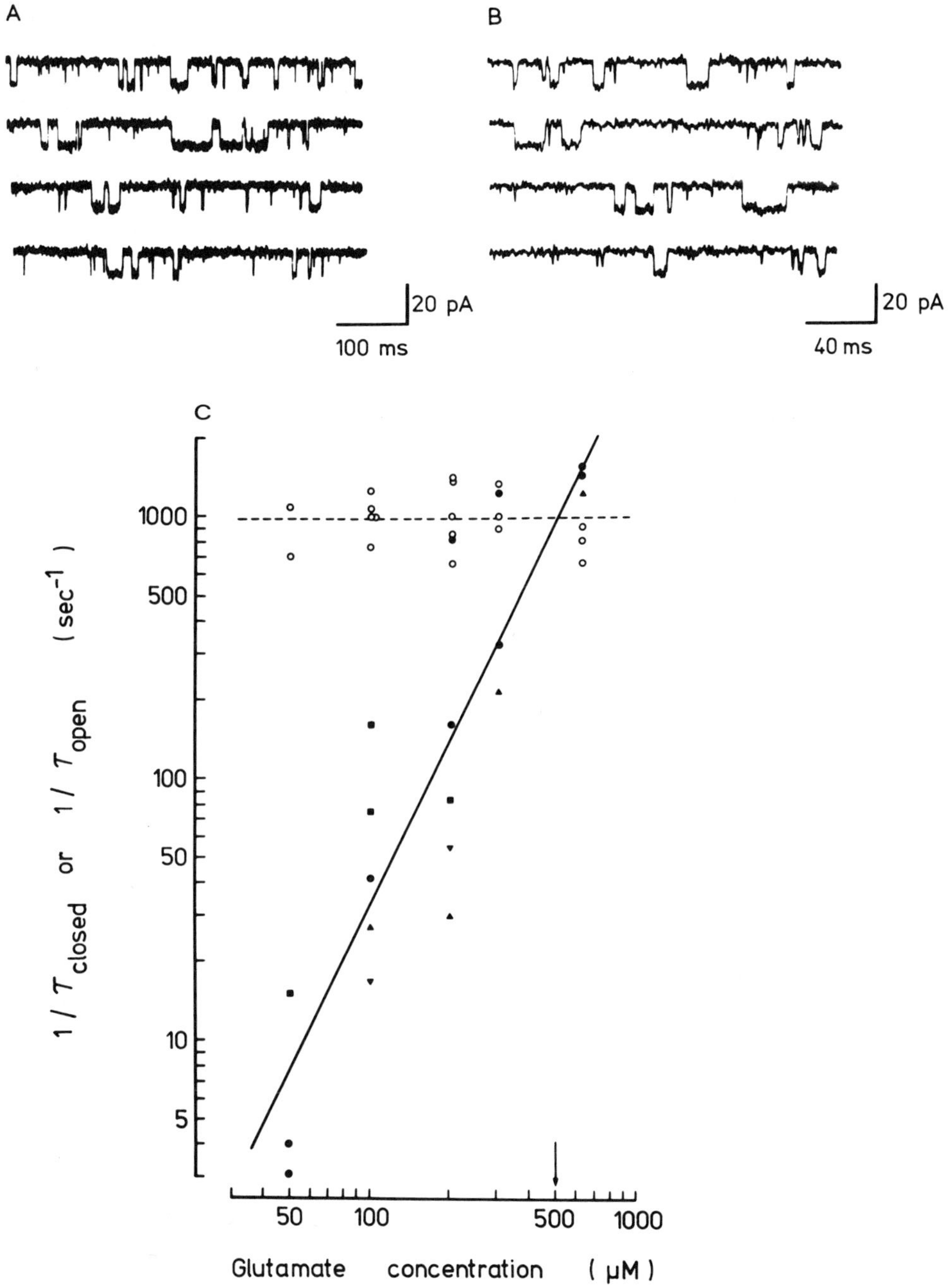

Figure 7. (*A,B*) Records of glutamate-activated, single-channel currents from two muscle fibers voltage-clamped at – 110 mV; note the difference in sweep speeds. In both experiments, the tip of the patch electrode contained 100 μM glutamate. Inward currents are seen as downward deflections. Bandwidth DC – 1 kHz. (*C*) Relationship between reciprocal closed time, $1/\tau_{closed}$, and glutamate concentrations obtained with internally perfused patch electrodes, $1/\tau_{closed}$ was measured in four patches (●, ■, ▲, ▼). Continuous line through the points has a slope of 2.1. Mean value of $1/\tau_{open}$ (○), obtained from the same records, is indicated by dashed line. Open and closed times are equal at the intercept of the lines, which occurs at 500 μM glutamate (arrow). (Reprinted, with permission, from Cull-Candy et al. 1981.)

concentrations. This seemed feasible as it is possible to find membrane patches where only a single receptor is active under the tip of the patch electrode. However, one limitation of conventional recording is that the concentration of drug contained in the patch electrode cannot be altered during recording. To overcome this problem, a system was devised that allows rapid internal perfusion of the tip of the patch electrode, with various known glutamate concentrations (for details, see Cull-Candy et al. 1981). In such experiments, no double-sized events were present even at high levels of glutamate when the channel was in its open state for a majority of the time.

Figure 7 is a plot of 1/mean closed time, $(1/\tau_{closed})$, which is equivalent to β', the net forward rate of the reaction involved in the opening of the receptor channel (closed $\underset{\alpha}{\overset{\beta'}{\rightleftharpoons}}$ open). The data points were obtained from four different membrane patches in which various glutamate concentrations were applied to individual receptors by internal perfusion of the patch-clamp electrode. Mean channel open-time is obtained from the same record. 1/open time, which is equivalent to the rate constant for channel closing α, was not appreciably dependent on concentration over the range examined. From such a plot it is possible to estimate the apparent dissociation constant of the receptor, i.e., the glutamate concentration that holds the channel in its open state for 50% of the time (i.e., when $\alpha = \beta'$, at the intercept of the lines in Fig. 7C). The apparent dissociation constant is about 500 μM glutamate, and the slope of the plot is close to 2. At the vertebrate end plate, roughly 30 μM ACh is required to hold the channel in its open state for 50% of the time (Sakmann et al. 1980), indicating that the affinity of the glutamate receptor for its neurotransmitter is very low by comparison.

Resolution of Fast Events: Channel Flicker

Recently, glutamate-receptor channels have been examined with a resolution that enables the detection of events as brief as 50-μsec duration (Cull-Candy and Parker 1982). This was possible partly because of the large current passing through the single glutamate-activated channel, especially when fibers are clamped at more hyperpolarized levels (i.e., − 140 mV to − 150 mV) and also by virtue of other technical improvements (described in Cull-Candy and Parker 1982). Figure 8, A, B, and C, shows single-channel currents obtained with wide-frequency-band recording, and it is apparent that brief closures (brief gaps) interrupt some channel openings (Fig. 8A). A similar phenomenon has been described for single AChR channels in vertebrate muscle fibers (see Colquhoun and Sakmann 1981). Because such gaps are brief and occur even when the frequency of channel openings is low, it is unlikely that a burst of openings results from several independent channel activations. This was apparently confirmed by forming histograms of channel-closed times, which have enabled a

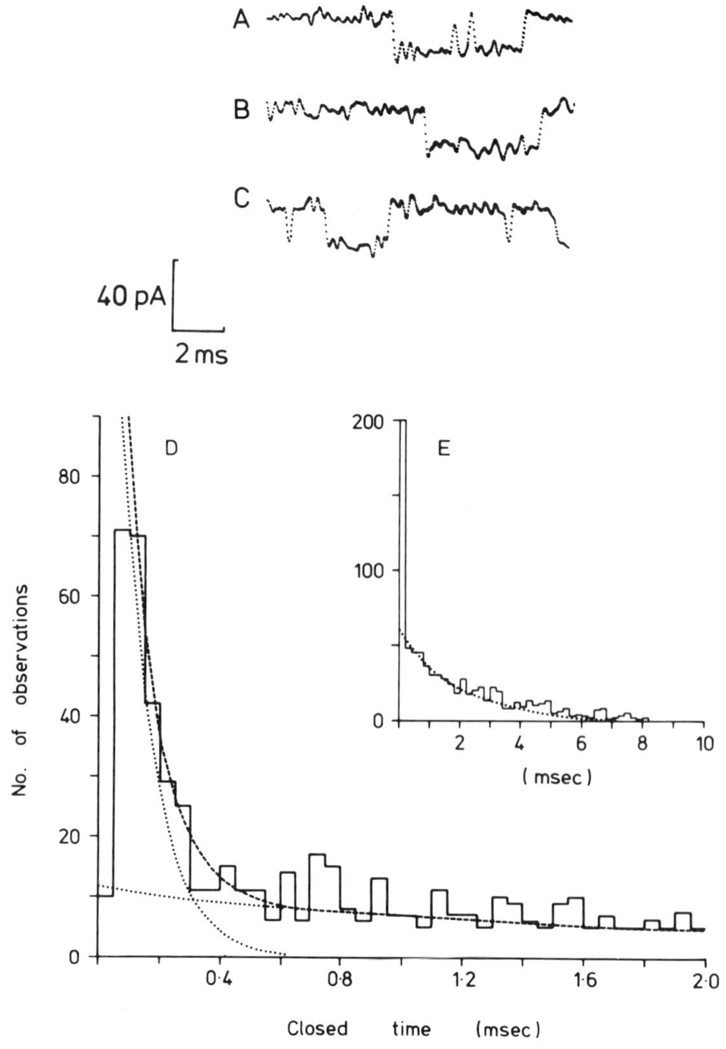

Figure 8. (*A,B,C*) Single-channel currents recorded with high-frequency resolution (DC − 3.2 kHz). Channel openings correspond to downward deflections. (*D,E*) Histograms (on differing time scales) of single-channel closed-time distributions, showing the excess of brief events. Data are measurements from 832 channel openings from a membrane patch where only one channel was active. Distribution in *D* has been fitted by the sum of two exponentials. Component exponentials are indicated by dotted curves, and the sum is indicated by a dashed curve; the slow exponential is shown in *E*. Time constants are $\tau_{fast} = 100$ μsec, $\tau_{slow} = 2$ msec; area under the exponentials is fast = 383, slow = 616. (Reprinted, with permission, from Cull-Candy and Parker 1982.)

quantitative analysis of the brief closures. As seen in Figure 8D,E, the closed times can be fitted with a single exponential, in agreement with earlier studies, except for the presence of a large excess of closings with durations less than 200 μsec (Cull-Candy and Parker 1982; but see also Gration et al. 1982). The lack of events in the 0–50 μsec bin probably represents the upper limit of the frequency resolution of the recording system. The histograms show the closed times on two different time scales, and the time constants of the exponential are $\tau_{fast} = 100$ μsec, $\tau_{slow} = 2$ msec. From the areas under the exponentials, it is estimated that there are roughly 0.6 closing per burst and hence 1.6 openings per burst. Since the average burst length is about 1.7 msec, the mean duration of each opening is roughly 1.0 msec.

An interesting possibility is that brief gaps, within a burst of openings, may result from multiple openings during a single occupancy, without dissociation of the agonist molecules (see Colquhoun and Hawkes 1981). In a simple reaction scheme of the type (del Castillo and Katz 1957; Magleby and Stevens 1972):

$$A + R \underset{k_{-1}}{\overset{k_1}{\rightleftharpoons}} AR_{closed} \underset{\alpha}{\overset{\beta}{\rightleftharpoons}} AR_{open}$$

the mean duration of the gaps will be $1/(\beta + k_{-1})$ and the mean number of gaps within a burst will be β/k_{-1}. Estimates of k_{-1} and β obtained from such analyses at AChR and glutamate-receptor channels do not seem compatible with the ideas that the agonist binding step is fast and that the conformation change is rate limiting (see Colquhoun and Sakmann 1981; Cull-Candy and Parker 1982). A further point of interest is that earlier estimates of channel lifetime, in which brief gaps were not detected, would be equivalent to the burst length of a burst of openings. Fairly good agreement has in fact been found between "mean burst length" and previous estimates of channel lifetime derived from both noise analysis and patch-clamp recording.

ACKNOWLEDGMENT

I am grateful to the Wellcome Trust and the Medical Research Council for support.

REFERENCES

ADAMS, P.R. 1976. Drug blockage of open end-plate channels. *J. Physiol.* **260:** 531.

ANDERSON, C.R. and C.F. STEVENS. 1973. Voltage clamp analysis of acetylcholine produced end-plate current fluctuations at frog neuromuscular junction. *J. Physiol.* **235:** 655.

ANDERSON, C.R., S.G. CULL-CANDY, and R. MILEDI. 1976. Glutamate and quisqualate noise in voltage-clamped locust muscle fibers. *Nature* **261:** 151.

———. 1978. Glutamate current noise: Post-synaptic channel kinetics investigated under voltage clamp. *J. Physiol.* **282:** 219.

COLQUHOUN, D. and A.G. HAWKES. 1981. On the stochastic properties of single ion channels. *Proc. R. Soc. Lond. B* **211:** 205.

COLQUHOUN, D. and B. SAKMANN. 1981. Fluctuations in the microsecond time range of the current through single acetylcholine receptor ion channels. *Nature* **294:** 464.

COLQUHOUN, D., F. DREYER, and R.E. SHERIDAN. 1979. The actions of tubocurarine at the frog neuromuscular junctions. *J. Physiol.* **293:** 247.

CRAWFORD, A.C. and R.N. MCBURNEY. 1976. The postsynaptic action of some putative excitatory transmitter substances. *Proc. R. Soc. Lond. B* **192:** 481.

CULL-CANDY, S.G. 1975. Effect of denervation and local damage on extrajunctional L-glutamate receptors in locust muscle. *Nature* **258:** 530.

———. 1976. Two types of extrajunctional L-glutamate receptors in locust muscle. *J. Physiol.* **255:** 449.

———. 1983a. New ways of looking at synaptic channels: Noise analysis and patch-clamp recording. In *Recent advances in physiology* (ed. P.F. Baker), vol. 10. Churchill-Livingstone.

———. 1983b. Miniature inhibitory junctional currents in voltage-clamped locust muscle fibres. *J. Physiol.* **342** (in press).

CULL-CANDY, S.G. and R. MILEDI. 1981. Junctional and extrajunctional channels activated by GABA in locust muscle fibres. *Proc. R. Soc. Lond. B* **211:** 527.

———. 1982. Properties of miniature excitatory juctional currents at the locust nerve-muscle junction. *J. Physiol.* **326:** 527.

———. 1983. Block of glutamate-activated synaptic channels by curare and gallamine. *Proc. R. Soc. Lond. B* **218:** 111.

CULL-CANDY, S.G. and I. PARKER. 1982. Rapid kinetics of single glutamate-receptor channels. *Nature* **295:** 410.

CULL-CANDY, S.G., R. MILEDI, and I. PARKER. 1981. Single glutamate-activated channels recorded from locust muscle fibres with perfused patch-clamp electrodes. *J. Physiol.* **321:** 195.

DEL CASTILLO, J. and B. KATZ. 1957. Interaction at end-plate receptors between different choline derivatives. *Proc. R. Soc. Lond. B* **146:** 369.

DUDEL, J. 1974. Nonlinear voltage dependence of excitatory synaptic current in crayfish muscle. *Pfluegers Arch. Gesamte Physiol.* **352:** 227.

———. 1977. Voltage dependence of amplitude and time course of inhibitory synaptic current in crayfish muscle. *Pfluegers Arch. Gesamte Physiol.* **371:** 167.

DUDEL, J., W. FINGER, and H. STETTMEIER. 1980. Inhibitory synaptic channels activated by γ-aminobutyric acid (GABA) in crayfish muscle. *Pfluegers Arch. Gesamte Physiol.* **387:** 143.

FINGER, W. and H. STETTMEIER. 1981. Analysis of miniature spontaneous inhibitory postsynaptic currents (sIPSCs) from current noise in crayfish opener muscle. *Pfluegers Arch. Gesamte Physiol.* **392:** 157.

GRATION, K.A.F., J.J. LAMBERT, R.L. RAMSEY, R.P. RAND, and P.N.R. USHERWOOD. 1982. Closure of membrane channels gated by glutamate receptors may be a two-step process. *Nature* **295:** 599.

HAMILL, O.P., A. MARTY, E. NEHER, B. SAKMANN, and F.J. SIGWORTH. 1981. Improved patch-clamp technique for high-resolution current recording from cells and cell-free membrane patches. *Pfluegers. Arch. Gesamte Physiol.* **391:** 85.

KATZ, B. and R. MILEDI. 1972. The statistical nature of the acetylcholine potential and its molecular components. *J. Physiol.* **224:** 665.

———. 1973a. The binding of acetylcholine to receptors and its removal from the synaptic cleft. *J. Physiol.* **231:** 549.

———. 1973b. The characteristics of "end-plate noise" produced by different polarizing drugs. *J. Physiol.* **230:** 707.

———. 1975. The effect of procaine on the action of acetylcholine at the neuromuscular junction. *J. Physiol.* **249:** 269.

——. 1978. A re-examination of curare action at the motor endplate. *Proc. R. Soc. Lond. B* **203:** 119.

Kordas, M. 1969. The effect of membrane polarization on the time course of the end-plate current in frog sartorius muscle. *J. Physiol.* **204:** 493.

Kriebel, M.E. and C.E. Gross. 1974. Multimodal distribution of frog miniature end-plate potentials in adult, denervated and tadpole leg muscle. *J. Gen. Physiol.* **64:** 85.

Lingle, C., J.S. Eisen, and E. Marder. 1981. Block of glutamatergic excitatory synaptic channels by chlorisondamine. *Mol. Pharmacol.* **19:** 349.

Magleby, K.L. and C.F. Stevens. 1972. A quantitative description of end-plate currents. *J. Physiol.* **223:** 173.

Manalis, R.S. 1977. Voltage-dependent effect of curare at the frog neuromuscular junction. *Nature* **267:** 366.

Marty, A., T.O. Neild, and P. Ascher. 1976. Voltage sensitivity of acetylcholine currents in *Aplysia* neurones in the presence of curare. *Nature* **261:** 50.

Mathers, D.A. and P.N.R. Usherwood. 1977. Concanavalin A blocks desensitization of glutamate receptors on insect muscle fibres. *Nature* **259:** 409.

Neher, E. and B. Sakmann. 1976. Single-channel currents recorded from membrane of denervated frog muscle fibres. *Nature* **260:** 799.

Onodera, K. and A. Takeuchi. 1976. Inhibitory postsynaptic current in voltage-clamped crayfish muscle. *Nature* **263:** 153.

——. 1978. Effects of membrane potential and temperature on the excitatory post-synaptic current in the crayfish muscle. *J. Physiol.* **276:** 183.

Patlak, J.B., K.A.F. Gration, and P.N.R. Usherwood. 1979. Single glutamate-activated channels in locust muscle. *Nature* **278:** 643.

Ruff, R.L. 1977. A quantitative analysis of local anaesthetic alteration of miniature endplate currents and end-plate current fluctuation. *J. Physiol.* **264:** 89.

Sakmann, B., J. Patlak, and E. Neher. 1980. Single acetylcholine-activated channels show burst-kinetics in presence of desensitizing concentrations of agonist. *Nature* **286:** 71.

Stevens, C.F. 1972. Interferences about membrane properties from electrical noise measurements. *Biophys. J.* **12:** 1028.

Usherwood, P.N.R., and S.G. Cull-Candy. 1975. Pharmacology of somatic nerve-muscle synapses. In *Insect muscle* (ed. P.N.R. Usherwood), p. 207. Academic Press, London.

Usherwood, P.N.R. and R. Grundfest. 1965. Peripheral inhibition in skeletal muscle of insects. *J. Neurophysiol.* **28:** 497.

Yamamoto, D. and H. Washio. 1979. Curare has a voltage-dependent blocking action on the glutamate synapse. *Nature* **281:** 372.

Chemical and Pharmacological Approaches to the Definition and Quantitation of Calcium Channels

P.J. GENGO, E. LUCHOWSKI, D.E. RAMPE, A. RUTLEDGE, A.M. TRIGGLE, D.J. TRIGGLE, and R.A. JANIS*

*Department of Biochemical Pharmacology, School of Pharmacy, State University of New York, Buffalo, New York 14260; *Miles Institute for Preclinical Pharmacology, New Haven, Connecticut 06509*

The role of Ca^{++} as a messenger species is made possible by the large, inwardly directed concentration gradient of free ionized Ca^{++}, by the presence of intracellular binding proteins with high affinity for Ca^{++}, and by the existence of chemically and electrically excitable Ca^{++} channels in the plasma membrane (Kretsinger 1977; Rosenberger and Triggle 1978; Cheung 1980).

Despite the importance of Ca^{++} as a current-carrying species and its fundamental role in excitation-contraction and stimulus-secretion coupling processes, knowledge of Ca^{++} channel function has lagged far behind that of Na^{+} or K^{+} channels (Hagiwara and Byerly 1981; Reuter 1983; Tsien 1983). Fortunately, this situation is being remedied with substantial progress in the electrophysiologic analysis of Ca^{++} currents and with the development of blocking drugs that inhibit Ca^{++} channel function in potent and specific fashion.

The blocking drugs are a chemically heterogeneous group of compounds variously referred to as Ca^{++} channel antagonists, Ca^{++} blockers, Ca^{++} entry blockers, or slow channel blockers (Fleckenstein 1971, 1977; Weiss 1981; Flaim and Zelis 1982; Sonnenblick and Lesch 1982; van Zwieten 1982). These compounds, which include the clinically available nifedipine, verapamil, and diltiazem (Fig. 1), are being employed increasingly for the treatment of a number of cardiovascular disorders including angina, supraventricular tachycardias, and hypertension, but it is likely that the spectrum of clinical use for existing and new agents of this category is much wider (Stone et al. 1980; Flaim and Zelis 1982; Janis and Triggle 1983). Thus, there is considerable basic and clinical interest in the definition of the sites and mechanisms of action of this group of agents.

The classification of these agents as Ca^{++} channel antagonists owes much to the pioneering investigations of Fleckenstein and his colleagues, who noted that these agents mimicked in cardiac tissue the electrophysiologic and mechanical effects of Ca^{++} withdrawal (Fleckenstein 1971, 1977, 1981). Many subsequent investigations have documented the ability of this group of agents to inhibit, with substantial selectivity, the Ca^{++} component of the cardiac current (inter alia, Kass and Tsien 1975; Bayer et al. 1982; Lee and Tsien 1983). Similarly, in smooth muscle, although comparable electrophysiologic data are largely lacking, the Ca^{++} channel antagonists inhibit $^{45}Ca^{++}$ uptake in response to depolarizing stimuli with a generally good correlation to their activities as inhibitors of excitation-contraction coupling (Godfraind and Dieu 1981; Meisheri et al. 1981; Janis and Triggle 1983).

This evidence, together with descriptions of stimulus selectivity (Triggle 1981; Flaim 1982) and structure-activity relationships including stereoselectivity (Triggle 1982; Janis and Triggle 1983), indicate that these agents are specific rather than nonspecific agents and that they have a selectivity of action for the voltage-dependent channel (Fig. 2). However, this evidence defines neither the sites nor the mechanisms of action. The chemical heterogeneity of this class of compounds suggests that they have multiple sites and mechanisms of action, and consistent with this there is substantial pharmacologic heterogeneity of action. Verapamil and diltiazem are approximately equieffective in cardiac and smooth muscle, whereas nifedipine and other 1,4-dihydropyridines are substantially more selective for smooth muscle. These differences parallel their relative therapeutic uses (Stone et al. 1980; Schwartz and Triggle 1984). Additionally, verapamil and D-600 show frequency and voltage dependence of action, a property expressed to a much lesser extent by the 1,4-dihydropyridines (Bayer et al. 1975; Ehara and Kaufmann 1978; Kass 1982; Lee and Tsien 1983). It is noteworthy also that neuronal Ca^{++} channels appear generally to be less sensitive to these agents than the

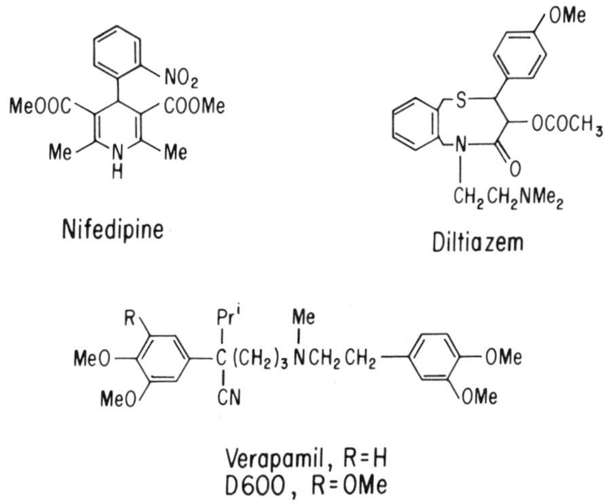

Figure 1. Structural formulae of Ca^{++} channel antagonists.

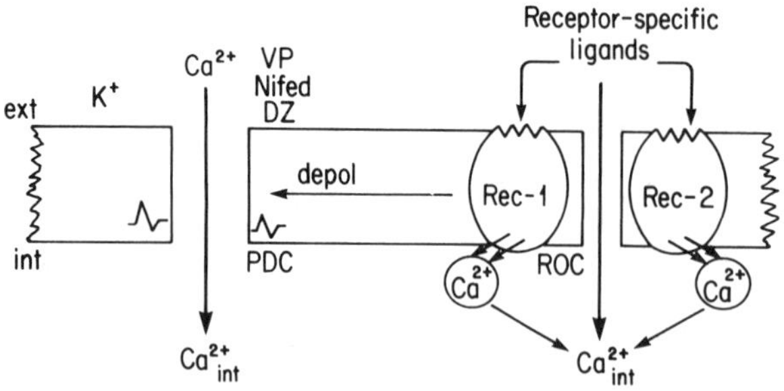

Figure 2. Schematic representation of Ca^{++} mobilization processes in excitable membranes.

channels in cardiac and smooth muscle (Nachshen and Blaustein 1979; Akaike et al. 1981; but see Toll 1982). The extent to which such apparent differences in sensitivity may reflect basic differences in the Ca^{++} channels as opposed to differences in associated regulatory sites, differences in the linkage between membrane excitation, channel activation, and response is not clear.

It appears likely, therefore, that the Ca^{++} channel antagonists represent compounds with substantial specificity for voltage-dependent and some receptor-operated Ca^{++} channels, but possessing a variety of sites and mechanisms of action. Delineation of these sites and mechanisms of action is important to rationalize the therapeutic use of these agents, to facilitate the development of new agents, and to provide probes for the investigation of Ca^{++} channels. Radioligand binding studies have been useful for the description of receptor binding and activation processes. We have used this approach in attempts to define the characteristics of Ca^{++} channel antagonist interaction in excitable tissues.

METHODS

Chemical synthetic procedures. 1,4-Dihydropyridines were synthesized by standard variations of the Hantzsch 1,4-dihydropyridine synthesis (Eisner and Kuthan 1972). Nitrendipine (ethylmethyl 1,4-dihydro-2,6-dimethyl-4[3-nitrophenyl]-3,5-pyridinecarboxylate), nisoldipine (isobutyl methyl 1,4-dihydro-2,6-dimethyl-4[2-nitrophenyl]-3,5-pyridinecarboxylate), nimodipine (isopropyl[2-methoxyethyl]1,4-dihydro-2,6-dimethyl-4[3-nitrophenyl]-3,5-pyridine carboxylate) were obtained from Miles Institute for Preclinical Pharmacology, New Haven, Conn. Bay 3628/3629 ([+]- and [−]-isopropyl methyl 1,4-dihydro-2,6-dimethyl-4[3-nitrophenyl]-3,5-pyridine carboxylate) was obtained from Bayer, AG, Wuppertal, F.R.G; verapamil and D-600 were from Knoll, AG, Ludwigshaften, F.R.G.; and diltiazem was from Marion Laboratories, Kansas City, Kansas.

Smooth Muscle Pharmacology

Smooth muscle preparations were prepared for recording of mechanical responses as previously described (Rosenberger et al. 1979; Nghiem et al. 1982).

Radioligand binding. Specific [^{3}H]nitrendipine binding to microsomal and other fractions of muscle and neuronal tissue was determined according to published protocols (Bolger et al. 1982, 1983; Triggle et al. 1982). [^{3}H]Nitrendipine and [^{3}H]2,6-dimethyl-3,5-dicarbomethoxy-4-(2-isothiocyanophenyl)-1,4-dihydropyridine were obtained from New England Nuclear.

Biochemical determinations. Determinations of molecular weight of the [^{3}H]nitrendipine-binding site in microsomal preparations were made by radiation inactivation (target size) analysis (Lo et al. 1982; Venter et al. 1983). Polyacrylamide gel electrophoresis of affinity-labeled membranes was performed as described (Venter et al. 1983).

RESULTS AND DISCUSSION

The inference, drawn from pharmacologic and physiologic data, that the Ca^{++} channel antagonists interact at a specific site to block channel function is confirmed by the existence of saturable, reversible, and high-affinity binding of [^{3}H]nitrendipine (inter alia, Bellemann et al. 1981, 1983; Bolger et al. 1982, 1983; Ehlert et al. 1982; Ferry and Glossmann 1982; Gould et al. 1982; Triggle et al. 1982) in membrane preparations from excitable tissues (Fig. 3; Table 1). Binding of [^{3}H]nitrendipine, and of the related compounds [^{3}H]nimodipine and [^{3}H]nisoldipine, shows an appropriate pharmacologic selectivity, being insensitive or weakly sensitive to a wide variety of receptor-directed ligands but sensitive to the several structural categories of Ca^{++} channel antagonists.

Correlations between binding and pharmacologic activities have been studied extensively in smooth muscle (Bolger et al. 1982, 1983; Triggle et al. 1982; Janis and Triggle 1983; Triggle and Janis 1983). For a series of 1,4-dihydropyridines, including enantiomeric pairs, an essentially 1:1 correlation exists over a wide potency range, between inhibition of [^{3}H]nitrendipine binding and inhibition of K^+-depolarization-induced contractions in guinea pig ileal longitudinal smooth muscle (Fig. 4A; Bolger et al. 1983). [^{3}H]Nitrendipine binding is sensitive to other classes of Ca^{++} channel antagonists, although, unlike the 1,4-dihydropyridines, these agents do not behave as simple competitive antagonists. Thus, D-600 (methoxyverapamil) serves as a stereoselective ([−]-D-600>[+]-D 600) partial antagonist producing a concentration-limited decrease in the affinity of [^{3}H]nitrendipine without change in B_{MAX}, whereas diltiazem increases B_{MAX} without change in K_d. [^{3}H]Nitrendipine binding in smooth muscle is also sensitive to organic di- and trivalent metal cations (Mn^{++}, Co^{++}, La^{3+}, Tm^{3+}, etc.) that serve as Ca^{++} channel antagonists. Additionally, [^{3}H]nitrendipine binding is Ca^{++} (and other divalent cation)-dependent, being abolished in the absence of Ca^{++} and gradedly restored with increasing Ca^{++} concentration, $EC_{50} \cong 5 \times 10^{-6}$ M (Bolger et al. 1983; E.M. Luchowski et al., in prep.).

These observations suggest that there are several

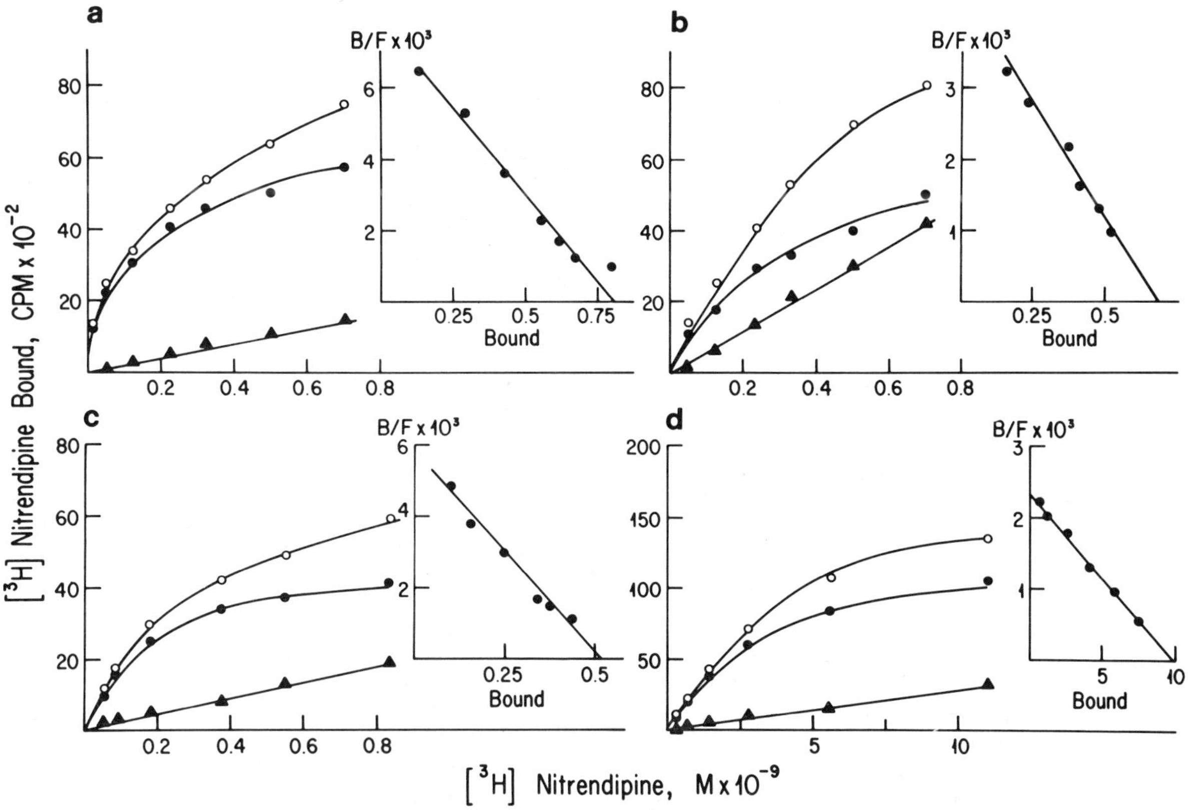

Figure 3. Binding of [^{3}H]nitrendipine to microsomal preparations from smooth, cardiac, and skeletal muscle and brain. Each plot is a representative example showing total binding (○), nonspecific binding (▲), and specific binding (●). The insets show the Scatchard transformations of specific binding. (*a*) Guinea pig ileal longitudinal smooth muscle, $B_{MAX} = 820$ fmoles/mg protein, K_d 1.1×10^{-10} M; (*b*) rat ventricle muscle, $B_{MAX} = 650$ fmoles/mg protein, $K_d = 1.8 \times 10^{-10}$ M; (*c*) guinea pig cerebral cortex synaptosomal preparations, $B_{MAX} = 505$ fmoles/mg protein, $K_d = 0.9 \times 10^{-10}$ M; (*d*) rat hindlimb skeletal muscle, $B_{MAX} = 9.2$ pmoles/mg protein, $K_d = 3.6 \times 10^{-9}$ M. Binding was carried out in 50 mM Tris buffer (pH 7.2) at 25°C (skeletal muscle at 10°C) and nonspecific binding was defined by 10^{-7}–10^{-6} M nitrendipine.

discrete sites at which Ca^{++} channel antagonists affect [^{3}H]nitrendipine binding and (by inference) Ca^{++} channel function. These include the binding site for the 1,4-dihydropyridines, at least one allosteric site with which other organic Ca^{++} antagonists interact, and a cation-binding site, which regulates 1,4-dihydropyridine interaction (Fig. 5). Although it is not yet possible to define the relationship of these several sites to Ca^{++} channel function, the existence of multiple sites and mechanisms of Ca^{++} channel antagonist interaction is consistent with the chemical and pharmacologic heterogeneity of these agents.

Table 1. Characteristics of Binding of [^{3}H]Nitrendipine in a Representative Series of Preparations

Preparation	K_d (nM)	B_{MAX} (fmoles mg^{-1})	References[a]
1. Smooth muscle			
Guinea pig ileum	0.16	1,100	Bolger et al. (1982, 1983)
Guinea pig bladder	0.15	500	G.T. Bolger et al. (unpubl.)
Guinea pig lung	0.18	40	G.T. Bolger et al. (unpubl.)
Rat myometrium	0.18	220	R.A. Janis et al. (unpubl.)
Rat vas deferens	0.20	230	R.A. Janis et al. (unpubl.)
Rat mesenteric artery	0.10	18	Triggle et al. (1982)
2. Cardiac muscle			
Rat ventricle	0.18	400	Bolger et al. (1982, 1983)
Canine ventricle	0.14	960	Sarmiento et al. (1983)
3. Neuronal preparation			
Rat cerebral cortex	0.10	102	Ehlert et al. (1982)
Rat synaptosomes	0.17	114	Yamamura et al. (1982)
PC12 cells	1.10	28	Toll (1982)
4. Skeletal muscle			
Rabbit skeletal muscle	1.80	2,000	Fosset et al. (1983)
T tubules	1.80	56,000	Fosset et al. (1983)

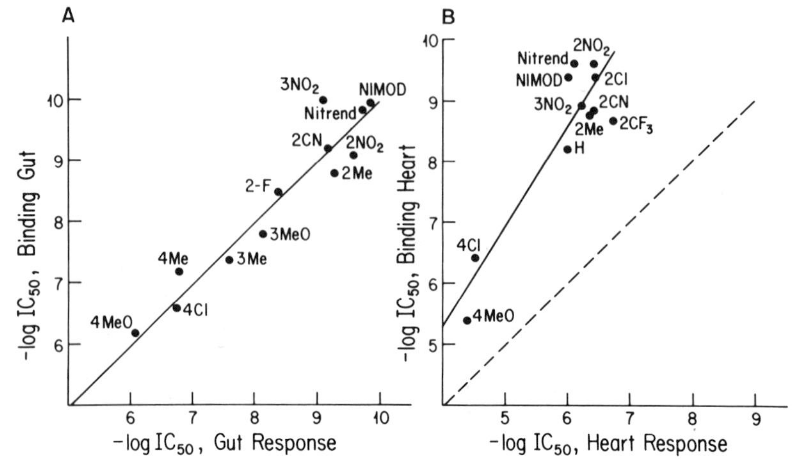

Figure 4. (*A,B*) Correlations between binding and pharmacologic response for a series of 1,4-dihydropyridines in smooth and cardiac muscle. The activities of a series of 2,6-dimethyl-3,5-dicarbomethoxy-4-(x-substituted phenyl)-1,4-dihydropyridines and nimodipine (Nimod) and nitrendipine (Nitrend) as competitive antagonists of [^{3}H]nitrendipine binding are plotted against the activities as inhibitors of mechanical response. (*A*) Guinea pig ileal longitudinal smooth muscle. Response is inhibition of the sustained contracture to K^+ depolarization. (*B*) Rabbit ventricle muscle. Response is inhibition of cat papillary muscle (data from Rodenkirchen et al. 1979).

The binding of [^{3}H]nitrendipine and [^{3}H]nimodipine to cardiac and brain tissue shows very similar characteristics to that revealed for smooth muscle. Binding is of similarly high affinity, is Ca^{++} dependent, and is inhibited by the several categories of Ca^{++} channel antagonist (Bellemann et al. 1981, 1983; Ehlert et al. 1982; Gould et al. 1982; Yamamura et al. 1982; Bolger et al. 1983; Sarmiento et al. 1983;

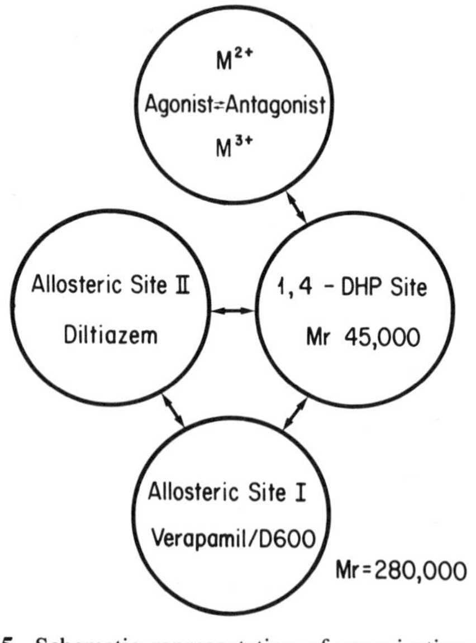

Figure 5. Schematic representation of organization of binding sites of Ca^{++} antagonists in smooth muscle. The molecular weight of the total nitrendipine-binding complex is 280,000, as determined by radiation inactivation analysis and the molecular weight of the dihydropyridine-binding site is 45,000, as determined by an affinity-labeling technique (Venter et al. 1983). Dihydropyridine binding is dependent upon Ca^{++} and other divalent cations (M^{++}/M^{3+} site), and the other structural categories of Ca^{++} channel antagonists are depicted as binding at two sites allosterically linked to the 1,4-dihydropyridine-binding site.

Triggle and Janis 1983). However, for these tissues, there exist significant discrepancies between the binding and pharmacologic activities of the 1,4-dihydropyridines. The discrepancy for cardiac tissue is shown in Figure 4B, in which the data reveal high-affinity binding and low-affinity pharmacology. Nonetheless, the 1,4-dihydropyridines exhibit qualitatively the same structure-activity relationship as inhibitors of excitation-contraction coupling in both cardiac and smooth muscle, consistent with a basic similarity in the 1,4-dihydropyridine-binding sites in both tissues. Similar discrepancies between 1,4-dihydropyridine binding and pharmacology are also apparent in neuronal and skeletal muscle tissues (Nachshen and Blaustein 1979; Triggle 1981; Gould et al. 1982; Yamamura et al. 1982; Bellemann et al. 1983; Kaplita and Triggle 1983).

Several hypotheses may be advanced to accommodate such discrepancies.

1. The properties of the binding site have been altered during tissue disruption procedures, e.g., loss of a phosphorylation-dephosphorylation cycle or other regulating factor(s). The extent of loss of such control would have to be tissue dependent.
2. The presence of endogenous ligands for the dihydropyridine-binding site may shift dose-response curves determined in intact tissues, but may not be a factor in binding studies in broken cell preparations.
3. There is a nonlinear relationship between binding site occupancy by a 1,4-dihydropyridine and inhibition of channel function. The existence of a "channel reserve" would shift the concentration-response inhibition curve to the right of the binding curve.
4. The high-affinity binding sites may not be uniquely related to functionally active Ca^{++} channels. This hypothesis suggests that there exists more than one class of voltage-dependent Ca^{++} channel, one sen-

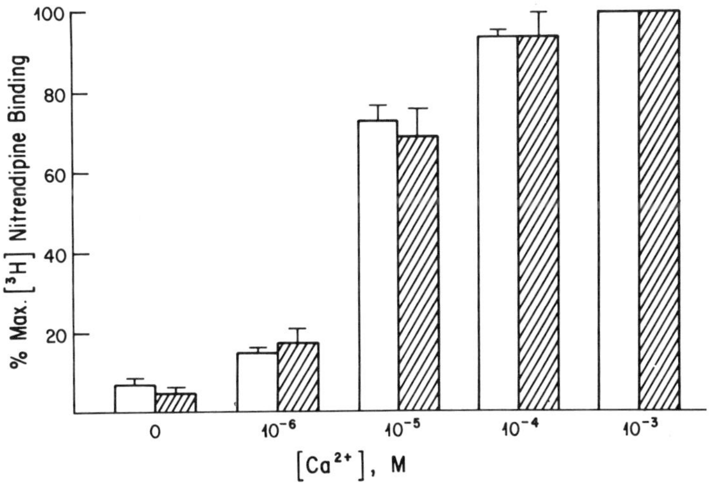

Figure 6. Dependence of binding of [^{3}H]nitrendipine (□) and of [^{3}H]2,6-dimethyl-3,5-dicarbomethoxy-4-(2-isothiocyanophenyl)-1,4-dihydropyridine (▨) to guinea pig ileal smooth muscle membranes upon the presence of Ca^{++}. Binding was determined at 25°C in 50 mM Tris buffer (pH 7.2) in membranes prepared in the presence of EDTA and incubated (zero Ca^{++}) in the presence of 5×10^{-5} M EDTA. Bars indicate mean ± S.E.M. ($n = 4$–8).

sitive to the Ca^{++} channel antagonists and one insensitive, and that the properties or extent of use of the different classes is tissue dependent.

5. Alternatively, the extent of association of [^{3}H]-nitrendipine binding with functional Ca^{++} channels may vary between tissues. This suggests that there may exist more than one category of [^{3}H]nitrendipine-binding site according to the state of association with the channel.
6. The affinity of the binding site may vary with the state (open, closed, inactivated) of the Ca^{++} channel, which, in turn, is dependent upon membrane potential, intracellular Ca^{++} activity, and other factors cited in items 2 and 3, above.
7. Finally, and perhaps least likely, the [^{3}H]1,4-dihydropyridine sites may be a cellular component of excitable tissues but are simply not related to Ca^{++} channels or Ca^{++} channel function.

Resolution of these several possibilities is of considerable importance since it will lead to a better appreciation of the relationship between [^{3}H]nitrendipine binding and functional Ca^{++} channels. The discrepancies between binding and pharmacologic activities of the 1,4-dihydropyridines should not, however, obscure the many fundamental similarities between ^{3}[H]dihydropyridine binding in different tissues. These similarities include, as noted previously, sensitivity to antagonism by the several structural classes of Ca^{++} antagonists, both organic and inorganic.

Of particular interest is the dependence of [^{3}H]1,4-dihydropyridine binding upon Ca^{++} and other divalent cations (Glossmann et al. 1982; Gould et al. 1982; Bolger et al. 1983). Treatment of membranes with chelating agents (EDTA, EGTA) virtually abolishes specific binding of [^{3}H]nitrendipine by reducing the number of available binding sites, and binding is restored in graded fashion by Ca^{++} (Fig. 6) and other cations in the sequence $Ca^{++} \cong Sr^{++} \cong Ba^{++} > Co^{++}$ Mg^{++} $Mn^{++} > Ni^{++} > Zn^{++}$ $Cd^{++} > La^{3+} \cong Sm^{3+}$ $\cong Tm^{3+}$. The least potent cations block the Ca^{++}-[^{3}H]nitrendipine interaction, and it is possible that the cations of this series interact at a common site to control the availability of [^{3}H]nitrendipine sites. Since cations of this series can serve as carriers or inhibitors of current flow through the Ca^{++} channel (Hagiwara and Byerly 1981), they may act at the cation coordination site of the Ca^{++} channel to regulate [^{3}H]nitrendipine binding (Gould et al. 1982). However, the high apparent affinity of Ca^{++} ($EC_{50} \cong 5 \times 10^{-6}$ M) in restoring [^{3}H]nitrendipine binding to smooth muscle membranes (Fig. 6) is not readily consistent with this proposal and, in fact, suggests interaction with an intracellular Ca^{++}-binding protein.

A role for a Ca^{++}-binding protein as a channel component is consistent with several observations that Ca^{++} channel inactivation may depend on Ca^{++} entry (Tsien 1983). Binding of Ca^{++} to an intracellularly

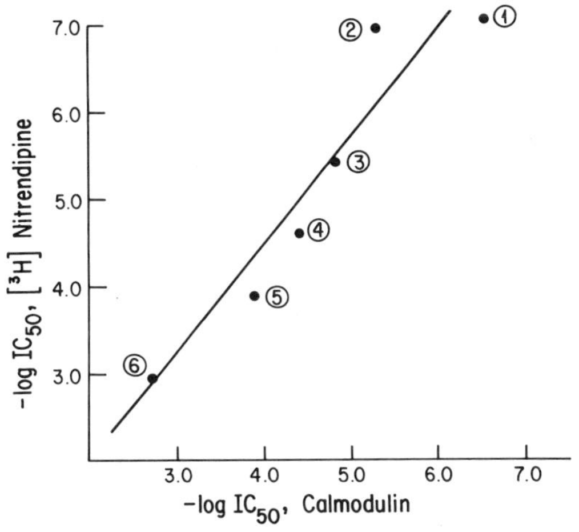

Figure 7. Correlation between activities of a series of compounds as inhibitors of specific [^{3}H]nitrendipine binding to guinea pig ileal smooth muscle membranes and of calmodulin-dependent phosphodiesterase activity. (Data from E.M. Luchowski et al., in prep.) (1) R24571; (2) pimozide; (3) trifluoperazine; (4) chlorpromazine; (5) imipramine; (6) chlorpromazine sulfoxide.

Table 2. Molecular Properties of [^{3}H]Nitrendipine-binding Sites

Property	Quantity	Reference
Guinea pig ileum		
K_d	0.15 nM	Bolger et al. (1982, 1983)
B_{MAX}	1100 fmole/mg protein[a]	
Guinea pig heart		
K_d	0.18 nM	Bolger et al. (1982)
B_{MAX}	400 fmole/mg protein	
Guinea pig heart and ileum		
M_r [^{3}H]2-NCS-DHP binding	45,000	Venter et al. (1983)
Guinea pig ileum		
M_r [^{3}H]nitrendipine (membrane)	278,000	Venter et al. (1983)

[a]Corresponds to a density of 2–10 binding sites μm^{-2} of membrane.

located Ca^{++}-binding protein component of the Ca^{++} channel could serve as the mechanism for this inactivation (Standen and Stanfield 1982). That calmodulin antagonists inhibit [^{3}H]nitrendipine binding with activities correlating with their inhibition of calmodulin-dependent phosphodiesterase (Fig. 7) offers further support to this suggestion.

The [^{3}H]nitrendipine data suggest the existence of several discrete binding sites on the Ca^{++} channel with which antagonists interact. A concept of the Ca^{++} channel as a multifunctional complex is quite consistent with the several separate molecular responsibilities of ion channels. Some evidence is available to support the view that the Ca^{++} channel may be an oligomeric assembly, and thus resemble in this respect the better characterized voltage-dependent Na^+ channel (Catterall 1980). The technique of radiation inactivation permits estimation of the molecular weight(s) of membrane-associated proteins (Kempner and Schlegel 1979). Application of this technique gives a value of 280,000 for the molecular weight of the [^{3}H]nitrendipine-binding site in smooth muscle membranes (Table 2; Venter et al. 1983). Furthermore, affinity labeling of smooth and cardiac muscle membranes with [^{3}H]2,6-dimethyl-3,5-dicarbomethoxy-4-(2-isothiocyano-phenyl)-1,4-dihydropyridine, a nifedipine analog whose binding behavior, including Ca^{++} dependence (Fig. 6), parallels that of [^{3}H]nitrendipine, gives an apparent subunit molecular weight of 45,000 on gel electrophoresis (Table 2; Venter et al. 1983).

Application of the radioligand binding approach has thus served to confirm the thesis, drawn from pharmacologic and physiologic data, that the Ca^{++} channel antagonists do indeed interact at specific sites of action. This approach further confirms that this chemically and pharmacologically heterogeneous group of compounds blocks Ca^{++} channel function at different sites and by different mechanisms. The Ca^{++} channel antagonists appear to be valuable tools for probing channel function, but major questions remain to be answered concerning quantitative discrepancies between binding and pharmacologic responses and the significance of apparently "silent" binding sites in brain and skeletal muscle.

ACKNOWLEDGMENTS

This work was supported by a grant from the National Institutes of Health (HL 16003).

REFERENCES

Akaike, N., A.M. Brown, K. Nishi, and Y. Tsuda. 1981. Actions of verapamil, diltiazem and other divalent cations on the calcium current of *Helix* neurones. *Br. J. Pharmacol.* **74:** 87.

Bayer, R., R. Hennekes, R. Kaufmann, and R. Mannhold. 1975. Inotropic and electrophysiological actions of verapamil and D600 in mammalian myocardium. I. Pattern of inotropic effects of the racemic compounds. *Naunyn-Schmiedebergs Arch. Pharmacol.* **290:** 49.

Bayer, R., R. Kaufmann, R. Mannhold, and R. Rodenkirchen. 1982. The action of specific Ca antagonists on cardiac electrical activity. *Prog. Pharmacol.* **5:** 53.

Bellemann, P., A. Schade, and R. Towart. 1983. Dihydropyridine receptor in rat brain labeled with [^{3}H]nimodipine. *Proc. Natl. Acad. Sci.* **80:** 2356.

Bellemann, P., D. Ferry, F. Lübbecke, and H. Glossmann. 1981. [^{3}H]Nitrendipine, a potent calcium antagonist, binds with high affinity to cardiac membranes. *Arzneim Forsch.* **31:** 2064.

Bolger, G.T., P.J. Gengo, E.M. Luchowski, H. Siegel, D.J. Triggle, and R.A. Janis. 1982. High affinity binding of a calcium channel antagonist to smooth and cardiac muscle. *Biochem. Biophys. Res. Commun.* **104:** 1604.

Bolger, G.T., P.J. Gengo, R. Klockowski, E. Luchowski, H. Siegel, R.A. Janis, A.M. Triggle, and D.J. Triggle. 1983. Characterization of binding of the Ca^{2+}-channel antagonist, [^{3}H]nitrendipine, to guinea pig ileal smooth muscle. *J. Pharmacol. Exp. Ther.* **225:** 291.

Catterall, W.A. 1980. Neurotoxins that act on voltage-sensitive sodium channels in excitable membranes. *Annu. Rev. Pharmacol. Toxicol.* **20:** 15.

Cheung, W.Y., ed. 1980. *Calcium and cellular function*, vol. 1. Academic Press, New York.

Ehara, T. and R. Kaufmann. 1978. The voltage- and time-dependent effects of (−)-verapamil on the slow inward current in isolated cat ventricular myocardium. *J. Pharmacol. Exp. Ther.* **207:** 49.

Ehlert, F.J., E. Itoga, W.R. Roeske, and H.I. Yamamura. 1982. The interaction of [^{3}H]nitrendipine with receptors for calcium antagonists in the cerebral cortex and heart of rats. *Biochem. Biophys. Res. Commun.* **104:** 937.

Eisner, V. and J. Kuthan. 1972. The chemistry of dihydropyridines. *Chem. Rev.* **72:** 1.

Ferry, D.R. and H. Glossmann. 1982. Identification of

putative calcium channels in skeletal muscle microsomes. *FEBS Lett.* **148:** 331.

FLAIM, S.F. 1982. Comparative pharmacology of calcium blockers based on studies of vascular smooth muscle. In *Calcium blockers. Mechanisms of action and clinical applications* (ed. S.F. Flaim and R. Zelis), p. 155. Urban and Schwarzenberg, Baltimore, Maryland.

FLAIM, S.F. and R. ZELIS, eds. 1982. *Calcium blockers. Mechanisms of action and clinical applications.* Urban and Schwarzenberg, Baltimore, Maryland.

FLECKENSTEIN, A. 1971. Specific inhibitors and promoters of calcium action in the excitation-contraction coupling of heart muscle and their role in the prevention or production of myocardial lesions. In *Calcium and the heart* (ed. P. Harris and L.H. Opie), p. 135. Academic Press, London.

——. 1977. Specific pharmacology of calcium in myocardium, cardiac pacemakers and vascular smooth muscle. *Annu. Rev. Pharmacol. Toxicol.* **17:** 149.

——. 1981. Fundamental actions of calcium antagonists on myocardial and cardiac pacemaker cell membranes. In *New perspectives on calcium antagonists* (ed. G.B. Weiss), p. 59. American Physiology Society, Bethesda, Maryland.

FOSSET, M., E. JAIMOVICH, E. DELPONT, and M. LAZDUNSKI. 1983. [^{3}H]Nitrendipine labelling of the Ca^{2+} channel in skeletal muscle. *Eur. J. Pharmacol.* **86:** 141.

GLOSSMANN, H., D.R. FERRY, F. LÜBBECKE, R. MEWES, and F. HOFMANN. 1982. Calcium channels: Direct identification with radioligand binding studies. *Trends Pharmacol. Sci.* **3:** 431.

GODFRAIND, T. and D. DIEU. 1981. The inhibition by flunarizine of the norepinephrine-evoked contraction and calcium influx in rat aorta and mesenteric arteries. *J. Pharmacol. Exp. Ther.* **217:** 510.

GOULD, R.J., K.M.M. MURPHY, and S.H. SNYDER. 1982. [^{3}H]Nitrendipine-labeled calcium channels discriminate inorganic calcium agonists and antagonists. *Proc. Natl. Acad. Sci.* **79:** 3656.

HAGIWARA, S. and L. BYERLY. 1981. Calcium channel. *Annu. Rev. Neurosci.* **4:** 69.

JANIS, R.A. and D.J. TRIGGLE. 1983. New developments in Ca^{2+} channel antagonists. *J. Med. Chem.* **26:** 775.

KAPLITA, P.V. and D.J. TRIGGLE. 1983. Actions of Ca^{2+} antagonists on the guinea pig ileal myenteric plexus preparation. *Biochem. Pharmacol.* **32:** 65.

KASS, R.S. 1982. Nisoldipine: A new, more selective calcium current blocker in cardiac Purkinje fibers. *J. Pharmacol. Exp. Ther.* **223:** 446.

KASS, R.S. and R.W. TSIEN. 1975. Multiple effects of calcium antagonists on plateau currents in cardiac Purkinje fibers. *J. Gen. Physiol.* **66:** 169.

KEMPNER, E.S. and W. SCHLEGEL. 1979. Size determination of enzymes by radiation inactivation. *Anal. Biochem.* **92:** 2.

KRETSINGER, R.H. 1977. Evaluation of the informational role of calcium in eukaryotes. In *Calcium binding proteins and calcium function* (ed. R.H. Wasserman et al.), p. 63. North-Holland, Amsterdam.

LEE, K.S. and R.W. TSEIN. 1983. Mechanism of calcium channel blockade by verapamil, D600, diltiazem and nitrendipine in single dialysed heart cells. *Nature* **302:** 790.

LO, M.M.S., E.A. BARNARD, and J.O. DOLLY. 1982. Size of acetylcholine receptors in the membrane. An improved version of the radiation inactivation method. *Biochemistry* **21:** 2210.

MEISHERI, K.D., O. HWANG, and C. VAN BREEMEN. 1981. Evidence for two separate Ca^{2+} pathways in smooth muscle plasmalemma. *J. Membr. Biol.* **59:** 19.

NACHSHEN, D.A. and M.P. BLAUSTEIN. 1979. The effects of some organic "calcium antagonists" on presynaptic nerve terminals. *Mol. Pharmacol.* **16:** 579.

NGHIEM, C., V.C. SWAMY, and D.J. TRIGGLE. 1982. Inhibition by D600 of norepinephrine- and clonidine-induced responses of the aortae from normotensive (WKY) and spontaneously hypertensive rats (SHR). *Life Sci.* **30:** 45.

REUTER, H. 1983. Calcium channel modulation by neurotransmitters, enzymes and drugs. *Nature* **301:** 569.

RODENKIRCHEN, R., R. BAYER, R. STEINER, F. BOSSERT, H. MEYER, and E. MÖLLER. 1979. Structure-activity studies on nifedipine in isolated cardiac muscle. *Naunyn-Schmiedebergs Arch. Pharmacol.* **310:** 69.

ROSENBERGER, L.B. and D.J. TRIGGLE. 1978. Calcium, calcium translocation and specific calcium antagonists. In *Calcium and drug action* (ed. G.B. Weiss), p. 3. Plenum Press, New York.

ROSENBERGER, L.B., M.K. TICKU, and D.J. TRIGGLE. 1979. The effects of Ca^{2+} antagonists on mechanical responses and Ca^{2+} movements in guinea pig ileal longitudinal smooth muscle. *Can. J. Physiol. Pharmacol.* **57:** 333.

SARMIENTO, J.G., R.A. JANIS, R.A. COLVIN, D.J. TRIGGLE, and A.M. KATZ. 1983. Binding of the calcium channel blocker nitrendipine to its receptor in purified sarcolemma from canine cardiac ventricle. *J. Mol. Cell. Cardiol.* **15:** 135.

SCHWARTZ, A. and D.J. TRIGGLE. 1984. Ca^{2+} channel antagonists. *Annu. Rev. Med.* (in press).

SONNENBLICK, E.H. and M. LESCH, eds. 1982. Calcium blockers I, II. *Prog. Cardiovasc. Dis.* **25:** 83.

STANDEN, N.B. and P.R. STANFIELD. 1982. A binding site model for calcium channel inactivation that depends on calcium entry. *Proc. R. Soc. Lond. B.* **217:** 101.

STONE, P.H., E.M. ANTMAN, J.E. MULLER, and E. BRAUNWALD. 1980. Calcium channel blocking agents in the treatment of cardiovascular disorders. II. Hemodynamic effects and clinical applications. *Ann. Intern. Med.* **93:** 886.

TOLL, L. 1982. Calcium antagonists. High affinity binding and inhibition of calcium transport in a clonal cell line. *J. Biol. Chem.* **257:** 13189.

TRIGGLE, D.J. 1981. Calcium antagonists: Basic chemical and pharmacological aspects. In *New perspectives on calcium antagonists* (ed. G.B. Weiss), p. 1. American Physiology Society, Bethesda, Maryland.

——. 1982. Chemical pharmacology of calcium antagonists. In *Calcium regulation by calcium antagonists* (ed. R.G. Rahwan and D.T. Witiak), p. 17. American Chemical Society, Washington, D.C.

TRIGGLE, D.J. and R.A. JANIS. 1983. Nitrendipine: Binding sites and mechanisms of action. In *Nitrendipine* (ed. A. Scriabine et al.). Urban and Schwarzenberg, Baltimore, Maryland. (In press)

TRIGGLE, C.R., D.K. AGRAWAL, G.T. BOLGER, E.E. DANIEL, C.Y. KWAN, E.M. LUCHOWSKI, and D.J. TRIGGLE. 1982. Calcium-channel antagonist binding to isolated vascular smooth muscle membranes. *Can. J. Physiol. Pharmacol.* **60:** 1738.

TSIEN, R.W. 1983. Calcium channels in excitable cell membranes. *Annu. Rev. Physiol.* **45:** 341.

VAN ZWIETEN, P.A., ed. 1982. Calcium antagonists—Recent developments. *Prog. Pharmacol.* **5:** 1.

VENTER, J.C., C.M. FRASER, J.S. SCHABER, C.Y. JUNG, G. BOLGER, and D.J. TRIGGLE. 1983. Molecular properties of the slow inward calcium channel: Molecular weight determinations by radiation inactivation and covalent affinity labelling. *J. Biol. Chem.* **258:** 9344.

WEISS, G.B., ed. 1981. *New perspectives on calcium antagonists.* American Physiology Society, Bethesda, Maryland.

YAMAMURA, H.I., H. SCHOEMAKER, R.G. BOLES, and W.R. ROESKE. 1982. Diltiazem enhancement of [^{3}H]nitrendipine binding to calcium channel associated drug receptor sites in rat brain synaptosomes. *Biochem. Biophys. Res. Commun.* **108:** 640.

Purification and Characterization of ATP-dependent Calcium Pumps from Synaptosomes

S.M. GOLDIN, S.Y. CHAN, D.M. PAPAZIAN, E.J. HESS, AND H. RAHAMIMOFF*
Department of Pharmacology, Harvard Medical School, Boston, Massachusetts 02115;
**Department of Biochemistry, Hadassah Medical School, Hebrew University, Jerusalem, Israel*

The primary event that initiates the release of neurotransmitter is an increase in the Ca^{++} concentration in the nerve terminal cytoplasm (for review, see Kelly et al. 1979). Neurotransmitter release is normally terminated by reduction in cytoplasmic Ca^{++} levels. Thus, the role of Ca^{++} in the initiation and termination of neurotransmitter release is analogous to its role in initiation and termination of muscle contraction (for review, see Carafoli and Crompton 1978). Ca^{++} fluxes responsible for muscle contraction are regulated by various mechanisms; e.g., the sarcoplasmic reticulum Ca^{++} pump, responsible for Ca^{++} removal from the myoplasm of cardiac muscle cells, can be regulated by cAMP (Tada et al. 1979) in response to catecholamines and, in parallel, by calmodulin (LePeuch et al. 1979). These forms of regulation involve phosphorylation by at least two distinct protein kinases of a regulatory protein associated with the Ca^{++} pump. Our understanding of the molecular basis of the regulation of muscle function has advanced significantly in the last decade, due in part to the identification and purification of the Ca^{++} pump involved (MacLennan and Holland 1975) and to the subsequent demonstrations in vitro of its regulation by calmodulin and cAMP.

Evidence exists that synaptosomes, a preparation enriched in nerve terminals derived from brain homogenates (Morgan 1976), contain ATP-dependent Ca^{++} pumps. Such pumps presumably remove Ca^{++} from the cytoplasm of the nerve terminal by extrusion across the plasma membrane and/or by sequestration of Ca^{++} within vesicular structures in the terminals. By analogy with the regulation of muscle contraction, regulation of neurotransmitter release by drugs and endogenous substances (e.g., the opiates and enkephalins) that act at presynaptic receptors may involve biochemical modification of the activity of proteins responsible for Ca^{++} transport in nerve terminals (Mudge et al. 1979; Ross and Cardenas 1979). The potential importance of the Ca^{++} pumps in the regulation of neurotransmitter release is a strong motive for their identification, purification, and characterization.

This paper summarizes the progress we have made in this area. We have been applying biochemical and immunological techniques to the study of ATP-dependent Ca^{++} transport in synaptosomes of mammalian brain, with the following questions in mind. How many molecularly distinct Ca^{++} pumps are there in nerve terminals? Are these pumps identical to the Ca^{++} pumps from other mammalian cells, such as the Ca^{++} pumps of sarcoplasmic reticulum (MacLennan and Holland 1975) and erythrocyte (Niggli et al. 1979), or are they specific to nerve cells? Is their function primarily to extrude Ca^{++} from the nerve terminal or to sequester Ca^{++} within it? Are they regulated by secondary mechanisms in addition to their primary regulation by the concentrations of Ca^{++} and ATP? If they are regulated, what is the physiological role of such mechanisms?

Purification of Synaptosome-derived Ca^{++} Pumps by Transport-specific Fractionation

Evidence suggesting that nerve terminals contain ATP-dependent Ca^{++} pumps has been obtained by several groups using vesicular membrane fractions obtained by osmotic lysis of mammalian brain synaptosomes. Such vesicular fractions accumulate Ca^{++} in an ATP-dependent manner (Blaustein et al. 1978a,b; Rahamimoff and Abramovitz 1978a,b). Another fraction, enriched in synaptic plasma membranes, catalyzes Ca^{++}-stimulated ATP hydrolysis (Sobue et al. 1979; Sorensen and Mahler 1981), an activity associated with Ca^{++} transport in other tissues (MacLennan and Holland 1975; Niggli et al. 1981). In both of the synaptosome-derived membrane fractions, the amount of enzymatic and/or transport activity is much lower than is found, for example, in muscle sarcoplasmic reticulum. This makes the identification and purification of such neuronal Ca^{++} pumps a relatively more difficult task.

We have developed a procedure—transport-specific fractionation—for the purification of membrane transport proteins (Goldin and Rhoden 1978) and have successfully applied it to the isolation of synaptosomal Ca^{++} pumps (described in detail in Papazian et al. 1979). This procedure uses reconstitution of the Ca^{++} pump into vesicles composed of exogenous phospholipid as a physical tool for its purification, in the following manner. The synaptosomal vesicle membrane fraction, derived from osmotically lysed synaptosomes (Rahamimoff and Abramovitz 1978a,b), is solubilized with the detergent, sodium cholate, in the presence of a large excess of soybean phospholipids. The subsequent removal of the detergent by dialysis randomly distributes individual protein molecules among small (400–700 Å) unilamellar vesicles so that one or at most a few membrane proteins, on the average, are present in each vesicle (Fig. 1). Since the membrane proteins responsible for ATP-dependent Ca^{++} transport are a dilute component of the synaptosomal membranes used

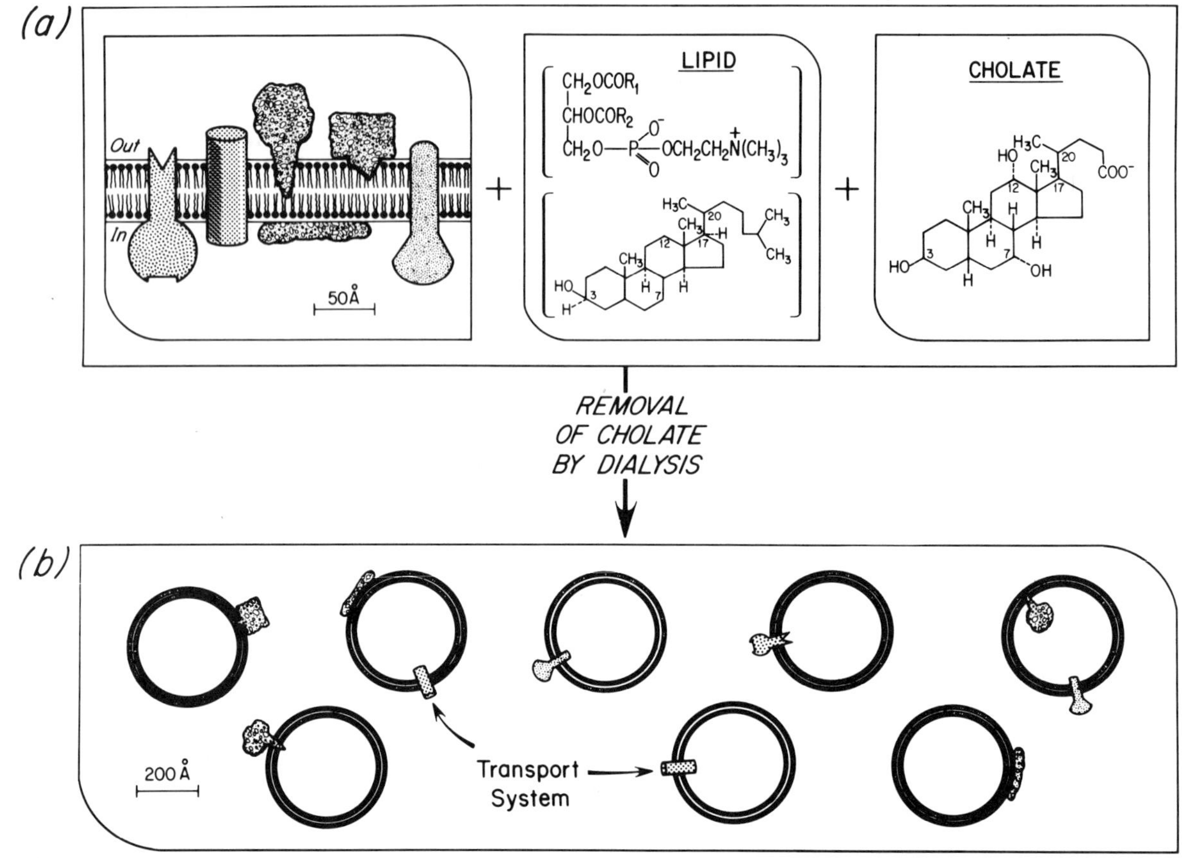

Figure 1. Schematic representation of the process of reconstitution of membrane proteins into artificial vesicles for their subsequent fractionation according to transport specificity. (*a*) The membrane fraction containing the transport system of interest is solubilized with cholate in the presence of a greater than 50-fold excess of exogenous lipid. After removing the cholate by hollow-fiber dialysis, small unilamellar vesicles result (*b*), containing, on the average, one protein molecule per vesicle. A fraction of these vesicles contains the transport system in biologically active form. A procedure is then devised for exploiting the specific ion transport properties of the reconstituted protein to create changes in the density of only those vesicles containing that transport protein. Details of the application of this technique to purify calcium pumps are described in the text.

for reconstitution, only a small fraction of the reconstituted vesicles contain a Ca^{++} pump.

The strategy for purification of the Ca^{++}-transporting vesicle fraction takes advantage of the observation that, as in isolated skeletal sarcoplasmic reticulum (Martonosi and Feretos 1964), oxalate greatly augments ATP-dependent Ca^{++} uptake, presumably due to the formation of an insoluble calcium oxalate complex within the vesicle. Formation of calcium oxalate would be expected to increase significantly the density of those vesicles exhibiting active uptake of Ca^{++}, enabling their separation from the rest of the vesicle population on isopycnic density gradients.

To achieve this purification, a two-stage gradient separation procedure has been devised. Reconstituted vesicles with oxalate trapped inside are incubated first with Mg^{++} and Ca^{++} but without ATP. The vesicles are applied to a linear density gradient. Centrifugation of the first gradient produces a well-defined major band of phospholipid vesicles above a diffuse region of nonspecifically permeable vesicles and unreconstituted material. The ATP-dependent Ca^{++} transport activity on this first gradient is aligned exactly with the major band of phospholipid vesicles (Fig. 2). The purpose of the first gradient is to remove the material that sediments to a position below the vesicles, since it would interfere with the subsequent purification. The phospholipid vesicle peak is pooled; incubated in the presence of Ca^{++}, Mg^{++}, and ATP; and resedimented on a second gradient. Figure 3 shows the transport activity and lipid phosphate profiles on the second gradient. Inclusion of ATP in the incubation before the vesicles are placed on the second gradient moves the peak of vesicles containing ATP-dependent Ca^{++} transport activity to a position of higher density; the density of the vast majority of the phospholipid vesicles is unaffected.

The two-stage gradient procedure results in an ~140-fold purification of the ATP-dependent Ca^{++} transport activity. Electrophoresis of the purified Ca^{++}-transporting vesicle fraction on an SDS gel (Fig. 4) shows that three components of the original synaptosomal vesicle fraction (of molecular weights ~230,000, 140,000, and 94,000, C230, C140, and C94, respectively) have been purified substantially.

Since reconstitution of transport activity is required

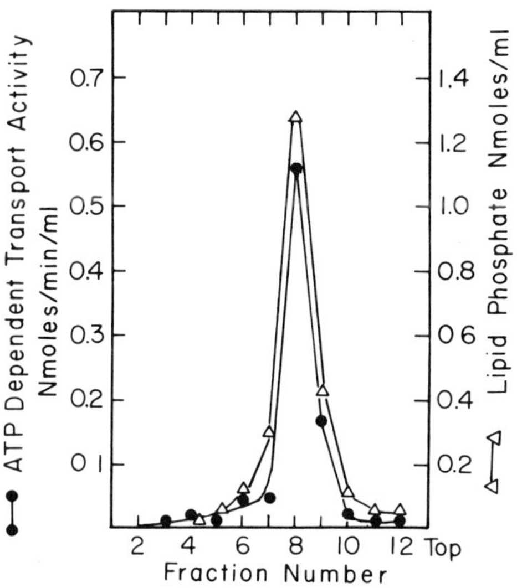

Figure 2. Isopycnic density gradient centrifugation of vesicles containing reconstituted synaptosomal membrane proteins. Vesicles are placed on a linear gradient formed from solutions containing 700 mM glycerol (upper phase) and 700 mM sucrose (lower phase). After centrifugation, gradient fractions are assayed for vesicle phospholipid (△) and ATP-dependent Ca^{++} uptake (●). (For further details, see Papazian et al. 1979.)

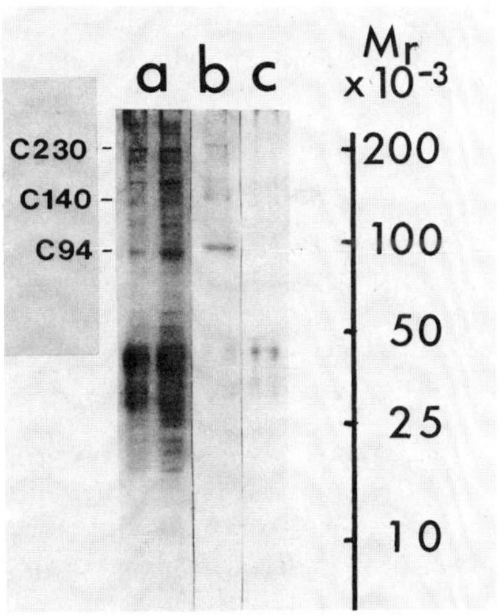

Figure 4. Coomassie-blue-stained SDS slab gels of the ATP-dependent Ca^{++} transport activity after its purification by transport-specific fractionation. (Lanes *a*) The protein profiles of the crude synaptosomal vesicle fraction; (lane *b*) the material from the region of the second gradient step containing the lower two thirds of the peak of purified Ca^{++} transport activity (fractions 8–14 of curve labeled □ in Fig. 3); (lane *c*) material obtained from fractions 8–14 of the control gradient of vesicles incubated in the absence of ATP (△ in Fig. 3). C230, C140, and C94 denote the three major polypeptides of M_r 230,000, 140,000, and 94,000 that have been purified by transport-specific fractionation. (For further details, see Papazian et al. 1979.)

for the transport-specific fractionation to work, we know that what we are purifying is a transport activity rather than merely an enzymatic or ligand-binding activity. However, further investigation was required to elucidate the relationship of the three polypeptides to one another: Are they subunits of a single Ca^{++} pump, or have we copurified more than one distinct Ca^{++} transporter from separate regions of the nerve terminal? The following studies address this question.

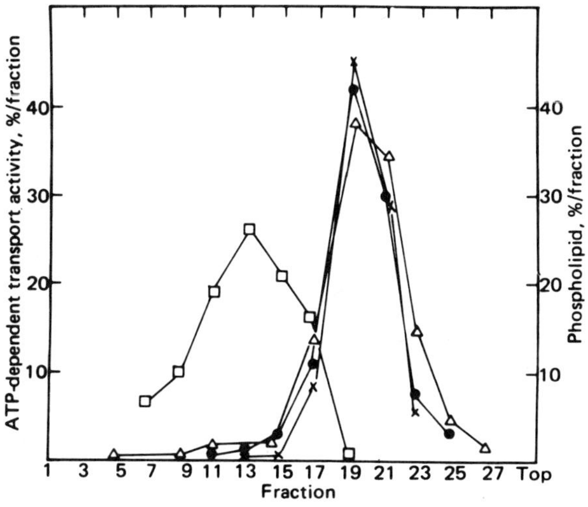

Figure 3. Results of the two-stage gradient purification of the reconstituted ATP-dependent Ca^{++} transport activity. Before being placed on the second gradient, a portion of the vesicle peak obtained from the first gradient (Fig. 2) is incubated with Mg^{++}, Ca^{++}, and ATP. After centrifugation, gradient fractions are assayed for vesicle phospholipid (●) and ATP-dependent Ca^{++} uptake (□). As a control, another portion of the same material obtained from the first gradient is preincubated with Mg^{++} and Ca^{++} but no ATP; in this case, the peaks of phospholipid (×) and ATP-dependent Ca^{++} uptake (△) coincide. (For further details, see Papazian et al. 1979.)

Purification of a Calmodulin-stimulated Ca^{++} Pump from Synaptic Plasma Membranes

A possible regulator of ATP-dependent Ca^{++} transport in nerve terminals is the Ca^{++}-binding protein, calmodulin. Calmodulin is a soluble protein, M_r ~17,000, that binds Ca^{++} with micromolar affinity. It regulates a number of cellular enzymes (for review, see Klee et al. 1980), including at least two distinct protein kinases in brain that in turn phosphorylate synaptosomal proteins (Kennedy and Greengard 1981). Calmodulin also directly stimulates the Ca^{++}-dependent ATPase of mammalian erythrocytes (Jarrett and Kyte 1979; Niggli et al. 1979). We determined that calmodulin does not stimulate ATP-dependent Ca^{++} uptake in the native synaptosomal vesicle fraction, after it has been depleted of endogenous calmodulin by repeated washing in Ca^{++}-free medium containing the calcium chelator EGTA (Papazian et al. 1981). In contrast, the synaptic plasma membrane fraction has a readily detectable Ca^{++}-stimulated ATPase activity activated more than twofold by calmodulin (Papazian et al. 1981). This latter result is consistent with the previous observations of Sobue et al. (1979) and Sorensen and Mahler (1981). We further showed that after reconstitution of synaptic plasma membranes into vesicles formed from egg phosphatidyl choline by the cholate dialysis procedure, a calmodulin-stimulated ATP-dependent Ca^{++} uptake activity can be resolved (Papazian et al. 1984).

These observations suggested that synaptic plasma membranes contain a calmodulin-stimulated Ca^{++} pump. We have used calmodulin-Sepharose affinity chromatography to purify the pump partially (described in detail in Papazian et al. 1984). As shown in the Coomassie-stained gel in Figure 5, the preponderant component that copurifies with the Ca^{++}-ATPase activity electrophoreses on an SDS gel as a band of M_r ~140,000 (lane 2). While this work was nearing completion, a similar result was obtained by Hakim et al. (1982). The affinity-purified component corresponds in molecular weight to C140 of Figure 4, the most minor of the three components purified from reconstituted synaptosomal vesicles by transport-specific fractionation. Since synaptosomal vesicles and synaptic plasma membranes are derived from the same synaptosomal lysate, the two membrane fractions are probably cross-contaminated with each other. It is therefore likely that C140 and the affinity-purified component, both implicated in Ca^{++} transport, are the same protein, which is found more abundantly in synaptic plasma membranes than synaptosomal vesicles. To determine whether C140 is indeed responsible for Ca^{++} transport, the following experiments were performed.

As shown in Figure 6, C140 is phosphorylated by [γ-^{32}P]ATP in a Ca^{++}-dependent manner. The phosphate is discharged by hydroxylamine: Hydroxylamine sensitivity is characteristic of the acyl phosphate bonds of the phosphorylated intermediates of the Na^+-K^+ ATPase (Hokin et al. 1965) and of the Ca^{++} ATPase of sarcoplasmic reticulum (Makinose 1969; Pang and Briggs 1973). These results implicate C140 functional-

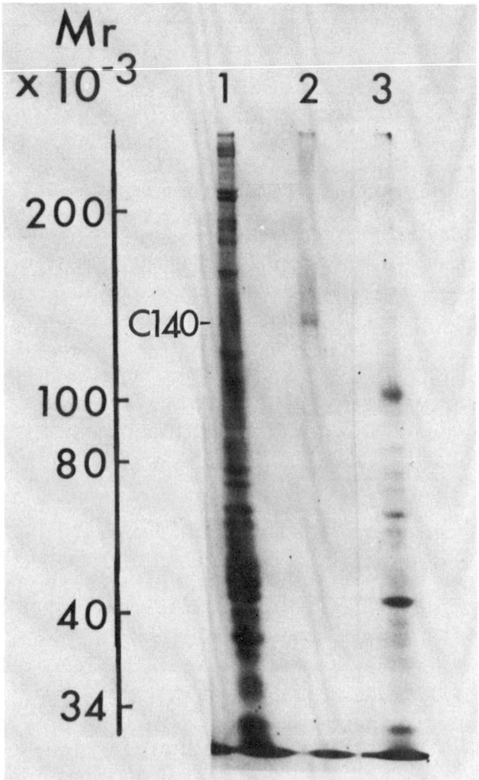

Figure 5. Coomassie-blue-stained SDS gel of the Ca^{++}-stimulated ATPase preparation purified from synaptic plasma membranes by calmodulin affinity chromatography. Synaptic plasma membranes are depleted of endogenous calmodulin and solubilized with the nonionic detergent Triton X-100. After addition of $CaCl_2$, the solubilized material is applied to a calmodulin-Sepharose 4B column. Unbound protein is washed through the column with a Ca^{++}-containing buffer. Protein bound to the column in a Ca^{++}-dependent manner is subsequently eluted with an EGTA-containing buffer. (Lane *1*) The synaptic plasma membrane fraction; (lane *2*) the affinity-purified protein eluted from the calmodulin column with an EGTA-containing buffer (the major component of the purified preparation corresponds in molecular weight to C140 of Fig. 4); (lane *3*) solubilized proteins from the synaptic plasma membrane fraction that do not bind to the calmodulin column, but pass through in the presence of Ca^{++}. (For further details, see Papazian et al. 1984.)

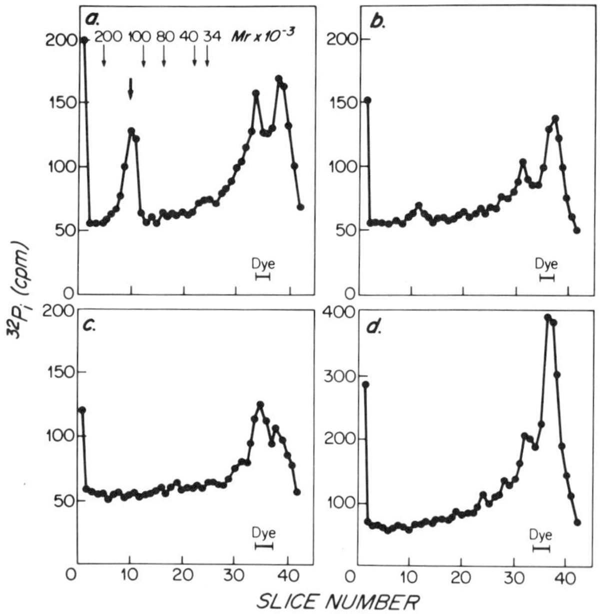

Figure 6. Ca^{++}-dependent, hydroxylamine-sensitive phosphorylation of the affinity-purified Ca^{++}-stimulated ATPase by [γ-^{32}P]ATP. The affinity-purified Ca^{++}-stimulated ATPase is incubated with [γ-^{32}P]ATP under various conditions, precipitated with trichloroacetic acid (TCA) in the presence of carrier bovine serum albumin (BSA), and subjected to electrophoresis at pH 2.4 and 15°C, conditions that preserve the unstable acyl enzyme intermediate. Phosphate incorporation is shown after: (*a*) incubation with ATP and $CaCl_2$ (the arrow denotes the position of the major protein component of the affinity-purified preparation, subjected to electrophoresis at pH 2.4 and 15°C, in the absence of carrier BSA, and stained with Coomassie blue); (*b*) incubation with ATP and EGTA; (*c*) incubation with ATP and $CaCl_2$, followed by treatment with hydroxylamine; (*d*) control incubation of buffer with ATP and $CaCl_2$, in the absence of affinity-purified protein, showing that the ^{32}P-containing material that comigrates with the tracking dye in *a*–*c* is unrelated to the Ca^{++}-stimulated ATPase. The variable number of counts present in the first slice of each gel represents material that did not enter the gel. (For further details, see Papazian et al. 1984.)

ly as a Ca^{++}-transporting ATPase. Reconstitution of the affinity-purified fraction, followed by transport-specific fractionation on density gradients using the Ca^{++} oxalate-loading procedure described above, provides more-direct evidence for the transport role of C140. Transport-specific fractionation reproducibly and specifically enriches only C140 in an ATP-dependent manner (Fig. 7, lane 1). The lower-molecular-weight components are largely protein contaminants of the soy lipid preparation (asolectin) used in the experiment (lanes 2 and 3). After reconstitution of the affinity-purified fraction into phosphatidyl choline vesicles, the ATP-dependent Ca^{++} transport activity is stimulated seven- to ninefold by the addition of calmodulin (Papazian et al. 1984). Thus, it can be concluded that C140 comprises a calmodulin-stimulated, ATP-dependent Ca^{++} pump.

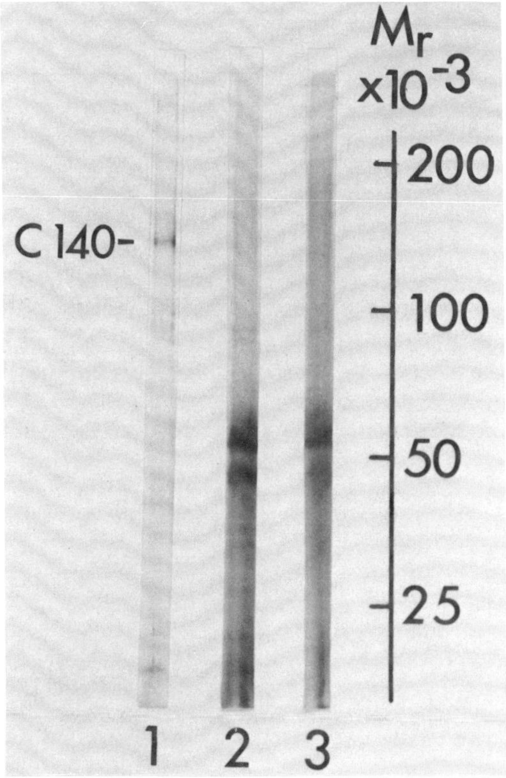

Figure 7. Silver-stained SDS gel of the affinity-purified, reconstituted ATP-dependent Ca^{++} pump after transport-specific fractionation. The affinity-purified preparation is reconstituted in asolectin. The reconstituted vesicles are incubated with Ca^{++} and Mg^{++}, with or without ATP, and subjected to equilibrium density-gradient centrifugation (as in Figs. 2 and 3). As in Fig. 3, material is collected from the region of the gradient below the position of the main phospholipid vesicle peak, pooled, and subjected to electrophoresis on an SDS gel. Proteins on the gel are visualized by silver staining, a more sensitive technique than Coomassie blue staining (Oakley et al. 1980). (Lane *1*) Vesicles have been incubated with Ca^{++}, Mg^{++}, and ATP before gradient centrifugation; (lane *2*) vesicles have been incubated with Ca^{++} and Mg^{++}, but without ATP before gradient centrifugation; (lane *3*) lipid vesicles composed of asolectin but containing no affinity-purified protein have been subjected to the same gradient centrifugation procedure. Material below the position of the main vesicle peak has been pooled and electrophoresed. Lanes *2* and *3* indicate that protein contaminants of asolectin vesicles contribute bands to the pertinent region of the gradient, but comparison with lane *1* indicates that only C140 is enriched in an ATP-dependent manner. (For further details, see Papazian et al. 1984.)

Neither C230 nor C94 is present in the calmodulin affinity-purified preparation. We have used immunologic techniques to study the involvement of these proteins in Ca^{++} transport (S.Y. Chan et al., in prep.).

Monoclonal Antibodies to a Ca^{++} Pump Purified from Synaptosomal Vesicles

When the transport-specific fractionation procedure is applied to Ca^{++} pumps of the synaptosomal vesicle fraction of bovine brain rather than rat brain, qualitatively similar results are obtained. Protein components corresponding to C230 and C94 are purified; however, in this case, minimal amounts of C140 are obtained (compare Fig. 8 with Fig. 4). C94 corresponds to the major protein of bovine brain synaptosomal vesicles phosphorylated by [γ-^{32}P]ATP in a hydroxylamine-sensitive manner (Fig. 9).

The material purified by transport-specific fractionation from bovine brain has been used as the antigen to immunize BALB/c mice for the production of antisera and monoclonal antibodies. By solid-phase radioimmunoassay, the 50% titer points of the antisera are typically 1/1000. The 50% titer points of ascites fluids prepared from our most active clones are typically 1/100,000. When employed in conjunction with the membrane-bound immunoglobulin-binding protein, protein A of *Staphylococcus aureus* (an adaptation of the method of Kessler 1975), monoclonal antibodies are capable of immunoprecipitating most of the ATP-dependent Ca^{++} uptake activity present in the native synaptosomal vesicle preparation (Fig. 10). After the synaptosomal vesicles are solubilized and the membrane proteins reconstituted by cholate dialysis into artificial vesicles at a multiplicity of one protein per vesicle, ~50% of the Ca^{++} transport activity can still be immunoprecipitated (Fig. 11). Immunoprecipitation is dependent on the presence of both antibody and protein A. As shown, the monoclonal antibody alone neither inhibited nor immunoprecipitated the Ca^{++} pump.

The Western blot procedure (Towbin et al. 1979) has been used to detect the specific interaction of antisera and monoclonal antibodies with synaptosomal vesicle proteins that have been subjected to SDS gel electrophoresis (Fig. 12). As shown, antisera bind to C230 strongly and to C94 to a much lower degree (lane b). Antisera also interact with several proteins of lower molecular weight. In contrast, all the monoclonal antibodies tested with positive titers bind primarily to C230 (lane c). C94 in isolated form has been obtained by eluting it from SDS gels of the material purified by transport-specific fractionation and has been used as the antigen to prepare an antiserum specific to it. The resulting antiserum to C94 binds predominantly to

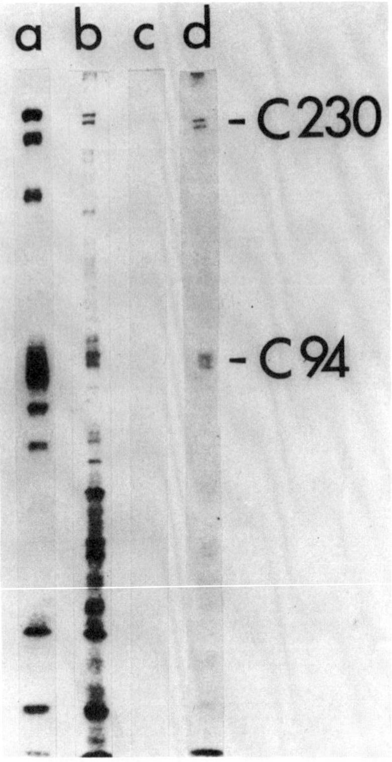

Figure 8. SDS slab gel electrophoresis of the ATP-dependent Ca^{++} transport component associated with bovine brain synaptosomal vesicles, after its purification by transport-specific fractionation. The reconstitution and purification procedures are essentially identical to those described in the text and Figs. 2–4 for the rat brain Ca^{++} transport component. Gels are silver-stained. (Lane *a*) Human red blood cell ghosts used as molecular weight standards; (lane *b*) protein profile of the unfractionated native bovine synaptosomal vesicles. The last two lanes are the results of electrophoresis of material collected from the dense region of the second gradient of the two-stage gradient purification procedure as described for rat brain. (Lanes *c* and *d*) The results if the vesicles are preincubated with Ca^{++} and Mg^{++} in the absence and presence of ATP, respectively. (For further details, see S.Y. Chan et al., in prep.)

C230 on Western blots (lane d), indicating that C230 and C94 are immunologically related.

The behavior of the antiserum to C94 on Western blots is somewhat paradoxical. It is possible that the antigenic sites primarily responsible for producing antisera in mice when C94 is injected in Freund's adjuvant are not exposed when C94 is bound to nitrocellulose, as required for the Western blot procedure. If C230 is a dimer of C94, its conformation on nitrocellulose may more closely resemble that of its native state; thus, more of these antigenic sites may be available for binding than is the case with the monomer. In this case, a stronger interaction of the antiserum with C230 than with C94 on the Western blot would be expected. The possible monomer-dimer relationship of C94 with C230 is under investigation. This explanation would also account for the selective affinity of the monoclonal antibodies for C230 on Western blots. In any case, even if

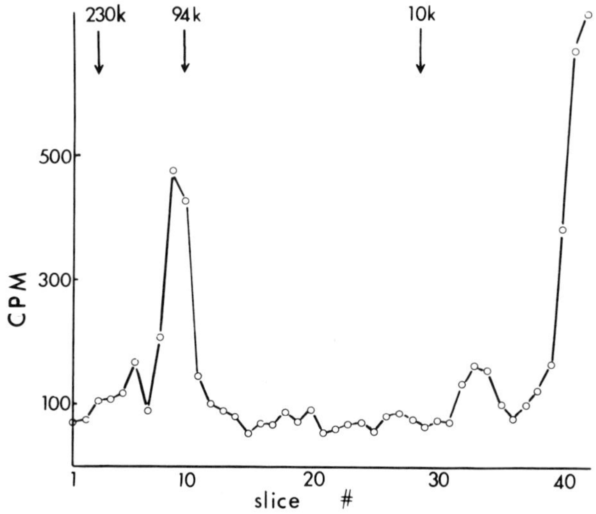

Figure 9. Mobility on acidic SDS gels of the phosphorylated intermediate of the bovine synaptosomal Ca^{++} pump. Native synaptosomal vesicles are incubated with [γ-^{32}P]ATP, TCA-precipitated, and subjected to gel electrophoresis using procedures similar to those described for Fig. 6.

C230 is not a dimer of C94, our results indicate that the two are immunologically related.

We have thus demonstrated that our antisera, and monoclonal antibodies that are capable of specifically immunoprecipitating the ATP-dependent Ca^{++} trans-

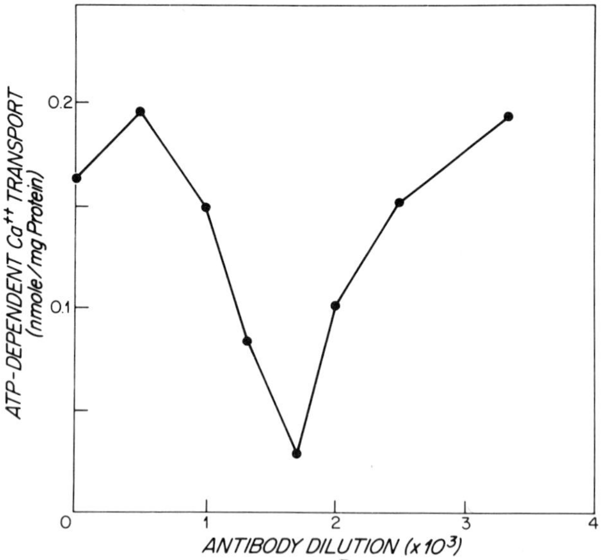

Figure 10. Immunoprecipitation of the ATP-dependent Ca^{++} uptake activity from the native bovine brain synaptosomal vesicle preparation. Synaptosomal vesicles are incubated with dilutions (1/∞–1/300) of ascites fluid prepared from monoclonal antibodies specific to the purified Ca^{++} transport component. A fixed concentration of membrane-associated protein A is then added; following additional incubation, the preparation is subjected to centrifugation. The supernatant is then assayed for ATP-dependent $^{45}Ca^{++}$ uptake. As shown, a 1/600 dilution of the antibody yields the optimal immunoprecipitation of the transport activity. (For further details, see S.Y. Chan et al., in prep.)

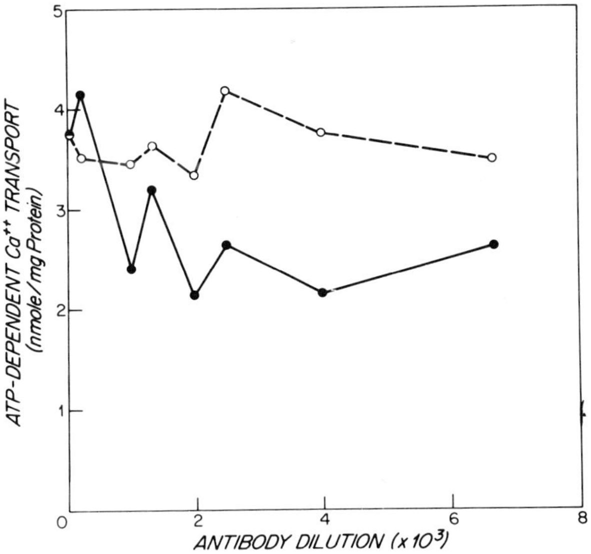

Figure 11. Immunoprecipitation of the ATP-dependent Ca^{++} uptake activity after its solubilization and reconstitution into artificial lipid vesicles. Native synaptosomal vesicle membrane proteins are solubilized and reconstituted into asolectin vesicles by the same method used to prepare such vesicles for transport-specific fractionation. Unfractionated, reconstituted vesicles were then incubated with varying dilutions of monoclonal antibody in the absence (○) and presence (●) of a fixed concentration of protein A. Following centrifugation the supernatant is assayed for ATP-dependent Ca^{++} uptake activity. As shown, antibody in the absence of protein A neither immunoprecipitates nor inhibits the transport activity. (For details, see S.Y. Chan et al., in prep.)

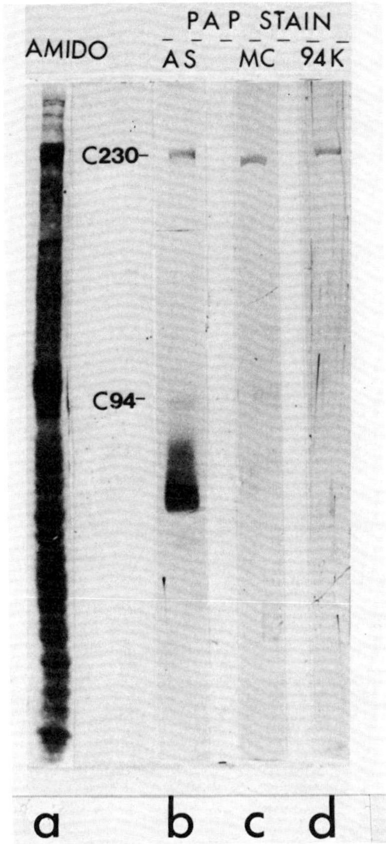

Figure 12. Specific interaction of antisera and monoclonal antibodies with native synaptosomal vesicle membrane proteins as detected by the Western blot procedure. (Lane *a*) An Amido-black-stained protein profile of crude synaptosomal vesicle proteins after they are electrophoretically transferred to the nitrocellulose sheet; (lanes *b–d*) specific binding of antibodies as detected by the peroxidase-antiperoxidase (PAP) staining procedure (Sternberger et al. 1970). (Lane *b*) Binding of mouse antiserum raised against the purified, reconstituted ATP-dependent Ca^{++} transport component; (lane *c*) results from binding of the monoclonal antibody that immunoprecipitates Ca^{++} transport activity (Figs. 10 and 11); (lane *d*) results from mouse antiserum prepared against C94 alone. (For further details, see text and S.Y. Chan et al., in prep.)

port activity, interact with C94 and C230, implicating these proteins in Ca^{++} transport.

The antisera and monoclonal antibodies have also been used to determine whether the synaptosomal vesicle-associated Ca^{++} pump is immunologically related to the Ca^{++} pumps of sarcoplasmic reticulum or erythrocytes. As shown in Figure 13, all monoclonal antibodies tested show no significant cross-reactivity for any component of bovine sarcoplasmic reticulum or bovine erythrocytes in solid-phase radioimmunoassay. This is despite the fact that the Ca^{++} pump of sarcoplasmic reticulum is a major membrane component (MacLennan and Holland 1975), whereas the Ca^{++} pump of synaptosomal vesicles is a comparatively minor membrane component. (The erythrocyte Ca^{++} pump is on the order of 0.1% of the erythrocyte membrane protein [from the data of Jarrett and Kyte 1979], levels roughly comparable to the concentration of the synaptosomal vesicle Ca^{++} pump.) The monoclonal antibodies cross-react with a component of axolemma, a membrane fraction derived from myelinated axons of bovine brain that is produced by dissociation of myelin and density gradient purification of the underlying excitable membrane (DeVries et al. 1978). Antisera to C94 and C230 behave similarly: They do not cross-react with sarcoplasmic reticulum or erythrocytes, but they do cross-react with axolemma (not shown). These results indicate that the synaptosomal Ca^{++} pump may be specific to nerve tissue and may be immunologically distinct from other well-characterized Ca^{++} pumps.

DISCUSSION

A principal conclusion derived from these studies is that there are at least two distinct ATP-dependent Ca^{++} pumps associated with mammalian brain synaptosomes. One, C140, is a calmodulin-stimulated Ca^{++} pump of M_r ~140,000 that is primarily associated with the synaptic plasma membrane fraction. The other, C94, is a Ca^{++} pump of M_r ~94,000, apparently specific for nerve tissue, that is primarily associated with synaptosomal vesicles. C94 does not appear to interact directly with calmodulin. On the basis of the preferential association of the two pumps with these different membrane fractions, we hypothesize that the probable function of C140 is to extrude Ca^{++} from the

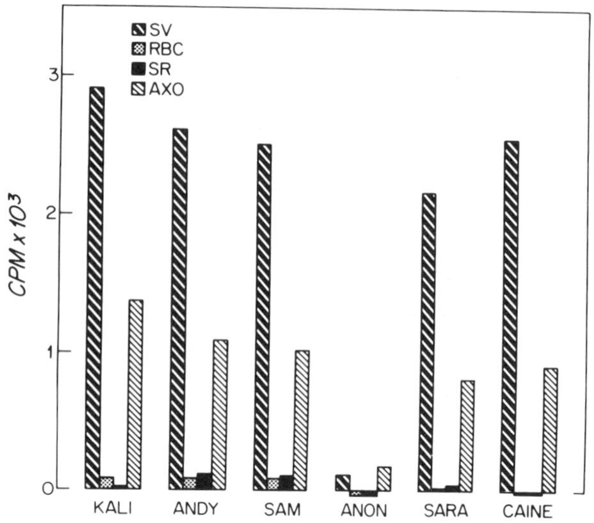

Figure 13. Interaction of monoclonal antibodies with bovine membrane fractions. Membrane fractions employed are synaptosomal vesicles (SV), erythrocyte ghosts (RBC), sarcoplasmic reticulum (SR), and axolemma (AXO). Each membrane fraction in separately insolubilized to 96-well microtiter plates and exposed, in turn, to antibody secreted into the cell culture medium and ^{125}I-labeled anti-mouse Fab. Names along the abscissa refer to individual clones. Further details are in the text and in S.Y. Chan et al. (in prep.). Other procedures used for determination of antibody cross-reactivity produce results consistent with the above and are described by S.Y. Chan et al. (in prep.).

nerve terminal; the probable role of C94 is to sequester Ca^{++} within the nerve terminal. Immunocytochemical localization studies to test the validity of this hypothesis are in progress. Although C230 copurifies with C94 by transport-specific fractionation, a phosphorylated intermediate of C230 has not been demonstrated and the details of its relationship to C94 and to Ca^{++} transport have not yet been resolved.

The physiological role of C140 in regulating Ca^{++} levels in nerve terminals may be analogous to the role of the Ca^{++} pump of erythrocytes. Erythrocytes must maintain Ca^{++} levels at or below $\sim 10^{-6}$ M to maintain the deformability of the cell membrane so that the cells can easily pass through narrow capillaries (Kirkpatrick et al. 1975). Because erythrocytes do not contain mitochondria and hence are metabolically deficient, economy in the utilization of ATP is important. Calmodulin binds Ca^{++} at micromolar levels and then activates the erythrocyte Ca^{++} pump, which extrudes Ca^{++} through the plasma membrane. This mechanism provides more dramatic on-off regulation of the pump than would be obtained if the only stimulation of the pump were by the substrate, Ca^{++}, according to Michaelis-Menten kinetics. Although nerve terminals are not metabolically deficient, they must consume a large amount of energy for synthesis, storage, and release of neurotransmitters; because termination of neurotransmitter release requires reduction of Ca^{++} concentrations to submicromolar levels (Kelly et al. 1979), a plasma membrane Ca^{++} pump stimulated by calmodulin is well suited for such a regulatory role.

Although C94 does not appear to be directly stimulated by calmodulin, calmodulin-stimulated protein kinases have been identified in brain (Kennedy and Greengard 1981). The possible regulation of ATP-dependent Ca^{++} transport in nerve terminals by calmodulin-stimulated and/or cyclic nucleotide-dependent protein kinases is under investigation.

Another question of immediate interest is the degree of homogeneity of C94 and C140. C140, purified by calmodulin affinity chromatography, appears as a broad band on SDS gels (Fig. 5); C94 (as well as C230) can, on occasion, be resolved as doublet bands (Fig. 8). This heterogeneity could be produced during the purification procedure or alternatively could reflect the true situation in vivo. Heterogeneity of the brain Na^+-K^+ ATPase has been established: There are at least two distinct molecular forms of the Na^+-K^+ ATPase in brain (Sweadner 1979), differing in their kinetics and in their affinity for cardiac glycosides, and differing slightly in their apparent molecular weights on SDS gels.

In conclusion, these studies have used biochemical and immunological techniques together with transport-specific fractionation—a procedure that uses reconstitution of membrane proteins into lipid vesicles as a physical tool for the purification of transport proteins—as a means of identifying and purifying proteins that function as Ca^{++} pumps in brain. Only a minority of neuronal ion transport proteins (for example, the action potential Na^+ channel and the Na^+-K^+ ATPase) contain specific toxin-binding sites or enzymatic activities that can be used as assays for their purification. Many (probably most) ion transport proteins, e.g., the action potential Ca^{++} channel, are also critical to the function of the nervous system but do not have high-affinity ligands or associated enzymatic activities. For such proteins, we may of necessity depend on reconstitution of their transport function as an assay and perhaps as a physical tool for their purification (Goldin et al. 1983).

ACKNOWLEDGMENTS

We wish to thank Colin Barnstable for invaluable guidance in the production of monoclonal antibodies. This work was supported by grants to S.M.G by the National Institutes of Health (NS-15236 and NS-16475), the McKnight Foundation, and the Searle Scholars Program. H.R. is the recipient of a U.S./Israel Binational Science Foundation Grant. S.Y.C. is a Pharmaceutical Manufacturers Association postdoctoral fellow. Stuart Klein provided excellent technical assistance.

REFERENCES

Blaustein, M.P., R.W. Ratzlaff, and E.S. Schweitzer. 1978a. Calcium buffering in presynaptic nerve terminals. II. Kinetic properties of the nonmitrochondrial Ca sequestration mechanism. *J. Gen. Physiol.* **72:** 43.

BLAUSTEIN, M.P., R.W. RATZLAFF, N.C. KENDRICK, and E.S. SCHWEITZER. 1978b. Calcium buffering in presynaptic nerve terminals. I. Evidence for involvement of a nonmitochondrial ATP-dependent sequestration mechanism. *J. Gen. Physiol.* **72:** 15.

CARAFOLI, E. and M. CROMPTON. 1978. The regulation of intracellular calcium. *Curr. Top. Membr. Transp.* **10:** 151.

DEVRIES, G.H., J. MATTHIEU, M. BENY, R. CHICHEPORTICHE, M. LAZDUNSKI, and M. DOLIVO. 1978. Isolation and partial characterization of rat CNS axolemma-enriched fractions. *Brain Res.* **147:** 339.

GOLDIN, S.M. and V. RHODEN. 1978. Reconstitution and "transport specific fractionation" of the human erythrocyte glucose transport system: A new approach for identification and isolation of membrane transport proteins. *J. Biol. Chem.* **253:** 2575.

GOLDIN, S.M., E.G. MOCZYDLOWSKI, and D.M. PAPAZIAN. 1983. Isolation and reconstitution of neuronal ion transport proteins. *Annu. Rev. Neurosci.* **6:** 419.

HAKIM, G., T. ITANO, A.K. VERMA, and J.T. PENNISTON. 1982. Purification of the Ca^{++}- and Mg^{++}-requiring ATPase from rat brain synaptic plasma membrane. *Biochem. J.* **207:** 225.

HOKIN, L.E., P.S. SASTRY, P.R. GALSWORTHY, and A. YODA. 1965. Evidence that a phosphorylated intermediate in a brain transport adenosine triphosphatase is an acyl phosphate. *Proc. Natl. Acad. Sci.* **54:** 177.

JARRETT, H.W. and J. KYTE. 1979. Human erythrocyte calmodulin: Further chemical characterization and the site of its interaction with the membrane. *J. Biol. Chem.* **254:** 8237.

KELLY, R.B., J.W. DEUTSCH, S.S. CARLSON, and J.A. WAGNER. 1979. Biochemistry of neurotransmitter release. *Annu. Rev. Neurosci.* **2:** 399.

KENNEDY, M.B. and P. GREENGARD. 1981. Two calcium/calmodulin-dependent kinases, which are highly concentrated in brain, phosphorylate protein I at distinct sites. *Proc. Natl. Acad. Sci.* **78:** 1293.

KESSLER, S.W. 1975. Rapid isolation of antigens from cells with a staphylococcal protein A-antibody adsorbant: Parameters of the interaction of antibody-antigen complexes with protein A. *J. Immunol.* **115:** 1617.

KIRKPATRICK, F.H., D.G. HILLMAN, and P.L. LACELLE. 1975. A23187 and red cells: Changes in deformability, K^+, Mg^{++}, Ca^{++} and ATP. *Experientia* **31:** 653.

KLEE, C.B., T.H. CROUCH, and P.G. RICHMAN. 1980. Calmodulin. *Annu. Rev. Biochem.* **49:** 489.

LEPEUCH, C.J., J. HAIECH, and J.G. DEMAILLE. 1979. Concerted regulation of cardiac sarcoplasmic reticulum calcium transport by cyclic adenosine monophosphate dependent and calcium-calmodulin-dependent phosphorylations. *Biochemistry* **18:** 5150.

MACLENNAN, D.H. and P.C. HOLLAND. 1975. Calcium transport in sarcoplasmic reticulum. *Annu. Rev. Biophys. Bioeng.* **4:** 377.

MAKINOSE, M. 1969. The phosphorylation of the membranal protein of the sarcoplasmic vesicles during active calcium transport. *Eur. J. Biochem.* **10:** 74.

MARTONOSI, A. and R. FERETOS. 1964. Sarcoplasmic reticulum. I. The uptake of Ca^{++} by sarcoplasmic reticulum fragments. *J. Biol. Chem.* **239:** 648.

MORGAN, I.G. 1976. Synaptosomes and cell separation. *Neuroscience* **1:** 159.

MUDGE, A.W., S.E. LEEMAN, and G.D. FISCHBACH. 1979. Enkephalin inhibits release of substance P from sensory neurons in culture and decreases action potential duration. *Proc. Natl. Acad. Sci.* **76:** 526.

NIGGLI, V., J.T. PENNISTON, and E. CARAFOLI. 1979. Purification of the (Ca^{++}-Mg^{++})-ATPase from human erythrocyte membranes using a calmodulin affinity column. *J. Biol. Chem.* **254:** 9955.

NIGGLI, V., E.S. ADUNYAH, J.T. PENNISTON, and E. CARAFOLI. 1981. Purified (Ca^{++}-Mg^{++})-ATPase of the erythrocyte membrane: Reconstitution and effect of calmodulin and phospholipids. *J. Biol. Chem.* **256:** 395.

OAKLEY, B.R., D.R. KIRSCH, and N.R. MORRIS. 1980. A simplified ultra-sensitive silver stain for detecting proteins in polyacrylamide gels. *Anal. Biochem.* **105:** 361.

PANG, D.C. and F.N. BRIGGS. 1973. Reaction mechanism of the cardiac sarcotubule calcium (II) dependent adenosine triphosphatase. *Biochemistry* **12:** 4905.

PAPAZIAN, D., H. RAHAMIMOFF, and S.M. GOLDIN. 1979. Reconstitution and purification by "transport specific fractionation" of an ATP-dependent calcium transport component from synaptosome-derived vesicles. *Proc. Natl. Acad. Sci.* **76:** 3708.

——. 1981. Synaptosomes contain two non-mitochondrial ATP-dependent calcium transport activities differing in their regulation by calmodulin. *Soc. Neurosci. Abstr.* **7:** 367.

——. 1984. Partial purification and functional identification of a calmodulin-activated ATP-dependent calcium pump from synaptic plasma membranes. *J. Neurosci.* (in press).

RAHAMIMOFF, H. and E. ABRAMOVITZ. 1978a. Calcium transport in a vesicular membrane preparation from rat brain synaptosomes. *FEBS Lett.* **89:** 223.

——. 1978b. Ca transport and ATPase activity of synaptosomal vesicles from rat brain. *FEBS Lett.* **92:** 163.

ROSS, D.H. and H.L. CARDENAS. 1979. Nerve cell calcium as a messenger for opiate and endorphin actions. *Adv. Biochem. Psychopharmacol.* **20:** 301.

SOBUE, K., S. ICHIDA, H. YOSHIDA, R. YAMAZAKI, and S. KAKIUCHI. 1979. Occurrence of a Ca^{++}- and modulator protein-activatable ATPase in the synaptic plasma membranes of brain. *FEBS Lett.* **99:** 199.

SORENSEN, R.G. and H.R. MAHLER. 1981. Calcium-stimulated adenosine triphosphatases in synaptic membranes. *J. Neurochem.* **37:** 1407.

STERNBERGER, L.A., P.H. HARDY, J.J. CUCULIS, and H.G. MYER. 1970. The unlabeled antibody enzyme method by immunohistochemistry: Preparation and properties of soluble antigen-antibody complex (horseradish peroxidase-antiperoxidase) and its use in identification of spirochetes. *J. Histochem. Cytochem.* **18:** 315.

SWEADNER, K.J. 1979. Two molecular forms of (Na^+-K^+)ATPase in brain: Separation and difference in affinity for strophanthidin. *J. Biol. Chem.* **254:** 6060.

TADA, M., F. OHMORI, M. YAMADA, and H. ABE. 1979. Mechanism of the stimulation of Ca^{++}-dependent ATPase of cardiac sarcoplasmic reticulum by adenosine 3':5'-monophosphate-dependent protein kinase: Role of the 22,000-dalton protein. *J. Biol. Chem.* **254:** 319.

TOWBIN, H., T. STAEHELIN, and J. GORDON. 1979. Electrophoretic transfer of proteins from polyacrylamide gels to nitrocellulose sheets: Procedure and some applications. *Proc. Natl. Acad. Sci.* **76:** 4350.

Studies on the Na^+-K^+ ATPase of Skeletal Muscle and Nerve

D.M. FAMBROUGH
Department of Embryology, Carnegie Institution of Washington, Baltimore, Maryland 21210

The sodium- and potassium-ion-stimulated adenosine triphosphatase, Na^+-K^+ ATPase (E.C. 3.6.1.3), catalyzes the transmembrane exchange of three Na^+ for two K^+. This major ion exchange process participates in maintenance of osmotic balance and preservation of an internal ionic milieu appropriate for intracellular processes. It is also the key transport mechanism in restoration of resting levels of intracellular Na^+ and K^+ in excitable cells. Even when intracellular ionic concentrations have been altered significantly by activity of nerve or muscle, the Na^+-K^+ ATPase, operating as an electrogenic ion pump, may preserve a high transmembrane potential and thus maintain electrical excitability.

The major focus of research on the Na^+-K^+ ATPase has been upon the mechanism of its action (Jørgensen 1982). Studies on the number and distribution of Na^+-K^+ ATPase molecules in excitable tissues and on the biosynthesis and turnover and metabolic regulation of the Na^+-K^+ ATPase have been relatively few. This was due in large measure to the lack of convenient ligands for such work. In the course of our study of cell-surface antigens on skeletal muscle, we found a monoclonal antibody that bound to an externally facing epitope on the Na^+-K^+ ATPase. This chapter reviews our research on the Na^+-K^+ ATPase and outlines some of our working hypotheses and directions for further research in this area.

Evidence That Monoclonal Antibody 24 Recognizes the Na^+-K^+ ATPase

In cloning hybridoma 24, which secretes an IgG, we screened for antibody binding to exposed sites on tissue-cultured chick spinal cord cells and myogenic cells. Monoclonal antibody (mAb) 24 bound with high affinity to neurons, myoblasts, and myotubes but showed insignificant binding to other cell types in these cultures. Similarly, the antibody recognized skeletal muscle fibers and peripheral nerve axons in sections of skeletal muscle (Fig. 1), but did not bind at detectable levels to other cell types such as capillary endothelial cells, Schwann cells, the interstitial fibroblasts of skeletal muscle, or the fibroblasts of the perineurium. The high level of binding of mAb 24 to excitable cells was the initial basis for our interest in the antigen it recognized. The antigen was first purified from [^{35}S]methionine-labeled myogenic cells. Its subunit structure, determined by SDS-polyacrylamide gel electrophoresis, immediately suggested that the antigen might be the Na^+-K^+ ATPase. Subsequently, we have purified fairly large amounts of the antigen from brain and kidney. It consists of two types of polypeptide chains of apparent molecular weights 105,000 (α) and 47,000 (β), held together by noncovalent interactions (Fig. 2). SDS-polyacrylamide gel electrophoretic analysis of antigen

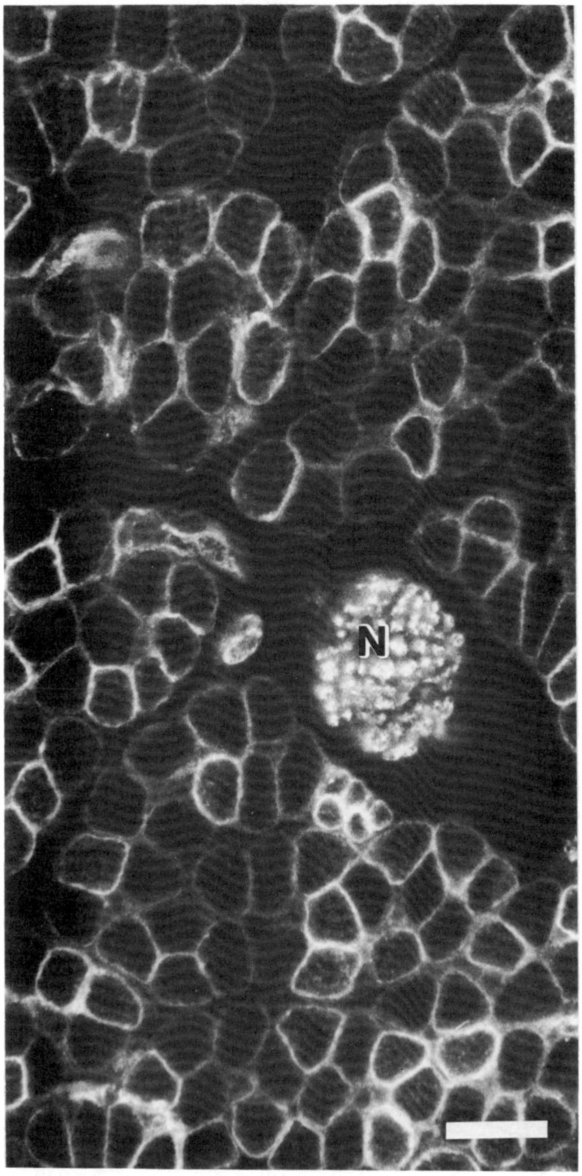

Figure 1. Fluorescence micrograph of cryosection through chick skeletal muscle, stained with a monoclonal antibody to the Na^+-K^+ ATPase followed by fluorescein-labeled second antibody. (N) Peripheral nerve bundle. Magnification bar, 20 μm. (Reprinted, with permission, from Fambrough and Bayne 1983.)

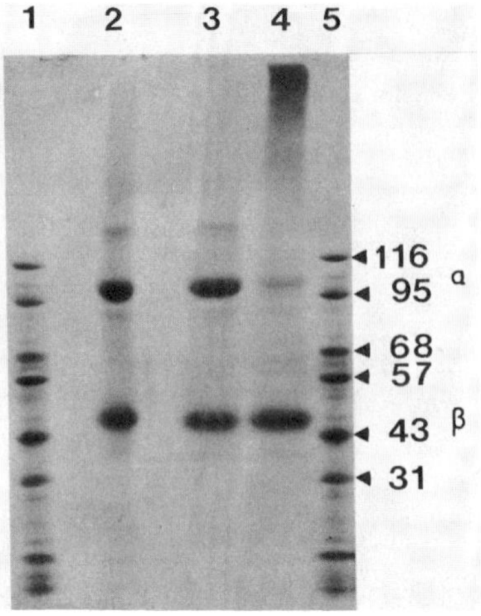

Figure 2. Electrophoretic analysis of antigen 24 on SDS-polyacrylamide gel. Coomassie-stained gel: (Lanes *1* and *5*) molecular weight standards; (lane *2*) unreduced antigen; (lane *3*) reduced antigen; (lane *4*) reduced antigen heated to 100°C in SDS sample buffer. (Reprinted, with permission, from Fambrough and Bayne 1983.)

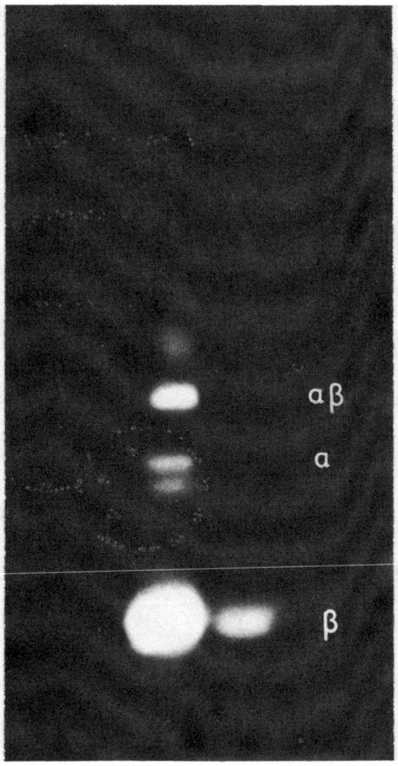

Figure 3. Autoradiograph (negative print) of chicken antigen 24 (*left lane*) and gull salt gland Na^+-K^+ ATPase (*right lane*) following SDS-polyacrylamide gel electrophoresis, transfer to nitrocellulose, and probing with rabbit anti-antigen-24 serum and ^{125}I-labeled protein A.

synthesized by cells treated with tunicamycin indicated that the unglycosylated polypeptide molecular weights are 105,000 and 32,000.

Because mAb 24 had no measurable effect on ATPase activity of the antigen and because solubilization of the antigen in detergent destroyed ATPase activity, identification of the antigen as the Na^+-K^+ ATPase involved making a large number of correlations of antigen with known characteristics of the Na^+-K^+ ATPase (Fambrough and Bayne 1983). In addition to subunit structure, the antigen resembled the Na^+-K^+ ATPase in sedimentation constant, in glycosylation of the β-subunit, and in having a basolateral distribution on kidney tubule cells and cells of the duck salt gland. Further, the number of ouabain-binding sites and antibody-binding sites on cultured myotubes was approximately equal at about 5×10^5 sites on the surface per myotube nucleus, and the antigen showed up-regulation when cultured myogenic cells were stressed by growth in medium with lowered K^+ concentration. mAb 24 was shown to recognize the only reasonable candidate for the Na^+-K^+ ATPase in chicken kidney microsomes enriched for Na^+-K^+ ATPase by the method of Jørgensen (1974). mAb 24 also was shown to recognize purified gull salt gland ATPase (a gift from Dr. J.A. Reynolds), and antisera raised against the purified antigen 24 were shown to bind to the β-subunit of gull salt gland Na^+-K^+ ATPase on protein blots (Fig. 3).

It was particularly important to establish beyond reasonable doubt that antigen 24 was the Na^+-K^+ ATPase because several of our observations are either at odds with or at least surprising in light of published information on the Na^+-K^+ ATPase. These observations include evidence for rapid assembly of subunits during biosynthesis, evidence for the occurrence of high concentrations of Na^+-K^+ ATPase in the internodal regions of myelinated nerve fibers, and evidence for antigenically different Na^+-K^+ ATPase molecules in different cell types.

Biosynthesis and Assembly

We established that the α- and β-subunits of the Na^+-K^+ ATPase recognized by mAb 24 existed as a stoichiometric complex after solubilization in nonionic detergent solution by showing that the two subunits sedimented together in sucrose gradient velocity sedimentation as a 7S complex with constant ratio of α to β across the antigen peak (Fambrough and Bayne 1983). We also demonstrated that unglycosylated antigen from cells grown in the presence of tunicamycin and also labeled antigen from cells pulse-labeled for a few minutes with [^{35}S]methionine could be immunoprecipitated with mAb 24. Thus, the antibody appears to recognize a proteinaceous aspect of the Na^+-K^+ ATPase rather than an epitope related to secondary modification of the molecule (Fambrough and Bayne 1983). Because the antibody is a monoclonal antibody and there is no reason to suspect any antigenic relatedness of the α- and β-subunits, it is nearly certain that the antibody recognizes a single portion of the molecule either on one of the subunits or generated at the interface of α- and β-subunits. Yet we find a constant ratio of α- to β-subunit immunoprecipitated by the monoclonal antibody. We conclude that the coprecipitation of α- and β-subunits reflects the fact that the subunits are together in a complex. Therefore, the coprecipitation of α- and β-subunits by monoclonal antibody represents an operational definition of subunit assembly.

We have examined the biosynthesis and assembly of Na^+-K^+ ATPase in tissue-cultured chick skeletal muscle, using standard pulse and pulse-chase strategies with [^{35}S]methionine. The recovered ^{35}S-labeled antigen represents about 0.02% of labeled protein solubilized by nonionic detergent extraction of the cultured cells. Analyzed by SDS gel electrophoresis, the Na^+-K^+ ATPase labeled for as little as 5 minutes with [^{35}S]methionine consists of α- and β-subunits with the typical ratio of incorporated radioactivity (α/β = 2.0–2.5). The incorporation of label into immunoprecipitable antigen is approximately linear with time, so that during a pulse-chase experiment label accumulates in the antigen over the course of the pulse and little additional label appears during the chase. These data suggest that there is not a significant pool of unassembled α- and/or β-subunits that later are assembled into antigenic complexes. Rather, the data are readily explained in terms of rapid association of α- and β-subunits into stoichiometric complexes. This conclusion appears contradictory to that reached by Sabatini and co-workers (1982), who proposed physically separated sites of synthesis for α- and β-subunits, with the α-subunit synthesized on free polysomes and delayed assembly of α- and β-subunits into complexes.

The assembly process is so rapid during Na^+-K^+ ATPase biosynthesis that it can even precede the addition of core oligosaccharide chains to asparaginyl residues of the β-chain. This is evidenced by the occurrence of a set of β-subunit forms associated with the α-subunit isolated from very short-term pulse-labeled cells. β-Subunits of apparent molecular weights 32,000, 35,000, 38,000, and 40,000 and the 105,000-molecular-weight α-subunit are seen (Fig. 4). Autoradiographs of these SDS gel electrophoresis patterns have been quantified by densitometry (Fig. 5). When the sum of the areas under the curve for all the β-subunit forms is used in determining the ratio of α- to β-subunit labeling, this ratio is a constant for a 10-minute pulse period and subsequent chase periods up to 8 hours. During the chase period, the β-subunit undergoes further posttranslational modification to the mature, 47-kD form. This form, unlike the lower-molecular-weight forms, is resistant to endoglycosidase H, whereas the lower-molecular-weight forms are converted to a 32-kD polypeptide by the enzyme (Fig. 4). The mature β-subunit is a receptor for wheat germ agglutinin.

In sum, our study of the biosynthesis and assembly of the Na^+-K^+ ATPase suggests that the assembly process is extremely rapid, that the α-subunit bears no major oligosaccharide side chains, and that the β-subunit bears three oligosaccharide chains of the complex, N-linked type.

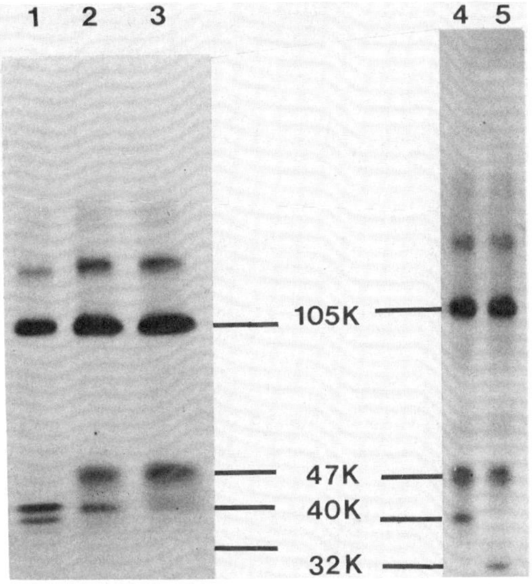

Figure 4. Pulse-chase labeling of Na^+-K^+ ATPase in myogenic cell cultures (*left lanes*) and endoglycosidase H digestion (*right lanes*), analyzed by SDS-polyacrylamide gel electrophoresis and autoradiography. Muscle cultures were labeled for 10 min with [^{35}S]methionine (lane *1*) and then chased for 40 min and 60 min (lanes *2* and *3*) before extraction and affinity purification of antigen on mAb 24 immunobeads. Lanes *4* and *5* illustrate control and endoglycosidase H digestion of [^{35}S]methionine-labeled Na^+-K^+ ATPase.

Regulation: Quantitative Aspects

The Na^+-K^+ ATPase is an exchange pump that can catalyze transport of Na^+ and K^+ ions only when both ionic species are available: Na^+ inside and K^+ outside the cell. We have adopted a strategy used by Cook and colleagues in their investigation of Na^+-K^+ ATPase regulation in HeLa cells (Vaughan and Cook 1972; Pollack et al. 1981) to stress myogenic cells in such a way that they might respond by up-regulation of their Na^+-K^+ ATPase. The K_m for K^+ uptake by HeLa cells is approximately 0.8 mM and for Na^+ extrusion about 10–15 mM. Thus, under normal conditions the Na^+-K^+ ATPase is maximally stimulated by the extracellular K^+ concentration (~5 mM) and ion transport is sensitive to the intracellular Na^+ concentration (~10–15 mM). By lowering external K^+ concentration to 0.5 mM, it becomes difficult for the Na^+-K^+ ATPase to transport ions. Na^+ accumulates in the cells, and the cells respond by increasing the number of Na^+-K^+ ATPase molecules in the plasma membrane. For HeLa cells, this up-regulation is accomplished by a marked slowing of the turnover rate of the Na^+-K^+ ATPase (Pollack et al. 1981). In pilot experiments, we found that overnight culture of myogenic cells in low-potassium medium (0.5 mM) resulted in an increase in Na^+-K^+ ATPase molecules in the plasma membrane of up to 74% compared with unstressed control cultures (Fambrough and Bayne 1983). We are currently studying the mechanism underlying this up-regulation. Formally, there are four types of mechanisms that could accomplish this up-regulation: (1) an increased rate of

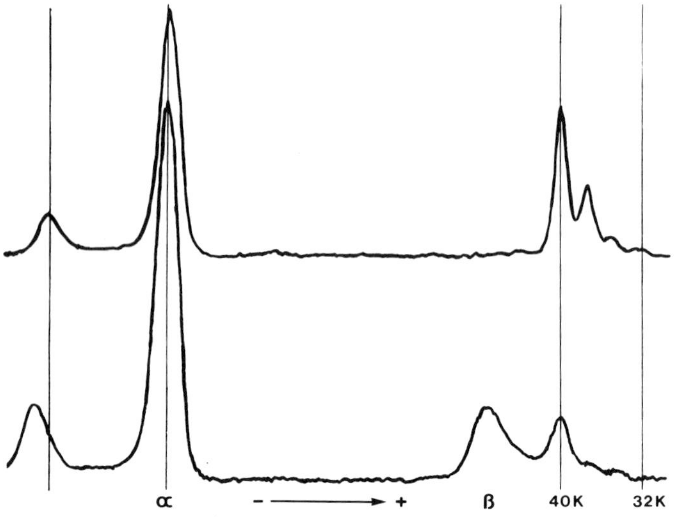

Figure 5. Densitometric tracings of autoradiograph of Na^+-K^+ ATPase isolated from myogenic cultures pulse-labeled for 10 min with [^{35}S]methionine (*upper trace*) and then chased for 40 min (*lower trace*) and subjected to SDS-polyacrylamide gel electrophoresis. Note conversion of immature β-subunit forms (M_r = 32,000, 35,000, 38,000, 40,000) toward mature β size (M_r = 47,000). Undissociated α-β-complex peaks at left also reflect maturation of β-subunit.

biosynthesis, (2) a decreased rate of degradation, (3) a transfer of Na^+-K^+ ATPase molecules from intracellular membranes to the plasma membrane, and (4) the revelation of masked Na^+-K^+ ATPase molecules. Fortunately, each of these possibilities can be investigated through the use of monoclonal antibodies to measure the number and distribution and kinetics of biosynthesis and turnover of the Na^+-K^+ ATPase.

Is up-regulation of the Na^+-K^+ ATPase in myogenic cell cultures relevant to situations in adult animals? The work of Norgaard et al. (1981) and Clausen et al. (1982) has suggested that up-regulation in response to lowered K^+ concentration may not occur in skeletal muscles of rats maintained on a low-potassium diet. On the other hand, an examination of the levels of Na^+-K^+ ATPase in skeletal muscle fibers and peripheral nerves (Figs. 1 and 6) suggests that the level of Na^+-K^+ ATPase may well be adjusted to the needs of the individual excitable cells. In the case of skeletal muscle fibers, the density of Na^+-K^+ ATPase molecules appears to be markedly different from one fiber to the next. This pattern of Na^+-K^+ ATPase levels in muscle tissue resembles the distribution of muscle fiber types determined by classical methods (Dubowitz and Pearse 1960; Guth and Samaha 1969). Such a distribution of Na^+-K^+ ATPase sites fits well with the motor unit structure of muscles. In each muscle, there are fibers that are virtually continuously active and contribute to the tone of the muscle and the maintenance of posture. Other muscle fibers are recruited to serve for shorter periods of time and others are only occasionally recruited into activity when short-term application of maximal or near maximal muscle force is required. As a working hypothesis, we propose that the level of Na^+-K^+ ATPase in a muscle fiber is set by the previous history of muscle activity as well as by constitutive expression at some basal level. We have begun to look at the relation of Na^+-K^+ ATPase level to muscle fiber types defined by physiological criteria and by pH sensitivity of the myosin ATPase. In the chicken anterior latissimus dorsi muscle, consisting of only slow, graded contraction fibers, the level of Na^+-K^+ ATPase is similar in all fibers, being slightly higher in those fibers that contain high pH-resistant myosin ATPase. There is a much larger variation in Na^+-K^+ ATPase level between muscle fibers of various leg muscles in which a variety of muscle fiber types are found.

In contrast to the variable level of expression in the sarcolemma of different skeletal muscle fibers, it appears that all peripheral nerve fibers display a very high level of axolemmal Na^+-K^+ ATPase (Fig. 1). This suggests the interesting possibility that up- and down-regulation of Na^+-K^+ ATPase may not occur in neurons. We speculate that up- and down-regulation, mediated by modulation of metabolism and distribution of Na^+-K^+ ATPase molecules, does not operate on a time scale suitable for neurons. Rather, we suggest that the large capacity for Na^+ extrusion and K^+ uptake needed by neuronal processes (in many cases far from the soma) can best be met by a constitutively high level of expression of the Na^+-K^+ ATPase in neurons. We plan to examine this possibility initially by studies on dorsal root ganglion cells in tissue culture.

Another interesting aspect of the distribution of Na^+-K^+ ATPase on peripheral nerve fibers is the obvious internodal location of a major portion of the Na^+-K^+ ATPase (Fig. 1). Access of antibody to internodal axolemma is afforded by the nature of the preparation—cryosection. Nodal axolemma represents an exceedingly small portion of total axolemma in myelinated nerve fibers and is not expected to contribute to the immunofluorescence observed in typical sections (which, even if they occasionally include a node of Ranvier, are at least five times thicker than the length of nodal membrane). This observation contrasts with the report of Wood et al. (1977) that the Na^+-K^+ ATPase located by immunohistochemistry in central myelinated nerves of the knifefish has a nodal localization. The difference between these observations can only be evaluated after additional quantitative study of

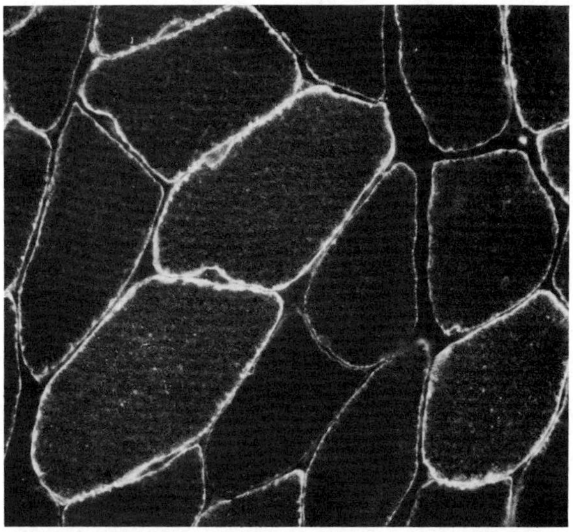

Figure 6. Fluorescence micrograph of cryosection through duck neck muscle stained with monoclonal antibody to the Na^+-K^+ ATPase followed by fluorescein-labeled second antibody. Note marked variation in intensity of staining of adjacent muscle fibers. Preliminary electron microscope immunohistochemical analysis suggests that the intracellular fluorescence is due to antigen sites in the T-tubule system.

both chicken and fish peripheral and central nervous systems.

Regulation: Evidence for Qualitatively Different Forms of Na^+-K^+ ATPase in Different Cell Types

Prior to our work with mAb 24, localization and quantification of Na^+-K^+ ATPase sites had been done primarily with [^{3}H]ouabain. In exploring the possible identity of antigen 24 and the Na^+-K^+ ATPase, we measured the number of antibody-binding sites and ouabain-binding sites on erythrocytes and on tissue-cultured fibroblasts and myotubes. In the case of erythrocytes, we found about 180 binding sites for divalent antibody per erythrocyte, corresponding well to the number of binding sites for ouabain. The numbers of antibody-binding sites and ouabain-binding sites on myotubes were the same within experimental error ($\sim 5 \times 10^5$ per myotube nucleus). However, there was a 25-fold discrepancy between number of ouabain-binding sites and antibody-binding sites on muscle-derived fibroblasts (Fambrough and Bayne 1983). The number of saturable, K^+-sensitive ouabain-binding sites on fibroblasts was similar to the number of sites on myotubes and was about half as large as the number of sites reported to be on HeLa cells. We concluded that there must be an antigenically different form of Na^+-K^+ ATPase on fibroblasts and probably also on other cell types that appeared unreactive with fluorescent mAb 24, such as capillary endothelial cells, Schwann cells, and connective tissue fibroblasts.

We are exploring the possible qualitative differences between Na^+-K^+ ATPase molecules in different cell types at two levels. First, we are preparing antisera and monoclonal antibodies against Na^+-K^+ ATPase purified with the use of monoclonal antibody. We expect that among the antibodies generated in response to this antigen as immunogen, we will find those that detect all forms of the Na^+-K^+ ATPase. These antibodies can be used for preparation and comparison of Na^+-K^+ ATPases from different source cells. Second, we are examining the possible genetic basis for multiple forms of Na^+-K^+ ATPase by attempting to clone cDNAs encoding subunits of the Na^+-K^+ ATPase. In collaboration with Dr. D. Helfman, we have screened an expression library of chicken smooth muscle cDNAs (Helfman et al. 1983), using antiserum prepared against the affinity-purified Na^+-K^+ ATPase and have found several clones of potential interest. As an initial strategy for verifying the pertinence of these clones to Na^+-K^+ ATPase, we are attempting to purify the expressed polypeptide from the bacterial clones and plan to use it to back-purify the antibodies from rabbit serum that bind to it. We then plan to test these for binding to each subunit of the Na^+-K^+ ATPase. Although there are now very limited amino acid sequence data on the Na^+-K^+ ATPase, such data as there are may also help in identification of the cDNA. An immediate goal of the research is determination of the amino acid sequences of the Na^+-K^+ ATPase subunits through sequencing the encoding cDNAs. Examination of the genetic basis of Na^+-K^+ ATPase structure will begin with Southern (1975) analysis of restricted genomic DNA, using probes derived from the expressed inserts of the expression clones.

Other evidence for multiple forms of Na^+-K^+ ATPase has come from binding studies (see Hansen 1976) and from the work of Sweadner (1979), which included evidence for a difference in electrophoretic mobility of α-subunits from myelinated central nervous system neurons compared with α-subunits from other sources, including neurons grown in tissue culture.

The existence of multiple forms of Na^+-K^+ ATPase will complicate the analysis of up-regulation. It is possible that some cell types express a mixture of forms, and it is possible that these respond differently to physiological signals. The problem might be manageable at the protein level through the use of monoclonal antibodies specific for each form and at the nucleic acid level through the use of cDNA probes specific for individually encoded Na^+-K^+ ATPase mRNAs.

Correlation of Na^+-K^+ ATPase and Intramembrane Particles

When fluorescein-conjugated mAb 24 was used to document the distribution of Na^+-K^+ ATPase molecules on tissue-cultured myotubes, a ubiquitous surface distribution was seen on live cells, with very little evidence of spatial organization (Fig. 7A). Some more intense striations in the long axis of the myotubes may be due to creases in the surface, which are also seen by

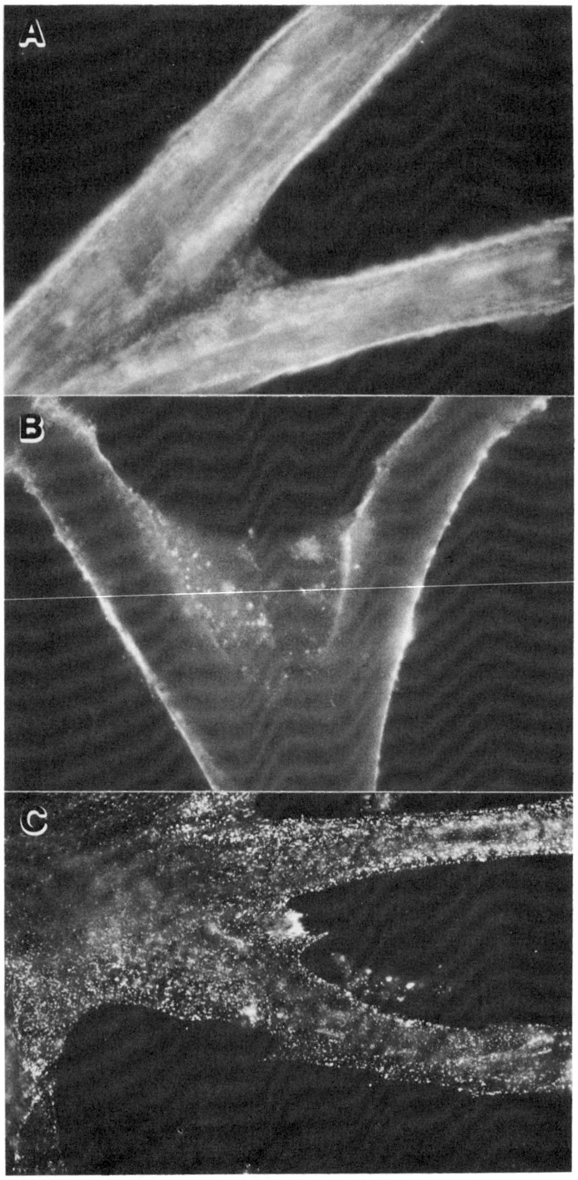

Figure 7. Immunofluorescence micrographs illustrating distribution of Na^+-K^+ ATPase on cultured chick myotubes. All myotubes stained with fluorescein-labeled mAb 24. (*A*) No further treatment. (*B*) Myotubes prefixed with formaldehyde fixative before incubation with monoclonal antibody followed by incubation with goat anti-mouse IgG. (*C*) Myotubes treated as in *B* but without prefixation.

scanning electron microscopy. This relatively uniform pattern of fluorescent staining can be altered to a pattern of small patches (Fig. 7C) by incubating the fluorescent-labeled myotubes with antibodies directed against mouse IgG. This alteration appears to be due to the cross-linking of bound antibody, and must depend upon some mobility of the Na^+-K^+ ATPase in the plasma membrane. The formation of patches can be prevented by fixation of the myotubes in formaldehyde fixative (McLean and Nakane 1974) prior to the binding of monoclonal antibody (Fig. 7B).

In collaboration with Dr. D. Pumplin, we have carried out a freeze-fracture study of myotubes treated as described above (Pumplin and Fambrough 1983). This study revealed that the fluorescent clusters generated by double-antibody labeling were spatially congruent with clusters of intramembrane particles (IMPs) seen in the freeze-fracture replicas of the same myotubes. The clustered particles represented up to 50% of the intramembrane particles in the replicas, or up to 400 IMPs/μm^2 of surface membrane. As shown in Figure 8, the IMP clusters were fairly loose and were often associated with slight indentation of the plasma membrane (not seen in control preparations without antibody cross-linking).

Several control experiments were carried out to establish that the IMP clusters were caused by clusters of Na^+-K^+ ATPase molecules. In these control experiments, several other antigens were clustered by monoclonal antibody and second antibody. In none of these cases did this cause the formation of IMP clusters nor were the Na^+-K^+ ATPase molecules clustered by the treatments. In other experiments, the Na^+-K^+ ATPase was clustered by double-antibody treatment and then the distribution of other surface antigens was explored with fluorescent-labeled monoclonal antibodies or with fluorescent α-bungarotoxin (for locating acetylcholine receptors). Clustering of Na^+-K^+ ATPase molecules did not cause co-clustering of any of the other surface components we monitored.

Further quantification of the IMPs clustered by mAb 24 and second antibody revealed that about 80% of the clustered particles were in the P-face replicas. The distribution of particle diameters was virtually identical to the distribution of particle diameters for the entire population of IMPs on the myotubes.

This correlation of IMPs and Na^+-K^+ ATPase on cultured myotubes may be relevant to observations made by Schotland et al. (1980, 1981) on the numbers of IMPs in freeze-fracture replicas of normal and dystrophic human muscle fibers. These investigators reported that dystrophic muscle fibers contained fewer IMPs than did normal fibers. Considering that the Na^+-K^+ ATPase may account for 20–50% of IMPs and that there may be rapid up- and down-regulation of Na^+-K^+ ATPase in response to physiological conditions, we can offer an explanation for the difference noted above: The difference in IMPs may reflect different levels of Na^+-K^+ ATPase expression in normal and dystrophic muscles due to differences in muscle usage.

SUMMARY

We have used monoclonal antibodies directed against skeletal muscle and neuronal plasma membranes to define surface antigens and have identified one of these antigens as the Na^+-K^+ ATPase. The monoclonal antibody to the Na^+-K^+ ATPase has facilitated an exploration of the biosynthesis and assembly of the Na^+-K^+ ATPase, the distribution of molecules on nerve and muscle, the regulation of expression of Na^+-K^+ ATPase, and the relation of the Na^+-K^+ ATPase to the intramembrane particles seen in freeze-fracture repli-

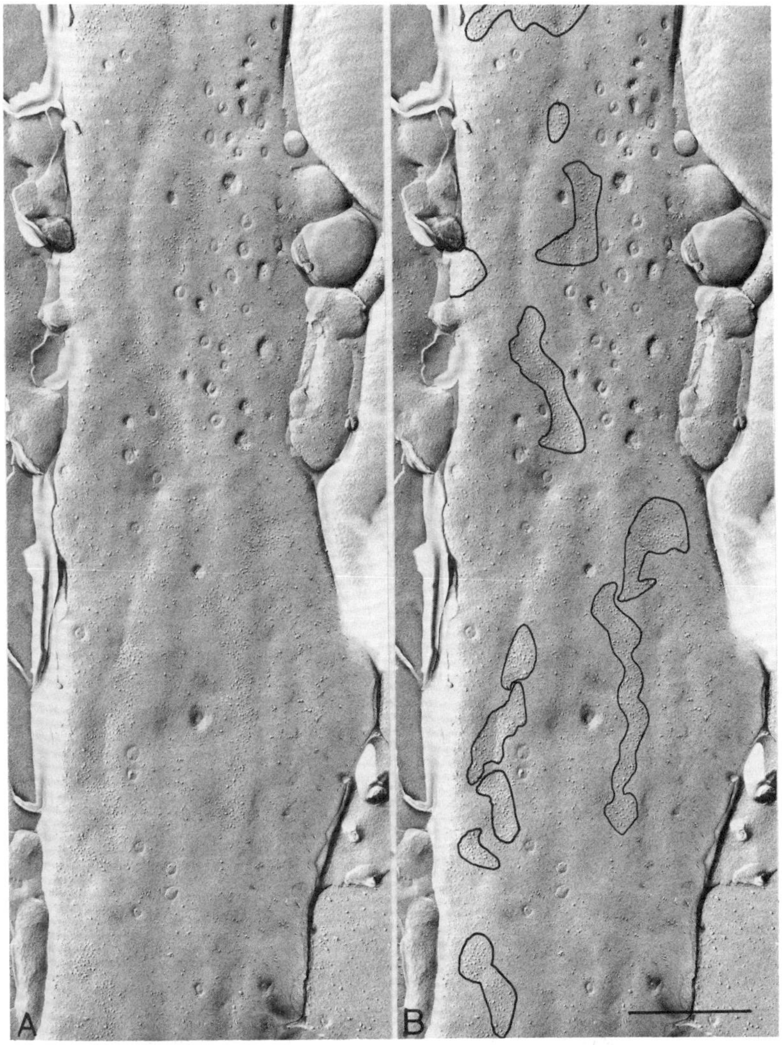

Figure 8. Freeze-fracture replica (P face) of very small myotube, made after cross-linking of Na^+-K^+ ATPase sites with mAb 24 followed by goat anti-mouse IgG. *A* and *B* are same field. Magnification bar, 0.5 μm. (Reprinted, with permission, from Pumplin and Fambrough 1983.)

cas. We have obtained new information on the glycosylation sites on the β-subunit and evidence for antigenically different forms of Na^+-K^+ ATPase related to differences in proteinaceous aspects of the molecule. Finally, we are attempting to move from the protein level to the nucleic acid level of analysis by using antiserum prepared against the Na^+-K^+ ATPase to search for encoding cDNAs in an expression library.

ACKNOWLEDGMENT

This research was supported by grant NS-16542 from the Department of Health and Human Services, U.S. Public Health Service.

REFERENCES

CLAUSEN, T., O. HANSEN, K. KJELDSEN, and A. NØRGAARD. 1982. Effect of age, potassium depletion and denervation on specific displaceable [^{3}H]-ouabain binding in rat skeletal muscle *in vivo*. *J. Physiol.* **333:** 367.

DUBOWITZ, V. and A.G.E. PEARSE. 1960. A comparative histochemical study of oxidative enzyme and phosphorylase activity in skeletal muscle. *Histochemie* **2:** 105.

FAMBROUGH, D.M. and E.K. BAYNE. 1983. Multiple forms of (Na^+ + K^+)-ATPase in the chicken. Selective detection of the major nerve, skeletal muscle, and kidney form by a monoclonal antibody. *J. Biol. Chem.* **258:** 3926.

GUTH, L. and F.J. SAMAHA. 1969. Qualitative differences between actomyosin ATPase of slow and fast mammalian muscle. *Exp. Neurol.* **25:** 138.

HANSEN, O. 1976. Non-uniform populations of *g*-strophanthidin binding sites of (Na^+ + K^+)-activated ATPase. *Biochim. Biophys. Acta* **433:** 383.

HELFMAN, D.M., J.R. FERAMISCO, J.C. FIDDES, G.P. THOMAS, and S.H. HUGHES. 1983. Identification of clones that encode chicken tropomyosin by direct immunological screening of a cDNA expression library. *Proc. Natl. Acad. Sci.* **80:** 31.

JØRGENSEN, P.L. 1974. Purification and characterization of (Na^+ + K^+)-ATPase. III. Purification from the outer medulla of mammalian kidney after selective removal of membrane components by sodium dodecylsulfate. *Biochim. Biophys. Acta* **356:** 36.

——. 1982. Mechanism of the Na^+,K^+ pump. Protein structure and conformations of the pure ($Na^+ + K^+$)-ATPase. *Biochim. Biophys. Acta* **694:** 27.

McLean, I.W. and P.K. Nakane. 1974. Periodate-lysine paraformaldehyde fixative: A new fixative for immunoelectron microscopy. *J. Histochem. Cytochem.* **22:** 1077.

Nørgaard, A., K. Kjeldsen, and T. Clausen. 1981. Potassium depletion decreases the number of ^{3}H-ouabain binding sites and the active Na-K transport in skeletal muscle. *Nature* **293:** 739.

Pollack, L.R., E.H. Tate, and J.S. Cook. 1981. Turnover and regulation of Na-K-ATPase in HeLa cells. *Am. J. Physiol.* **241:** C173.

Pumplin, D.W. and D.M. Fambrough. 1983. ($Na^+ + K^+$)-ATPase correlated with a major group of intramembrane particles in freeze fracture replicas of cultured chick myotubes. *J. Cell Biol.* (in press).

Sabatini, D., D. Colemen, E. Sabban, J. Sherman, T. Morimoto, G. Kreibish, and M. Adesnik. 1982. Mechanisms for the incorporation of proteins into the plasma membrane. *Cold Spring Harbor Symp. Quant. Biol.* **46:** 807.

Schotland, D.L., E. Bonilla, and Y. Wakayama. 1980. Application of freeze-fracture technique to the study of human neuromuscular disease. *Muscle Nerve* **3:** 21.

——. 1981. Freeeze-fracture studies of human muscular dystrophy. *Acta Neuropathol.* **54:** 189.

Southern, E. 1975. Detection of specific sequences among DNA fragments separated by gel electrophoresis. *J. Mol. Biol.* **98:** 503.

Sweadner, K.J. 1979. Two molecular forms of ($Na^+ + K^+$)-stimulated ATPase in brain. Separation, and difference in affinity for strophanthidin. *J. Biol. Chem.* **254:** 6060.

Vaughan, G.L. and J.S. Cook. 1972. Regeneration of cation-transport capacity in HeLa cell membranes after specific blockade by ouabain. *Proc. Natl. Acad. Sci.* **69:** 2627.

Wood, J.G., D.H. Jean, J.N. Whitaker, B.J. McLaughlin, and R.W. Albers. 1977. Immunocytochemical localization of the sodium, potassium activated ATPase in knifefish brain. *J. Neurocytol.* **6:** 571.

Detection and Regulation of the Tyrosine Hydroxylase mRNA Levels in Rat Adrenal Medulla and Brain Tissues

J. MALLET,* N. FAUCON BIGUET,* M. BUDA,† A. LAMOUROUX,* AND D. SAMOLYK*
*Institut de Microbiologie, Université Paris-Sud, 91405 Orsay, France; †Institut National de La Santé et de la Recherche Médicale, Hôpital Ste-Eugénie, 69230 Saint-Genis Laval, France

The catecholamines are an important class of neurotransmitters present both in the central and peripheral nervous systems. They constitute a useful model for the analysis of the mechanisms involved in the regulation of gene expression during development. Tyrosine hydroxylase (tyrosine 3-monooxygenase; L-tyrosine, tetrahydropterine:oxygen oxidoreductase [3-hydroxylating], [EC 1.14.16.2]), the rate-limiting enzyme in the synthesis of catecholamine, has been intensively investigated because of its key role in the physiology of adrenergic neurons. The regulation of its expression has been shown to be influenced by transsynaptic humoral orthograde and retrograde processes during development (Edgar and Thoenen 1978; Black and Patterson 1980; Le Douarin 1980).

The expression of information encoded in the neuronal genome is also regulated during the normal functioning of the mature nervous system. Nervous brain stimulations such as cold, immobilization stress, or administration of drugs such as reserpine cause an increase of the enzymes tyrosine hydroxylase and dopamine-β-hydroxylase (DBH) with a latency of several days (Thoenen et al. 1969; Joh et al. 1973; Reis et al. 1974; Zigmond and Chalazonitis 1979). This process is referred to as transsynaptic induction and the underlying mechanism may provide a molecular basis for neural adaptation to functional requirement.

Recently, we reported the isolation of recombinant plasmids carrying DNA sequences complementary to a mRNA coding for rat tyrosine hydroxylase antigen (Lamouroux et al. 1982). This probe provides a tool for analysis of the tyrosine hydroxylase gene and the transcriptional and posttranscriptional regulation of its expression during development and electrical activity of catecholaminergic neurons.

We describe here a very sensitive dot-blot assay for tyrosine hydroxylase mRNA and its use to study the transsynaptic regulation of the tyrosine hydroxylase gene by reserpine. It has also allowed us to localize tyrosine hydroxylase mRNA expression in various brain regions.

MATERIALS AND METHODS

Materials. Restriction enzymes and DNA polymerase I were from Boehringer Mannheim. DNase was from Worthington. Nitrocellulose paper (BA 85) was purchased from Schleicher and Schuell. α-^{32}P-labeled deoxynucleotides (>600 Ci/mmoles) were from the Radiochemical Centre (Amersham, England).

Reserpine treatment of rats. Male Sprague-Dawley rats (200 g; Iffa Credo Breeding Laboratories) were housed five per cage in an animal room kept at 21°C ± 1°C and maintained on a 12-hour light cycle. Food and water were given ad libitum. Reserpine was dissolved in a few drops of glacial acetic acid, diluted with saline, and neutralized to pH 4 with 1 N NaOH. Animals were injected intraperitoneally with a single dose of reserpine (10 mg/kg) at a concentration of 2.4 mg/ml. Control rats were injected with saline. The animals were sacrificed 4 days after the drug administration.

Growth of cells, tissue dissection, and RNA extractions. Rat pheochromocytoma PC12 cells, originally obtained through the generosity of L.A. Greene, were grown and harvested as described (Mallet et al. 1981). Reserpine-treated animals were sacrificed by decapitation. Each brain was quickly removed from the skull and a frontal section was made with a razor blade at the level of the anterior hypothalamus. The two brain parts were placed on their cut faces and frozen on dry ice. Simultaneously, the two adrenals were dissected and rapidly frozen. The posterior part of each brain was then cut into 500-μm-thick serial slices. The locus coeruleus (two sides) was punched using a 1.0-mm inner diameter needle from three adjacent frozen slices, as previously described (Buda et al. 1975). The median part of the substantia nigra was removed in a similar manner from the four adjacent slices containing this structure. Septum, caudate nuclei, and cerebella were separated from 14-day-old rat brain and stored frozen (−70°C) until use. Livers were dissected from adult rats.

Total RNA, from brain samples, adrenal medulla, and PC12 cells, was extracted as described by Feramisco et al. (1982), using a Dounce homogenizer. Liver RNA was purified by the method of Auffray and Rougeon (1980).

Preparation of cDNA probe. The plasmid pTH-1 contains 742 bases of the tyrosine hydroxylase mRNA sequence cloned into the *Pst*I site of pBR322. The plasmid was propagated in *Escherichia coli* C600. Following isolation, it was purified twice by cesium chloride-ethidium bromide gradients. The 742-bp insertion was cut from the plasmid by digestion with *Pst*I, separated on low-melt agarose (BRL), and purified by passage over a DEAE-cellulose column. The probe was nick-translated as previously described (Rigby et al. 1977) to a specific activity of 2×10^8 to 5×10^8 cpm/μg with [α-^{32}P]dATP and [α-^{32}P]dCTP as labeled nucleotides.

Dot-blot analysis. RNA samples were resuspended in 100 μl of 3 M NaCl/0.3 M sodium citrate (20× SSC). Aliquots of 50 μl were spotted on a nitrocellulose paper (Scheicher and Schull, BA 85) which had been equilibrated in 20× SSC. This was conveniently done by using a microfold dot blot (Scheicher and Schull) and applying a slight vacuum by means of an aspirator. The filtration time was about 1 nm for each set of samples. After sample application, the nitrocellulose paper was dried and baked under reduced pressure at 80°C for 2 hours. Filters were prehybridized and also hybridized at 42°C according to Thomas (1980), except that the mixture included 10% dextran sulfate. The nick-translated probes were denatured at 100°C for 2 minutes, cooled, and added to the hybridization buffer, and the blots were hybridized for about 48 hours. They were then washed successively for 20 minutes at 65°C with solutions of 2× SSC, 0.3× SSC, and 0.1× SSC containing 0.1% SDS. The blots were exposed on Fuji RX X-ray films. After different times of exposures, the developed films were scanned with a Vernon microdensitometer.

RESULTS

Determination of Tyrosine Hydroxylase mRNA Relative Concentrations by the Dot-blot Method

A prerequisite for using the dot-blot method to study changes in tyrosine hydroxylase mRNA levels between different RNA samples was to show that the tyrosine hydroxylase mRNA concentrations can be determined accurately and reproducibly in very small amounts of total RNA isolated from the tissues of interest.

To quantitate tyrosine hydroxylase mRNA, serial dilutions of total RNA from PC12 cells were spotted onto nitrocellulose filters. The amount of tyrosine hydroxylase mRNA in each spot was measured by hybridization to radioactive, nick-translated rat tyrosine hydroxylase cDNA following exposure of the nitrocellulose blot to X-ray film. The intensity of each spot on the autoradiogram, which reflects the amount of tyrosine hydroxylase mRNA, can be quantified by densitometry. Figure 1 indicates that there is a linear relationship between the peak height (cm) of the dots by densitometric scanning and the amount of RNA spotted. This method is extremely sensitive; tyrosine hydroxylase mRNA can be detected in as little as 20 ng of PC12 total RNA. Liver RNA never gave a positive signal for tyrosine hydroxylase mRNA and was therefore used as a negative control on all subsequent dot blots.

Effect of Reserpine Treatment on Tyrosine Hydroxylase mRNA in Rat Adrenal, Locus Coeruleus, and Substantia Nigra

Adult rats were injected with a single dose of reserpine (10 mg/kg), and 4 days later, adrenals and areas of locus coeruleus and substantia nigra were dissected out and total RNA was extracted. The relative levels of tyrosine hydroxylase mRNA were determined by dot-blot hybridization to the tyrosine hydroxylase probe, as described in Materials and Methods. Reserpine caused 65% and 130% increases in tyrosine hydroxylase mRNA content in locus coeruleus and adrenal medulla,

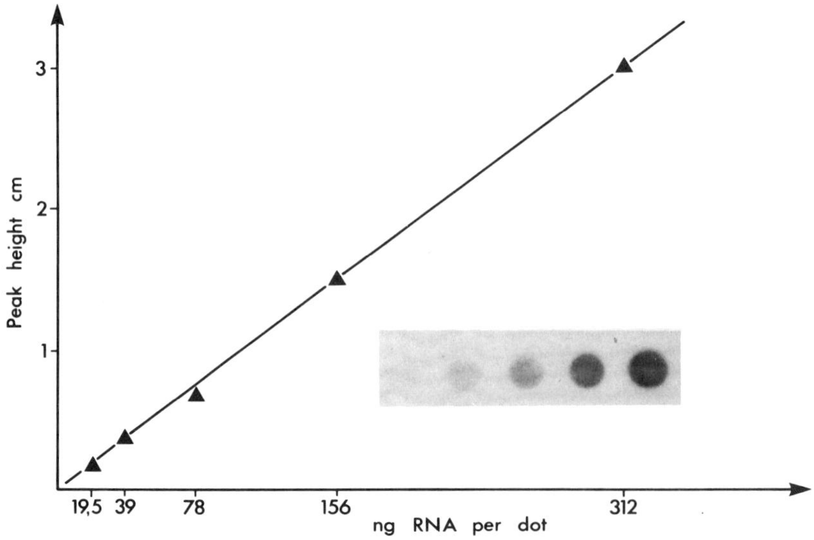

Figure 1. Quantitation of mRNA by scanning densitometry. Varying amounts of total RNA from PC12 cells were spotted on nitrocellulose paper, and tyrosine hydroxylase mRNA was identified by hybridization to a specific cDNA probe and autoradiography for 36 hr. The graph represents the correlation between the amount of RNA spotted and the heights of the corresponding peaks of the densitometry scan.

respectively, whereas a slight decrease was observed in substantia nigra (Fig. 2). These variations in tyrosine hydroxylase mRNA correlate well with the variations in tyrosine hydroxylase enzyme activity (results not shown).

Presence of Tyrosine Hydroxylaselike mRNA in Various Brain Tissues

Three brain regions that do not contain dopaminergic and noradrenergic cell bodies were analyzed for the presence of tyrosine hydroxylaselike mRNA. Cerebella, caudate nuclei, and septum were dissected from 15-day-old rats. Total RNA was purified and blotted onto nitrocellulose paper. Hybridization with the tyrosine hydroxylase probe at high stringency revealed the presence of tyrosine hydroxylaselike mRNA in these tissues (Fig. 3, a, b, and c). Hybridization with RNA from substantia nigra and from PC12 cells is also shown for comparison (Fig. 3, d and e). Finally, as a control, liver RNA was also spotted onto the nitrocellulose paper and no signal could be detected (Fig. 3, f). The tyrosine hydroxylaselike mRNA levels in cerebellum, caudate nucleus, and septum were, respectively, 80, 125, and 200 times lower than in PC12 cells.

DISCUSSION

In this paper, we report on a study of the transsynaptic regulation of tyrosine hydroxylase and the expression of its gene in various brain regions using dot hybridization to compare relative levels of tyrosine hydroxylase mRNA. This technique is rapid, sensitive, and reproducible. Tyrosine hydroxylase mRNA can easily be detected from samples containing nanogram amounts of total RNA.

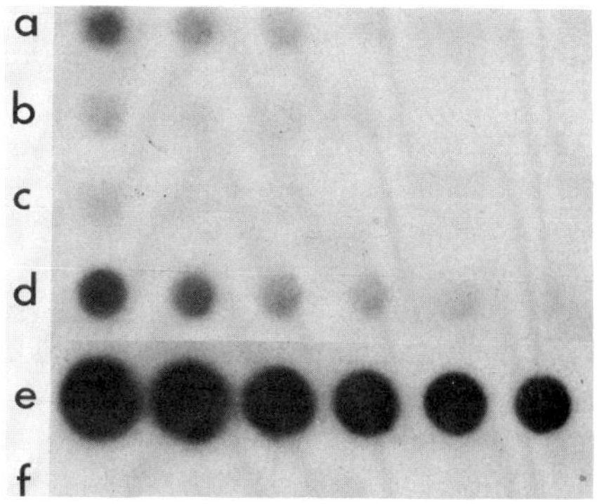

Figure 3. Tyrosine hydroxylase mRNA in various tissue. RNA was extracted and dotted on nitrocellulose paper as described in Materials and Methods. The following tissues were analyzed: (*a*) cerebellum; (*b*) caudate nucleus; (*c*) septum, (*d*) substantia nigra; (*e*) PC12 cells; and (*f*) liver. Cerebellum, caudate nucleus, and septum were from 14-day-old rats; the other tissues were from adult rats. Spots on each horizontal line correspond to 1:2 serial dilution of RNA starting with 10 μg in the first spot. Autoradiographic exposure time, 2 days.

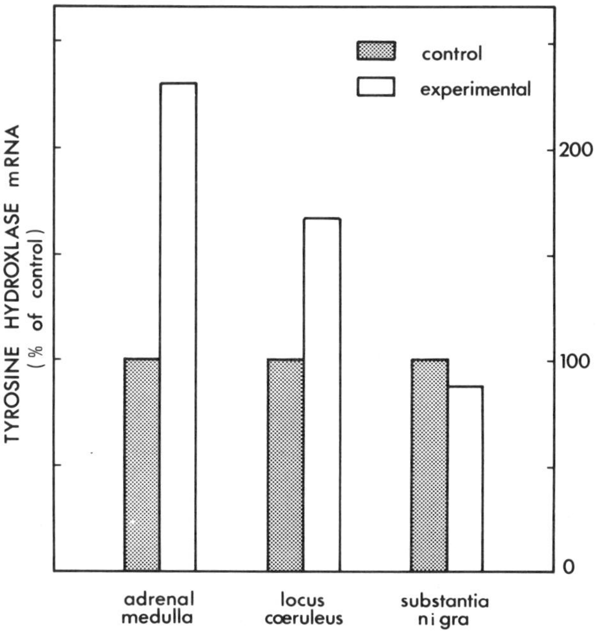

Figure 2. Effect of reserpine on tyrosine hydroxylase mRNA in adrenal medulla, locus coeruleus, and substantia nigra. Reserpine treatment, RNA purification, and dot-blot assays were as described in Materials and Methods. In each case, 10 μg, 5 μg, and 2.5 μg of RNA was spotted onto the nitrocellulose paper. For each tissue, an mRNA ratio between treated and control samples was determined for each mRNA amount and these ratios were compared.

In a first series of experiments, we have analyzed the effect of reserpine on tyrosine hydroxylase mRNA in adrenal glands, locus coeruleus, and substantia nigra. The administration of this drug has been shown to increase the activity of tyrosine hydroxylase (and also of DBH) (up to two- to threefold) in both peripheral and central noradrenergic neurons (Thoenen et al. 1969; Joh et al. 1973; Reis et al. 1974; Zigmond et al. 1974). The enzyme activity reaches a peak 48–72 hours after injection of reserpine and returns to control values after about 3 weeks. This effect can be abolished by interruption of the preganglionic sympathetic fibers and is therefore referred to as transsynaptic induction (Thoenen et al. 1969). Joh et al. (1973) and Reis et al. (1974) have shown that this induction is entirely attributable to synthesis of specific tyrosine hydroxylase enzyme protein and not to activation of preexisting enzyme molecules.

The results of our preliminary study of the mechanism underlying this long-term transsynaptic induction show that reserpine elicits an increase in tyrosine hydroxylase mRNA in adrenal gland and locus coeruleus. This increase correlates well with the increase in tyrosine hydroxylase enzymatic activity. Further investigation is necessary to determine whether the increase in tyrosine hydroxylase mRNA is due to increased synthesis or to decreased degradation.

In contrast to noradrenergic neurons, we found that reserpine induced a slight decrease of tyrosine hydroxylase mRNA in substantia nigra. This absence of an effect by reserpine on tyrosine hydroxylase in dopaminergic neurons confirms and extends previous reports by Reis et al. (1974). It is at present not clear whether the different response to various modulating factors of the tyrosine hydroxylase gene in different tyrosine hydroxylase-producing cells lies in the ability of the cell to recognize the modulating factor rather than in the tyrosine hydroxylase gene itself being nonresponsive. In the rat genome, initial evidence (C. Boni et al., unpubl.) indicates that there is only one tyrosine hydroxylase gene and the first hypothesis is then more likely.

Using the dot-blot method, we also have demonstrated the presence of tyrosine hydroxylaselike mRNA in brain tissues that have not previously been found to contain dopaminergic or noradrenergic cell bodies. Under highly stringent conditions, the tyrosine hydroxylase probe hybridizes to RNA purified from caudate nucleus, septum, and cerebellum. Under such conditions, only mRNA with a high degree of homology to the tyrosine hydroxylase gene would be detected. The relative amount of these mRNA species is low compared with tyrosine hydroxylase mRNA contained in PC12 cells; but considering these tissues are heterogeneous, it is conceivable that this RNA represents a high percentage of total mRNA in certain cells. In situ hybridization on tissue slices will allow us to address these questions. The significance of the presence of tyrosine hydroxylaselike mRNA in nonadrenergic tissue remains to be determined. These tyrosine hydroxylaselike mRNAs might represent tyrosine hydroxylase mRNAs that have not been fully processed and are not translated. Such processing might occur under certain conditions and could provide a molecular basis for neural regulation and response to perturbation. It should also be noted that adrenergic terminals exist in caudate nucleus, septum, and cerebellum (Lindvall and Björklund 1978) and the possibility that tyrosine hydroxylase mRNA is transported should not be excluded at this point.

The tyrosine hydroxylase cDNA probe is also being used to study tyrosine hydroxylase nuclear gene organization and structure. This probe has also allowed us to isolate human tyrosine hydroxylase cDNA clones (J.F. Julien et al., unpubl.). These clones will permit us to study the biochemical genetics of the catecholamine system in humans.

ACKNOWLEDGMENTS

We thank C. Boni, B. Grima, J.F. Julien, A. Kahn, and J.L. Nahon for helpful discussions. We also thank J. Powell and S. Farrand for critical reading of the manuscript. This work was supported by grants from the Centre National de la Recherche Scientifique, the Institut National de la Santé et de la Recherche Médicale, and the Fondation pour la Recherche Médicale Française. A. Lamouroux received a fellowship from the Fondation pour la Recherche Médicale Française.

REFERENCES

Auffray, C. and F. Rougeon. 1980. Purification of mouse immunoglobulin heavy-chain messenger RNAs from total myeloma tumor RNA. *Eur. J. Biochem.* **107:** 303.

Black, I.B. and P.H. Patterson. 1980. Developmental regulation of neurotransmitter phenotype. *Curr. Top. Dev. Biol.* **15:** 27.

Buda, M., B. Roussel, B. Renaud, and J.F. Pujol. 1975. Increase in tyrosine hydroxylase activity in the locus coeruleus of the rat brain after contralateral lesioning. *Brain Res.* **93:** 564.

Edgar, D.E. and H. Thoenen. 1978. Selective enzyme induction in a nerve growth factor—Responsive pheochromocytoma cell line (PC 12). *Brain Res.* **154:** 186.

Feramisco, J.R., J.E. Smart, K. Burridge, D.M. Helfman, and G.P. Thomas. 1982. Co-existence of vinculin and a vinculin-like protein of higher molecular weight in smooth muscle. *J. Biol. Chem.* **257:** 11024.

Joh, T.H., C. Geghman, and D. Reis. 1973. Immunochemical demonstration of increase accumulation of tyrosine hydroxylase protein in sympathetic ganglia and adrenal medulla elicited by reserpine. *Proc. Natl. Acad. Sci.* **70:** 2767.

Lamouroux, A., N. Faucon Biguet, D. Samolyk, A. Privat, J.C. Salomon, J.F. Pujol, and J. Mallet. 1982. Identification of cDNA clones coding for rat tyrosine hydroxylase antigen. *Proc. Natl. Acad. Sci.* **79:** 3881.

Le Douarin, N. 1980. The ontogeny of the neural crest in avian embryo chimaeras. *Nature* **286:** 663.

Lindvall, O. and A. Björklund. 1978. Organization of catecholamine neurons in the rat central nervous system. In *Handbook of psychopharmacology* (ed. L.L. Iversen et al.), vol. 9, p. 139. Plenum Press, New York.

Mallet, J., A. Privat, D. Samolyk, A. Bérod, A. Keller, and J.F. Pujol. 1981. Synthèse *in vitro* de la tyrosine hydroxylase. *C.R. Hebd. Séances Acad. Sci. Ser. D* **292:** 455.

Reis, D., T.H. Joh, R.A. Ross, and V.M. Pickel. 1974. Reserpine selectivity increases tyrosine hydroxylase and dopamine-β-hydroxylase enzyme protein in central noradrenergic neurons. *Brain Res.* **81:** 380.

Rigby, P.W., M. Dieckmann, C. Rhodes, and P. Berg. 1977. Labeling deoxyribonucleic acid to high specific activity *in vitro* by nick-translation with DNA polymerase I. *J. Mol. Biol.* **113:** 237.

Thoenen, H., R.A. Mueller, and J. Axelrod. 1969. Trans-synaptic induction of adrenal tyrosine hydroxylase. *J. Pharmacol. Exp. Ther.* **169:** 249.

Thomas, P.S. 1980. Hybridization of denatured RNA and small DNA fragments transferred to nitrocellulose. *Proc. Natl. Acad. Sci.* **77:** 5201.

Zigmond, R.E. and A. Chalazonitis. 1979. Long-term effects of preganglionic nerve stimulation on tyrosine hydroxylase activity in the rat superior cervical ganglion. *Brain Res.* **164:** 137.

Zigmond, R.E., F. Schon, and L.L. Iversen. 1974. Increased tyrosine hydroxylase activity in the locus coeruleus of rat brain stem after reserpine treatment and cold stress. *Brain Res.* **70:** 547.

Cloning and Characterization of Rat-brain-specific Transcripts: Rare, Brain-specific Transcripts and Tyrosine Hydroxylase

D.M. CHIKARAISHI, M.H. BRILLIANT, AND E.J. LEWIS
Division of Molecular and Cellular Biology, National Jewish Hospital and Research Center/National Asthma Center, Denver, Colorado 80206

The sequence complexity of RNA from adult tissues of several rodents has been measured by saturation hybridization in which excess RNA is annealed to radioactive, single-copy DNA. In the rat (Shearer 1977; Chikaraishi et al. 1978; Grouse et al. 1978; Kaplan et al. 1978; Chikaraishi 1979; Lee et al. 1979; Wilkes et al. 1979; Beckmann et al. 1981; Kaplan and Finch 1982), mouse (Hahn and Laird 1971; Brown and Church 1972; Grouse et al. 1972; Bantle and Hahn 1976; Grady et al. 1978; Kuroiwa and Natori 1979; Van Ness et al. 1979), and rabbit (Brown and Church 1972; Church and Schultz 1974), the brain has consistently shown a higher complexity in both its nuclear and cytoplasmic RNAs compared with other tissues. Assuming asymmetric transcription, brain nuclear RNA possesses sequence complexity equivalent to between 31% (5.9×10^8 nucleotides in rat; Chikaraishi et al. 1978) and 42% (5.7×10^8 nucleotides in mouse; Bantle and Hahn 1976) of the single-copy coding capacity of the rodent genome. The levels of complexity of other adult tissues, including melanoma, lung, mammary gland, liver, kidney, spleen, and thymus, range from 10% to 26%, depending on the tissue (Fig. 1) (Chikaraishi et al. 1978; N. Sueoka et al., pers. comm.).

Unlike liver nuclear RNA, whose complexity is unchanged in newborn and adult animals, rat brain nuclear RNA complexity gradually increases from an estimated 21% in the newborn to the adult level of 31% at 2 months of age (N. Sueoka et al., pers. comm.). A parallel increase has been reported for the total poly(A)$^+$ RNA (Kaplan and Finch 1982).

The highly complex nuclear RNAs are present at low abundance (less than 1 copy/cell) and form a nested hierarchy such that the sequences expressed in a given tissue are also found in the nuclear RNA of any other tissue having a higher absolute complexity (see Fig. 1) (Chikaraishi et al. 1978; Beckmann et al. 1981; N. Sueoka et al., unpubl.). Therefore, by additive or sequential hybridization, the sequences in kidney nuclear RNA are also found in liver and those in liver are present in brain. This suggests an extensive overlap exists among the nuclear RNAs of various rat tissues, particularly among tissues like brain (31%) and melanoma (26%), which have close absolute complexities.

The distinguishing characteristic of brain RNA is its high polysomal complexity as measured by saturation hybridization. Both rat (Chikaraishi 1979) and mouse (Van Ness et al. 1979) brain polysomal RNAs have complexities of 15–20% of the single-copy DNA, equivalent to 120,000–180,000 average-sized mRNAs. We estimate the polysomal and cytoplasmic complexities of rat brain RNA to be 16% and 29%, respectively, of the unique-sequence coding capacity. We were surprised that half of the cytoplasmic complexity resided in nonpolyadenylated RNAs (Chikaraishi 1979), a finding that was also confirmed for mouse brain mRNAs (Van Ness et al. 1979). The vast majority of these complex poly(A)$^-$ mRNAs are not found among the poly(A)$^+$ mRNAs and may constitute a unique class of messages. That these poly(A)$^-$ RNAs actually function as genuine mRNAs is suggested by the fact that they can be released from large polysomes (>150S) by puromycin and by the fact that they can be translated in vitro with the same molar efficiency as poly(A)$^+$ mRNAs (Chikaraishi 1979). Hahn et al. (1982) have shown that 90% of these poly(A)$^-$ mRNAs are found only in the brain and may be postnatally activated in the mouse.

To determine if the high sequence complexity in brain results from high complexity in every brain cell or from the summation of lower, incompletely overlapping complexities from different cells, we investigated clonal cell lines and various brain sections (Table 1). The nuclear RNA complexities of the brain-derived lines B103 and B27 (Schubert et al. 1974) were 29.0% and 27.0%, respectively, whereas that of a peripheral neurotumor line RT4-D (Imada and Sueoka 1978) was 22.2% (Beckmann et al. 1981). Within the error of the measurement, all of the brain sections had nuclear complexities equal to that of whole brain. Therefore, with the exception of the peripheral nervous system line, the cell lines and brain sections contain at least 85% of the nuclear sequences found in total brain. However, B103 and B27 had polysomal mRNA complexities of 6.4% and 7.8%, respectively, whereas total brain complexity was 16%. Cytoplasmic RNA values also reflected the same difference, with B103 and B27 equal to 6.2% and 8.6%, respectively, whereas brain was equal to 20%. As shown in Table 1, some of the brain sections had reduced poly(A)$^+$ cytoplasmic complexity, with the cerebellum having about half the complexity of whole brain. In general, these data support the view that in the brain most cells make a very similar set of nuclear RNAs from which a limited subset of mRNAs are selected that vary between cell types. The low concentration (0.15 copy/cell) estimated for the average complex mRNA may reflect its complete absence as mRNA in

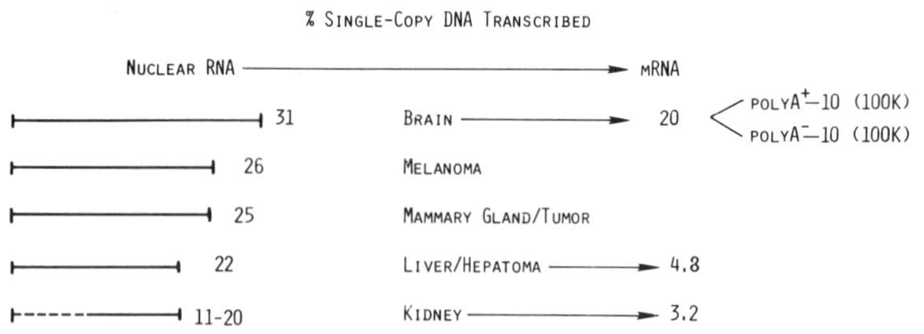

Figure 1. RNA sequence complexity of adult rat tissues. Nuclear RNA complexities are shown as lines representing the percentages of single-copy DNA transcribed, assuming asymmetric transcription. The kidney nuclear value may be higher than that originally published and is shown as a dotted line. Overlapping lines reflect sequences held in common between nuclear RNAs. Values given in parentheses are the number of different mRNAs estimated by assuming an average mRNA is 2000 nucleotides in length. Data are from Chikaraishi et al. (1978), Chikaraishi (1979), and N. Sueoka et al. (pers. comm.).

most cells, although it may be quite abundant in other cells. Therefore, the mechanism by which posttranscriptional processing of mRNAs appropriate for a given cell type is accomplished may be an important factor in regulating neural gene expression.

Saturation hybridization overestimates complexity if a significant number of transcribed genes have homologous pseudogenes in the genome. By definition, a pseudogene would not be transcribed into functional mRNA, but, by close homology with a functional gene, would cross-anneal with mRNA and be counted as "transcribed." Another limitation of complexity measurements is that they characterize entire populations of diverse molecules and therefore reflect only the overall pattern of transcription. To determine how well individual RNA species fit the pattern, we have prepared a single-copy library of rat DNA in the single-stranded phage M13, and individual clones selected at random have been tested for their presence in brain RNA. To compare with these random clones (for which there is no known function), we have prepared a partial cDNA clone to tyrosine hydroxylase, a gene whose expression is limited to catecholamine-containing neurons and to certain cells derived from the neural crest and whose product is the rate-limiting enzyme in catecholamine synthesis.

MATERIALS AND METHODS

Materials, cDNA clones, antisera, and cell lines. Restriction enzymes were purchased from Bethesda Research Laboratories or New England BioLabs; T4 polynucleotide kinase was obtained from P-L Biochemicals or New England Nuclear; and S1 nuclease was purchased from P-L Biochemicals or Bethesda Research Laboratories. [γ-^{32}P]ATP (9000 c/mmole) was prepared by a modification of the technique of Johnson and Walseth (1979). Other isotopes were purchased from New England Nuclear.

cDNA clones for rat β-casein, rat albumin, and a rat prostate-specific glycoprotein were generously provided by Dr. J. Rosen (Baylor College of Medicine, Houston, Texas), Dr. J. Boulter (University of California, San Diego, Calif.), and Dr. M. Parker (Imperial Cancer Research Fund, London), respectively. These cDNA clones in pBR322 were recloned into M13mp7 for use in S1 nuclease transcription mapping.

Rabbit antiserum to tyrosine hydroxylase (Meligeni et al. 1981) was generously provided by Dr. A.W. Tank and Dr. N. Weiner (University of Colorado, Denver, Colo.). A sheep antiserum to rat phenylalanine hydroxylase (Shiman et al. 1979) was fractionated and a purified immunoglobulin fraction (Steinbuch and

Table 1. Percentage of Single-copy DNA Transcribed, Assuming Assymetric Transcription

Tissue	Nuclear RNA	mRNA	Poly(A)$^+$ cytoplasmic RNA
Brain			
Anterior cerebral cortex	32		7.6
Parietal cerebral cortex	32		7.2
Occipital cerebral cortex	29		10.1
Regions underlying cortex	31		10.1
Cerebellum	30		4.6
Whole brain	31	20.0	10.1
Cell lines			
B103	29.0	6.1	
B27	27.0	8.6	
RT4-D	22.2		

Data are from Beckmann et al. (1981).

Audran 1969) was used for the experiments described. This was the kind gift of Dr. R. Shiman (Pennsylvania State University, Hershey, Penna.).

The RT4-D and RT4-B8 cells were provided by Dr. N. Sueoka (University of Colorado, Boulder, Colo.), (Imada and Sueoka 1978); Dr. A.W. Tank and Dr. N. Weiner provided the PC8b cells, and Dr. R. Goodman (Children's Hospital, Philadelphia, Penna.) provided the PC-G2 and PC-GIS cells. The B27 and B103 lines were from Dr. D. Schubert (Salk Institute, San Diego, Calif.) (Schubert et al. 1974), and the rat melanoma was provided as a transplantable tumor (Iglisias 1960) carried in AC rats.

Preparation of RNAs. Total RNA from cells or tissue was extracted with guanidinium thiocyanate and purified by centrifugation through 5.7 M CsCl as described (E.J. Lewis et al., in prep.). Polysomal RNA was prepared as described (Sala-Trepat et al. 1978).

Preparation and screening of the single-copy rat DNA library in M13. Total rat DNA was digested to completion with the restriction endonuclease *Mbo*I. Unique-sequence DNA was isolated from the total digest by three cycles of reassociation to $C_0t = 200$ followed by hydroxyapatite chromatography. The unique-sequence DNA (~200–400 bases in length) was ligated into the *Bam*HI site of the replicative form of the bacteriophage M13mp7, and recombinant (clear) plaques were selected from indicator plates. Single-stranded DNA from individual, recombinant M13mp7 clones was prepared as described (Harrington and Chikaraishi 1983) and digested with *Sau*3A, which restricts at the same site as *Mbo*I. The rat insert is preferentially cleaved out because it is flanked by a 14-bp duplex formed by the symmetrical sequence between the *Bam*HI site and the *Eco*RI site in mp7, which includes the *Sau*3A site. The restricted DNA was labeled with $[\gamma\text{-}^{32}P]$ATP and polynucleotide kinase either by the forward reaction (Maxam and Gilbert 1980) or by the exchange reaction (Berkner and Folk 1980), which we have found to be simpler and more reproducible. The labeled DNA is a mixture of M13 fragments and the rat insert fragment that has an estimated specific activity of between 1×10^7 to 5×10^7 cpm/μg.

Transcription mapping on single-stranded DNA. Labeled, single-stranded DNA (2–5 ng) and the indicated amount of cold RNA were first boiled and then annealed in 2× SSC (1× SSC = 0.15 M NaCl, 0.015 M Na citrate, pH 7.0) for 12–18 hours at 65°C in a final volume of 50 μl overlaid with mineral oil. After hybridization, the mixture was diluted to 350 μl with cold S1 buffer (0.3 M NaCl, 0.03 M Na acetate, pH 4.5, 3 mM $ZnCl_2$, 50 mM NaF) and 200–500 units/ml of S1 added. The digestion was incubated at 37°C for 1 hour and terminated with addition of 50 μl of 2.5 M NH_4 acetate, 50 mM EDTA. After transfer to a fresh tube, the digestions were extracted twice with phenol and once with $CHCl_3$-isoamyl alcohol. Twenty micrograms of carrier *Escherichia coli* tRNA were added to the aqueous supernatant, and the nucleic acids were precipitated with ethanol. The size of the pelleted S1-resistant DNA was resolved by electrophoresis on 8% polyacrylamide–8 M urea gels in Tris-borate (pH 8.3) as described (D.M. Chikaraishi et al., in prep.).

Tyrosine hydroxylase cDNA clone. A partial cDNA clone containing sequences homologous to tyrosine hydroxylase (TH) mRNA was selected as described (E.J. Lewis et al., in prep.). In brief, a cDNA library was made from poly(A)$^+$ RNA extracted from a subclone of the PC12 cell line (PC8b) that had been treated with dexamethasone and 8-bromo cyclic AMP (cAMP), agents known to increase TH enzyme activity in PC8b cells. Double-stranded cDNA was inserted into the *Pst*I site of pBR322 using GC tailing, and the recombinant plasmids were used to transform *E. coli* strain HB101. Colonies containing plasmids with inserted sequences were initially screened by DNA dot hybridization with cDNA preparations enriched in TH sequences from induced cells and cDNAs from uninduced cells. Plasmid DNAs that preferentially hybridized cDNA sequences from the induced cells were used to select TH mRNA. The hybrid-selected RNA was translated in vitro, and TH was identified by immunoprecipitation of a 60-kD translation product with antiserum against TH. One plasmid, pTH.4, was identified as containing a 400-bp sequence complementary to TH mRNA.

For transcription mapping, the TH sequences in pBR322 were excised with *Pst*I and *Eco*RI and recloned in the *Eco*RI and *Pst*I sites of M13mp8 and M13mp9. An additional 0.75 kb of pBR322 sequences from the *Pst*I site to the *Eco*RI site were cloned into M13 along with the TH sequences. By RNA dot hybridization (Thomas 1980), it was determined that the TH insert in M13mp8 annealed to TH mRNA, whereas the opposite orientation of TH sequences in M13mp9 did not.

Immunoprecipitation and in vitro translation. Immunoprecipitation with whole *Staphylococcus aureus* bacteria as adsorbent following in vitro translation of total RNA in rabbit reticulocyte lysates (Pelham and Jackson 1976) was as described by E.J. Lewis et al. (in prep.).

Southern blot hybridization. Southern blot hybridization of restricted genomic DNA was as described (D.M. Chikaraishi et al., in prep.) except 20 μg of DNA was used in each lane. After hybridization in 3× SSC, 0.2% bovine serum albumin, 0.2% Ficoll, 0.2% polyvinylpyrrolidone, 0.1% SDS, and either 50 μg/ml denatured salmon sperm DNA or 100 μg/ml denatured *E. coli* DNA for 12–18 hours at 65°C, the filters were washed for 15 minutes in 2× SSC at room temperature, 2× SSC at 60°C, 1× SSC at 60°C, 0.5× SSC at 60°C, and twice in 0.2× SSC at 60°C.

RESULTS

Screening of Clones Corresponding to Rare-abundance RNA

Based on RNA complexity measurements, 15% of the single-copy DNA clones should be represented as brain nuclear RNA. However, these complex transcripts are so rare in abundance (0.1–1.0 copy/cell, equivalent to 10^{-6} to 10^{-7} of the total RNA per cell) that their detection is impossible using techniques employed for screening cDNA clones. After investigating a number of techniques, the S1 transcription mapping method of Berk and Sharp (1978) was modified as diagramed in Figure 2A to be sufficiently sensitive to detect RNAs present at 0.05–0.1 copy/cell. In brief, restricted, end-labeled, single-stranded cloned DNA was annealed to total cellular RNA (~50 μg). The cloned insert DNA was in excess compared with the estimated abundance of the homologous RNA used in the annealing, such that the rate of hybridization was dependent on the concentration of probe DNA. Those rat DNA fragments hybridized with RNA were resistant to digestion by the single-strand-specific S1 nuclease and were resolved by gel electrophoresis. To produce a detectable band, at least 30 nucleotides of DNA, including the end-labeled nucleotide, must be engaged in an S1-resistant duplex with RNA. Short stretches of DNA (10–20 nucleotides in length), which might adventitiously anneal, resulting in high nonspecific backgrounds in RNA dot or Northern blot hybridization experiments, are not scored by S1 transcription mapping.

In reconstruction experiments, in which known amounts of β-casein or albumin mRNA were annealed to their respective cDNA clones in M13, mRNA levels equivalent to 0.1 copy/cell were detected following a 3-day exposure of the resulting gel to X-ray film.

Characterization of Clones Representing Rare-abundance RNA

Our preliminary screen indicated that 8 of 50 (15%) of the randomly selected, single-copy clones annealed to total brain RNA, close to the predicted value of 15.5%. (The complexity value of 31% for brain nu-

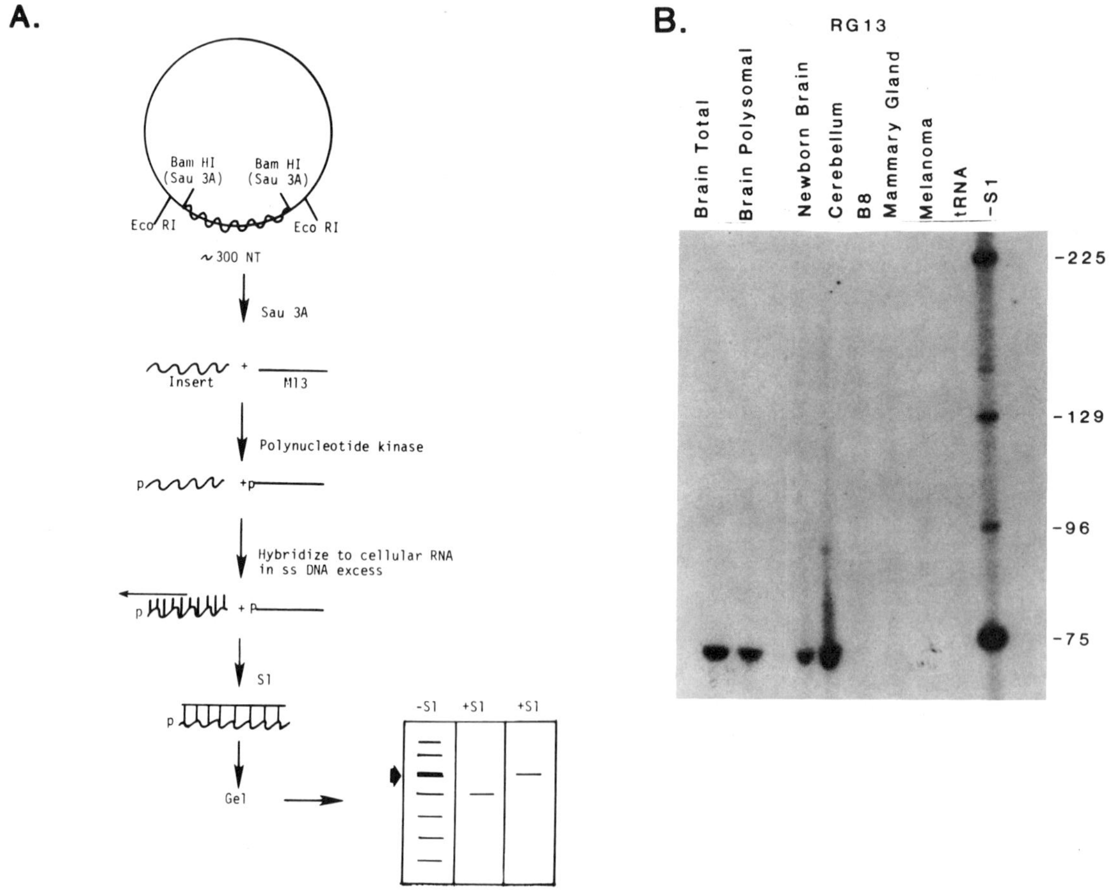

Figure 2. Transcription mapping using single-copy DNA clones in M13. (*A*) Schematic diagram of the S1-transcription-mapping technique. The resulting gel depicts from left to right the bands obtained with (1) no digestion of the labeled DNA probe where the arrow indicates the rat insert fragment, (2) S1 digestion of a RNA-DNA hybrid in which the rat DNA fragment contained some sequences not transcribed into RNA and hence the S1-resistant band is shorter than the original insert band, and (3) S1 digestion of an RNA-DNA hybrid in which all the DNA sequences were present in the RNA. (*B*) Transcription mapping of clone rg13 using 50 μg of total RNA extracted from the indicated source, as described in Materials and Methods. B8 is a cell line derived from the RT4 neurotumor family (Imada and Sueoka 1978). Sizes of labeled fragments are given in nucleotides.

clear RNA was based on double-stranded DNA and must be halved since only one strand of the DNA is present in the M13 clones.) Because total RNA includes nuclear RNA, total RNA reflects the complexity of nuclear RNA.

Figure 2B depicts the hybridization of one of the randomly selected, single-copy DNA clones with various RNAs. This clone, called rg13 (rg for "rat genomic"), appears to be brain specific, giving an S1-resistant band, 75 nucleotides in length, when annealed with total or polysomal RNA from brain, whereas no band was detected when rg13 was annealed to melanoma, mammary gland, liver, and kidney total RNAs. It is, however, transcribed in cerebellum and newborn brain, but not in the peripheral neurotumor line RT4-B8.

In the preliminary analysis of the other transcribed clones (see Fig. 3), six of the eight clones represent polysomal RNAs in brain, consistent with the observation that about two-thirds of the brain nuclear RNA sequences are cytoplasmic, whereas half ($^{3}/_{6}$) of the clones expressed in liver nuclear RNA are on liver polysomes. Some clones (rg3, rg36, and rg51) are on the polysomes of both liver and brain (the only two tissues assayed for polysomal RNA) and may encode common "housekeeping" proteins; others (rg18 and 26) are only found in nuclear RNA and are not mRNA in either brain or liver. One clone, rg34, appears to be present in the nuclear RNA of both tissues, but polysomal only in brain, and may qualify as encoding a brain-specific protein. Therefore, rg34 is of interest in that its sequences may be posttranscriptionally selected for messenger expression.

In line with the nested model for nuclear RNAs, most ($^{6}/_{8}$) of the clones are expressed in all tissues, the exceptions being rg13, which is found only in brain, and

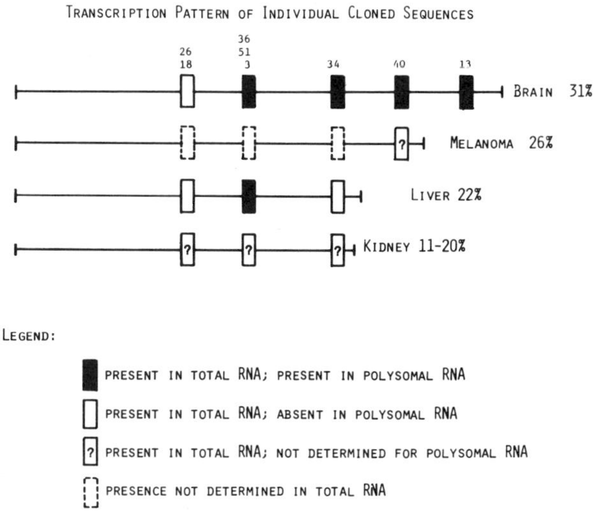

Figure 3. Transcriptional patterns of genomic clones encoding rare RNA. The lines represent the overlapping nuclear RNA complexities of various tissues. The boxes (not drawn to scale) represent individual transcribed clones, with the number of the clone above the box. (■) Present in total RNA and in polysomal RNA; ([?]) present in total RNA, absent in polysomal RNA; (□) present in total RNA, not determined for polysomal RNA; ([ ]) presence not determined in total RNA.

rg40, transcribed only in brain and melanoma. Since we do not know if rg40 is polysome associated in melanoma, it, too, may be brain specific for messenger expression. In accord with their higher complexity values, brain and, to a lesser extent, melanoma, would be expected to express sequences, such as rg13 and rg40, not shared with other tissues.

The limited spectrum of single-copy DNA clones we have so far investigated fit the pattern predicted by the overall complexity measurements and encode RNAs that fall into the rare-abundance class (0.5–10 copies/cell). These, however, may be quite different than the well-characterized sequences corresponding to abundant tissue-specific RNAs. For example, using cDNA clones specific for β-casein, albumin, or a prostate-specific glycoprotein, we could not detect at the level of 0.1 copy/cell (data not shown) these RNAs in any tissue other than that appropriate for the gene product, suggesting that these genes may not participate in the nested hierarchy predicted for the rare RNAs.

Isolation of a cDNA Clone for TH mRNA

In addition to identifying clones representing the large number of rare brain mRNAs, whose function is unknown, we have begun cloning known nerve-specific genes. To this end, we have prepared a partial cDNA clone to rat TH (E.J. Lewis et al., in prep.), the rate-limiting enzyme in catecholamine biosynthesis. TH, which converts tyrosine to dihydroxyphenylalanine (DOPA), is found only in catecholamine-producing neurons and the adrenal medulla. The cloning of TH was facilitated by the availability of clonal cell lines derived from pheochromocytomas (tumors of the adrenal medulla), which contain high levels of TH mRNA (Baetge et al. 1981; Thibault et al. 1981). Poly(A)$^+$ RNA was prepared from one such line, PC8b, a derivative of the PC12 line (Greene and Tischler 1976), after the cells had been induced by simultaneous treatment with 8-bromo cAMP and dexamethasone, under conditions in which both inducers increase TH mRNA activity in an additive manner. Approximately 1% of the mRNA activity, assayed by in vitro translation and immunoprecipitation, was specific for TH. Double-stranded cDNA was prepared from this enriched RNA and cloned into the *Pst*I site of pBR322 by GC tailing. A TH cDNA clone was selected by differential hybridization and by hybrid-selected translation as described by E.J. Lewis et al. (in prep.).

Our cDNA insert is approximately 400 nucleotides in length and represents 20% of the TH mRNA sequences. By Northern analysis, the length of TH mRNA is about 1800 nucleotides, although minor, presumably precursor, RNAs can be detected up to 6 kb in length (E.J. Lewis et al., in prep.). In addition, we have isolated a number of genomic clones, spanning at least 20 kb and containing sequences homologous to the original cDNA, that will be used in determining the nature of these precursor transcripts as well as the genomic organization of the TH gene.

TH mRNA Levels

To quantitate TH RNA sequences by S1 transcription mapping, we recloned the TH cDNA into the single-stranded vectors M13mp8 and M13mp9, which accommodate opposite strands of the TH cDNA. As expected, only one strand of the TH DNA, that in mp8, annealed to TH mRNA as determined by RNA dot hybridization (data not shown). This fact also allows determination of the direction of transcription with respect to the cDNA.

Figure 4 is a transcription map in which the TH-mp8 recombinant DNA was cleaved with the restriction enzyme *Sau*3A, end-labeled, and annealed to various cellular RNAs. Multiple TH-specific bands reflect multiple *Sau*3A sites within the TH sequences and the inability of S1 to cleave homopolymeric stretches created by the GC tailing during the original cloning.

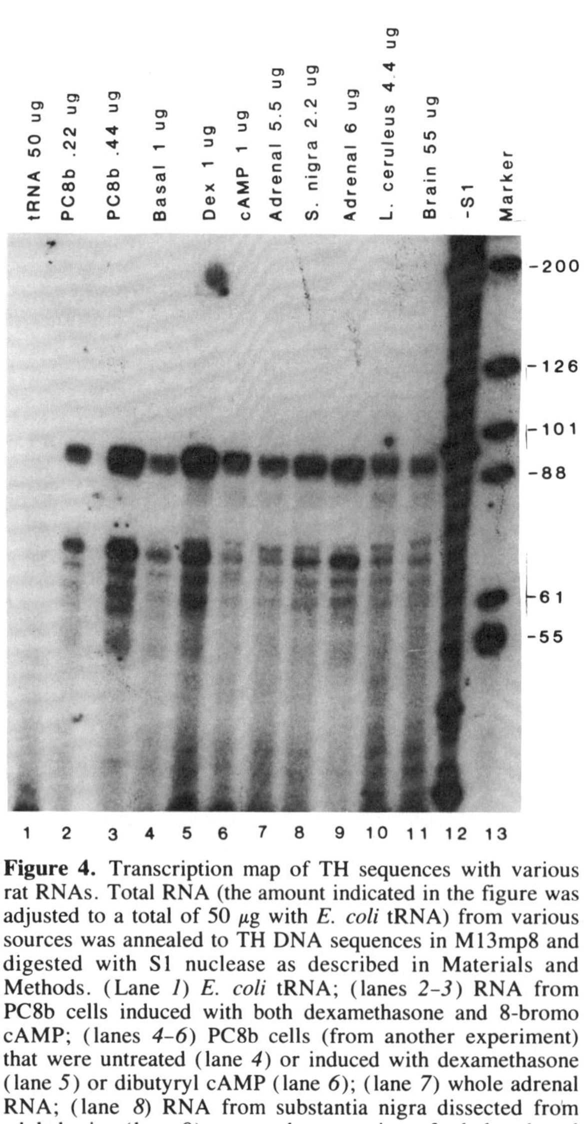

Figure 4. Transcription map of TH sequences with various rat RNAs. Total RNA (the amount indicated in the figure was adjusted to a total of 50 μg with *E. coli* tRNA) from various sources was annealed to TH DNA sequences in M13mp8 and digested with S1 nuclease as described in Materials and Methods. (Lane *1*) *E. coli* tRNA; (lanes *2–3*) RNA from PC8b cells induced with both dexamethasone and 8-bromo cAMP; (lanes *4–6*) PC8b cells (from another experiment) that were untreated (lane *4*) or induced with dexamethasone (lane *5*) or dibutyryl cAMP (lane *6*); (lane *7*) whole adrenal RNA; (lane *8*) RNA from substantia nigra dissected from adult brain; (lane *9*) a second preparation of whole adrenal RNA; (lane *10*) RNA from the locus coeruleus dissected from adult brain; (lane *11*) whole Sprague-Dawley brain RNA: (lane *12*) no S1 digestion of probe; (lane *13*) marker; ^{32}P-labeled SV40 DNA fragments generated by *Eco*RII digestion. Lengths are given in nucleotides.

The PC8b cell line contains a higher level of TH mRNA than the other tissues assayed since 5-fold more adrenal RNA and 50-fold more brain RNA were needed to yield a comparable signal. When the PC8b cells are exposed to dexamethasone (1 μM for 48 hr), there is a substantial increase in TH mRNA (lane 5) whereas treatment with dibutyryl cAMP (1 mM for 48 hr) gives a more modest induction (lane 6).

Total brain has a low overall concentration of TH RNA, which is substantially increased in RNA extracted from localized areas such as the substantia nigra or locus coeruleus, which are known to have high levels of catecholamine synthetic enzymes. In contrast, no detectable TH RNA (at the level of 0.1 copy/cell) was found in RNA extracted from liver, prostate, mammary gland, or RT4-B8 peripheral neurotumor cells (data not shown).

PC8b, adrenal, and brain RNA give the same sized S1-resistant bands, suggesting that these tissues have TH RNAs that are probably identical in the region homologous to the end-labeled probe. In addition, by Northern analysis, both adrenal and cell-line RNA have the same size bands (E.J. Lewis et al., in prep.). However, further analysis using different probes is necessary to exclude tissue-specific differences among the various TH mRNAs.

Immunoprecipitation of TH by Antiserum to Phenylalanine Hydroxylase

Preliminary data (Joh et al. 1981; Ross et al. 1982) suggested that three of the four enzymes involved in catecholamine biosynthesis may be structurally related. From cross-reactivity of antisera and similarities in amino acid compositions of peptide fragments from TH, dopamine β-hydroxylase, and phenylethanolamine-*N*-methyltransferase, it has been suggested that these enzymes share common protein domains and may belong to a multigene family.

Antibodies to purified rat liver phenylalanine hydroxylase (PH), which converts phenylalanine to tyrosine, cross-react with what appears to be TH, suggesting that PH may also be related to the three catecholamine biosynthetic enzymes. Figure 5 shows that PH antibodies immunoprecipitate a 60-kD protein from the in vitro translation products programmed by PC12 RNA that comigrates with that immunoprecipitated by TH antiserum. If the PC12 RNA translation products are first precipitated with TH antiserum, no precipitation of the 60-kD protein can be detected upon subsequent incubation of the supernatant with PH antiserum (data not shown). As shown in Figure 5, the 60-kD band precipitated by PH antibodies is actually stronger than the homologous precipitation of PH from in vitro translation of liver RNA (lane 3). This, however, could be because the percentage of TH-specific mRNA in PC12 cells is higher than the per-

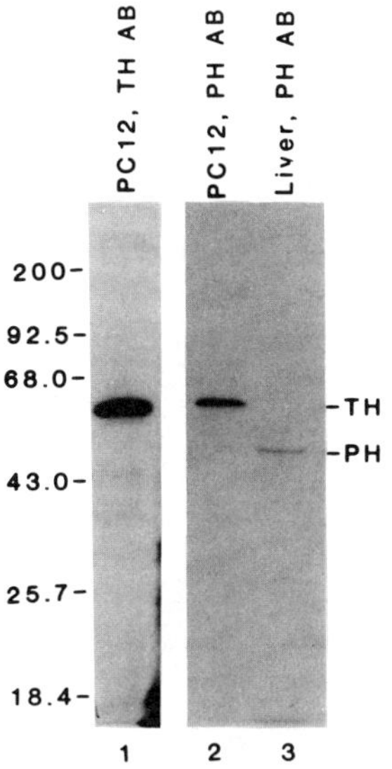

Figure 5. Immunoprecipitation of in vitro translation products with antiserum to TH and PH. Translations and immunoprecipitations were performed as described in Materials and Methods. Total RNA from PC12 cells was translated and the product was immunoprecipitated with antiserum against TH (lane *1*) or with antibodies against PH (lane *2*). Total liver RNA was translated and the products immunoprecipitated with antibodies against PH (lane *3*). Molecular-weight markers were run in a separate lane, and their sizes in kilodaltons are indicated.

centage of PH mRNA in liver cells (Robson et al. 1982).

Southern Blot Analysis of Rat TH DNA Sequences

Figure 6 (lanes 1–4) shows that the pTH.4 sequences anneal to a single band in rat DNA that has been digested with various restriction enzymes, fractionated on agarose gels, and transferred to nitrocellulose by the method of Southern (1975). A single band at 22 kb was also observed using *Eco*RI-digested DNA from rat liver, brain, RT4-D neurotumor cells, PC8b, and the PC-G2 and PC-G1S lines (Goodman et al. 1979), which are independent clonal isolates from a primary pheochromocytoma (data not shown). Since the intensity of the bands was equal in all cases, there is no evidence of rearrangement or gene amplification in any of the DNAs examined.

We presume the bands in Figure 6, lanes 1–4, are genomic TH sequences homologous to those in pTH.4 and have cloned a 9.5-kb *Eco*RI–*Bam*HI fragment corresponding to that shown in lane 3 to confirm its identity as part of the TH gene.

The Southern analyses shown in lanes 1–4 were performed under conditions that would select against poorly matched duplexes, using salmon sperm DNA as a competitor to preempt nonspecific sites. When, however, the Southern blot hybridizations were washed less stringently (i.e., 0.5× SSC at 55°C instead of 0.2× SSC at 60°C) and a prokaryotic DNA (*E. coli*) was used as competitor (rather than salmon sperm DNA), many minor bands could be observed (lanes 5 and 6). Lanes 7 and 8 show the annealing to *Bam*HI-digested calf thymus and salmon sperm DNAs, respectively. Although there is little cross-annealing to bovine DNA, the rat TH homologous sequences appear to be ubiquitous in the salmon sperm genome and probably cross-hybridize to a repetitive DNA element.

DISCUSSION

Rat DNA Clones Corresponding to Rare RNAs

Using a modified S1-mapping technique, we have identified single-copy genomic clones that are homologous to RNAs belonging to the high-complexity, rare-abundance class. These clones differ from those obtained by cDNA cloning, which selects sequences encoding high-abundance and low-complexity RNAs.

Eight of 50 randomly selected unique-sequence clones encode rare transcripts present in the nuclear RNA of brain and other tissues. Their frequency and pattern of expression among various tissues follows that determined for the overall pattern of complex transcripts suggested by earlier complexity measurements. This is in contrast to the expression of high-abundance, tissue-specific genes (β-casein, albumin, and prostate glycoprotein) that are found only in their appropriate tissues and do not follow the nested hierarchy predicted for the rare nuclear RNAs.

At least one of the clones we have investigated is expressed only in brain and may be particularly valuable for investigating neural gene expression, since the mammalian brain expresses 150,000 different mRNAs, almost all of which fall into the rare-abundance class. Neural-specific clones may represent RNAs expressed at low levels in all neural cells, or their low average copy number may reflect high abundance in a few cells but total absence in most cells, as may be the case for TH. To distinguish between these two possibilities, the presence of these sequences is being assayed in RNA extracted from various regions of the brain, from preparations of separated glia and neurons, and from various neural cell lines. Eventually, precise localization may be possible by using in situ hybridization, as has been described for pro-opiomelanocortin (Gee et al. 1982).

The nature of the proteins encoded by the 150,000 different brain mRNAs is unknown, as are their functions. Grouse et al. (1978, 1980) have suggested that the number of complex RNAs increases with environmental stimuli and may constitute changes due to learning and memory. The availability of clones may help to identify the proteins encoded by these complex mRNAs. However, techniques used with cDNA clones, such as hybrid-selected translation and two-dimensional electrophoresis, are not yet feasible with RNAs present at levels as low as 10^{-6} to 10^{-7} of the total RNA. Instead, peptides could be synthesized, based on DNA sequence, that may be used to elicit antiserum that

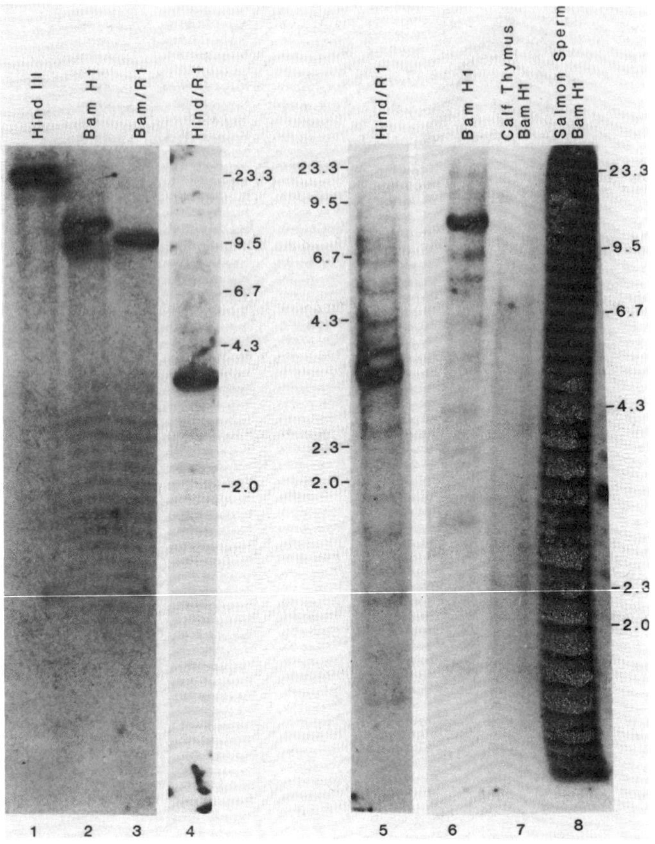

Figure 6. Southern blot hybridization probed with nick-translated pTH.4 cDNA. Sprague-Dawley rat liver DNA (lanes *1–6*) was digested with the restriction enzymes indicated above each lane; calf thymus DNA (lane *7*) and salmon sperm DNA (lane *8*) were digested with *Bam*HI. The DNA was fractionated on 1.1% agarose gels, transferred to nitrocellulose filters, and annealed as described in Materials and Methods. Lanes *1–4* were preannealed and annealed with denatured salmon sperm DNA (50 μg/ml) as competitor, and those blots were washed as indicated in Materials and Methods. Lanes *5–8* were preannealed and annealed in the presence of 100 μg/ml denatured *E. coli* DNA as competitor and washed at 55°C as described in Materials and Methods, except that the 0.2 × SSC washes were omitted. Phage λ DNA restricted with *Hin*dIII was run as marker on a parallel lane in each gel; sizes are given in kilobase pairs.

cross-reacts with the natural protein, as described for prosomatostatin (Lechan et al. 1983). Possibly, some of these proteins may fall into multigene families, which may be discerned by Southern blot analysis.

TH RNA

Using a partial cDNA clone to TH, we have assayed TH-specific RNA in various tissues and cells by S1 mapping as well as RNA dot and Northern blot analysis (E.J. Lewis et al., in prep.).

The overall concentration of TH RNA in brain is on the order of 1–10 copies/cell, a figure estimated by comparison to the signal obtained between a known amount of specific mRNA and its homologous probe. TH RNA is enriched at least 10-fold in RNA from the substantia nigra or locus coeruleus and is about 50-fold higher in the PC8b cell line when induced by the synthetic glucorticoid, dexamethasone, or by cAMP analogs.

Possible Multigene Family That Includes TH and PH

Others have suggested that TH, dopamine β-hydroxylase, and phenylethanolamine-*N*-methyltransferase are members of a multigene family based upon cross-reacting antisera and partial peptide mapping (Joh et al. 1981; Ross et al. 1982; O'Malley et al. 1983). Using specific antibodies against PH, a 60-kD protein that comigrates with TH was immunoprecipitated from translation products programmed with PC12 RNA. This precipitation is precluded if the translation products are first precipitated with anti-TH serum, suggesting that both sera react with the same protein. We were unable, however, to show that TH antiserum precipitated PH from liver RNA-programmed translations.

Because the PH antigen was prepared from rat liver, it is unlikely to be contaminated with even trace amounts of TH, since TH has never been detected in liver. Therefore, the reactivity of anti-PH antibodies with TH cannot be accounted for by contamination of the original antigen, which could be a problem with TH, dopamine β-hydroxylase, and phenylethanolamine-*N*-methyltransferase, which are often prepared from the same tissue.

Because TH and PH hydroxylate by similar enzymatic mechanisms and require the same tetrahydropteridine cofactor, it seems possible that both proteins might share structural similarities, which might also be shared with tryptophenyl hydroxylase. However, positive identification of the 60-kD protein as TH requires peptide mapping and enzymatic analysis. In this regard, Friedman et al. (1972) showed that PH antiserum blocked TH enzyme activity, although their antiserum was unable to precipitate TH from extracts of rat brain or adrenal. In addition, Robson et al. (1982) suggested that nucleic acid homology might exist between PH and TH mRNAs by demonstrating that PH cDNA annealed to rat adrenal RNA, which is not known to express PH, but does have TH RNA. However, the size of the cross-hybridizing RNA (21S) was significantly larger than that measured for TH (18S) (Lamouroux et al. 1982; E.J. Lewis et al., in prep.).

By Southern analysis, using *E. coli* DNA as competitor, multiple rat genomic DNA sequences cross-anneal to TH cDNA under conditions that favor the detection of weak (imperfect or short) hybrids. In contrast, with salmon sperm DNA as competitor during hybridization, only a single, strong band containing TH-homologous sequences was observed. The substitution of *E. coli* DNA in place of salmon sperm DNA was important in facilitating detection of minor bands since salmon sperm DNA (but not mammalian DNA) has sequences (probably repetitive elements) that anneal to TH cDNA (Fig. 6, lane 8).

It is unclear if the minor TH bands in rat DNA are adventitious, cross-annealing between functionally unrelated sequences, or if they define members of a TH multigene family. The availability of a full-length TH cDNA clone and genomic TH DNA clones may help decide this question; however, cross-hybridization of the available rat cDNA clones for TH (Lamouroux et al. 1982; E.J. Lewis et al., in prep.), for PH (Robson et al. 1982), and for dopamine β-hydroxylase (O'Malley et al. 1983) may provide a direct answer.

If PH, TH, dopamine β-hydroxylase, and phenylethanolamine-*N*-methyltransferase do comprise a multigene family, it would be unique in that the members of this family catalyze sequential steps along a single biosynthetic pathway.

ACKNOWLEDGMENTS

We would like to thank Drs. A.W. Tank, N. Weiner, and R. Shiman for antiserum and helpful discussions, and Dr. J. Masserano for the dissected brain tissues. We want to thank Ms. D.M. Krzemien for excellent technical assistance and Ms. J. Derry, Ms. C. Breibart, and Ms. N. de Stackleberg for aid in preparation of the manuscript. We would like to acknowledge Dr. C. Harrington, Mr. H. Tomasiewicz, and Dr. N. Sueoka for stimulating discussions. This work was supported by National Institutes of Health grant GM-29302 to D.M.C.

REFERENCES

Baetge, E.E., B.B. Kaplan, D.J. Reis, and T.H. Joh. 1981. Translation of tyrosine hydroxylase from poly(A)-mRNA in pheochromocytoma cells is enhanced by dexamethasone. *Proc. Natl. Acad. Sci.* **78:** 1269.

Bantle, J.A. and W.E. Hahn. 1976. Complexity and characterization of polyadenylated RNA in mouse brain. *Cell* **8:** 139.

Beckmann, S.L., D.M. Chikaraishi, S.S. Deeb, and N. Sueoka. 1981. Sequence complexity of nuclear and cytoplasmic ribonucleic acids from clonal neurotumor cell lines and brain sections of the rat. *Biochemistry* **20:** 2684.

Berk, A.J. and P.A. Sharp. 1978. Spliced early mRNA of simian virus 40. *Proc. Natl. Acad. Sci.* **75:** 1274.

Berkner, K.L. and W.R. Folk. 1980. Polynucleotide kinase exchange as an assay for class II restriction endonucleases. *Methods Enzymol.* **65:** 28.

Brown, I.R. and R.B. Church. 1972. Transcription of nonrepeated DNA during mouse and rabbit development. *Dev. Biol.* **29:** 73.

Chikaraishi, D.M. 1979. Complexity of cytoplasmic polyadenylated and nonpolyadenylated rat brain RNAs. *Biochemistry* **18:** 3249.

Chikaraishi, D.M., S. Deeb, and N. Sueoka. 1978. Sequence complexity of nuclear RNAs in adult rat tissues. *Cell* **13:** 111.

Church, R.B. and G.A. Schultz. 1974. Differential gene activity in the pre- and postimplantation mammalian embryo. *Curr. Top. Dev. Biol.* **8:** 179.

Friedman, P.A., T.G. Lloyd, and S. Kaufman. 1972. Production of antibodies to rat liver phenylalanine hydroxylase; cross-reactivity with other pterin-dependent hydroxylases. *Mol. Pharmacol.* **8:** 501.

Gee, C., J.L. Roberts, and S.J. Watson. 1982. Application of cellular level hybridization histochemistry to the study of neuroendocrine gene expression. *Soc. Neurosci. Abstr.* **8:** 703.

Goodman, R., C. Chandler, and H.R. Herschman. 1979. Pheochromocytoma cell lines as models of neuronal differentiation. *Cold Spring Harbor Conf. Cell Proliferation* **6:** 653.

Grady, L.J., A.B. North, and W.D. Campbell. 1978. Complexity of poly (A^+) and poly (A^-) polysomal RNA in mouse liver and cultured mouse fibroblasts. *Nucleic Acids Res.* **5:** 697.

Greene, L.A. and A.S. Tischler. 1976. Establishment of a noradrenergic clonal line of rat adrenal pheochromocytoma cells which respond to nerve growth factor. *Proc. Natl. Acad. Sci.* **73:** 2424.

Grouse, L., M.D. Chilton, and B. McCarthy. 1972. Hybridization of ribonucleic acid with unique sequences of mouse deoxyribonucleic acid. *Biochemistry* **11:** 798.

Grouse, L.D., B.K. Schrier, C.H. Letendre, and P.G. Nelson. 1980. RNA sequence complexity in central nervous system development and plasticity. *Curr. Top. Dev. Biol.* **16:** 381.

Grouse, L.D., B.K. Schrier, E.L. Bennett, M.R. Rosenzweig, and P.G. Nelson. 1978. Sequence diversity studies of rat brain RNA: Effects of environmental complexity on rat brain RNA diversity. *J. Neurochem.* **30:** 191.

Hahn, W.E. and C.D. Laird. 1971. Transcription of nonrepeated DNA in mouse brain. *Science* **173:** 158.

Hahn, W.E., J. Van Ness, and N. Chaudhari. 1982. Overview of the molecular genetics of mouse brain. In *Molecular genetic neuroscience* (ed. F.O. Schmitt et al.), p. 323. Raven Press, New York.

Harrington, C.A. and D.M. Chikaraishi. 1983. Identification and sequence of the initiation site for rat 45S ribosomal RNA synthesis. *Nucleic Acids Res.* **10:** 3317.

Iglisias, R. 1960. Spontaneous transplantable melanoma of the rat. *Proc. Am. Assoc. Cancer Res.* **3:** 122.

Imada, M. and N. Sueoka. 1978. Clonal sublines of rat neurotumor RT4 and cell differentiation. I. Isolation and characterization of cell lines and cell type conversion. *Dev. Biol.* **66:** 97.

Joh, T.H., E.E. Baetge, B.B. Kaplan, M.E. Ross, M.J. Brodsky, V.R. Albert, D.H. Park, and D.J. Reis. 1981. Do catecholamine synthesizing enzymes share common gene coding sequences. *Soc. Neurosci. Abstr.* **1:** 206.

Johnson, R.A. and T.F. Walseth. 1979. The enzymatic preparation of [α^{32}P]ATP, [α^{32}P]GTP, [^{32}P]cAMP, and [^{32}P]GMP and their use in the assay of adenylate and guanylate cyclases and cyclic nucleotide phosphodiesterases. *Adv. Cyclic Nucleotide Res.* **10:** 135.

Kaplan, B.B. and C.E. Finch. 1982. The sequence complexity of brain ribonucleic acids. In *Molecular ap-*

proaches to neurobiology (ed. I. Brown). Academic Press, New York.

KAPLAN, B.B., B.S. SCHACHTER, H.H. OSTERBURG, J.S. DE VELLIS., and C.E. FINCH. 1978. Sequence complexity of polyadenylated RNA obtained from rat brain regions and cultured rat cells of neural origin. *Biochemistry* **17:** 5516.

KUROIWA, A. and S. NATORI. 1979. Preferential expression of unique sequences adjacent to middle repetitive sequences in mouse cytoplasmic RNA. *Nucleic Acids Res.* **7:** 751.

LAMOUROUX, A., N.F. BIGUET, D. SAMOLYK, A. PRIVAT, J.C. SALOMON, J.F. PUJOL, and J. MALLET. 1982. Identification of cDNA clones coding for rat tyrosine hydroxylase antigen. *Proc. Natl. Acad. Sci.* **79:** 3881.

LECHAN, R.M., R.H. GOODMAN, M. ROSENBLATT, S. REICHLIN, and J.F. HABENER. 1983. Prosomatostatin-specific antigen in rat brain: Localization by immunocytochemical staining with an antiserum to a synthetic sequence of preprosomatostatin. *Proc. Natl. Acad. Sci.* **80:** 2780.

LEE, D., C.E. CASTRO, L.L. JAGODZINSKI, and J.S. SEVALL. 1979. Functional significance of rat liver nonhistone protein-DNA interactions: RNA hybridization of protein-bound DNA. *Biochemistry* **18:** 3160.

MAXAM, A.M. and W. GILBERT. 1980. Sequencing end-labeled DNA with base-specific chemical cleavage. *Methods Enzymol.* **65:** 499.

MELIGENI, J., A.W. TANK, J.K. STEPHENS, E. DREYER, and N. WEINER. 1981. In vivo phosphorylation of rat adrenal tyrosine hydroxylase during acute decapitation stress. *Cold Spring Harbor Conf. Cell Proliferation* **8:** 1377.

O'MALLEY, K.L., A. MAURON, J. RAESE, J.D. BARCHAS, and L. KEDES. 1983. Genes for catecholamine biosynthesis: Cloning by expression and identification of the cDNA for rat dopamine β-hydroxylase. *Proc. Natl. Acad. Sci.* **80:** 2161.

PELHAM, H.R.B. and R.J. JACKSON. 1976. An efficient mRNA-dependent translation system from reticulocyte lysates. *Eur. J. Biochem.* **67:** 247.

ROBSON, K.J.H., T. CHANDRA, R.T.A. MACGILLIVRAY, and S.L.C. WOO. 1982. Polysome immunoprecipitation of phenylalanine hydroxylase mRNA from rat liver and cloning its cDNA. *Proc. Natl. Acad. Sci.* **79:** 4701.

ROSS, M.E., C.Y. LAI, D.J. REIS, and T.H. JOH. 1982. Evidence for homologous protein domains among catecholamine synthesizing enzymes revealed by fingerprinting on high performance liquid chromatography. *Soc. Neurosci. Abstr.* **8:** 885.

SALA-TREPAT, J.M., M.J. SAVAGE, and J. BONNER. 1978. Isolation and characterization of poly(adenylic acid)-containing messenger ribonucleic acid from rat liver polysomes. *Biochim. Biophys. Acta* **519:** 173.

SCHUBERT, D., S. HEINEMANN, W. CARLISLE, H. TARIKAS, B. KIMES, J. PATRICK, J.H. STEINBACH, W. CULP, and B.L. BRANDT. 1974. Clonal cell lines from rat nervous system. *Nature* **249:** 224.

SHEARER, R.W. 1977. Altered control of transcription during tumor progression. In *Prevention and detection of cancer* (ed. H.E. Nieburgs), part I, vol. 1, p. 53. Marcel Dekker, New York.

SHIMAN, R., D.W. GRAY, and A. PATER. 1979. A simple purification of phenylalanine hydroxylase by substrate-induced hydrophobic chromatography. *J. Biol. Chem.* **254:** 11300.

SOUTHERN, E.M. 1975. Detection of specific sequences among DNA fragments separated by gel electrophoresis. *J. Mol. Biol.* **98:** 503.

STEINBUCH, M. and R. AUDRAN. 1969. The isolation of IgG from mammalian sera with the aid of capryl acid. *Arch. Biochem. Biophys.* **134:** 179.

THIBAULT, J., D. VIDAL, and F. GROS. 1981. In vitro translation of mRNA from rat pheochromocytoma tumors, characterization of tyrosine hydroxylase. *Biochem. Biophys. Res. Commun.* **99:** 960.

THOMAS, P.S. 1980. Hybridization of denatured RNA and small DNA fragments transferred to nitrocellulose. *Proc. Natl. Acad. Sci.* **77:** 5201.

VAN NESS, J., I.H. MAXWELL, and W.E. HAHN. 1979. Complex population of nonpolyadenylated messenger RNA in mouse brain. *Cell* **18:** 1341.

WILKES, P.R., G.D. BIRNIE, and J. PAUL. 1979. Changes in nuclear and polysomal polyadenylated RNA sequences during rat liver regeneration. *Nucleic Acids Res.* **6:** 2193.

Dopamine β-Hydroxylase Rat mRNA: Structure, Regulation, and Tissue Localization

K. O'MALLEY,* A. MAURON,* G. MAKK,* D.L. WONG,† R.D. CIARANELLO,† J.D. BARCHAS,† AND L. KEDES*

*The MEDIGEN Project, Department of Medicine and the †Nancy Pritzker Laboratory of Behavioral Neurochemistry, Department of Psychiatry and Behavioral Sciences, Stanford University School of Medicine, Stanford, California 94305 and Veterans Administration Medical Center, Palo Alto, California 94304

To study the organization and expression of the genes involved in catecholamine biosynthesis, we have begun to synthesize and clone corresponding cDNAs from polyadenylated mRNA enriched in template activity for catecholamine enzymes. Reported structural homologies and antisera cross-reactivity (Joh et al. 1981) allowed us to devise a plan whereby available antisera for one enzyme could be used to purify not only its own mRNA but also that of a cross-reacting mRNA species.

We used available tyrosine hydroxylase (TH; E.C. 1.14.16.20) antisera to identify products produced by the in vitro protein translation of fractionated mRNAs prepared from a noradrenergic clonal cell line. Fractions enriched in TH, dopa-decarboxylase (DDC; E.C. 4.1.1.26), and dopamine β-hydroxylase (DBH; 3,4-dihydroxyphenylethylamine, ascorbate:oxygen oxidoreductase [β-hydroxylating]; E.C. 1.14.17.1) mRNA served as templates in the preparation of double-stranded complementary DNA clones. We recently reported the identification of such a recombinant clone coding for DBH from the rat pheochromocytoma cell line PC12 (O'Malley et al. 1983). This clone, pDBH-1, contained only a portion of the coding region calculated for the native polypeptide.

This paper describes the isolation of nearly full-length clones from a new PC12 library and an additional rat brain library. We also describe the tissue distribution of the DBH message and its frequency in the rat genome. Preliminary results of the regulation of DBH gene expression in the rat pheochromocytoma cell lines PC12 and PCG2 are presented.

METHODS AND MATERIALS

Materials. Avian myeloblastosis virus reverse transcriptase and terminal deoxyribonucleotidyltransferase were purchased from PB Biochemicals. Oligo(dT)-cellulose was obtained from Collaborative Research. Restriction endonucleases were purchased from New England BioLabs. Rat pheochromocytoma cell lines PC12 and PCG2 were maintained in Dulbecco's modified tissue culture medium supplemented with 5% horse serum and 10% fetal calf serum. Cultures were maintained in 12% CO_2/88% O_2 at 37°C.

Isolation of poly(A)⁺ mRNA. Rat tissue was dissected immediately upon death, various areas removed, frozen on dry ice, and stored at −70°C until use. Total cytoplasmic RNA from either brain tissue or PC12 cells was prepared by a modification of the Chirgwin et al. (1979) procedure. Following extraction in guanidine thiocyanate, the homogenate was layered onto a 7.5 M CsCl gradient and sedimented for 24 hours at 23,000*g*. The pelleted RNA was recovered and precipitated with ethanol. Poly(A)⁺ mRNA was prepared by two cycles of adsorption and elution from an oligo(dT)-cellulose column (Aviv and Leder 1972).

Construction of cDNA clones. A cDNA library was prepared from 18–20S PC12 mRNA using fractions enriched for the catecholamine enzymes TH, DDC, and DBH. An additional library was prepared using unfractionated rat brain caudate nucleus poly(A)⁺ mRNA. Both libraries were synthesized by the procedure of Okayama and Berg (1982). *E. coli* HB101 was transformed to ampicillin resistance with a frequency of ~2×10^5/μg starting RNA. Recombinant clones (2000 from the PC12 library; 10,000 from the rat library) were picked onto nitrocellulose filters, grown for 6 hours on media containing 50 μg/ml ampicillin, and then transferred to LB plates containing 240 μg/ml chloramphenicol. Colonies were amplified for 24 hours and then prepared for hybridization as described by Grunstein and Hogness (1975). Remaining transformants (20,000) were scraped from their plates and used to prepare plasmid DNA by the cleared lysate procedure (Konkel et al. 1979) without amplification.

The plasmid pDBH-1 and subsequent isolated clones were mapped by the Smith and Birnstiel procedure (1978). An internal *Hin*f fragment of pDBH-1 was end-labeled using T4 polynucleotide kinase and [γ-^{32}P]ATP to a specific activity of 1×10^6 cpm/pmole (Wallace et al. 1981). The labeled probe was separated from free nucleotides by Sephadex G-50 chromatography. Nick-translation of DNA fragments was performed as described (Rigby et al. 1977). All recombinant DNA procedures followed the pertinent NIH guidelines.

Preparation of rat genomic blots. High-molecular-weight rat thymus was prepared by the following procedure. Thymus cells of young rats (120 g) were homogenized in 30 mM Tris (pH 7.5), 120 mM KCl, and 10 mM mercaptoethanol. After adjusting the NaCl concentration to 100 mM and SDS to 0.05%, the lysate was digested with RNase (heat-pretreated, 50 μg/ml, 1.5 hr at 37°C) followed by proteinase K (Merck, 2 mg/ml, 3 hr at 37°C). The resulting DNA was extracted twice with phenol-chloroform-isoamyl alcohol (10:10:1), once with chloroform, and dialyzed in 10 mM Tris (pH 7.5) and 1 mM EDTA. Purified DNA was digested with a two- to threefold excess of the restriction endonuclease *Eco*RI under conditions specified by the supplier. DNA fragments were separated by electrophoresis in 0.8 agarose gels in 89 mM Tris, 89 mM borate, 2.5 mM EDTA and transferred onto nitrocellulose (Southern 1975). After prehybridization for 6 hours at 65°C in 0.6 M NaCl, 0.06 M sodium citrate (pH 7.0), 0.05% polyvinylpyrrolidone, 0.05% Ficoll, 0.05% bovine serum albumin, and 50 mM NaP_i (pH 7.0), the blots were hybridized for 18 hours with 10^6 cpm/ml nick-translated DNA probe in the above buffer plus 10% dextran sulfate. Filters were washed extensively in 0.06 M NaCl, 0.006 M sodium citrate (pH 7.0), and 0.1% SDS at 65°C.

Preparation of RNA blots. Either total cytoplasmic (20 μg) or poly(A)$^+$ mRNA (3 μg) was denatured in formaldehyde and formamide and electrophoresed in 1.2% agarose gels, as described by Lehrach et al. (1977). Rat ribosomal RNA was used for size calibration. Fractionated RNA was transferred to nitrocellulose and hybridized with nick-translated pDBH-2 or pDBH-3. Hybridization buffer was prepared as described above with the addition of 50% formamide, 0.1% SDS, and 50 μg/ml denatured calf thymus DNA. Hybridizations and filter washing (0.075 M NaCl, 0.0075 M sodium citrate, 0.1% SDS) were carried out at 42°C. Following autoradiography, blots were scanned using an Optronics P-1000 digitizer driven by a PDP-10 computer (graciously put at our disposal by Jerry Latter, The Linus Pauling Institute for Science and Medicine, Palo Alto, Calif.).

RESULTS

Isolation of Longer cDNA Clones

We recently reported the molecular cloning of a cDNA for rat DBH (O'Malley et al. 1983). The plasmid containing nucleotide sequences specific for DBH, pDBH-1, had an insert size of 330 bp. Since the native enzyme has a subunit size of 75,000, pDBH-1 could code for only a portion of the molecule. Therefore, using the high-efficiency cloning system developed by Okayama and Berg (1982), we constructed a new PC12 cDNA library. Our starting RNA was the same 18S–20S poly(A)$^+$ mRNA enriched in TH, DDC, and DBH messages that we used to construct the expression library (O'Malley et al. 1983). Starting with 1 μg of enriched PC12 RNA, we obtained ~22,000 transformants with an average insert size of 2100 bp. Two thousand transformants were picked onto nitrocellulose and processed as described in Methods and Materials.

In the process of cloning pDBH-1, the *Pst*I cleavage site in pBR322 had not been regenerated. Therefore, to obtain an internal fragment free of vector, we mapped this clone using double-restriction digests and Smith and Birnstiel mapping procedures (1978). We were able to isolate an 80-bp DNA fragment from pDBH-1 fragment generated by digestion with the restriction endonuclease *Hin*fI. This was end-labeled and used to screen the ordered clones. Due to the short length of the probe, hybridization and washing were done at 60°C. Six clones strongly hybridizing to the kinased probe were identified after autoradiography. These six clones were prepared as minilysates (Birnboim and Doly 1979), cleaved with restriction nuclease *Hin*fI, electrophoresed in a 2% minigel, blotted onto nitrocellulose, and rehybridized with the labeled probe. All six transformants contain an 80-bp *Hin*fI-generated DNA fragment comigrating and cohybridizing with the isolated fragment from pDBH-1.

One advantage of the Okayama-Berg cloning vector is that the insert is oriented within it. Therefore, it was easy to compare the DNA structure of the individual hybridization-positive clones. Based on double-digest restriction endonuclease mapping, all six clones represented varying lengths of the same mRNA species ranging from 800 to 1800 bp. The 1800-bp clone was designated pDBH-2.

Recently, Gunning et al. (1982) have devised a procedure for increasing the yield of specific sequences at or near full length from cDNA libraries. First, we determined from Southern blots of the cDNA library cut with enzymes that are known to cut the Okayama-Berg vector only once the distribution of cDNA sequences that hybridized to our probe. From the autoradiogram, we could see that restriction endonucleases *Eco*RI and *Cla*I cut outside our insert, whereas *Pvu*II and *Hin*dIII gave multiple lower-molecular-weight bands. Therefore, we digested 50 μg of the PC12 cleared lysate library with both *Cla*I and *Eco*RI. After fractionation on an agarose gel, we cut out the region containing the highest-molecular-weight components. In this gel fragment, the concentration of specific DBH cDNAs should be greatly increased relative to the library as a whole, since many other inserts of the same length will have been cut by these enzymes. The DNA was electroeluted, religated, and used to transform HB101.

When this size-selected portion of the PC12 library was screened with a 5′ fragment from pDBH-2, 12 additional clones were detected. Minilysates from nine of these clones showed the same size and restriction pattern as did pDBH-2; the remaining three clones were shorter. The fact that so many independent cloning events each generated a truncated cDNA of 1800 bp suggests the presence of considerable mRNA secondary structure such that the reverse transcriptase molecule was unable to proceed beyond that point on the template.

Data (described below) suggestive of the presence of DBH-like mRNA in rat caudate nucleus convinced us to screen a caudate cDNA library for clones hybridizing to a pDBH-2 probe. We screened 10,000 rat caudate transformants and obtained two hybridization-positive clones. The longest clone was 2300 bp and was designated pDBH-3.

The arrangement and orientation of these clones is shown in Figure 1. Restriction mapping and subsequent sequence data (K. O'Malley et al., unpubl.) indicate that the 1800-bp overlapping region between pDBH-2 and pDBH-3 is identical. The size of the message hybridizing to either of these clones is 2500–2700 bp (data not shown). By subtraction, pDBH-3 is 200–400 bp short of the full-length message. However, based on amino acid composition data, pDBH-3 should include 90–95% of the DBH coding region.

Rat Genomic Blots

The DBH cDNA clones allowed us to study the corresponding gene structure using genomic Southern blots. To assess the number of DBH gene fragments contained in the rat genome, a radiolabeled 600-bp fragment centrally located in pDBH-2 was hybridized to nitrocellulose blots of *Eco*RI-digested rat thymus DNA. The results from one such experiment are shown in Figure 2. It is apparent from this autoradiogram that the rat has 20–25 genes and/or pseudogenes hybridizing to the cDNA probe under stringent conditions. We have determined (results not shown) that the hybridizing genomic fragments are present once per genome and exhibit varying degrees of homology to the probe.

Tissue Distribution

To examine the tissue distribution of the DBH mRNA, we isolated total cytoplasmic RNA from heart, liver, kidney, and adrenal glands. Equal amounts of RNA were electrophoresed in denaturing gels and transferred to nitrocellulose. The same nick-translated probe described above was used to probe the RNA blots. The resulting autoradiograms were scanned by computer and relative amounts of RNA determined. Data from several experiments are shown in Table 1. Relative amounts of DBH-like mRNA are shown as percentages in comparison to the level of DBH mRNA from PC12 cells. Using this methodology, we detected a DBH-like mRNA present to some extent in all of these tissues, although nowhere near the levels found in the overexpressing noradrenergic cell line.

We also examined total cytoplasmic RNA isolated from various brain regions and blotted onto nitrocellulose as described above. These results are also shown in Table 1. It is apparent that RNA from every brain area examined shows some hybridization, with the locus coeruleus area, the hypothalamus, and cerebellum being more enriched in the reacting species.

Regulation of DBH Gene Expression in PC12 and PCG2 Cell Lines

The PC cell lines have been established from a transplantable rat pheochromocytoma first described by Warren and Chute (1972). The best known cell line, PC12, was cloned from this pheochromocytoma by Greene and Tischler (1976) and shown to contain high levels of all the catecholamine biosynthetic enzymes except phenylethanolamine-*N*-methyltransferase (PNMT). Furthermore, these cells have been shown by many investigators to exhibit several of the properties of noradrenergic neurons with respect to catecholamine synthesis, storage, and release. In particular, PC12 and the closely related clones PCG1 and PCG2 were shown to exhibit varying degrees of TH activity regulation by glucocorticoids (Edgar and Thoenen 1978; Otten and

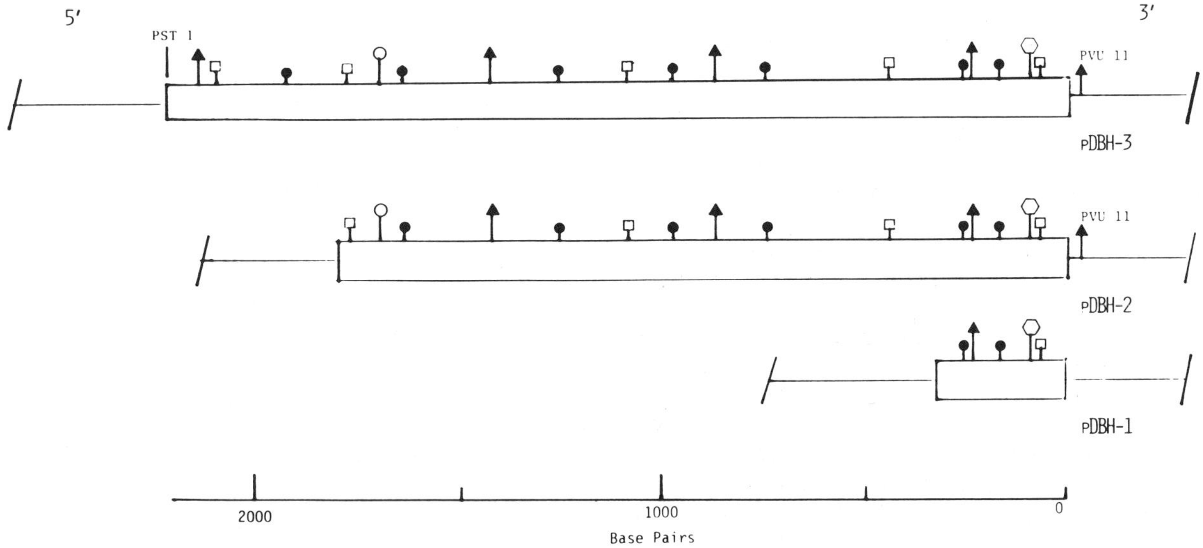

Figure 1. Restriction endonuclease map of inserts from cDNA clones coding for rat DBH. Symbols represent various restriction endonucleases used in determining the map. The clones pDBH-1 and pDBH-2 were obtained from a PC12 cDNA library. Clone pDBH-3 was found by screening a rat brain cDNA library.

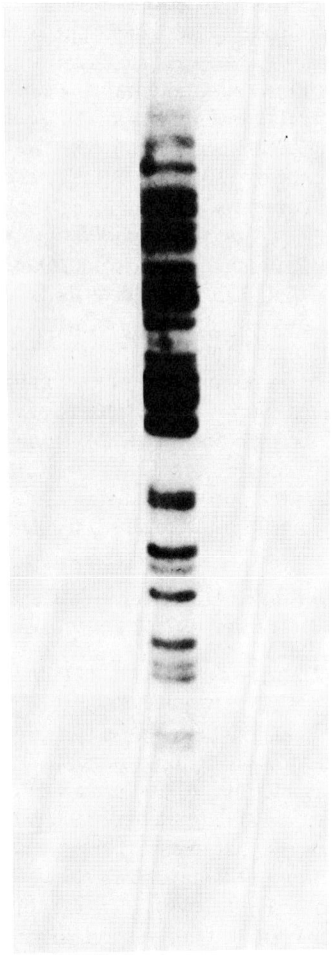

Figure 2. Determination of the number of *Eco*RI fragments in the rat genome homologous to DBH cDNA.

Towbin 1980; Schubert et al. 1980), thus duplicating, at least in part, physiological processes observed in the normal ganglia. Since TH and DBH seem often to be subject to coordinated regulation in adrenergic neurons (Otten and Thoenen 1976), we asked whether substances known to affect catecholamine biosynthetic enzymes in these cells and/or normal adrenergic neurons regulate DBH at the mRNA level. PC12 and PCG2 cultures were grown in the presence of 10 μM dexamethasone for 1, 2, or 3 days and harvested. Untreated cells from the same passage were harvested at the same times to control for the effect of cell density. RNA was prepared and analyzed on formaldehyde-agarose gels as described above. The relative quantity of DBH mRNA was determined from the intensity of the autoradiographic signals. Figure 3 shows the result of this experiment: The effect of dexamethasone is expressed as the ratio of DBH in treated over control culture at any given day ($\times$100). In both PC12 and PCG2, dexamethasone does not cause any increase in the abundance of DBH mRNA but rather a slight, but possibly significant drop. We can therefore conclude that glucocorticoids show no inductive effect on DBH

Table 1. Abundance of DBH-like mRNA in PC12 Cells and Different Rat Tissues and Brain Parts

Origin of RNA	Relative abundance of DBH-like mRNA (PC12 = 100)
Adrenal	33
Heart	29
Kidney	26
Liver	9
Locus coeruleus	40
Hypothalamus	27
Olfactory bulb	19
Cerebellum	44
Brain stem	23
Caudate	26
Cortex	31

A single RNA gel containing equal amounts of RNA from PC12 and the various tissues was quantitated as explained in the text. Results are relative mRNA quantities referred to DBH-like mRNA content in PC12.

mRNA levels in the pheochromocytoma lines that we have examined.

PC cell lines are commonly used as a model system for the normal adrenergic neuron, but they really originate from a rather specialized such cell, the adrenal medullary cell. The regulation of DBH activity in the adrenal medulla has been extensively studied by the groups of Axelrod and Ciaranello (Ciaranello et al. 1975, 1976) who showed that de novo synthesis of the enzyme occurs upon stimulation of the presynaptic nerve. This constitutes a well-documented instance of transsynaptic induction, a phenomenon thought to occur in other adrenergic neurons as well (Thoenen 1975). We wish to assess whether transsynaptic induction entails the specific induction of DBH mRNA. Our first attempt was to try and model this situation in the PC system by growing PCG2 cells in the presence of the acetylcholine analog carbachol. Cells were plated out and grown for 1, 2, and 3 days in concentrations of

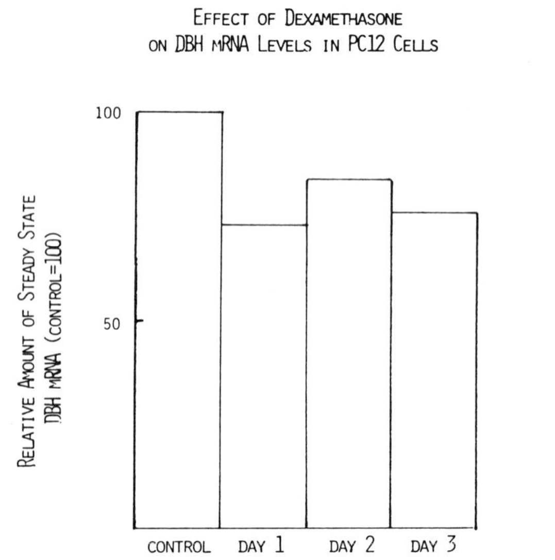

Figure 3. Effect of dexamethasone on DBH mRNA levels in PC12 cells.

carbachol of 0.1, 1, 10, and 50 μM. They were harvested and RNA was analyzed as before. Again, as with dexamethasone, we do not detect any elevation of DBH mRNA but a slight decrease instead.

This initial failure to set up in culture a situation similar to the in vivo induction can probably be explained, considering our data on the relative abundance of DBH mRNA in different tissues. These experiments (Table 1) show that the PC cultures contain abnormally high levels of DBH mRNA compared with the normal adrenal tissue. This suggests that the DBH mRNA is somehow derepressed in the cultured cells. This could be due to a defective control mechanism in these abnormal cells. On the other hand, it could be that some element of the transsynaptic induction mechanism is being chronically stimulated in our conditions of culture.

DISCUSSION

We recently isolated a clone, pDBH-1, coding for a rat cDNA copy of DBH mRNA. This transformant was isolated by a sensitive colony immunoassay in which several bacterial promoters enhanced the expression of a foreign insert. Since pDBH-1 had an insert size of 330 bp, it is apparent that even a small insert can be effective as a fusion peptide carrying immunoreactive determinants. Subsequent sequence information (K. O'Malley et al., unpubl.) indicates that pDBH-1 would include ~200 bp of the protein-coding region. Therefore, 60–70 amino acids were sufficient for antisera recognition. The effectiveness of this technique was also shown by Ruther et al. (1982). Using an analogous colony immunoassay, this group described the isolation of a clone 33 bp long, i.e., 10 amino acids, that was identified by antibodies against the gene product.

We obtained clone pDBH-2 by using the short expression transformant to screen libraries prepared by the Okayama and Berg procedure (1982). Initially, we screened a library made from size-fractionated poly(A)$^+$ mRNA isolated from PC12 cells. From this library we obtained a clone encoding ~60% of the predicted protein structure. Analysis of additional hybridization-positive transformants suggested an area of considerable secondary structure at ~1800 bp upstream from the poly(A) tract. After screening 22,000 transformants, we were unable to find any DBH-like clones exceeding 1800 bp; 60% of the DBH clones stopped at this point.

Results from the tissue-specificity experiments indicated the presence of a DBH-like mRNA present in various brain regions. Therefore, we screened a rat brain library using a coding-region fragment from pDBH-2. From the results shown in Figure 1 and subsequent sequence data, we conclude that the rat adrenal and caudate DBH mRNA are identical. Although posttranslational modifications might be tissue-specific, presumably the DBH protein is similar as well.

Recently, Matsui et al. (1982) reported the purification of bovine brain DBH. Their data showed a very close similarity between the bovine brain and adrenal enzymes based on immunological titration studies. By gel filtration studies, they estimated the molecular weight of the brain enzyme at 400,000. This molecular weight is considerably larger than that reported for its adrenal counterpart. Several groups have reported the molecular weight of bovine adrenal DBH to be 290,000 composed of four identical subunits (Friedman and Kaufman 1965; Craine et al. 1973; Wallace et al. 1973). Matsui et al. proposed additional subunits, either identical or different, and/or larger precursor DBH subunits to account for the difference in size between tissues. Apparent size differences might also be due to posttranslational modifications, particularly glycosylation.

The DBH cDNA clone allowed us to examine the rat genome for the presence of DBH coding sequences. By Southern blot analysis, we observed between 20 and 25 distinct *Eco*RI fragments hybridizing to the PC12 DBH probe under stringent conditions. This result lends credence to the hypothesis that the shared antigenic determinants and peptide patterns among the catecholamine enzymes (Joh et al. 1981) could be due to structural homologies at the nucleotide level. At this point, we could not rule out the possibility that DBH is a member of its own multigene family excluding the other catecholamine enzymes. If this is the case, then reported similarities would be due to factors affecting antigenic determinants such as glycosylation or conformational traits, rather than the protein's primary structure. Gene probes coding for TH, DDC, and PNMT should soon answer this question.

Rat DBH has been isolated not only from brain and adrenal glands, but also from heart, salivary glands, vas deferens, and pineal (Potter and Axelrod 1963; Kato et al. 1978). This would be consistent with our rather surprising finding that DBH-like mRNA is present in nonneural tissue. Considering, however, that our DBH cDNA clone is clearly encoded by a multigene family, we must assess the possibility that in nonneural tissue, other genes, coding for proteins distinct from, but related to, DBH, are transcribed into mRNA species of similar size and show extensive cross-hybridization to the DBH cDNA probe.

In the brain, noradrenergic cell bodies appear to be restricted primarily in the pons/medulla area from which their axons project both to the spinal cord and rostrally (Ungerstedt 1971; Dahlstrom and Fuxe 1974). The locus coeruleus contains the most abundant noradrenergic cell population and is responsible for noradrenergic innervation of all of the cortices, geniculate bodies, colliculi, thalamus, and hypothalamus (Ungerstedt 1971; Dahlstrom and Fuxe 1964). Strong DBH immunoreactivity can be found in all areas innervated by the diffuse noradrenergic pathway (Fuxe et al. 1971). Based on these studies, we expected to find DBH mRNA in the hindbrain (pons/medulla), locus coeruleus area, and hypothalamus (Table 1). Unexpectedly, we also find the presence of a DBH-like mRNA in the cerebellum, olfactory bulb, striatum, and cortex. We conclude either that the use of gene probes in distribution studies is a more sensitive approach in revealing aspects of neuronal distribution (i.e., the presence of local populations of cell bodies) or that we are seeing a cross-hybridizing species of mRNA. To examine the latter possibility, we are now in the process of performing S1 nuclease experiments on RNA/DNA hybrids formed between cytoplasmic DNA isolated from various brain regions and the clone pDBH-3.

In our examination of DBH mRNA levels in the PC cells, we were unable to identify any substances inducing DBH mRNA. We relate this to the abnormally high constitutive level of DBH mRNA in these cells and have therefore shifted our priorities to substantiating the existence of a regulation of DBH mRNA in vivo. The experimental preparation used to demonstrate transsynaptic induction of DBH activity has classically been the reserpinized rat (Ciaranello et al. 1975) and in collaboration with D. Wong and R. Ciaranello, we have begun to investigate it at the mRNA level. Preliminary results indicate that chronic stimulation of the splanchnic nerve by the reflex action of reserpine causes a fourfold elevation of the relative abundance of DBH mRNA in the adrenal medulla. Interestingly, this would bring the DBH mRNA level in line with its value in the PC cultures. However, further experiments in a format amenable to statistical evaluation are needed to substantiate this conclusion.

ACKNOWLEDGMENTS

This work was supported in part by grants from the National Institutes of Health and the Veterans Administration to L.K. and by National Institutes of Mental Health grant MH23861 and Office of Naval Research grant SRO-001 to J.D.B.

REFERENCES

Aviv, H. and P. Leder. 1972. Purification of biologically active globin messenger RNA by chromatography on oligo thymidylic acid-cellulose. *Proc. Natl. Acad. Sci.* **69:** 1408.

Birnboim, H.C. and J. Doly. 1979. A rapid alkaline extraction procedure for screening recombinant plasmid DNA. *Nucleic Acids Res.* **7:** 1513.

Chirgwin, J.M., A.E. Pryzbyla, R.J. MacDonald, and W.J. Rutter. 1979. Isolation of biologically active ribonucleic acid from sources enriched in ribonuclease. *Biochemistry* **18:** 5294.

Ciaranello, R.W., G.F. Wooten, and J. Axelrod. 1975. Regulation of dopamine-beta-hydroxylase in rat adrenal glands. *J. Biol. Chem.* **250:** 3204.

———. 1976. Regulation of dopamine-beta-hydroxylase. II. Receptor interaction in the regulation of enzyme synthesis and degradation. *Brain Res.* **113:** 349.

Craine, J.E., G.H. Daniels, and S. Kaufman. 1973. Dopamine-beta-hydroxylase. The subunit structure and anion activation of the bovine adrenal enzyme. *J. Biol. Chem.* **248:** 7838.

Dahlstrom, A. and K. Fuxe. 1964. Evidence for the existence of monoamine containing neurons in the central nervous system. *Acta Physiol. Scand.* (Suppl. 232) **62:** 1.

Edgar, D.H. and H. Thoenen. 1978. Selective enzyme induction in a nerve growth factor responsive pheochromocytoma cell line (PC12). *Brain Res.* **154:** 186.

Friedman, S. and S. Kaufman. 1965. 3,4-Dihydroxyphenylethylamine beta-hydroxylase. Purification and properties. *J. Biol. Chem.* **240:** 4763.

Fuxe, K., M. Goldstein, T. Hokfelt, and T.H. Joh. 1971. Cellular localization of dopamine beta-hydroxylase and phenylethanolamine-*N*-methyltransferase as revealed by immuno-histochemistry. *Prog. Brain Res.* **34:** 127.

Greene, L.A. and A.S. Tischler. 1976. Establishment of a neuradrenergic clonal line of rat adrenal pheochromocytoma cells which respond to nerve growth factor. *Proc. Natl. Acad. Sci.* **73:** 2424.

Grunstein, M. and D. Hogness. 1975. Colony hybridization: A method for the isolation of cloned DNAs that contain a specific gene. *Proc. Natl. Acad. Sci.* **72:** 3901.

Gunning, P., P. Ponte, H. Okayama, J. Engel, H. Blau, and L. Kedes. 1982. Isolation and characterization of full-length cDNA clones for human alpha- beta- and gamma-actin mRNAs: Skeletal but not cytoplasmic actins have an amino terminal cysteine that is subsequently removed. *Mol. Cell. Biol.* **3:** 787.

Joh, T.H., E.E. Baetge, B.B. Kaplan, M.E. Ross, M.J. Brodsky, V.R. Albert, D.H. Park, and D.J. Reis. 1981. Do catecholamine synthesizing enzymes share common gene coding sequences? *Soc. Neurosci. Abstr.* **7:** 206.

Kato, T., Y. Wakui, T. Nagatsu, and T. Ohnishi. 1978. An improved dual-wavelength spectrophotometric assay for dopamine-beta-hydroxylase activity. *Biochem. Pharmacol.* **27:** 829.

Konkel, D.A., J.V. Maizel, and D. Leder. 1979. The evolution and sequence comparison of a recently diverged mouse chromosomal beta-globin gene. *Cell* **18:** 865.

Lehrach, H.D., D. Diamond, J.M. Wozney, and H. Boedtker. 1977. RNA molecular weight determinations by gel electrophoresis under denaturing conditions, a critical reexamination. *Biochemistry* **16:** 4743.

Matsui, H., C. Yamamoto, and T. Nagatsu. 1982. Purification and properties of bovine brain dopamine beta-hydroxylase. *J. Neurochem.* **39:** 1066.

Okayama, H. and P. Berg. 1982. High efficiency cloning of full-length cDNA. *Mol. Cell. Biol.* **2:** 161.

O'Malley, K., A. Mauron, J. Raese, J.D. Barchas, and L. Kedes. 1983. Genes for catecholamine biosynthesis: Cloning by expression and identification of the cDNA for rat dopamine beta-hydroxylase. *Proc. Natl. Acad. Sci.* **80:** 2161.

Otten, U. and H. Thoenen. 1976. Selective induction of tyronic hydroxylase and dopamine-beta-hydroxylase in sympathetic ganglia in organ culture: Role of glucocorticoids as modulators. *Mol. Pharmacol.* **12:** 353.

Otten, U. and M. Towbin. 1980. Permissive action of glucocorticoids in induction of tyronine hydroxylase by nerve growth factor in a pheochromocytoma cell line. *Brain Res.* **193:** 304.

Potter, C.T. and J. Axelrod. 1963. Subcellular localization of catecholamines in tissue of the rat. *J. Pharmacol.* **142:** 291.

Rigby, P.W.J., M. Dieckmann, C. Rhodes, and P. Berg. 1977. Labeling deoxyribonucleic acid to high specific activity in vitro by nick translation cDNA polymerase I. *J. Mol. Biol.* **113:** 237.

Ruther, U., M. Koenen, A.E. Sippel, and B. Muller-Hill. 1982. Exon cloning: Immunoenzymatic identification of exons of the chicken lysozyme gene. *Proc. Natl. Acad. Sci.* **79:** 6852.

Schubert, D., M. LaCorbiere, F.G. Klier, and J.M. Steinbach. 1980. The modulation of neurotransmitter synthesis by steroid hormone and insulin. *Brain Res.* **190:** 67.

Smith, H.O. and M. Birnstiel. 1978. A simple method for

DNA restriction site mapping. *Nucleic Acids Res.* **3:** 2387.

SOUTHERN, E. 1975. Detection of specific sequences among DNA fragments separated by gel electrophoresis. *J. Mol. Biol.* **98:** 503.

THOENEN, H. 1975. Trans-synaptic regulation of the synthesis of specific neuronal proteins. *J. Neural Transm.* (Suppl.) **12:** 127.

UNGERSTEDT, U. 1971. Stereotoxic mapping of the monoamine pathways in the rat brain. *Acta Physiol. Scand.* (Suppl.) **367:** 1.

WALLACE, E.F., M.J. KRANTZ, and W. LOVENBERG. 1973. Dopamine-beta-hydroxylase: A tetrameric glycoprotein. *Proc. Natl. Acad. Sci.* **70:** 2253.

WALLACE, R.B., M. SCHOLD, M.J. JOHNSON, P. DEMBEK, and K. ITAKURA. 1981. Oligonucleotide directed mutagenesis of the human beta-globin gene: A general method for producing specific point mutations in cloned DNA. *Nucleic Acids Res.* **9:** 3647.

WARREN, S. and R.N. CHUTE. 1972. Pheochromocytoma. *Cancer* **29:** 327.

Evidence for the Existence of Homologous Gene Coding Regions for the Catecholamine Biosynthetic Enzymes

T.H. JOH, E.E. BAETGE, M.E. ROSS, AND D.J. REIS
Laboratory of Neurobiology, Cornell University Medical College, New York, New York 10021

The classical catecholamine (CA) biosynthetic pathway of Blaschko (1939) involves the activity of four enzymes: tyrosine hydroxylase (TH), which catalyzes the first step in the pathway of the conversion of tyrosine to L-dopa; aromatic L-amino acid decarboxylase (dopadecarboxylase, DDC), which catalyzes the conversion of L-dopa to dopamine (DA); dopamine-β-hydroxylase (DBH), which converts DA to norepinephrine (NE); and phenylethanolamine-*N*-methyltransferase (PNMT), the last enzyme in the chain, which catalyzes the biosynthesis of epinephrine (E) from NE.

An important feature of the CA biosynthetic pathway, however, is that all four enzymes are not expressed together in all cells. Thus, whereas all are expressed in neurons or chromaffin cells synthesizing adrenaline, PNMT is not expressed in neurons, ganglia, or chromaffin cells synthesizing NE. In cells producing DA as their final neurotransmitter, DBH as well as PNMT are not present. Based on the classical one gene for one enzyme hypothesis, these facts would suggest that separate genes code for each enzyme and that these genes are probably independently regulated. Thus, one would predict from this concept that neurons and cells containing still other combinations of the CA biosynthetic enzymes should exist. That this may be the case has been suggested by recent studies in our laboratory identifying neurons in the CNS that contain only PNMT without DDC, TH, or DBH (C. Ross et al., in prep.) and other neurons of the CNS that contain DDC alone (Jaeger et al. 1983).

On the other hand, studies on the regulation of the catecholamine biosynthetic enzymes in adult mammals have demonstrated a high degree of coordinated regulation. Two examples may be cited. It has long been known that the activity of TH and DBH can be increased in response to nerve impulse activity or by the administration of drugs in sympathetic ganglia, noradrenergic neurons of the locus coeruleus, or in chromaffin cells of the adrenal medulla where PNMT is also coordinately regulated along with TH and DBH. This enzyme induction phenomenon is most easily studied following administration of the drug reserpine. We have demonstrated that following reserpine, the increase of TH and DBH in noradrenergic neurons of the locus coeruleus is due to accumulation of more enzyme protein (Reis et al. 1975) and is a consequence of the increase in the relative rate (rate of enzyme synthesis/rate of the total protein synthesis) of biosynthesis of these enzymes (Ross et al. 1978). A reduction of TH and DBH following an identical pattern also can be initiated in cells of the locus coeruleus following axotomy (Ross et al. 1975). The reversible reduction of enzyme activity that follows damage to axons and lasts from 14 to 21 days is identical for the two enzymes and in both instances is due to a relative reduction in the rate of their biosynthesis (Ross et al. 1979).

The above results suggest that phenotypic expression of CA-synthesizing enzymes in neurons and chromaffin cells may be independently regulated and that one cell type can express a single CA enzyme. Alternatively, when more than one enzyme is expressed in a CA cell type, the reduction or induction of their synthesis occurs in a linked coordinated fashion. We have recently reported our preliminary results for the existence of similar protein domains in the primary structure of these enzymes, and have postulated that these enzymes are probably coded for by genes containing similar coding sequences (Joh et al. 1981).

In the present studies, we analyze the structural relationship of genes coding for these enzymes and present evidence for the existence of homologous gene coding regions. The present results suggest that the CA-synthesizing enzymes may be coded for by genes that have evolved through duplication of a common ancestral precursor.

METHODS

Purification of and production of antibodies to CA-synthesizing enzymes. CA-synthesizing enzymes, TH, DBH, and PNMT, were purified, and antibodies against purified enzymes were raised in rabbits as described elsewhere (Joh and Ross 1983). The molecular weight of each enzyme was determined by SDS-polyacrylamide gel electrophoresis. Molecular weights of TH, DBH, and PNMT were 60, 75, and 31 kD, respectively. It is known that when TH is treated with trypsin for a short period of time (10–30 min at room temperature), a smaller (35 kD) species can be obtained. This trypsin-digested form of TH (tTH) usually has higher specific activity than the native form of TH.

Antibodies to each enzyme, including tTH, do not cross-react with any other protein except their own antigen in crude, partially purified, or purified enzyme preparations (Joh and Ross 1983). No cross-reactivity with other CA-synthesizing enzymes has been demonstrated by immunohistochemistry and immunochemical precipitation.

Isolation of homogeneous enzymes on SDS-polyacrylamide gel electrophoresis. Amino acid analyses and peptide mapping for comparison of primary structures require the use of chemically pure enzyme proteins. Final isolation of homogeneous enzyme was achieved by separating enzyme purified by conventional purification methods (Joh and Ross 1983) on SDS-polyacrylamide gel electrophoresis and eluting each enzyme as identified by its characteristic molecular weight (Ross 1982).

SDS-slab gel electrophoresis (Laemmli 1970) of enzymes was carried out on a 12.5% polyacrylamide gel in Tris-Cl buffer (pH 8.7) containing 0.1% SDS. Enzyme proteins are first linked with fluorescamine and layered on 5% acrylamide stacking gels. When the dye front migrates 15 cm or more, electrophoresis is stopped. The slab gel is then placed under a UV lamp to visualize the fluorescent protein bands. The appropriate enzyme protein of characteristic molecular weight (TH = 60; DBH = 75, PNMT = 31) is then cut from its slab gel and eluted in the following manner.

Gel strips containing the enzyme of interest are minced with 10 ml of 1 M NH_4OH containing 1% SDS in a clean polycarbonate centrifuge tube, and the mixture is allowed to stand at room temperature for 5 hours with occasional shaking. After filtering the mixture through a 0.2-μm Nalgene filter, the solution is dialyzed against a large volume of water to remove as much NH_4OH and SDS as possible from the solution. The dialysate is then lyophilized. The resulting powder is suspended in 1–1.5 ml of water, and enzyme in the solution is precipitated in four volumes of 100% acetone, which is cooled to −20°C in a dry ice–ethanol bath. The mixture is allowed to stand for 30 minutes at −20°C, and the precipitate is collected by centrifugation. For peptide mapping, the precipitate is dissolved in Tris-Cl buffer containing SDS, whereas for amino acid analysis it is suspended in ammonium carbonate (pH 8.0).

Peptide mapping of TH, DBH, and PNMT. Two procedures were used to identify similar protein domains in these enzymes: (1) a limited proteolysis of enzymes at specific cleavage points by digestion with chymotrypsin, CNBr, or *Staphylococcus aureus* V8 protease, identification of similar peptide on SDS-polyacrylamide gel electrophoresis gels, and elution for subsequent analysis of their amino acid compositions; (2) an extensive proteolysis of enzymes with trypsin, isolation of peptides by HPLC, and subsequent amino acid analysis.

For limited protein digestion, enzyme protein eluted from SDS-polyacrylamide gel electrophoresis gels and precipitated with acetone was suspended in 80 mM Tris-Cl buffer (pH 6.8) containing 0.5% SDS and boiled for 3 minutes. Samples were then digested with the proteases at 30°C for 30 minutes. The reactions were stopped by boiling samples for 5 minutes. For CNBr digestion, a large excess of CNBr suspended in 70% formic acid is added to eluted enzymes, and samples are incubated at room temperature for 24 hours. CNBr and formic acid were removed by repeated dilution of samples with water and lyophilization. Digested peptides were then suspended in Tris-Cl buffer (pH 8.6) containing 1% SDS for further analysis by SDS-polyacrylamide gel electrophoresis.

Peptide products of the above limited digestions were separated by SDS-polyacrylamide gel electrophoresis. Identification of peptides on SDS-gels was carried out either by staining peptides with Coomassie blue or by elution of peptides from gels and subsequent analysis of their amino acid compositions.

For an extensive proteolysis, the enzymes, eluted from SDS-gels and precipitated with acetone, were resuspended in buffer containing 0.3 Tris (pH 8.5), 0.015 M EDTA, and 6 M guanidine hydrochloride. Each enzyme was reduced with a tenfold molar excess of dithiothreitol (DTT) for 4 hours at room temperature. Once reduced, sulfhydryl groups in the protein were carboxymethylated (Lai et al. 1976) with an equimolar (to DTT) amount of recrystalized iodoacetic acid (IAA) containing 4×10^7 cpm per sample of 2-[^{14}C]IAA. After a 20-minute incubation at room temperature, an excess of mercaptoethanol was added, and samples were dialyzed overnight against 2000 volumes of distilled water. The protein content of each sample was then approximated using a Coomassie blue microprotein assay (Bio-Rad), and protein concentrations were adjusted to 0.9 mg/ml. The pH of each sample was adjusted to 8.5 by bringing them to 50 mM Tris-HCl with a 2 M stock buffer. Proteins were then digested with trypsin that had been treated with tosylphenyl-carboxymethyl-ketone (TPCK trypsin) to inactivate traces of exogenous chymotrypsin activity. A 1:50 ratio (w/w) of TPCK trypsin to protein was used. Samples were incubated at room temperature for 16 hours, and the digestion was stopped with a one-tenth volume of trifluoroacetic acid (TFA). Solid guanidine HCl was added to a final concentration of 6 M salt to solubilize any material that precipitated at low pH.

The trypsin digests (3 nmoles each) were then analyzed by reversed-phase HPLC using an RP-18 column (LiChrosorb, Alltech Associates). Peptides were separated over 3 hours using a 5–40% propanol gradient in 0.1% TFA. Peptides were detected by their absorbance at 206 nm and peak fractions were collected. One-tenth volume of each sample was scintillation counted to identify peaks containing carboxymethyl cysteine residues. Peptide fractions held in common among TH, DBH, and PNMT were concentrated on a Speed Vac apparatus and hydrolyzed for amino acid analysis.

Determination of amino acid composition. Amino acid compositions of peptides obtained by chemical or enzymatic digestion were determined as described by Spackman et al. (1958) and Stein et al. (1975).

In vitro mRNA translation and enzyme immunoprecipitation. Polysomal poly(A) mRNA was purified from bovine adrenal medulla, and the mRNA was translated in vitro in a reticulocyte lysate translation system as described previously (Baetge et al. 1981; Park et al. 1982). Total bovine adrenal mRNA was isolated as described by Seeburg et al. (1977).

Upon completion of translation reactions, the lysates were centrifuged at 100,000*g* for 1 hour at 2°C to remove polysomes. Excess amounts of antibody to TH, DBH, tTH, or PNMT were added to 100,000*g* supernatants from mRNA translation reactions. The mixture was allowed to stand at room temperature for 30–60 minutes and at 4°C for 16 hours. An excess of Protein-A-coated *Staphylococcus aureus* cells was added, and the mixture was incubated for 20 minutes at room temperature. The immune-complexed cells were pelleted and washed five times with 0.5–1.0 ml 50 mM Tris-HCl (pH 7.4), containing 150 mM NaCl, 10 mM L-methionine, 0.05% NP-40, and 0.02% Na-azide. The washed pellets were boiled in SDS-polyacrylamide gel electrophoresis sample buffer (62.5 mM Tris-HCl [pH 6.8], 5% 2-mercaptoethanol, 2% SDS, 10% glycerol, and 0.01% bromophenol blue), the cells were pelleted, and the supernatant was removed.

The boiled immunoprecipitates were subjected to SDS-slab gel electrophoresis (Laemmli 1970). For fluorography, slabs were fixed in 10% acetic acid and 45% methanol, impregnated with NEN En^{3}Hance, and dried by heat vacuum. The dried gels were exposed to Kodak XAR X-ray film at −70°C for 1–7 days.

Partial purification of PNMT and DBH mRNAs. PNMT mRNA was purified by immunoprecipitation of bovine adrenal medullary polysomes with affinity-purified PNMT antibodies, followed by oligo(dT)-cellulose column chromatography, as detailed elsewhere (Baetge et al. 1981).

DBH mRNA was partially purified from total bovine adrenal mRNA by sucrose gradient fractionation; 5–20% neutral sucrose gradients were poured into SW 41 tubes and loaded with 500 μg total poly(A) mRNA. Centrifugation proceeded at 22,000 rpm for 16 hours at 10°C. The gradients were separated into 0.35-ml fractions and monitored for absorbance at 260 nm. RNA-containing fractions were precipitated with ethanol and ammonium acetate, and the pelleted RNA was translated in the reticulocyte lysate translation system as described. Translated fractions were precipitated with DBH antibody, and precipitates were visualized on SDS-polyacrylamide gel electrophoresis. Fractions containing the highest levels of DBH mRNA were utilized for cDNA synthesis and cloning.

Synthesis of cDNA. Production of double-stranded cDNA for immune-purified PNMT mRNA and sucrose-gradient-purified DBH mRNA was by two separate methods. PNMT cDNA synthesis was detailed in our previous publication (Baetge et al. 1983). Sucrose-gradient-enriched DBH mRNA was used as a template for the synthesis of double-stranded cDNA using the reaction conditions of Ohno et al. (1980). Tailing of double-stranded cDNA and pBR322, transformation, and cDNA clone identification were detailed elsewhere (Baetge et al. 1983).

Enzyme cDNA probe production. PNMT or DBH enzyme plasmid DNA was nick-translated to specific activities of 1×10^8 to 2×10^8 cpm/mg as described previously (Baetge et al. 1983).

Northern blot hybridization. Poly(A) mRNA was denatured at 50°C for 1 hour in 1 M glyoxal, 50% DMSO, and 10 mM sodium phosphate (pH 7.0) and subjected to electrophoresis in 1.1–1.5% agarose gels in 10 mM sodium phosphate running buffer at 3–6 V/cm with constant buffer recirculation (Charmichael and McMaster 1980). The glyoxylated RNA was transferred to nitrocellulose filters with 3 M NaCl and 0.3 M sodium citrate (20× SSC) by capillary flow for 16 hours (Southern 1975). Baking, hybridization, and washing of the filters were carried out as described by Thomas (1980).

Southern blot hybridization. High-molecular-weight DNA was isolated from crude bovine adrenal nuclei as described by Blin and Stafford (1976). DNA was digested with appropriate restriction enzymes according to the conditions suggested by the manufacturer, and restriction digests were run in 0.8–1.0% agarose gels in 40 mM Tris-acetate, 2 mM EDTA (pH 8.0) at 4–6 V/cm. DNA gels were stained for 20 minutes in 1 μg/ml ethidium bromide and photographed under UV illumination. The gels were then processed essentially as described by Wahl et al. (1979) with omission of the initial depurination step. The blot was baked, then prehybridized in 50% deionized formamide, 5× SSC, 5× Denhardt's solution, 50 mM sodium phosphate (pH 6.5), 1% glycine, and 100 μg/ml sonicated and boiled salmon sperm DNA at 42°C for 4–6 hours. The filter was hybridized in 50% formamide, 5× SSC, 1× Denhardt's solution, 20 mM sodium phosphate (pH 6.5), 100 μg/ml sonicated, denatured salmon DNA, 10% sodium dextran sulfate, and 5×10^6–5×10^7 cpm ^{32}P-labeled probe at 42°C for 12–16 hours. The nitrocellulose filter was washed exactly as described above for Northern blot hybridization.

RESULTS

Peptide Mapping of TH, DBH, and PNMT: Comparison of Large Fragments

Digestion of CA-synthesizing enzymes with CNBr and chymotrypsin produced 55K proteins from both TH and DBH, and amino acid composition of the 55K peptides was very similar (Joh et al. 1981; Ross 1982). Digestion of these enzymes with *S. aureus* V8 protease produced patterns of SDS-polyacrylamide gel electrophoresis shown in Figure 1. Major peptides included a 10K moiety common to DBH and TH and a 20K peptide common to DBH and PNMT. A somewhat diffuse minor band appeared among all three enzymes and a minor 15K peptide was common to PNMT and DBH.

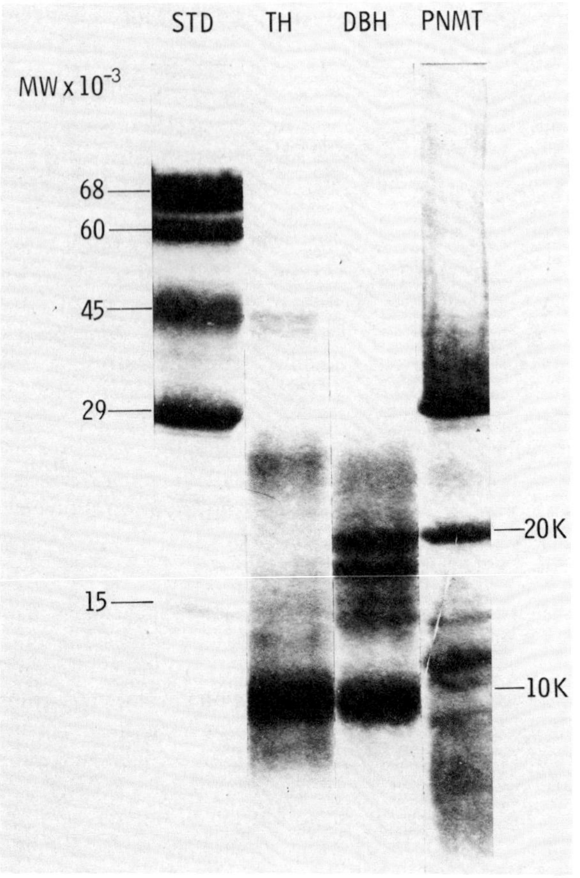

Figure 1. SDS-polyacrylamide gel electrophoresis of *Staphylococcus aureus* V8 protease digests of eluted CA-synthesizing enzymes. Proteins are stained with Coomassie blue (STD = molecular weight standards). The fluorescamine-labeled 20-kD peptides generated from DBH and PNMT and the 10-kD peptides generated from DBH and TH were cut from a duplicate unstained gel slab and eluted, and amino acid compositions were determined.

The 10K and 20K peptides were eluted from an identical unstained gel and were analyzed for amino acid content. An amino acid analysis revealed a striking correlation, indicating that the 10K peptides in TH and DBH as well as the 20K peptides in DBH and PNMT are nearly identical (results not shown).

Peptide Mapping of TH, DBH, and PNMT: Comparison of Small Fragments

In the 5–40% propanol gradient, 32 peptides were separated from TH and PNMT digests, whereas 37 peptides were separated from DBH. Of those, a maximum of 11 peaks in TH, 13 peaks in DBH, and 10 peaks in PNMT contained radioactivity, indicating the presence of cysteine residues (data not shown). Comparing retention times of the peaks, 10–12 peptides in PNMT were comparable to those found in either TH or DBH (data not shown). Of those, two sets of peptides eluting with similar retention times among TH, DBH, and PNMT were tagged with ^{14}C, indicating the presence of cysteine residues in their composition.

The retention times of peptides separated from TH and DBH digests show a striking similarity (Fig. 2). Of the 32 peaks in TH, 21 elute at times comparable with the appearance of peptides in DBH. Furthermore, many of these peaks display a rather characteristic morphology as seen in the superimposition of the two elution patterns presented in Figure 2. Amino acid composition of the 21 peptides common to TH and DBH revealed a high degree of homology between these two enzymes.

Similar analyses of peptides and amino acid composition of the peptides revealed that homology between PNMT and TH, and PNMT and DBH is less than that of TH and DBH. However, approximately 25–30% homologies exist among these enzymes.

Cross-reactivity of Antibodies with mRNA Translation Products

Antibody cross-reactivity of the proteins obtained from in vitro mRNA translation products was different from in vivo translation products (Fig. 3). Antibodies raised against tTH reacted predominantly with a 31-kD protein, identical in size to PNMT. When an antibody raised against the native form of rat adrenal TH was used, coprecipitation of a small amount of a 72-kD protein in addition to the 60-kD TH was observed. This 72-kD protein is identical in size to the protein subunit precipitated with DBH antibody. In addition, the DBH antibody precipitated a small amount of a 60-kD protein, identical in size to TH.

Enzyme Clone Identification through Hybrid Selection

Identification of PNMT clones was detailed elsewhere (Baetge et al. 1983), and similar procedures were used to identify DBH clones.

Preliminary hybrid selection of prospective DBH ampicillin-sensitive clones was carried out on pools of six clones. Positive pools were then analyzed separately as single clones. Pools 1 and 3 contain a plasmid that hybridized to an mRNA species coding for a 72-kD protein as determined by in vitro translation. DBH antibody immunoprecipitation of these same hybrid-selected translation products revealed that pools 1 and 3 each contain one or more recombinant clones that specifically select an mRNA coding for DBH. Pool 1 was rescreened as separate clones. Clones 8, 103, and 335 specifically hybridized to an mRNA that upon translation and immunoprecipitation was demonstrated to code for DBH. Following immunoprecipitation with DBH antibody, as shown in Figure 4, clones 8, 103, and 335 specifically hybrid-selected mRNAs coding for immunoprecipitable DBH protein and were considered to be DBH cDNA clones.

Cross-hybridization of DBH and PNMT mRNAs

While carrying out hybrid selection analysis with suspected DBH cDNA clones, the PNMT cDNA clone

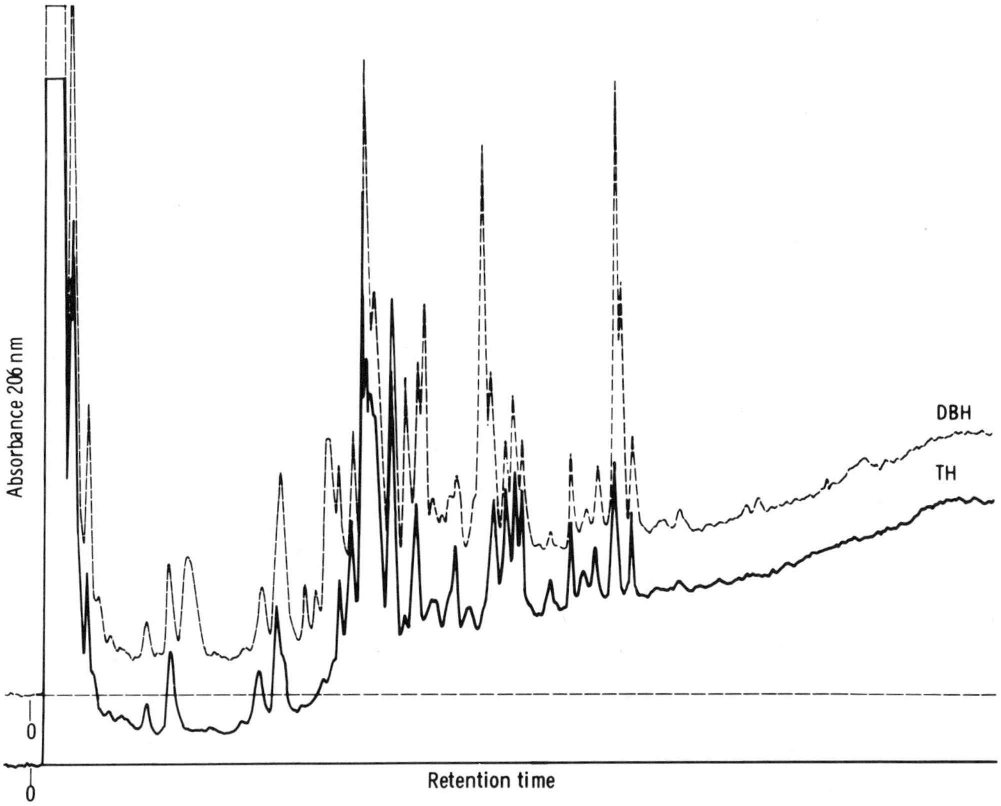

Figure 2. Comparison of HPLC separations of TPCK trypsin digests of TH and DBH. Patterns are superimposed to underscore the similarity between the elution patterns of these two proteolytic digests.

22 was used as an internal control. Unexpectedly, this PNMT cDNA clone 22 appeared to hybrid-select an mRNA that coded for a 72K protein. Following immunoprecipitation with DBH antibody, it became apparent, as shown in Figures 5 and 6, that PNMT clone 22 hybrid-selected DBH mRNA in addition to PNMT mRNA. The results of immunoprecipitation of these same hybrid-selected mRNA translation products with PNMT antibody are shown in Figure 6. Here it is evident that DBH clones 71, 78, and 335 significantly cross-hybrid-select PNMT mRNA. Indeed, both DBH clones 78 and 335 were selected by replicate screening of a cDNA library with a PNMT cDNA clone 22.

Northern Blot Hybridization

^{32}P-labeled PNMT cDNA probe 22 and DBH probe 71 were used for hybridization with poly(A) mRNA bound to nitrocellulose. PNMT probe 22 hybridized to an mRNA from bovine adrenal medulla of approximately 1100 nucleotides in length, using hybridization conditions described in Methods. Under slightly less stringent hybridization conditions (30% formamide, 5× SSC, 50 mM sodium phosphate [pH 6.5], 10 μg/ml denatured salmon sperm DNA, 0.02% Denhardt's solution, and 10% dextran sulfate at 42°C for 16–24 hr), the PNMT probe hybridized weakly to an mRNA of unknown specificity having a length of 5500 nucleotides (Fig. 7). When DBH probe 103 was used as a probe in Northern blot hybridization, an mRNA of approximately 5500 nucleotides was hybridized (Fig. 7), which was identical in size to the cross-hybridizing mRNA species found when PNMT cDNA probe 22 was used under less stringent conditions.

Southern Blot Hybridization

High-molecular-weight DNA (>30 kb) isolated from bovine adrenal medulla was digested with various restriction enzymes and the digests dispersed on agarose gels and transferred to nitrocellulose. The filters were hybridized with DBH cDNA probe 71, and the results are shown in Figure 8. It can be seen that the probe hybridized to one to two bands per lane. Comparison of DBH and PNMT Southern blots is shown in the same figure. PNMT cDNA probe 22 hybridizes to a 4.2-kb molecular weight band in *Bam*HI-digested DNA identical in size to the 4.2-kb molecular weight band found with the DBH cDNA probe. In addition, PNMT 22 hybridized to a 14-kb *Eco*RI DNA band, which is identical in molecular weight to the top *Eco*RI band that hybridized to the DBH 71 probe.

DISCUSSION

In the present studies, we discovered that the CA biosynthetic enzymes are related to one another not only by being sequential enzymes in a pathway, but also by what appeared to be similar protein structures. This is consistent with our biochemical analysis of

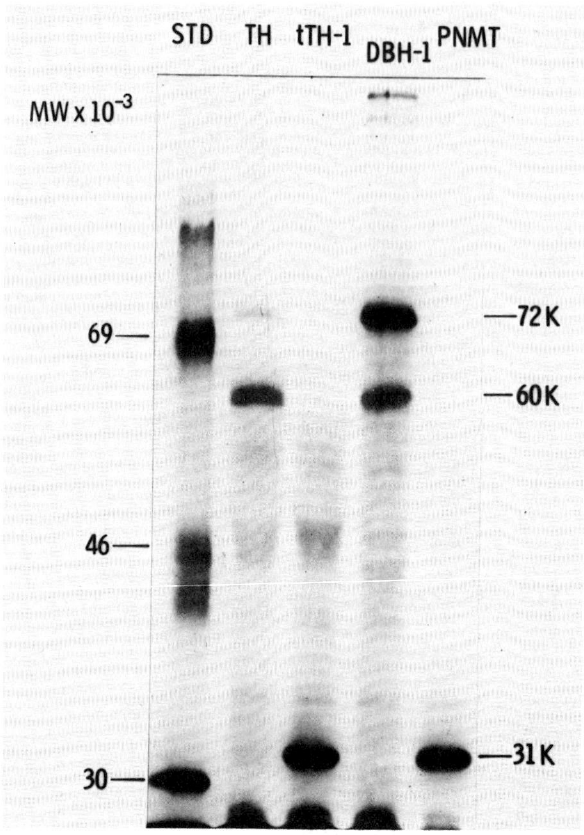

Figure 3. SDS-polyacrylamide gel electrophoresis fluorograph of bovine adrenal medulla poly(A) mRNA translation products immunoprecipitated with CA enzyme antibodies. (TH) Rat adrenal TH immunoprecipitation showing major immunoprecipitation of the 60-kD (K) TH subunit with slight cross-reactivity with a 72K protein; (tTH-1) bovine tTH antibody precipitates a 31K protein identical in size to PNMT; (PNMT) bovine PNMT antibody precipitation of 31K PNMT. DBH-1 antibody precipitates predominantly the 72K DBH but also cross-reacts with a 60K protein, identical in size to TH.

these enzymes, which showed that the amino acid composition of peptides of identical molecular weight produced by proteolytic digestion of the enzymes is nearly identical, and that the amino acid composition of peptides obtained from extensive proteolytic digestion of these enzymes separated by HPLC showed high degrees of homology.

The establishment of an mRNA-dependent in vitro translation assay and immunoprecipitation of the enzyme protein from mRNA translation products revealed that the CA enzymes, TH, DBH, and PNMT, share common antigenic sites when they are in the form of newly translated polypeptide chains, but before any posttranslational modification has occurred. The common antigenic determinants may reflect similarities in specific portions of their primary or folded structures as they exist immediately upon release from the ribosome.

It has been demonstrated in our laboratory that the antibodies to TH, DBH, and PNMT are highly specific in their ability to recognize soluble or immobilized antigen in vivo (Joh and Ross 1983). Indeed, these antibodies have been used for immunocytochemical localization of the specific CA enzymes expressed in dopaminergic and noradrenergic, as well as adrenergic, cell types with no demonstrated cross-reactivity. They are also absolutely specific in their ability to remove enzyme activity from crude or purified enzyme preparations by immunoprecipitation (Joh et al. 1973; Reis et al. 1975; Baetge et al. 1981; Park et al. 1982). DBH antibody does not inhibit TH activity and tTH antibody does not inactivate PNMT. In addition, when tTH, DBH, and PNMT antibodies are used to stain native proteins transferred to nitrocellulose paper (Western blots), no cross-reactivity is observed (T.H. Joh et al., in prep.). tTH antibody only reacts with 60-kD TH, DBH antibody with 75-kD DBH, and PNMT with 31-kD PNMT.

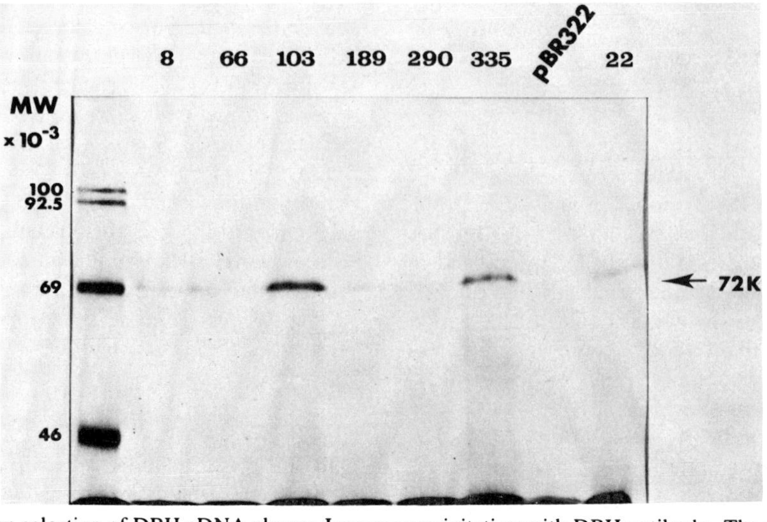

Figure 4. Hybridization selection of DBH cDNA clones: Immunoprecipitation with DBH antibody. The SDS-polyacrylamide gel electrophoresis fluorograph shows DBH immunoprecipitated from the translation products.

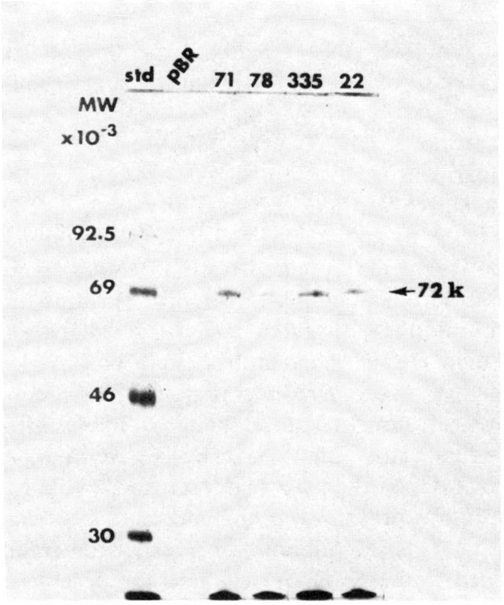

Figure 5. Hybridization selection of DBH cDNA clones: Immunoprecipitation with DBH antibody. PNMT clone 22 hybrid-selects mRNA coding for DBH.

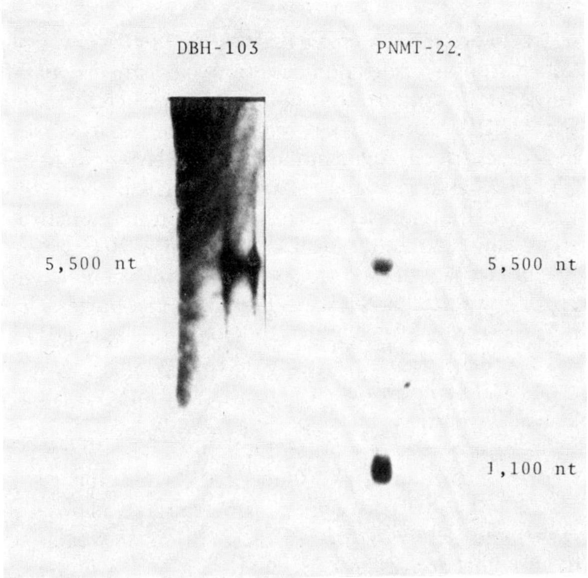

Figure 7. Northern blot hybridization: Comparison of DBH and PNMT cDNA probes. DBH probe 103 hybridizes to a 5500-nucleotide mRNA. PNMT-22 hybridizes predominantly to the 1100-nucleotide mRNA species but cross-hybridizes with the 5500-nucleotide mRNA which hybridizes to DBH probe 103. DBH probes were hybridized in 50% formamide. PNMT probes were hybridized in 30% formamide with remaining conditions exactly as described in Methods; 6 μg bovine adrenal medulla poly(A) mRNA per lane.

Cross-reactivity of the CA enzyme antibodies with in-vitro-synthesized enzymes may be understood if one considers the inherent nature of the polyclonal antibodies produced against these enzymes, as well as the posttranslationally unmodified state of the in-vitro-translated enzyme. For example, since only active enzyme is injected for antibody production, the majority of the antibodies will recognize antigenic sites present on enzyme molecules as they exist in their tertiary, posttranslationally modified, and catalytically active conformation. A major antigenic site present on a catalytically active posttranslationally modified enzyme may be buried or unexposed in vivo on a second enzyme. When both are translated in vitro, the buried antigenic site becomes exposed through lack of posttranslational modification and the antibody to one enzyme now cross-reacts with a second enzyme.

The first evidence for nucleic acid sequence homol-

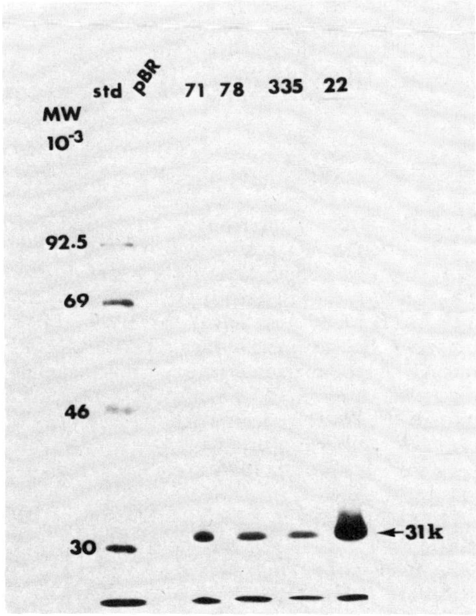

Figure 6. Hybridization selection of suspected DBH cDNA clones: Immunoprecipitation with PNMT antibody. DBH clones 71, 78, and 337 hybrid-select mRNA coding for PNMT.

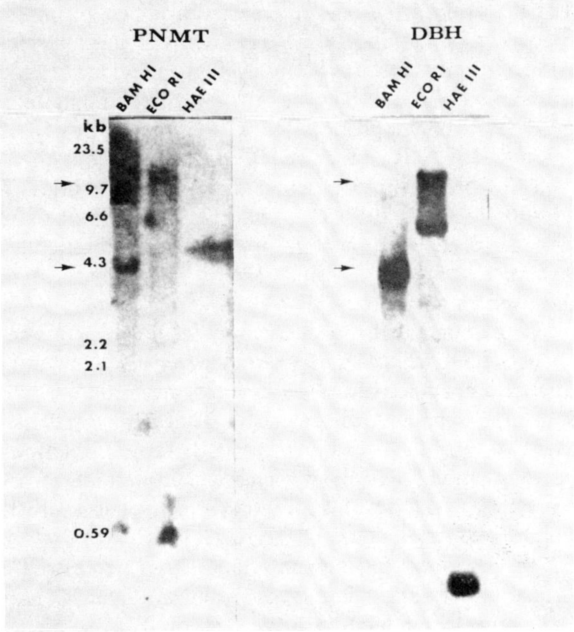

Figure 8. Southern blot hybridization: Comparison of DBH and PNMT cDNA probes. Bovine adrenal medulla DNA was digested and hybridized as described in Methods. The arrows indicate the similar-sized bands; 10 μg bovine adrenal medulla DNA per lane.

ogy between the CA enzymes DBH and PNMT was obtained from hybridization selection experiments. In addition to predominantly hybridizing to mRNA coding for DBH, several DBH clones specifically cross-hybridized to a small amount of mRNA coding for PNMT. Similarly, the PNMT clone 22, while selecting mostly PNMT message, also hybridized to a small but significant amount of mRNA coding for DBH.

The fact that cDNA probes specific for each enzyme cross-hybridize on Northern blots and co-hybrid-select both DBH and PNMT mRNAs from total adrenal medulla mRNA (Figs. 5 and 6) strongly suggests that regions of homology exist in the nucleic acid sequences as well as in the primary protein structure.

There are several similar findings in the literature. Griffin-Shea et al. (1980) found that cDNA clones corresponding to two distinct chorion genes, A_1 and A_2, in *Drosophila melanogaster*, cross-hybridized weakly in hybrid-selection assays. Similarly, it was previously found that the antisera produced against these proteins also cross-reacted (Thireos et al. 1979). From these data, it was suggested that some chorion genes are evolutionarily homologous and may be members of a multigene family.

In comprehensive studies conducted by Scheller et al. (1982, 1983), the structures of three genes encoding the A and B peptide neurotransmitters and egg laying hormone (ELH) in *Aplysia* have been determined. These three genes directly mediate the behavior associated with the egg laying process in *Aplysia*. The three genes are 90% homologous, as determined by blot hybridization, restriction mapping, and sequencing. In addition to determining that the cDNAs to A and B peptide and ELH all cross-hybridize with one another, it was further determined by in vitro translation that the protein precursor containing the A peptide was immunoprecipitated with both the antibody to A peptide and the antibody to ELH. Likewise, the precursor containing ELH was immunoprecipitated with antibodies directed against ELH or B peptide. Each of the three genes contains regions homologous to either A or B peptide in addition to ELH, suggesting that the evolution of three separate genes with separate functions and specific patterns of expression was the result of gene duplication followed by divergence.

Using amino acid sequence analysis, homologous sequences have been identified in members of a bacterial catabolic enzyme pathway known as the β-ketoadipate pathway (Yeh and Ornston 1980). These enzymes, like the CA enzymes, catalyze very different reactions (isomerization, hydrolysis, hydroxylation, and decarboxylation), and yet this amino acid sequence analysis suggests common ancestral origin.

The CA enzyme genes may indeed have followed a similar pattern of evolution consisting of duplication or triplication of an ancestral gene, followed by divergent evolution with conservation of sequences essential for catalytic activity, substrate, and cofactor binding, etc., to produce the unique enzyme functionalities.

To address further the question of CA enzyme gene organization in chromosomal DNA, high-molecular-weight genomic DNA isolated from bovine adrenal medulla was treated with various restriction endonucleases and analyzed by Southern hybridization using cDNA probes to DBH and PNMT.

Comparison of Southern blots performed with both DBH and PNMT probes revealed that both PNMT and DBH cDNA probes hybridized to several common restriction fragments in paired restriction enzyme digests of bovine adrenal medulla DNA (Fig. 8). For example, a 4.2-kb *Bam*HI fragment is hybridized to both by PNMT and DBH cDNA probes. Both probes also hybridize to a 14-kb *Eco*RI fragment. These common restriction fragments could be completely different gene segments of similar size or the result of cross-hybridization of the two cDNA probes with homologous regions present in each gene.

Finally, if these common-sized restriction fragments are indeed the same DNA segment, the possibility exists either that these enzyme genes are linked or that these enzymes are each coded for by a colinear gene segment that can differentially or coordinately produce DBH and PNMT enzyme mRNAs.

In conclusion, the evidence presented suggests the existence of homologous gene coding regions in PNMT and DBH cDNAs. These homologies may be the result of duplication of a common ancestral gene. Southern blot analysis suggests that the two enzymes are coded for by single genes that may be located in close proximity to each other in the DNA, and points to the possible existence of either a single gene or linked genes coding for the CA enzymes PNMT and DBH.

ACKNOWLEDGMENTS

We are deeply in debt to Dr. Hong Mo Moon, NYS Institute of Basic Research for Developmental Disabilities, for his excellent technical advice. This work is supported by National Institutes of Health grants HM24285 and HL18974.

REFERENCES

Baetge, E.E., B.B. Kaplan, D.J. Reis, and T.H. Joh. 1981. Translation of tyrosine hydroxylase from poly(A)-mRNA in pheochromocytoma cells is enhanced by dexamethasone. *Proc. Natl. Acad. Sci.* **78:** 1269.

Baetge, E.E., H.M. Moon, B.B. Kaplan, D.H. Park, D.J. Reis, and T.H. Joh. 1983. Identification of clones containing DNA complementary to phenylethanolamine N-methyltransferase mRNA. *Neurochem. Int.* (in press).

Blaschko, H. 1939. The specific action of L-dopa decarboxylase. *J. Physiol.* **96:** 50.

Blin, N. and D. Stafford. 1976. Isolation of high molecular weight DNA. *Nucleic Acids Res.* **3:** 2303.

Charmichael, G.G. and G.K. McMaster. 1980. The analysis of nucleic acids in gels using glyoxal and acridine orange. *Methods Enzymol.* **65:** 380.

Griffin-Shea, R.G. Thireos, and F. Kafatos. 1980. Chorion cDNA clones of *D. melanogaster* and their use in studies of sequence homology and chromosomal location of chorion genes. *Cell* **19:** 915.

Jaeger, C.B., G. Teitelman, T.H. Joh, V.R. Albert, D.H. Park, and D.J. Reis. 1983. Some neurons of rat

CNS contain aromatic L-amino acid decarboxylase but not monoamines. *Science* **219:** 1233.

JOH, T.H. and M.E. ROSS. 1983. Preparation of catecholamine synthesizing enzymes as immunogens for immunohistochemistry in immunocytochemistry. In *Neuroimmunocytochemistry. Methods in Neurosciences* (ed. A.C. Cuello). Wiley, New York. (In press)

JOH, T.H., C. GEGHMAN, and D.J. REIS. 1973. Immunochemical demonstration of increased accumulation of tyrosine hydroxylase protein in sympathetica ganglia and adrenal medulla elicited by reserpine. *Proc. Natl. Acad. Sci.* **70:** 2767.

JOH, T.H., E.E. BAETGE, B.B. KAPLAN, M.E. ROSS, M.J. BRODSKY, V.R. ALBERT, D.H. PARK, and D.J. REIS. 1981. Do catecholamine enzymes share common gene coding sequences? *Soc. Neurosci. Abstr.* **7:** 206.

LAEMMLI, U.K. 1970. Cleavage of structural proteins during the assembly of the head of bacteriophage T4. *Nature* **227:** 3881.

LAI, C.Y., E. MENDEZ, and D. CHANG. 1976. Chemistry of cholera toxin: The subunit structure. *J. Infect. Dis.* **133:** S23.

OHNO, T., P.J. COZENS, A.C.B. CATO, and J.P. JOST. 1980. Recombinant plasmids containing avian vitellogenin structural gene sequences derived from complementary DNA. *Biochim. Biophys. Acta* **79:** 409.

PARK, D.H., E.E. BAETGE, B.B. KAPLAN, V.R. ALBERT, D.J. REIS, and T.H. JOH. 1982. Different forms of adrenal phenylethanolamine *N*-methyltransferase: Species specific post-translational modification. *J. Neurochem.* **38:** 410.

REIS, D.J., T.H. JOH, and R.A. ROSS. 1975. Effects of reserpine on activities and amounts of tyrosine hydroxylase and dopamine β-hydroxylase in catecholaminergic neuronal systems in rat brain. *J. Pharmacol. Exp. Ther.* **193:** 775.

ROSS, M.E. 1982. "Studies of neurotransmitter synthesizing enzymes. I. Production of monoclonal antibodies to tyrosine hydroxylase and choline acetyltransferase. II. Comparative structural analyses of catecholamine synthesizing enzymes: Are they derived from the same gene?" Ph.D. thesis, Cornell University Medical College, New York, New York.

ROSS, R.A., T.H. JOH, and D.J. REIS. 1975. Reversible changes in the accumulation and activities of tyrosine hydroxylase and dopamine β-hydroxylase in neurons of the nucleus locus coeruleus during the retrograde reaction. *Brain Res.* **92:** 57.

——. 1978. Increase in the relative rate of synthesis of dopamine β-hydroxylase in the nucleus locus coeruleus elicited by reserpine. *J. Neurochem.* **31:** 1491.

——. 1979. Reduced rate of biosynthesis of dopamine β-hydroxylase in the nucleus locus coeruleus during the retrograde reaction. *Brain Res.* **169:** 174.

SCHELLER, R.H., J.F. JACKSON, L.B. MCALLISTER, B.S. ROTHMAN, E.R. MAYERI, and R. AXEL. 1983. A single gene encodes multiple neuropeptides mediating a stereotyped behavior. *Cell* **28:** 32.

SCHELLER, R.H., J.F. JACKSON, L.B. MCALLISTER, J.H. SCHWARTZ, E.R. KANDEL, and R. AXEL. 1982. A family of genes that codes for ELH, a neuropeptide eliciting a stereotyped pattern of behavior in *Aplysia*. *Cell* **28:** 707.

SEEBURG, P.H., H. SHINE, J.A. MARTIAL, A. ULLRICH, J.D. BAXTER, and H.M. GOODMAN. 1977. Nucleotide sequence of part of the gene for human chorionic somatomatropin: Purification of DNA complementary to predominant mRNA species. *Cell* **12:** 157.

SOUTHERN, E. 1975. Detection of specific sequences among DNA fragments separated by gel electrophoresis. *J. Mol. Biol.* **98:** 503.

SPACKMAN, D.H., W.H. STEIN, and S. MORE. 1958. Automatic recording apparatus for use in the chromatography of amino acids. *Anal. Chem.* **30:** 1190.

STEIN, S., A. SJOERDSMA, and S. UDENFRIEND. 1975. Amino acid analysis with fluorescamine at the picomole level. *Arch. Biochem. Biophys.* **155:** 202.

THIREOS, G., R.G. SHEA, and F.C. KAFATOS. 1979. Identification of chorion protein precursors and the mRNAs that encode them in *Drosophila melanogaster*. *Proc. Natl. Acad. Sci.* **76:** 6279.

THOMAS, P.S. 1980. Hybridization of denatured RNA and small DNA fragments transferred to nitrocellulose. *Proc. Natl. Acad. Sci.* **77:** 5201.

WAHL, G.M., R.A. PADGETT, and G.R. STARK. 1979. Gene amplification causes overproduction of the first three enzymes of UMP synthesis in *N*-(phosphonacetyl)-L-aspartate-resistant hamster cells. *J. Biol. Chem.* **254:** 8679.

YEH, W.K. and L.N. ORNSTON. 1980. Origins of metabolic diversity: Substitution of homologous sequences into genes for enzymes with different catalytic activities. *Proc. Natl. Acad. Sci.* **77:** 5365.

Neuron-specific Phosphoproteins as Models for Neuronal Gene Expression

L.J. DeGennaro,* S.D. Kanazir,† W.C. Wallace,† R.M. Lewis,† and P. Greengard†
*Department of Neurochemistry, Max Planck Institute for Psychiatry, Martinsried, Federal Republic of Germany;
†The Rockefeller University, New York, New York 10021

Recent advances in molecular biology have led to the determination of the general features of eukaryotic genes and to an understanding of the processes required for the expression of these genes as mature polypeptide-coding mRNAs. It is now possible to determine the level at which the regulation of gene expression occurs—at transcription, processing, mRNA stability, or translation. A topic of much current interest in developmental neurobiology is the investigation of the molecular mechanisms that regulate the expression of genes coding for neuron-specific proteins. This interest arises from the observation that the nervous system is composed of a great number of different cell types that exhibit a complex array of interconnections and functions and from the inference that this diversity must require the modulation of gene expression.

The developing nervous system provides a particularly challenging model for studying cell differentiation because of the large number of different cell types and the complex array of interconnections that are necessary for the proper functioning of the adult nervous system. This complexity can be exploited for the study of the regulation of gene expression since the diversity of cell types arises from the expression of genes coding for a specialized subset of proteins unique to the nervous system. Determining the mechanisms by which individual cells in the nervous system control the expression of specific genes during development would represent a major step toward understanding how cell differentiation occurs.

Neuron-specific phosphoproteins may provide a model system for studying control of gene expression during neuronal development. Over 70 nervous system-specific phosphoproteins have been identified (Walaas et al. 1983a), and evidence has been presented that indicates that protein phosphorylation appears to regulate many responses of neurons to neurotransmitter stimulation (Greengard 1981). The physicochemical properties and neuronal distribution of two of these proteins, Synapsin I and Protein III, have been extensively characterized. Synapsin I is a neuron-specific phosphoprotein found throughout the central and peripheral nervous system. It is composed of two polypeptides of 86 kD (Ia) and 80 kD (Ib). These polypeptides of Synapsin I exhibit many similar properties, including substantial immunological cross-reactivity (Browning et al. 1982). Immunohistochemistry by light and electron microscopy has localized Synapsin I specifically to the synaptic terminal region of most, if not all, neurons in association with the membranes of synaptic vesicles (DeCamilli et al. 1983a,b; Huttner et al. 1983). Synapsin I comprises approximately 0.4% of the total protein of a cortical homogenate of rat brain and 6% of the total protein in a highly purified fraction of synaptic vesicles (Goelz et al. 1981; Huttner et al. 1983). The protein is phosphorylated by both cAMP-dependent and Ca^{++}/calmodulin-dependent protein kinases, and the state of phosphorylation is altered by conditions that affect neuronal activity (Greengard 1981).

Protein III is also composed of two polypeptides of 74 kD and 55 kD, respectively (Huang et al. 1982). Both polypeptides of Protein III are substrates for the cAMP-dependent protein kinase. These polypeptides exhibit substantial immunological cross-reactivity between themselves, as well as with the polypeptides of Synapsin I.

Synapsin I and Protein III represent distinct, well-characterized, neuron-specific biochemical markers. The analysis of the regulation of the expression of the genes that code for these proteins may therefore serve as a model system for the study of neuronal gene expression. In the present study, we describe the application of an in vitro translation system to assay Synapsin I and Protein III mRNA activities. Using this system we have measured the activity of polysome-associated Synapsin I and Protein III mRNAs during postnatal development of the rat brain and have partially purified the mRNA coding for these neuron-specific phosphoproteins.

MATERIALS AND METHODS

Preparation of polysomes. Rats were killed by decapitation, and the brains, including cerebella, were removed, immediately frozen in liquid nitrogen, and stored at −70°C until use. Polysomes were prepared according to the procedures of MacDonald et al. (1977) and Gough and Adams (1978). All buffers were autoclaved for 30 minutes, and glassware was heat-treated at 180°C for 16 hours. Polysomes were resuspended in Buffer B (0.3 M NaCl, 5 mM $MgCl_2$, 50 mM Tris-HCl [pH 7.4]) and adjusted to a final concentration of approximately 100 A_{260} units/ml. The polysomes were then rapidly frozen and stored at −70°C. Typical yields were approximately 10 A_{260} units of free polysomes/g and 1.6 A_{260} units of bound polysomes/g (wet weight) of brain.

Preparation of anti-Synapsin I and anti-Protein III immunoglobulin. Rabbit anti-Synapsin I serum raised against purified bovine Synapsin I was a gift of Dr. E.J. Nestler of this laboratory. Rabbit anti-Protein III serum raised against purified bovine Protein III was a gift of Dr. M.D. Browning of this laboratory. Specific immunoglobulins were purified from the sera by antigen-Sepharose affinity chromatography. Purified bovine Synapsin I or Protein III was covalently linked to 0.5 ml (packed volume) of CNBr-activated Sepharose as suggested by the supplier (Pharmacia). Purification of the specific immunoglobulin was then performed following the MCPII immunoglobulin purification procedure of Gough and Adams (1978).

In vitro translation. Rabbit reticulocytes were prepared and lysed by the method of Evans and Lingrel (1966) and stored in 0.5-ml aliquots in liquid nitrogen. Endogenous mRNA was inactivated by the procedure of Pelham and Jackson (1976), and the lysate was used for in vitro translation directed by added polysomes or partially purified mRNA as described previously (DeGennaro et al. 1983).

Immunoprecipitation and visualization of translation products. After termination of the reaction, translation reaction mixtures were adjusted to 0.3 M NaCl, 0.5% NP-40, and 5 mg/ml bovine serum albumin. Immediately before immunoprecipitation, the reaction mixtures were precleared by incubation for 5 minutes with 10 μl *Staphylococcus aureus*, Cowan I, followed by centrifugation for 2 minutes in a Beckman microfuge. The precleared supernate was incubated with anti-Synapsin I immunoglobulin or anti-Protein III immunoglobulin for 20 minutes and then treated with 10 μl of *S. aureus*, Cowan I, for 10 minutes. Bacteria-bound immunocomplexes were pelleted by centrifugation as above. Pellets were washed three times by resuspension in 0.8 ml of NET buffer containing NP-40 (Ivarie and Jones 1979) followed by centrifugation. The final pellets were suspended in 100 μl of SDS-sample buffer (Laemmli 1970) and heated at 100°C for 2 minutes before electrophoresis. SDS-slab polyacrylamide gel electrophoresis (PAGE) was performed as described by Laemmli (1970). Following electrophoresis, the gels were fixed and processed for fluorography with En^3-Hance (New England Nuclear). The gels were exposed to Kodak XAR-5 film at −70°C.

Translation assay during development. Rat pups were taken at various times after birth. Four pups (each from a different litter) were killed by decapitation, and the entire brain was removed and immediately frozen in liquid nitrogen. The brains were stored at −70°C until all the brains for the entire experiment were obtained so that all the polysomes were collected under identical conditions. The four brains for each time point were pooled. Polysomes were isolated as described above. To overcome the variation in polysome recovery from day-to-day during development, equal amounts of polysomes from rat pups of different ages (as determined by the absorbance at 260 nm) were translated in the rabbit reticulocyte lysate system. Generally, 0.2–0.4 absorbance units of polysomes were translated per assay.

After the translation incubation, total incorporation of [^{35}S]methionine into newly synthesized peptides was determined by the filter paper assay (DeGennaro et al. 1983). A 5-μl aliquot of the translation mixture was analyzed for total newly synthesized protein by SDS-polyacrylamide gel electrophoresis and fluorography. The remainder of the translation mixture was analyzed for Synapsin I or Protein III by immunoprecipitation, SDS-polyacrylamide gel electrophoresis, and fluorography as described above. In the case of Synapsin I, the immunoprecipitated [^{35}S]methionine-containing bands, which had the same electrophoretic mobility as standard [^{32}P]phospho-Synapsin I, were cut from the gel. These gel pieces were reswollen in water and then incubated with 1 ml of Protosol overnight at 65°C in large glass scintillation vials, after which 15 ml of Liquiscint scintillation fluid was added. The amount of [^{35}S]methionine was determined by liquid scintillation spectrophotometry, and the [^{35}S]methionine content of the gel band was considered to be a measure of the amount of Synapsin I synthesized during the translation incubation. In the case of Protein III, the relative incorporation of [^{35}S]methionine was determined by densitometric analysis of fluorograms. The peak areas were normalized per gram wet weight of brain tissue.

mRNA purification. Synapsin I and Protein III mRNAs were partially purified by specific immunoadsorption of polysomes followed by oligo(dT)-cellulose affinity chromatography, as described by Shapiro and Young (1981). Polysomes (100–200 A_{260} units) suspended in 2 ml of Buffer B containing 0.1% NP-40 and 20 units/ml heparin were incubated with 250 μg of purified immunoglobulin for 60 minutes at 4°C. The antibody-bound polysomes were isolated by affinity chromatography using Protein A-Sepharose resin (Pharmacia). RNA was released from specifically bound polysomes with buffer containing 20 mM EDTA. The released RNA was fractionated on oligo(dT)-cellulose. The eluted poly(A)$^+$ RNA fraction was precipitated with 2.5 volumes of ethanol in the presence of 0.2 M sodium acetate. The precipitate was collected by centrifugation at 0°C in an HB-4 rotor at 16,300*g* for 30 minutes. The pellet was vacuum-dried and resuspended in 0.2 ml of water or 1 mM EDTA and 10 mM Tris-HCl (pH 8.0). This partially purified mRNA was assayed by the in vitro translation system described above.

RESULTS

In Vitro Translation of Synapsin I and Protein III Polysome-associated mRNAs

When polysomes isolated from rat brain were used to direct protein synthesis in the rabbit reticulocyte lysate translation reaction mixture, numerous polypeptides ranging from M_r 18,000 to 250,000 were synthesized. Synapsin I and Protein III were tentatively identified by immunoprecipitation to be among those proteins synthesized by cytosolic polysomes isolated from total rat brain (Figs. 1 and 2). Incorporation of [^{35}S]methionine into newly synthesized proteins ceased after 30 minutes of incubation under the experimental conditions employed. The addition of a second aliquot of polysomes, but not of additional reticulocyte lysate, stimulated further protein synthesis. Protein synthesis in this system, therefore, probably resulted from the elongation of existing nascent polypeptide chains without reinitiation (DeGennaro et al. 1983).

The proteins immunoprecipitated by the appropriate immunoglobulins were identical to authentic standard phosphoproteins by several other criteria. For both proteins, the newly synthesized protein exhibited the same electrophoretic mobility as the mature protein isolated from adult brains (Figs. 1 and 2).

In the case of Synapsin I, the two [^{35}S]methionine-containing polypeptides appeared to be synthesized in the same 1:2 ratio characteristic of mature Synapsin Ia and Synapsin Ib. (Synapsin Ia and Synapsin Ib contain an equal number of methionine residues [Ueda and Greengard 1977].) Newly synthesized Protein III also appeared as two polypeptides, Protein IIIa and Protein IIIb. In addition, the immunoprecipitation of each of these proteins was selectively inhibited by the addition of the corresponding nonradioactive purified protein to the reaction mixture just prior to the addition of immunoglobulin (Figs. 1 and 2). Finally, when cytosolic polysomes isolated from rat liver were used to program protein synthesis, neither Synapsin I nor Protein III, both being neuron-specific proteins, was detected among the numerous proteins synthesized (Figs. 1 and 2).

We have also investigated the intracellular location of the synthesis of these two proteins. For the study of Synapsin I synthesis, both cytosolic and membrane-bound polysomes were isolated from rat whole brain. When each fraction was incubated with the in vitro translation mixture and the resulting newly synthesized proteins immunoprecipitated with anti-Synapsin I immunoglobulin, Synapsin I was detected only among the proteins synthesized by the cytosolic polysomes and not among those synthesized by the membrane-bound polysomes (Fig. 3). Using this same procedure, we found that Protein III was also synthesized by cytosolic polysomes. This location of synthesis is consistent with the intracellular location of these proteins: Synapsin I and Protein III are extrinsically associated with the cytoplasmic surface of the synaptic vesicle membrane. After immunoprecipitation from in vitro translation reaction mixtures, newly synthesized Synapsin I and Protein III were shown to act as substrates for cAMP-dependent protein kinase, indicating another similarity with the mature phosphoprotein (Fig. 4).

Translation of Polysome-associated Synapsin I and Protein III mRNAs during Development

Cytosolic polysomes isolated from total rat brain at various stages of postnatal development were used to

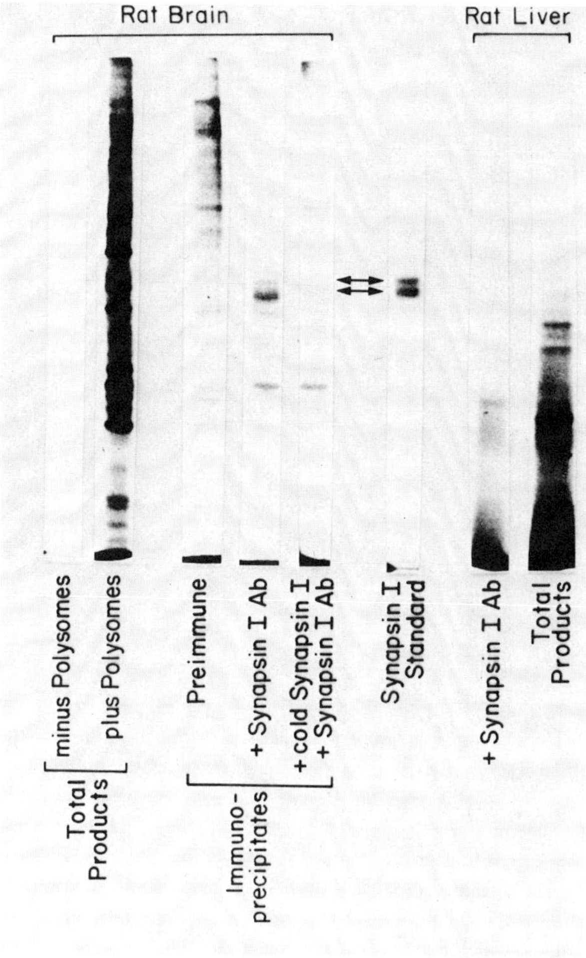

Figure 1. In vitro translation of polysome-associated Synapsin I mRNA. Polysomes were purified from the brains of 14-day-old rats and translated in a reticulocyte lysate translation reaction in the presence of [^{35}S]methionine (DeGennaro et al. 1983). Total translation products (Total Products) or translation products precipitated by Synapsin I immunoglobulin (Immunoprecipitates) were separated by SDS-PAGE, and the [^{35}S]methionine incorporated into protein was visualized by fluorography. The amount of translation reaction mixture used to analyze total products represented 2% of that used for immunoprecipitation. (minus Polysomes) Translation products after the reticulocyte lysate mixture had been treated with micrococcal nuclease, but without the addition of polysomes; (plus Polysomes) translation products of the reticulocyte lysate mixture after treatment with micrococcal nuclease, followed by the addition of cytosolic polysomes; (Preimmune serum) newly synthesized polypeptides immunoprecipitated from total translation products using preimmune serum; (+ Synapsin I Ab) newly synthesized polypeptides immunoprecipitated from total translation products using affinity-purified rabbit anti-Synapsin I immunoglobulin; (+ cold Synapsin I + Synapsin I Ab) newly synthesized polypeptides immunoprecipitated from total translation products using affinity-purified anti-Synapsin I immunoglobulin in the presence of 0.5 μg of nonradioactive, purified Synapsin I. Cytosolic polysomes purified from rat liver were translated in a reticulocyte lysate translation reaction. (+ Synapsin I Ab) Newly synthesized polypeptides immunoprecipitated from total translation products using affinity-purified anti-Synapsin I immunoglobulin; (Total Products) total translation products. Synapsin I standard is purified rat Synapsin I that had been phosphorylated by the catalytic subunit of cAMP-dependent protein kinase. Arrows indicate the positions of Synapsin Ia (*upper*) and Synapsin Ib (*lower*).

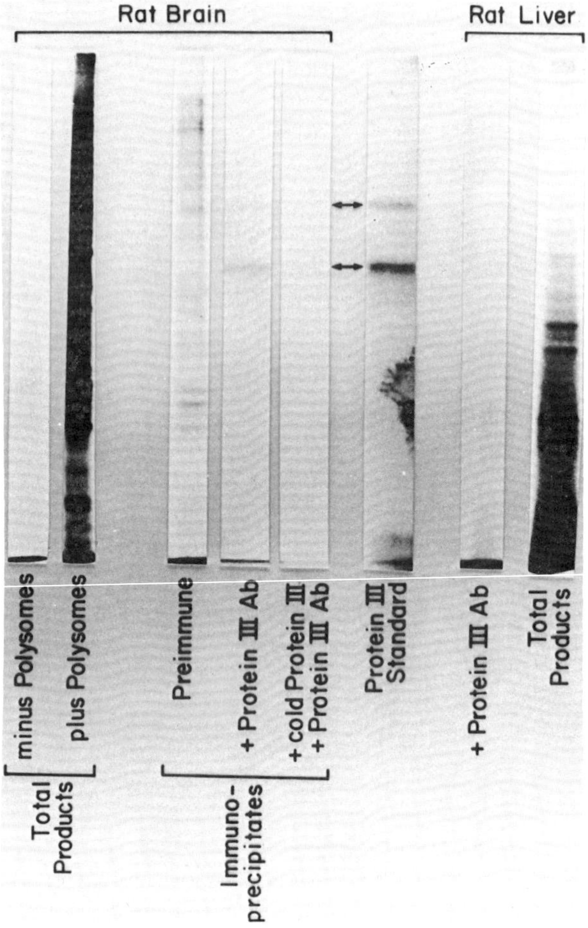

Figure 2. In vitro translation of polysome-associated Protein III mRNA. Polysomes were purified from the brains of 14-day-old rats and translated in a reticulocyte lysate translation reaction in the presence of [^{35}S]methionine (DeGennaro et al. 1983). Total translation products (Total Products) or translation products precipitated by Protein III immunoglobulin (Immunoprecipitates) were separated by SDS-PAGE, and the [^{35}S]methionine incorporated into protein was visualized by fluorography. The amount of translation mixture used to analyze total products represented 2% of that used for immunoprecipitation. (minus Polysomes) Translation products after the mixture had been treated with micrococcal nuclease, but without the addition of polysomes; (plus Polysomes) translation products of the reticulocyte lysate mixture after treatment with micrococcal nuclease, followed by the addition of cytosolic polysomes; (Preimmune serum) newly synthesized polypeptides immunoprecipitated from total translation products using preimmune serum; (+ Protein III Ab) newly synthesized polypeptides immunoprecipitated from total translation products using affinity-purified rabbit anti-Protein III immunoglobulin; (+ cold Protein III + Protein III Ab) newly synthesized polypeptides immunoprecipitated from total translation products using affinity-purified anti-Protein III immunoglobulin in the presence of 0.3 μg of nonradioactive, purified Protein III. Cytosolic polysomes purified from rat liver were translated in a reticulocyte lysate translation reaction. (+ Protein III Ab) Newly synthesized polypeptides immunoprecipitated from total translation products using affinity-purified anti-Protein III immunoglobulin; (Total Products) total translation products. Protein III standard is purified rat Protein III that had been phosphorylated by the catalytic subunit of cAMP-dependent protein kinase. Arrows indicate positions of Protein IIIa (*upper*) and Protein IIIb (*lower*).

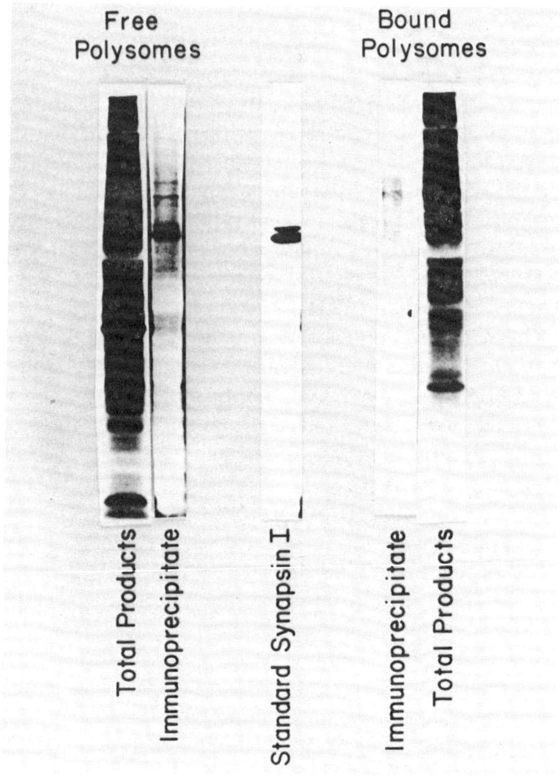

Figure 3. In vitro translation of Synapsin I mRNA: Comparison of cytosolic and membrane-bound polysomes. Cytosolic and membrane-bound polysomes were purified from the brains of 14-day-old rats and translated in a reticulocyte lysate translation reaction in the presence of [^{35}S]methionine (DeGennaro et al. 1983). The total translation products (Total Products) or translation products precipitated by Synapsin I immunoglobulin (Immunoprecipitates) were separated by SDS-polyacrylamide gel electrophoresis, and the [^{35}S]methionine incorporated into protein was visualized by fluorography. (Free Polysomes) Translation products of the reticulocyte lysate mixture after treatment with micrococcal nuclease, followed by the addition of cytosolic polysomes; (Bound Polysomes) translation products of the reticulocyte lysate mixture after treatment with micrococcal nuclease, followed by the addition of membrane-bound polysomes. Synapsin I standard is purified rat Synapsin I that had been phosphorylated by cAMP-dependent protein kinase.

direct protein synthesis in the in vitro system described above. Fluorograms of the translation products present in anti-Synapsin I and anti-Protein III immunoprecipitates are shown in Figures 5 and 7, respectively. Synapsin I mRNA activity measured by this assay increased more than eightfold from birth to postnatal day 14, and then decreased to the adult level (Figs. 5 and 6A,B). The activity of polysome-associated Protein III mRNA was also measured during postnatal rat brain development. Polysome-associated Protein III mRNA activity exhibited developmental changes similar to those observed for Synapsin I (Fig. 7). During the same period, maximal total polysome-associated mRNA activity was observed on postnatal day 2 and steadily decreased with development until it reached a minimum

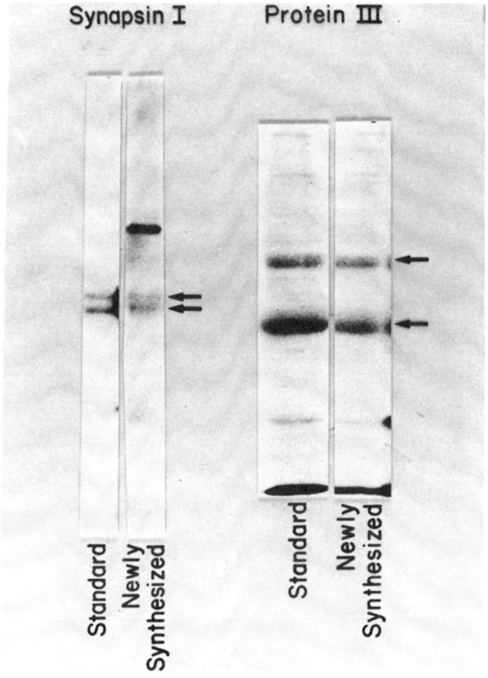

Figure 4. Phosphorylation of newly synthesized Synapsin I and Protein III. Polysomes from rat brain were translated in the rabbit reticulocyte lysate translation system containing nonradioactive methionine. The newly synthesized Synapsin I or Protein III was immunoprecipitated from the translation mixture with the corresponding immunoglobulin. The antigen was dissociated from the antibody and phosphorylated using the catalytic subunit of cAMP-dependent protein kinase and [γ-^{32}P]ATP as described (Nestler and Greengard 1982). The phosphorylation reaction mixture was separated by SDS-polyacrylamide gel electrophoresis and the proteins that incorporated ^{32}P were visualized by autoradiography. (Standard) Purified Synapsin I or Protein III added to the translation reaction mixture, immunoprecipitated, and phosphorylated; (Newly Synthesized) Synapsin I or Protein III synthesized by polysomes, immunoprecipitated, and phosphorylated. The arrows indicate the positions of the appropriate phosphoproteins. The higher-molecular-weight phosphorylated band, which has only appeared in translation mixtures containing polysomes that have been immunoprecipitated with rabbit anti-Synapsin I immunoglobulin and goat anti-rabbit immunoglobulin, has not been identified.

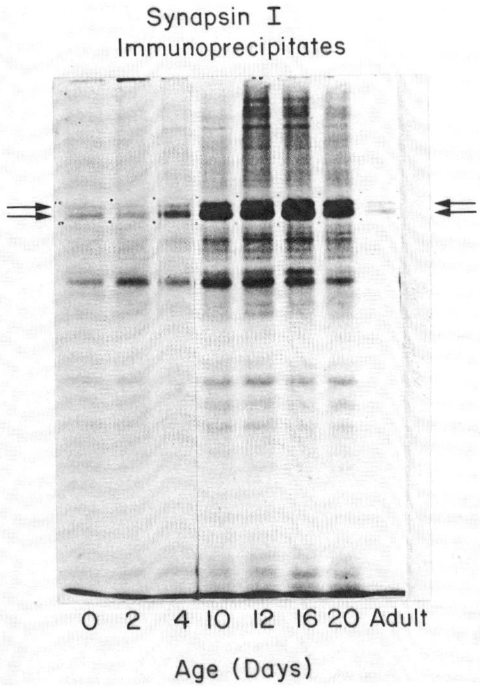

Figure 5. Synapsin I synthesized by polysomes isolated from brains of rats of various ages. Equal amounts of cytosolic polysomes were translated in a micrococcal-nuclease-treated rabbit reticulocyte lysate translation reaction in the presence of [^{35}S]methionine (DeGennaro et al. 1983). Synapsin I was immunoprecipitated from the translation mixture with affinity-purified anti-Synapsin I immunoglobulin. The immunoprecipitates were separated by SDS-polyacrylamide gel electrophoresis and visualized by fluorography. A representative experiment showing the translation products present in Synapsin I immunoprecipitates is presented. The arrows indicate the positions of Synapsin Ia and Synapsin Ib.

in translations directed by polysomes prepared from adult brain (Fig. 6A). These differences in developmental patterns of Synapsin I and Protein III, relative to those of the total polysome-associated mRNA activities, indicate that the changes in Synapsin I and Protein III mRNA activities are specifically controlled changes.

Purification of Synapsin I and Protein III mRNAs

The mRNAs for Synapsin I and Protein III were partially purified by specific immunoadsorption of polysomes binding anti-Synapsin I or -Protein III immunoglobulin and were separated from the remaining polysomes by chromatography on Protein A-Sepharose. The polyadenylated fraction of the immunoadsorbed polysomal RNA was isolated by oligo(dT)-cellulose affinity chromatography. Results of the in vitro translation of these RNA fractions are shown in Figure 8. Minus RNA lanes show the total proteins and immunoprecipitated products synthesized by the micrococcal-nuclease-treated reticulocyte lysate alone. Plus RNA lanes show the total proteins and immunoprecipitated products synthesized by the lysate in the presence of polyadenylated RNA purified as described above. The partially purified poly(A)$^+$ RNA directed the synthesis of new polypeptides that had the same electrophoretic mobilities as the polypeptides of purified rat Synapsin I and Protein III. The identities of the new translation products were confirmed by specific immunoprecipitation from the translation reaction mixtures. Based on the amount of [^{35}S]methionine incorporated into the immunoprecipitated Synapsin Ia and Synapsin Ib (or Protein IIIa and Protein IIIb) relative to the amount of [^{35}S]methionine incorporated into the total translation products, the purity of these mRNAs varied from 50% to 80%. This estimation is based on the assumption that Synapsin I and Protein III contain the same number of methionine residues as the

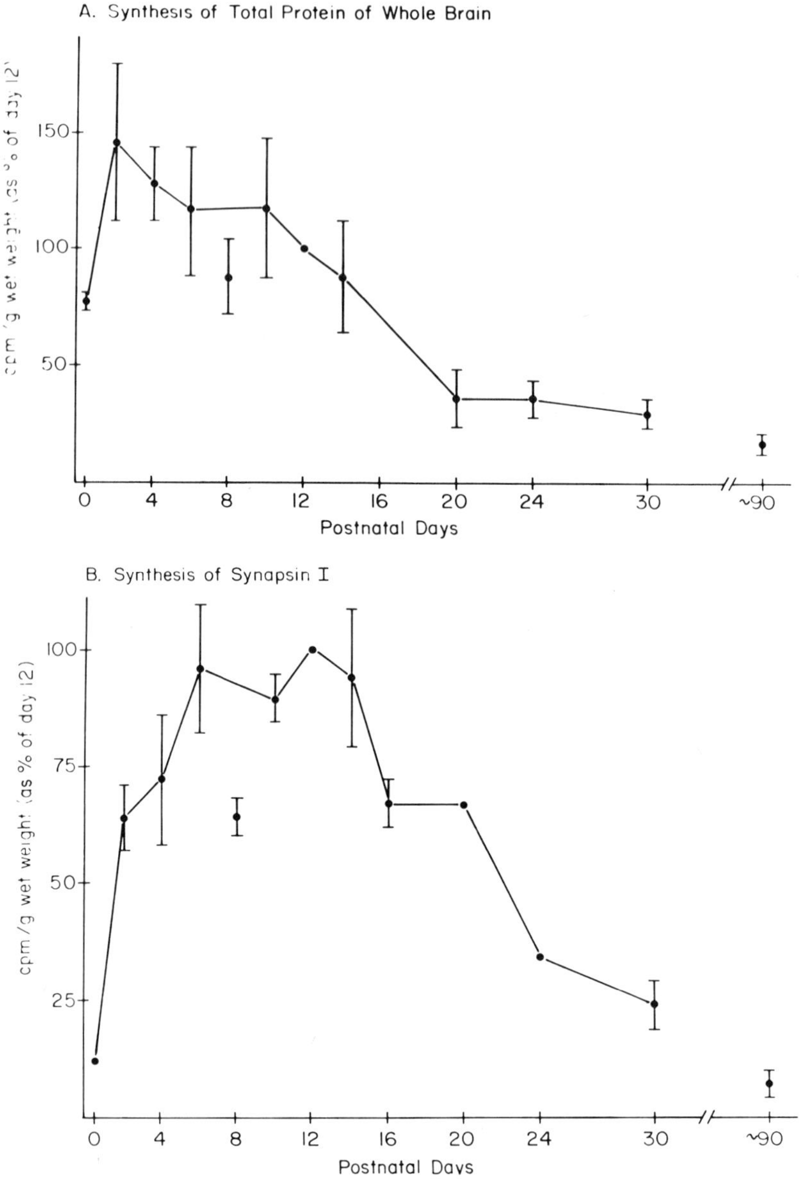

Figure 6. Incorporation of [^{35}S]methionine into total translation products and into Synapsin I as directed by polysomes isolated from brains of rats of various ages. Cytosolic polysomes were purified and translated in a rabbit reticulocyte lysate translation reaction system in the presence of [^{35}S]methionine. (*A*) The total incorporation of [^{35}S]methionine into TCA-insoluble protein was determined by a filter paper assay. (*B*) The reaction mixture was incubated with anti-Synapsin I immunoglobulin, and the proteins present in the immunoprecipitate were separated by SDS-polyacrylamide gel electrophoresis and visualized by fluorography. The [^{35}S]methionine-containing band corresponding to Synapsin I was cut from the gel and analyzed by liquid scintillation spectrometry to determine the amount of [^{35}S]Synapsin I. The incorporated counts of [^{35}S]methionine, for both total translation products and Synapsin I, were calculated per gram wet weight of brain tissue and are expressed as the percentage of the value determined for postnatal day 12. Data are presented as means ± S.D. of the means for three separate experiments.

average for all other polypeptides synthesized by the partially purified mRNA. This purity represents approximately a 200-fold enrichment of Synapsin I and Protein III mRNAs.

The anti-Synapsin I immunoglobulin used as the immunoaffinity adsorbant in the purification of the mRNA fraction whose translation is illustrated in Figure 8A showed substantial cross-reactivity with purified Protein III. As a consequence, mRNA coding for Protein III is also enriched in the partially purified RNA fraction. This is indicated by the in vitro synthesis of Protein IIIb observed in translation reactions directed by this preparation of poly(A)$^+$ RNA (Fig. 8A).

DISCUSSION

We have used an in vitro translation system to assay Synapsin I and Protein III mRNA activities in the brain.

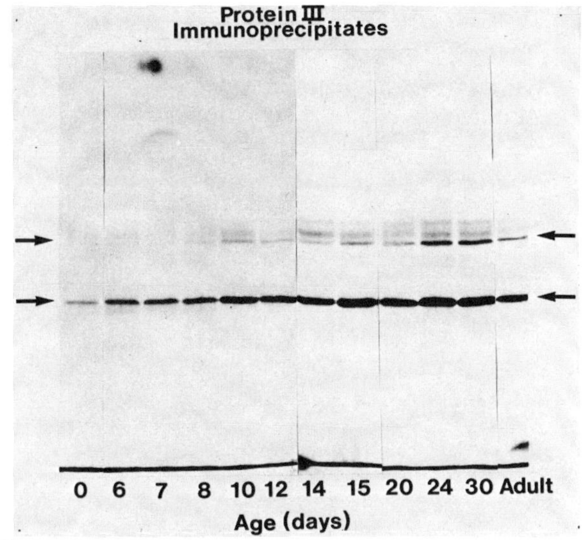

Figure 7. Protein III synthesized by polysomes isolated from brains of rats of various ages. Equal amounts of cytosolic polysomes were translated in a micrococcal-nuclease-treated rabbit reticulocyte lysate translation reaction in the presence of [^{35}S]methionine (DeGennaro et al. 1983). Protein III was immunoprecipitated from the translation mixture with affinity-purified rabbit anti-Protein III immunoglobulin. The immunoprecipitates were separated by SDS-polyacrylamide gel electrophoresis and visualized by fluorography. A representative experiment showing the translation products present in Protein III immunoprecipitates is presented. The arrows indicate the positions of Protein IIIa and Protein IIIb.

The results indicate that these two neuron-specific proteins can be synthesized by brain polysomes in a reticulocyte lysate translation system and that the newly synthesized proteins can be detected by immunoprecipitation. The identities of de novo-synthesized Synapsin I and Protein III have been confirmed on the basis of similarities to the standard purified proteins in antigenicity, electrophoretic mobility, and absence in translations directed by liver polysomes (Figs. 1 and 2). Further confirmation of the similarity of the newly synthesized Synapsin I and Protein III to their native counterparts derives from evidence of the intracellular location of their synthesis on the cytosolic fraction of polyribosomes (Fig. 3) and by their ability to undergo phosphorylation by the cAMP-dependent protein kinase (Fig. 4).

The similarity between the newly synthesized and the mature forms of the polypeptides of Synapsin I and of Protein III, including the molar ratios of the constituent polypeptides, may be due to the translation of mRNAs that are unique to each polypeptide. For example, the mRNA coding for Synapsin Ib may occur at twice the concentration of that coding for Synapsin Ia. Alternatively, either a larger precursor could be processed to yield the polypeptides in the appropriate ratios or the selective posttranslational modification of one polypeptide could produce the proper amount of the other. We have not detected the synthesis of any large precursor. Precursor processing and posttranslational modification would each require either that specific modifying enzymes are already present in the reticulocyte lysate or that they are synthesized from partially purified mRNA during the translation reaction. In addition, we have observed that there was no interconversion between purified mature Synapsin Ia, Synapsin Ib, Protein IIIa, and Protein IIIb added to the in vitro translation mixture (data not shown). These results argue against any specific processing or posttranslational modification to yield the final, mature form of each of these proteins.

The in vitro translation of Synapsin I and Protein III has been used to assay changes in their polysome-associated mRNA activities during postnatal development in the rat brain (Figs. 5, 6, and 7). The changes in Synapsin I and Protein III mRNA activities measured by the translation assay correlated with morphological and biochemical evidence of synaptogenesis. The time during development when these mRNA activities were highest (between 1 and 4 weeks of postnatal life) coincided with a period of rapid formation of recognizable synapses (Aghajanian and Bloom 1967).

Purification of Synapsin I and Protein III mRNAs by specific immunoadsorption of polysomes followed by oligo(dT)-cellulose affinity chromatography resulted in an enrichment of approximately 200-fold in these mRNA species. Both Synapsin I and Protein III could be synthesized when their corresponding partially purified mRNAs were used to direct translation in the rabbit reticulocyte lysate translation system (Fig. 8A,B). These in vitro translation results suggest that no neuron-specific factors are absolutely required for the translation of these mRNAs. However, we have not determined whether the efficacy of translation may be affected by such factors. The translation of partially purified Synapsin I and Protein III mRNAs resulted in the synthesis of the corresponding polypeptides of each protein in the same molar ratio in which they occur in vivo. Possible explanations for this constant ratio have been discussed above. Finally, the retention of Synapsin I and Protein III mRNAs by oligo(dT)-cellulose suggests that the messages are polyadenylated.

We are currently using the partially purified Synapsin I and Protein III mRNA preparations as templates for reverse transcriptase and the production and cloning of complementary DNA molecules. These clones will be used to study the structure and function of the genes coding for these neuron-specific phosphoproteins.

ACKNOWLEDGMENTS

We are indebted to several people for their contributions to this work. Annette E. Gwardyak provided editorial assistance, Karen Hageman prepared the figures, Philip L. Mobley and Wilson C.-S. Wu offered critical comments about the manuscript, Wilson C.-S. Wu provided purified Synapsin I, and Michael D. Browning provided purified Protein III.

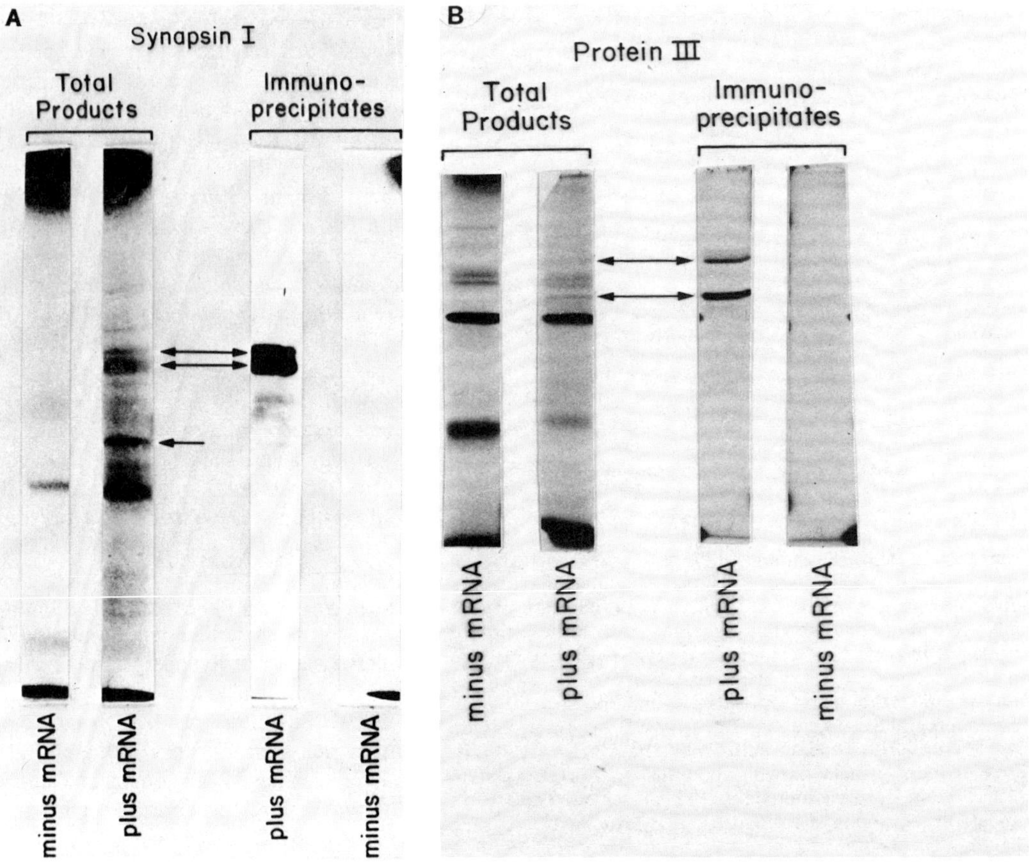

Figure 8. Translation products of partially purified mRNA for Synapsin I (*A*) and Protein III (*B*). mRNAs for Synapsin I and Protein III were purified by immunoadsorption of polysomes followed by oligo(dT)-cellulose affinity chromatography (DeGennaro et al. 1983). [^{35}S]Methionine was incorporated into translation products in a translation reaction (using a reaction mixture that had been treated with micrococcal nuclease) in the absence (minus mRNA) or presence (plus mRNA) of partially purified mRNA for Synapsin I or Protein III. The translation products were separated by SDS-polyacrylamide gel electrophoresis, either before (Total Products) or after (Immunoprecipitates) immunoprecipitation with the corresponding immunoglobulin, and visualized by fluorography. (*A*) The upper two arrows indicate the positions of Synapsin Ia and Synapsin Ib; the lower arrow indicates the position of Protein IIIb, which is antigenically similar to Synapsin I. (*B*) The arrows indicate the positions of Protein IIIa and Protein IIIb.

REFERENCES

AGHAJANIAN, G.K. and F.E. BLOOM. 1967. The formation of synaptic junctions in developing rat brain: A quantitative electron microscopic study. *Brain Res.* **6:** 716.

BROWNING, M.D., C.-K. HUANG, and P. GREENGARD. 1982. Biological similarity of two phosphoproteins (Protein IIIa and IIIb) associated with synaptic vesicles in rat brain. *Soc. Neurosci. Abstr.* **8:** 794.

DECAMILLI, P., R. CAMERON, and P. GREENGARD. 1983a. Synapsin I (Protein I), a nerve terminal-specific phosphoprotein. I. Its general distribution in synapses of the central and peripheral nervous system demonstrated by immunofluorescence in frozen and plastic sections. *J. Cell Biol.* **96:** 1337.

DECAMILLI, P., S.M. HARRIS, JR., W.B. HUTTNER, and P. GREENGARD. 1983b. Synapsin I (Protein I), a nerve terminal-specific phosphoprotein. II. Its specific association with synaptic vesicles demonstrated by immunocytochemistry in agarose embedded synaptosomes. *J. Cell Biol.* **96:** 1355.

DEGENNARO, L.J., W.C. WALLACE, and P. GREENGARD. 1983. *In vitro* translation of Synapsin I, a neuron-specific protein: Developmental studies and messenger RNA purification. *J. Biol. Chem.* (in press).

EVANS, M.J. and J.B. LINGREL. 1966. Hemoglobin messenger ribonucleic acid. Distribution of the 9S ribonucleic acid in polysomes of different sizes. *Biochemistry* **8:** 829.

GOELZ, S.E., E.J. NESTLER, B. CHEHRAZI, and P. GREENGARD. 1981. Distribution of Protein I in mammalian brain as determined by a detergent-based radioimmunoassay. *Proc. Natl. Acad. Sci.* **78:** 2130.

GOUGH, N.M. and J.M. ADAMS. 1978. Immunoprecipitation of specific polysomes using *Staphylococcus aureus*: Purification of the immunoglobulin K chain messenger RNA from the mouse myeloma MPC11. *Biochemistry* **17:** 5560.

GREENGARD, P. 1981. Intracellular messengers in the brain. *Harvey Lect.* **75:** 277.

HUANG, C.-K., M.D. BROWNING, and P. GREENGARD. 1982. Purification and characterization of Protein IIIb, a mammalian brain phosphoprotein. *J. Biol. Chem.* **257:** 6524.

HUTTNER, W.B., W. SCHIEBLER, P. GREENGARD, and P. DECAMILLI. 1983. Synapsin I (Protein I), a nerve terminal-specific phosphoprotein. III. Its association with synaptic vesicles studied in a highly-purified synaptic vesicle preparation. *J. Cell Biol.* **96:** 1374.

IVARIE, R.D. and R.P. JONES. 1979. A rapid sensitive assay for specific protein synthesis in cells and in cell-free translations: Use of *Staphylococcus aureus* as an absorbent for immune complexes. *Anal. Biochem.* **97:** 24.

LAEMMLI, U.K. 1970. Cleavage of structural proteins during

the assembly of the head of bacteriophage T4. *Nature* **227:** 680.

MacDonald, R.J., A.E. Przybyla, and W.J. Rutter. 1977. Isolation and *in vitro* translation of the messenger RNA coding for pancreatic amylase. *J. Biol. Chem.* **252:** 5522.

Nestler, E.J. and P. Greengard. 1982. Distribution of Protein I and regulation of its state of phosphorylation in the superior cervical ganglion. *J. Neurosci.* **2:** 1011.

Pelham, H.R. and R.I. Jackson. 1976. An efficient mRNA-dependent translation system from reticulocyte lysates. *Eur. J. Biochem.* **67:** 247.

Shapiro, S.Z. and J.R. Young. 1981. An immunochemical method for mRNA purification. *J. Biol. Chem.* **256:** 1495.

Ueda, T. and P. Greengard. 1977. Adenosine 3′:5′-monophosphate regulated phosphoprotein system of neuronal membranes. I. Solubilization, purification, and some properties of an endogenous phosphoprotein. *J. Biol. Chem.* **252:** 5155.

Walaas, S.I., A.C. Nairn, and P. Greengard. 1983a. Regional distribution of calcium- and cyclic AMP-regulated protein phosphorylation systems in mammalian brain. I. Particulate systems. *J. Neurosci.* **3:** 291.

———. 1983b. Regional distribution of calcium- and cyclic AMP-regulated protein phosphorylation systems in mammalian brain. II. Soluble systems. *J. Neurosci.* **3:** 302.

Expression of Cell-type-specific Neuronal Phosphoproteins

R.M. LEWIS, W.C. WALLACE, S.D. KANAZIR, AND P. GREENGARD
The Rockefeller University, New York, New York 10021

Until recently, the investigation of cell differentiation in the nervous system has been largely restricted to qualitative descriptions of anatomical changes or measurements of accumulated cell products. Recent advances in techniques developed for assaying gene expression are now being applied to studies of neuronal systems. In vitro translation assays have been used to characterize mRNA translation products in the brain (Fando and Wasterlain 1980) and in neuroblastoma cell lines (Morrison et al. 1979). Some specific examples are the acetylcholine receptor (AChR) (Merlie et al. 1981), neuron-specific enolase (Walker and Zomzely-Neurath 1980), and tyrosine hydroxylase (Baetge et al. 1981). Complementary DNA probes have been used to measure the levels of mRNAs coding for the neural hormones pro-opiomelanocortin (Civelli et al. 1982), proenkephalin (Gubler et al. 1981), and the egg laying hormone (ELH) of *Aplysia* (Scheller et al. 1982). Levels of preprolactin mRNA have been measured during the development of the fetal bovine pituitary (Nilson et al. 1980), and the mRNAs coding for microtubule-associated proteins during brain development have been analyzed (Ginzburg et al. 1982).

Two neuron-specific phosphoproteins that appear to be present in all nerve cells are Synapsin I (DeCamilli et al. 1983a,b) and Protein III (Browning et al. 1982). An in vitro translation assay has been utilized to measure the synthesis of these two phosphoproteins in the developing postnatal rat brain (DeGennaro et al. 1983 and this volume). Two neuron-specific phosphoproteins that appear to be present in only certain types of nerve cells are DARPP-32 and G substrate. DARPP-32 and G substrate are promising candidates as model proteins for studying cell-type-specific gene expression in the nervous system.

DARPP-32 (dopamine and adenosine 3′:5′-monophosphate-regulated phosphoprotein of M_r 32,000) is composed of a single peptide (Hemmings et al. 1983) which is phosphorylated in intact cells from the caudatoputamen in the presence of dopamine or 8-bromo-cAMP, and in cell-free preparations by cAMP-dependent protein kinase (Walaas and Greengard 1983). DARPP-32 is enriched in the caudatoputamen, nucleus accumbens, olfactory tubercle, bed nucleus of the stria terminalis, and portions of the amygdaloid complex, and appears to be localized to neurons that contain D-1 receptors (dopamine receptors coupled to adenylate cyclase) (Ouimet et al. 1983; Walaas et al. 1983; Walaas and Greengard 1983). G substrate is composed of a single peptide of M_r 23,000 that is phosphorylated on two threonine residues by cGMP-dependent protein kinase (Schlichter et al. 1978; Aswad and Greengard 1981a,b). In its phosphorylated form, G substrate can function as a specific inhibitor of cerebellar phosphatase III (P.F. Simonelli et al., in prep.). G substrate is found exclusively in the cytoplasm of the Purkinje cells of the cerebellum (Schlichter et al. 1978, 1980; Dolphin et al. 1983; Detre et al., in prep.).

We have chosen DARPP-32 and G substrate as model proteins for studying cell-type-specific gene expression in the nervous system because they exhibit a high degree of cell-type specificity and are very likely involved in functions characteristic of these specialized cells. We have used an in vitro translation system to study the synthesis of DARPP-32 and G substrate by polysomes isolated from specific regions of the brain and to examine G substrate synthesis during the postnatal development of the rat cerebellum. In addition, the mRNAs coding for DARPP-32 and G substrate have been partially purified.

MATERIALS AND METHODS

Preparation of polysomes. Bovine caudate nuclei were dissected from the brains of calves immediately after death and frozen in liquid nitrogen. Rats were killed by decapitation and the cerebella were rapidly removed and frozen in liquid nitrogen. Frozen rabbit cerebella were purchased from Pel-Freeze. Fresh rabbit cerebella were isolated immediately from animals killed by air embolism. Virtually identical results were obtained with the two types of preparations. Polysomes were purified by centrifugation through sucrose (MacDonald et al. 1977; Gough and Adams 1978), as previously described (DeGennaro et al. 1983), with the following modifications. Buffer A did not contain sodium iodoacetate; cytosolic polysomes were routinely isolated by centrifugation through 20 ml of 2 M sucrose in Buffer A; and polysomes were stored in liquid nitrogen.

Preparation of anti-DARPP-32 immunoglobulin. Rabbit anti-DARPP-32 serum raised against purified bovine DARPP-32 (Hemmings et al. 1983) was obtained from H.C. Hemmings, Jr. of this laboratory. Anti-DARPP-32 immunoglobulin was purified from the serum by DARPP-32-Sepharose affinity chromatography. The DARPP-32-Sepharose affinity resin was prepared by H.C. Hemmings, Jr. according to the procedure suggested by the supplier (Pharmacia). Serum was loaded on the column and the column was washed with 0.15 M NaCl. Fractions containing antibody were eluted with 4 M $MgCl_2$ and dialyzed against 0.1 M NaCl and 0.05 M Tris (pH 7.4).

Preparation of anti-G-substrate immunoglobulin. Rabbit anti-dephospho-G-substrate serum, raised against a seven-amino-acid synthetic peptide corresponding to the site of phosphorylation in G substrate (Nairn et al. 1982), was obtained from A.C. Nairn of this laboratory. Anti-dephospho-G-substrate immunoglobulin was purified from the serum by dephospho-G-substrate synthetic peptide-Sepharose affinity chromatography. The G-substrate-Sepharose affinity resin was prepared by A.C. Nairn according to the procedure suggested by the supplier (Pharmacia). Serum was loaded on the column and the column was washed with 0.15 M NaCl. Fractions containing antibody were eluted with 4 M $MgCl_2$ and dialyzed against 0.1 M NaCl and 0.05 M Tris (pH 7.4).

In vitro translation. Rabbit reticulocyte lysate was prepared (Evans and Lingrel 1966), endogenous mRNA was inactivated (Pelham and Jackson 1976), and the lysate was used for in vitro translation directed by added polysomes or partially purified mRNA as previously described (DeGennaro et al. 1983) with some modifications. The reaction mixture was incubated at 30°C for 60 minutes and terminated by pelleting the polysomes by centrifugation at 95,000*g* for 5 minutes in a Beckman airfuge. Less than 5% of the incorporated [^{35}S]methionine was found in this pellet. The incorporation of [^{35}S]methionine into total TCA-insoluble protein was determined as previously described (DeGennaro et al. 1983).

Immunoprecipitation and visualization of translation products. The immunoprecipitation technique was a modification of previously described protocols (Ivarie and Jones 1979; DeGennaro et al. 1983). The final buffer used for resuspending SAC (*Staphylococcus aureus*, Cowan I) contained 25 mg/ml BSA. After termination of the translation reaction, the supernatants were adjusted to 0.2 M NaCl, 10 mM EDTA, 10 mM NaH_2PO_4, and precleared for 10 minutes with 20 μl of SAC. The precleared supernate was then incubated with the appropriate immunoglobulin for at least 60 minutes and treated with 20 μl of SAC for 10 minutes. After washing, the pellets were resuspended in 100 μl of SDS-sample buffer and heated at 100°C for 2 minutes before electrophoresis in 12% acrylamide resolving gels (Laemmli 1970). The gels were fixed, processed for fluorography with Autofluor (National Diagnostics), and exposed at −70°C to Kodak XAR-5 film which had been preflashed to obtain linear exposures (Laskey and Mills 1975).

Phosphorylation of newly synthesized DARPP-32. Cytosolic polysomes from bovine caudate nucleus were used to direct the synthesis of DARPP-32 in the in vitro translation system in the absence of [^{35}S]methionine. After termination of the reaction, the lysate supernate was adjusted to 250 mM NaCl, 50 mM NaF, 14 mM EDTA, 10 mM NaH_2PO_4 (pH 7.5), 1.5% NP-40, and purified anti-DARPP-32 immunoglobulin was added. After a 4-hour incubation at 4°C, 6 μg of purified rabbit immunoglobulin G (Sigma) and 40 μl of goat antiserum against rabbit γ-globulin (Sigma) were added. After an overnight incubation at 4°C, the precipitate was collected by centrifugation and washed twice by suspension in 150 mM NaCl, 10 mM NaH_2PO_4 (pH 7.5) followed by centrifugation at 8000*g* for 2 minutes in a Beckman microfuge. The washed pellet was resuspended in 11 mM citric acid (pH 2.5) and this solution was centrifuged at 8000*g* for 5 minutes. The supernate was adjusted to pH 7 with NaH_2PO_4 and phosphorylated using [γ-^{32}P]ATP and the catalytic subunit of cAMP-dependent protein kinase as previously described (Walaas et al. 1983). Proteins in the sample were separated on 12% acrylamide gels and exposed for autoradiography.

Translation assay for polysome-associated G-substrate mRNA during development. Rat pups were analyzed at various times after birth. Four to eight pups (each from a different litter) for each time point were killed by decapitation and the cerebella were removed and immediately frozen in liquid nitrogen. The cerebella were stored at −70°C until all cerebella for each experiment were collected. Isolation of polysomes, in vitro translation, immunoprecipitation with anti-dephospho-G-substrate immunoglobulin, separation on SDS-polyacrylamide gels, and fluorography were carried out as described above. The amount of polysomes used for each assay was adjusted such that an equal number of total counts incorporated into TCA-precipitable protein was obtained for each time point. The fluorographs were scanned by a densitometer. The values for the G-substrate peak were normalized to the total number of counts loaded on the gel to adjust for any variability in recovery that might have occurred during the immunoprecipitation and washing procedures. The value for each time point was normalized to the value for postnatal day 20. The data are expressed as the means ± S.E. of the means for eight experiments.

mRNA purification. The mRNAs for DARPP-32 and G substrate were partially purified by specific immunoadsorption of polysomes followed by oligo(dT)-cellulose affinity chromatography (Shapiro and Young 1981) as previously described (DeGennaro et al. 1983) with some modifications. The oligo(dT)-cellulose resin was equilibrated and washed with 0.5 M NaCl, 0.1% SDS, 1 mM EDTA, and 20 mM Tris-HCl (pH 7.4). The mRNA was precipitated without the addition of yeast tRNA.

RESULTS

In vitro translation of polysome-associated DARPP-32 mRNA. Cytosolic polysomes isolated from bovine caudate nucleus directed the in vitro synthesis of numerous polypeptides ranging from M_r 18,000 to M_r 250,000 (Fig. 1, "plus Polysomes"). A number of polypeptides could be nonspecifically precipitated from this mixture with rabbit preimmune serum and SAC (Fig. 1, "Preimmune"). When anti-DARPP-32 rabbit

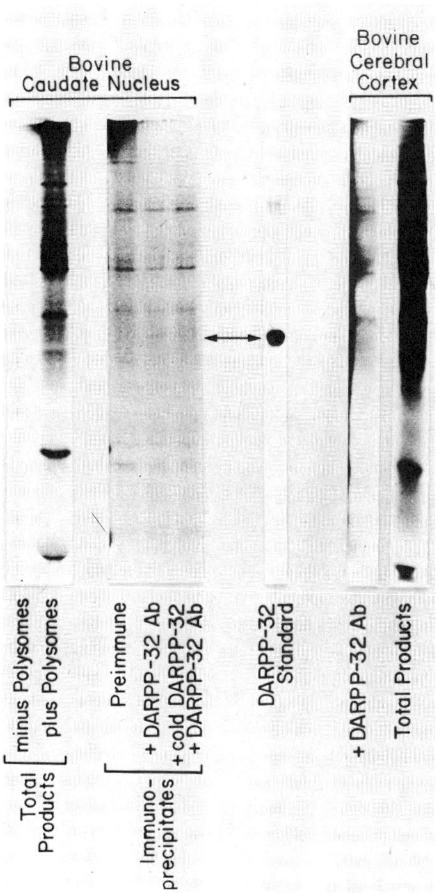

Figure 1. In vitro translation of polysome-associated DARPP-32 mRNA. Polysomes were purified from the caudate nucleus of bovine calves and translated in a reticulocyte lysate translation reaction in the presence of L-[^{35}S]methionine. The total translation products (Total Products) or translation products precipitated by DARPP-32 immunoglobulin (Immunoprecipitates) were separated by SDS-polyacrylamide gel electrophoresis and the [^{35}S]methionine incorporated into protein was visualized by fluorography. The amount of translation reaction mixture used to analyze total products represented 2% of that used for immunoprecipitation. (minus Polysomes) Translation products after the reticulocyte lysate mixture had been treated with micrococcal nuclease, but without the addition of polysomes; (plus Polysomes) translation products after the reticulocyte lysate mixture had been treated with micrococcal nuclease, followed by the addition of cytosolic polysomes; (Preimmune) newly synthesized polypeptides that were precipitated from total translation products using preimmune serum; (+ DARPP-32 Ab) newly synthesized polypeptides that were precipitated from total translation products using affinity-purified rabbit anti-DARPP-32 immunoglobulin; (+ cold DARPP-32 + DARPP-32 Ab) newly synthesized polypeptides that were precipitated from total translation products using affinity-purified anti-DARPP-32 immunoglobulin in the presence of nonradioactive, purified DARPP-32. Cytosolic polysomes purified from the cerebral cortex of bovine calves were translated in a reticulocyte lysate translation reaction. (+ DARPP-32 Ab) Newly synthesized polypeptides that were precipitated from total translation products with affinity-purified anti-DARPP-32 immunoglobulin; (Total Products) total translation products. DARPP-32 standard is purified bovine DARPP-32 which had been phosphorylated by the catalytic subunit of cAMP-dependent protein kinase. The arrow indicates the position of DARPP-32.

serum (not shown) or affinity-purified immunoglobulin (Fig. 1, " + DARPP-32 Ab") was added to the in vitro translation mixture and precipitated with SAC, a single additional band of M_r 32,000, which comigrated with mature DARPP-32 isolated from bovine caudate nucleus, could be detected. The immunoprecipitation of this polypeptide could be selectively inhibited by the addition of nonradioactive, purified DARPP-32 to the reaction mixture prior to the addition of anti-DARPP-32 immunoglobulin (Fig. 1, " + cold DARPP-32 + DARPP-32 Ab"). No polypeptide corresponding to DARPP-32 could be immunoprecipitated from translation reactions programmed with cytosolic polysomes isolated from bovine cerebral cortex (Fig. 1, "Bovine Cerebral Cortex, + DARPP-32 Ab") or programmed with the microsomal (membrane-bound) polysome fraction of bovine caudate nucleus (data not shown). Newly synthesized DARPP-32, immunoprecipitated from the in vitro translation mixture programmed with cytosolic polysomes from bovine caudate nucleus, could be phosphorylated by the catalytic subunit of cAMP-dependent protein kinase (Fig. 2).

Purification of DARPP-32 mRNA. The mRNA for DARPP-32 was partially purified from the total cyto-

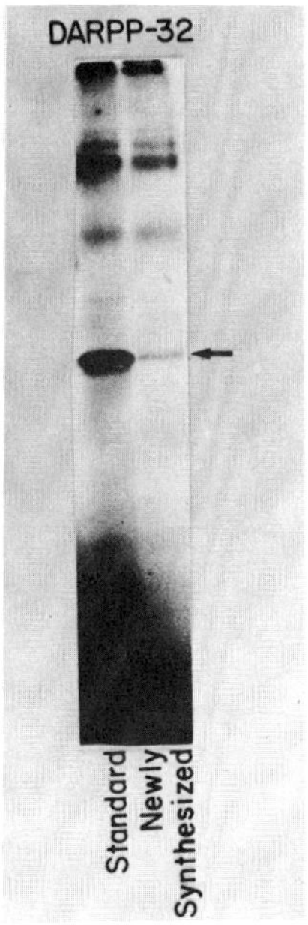

Figure 2. Phosphorylation of newly synthesized DARPP-32. Cytosolic polysomes from bovine caudate nucleus were translated in the rabbit reticulocyte lysate translation system containing nonradioactive methionine. The newly synthesized DARPP-32 was immunoprecipitated from the translation mixture, dissociated from the antibody, and phosphorylated using the catalytic subunit of cAMP-dependent protein kinase and [γ-^{32}P]ATP. Proteins in the phosphorylation reaction mixture were separated by SDS-polyacrylamide gel electrophoresis and the proteins that incorporated ^{32}P were visualized by autoradiography. (Standard) Purified DARPP-32 added to the translation reaction mixture, immunoprecipitated, and phosphorylated; (Newly Synthesized) DARPP-32 synthesized by polysomes, immunoprecipitated, and phosphorylated. The arrow indicates the position of DARPP-32.

solic polysomes of bovine caudate nucleus by specific immunoadsorption of DARPP-32-synthesizing polysomes. The polyadenylated mRNA from these specific polysomes was isolated by oligo(dT)-cellulose affinity chromatography. The polyadenylated fraction was translated in vitro. The newly synthesized DARPP-32 was immunoprecipitated with anti-DARPP-32 immunoglobulin. A single polypeptide comigrating with purified DARPP-32 could be detected (Fig. 3).

In vitro translation of polysome-associated G-substrate mRNA. Cytosolic polysomes isolated from rat cerebella directed the in vitro synthesis of numerous polypeptides ranging from M_r 18,000, to M_r 250,000 (Fig. 4, "plus Polysomes"). A number of polypeptides could be nonspecifically precipitated from this mixture with rabbit preimmune serum and SAC (Fig. 4, "Preimmune"). When anti-G-substrate rabbit serum (Fig. 4, "+G-Substrate Ab") or affinity-purified immunoglobulin (not shown) was added to the in vitro translation mixture and precipitated with SAC, a single additional band of M_r 23,000, which comigrated with mature G substrate isolated from rat cerebella, could be

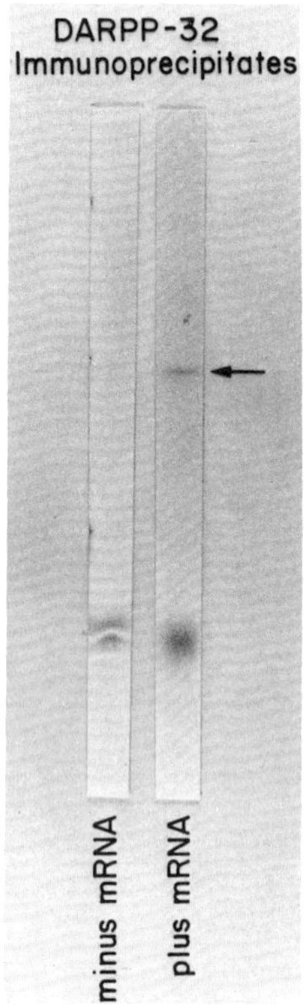

Figure 3. Translation products of partially purified mRNA for DARPP-32. mRNA for DARPP-32 was purified by immunoadsorption of polysomes followed by oligo(dT)-cellulose affinity chromatography. [^{35}S]methionine was incorporated into translation products in a translation reaction mixture treated with micrococcal nuclease which had been in the absence (minus mRNA) or presence (plus mRNA) of the purified mRNA. The translation products were immunoprecipitated, separated by SDS-polyacrylamide gel electrophoresis, and visualized by fluorography. The arrow indicates the position of DARPP-32.

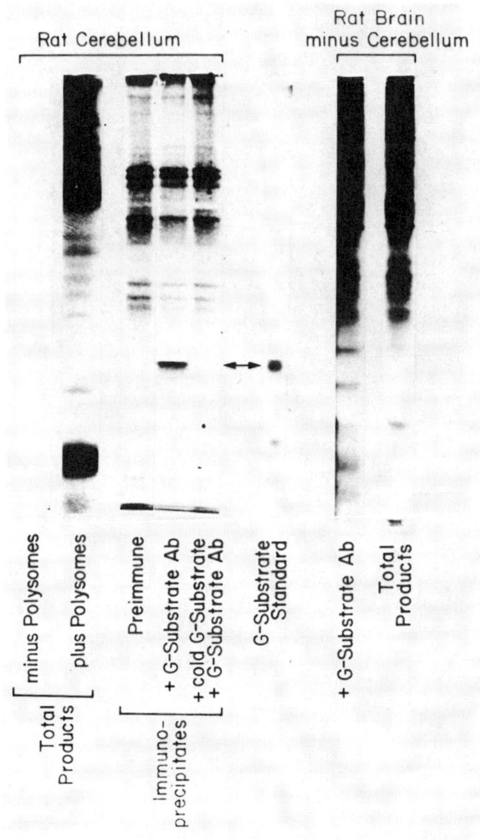

Figure 4. In vitro translation of polysome-associated G-substrate mRNA. Polysomes were purified from the cerebella of 14-day-old rats and translated in a reticulocyte lysate translation reaction in the presence of L-[^{35}S]methionine. The total translation products (Total Products) or translation products precipitated by G-substrate immunoglobulin (Immunoprecipitates) were separated by SDS-polyacrylamide gel electrophoresis and the [^{35}S]methionine incorporated into protein was visualized by fluorography. The amount of translation reaction mixture used to analyze total products represented 2% of that used for immunoprecipitation. (minus Polysomes) Translation products after the reticulocyte lysate mixture had been treated with micrococcal nuclease, but without the additon of polysomes; (plus Polysomes) translation products after the reticulocyte lysate mixture had been treated with micrococcal nuclease, followed by the addition of cytosolic polysomes; (Preimmune) newly synthesized polypeptides that were precipitated from total translation products using preimmune serum. (This preimmune serum precipitated an unidentified protein not precipitated by the immune serum.) (+G-substrate Ab) Newly synthesized polypeptides that were precipitated from total translation products using anti-G-substrate rabbit serum; (+cold G-substrate, +G-substrate Ab) newly synthesized polypeptides that were precipitated from total translation products using affinity-purified anti-G-substrate immunoglobulin in the presence of nonradioactive, purified G substrate. Cytosolic polysomes purified from rat brain minus cerebellum were translated in a reticulocyte lysate translation reaction. (+G-substrate Ab) Newly synthesized polypeptides that were precipitated from total translation products with affinity-purified anti-G-substrate immunoglobulin; (Total Products) total translation products. G-substrate standard is partially purified rabbit G substrate which had been phosphorylated by cGMP-dependent protein kinase. The arrow indicates the position of G substrate.

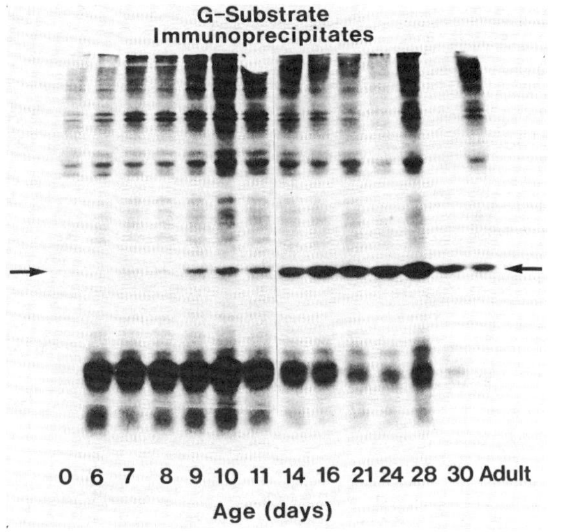

Figure 5. G substrate synthesized by polysomes isolated from cerebella of rats of various ages. Cytosolic polysomes were translated in a micrococcal nuclease-treated rabbit reticulocyte lysate translation reaction in the presence of L-[^{35}S]methionine. G substrate was immunoprecipitated from the translation mixture with rabbit serum containing anti-G-substrate immunoglobulin. The immunoprecipitates were separated by SDS-polyacrylamide gel electrophoresis and visualized by fluorography. A representative experiment showing the translation products present in G-substrate immunoprecipitates is presented. The arrows indicate the position of G substrate.

detected. The immunoprecipitation of this polypeptide could be selectively inhibited by the addition of nonradioactive purified G substrate to the reaction mixture prior to the addition of anti-G-substrate rabbit serum (Fig. 4, "+ cold G-Substrate + G-Substrate Ab"). No polypeptide corresponding to G substrate could be immunoprecipitated from translation reactions programmed with cytosolic polysomes isolated from rat brain minus cerebellum (Fig. 4, "Rat Brain minus Cerebellum, + G-Substrate Ab") or programmed with the microsomal (membrane-bound) polysome fraction of rat cerebellum (data not shown). Studies with rabbit cerebella gave results similar to those described for rat cerebella. Newly synthesized G substrate, immunoprecipitated from the in vitro translation mixture programmed with cytosolic polysomes from rabbit cerebella, could be phosphorylated by cGMP-dependent protein kinase (data not shown).

Translation of polysome-associated G-substrate mRNA during development. Cytosolic polysomes isolated from rat cerebella at various stages of postnatal development were used to direct the in vitro synthesis of cerebellar proteins in the presence of L-[^{35}S]methionine. G substrate was immunoprecipitated, separated on SDS-polyacrylamide gels, and visualized by fluorography (Fig. 5). Synthesis of G substrate could not be detected before postnatal day 6 in the rat cerebellum.

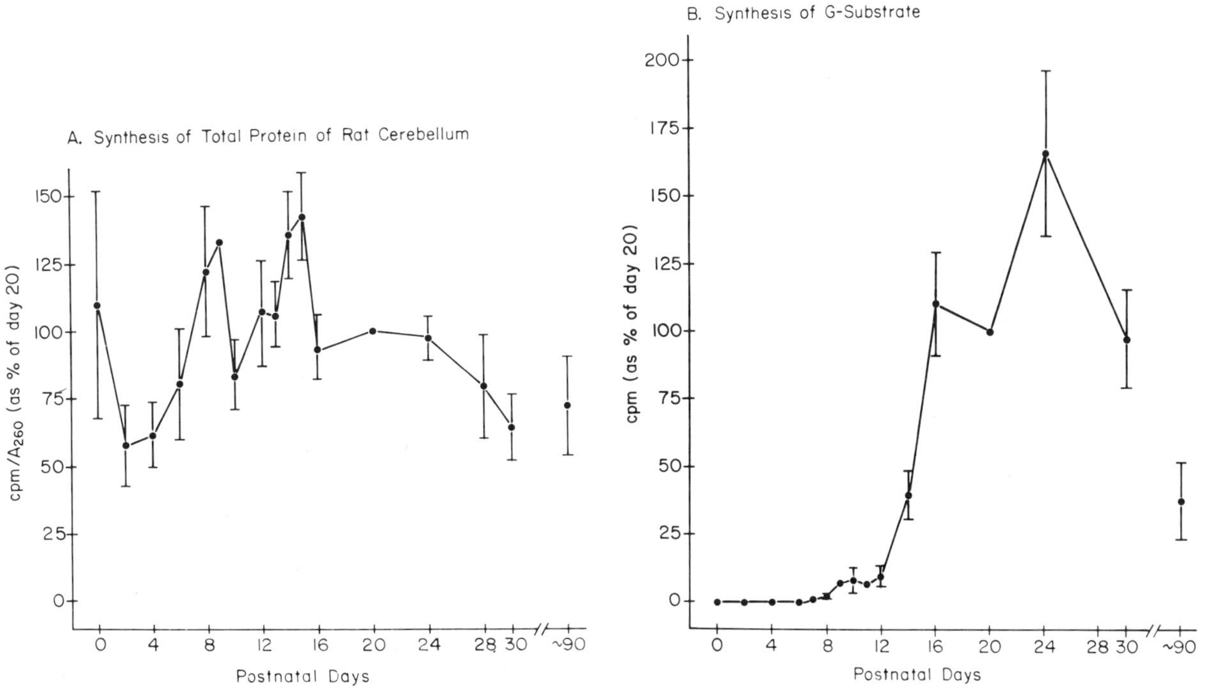

Figure 6. Incorporation of [^{35}S]methionine into total translation products and into G substrate as directed by polysomes isolated from cerebella of rats of various ages. Cytosolic polysomes were purified and translated in a rabbit reticulocyte lysate translation reaction system in the presence of L-[^{35}S]methionine. (*A*) The total incorporation of [^{35}S]methionine into TCA-insoluble protein was determined by a filter paper assay and is expressed as cpm per A_{260} unit of polysomes, normalized to the value for day 20. (*B*) The reaction mixture was incubated with anti-G-substrate immunoglobulin and the proteins present in the immunoprecipitate were separated by SDS-polyacrylamide gel electrophoresis and visualized by fluorography. The band corresponding to G substrate was scanned by a densitometer. The value for each peak was calculated per unit of total translation products and normalized to the value for day 20. Data are presented as means ± S.E. of the means for eight separate experiments.

The level of synthesis increased slowly from postnatal day 6 to 12, increased rapidly from postnatal day 12 to 24, and then decreased to the adult level (Fig. 6B). There was no correlation between the relative level of G-substrate synthesis (Fig. 6B) and the relative total protein-synthesizing activity of the cerebellar polysomes (Fig. 6A) during development.

Purification of G-substrate mRNA. The mRNA for G substrate was partially purified from the total cytosolic polysomes of rabbit cerebella by specific immunoadsorption of G-substrate-synthesizing polysomes. The polyadenylated mRNA from these specific polysomes was isolated by oligo(dT)-cellulose affinity chromatography. The polyadenylated fraction was translated in vitro. The newly synthesized G substrate was immunoprecipitated with anti-G-substrate immunoglobulin. A polypeptide comigrating with purified G substrate was detected (Fig. 7, lower arrow). A second band, of higher molecular weight, was also detected under these conditions (Fig. 7, upper arrow).

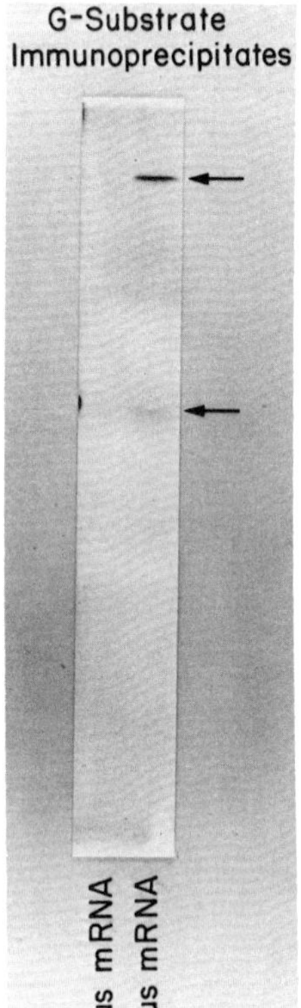

Figure 7. Translation products of partially purified mRNA for G substrate. mRNA for G substrate was purified by immunoadsorption of polysomes followed by oligo(dT)-cellulose affinity chromatography. [^{35}S]methionine was incorporated into translation products in a translation reaction mixture, which had been treated with micrococcal nuclease, in the absence (minus mRNA) or presence (plus mRNA) of the purified mRNA. The translation products were immunoprecipitated, separated by SDS-polyacrylamide gel electrophoresis, and visualized by fluorography. The lower arrow indicates the position of G substrate; the upper arrow indicates the position of a coimmunoprecipitating polypeptide (see text).

DISCUSSION

We have investigated the in vitro translation of two cell-type-specific, neuronal phosphoproteins, DARPP-32 and G substrate. Cytosolic polysomes isolated from the bovine caudate nucleus were used to direct the synthesis of DARPP-32. The newly synthesized polypeptide was identified as DARPP-32 by a number of criteria: (1) upon immunoprecipitation with anti-DARPP-32 immunoglobulin, a newly synthesized polypeptide was found to comigrate with purified DARPP-32 on SDS polyacrylamide gels, (2) immunoprecipitation of the newly synthesized polypeptide could be inhibited by the addition of purified DARPP-32, (3) no polypeptide corresponding to DARPP-32 was synthesized by polysomes from bovine cerebral cortex, a region of the brain in which little DARPP-32 has been detected, (4) the polypeptide was synthesized by cytosolic polysomes, but not by microsomal polysomes, which is consistent with the cytosolic location of DARPP-32, (5) the polypeptide was phosphorylated by added catalytic subunit of cAMP-dependent protein kinase. G substrate was synthesized by cytosolic polysomes isolated from rat or rabbit cerebella and identified by criteria analogous to those described above for DARPP-32.

Both DARPP-32 and G substrate could be synthesized in vitro when their corresponding partially purified mRNAs were used to program the reticulocyte lysate translation system, which indicates that there is no absolute requirement for neuron-specific factors for translating these neuron-specific mRNAs. We have not yet tested the possibility that the efficiency of translation could be affected by neuron-specific factors. Since both DARPP-32 and G-substrate mRNAs bind to oligo(dT)-cellulose, they most likely contain polyadenylated sequences. The newly synthesized DARPP-32 comigrates with purified DARPP-32, and no other peptides can be detected. This suggests that DARPP-32 is synthesized without any major posttranslational modifications that greatly affect its molecular weight, but does not definitively rule out the possibility of a precursor that could not be detected by the assay conditions used. The partially purified mRNA for G substrate directed the synthesis of a polypeptide comigrating with G substrate as well as a higher-molecular-weight polypeptide. We are currently investigating the possibility that this larger polypeptide is a precursor for G substrate. The structure of the gene for G substrate and the processing of its mRNA is especially interesting in view of the high degree of similarity between the amino acid sequences surrounding the two phosphothreonine residues. In a sequence of 26 residues, there are 18 identical residues and 4 conservative substitutions (Aitken et al. 1981), which suggests that G substrate contains an internal gene duplication, as is found, for example, in the calcium-binding proteins calmodulin and troponin (Watterson et al. 1980).

G-substrate synthesis was examined in the developing postnatal rat cerebellum. The increase in G-substrate synthesis from postnatal day 12 to day 24 was concurrent with an increase in the total activity of cGMP-dependent protein kinase in the cerebellum (Bandle and Guidotti 1979). By postnatal day 3 in the rat, Purkinje cells have formed in a monolayer arrangement (Woodward et al. 1969). By postnatal day 12, the Purkinje cells have started to form apical dendritic trees (Berry and Bradley 1976), contacts with parallel fibers (Woodward et al. 1969), and temporary synapses with climbing fibers (Shimono et al. 1976). Although the Purkinje cells are well established as a definitive cell type by postnatal day 12, they still do not synthesize G substrate at a very high rate. It is only after postnatal day 12, when the Golgi, basket, stellate, and granule cells begin to show activity, and the Purkinje cells have begun to establish final contact with climbing fibers (Shimono et al. 1976), that rapid synthesis of G substrate begins. The observation that the synthesis of G substrate occurs late in the development of Purkinje cells supports the interpretation that this phosphoprotein is involved in a function characteristic of these specialized cells.

We are currently using the partially purified mRNAs for DARPP-32 and G substrate to construct complementary DNA molecular clones. These clones will be used to study the structure and expression of the genes coding for DARPP-32 and G substrate.

ACKNOWLEDGMENTS

This work was supported by U.S. Public Health Service Grants MH-17387 and NS-08440 to P.G. and by Grant #PF2052 from the American Cancer Society to R.M.L. We thank Hugh C. Hemmings, Jr. for supplying bovine caudate, purified DARPP-32, DARPP-32-Sepharose affinity resin, and anti-DARPP-32 serum; Angus C. Nairn for supplying purified G substrate, G-substrate-Sepharose affinity resin, and anti-G-substrate serum; Annette E. Gwardyak for providing editorial assistance; and Karen J. Hageman for preparing the figures.

REFERENCES

Aitken, A., T. Bilham, P. Cohen, D. Aswad, and P. Greengard. 1981. A specific substrate from rabbit cerebellum for guanosine 3′:5′-monophosphate-dependent protein kinase. III. Amino acid sequences at the two phosphorylation sites. *J. Biol. Chem.* **256:** 3501.

Aswad, D.W. and P. Greengard. 1981a. A specific substrate from rabbit cerebellum for guanosine 3′:5′-monophosphate-dependent protein kinase. I. Purification and characterization. *J. Biol. Chem.* **256:** 3487.

———. 1981b. A specific substrate from rabbit cerebellum for guanosine 3′:5′-monophosphate-dependent protein kinase. II. Kinetic studies on its phosphorylation by guanosine 3′:5′-monophosphate-dependent and adenosine 3′:5′-monophosphate-dependent protein kinases. *J. Biol. Chem.* **256:** 3494.

Baetge, E.E., B.B. Kaplan, D.J. Reis, and T.H. Joh. 1981. Translation of tyrosine hydroxylase from poly(A)-mRNA in pheochromocytoma cells is enhanced by dexamethasone. *Proc. Natl. Acad. Sci.* **78:** 1269.

Bandle, E. and A. Guidotti. 1979. Ontogenic studies of cGMP-dependent protein kinase in rat cerebellum. *J. Neurochem.* **32:** 1343.

Berry, M. and P. Bradley. 1976. The growth of the dendritic trees of Purkinje cells in the cerebellum of the rat. *Brain Res.* **112:** 1.

Browning, M.D., C.-K. Huang, and P. Greengard. 1982. Biological similarity of two phosphoproteins (Proteins IIIa and IIIb) associated with synaptic vesicles in rat brain. *Soc. Neurosci. Abstr.* **8:** 794.

Civelli, O., N. Birnberg, and E. Herbert. 1982. Detection and quantitation of pro-opiomelanocortin mRNA in pituitary and brain tissues from different species. *J. Biol. Chem.* **257:** 6783.

DeCamilli, P., R. Cameron, and P. Greengard. 1983a. Synapsin I (Protein I), a nerve terminal-specific phosphoprotein. I. Its general distribution in synapses of the central and peripheral nervous system demonstrated by immunofluorescence in frozen and plastic sections. *J. Cell Biol.* **96:** 1337.

DeCamilli, P., S.M. Harris, Jr., W.B. Huttner, and P. Greengard. 1983b. Synapsin I (Protein I), a nerve terminal-specific phosphoprotein. II. Its specific association with synaptic vesicles demonstrated by immunocytochemistry in agarose-embedded synaptosomes. *J. Cell Biol.* **96:** 1355.

DeGennaro, L.J., W.C. Wallace, and P. Greengard. 1983. *In vitro* translation of Synapsin I, a neuron-specific protein: Developmental studies and messenger RNA purification. *J. Biol. Chem.* (in press).

Dolphin, A.C., J.A. Detre, D.J. Schlichter, A.C. Nairn, H.H. Yeh, D.J. Woodward, and P. Greengard. 1983. Cyclic nucleotide-dependent protein kinases and some major substrates in the rat cerebellum after neonatal X-irradiation. *J. Neurochem.* **40:** 577.

Evans, M.J. and J.B. Lingrel. 1966. Hemoglobin messenger ribonucleic acid. Distribution of the 9S ribonucleic acid in polysomes of different sizes. *Biochemistry* **8:** 829.

Fando, J.L. and C.G. Wasterlain. 1980. A simple, reproducible cell-free system for measuring brain protein synthesis. *Neurochem. Res.* **5:** 197.

Ginsburg, I., T. Scherson, D. Giveon, L. Behar, and U.Z. Littauer. 1982. Modulation of mRNA for microtubule-associated proteins during brain development. *Proc. Natl. Acad. Sci.* **79:** 4892.

Gough, N.M. and J.M. Adams. 1978. Immunoprecipitation of specific polysomes using *Staphylococcus aureus:* Purification of the immunoglobulin κ chain messenger RNA from the mouse myeloma MPC11. *Biochemistry* **17:** 5560.

Gubler, U.D., L. Kilpatrick, P.H. Seeburg, L.P. Gage, and S. Udenfriend. 1981. Detection and partial characterization of proenkephalin mRNA. *Proc. Natl. Acad. Sci.* **78:** 5484.

Hemmings, H.C., Jr., A.C. Nairn, D.W. Aswad, and P. Greengard. 1983. DARPP-32, a dopamine and adenosine 3′:5′-monophosphate-regulated phosphoprotein enriched in dopamine-innervated brain regions. II. Purification and characterization of the phosphoprotein from bovine caudate nucleus. *J. Neurosci.* (in press).

Ivarie, R.D. and P.P. Jones. 1979. A rapid sensitive assay for specific protein synthesis in cells and in cell-free translations: Use of *Staphylococcus aureus* as an adsorbent for immune complexes. *Anal. Biochem.* **97:** 24.

Laemmli, U.K. 1970. Cleavage of structural proteins during the assembly of the head of bacteriophage T4. *Nature* **227:** 680.

Laskey, R.A. and A.D. Mills. 1975. Quantitative film detection of ^{3}H and ^{14}C in polyacrylamide gels by fluorography. *Eur. J. Biochem.* **56:** 335.

MacDonald, R.J., A.E. Przybyla, and W.J. Rutter.

1977. Isolation and *in vitro* translation of the messenger RNA coding for pancreatic amylase. *J. Biol. Chem.* **252:** 5522.

MERLIE, J.P., J.G. HOFLER, and R. SEBBANE. 1981. Acetylcholine receptor synthesis from membrane polysomes. *J. Biol. Chem.* **256:** 6995.

MORRISON, M.R., R. BRODEUR, S. PARDUE, F. BASKIN, C.L. HALL, and R.N. ROSENBERG. 1979. Differences in the distribution of poly(A) size classes in individual messenger RNAs from neuroblastoma cells. *J. Biol. Chem.* **254:** 7675.

NAIRN, A.C., J.A. DETRE, J.E. CASNELLIE, and P. GREENGARD. 1982. Serum antibodies that distinguish between the phospho- and dephospho-forms of a phosphoprotein. *Nature* **299:** 734.

NILSON, J.H., K.J. BARRINGER, E.M. CONVEY, K. FRIDERICI, and F.M. ROTTMAN. 1980. Ontogeny of pituitary hormone mRNAs in the bovine fetus. Quantitation of pre-prolactin mRNA as a function of gestation. *J. Biol. Chem.* **255:** 5871.

OUIMET, C.C., P.E. MILLER, H.C. HEMMINGS, JR., S.I. WALAAS, and P. GREENGARD. 1983. DARPP-32, a dopamine and adenosine 3′:5′-monophosphate-regulated phosphoprotein enriched in dopamine-innervated brain regions. III. Immunocytochemical localization. *J. Neurosci.* (in press).

PELHAM, H.R. and R.I. JACKSON. 1976. An efficient mRNA-dependent translation system from reticulocyte lysates. *Eur. J. Biochem.* **67:** 247.

SCHELLER, R.H., J.F. JACKSON, L.B. MCALLISTER, J.H. SCHWARTZ, E.R. KANDEL, and R. AXEL. 1982. A family of genes that codes for ELH, a neuropeptide eliciting a stereotyped pattern of behavior in *Aplysia. Cell* **28:** 707.

SCHLICHTER, D.J., J.E. CASNELLIE, and P. GREENGARD. 1978. An endogenous substrate for cGMP-dependent protein kinase in mammalian cerebellum. *Nature* **273:** 61.

SCHLICHTER, D.J., J.A. DETRE, D.W. ASWAD, B. CHEHRAZI, and P. GREENGARD. 1980. Localization of cyclic GMP-dependent protein kinase and substrate in mammalian cerebellum. *Proc. Natl. Acad. Sci.* **77:** 5537.

SHAPIRO, S.Z. and J.R. YOUNG. 1981. An immunochemical method for mRNA purification. *J. Biol. Chem.* **256:** 1495.

SHIMONO, T., S. NOSAKA, and K. SASAKI. 1976. Electrophysiological study on the postnatal development of neuronal mechanisms in the rat cerebellar cortex. *Brain Res.* **108:** 279.

WALAAS, S.I. and P. GREENGARD. 1983. DARPP-32, a dopamine and adenosine 3′:5′-monophosphate regulated phosphoprotein enriched in dopamine-innervated brain regions. I. Regional and cellular distribution in the rat brain. *J. Neurosci.* (in press).

WALAAS, S.I., D.W. ASWAD, and P. GREENGARD. 1983. A dopamine- and cyclic AMP-regulated phosphoprotein enriched in dopamine-innervated brain regions. *Nature* **301:** 69.

WALKER, W.A. and C. ZOMZELY-NEURATH. 1980. *De novo* synthesis of neuron-specific enolase in a rabbit reticulocyte translation system programmed by poly(A)-RNA from rat brain. *Neurochem. Res.* **5:** 361.

WATTERSON, D.M., F. SHARIEF, and T.C. VANAMAN. 1980. The complete amino acid sequence of the Ca^{2+}-dependent modulator (calmodulin) of bovine brain. *J. Biol. Chem.* **255:** 962.

WOODWARD, D.J., B.J. HOFFER, and L.W. LAPHAM. 1969. Correlative survey of electrophysiological, neuropharmacological and histochemical aspects of cerebellar maturation in rat. In *Neurobiology of cerebellar evolution and development* (ed. R. Llinas), p. 725. American Medical Association, Chicago, Illinois.

Studies on Voltage-operated Calcium Channels Using Radioligands

R.J. Gould, K.M.M. Murphy, and S.H. Snyder

Departments of Neuroscience, Pharmacology, and Experimental Therapeutics, Psychiatry and Behavioral Sciences, The Johns Hopkins University School of Medicine, Baltimore, Maryland 21205

Since the pioneering work of Hodgkin and Huxley, the study of ion channels has been carried forward primarily by electrophysiologists (Hagiwara and Byerly 1981; Edwards 1982; Tsien 1983). The advent of specific and potent neurotoxins for Na^{+} channels has allowed for the biochemical analysis of these channels at the molecular level (Catterall 1980). Yet, our knowledge about the Ca^{++} channel has proceeded at a relatively slow pace, both biophysically and biochemically. The experimental difficulties in studying Ca^{++} channels have centered around two aspects. First, a suitable preparation for electrophysiological and biophysical studies has not been available (Hagiwara and Byerly 1981). Such a preparation should have a strong Ca^{++} current that can be resolved from Na^{+} and K^{+} currents and whose reverse potential can be measured. Recently, significant advances in these problems have been made, and subclassifications of Ca^{++} channels have been made (Hagiwara and Byerly 1981; Tsien 1983). The second problem has been the lack of specific and potent pharmacological agents to employ as tools for biochemical analysis. The advent of the class of compounds known variously as Ca^{++} antagonists, Ca^{++} blockers, Ca^{++} channel inhibitors, Ca^{++} entry blockers, or Ca^{++} channel antagonists may provide such a biochemical and pharmacologic tool.

The Ca^{++} channel antagonists represent a structurally diverse class of compounds (Fig. 1). The most specific and potent of these agents are the 1,4-dihydropyridines, such as nitrendipine (1,4-dihydro-2,6-dimethyl-4-[3-nitrophenyl]-3,ethyl,5 methyl ester-pyridine dicarboxylate) (Fleckenstein 1977; Henry 1980; Triggle 1982). A large number of these compounds have been made and the structure-activity relationship for the 1,4-dihydropyridines have been studied in some detail (Loev et al. 1974; Rodenkirchen et al. 1978; Bolger et al. 1983). The 1,4-dihydropyridine ring is essential for activity, as the pyridine derivative is inactive. A substituted phenyl ring in the 4 position of dihydropyridine ring is necessary for high potency. Compounds with *ortho-* and *meta*-substituents on the phenyl ring are the most active. Ester substituents of a methyl group or larger in the 3 and 5 positions cause increased activity (Janis and Triggle 1983). Structure-activity relationships for the other classes of compounds such as verapamil, prenylamine, and lidoflazine have been less well studied.

The evidence for these agents as Ca^{++} channel antagonists comes primarily from smooth muscle and cardiac muscle. These agents inhibit $^{45}Ca^{++}$ influx into vascular smooth muscle at concentrations that inhibit contractions initiated under similar depolarizing conditions (Godfraind and Dieu 1981; Van Breemen et al. 1981; Cauvin et al. 1983; Godfraind 1983; Janis and Triggle 1983). Moreover, if one increases the Ca^{++} concentration of the surrounding buffer, the inhibition by these agents can be overcome (Godfraind 1982; Triggle 1982). As skinned smooth muscle preparations are insensitive to verapamil, diltiazem, and nifedipine, even at 100 μM, the site of action of the Ca^{++} channel antagonists appears to be at the level of the membrane rather than the contractile apparatus (Fleckenstein 1977; Diecke 1983).

In cardiac preparations, Ca^{++} channel antagonists inhibit the slow inward current that is dependent upon extracellular Ca^{++} (Fleckenstein 1977, 1982). The result of such inhibition is a depressant effect on the rate of sinus node discharge leading to a negative chronotropic effect. The agents can also relax papillary muscle much as they do smooth muscle (Fleckenstein 1977, 1982). Much of the evidence from smooth and cardiac muscle for these agents acting as Ca^{++} channel antagonists has recently been reviewed (Fleckenstein 1977; Janis and Triggle 1983). They appear to be potent and specific inhibitors and thus may prove useful in defining Ca^{++} channels biochemically and pharmacologically.

There have been a few reported effects of these agents in neural tissue. The dihydropyridines inhibit $^{45}Ca^{++}$ uptake in depolarized PC12 cells as potently as they inhibit Ca^{++} channels in smooth and cardiac muscle (Toll 1982). A number of investigators have reported that this class of Ca^{++} channel antagonists does not affect Ca^{++} channels in neuronal tissue. For example, they do not affect electrophysiological measurements of voltage-operated Ca^{++} channels, $^{45}Ca^{++}$ uptake into synaptosomes, or Ca^{++}-dependent neurotransmitter release at low concentrations (Starke and Schumann 1973; Blakeley et al. 1981; Douglas and Taraskavitch 1982; Hogestatt et al. 1982). Thus, the properties of the Ca^{++} channels in neural systems appear to be different from those in smooth and cardiac muscle. Whether this difference is due to a different protein, or due to differential regulation of the channellike phosphorylation is unknown. The recent development of [^{3}H]nitrendipine and [^{3}H]verapamil of high specific activity has facilitated the biochemical characterization of binding sites associated with Ca^{++} chan-

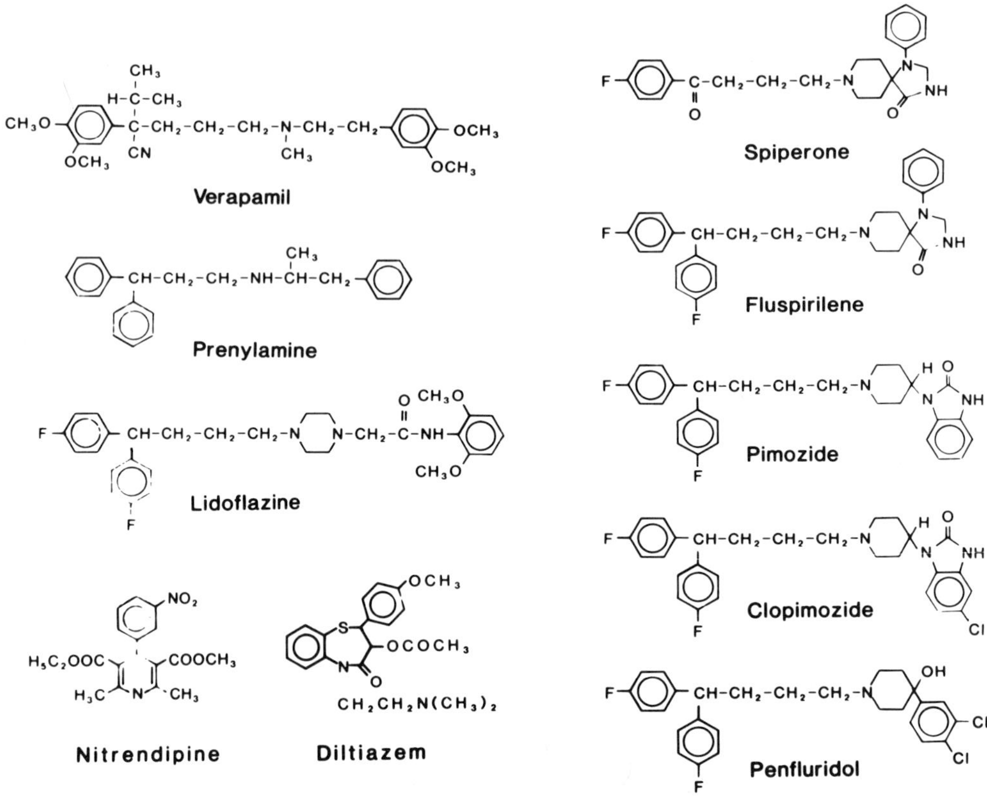

Figure 1. Structures of Ca^{++} channel antagonists.

nels in smooth and cardiac muscle and shown the presence of similar binding sites in neural tissue.

METHODS

Male Sprague-Dawley rats were killed by cervical dislocation and decapitation. Brains were removed immediately, washed with 0.9% ice-cold NaCl, and the cerebral cortices separated. Tissue was homogenized using a Brinkmann Polytron in 10 volumes of 50 mM Tris-HCl (pH 7.7) and centrifuged at 20,000 rpm for 10 minutes in a Sorvall SS-34 rotor. The pellet was washed three times with ice-cold 50 mM Tris-HCl (pH 7.7) and resuspended to a final concentration of 1–4 mg wet weight/ml in 50 mM Tris-HCl. In those experiments exploring the ionic regulation of [^{3}H]nitrendipine binding, cerebral cortex or heart was homogenized in 10 volumes of 50 mM Tris-HCl (pH 7.7) containing 10 mM EDTA and immediately centrifuged at 20,000 rpm for 10 minutes. The pellet was resuspended into 50 mM Tris-HCl (pH 7.7) and 10 mM EDTA and incubated on ice for 30 minutes. Following centrifugation as before, the pellet was resuspended into 50 mM Tris-HCl (pH 7.7) containing 10 mM EGTA and again incubated on ice for 30 minutes. The pellet obtained after centrifuging was then washed three times with 10 μM EGTA in 50 mM Tris-HCl (pH 7.7) and finally resuspended at 4 mg wet weight in this EGTA containing Tris buffer. Incubations were performed at 25°C as described in the appropriate figure and table legends. Incubations were terminated by rapidly filtering through GF/B glass fiber filters (Whatman). The filters were washed three times with 3 ml each of 50 mM Tris-HCl (pH 7.7) and the radioactivity retained was determined by liquid scintillation spectrometry using NEN-947.

[^{3}H]nitrendipine (85 Ci/mmole) and oxidized [^{3}H]nitrendipine (85 Ci/mmole) were obtained from New England Nuclear. Nifedipine (Pfizer), nimodipine, nisoldipine, nitrendipine (Miles), SKF24260 (Smith Kline & French), PY108-068 (Sandoz), diltiazem (Tanabe and Marion Laboratories), cinnarizine, flunarizine, lidoflazine, pimozide, clopimozide, fluspirilene (Janssen Pharmaceutical), penfluridol (Marion Laboratories), verapamil, and D-600 (gifts of Dr. John Daly and Knoll Pharmaceutical) were initially dissolved in absolute ethanol at 10^{-3} M and subsequently diluted to the appropriate concentrations with 50 mM Tris-HCl (pH 7.7). All other chemicals were obtained from standard commercial sources.

RESULTS

The binding of [^{3}H]nitrendipine appears to be identical in ileum, heart, and brain, with the only difference being in the number of binding sites in these tissues. Thus, a wide variety of dihydropyridine derivatives show virtually identical affinities for [^{3}H]nitrendipine-binding sites in these three preparations (Table 1). In heart, cerebral cortex, and ileum, a similar order of potency of SKF24260 > nitrendipine > PY108-068 > nisol-

Table 1. Characteristics of [^{3}H]Nitrendipine Binding

	Heart	Cerebral cortex	Ileum
K_d (nM)	0.16	0.19	0.14
B_{max} (pmoles/g)	6	8	0.9
	a. Dihydropyridine Drugs (K_i, nM)		
Nifedipine	0.49±0.12	0.90±0.15	0.80 ±0.10
Nitrendipine	0.12±0.03	0.18±0.05	0.12 ±0.08
Nimodipine	0.32±0.16	0.28±0.11	0.12 ±0.03
Nisoldipine	0.17±0.08	0.13±0.04	0.07 ±0.001
SKF24260	0.07±0.04	0.08±0.03	0.038±0.019
PY108-068	0.15±0.06	0.12±0.03	0.102±0.047
	b. Nondihydropyridine Drugs (EC_{50}, nM)		
Verapamil	29±6.0	27± 5	23±5
D-600	1.2 ±0.6	6± 2	1.9±0.6
Tiapamil	1200±0.200	1000±100	1060±230
Flunarizine	360± 80	550± 80	260± 90
Cinnarizine	1140±260	1280±260	1050±270
Lidoflazine	1120±320	660± 90	200± 20
Prenylamine	280± 50	250± 60	220± 20
Diltiazem	400±300	500±200	n.d.

dipine>nimodipine>nifedipine is seen. In addition, a variety of nondihydropyridine derivatives also inhibit [^{3}H]nitrendipine binding and the concentrations of these drugs necessary to decrease binding to 50% of control levels are also virtually identical (Table 1). This binding appears to be to plasma membranes of neural cells, as the subcellular distribution of this binding parallels that of [^{3}H]dihydroalprenolol binding. Autoradiographic studies have demonstrated that the binding is localized to regions rich in synaptic connections (Murphy et al. 1982; Gould et al. 1983a). Thus, pharmacologically and anatomically, [^{3}H]nitrendipine-binding sites in neural tissue appear to be associated with voltage-dependent Ca^{++} channels.

[^{3}H]nitrendipine binding is regulated in a complex manner by multivalent cations. The binding is absolutely dependent on Ca^{++} (Gould et al. 1982; Marangos et al. 1982). If membranes prepared from rat cerebral cortices are washed as described in Methods with EDTA and EGTA, 90% of the [^{3}H]nitrendipine binding is abolished (Table 2). A detailed analysis of the effects of the multivalent ions indicates that ions which are antagonists of Ca^{++} at voltage-dependent channels (Hagiwara and Byerly 1981) inhibit binding at higher concentrations. This suggests that the nitrendipine-binding site is sensitive to cations that are agonists and antagonists at voltage-dependent channels.

As mentioned previously, the Ca^{++} channel antagonists are a chemically diverse group of compounds (Fig. 1). Do all these compounds act at the same binding site within the Ca^{++} channel, or are there multiple drug-binding sites? Recently, we and others have shown that there are at least two drug-binding sites within the Ca^{++} channel complex (Ehlert et al. 1982; Ferry and Glossmann 1982; Bolger et al. 1983; Murphy et al. 1983a,b). Although drugs such as verapamil and lidoflazine inhibit [^{3}H]nitrendipine binding, the fashion in which they do so is not consistent with a competitive interaction (Ehlert et al. 1982; Murphy et al. 1983a,b). Verapamil is unable to inhibit [^{3}H]nitrendipine binding completely except at very high concentrations (Fig. 2). Whereas lidoflazine and related drugs can inhibit completely at low [^{3}H]nitrendipine concentrations, they do not at higher concentrations (Murphy et al. 1983a). Moreover, the Ca^{++} channel antagonist, diltiazem, actually stimulates [^{3}H]nitrendipine binding (Fig. 2). Although inhibition of binding can be explained by competition of drugs for a single site, it is difficult to explain enhancement of binding at a single site without invoking a complex model involving positive cooperativity.

As shown in Figure 3, these nondihydropyridine Ca^{++} channel antagonists affect the dissociation rate of [^{3}H]nitrendipine from its binding site in cerebral cortex. Thus, while nimodipine, a dihydropyridine, does not affect the dissociation rate, verapamil and lidofla-

Table 2. Effect of Cation Replacement on [^{3}H]Nitrendipine Binding in Cerebral Cortex

	K_d (pM)	B_{max} (pmoles/g wet wt.)
Tris-washed membranes	137	11.5
EGTA-washed membranes	176	1.8
+ 1 mM $CaCl_2$	188	8.0
+ 1 mM $SrCl_2$	267	9.0
+ 1 mM $MnCl_2$	219	7.9
+ 1 mM $CoCl_2$	199	4.2
+ 1 mM $MgCl_2$	198	4.7
+ 1 mM $BaCl_2$	129	2.0
+ 1 mM $CaCl_2$ + 1 mM $CoCl_2$	174	5.0
+ 1 mM $CaCl_2$ + 30 μM $LaCl_3$	180	6.0
+ 1 mM $CaCl_2$ + 100 μM $LaCl_2$	196	3.0

Saturation binding experiments were performed on Tris- or EGTA-washed membranes prepared as described in the Materials and Methods in the absence or presence of the ions indicated. Total and nonspecific [^{3}H]nitrendipine binding were determined in triplicate at nine concentrations of [^{3}H]nitrendipine between 0.005 nM and 2.0 nM in absence or presence of 60 nM nifedipine. The dissociation constants (K_d) and maximal binding capacity (B_{max}) were determined by Scatchard analysis. Values presented are the means of two to four independent determinations.

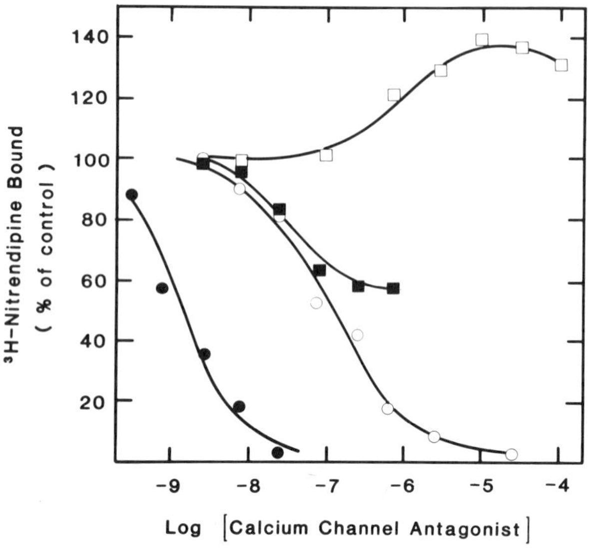

Figure 2. Effect of Ca^{++} channel blockers on [^{3}H]nitrendipine binding. Increasing concentrations of nifedipine (●), prenylamine (○), verapamil (■), or diltiazem (□) were added to [^{3}H]nitrendipine-binding assays performed as described in Methods.

zine accelerate dissociation. By contrast, diltiazem slows the dissociation rate (Fig. 3). These changes in dissociation rate are not consistent with these drugs acting at the dihydropyridine site. At fixed concentrations of [^{3}H]nitrendipine, such alterations in dissociation rates result in inhibition or stimulation of binding. This would suggest possibly three sites of drug action within the Ca^{++} channel: (1) a dihydropyridine site, (2) a site responsible for negative heterotropic interactions (the verapamil site), and (3) a site responsible for positive heterotropic interactions (the diltiazem site).

We have analyzed the interactions of verapamil and diltiazem and concluded that they act at the same site (Murphy et al. 1983a). As shown in Figure 4a, diltiazem can reverse the inhibition by D-600, prenylamine, or tiapamil, a verapamil analog. This reversal is stereospecific, as *trans*-diltiazem is ineffective in reversing tiapamil inhibition (Fig. 4b). Similar experiments have shown that methoxyverapamil (D-600), prenylamine, flunarizine, lidoflazine, and tiapamil all act at the same site (Murphy et al. 1983a,b). Methoxyverapamil reverses the inhibition of binding induced by prenylamine, flunarizine, lidoflazine, and tiapamil. Although one could invoke alternative explanations, the simplest, most straightforward explanation is that these drugs all act at a similar site. Consequently, it would appear that verapamil and the other drugs that cause negative heterotropic effects act at the same site as diltiazem. Thus, there appear to be only two drug receptor sites within the Ca^{++} channel—the dihydropyridine site and the verapamil/diltiazem site.

The structure-activity relationship of a variety of drugs for this site suggests a hydrophobic site similar to the hydrophobic trifluoperazine site on calmodulin (R.J. Gould et al., in prep.). Several drugs that inhibit calmodulin-stimulated phosphodiesterase drugs also inhibit [^{3}H]nitrendipine binding with similar potencies (Table 3). The diphenylbutylpiperidines (DPBP), which include pimozide, clopimozide, fluspirilene, and penfluridol (Fig. 1), are a notable exception to this. These drugs are much more potent at inhibiting [^{3}H]nitrendipine binding than at inhibiting calmodulin effects (Table 3).

Moreover, the DPBP neuroleptics have similar potencies at [^{3}H]nitrendipine sites and at D_2 dopamine receptors labeled with [^{3}H]spiperone (Table 3). By contrast, with the exception of thioridazine, other neuroleptics are at least 100 times more potent at dopamine receptors than at [^{3}H]nitrendipine sites (Gould et al. 1983b).

Earlier we showed that Ca^{++} antagonists of the verapamil class influence [^{3}H]nitrendipine binding in an allosteric fashion, even though some of these, such as lidoflazine and prenylamine, seem competitive because they displace [^{3}H]nitrendipine binding completely with a pseudo Hill coefficient of 1.0 (Murphy et al. 1983a,b). The DPBP neuroleptics also inhibit [^{3}H]nitrendipine binding completely with competition curves that are parallel to those of dihydropyridines and pseudo Hill coefficients of 0.8–1.0 (Gould et al. 1983b).

Lidoflazine and prenylamine allosterically modulate [^{3}H]nitrendipine binding by competing with verapamil

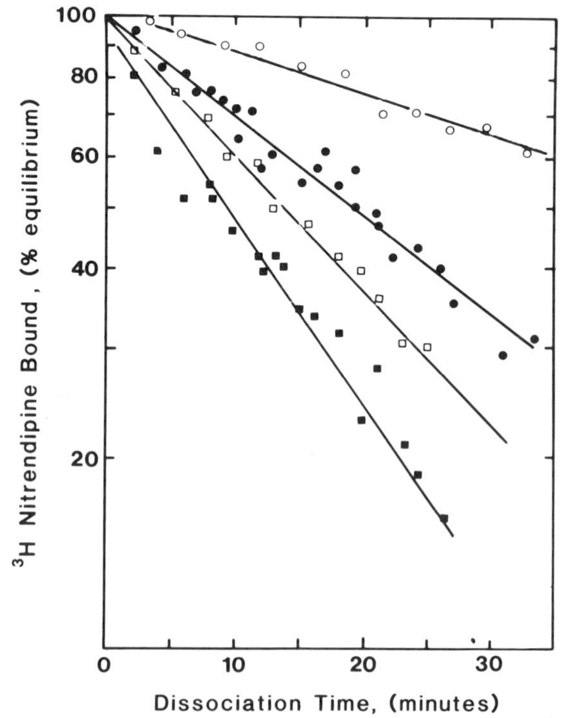

Figure 3. Ca^{++} channel blockers alter dissociation of [^{3}H]nitrendipine. Rat cerebral cortex membranes were incubated with 0.5 nM [^{3}H]nitrendipine for 1–1.5 hr. Two-ml aliquots were added to tubes containing 100 nM nifedipine alone (●), 100 nM nifedipine and 0.1 μM verapamil (■), 100 nM nifedipine and 2.5 μM prenylamine (□), or 100 nM nifedipine and 25 μM diltiazem (○).

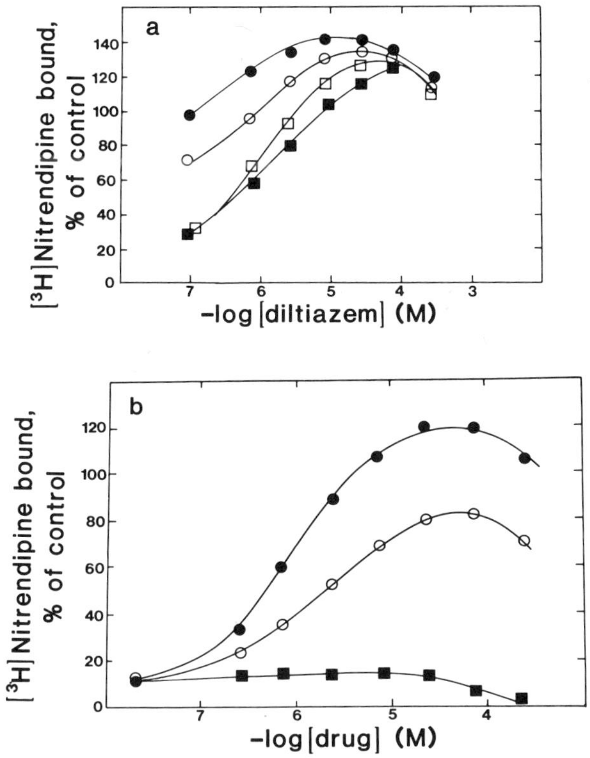

Figure 4. Diltiazem reverses the effects of D-600, prenylamine, and tiapamil on [^{3}H]nitrendipine binding. (*a*) Concentration-response curves for the effects of diltiazem on [^{3}H]nitrendipine binding were determined in the absence of any inhibiting agent (●) or in the presence of 0.25 μM D-600 (○), 0.75 μM prenylamine (■), or 2.5 μM tiapamil (□). Values for each condition are presented as the percent of specific nitrendipine binding determined in the absence of all inhibiting agents. Results are from one of three experiments carried out for each drug with similar results. (*b*) Concentration-response curves for the effects of *cis*-diltiazem (●), desmethyl-*cis*-diltiazem (○), or *trans*-diltiazem (■) were carried out using at least six concentrations of each of these agents in the presence of 0.2 nM [^{3}H]nitrendipine and 2.5 μM tiapamil incubated with brain membranes. Values are presented as the percentage of [^{3}H]nitrendipine binding determined in the absence of tiapamil, diltiazem, or diltiazem analogs. (Reprinted, with permission, from Murphy et al. 1983b.)

and related drugs at a single site. Thus, the inhibition of [^{3}H]nitrendipine binding by these drugs can be reversed by methoxyverapamil (D-600) (Murphy et al. 1983b). To ascertain if the DPBP neuroleptics act at the verapamil site or the dihydropyridine binding site, we explored the influence of methoxyverapamil upon the inhibition of [^{3}H]nitrendipine binding by DPBPs (Fig. 5a,b). Inhibition of [^{3}H]nitrendipine binding by fluspirilene and pimozide (Fig. 5a,b) as well as clopimozide and penfluridol (data not shown) is reversed by methoxyverapamil. Methoxyverapamil alone maximally inhibits [^{3}H]nitrendipine binding about 40%. However, in the presence of concentrations of DPBPs, which themselves greatly reduce or abolish [^{3}H]nitrendipine binding, methoxyverapamil (D-600) restores binding to a level comparable to the inhibition observed from methoxyverapamil alone. These findings indicate that methoxyverapamil and the DPBPs act at the same site, but that the DPBPs are more efficacious in allosterically reducing the affinity of the [^{3}H]nitrendipine-binding site. Numerous Ca^{++} channel antagonist drugs act at the verapamil site but with varying efficacies in allosterically reducing [^{3}H]nitrendipine binding (Murphy et al. 1983b). The Ca^{++} channel antagonist diltiazem also acts at this site but allosterically enhances [^{3}H]nitrendipine binding (Murphy et al. 1983b). Thus, the pharmacologic profile of the [^{3}H]nitrendipine binding site associated with Ca^{++} channels suggests a complex consisting of at least two binding sites. One of these binding sites shows a great deal of specificity for the dihydropyridine structure. The second site, the verapamil-diltiazem site, is more promiscuous and will accept a variety of drugs, showing a structure-activity profile similar to the hydrophobic binding site in calmodulin.

Biochemical characterization using a variety of enzymes has shown that the nitrendipine-binding site is a protein (Table 4). Trypsin reduces binding, as does heat treatment. As discussed earlier, it is probably a metalloprotein, possibly containing Ca^{++} or similar divalent cations. Phospholipase A treatment decreases binding, but not due to the detergent effect of the en-

Table 3. Calmodulin Inhibitors as Ca^{++} Channel Blockers (EC_{50}, nM)

	PDE activation[a,b]	[^{3}H]nitrendipine binding	[^{3}H]spiperone
Pimozide	7,000	13	4
Clopimozide	—	17	10
Fluspirilene	—	21	22
Penfluridol	10,000	30	16
Haloperidol	60,000	2,100	8
Sulpiride	> 100,000	>250,000	600
Spiperone	17,000	2,500	2
Trifluperazine	10,000	25,000	10
Thiothixene	30,000	25,000	n.d.
cis-Flupenthixol	25,000	2,500	n.d.

[a]Data taken from Weiss et al. (1980). Other values reported are concentrations necessary to reduce specific binding by 50%.
[b]PDE, phosphodiesterase.

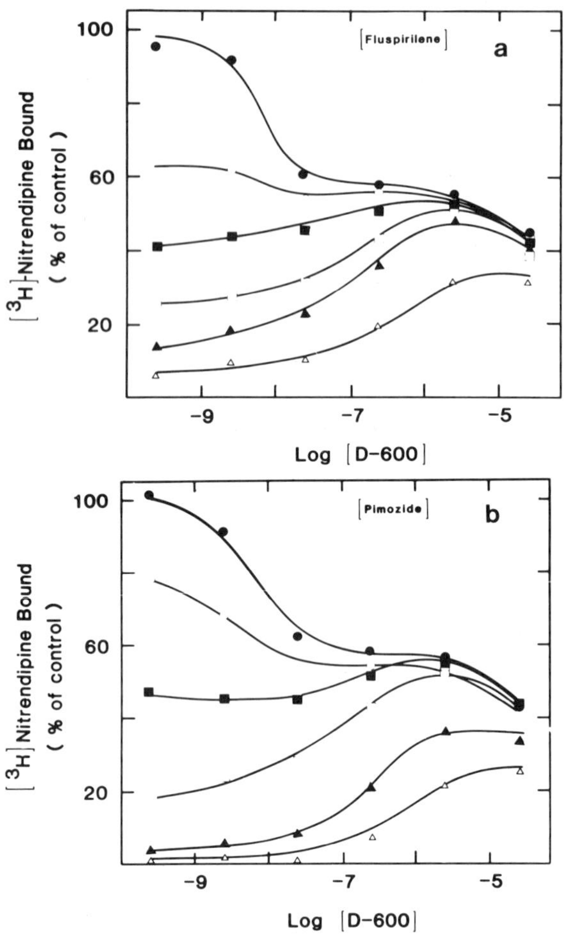

Figure 5. Methoxyverapamil (D-600) reverses the inhibition of [^{3}H]nitrendipine binding produced by fluspirilene or pimozide. Concentration-response curves for the influence of (D-600) on [^{3}H]nitrendipine binding to membranes were obtained with six concentrations of D-600 and 0.2 nM [^{3}H]nitrendipine. [^{3}H]nitrendipine binding was determined at each concentration of (D-600) in the absence (●) or presence of 7.5 nM (○), 25 nM (■), 75 nM (□), 250 nM (▲), or 750 nM (△) fluspirilene (*a*) or pimozide (*b*). (Reprinted, with permission, from Gould et al. 1983b.)

zymic products, as bovine serum albumin does not protect against the binding decrease (Table 4). Likewise phospholipase C treatment of the membrane decreases binding.

Table 4. Biochemical Characteristics of [^{3}H]Nitrendipine Binding

	Percent of binding
Trypsin	8
Trypsin + trypsin inhibitor	67
Phospholipase A_2	46
Phospholipase A_2 and BSA	53
Phospholipase C	29
Phorbolipase C and BSA	10
N-ethylmaleimide	58
2-Mercaptoethanol	68
Heat (60°C, 15 min)	0

DISCUSSION

We have proposed a simple model of the Ca^{++} channel (Fig. 6). The dihydropyridine-binding site appears to be a metalloprotein of radiation inactivation target size of 250,000 (Norman et al. 1983). There appears to be a second drug binding site, the verapamil-diltiazem site. There is no evidence as to whether this site is on the same polypeptide chain as the dihydropyridine site. The binding site for dihydropyridines requires Ca^{++} or a divalent cation of similar ionic radius for full activity (Fig. 7). By contrast, binding at the verapamil-diltiazem site appears to be very sensitive to inhibition by Ca^{++} (I.J. Reynolds et al., unpubl.). This would also suggest two sites of action of these drugs. Moreover, an nonpermeable, charged quaternary-ammonium derivative of verapamil, D-890, is inactive when administered to the exterior of right ventricular tubiculae or isolated heart cells. If it is microinjected into the interior of the cells, it is very potent and of rapid onset (Hescheler et al. 1982). This suggests an intracellular site of action for these drugs, and is consistent with these drugs acting at a different site than the dihydropyridines, which have been assumed to act at the exterior of the cell. This second site shows a similar structure-activity relationship as the hydrophobic binding site of calmodulin. It is, however, more specific for verapamil, pimozide, fluspirilene, and related drugs than calmodulin is.

The ionic effects suggest that the divalent cation site responsible for dihydropyridine binding is different from the metal cation coordination site responsible for movement through the channel. Whereas Ba^{++} is the most efficacious ion at this latter site, Ca^{++}, Sr^{++}, and Mn^{++} are approximately equally efficacious at the dihydropyridine site (Hagiwara and Byerly 1981; Gould et al. 1982). Ba^{++} can only restore 50% of the binding (Table 2). Interestingly, nitrendipine is much less effective in blocking currents carried by Ba^{++} than in blocking Ca^{++} currents in single dialyzed heart cells (Lee and Tsien 1983). This may suggest that the cation site involved in dihydropyridine binding is involved in regulating current at a site distinct from cation coordination sites described previously. By substi-

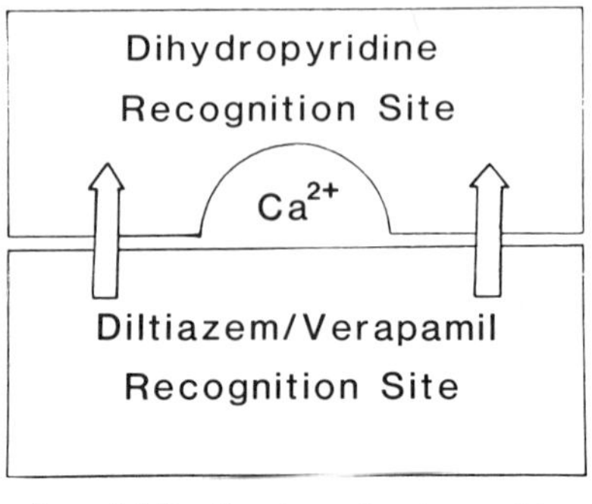

Figure 6. A model for the sites of action and interactions for inorganic and organic Ca^{++} channel antagonists. (Reprinted, with permission, from Murphy et al. 1983b.)

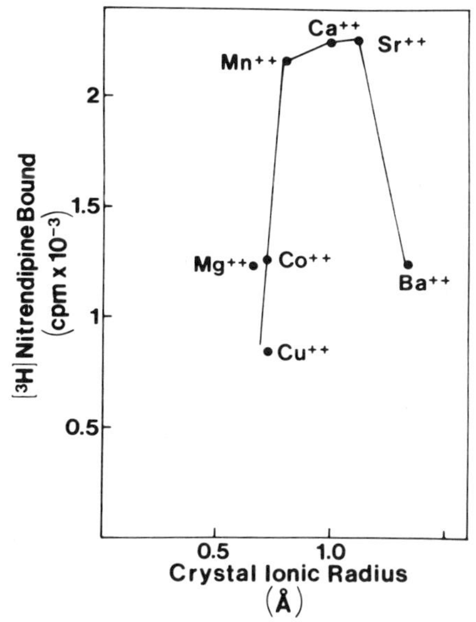

Figure 7. Maximal stimulation of [^{3}H]nitrendipine binding versus ionic crystal radii of divalent cations. (Reprinted, with permission, from Gould et al. 1982.)

tuting Ba^{++} for Ca^{++}, nitrendipine binds to fewer sites and is thus less effective. This is consistent with several cation sites within the Ca^{++} channel.

Do Ca^{++} channel antagonists have any psychopharmacologic effects? To date, the only reported effect is prevention of amnesia following anoxia or electroconvulsive shock (Hoffmeister et al. 1982). However, the results with diphenylbutylpiperidines may suggest a role for these drugs acting at Ca^{++} channels in schizophrenia. Pimozide, penfluridol, and fluspirilene are about as potent at nitrendipine sites as they are at dopamine (D_2) sites (Table 3). Thus, at clinical doses that relieve symptoms believed to be due to actions at D_2 sites, dihydropyridine sites would be similarly occupied. The diphenylbutylpiperidines appear to be somewhat unique in the symptoms they relieve (Table 5). In addition to the positive symptoms of acute schizophrenia (hallucinations, delusions), they relieve the negative symptoms of chronic schizophrenia, such as flattened effect and social withdrawal (Crow 1980). By contrast, classic neuroleptics, such as phenothiazines and butyrophenones, are much less effective at the negative symptoms. These classic neuroleptics are much less potent at nitrendipine-associated sites (Gould et al. 1983b). Thus, part of the action of the diphenylbutylpiperidines may be mediated via the nitrendipine-associated sites. If so, other Ca^{++} channel antagonists may have similar efficacy at relieving the negative symptoms of schizophrenia and may imply an important role of Ca^{++} channels in emotions.

ACKNOWLEDGMENTS

This work was supported by USPHS grants DA-00266, MH-18501 and NS-16375, RSA DA-00074 to S.H.S., a grant of the McKnight Foundation, training grant MH-15330 to R.J.G., training grant GM-07309 to K.M.M.M., and a grant from Miles Laboratories. We thank Dawn C. Dodson for manuscript preparation.

REFERENCES

Blakely, A.G.H., D.A. Brown, T.C. Cunnane, A.M. French, J.C. McGrath, and N.C. Scott. 1981. Effects of nifedipine on electrical and mechanical responses of rat and guinea pig vas deferens. *Nature* **294:** 759.

Bolger, G.T., P. Gengo, R. Klockowski, E. Luchowski, H. Siegel, R.A. Janis, A.M. Triggle, and D.J. Triggle. 1983. Characterization of binding of the Ca^{++} channel antagonist, [^{3}H]nitrendipine, to guinea pig ileal smooth muscle. *J. Pharmacol. Exp. Ther.* **225:** 291.

Catterall, W.A. 1980. Neurotoxins that act on voltage-sensitive sodium channels in excitable membranes. *Annu. Rev. Pharmacol. Toxicol.* **20:** 15.

Cauvin, C., R. Loutzenhiser, and C. Van Breemen. 1983. Mechanisms of calcium antagonist-induced vasodilation. *Annu. Rev. Pharmacol. Toxicol.* **23:** 373.

Crow, T.J. 1980. Molecular biology of schizophrenia: More than one disease process? *Br. Med. J.* **280:** 66.

Diecke, F.P.J. 1983. Studies with nitrendipine on chemically skinned smooth muscle strips. In *Nitrendipine* (ed. A. Scriabine et al.). Urban and Schwarzenberg, Baltimore, Maryland. (In press.)

Douglas, W.W. and P.S. Taraskavitch. 1982. Slowing effects of dopamine and calcium-channel blockers on fre-

Table 5. Diphenylbutylpiperidines as Antischizophrenic Agents

Agent	Clinical efficacies			Reference
Pimozide	decreased anergia	vs.	haloperidol	Haas and Beckman (1982)
	decreased emotional withdrawal			
	improved "expression of feeling"	vs.	carpipramine	Kudo (1972)
	improved "contact"			
	improved "work or recreation"			
	reduced emotional withdrawal	vs.	chlorpromazine	Kolivakis et al. (1974)
Penfluridol	improved anergia	vs.	fluphenazine	Lapierre (1978)
	improved emotional withdrawal			
	improved social competence	vs.	thioridazine	Jus et al. (1974)
Fluspirilene	improved uncooperativeness	vs.	fluphenazine decanoate	Frangos et al. (1978)
	improved blunted affect			
	improved depressive mood			
	improved sociability	vs.	fluphenazine	Faltus (1974)
	improved affectivity			
	improved blunted affect	vs.	fluphenazine enanthate	Villeneuve et al. (1970)

quency of sodium spikes in rat pars intermedia cells. *J. Physiol.* **326:** 201.

EDWARDS, C. 1982. The selectivity of ion channels in nerve and muscle. *Neuroscience* 7: 1335.

EHLERT, F.J., W.R. ROESKE, E. ITOGA, and H.I. YAMAMURA. 1982. The binding of [^{3}H]nitrendipine to receptors for calcium channel antagonists in the heart, cerebral cortex and ileum of rats. *Life Sci.* **30:** 2191.

FALTUS, F. 1974. The depot compounds Moditen (fluphenazine) and Imap (fluspirilene) in clinical practice. *Activ. Nerv. Sup.* **16:** 167.

FERRY, D.R. and H. GLOSSMANN. 1982. Evidence for multiple receptor sites within the putative calcium channel. *Naunyn-Schmeidebergs Arch. Pharmacol.* **320:** 80.

FLECKENSTEIN, A. 1977. Specific pharmacology of calcium in myocardium cardiac pacemakers and vascular smooth muscle. *Annu. Rev. Pharmacol. Toxicol.* **17:** 149.

——. 1982. Fundamental actions of calcium antagonists on myocardial and cardiac pacemaker cell membranes. In *New perspectives on calcium antagonists* (ed. G.B. Weiss), p. 59. American Physiological Society, Bethesda, Maryland.

FRANGOS, H., N.P. ZISSIS, I. LEONTOPOULOS, N. DIAMANTAS, S. TSITOURIDIS, I. GAVRIL, and K. TSOLIS. 1978. Double-blind therapeutic evaluation of fluspirilene compared with fluphenazine decanoate in chronic schizophrenics. *Acta Psychiatr. Scand.* **57:** 436.

GODFRAIND, T. 1982. Calcium influx and receptor-response coupling. In *New perspectives on calcium antagonists* (ed. G.B. Weiss), p. 95. American Physiological Society, Bethesda, Maryland.

——. 1983. Actions of nifedipine on calcium fluxes and contraction in isolated rat arteries. *J. Pharmacol. Exp. Ther.* **224:** 443.

GODFRAIND, T. and D. DIEU. 1981. The inhibition by flunarizine of the norepinephrine-evoked contraction and calcium influx in rat aorta mesenteric arteries. *J. Pharmacol. Exp. Ther.* **217:** 510.

GOULD, R.J., K.M.M. MURPHY, and S.H. SNYDER. 1982. [^{3}H]Nitrendipine labeled calcium channels discriminate inorganic calcium agonists and antagonists. *Proc. Natl. Acad. Sci.* **79:** 3656.

——. 1983a. *In vitro* autoradiography of [^{3}H]nitrendipine localizes calcium channels to synaptic rich zones. In *Nitrendipine* (ed. A. Scriabine et al.). Urban and Schwarzenberg, Baltimore, Maryland. (In press.)

GOULD, R.J., K.M.M. MURPHY, I.J. REYNOLDS, and S.H. SNYDER. 1983b. Antischizophrenic drugs of the diphenylbutylpiperidine type act as calcium channel antagonists. *Proc. Natl. Acad. Sci.* **80:** 5122.

HAAS, S. and H. BECKMANN. 1982. Pimozide versus haloperidol in acute schizophrenia. A double-blind controlled study. *Pharmacopsychiatria* **15:** 70.

HAGIWARA, S. and L. BYERLY. 1981. Calcium channel. *Annu. Rev. Neurosci.* **4:** 69.

HENRY, P.D. 1980. Comparative pharmacology of calcium antagonists: Nifedipine, verapamil and diltiazem. *Am. J. Cardiol.* **46:** 1047.

HESCHELER, J., D. PELZER, G. TRUBE, and W. TRAUTWEIN. 1982. Does the organic calcium channel blocker D-600 act from inside or outside on the cardiac cell membrane. *Pfluegers Arch.* **393:** 287.

HOFFMEISTER, F., U. BENZ, A. HEISE, H.P. KRAUSE, and V. NEUSER. 1982. Behavioral effects of nimodipine in animals. *Arzneim. Forsch. Drug Res.* **32:** 347.

HOGESTATT, E.D., K.-E. ANDERSSON, and L. EDVINSSEN. 1982. Effects of nifedipine on potassium-induced contraction and noradrenaline release in cerebral and extracranial arteries from rabbit. *Acta Physiol. Scand.* **114:** 283.

JANIS, R.A. and D.J. TRIGGLE. 1983. New developments in Ca^{2+} channel antagonists. *J. Med. Chem.* **26:** 775.

JUS, A., R. PINEAU, K. JUS, A. VILLENEUVE, J. GAUTIER, P. PIRES, A. DROLET, M. COTE, and R. VILLENEUVE. 1974. Penfluridol: A long-acting oral neuroleptic as therapeutic agent in chronic schizophrenia. *Curr. Ther. Res. Clin. Exp.* **16:** 1041.

KOLIVAKIS, T., H. AZIM, and E. KINGSTONE. 1974. A double-blind comparison of pimozide and chlorpromazine in the maintenance care of chronic schizophrenic outpatients. *Curr. Ther. Res. Clin. Exp.* **16:** 998.

KUDO, Y. 1972. A double-blind comparison of pimozide with carpipramine in schizophrenic patients. *Acta Psychiatr. Belg.* **72:** 685.

LAPIERRE, Y.D. 1978. A controlled study of penfluridol in the treatment of chronic schizophrenia. *Am. J. Psychiatry* **135:** 956.

LEE, K.S. and R.W. TSIEN. 1983. Mechanism of calcium channel blockade by verapamil, D-600, diltiazem and nitrendipine in single, dialyzed heart cells. *Nature* **302:** 790.

LOEV, B., M.M. GOODMAN, K.M. SNADER, R. TEDESCHI, and E. MACKO. 1974. "Hantzch-type" dihydropyridine hypotensive agents. *J. Med. Chem.* **17:** 956.

MARANGOS, P.J., J. PATEL, G. MILLER, and A.M. MARTINO. 1982. Specific calcium antagonist binding sites in brain. *Life Sci.* **31:** 1575.

MURPHY, K.M.M., R.J. GOULD, and S.H. SNYDER. 1982. Autoradiographic visualization of [^{3}H]nitrendipine binding sites in rat brain: Localization to synaptic zones. *Eur. J. Pharmacol.* **81:** 517.

——. 1983a. Regulation of [^{3}H]nitrendipine binding: A single allosteric site for verapamil, diltiazem and prenylamine. In *Nitrendipine* (ed. A. Scriabine et al.). Urban and Schwarzenberg, Baltimore, Maryland. (In press.)

MURPHY, K.M.M., R.J. GOULD, B.L. LARGENT, and S.H. SNYDER. 1983b. A unitary mechanism of calcium antagonist drug action. *Proc. Natl. Acad. Sci.* **80:** 860.

NORMAN, R.I., M. BORSOTTO, M. FOSSETT, M. LAZDUNSKI, and J.C. ELLORY. 1983. Determination of the molecular size of the nitrendipine-sensitive Ca^{2+} channel by radiation inactivation. *Biochem. Biophys. Res. Commun.* **111:** 878.

RODENKIRCHEN, R., R. BAYER, R. STEINER, F. BOSSERT, H. MEYER, and E. MOLLER. 1978. Structure-activity studies on nifedipine in isolated cardiac muscle. *Naunyn-Schmiedebergs Arch. Pharmacol.* **310:** 69.

STARKE, K. and H.J. SCHUMANN. 1973. Effect of nifedipine on the sympathetic cardiac nerves. *Arzneim. Forsch.* **23:** 193.

TOLL, L. 1982. Calcium antagonists. High-affinity binding and inhibition of calcium transport in a clonal cell line. *J. Biol. Chem.* **257:** 13189.

TRIGGLE, D.J. 1982. Calcium antagonists: Basic chemical and pharmacological aspects. In *New perspectives on calcium antagonists* (ed. G.B. Weiss), p. 1. American Physiological Society, Bethesda, Maryland.

TSIEN, R. 1983. Calcium channels in excitable cell membranes. *Annu. Rev. Physiol.* **45:** 341.

VAN BREEMAN, C., O. HWANG, and K.D. MEISHERI. 1981. The mechanism of inhibitory action of diltiazem on vascular smooth muscle contractility. *J. Pharmacol. Exp. Ther.* **218:** 459.

VILLENEUVE, A., K. DOGAN, R. LACHANCE, and C. PROULX. 1970. A controlled study of fluspirilene in chronic schizophrenia. *Curr. Ther. Res. Clin. Exp.* **12:** 819.

WEISS, B., W. PROZIALECK, M. CIMINO, M.S. BARNETTE, and T.L. WALLACE. 1980. Pharmacological regulation of calmodulin. *Ann. N.Y. Acad. Sci.* **356:** 319.

Peptides in Neuronal Function: Studies Using Frog Autonomic Ganglia

Y.N. JAN, C.W. BOWERS, D. BRANTON, L. EVANS, AND L.Y. JAN
Department of Physiology, School of Medicine, University of California, San Francisco, California 94143

At present, more than 20 peptides are considered as putative neurotransmitters, based primarily on their localization in nerve terminals and their actions on nerve cells (Hökfelt et al. 1980). The list is almost certainly not exhaustive and it is likely that more will be found in the future. If many peptides turn out to be neurotransmitters, one might wonder why the nervous system utilizes so many different transmitters. A detailed analysis of the action of a number of neuropeptides may reveal characteristics unique to peptidergic transmission and thus shed light on the functional significance of peptide transmitters and perhaps explain the need for multiple transmitters.

Frog sympathetic ganglia and parasympathetic cardiac ganglia are excellent preparations for this purpose (Langley and Orbeli 1910; McMahan and Kuffler 1971). Structurally, they are very simple. Neurons in these ganglia are unipolar and have no dendrites. Preganglionic fibers make synaptic contacts almost exclusively on the ganglionic cell bodies. This arrangement facilitates analysis of the physiological effects of neurotransmitters and allows accurate anatomical description of synaptic connections. Analysis of possible actions of peptides in these two preparations has already revealed interesting features. In the frog sympathetic ganglia, a luteinizing hormone releasing hormone (LHRH)-like peptide is almost certain to be the transmitter that mediates a slow excitatory postsynaptic potential, the late slow EPSP. Certain properties of this peptidergic transmission appear to be strikingly different from those of conventional cholinergic transmission. For instance, although the LHRH-like peptide and acetylcholine (ACh) seem to be released from the same nerve terminals that form synapses of conventional morphology, the peptide can diffuse far greater distances than ACh and activate sympathetic neurons nonsynaptically. In the frog heart septum, all of the terminals on the parasympathetic neurons are positively stained by antibodies to substance P (SP). However, little or no electrical effects on those neurons have been detected that may be attributed to the SP-like immunoreactivity.

Our preliminary results indicate that one target of the SP-like material may be the mast cells situated near the neurons. These results prompted us to propose a hypothesis about the general mode of actions of neuropeptides and to suggest why the nervous system uses so many different chemical messengers.

METHODS

Both preparations used here, the paravertebral sympathetic ganglia and the parasympathetic ganglia of the interartrial septum of the heart, were isolated from bullfrogs (*Rana catesbiana*) as previously described (McMahan and Kuffler 1971; Jan and Jan 1982).

Electrophysiological experiments. The isolated ganglia were pinned to a thin layer of Sylgard in a petri dish and viewed with Nomarski optics. The bath was continuously perfused at room temperature with Ringer's solution containing 115 mM NaCl, 2 mM KCl, 3.6 mM $CaCl_2$, and 5 mM HEPES (pH 7.2). Conventional recording techniques were used (Jan and Jan 1982). For the cardiac septum physiology, glucose (5 mM) was added to the perfusion solution.

Histochemical experiments—Immunocytochemistry. Bullfrog atrial septa were removed and fixed by immersion in 4% paraformaldehyde and 0.09 M sodium phosphate (pH 7.2) for 45 minutes at room temperature. The tissue was processed for immunofluorescence in 0.3% Triton, 0.05 M Tris buffer (pH 7.2), 0.65% NaCl, and optimal dilutions of antibody (1:200 for anti-SP mouse monoclonal) overnight (4°C). The tissue was then treated with a second antibody, rabbit anti-globulin (rhodamine-labeled) or mouse anti-globulin (fluorescein-labeled) for 2 hours. The preadsorption for SP staining was achieved by preincubating the anti-SP monoclonal (1:100) with 140 μg/ml mammalian SP overnight (4°C) and then diluting the solution 1:1 with the appropriate Triton-buffer solution. Tissue treated with this preadsorbed antibody showed no staining, whereas similarly treated control antibody solutions (without SP) gave normal fluorescent labeling.

Mast cell staining. The procedure of Blaies and Williams (1981) was used with little alteration to stain mast cells in septal whole mounts. Fixed tissue was exposed to 0.5% Astra Blue, 2% $MgCl_2$ $6H_2O$, and 0.1% pararosaniline hydrochloride in 76% ethanol for 2 hours. Tissue was rinsed in Ringer's and not counterstained. Tissue was cleared with xylene and mounted on slides.

PEPTIDERGIC TRANSMISSION IN FROG SYMPATHETIC GANGLIA

It has been known for many years that ACh is a transmitter contained in the preganglionic fibers. Three

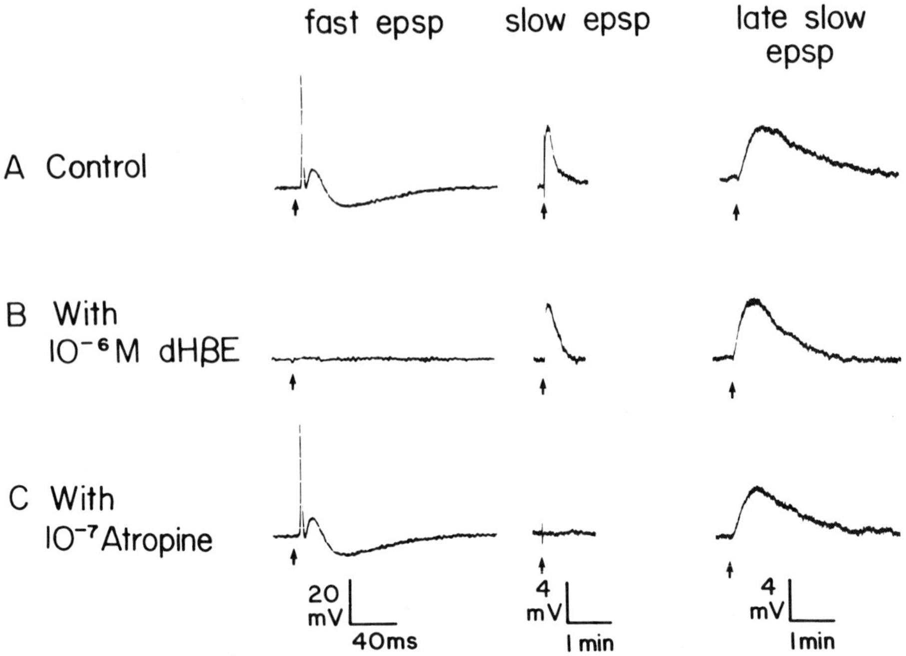

Figure 1. Effects of nicotinic cholinergic blocker, dihydro-β-erythroidine (dHβE), and a muscarine cholinergic blocker, atropine, on the fast EPSP and late slow EPSP recorded from the same B cell. Arrows indicate time of nerve stimulation. The fast EPSP was initiated by a single stimulus of the sympathetic chain above the 7th ganglion. The slow EPSP was induced by stimulating the chains at 30 Hz for 0.7 sec. Two separate nerves, the 7th and 8th spinal nerves, were stimulated at 5 Hz for 20 sec to induce the late slow EPSP. (Reprinted, with permission, from Jan and Jan 1982.)

types of synaptic responses are mediated by ACh, namely, (1) the nicotinic fast EPSPs, which last for about 30–50 msec; (2) the muscarinic slow EPSPs lasting 30–60 seconds; and (3) the muscarinic slow inhibitory postsynaptic potential (IPSP) of 1–2 seconds duration. In 1968, a fourth synaptic potential was discovered by Nishi and Koketsu and named the late slow EPSP. This response lasts for several minutes and is not mediated by ACh (Fig. 1). From studies done in the last few years (Jan et al. 1979, 1980a,b; Jan and Jan 1982), we believe that the late slow EPSP is mediated by a peptide transmitter that resembles mammalian LHRH. Briefly, we will outline the evidence as follows.

An LHRH-like Peptide Is the Transmitter Mediating the Late Slow EPSP in Bullfrog Sympathetic Ganglia

Presence. An LHRH-like peptide is contained in the appropriate preganglionic fibers. Radioimmunoassays using an antiserum highly specific for LHRH revealed that the bullfrog sympathetic ganglia contained a high level of LHRH-like peptide (Jan et al. 1979). The LHRH-like peptide is localized to preganglionic nerve terminals by immunohistochemistry (Jan et al. 1980a). Cutting appropriate preganglionic fibers greatly reduces the staining in nerve terminals and removes from ganglia almost all the LHRH-like peptide detectable by radioimmunoassays, demonstrating that the LHRH-like peptide is contained in those preganglionic fibers that generate the late slow EPSP.

Release. The LHRH-like peptide is released into the extracellular medium when the appropriate preganglionic fibers are stimulated. By either stimulating the appropriate spinal nerves electrically or raising the extracellular K^+ concentration, one can detect the release of the LHRH-like peptide by radioimmunoassays (Jan et al. 1979; Jan and Jan 1982). This release requires the presence of extracellular Ca^{++}.

Mimicry. The physiological effects of the late slow EPSP can be mimicked by applying LHRH to the surface of sympathetic neurons. A depolarizing response in sympathetic neurons is induced by synthetic LHRH, which may be applied either in the bath or locally through a micropipette via a brief pulse of pressure. This depolarizing response persists in media devoid of Ca^{++}, indicating that LHRH acts directly on sympathetic neurons. The LHRH-induced depolarization resembles the late slow EPSP in considerable detail (Jan and Jan 1982; Jan et al. 1979, 1980b; Kuffler 1980): (1) Both are associated with similar conductance changes; (2) their amplitudes vary in parallel as the membrane potential is shifted over a wide range, suggesting that similar ionic mechanisms are involved; (3) both responses increase the excitability of the neurons; and (4) the responses interact with the cholinergic responses in a parallel manner. The similarities between physiological changes during the LHRH-induced depolarization and those during the late slow EPSP suggest that LHRH acts on the same receptors as the natural transmitter for the late slow EPSP.

Pharmacology. The late slow EPSP is blocked by antagonists of LHRH. Several analogs of LHRH that function as antagonists in mammalian anterior pituitaries (Rivier and Vale 1978) also block the LHRH-induced depolarization in frog sympathetic neurons. Furthermore, they also block the nerve-evoked late slow EPSP without affecting the resting membrane potential, membrane resistance, or the cholinergic synaptic potentials (Jan et al. 1980b; Jan and Jan 1982) (Figs. 2 and 3). Therefore, pharmacologically, the receptor for the late slow EPSP behaves like a LHRH receptor.

Based on the above four criteria, there seems to be very little doubt that an LHRH-like peptide is the transmitter mediating the late slow EPSP. However, several issues are yet to be resolved completely.

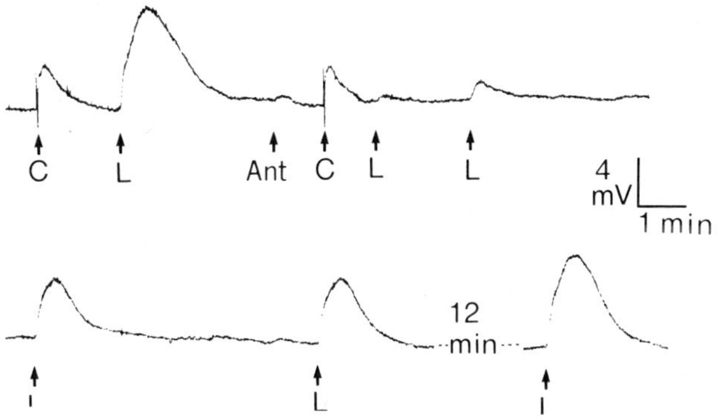

Figure 3. Blockade and recovery of LHRH response (L) following the application of a brief pulse (10 sec) of an LHRH antagonist. Notice that the antagonist has no effect on cholinergic responses (C). This is a recording from another B cell. Nerve stimulation and peptide application were done as in Fig. 2.

The Identity of the LHRH-like Peptide

The LHRH-like transmitter in frog sympathetic ganglia is immunologically and chromatographically similar to, but not identical to, mammalian LHRH (Jan et al. 1979; Eiden and Eskay 1980). To date, three forms of LHRH are known. They are from the hypothalami of mammal, chicken, and fish (King and Millar 1982; Miyamoto et al. 1982; Sherwood et al. 1983). A recent report suggested that fish LHRH is immunologically and chromatographically most similar to the frog ganglion peptide (Eiden et al. 1982). We have examined the relative physiological potencies of mammalian, chicken, and fish LHRH on frog ganglion cells (Table 1). Fish LHRH is much more potent than mammalian LHRH, which is much more potent than chicken LHRH. The threshold for a detectable depolarization produced by mammalian LHRH is about 10^{-6} M, whereas the threshold for fish LHRH is about 10^{-7} M. Fish LHRH-derived antagonists and superagonists are also more potent than mammalian analogs. By contrast, it has been reported that fish LHRH is only 2–3% as potent as mammalian LHRH in mammalian assay systems (Sherwood et al. 1983). The relative potency of fish LHRH in the frog ganglion supports the idea that fish LHRH is more similar than mammalian LHRH to the ganglion LHRH-like peptide. However, caution must be used in drawing conclusions about the structure of the ganglion peptide. Fish LHRH differs from mammalian LHRH at positions 7 and 8. Fish LHRH is Trp-7, Leu-8, whereas mammalian LHRH is Leu-7, Arg-8. We, for example, found that the Leu-7, Trp-8 analog, which is extremely difficult to distinguish from fish LHRH, even by HPLC (Sherwood et al. 1983), is approximately equipotent with fish LHRH on frog ganglion cells. Determination of the structure of the frog ganglion LHRH-like peptide must await direct determination of its amino acid sequence.

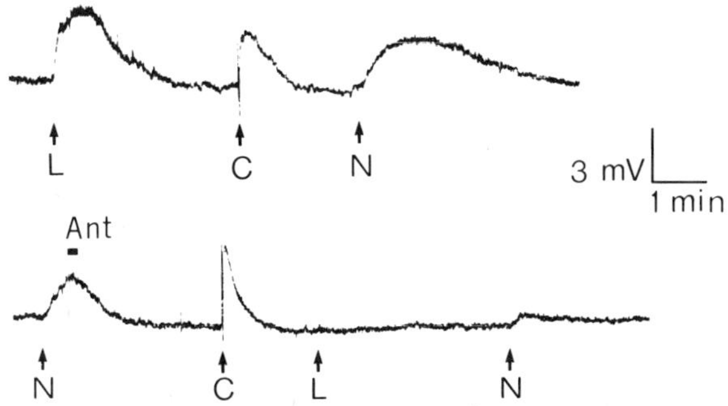

Figure 2. Effect of an antagonist of LHRH on the late slow EPSP (N), the LHRH-induced response (L), and the cholinergic response (C). On the lower trace, after a late slow EPSP was initiated, a brief pulse of an antagonist of LHRH (acetyl-D-NAL(2)[1],4-FDPhe[2],DTrp[3],DArg[6],Trp[7],Leu[8]LHRH) was delivered via pressure ejection onto the cell from a micropipette positioned nearby. Notice that late slow EPSP was truncated. Subsequently, both fish LHRH-induced response and the late slow EPSP were blocked, the nicotinic cholinergic response was not affected, whereas the muscarinic cholinergic response was actually enhanced. This recording is from a B cell. Cholinergic responses and the late slow EPSP were elicited by stimulating the sympathetic chain above the 7th ganglion and the 7th and 8th spinal nerves, respectively. LHRH response was elicited with a brief (3 sec) pulse of fish LHRH via pressure ejection.

The Ionic Basis of the Late Slow EPSP

Adams and Brown and their colleagues discovered that the late slow EPSP causes inactivation of a novel K^+ channel, the M channel (Adams and Brown 1980; Adams et al. 1982). Although the M channel probably is an important component in generating the late slow EPSP, it cannot fully account for the ionic basis of the late slow EPSP. In addition to the M channel, at least one more ionic current must be involved, because the late slow EPSP and the LHRH-induced depolarization usually persist at the K^+ equilibrium potential (Jan and Jan 1982; Katayama and Nishi 1982; Kuffler and Sejnowski 1983). This additional component has not yet been fully characterized.

Removal of the LHRH-like Peptide after Release

In the classical picture of synaptic transmission, the transmitter molecules are quickly removed after release either by enzyme degradation or by an efficient re-

Table 1. Threshold Concentrations of LHRH Analogs to Depolarize Frog Ganglion Cells

LHRH analog		Threshold on frog ganglion
Fish LHRH	pGlu^1-His^2-Trp^3-Ser^4-Tyr^5-Gly^6-Trp^7-Leu^8-Pro^9-Gly^{10}NH$_2$	10^{-7}–10^{-8} M
Reverse fish LHRH	-Gly^6-Leu^7-Trp^8-Pro^9-Gly^{10}NH$_2$	10^{-7}–10^{-8} M
Fish superagonist	-DArg^6-Trp^7-Leu^8-Pro^{9}NHET	10^{-9}–10^{-10} M
Mammalian LHRH	-Gly^6-Leu^7-Arg^8-Pro^9-Gly^{10}NH$_2$	10^{-6}–10^{-7} M
Mammalian superagonist	-DLys^6-Leu^7-Arg^8-Pro^{9}NHET	10^{-7}–10^{-8} M
Mammalian superagonist	-DTrp^6-Leu^7-Arg^8-Pro^{9}NHET	10^{-8}–10^{-9} M
Chicken LHRH	-Gly^6-Leu^7-Gln^8-Pro^9 Gly^{10}NH$_2$	10^{-4}–10^{-5} M

Analogs were applied via pressure ejection from 5–8-μm micropipettes, placed about 30 μm from the cell, at a pressure of 2 lb/in^2 for 5 sec. BSA (0.25%) was added as a carrier when concentrations were nanomolar or lower. The values given for each analog represent the range within which a depolarization was first detectable.

uptake system. This does not seem to be the case for the LHRH-like peptide. As will be discussed later, the LHRH-like peptide appears to have a fairly long lifetime (many seconds) after they are released; apparently, there are no efficient removal systems. Whether removal mechanisms other than simple diffusion are operative at all is not yet clear.

Features of Peptidergic Transmission

Given that a peptide can function as a transmitter for a slow synaptic potential, our next question is: Do peptide transmitters serve functions qualitatively different from those of "classical" transmitters? One striking difference is that the peptidergic responses in autonomic neurons are 10–10^4 times slower than cholinergic responses (Dun and Karczmar 1979; Jan et al. 1979; Konishi et al. 1979; Tsunoo et al. 1982). It turns out that in frog ganglia the slow time course of the late slow EPSP is partially due to the long lifetime of the nerve-released peptide transmitter. Further, the LHRH-like peptide appears to be contained and released from the same preganglionic fibers as ACh. Whereas ACh acts only synaptically, the LHRH-like peptide can diffuse and activate sympathetic neurons that are not in synaptic contact with LHRH-positive nerve terminals. Experimental results in support of the above statements are reviewed in the following sections.

The LHRH-like Peptide and ACh Are Probably Contained and Released from the Same Preganglionic Fibers

In the last two ganglia of the lumbar chain, the 9th and 10th ganglia, there are two classes of sympathetic neurons—the larger B cells are in synaptic contact with preganglionic B fibers arising from the 3rd, 4th, and 5th spinal nerves, whereas the smaller C cells are in synaptic contact with preganglionic C fibers arising from the 7th and 8th spinal nerves. As summarized in Figure 4, stimulation of the 3rd, 4th, and 5th spinal nerves generates cholinergic synaptic potentials in B cells only. On the other hand, stimulation of the 7th and 8th spinal nerves generates the late slow EPSP in both B cells and C cells, as well as cholinergic responses in C cells. This raises the following question: Does the same preganglionic C fiber generate both cholinergic responses and the peptidergic late slow EPSP? In other words, does the same preganglionic C fiber contain and release both ACh and the LHRH-like peptide?

Coexistence? In the absence of any known specific markers for ACh or cholineacetyltransferase in frog sympathetic ganglia, we have approached this question with indirect means. In double-labeling experiments, we marked all terminals of preganglionic C fibers with horseradish peroxidase (HRP) or a monoclonal antibody specific for a 65-kD vesicle membrane protein (Matthew et al. 1981) and the appropriate fluorescein-labeled second antibodies. The same terminals were then treated with rabbit antibodies against LHRH and rhodamine-labeled second antibodies that bind rabbit antibodies so that we could estimate the proportion of preganglionic C fiber terminals that contain the LHRH-like peptide. It was found that almost all terminals of preganglionic C fibers contained the LHRH-like peptide (Jan and Jan 1983), as shown in Figure 5. Since some of these terminals must also contain ACh, which generates in C cells the fast EPSP of only 1–2-msec latency, at least some of the terminals of preganglionic C fibers must contain both ACh and the LHRH-like peptide. Morphologically, terminals of preganglionic C fibers form conventional synaptic contacts with C cells and are uniform in appearance under an electron microscope. Therefore, conceivably all terminals of preganglionic C fibers may contain both ACh and the LHRH-like peptide.

Co-release? Physiological experiments indicate that each preganglionic C fiber probably releases both ACh and the LHRH-like peptide: A sympathetic C neuron typically receives several cholinergic inputs with different thresholds for stimulation, so that one may raise the stimulation strength gradually to recruit cholinergic preganglionic fibers one by one. In doing so, we found that each time a cholinergic fiber is recruited, repetitive stimulation at that stimulation strength also resulted in a larger late slow EPSP, indicating that the peptidergic inputs to a C cell have thresholds similar to the thresholds of its cholinergic inputs (Jan and Jan 1983). Most likely, the same preganglionic fibers are responsible for

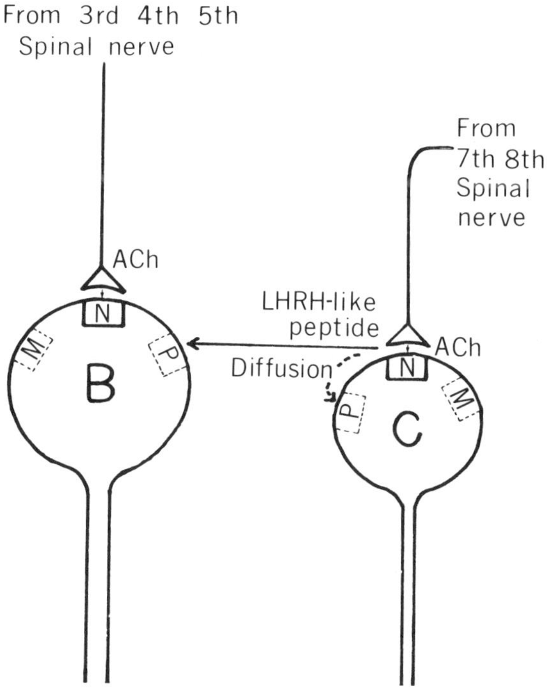

Figure 4. Scheme of innervation of neurons in the 9th or 10th ganglia, the last two ganglia in the sympathetic chain. B cells are larger, whereas C cells are smaller. Cholinergic axons for B neurons arise from the 3rd, 4th, and 5th spinal nerves, whereas preganglionic fibers for C cells come through the 7th and 8th spinal nerves. LHRH-positive nerve terminals are present only on C cells. Most likely, both the LHRH-like peptide and ACh are contained and released from the same preganglionic fibers for C cells. (N) Nicotinic cholinergic receptors; (M) muscarinic cholinergic receptors; (P) LHRH receptors. Of the three types of receptors, only the nicotinic cholinergic receptors have been localized. They are situated right in opposition to the synaptic boutons. (Reprinted, with permission, from Jan and Jan 1982.)

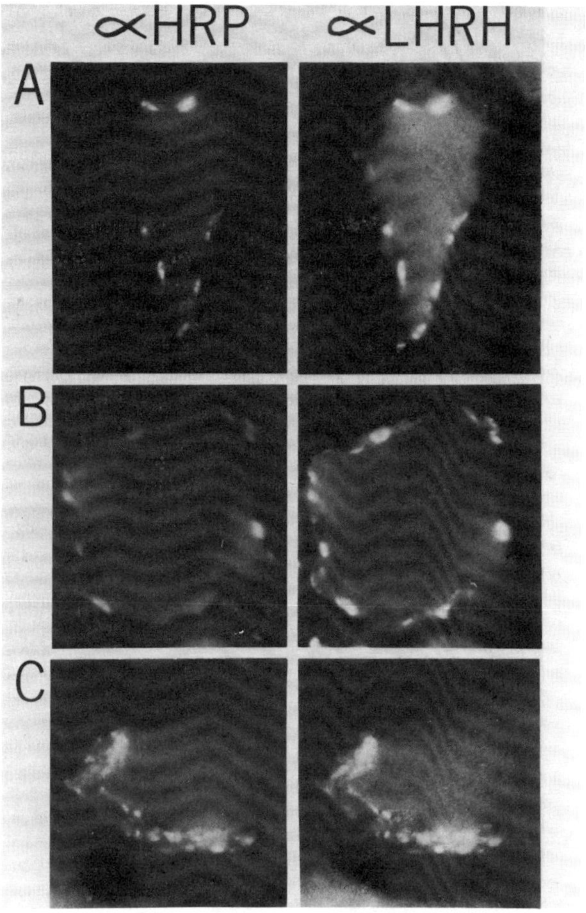

Figure 5. Double labeling of the same terminals of preganglionic C fibers with antibodies to LHRH and horseradish peroxidase (HRP). Preganglionic C fibers were first filled with HRP, then treated with fluorescein-coupled goat antibodies to HRP, rabbit antibodies to LHRH, and rhodamine-labeled goat antibodies to rabbit antibodies. Note that all terminals filled with HRP contain the LHRH-like peptide. (Reprinted, with permission, from Jan and Jan 1983.)

both the cholinergic fast EPSP and the peptidergic late slow EPSP. At present, there is much circumstantial evidence suggesting that a neuron may use more than one transmitter (Hökfelt et al. 1980). However, to prove it, one needs to demonstrate first that the substances in question actually function as transmitters. Frog sympathetic ganglia lends itself to a rigorous test of the idea of coexistence and co-release of more than one transmitter. The results described above provide strong evidence that suggests this is the case.

The LHRH-like Peptide Can Diffuse for Tens of Microns before Activating Sympathetic Neurons

Although stimulating preganglionic C fibers arising from the 7th and 8th spinal nerves generates the late slow EPSP in both B cells and C cells, preganglionic C fibers were found to make synaptic contacts on C cells exclusively. Therefore, the late slow EPSP recorded in B cells must be due to peptide transmitters released from preganglionic C fibers that are at a distance from B cells. Evidence in support of these statements is reviewed as follows.

Preganglionic C fibers make synaptic contacts with C cells but not B bells. When we cut all preganglionic B fibers by cutting the sympathetic chain between the 6th and 7th ganglion and allowed 5–10 days for their terminals to degenerate, electron microscopy revealed no synaptic boutons on the larger B cells, even though many synaptic boutons were found on smaller C cells. Similarly, filling the remaining preganglionic C fibers with HRP marked 30–50 nerve terminals around each C cell, but no terminals at all on the larger B cells (Jan et al. 1980a). These observations correlate well with immunohistochemistry, which showed that LHRH-positive terminals are found on the smaller C cells, but not on the larger B cells (Jan et al. 1980a; Jan and Jan 1982). Taken together, they showed that no synaptic contacts exist between preganglionic C fibers and B cells.

The late slow EPSPs recorded in B cells are due to nonsynaptic action of the peptide transmitter. If we cut all preganglionic B fibers arising from the 3rd, 4th, and 5th spinal nerves as described in the preceding paragraph and recorded from neurons in the denervated ganglia 5–10 days after the operation, we found no cholinergic synaptic potentials in B cells with nerve stimulation, whereas the late slow EPSP was still generated in those B cells when we stimulated the preganglionic C fibers arising from the 7th and 8th spinal nerves. Three of those B cells that showed the late slow EPSP but no longer showed the cholinergic synaptic potentials were serial-sectioned. Electron microscopy revealed no synaptic boutons on the surfaces of these B cells (Jan et al. 1980a). Therefore, the late slow EPSP recorded in these B cells must be generated by preganglionic C fibers that are not in synaptic contact with their cell surface.

The LHRH-like peptide remains active for at least 30 seconds after release and can therefore diffuse for many microns. A late slow EPSP normally lasts for several minutes, as shown in Figure 2. Thirty seconds after the initiation of the late slow EPSP, applying antagonist of LHRH through a micropipette caused the response to be truncated (Jan and Jan 1982) (Fig. 2), suggesting that the LHRH analog can compete with the peptide transmitter for receptor binding 30 seconds after it is released. This is probably due to long lifetime of the peptide transmitter after release, rather than prolonged release of peptide transmitters after nerve stimulation, because applying antagonists 30 seconds after a brief pulse of LHRH also caused the LHRH-induced response to be truncated (Jan and Jan 1982). Thus, the slow time course of the late slow EPSP is partly caused by the long lifetime of the peptide transmitter. As expected, the rise time and duration of the late slow EPSP in B cells are on the average greater than those of the late slow EPSP in C cells (Jan and Jan 1982).

In summary, the frog sympathetic ganglion is composed of two cell types, B cells (30–70 μm in diameter) and C cells (20–45 μm in diameter). LHRH-positive preganglionic C fibers make synaptic contacts exclusively with C cells, but are capable of generating late slow EPSP in B cells. This nonsynaptic activation of B cells is possible because the LHRH-like peptide transmitter remains active for at least 30 seconds after release and can diffuse some distance before activating receptors and generating the late slow EPSP. Since B cells and C cells are juxtaposed within the ganglion and most of the LHRH-positive synaptic boutons are clustered around the axon hillock region of C cells, the LHRH-like peptide probably has to diffuse for tens of microns before reaching the surface of B cells. On the other hand, although almost every B cell tested responded to LHRH with a depolarization, B cells from different regions of a sympathetic ganglion often showed late slow EPSPs of vastly different rise time and amplitude. Most likely, the peptide transmitter does not diffuse for distances comparable with the diameter of the ganglion ($\sim$1 mm). Thus, the action of this peptide transmitter is analogous to that of paracrine hormones.

ACTIONS OF SP IN FROG PARASYMPATHETIC CARDIAC GANGLIA

Because the caudal sympathetic system of the bullfrog had proven useful in characterizing an example of peptidergic transmission, other simple nervous systems within the frog were examined with immunohistochemical methods to determine possible candidates for peptide neurotransmitters. Recently, Campbell et al. (1982) reported evidence that the parasympathetic cardiac neurons in the toad (*Bufo marinus*) contain a somatostatinlike substance that is released during neural activity and alters atrial contractions. This prompted us to examine the cardiac ganglion in bullfrogs. This ganglion is located in the atrial septum of the frog and is thin enough to allow immunofluorescence in whole mounts. In addition, the bullfrog cardiac ganglion is accessible to intracellular electrophysiological recording techniques, and nicotinic cholinergic transmission has been well characterized in the ganglia of *Rana pipiens* by Dennis et al. (1971). We have recently examined the bullfrog ganglion with routine immunofluorescence techniques using antibodies against several different peptides, including LHRH, somatostatin, SP, bombesin, Leu-enkephalin, cholecystokinin, and physalaemin. Of these antibodies, only the anti-SP antibodies resulted in specific staining. We have begun to explore the possible physiological relevance of the SP immunoreactivity using electrophysiological and anatomical techniques.

Every Parasympathetic Neuron in the Septum Is in Synaptic Contact with SP-positive Terminals

The preganglionic parasympathetic nerve terminals of the bullfrog cardiac ganglion exhibit intense SP-positive staining when treated with routine immunohistochemical and immunofluorescence techniques (Fig. 6A,C). The SP-positive staining was apparent with two different anti-SP antibodies, a monoclonal as well as a complex serum, and the staining was effectively abolished by preincubating the monoclonal antibody with SP (70 μg/ml) (Fig. 6B–C). Every neuron so far examined in this tissue has had terminal boutons that were positive for SP immunofluorescence. To estimate the percentage of nerve terminals that are SP-positive, the nerve terminals were visualized with a monoclonal antibody against synaptic vesicles (Matthew et al. 1981). In preliminary studies, the vast majority, if not all, of the preganglionic terminals identified with this vesicle-specific antibody were also SP-positive. The vagal origin of these SP-positive terminals was verified by bilaterally transecting the vagi as they exited the cranium. Two weeks after such surgery, no SP-positive terminals were present on the neurons of the cardiac ganglia.

SP Caused Little Electrophysiological Change in Parasympathetic Neurons

Since SP has been shown to cause slow (seconds to minutes) electrophysiological changes in neurons of several different systems (Otsuka et al. 1975; Dun and Karczmar 1979; Tsunoo et al. 1982), including bullfrog sympathetic ganglia (Jan and Jan 1982), the presence of an SP-like substance in parasympathetic preganglionic terminals raised the possibility that similar SP-mediated electrical events may be present in this system. However, local administration of SP (0.1 mg/ml) by a pressure ejection system while recording intracellularly from septal neurons produced no reliable effects on membrane potential (Fig. 7C,E). Preliminary results also indicate that the peptide causes no change in membrane conductance nor does SP alter the amplitude or time course of orthodromically elicited EPSPs in low (0.6 mM) Ca^{++}. Thus, at the present time, no direct effect of mammalian SP has been observed on the postsynaptic membrane properties or on the amount of ACh released from the presynaptic terminals.

If an SP-like substance in the preganglionic terminals is not identical to the mammalian SP used in the pressure ejection studies, the endogenous substance, if re-

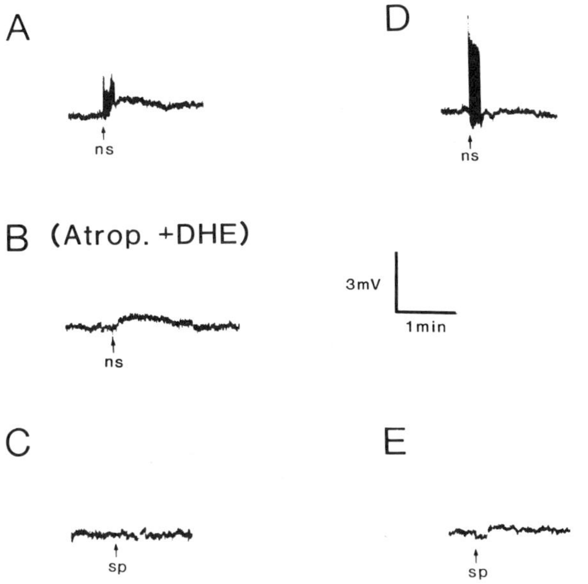

Figure 7. Intracellular recordings of membrane potential in two cells of the bullfrog cardiac ganglion. Arrows indicate onset of nerve stimulation (ns) or SP application (sp), respectively. (*A*) Small slow depolarization in response to supramaximal orthodromic stimulation at 10 Hz for 10 sec. (*B*) Small depolarization remains in presence of 5×10^{-6} M atropine and 3×10^{-5} M dihydro-β-erythroidine (dHβE). Same cell as *A*. (*C*) No effect of SP (0.1 mg/ml) applied by local pressure ejections for 5 sec. Same cell as *A*. (*D*) Nerve cell exhibiting no detectable change in membrane potential in response to nerve stimulation (10 Hz, 10 sec). (*E*) Nerve cell exhibiting no detectable change in membrane potential in response to SP (0.1 mg/ml), 5-sec pulse. Same cell as *D*.

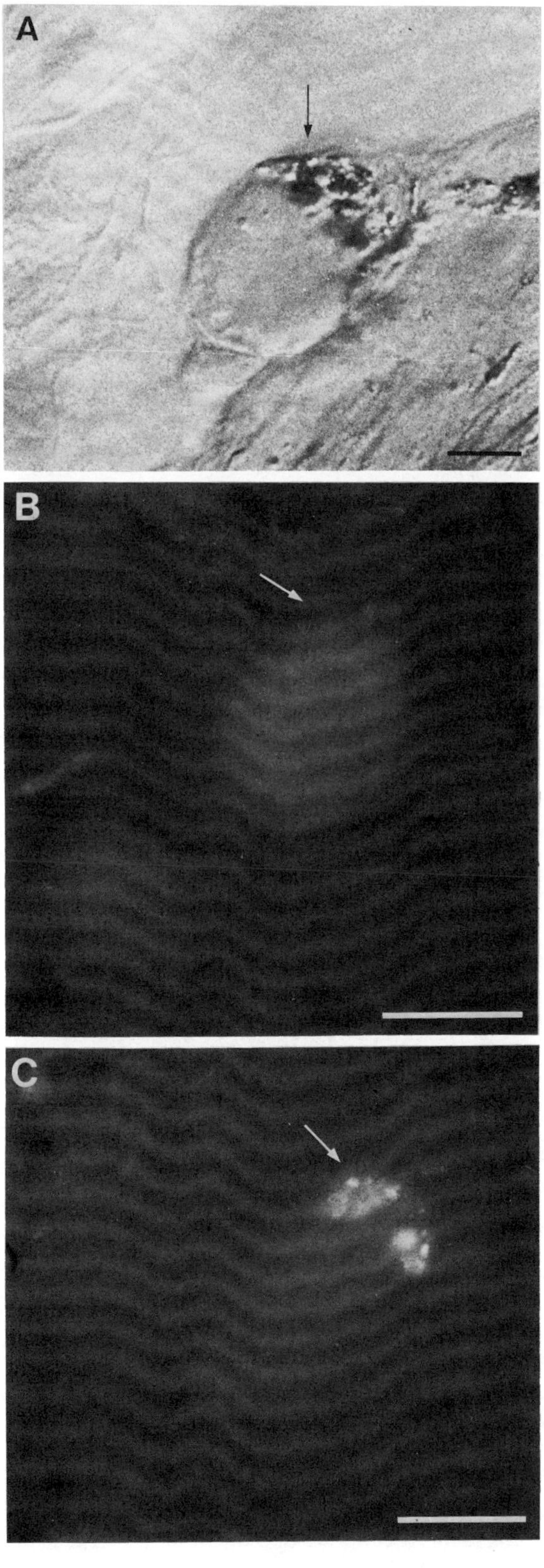

Figure 6. SP-positive staining in preganglionic nerve terminals of the bullfrog cardiac ganglion. (*A*) Light micrograph of terminal boutons (arrow) on neuronal cell body adjacent to nerve trunk treated with anti-SP antibody and visualized with biotinylated second antibody and avidin-biotinylated HRP complex with subsequent diamino benzidine-peroxide reaction (Vectastain Kit, Vector Laboratories, Inc.). Bar, 20 μm. (*B*) Absence of immunofluorescent nerve terminals on cell body exposed to anti-SP antibody preadsorbed with SP (see Methods). Arrow points at axon hillock region of nerve cell. Bar, 40 μm. (*C*) Terminals fluoresce brightly (arrow) in tissue treated similarly to *B* except with no SP in preadsorption solution. Bar, 40 μm.

leased, could mediate events that are not mimicked by the mammalian SP. We tested this possibility by stimulating ganglionic neurons orthodromically and monitoring their responses intracellularly (Fig. 7A,D). A significant proportion of the neurons (~40%) exhibited long-lasting (0.5–3 min) changes in membrane potential in response to bursts of orthodromic nerve stimulation (e.g., 10 Hz for 10 sec). Both depolarizations and hyperpolarizations were observed but always of low amplitude (0.5–3 mV) (Fig. 7A). Even in neurons exhibiting these small slow responses to nerve stimulation, mammalian SP (0.1 mg/ml) did not produce similar effects (Fig. 7A,C). The slow responses of the cardiac ganglion were not blocked by dihydro-β-erythroidine (5×10^{-5} M) and atropine sulfate (10^{-5} M) in the perfusion bath, indicating that cholinergic mechanisms are not involved (Fig. 7A,B). Therefore, if the slow synaptic potentials in parasympathetic neurons are caused by the SP-like material released from preganglionic terminals, they are generated by receptors that do not respond to substance P. In any case, the amplitudes of the noncholinergic responses are very small.

Actions of SP on Mast Cells

Assuming that an SP-like substance is released from preganglionic terminals, it is possible that the primary effect of the substance is not an electrical event. For instance, neurally evoked metabolic events that do not depend on measurable changes in membrane potential have recently been suggested in brown adipose tissue (Girardier and Schneider-Picard 1983). It is also possible, in analogy to the LHRH system in sympathetic ganglia, that even though an SP-like substance is in terminals apposed to parasympathetic neurons, these neurons are not the physiological target for the peptide. One possible target for SP is mast cells, which are abundant in the endoneurium of the cardiac nerves (Fig. 8A,B,C). Using Nomarski optics or specific histological stains, these mast cells in the bullfrog can be observed to degranulate within seconds in response to exogenously applied SP (0.01 mg/ml) (Fig. 8D). Such an effect of SP has also been characterized with mammalian mast cells based on histamine release (Johnson and Erdos 1973; Fewtrell et al. 1982). Lembeck and Holzer (1979) have provided pharmacological evidence that SP, released from peripheral sensory nerve endings, may result in histamine release from mast cells and they suggest that this interaction is partially responsible for sensory nerve-dependent neurogenic inflammation. Because of the sensitivity of the bullfrog mast cell to SP and the relatively close proximity of these mast cells to nerve cells in the cardiac ganglion (Fig. 8B,C), we are now examining the possible interaction of nerve stimulation on endoneurial mast cells in this system.

Discussion

The localization of SP immunoreactivity to all or most of the nerve terminals in the cardiac ganglia of the bullfrog suggests a transmitter role for a SP-like substance. Since at least some of the terminals on every neuron should contain and release ACh (Dennis et al. 1971), it is very likely that the SP immunoreactivity is present in the same terminals as ACh. However, at the present time, the exact nature of the substance responsible for the immunohistochemical staining has not been characterized, nor is it known whether the SP-like substance is released during neural activity. It is interesting that sympathetic neurons in the bullfrog are sensitive to exogenously applied SP (Jan and Jan 1982), although they are generally not innervated by SP-containing terminals, whereas cardiac ganglion neurons in the bullfrog appear to be innervated by SP-containing terminals but do not seem to respond to SP. The preganglionic terminals of another parasympathetic ganglion, the avian ciliary ganglion, have recently been reported to contain SP as well as enkephalin (Erichsen et al. 1982). As yet, no function has been established for the SP or enkephalin immunoreactivity in the ciliary ganglion, as is the case with most reports of immunohistochemical localization of different peptides in the CNS and peripheral nervous system. Similarly, even though the bullfrog cardiac ganglion exhibits distinct SP-positive terminals, and this system is accessible to detailed electrophysiological measurements, no phenomena that are normally associated with neural transmission and that can be attributed to an SP-like substance have thus far been observed (i.e., changes in membrane potential, membrane conductance, or synaptic efficacy). This situation may be common in systems exhibiting peptiergic immunoreactivity, and the cardiac ganglion may prove to be useful in elucidating novel types of interactions between neuropeptides and their targets.

NEUROPARACRINE HYPOTHESIS AND ITS IMPLICATIONS IN ORGANIZATION OF THE NERVOUS SYSTEM

In frog sympathetic ganglia, we have seen that an LHRH-like peptide serves as a transmitter that mediates a slow synaptic potential. Although this peptide transmitter is released from nerve terminals making close synaptic contact with a sympathetic neuron, the peptide transmitter can diffuse at least tens of microns and act on cells that are not in synaptic contact with the nerve terminals from which the peptide transmitters are released.

Although the extent of diffusion of peptides from nerve terminals has not been well characterized in most vertebrate systems, as described in the next section, the slow time course of phenomena associated with peptidergic transmission is consistent with considerable diffusion. If the LHRH system of the bullfrog sympathetic system is representative, it is possible that a general property of neurally released peptides is the capacity to influence targets relatively distant from the nerve terminal. The suggestion of an interaction between sensory nerves and mast cells by Lembeck and Holzer (1979) is almost certainly dependent on this type of dif-

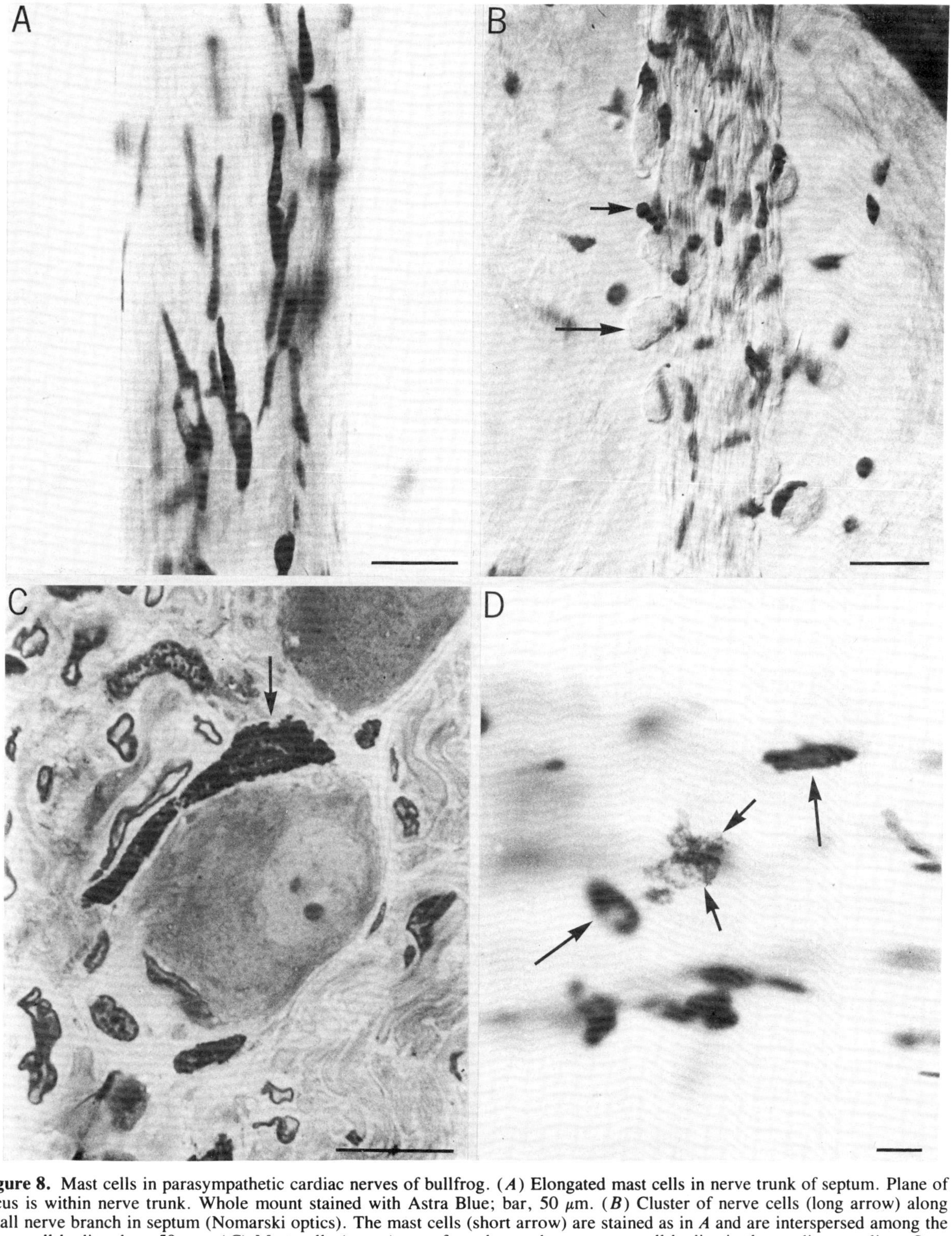

Figure 8. Mast cells in parasympathetic cardiac nerves of bullfrog. (*A*) Elongated mast cells in nerve trunk of septum. Plane of focus is within nerve trunk. Whole mount stained with Astra Blue; bar, 50 μm. (*B*) Cluster of nerve cells (long arrow) along small nerve branch in septum (Nomarski optics). The mast cells (short arrow) are stained as in *A* and are interspersed among the nerve cell bodies; bar, 50 μm. (*C*) Mast cells (arrow) are often observed near nerve cell bodies in the cardiac ganglion. One-micron plastic section, toluidine; bar, 28 μm. (*D*) Two mast cells (short arrows) on surface of nerve trunk that were exposed to a 1-sec pulse of 0.1 mg/ml SP 4 min before fixation. Cells visibly degranulated under Nomarski optics within seconds. Note the greatly reduced staining as compared with nearby mast cells (long arrows), located more deeply within the nerve trunk and just out of the focal plane. Stained as in *A*. Bar, 20 μm.

fusion, since it is unlikely that close synaptic contacts occur on mast cells. Although a neural influence on mast cells in the frog cardiac ganglion has not been established, the possibility exists if a substance, perhaps SP, can diffuse at least several microns away from the terminals. If such diffusion is typical of neurally released peptides, peptidergic transmission will likely exhibit the following properties.

The responses mediated by peptides should be considerably slower than those caused by the focal action of transmitters such as ACh. Let $\bar{x}$ be the average distance a peptide molecule has to travel to reach its receptors. The average time, t, it takes for a molecule to reach its receptor can be roughly estimated from $t = \bar{x}/2D$, where D is the diffusion constant. A value of 10^{-6} cm^2/sec for D should be a reasonable estimate. For $\bar{x} = 50$ μm (about the diameter of a large cell), $t = 10$ seconds. For $\bar{x} = 100$ μm (about two cell diameters), $t = 50$ seconds, which is comparable to the rise time of a typical late slow EPSP. For $\bar{x} = 1000$ μm (about the diameter of a ganglion), $t = 100$ minutes, which is much longer than the time course of the late slow EPSP. This estimate implies that the radius of influence from a peptide containing terminal is of the order of 100 μm. In a number of vertebrate preparations (mostly autonomic ganglia) where peptides are suspected to mediate some of the synaptic potentials, almost without exception those responses are slow in onset (hundreds of msec) and take seconds or up to a minute to reach their peak value (Otsuka et al. 1975; Katayama and North 1978; Dun and Karczmar 1979; Konishi et al. 1979; Tsunoo et al. 1982). This would translate into an effective radius of 10–100 μm.

The peptide receptors on distant targets should have high affinity for the ligands. At a synapse such as the skeletal neuromuscular junction, where the sites of transmitter release are in close apposition to the receptors, the peak concentration of the transmitter during a synaptic response could be as high as 10^{-3} M in the vicinity of the receptors. In this case, the receptors need not have a very high affinity for the transmitter molecules. In contrast, for peptides to act on receptors tens of microns away from the release sites, their concentration will be greatly reduced before they reach the receptors. Thus, one would expect peptide receptors to have a considerably higher affinity so that they can detect low levels of peptides. Results of receptor binding studies thus far indicate that peptide receptors do have high affinity for peptides. The dissociation constants were found to be in the range of 10^{-10} M to 10^{-9} M, which is 10–10^4 times lower than that of classical transmitters such as ACh and amino acids (e.g., Moody et al. 1978; Taylor and Pert 1979; Hanley et al. 1980).

Based on studies of bullfrog sympathetic ganglia plus the general findings of slow peptidergic responses and high-affinity peptide receptors, we would like to state explicitly a neuroparacrine hypothesis concerning the actions of neuropeptides in general.

Peptides in the nervous system act as paracrine hormones. In other words, peptides can diffuse in the extracellular space and affect cells many microns distant from the release sites.

The hypothesis that neuropeptides generally act as paracrine hormones has been a part of the literature almost since peptides were discovered in neurons (see, for instance, Guillemin 1978). The suggestion was based primarily on the fact that these peptides were first characterized as hormones and might therefore retain their hormonelike character even if released from neurons. Moreover, neuroparacrine function does not appear to be unique to peptides. For instance, norepinephrine probably diffuses for microns before acting on its target cells because some adrenergic varicosities and terminals do not appear to be in synaptic contact with any target tissues (e.g., see Burnstock and Costa 1975). The unique feature discovered for peptidergic transmission in frog ganglia is that peptides diffuse from terminals that make morphologically defined close synaptic contacts and act on cells that are not in direct contact with these terminals. This leads to the second point of the hypothesis.

Peptides, including those released from nerve terminals that are in synaptic contact with other cells, can diffuse many microns (perhaps up to about a hundred microns) and act upon cells that are not in synaptic contact with those peptidergic terminals. If this is generally true, the conventional morphological criteria for synapses may actually be misleading in identifying target neurons for a peptide transmitter, whereas localization of peptide receptors might be the most important criterion.

Selective communication using diffusible peptides as chemical messengers thus will not necessarily involve precise circuitry of synaptic contacts. Instead, different neurons in a given region may express different subsets of receptors, so that a peptide transmitter released in

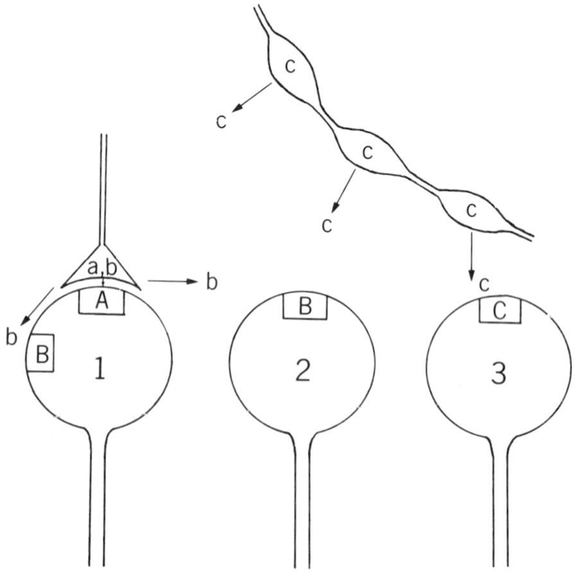

Figure 9. Hypothetical scheme for selective, long-range communication between neurons. The transmitter may be released either from varicosities or from nerve terminals that form synaptic contacts with other neurons. If the transmitter has long lifetime, it may diffuse and activate those neurons in the vicinity that express the appropriate receptors. (a,b,c) Different transmitter molecules; (A,B,C) different receptors for the respective transmitters.

the vicinity will influence only those neurons that have the appropriate receptors on their surfaces (Fig. 9). For this type of interneuronal communication to be used extensively without cross talks between parallel pathways, a necessary requirement is that many different molecules are used as transmitters. Perhaps this is one reason for the multiplicity of putative peptide transmitters. Intuitively, in situations where the speed of action is not crucial or perhaps a slower and prolonged influence is more desirable, this arrangement seems more economical. A terminal may release a multitude of transmitters, each influencing a different subset of cells in the general vicinity. Furthermore, plasticity may evoke merely an alteration in the type of receptors expressed by the "postsynaptic" cell.

ACKNOWLEDGMENTS

We are grateful to Drs. J. Rivier, W. Vale, and M. Brown of the Salk Institute for peptides and antisera. We thank Ms. Bonnie Iwamasa for her unfailing help in preparing this manuscript. This research was supported by grants from the National Institutes of Health (NS-15757, NS-16033) and a McKnight Award to Y.N.J. C.W.B. is a recipient of a Muscular Dystrophy fellowship and D.B. is a NIH postdoctoral fellow.

REFERENCES

Adams, P.R. and D.A. Brown. 1980. Luteinizing hormone-releasing factor and muscarinic agonists act on the same voltage-sensitive K$^+$-current in bullfrog sympathetic neurons. *Br. J. Pharmacol.* **68:** 353.

Adams, P.R., D.A. Brown, and A. Constanti. 1982. Pharmacological inhibition of the M-current. *J. Physiol.* **332:** 223.

Blaies, D.M. and J.F. Williams. 1981. A simplified method for staining mast cells with astra blue. *Stain Technol.* **56:** 91.

Burnstock, G. and M. Costa. 1975. *Adrenergic neurons*, p. 52. Wiley, New York.

Campbell, G., I.L. Gibbins, J.L. Morris, J.B. Furness, M. Costa, J.R. Oliver, A.M. Beardsley, and R. Murphy. 1982. Somatostatin is contained in and released from cholinergic nerves in the heart of the toad *Bufo marinus*. *Neuroscience* **7:** 2013.

Dennis, M.J., A.J. Harris, and S.W. Kuffler. 1971. Synaptic transmission and its duplication by focally applied acetylcholine in parasympathetic neurons in the heart of the frog. *Proc. R. Soc. Lond. B* **177:** 509.

Dun, N.J. and A.G. Karczmar. 1979. Actions of substance P on sympathetic neurons. *Neuropharmacology* **18:** 215.

Eiden, L.E. and R.L. Eskay. 1980. Characterization of LRF-like immunoreactivity in the frog sympathetic ganglia: Non identity with LRF decapeptide. *Neuropeptides* **1:** 29.

Eiden, L.E., E. Loumaye, N. Sherwood, and R.L. Eskay. 1982. Two chemically and immunologically distinct forms of lutenizing hormone-releasing hormone are differentially expressed in frog neural tissue. *Peptides* **3:** 323.

Erichsen, J.T., H.J. Karten, W.D. Eldred, and N.C. Brecha. 1982. Localization of substance P-like and enkephalin-like immunoreactivity within preganglionic terminals of the avian ciliary ganglion: Light and electron microscopy. *J. Neurosci.* **2:** 994.

Fewtrell, C.M.S., J.C. Foreman, C.C. Jordan, P. Oehme, H. Renner, and J.M. Stewart. 1982. The effects of substance P on histamine and 5-hydroxytryptamine release in the rat. *J. Physiol.* **330:** 393.

Girardier, L. and G. Schneider-Picard. 1983. Alpha and beta-adrenergic mediation of membrane potential changes and metabolism in rat brown adipose tissue. *J. Physiol.* **335:** 629.

Guillemin, R. 1978. Peptides in the brain: The new endocrinology of the neuron. *Science* **202:** 390.

Hanley, M.R., B.E.B. Sandberg, C.M. Lee, L.L. Iversen, D.E. Brundish, and R. Wade. 1980. Specific binding of ^{3}H-Substance P to rat brain membranes. *Nature* **286:** 810.

Hökfelt, T., O. Johannson, Å. Ljungdahl, J.M. Lundberg, and M. Schultzberg. 1980. Peptidergic neurons. *Nature* **284:** 515.

Jan, L.Y. and Y.N. Jan. 1982. Peptidergic transmission in sympathetic ganglia of the frog. *J. Physiol.* **327:** 219.

Jan, L.Y., Y.N. Jan, and M.S. Brownfield. 1980a. Peptidergic transmitters in synaptic boutons of sympathetic ganglia. *Nature* **288:** 380.

Jan, Y.N. and L.Y. Jan. 1983. Coexistence and co-release of acetylcholine and the LHRH-like peptide from the same preganglionic fibers in frog sympathetic ganglia. *Fed. Proc.* **42:** 2929.

Jan, Y.N., L.Y. Jan, and S.W. Kuffler. 1979. A peptide as a possible transmitter in sympathethic ganglia of the frog. *Proc. Natl. Acad. Sci.* **76:** 1501.

——. 1980b. Further evidence for peptidergic transmission in sympathetic ganglia. *Proc. Natl. Acad. Sci.* **77:** 5008.

Johnson, A.R. and E.G. Erdos. 1973. Release of histamine from mast cells by vasoactive peptides. *Proc. Soc. Exp. Biol. Med.* **142:** 1252.

Katayama, Y. and S. Nishi. 1982. Voltage-clamp analysis of peptidergic slow depolarization in bullfrog sympathetic ganglion cells. *J. Physiol.* **333:** 305.

Katayama, Y. and R.A. North. 1978. Does substance P mediate slow synaptic excitation within the myenteric plexus? *Nature* **274:** 387.

King, J.A. and R.P. Millar. 1982. Structure of chicken hypothalamic luteinizing hormone-releasing hormone. I. Structural determination on partially purified material. *J. Biol. Chem.* **257:** 10722.

Konishi, S., A. Tsunoo, and M. Otsuka. 1979. Enkephalin presynaptically inhibits cholinergic transmission in sympathetic ganglia. *Nature* **282:** 515.

Kuffler, S.W. 1980. Slow synaptic responses in autonomic ganglia and the pursuit of a peptidergic transmitter. *J. Exp. Biol.* **89:** 257.

Kuffler, S.W. and T.J. Sejnowski. 1983. Peptidergic and muscarinic excitation at amphibian sympathetic synapses. *J. Physiol.* (in press).

Langley, J.N. and L.A. Orbeli. 1910. Observations on sympathetic ganglia and sacral autonomic system of the frog. *J. Physiol.* **41:** 450.

Lembeck, F. and P. Holzer. 1979. Substance P as a neurogenic mediator of antidromic vasodilatation and neurogenic plasma extravasation. *Naunyn-Schmeidebergs Arch. Pharmakol.* **310:** 175.

Matthew, W., L. Tsavaler, and L.I. Reichardt. 1981. Identification of a synaptic vesicle specific membrane protein that is widely distributed in neuronal and neurosecretory tissue. *J. Cell Biol.* **91:** 257.

McMahan, U.J. and S.W. Kuffler. 1971. Visual identification of synaptic boutons on living ganglion cells and of varicosities in postganglionic axons in the heart of the frog. *Proc. R. Soc. Lond. B* **177:** 485.

Miyamoto, K., Y. Hasegawa, T. Minegishi, M. Nomura, Y. Takahashi, M. Igarashi, K. Kanagawa, and H. Matsuo. 1982. Isolation and characterization of chicken hypothalamic luteinizing hormone-releasing hormone. *Biochem. Biophys. Res. Commun.* **107:** 820.

Moody, T.W., C.B. Pert, J. Rivier, and M. Brown. 1978. Bombesin: Specific binding to rat brain membranes. *Proc. Natl. Acad. Sci.* **75:** 5372.

Nishi, S., and K. Koketsu. 1968. Early and late after-discharges of amphibian sympathetic ganglion cells. *J. Neurophysiol.* **31:** 109.

Otsuka, M., S. Konishi, and T. Takahashi. 1975. Hypothalamic substance P as a candidate for transmitter of primary afferent neurons. *Fed. Proc.* **34:** 1922.

Rivier, J. and W. Vale. 1978. [D-pGlu^{1},D-phe^{2}-Trp^{3,6}]-LRF. A potent luteinizing hormone releasing factor antagonist in vitro and inhibitor of ovulation in the rat. *Life Sci.* **23:** 869.

Sherwood, N., L. Eiden, M. Brownstein, J. Spiess, J. Rivier, and W. Vale. 1983. Characterization of a teleost gonadotropin-releasing hormone. *Proc. Natl. Acad. Sci.* **80:** 2794.

Taylor, D.P. and C.B. Pert. 1979. Vasoactive intestinal polypeptide: Specific binding to rat brain membranes. *Proc. Natl. Acad. Sci.* **76:** 660.

Tsunoo, A., S. Konishi, and M. Otsuka. 1982. Substance P as an excitatory transmitter of primary afferent neurons in guinea-pig sympathetic ganglia. *Neuroscience* **7:** 2025.

Generation of Diversity and Evolution of Opioid Peptides

E. HERBERT,* E. OATES,* G. MARTENS,* M. COMB,* H. ROSEN,* AND M. UHLER†
*Department of Chemistry, University of Oregon, Eugene, Oregon 97403; †Department of Genetics and Medicine, Stanford University Medical School, Stanford, California 94305

Control of behavior in animals involves the transfer of information at chemical synapses and the conversion of neural signals into hormonal messages by the neuroendocrine system. The hormonal messengers activate the appropriate target systems in the body to produce specific responses. To understand how these responses are regulated, it is necessary to identify the transmitter molecules in specific pathways in the brain and the hormonal messenger molecules released by the neuroendocrine system. It is also necessary to understand how the synthesis and release of these molecules are regulated.

In the past 15 years a number of neuroactive peptides have been discovered that mediate specific behavioral responses in animals. Some of these peptides may be the neurotransmitter and neuromodulator molecules we have been searching for in the brain. For example, substance P and enkephalins function as neurotransmitters and neuromodulators in the pain pathways in animals. Egg-laying behavior in the marine invertebrate *Aplysia* is regulated in stepwise fashion by a related group of peptides known as egg-laying hormones (Scheller et al. 1983). Complex stress responses are coordinated by adrenocorticotropin (ACTH), β-endorphin, corticotropin-releasing factor, and γ-melanocyte-stimulating hormone (MSH). Other peptides have been implicated in temperature regulation and induction of sleep.

Diverse Nature of Neuroactive Peptides as Illustrated by Opioid Peptides

The variety of neuroactive peptides in nature is astonishing. For example, there are now 18 distinct peptides that are known to exhibit opioid activity when injected into animals. The pentapeptides, methionine (Met)- and leucine (Leu)-enkephalin (Hughes et al. 1975), are the simplest of these substances. The other opioid peptides are carboxyterminal extensions of either Met- or Leu-enkephalin, as shown in Figure 1 (a partial list of opioid peptides).

In the past 4 years it has been shown through the use of recombinant DNA technology that all of the known opioid peptides are derived from three different precursor proteins. The first opioid precursor protein to be characterized was pro-opiomelanocortin (POMC), which gives rise to the opioid β-endorphin and a variety of other peptides, including ACTH and α-, β-, and γ-MSH (Nakanishi et al. 1979). The second precursor protein to be characterized was proenkephalin, which contains six copies of Met-enkephalin and one copy of Leu-enkephalin (Comb et al. 1982; Gubler et al. 1982; Noda et al. 1982b). In the adrenal medulla, this precursor gives rise to a number of different opioid peptides, including Met- and Leu-enkephalin, peptide E (which contains one copy of Met-enkephalin and one copy of Leu-enkephalin), peptide F (which contains two copies of Met-enkephalin), and Met-enkephalin-Arg-Gly-Leu (Kilpatrick et al. 1981). The third precursor is prodynorphin, which contains three copies of Leu-enkephalin and gives rise to the opioids dynorphin and α-neo-endorphin (Kakadani et al. 1982) (Figs. 1 and 2). The integrity of the enkephalin sequence is essential for the morphinomimetic activity of all these peptides. Some remarkable similarities in the structures of the opioid peptide precursors are revealed in Figure 2. First, the precursors are almost all of the same length, and the sequences of the biologically active peptides are confined almost exclusively to the carboxyterminal half of the precursors. The aminoterminal region of each precursor has a high concentration of cysteine residues, and the distribution of these residues is similar in each case, indicating that formation of disulfide bridges may be essential for stabilizing the protein in conformations required for correct processing. Finally, the biologically active domains in each precursor are flanked on both sides by pairs of basic amino acid residues, creating potential cleavage sites for trypsinlike enzymes (Fig. 2).

The precursors of the opioid peptides belong to a growing class of proteins known as polyfunctional proteins or polyproteins because they give rise to more than one biologically active peptide or to multiple copies of the same peptide. Some of the known polyfunctional proteins are depicted in Figure 3. Polyproteins are found in organisms as diverse as yeast and humans. As shown in Figure 3 the precursor of α-factor, a pheromone that controls yeast mating behavior, contains four copies of this pheromone (Kurjan and Herskowitz 1982). The precursors of the posterior pituitary hormones, oxytocin and vasopressin, also contain their binding proteins, the neurophysins (Brownstein et al. 1980; Land et al. 1982), and the precursor of glucagon contains two copies of this hormone (Lund et al. 1982).

Tissue Diversity and Sites of Synthesis of Opioid Peptides

Each class of opioid peptides is expressed in a wide variety of tissues. For example, peptides of the

β-endorphin (31 amino acids)	Tyr·Gly·Gly·Phe·Met	Thr·Ser·Glu·Lys·Ser·Gln·Gln·Thr·Pro·Leu·Val·Thr·Leu·Phe·Lys·Ana·Ala·Ile·Ile·Lys^{31}
Peptide E	Tyr·Gly·Gly·Phe·Met	Arg·Arg·Val·Gly·Arg·Pro·Glu·Trp·Trp·Met·Asp·Tyr·Gln·Lys·Arg·Tyr·Gly·Gly·Phe·Leu^{25}
Met-enk·arg·phe	Tyr·Gly·Gly·Phe·Met	Arg·Phe^{7}
Met-enk·arg·gly·leu	Tyr·Gly·Gly·Phe·Met	Arg·Gly·Leu^{8}
Met-enkephalin	Tyr·Gly·Gly·Phe·Met	
Leu-enkephalin	Tyr·Gly·Gly·Phe·Leu	
Dynorphin 1-17	Tyr·Gly·Gly·Phe·Leu	Arg·Arg·Ile·Arg·Pro·Lys·Leu·Lys·Trp·Asp·Asn·Gln^{17}
Dynorphin 1-8	Tyr·Gly·Gly·Phe·Leu	Arg·Arg·Ile^{8}
α-Neo-endorphin	Tyr·Gly·Gly·Phe·Leu	Arg·Lys·Tyr·Pro·Lys^{10}
β-Neo-endorphin	Tyr·Gly·Gly·Phe·Leu	Arg·Lys·Tyr·Pro^{9}
Dynorphin B or Rimorphin	Tyr·Gly·Gly·Phe·Leu	Arg·Arg·Gln·Phe·Lys·Val·Val·Thr^{13}

Figure 1. Structure of the opioid peptides depicting common functional sequences of Met-enkephalin (upper box) and Leu-enkephalin (lower box). Only 25 of the 31 amino acids of β-endorphin (bovine) are shown.

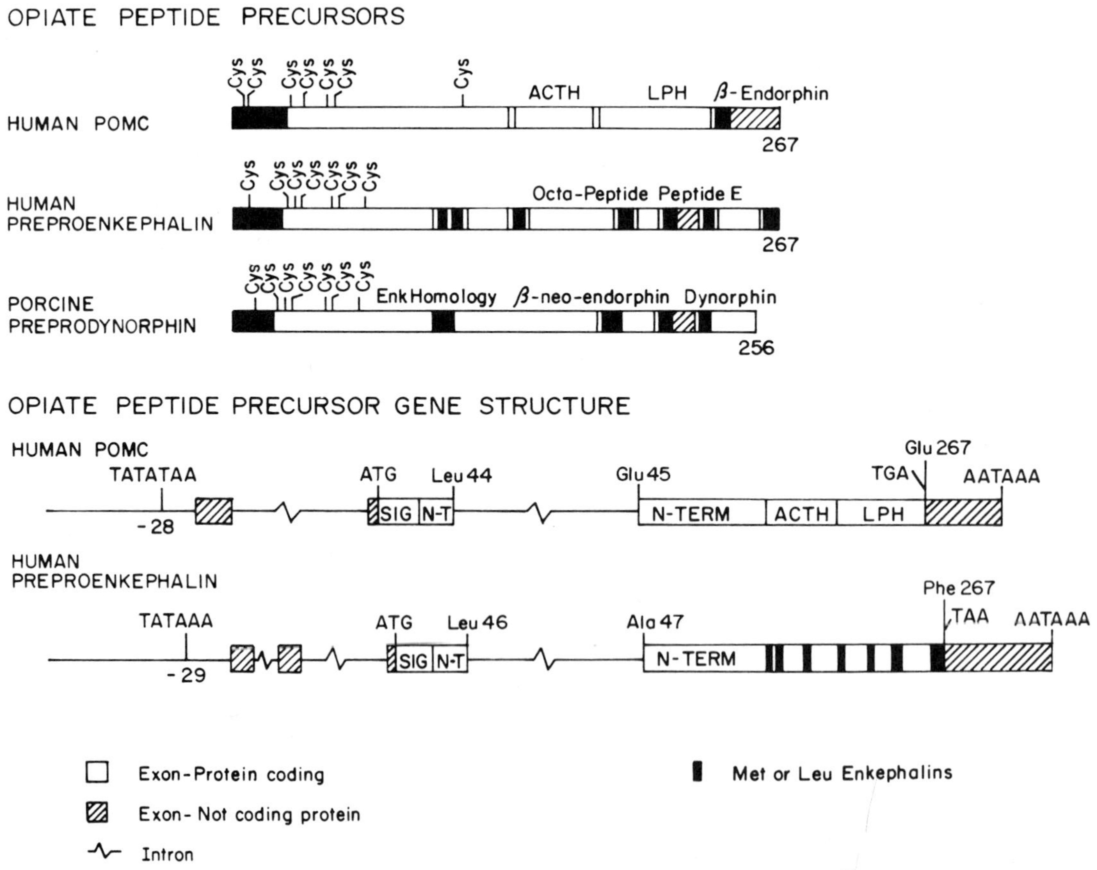

Figure 2. Schematic representation of the three opiate peptide precursors and of two opiate peptide genes. In the opiate peptide precursors, the black box at the amino terminus represents the signal peptide. The enkephalin homology sequence in the preprodynorphin precursor indicates a sequence found highly homologous to an enkephalin unit of the preproenkephalin sequence. This enkephalin homology sequence, however, does not code for an enkephalin peptide. In the gene structures, SIG and N-T or N-TERM represent, respectively, the signal peptide and the aminoterminal part of the precursors.

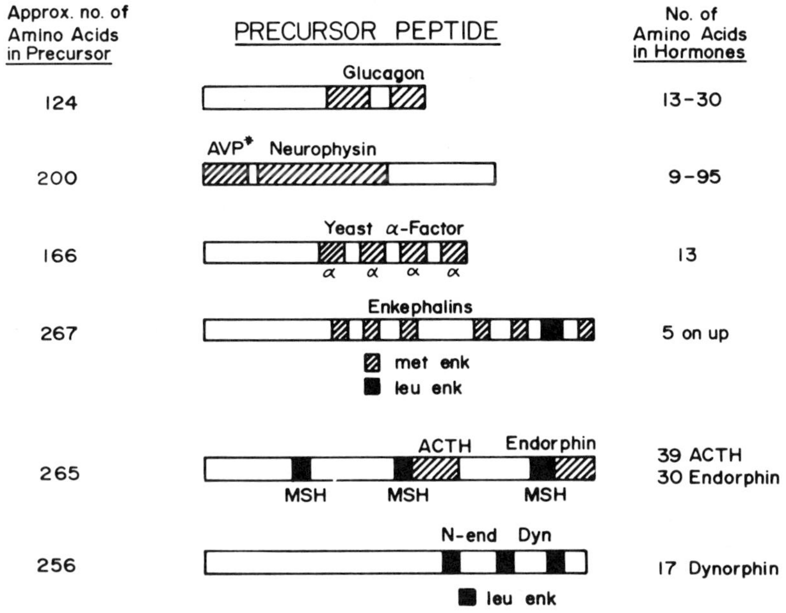

Figure 3. Large polyprotein precursors of small bioactive peptides.

ACTH/endorphin family are found in the anterior and intermediate lobes of the pituitary (Krieger et al. 1980), the hypothalamus, and several other regions of the brain (Krieger et al. 1977), the placenta, and intestine. Civelli et al. (1982) have shown that POMC mRNA is present in each of these sites at a level comparable to that of the POMC peptides, indicating that each of these tissues synthesizes its own peptides from a form of POMC very similar or identical to that found in the pituitary.

Peptides of the enkephalin family are also found in a wide variety of tissues, including the adrenal medulla (Kilpatrick et al. 1982) and a number of regions of the brain. Again, studies of proenkephalin mRNA levels in these tissues (M. Comb et al., unpubl.) indicate that each tissue synthesizes its own enkephalin peptides from the same or a very similar form of proenkephalin.

Dynorphin and α-neo-endorphin are also expressed in many different sites in the brain, but not enough is known about prodynorphin mRNA levels in these sites to draw any conclusions about the sites of synthesis of these peptides.

Generation of Diversity and Tissue-specific Expression of Neuropeptides

One of the most fascinating phenomena that has been observed in the study of neuropeptides is the tissue-specific expression of these peptides. Detailed studies of the synthesis and processing of POMC in pituitary-cell-culture systems by pulse-label and pulse-chase methods have shown that this precursor is processed to different hormones in the anterior and neurointermediate lobes of mouse and rat pituitary (Fig. 3) (Crine et al. 1978; Mains and Eipper 1979; Hinman and Herbert 1980). In the anterior pituitary, POMC is processed to $ACTH_{1-39}$, β-lipotropin (β-LPH), and aminoterminal fragments (Fig. 4). These peptides are processed further in the neurointermediate lobe. ACTH is converted to α-MSH (the first 13 amino acids of ACTH with an acetylated amino terminus and an amidated carboxyl terminus) and to the corticotropinlike intermediate-lobe peptide (CLIP) from the carboxyterminal half of ACTH (Scott et al. 1974). β-LPH is processed to γ-LPH and β-endorphin and acetylated derivatives of β-endorphin (Smyth and Zakarian 1979).

Thus, POMC is converted to two completely different sets of hormones in the two lobes of the pituitary (Fig. 4). Furthermore, the synthesis of POMC and the release of POMC peptides are regulated differently in the anterior and intermediate lobes of the pituitary. In the anterior lobe, these processes are stimulated by corticotropin-releasing factor, a hypothalamic peptide, and inhibited by glucocorticoids (Yates and Maran 1974). Release and synthesis of POMC peptides in the neurointermediate lobe, on the other hand, are regulated in a negative manner by catecholamines that are released by nerve endings that have their cell bodies in the hypothalamus. The processing of POMC in the hypothalamus may be different from that in either lobe of the pituitary.

Proenkephalin is also processed differently in different tissues. The predominant products of proenkephalin processing in bovine and human adrenal medulla are much higher in molecular weight than the products observed in the brain, where one observes mostly enkephalins (E. Weber et al., in prep.).

How Do Tissue-specific Differences in Processing Arise?

There are at least three different ways in which this kind of tissue diversity could come about. First, dif-

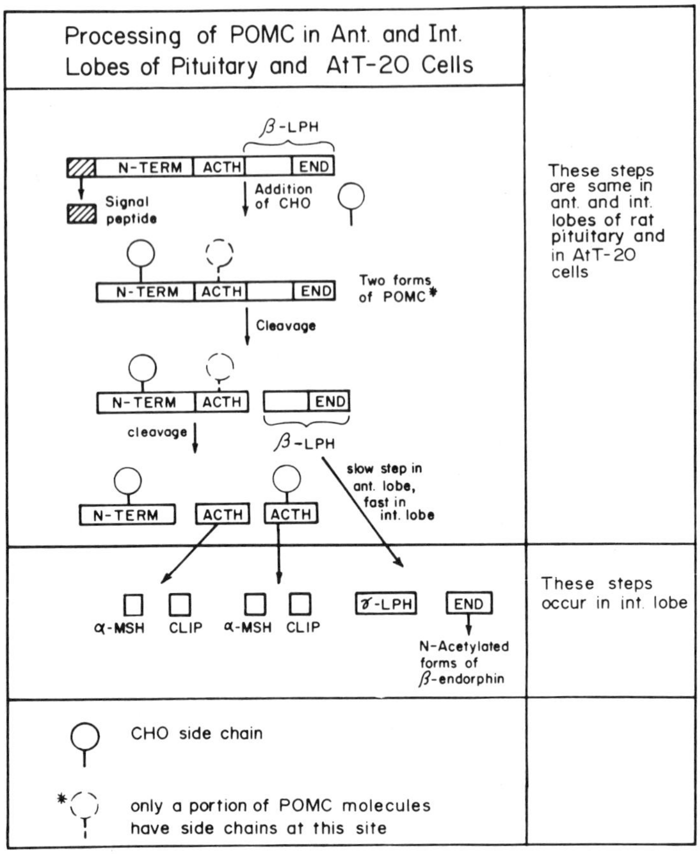

Figure 4. Processing of pro-opiomelanocortin in anterior and intermediate lobes of mouse and rat pituitary and mouse pituitary tumor cells (AtT-20-D_{16V} cells).

ferent tissues could have different sets of processing enzymes. Second, there could be a family of closely related POMC genes that are expressed selectively in different tissues to produce slightly different forms of POMC in each tissue. The structure of POMC would then dictate how it is processed in each tissue. A third possibility is the production of more than one mRNA molecule from a single gene by alternate modes of splicing of the precursor mRNA. Generation of diversity by alternate splicing of RNA has been observed for the calcitonin gene (Amara et al. 1982) and the growth hormone gene (DeNoto et al. 1981). The same gene that codes for calcitonin in the thyroid codes for another peptide in the hypothalamus. In the case of rat growth hormone, two different mRNA species have been detected. One species has 45 bases deleted from the region of an RNA splice junction in the coding portion of the mRNA. This mRNA species codes for a variant form of growth hormone that is smaller than the predominant form of growth hormone. The variant form of growth hormone has a different biological action than the other form (Lewis et al. 1980).

There is very little information available about the nature of the enzymes in different parts of the pituitary and brain that might be involved in processing of POMC and proenkephalin. Therefore, it is difficult to say whether differential processing is related to different sets of processing enzymes, although this seems like a good possibility.

If the amino acid sequence of POMC dictates how it is processed in each tissue, one would expect to find more than one POMC gene in the mouse and rat, making possible differential tissue expression. Thus far, published reports have described the structure of a single POMC gene in the cow (Nakanishi et al. 1980), human (Chang et al. 1980), and rat (Drouin and Goodman 1980). These genes are similar in structure. Uhler et al. (1983) have recently found two POMC genes in the mouse. The structure of one of these genes (α-POMC gene) is similar to the genes described in rat, cow, and human, with introns in the same places. The other POMC gene (β-POMC gene) is homologous to the α-POMC gene in the region that encodes β-LPH and ACTH but quite different in the region that encodes the aminoterminal domain of POMC. The β gene is also located on a different chromosome than the α gene and has a translation stop codon in place of the codon for the first amino acid in β-endorphin (Tyr). β-POMC appears to be a pseudogene, and no mRNA or protein product of the β-POMC gene has been detected in pituitary or hypothalamic tissue. Therefore, since the mouse has only one functional POMC gene, diversity appears to be generated at another level of gene expression.

As mentioned above, diversity can also be generated during gene expression by alternate modes of splicing of RNA transcribed from a single gene. Since we had observed several species of POMC in reticulocyte lysates following cell-free translation of rat pituitary mRNA (E. Oates and E. Herbert, unpubl.), we decided to examine the possibility that two or more species of POMC mRNA might exist in rat pituitary, possibly as a result of alternate modes of splicing of the RNA. Because we had shown previously that the heterogeneity resides in the aminoterminal region of POMC (E. Oates et al., unpubl), we analyzed the nucleotide sequences in the 5′ region of POMC mRNA in rat pituitary.

We used a specific cDNA-priming technique to determine these sequences. Two oligodeoxynucleotides, a dodecamer and a tetradecamer, were synthesized complementary to two different sequences in the 5′ region of POMC mRNA by either the phosphite method (Matteucci and Caruthers 1981) or the phosphotriester method (Narang et al. 1980). The locations of the mRNA sequences complementary to the dodecamer and the tetradecamer are shown at the bottom of Figure 5. The primers were labeled with ^{32}P at the 5′ end, hybridized to rat neurointermediate lobe poly(A) RNA (cytoplasmic RNA), and elongated with reverse transcriptase. The cDNA products of these reactions were fractionated by gel electrophoresis, eluted from the gels, and sequenced (Maxam and Gilbert 1977).

The autoradiograph in Figure 6A shows the array of products generated when cDNA is copied from rat neurointermediate lobe mRNA primed with the dodecamer (E. Oates and E. Herbert, in prep.). Sequencing of the cDNA in the bands eluted from the gel indicates that the cDNA products labeled R_1 and R_2 are copied from two different forms of POMC mRNA. These products differ in size by 30 bases (Fig. 6B). The labeled intermediately sized products below R_1 and R_2 are prematurely terminated POMC cDNAs (Fig. 6A). R_1 cDNA has the same sequence as R_2 except that it contains a 30-base insert in the 5′ untranslated region of the mRNA. Thus, R_1 and R_2 are DNA sequences copied from two different mRNA species present in the neurointermediate lobe of the pituitary. Synthesis of cDNA using the dodecamer as primer consistently gives an R_1/R_2 of approximately 0.25. The same ratio of products is found when rat anterior pituitary mRNA is used as template, suggesting that the anterior and neurointermediate lobes contain the same mRNA species.

The differences in the sequences of R_1 and R_2 cDNAs are highlighted in Figure 6B. The 30-base insertion aligns well with the first intron splice junction in POMC mRNA from other species (Nakanishi et al. 1979; Whitfeld et al. 1982; Uhler et al. 1983). Moreover, it is identical (with the exception of 2 bases) with 30 bases of the 3′ end of intron A of the human POMC gene (Whitfeld et al. 1982) (Fig. 6B). (We have very recently been informed that the 30-base insert in R_1 cDNA is identical with the sequence at the 3′ end of intron A of the rat POMC gene just sequenced by J. Drouin, University of Montreal.)

Thus, the insertion sequence location and the sequence homologies strongly suggest that the two rat POMC mRNA species are a result of alternate modes of splicing of RNA. This explanation is further strengthened by the fact that only one POMC gene has been detected in the rat (Drouin et al. 1983).

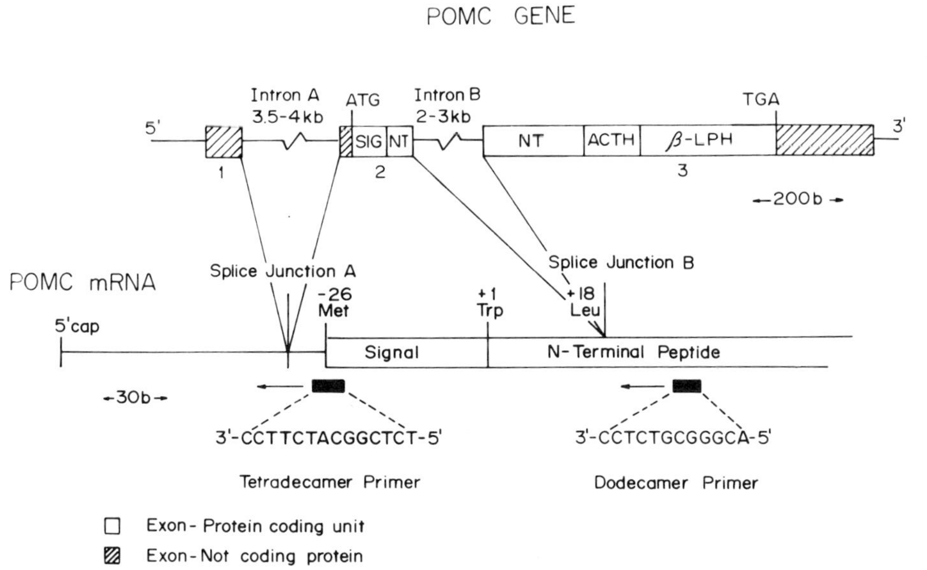

Figure 5. A schematic of the generalized POMC gene (*top*) and the 5′ end of its mRNA (*bottom*). Two introns (A and B) separate the gene into three exons (1–3). Introns A and B are removed to produce mature mRNA by a process of RNA splicing; the locations of these introns are shown as splice junctions A and B in the mRNA. The synthetic DNA primers are positioned where they hybridize to the mRNA. The primers were synthesized by either the phosphite or triester methods and purified by electrophoresis on 1.2-mm × 40-cm 24% acrylamide sequencing-type gels, using UV_{260} imaging to detect the DNA. Eluted oligomers were desalted on a Sephadex G-25 (fine) column run in 100 mM NH_4HCO_3, concentrated by lyophilization, and characterized by Maxam-Gilbert sequencing techniques. Arrows indicate the direction that reverse transcriptase elongates these primers.

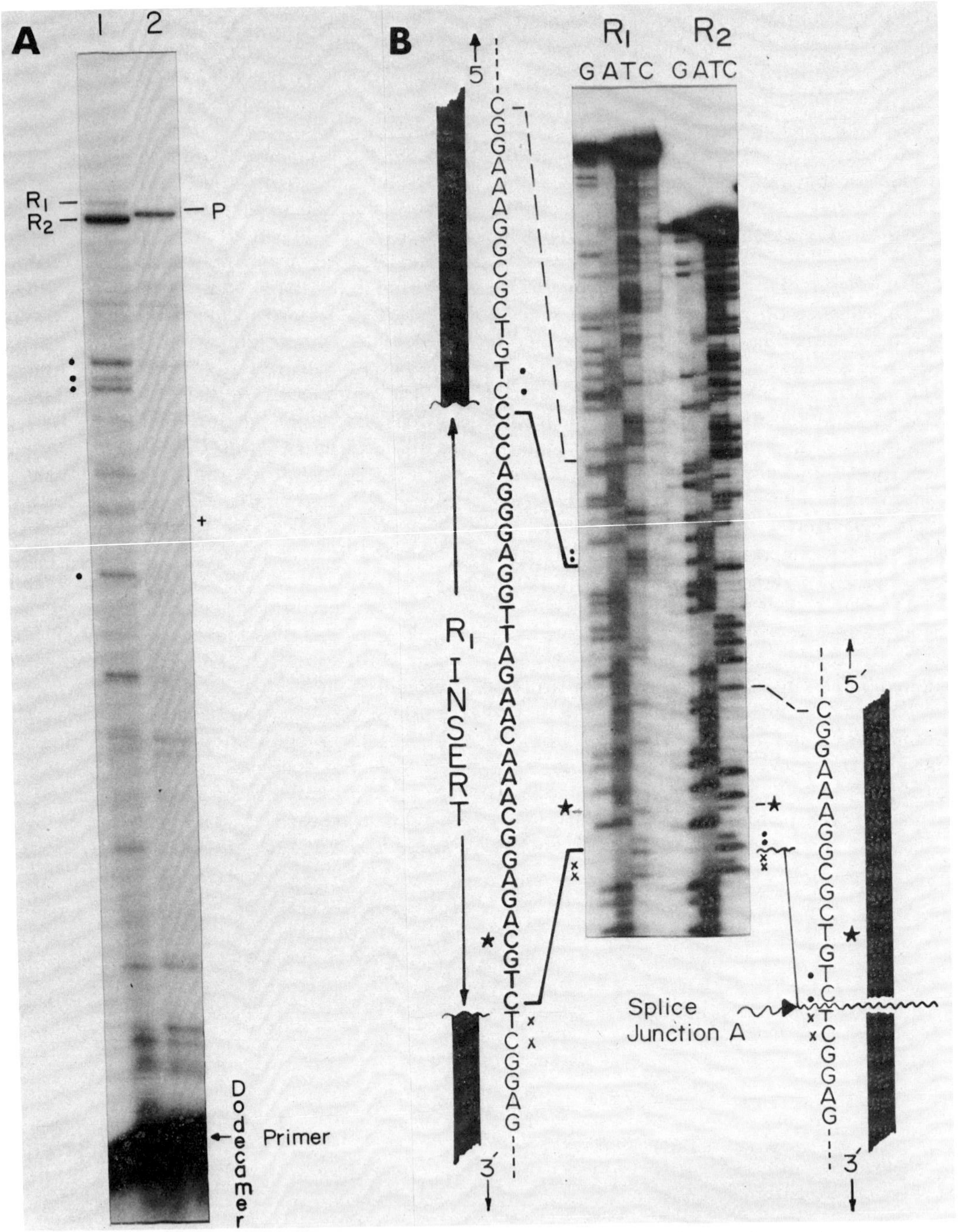

Figure 6. Sequence of the two rat pituitary POMC mRNAs. Poly(A) mRNA isolated from Sprague-Dawley rat neurointermediate pituitaries was hybridized with a synthetic, radioactively labeled oligodeoxynucleotide probe complementary to amino acids 57–60 of the precursor. Primer extension was achieved by the action of reverse transcriptase in the presence of nonlabeled nucleotides. The products of this reaction were subjected to electrophoresis on an 8% urea-polyacylamide gel (*A*). The two bands, R_1 and R_2, were further sequenced by the Maxam-Gilbert technique (*B*). (*A*) Lane *1* shows the results of priming rat pituitary mRNA; lane *2* shows a similar result obtained with porcine pituitary mRNA. (●) Prematurely terminated POMC products; (+) a non-POMC-related product. (*B*) The sequences of the two rat products. A 30-base addition is found in R_1, inserted at splice junction A, which indicates the junction site between exons I and II in the rat POMC gene.

Similar priming experiments with mRNA from mouse anterior pituitary cells (AtT-$20D_{16V}$ cell line) show that there are two POMC mRNA species in the mouse pituitary that differ from one another in the same way as R_1 and R_2 in the rat.

It should also be noted in Figure 6A that only one form of cDNA is made from porcine pituitary mRNA (labeled P in Fig. 6A). The reason for the existence of two forms of POMC mRNA in mouse and rat and only one in the pig is not known.

S. Amara et al. (in prep.) have used similar priming and cDNA extension techniques to demonstrate a 24-base insertion in the 5′ region of rat calcitonin mRNA. The insertion aligns with a splice junction located 9 bases before the initiator AUG. This insertion appears to result from alternate RNA splicing, as in the case of rat and mouse POMC mRNA.

The diversity of rat POMC mRNA described here does not account for the heterogeneity observed in rat POMC protein, described above. However, it is possible that the two POMC mRNA species also differ in regions 3′ to the ones sequenced here and that these differences account for the protein heterogeneity observed.

The significance of the 30-base insertion in the 5′ untranslated region of rat POMC mRNA is not yet known. The proximity of the insertion to the initiator AUG is compatible with an effect on translation initiation, or the insertion may be a reflection of a particular process involved in RNA splicing. Whatever the reason for the insertion, it can be concluded unequivocally that neither species of rat POMC mRNA is an intermediate in nuclear mRNA processing, because both species are present in cytoplasmic ribosomal fractions (E. Oates et al., unpubl.).

Structure of Human POMC and Proenkephalin Genes

The similarities noted in the structures of the three opioid precursors in the upper part of Figure 2 are reflected at the level of gene organization. The structures of the human POMC (Cochet et al. 1982) and proenkephalin genes (Noda et al. 1982a; Comb et al. 1983) are shown in the lower part of Figure 2. At the 3′ end of the genes is a large exon that contains all of the bioactive domains and most of the aminoterminal portions of these precursors. On the 5′ side of the major exon is a large intron of 3000 bases followed by a small exon containing the region coding for the amino terminus of the protein. The region coding for the remainder of the 5′ untranslated portion of the mRNA is further upstream in the same exon in the POMC gene or in another exon in the proenkephalin gene. Hence, it is clear from the organization of introns and exons that the POMC and proenkephalin genes belong to a closely related family of genes. Southern analysis of human DNA using cloned human proenkephalin cDNA as a probe indicates that there is only one proenkephalin gene in the human. This gene is located on chromosome 12 (M. Comb et al., unpubl., done in collaboration with F. Ruddle and P. Barker at Yale University), whereas the POMC gene is located on chromosome 2 (Owerbach et al. 1981). Hence, the POMC and proenkephalin genes are not linked in the human.

A more detailed analysis of the proenkephalin gene shows that it shares a number of sequence homologies with the POMC gene. One common feature of great interest is the clustering of C_pG sequences in the 5′ and 3′ regions of the POMC and proenkephalin genes. The frequency of occurrence of this dinucleotide sequence is 10 times as high in the 5′ region of the human proenkephalin gene (Fig. 2, exons I, II, and III) as in the human genome as a whole. Clustering of C_pG sequences is of particular importance because methylation of cytosine residues in this sequence has been implicated in the control of gene expression during development. An inverse relationship has been observed between the level of methylation of cytosine in C_pG and the transcriptional activity of a gene (Felsenfeld and McGee 1982).

Methylation patterns in a gene can be analyzed by using methylation-sensitive restriction enzymes, i.e., restriction enzymes whose action is prevented by methylation of cytosine in C_pG. The failure of such an enzyme to cut a particular restriction site containing a C_pG indicates the presence of a methylated cytosine at that site. Hence, one can correlate the ability of the restriction enzyme to cut a fragment of DNA and the methylation level of C_pG in the fragment.

DNA from different human tissues was cleaved with methylation-sensitive enzymes, and the fragments were analyzed by Southern hybridization using proenkephalin genomic probes (Comb et al. 1983). The only regions of the proenkephalin gene that exhibited tissue-specific methylation patterns were those in exon I and between exon IV and the 3′ end of the gene. However, it is not possible at this time to conclude that low levels of methylation of C_pG sequences in the latter regions of the proenkephalin gene are correlated with transcriptional activity of this gene. Further experiments to clarify this point are in progress.

Evolutionary Origin of the Opioid Peptide Genes

Peptides of the opioid family have been reported to occur in species ranging from man to tetrahymena (Le Roith et al. 1982). Comparison of the POMC genes in the human (Cochet et al. 1982), cow (Nakanishi et al. 1981), rat (Drouin and Goodman 1980), and mouse (Uhler et al. 1983) indicates that in mammals the structural organization of this gene is highly conserved. We were interested in knowing if the organization of the opioid genes is the same in lower organisms as in mammals. Hence, we determined the structure of a proenkephalin gene in *Xenopus laevis*, a toad that diverged from the principal line of vertebrate evolution about 350 million years ago (G. Martens et al., in prep.). The major exon codes for all seven enkephalin sequences in the gene (Fig. 7). Figure 7 shows that the major exon of the *X. laevis* gene is remarkably similar in organization to the human gene, with seven enkephalin sequences separated by spacer regions almost identical in length to those in the human gene. The toad precursor has pairs of basic amino acids flanking the biologically active peptides, suggesting that trypsinlike enzymes are involved in cleaving the peptides out of the precursor, as in the case of human proenkephalin. However, enkephalin units 4 and 6 are flanked carboxyterminally by Arg-Arg sites (instead of Lys-Lys or Lys-Arg sites),

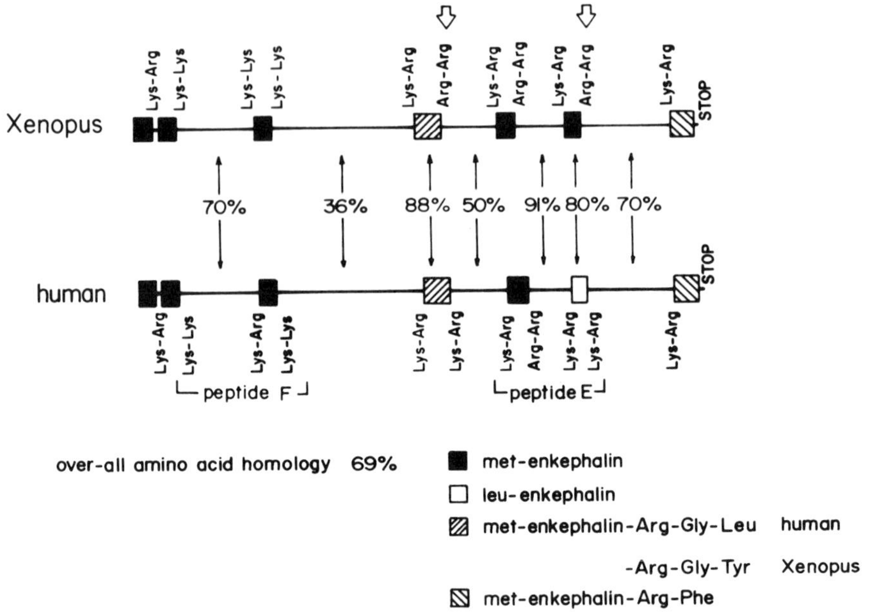

Figure 7. Sequence comparison of the proenkephalin precursors of human and *Xenopus laevis*. The *X. laevis* peptide sequence has been derived from the nucleic acid sequence of a gene isolated from a bacteriophage λ-*Xenopus* genomic library. The numbers represent the percentage of amino acid homology between the spacer regions of the enkephalin units.

which in mammalian precursors are not readily cleaved. Hence, it is possible that different sets of enkephalin end products are produced in the amphibian than those in the mammal. It should also be noted that both precursors have Met-enkephalin-Arg-Phe in carboxyterminal positions, but that the toad has the octapeptide Met-enkephalin-Arg-Gly-Tyr in place of Met-enkephalin-Arg-Gly-Leu in the human.

Perhaps the most interesting feature of toad proenkephalin is the substitution of Met-enkephalin for Leu-enkephalin at the carboxyl terminus of the peptide E sequence. This means that there is no Leu-enkephalin sequence in the toad proenkephalin gene, suggesting that the Met-enkephalin sequence evolved first and that the switch from Met-enkephalin to Leu-enkephalin occurred relatively recently (less than 350 million years ago). Obviously, proenkephalin genes in other nonmammalian species have to be analyzed to test this idea more thoroughly.

Finally, it is important to point out that despite the similarity in length of the spacer regions between enkephalins, the sequences are not very well conserved in these regions except for peptide E (Fig. 7). The amino acid homologies in the spacer regions range from 91% (peptide E spacer) to 36% (for the longest spacer). Hence, except for peptide E, most of the sequences in these regions do not appear to be important for processing of the precursor or for other functions.

The repeated sequences of biologically active peptides in the opioid peptide precursors suggest that these peptides arose by a series of duplication events (Fig. 8, top line). Evidence for duplication is even more apparent at the nucleic acid level than at the peptide level. A computer search for internal homologies in nucleic acid sequences in each opioid gene and for sequence homologies among these genes has provided striking evidence for the occurrence of duplication events during the evolution of the proenkephalin gene. In the toad proenkephalin gene, the regions coding for the first two enkephalin sequences and the flanking basic amino acids could constitute the repeat unit (48 bp) since it shows 73% homology with a second unit about 70 bp downstream (Fig. 8, line 2) and somewhat lower homology with three other units further downstream. Note that in each of units 3, 4, and 5, there is one sequence coding for enkephalin and one exhibiting enkephalin-sequence homology (pseudoenkephalin) resulting from a phase shift or frameshift. This suggests that units 3, 4, and 5 arose by duplication of unit 1 and then diverged further during evolution.

The organization and spacing of enkephalin units in the prodynorphin gene is similar to that in the proenkephalin gene in the carboxyterminal portion of the precursor (Fig. 8, line 3). Similarity between the two precursors is even more apparent at the level of nucleic acid sequence (cDNA sequence from Kakidani et al. 1982), as shown in lines 2 and 4 of Figure 8. There are five homologous repeat units in the prodynorphin gene (Fig. 8, line 4). Two of these are pseudoenkephalin units exhibiting phase shifts as the comparable units in the proenkephalin gene. The phase shifts in these sequences in units 3, 4, and 5 suggest that these units may have originated as parts of introns where frameshifts appear to occur more frequently than in exons.

It is intriguing to speculate that the enkephalin repeat unit is the primordial building block of the gene. Duplication and rearrangements of these building blocks led to the proenkephalin genes that exist today.

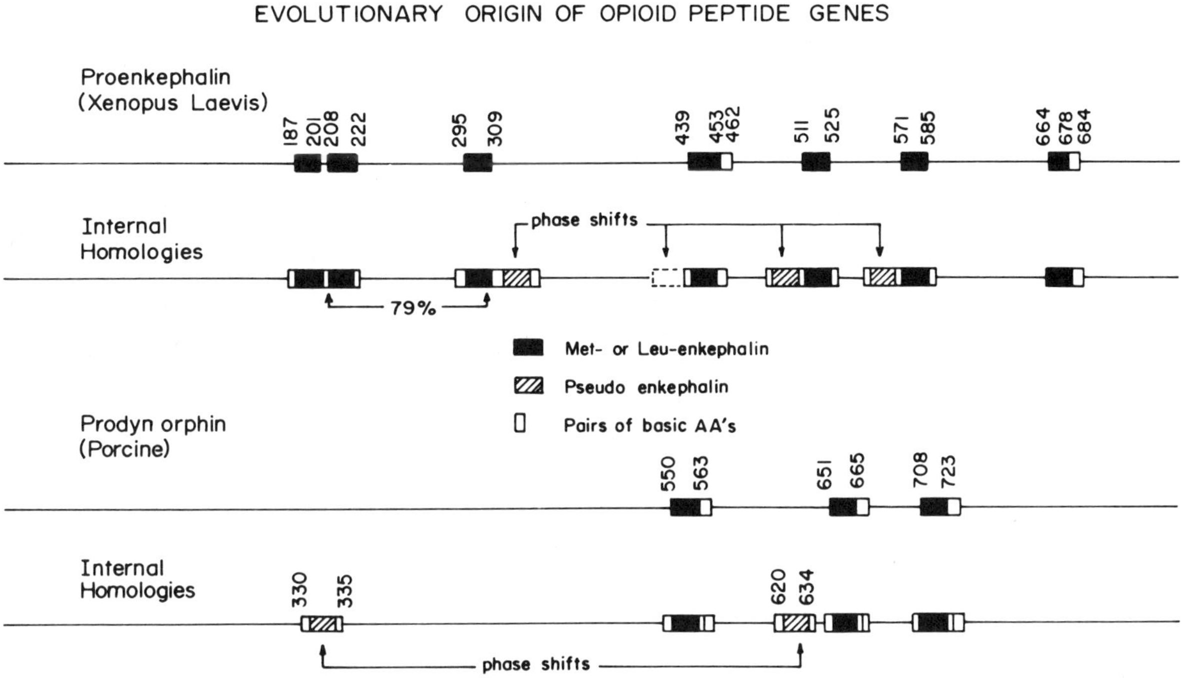

Figure 8. Primordial duplication units in the evolution of proenkephalin and prodynorphin genes. The numbering system starts with the first nucleotide in the nucleic acid sequence (G. Martens et al., in prep.).

It is of interest that a similarly sized repeat unit has been reported as the building block of the immunoglobulin family of genes (Yazaki and Ohno 1983), implying that repeat units of this size may have been widely used in nature to construct multigene families.

The duplication events that generated the multiple copies of the enkephalin and MSH units, a feature common to all three opioid precursors, are likely to be an important theme in the evolution of neuroendocrine genes, as evidenced by the growing number of precursors containing multiple, homologous bioactive peptides (opioid precursors, ELH precursor [Scheller et al. 1983], α-factor precursor [Kurjan and Herskowitz 1982], glucogon precursor [Lund et al. 1982]). Duplication of the bioactive units within the ancestral opioid precursor gene provides a fertile template upon which mutation could act to produce related yet functionally diversified sets of peptides. Duplication of the entire ancestral opioid peptide precursor gene, or the opioid peptide receptor gene, to form a related gene family would also serve a similar function.

ACKNOWLEDGMENTS

This work is supported by the following grants (E.H., principal investigator): National Institute of Arthritis, Diabetes, Digestive and Kidney Diseases grants AM-16879 and AM-30155 and National Institute of Drug Abuse grant DA-02736. G.M. is supported by a postdoctoral fellowship from the Netherlands Organization for the Advancement of Pure Research (ZWO). M.C. is supported by a Molecular Biology Training Grant (GM-07759) from the National Institutes of Health. H.R. is supported by a Chaim Weizman postdoctoral fellowship.

REFERENCES

Amara, S.G., V. Jones, M.G. Rosenfeld, E.S. Ong, and R.M. Evans. 1982. Alternative RNA processing in calcitonin gene expression generates mRNAs encoding different polypeptide products. *Nature* **298:** 240.

Brownstein, M.J., J.F. Russell, and H. Gainer. 1980. Synthesis, transport, and release of posterior pituitary hormones. *Science* **207:** 373.

Chang, A.C.Y., M. Cochet, and S. Cohen. 1980. Structural organization of human genomic DNA encoding the pro-opiomelanocortin peptide. *Proc. Natl. Acad. Sci.* **77:** 4890.

Civelli, O., N. Birnberg, and E. Herbert. 1982. Detection and quantitation of pro-opiomelanocortin mRNA in pituitary and brain tissues from different species. *J. Biol. Chem.* **257:** 6783.

Cochet, M., A.C.Y. Chang, and S. Cohen. 1982. Characterization of the structural gene and putative 5′-regulatory sequences for human pro-opiomelancortin. *Nature* **297:** 335.

Comb, M., H. Rosen, P. Seeburg, J. Adelman, and E. Herbert. 1983. Primary structure of the human proenkephalin gene. *DNA* **2:** (in press).

Comb, M., P. Seeburg, J. Adelman, L. Eiden, and E. Herbert. 1982. Primary structure of the human Met- and Leu-enkephalin precursor and its mRNA. *Nature* **295:** 663.

Crine, P., C. Gianoulakis, N.G. Seidah, F. Gosard, P.D. Rezalla, M. Lis, and M. Chretien. 1978. Biosynthesis of β-lipotropin and a larger molecular weight precursor in rat pars intermedia. *Proc. Natl. Acad. Sci.* **75:** 4719.

DeNoto, J.M., D.D. More, and H.M. Goodman. 1981. Human growth hormone DNA sequence and mRNA structure: Possible alternative splicing. *Nucleic Acids Res.* **9:** 3719.

DROUIN, J. and H.M. GOODMAN. 1980. Most of the coding region of rat ACTH-β-LPH precursor gene lacks intervening sequences. *Nature* **288:** 610.

DROUIN, J., P. BURBACK, J. SHARRON, and J.P. GAYNER. 1983. Expression of the pro-opiomelanocortin (POMC) gene. (UCLA Symposia on Molecular and Cellular Biology, Abstracts). *J. Cell Biochem. Suppl.* **7A:** 107.

FELSENFELD, G. and J. MCGHEE. 1982. Methylation and gene control. *Nature* **296:** 602.

GUBLER, U., P. SEEBURG, B.J. HOFFMAN, L.P. GAGE, and S. UDENFRIEND. 1982. Molecular cloning establishes proenkephalin as precursor of enkephalin-containing peptides. *Nature* **295:** 206.

HINMAN, M.B. and E. HERBERT. 1980. Processing of the precursor to adrenocorticotropic hormone and β-lipotropin in monolayer cultures of mouse anterior pituitary. *Biochemistry* **19:** 5395.

HUGHES, J., T.W. SMITH, H.W. KOSTERLITZ, L.A. FORTHEGILL, B.A. MORAN, and H.R. MORRIS. 1975. Identification of two related pentapeptides from the brain with potent opiate agonist activity. *Nature* **258:** 577.

KAKIDANI, H., Y. FURUTANI, H. TAKAHASHI, M. NODA, Y. MORIMOTO, T. HIROSE, M. ASAI, S. INAYAMA, S. NAKANISHI, and S. NUMA. 1982. Cloning and sequence analysis of cDNA for porcine β-neo-endorphin/dynorphin precursor. *Nature* **298:** 245.

KILPATRICK, D.L., B.N. JONES, R.V. LEWIS, A.S. STERN, K. KOJIMA, J.E. SHIVELY, and S. UDENFRIEND. 1982. A 18,200-dalton adrenal protein that contains four met-enkephalin sequences. *Proc. Natl. Acad. Sci.* **79:** 3057.

KILPATRICK, D.L., T. TANIGUCHI, B.N. JONES, A.S. STERN, J.E. SHIVELY, J. HULLIHAN, S. KIMURA, S. STEIN, and S. UDENFRIEND. 1981. A highly potent 3200-dalton adrenal opioid peptide that contains both a met- and leu-enkephalin sequence. *Proc. Natl. Acad. Sci.* **78:** 3265.

KRIEGER, D.T., A.S. LIOTTA, and M.J. BROWNSTEIN. 1977. Presence of corticotropin in brain of normal and hypophysectomized rats. *Proc. Natl. Acad. Sci.* **74:** 648.

KRIEGER, D.T., A.S. LIOTTA, M.J. BROWNSTEIN, and E.A. ZIMMERMAN. 1980. ACTH, β-lipotropin and related peptides in brain, pituitary and blood. *Recent Prog. Horm. Res.* **36:** 272.

KURJAN, J. and I. HERSKOWITZ. 1982. Structure of a yeast pheromone gene (MFα): A punitive α-factor precursor containing four tandem copies of mature α-factor. *Cell* **30:** 933.

LAND, H., G. SCHÜTZ, H. SCHMALE, and D. RICHTER. 1982. Nucleotide sequence of cloned cDNA encoding bovine arginine vasopressin-neurophysin II precursor. *Nature* **295:** 299.

LEROITH, D., A.S. LIOTTA, J. ROTH, J. SHILOACH, M.E. LEWIS, C.B. PERT, and D.T. KRIEGER. 1982. Corticotropin and β-endorphin-like materials are native to unicellular organisms. *Proc. Natl. Acad. Sci.* **79:** 2086.

LEWIS, U.L., L.F. BONEWALD, and L.J. LEWIS. 1980. The 20,000-dalton variant of human growth hormone: Location of the amino acid deletions. *Biochem. Biophys. Res. Commun.* **92:** 511.

LUND, P.K., R.H. GOODMAN, P.C. DEE, and J.F. HABENER. 1982. Pancreatic preglucagon DNA contains two glucagon-related coding sequences arrayed in tandem. *Proc. Natl. Acad. Sci.* **79:** 345.

MAINS, R.E. and B.A. EIPPER. 1979. Synthesis and secretion of corticotropins, melanotropins and endorphins by rat intermediate pituitary cells. *J. Biol. Chem.* **254:** 7885.

MATTEUCCI, M.D. and M.H. CARUTHERS. 1981. Synthesis of deoxyoligonucleotides on a polymer support. *J. Am. Chem. Soc.* **103:** 3185.

MAXAM, A.M. and W. GILBERT. 1977. A new method for sequencing DNA. *Proc. Natl. Acad. Sci.* **74:** 560.

NAKANISHI, S., A. INOUE, T. KITA, M. NAKAMURA, A.C.Y. CHANG, S. COHEN, AND S. NUMA. 1979. Nucleotide sequence of cloned cDNA for bovine corticotropin-β-lipotropin precursor. *Nature* **278:** 423.

NAKANISHI, S., Y. TERANISHI, M. NODA, M. NOTAKE, Y. WATANABE, H. KAKIDANI, H. JINGAMI, and S. NUMA. 1980. The protein-coding sequence of the bovine ACTH-β-LPH precursor gene is split near the signal peptide region. *Nature* **287:** 752.

NAKANISHI, S., Y. TERANISHI, Y. WATANABE, M. NOTAKE, M. NODA, H. KAKIDANI, H. JINGAMI, and S. NUMA. 1981. Isolation and characterization of the bovine corticotropin/β-lipotropin precursor gene. *Eur. J. Biochem.* **115:** 429.

NARANG, S.A., R. BROUSSEAU, H.M. HSIUNG, and J.J. MICHNIEWICZ. 1980. Chemical synthesis of deoxyoligonucleotides by the modified triester method. *Methods Enzymol.* **65:** 6100.

NODA, M., Y. TERAHASHI, H. TAKAHASHI, M. TOYOSATO, M. NOTAKE, S. NAKANISHI, and S. NUMA. 1982a. Isolation and structural organization of the human preproenkephalin gene. *Nature* **297:** 431.

NODA, M., Y. FURUTANI, H. TAKAHASHI, M. YOYOSATO, T. HIROSE, S. INAYAMA, S. NAKANISHI, and S. NUMA. 1982b. Cloning and sequence analysis of cDNA for bovine adrenal preproenkephalin. *Nature* **295:** 202.

OWERBACH, D., W.J. RUTTER, J.R. ROBERTS, P. WHITFELD, J. SHINE, P.H. SEEBURG, and T.B. SHOWS. 1981. The pro-opiocortin gene is located on chromosome 2 in humans. *Somatic Cell Genet.* **7:** 359.

SCHELLER, R.H., J.F. JACKSON, L.B. MCALLISTER, B.S. ROTHMAN, M. EARL, and R. AXEL. 1983. A single gene encodes multiple neuropeptides mediating a stereotyped behavior. *Cell* **32:** 7.

SCOTT, A.P., P.J. LOWRY, J.G. RATCLIFF, L.H. REES, and J. LANDON. 1974. Corticotropin-like peptides in the rat pituitary. *J. Endocrinol.* **61:** 355.

SMYTH, D. and S. ZAKARIAN. 1979. Isolation and characterization of endogenous [α]-N-acyl derivatives of the C- and C′-fragment of lipotropin. In *Endorphins in mental health* (ed. E. Usdin et al.), p. 84. Macmillan, New York.

UHLER, M., E. HERBERT, P. D'EUSTACHIO, and F. RUDDLE. 1983. The mouse genome contains two nonallelic pro-opiomelanocortin genes. *J. Biol. Chem.* **258:** 9444.

WHITFELD, P.L., P. SEEBURG, and J. SHINE. 1982. The human pro-opiomelanocortin gene: Organization, sequence and interspersion with repetitive DNA. *DNA* **1:** 133.

YATES, F.F. and J.W. MARAN. 1974. Stimulation and inhibiton of adrenocorticotropin release. *Handbook Physiol.* **4:** 367.

YAZAKI, A. and S. OHNO. 1983. Recurrence of 49-base decamers, nonamers and octamers within mouse Cμ gene Ig heavy chain and its primordial building block. *Proc. Natl. Acad. Sci.* **80:** 2337.

Analysis of POMC Gene Expression by Transcription Assay and In Situ Hybridization Histochemistry

J.L. ROBERTS, J.H. EBERWINE, AND C.E. GEE
Center for Reproductive Sciences and Department of Biochemistry, Columbia University, College of Physicians and Surgeons, New York, New York 10032

Until recently, interest was focused primarily on the synthesis and processing of neuropeptide hormones, and not on the regulation of peptide hormone gene expression. With the development of recombinant DNA technology, a better understanding of the regulation of genes encoding these neuromodulators will be forthcoming. The pathway of expression of a peptide hormone gene involves many steps (see Fig. 1). Under appropriate hormonal and environmental conditions, the peptide hormone gene is transcribed into a heterogeneous nuclear RNA (hnRNA), which is then processed to yield the mature mRNA within the nucleus. After transport to the cytoplasm, the mRNA is either translated into protein, stored, or degraded. This protein can then undergo tissue-specific cotranslational and/or posttranslational modifications, such as glycosylation and proteolysis, to give the individual mature peptide hormone. These peptides are stored in secretory granules that await the proper stimulus to signal secretion. It is the interest of our laboratory to study the early events in peptide hormone gene expression, such as quantitating the levels of the mRNA or analyzing the levels of nuclear transcription for specific neuropeptides.

The proopiomelanocortin (POMC) peptide hormone system is an excellent model for studying neuropeptide gene regulation. POMC is a 30,000-molecular-weight polyhormone protein that can be processed posttranslationally to several peptide hormones, including β-endorphin, adrenocorticotropin hormone (ACTH), and the melanocyte-stimulating hormones (MSHs) (for review, see Eipper and Mains 1980). It is expressed in a variety of tissues, the primary sites being the intermediate and anterior lobes of the pituitary and the periarcuate region of the hypothalamus (Liotta et al. 1979).

Although it appears that the pituitary and hypothalamus are producing a similar, if not identical, POMC protein, the posttranslational modifications in these tissues vary, resulting in a different set of hormones. For example, in the anterior pituitary ACTH is produced, whereas in the intermediate lobe all of the ACTH is processed to yield the α-MSH peptide. The POMC neurons of the arcuate region appear to produce a mixture of both ACTH and α-MSH. These alternate pathways in processing the POMC peptides and the resultant biological activities may necessitate the differential regulation of gene expression seen in these tissues.

Because POMC is expressed in both a neural tissue and a nonneural (the pituitary) tissue, it becomes a very useful model for the study of neuropeptide hormone gene regulation. Indeed, what we have tried to do is to study POMC gene regulation in the pituitary, a much more experimentally accessible tissue, and then make inference as to how this regulation may be related to the expression of the POMC peptide hormone gene in the hypothalamus. In this paper, we discuss some of the studies we have performed along these lines and try to show how we have been able to apply the techniques used in the pituitary to a direct analysis of POMC gene regulation in the periarcuate region of the hypothalamus.

Heterogeneity of Regulation in Anterior and Intermediate Pituitary Lobes

A variety of hormonal and environmental factors selectively modulate the synthesis and secretion of POMC peptides in the anterior and neurointermediate lobes of the pituitary. Several of these modulators originate in the brain. For example, corticotropin-releasing factor (CRF) and vasopressin, two hypothalamic peptides, stimulate the synthesis and release of ACTH and β-endorphin in the anterior lobe (Vale et al. 1978). Neurotransmitters such as norepinephrine, serotonin, and dopamine, which are transported to the pituitary either by the portal vasculature or directly via innervation of the intermediate lobe, similarly play important roles in the regulation of POMC peptide production in the anterior and intermediate lobes (Chen et al. 1982).

In the anterior lobe of the pituitary, the synthesis and release of ACTH and β-endorphin are subject to both positive regulation by CRF and negative feedback regulation by glucocorticoids. However, in the intermediate lobe, CRF and glucocorticoids have no effect on POMC synthesis and secretion. In this tissue, catecholamines are involved in regulation. Specifically, dopamine administered at physiological concentrations inhibits the release of POMC-derived peptides from the intermediate lobe. A major point of interest now becomes the determination of whether expression of the POMC gene itself is influenced by any of those factors that modulate the secretion of its peptide. If indeed modulators of secretion also have effects on expression of the gene, it will be of interest to learn how the cell elicits the differential expression of the POMC gene

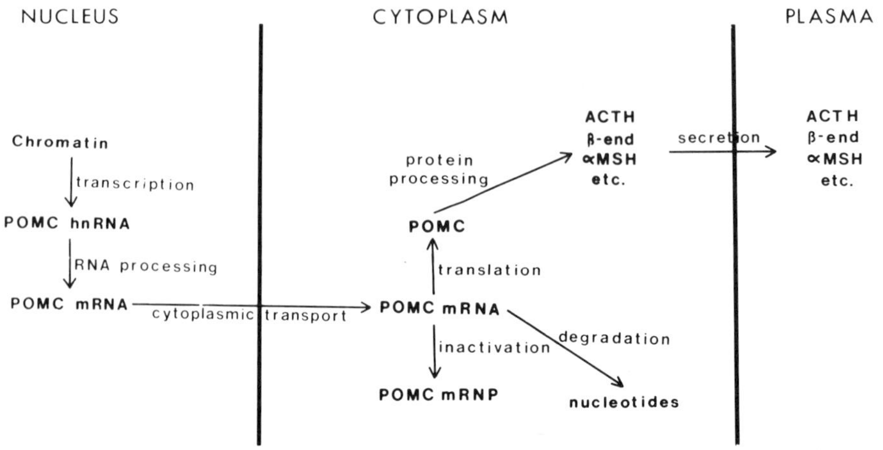

Figure 1. Pathway of expression of the POMC gene, from transcription (*left*) to secretion (*right*).

observed for many of these compounds on the two lobes of the pituitary.

Investigators in many laboratories have been involved in the study of POMC gene expression in the pituitary. Using either indirect cell-free protein translation/immunoprecipitation or RNA:DNA hybridization assays, several groups have demonstrated that glucocorticoids lower the POMC mRNA level in rat pituitary. Nakanishi et al. (1977) have reported a twofold increase in total pituitary POMC mRNA after adrenalectomy and a fourfold decrease following dexamethasone treatment of adrenalectomized rats. The half-time of this deinduction was 24 hours. Schachter et al. (1982) assayed POMC mRNA in separated anterior and neurointermediate lobes (the intermediate lobe with the posterior lobe still attached) from pituitaries of dexamethasone-treated adrenalectomized rats. They found that the glucocorticoids decreased 30–50-fold the level of anterior lobe POMC mRNA induced after 7 days of adrenalectomy. The same treatment has resulted in only a threefold suppression of the level of POMC mRNAs in the intermediate lobe. Herbert et al. (1981) obtained similar data, with the exception that they saw no change in the level of intermediate lobe POMC mRNA. They also demonstrated that adrenalectomy after a lag period of 6–10 hours progressively increased mRNA levels in the anterior lobe tenfold after 4 weeks. The response of POMC mRNA levels to the different steroids at various concentrations has also been studied in the mouse POMC-producing ATT20 cells (Roberts et al. 1979). On the basis of these results, it is believed that the inhibitory effect of dexamethasone is mediated by the classic cytoplasmic/nuclear glucocorticoid receptor (for review, see Roberts et al. 1982).

It has also been shown that dopaminergic compounds exert an inhibitory effect on POMC gene expression in the rat pituitary. Chen et al. (1982), using the dopamine antagonist haloperidol, were able to show a fourfold to sixfold time- and dose-dependent increase in the intermediate lobe levels of POMC mRNA. This effect was detectable after 6 hours and reached a maximum after 3 days. Ergocryptine, a dopamine agonist, decreased the POMC mRNA level twofold to threefold in the intermediate lobe. There is no effect of either compound on POMC mRNA levels in the anterior lobe. Recently, Hollt et al. (1982), using a translation/immunoprecipitation assay, have shown a similar stimulatory effect of haloperidol on POMC mRNA in the intermediate lobe of rats.

These studies show a differential effect of these various hormonal compounds on the regulation of POMC gene expression at the mRNA level in these different tissues. But such studies only measure the steady-state levels of mRNA and do not yield information as to how the changes came about. It is possible that there are altered rates of transcription of the POMC gene yielding the changed levels in mRNA. It is also possible that the stability of the mRNA has changed with the hormonal treatment. Indeed, examples of both types of controls yielding changed mRNA levels abound in the literature (McKnight and Palmiter 1979; Spindler et al. 1982). To probe further into the mechanism of hormonal regulation of gene expression, we initiated studies that would determine the effects of the various hormonal and environmental factors on POMC gene transcription.

We developed a POMC gene transcription assay using a modification of several published procedures (McKnight and Palmiter 1979; Spindler et al. 1982). In this assay, nuclei are isolated from hormonally manipulated animals (see Fig. 2). These nuclei are then incubated with radioactive RNA precursors, allowing those RNA chains initiated in vivo to be further elongated, with no new initiation of transcription. POMC-specific transcripts are analyzed by incubating the newly synthesized radioactive RNA with POMC gene DNA denatured and bound to nitrocellulose filters. The POMC-specific sequences hybridize to the filter and then can be separated from the large majority of non-POMC transcripts by extensive washing of the filter. The counts per minute of labeled RNA and the amount of POMC-specific transcripts present in the nuclear transcription mixture are determined using a scintillation counter. Corrections are made for a non-POMC DNA-

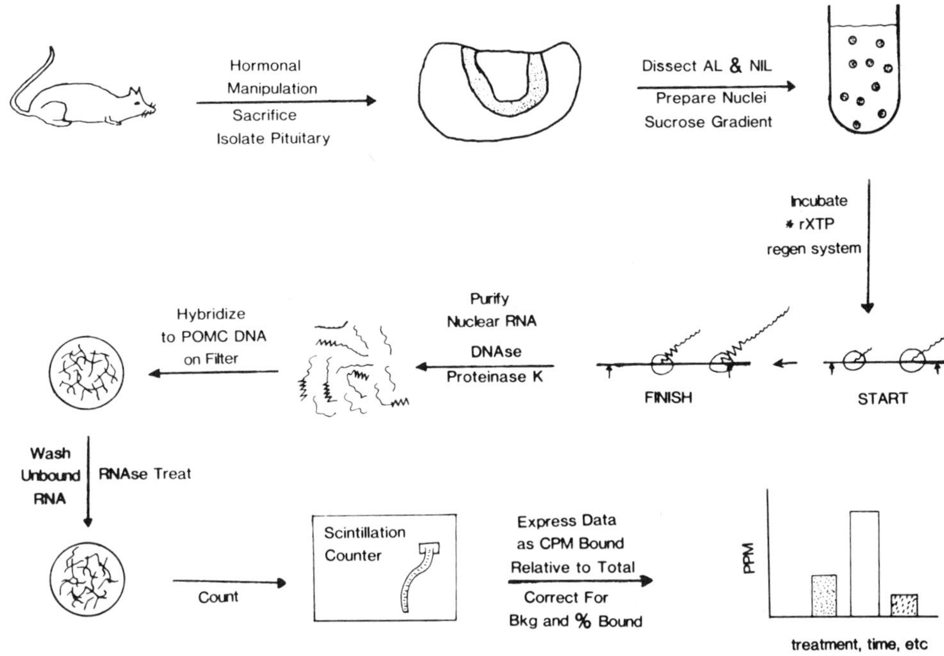

Figure 2. Diagram above "START" depicts the RNA polymerases present on the POMC gene within the nuclear chromatin isolated following specific hormonal treatment. In the presence of ^{32}P-labeled rXTP, the polymerases are permitted to continue to transcribe the previously initiated hnRNA transcripts as depicted in the diagram above "FINISH." These elongated transcripts are then isolated, and the specific POMC transcripts are quantitated. (AL) Anterior lobe; (NIL) neurointermediate lobe; (rXTP) ribose nucleotide triphosphates; (PPM) parts per million.

binding background as well as percentage of hybridization. These data are generally presented in parts per million of total transcription.

The effect of glucocorticoids on POMC gene transcription was measured at various time periods after administration of the steroid to the animal. The results of such studies are shown in Figure 3. POMC gene transcription is rapidly affected by glucocorticoids, with a maximal effect between 15 and 60 minutes. The synthetic glucocorticoid, dexamethasone, elicited a fourfold suppression, whereas the natural glucocorticoid, corticosterone, elicited only a threefold suppression. After this maximal effect of the glucocorticoid, the system slowly returns toward normal levels. It is not clear whether this is due to some inherent mechanism of the steroid washing out of the cell or whether it is due to some refractory mechanism in the transcription itself. One point does address this question: When the natural glucocorticoid corticosterone was used, which is metabolically cleared more rapidly than dexamethasone, POMC transcription returned to the normal state more quickly than when dexamethasone was used, suggesting that this return to normal is due to a washout of the steroid. Interestingly, in the intermediate lobe, glucocorticoids had no effect on POMC gene transcription. This is in agreement with the minimal effect of glucocorticoids on POMC mRNA levels in this tissue. Thus, the conclusion is that, at least in part, the lowering of the POMC mRNA level in response to glucocorticoid treatment comes from suppression of transcription of the POMC gene by glucocorticoids. Whether this is a direct effect of the glucocorticoid receptor on the gene or indirect through another gene system is not clear, although the rapidity of the effect would argue for direct effects.

The effect of adrenalectomy on POMC transcription was also measured. As shown in Figure 4, bilateral adrenalectomy caused a significant increase in transcription of the POMC gene by 4 days. With time, this stimulation gradually returns toward the control values. Thus, the increase seen by various research groups in pituitary POMC mRNA following adrenalectomy is accounted for in part by the increased synthesis of the POMC mRNA.

Another interesting feature of this study comes from the sham injection control animals. When animals were injected intraperitoneally with saline and then analyzed at varying times for POMC gene transcription, a reproducible stimulation at short time periods, inhibition at intermediate time periods, and return to control levels within 2 hours were seen. We interpret these data in light of the well-characterized physiological responses involving ACTH or glucocorticoids in response to stress. The short-time-period stimulation of POMC transcription occurs at the same time POMC peptides are being released by CRF. This CRF is being released by the hypothalamus in response to the stressful stimulus. At intermediate time periods of 30–60 minutes, the ACTH released from the anterior pituitary has elicited a stimulatory response of glucocorticoid secretion from the adrenal gland. These glucocorticoids are having a feedback inhibitory effect on the anterior pituitary,

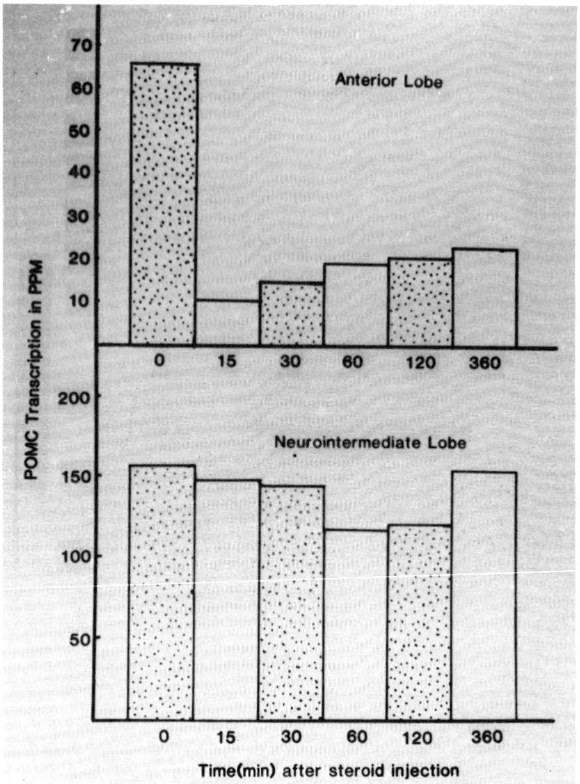

Figure 3. Three adult male Sprague-Dawley rats were injected intraperitoneally with 100 μg of dexamethasone at various times prior to sacrifice. The pituitary anterior lobes were separated from interior lobes and were processed through the nuclear transcription assay (see Fig. 2). PPM indicates parts per million and is a direct measurement of the amount of transcription of the POMC gene under assay conditions. PPM was corrected for amount of hybridization using a POMC [^{3}H]cRNA standard and background nonspecific DNA counts per minute. Background was generally 5–20% of specific counts bound by POMC DNA-containing filter.

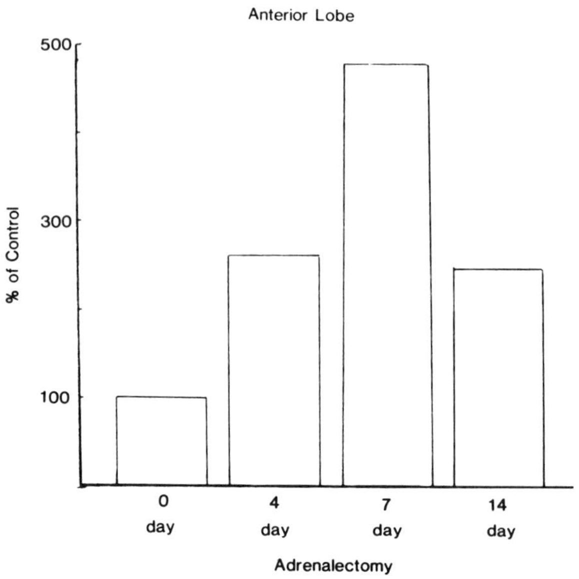

Figure 4. Rats in groups of three were bilaterally adrenalectomized for various lengths of time. They were then sacrificed, and their anterior lobes were isolated and processed through the nuclear transcription assay. Control values were from sham-operated animals.

blocking CRF-stimulated POMC peptide secretion. The glucocorticoid and ACTH hormone levels return to control values, as do POMC transcription levels after 2 hours, since there are no further stressful stimuli. These observations would suggest that POMC gene transcription is tightly coupled with POMC secretion in the anterior lobe of the pituitary. Such a mechanism is appealing in that the general phenomenon of secretion can be directly related then to stimulating overall transcription of all genes required in the peptide hormone biosynthetic process prior to secretion of the peptide hormone. It is important to point out that these rapid, acute changes of POMC hnRNA in response to glucocorticoids can readily be observed using the nuclear runoff transcription assay, whereas changes in cytoplasmic levels of POMC mRNA take 6 hours to detect. This difference can be attributed to the relative sizes of the pools of cytoplasmic and nuclear POMC RNAs. It is possible that short-lived acute changes may go undetected because cytoplasmic mRNA levels do not change rapidly enough to reflect the changes in hnRNA. This point is elaborated on in the legend to Figure 5.

The ability of the POMC gene to respond differently to various hormonal stimuli in the two different lobes of the POMC pituitary is also interesting. Extensive studies have shown that there is only one POMC gene present in the rat haploid genome complement (for review, see Eberwine and Roberts 1983). This rules out the possibility of a different gene being expressed that would code for the same structural mRNA but be subject to different hormonal control. Thus the question is, How can the same gene respond differently to the same stimulus in the different expressing tissues? The answer to the question appears to be an emerging idea of the generation of biological complexity from a defined genomic complement, i.e., that the expressing cells can differentially recognize specific effector compounds. They do this either by blocking or by not expressing at all the receptor mechanism necessary to recognize that compound. In the case described here, the intermediate lobe, which is not responding to glucocorticoids, does not have a normal glucocorticoid receptor. Autoradiographic studies have shown that the intermediate lobe cells lack the ability to concentrate radioactive glucocorticoid in their nuclei. Thus, the lack of an effect of glucocorticoids probably does not reside in some inherent inability of the gene to respond to glucocorticoids; the cell simply does not recognize the glucocorticoids because there is no receptor. This example of tissue specificity in the anterior and interior lobes of the pituitary control of transcription of the POMC gene is reminiscent of the cotranslational/posttranslational processing specificity exhibited by these same tissues. It is probable that specificity of cell type does not engender solely the production of a specific protein product. Cell-type specificity is composed of the sum total of

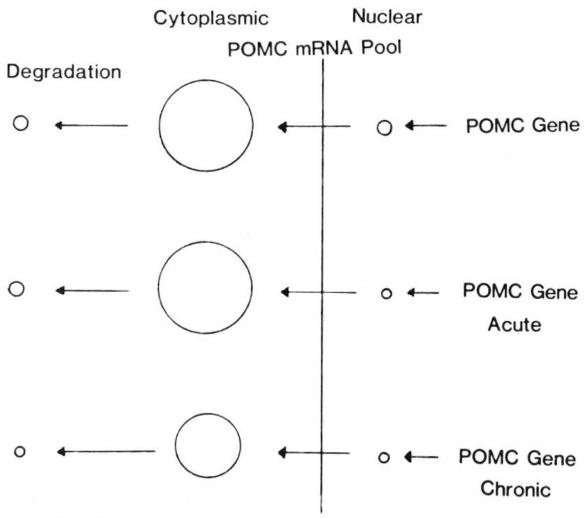

Figure 5. Scheme depicting the value of looking at changes in transcription to study hormonal modulation of gene expression. The open circles represent relative POMC mRNA pool sizes. Rapid changes in POMC transcription from acute stimuli can go undetected in the cytoplasm because of the large pool of POMC mRNA. To detect changes in cytoplasmic mRNA, one must perturb the pool significantly, as with chronic stimuli. Hence, rapid changes in transcription that are short-lived may not be reflected in a change in the large cytoplasmic POMC mRNA pool.

many processes that affect specific protein production. These processes range from direct transcriptional and posttranscriptional processing events to the indirect consequences of receptor presence or absence.

In Situ Hybridization Histochemistry

Although the primary site of synthesis of POMC is in the pituitary, POMC-derived peptides have been identified in a variety of tissues, most notably in the brain. A major question arises concerning these brain POMC peptides: Are they synthesized within the CNS itself or are they taken up from the plasma by retrograde circulatory transport from the pituitary? POMC peptides have been detected immunohistochemically and biochemically in the medial basal hypothalamus, in the amygdala, and throughout the brain stem. POMC-peptide-containing cell bodies, however, have been identified in only two cell groups: principally in the periarcuate region of the hypothalamus and, to a lesser extent, in the nucleus of the tractus solitarius. Such observations would suggest that POMC peptides may originate in the medial basal hypothalamus and reach other regions of the CNS by axonal transport.

Several other studies support this concept of localized brain synthesis of POMC peptides. In a biosynthetic study, Liotta et al. (1979) incubated bovine hypothalamus with radioactive amino acids and were able to show de novo synthesis of POMC. In a different type of study, Civelli et al. (1982) identified POMC mRNAs in nucleic acid extracts of rat and bovine hypothalami by solution hybridization as well as Northern gel blot analysis. These researchers also found a much smaller amount of a POMC-like mRNA per microgram of total RNA in the rat amygdala and cortex; but this mRNA was about 100 nucleotides shorter than the authentic pituitary and hypothalamic POMC mRNAs, so its identity is not clear. These studies strongly argue in favor of localized synthesis of POMC in the hypothalamus, but because of the close proximity of a rich POMC-synthesizing tissue, the pituitary, there always exists the possibility of tissue contamination during dissection. Thus, a technique capable of localizing specific sequences of mRNA in individual cells, a prerequisite for synthesis of the protein for which the RNA encodes, would be very good proof of localized synthesis of that particular peptide. The presence of POMC-specific mRNA in a POMC-peptide-containing cell in the brain would therefore be taken as strong evidence for POMC biosynthesis in brain tissue.

To localize the POMC-specific mRNA in individual cells, we have employed the newly refined technique of in situ cDNA : mRNA hybridization on fixed sections of brain tissue (Gee et al. 1983). This technique is conceptually very similar to immunohistochemistry, except that a DNA or RNA probe is used to detect the presence of a specific nucleic acid sequence. In our experiment we fixed the tissue by perfusing the animal through the aorta with 4% formaldehyde in phosphate-buffered saline (PBS) for 20 minutes. This particular fixation method preserves both the peptide and the mRNA in the cytoplasm of cells. The brain was partitioned into three blocks by coronal cuts, one rostral and one caudal to the hypothalamus. These blocks of tissue were then placed into 15% sucrose in PBS at 4°C overnight to rinse out excess fixative. The tissue was embedded in OTC compound and frozen in liquid nitrogen. Sections (10 μm) were cut with a cryostat microtome and placed onto gelatin-coated glass microscope slides. Tritiated POMC cDNA probes were hybridized to the section, the unhybridized probe was washed off, and the section was processed for autoradiography. The presence of autoradiographic grains over cytoplasm therefore represents the presence of the mRNA within the cytoplasm of that cell. We have previously published a more detailed account of this protocol, including a variety of controls to ensure that the signal we are seeing is actually derived from POMC-specific mRNA sequences (Gee and Roberts 1983).

Although this technique is very similar in principle to immunohistochemistry, there are several characteristics that set it apart. A major difference is that the mRNA, intrinsic to its function, is covered up by ribosomes. Thus, it is often difficult after fixation of the tissue for the cDNA probe to recognize specific sequences in the mRNA. This may be one of the reasons we, as well as other investigators, have observed that probes 15–50 bases in length are much better at giving a positive hybridization signal relative to longer probes (Brahic and Haase 1978; Gee and Roberts 1983).

An application of the in situ hybridization histochemistry technique to the POMC system in the rat hypothalamus is shown in Figure 6. In this case, POMC-specific

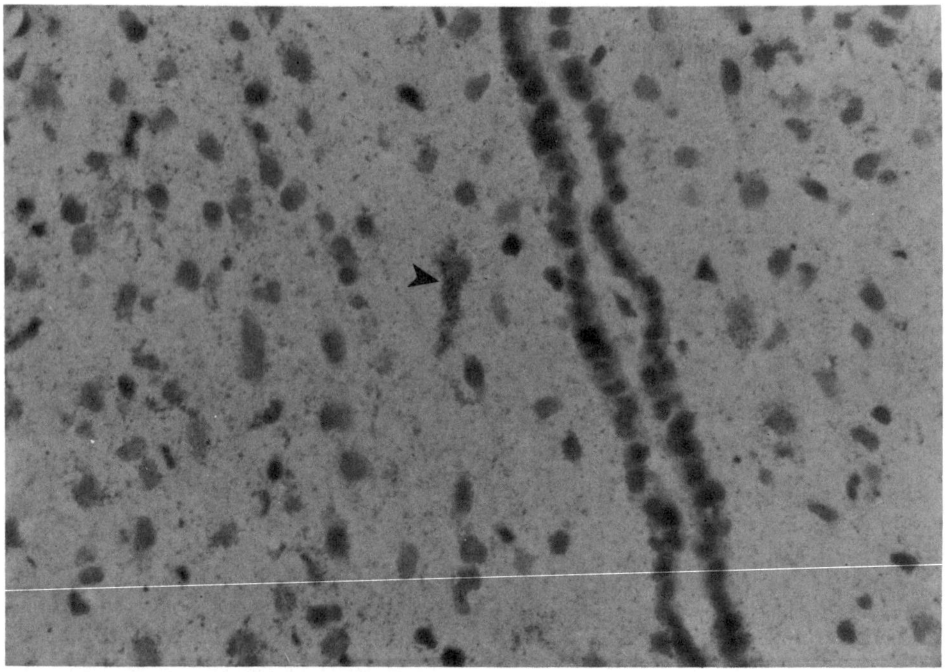

Figure 6. POMC in situ hybridization histochemistry in the periarcuate region of the rat hypothalamus. Arrow indicates a neuron that has autoradiographic grains over the cell body, including a cell process, possibly dendritic. The two rows of cells at the right are the epithelial cells lining the third ventricle of the medial basal hypothalamus.

mRNA sequences were identified in several neuronal cell bodies in the periarcuate region of the rat medial basal hypothalamus. Not only does one see cytoplasmic localization of the mRNA, but one may also see autoradiographic grains over neuronal processes, most likely dendritic. By performing POMC immunohistochemistry on these serial sections to those in Figure 6, it is shown that the same cells that record positive for POMC mRNA also have POMC-derived peptides in them (Gee et al. 1983). This observation, in light of the biosynthetic studies described above, strongly suggests that POMC is synthesized in neurons in the mammalian CNS.

The in situ hybridization histochemistry technique will have great value in the study of specific gene expression in the CNS. Indeed, this technique has already been used to identify and to study the biosynthesis of many different gene products in a variety of different tissues (Mohen and Palmiter 1980; Hudson et al. 1981). A great promise that this technique holds forth is that it may allow the researcher to analyze and study the regulation of expression of specific gene products in individual cells. Using quantitative autoradiography, it may be possible to actually quantitate the amount of mRNA in an individual cell in response to various hormonal, behavioral, and environmental manipulations. Because of the extreme complexity of the mammalian brain, the possibility of analyzing gene expression in individual cells is exciting. For example, in the many different cells of the medial basal hypothalamus that express the POMC gene, it may be that there are a variety of different functional classes of cells, each one responsive to a different set of hormonal, environmental, or behavioral stimuli. Additionally, if an intervening sequence probe derived from the gene of interest is available, then it may be possible to quantitate hnRNA levels within the nuclei of expressing cells using in situ hybridization histochemistry. This would allow for a distinction to be made between specific nuclear and cytoplasmic regulatory events. Thus, only through a technique that allows the study of mRNA levels in individual cells will it be possible to ask questions of how neurospecific genes are regulated in the CNS.

ACKNOWLEDGMENTS

We thank Vilma Lorenzo and Edith Kupsaw for typing the manuscript and Pamela Ross for the photography. This work was supported by National Institutes of Health grant AM-27484 to J.L.R.

REFERENCES

Brahic, M. and A.T. Haase. 1978. Detection of viral sequences of low reiteration frequency by in situ hybridization. *Proc. Natl. Acad. Sci.* **75:** 6125.

Chen, C.L.C., F.T. Dionne, and J.L. Roberts. 1982. Regulation of proopiomelanocortin mRNA levels in rat pituitary by dopaminergic compounds. *Proc. Natl. Acad. Sci.* **80:** 2211.

Civelli, O., N. Birnberg, and E. Herbert. 1982. Expression of POMC mRNA in neural tissues. *J. Biol. Chem.* **257:** 6783.

Eberwine, J. and J.L. Roberts. 1983. Analysis of proopiomelanocortin gene structure and function. *DNA* **2:** 1.

Eipper, B.A. and R.E. Mains. 1980. Structure and biosynthesis of pro-adrenocorticotropin/endorphin and related peptides. *Endocr. Rev.* **1:** 1.

GEE, C.E. and J.L. ROBERTS. 1983. In situ hybridization histochemistry: A technique for the study of gene expression in single cells. *DNA* **2:** 155.

GEE, C.E., C.-L.C. CHEN, J.L. ROBERTS, R. THOMPSON, and S.J. WATSON. 1983. Identification of POMC neurons in the rat hypothalamus by in situ hybridization. *Nature* (in press).

HERBERT, E., N. BIRNBERG, J.C. LISSITSKY, O. CIVELLI, and M. UHLER. 1981. Pro-opiomelanocortin: A model for the regulation of expression of neuropeptides in pituitary and brain tissue. *Neurosci. Lett.* **12:** 16.

HOLLT, V.I., B.R. HAARMANN, and A. HERTY. 1982. Chronic haloperidol treatment increases the level of in vitro translatable mRNA coding for POMC in the rat pars intermedia of the pituitary. *Endocrinology* **111:** 1885.

HUDSON, P., J. PENSCHOW, J. SHINE, G. RYAN, H. NIALL, and J. COGHLAN. 1981. Hybridization histochemistry: Use of recombinant DNA as a "homing probe" for tissue localization of specific mRNA populations. *Endocrinology* **108:** 353.

LIOTTA, A.S., D. GILDERSLEEVE, M.J. BROWNSTEIN, and E.T. KRIEGER. 1979. Biosynthesis in vitro of immunoreactive 31,000-dalton corticotropin/endorphin-like material by bovine hypothalamus. *Proc. Natl. Acad. Sci.* **76:** 1448.

MCKNIGHT, G.S. and R.D. PALMITER. 1979. Transcriptional regulation of the ovalbumin and conalbumin genes by steroid hormones in chick oviduct. *J. Biol. Chem.* **254:** 9050.

MOHEN, R.C. and R.D. PALMITER. 1980. Changes in hormone responsiveness of chick oviduct during primary stimulation with estrogen. *Dev. Biol.* **78:** 450.

NAKANISHI, S., T. KITA, S. TAII, H. IMURA, and S. NUMA. 1977. Glucocorticoid effect on the level of corticotropin messenger RNA activity in rat pituitary. *Proc. Natl. Acad. Sci.* **74:** 3283.

ROBERTS, J.L., M.L. BUDARF, J.D. BAXTER, and E. HERBERT. 1979. Selective reduction of pro-ACTH/endorphin proteins and mRNA by glucocorticoids in pituitary tumor cells. *Biochemistry* **18:** 3907.

ROBERTS, J.L., C.-L. CHEN, J.H. EBERWINE, M.J.Q. EVINGER, C. GEE, E. HERBERT, and B.S. SCHACHTER. 1982. Glucocorticoid regulation of pro-opiomelanocortin gene expression in rodent pituitary. *Recent Prog. Horm. Res.* **38:** 227.

SCHACHTER, B.S., L.K. JOHNSON, J.D. BAXTER, and J.L. ROBERTS. 1982. Differential regulation by glucocorticoids of pro-opiomelanocortin mRNA levels in the anterior and intermediate lobes of rat pituitary. *Endocrinology* **110:** 1442.

SPINDLER, S.R., S.H. MELLON, and J.D. BAXTER. 1982. Growth hormone gene transcription is regulated by thyroid and glucocorticoid hormones in cultured rat pituitary tumor cells. *J. Biol. Chem.* **257:** 11627.

VALE, W., C. RIVIER, L. YANG, S. MINICK, and R. GUILLEMIN. 1978. Effects of purified hypothalamic corticotropin-releasing factor and other substances on the secretion adrenocorticotropin and β-endorphin-like immunoactivities *in vitro*. *Endocrinology* **103:** 1910.

The Enkephalinergic Neuron: Implications of a Polyenkephalin Precursor

J. ROSSIER, D. LISTON, G. PATEY, M. CHAMINADE, A.S. FOUTZ, A. CUPO,* P. GIRAUD,† M.P. ROISIN,‡ J.P. HENRY,‡ P. VERBANCK,§ AND J.-J. VANDERHAEGHEN§

Laboratoire de Physiologie Nerveuse, Centre National de la Recherche Scientifique, 91190 Gif-sur-Yvette, France

Neurons and neurosecretory cells that release peptides as intercellular messengers employ biosynthetic strategies that differ substantially from those used by neurons that release the more classical neurotransmitters. The peptide hormones/transmitters are first synthesized as part of a larger precursor protein that is then cleaved enzymatically to produce the active peptides (for reviews, see Steiner et al. 1980; Chrétien et al. 1983). In recent years the primary structures of a number of peptide neurohormone precursors have been established. From these sequences, several general features of peptide neurohormone biosynthesis are emerging. The sequence of the active peptide is usually bounded by two pairs of basic amino acid residues (lysine, arginine), which appear to act as signals for cleavage by the processing enzymes. The precursors frequently contain multiple copies of a particular peptide sequence and/or multiple, distinct peptides with varying physiological functions. Tissue-specific processing of the precursor can determine the nature and actions of the final products. Thus, the cell displays a remarkable economy in the production of these potent agents.

An outstanding example of this biological economy is the biosynthesis of the opioid peptides, methionine (Met)-enkephalin (Tyr-Gly-Gly-Phe-Met) and leucine (Leu)-enkephalin (Tyr-Gly-Gly-Phe-Leu). Following the pioneering work of Schultzberg et al. (1978) demonstrating that the adrenal glands of several species contain enkephalinlike materials, several groups have studied the pathway of enkephalin biosynthesis in the bovine adrenal medulla (for review, see Rossier 1981). Recently, the primary sequence of the initial precursor, proenkephalin, has been deduced from the sequence of the cloned cDNA that codes for this protein (Comb et al. 1982; Gubler et al. 1982; Noda et al. 1982). Bovine adrenal glands or human pheochromocytoma were used as a souce of mRNA in these cloning studies. The structures of adrenal proenkephalin from these two species display a high degree of homology. The protein contains one copy of the sequence of Leu-enkephalin and six copies of the sequence of Met-enkephalin. Two copies of the Met-enkephalin sequence possess carboxyterminal extensions of two or three amino acids, giving rise to the octapeptide Met-enkephalin-Arg-6–Gly-7–Leu-8 and the heptapeptide Met-enkephalin-Arg-6–Phe-7 (Rossier et al. 1980a; Kilpatrick et al. 1981a; Ikeda et al. 1982). Each of these sequences is sandwiched between pairs of basic amino acid residues.

Numerous enkephalin-containing peptides with molecular weights of up to 18,200 have been purified from the adrenal medulla and characterized (Fig. 1) (Jones et al. 1980, 1982a,b; Mizuno et al. 1980; Stern et al. 1981; Kilpatrick et al. 1982). Pulse-labeling studies with adrenal medullary cells in culture have established a precursor-product relationship between high-molecular-weight enkephalin-containing peptides and the enkephalins (Rossier et al. 1980b). Many of these peptides isolated from the adrenal medulla contain multiple copies of the enkephalin sequence and appear to represent intermediates in the processing of proenkephalin into the active peptide products. Some of these peptides are themselves potent opiate agonists (Mizuno et al. 1980; Kilpatrick et al. 1981b).

The pathway of enkephalin biosynthesis in the brain has been more difficult to study, since enkephalin-containing proteins similar to those found in the adrenal gland are not present in high concentrations in brain. However, biochemical (Baird et al. 1982; Dandekar and Sabol 1982; Liston et al. 1983; Sabol et al. 1983) and immunohistochemical (Rossier et al. 1980a; Höllt et al. 1982; Watson and Akil 1982; Weber et al. 1982; Khachaturian et al. 1983) studies have suggested that the biosynthetic precursor is similar in these two tissues. By the use of specific antisera that recognize various regions of the precursor and daughter peptides, we have studied the distribution and release of proenkephalin-derived peptides in the adrenal gland and in the brain. The biochemical features of this polyenkephalin precursor present interesting physiological consequences for the functions of peptidergic neurons.

EXPERIMENTAL PROCEDURES

Radioimmunoassays. Radioimmunoassays (RIA) for Met-enkephalin, Leu-enkephalin, and the heptapeptide were performed as described by Giraud et al. (1983). RIA for synenkephalin, the aminoterminal fragment of proenkephalin, was performed as described by Liston et al. (1983). This assay also recognizes larger intermediates in the processing of proenkephalin that contain the aminoterminal fragment.

Present addresses: *Centre d'Immunologie INSERM/CNRS, Marseille, France; †Médecine expérimentale, Faculté de Médecine de Marseille Nord, France; ‡Institut de Biologie Physico-chimique, Paris 5, France; §Laboratoire de Neuropathologie, Université libre de Bruxelles, Belgique.

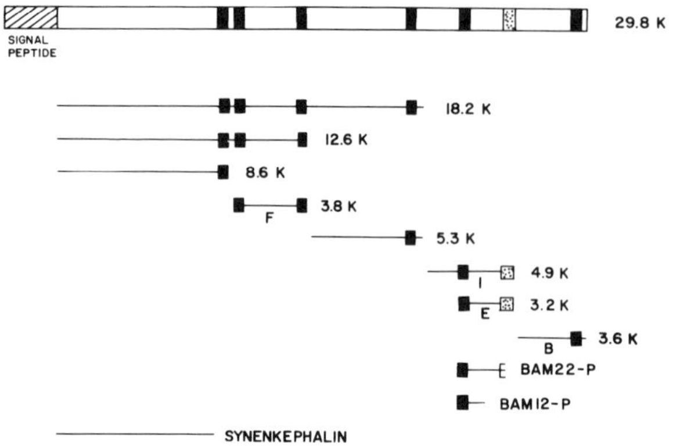

Figure 1. Schematic representation of the family of peptides derived from adrenal preproenkephalin. The primary sequence of adrenal proenkephalin has been deduced from the nucleotide sequence of the cloned cDNA for the prohormone. The putative signal peptide characteristic of secreted proteins occupies residues 1–24 of preproenkephalin. The daughter peptides illustrated have all been purified from bovine adrenal medulla and characterized by several groups of investigators. Sequences of active peptides are bounded by pairs of basic amino acids (lysine, arginine). (■) Sequence of Met-enkephalin; (⊡) sequence of Leu-enkephalin.

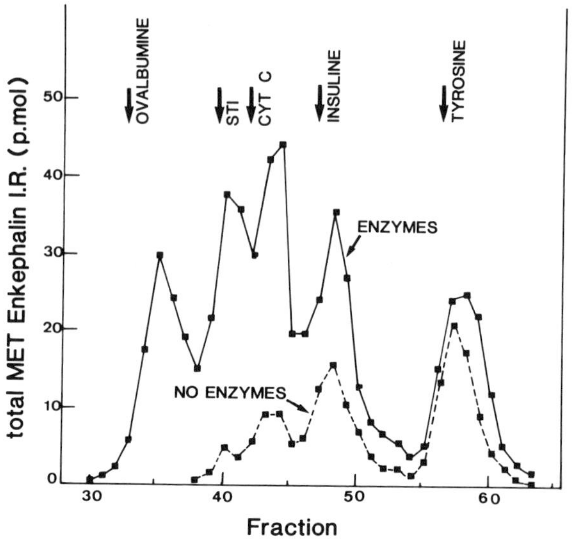

Figure 2. Gel filtration of chromaffin granule lysate. Granules isolated from bovine adrenal medullas were lysed in hypotonic solution and centrifuged for 1 hr at 105,000*g*. The supernatant was applied to a column (1.9 × 90 cm) of Ultrogel AcA 44 (LKB Instruments) equilibrated with 0.5 M NaCl, 0.05 M Tris-HCl, pH 7.4. Aliquots of each fraction were assayed for Met-enkephalin IR before and after enzymatic digestion with trypsin and carboxypeptidase B. The enzymatic treatment greatly increased the amount of Met-enkephalin IR in fractions containing proteins larger than M_r 2000. The Met-enkephalin IR recovered without enzymatic digestion represented only 12% of the total Met-enkephalin IR, showing that the major part of Met-enkephalin sequences appears in larger peptides in the adrenal medulla.

Determination of total Met-enkephalin immunoreactivity (IR). Sequences of Met-enkephalin that appear in the form of larger peptides are frequently not recognized by our Met-enkephalin RIA. To estimate the amount of this cryptic Met-enkephalin present in the various preparations, a sequential digestion with trypsin and carboxypeptidase B was performed. This treatment is described in detail by Liston et al. (1983).

Immunohistochemistry. For immunohistochemical studies on the localization of proenkephalin-derived peptides, bovine brains and pituitaries were obtained fresh from a local slaughter house. Tissues were cut into slices 1 cm thick within 1 hour after death and fixed in "Bouin Hollande sublimé" for 2 days (Ganter and Jolles 1970). Tissues were then embedded in paraffin, and sections 5 μm thick were prepared for immunohistochemical staining. Peptides were visualized by use of the unlabeled peroxidase-antiperoxidase technique of Sternberger et al. (1970) as modified by Vandesande et al. (1977).

Adsorbed sera against oxytocin and vasopressin were generous gifts from Dr. F. Vandesande. Briefly, antivasopressin serum was adsorbed by treatment with oxytocin coupled to Sepharose 4B beads. Antioxytocin serum was similarly treated with vasopressin-Sepharose 4B. Complete details on the preparation and specificity of these sera have been presented elsewhere (Vandesande et al. 1977).

RESULTS AND DISCUSSION

Studies in the Adrenal Gland

Analysis by gel filtration of enkephalin IR in bovine adrenal medulla. Figure 2 shows the distribution of Met-enkephalin IR in the lysate of bovine adrenal chromaffin granules analyzed by gel filtration. In the absence of any enzymatic treatment, the main peak of Met-enkephalin IR is seen in the total volume of the column. Nevertheless, 59% of the immunoreactivity is associated with molecules having an apparent M_r ranging from 5,000 to 17,000. The immunoreactivity associated with these large molecules is greatly enhanced by sequential enzymatic treatment with trypsin and carboxypeptidase B. This work confirms previous observations indicating that most of the enkephalin sequences found in the bovine adrenal were in the form of large precursors of various sizes (Lewis et al. 1980). Particular attention was paid to the void volume fractions and fractions containing proteins with apparent M_r greater than 20,000. These fractions were assayed for Leu-enkephalin IR before and after enzymatic digestion. Detectable Leu-enkephalin IR was not present in these fractions.

Crude extracts from bovine adrenal medullas were also prepared and analyzed by gel filtration chromatography. In these crude extracts, no Leu-enkephalin IR was present in fractions containing proteins larger than M_r 20,000. It thus appears that the unprocessed proenkephalin molecule does not accumulate to a detectable level in the adrenal medulla, contrary to an early report (Lewis et al. 1980). Indeed, the largest enkephalin-containing protein reported has an M_r of 18,200 and does not contain the sequence of Leu-enkephalin (Kilpatrick et al. 1982).

Subcellular localization of enkephalins in the adrenal medulla. Morphological studies of the adrenal medulla have revealed the existence of two types of chromaffin cells, containing either adrenaline or noradrenaline (Verhofstad et al. 1982). Recently, Livett et al. (1982) and Pelto-Huikko et al. (1982) have shown that enkephalin IR was present in the adrenergic cells but absent from the noradrenergic cells.

We have been able to confirm these immunohistochemical studies by subcellular fractionation and biochemical assay, showing that enkephalins are associated mainly with adrenergic chromaffin granules (Roisin et al. 1983). Adrenergic and noradrenergic granules of bovine adrenal medulla were purified by two successive self-generating Percoll-sucrose gradients as described by Terland et al. (1979). The large-chromaffin-granule fraction was mixed with Percoll-sucrose and centrifuged. Dense fractions were collected and centrifuged a second time with Percoll-sucrose. The first centrifugation resolved the large-granule fraction into two peaks of protein. The lighter protein peak corresponded to mitochondria. The more dense peak contained noradrenergic and adrenergic chromaffin granules. In the chromaffin peak, the most dense fractions (Fig. 3a, pool b) contained 25% of the total noradrenaline but very little adrenaline. Noradrenaline represented 81% of the total catecholamines in these fractions, compared with only 27% in the original large-granule fraction. The total Met-enkephalin IR closely parallelled the profile of adrenaline in this initial Percoll-sucrose gradient (Fig. 3).

Fractions enriched in noradrenaline (Fig. 3a, pool b) were subjected to centrifugation on a second Percoll-sucrose gradient (Fig. 3b), which separated the dense, noradrenaline-rich granules from the lighter, adrenaline-rich granules. In fraction 6, noradrenaline represented 90% of the total catecholamines. Thus, the two self-generating gradients gave a 15-fold enrichment of noradrenaline. Met-enkephalin IR was absent from the dense, noradrenergic granules but was found associated with the adrenergic granules present in the lighter fractions (maximum in fraction 11).

The association of total Met-enkephalin IR with adrenaline was clearly shown in the density gradient obtained with fractions 6–8 (Fig. 3a, pool c) of the first gradient (Fig. 3c). The distributions of total Met-

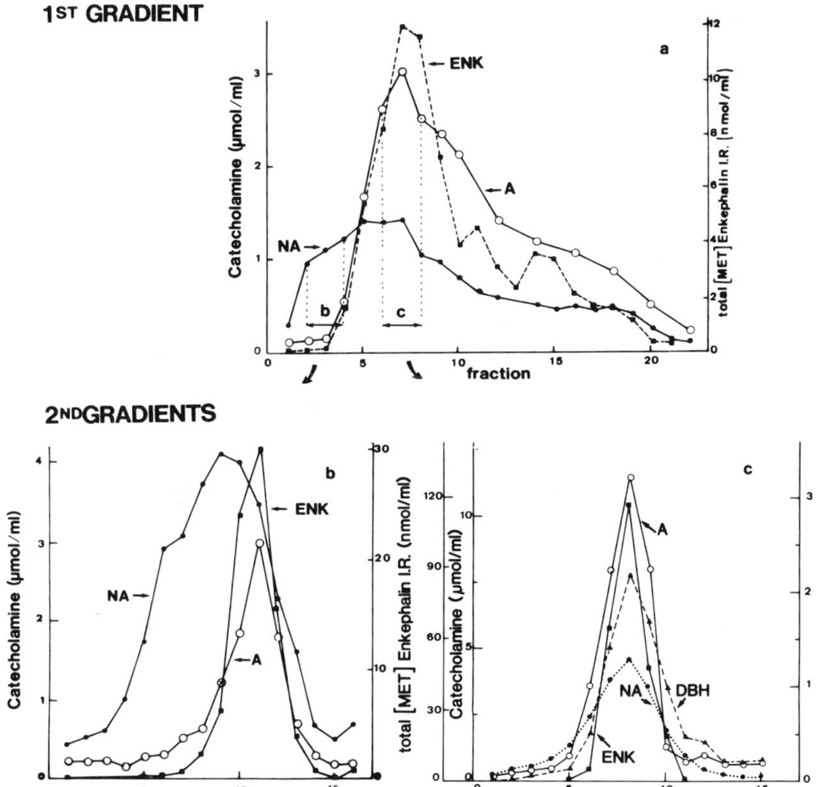

Figure 3. Enkephalin-IR materials in the bovine adrenal medulla are associated with adrenergic chromaffin granules. The dense, noradrenergic granules and the lighter, adrenergic granules were separated by two successive centrifugations on self-generating Percoll gradients (Terland et al. 1979). In brief, a 0.5-ml aliquot (containing 10–20 mg protein) of the large-granule fractions, prepared according to Smith and Winkler (1967), was mixed with 7.5 ml of a mixture of Percoll (Pharmacia; Uppsala, Sweden) and 0.5 M sucrose in 20 mM HEPES buffer, pH 7.0 (1:1, v/v). The tubes were centrifuged at 58,000*g* for 30 min, and fractions were collected from the bottom of the tubes (*a*). Homologous noradrenergic fractions (2–4) (total vol. 7.5 ml) of eight gradients similar to that described above were mixed with 2.8 ml of the same Percoll-sucrose solution and centrifuged again (*b*). A similar protocol was applied to the adrenergic fractions (6–8) (*c*).

enkephalin IR, adrenaline, and dopamine-β-hydroxylase (DBH) were parallel, with a maximum in fraction 8. For each mole of Met-enkephalin present in these fractions, 97 moles of adrenaline and 136 moles of total catecholamines were also present.

Corelease of catecholamines and enkephalins by the adrenal gland. The colocalization of catecholamines and enkephalins in adrenergic granules suggested that these molecules might be coreleased in vivo. Several studies have indicated that opiatelike products are coreleased with catecholamines from the adrenal medulla (Kilpatrick et al. 1980; Stine et al. 1980; Livett et al. 1981). These studies have employed adrenal cell cultures or the perfused bovine adrenal gland in order to demonstrate the release of enkephalins. These in vitro models are restricted to the use of chemical stimuli to induce release. Other investigators have used cannulation of the dog adrenal in vivo (Hexum et al. 1980; Govoni et al. 1981). In this model, released materials were assayed after extraction from the blood. The half-life of the enkephalins in the blood is known to be very short (1–2 min) (Hambrook et al. 1976); thus, absolute amounts of enkephalins released are difficult to ascertain.

To circumvent these limitations, we chose to study enkephalin release from the perfused cat adrenal in situ. By use of gel exclusion chromatography and reverse-phase high-performance liquid chromatography (HPLC), we have characterized the nature of the enkephalinlike material released into the perfusate following splanchnic nerve stimulation. This system offers the advantages of allowing nerve stimulation while using a perfusing medium that does not contain blood elements that may interfere with our assays. The in situ model thus allows identification and quantification of the released materials.

As has been reported by others (Livett et al. 1981), we have found that stimulation of the adrenal gland leads to concurrent release of catecholamines and enkephalin IR. We have examined the nature of the enkephalinlike materials released following splanchnic nerve stimulation. Perfusate collected during a 10-minute stimulation was fractionated on a Sephadex G-100 column (1.6 × 90 cm). Fractions containing low-molecular-weight materials (<2000) were analyzed by reverse-phase HPLC (Fig. 4). More than 90% of the low-molecular-weight Met-enkephalin IR material eluted with the same retention time as authentic Met-enkephalin. Digestion of the HPLC column fractions with trypsin and carboxypeptidase B revealed two additional peaks of Met-enkephalin IR. These two peaks eluted in the same positions as the octapeptide and the heptapeptide. The relative proportions of Met-enkephalin IR found in the three peaks after enzymatic digestion were Met-enkephalin, 70%; octapeptide, 16%; and heptapeptide, 14%. Thus, these peptides are released from the adrenal gland following nerve stimulation in the same ratio (4:1:1) as is present in the proenkephalin molecule.

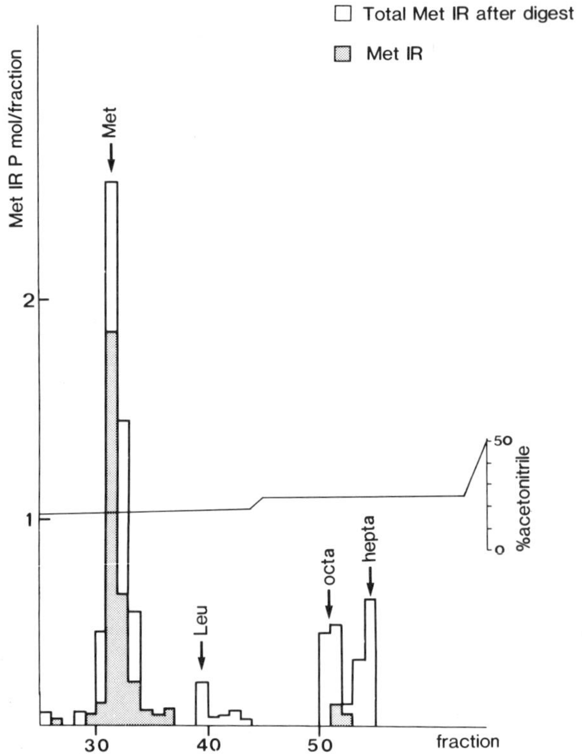

Figure 4. Reverse-phase HPLC analysis of the low-molecular-weight enkephalinlike materials released by splanchnic nerve stimulation of the perfused cat adrenal in situ. The gland was perfused with bicarbonate-buffered Krebs-Ringer solution at a constant pressure of 70 mm Hg. Perfusate was collected during a 10-min electrical stimulation (15 Hz, 15 V, 1 msec square pulse) of the splanchnic nerve and fractionated on a column of Sephadex G-100. Fractions containing low-molecular-weight materials were injected onto an Altex RP-8 column and eluted with a discontinuous gradient of acetonitrile in 0.25 M triethylamine formate, pH 3.1. Fractions were evaporated under reduced pressure and resuspended in 50 mM Tris, 2 mM $CaCl_2$, pH 8.4. Aliquots were assayed for Met-enkephalin-IR before (▨) and after (□) enzymatic treatment. Without enzymatic digestion, more than 90% of the Met-enkephalin IR eluted in the same position as authentic Met-enkephalin. Following digestion, additional peaks of Met-enkephalin IR appeared in the elution positions corresponding to the octapeptide and the heptapeptide. The molar ratio of Met-enkephalin:octapeptide:heptapeptide after enzymatic digestion was 4:1:1, the same as that found in the sequence of proenkephalin. Some immunoreactivity is present in the elution position of Leu-enkephalin due to cross-reactivity (4%) with the anti-Met-enkephalin RIA.

In experiments utilizing more intense stimuli, i.e., perfusion with solutions containing 50 mM K^+ or 0.1 mM acetylcholine, we found a much greater release of total Met-enkephalin IR than could be accounted for by the stoichiometric release of the octapeptide and the heptapeptide. Perfusates obtained during these more intense stimuli were analyzed by gel exclusion chromatography (Fig. 5). Following perfusion of the gland with 50 mM K^+, 88% of the total Met-enkephalin IR occurred in peptides with molecular weights greater than 5000. Electrical stimulation of the splanchnic nerve resulted in the release of only 53% of the total

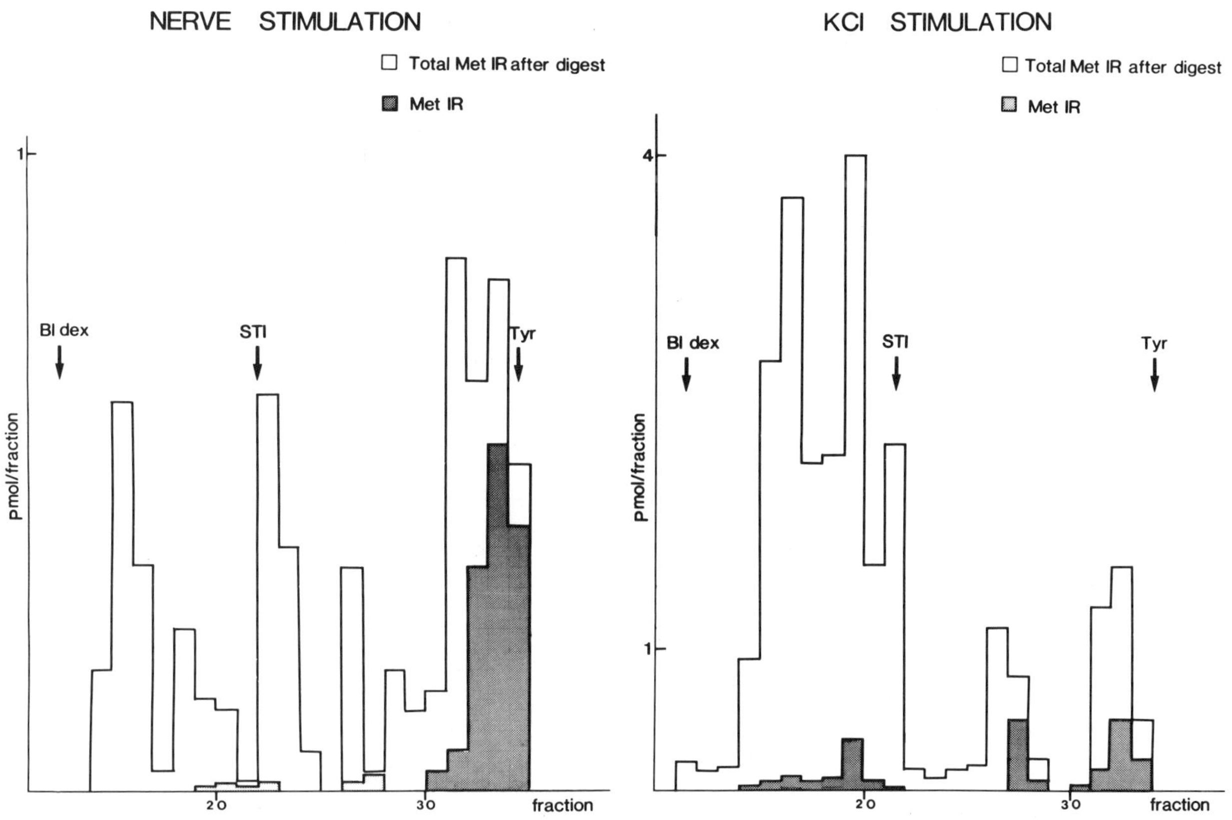

Figure 5. Gel-filtration chromatography of perfusates of cat adrenal gland following splanchnic nerve stimulation or 50 mM KCl stimulation. Perfusates were collected and fractionated on a column of Sephadex G-100 as described in the legend to Fig. 4. Perfusion with medium containing 50 mM K^+ (*right*) resulted in a much greater release of high-molecular-weight enkephalin-containing peptides than did electrical stimulation of the nerve (*left*).

Met-enkephalin IR in the form of larger peptides. The two stimuli thus resulted in markedly different profiles of the nature of Met-enkephalin-containing materials released.

We have found that, in the cat adrenal medulla, large enkephalin-containing peptides represent 80–90% of the total Met-enkephalin present in the gland, in agreement with the findings of others (Viveros et al. 1979). It was expected that these large enkephalin-containing peptides would be released simultaneously with catecholamines in response to physiological stimuli. However, the present results indicate that the nature of the released peptides depends on the nature of the stimuli. In response to an intense stimuli (K^+), large quantities of high-molecular-weight peptides are indeed released. When a more gentle stimuli is used (nerve stimulation), a much greater proportion of the released material resides in small peptides.

This difference in the nature of the liberated enkephalin IR may indicate that under physiological conditions most of the material released represents final products obtained by complete processing of proenkephalin. When drastic stimuli are used, partially processed peptides are also released. This suggests a heterogeneity in the population of chromaffin granules, similar to that described for neurosecretory granules in the posterior pituitary (Nordmann and Labouesse 1981). Granules may exist in a fully matured form, wherein processing of proenkephalin is complete with Met-enkephalin, the octapeptide, and the heptapeptide being the major products. Immature granules would contain larger quantities of the unprocessed intermediate forms of the proenkephalin molecule, many of which have been purified and sequenced (Fig. 1). Physiological stimulation of the splanchnic nerve would preferentially cause the release of the mature granules, which contain Met-enkephalin, the octapeptide, and the heptapeptide in a molar ratio of 4:1:1. When intense stimuli are used, the pool of mature granules is rapidly depleted and immature granules containing larger enkephalin-containing peptides are also released. Although such a scheme is somewhat speculative, it is consistent with the present results.

Studies in Brain

Nature of enkephalinlike materials released from brain slices in vitro. Unlike the adrenal gland, most brain regions contain relatively small amounts of cryptic Met-enkephalin (see Table 1). Moreover, large enkephalin-containing peptides have not yet been isolated from brain tissue. This suggests that the large precursor intermediates found in the adrenal medulla do

Table 1. Regional Distribution of Proenkephalin IR and Met-enkephalin IR in the Bovine Brain

Region	Proenkephalin IR (fmoles/mg tissue)	Met-enkephalin IR (fmoles/mg tissue)		Met/pro ratio	
		before digestion	after digestion	before	after
Globus pallidus (4)	1098 ± 201	4128 ± 695	4774 ± 995	3.8	4.3
Caudate (9)	375 ± 17	1844 ± 214	2650 ± 351	4.9	7.1
Putamen (9)	256 ± 34	913 ± 91	1335 ± 116	3.6	5.2
Hypothalamus (8)	124 ± 14	576 ± 57	1547 ± 177	4.6	12.5
Thalamus (9)	42 ± 5	176 ± 33	—	4.2	—
Hippocampus (9)	10 ± 1	26 ± 4	—	2.6	—
Cerebral cortex (9)	11 ± 2	65 ± 11	—	5.9	—
Cerebellum (9)	4 ± 1	14 ± 4	—	3.5	—
Brain stem (9)	53 ± 3	143 ± 9	—	2.7	—
Posterior pituitary (7)	1806 ± 417	6405 ± 223	6979 ± 1649	3.5	3.9

Proenkephalin-IR and Met-enkephalin IR before digestion were determined directly on aliquots of buffered acetic acid extracts of tissues. Separate aliquots of the acetic acid extracts were lyophilized and subjected to the sequential enzymatic digestion described in Experimental Procedures. Digestion controls consisted of samples carried through the digestion procedure without the addition of trypsin and carboxypeptidase B. Controls were within 20% of the values found in the untreated extracts. Values are mean ± S.E.M. of the number of animals given in parentheses.

not accumulate in brain and that proenkephalin is rapidly processed into the final products. We have investigated this hypothesis through analysis of the materials released from rat striatal slices in vitro following stimulation with K^+.

One of the major problems encountered when working with brain slices is the presence of numerous peptidases capable of rapid degradation of the released peptides (for review, see Schwartz 1983). Now at least three enzymes have been implicated in the inactivation of the enkephalins and/or of the heptapeptide: (1) "enkephalinase," a dipeptidyl carboxypeptidase that cleaves the Gly-Phe bond of the enkephalins and of the heptapeptide and is specifically blocked by 0.1 μM thiorphan; (2) angiotensin-converting enzyme, also a dipeptidyl-carboxypeptidase, which successively releases the dipeptides Arg-Phe and Phe-Met from the heptapeptide and is inhibited by 1 μM captopril; (3) aminopeptidase(s), which cleaves the Tyr-Gly bond of the enkephalins and probably also of the heptapeptide and is inhibited by 20 μM bestatin.

Slices perfused with a physiological medium containing this cocktail of inhibitors were subjected to depolarization by 50 mM K^+ (Fig. 6). Under these conditions, the ratio of Met-enkephalin IR to Leu-enkephalin IR released into the perfusate was found to be close to 4:1, as would be expected if processing of the precursor forms was complete. The ratio of Met-enkephalin IR to heptapeptide IR in the perfusate was between 8:1 and 9:1. This ratio is not the same as that found in crude tissue extracts (Giraud et al. 1983) or in the slices themselves following K^+-induced depolarization (Patey et al. 1983), where the ratio of Met-enkephalin IR to heptapeptide-IR was close to 4:1. It is possible that the cocktail of inhibitors used to prevent proteolytic degradation of the released peptides was not sufficient to protect the heptapeptide. Indeed, perfusion of rat striatal slices with a solution containing the tritiated heptapeptide (5 nM) indicated that as much as 25% of the peptide was degraded in the presence of the inhibitors.

Work is currently in progress to investigate whether the octapeptide is also released with Met-enkephalin and the heptapeptide in a stoichiometric ratio similar to that found in proenkephalin.

The relative abundance in brain of Met-enkephalin, Leu-enkephalin, the heptapeptide, and the octapeptide suggests that the brain utilizes a precursor similar to adrenal proenkephalin. Several recent immunohistochemical studies have demonstrated the presence in brain of substances that resemble the intermediates in the processing of adrenal proenkephalin that have been isolated from the adrenal medulla (Rossier et al. 1980a; Baird et al. 1982; Höllt et al. 1982; Watson et al. 1982a; Weber et al. 1982; Khachaturian et al. 1983).

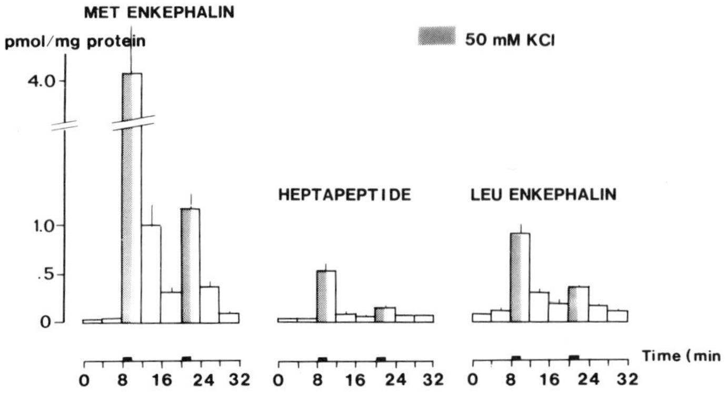

Figure 6. Release of enkephalinlike peptides from rat striatal slices in vitro. Slices (350 μm) were washed and placed in chambers (250 μl) of a superfusion apparatus. Bicarbonate-buffered Krebs-Ringer medium containing bovine serum albumin (0.01%), thiorphan (0.1 μM), captopril (1 μM), and bestatin (20 μM) was pumped at a rate of 250 μl/min. Following a 30-min stabilization period, fractions (1 ml) were collected and incubated at 100°C for 10 min and centrifuged to remove insoluble carbonates before assay for Met-enkephalin, Leu-enkephalin, and the heptapeptide. Results are expressed in terms of the protein content of the slices and represent the mean + S.E. of five independent determinations.

This hypothesis is further supported by our recent work concerning synenkephalin.

Purification from brain of synenkephalin, the aminoterminal fragment of proenkephalin. Whereas the adrenal medulla has been shown to be rich in large enkephalin-containing peptides, similar compounds have not yet been isolated from brain. We have recently reported the development of an RIA that recognizes the aminoterminal region of adrenal proenkephalin (Liston et al. 1983). This RIA cross-reacts with an antigen in brain that does not give rise to Met- or Leu-enkephalin following digestion with trypsin and carboxypeptidase B. We have used the RIA to follow the purification of this antigen, which we have called synenkephalin.

Synenkephalin IR was purified to homogeneity approximately 13,000-fold from an acetic acid extract of bovine caudate nucleus as outlined in Table 2. The recovery of immunoreactivity from each of the chromatographic procedures employed was 61–98%, with an overall yield of 20%. Amino acid analysis of the pure protein indicates that synenkephalin corresponds to residues 1–70 of adrenal proenkephalin and is thus the aminoterminal fragment of the prohormone. Further cleavage of synenkephalin appears to be limited, as seen by its accumulation in brain tissue in a molar ratio with Met-enkephalin close to that found in the sequence of proenkephalin (Table 1).

Distribution of synenkephalin IR in brain. The synenkephalin RIA was used to compare the qualitative and quantitative distribution in brain of this molecule with the distribution of Met-enkephalin. A primary goal of this study was to determine the molar ratio of Met-enkephalin IR to synenkephalin IR in the brain.

The general distribution of synenkephalin IR was found to be similar to the distribution of Met-enkephalin (Table 1). High concentrations of synenkephalin IR were present in the basal ganglia, especially the globus pallidus. The highest levels of both synenkephalin IR and Met-enkephalin IR were found in the posterior pituitary.

The molar ratio of Met-enkephalin IR to synenkephalin IR ranged between 2.6 and 5.9 in the regions tested. The five regions containing the highest concentrations of synenkephalin IR and Met-enkephalin IR (posterior pituitary, globus pallidus, caudate and putamen nucleus, and hypothalamus) were further examined for total Met-enkephalin IR after treatment with trypsin and carboxypeptidase B. In the globus pallidus and posterior pituitary, the increase in Met-enkephalin IR after digestion was small, between 9% and 16%. In the caudate and putamen, digestion led to a 45% increase in Met-enkephalin IR. The most striking effect of digestion was observed in the hypothalamus, with an increase in Met-enkephalin IR of 169% over the undigested samples. These results indicate that in some tissues, especially the hypothalamus, there is a considerable amount of Met-enkephalin residing in larger peptides.

Immunocytochemical studies: Colocalization of synenkephalin IR and Met-enkephalin IR in oxytocinergic magnocellular neurons. Initial immunocytochemical studies indicated that synenkephalin IR and Met-enkephalin IR had the same general distribution throughout the bovine brain (Liston et al. 1983). Recently, we have examined more closely the distribution of these peptides in the hypothalamus (Fig. 7) (Vanderhaeghen et al. 1983). In the bovine hypothalamo-neurohypophyseal system, synenkephalin IR and Met-enkephalin IR were found in magnocellular neuronal cell bodies in the dorsal part of the supraoptic nuclei and in the peripheral part of the paraventricular nuclei. A densely staining network of nerve terminals was found in the external part of the median eminence and in the posterior hypophysis. This general distribution is identical with that of the neurohypophyseal hormone, oxytocin. The precise localization of synenkephalin IR and Met-enkephalin IR was then compared with the distribution of oxytocin and vasopressin in serial 5-μm sections through the magnocellular nuclei (Fig. 8). Oxytocin IR was nearly always present in cells that stained with synenkephalin and Met-enkephalin antisera. The vasopressin-immunoreactive cells were never stained with either the synenkephalin or the Met-enkephalin antisera. These results extend the observations of Martin and Voigt (1980) and Martin et al. (1983), who have shown that in the rat neurohypophysis, Met-enkephalin IR coincides with oxytocin IR in neuronal terminals.

Table 2. Summary of the Purification of Synenkephalin from Bovine Caudate Nucleus

Step	Total nmoles[a]	Specific activity (nmoles/mg protein)	Yield of step (%)	Total yield (%)
Acetic acid extraction	27.3	0.009	100	100
Reextraction in 30% 1-propanol	30.5	0.033	111	111
G-100 chromatography	20.0	0.046	66	73
Chromatofocusing	13.5	4.5	68	49
Reverse-phase HPLC	8.2	—[b]	61	30
Gel-permeation HPLC	5.6	—[b]	68	21
Reverse-phase HPLC	5.5	120[c]	98	20

[a]Total nmoles of synenkephalin-IR recovered at the end of the step.
[b]Protein assay was not performed due to the small amount of protein present in the later steps of the purification.
[c]Specific activity was calculated assuming M_r 7800, based on amino acid analysis.

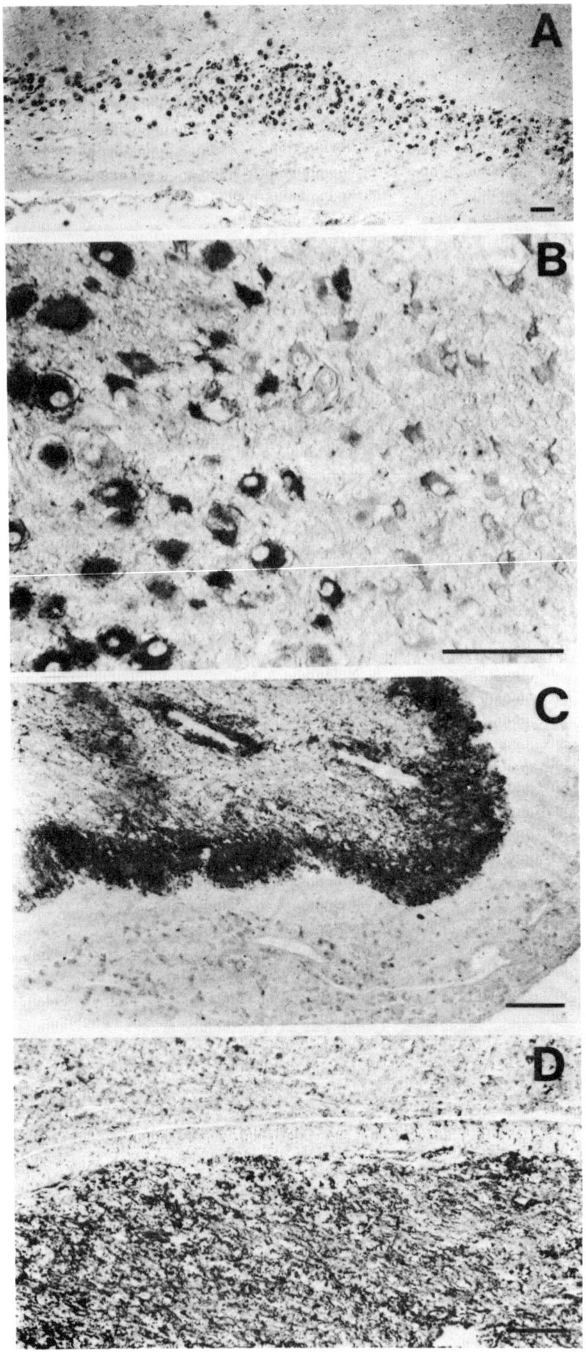

Figure 7. Immunohistochemical localization of synenkephalin IR in neuronal elements of the bovine hypothalamo-neurohypophyseal system. As shown in *A*, synenkephalin IR was restricted to neuronal cell bodies in the dorsal part of the supraoptic nucleus, similar to oxytocin. Most of the ventrally located unstained cells stained positively for vasopressin. (*B*) Higher magnification of *A* (Nomarski optics). Synenkephalin-IR nerve terminals were present in the external median eminence (*C*) and the neurohypophysis (*D*). The adenohypophysis (top of *D*) and intermediate lobe do not contain immunoreactive materials. Calibration bars for *A–D* are 100 μm.

It has been reported that another opioid peptide, dynorphin, is present in vasopressin-containing magnocellular neurons (Watson and Akil 1982; Watson et al 1982a,b; Weber et al. 1982). The dynorphins and neoendorphins arise from a precursor protein, prodynorphin (proenkephalin B), which does not contain the sequence of Met-enkephalin (Kakidani et al. 1982). Thus, it appears that the proenkephalin and prodynorphin systems are present in different subpopulations of magnocellular neurons, the enkephalin family being present in most oxytocin IR cells and the dynorphin family in some vasopressin IR cells.

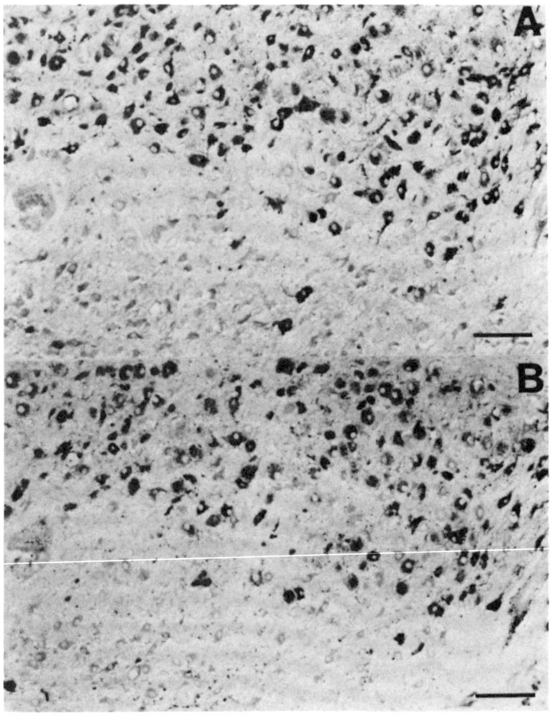

Figure 8. Serial, 5-μm frontal sections through the bovine supraoptic nucleus showing colocalization of synenkephalin IR and oxytocin IR. Most cells staining positively with the antisynenkephalin serum (*B*) were also stained with the antioxytocin serum (*A*). Most of ventrally located unstained cells stained positively for vasopressin. Bar, 100 μm.

The physiological significance of this anatomical separation of opiate peptides may be related to the differential potency that they exhibit at the various subclasses of opiate receptors. Peptides derived from proenkephalin are highly potent ligands for the μ- and δ-receptors (Corbett et al. 1982), and dynorphin has been shown to be much more potent at the κ-receptor (Chavkin et al. 1982). These peptides are likely candidates as modulators of oxytocin and vasopressin release at the neurohypophyseal level (Clarke et al. 1979; Rossier et al. 1979). A distinction between the receptors activated by concurrent release of oxytocin/enkephalin or vasopressin/dynorphin would minimize "cross-talk" between the two systems, allowing independent regulation of oxytocin and vasopressin release. This hypothesis is supported by the recent work of Leander (1983), who has proposed that dynorphin inhibits the release of vasopressin via autoreceptors.

Processing of proenkephalin during axonal transport. The immunocytochemical colocalization of proenkephalin-derived peptides in magnocellular oxytocinergic neurons allowed us to exploit the unique anatomical features of this neurosecretory system. Oxytocinergic cell bodies are present in the supraoptic and paraventricular nuclei of the hypothalamus, sending numerous axons coursing through the pituitary stalk to terminate in the posterior pituitary. Gainer et al. (1977) have shown that synthesis of neurohypophyseal hormones begins in the cell bodies with the larger prohormones, which are packaged into granules and transported down the axon to the terminals. During this axonal transport, the prohormone is processed into the active peptide products that are ultimately released during nerve activity.

The intraneuronal processing of proenkephalin in the hypothalamo-neurohypophyseal system appears to follow this general model. By dissecting the system into three components, consisting of the cell bodies (supraoptic nucleus), axons (pituitary stalk), and terminals (posterior pituitary), the processing of proenkephalin during axonal transport becomes apparent (Table 3 and Fig. 9). In the cell bodies, cryptic Met-enkephalin represents 75% of the total Met-enkephalin content present in acetic acid extracts. Analysis of these extracts by gel filtration chromatography demonstrates that 60% of the total Met-enkephalin IR resides in peptides larger than M_r 5000. The amount of cryptic Met-enkephalin present in the axons of the pituitary stalk is reduced to 56% of total Met-enkephalin. In the neuronal terminals of the posterior pituitary, only 11% of the total Met-enkephalin IR is contained in larger peptides, nearly all of which exhibits molecular weights below 5000 on gel exclusion chromatography. These results indicate that in the hypothalamo-neurohypophyseal enkephalinergic neuron, processing of the prohormone occurs during axonal transport from the site of synthesis in the cell body to the site of release of active peptides in the terminals.

The enzymes responsible for processing of proenkephalin have not yet been characterized. Both trypsin-like (Evangelista et al. 1982; Lindberg et al. 1982) and carboxypeptidase-B-like (Fricker and Snyder 1982; Hook et al. 1982; Wallace et al. 1982) enzymes have been found in adrenal chromaffin granules. Several endopeptidase activities that may play a role in proenkephalin processing have been described in the posterior pituitary (Wilk and Orlowski 1980; Loh and Gainer 1982; Mizuno et al. 1982). We have recently shown that enzymes of the kallikrein family of proteases are also candidates for this action (Prado et al. 1983).

SUMMARY AND CONCLUSIONS

The study of the biochemical and physiological functions of the enkephalinergic cell has greatly extended our understanding of peptidergic cells in general. In the adrenal gland, the major part of the proenkephalin-derived peptides is present as intermediates in the processing of the precursor. These peptides are contained within the adrenergic chromaffin granules, from which they are released in response to stimulation of the cell. The nature of the products released depends on the nature of the stimulus, but it appears that mature granules containing completely processed peptides are preferentially released under physiological conditions.

In the brain, the presence and release of the heptapeptide that comprises the carboxyl terminus of adrenal proenkephalin suggest that similar mechanisms are operating centrally. The identity of brain and adrenal proenkephalin is further supported by the purification from brain of a large fragment of the proenkephalin molecule, synenkephalin, and the occurrence in brain of this and the other proenkephalin-derived peptides in a molar ratio close to that found in the sequence of the adrenal precursor. The processing of proenkephalin in brain appears to follow the classical models first proposed for peptide hormones (Steiner et al. 1980), which may thus be generalized to include peptide neurotransmitters/neuroregulators.

In addition, the results presented in this paper indicate that enkephalins may be cotransmitters in at least two diverse systems. Enkephalins and catecholamines are colocalized in the adrenergic granules of the adrenal gland. In the brain, enkephalins and oxytocin are colocalized in the magnocellular neurons of the hypothalamo-neurohypophyseal oxytocinergic pathway. In both of these systems, the enkephalins are present in a molar concentration that is less than 1% of the concentration of the principal messenger. Such colocalization, cou-

Table 3. Met-enkephalin Concentrations in the Bovine Hypothalamo-neurohypophyseal System

Region	Met-enkephalin (pmoles/mg tissue) before digest	Met-enkephalin (pmoles/mg tissue) after digest	Increase (%)
Supraoptic nucleus	0.565 ± 0.103	2.25 ± 0.44	297
Stalk	4.56 ± 1.01	10.26 ± 2.09	125
Posterior pituitary	28.78 ± 7.77	32.32 ± 10.12	13

Tissues were homogenized in 4 volumes of 1 M acetic acid that had been adjusted to pH 1.9 with HCl. Following centrifugation, the supernatants were lyophilized and assayed for Met-enkephalin before and after digestion with trypsin and carboxypeptidase B. Values are the mean ± S.E.M. of tissues from 6–7 animals.

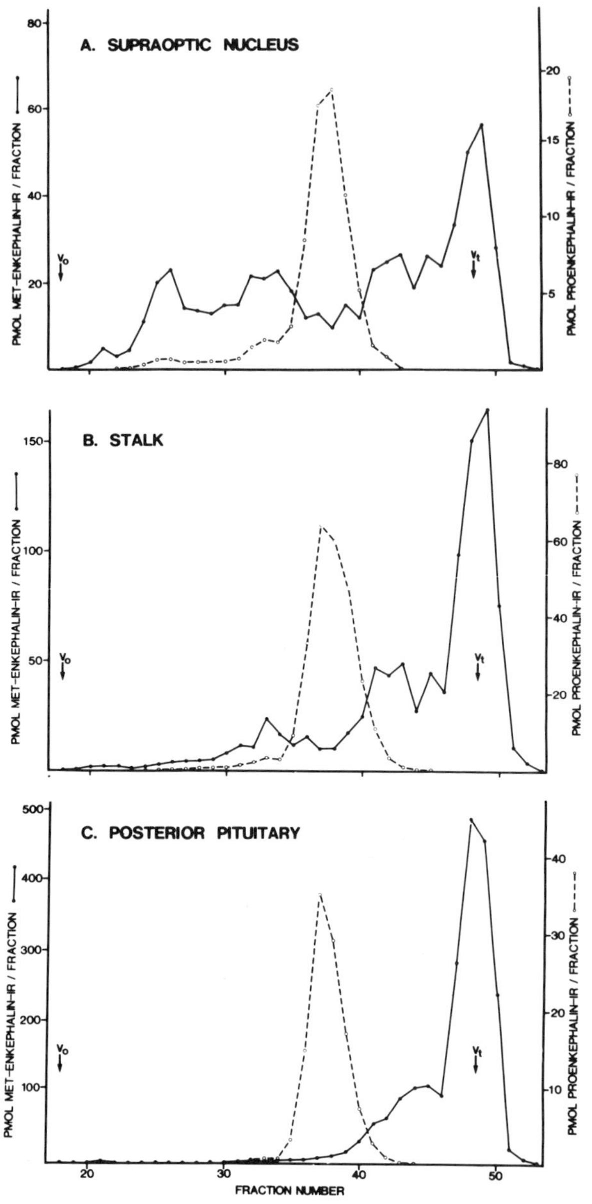

Figure 9. Molecular forms of proenkephalin IR in the bovine hypothalamo-neurohypophyseal system. The supraoptic nucleus, pituitary stalk, and posterior pituitary were dissected from 6–8 cows and homogenized in 4 volumes of 1 M acetic acid adjusted to pH 1.9 with HCl. Soluble proteins were fractioned on a column (1.6 × 90 cm) of Sephadex G-100 equilibrated with 1 M acetic acid, pH 1.9. Aliquots of each fraction were lyophilized and assayed for total Met enkephalin IR after enzymatic digestion (——) and for proenkephalin IR (-----). The amount of high-molecular-weight enkephalin-containing peptides associated with proenkephalin IR decreased as the distance from the cell body increased.

pled with the numerous active peptides that may arise from proenkephalin, suggests many elegant but complex schemes of neurotransmitter interactions. For example, release of enkephalins in the neurohypophysis may regulate oxytocin release through an action on autoreceptors of the oxytocinergic terminals. In the adrenal the coreleased enkephalins may act by regulating presynaptically the cholinergic output of the splanchnic nerve. However, further studies are needed to define clearly the physiological roles of such cotransmission.

From the abundance of proenkephalin-derived peptides in the basal ganglia, it appears that enkephalins may represent the principal transmitter in some central neurons. Assuming that enkephalinergic cells comprise 1% of the volume of the bovine caudate nucleus, it can be calculated that proenkephalin may account for as much as 1% of the total protein of these cells. This is in sharp contrast to the much lower concentration of proteins that play a catalytic role in the biosynthesis of the more classical neurotransmitters, such as acetylcholine (Rossier 1977). The organization of proenkephalin into a polyenkephalin precursor may thus be interpreted as a very economical way to reduce the demands on protein synthesis in the active enkephalinergic neuron.

In addition to the seven copies of the enkephalin sequence, other messages may be present in the prohormone, as well. The abundance of synenkephalin in the caudate nucleus, where it is present in a concentration approximately one-fifth of that of Met-enkephalin, implies an important function for this protein. This function may be limited to service as a template for the processing of the prohormone, positioning the enkephalin-containing sequences of the molecule in an orientation that directs and facilitates cleavage at the pairs of basic amino acid residues that encompass the enkephalin sequences. However, other regulatory functions such as modulation of the actions of the daughter peptides cannot be ruled out.

ACKNOWLEDGMENTS

The authors express their sincere thanks for the outstanding professional photographic assistance of Mr. Jean Paul Bouillot and for the reliable clerical help of Mrs. Hélène Hryn. This work was supported by the Esther A. and Joseph Klingenstein Fund (New York), CNRS-NSF exchange program, INSERM (CRL-81-6025), Fondation pour la Recherche Médicale, the Belgian Medical Scientific Research Fund (3.45.21.82), and the Queen Elisabeth Medical Center Foundation.

REFERENCES

Baird, A., N. Ling, P. Bohlen, R. Benoit, R. Klepper, and R. Guillemin. 1982. Molecular forms of the putative enkephalin precursor BAM-12P in bovine adrenal, pituitary, and hypothalamus. *Proc. Natl. Acad. Sci.* **79:** 2023.

Chavkin, C., I.F. James, and A. Goldstein. 1982. Dynorphin is a specific endogenous ligand of the κ opioid receptor. *Science* **215:** 413.

Chrétien, M., G. Boileau, C. Lazure, and N.G. Seidah. 1983. Processing of peptide hormone and neuropeptide precursors. In *The secretory process* (ed. A.M. Poisner and J.M. Trifaro), vol. 2. Elsevier Biomedical Press, Amsterdam. (In press)

Clarke, G., P. Wood, L. Merrick, and D.W. Lincoln. 1979. Opiate inhibition of peptide release from the neurohumoral terminals of hypothalamic neurons. *Nature* **282:** 746.

Comb, M., P.H. Seeburg, J. Adelman, L. Eiden, and E. Herbert. 1982. Primary structure of the human Met- and Leu-enkephalin precursor and its mRNA. *Nature* **295:** 633.

Corbett, A.D., S.J. Paterson, A.T. McKnight, J. Magnan, and H.W. Kosterlitz. 1982. Dynorphin 1-8 and dynorphin 1-9 are ligands for the κ-subtype of receptor. *Nature* **299:** 79.

Dandekar, S. and S.L. Sabol. 1982. Cell-free translation and partial characterization of proenkephalin messenger RNA from bovine striatum. *Biochem. Biophys. Res. Commun.* **105:** 67.

Evangelista, R., P. Ray, and R.V. Lewis. 1982. A "trypsin-like" enzyme in adrenal chromaffin granules: A proenkephalin processing enzyme. *Biochem. Biophys. Res. Commun.* **106:** 895.

Fricker, L.D. and S.H. Snyder. 1982. Enkephalin convertase: Purification and characterization of a specific enkephalin-synthesizing carboxypeptidase localized to adrenal chromaffin granules. *Proc. Natl. Acad. Sci.* **79:** 3886.

Gainer, H., Y. Sarne, and J. Brownstein. 1977. Biosynthesis and axonal transport of rat neurohypophysial proteins and peptides. *J. Cell Biol.* **73:** 366.

Ganter, P. and G. Jolles. 1970. *Histochimie normale et pathologique.* Gauthier-Villars, Paris.

Giraud, P., E. Castanas, G. Patey, C. Oliver, and J. Rossier. 1983. Regional distribution of methionine-enkephalin-Arg6-Phe7 in the rat brain: Comparative study with the distribution of other opioid peptides. *J. Neurochem.* **41:** 154.

Govoni, S., I. Hanbauer, T.D. Hexum, H.-Y.T. Yang, G.D. Kelly, and E. Costa. 1981. In vivo characterization of the mechanisms that secrete enkephalin-like peptides stored in dog adrenal medulla. *Neuropharmacology* **20:** 639.

Gubler, U., P. Seeburg, B.J. Hoffman, L.P. Gage, and S. Udenfriend. 1982. Molecular cloning establishes proenkephalin as precursor of enkephalin-containing peptides. *Nature* **295:** 206.

Hambrook, J.M., B.A. Morgan, M.J. Rance, and C.F.C. Smith. 1976. Mode of deactivation of the enkephalins by rat and human plasma and rat brain homogenates. *Nature* **262:** 782.

Hexum, T.D., I. Hanbauer, S. Govoni, H.-Y.T. Yang, and E. Costa. 1980. Secretion of enkephalin-like peptides from canine adrenal gland following splanchnic nerve stimulation. *Neuropeptides* **1:** 137.

Höllt, V., I. Haarmann, C. Grimm, A. Herz, F.C. Tulunay, and H.H. Loh. 1982. Pro-enkephalin intermediates in bovine brain and adrenal medulla: Characterization of immunoreactive peptides related to BAM-22P and peptide F. *Life Sci.* **31:** 1883.

Hook, V.Y.H., L.E. Eiden, and M.J. Brownstein. 1982. A carboxypeptidase processing enzyme for enkephalin precursors. *Nature* **295:** 341.

Ikeda, Y., K. Nakao, T. Yoshimasa, N. Yanaihara, S. Numa, and H. Imura. 1982. Existence of met-enkephalin-Arg6-Gly7-Leu8 with met-enkephalin, leu-enkephalin and met-enkephalin-Arg6-Phe7 in the brain of guinea pig, rat and golden hamster. *Biochem. Biophys. Res. Commun.* **107:** 656.

Jones, B.N., A.S. Stern, R.V. Lewis, S. Kimura, S. Stein, S. Udenfriend, and J.E. Shively. 1980. Structure of two adrenal polypeptides containing multiple enkephalin sequences. *Arch. Biochem. Biophys.* **204:** 392.

Jones, B.N., J.E. Shively, D.L. Kilpatrick, K. Kojima, and S. Udenfriend. 1982a. Enkephalin biosynthetic pathway: A 5300-dalton adrenal polypeptide that terminates at its COOH end with the sequence (Met)enkephalin-Arg-Gly-Leu-COOH. *Proc. Natl. Acad. Sci.* **79:** 1313.

Jones, B.N., J.E. Shively, D.L. Kilpatrick, A.S. Stern, R.V. Lewis, K. Kojima, and S. Udenfriend. 1982b. Adrenal opioid proteins of 8,600 and 12,600 daltons: Intermediates in proenkephalin processing. *Proc. Natl. Acad. Sci.* **79:** 2096.

Kakidani, H., Y. Furutani, H. Takahashi, M. Noda, Y. Morimoto, T. Hirose, M. Asai, S. Inayama, S. Nakanishi, and S. Numa. 1982. Cloning and sequence analysis of cDNA for porcine β-neo-endorphin/dynorphin precursor. *Nature* **298:** 245.

Khachaturian, H., M.E. Lewis, V. Höllt, and S.J. Watson. 1983. Telencephalic enkephalinergic systems in the rat brain. *J. Neurosci.* **3:** 844.

Kilpatrick, D.L., B.N. Jones, K. Kojima, and S. Udenfriend. 1981a. Identification of the octapeptide (Met)enkephalin-Arg6-Gly7-Leu8 in extracts of bovine adrenal medulla. *Biochem. Biophys. Res. Commun.* **103:** 698.

Kilpatrick, D.L., R.V. Lewis, S. Stein, and S. Udenfriend. 1980. Release of enkephalins and enkephalin-containing polypeptides from perfused beef adrenal glands. *Proc. Natl. Acad. Sci.* **77:** 7473.

Kilpatrick, D.L., B.N. Jones, R.V. Lewis, A.S. Stern, K. Kojima, J.E. Shively, and S. Udenfriend. 1982. An 18,200 dalton adrenal protein that contains four (Met) enkephalin sequences. *Proc. Natl. Acad. Sci.* **79:** 3057.

Kilpatrick, D.L., T. Tanigushi, B.N. Jones, A.S. Stern, J.E. Shively, J. Hullihan, S. Kimura. S. Stein, and S. Udenfriend. 1981b. A highly potent 3,200-dalton adrenal opioid peptide that contains both a [Met]- and [Leu]-enkephalin sequence. *Proc. Natl. Acad. Sci.* **78:** 3265.

Leander, J.D. 1983. A kappa opioid effect: Increased urination in the rat. *J. Pharmacol. Exp. Ther.* **224:** 89.

Lewis, R.V., A.S. Stern, S. Kimura, J. Rossier, S. Stein, and S. Udenfriend. 1980. An about 50,000 dalton protein in adrenal medulla: A common precursor of (Met)- and (Leu)-enkephalin. *Science* **208:** 1459.

Lindberg, I., H.-Y.T. Yang, and E. Costa. 1982. An enkephalin-generating enzyme in bovine adrenal medulla. *Biochem. Biophys. Res. Commun.* **106:** 186.

Liston, D.R., J.J. Vanderhaeghen, and J. Rossier. 1983. Presence in brain of synenkephalin, a proenkephalin-immuno-reactive protein which does not contain enkephalin. *Nature* **302:** 62.

Livett, B.G., R. Day, R.P. Elde, and P.R.C. Howe. 1982. Co-storage of enkephalins and adrenaline in the bovine adrenal medulla. *Neuroscience* **7:** 1323.

Livett, B.G., D.M. Dean, L.G. Whelan, S. Udenfriend, and J. Rossier. 1981. Co-release of enkephalin and catecholamines from cultured adrenal chromaffin cells. *Nature* **289:** 317.

Loh, Y.P. and H. Gainer. 1982. Characterization of pro-opiocortin-converting activity in purified secretory granules from rat pituitary neurointermediate lobe. *Proc. Natl. Acad. Sci.* **79:** 108.

Martin, R. and K.H. Voigt. 1980. Enkephalins co-exist with oxytocin and vasopressin in nerve terminals of rat neurohypophysis. *Nature* **289:** 502.

Martin, R., R. Geis, R. Holl, M. Schafer, and K.H. Voigh. 1983. Co-existence of unrelated peptides in oxytocin and vasopressin terminals of rat neurohypophyses: Immunoreactive methionine-enkephalin-, leucine-enkephalin- and cholecystokinin-like substances. *Neuroscience* **8:** 213.

Mizuno, K., N. Minamino, K. Kangawa, and H. Matsuo. 1980. A new family of endogenous "big" met-enkephalins from bovine adrenal medulla: Purification and structure of docosa-(BAM-22P) and eicosapeptide (BAM-20P) with very potent opiate activity. *Biochem. Biophys. Res. Commun.* **97:** 1283.

Mizuno, K., A. Miyata, K. Kangawa, and H. Matsuo. 1982. A unique proenkephalin-converting enzyme purified from bovine adrenal chromaffin granules. *Biochem. Biophys. Res. Commun.* **108:** 1235.

Noda, M., Y. Furutani, H. Takahashi, M. Toyosato, T. Hirose, S. Inayama, S. Nakanishi, and S. Numa. 1982.

Cloning and sequence analysis of cDNA for bovine adrenal prepro-enkephalin. *Nature* **295:** 202.

NORDMANN, J.J. and J. LABOUESSE. 1981. Neurosecretory granules: Evidence for an aging process within the neurohypophysis. *Science* **211:** 595.

PATEY, G., A. CUPO, P. GIRAUD, M. CHAMINADE, and J. ROSSIER. 1983. The heptapeptide Met-enkephalin-Arg6-Phe7 is released from rat striatum in vitro by high potassium. *Neurosci. Lett.* **37:** 267.

PELTO-HUIKKO, M., T. SALMINEN, and A. HERNOVEN. 1982. Enkephalin-like immunoreactivity is restricted to the adrenaline cells in the hamster adrenal medulla. *Histochemistry* **73:** 493.

PRADO, E.S., L. PRADO DE CARVALHO, M.S. ARAUJO-VIEL, N. LING, and J. ROSSIER. 1983. A Met-enkephalin-containing-peptide, BAM 22P, as a novel substrate for glandular kallikreins. *Biochem. Biophys. Res. Commun.* **112:** 366.

ROISIN, M.P., A. ARTOLA, J.P. HENRY, and J. ROSSIER. 1983. Enkephalins are associated with adrenergic granules in bovine adrenal medulla. *Neuroscience* (in press).

ROSSIER, J. 1977. Choline acetyltransferase: A review with special reference to its cellular and subcellular localization. *Int. Rev. Neurobiol.* **20:** 283.

——. 1981. Enkephalin biosynthesis: The search for a common precursor. *Trends Neurosci.* **4:** 94.

ROSSIER, J., E. BATTENBERG, Q. PITTMAN, A. BAYON, L. KODA, R. MILLER, R. GUILLEMIN, and F. BLOOM. 1979. Hypothalamic enkephalin neurones may regulate the neurohypophysis. *Nature* **277:** 653.

ROSSIER, J., Y. AUDIGIER, N. LING, J. CROS, and S. UDENFRIEND. 1980a. Met-enkephalin-Arg6-Phe7, present in high amounts in brain of rat, cattle and man, is an opioid agonist. *Nature* **288:** 88.

ROSSIER, J., J.M. TRIFARO, R.V. LEWIS, R.W.H. LEE, A. STERN, S. KIMURA, S. STEIN, and S. UDENFRIEND. 1980b. Studies with [^{35}S]methionine indicate that the 22,000-dalton (Met)enkephalin-containing protein in chromaffin cells is a precursor of (Met)enkephalin. *Proc. Natl. Acad. Sci.* **77:** 6889.

SABOL, S.L., C.M. LIANG, S. DANDEKAR, and L.S. KRNZLER. 1983. In vitro biosynthesis and processing of immunologically identified methionin-enkephalin precursor protein. *J. Biol. Chem.* **258:** 2697.

SCHULTZBERG, M., J.M. LUNDBERG, T. HOKFELT, L. TERENIUS, J. BRANDT, R.P. ELDE, and M. GOLDSTEIN. 1978. Enkephalin-like immunoreactivity in gland cells and nerve terminals of the adrenal medulla. *Neuroscience* **3:** 1169.

SCHWARTZ, J.C. 1983. Metabolism of enkephalins and the inactivating neuropeptide concept. *Trends Neurosci.* **6:** 45.

SMITH, A.D. and H. WINKLER. 1967. A simple method for the isolation of adrenal chromaffin granules on a large scale. *Biochem. J.* **103:** 480.

STEINER, D.F., P.S. QUINN, S.J. CHAN, J. MARSH, and H.S. TAGER. 1980. Processing mechanisms in the biosynthesis of proteins. In *Precursor processing in the biosynthesis of proteins* (ed. M. Zimmermann et al.), p. 1. New York Academy of Sciences, New York, New York.

STERN, A.S., B.N. JONES, J.E. SHIVELY, S. STEIN, and S. UDENFRIEND. 1981. Two adrenal opioid polypeptides: Proposed intermediates in the processing of proenkephalin. *Proc. Natl. Acad. Sci.* **78:** 1962.

STERNBERGER, L.A., P.H.J. HARDY, J.J. CUCULIS, and H.G. MEYER. 1970. The unlabeled antibody-enzyme method of immunohistochemistry. Preparation and properties of soluble antigen-antibody complex (horseradish peroxidase-antihorseradish peroxidase) and its use in identification of spirochetes. *J. Histochem. Cytochem.* **18:** 315.

STINE, S.M., H.-Y.T. YANG, and E. COSTA. 1980. Release of enkephalin-like immunoreactive material from isolated bovine chromaffin cells. *Neuropharmacology* **19:** 683.

TERLAND, O., T. FLATMARK, and H. KRIVI. 1979. Isolation and characterization of noradrenaline storage granules of bovine adrenal medulla. *Biochim. Biophys. Acta* **553:** 460.

VANDERHAEGHEN, J.J., F. LOSTRA, D. LISTON, and J. ROSSIER. 1983. Proenkephalin, met-enkephalin and oxytocin immunoreactivities are co-localized in bovine hypothalamic magnocellular neurons. *Proc. Natl. Acad. Sci.* **80:** 5139.

VANDESANDE, F., K. DIERICKX, and J. DE MEY. 1977. The origin of the vasopressinergic and oxytocinergic fibres of the external region of the median eminence of the rat hypophysis. *Cell Tissue Res.* **180:** 443.

VERHOFSTAD, A.A.J., H.W.N. STEINBUSCH, H.W.J. JOOSTEN, B. PENKE, and J. VARGA. 1982. Demonstration of NA and A storing cells in the adrenal medulla by immunohistochemistry. In *Proceedings of the 11th Meeting of the Anatomy Association.* p. 180.

VIVEROS, O.H., E.J.J. DILIBERTO, E. HAZUM, and K.-J. CHANG. 1979. Opiate-like materials in the adrenal medulla: Evidence for storage and secretion with catecholamines. *Mol. Pharmacol.* **16:** 1101.

WALLACE, E.F., C.J. EVANS, S.M. JURIK, I.N. MEFFORD, and J.D. BARCHAS. 1982. Carboxypeptidase B activity from adrenal medulla—Is it involved in the processing of proenkephalin? *Life Sci.* **31:** 1793.

WATSON, S.J. and H. AKIL. 1982. Recent studies on dynorphin and enkephalin precursor fragments in central nervous system. *Adv. Biochem. Psychopharmacol.* **33:** 35.

WATSON, S.J., H. KHACHATURIAN, H. AKIL, D.H. COY, and A. GOLDSTEIN. 1982a. Comparison of the distribution of dynorphin systems and enkephalin systems in brain. *Science* **218:** 1134.

WATSON, S.J., H. AKIL, W. FISCHLI, A. GOLDSTEIN, E. ZIMMERMAN, G. NILAVER, and T.B. VAN WIMERSMA GREID. 1982b. Dynorphin and vasopressin: Common localization in magnocellular neurons. *Science* **216:** 85.

WEBER, E., K.A. ROTH, C.J. EVANS, J.K. CHANG, and J.D. BARCHAS. 1982. Immunohistochemical localization of dynorphin (1-8) in hypothalamic magnocellular neurons: Evidence for absence of proenkephalin. *Life Sci.* **31:** 1761.

WILK, S. and M. ORLOWSKI. 1980. Cation-sensitive neutral endopeptidase: Isolation and specificity of the bovine pituitary enzyme. *J. Neurochem.* **35:** 1172.

The Molecular Basis of a Neuroendocrine Fixed Action Pattern: Egg Laying in *Aplysia*

A.C. MAHON AND R.H. SCHELLER
Department of Biological Sciences, Stanford University, Stanford, California 94305

The egg-laying behavior of the marine snail, *Aplysia*, is a stereotyped, fixed action pattern. As such, it exhibits a number of fixed behavioral responses accompanied by distinct physiological changes. These include a cessation of walking and feeding and increased respiratory activity, followed by head waving and egg deposition (Arch et al. 1978; Strumwasser et al. 1980). The appropriate stimulus induces the full repertoire in an invariant, all-or-none fashion.

The distinct behavioral and physiological changes that constitute the egg-laying behavior occur as fixed acts. These are simple behaviors with characteristically steep stimulus-response curves. They are mediated by a defined set of neurons, or neural circuit, which acts on a target tissue (muscle or gland) to elicit an observable response. Stimulation of this circuit directly results in a change in the effector tissue that leads to movement or a physiological response. A simple working hypothesis proposes that the neural circuitry for stereotyped behavior patterns, such as the egg-laying behavior, is a composite of those for each fixed act (Kandel 1976). The simultaneous generation of each fixed act is accomplished by the stimulation of a single control element which then activates each neural circuit.

The generation of the egg-laying behavior and its coordination is directly linked to the bag cells—two clusters of electrically coupled neurosecretory cells located on the rostral margin of the abdominal ganglion (Fig. 1) (Frazier et al. 1967; Kupfermann 1970; Kupfermann and Kandel 1970; Arch et al. 1978; Strumwasser et al. 1980). Egg-laying is always preceded by bag cell firing (Pinsker and Dudek 1977). Furthermore, the entire behavioral pattern can be elicited by injecting bag cell extracts into the hemolymph of the animal (Kupfermann 1970).

The bag cells send granule-filled processes into the vascularized sheath surrounding the abdominal ganglion. Stimulation produces a prolonged, synchronous discharge and release of peptides from the bag cell processes (Frazier et al. 1967; Arch 1972; Loh et al. 1975; Stuart and Strumwasser 1980). The released peptides may act directly on neurons of the abdominal ganglion (Branton et al. 1978a,b; Stuart and Strumwasser 1980; Mayeri and Rothman 1982; Scheller et al. 1983a) or may diffuse into the hemolymph and be dispersed throughout the animal (Kupfermann 1970).

A number of bag cell peptides have been isolated. The most fully characterized of these is the egg-laying hormone (ELH) (Arch 1972). This peptide promotes release of eggs from the ovotestis by inducing smooth muscle contraction (Dudek and Tobe 1979; Rothman et al. 1982). ELH also acts locally to activate neurons in

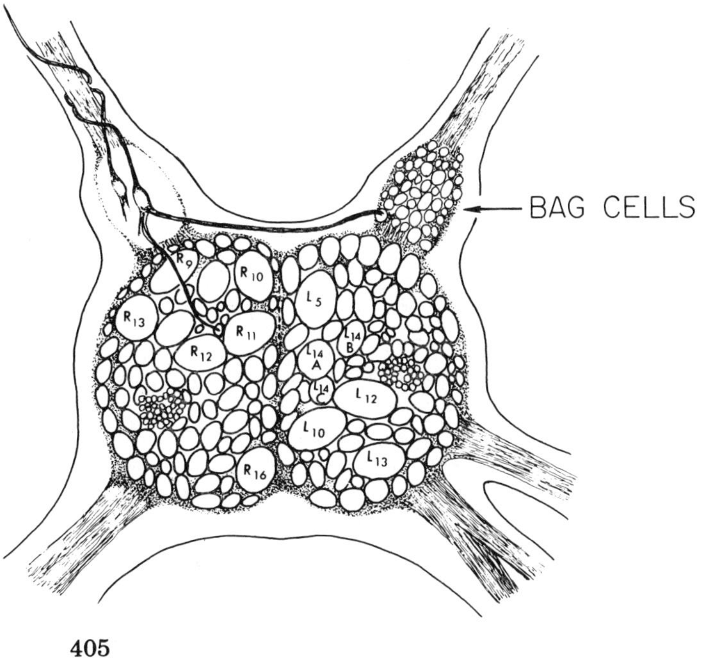

Figure 1. Schematic representation of the ventral surface of the abdominal ganglion. Cells are labeled L or R (designating left or right hemiganglion) and with identifying number (Frazier et al. 1967). The two clusters of bag cell neurons are diagramed on the rostral side of the ganglion. Two cells of the right cluster are shown to illustrate the relationship of bag cell processes to the connective and the main part of the ganglion.

the abdominal ganglion (Branton et al. 1978a; Stuart and Strumwasser 1980; Mayeri and Rothman 1982; Scheller et al. 1983a) that are thought to control the autonomic functions associated with egg laying (Kandel 1976). In addition, ELH modifies the activity of neurons in the head ganglia (Stuart and Strumwasser 1980), where locomotion, feeding, and head movements are controlled (Kandel 1976). Thus, ELH acts both locally and at a distance to promote aspects of the egg-laying behavior.

Additional bag cell peptides that are active on identified neurons have been isolated (Rothman et al. 1981; Scheller et al. 1983a). None of these peptides alone, including ELH, can generate the complete egg-laying behavior. This suggests that a combination of bag cell peptides are involved in egg laying. It is thought that these peptides modify the activity of both neural and nonneural targets to generate the fixed acts and physiological changes associated with this behavior. Coordination of these responses presumably is achieved by simultaneous release from one site, the bag cells.

In addition to the bag cells, the atrial gland, which is a secretory organ of the reproductive tract, may also have a role in the egg-laying behavior. Extracts of this gland initiate egg laying when injected into mature animals (Arch et al. 1978). The major secretory products of this gland are two closely related 34-amino-acid proteins, the A and B peptides (Heller et al. 1980). When applied to the bag cells in vitro, the atrial peptides induce a burst discharge of these neurons and release of bag cell peptides (Heller et al. 1980).

The peptides controlling the egg-laying behavior in *Aplysia* are encoded by a small multigene family (Scheller et al. 1982, 1983b). These genes are differentially expressed in the bag cells and the atrial gland to give distinct proteins. In what follows, the nature of these proteins and their role in the egg-laying behavior are discussed.

EXPERIMENTAL PROCEDURES

Detailed procedures for the experimental operations utilized in the study were presented previously (Scheller et al. 1982, 1983b).

RESULTS AND DISCUSSION

The Gene Encoding ELH Is Part of a Small Gene Family

In sexually mature *Aplysia*, ELH accounts for approximately half of the bag cell protein (Arch 1972), suggesting that a large portion of bag cell poly(A)$^+$ mRNA contains ELH-coding sequences. This RNA is rare in other neural cells (McAllister et al. 1983) and presumably absent from nonneural tissue. This tissue specificity was used to identify a gene encoding ELH. A preliminary screen of an *Aplysia* genomic library revealed clones that hybridized to radiolabeled cDNA synthesized from bag cell and atrial gland poly(A)$^+$ RNAs, but not to cDNA from hepatopancreas or abdominal ganglion poly(A)$^+$ RNA. Partial nucleotide sequence analysis of one differentially selected clone identified ELH-coding regions. Subsequently, this sequence was used to identify additional recombinants (Scheller et al. 1982).

Restriction endonuclease maps of six independently isolated cloned DNA sequences are presented in Figure 2. Nine discrete regions bearing homology to ELH-coding sequences were identified (dark arrows, Fig. 2). Annealing experiments with cDNA synthesized from poly(A)$^+$ RNA to numerous tissues indicate that these genes are homologous to messages prevalent only in the bag cells and atrial gland. Under very stringent hybridization conditions, the ELH genes hybridize exclusively to bag cell mRNA, whereas the *A* and *B* genes hybridize only to atrial gland mRNA (Fig. 2) (Scheller et al. 1983b).

Clones ELH-1, ELH-3, and ELH-8 each contain two genes, one homologous to bag cell RNA and one homologous to atrial gland RNA. DNA sequence analysis of ELH-1 (below) indicates that the gene expressed in the atrial gland encodes the B peptide, whereas the bag cell gene encodes ELH. By heteroduplex analysis, the ELH-1 clone was shown to contain a large inverted repeat of at least 6 kb (Scheller et al. 1982). The three clones containing the linked *B* and ELH genes (ELH-1, ELH-3, and ELH-8) are identical in their restriction enzyme maps except for some DNA inserts. These include approximately 1.7-kb DNA sequences which appear 5′ to the ELH-coding region on ELH-1 and 5′ to the B-coding region on ELH-3. Clone ELH-8 contains an additional 200 nucleotides 4 kb downstream from the ELH gene.

The restriction maps of ELH-13 and ELH-18 are clearly distinct from the cloned DNA sequences containing the linked *B* and ELH genes. ELH-13 and ELH-18 contain one gene that hybridizes to atrial gland mRNA. Partial nucleotide sequencing of ELH-18 predicts that this gene encodes the A peptide. The restriction enzyme maps of ELH-18 and ELH-13 are identical except for a 1.7-kb insert on ELH-18. Clone ELH-1, which contains the *B* and ELH genes, and clone ELH-18 form a heteroduplex that extends past the coding regions (McAllister et al. 1983).

Taken together, these results suggest that the six clones contain at least three nonallelic members of the ELH gene family. These genes are expressed in a tissue-dependent manner in either the bag cells to produce ELH or the atrial gland to give the A and B peptides. Since the genes exhibit considerable homology that extends past the coding regions, it is probable that the gene family arose by duplication from a common ancestral DNA sequence.

The Genes of the ELH Family Encode Protein Precursors

DNA sequence analysis reveals three distinct proteins encoded by the ELH gene family (Fig. 3). Each is a

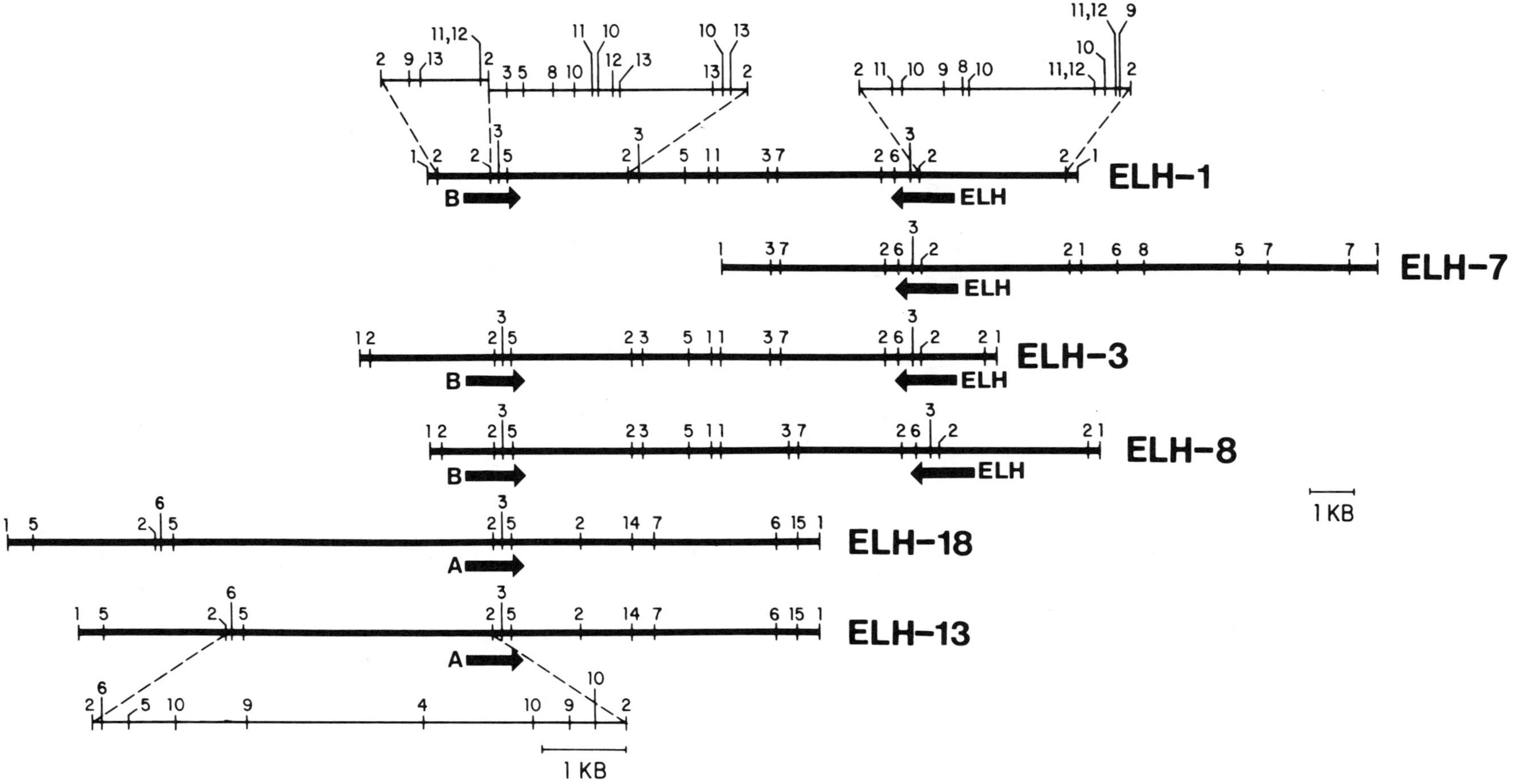

Figure 2. Restriction enzyme maps of ELH recombinant clones. Restriction enzyme maps were determined by a combination of single, partial, and double digests on intact clones and isolated fragments. The arrows indicate the position of mRNA homologous sequences and point in the direction of transcription as determined from DNA sequencing. Restriction enzymes: (1) *Eco*RI; (2) *Pst*I; (3) *Xho*I; (4) *Stu*I; (5) *Pvu*II; (6) *Hin*dIII; (7) *Bgl*I; (8) *Xba*I; (9) *Ava*II; (10) *Hinc*II; (11) *Hae*II; (12) *Hha*I; (13) *Hpa*II; (14) *Bam*HI; (15) *Sal*I.

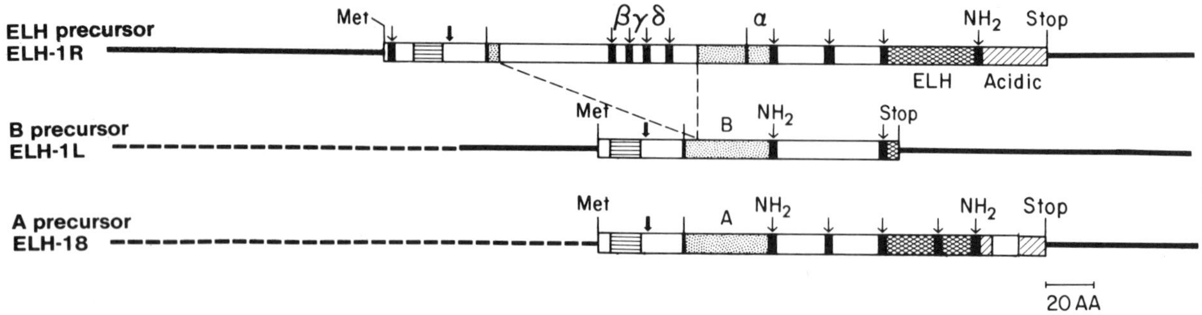

Figure 3. Comparison of the precursors encoding the ELH and A and B peptides. Each of the three proteins is initiated by a methionine followed by a hydrophobic region (▤). Thick arrows represent the putative cleavage site of the signal sequence. Line above the sequence represents potential cleavages at single arginine residues, whereas arrows represent potential or known cleavages at dibasic, tribasic, or tetrabasic residues. If carboxyterminal amidation is believed to occur, an NH_2 appears above the arrow. The A/B peptide homology is represented by stippled boxes, the ELH homology by crosshatched boxes, and the acidic peptide homology by parallel lines enclosed in boxes. Solid lines symbolize sequenced noncoding regions, and dashed lines depict regions not sequenced.

precursor protein containing either ELH or the A or B peptides (Scheller et al. 1983b). These proteins exhibit features in common with other similarly characterized neuropeptide precursors (Herbert et al. 1981; Amara et al. 1982; Craig et al. 1982; Kakidani et al. 1982; Land et al. 1982). All begin with a short hydrophobic signal sequence essential for membrane association, vesiculation, and secretion (Blobel and Dobberstein 1975; Blobel 1980; Sabatini et al. 1982). All contain one or more biologically active peptides flanked by proteolytic cleavage sites (Steiner et al. 1980). These may be dibasic amino acid residues (Herbert et al. 1981) or single arginine residues (Amara et al. 1982). The precursor proteins can release peptides with either amidated or unmodified carboxyterminal ends. In all cases where there is a blocked carboxy end, the peptide sequence in the precursor is followed by Gly-Lys-Arg. This sequence signals cleavage followed by destruction of the glycine and transamidation of the amino acid preceding the glycine. The peptide released thus contains a blocked carboxyl terminus, which presumably confers protection from degradation by carboxypeptidases.

Our data suggest that genes *A* and *B* encode two abundant atrial gland proteins. First, the nucleotide sequences predict that these genes encode proteins of 21 kD and 13.8 kD, respectively (Scheller et al. 1983b). Processing of these proteins at predicted proteolytic cleavage sites would release two major secretory products of the atrial gland, the A and B peptides. Second, the *A* and *B* genes form stable hybrids with atrial gland poly(A)$^+$ RNA (above). In vitro, this RNA directs the synthesis of two prominent proteins of 21 kD and 13 kD that are immunoprecipitated with antibody directed against the A peptide (Scheller et al. 1982, 1983b). These data provide strong evidence that the ELH family genes, *A* and *B*, are transcribed and translated in vivo.

The relationship of the ELH gene to bag cell mRNA appears more complex. When an ELH-1 coding region probe is hybridized to bag cell mRNA, RNA species of about 1.2, 1.45, and 1.6 kb in length are identified (Scheller et al. 1982). Several cDNA clones have been isolated from an abdominal ganglion cDNA library. The longest of these are 1.2 kb in length. We have compared the nucleotide sequences of these clones to the ELH genomic DNA. In all cDNAs examined, the genomic DNA immediately 5′ to the coding region is missing. Thus, either the ELH gene expressed in the bag cells is different from ELH-1 or the sequences 5′ to the ELH-1 gene have been removed by RNA splicing. Preliminary evidence suggests that the protein-coding regions of ELH-1 and of the cDNAs are identical or differ by only a few base changes. These results suggest that ELH-1 or a gene that encodes a very similar protein is expressed in the bag cells.

Divergence of the ELH Gene Family Has Generated Functionally Distinct Proteins

Direct comparison of the nucleotide sequences encoding the A, B, and ELH proteins suggests that these genes are derived from a common ancestor. The genes are greater than 90% homologous. Further, the DNA flanking the genes forms stable intermolecular hybrids (Scheller et al. 1982; 1983b; McAllister et al. 1983). Each precursor contains either a full ELH, A or B peptide, or additional unique peptides of possible physiological significance. Regions of divergence are attributed to single base changes and insertions or deletions in the genomic DNA. These alterations give rise to distinct precursor proteins. The consequence of specific DNA alterations on precursor structure is discussed below.

The A and B peptides are released from amino acid positions 36 to 70 in their respective precursors (Fig. 3). Point mutations in the DNA encoding these regions have caused the peptides to diverge at 4 out of 34 amino acids. Similarly, an overlapping but distinct set of base differences has caused the *A* and *B* genes to diverge from the ELH gene at 13 amino acids. The ELH gene also has two additional amino acids that are not present in the *A* and *B* genes in this coding block. Of considerable significance, a single base difference in the ELH gene generates an arginine codon ten amino acids from the carboxyterminal end of the A and B peptides. This signals the aminoterminal cleavage site for a small peptide, α-BCP (bag cell peptide). This peptide is homologous in six out of nine amino acids to the carboxyterminal end of the A and B peptides. Mutation has caused the carboxyterminal cleavage sites of α-BCP and the A and B peptides to diverge so that, in contrast to the atrial peptides, α-BCP is not amidated.

In addition to these single base changes, the region homologous to the A and B peptide nucleotide sequences within the ELH gene is interrupted by a DNA fragment encoding 80 amino acids not present on the A or B precursors (Fig. 3). This segment interrupts the ELH gene at a site corresponding to the sequence TCATCA in the *B* gene. The segment is flanked at both ends by an imperfect repeat of this sequence reading 5′-TCATCA . . . TCAGCA-3′. From this, we are not able to tell if this DNA region was inserted or deleted. However, multiple copies of the insert are not found in the genome (below), which suggests that it is not a transposable element. The insert encodes four dibasic cleavage sites. Processing at these sites would release three small peptides: β-BCP, γ-BCP, and δ-BCP. Due to the insert and base changes, the ELH precursor may produce as many as six distinct peptides from the region homologous to the DNA block encoding the A and B peptides.

Similarly, the region within the atrial genes that is homologous to the nucleotide sequences encoding ELH has diverged so that ELH cannot be expressed. A single base deletion after the sixth amino acid introduces a stop codon in the B peptide. This deletion is not present in the A precursor. However, in this region, eight amino acids of the ELH and A precursors have diverged. A single base change introduces a potential proteolytic cleavage site in the region of the *A* gene homologous to ELH. If utilized, this precursor would release two peptides rather than ELH.

Additional divergence is observed in regions not encoding ELH or the A or B peptides (Fig. 3). An acidic protein with an amino acid sequence identical to the carboxyterminal end of the ELH precursor has been isolated from bag cell extracts (Scheller et al. 1983b). The role of the acidic peptide in egg laying is unknown. This peptide contains five acidic residues that may enable it to serve as a carrier protein for the basic α- and β-BCP (Breslow 1979; Brownstein et al. 1980). The region in the A precursor homologous to the acidic peptide has diverged considerably.

The Size of the ELH Gene Family

Expression of the ELH gene family is not limited to the bag cells and the atrial gland. A small number of neurons in the central ganglia produce RNA homologous to ELH-coding regions (L.B. McAllister et al., in prep.). These neurons may have a role in the egg-laying behavior. To characterize the processes that generate this behavior fully, it is necessary to identify and characterize all ELH genes. This will be useful in predicting protein structure and in determining the expression and processing of ELH genes in all tissues.

We have characterized three nonallelic members of the ELH gene family that encode proteins very similar or identical to the major secretory products of the bag cells and the atrial gland. Southern analysis of genomic DNA from five individuals reveals at least nine distinct *Eco*RI fragments homologous to ELH-coding regions (Scheller et al. 1982). Three of these correspond in size to the fragments containing sequenced genes. No individual has all nine fragments, indicating a substantial amount of polymorphism. Thus, the ELH gene family has at least three and probably less than nine nonallelic members.

We have isolated DNA sequences from coding and noncoding regions of clone ELH-1 to analyze the genomic representation of this gene and its flanking regions (Fig. 4). Probe A contains DNA 2 kb upstream from the ELH gene. Probe B contains the 1.7-kb insert which is 5′ to the ELH gene in clone ELH-1. Probe C spans the region immediately 5′ to the ELH-coding

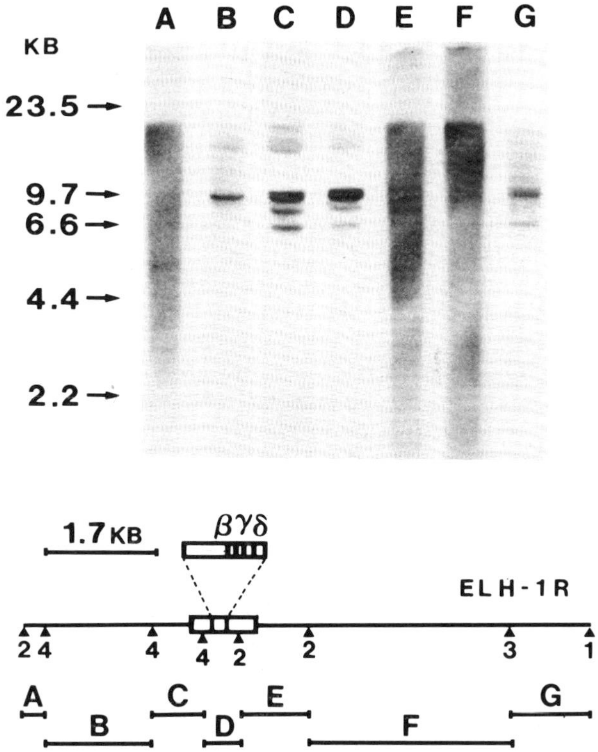

Figure 4. The distribution in genomic DNA of sequences from clone ELH-1. (*Top*) The *Eco*RI-digested sperm DNA from one individual was size-fractionated by gel electrophoresis, transferred to nitrocellulose (Southern 1975), and hybridized to DNA fragments from clone ELH-1. Lanes *A–G*: Hybridization to probes A–G. (*Bottom*) Probes A–G were generated by restriction endonuclease enzyme digestion of the right half of clone ELH-1 (solid line) with *Eco*RI (1), *Pst*I (2), *Xho*I (3), and *Hinc*II (4). (Open box) ELH-coding region. The 240-nucleotide insert encoding β-, γ-, and δ-BCP and the 1.7-kb insert are shown above the ELH-1 clone.

block. Analysis of ELH cDNA clones reveals that this sequence may be part of an intron. Probe D contains sequences encoding the aminoterminal portion of the ELH precursor. Included is the entire insert encoding β-, γ-, and δ-BCP. In addition, this fragment contains regions homologous to the A and B peptides. Probe E encodes ELH and the acidic peptide. Noncoding sequences 3′ to the ELH gene are also included in this fragment. Fragments F and G contain sequences that are 3′ to the gene.

Radiolabeled ELH-1 probes were hybridized to *Eco*RI digests of DNA from a single animal (Fig. 4). Seven bands homologous to ELH-coding regions were distinguished (lanes C and D). These correspond to seven of the nine fragments previously identified. The lower band in the dark doublet at 8.5 kb (lanes C and D) is identical to the right *Eco*RI fragment of ELH-1 which encodes ELH. This band hybridizes strongly to the bag cell peptide insert (probe D), which is present on the ELH precursor only, and to the region 5′ to ELH-1 (probe C). The upper band in the 8.5-kb doublet also hybridizes strongly to these probes, suggesting that both fragments encode β-, γ-, and δ-BCP. However, in contrast to the ELH-1 fragment, the upper band is not associated with the 1.7-kb insert (lane B). The existence of an 8.5-kb doublet in all six animals examined (Scheller et al. 1982) suggests that there are two nonallelic ELH genes. Analysis of cDNA clones from an abdominal library provides further evidence for two ELH genes (data not presented).

The remaining five restriction fragments have similar hybridization patterns, which are distinct from those of the ELH genes. The band at 6.5 kb (lanes C and D) is identical to the left *Eco*RI fragment of ELH-1 which encodes the B peptide (Scheller et al. 1982). The 18-kb band is identical to the A clone, ELH-18 (Scheller et al. 1982). Weak hybridization of these restriction fragments to probe D is expected since the *A* and *B* genes do not contain the bag cell peptide insert. Stronger hybridization to probe C is expected, since homology to ELH in the *A* and *B* genes extends past the coding regions in both the 5′ and 3′ directions. The remaining three *Eco*RI restriction enzyme fragments have not been identified. The similarity of their hybridization patterns to the *A* and *B* genes suggests that these contain DNA encoding atriumlike genes. The 7.7-kb restriction fragment is present in all animals examined. In contrast, different combinations of the large (>18 kb) restriction fragments are evident. Thus, in addition to the *A* and *B* genes, at least one and possibly two additional nonallelic atriumlike genes probably exist. We have not identified restriction enzyme fragments corresponding in length to those of ELH-3 in any animal examined. Thus, we conclude that some DNA rearrangement has occurred in this recombinant during the cloning procedure.

Approximately half of *Aplysia* DNA is repetitive (Angerer et al. 1975). Within clone ELH-1 there are at least three families of short (200–400 bp each) sequences that are repeated several times (McAllister et al. 1983). In addition, fragments of this clone are highly repeated in the genome (Fig. 4, lanes E–G). These repeated sequences are both downstream from (Fig. 4, lanes E–G) and upstream of (Fig. 4, lane A) ELH-coding regions. The DNA sequence encoding β-, γ-, and δ-BCP (probe D) and the 1.7-kb element in ELH-1 (probe B) have only two and one copies, respectively, in the genome. Thus, these inserts are probably not transposable elements that have been dispersed throughout the genome.

The Expression of the ELH Gene Family Throughout Development

Members of the ELH gene family are expressed in the bag cells, the atrial gland, and in neurons of the central ganglia (Chiu and Strumwasser 1981; L.B. McAllister et al., in prep.). Two techniques have been used to identify the cells that express members of the ELH gene family: (1) immunocytochemistry and (2) in situ hybridization of radiolabeled ELH DNA probes to fixed *Aplysia* tissue (L.B. McAllister et al., in prep.). An example of the latter technique is shown in Figure 5, where an ELH DNA probe clearly distinguishes the bag cells from other abdominal ganglion cells. Using these techniques, we have identified approximately 20 nonbag cell neurons that express the ELH gene family. These are distributed throughout the nervous system in the abdominal, cerebral, buccal, and pleural ganglia. The processes of these neurons terminate on cell bodies and in the neuropil (L.B. McAllister et al., in prep.). This suggests that the ELH-producing cells function as interneurons.

The expression of ELH genes has been followed throughout development using in situ hybridization (L.B. McAllister et al., in prep.). The development of *Aplysia* occurs in four stages: the larval (veliger), metamorphic, juvenile, and adult (Kandel 1976). The veliger has a rudimentary nervous system in which all ganglia, but not the bag cells, are represented. During metamorphosis (days 34–41), the free-swimming veliger gradually takes on the appearance and behavior of the benthic juvenile. By this time, the CNS is well developed. In contrast, the reproductive system develops late in the juvenile stage. Bag cells are identified 20 days postmetamorphosis. The cells increase in number from 10 to 400 in the next 200 days. The atrial gland is first discernible 50 days postmetamorphosis. The sequence of development of the ganglia, bag cells, and atrial gland is paralleled by ELH gene family expression in these tissues. Thus, in situ hybridization detects ELH-producing neurons in the young juvenile. Hybridization of ELH probes to the bag cells and atrial gland occurs 20 and 40 days, respectively, postmetamorphosis (L.B. McAllister et al., in prep.).

In situ studies suggest that ELH-producing cells originate in proliferative zones on the body wall. In these areas, ELH-containing cells can be seen from day 25 of the larval stage to at least 35 days postmetaphorphosis. At about 10 days postmetamorphosis, cells are

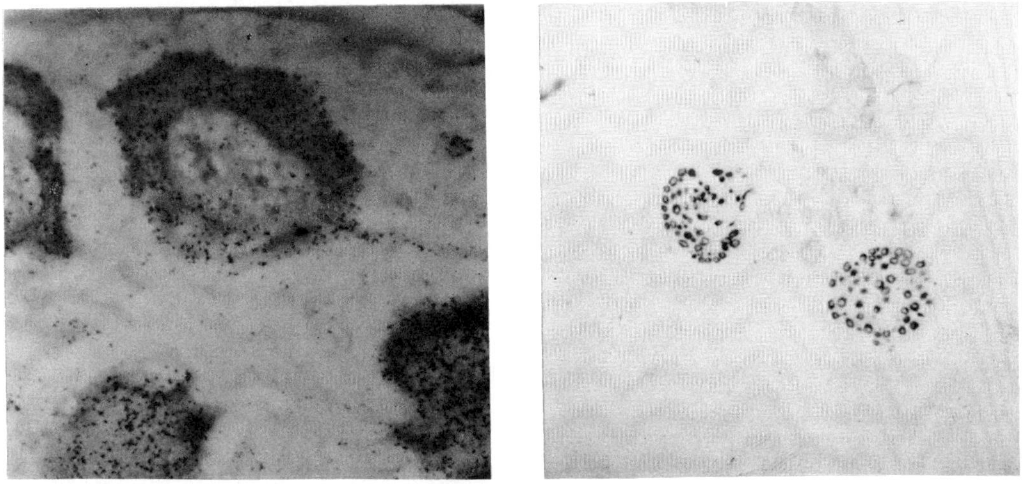

Figure 5. In situ hybridization of DNA encoding ELH to the bag cells. Radiolabeled DNA was hybridized to fixed abdominal ganglion tissue sections following the procedure of L.B. McAllister et al. (in prep.). (*Left*) Autoradiogram of a section through the bag cells and abdominal ganglion. Magnification, 20×. (*Right*) A single bag cell from the section at the left. Magnification, 600×.

seen extending into the body cavity. In addition, ELH-positive cells are identified on fibrous connective tissue strands at this time (L.B. McAllister et al., in prep.). These observations are consistent with a model of bag cell development in which these cells originate in the body wall and migrate to the CNS along fibrous connective tissue tracts.

The Egg-laying Peptides and Behavior

The amino acid sequence of the bag cell precursor can be used to predict and experimentally test the precise nature of the bag cell peptides. If each cleavage site were recognized, at least 11 distinct peptides could be released from this protein. Although we do not know the precise processing events, indirect evidence suggests that a number of cleavage sites are utilized. Peptides with sequences identical to ELH, α-BCP, and the acidic protein have been isolated in roughly equimolar amounts from bag cell extracts (Chiu et al. 1979; Mayeri and Rothman 1982). In addition, a pentapeptide with the amino acid composition of β-BCP has been isolated. These proteins have been synthesized or purified from bag cell extracts, and their activity on identified neurons has been tested (Mayeri and Rothman 1982; Scheller et al. 1983a). In this way the mechanisms involved in the coordination of the egg-laying behaviors are being probed.

ELH has a prolonged effect on the activity of several abdominal ganglion neurons. It mediates burst augmentation in R15. In addition, it induces the normally quiescent cells of the left lower quadrant to fire tonically. One of these cells, L10, directs the neural circuit for the stereotyped increase in heart beat observed during egg laying (Mayeri et al. 1979a,b). Like ELH, the bag cell peptide, α-BCP, has several sites of action in the abdominal ganglion. It hyperpolarizes the left upper-quadrant cells, L2, L3, L4, and L6, thus causing prolonged inhibition. This peptide also depolarizes the bag cells, a form of autoregulation that may contribute to their prolonged discharge. The amino acid sequence of α-BCP is homologous to the carboxyterminal end of the A and B peptides. In contrast to the atrial peptides, α-BCP is not amidated. This is consistent with the function of these proteins. The blocked carboxyl terminal may provide stability to the atrial peptides, which exert long-range action. This added stability may not be necessary for α-BCP, which is released directly onto neurons. In addition to ELH and α-BCP, the pentapeptide, β-BCP, may contribute to the egg-laying behavior. Bag cell discharge results in transient excitation of L1 and R1. This effect is thought to be mediated by β-BCP. These studies provide evidence that the ELH precursor is processed to several biologically active peptides. These may then generate different aspects of the egg-laying behavior.

SUMMARY

We describe a gene family that encodes the proteins controlling the egg-laying behavior of *Aplysia*. The family evolved by duplication and divergence from a common ancestral gene. The ELH gene family is expressed in the atrial gland, in the bag cells, and in a small network of neurons in the central ganglia. The bag cells and the atrial gland express distinct members of the family that encode different precursor proteins. These contain one or more biologically active peptides that can be released by proteolytic cleavage. The bag cell precursor releases several peptides that have multiple sites of action and may generate different components of the egg-laying behavior. Coordination of the full stereotyped behavior is achieved by simultaneous release of these peptides from the bag cell processes.

ACKNOWLEDGMENTS

This work was supported by funds from the Stanford Department of Biological Sciences and by a National Institutes of Health postdoctoral fellowship (to A.C.M.). We would like to thank Linda Harwood for secretarial assistance.

REFERENCES

AMARA, S.G., V. JONAS, M.G. ROSENFELD, E.S. ONG, and R.M. EVANS. 1982. Alternative RNA processing in calcitonin gene expression generates mRNAs encoding different polypeptide products. *Nature* **298:** 240.

ANGERER, R.C., E.H. DAVIDSON, and R.J. BRITTEN. 1975. DNA sequence organization in the mollusc *Aplysia californica*. *Cell* **6:** 29.

ARCH, S. 1972. Biosynthesis of the egg-laying hormone (ELH) in the bag cell neurons of *Aplysia californica*. *J. Gen. Physiol.* **60:** 102.

ARCH, S., T.R. SMOCK, R. GURVIS, and C. MCCARTHY. 1978. Atrial gland induction of the egg-laying response in *Aplysia californica*. *J. Comp. Physiol.* **128:** 67.

BLOBEL, G. 1980. Intracellular membrane topogenesis. *Proc. Natl. Acad. Sci.* **77:** 1496.

BLOBEL, G. and B. DOBBERSTEIN. 1975. Transfer of proteins across membranes. I. Presence of proteolytically processed and unprocessed nascent immunoglobulin light chains on membrane-bound ribosomes of murine myeloma. *J. Cell Biol.* **67:** 835.

BRANTON, W.D., S. ARCH, T. SMOCK, and E. MAYERI. 1978a. Evidence for mediation of a neuronal interaction by a behaviorally active peptide. *Proc. Natl. Acad. Sci.* **75:** 5732.

BRANTON, W.D., E. MAYERI, P. BROWNELL, and S. SIMON. 1978b. Evidence for local hormonal communication between neurons in *Aplysia*. *Nature* **274:** 70.

BRESLOW, E. 1979. Chemistry and biology of the neurophysins. *Annu. Rev. Biochem.* **48:** 251.

BROWNSTEIN, M.J., J.T. RUSSELL, and H. GAINER. 1980. Biosynthesis, axonal transport and release of posterior pituitary hormones. *Science* **207:** 373.

CHIU, A.Y. and F. STRUMWASSER. 1981. An immunohistochemical study of the neuropeptidergic bag cells of *Aplysia*. *J. Neurosci.* **1:** 812.

CHIU, A.Y., M.W. HUNKAPILLER, E. HELLER, D.K. STUART, L.E. HOOD, and F. STRUMWASSER. 1979. Neuropeptide egg-laying hormone of *Aplysia californica*: Purification and primary structure. *Proc. Natl. Acad. Sci.* **76:** 6656.

CRAIG, R.K., L. HALL, M.R. EDBROOKE, J. ALLISON, and I. MACINTYRE. 1982. Partial nucleotide sequence of human calcitonin precursor mRNA identifies flanking cryptic peptides. *Nature* **295:** 345.

DUDEK, F.E. and S.S. TOBE. 1979. Bag cell peptides act directly on ovotestis of *Aplysia californica*: Basis for an *in vivo* bioassay. *Gen. Comp. Endocrinol.* **36:** 618.

FRAZIER, W.T., E.R. KANDEL, I. KUPFERMANN, R. WAZIRI, and R.E. COGGESHALL. 1967. Morphological and functional properties of identified neurons in the abdominal ganglion of *Aplysia californica*. *J. Neurophysiol.* **30:** 1288.

HELLER, E., L.K. KACZMAREK, M.W. HUNKAPILLER, L.E. HOOD, and F. STRUMWASSER. 1980. Purification and primary structure of two neuroactive peptides that cause bag cell afterdischarge and egg-laying in *Aplysia*. *Proc. Natl. Acad. Sci.* **77:** 2328.

HERBERT, E., N. BIRNBERG, J.-C. LISSITSKY, O. CIVELLI, and M. UHLER. 1981. Pro-opiomelanocortin: A model for the regulation of expression of neuropeptides in pituitary and brain. *Neurosci. Comment.* **1:** 16.

KAKIDANI, H., Y. FURUTANI, H. TAKAHASHI, M. NODA, Y. MORIMOTO, T. HIROSE, M. ASAI, S. INAYAMA, S. NAKANISHI, and S. NUMA. 1982. Cloning and sequence analysis of cDNA for porcine β-neo-endorphin/dynorphin precursor. *Nature* **298:** 245.

KANDEL, E.R. 1976. *Cellular basis of behavior*. W.H. Freeman, San Francisco.

KUPFERMANN, I. 1970. Stimulation of egg-laying by extracts of neuroendocrine cells (bag cells) of abdominal ganglion of *Aplysia*. *J. Neurophysiol.* **33:** 877.

KUPFERMANN, I. and E.R. KANDEL. 1970. Electrophysiological properties and functional interconnections of two symmetrical neurosecretory clusters (bag cells) in abdominal ganglion of *Aplysia*. *J. Neurophysiol.* **33:** 865.

LAND, H., G. SCHUTZ, H. SCHMAK, and D. RICHTER. 1982. Nucleotide sequence of cloned cDNA encoding bovine arginine vasopressin-neurophysin II precursor. *Nature* **295:** 299.

LOH, Y.P., Y. SARNE, and H. GAINER. 1975. Heterogeneity of proteins synthesized, stored and released by the bag cells of *Aplysia californica*. *J. Comp. Physiol.* **100:** 283.

MAYERI, E. and B. ROTHMAN. 1982. Nonsynaptic peptidergic neurotransmission in the abdominal ganglion of *Aplysia*. In *Neurosecretion—Molecules, cells and systems* (ed. D.S. Farner and K. Lederis), p. 307. Plenum Press, New York.

MAYERI, E., P. BROWNELL, and W.D. BRANTON, 1979a. Multiple, prolonged actions of neuroendocrine bag cells on neurons in *Aplysia*. II. Effects on beating pacemaker and silent neurons. *J. Neurophysiol.* **42:** 1185.

MAYERI, E., P. BROWNELL, W.D. BRANTON, and S.B. SIMON. 1979b. Multiple, prolonged actions of neuroendocrine bag cells on neurons in *Aplysia*. *J. Neurophysiol.* **42:** 1165.

MCALLISTER, L.B., A.C. MAHON, and R.H. SCHELLER. 1983. Evolution of egg-laying behavior in *Aplysia*. In *Symposia on the evolution of invertebrate behavior* (ed. H. Huddle). (In press)

PINSKER, H.M. and F.E. DUDEK. 1977. Bag cell control of egg-laying in freely behaving *Aplysia*. *Science* **197:** 490.

ROTHMAN, B.S., L. PADGETT, and E. MAYERI. 1981. More than one peptide neurotransmitter may mediate bag cell actions on central neurons of *Aplysia*. *Neuroscience* **7:** 636 (Abstr.).

ROTHMAN, B.S., G. WEIR, and F.E. DUDEK. 1982. Direct action of egg-laying hormone on ovotestis of *Aplysia*. *Science* **197:** 490.

SABATINI, D.D., G. KREIBICH, T. MORIMOTO, and M. ADESNIK. 1982. Mechanisms for the incorporation of proteins in membranes and organelles. *J. Cell Biol.* **92:** 1.

SCHELLER, R.H., B.S. ROTHMAN, and E. MAYERI. 1983a. A single gene encodes multiple peptide transmitter candidates involved in a stereotyped behavior. *Trends Neurosci.* **6:** 340.

SCHELLER, R.H., J.F. JACKSON, L.B. MCALLISTER, B.S. ROTHMAN, E. MAYERI, and R. AXEL. 1983b. A single gene encodes multiple neuropeptides mediating a stereotyped behavior. *Cell* **32:** 7.

SCHELLER, R.H., J.F. JACKSON, L.B. MCALLISTER, J.H. SCHWARTZ, E.R. KANDEL, and R. AXEL. 1982. A family of genes that codes for ELH, a neuropeptide eliciting a stereotyped pattern of behavior in *Aplysia*. *Cell* **28:** 707.

Southern, E.M. 1975. Detection of specific sequences among DNA fragments separated by gel electrophoresis. *J. Mol. Biol.* **98:** 503.

STEINER, D.F., P.S. QUINN, S.J. CHAN, J. MAKH, and H.S. TAGER. 1980. Processing mechanisms in the biosynthesis of protein. *Ann. N.Y. Acad. Sci.* **343:** 1.

STRUMWASSER, F., L.K. KACZMAREK, A.Y. CHIU, E. HELLER, K.R. JENNINGS, and D.P. VIELE. 1980 . Peptides controlling behavior in *Aplysia*. *Soc. Gen. Physiol. Ser.* **35:** 197.

STUART, D.K. and F. STRUMWASSER. 1980. Neuronal sites of action of a neurosecretory peptide, egg-laying hormone, in *Aplysia californica*. *J. Neurophysiol.* **43:** 399.

Molecular Events in Developmental Regulation of Neuroendocrine Genes: Characterization of the Novel Neuropeptide CGRP

R.M. EVANS,* S. AMARA,† AND M.G. ROSENFELD†

*The Salk Institute for Biological Studies, San Diego, California 92138;
†University of California, San Diego, School of Medicine, La Jolla, California 92093

The complexity of CNS function is attributable to the extraordinarily intricate pattern of neural interconnections, the large number of regulatory molecules that are produced and secreted by neurons, and the diversity of the neuronal receptor systems. What remains unclear is whether there exist any biological processes unique to neural tissues that are responsible for the requisite functional complexity of the brain. The challenge, then, is to define the developmental mechanisms responsible for the neuronal migration and interconnection patterns and for neuron-specific patterns of gene expression. These events are likely to be critical for specific functions such as memory, learning, and behavior. In this paper we shall try to describe an approach used to identify novel regulatory neuropeptides in the absence of prior structural or functional information. The analysis of calcitonin gene expression in brain revealed the operation of a new developmental mechanism involved in the generation of diversity of brain regulatory peptides and provided new insights into the regulation of brain-specific gene expression. As a consequence of these studies, we discovered a novel neuropeptide exhibiting a unique pattern of production and distribution in the central and peripheral nervous systems.

A Single Calcitonin Gene Generates Several Structurally Discrete mRNAs

We have recently reported that the calcitonin gene is capable of generating at least two different mRNA and protein products by an alternative RNA processing mechanism (Amara et al. 1982a; Evans et al. 1982; Rosenfeld et al. 1982, 1983a). In the thyroid, the calcitonin gene gives rise to the mRNA product that is translated into the precursor for the calcium-regulating hormone, calcitonin, and two other peptides (Amara et al. 1982b). The alternative mRNA product has been cloned from a medullary thyroid carcinoma tumor cell, and its sequence predicts the production of a protein precursor that would be proteolytically processed to generate three peptides, including a 37-amino-acid product termed calcitonin gene-related peptide (CGRP) (see Fig. 1). The structural basis for the alternative RNA processing events responsible for the generation of calcitonin and CGRP mRNAs has been elucidated. The calcitonin gene (see Fig. 2) contains all of the sequences found in both calcitonin and CGRP cDNA clones. Five introns separate the gene into functional domains. The fourth exon encodes the entire calcitonin sequence followed by a glycine residue to signal the amidation of the carboxyterminal proline of calcitonin and flanked by paired basic residues to signal proteolytic excision of the hormone. The cleavage events generate the 32-amino-acid hormone, calcitonin, and a carboxyterminal cleavage peptide (CCP) (Amara et al. 1982a; Jacobs et al. 1981). The exon also contains the translational stop signal and the entire 3′-untranslated region of calcitonin mRNA, including the polyadenylation signal (AATAAA). The fifth exon encodes the carboxyl 52 amino acids of CGRP precursor, including the predicted 37-amino-acid excision product (CGRP)-associated processing signals, and a translation stop signal. The remainder of the fifth and the entire sixth exon encodes the 3′-untranslated region of CGRP. The first three exons are utilized in both calcitonin and CGRP

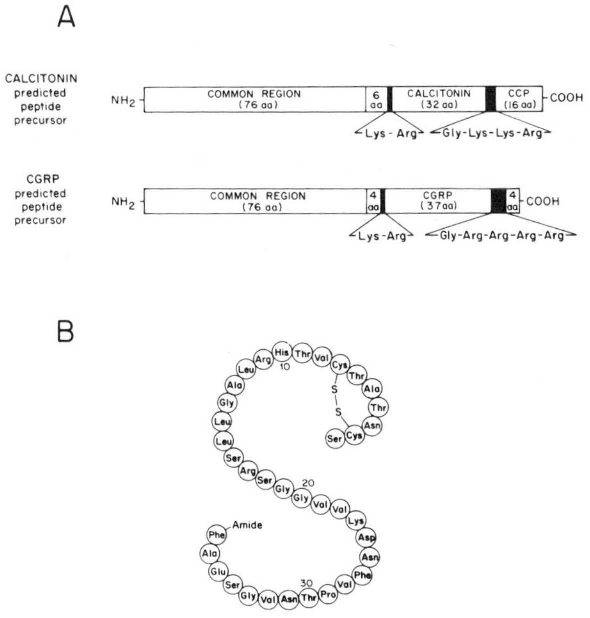

Figure 1. (*A*) Structure of the predicted peptide products of calcitonin and CGRP mRNAs. The products of the calcitonin gene have been structurally confirmed. The nucleotide structures of calcitonin and CGRP mRNAs that encode the 76-amino-acid "common region" are identical; the nucleotide sequences are completely divergent thereafter. (*B*) Predicted structure of CGRP following processing from the CGRP precursor protein.

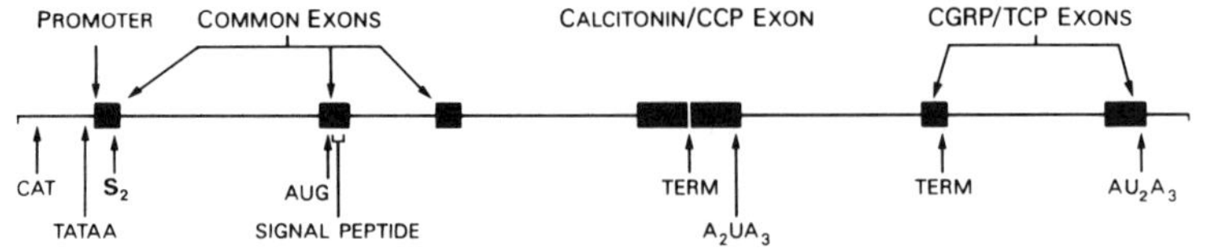

Figure 2. Structure of the rat calcitonin gene. The terminator codons of the CGRP and calcitonin exons are referred to as TERM. The carboxyl pentapeptide predicted by the CGRP cDNA structure is referred to as TCP.

mRNAs and encode the hydrophobic leader sequence and an aminoterminal protein common to both calcitonin and CGRP protein precursors. The fourth exon is utilized exclusively in the formation of calcitonin mRNA, whereas the fifth and sixth exons are utilized only in the assembly of CGRP mRNA (Amara et al. 1982a). Sequence analyses of the calcitonin and CGRP genomic and cDNA clones have thus established that one neuroendocrine gene can generate two discrete mRNAs encoding entirely divergent carboxyterminal peptides.

CGRP: A Novel Neuropeptide

The specific hybridization of CGRP exon-specific probes to poly(A)-rich RNA from rat trigeminal ganglion and hypothalamic RNA preparations, but not from thyroid, suggested that CGRP mRNA may be specifically synthesized in the brain (Amara et al. 1982a). The structure of CGRP mRNA predicts a precursor peptide that will be cleaved to generate three peptides, including a 37-amino-acid peptide with a disulfide bridge at the aminoterminus and a phenylalanine amide as the carboxyterminal amino acid. To provide evidence for the production of this predicted peptide in the brain and to determine its precise sites of synthesis, a strategy was devised to utilize both immunohistochemical and molecular biological approaches. Antibody to a synthetic peptide containing the predicted 14 carboxyterminal amino acids of CGRP was generated, and this antiserum was used for immunohistochemical studies designed to identify and to determine the distribution of CGRP in neural tissues (Rosenfeld et al. 1983b). Immunofluorescent localization revealed that CGRP was present in a large number of cell groups and pathways throughout the CNS and has a distribution distinct from that of any other known neuropeptide. The distribution of CGRP is schematically summarized in Figure 3 and presented in detail elsewhere (Rosenfeld et al. 1983b). Staining in sensory systems includes a subpopulation of small neurons in the sensory ganglia, including both the trigeminal ganglion and the spinal sensory ganglia, as well as their terminals in the brain stem and spinal cord. This system is believed to relay somatic and cutaneous pain and/or temperature information. CGRP is present in most components of the neuronal pathways subserving the sensory and motor aspects of ingestive behavior, including sensory endings in the taste buds and central endings of these fibers in the rostral part of the nucleus of the solitary tract (NTS) and in the relay systems to the taste area of the cerebral cortex. CGRP is present in motor nuclei (hypoglossal, facial) and fibers innervating musculature of the tongue and pharynx (swallowing). CGRP is also present in one of the two motor nuclei of the vagus nerve, particularly in the rostral portion of the nucleus ambiguus, which serves motor functions in the heart and/or gastrointestinal tract. In addition, CGRP is present in motor end plates in the striated muscle of the upper esophagus, the first neuropeptide thus identified at neuromuscular junctions. CGRP distribution also suggests that it plays a part in the relay of visceral sensory information in the vagus and glossopharyngeal nerves. CGRP is also present in specific cells in the olfactory bulb and in the nucleus of the eighth cranial nerve.

CGRP staining was also observed in thin, beaded fibers in virtually every organ, including the heart, lung, gastrointestinal tract, and epidermis. These fibers

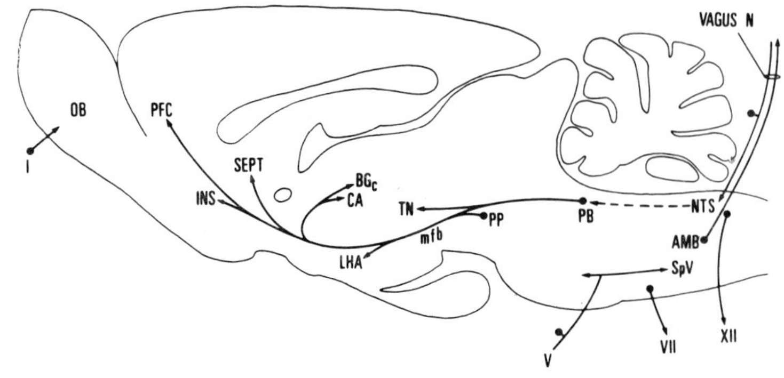

Figure 3. Partial map of CGRP immunoreactive nuclei and pathways in the brain. The map is based on detailed analyses of L.W. Swanson and P.E. Sawchencko, reported, in part, elsewhere (Rosenfeld et al. 1983b). (NTS) Nucleus tratus solitarius; (AMB) nucleus ambiguus; (SpV) spinal tract fifth cranial nerve; (XII) nucleus twelfth cranial nerve; (VII) motor nucleus seventh cranial nerve; (V) trigeminal ganglion fifth cranial nerve; (PB) parabranchial nucleus; (PP) peripeduncular nucleus; (mfb) medial forebrain bundle; (LHA) projections to lateral hypothalamic area; (TN) hypothalamic taste nucleus; (CA) projections to central nucleus of the amygdala; (BG_c) projection to patch of caudoputamen and globus pallidus; (SEPT) lateral ceptal nucleus and bed nucleus of the stria terminalis; (PFC) projections to layer III of several central areas including INS and infralimbic prefrontal area; (OB) occasional positive cell in olfactory bulb.

were often associated with the smooth muscle of blood vessels. Antisera against calcitonin failed to stain any region of the brain or sensory ganglia, although bright cells and beaded fibers were observed in the thyroid; conversely, there is little, if any, CGRP in the thyroid gland. CGRP is present in the endocrine system, as suggested by immunohistochemical staining of a subset of isolated chromaffin cells in the adrenal medulla and by CGRP-stained fibers that project through the adrenal cortex to a subcapsular zone, and in peripheral pancreatic islet cells (Rosenfeld et al. 1983b).

To ensure that the immunohistochemically stained material was, in fact, CGRP and to verify the sites of CGRP biosynthesis, it was shown that CGRP mRNA is synthesized in brain. S1-nuclease mapping analysis (Fig. 4) shows that the content of authentic CGRP mRNA in various regions of the brain reflects the relative content of CGRP neurons, e.g., approximately 1 in 50–100 cells in the trigeminal ganglion and 1 in 10^6 cells in midbrain. No immunoreactive calcitonin or calcitonin mRNA was detected in brain. The unusually large amount of immunoreactive peptide present in CGRP terminal fields reflects the very high transcription rate of the gene in neural tissues. The production of CGRP was quantitated using radioimmunoassay (e.g., 2–5 mg/g tissue in trigeminal ganglion), and the immunoreactive material was shown by gel filtration to have the molecular weight corresponding to the predicted processing product shown in Figure 1. These data together suggest that the calcitonin gene is expressed in a highly specific fashion in the central and peripheral nervous systems, as well as in specific components of the endocrine system, to generate a novel 37-amino-acid neuropeptide.

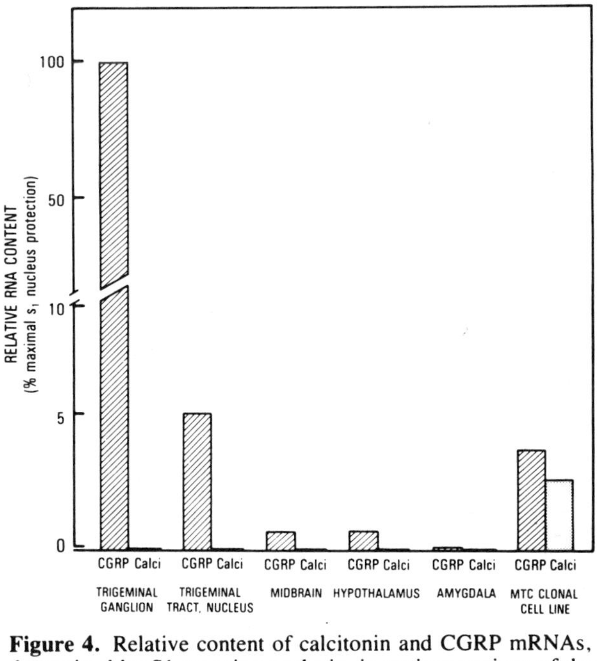

Figure 4. Relative content of calcitonin and CGRP mRNAs, determined by S1 mapping analysis, in various regions of the brain. A rat MTC tumor line provides the assay control (*right panel*).

The distribution of CGRP in specific sensory, integrative, and motor systems suggests several possible functions for the peptide. The distribution of CGRP in the olfactory and gustatory systems; in the hypoglossal, facial, and vagal nuclei; and in the hypothalamus and limbic regions strongly suggests that it may have a particularly important role in the sensory, integrative, and motor components of ingestive behavior. A role is also suggested in the processing of painful stimuli. The widespread presence in fibers in vascular musculature and in a vagal motor nucleus is suggestive of effects on cardiovascular homeostasis. The availability of synthetic CGRP has permitted demonstration of unique effects on blood pressure and catecholamine release (Fisher et al. 1983).

Mechanism of Tissue-specific Gene Expression

Nucleic acid hybridization analysis has demonstrated that the production of CGRP and calcitonin mRNAs is restricted to specific tissues (Rosenfeld et al. 1983b), suggesting that the production of these two mRNAs may be determined by a developmental event, such as the expression of tissue-specific regulatory factors. Structural analysis of the gene and its mRNA products has provided insights into the mechanism associated with the regulation of these two mRNAs. Although several models are possible (for review, see Darnell 1978, 1982; Evans et al. 1982; Rosenfeld et al. 1983a), we propose that selective utilization of alternative poly(A) sites (confirmed by S1 and DNA sequence analysis) (S.G. Amara et al., in prep.) provides a structural basis for the production of either calcitonin- or CGRP-specific sequences (see Fig. 4). Accordingly, the poly(A) site that is part of the calcitonin exon will be utilized in the thyroid, whereas the second poly(A) site that is adjacent to a CGRP exon will be utilized in the brain (Fig. 5). Several events could account for this selective polyadenylation. First, there could be two cap sites, such that CGRP and calcitonin mRNAs are the products of separate transcription units. The 5′ cap terminus of calcitonin and CGRP mRNAs have been determined by S1-nuclease mapping and by primer extension and sequencing techniques (S.G. Amara et al., in prep.) to be unique and identical for each mRNA. According to the cap-promoter hypothesis (Ziff and Evans 1978), this defines one transcription unit in the case of the calcitonin gene (see Fig. 2). Immediately preceding the established cap site is the promoter sequence TATAA, which is then preceded by a "CAT" box sequence (Benoist et al. 1980). Interestingly, the first exon contains a second utilized splice site (indicated S_2 in Fig. 2); however, the utilization of the alternative splice sites is unrelated to the exon selection events in calcitonin gene transcription. The second possibility is that use of alternative sites of termination of gene transcription provides the critical, regulated event. To evaluate this possibility, specific genomic fragments free of any repeat sequences were subcloned and used to perform DNA-excess hybridization of nascent RNA transcripts

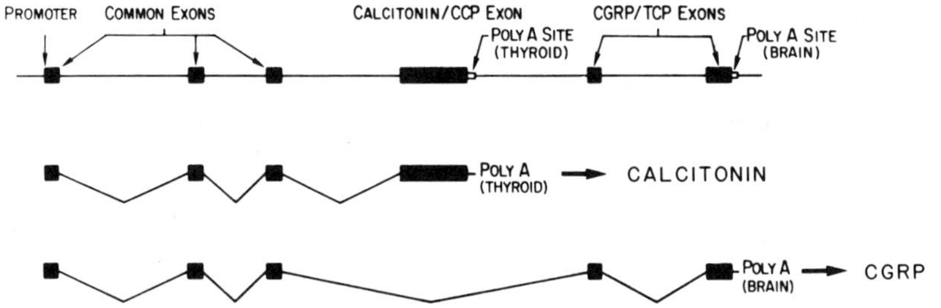

Figure 5. Exon inclusion due to poly(A)-site selection in the products of calcitonin gene expression in thyroidal "C" cells and in brain.

labeled in vitro by permitting elongation in isolated nuclei. These experiments demonstrate that, irrespective of whether calcitonin or CGRP mRNA is the mRNA product, transcription proceeds through the CGRP exons. It remains unknown whether an identical site of termination is used in each case. Next, the possibility of genomic rearrangement was considered, using restriction endonuclease mapping; however, no genomic structural alterations were detected. Therefore, it is presumed that a cellular mechanism exists that permits a direct recognition of one of the two alternative poly(A) sites and that dictates where the critical endonucleolytic cleavage will occur. One possible mechanism is that a sequence in the calcitonin gene or nuclear RNA transcript is recognized by a critical positive or negative factor that determines whether or not the first poly(A) site is utilized. The poly(A)-addition signals are AAUAAA, in the case of the calcitonin exon, and AUUAAA, in the case of the CGRP exon; however, it is unlikely that this single nucleotide difference is sufficient to specify the action of a putative regulatory factor. The mechanism of proposed poly(A)-site selection in the case of immunoglobulin heavy-chain gene and viral gene expression (Alt et al. 1980; Early et al. 1980; Ziff 1980; Maki et al. 1981) is also unknown. However, the similarities of alternative RNA processing events in the neuroendocrine and immune system suggest that the mechanism involved in calcitonin gene expression is prototypic for important and commonly utilized strategies in developmental regulation of gene expression.

Speculations Concerning the Origin of the Calcitonin/CGRP Gene

There are many possible models to explain the origin of the mammalian calcitonin gene structure. One possibility is that early in evolution there were separate calcitonin and CGRP genes each encoding only a single mRNA. Possibly at some point during evolution, the exon encoding either calcitonin or CGRP translocated, creating a gene that would be the evolutionary precursor calcitonin-CGRP gene. If this were true, then at least one of the original primordial genes might still exist in the mammalian genome and would either encode a peptide related to calcitonin or CGRP or, if no longer expressed, be present as a pseudogene. Alternatively, the poly(A)-site selection could have been the result of some selective advantage for the use of two endonucleolytic sites in a single transcription unit. There is no evidence to distinguish between these alternative possibilities. It is tempting to speculate that genomic rearrangements could provide a means for a particular peptide-encoding domain to be expressed as a component of different transcription units expressed in different tissues and under different regulatory mechanisms. If these events did occur, they might have occurred quickly if the translocated regions contained transposable genetic elements (McClintock 1950). Whatever molecular mechanisms account for the events observed in calcitonin gene expression, their complete elucidation will provide insights into specific neural developmental events.

ACKNOWLEDGMENTS

We thank our colleagues Dr. Jean-Jacques Mermod, Vivian Jonas, Estellita Ong, and Lana Stolarsky for their contributions to these investigations; our collaborators, Drs. Larry Swanson and Paul Sawchenko for their extensive immunohistochemical analysis; Drs. Jean Rivier and Wylie Vale for synthetic oligonucleotide and CGRP antisera; and Dr. Bernard Roos for providing fully characterized rat MTC tissues for these analyses. Dr. J. Pollak identified CGRP at neuroendocrine junctions. Valerie Jameson and Margaret Richards provided secretarial assistance in preparation of the manuscript. This work was supported by grants from the American Cancer Society and the National Institutes of Health.

REFERENCES

Alt, F.W., A.L. Bothwell, M. Kapp, E. Siden, G. Mather, M. Roshland, and D. Baltimore. 1980. Synthesis of secreted and membrane-bound immunoglobulin in Mu heavy chains is directed by mRNAs that differ at their 3′ ends. *Cell* **20:** 293.

Amara, S.G., V. Jonas, M.G. Rosenfeld, E.S. Ong, and R.M. Evans. 1982a. Alternative RNA processing in calcitonin gene expression generates mRNAs encoding different polypeptide products. *Nature* **298:** 240.

Amara, S.G., V. Jonas, J.A. O'Neil, W. Vale, J. Rivier, B.A. Roos, R.M. Evans, and M.G. Rosenfeld. 1982b. Calcitonin COOH-terminal cleavage peptide as a model for identification of novel neuropeptides predicted by recombinant DNA analysis. *J. Biol. Chem.* **257:** 2129.

BENOIST, C., K. O'HARE, R. BREATHNACH, and P. CHAMBON. 1980. The ovalbumin gene sequence of putative control regions. *Nucleic Acids Res.* **8:** 127.

DARNELL, J.E. 1978. Implications of RNA-RNA splicing in evolution of eukaryotic cells. *Science* **202:** 1257.

——. 1982. Variety in the level of gene control in eukaryotic cells. *Nature* **297:** 365.

EARLY, P., J. ROGERS, M. DAVIS, K. CALAME, M. BOND, R. WALL, and L. HOOD. 1980. Two RNAs can be produced from a single immunoglobulin gene by alternative RNA processing pathways. *Cell* **20:** 313.

EVANS, R.M., S.G. AMARA, and M.G. ROSENFELD. 1982. RNA processing regulation of neuroendocrine gene expression. *DNA* **1:** 323.

FISHER, L.A., D.O. KIKKAWA, J.E. RIVIER, S.G. AMARA, R.M. EVANS, M.G. ROSENFELD, W.W. VALE, and M.R. BROWN. 1983. Biological actions of calcitonin gene-related peptide. *Nature* (in press).

JACOBS, J.W., R.H. GOODMAN, W.W. CHIN, P.C. DEE, J.F. HABENER, N.H. BELL, and J.T. POTTS, JR. 1981. Calcitonin messenger RNA encodes multiple polypeptides in a single precursor. *Science* **213:** 457.

MAKI, R., W. ROEDER, A. TRAUNECKER, C. SIDMAN, M. WABI, W. RASCHKE, and S. TONEGAWA. 1981. The role of DNA rearrangement and alternative RNA processing in the expression of immunoglobulin delta genes. *Cell* **24:** 353.

MCCLINTOCK, B. 1950. The origin and behavior of mutable loci in maize. *Proc. Natl. Acad. Sci.* **36:** 344.

ROSENFELD, M.G., S. AMARA, N. BIRNBERG, J.-J. MERMOD, G.H. MURDOCH, and R.M. EVANS. 1983a. Calcitonin, prolactin and growth hormone gene expression as model systems for the definition of neuroendocrine regulation. *Recent Prog. Horm. Res.* **29:** 305.

ROSENFELD, M.G., J.-J. MERMOD, S.G. AMARA, L.W. SWANSON, P.E. SAWCHENKO, J. RIVIER, W. VALE, and R.M. EVANS. 1983b. Production of a novel neuropeptide encoded by the calcitonin gene via tissue-specific RNA processing. *Nature* **304:** 129.

ROSENFELD, M.G., C.R. LIN, S.G. AMARA, L. STOLARSKY, B.A. ROOS, E.S. ONG, and R.M. EVANS. 1982. Calcitonin mRNA polymorphism: Peptide switching associated with alternative RNA splicing events. *Proc. Natl. Acad. Sci.* **79:** 1717.

ZIFF, E.B. 1980. Transcription and RNA processing by the DNA tumour viruses. *Nature* **287:** 491.

ZIFF, E.B. and R.M. EVANS. 1978. Coincidence of the promoter and capped 5′ terminus of RNA from the adenovirus 2 major late transcription unit. *Cell* **15:** 1463.

The Kallikrein Multigene Family: Specific Processing of Biologically Active Peptides

J. SHINE, A.J. MASON, B.A. EVANS, AND R.I. RICHARDS
Centre for Recombinant DNA Research, Research School of Biological Sciences, Australian National University, Canberra, A.C.T. 2601, Australia

Since the initial discovery that the proteolytic digestive enzymes such as trypsin and chymotrypsin are activated by the selective cleavage of a peptide bond, it has become apparent that similar mechanisms operate in many fundamental biological processes. Limited proteolysis plays an important role in blood coagulation, blood pressure regulation, complement reaction, and polypeptide hormone biosynthesis (Neurath and Walsh 1976). The activity of many biologically active peptides, including neuropeptides, can thus be specifically regulated at the posttranslational level by at least two control mechanisms—the intracellular or intercellular compartmentalization of the substrate hormone and the processing enzyme, and the high degree of specificity exhibited by these enzymes. Most of the polypeptide hormones are initially translated as a larger precursor containing both a prepeptide or "signal" sequence needed for translocation through the endoplasmic reticulum and extra "propeptide" sequences, which must be specifically removed before activation of the hormone sequences. Such hormones include adrenocorticotropin, β-endorphin, insulin, relaxin, vasopressin, oxytocin, and the enkephalin precursors.

As a result of molecular cloning studies, it has recently become clear that many of the smaller peptide hormones are initially synthesized as part of a common precursor containing several different hormone sequences—often peptides that are regulated by the same physiological stimuli, e.g., adrenocorticotropic hormone (ACTH)-β-endorphin precursor (Whitfeld et al. 1982). An examination of the primary structure of hormone precursors reveals that cleavage at one or a number of peptide bonds is required to release the biologically active component of the precursor. It would therefore appear that limited proteolysis plays an important role in hormone biosynthesis, and thus may be an important step in regulating biological activity.

It appears that most such propolypeptides are processed by a group of intracellular proteases located in the Golgi apparatus and developing secretory granules. For several neuropeptides, processing from the precursor occurs mainly during transport of the secretory granule along the axon. In general, there appears to be a requirement for dibasic residues as a signal for proteolytic processing, although other structural features must be required for the highly specific cleavages involved (Docherty and Steiner 1982).

Although very little is known about the enzyme(s) responsible for the specific processing of many hormonal peptides, a considerable amount of data has accumulated concerning the processing of nerve growth factor (NGF) and epidermal growth factor (EGF). In these particular cases, the processing enzyme remains bound to its specific substrate after the cleavage event. This phenomenon of complex formation has greatly aided the detailed study of these particular processing enzymes (Berger and Shooter 1977; Frey et al. 1979). Detailed biochemical studies have revealed that these enzymes are highly homologous serine proteases closely related to glandular kallikrein (Bothwell et al. 1979).

We describe here the molecular cloning of the genes encoding this group of enzymes and demonstrate that they constitute a large highly homologous multigene family closely linked in the genome and highly conserved in their DNA sequence and structural organization. The high degree of functional specificity of different kallikreins is accounted for by the finding that small differences in the coding potential of individual members of this gene family are centered on residues known to be involved in determining the substrate specificity of serine proteases.

EXPERIMENTAL PROCEDURES

Isolation of nucleic acids. High-molecular-weight mammalian genomic DNA was prepared from liver essentially as described by Gross-Bellard et al. (1973). All manipulations were carried out at 4°C unless otherwise stated. Liver (10 g) in NKM buffer (0.15 M NaCl, 5 mM KCl, 2 mM $MgCl_2$) was broken up by 10–12 strokes in a tissue homogenizer. The cells were washed twice in NKM buffer and lysed by the addition of 2 mM $MgCl_2$. An equal volume of TKM buffer (10 mM Tris-HCl [pH 7.4], 0.2 M KCl, 2 mM $MgCl_2$) was added, and the cell debris, containing intact nuclei, was collected by centrifugation at 5000*g* for 10 minutes. The precipitate was resuspended in NKM buffer, washed twice, and added dropwise to a slowly stirring solution of 10 mM NaCl, 10 mM Tris-HCl (pH 8.0), 10 mM EDTA, 0.5% SDS, and 100 μg/ml of proteinase K at 37°C over a 12-hour period. A further 100 μg/ml of proteinase K was added, and the solution was incubated overnight at 37°C. The solution was gently extracted with an equal volume of phenol, saturated with TE (10 mM Tris-HCl [pH 8.0], 1 mM EDTA) and chloroform. The aqueous phase was then dialyzed against three changes of TE for 16 hours. The viscous solution was treated with ribonuclease A (20 μg/ml, 37°C, 3 hr), extracted with phenol-chloroform, and dialyzed against TE for a further 16 hours. The high-molecular-weight DNA (200–500 μg/ml) was stored at 4°C.

Bacteriophage λ DNAs (10^5–10^6 pfu) stored in PSB (0.01 M Tris-HCl [pH 7.4], 0.1 M NaCl, 0.01 M $MgCl_2$, 0.05% gelatin) were absorbed to 100 μl of a stationary culture of *Escherichia coli* LE 392 in 1 ml of LM broth at 37°C for 30 minutes. The infected culture was added to 40 ml of LM broth and incubated at 37°C with vigorous shaking, until lysis had occurred.

After cell debris had been removed by centrifugation (10,000 rpm, 15 min), the supernatant was treated with deoxyribonuclease I (10 μg/ml) and ribonuclease A (10 μg/ml) for 30 minutes at 37°C. Phage particles were precipitated by the addition of a one-fifth volume of 25% polyethyleneglycol 6000 and 2.5 M NaCl and incubated at 4°C overnight. The precipitated phage (8000 rpm, 20 min) were resuspended in 0.01 M Tris-HCl (pH 7.4) and 0.1 M NaCl and incubated at 37°C with SDS (0.5%) and Pronase (10 μg/ml) for 30 minutes. DNA was purified by phenol-chloroform extraction, adjusted to 0.3 M with NaOAC (pH 6.0), and precipitated with 2.5 volumes of ethanol.

Restriction endonuclease digestion. DNA was restricted at concentrations of 50–200 μg/ml in the presence of *Hae*III buffer (6.6 mM $MgCl_2$, 6.6 mM β-mercaptoethanol, 6.6 mM Tris-HCl at pH 7.4). NaCl was added to this reaction as specified by the manufacturer's conditions for each enzyme. Generally, a two-fold excess of enzyme was used, and the digest was allowed to continue for 2 hours at 37°C. The completeness of the digest was monitored by agarose gel electrophoresis.

Synthesis of hybridization probes. DNA hybridization probes were prepared by primed synthesis with the Klenow fragment of *E. coli* DNA polymerase using denatured random herring sperm DNA primers. Random primers of 8–12 nucleotides in length were prepared by treating herring sperm DNA with DNase I (Taylor et al. 1976). Plasmid DNA or isolated insert DNA (100–200 ng) was denatured in the presence of 100 μg of primers by boiling for 2 minutes and cooling on ice for 1 minute. The reaction mixture was adjusted to 30 μl by the addition of deoxyribonucleotide triphosphates (dATP, dGTP, and dTTP to 0.75 mM), 30 μCi of [α-^{32}P]dCTP (3000 Ci/mmole), reverse transcriptase buffer (final concentration, 50 mM Tris-HCl [pH 8.3], 7 mM $MgCl_2$, 20 mM KCl, 20 mM β-mercaptoethanol), and 1 unit of DNA polymerase I (Klenow fragment). After a 30-minute incubation at 37°C, the reaction was terminated by the addition of EDTA (to 12.5 mM) and extracted with phenol-chloroform. Following chromatography on Sephadex G-50, the specific activity was measured as between 10^7 and 10^8 cpm μg/DNA.

Hybridization. Bacteriophage λ libraries were screened by in situ hybridization (Benton and Davis 1979). Phage on nitrocellulose filters were lysed, and the DNA was denatured by placing the filter on Whatman 3MM paper saturated with 0.5 M NaOH and 0.5 M NaCl for 1–2 minutes. Filters were neutralized on 0.5 M Tris-HCl (pH 7.2) and 2 M NaCl, rinsed in 2× SSC, and allowed to air dry before being baked at 80°C in vacuo for 2 hours.

For Southern blots (Southern 1975), DNA fractionated on agarose gels was depurinated in 0.2 M HCl (5–10 min), rinsed in water, and denatured in 0.5 M NaOH and 0.5 M NaCl (30–60 min). The gel was neutralized by washing in 0.5 M Tris-HCl (pH 7.2) and 2 M NaCl (30–60 min). The denatured DNA was transferred to a 0.45-μm nitrocellulose filter by blotting the gel in 20× SSC (3 M NaCl, 3 M trisodium citrate) for 12–16 hours. The filter was washed in 2× SSC, air-dried, and baked at 80°C in vacuo for 2–3 hours.

Nitrocellulose filters were prehybridized in a solution of 3×SSC, 0.1% SDS, denatured herring sperm DNA (10 μg/ml), *E. coli* tRNA (20 μg/ml), 0.2% Ficoll, 0.2% bovine serum albumin, Fraction V, 0.2% polyvinylpyrrolidone, and 0.05 M HEPES (pH 7.0) for at least 2 hours at 65°C. Hybridization to a denatured DNA probe was carried out at 65°C for 16 hours. The filter was then washed in varying concentrations of SSC (0.1– 2×) and at different temperatures (20–65°C) depending on the stringency of hybridization required. Filters were then air-dried and exposed for autoradiography at −70°C with intensifying screens for 1–7 days.

RESULTS

Isolation of a Kallikrein cDNA Clone

The molecular cloning of complementary DNA (cDNA) synthesized from an mRNA population is a well-established method for obtaining purified gene fragments for sequence analysis and use as hybridization probes. During a study on the isolation of cDNA clones encoding bioactive peptides synthesized in the male mouse submaxillary gland, cDNA sequences homologous to pig pancreatic kallikrein were identified (Richards et al. 1982). The submaxillary gland of the male mouse was chosen as a source of mRNA for these studies, since it contains many different growth factors and other bioactive peptides as well as kallikrein activities. The latter comprise approximately 5% of the total soluble protein found in this tissue (Bothwell et al. 1979; Wilson and Shooter 1979). Analysis of recombinant clones containing cDNA inserts corresponding to mRNA of 1–5% abundance resulted in the isolation of one clone (pMK-1) that encoded a protein possessing 65% homology with pig pancreatic kallikrein. This plasmid contains cDNA coding for a peptide sequence with homology to all of the known kallikreins, including γ-NGF, EGF-binding protein (EGF-BP), and γ-renin. Thus, the mRNA from which pMK-1 was derived appears to encode a previously uncharacterized member of the kallikrein subgroup of the serine protease gene family. This suggests that there are at least five members of this enzyme family expressed in the male mouse submaxillary gland. This is in agreement with the data of Wilson and Shooter (1979), who found at least four enzyme activities in addition to EGF-BP, γ-NGF, and β-NGF endopeptidase in the submaxillary gland. Given the known roles of γ-NGF and EGF-BP, one may speculate that the pMK-1-encoded protein proteolytically modifies the precursor of a yet uncharacterized hormone or growth factor.

Genomic Organization of the Glandular Kallikrein Gene Family

High-molecular-weight mouse DNA was digested with the restriction endonuclease *Eco*RI, electrophoresed in a 1% agarose gel, and transferred to a nitrocellulose filter according to the method of Southern (1975). The denatured DNA immobilized on the filter was then hybridized with ^{32}P-labeled pMK-1 sequences. Figure 1 shows the hybridization pattern obtained. In mouse DNA there are 12 different-sized restriction fragments that hybridize with pMK-1. The variable intensity of hybridization has been shown to result from the conservation of particular restriction sites around several different genes (see below). Analysis of bacteriophage λ genomic clones and the data from Southern blot analysis suggest that there are 25–30 distinct kallikrein genes.

The distribution of kallikrein genes within the mouse genome was investigated by using pMK-1 as a hybridization probe on DNA isolated from a variety of Chinese hamster–mouse hybrid cell lines. These cell lines were constructed by D. Cox (University of California, San Francisco) and contained differing numbers and combinations of mouse chromosomes that proved ideal for mapping sequences to particular chromosomes. When the hybrid DNAs were cleaved with *Eco*RI and subjected to Southern blot analysis, it was observed that when hybridization to pMK-1 did occur, the complete set of *Eco*RI fragments was also present (Mason et al. 1983). Chromosome 7 was the only chromosome common to the DNAs that hybridized, and therefore all of the kallikrein genes detectable by hybridization to pMK-1 are present on this chromosome (Mason et al. 1983).

The identification of a large number of fragments that cross-react with pMK-1 and the demonstration of their presence only in hybrid Chinese hamster–mouse cells containing the mouse chromosome 7 indicate that the members of the mouse kallikrein gene family are both highly homologous in primary structure and closely linked on chromosome 7.

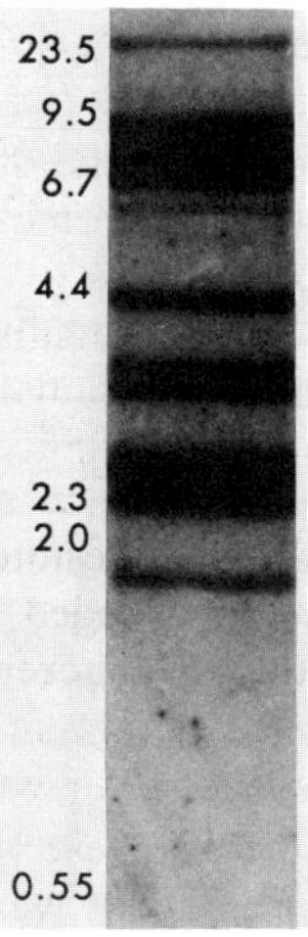

Figure 1. Genomic blot analysis of the kallikrein genes. Autoradiograph of an *Eco*RI digest of mouse DNA hybridized to the kallikrein cDNA clone pMK-1. Numbers refer to the size of DNA fragments in kilobase pairs.

Molecular Anatomy of a Kallikrein Gene

A bacteriophage λ genomic library of mouse DNA was constructed by ligating a partial *Mbo*I digest of high-molecular-weight DNA (10–20 kb) into *Bam*HI-cleaved Charon-28 DNA (Liu et al. 1980), and the resultant recombinant DNA was packaged in vitro (Hohn 1979). Bacteriophage from the cloned mouse DNA library were screened with the pMK-1 hybridization probe. One of the clones isolated in this fashion was characterized in detail by restriction endonuclease analysis and DNA sequencing. These data enabled the construction of a physical map of this genomic clone (λMSP-1) as shown in Figure 2. λMSP-1 contains 13.5 kb of mouse genomic DNA and is composed of two closely linked kallikrein genes having the same direction of transcription. The first gene (*mGK-1*) is complete in this clone and consists of five exons and four introns, together spanning 4.5 kb. The 3′ end of another clone (*mGK-2*) is located 3.7 kb upstream of the 5′ end of *mGK-1*. In λMSP-1, no other kallikrein gene sequences could be detected downstream from *mGK-1*.

The sequences present in the cDNA clone pMK-1 were found to hybridize to corresponding sequences in *mGK-1* and *mGK-2*, although they were interrupted by the presence of two intervening sequences. As pMK-1 is only a partial cDNA clone and contains the coding information for only approximately half of a kallikrein mRNA, the exons comprising each of the two genes were also mapped by hybridization to mRNA and confirmed by DNA sequence analysis (Mason et al. 1983).

The coding potential of *mGK-1* predicts an active 170-amino-acid serine protease with histidine, serine, and aspartic acid residues in the correct spatial configuration necessary for the formation of a serine protease catalytic site (Young et al. 1978). The presence of aspartic acid at position 207 also gives the protease the specificity needed for cleavage at basic residues (Krieger et al. 1974).

A comparison of the predicted amino acid sequences of *mGK-1*, *mGK-2*, and pMK-1 with the known amino acid sequences of γ-NGF, EGF-BP, and γ-renin is shown in Figure 3. As all of these proteases share extensive amino acid sequence homologies in the ranges of 73–80%, it is clear that γ-NGF, EGF-BP, and γ-

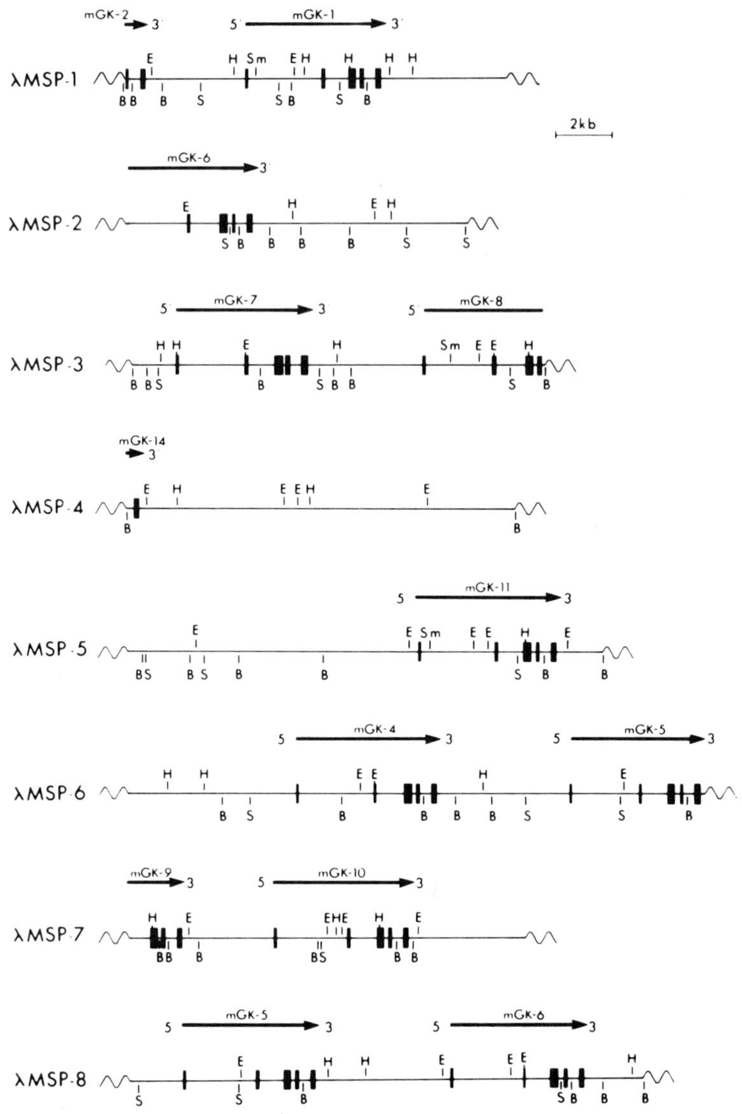

Figure 2. Physical map of bacteriophage λ clones containing kallikrein gene sequences. (■) Coding sequences (exons) of the kallikrein genes. Arrows indicate the direction of transcription. (∿∿∿) Arms of the Charon-28 vector. (B) *Bam*HI; (E) *Eco*RI; (H) *Hin*dIII; (S) *Sac*I: (Sm) *Sma*I. (Reprinted, with permission, from Mason et al. 1983.)

renin are encoded by members of the same kallikrein multigene family as the clones isolated here. All these proteases are structurally very similar, including the same five disulfide bridges, as previously noted for EGF-BP and γ-NGF (Server and Shooter 1976). Even though EGF-BP and γ-NGF are highly homologous, they differ in their substrate specificities (Frey et al. 1979), a function attributed to alterations in the residues that make up the substrate binding pocket of serine proteases (Tschesche et al. 1979). By analogy with the overall structure of trypsin and elastase, the residues that line the substrate binding pocket can be predicted (Tschesche et al. 1979). The kallikreins γ-NGF, EGF-BP, and γ-renin, as well as those encoded by *mGK-1*, *mGK-2*, and pMK-1, show a significant variation in these particular residues compared with the rest of the sequence (Fig. 3). It is therefore tempting to speculate that each of these kallikreins exhibits a different substrate specificity.

The Kallikrein Gene Locus

The demonstration that in the mouse all of the kallikrein genes are located on chromosome 7 (Mason et al. 1983), and that two members of this gene family are present in a single λ clone spanning only 13.5 kb, suggests that perhaps all the individual members of the kallikrein gene family may be tightly linked in one locus on chromosome 7. In an attempt to determine whether other members of the family were also as closely linked as *mGK-1* and *mGK-2*, further bacteriophage λ genomic isolates were analyzed in detail. The previously determined information on the detailed structure of the *mGK-1* gene was used in the construction of exon-specific hybridization probes. These were used to ascertain whether other kallikrein genes were organized into the same exon-intron arrangement as that of *mGK-1*. A Charon-28 BALB/c mouse DNA library (supplied by P. Leder, Harvard Medical School) was therefore

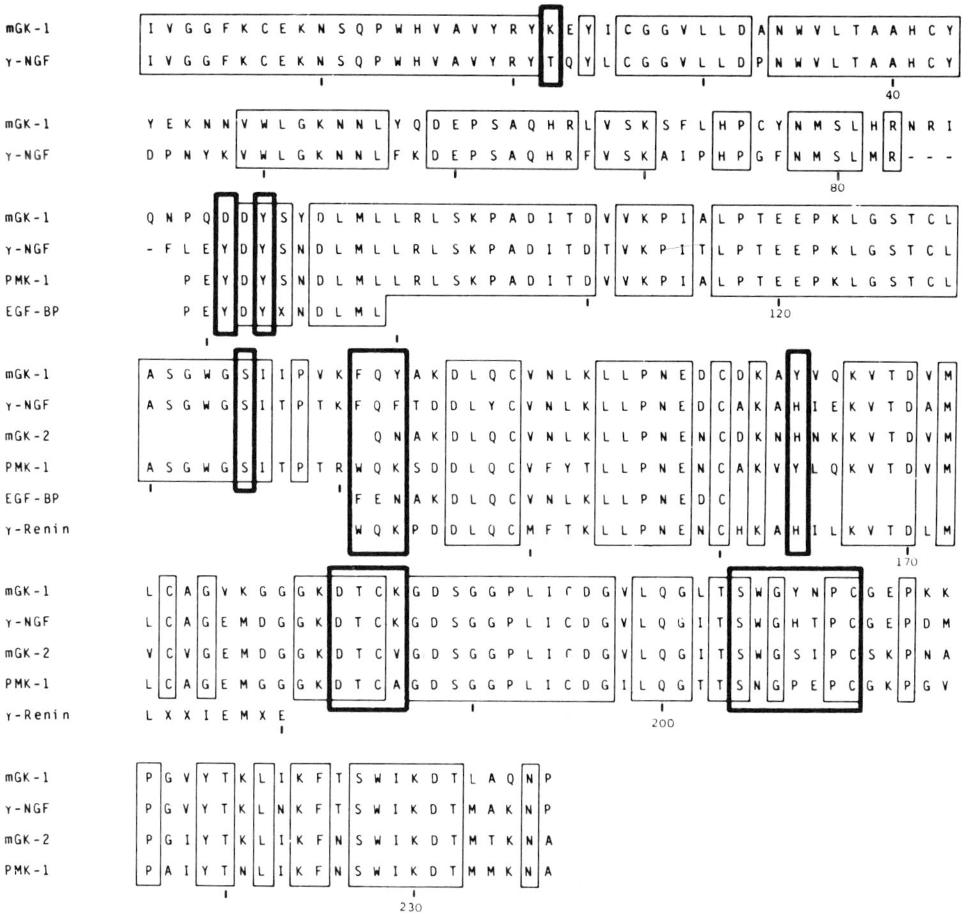

Figure 3. Comparison of the amino acid sequence of different kallikreins. Lightly boxed areas indicate regions of the sequences with identical residues. Heavily boxed areas indicate regions believed to line the substrate binding pocket. (Reprinted, with permission, from Mason et al. 1983.)

screened for kallikrein sequences utilizing pMK-1 as a hybridization probe. Approximately 4×10^5 phage were screened, and of the 65 clones initially picked, 30 were purified to homogeneity and analyzed in detail using the exon-specific probes derived from *mGK-1* or pMK-1.

The restriction maps and the location of the kallikrein genes within eight of the clones, including λMSP-1, are illustrated in Figure 2. The use of exon-specific probes enabled the position of each exon to be determined and the direction of transcription of each gene to be established. It should be noted that the lengths of certain regions of DNA, particularly those of the first intervening sequence, are only approximate. More precise restriction mapping and limited DNA sequence analysis are needed to clarify the exact location of some exons.

Several of these additional clones (λMSP-3, 6, 7, and 8) contain part, or all, of two kallikrein genes. These linked genes are transcribed from the same strand and are separated by a spacer region of between 3.5 kb and 5.0 kb.

Sequence analysis of one the genes present on λMSP-8 (B. Evans, unpubl.) indicates that it is identical with the *mGK-6* kallikrein gene contained on λMSP-2. This assignment was also based on the similarity of the restriction maps of the two clones in this region—the only difference being the presence of an additional *Eco*RI site in intron A of the *mGK-6* gene contained on λMSP-8. Since λMSP-2 was obtained from a Quackenbush mouse genomic library (Mason et al. 1983), this difference is probably reflective of the slight restriction site polymorphisms observed between different mice strains. Furthermore, it has been possible to map a series of overlapping λ clones and demonstrate that at least four kallikrein genes (*mGK-3, mGK-4, mGK-5,* and *mGK-6*) are tightly linked in the genome (Fig. 4). These four genes are encoded on the same strand of DNA and are roughly 4–5 kb apart. However, examination of other clusters of kallikrein genes shows that some intergenic spaces are considerably larger than that observed for this group of four genes (B. Evans, unpubl.).

Gene-specific Expression

All of the kallikrein gene sequences so far analyzed possess sequence homology of approximately 84%. Hybridization data (Fig. 1) indicate that all the mouse kallikrein genes share such extensive sequence homol-

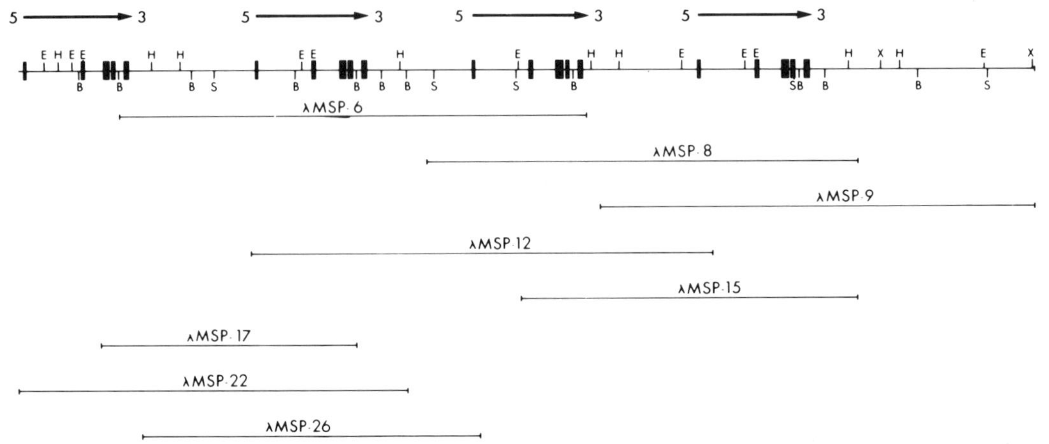

Figure 4. Physical map of a region of the mouse genome containing four kallikrein genes. (■) Coding regions (exons). Arrows indicate direction of transcription. The overall restriction map was generated by analysis of the illustrated overlapping λ clones.

ogy. If the expression of individual kallikrein genes is to be studied, a means of distinguishing between different members of this highly homologous gene family is essential. The use of specific oligonucleotides as hybridization probes representing the more variable regions of the coding sequence was therefore investigated. From the gene sequences available, a 19-bp stretch from the 3′ end of the coding region was selected as the most divergent in nucleotide sequence. The sequence from this region of three genes is compared in Figure 5a. Of the 19 bases, both *mGK-2* and pMK-1 have a total of 7 base mismatches with the corresponding sequence from *mGK-1*. A synthetic oligonucleotide of sequence 5′ CAGGGGTTTTGGGCCAAAG 3′, which is complementary to this region of *mGK-1*, was chemically synthesized by the phosphoramidite method. Approximately 10 ng of the 19-mer (a gift from J. Coghlan, Howard Florey Institute, Melbourne) was labeled at the 5′ terminus to high specific activity with [γ-^{32}P]ATP and T4 polynucleotide kinase. This probe was hybridized to a nitrocellulose filter on which restricted DNA, containing 14 distinct kallikrein genes, had been immobilized (Fig. 5b). Only sequences from *mGK-1* hybridized with the labeled oligonucleotide (Fig. 5c). Under the hybridization conditions used, this oligonucleotide is therefore a specific probe able to detect only *mGK-1* kallikrein sequences. To demonstrate that the *mGK-1* gene is expressed in the male submaxillary gland, the 19-mer was shown by Northern blot analysis to hybridize specifically to *mGK-1* mRNA present in this tissue.

These experiments demonstrate that by using synthetic oligonucleotides homologous to short stretches of a particular kallikrein gene, the tissue-specific expression of that gene can be studied.

DISCUSSION

The assignment of all kallikrein genes to chromosome 7 and the tight linkage of cloned genes are consistent with the clustering of these genes into a single locus. Of the 30 λ clones analyzed to date, a total of 14 distinct kallikrein genes have been identified. The molecular anatomy of each of these genes is identical with that determined for *mGK-1*. Many of the kallikrein genes are tightly clustered in a head-to-tail arrangement, each separated by an intergenic space of between 3.5 kb and 5.0 kb (Figs. 2 and 4).

The strong conservation of restriction sites observed between different genes and their high degree of cross-reactivity suggest that sequences in the coding and noncoding regions are highly conserved. The location of *Eco*RI restriction sites around different kallikrein genes is strongly conserved. The different *Eco*RI fragments from the 14 genes so far analyzed, which hybridize to the pMK-1 probe, are all represented among the *Eco*RI fragments detected by genomic blot analysis (Fig. 1). Three of these genes contain hybridizing sequences on 6.7-kb *Eco*RI fragments and another three contain hybridizing sequences on 2.4-kb *Eco*RI fragments. Therefore, it is apparent that the increased intensity of hybridization of certain *Eco*RI fragments (e.g., 2.4 kb and 6.7 kb) observed by genomic blot analysis can be explained by the conservation of *Eco*RI restriction sites around several distinct genes. By a comparison of the intensity of hybridization of different-sized *Eco*RI restriction fragments (Fig. 1) with the prevalence of the same-sized *Eco*RI fragments among different kallikrein gene isolates, it is possible to estimate that there are between 25–30 distinct kallikrein genes in the mouse.

The high degree of sequence homology observed in the coding and noncoding regions of different kallikrein genes suggests either that these 25–30 genes have recently diverged from a common ancestor or that the homology is maintained by a mechanism of gene conversion, as observed in the immunoglobulin genes (Baltimore et al. 1981).

The kallikrein genes each comprise five exons spanning approximately 4.5 kb. The introns separate the coding regions of a preprokallikrein as shown in Figure 6. As illustrated, the histidine, serine, and aspartic acid residues that are essential for the formation of a serine-

a

```
mGK-2     5'-CTATGACGAAAAATGCCTA-3'
             *  * **    **   *
mGK-1     5'-CTTTGGCCCAAAACCCCTG-3'
             *  ****    **
pMK-1     5'-CTATGATGAAAAATGCCTG-3'


Primer    5'-CAGGGGTTTTGGGCCAAAG-3'
```

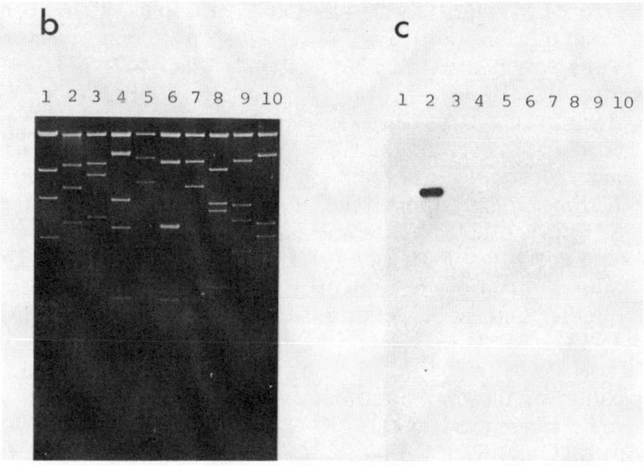

Figure 5. *mGK-1*-specific hybridization probe. (*a*) Comparison of the nucleotide sequence of *mGK-2*, *mGK-1*, and pMK-1 across the 19-bp region used to design an *mGK-1*-specific oligonucleotide. This region includes the last two nucleotides of the codon for amino acid residue 256, through to the first two nucleotides of the termination codon. Asterisks indicate the positions at which the sequence of *mGK-2* or pMK-1 differ from that of *mGK-1*. The sequence of the synthetic 19-mer, which is complementary to this region of *mGK-1*, is shown. (*b*) Ethidium-bromide-stained 1% agarose gel after electrophoresis of *Eco*RI-cleaved DNA from λ clones: (Lane *2*) λMSP-1; (lane *3*) λMSP-22; (lane *4*) λMSP-8; (lane *5*) λMSP-25; (lane *6*) λMSP-7; (lane *7*) λMSP-5; (lane *8*) λMSP-11; (lane *9*) λMSP-20; (lane *10*) λMSP-4. These clones contain 14 different kallikrein genes (see Fig. 2). The marker DNA is λ*c*I857 DNA cut with *Hin*dIII. Sizes in kilobases are 23.5, 9.5, 6.7, 4.4, 2.3, and 2.0. (*c*) Autoradiograph of the Southern blot of *b* hybridized with the 5′-end-labeled *mGK-1* primer. The nitrocellulose filter was hybridized with the labeled primer for 16 hr in 50% formamide, 3 × SSC at 4°C. The filter was washed twice at 21°C in two changes of 2 × SSC. The only fragment that hybridized was the 8.0-kb *Eco*RI fragment containing *mGK-1* from λMSP-1.

protease-active site are located on separate exons. Furthermore, the aspartic acid residue required for cleavage at basic amino acids, a characteristic of these enzymes, is also located on a separate exon. The residues lining the substrate specificity pocket, which tend to be the more variable regions of the protein, are widely distributed throughout the exons (see Fig. 3). These data are consistent with the evolution of a serine protease by the formation of an active site through the joining of different exons—each encoding an amino acid sequence crucial to the final tertiary structure of the catalytic pocket.

Since functional differences exist between the kallikreins so far analyzed, it would appear that such differences also exist between all kallikreins and that each

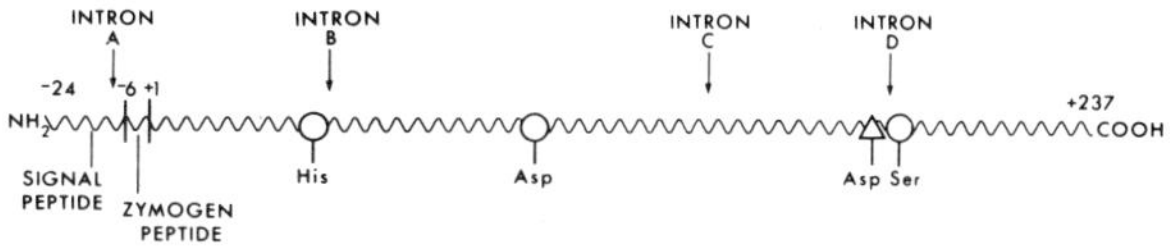

Figure 6. Molecular anatomy of a kallikrein gene. Positions of introns are shown in relation to the structure of a preprokallikrein. (○) Residues involved in the formation of the active site. Aspartic acid (△) confers the ability to cleave at basic residues. (Reprinted, with permission, from Mason et al. 1983.)

individual member is involved in processing a specific peptide precursor. However, at present, it is not known whether all of the 25–30 kallikrein genes are transcriptionally active. In the submaxillary gland there are at least six members known to be expressed (γ-NGF, EGF-BP, γ-renin, β-NGF endopeptidase, pMK-1, and *mGK-1*), and it has been demonstrated by immunocytological techniques that kallikreins are expressed in the submaxillary gland, kidney, spleen, pancreas, and brain cells (Orstavik 1980; Schachter 1980). However, due to the immunological cross-reactivity of many members of this family, it is impossible to ascertain which particular kallikrein protease is being detected. The use of oligonucleotides as gene-specific probes should resolve this problem.

Since EGF-BP and γ-NGF are members of the kallikrein family, it is tempting to speculate that the other kallikreins each act to process a specific bioactive peptide. The most challenging aspect of this work is to determine the specific roles played by each kallikrein gene in the activation and processing of a variety of physiologically important polypeptides. The results gained from the tissue studies should give clues as to what polypeptide each member is processing, since presumably the processed polypeptide would be coordinately expressed with its respective processing enzyme. Furthermore, the availability of cloned kallikrein genes will now permit the introduction of these genes into tissue-culture cells in experiments aimed at understanding the coordinate regulation of substrate (e.g., β-NGF) and processing (e.g., γ-NGF) genes.

ACKNOWLEDGMENTS

A.J.M. is supported by a Commonwealth postgraduate research award and B.A.E. is the recipient of a National Health and Medical Research Council postdoctoral fellowship.

REFERENCES

Baltimore, D. 1981. Gene conversion: Some implications for immunoglobulin genes. *Cell* **24:** 592.

Benton, W.D. and R.W. Davis. 1979. Screening λgt recombinant clones by hybridisation to single plaques *in situ*. *Science* **196:** 180.

Berger, E.A. and E.M. Shooter. 1977. Evidence for pro-B-nerve growth factor, a biosynthetic precursor to B-nerve growth factor. *Proc. Natl. Acad. Sci.* **74:** 3647.

BOTHWELL, M.A., W.H. WILSON, and E.M. SHOOTER. 1979. The relationship between glandular kallikrein and growth factor-processing proteases of mouse submaxillary gland. *J. Biol. Chem.* **254:** 7287.

DOCHERTY, K. and D.F. STEINER. 1982. Post-translational proteolysis in polypeptide hormone biosynthesis. *Annu. Rev. Physiol.* **44:** 625.

FREY, P., R. FORAND, T. MACIAG, and E.M. SHOOTER. 1979. The biosynthetic precursor of epidermal growth factor and the mechanism of its processing. *Proc. Natl. Acad. Sci.* **76:** 6294.

GROSS-BELLARD, M., P. OUDET, and P. CHAMBON. 1973. Isolation of high molecular weight DNA from mammalian cells. *Eur. J. Biochem.* **36:** 32.

HOHN, B. 1979. *In vitro* packaging of lambda and cosmid DNA. *Methods Enzymol.* **68:** 299.

KRIEGER, M., L.M. KAY, and R.M. STROUD. 1974. Structure of specific binding of trypsin: Comparison of inhibited derivatives and a model for substrate binding. *J. Mol. Biol.* **83:** 209.

LIU, C.-P., P.W. TUCKER, J.F. MUSHINSKI, and F.R. BLATTNER. 1980. Mapping of heavy chain genes for mouse immunoglobulins M and D. *Science* **209:** 1348.

MASON, A.J., B.A. EVANS, D.R. COX, J. SHINE, and R.I. RICHARDS. 1983. Structure of mouse kallikrein gene family suggests a role in specific processing of biologically active peptides. *Nature* **303:** 300.

NEURATH, H. and K.A. WALSH. 1976. Role of proteolytic enzymes in biological regulation: A review. *Proc. Natl. Acad. Sci.* **73:** 3825.

ORSTAVIK, T.B. 1980. The kallikrein-kinin system in exocrine organs. *J. Histochem. Cytochem.* **28:** 881.

RICHARDS, R.I., D.F. CATANZARO, A.J. MASON, B.J. MORRIS, J.D. BAXTER, and J. SHINE. 1982. Mouse glandular kallikrein genes. *J. Biol. Chem.* **257:** 2758.

SCHACHTER, M. 1980. Kallikreins (kininogenases)—A group of serine proteases with bioregulatory actions. *Pharmacol. Rev.* **31:** 1.

SERVER, A.C. and E.M. SHOOTER. 1976. Comparison of the arginine esteropeptidases associated with the nerve and epidermal growth factors. *J. Biol. Chem.* **251:** 165.

SOUTHERN, E.M. 1975. Detection of specific sequences among DNA fragments separated by gel electrophoresis. *J. Mol. Biol.* **98:** 503.

TAYLOR, J.M., R. ILLMENSEE, and J. SUMMERS. 1976. Efficient transcription of RNA into DNA by avian sarcoma virus polymerase. *Biochim. Biophys. Acta* **442:** 324.

TSCHESCHE, H., G. MUIR, G. GODFEC, F. FIELDER, W. EHRET, C. HIRSCHAUER, M. LEMON, H. FRITZ, G. SCHMIDT-KASTNER, and C. KUTZBACH. 1979. The primary structure of porcine glandular kallikreins. In *Kinins II: Biochemistry, pathophysiology and clinical aspects* (ed. S. Fujii), p. 245. Plenum Press, New York.

WHITFELD, P.L., P.H. SEEBURG, and J. SHINE. 1982. The human pro-opiomelanocortin gene: Organisation, sequence and interspersion with repetitive DNA. *DNA* **1:** 133.

WILSON, W.H. and E.M. SHOOTER. 1979. Structural modification of the amino-terminus of nerve growth factor. *J. Biol. Chem.* **254:** 6002.

YOUNG, C.L., W.C. BARKER, C.M. TOMASELLI, and M.O. DAYHOFF. 1978. In *Atlas of protein sequence and structure* (ed. M.O. Dayhoff), vol. 5, p. 73. National Biomedical Research Foundation, Silver Spring, Maryland.

The Biosynthesis and Processing of Proteins in the Mouse 7S Nerve Growth Factor Complex

T.L.J. DARLING,* P.E. PETRIDES,* P. BEGUIN,* P. FREY,* E.M. SHOOTER,* M. SELBY,† AND W.J. RUTTER†

*Department of Neurobiology, Stanford University School of Medicine, Stanford, California 94305;
†Department of Biochemistry and Biophysics, University of California, San Francisco, California 94143

Although a great deal is known about the structure of the nerve growth factor (NGF) and its role in the development and maintenance of sensory and sympathetic neurons, a number of significant questions remain to be answered concerning both the physiological sources of the factor and the mechanisms of its actions. Two main approaches to the first question, the identity of the tissues or cells that produce NGF, have involved the development of specific and highly sensitive radioimmunoassays for NGF (Thoenen et al., this volume) and the isolation of the cDNA for the NGF mRNA (Scott et al. 1983; Ullrich et al. 1983). It is reasonable to assume that the application of these two methodologies will lead quite rapidly to a better understanding of the sites of biosynthesis of NGF. A considerable amount of indirect evidence suggests that the target tissues of both sensory and sympathetic neurons are principal sites of NGF synthesis (Thoenen and Barde 1980). The evidence includes the demonstration that developing sensory and sympathetic axons are dependent on NGF (Hendry 1975; Levi-Montalcini et al. 1975) and that at least one target organ, the submaxillary gland, can be replaced by a slowly releasing source of NGF without impairing sympathetic innervation (Dibner et al. 1977). A specific and physiologically significant retrograde flow of NGF from target organ to sensory and sympathetic cell bodies has been demonstrated (Johnson et al. 1978; Dumas et al. 1979), and sympathetic neurons have been shown to survive and to continue axonal growth when only the growing tips are exposed to NGF (Campenot 1977). The findings that rat irises produce significant amounts of NGF when placed in culture or when they are denervated in situ and that this production falls to undetectable levels (at least as judged by bioassay) when reinnervation occurs also support the idea (Ebendal et al. 1980). Whether target organs are the sources of NGF for the sensory and sympathetic neurons as they acquire dependency on NGF in embryonic or early postnatal development remains to be determined.

The successful production of the cDNA for NGF mRNA has been based on the structure of NGF isolated from one of its most prolific sources, the submaxillary gland of the adult male mouse (Scott et al. 1983; Ullrich et al. 1983). The end product of NGF biosynthesis in this gland is the complex 7S NGF, which contains the β-NGF dimer (described here as NGF), the trypsinlike γ-subunit, the acidic α-subunit, and zinc ions (Varon et al. 1968; Pattison and Dunn 1975). The two peptide chains in the NGF dimer are identical (Angeletti et al. 1971; Greene et al. 1971), containing 118 amino acids and 3 intrachain disulfide bridges and ending in carboxyterminal arginine residues (Angeletti and Bradshaw 1971). Earlier work, noting the presence of a trypsinlike enzyme in the complex with the NGF dimer (Greene et al. 1969), led to the hypothesis that the γ-subunit functions to produce the NGF peptide chain by cleavage of a larger precursor at the arginyl residues, which become the carboxyterminal residues of NGF (Angeletti and Bradshaw 1971; Perez-Polo et al. 1972). In support of this hypothesis, Server and Shooter (1976) noted that other homologous trypsinlike enzymes could not replace the γ-subunit in the 7S NGF complex, whereas Bothwell and Shooter (1978), finding that the enzymatic activity of the γ-subunit is suppressed almost entirely in 7S NGF and that NGF is a competitive inhibitor of the γ-subunit, provided evidence that the carboxyterminal arginyl residue of NGF interacts with the active site of the γ-subunit. Indeed, removal of the carboxyterminal arginyl residues from NGF produces a protein that no longer forms a 7S NGF complex (Moore et al. 1974).

The synthesis of NGF in the mouse submaxillary gland was first demonstrated by Berger and Shooter (1978), who also provided evidence from the kinetics of the labeling in gland slices and by peptide mapping of a larger precursor with a molecular weight of 22,000, which is cleaved by the γ-subunit to yield a peptide chain of the size of NGF (Berger and Shooter 1977). The determination of the nucleotide sequence of the NGF mRNA leads to predictions about the biosynthesis of NGF that confirm and extend these observations as reported here.

EXPERIMENTAL PROCEDURES

Nerve growth factor and its antiserum. 7S NGF was isolated from the submaxillary glands of 60-day-old male Swiss Webster mice by the procedure of Varon et al. (1972) as modified by Darling and Shooter (1983). The modifications include addition of 1 μM $ZnCl_2$ to all buffers to increase the stability of the 7S NGF complex. The NGF protein was prepared from 7S NGF according to Smith et al. (1969) and further purified as follows. A 1-ml sample of NGF (2–10 mg/ml) was adjusted to 2 N in acetic acid by the addition of

0.136 ml of glacial acetic acid and chromatographed on a column of Sephadex G-75 (0.8 × 120 cm) equilibrated with 2 N acetic acid. The flow rate was 7.5 ml/hr and the NGF eluted as a major peak with an apparent molecular weight of 22,000. The pooled eluate was concentrated in ¼-inch dialysis tubing (Union Carbide) against 0.02% acetic acid.

The α-subunit was prepared from 7S NGF by the method of Smith et al. (1969) and further purified by gel filtration on a Bio-Gel P-60 column (2.5 × 40 cm in 50 mM phosphate buffer, pH 7.2) and by passage through an affinity column (1 × 10 cm in 50 mM phosphate buffer, pH 7.2) of Aprotinin (Sigma) coupled to cyanogen-bromide-activated Sepharose 4B (Pharmacia).

Antisera were raised in New Zealand white rabbits according to Ishii and Shooter (1975). The IgG fraction was prepared by the method of Reif (1969) and after concentration through an Amicon X-100 membrane and dialysis against 10 mM sodium phosphate buffer, pH 7.2, containing 0.15 M NaCl stored at −20°C. For some experiments, the IgG fraction was purified on an NGF-affinity column. Approximately 5 mg of NGF was filtered through a 3.5-ml column of Affigel (Bio-Rad) according to the manufacturer's instructions. Unreacted sites were blocked with 0.1 M glycine; the capacity of the column was 1 mg NGF/ml settled gel. About 20 ml of the anti-NGF antiserum was applied to the column followed by washing with 0.1 M Tris-Cl buffer, pH 7.6, and 0.1 M sodium acetate buffer, pH 4.0. The IgG fraction was eluted with 0.1 N HCl with the fractions being collected in 1 M Tris-Cl buffer, pH 8.0. Flow rate was 0.5 ml/hr. The IgG pool was dialyzed against 0.1 M sodium acetate buffer, pH 4.0, and stored at −20°C.

In vitro incorporation experiments in submaxillary gland slices. Male mice were castrated and kept for 21 days to reduce endogenous levels of NGF in the submaxillary gland. They were then injected daily with testosterone for 2 to 4 days as described by Ishii and Shooter (1975). Glands were removed immediately, sliced, and weighed, and samples (200–400 mg tissue) were put into organ culture with 1 ml of Dulbecco's modified Eagle's medium (DME) (GIBCO) containing [^{3}H]valine (2.5 mCi/ml), [^{35}S]cystine (6 mCi/ml), or [^{35}S]methionine (0.1 Ci/ml) at 37°C and 5% CO_2. After an appropriate time, the slices were rinsed with DME containing 1 mg/ml valine, cysteine, or methionine and homogenized in sodium acetate buffer (0.41 g sodium acetate, 2.92 g NaCl per 100 ml and brought to pH 4.0 with acetic acid) containing 1% Triton X-100, 1 mg/ml bovine serum albumin, 1 mg/ml L-valine, L-cystine, or L-methionine, 200 μg/ml sodium azide, 50 μg/ml pancreatic trypsin inhibitor, and 5 mM phenylmethanesulfonyl fluoride. After centrifugation (105,000*g* for 60 min), the supernatant was stored on ice while an aliquot was subjected to immunotitration as described by Berger and Shooter (1977). A further 0.32 ml of the supernatant was then used for single-step immunoprecipitation by addition of antiserum to the equivalence point. After incubation overnight at 4°C, the immunoprecipitate was collected by centrifugation, washed with sodium acetate buffer, pH 4, and, after dissolving in 0.1 N NaOH, subjected to a second immunoprecipitation (Berger and Shooter 1977). The labeled proteins in the washed, second immunoprecipitate were solubilized by boiling for 5 minutes in sample buffer containing SDS (Laemmli 1970) and analyzed by electrophoresis in 14% (tube) or 16.5% (slab) acrylamide gels.

Cell-free translation of submaxillary gland RNA. RNA from male submaxillary glands was isolated by the method of Chirgwin et al. (1979) except that sodium *N*-lauroylsarcosine and antifoam were omitted and translated at a concentration of 0.7 mg/ml in a rabbit reticulocyte lysate (Pelham and Jackson 1976). Poly(A) mRNA was isolated by chromatography on oligo(dT) cellulose (Aviv and Leder 1972) and used at concentrations of 50–70 μg. Reaction mixtures (0.3–1.0 ml) contained 0.5–2.0 mCi L-[^{35}S]methionine or L-[^{35}S]cystine and were incubated for 1 or 2 hours. The translation of total gland RNA was carried out both in the absence and presence of anti-NGF IgG (100 μg/ml). At the end of the reaction, anti-NGF antiserum (10 μl) was added, followed 1 hour later by 200 μl of a 10% suspension of fixed *Staphylococcus aureus* cells (Enzyme Center, Boston). The immunoprecipitate was washed with buffer containing 1% Triton X-100 and then with a solution of 10 μg/ml trypsin inhibitor in water prior to solubilization in sample buffer. For the precipitation of intermediates in α-subunit biosynthesis, the *S. aureus* cells were prewashed with 1 mg/ml protamine sulfate and with 15% serum from a nonimmunized rabbit. A volume (2.5–5.0 ml) of the anti-α-subunit antiserum was added to the reaction mixture, the mixture was incubated overnight at 4°C, and 15–30 μl of a 10% suspension of the washed *S. aureus* cells was added. The cells with adsorbed immune complexes were washed several times with 2% Sarkosyl in 50 mM Tris buffer, pH 7.4, prior to solubilization in sample buffer.

Peptide maps. [^{3}H]Valine-containing tryptic peptides were obtained from the intermediates in NGF biosynthesis after separation in acrylamide gel. Prior to electrophoresis, the immunoprecipitate was oxidized with performic acid (Glazer et al. 1975). Appropriate gel sections were washed sequentially in 25% isopropanol and 10% methanol and dried under a heat lamp. A volume of 1 ml of trypsin (50 μg) in 50 mM ammonium bicarbonate was added, and the solution was incubated overnight at 37°C. The supernatant was removed and counted, and aliquots (0.3 ml) were diluted with an equal volume of 0.5% trifluoracetic acid prior to loading directly onto a reverse-phase column (Supelcosil RP18, 5-μm particle size). After 30 minutes of equilibration with Buffer A (0.5% trifluoracetic acid), the column was developed with a 210-minute gradient to 35% Buffer B (80% acetonitrile in 0.5% trifluoracetic acid) at a flow rate of 0.5 ml/min. Fractions were collected at 90-second intervals.

The L-[^{35}S]cysteine-labeled α-subunit precursor was extracted from appropriate gel slices, carboxymethylated, and digested with trypsin. Purified α-subunit was carboxymethylated with [^{3}H]iodocetate and trypsinized as described earlier (Berger and Shooter 1977). The ^{35}S- and ^{3}H-labeled peptides were mixed and analyzed by isoelectric focusing in acrylamide gel in 2% Ampholines (LKB), pH 3–6.

RESULTS

Predictions from the cDNA Nucleotide Sequence for NGF

The predicted NGF mRNA sequence has the longest open reading frame beginning at a methionine residue 187 residues upstream from the aminoterminal serine residue of NGF (Fig. 1). Translation of this mRNA would produce an initial gene product, prepro-NGF, comprising 307 amino acids and a molecular weight of 33,800. If either of the two other methionine codons nearer to the start of the NGF structural sequence is used for initiation, the weight of the prepro-NGF is reduced by approximately 7300 and 7500, respectively. The termination of the signal peptide sequence for the longest prepro-NGF is not entirely clear, but it is assumed in Figure 1 that release of the signal sequence removes 25 amino acids resulting in a pro-NGF with a molecular weight of 30,800. Because the pro sequence of pro-NGF contains both a pair of arginine residues and a group of four contiguous basic amino acid residues, intermediates in the processing of pro-NGF may occur. These potential intermediates have molecular weights of 22,000 and 18,500, respectively. The alignment of intermediates in Figure 1 is not meant to imply any particular order in the appearance of these intermediates nor of the stage in the processing when the carboxyterminal dipeptide is released. The predicted amino acid sequence of prepro-NGF has three potential *N*-glycosylation sites, two of which, in the pro sequence, are marked by asterisks in Figure 1, and a third is in the NGF sequence; the latter is not glycosylated in NGF (Angeletti et al. 1973).

Intermediates in NGF Biosynthesis Identified by the In Vitro Labeling

A relatively stable NGF precursor with a molecular weight of 22,000 was identified by Berger and Shooter (1977). This 22K species possessed all the characteristic cysteine-containing tryptic peptides of NGF, and on the basis of the kinetics of its continuous or pulse-chase labeling, these authors identified it as pro-NGF. However, this species clearly arises by cleavage of the longer pro-NGF after the Arg-Arg site in the pro sequence (Fig. 1). Whether it retains the extra carboxyterminal dipeptide is not known. A second intermediate of smaller molecular weight was also observed in small amounts in the immunoprecipitate, presumably the 18.5K species.

Precipitation of an in vitro, [^{3}H]valine-labeled incubation mixture with a different anti-NGF antiserum revealed an NGF species of larger molecular weight than the 22K species (Fig. 2). The kinetics of the chase of label from this 29K species was similar to that from the 22K species (Fig. 3). The incorporation of radioactivity into trichloroacetic-acid-precipitable material was blocked after about 1 hour of pulse. Thus, the continued labeling of NGF after this time interval at the expense of both 22K and 29K species suggests that the 29K is being converted into NGF and is an NGF precursor. Its molecular weight is close to that anticipated for pro-NGF (Fig. 1).

Yet another anti-NGF antiserum precipitated a species with a molecular weight of 34,000–35,000 as well as the lower-molecular-weight species (Fig. 4). In this experiment an additional protease inhibitor, leupeptin, was added, and the immunoprecipitation was carried out immediately at the end of the incubation of the slices. The preponderance of the two higher-molecular-weight components in this analysis compared with the amounts of 22K, 20K, and 13K (NGF) species suggests that processing of these two species to lower-molecular-weight NGF precursors was, as a consequence, inhibited. The corresponding species were observed when an affinity-purified IgG fraction was used (Fig. 5), although in this experiment the longer time before immunoprecipitation resulted in a different distribution of these components.

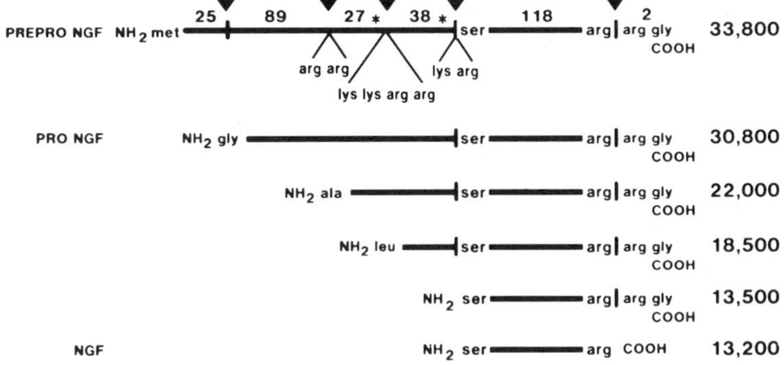

Figure 1. Intermediates in NGF biosynthesis, based on the nucleotide sequence of NGF mRNA (Scott et al. 1983; Ullrich et al. 1983).

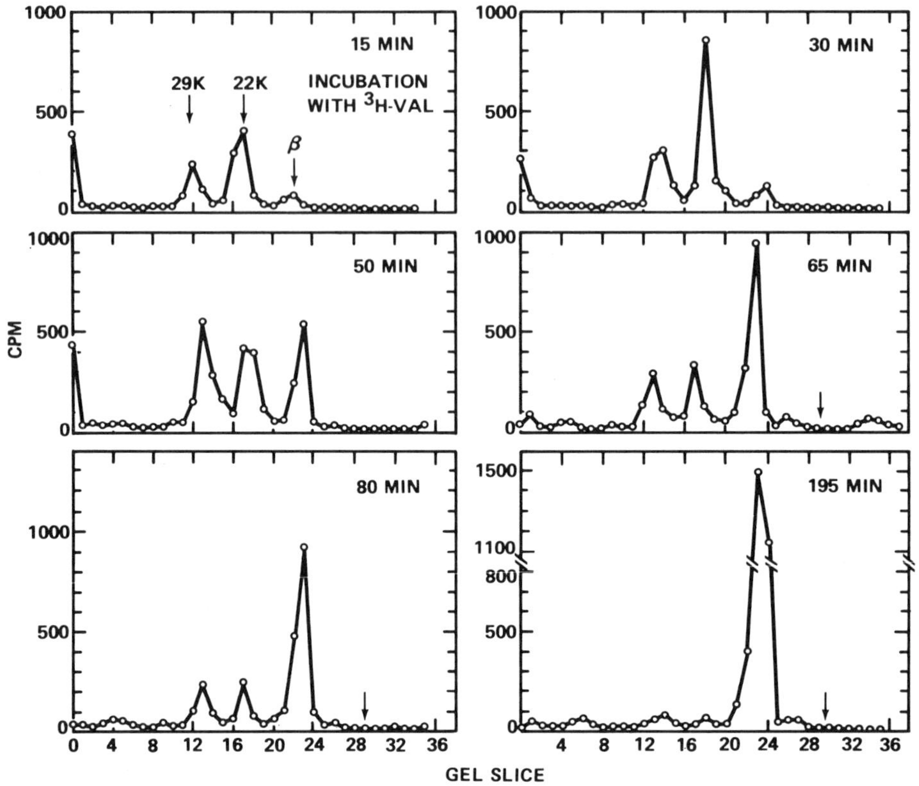

Figure 2. Pulse-chase labeling of NGF species in slices of the mouse submaxillary gland. Tissue slices (200 mg) were obtained as described in Experimental Procedures and incubated for 15 min with [^{3}H]valine before addition of excess unlabeled valine. The times indicated in each box include the 15-min pulse. Tissue extracts were immunoprecipitated and analyzed by electrophoresis under denaturing conditions as described in Experimental Procedures.

Peptide Mapping of the Intermediates in NGF Biosynthesis

The three major 34K–35K, 29K–32K, and 22K species were isolated, and their tryptic peptides were compared with those of NGF after separation by reverse-phase high-performance liquid chromatography (HPLC). NGF contains five [^{3}H]valine-containing tryptic peptides. More than five peptides were observed in the analysis (Fig. 6), the extra peptides presumably resulting from side reactions during the performic acid oxidation. Peptides eluting in fractions 53, 57, 71, 73, 82, and 107 were common to NGF and the three higher-molecular-weight species, providing evidence that the latter peptides contain the NGF sequence. Thus, the molecular weights and peptide maps of the four labeled proteins precipitated by anti-NGF antiserum are consistent with predictions outlined in Figure 1 for intermediates in NGF biosynthesis.

Cell-free Translation of Submaxillary Gland RNA

Preliminary results with the cell-free translation of total submaxillary gland RNA identified NGF species with molecular weights similar to those observed in the in vitro-labeling experiments. Addition of the purified NGF IgG to the incubation mixture at the end of the incorporation of label produced relatively few specifically precipitated counts, and analyses of the precipitate showed only a broad band of label migrating between the position of NGF and the dye front. However, if the NGF IgG was added to the cell-free incubation during incorporation of the labeled amino acid, significant incorporation of label into the immunoprecipitate was observed, and subsequent analyses revealed species with molecular weights of 35,000, 29,000, 22,000, 19,000, and 13,000. The appearance of the lower-molecular-weight NGF species suggests that a number of active trypsinlike enzymes are also being produced in the cell-free translation of the total gland RNA, which in turn results in the processing of prepro-NGF. It is significant that the size of the highest-molecular-weight species agrees with the prediction in Figure 1 for a prepro-NGF initiated at the first methionine residue, but it is unexpected to observe the apparent release of the signal peptide sequence and the appearance of the putative 29K pro-NGF.

Precursors in α-Subunit Biosynthesis

Similar in vitro incorporation experiments to those reported above but using anti-α-subunit antiserum to precipitate the products identified the precursor in α-subunit biosynthesis as a 32K single peptide chain. During continuous labeling, increasing amounts of radioactivity appeared in the 32K species over the first

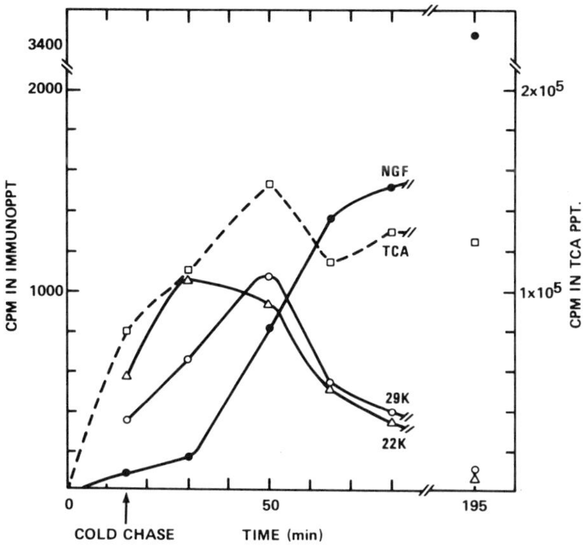

Figure 3. Time course of β-NGF biosynthesis in the mouse submaxillary gland; the kinetics of the pulse-chase labeling of the 29K and 22K NGF intermediates. Data were obtained from the analyses shown in Fig. 2.

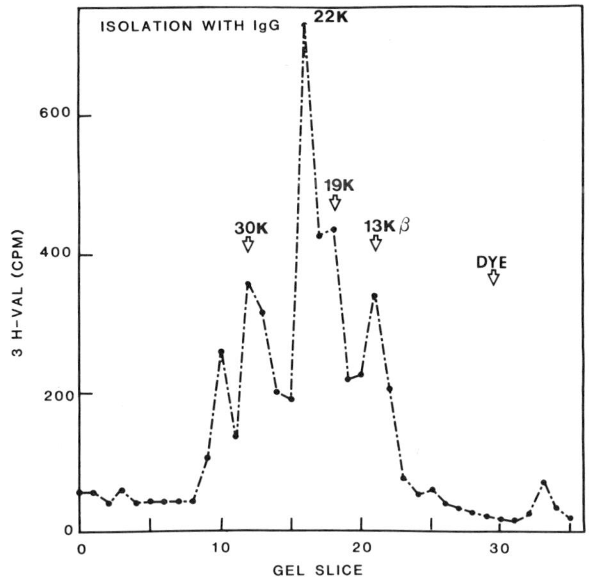

Figure 5. Intermediates in NGF biosynthesis in slices of submaxillary gland precipitable by NGF IgG. The experiment was performed as described in the legend to Fig. 2 except that affinity-purified NGF IgG was used instead of antiserum.

2 hours (Fig. 7). After this time the amount of the 32K species remained constant while label in the α-subunit increased at a linear rate. When the electrophoretic analyses were carried out under reducing conditions, it was observed that radioactivity appeared in both the 16K and 10K peptide chains, which are joined by disulfide bridges in the intact α-subunit, and that the rate of increase of label in both chains was similar. In a pulse-chase experiment, addition of unlabeled L-cysteine caused a continuous decrease in the trichloroacetic-acid-precipitable protein as well as in the 32K species but an increase in the label associated with the α-subunit. These experiments together with the identity of the tryptic peptides from the α-subunit and the 32K species confirmed that the 32K species is a precursor of the α-subunit and that its processing involves two simultaneous proteolytic cleavages to generate the two-peptide-chain α-subunit.

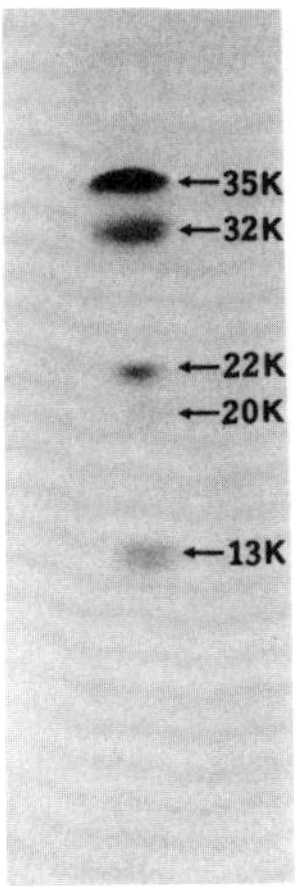

Figure 4. Intermediates in NGF biosynthesis in slices of submaxillary gland. Tissue slices (0.67 g in 2 ml of medium) were incubated with [^{3}H]valine (7 mCi) for 1 hr and then homogenized as described in Experimental Procedures, but with the addition of leupeptin (100 μg/ml). Immunoprecipitation was carried out immediately by adding anti-NGF antiserum to 70% excess beyond the equivalence point. Immunoprecipitates were washed with trypsin inhibitor (50 μg/ml) and leupeptin (100 μg/ml) and analyzed as described in Experimental Procedures. The gel was stained, treated with En^{3}Hance, and fluorographed.

In cell-free translation, anti-α-subunit antiserum precipitated a single component with a molecular weight of 30,000, which migrated slightly faster than the 32K precursor identified in the in vitro labeling. These results are consistent with the identification of the cell-free translation product as prepro-α-subunit and the in vitro-labeled product as a glycosylated pro-α-subunit.

DISCUSSION

The identification of the initial gene product for NGF has been hampered by the lack of an NGF antibody that recognizes the longer NGF precursor (Thoenen and Barde 1980). One of the antibodies used in this work has this specificity, and the sizes of the four labeled species precipitated by this antibody, in both in vitro and cell-free experiments, are consistent with one set of predictions from the nucleotide sequence of the NGF mRNA. The presence of two potential *N*-glycosylation sites in the precursor would lead to ambiguities in the molecular weights of pro-NGF (31K) and of the 22K and 19K species, and thus the question as to whether these sites are modified needs to be resolved. In spite of the uncertainty of predicting the signal peptide from the amino acid sequence of the precursor, the present data do appear to rule out initiation of translation at the sec-

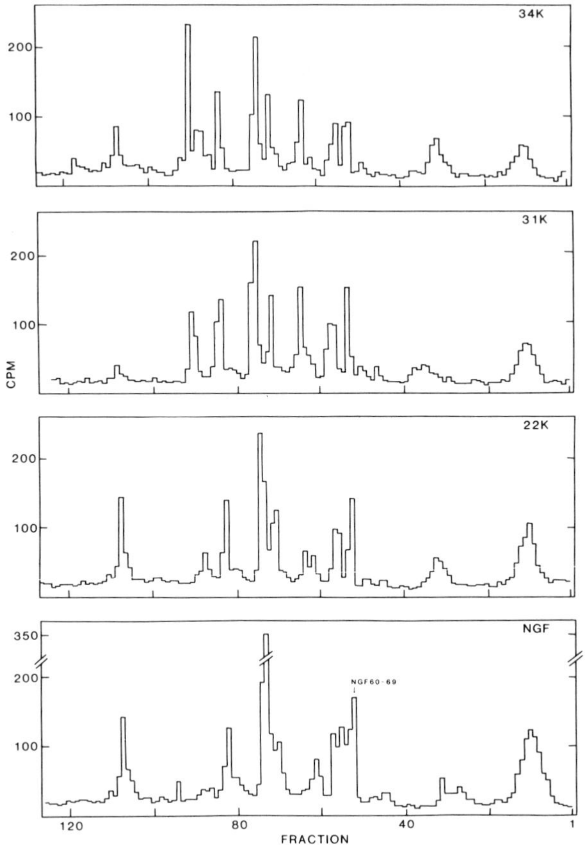

Figure 6. [^{3}H]Valine-containing tryptic peptides of the intermediates in NGF biosynthesis. Analysis was by HPLC as described in Experimental Procedures.

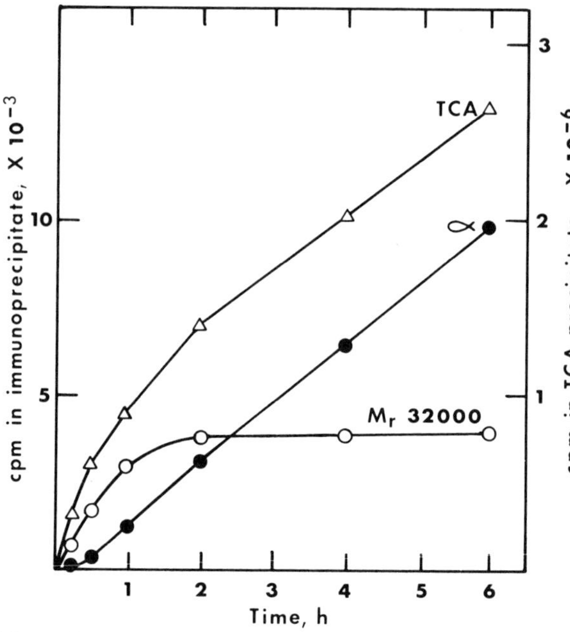

Figure 7. Time course of the labeling of the α-subunit and its 32K precursor in slices of mouse submaxillary glands. Tissue (400 mg) was incubated in 0.8 ml of cysteine-free medium with [^{35}S]cysteine (1.1 mCi). At the indicated times, aliquots (60–80 μg) of tissue were removed and homogenized in 2.5 ml of extraction buffer, and immunoprecipitates were obtained with anti-α-subunit antiserum and analyzed as described in Experimental Procedures.

ond or third methionine codons in the mRNA. If this were to occur, the largest precursor would have a molecular weight of only 27,000, and the molecular-weight difference between this and the 34K species is considerably greater than any apparent molecular-weight differences produced by glycosylation.

It is not possible to identify from these results the exact pathway used in the in vivo processing of the NGF precursor, although all potential proteolytic cleavage sites in the precursor (outside of the NGF protein) are used. As noted earlier, the 22K species is a significant intermediate in the processing of pro-NGF (Berger and Shooter 1977), whereas the smaller 19K species only appears in relatively small amounts. Earlier models had suggested that some or all of the extrapeptide material in the NGF precursor was an extension at the carboxyl terminus (Angeletti and Bradshaw 1971; Server and Shooter 1977). This is confirmed by the structure of the NGF mRNA, although the extension at the carboxyl terminus is only two amino acids with the bulk of the pro sequence at the amino terminus of NGF. The resolution of the electrophoretic system is not sufficient to determine when the carboxyterminal dipeptide is removed or, indeed, whether it is removed as a dipeptide rather than sequentially in amino acid residues. What is clear, however, is that the enzyme that removes the dipeptide (or the last arginine residue before the NGF carboxyterminal arginine) is the γ-subunit, for it is this enzyme that appears in the 7S NGF complex and that cannot be replaced in the complex by at least some of the other proteases in the gland (Server and Shooter 1976). It will be of interest to demonstrate the specificity of the cleavage at the carboxyl terminus directly on the isolated pro-NGF.

The γ-subunit is also capable of cleaving after the Lys-Arg sequence that determines the amino terminus of NGF (Berger and Shooter 1977), but whether this processing is highly specific is not yet known. It was observed that other proteases besides the γ-subunit were capable of processing the 22K species to NGF, but these experiments were carried out either on partially denatured 22K species or on the latter adsorbed to the NGF antibody (Berger and Shooter 1977). Seven different trypsinlike (kallikrein) enzymes have been identified in the male mouse submaxillary gland (Schachter 1969; Bothwell et al. 1979). They share extensive immunological cross-reactivity and amino acid sequence homology (Thomas et al. 1981; Bothwell et al. 1979). The cDNA for one of those kallikrein enzymes has been isolated and characterized (Richards et al. 1982), and hybridization experiments with restriction fragments of this cDNA have revealed a large gene family (>40 members; Shine et al., this volume) of kallikreins having approximately 75% amino acid sequence homology with each other (Mason et al. 1983). Since each of the enzymes thus far examined display amino acid differences in the active site region, it is likely that each member of this family will have rather specific substrates.

When cell-free translation of submaxillary gland mRNA is carried out with L-[^{35}S]cysteine and the incubation mixture is precipitated with anti-γ-subunit antiserum, a broad, labeled band is observed in the molecular-weight region of 29K–31K; precipitation of this band is inhibited when unlabeled γ-subunit is added along with the antiserum (L. Guidice and E. Shooter, unpubl.). This result is consistent with the average length of the family of preprokallikreins (261 amino acid residues) reported recently (Shine et al., this volume).

The biosynthesis of the α-subunit involves a prepro-α-subunit and pro-α-subunit, with the latter being cleaved simultaneously at two sites to produce the α-subunit. In the experiments reported here, the anti-α-subunit antiserum precipitated one species for both the prepro and pro forms. Other antisera precipitated two bands from cell-free translation mixtures, the one with a molecular weight of 30,000 noted here and another of slightly smaller molecular weight (29,000). Neither of these two species was completely chased by addition of unlabeled α-subunit, but the 29K species was eliminated by addition of unlabeled γ-subunit prior to immunoprecipitation (L. Giudice and E. Shooter, unpubl.). This apparent cross-reactivity of the precursors for the α- and γ-subunits is of interest in terms of reports that the α-subunit contains two closely related proteins, possibly arising from two separate genes (Isackson et al. 1983).

Finally, the characterization of the NGF mRNA clears up a controversy concerning the form of NGF in the mouse submaxillary gland. Young et al. (1978) suggested that the NGF protein was not present in the gland in the 7S NGF complex as originally proposed (Varon et al. 1967). Instead it was argued that the major form of NGF in the gland (HMW-NGF) had a molecular weight of 116,000 (Young et al. 1978) and that the NGF protein was released from this larger precursor form by autocatalytic proteolysis (Young 1979). The defined size of the initial gene product (34K) disproves this latter hypothesis. The HMW-NGF species isolated by Young et al. (1978) is, in fact, a 7S NGF complex that, because of the way it was isolated, contains a preponderance of the γ^3-subunit (D. Estell and E.M. Shooter, unpubl.). The γ^3-subunit, one of the several posttranslational modifications of the γ-subunit (Burton and Shooter 1981; Thomas et al. 1981), makes the most stable 7S NGF complexes (Nichols and Shooter 1983), and it is this type of 7S NGF complex that survives the isolation procedure devised by Young et al. (1978).

ACKNOWLEDGMENTS

This research was supported by grants from the National Institutes of Health (NS-04270 to E.M.S. and GM-21344 to W.J.R.), the National Science Foundation (BNS-79-14088), and the Dysautonomia Foundation. T.D. was supported by a National Institutes of Health fellowship (5-T32-AM-07217), and P.F. was supported by a fellowship from the Swiss National Science Foundation.

REFERENCES

Angeletti, R.A. and R.A. Bradshaw. 1971. Nerve growth factor from mouse submaxillary gland: Amino acid sequence. *Proc. Natl. Acad. Sci.* **68:** 2417.

Angeletti, R.H., R.A. Bradshaw, and R.D. Wade. 1971. Subunit structure and amino acid composition of mouse submaxillary gland nerve growth factor. *Biochemistry* **10:** 463.

Angeletti, R.H., M.A. Hermodson, and R.A. Bradshaw. 1973. Amino acid sequences of mouse 2.5S nerve growth factor II. Isolation and characterization of the thermolytic and peptic peptides and the complete covalent structure. *Biochemistry* **12:** 100.

Aviv, H. and P. Leder. 1972. Purification of biologically active globin messenger RNA by chromatography on oligothymidylic acid-cellulose. *Proc. Natl. Acad. Sci.* **69:** 1408.

Berger, E.A. and E.M. Shooter. 1977. Evidence for pro-βNGF, a biosynthetic precursor to β nerve growth factor. *Proc. Natl. Acad. Sci.* **74:** 3647.

——. 1978. The biosynthesis of β nerve growth factor in mouse submaxillary gland. *J. Biol. Chem.* **243:** 804.

Bothwell, M.A. and E.M. Shooter. 1978. Thermodynamics of the interaction of the subunits of 7S nerve growth factor. *J. Biol. Chem.* **253:** 8458.

Bothwell, M.A., W.H. Wilson, and E.M. Shooter. 1979. The relationship between glandular kallikrein and growth factor-processing proteases of mouse submaxillary gland. *J. Biol. Chem.* **254:** 7287.

Burton, L.E. and E.M. Shooter. 1981. The molecular basis of the heterogeneity of the γ-subunit of 7S nerve growth factor. *J. Biol. Chem.* **256:** 1011.

Campenot, R.B. 1977. Local control of neurite development by nerve growth factor. *Proc. Natl. Acad. Sci.* **74:** 4516.

Chirgwin, J., A. Przybyla, R. Macdonald, and W.J. Rutter. 1979. Isolation of biologically active ribonucleic acid from sources enriched in ribonuclease. *Biochemistry* **18:** 5294.

Darling, T.L. and E.M. Shooter. 1983. Methods for the preparation and assay of nerve growth factor. In *Methods of cell culture* (ed. G. Sato et al.). Alan R. Liss, New York. (In press.)

Dibner, M.D., C. Mytilineou, and I.B. Black. 1977. Target organ regulation of sympathetic neuron development. *Brain Res.* **123:** 301.

Dumas, M., M.E. Schwab, and H. Thoenen. 1979. Retrograde axonal transport of specific macromolecules as a tool for characterizing nerve terminal membranes. *J. Neurobiol.* **10:** 179.

Ebendal, T., L. Olson, A. Seiger, and K.O. Hedlund. 1980. Nerve growth factors in the rat iris. *Nature* **286:** 25.

Greene, L.A., E.M. Shooter, and S. Varon. 1969. Subunit interaction and enzymatic activity of mouse 7S nerve growth factor. *Biochemistry* **8:** 3735.

Greene, L.A., S. Varon, A. Piltch, and E.M. Shooter. 1971. Substructure of the β subunit of mouse 7S nerve growth factor. *Neurobiology* **1:** 37.

Glazer, A.N., R.J. Delange, and D.S. Sigman. 1975. *Chemical modification of proteins* (ed. T.S. Work and E. Work), p. 23. Elsevier, New York.

Hendry, I.A. 1975. The response of adrenergic neurons to axotomy and nerve growth factor. *Brain Res.* **94:** 87.

Isackson, P.J., D.K. Hopson, S.J. Nisco, and R.A. Bradshaw. 1983. Characterization of two forms of the α subunit of mouse submandibular gland 7S nerve growth factor. *Fed. Proc.* **42:** 1831.

Ishii, D.N. and E.M. Shooter. 1975. Regulation of nerve growth factor synthesis in mouse submaxillary glands by testosterone. *J. Neurochem.* **25:** 843.

Johnson, E.M., Jr., R.Y. Andres, and R.A. Bradshaw. 1978. Characterization of retrograde transport of nerve growth factor (NGF) using high specific activity (^{125}I) NGF. *Brain Res.* **150:** 319.

Laemmli, U.K. 1970. Cleavage of structural proteins during assembly of the head of bacteriophage T4. *Nature* **227:** 680.

Levi-Montalcini, R., L. Aloe, E. Mugnaini, F. Oesch, and H. Thoenen. 1975. Nerve growth factor induces volume increase and enhances tyrosine hydroxylase synthesis in chemically axotomized sympathetic ganglia of newborn rats. *Proc. Natl. Acad. Sci.* **72:** 595.

Mason, A.J., B.A. Evans, D.R. Cox, J. Shine, and R.I. Richards. 1983. Structure of mouse kallikrein gene family suggests a role in specific processing of biologically active peptides. *Nature* **303:** 300.

Moore, J.B., Jr., W.C. Mobley, and E.M. Shooter. 1974. Proteolytic modification of the β nerve growth factor protein. *Biochemistry* **13:** 833.

Nichols, R.A. and E.M. Shooter. 1983. Characterization of the differential interaction of the microheterogeneous forms of the γ subunit of 7S nerve growth factor with natural and synthetic ligands. *J. Biol. Chem.* (in press).

Pattison, S.E. and M.F. Dunn. 1975. On the relationship of zinc ion to the structure and function of the 7S nerve growth factor protein. *Biochemistry* **14:** 2733.

Pelham, H.R.B. and R.J. Jackson. 1976. An efficient mRNA-dependent translation system from reticulocyte lysates. *Eur. J. Biochem.* **67:** 247.

Perez-Polo, J.R., J.R. Bamburg, W.W.W. De Jong, D. Straus, M. Baker, and E.M. Shooter. 1972. Nerve growth factors of the mouse submaxillary gland. In *Nerve growth factor and its antiserum* (ed. E. Zaimis and J. Knight), p. 19. Athlone Press, London.

Richards, R.I., D.F. Catanzaro, A.J. Mason, B.J. Morris, J.D. Baxter, and J. Shine. 1982. Mouse glandular kallikrein genes. Nucleotide sequence of cloned cDNA coding for a member of the kallikrein arginyl esteropeptidase group of serine proteases. *J. Biol. Chem.* **257:** 2758.

Schachter, M. 1969. Kallikreins and kinins. *Physiol. Rev.* **49:** 509.

Scott, J., M. Selby, M. Urdea, M. Quiroga, G.I. Bell, and W.J. Rutter. 1983. Isolation and nucleotide coding sequence of a cDNA encoding the precursor of mouse nerve growth factor. *Nature* **302:** 538.

Server, A.C. and E.M. Shooter. 1976. A comparison of the arginine esteropeptidases associated with the nerve and epidermal growth factor. *J. Biol. Chem.* **251:** 165.

———. 1977. Nerve growth factor. *Adv. Protein Chem.* **31:** 339.

Smith, A.P., L.A. Greene, H.R. Fish, S. Varon, and E.M. Shooter. 1969. Subunit equilibria of the 7S nerve growth factor protein. *Biochemistry* **8:** 4918.

Thoenen, H. and Y.A. Barde. 1980. Physiology of nerve growth factor. *Physiol. Rev.* **60:** 1284.

Thomas, K.A., R.E. Silverman, I. Jeng, N.C. Baglan, and R.A. Bradshaw. 1981. Electrophoretic heterogeneity and polypeptide chain structure of the γ-subunit of mouse submaxillary 7S nerve growth factor. *J. Biol. Chem.* **256:** 9147.

Ullrich, A., A. Gray, C. Berman, and J.J. Dull. 1983. Human β nerve growth factor gene sequence highly homologous to that of mouse. *Nature* **303:** 821.

Varon, S., J. Nomura, and E.M. Shooter. 1967. The isolation of the mouse nerve growth factor protein in a high molecular weight form. *Biochemistry* **6:** 2202.

———. 1968. Reversible dissociation of the mouse nerve growth factor protein into different subunits. *Biochemistry* **7:** 1296.

Varon, S., J. Nomura, J.R. Perez-Polo, and E.M. Shooter. 1972. The isolation and assay of the nerve growth factor proteins. In *Methods of neurochemistry* (ed. R. Fried), vol. 3., p. 203. Marcel Dekker, New York.

Young, M. 1979. Proteolytic activity of nerve growth factor—A case of autocatalytic activation. *Biochemistry* **18:** 3050.

Young, M., J.D. Saide, R.A. Murphy, and M.H. Blanchard. 1978. Nerve growth factor: Multiple dissociation products in homogenates of the mouse submandibular gland. Purification and molecular properties of the intact undissociated form of the protein. *Biochemistry* **17:** 1490.

Sequence Homology of Human and Mouse β-NGF Subunit Genes

A. ULLRICH, A. GRAY, C. BERMAN, L. COUSSENS, AND T.J. DULL
Department of Molecular Biology, Genentech, Inc., South San Francisco, California 94080

Various features of the nerve growth factor (NGF) gene system (Carpenter and Cohen 1979; Levi-Montalcini 1982; Yankner and Shooter 1982) are of particular interest to the molecular biologist studying gene regulation in mammalian cells. The expression of the three polypeptide chains (α, β, γ) that comprise the NGF complex is under androgen control in the mouse submaxillary gland (MSG), resulting in about 10–30-fold higher levels of NGF in the male MSG than in the female. In other tissues of the mouse, as well as in most other species, the genes of the NGF complex are subject to other regulatory mechanisms, which are not at present understood. Because of the importance of NGF in the formation and maintenance of sympthetic nerve contacts, the regulation of genes coding for the NGF-complex polypeptides during development is of great interest. Formation and secretion of the NGF complex in the male MSG include several levels of regulation: androgen-simulated transcription of three genes (α, β, γ), formation of mature mRNAs from primary transcripts, translation, processing of the β-NGF precursor protein by the γ-NGF arginyl esteropeptidase, assembly of the three subunits into the 7S NGF complex, and transport into the lumen of secretory ducts of the submaxillary gland.

As a first step in the investigation of these cellular events, we have isolated mouse β-NGF complementary DNA (cDNA) and determined its nucleotide sequence. A mouse β-NGF cDNA clone was then used to isolate the human β-NGF gene, the coding regions of which are highly homologous to the mouse prepro-β-NGF nucleotide and amino acid sequences.

METHODS

Procedures employed in our experiments were carried out basically as described in detail elsewhere (Southern 1975; Lehrach et al. 1977; Maxam and

```
    1                                   10                                      20                                      30
    Ser Ser Thr His Pro Val Phe His Met Gly Glu Phe Ser Val Cys Asp Ser Val Ser Val Trp Val Gly Asp Lys Thr Thr Ala Thr Asn
    TCG TCG ACG CAT CCG GTG TTT CAT ATG GGG GAA TTT TCG GTG TGT GAT TCG GTG TCG GTG TGG GTG GGG GAT AAA ACG ACG GCG ACG AAT
5'    A   A   A   C   A   A   C   C       A   G   C   A   A   C   C   A   A   A   A       A   A   C   G   A   A   A   A   C
    AGT AGT   T       T   T               T         AGT   T         AGT   T AGT   T       T   T           T   T   T   T
      C   C   C       C   C               C           C   C           C   C   C   C       C   C           C   C   C   C

                                        40                                      50                                      60
    Ile Lys Gly Lys Glu Val Thr Val Leu Ala Glu Val Asn Ile Asn Asn Ser Val Phe Arg Gln Tyr Phe Phe Glu Thr Lys Cys Arg Ala
    ATT AAA GGG AAA GAA GTG ACG GTG CTG GCG GAA GTG AAT ATT AAT AAT TCG GTG TTT CGG CTA TAT TTT TTT GAA ACG AAA TGT CGG GCG
      C   G   A   G   G   A   A   A T A   A   G   A   C   C   C   C   A   A   C A A   G   C   C   C   G   A   G   C A A   A
      A       T           T   T   T   T   T       T       A         AGT   T       T                       T           T   T
              C           C   C   C   C   C       C                   C   C       C                       C           C   C
                                                                              -----------------------
                                                                             | 3' ATA AAA AAA CTT TG 5' |
                                                                             |      G   G   G   C       |
                                                                             |         16 x 14          |
                                                                              -----------------------

                                        70                                      80                                      90
    Ser Asn Pro Val Glu Ser Gly Cys Arg Gly Ile Asp Ser Lys His Trp Asn Ser Tyr Cys Thr Thr Thr His Thr Phe Val Lys Ala Leu
    TCG AAT CCG GTG GAG TCG GGG TGT CGG GGG ATT GAT TCG AAA CAT TGG AAT TCG TAT TGT ACG ACG ACG CAT ACG TTT GTG AAA GCG CTG
      A   C   A   A   A   A   A   C A A   A   C   C   A   G   C       C   A   C   C   A   A   A   C   A   C   A   G   A T A
    AGT       T   T     AGT   T       T   T   A     AGT                 AGT           T   T   T       T       T       T   T
      C       C   C       C   C       C   C           C                   C           C   C   C       C       C       C   C
                                                  -----------------------
                                                 | 3' TTT GTA ACC TTA 5' |
                                                 |      C   G       G    |
                                                 |        8 x 12         |
                                                  -----------------------

                                      100                                     110                             118
    Thr Thr Asp Glu Lys Gln Ala Ala Tyr Arg Phe Ile Arg Ile Asn Thr Ala Cys Val Cys Val Leu Ser Arg Lys Ala Thr Arg
    ACG ACG GAT GAA AAA CAA GCG GCG TAT CGG TTT ATT CGG ATT AAT ACG GCG TGT GTG TGT GTG CTG TCG CGG AAA GCG ACG CGG
      A   A   C   G   G   G   A   A   C A A   C   C A A   C   C   A   A   C   A   C   A T A   A A A   G   A   A A A     3'
      T   T                   T   T       T       A   T   A       T   T       T       T   T AGT   T       T   T   T
      C   C                   C   C       C           C           C   C       C       C   C   C   C       C   C   C
       -----------------------
      | 3' CTA CTT TTT GTT CG 5' |
      |      G   C   C   C       |
      |         16 x 14          |
       -----------------------
```

Figure 1. Mouse β-NGF amino acid and possible nucleotide sequences. Boxed areas indicate the synthetic oligodeoxyribonucleotide pools used for cDNA priming and/or for colony hybridization Pools are sequentially numbered in the text starting with 1 for the 3′ most one. Numbers indicated below oligonucleotide sequences refer to the number of oligonucleotides per pool and oligonucleotide length.

Gilbert 1977; Wallace et al. 1981; Maniatis et al. 1982).

RESULTS

Isolation of cDNA Clones Coding for Mouse Pro-β-NGF

The cDNA cloning approach took advantage of the known 118-amino-acid-long sequence of the mouse β-NGF subunit, employing synthetic oligonucleotide primers; the difference in NGF levels between male and female MSGs (Ishii and Shooter 1975) was used as an additional means of identification. Three small portions of the β-NGF amino acid sequence were chosen, and oligonucleotide pools complementary to all possible sequences coding for them were synthesized (Fig. 1) (Crea and Horn 1980). Initial attempts to identify or isolate β-NGF cDNA clones from an oligo(dT)-primed, male MSG cDNA bank (~10,000 clones) failed, using the synthetic oligonucleotides as hybridization probes (Wallace et al. 1981). This result indicated that although β-NGF comprised approximately 0.1% of the protein in the male MSG (Harper and Thoenen 1980), its mRNA was not of equal abundance. Therefore, the primer pool representing sequences closest to the carboxyl terminus of the protein (Fig. 1) was used to prime specifically reverse transcription of male MSG poly(A)$^+$ RNA, in order to first enrich for β-NGF-specific nucleotide sequences. Oligonucleotide-primed cDNA molecules more than 200 bp in length were cloned into pBR322; 10,000 clones were screened using the 5′-^{32}P-labeled β-NGF primer pool originally used in the cDNA priming as hybridization probe. Approximately 8% of the clones gave a positive signal under high-stringency hybridization conditions.

Positive clones were rescreened using radiolabeled primer pools 2 and 3, derived from sequences upstream of oligonucleotide pool 1, as hybridization probes (Fig 1). In addition, [^{32}P]cDNAs primed with pool 1 from male or female poly(A)$^+$ MSG RNA were used as probes on duplicate filters. Nine male-specific clones that hybridized with oligonucleotide pools 2 and 3 were identified. Restriction enzyme analyses demonstrated

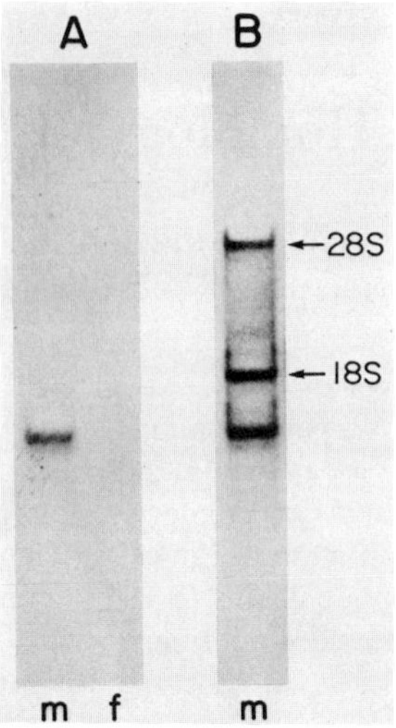

Figure 2. Northern blot hybridization analysis of male and female MSG poly(A)$^+$ RNAs. Male and female MSG poly(A)$^+$ RNAs, purified (Ullrich et al. 1977) from adult (~30 g) Swiss Webster male or female mice, were separated on a 1.2% agarose-formaldehyde gel, and Northern analysis was carried out as described by Lehrach et al. (1977). (Lane *A*/m) 2 μg male MSG poly(A)$^+$ RNA; (lane *A*/f) 2 μg female MSG poly(A)$^+$ RNA. A radiolabeled (Taylor et al. 1976), 665-bp *Hga*I–*Hinc*II fragment of clone pmβN-9G1 was used as hybridization probe. Lane *B* was hybridized with a mixture of the probe used in *A* and reverse-transcribed, radiolabeled (Taylor et al. 1976) rat ribosomal RNA. The detected hybridization reflects the presence of small amounts of ribosomal RNA in our poly(A)$^+$ RNA preparation and provides an internal standard. 28S and 18S rat ribosomal RNAs are 4718 and 1874 nucleotides in length, respectively (I. Wool, pers. comm.).

that all nine clones had common *Hae*III and *Hinf*I fragments. The clone containing the longest DNA insert (pmβN-9G1; 716 bp) was sequenced in its entirety. The amino acid sequence deduced from the nucleotide sequence contained the expected β-NGF sequence in one translational frame in addition to an aminoterminal pro-sequence (Fig. 4).

To determine the size of the complete β-NGF mRNA, we used Northern blot hybridization (Lehrach et al. 1977) as well as primer extension analysis. A 665-bp-long [^{32}P]DNA fragment that included β-NGF

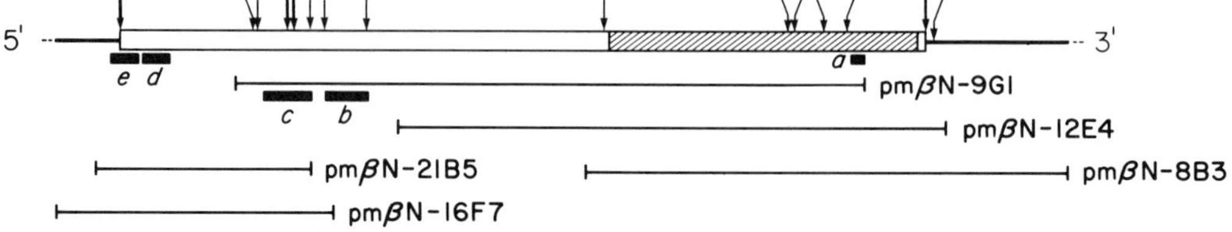

Figure 3. Bacterial clones containing overlapping mouse β-NGF cDNA sequences. Solid bars *a*, *b*, and *c* designate the positions of DNA fragments used to prime cDNA synthesis. The upper line and boxed region represent the entire region of mouse β-NGF cDNA covered by our clones. The base-pair scale begins arbitrarily at zero, which depicts the beginning of our nucleotide sequence. The boxed region encloses the open reading frame, and the hatched area codes for mature β-NGF. The three potential sites of translation initiation are indicated (ATG), as is the TGA termination codon.

as well as propeptide sequences (*Hga*I–*Hinc*II, Fig. 3) hybridized to a male MSG-specific RNA species of approximately 1300 nucleotides in length (Fig. 2). A primer extension experiment using two short, double-stranded, 5′-end-labeled restriction fragments localized the 5′ end of the β-NGF mRNA to about 230 bases upstream of the 5′ end of the pmβN-9G1 cDNA fragment, leaving approximately 370 bases downstream from the 3′ end of our clone (data not shown). All but about 30 nucleotides of the missing 5′ sequences are contained in clones pmβN-16F7 and pmβN-21B5, which overlap each other, and clone pmβN-9G1 (Fig. 3).

To obtain cloned cDNA that included the sequences from the 3′ end of the primer sequence downstream to the 3′ poly(A) sequence, we first enriched for β-NGF mRNA by fractionating total poly(A)$^+$ male MSG RNA on a preparative urea agarose gel (Lehrach et al. 1977). The largest-size fraction containing sequences hybridizing to a β-NGF cDNA probe was used for oligo(dT)-primed cDNA synthesis and cloning. The screening of 4400 clones resulted in four positive hybridization signals. Nucleotide sequence analysis of clones pmβN-12E4 and pmβN-8B3 added 239 nucleotides to the 3′-coding and untranslated sequences (Fig. 4).

Isolation and Characterization of the Human Chromosomal β-NGF Gene

A human gene library (Lawn et al. 1978) was screened using a 665-bp mouse β-NGF cloned cDNA fragment (pmβN-9G1 *Hga*I–*Hinc*II; Fig. 3) as a radioactive hybridization probe. Twenty-seven hybridization-positive recombinant phage were plaque-purified and partially characterized by *Eco*RI digestion; six different types of restriction patterns were identified. Each pattern category shared restriction fragments and thus appeared to overlap the same genomic region. λhβN8 was further characterized by multiple restriction digests, Southern blotting, and nucleotide sequencing; Figure

```
                                                                          -187                        -180
                                                                          met leu cys leu lys pro val lys leu gly ser
AGCGCATCGAGTGACTTTGGAGCTGGCCTTATATTTGGATCTCCCGGGCAGCTTTTTGGAAACTCCTAGTGAAC ATG CTG TGC CTC AAG CCA GTG AAA TTA GGC TCC
1       10        20        30        40        50        60        70              80                 90          100

                -170                            -160                            -150
leu glu val gly his gly gln his gly gly val leu ala cys gly arg ala val gln gly ala gly trp his ala gly pro lys leu thr
CTG GAG GTG GGA CAC GGG CAG CAT GGT GGA GTT TTG GCC TGT GGT CGT GCA GTC CAG GGG GCT GGA TGG CAT GCT GGA CCC AAG CTC ACC
 110        120        130        140        150        160        170        180        190

                -140                            -130                            -120
ser val ser gly pro asn lys gly phe ala lys asp ala ala phe tyr thr gly arg ser glu val his ser val met ser met leu phe
TCA GTG TCT GGG CCC AAT AAA GGT TTT GCC AAG GAC GCA GCT TTC TAT ACT GGC CGC AGT GAG GTG CAT AGC GTA ATG TCC ATG TTG TTC
 200        210        220        230        240        250        260        270        280

                -110                            -100                            -90
tyr thr leu ile thr ala phe leu ile gly val gln ala glu pro tyr thr asp ser asn val pro glu gly asp ser val pro glu ala
TAC ACT CTG ATC ACT GCG TTT TTG ATC GGC GTA CAG GCA GAA CCG TAC ACA GAT AGC AAT GTC CCA GAA GGA GAC TCT GTC CCT GAA GCC
 290        300        310        320        330        340        350        360        370

                -80                             -70                             -60
his trp thr lys leu gln his ser leu asp thr ala leu arg arg ala arg ser ala pro thr ala pro ile ala ala arg val thr gly
CAC TGG ACT AAA CTT CAG CAT TCC CTT GAC ACA GCC CTC CGC AGA GCC CGC AGT GCC CCT ACT GCA CCA ATA GCT GCC CGA GTG ACA GGG
 380        390        400        410        420        430        440        450        460

                -50                             -40                             -30
gln thr arg asn ile thr val asp pro arg leu phe lys lys arg arg leu his ser pro arg val leu phe ser thr gln pro pro pro
CAG ACC CGC AAC ATC ACT GTA GAC CCC AGA CTG TTT AAG AAA CGG AGA CTC CAC TCA CCC CGT GTG CTG TTC AGC ACC CAG CCT CCA CCC
 470        480        490        500        510        520        530        540        550

                -20                             -10                                 1
thr ser ser asp thr leu asp leu asp phe gln ala his gly thr ile pro phe asn arg thr his arg ser lys arg ser ser thr his
ACC TCT TCA GAC ACT CTG GAT CTA GAC TTC CAG GCC CAT GGT ACA ATC CCT TTC AAC AGG ACT CAC CGG AGC AAG CGC TCA TCC ACC CAC
 560        570        580        590        600        610        620        630        640

                10                              20                              30
pro val phe his met gly glu phe ser val cys asp ser val ser val trp val gly asp lys thr thr ala thr asp ile lys gly lys
CCA GTC TTC CAC ATG GGG GAG TTC TCA GTG TGT GAC AGT GTC AGT GTG TGG GTT GGA GAT AAG ACC ACA GCC ACA GAC ATC AAG GGC AAG
 650        660        670        680        690        700        710        720        730

                40                              50                              60
glu val thr val leu ala glu val asn ile asn asn ser val phe arg gln tyr phe phe glu thr lys cys arg ala ser asn pro val
GAG GTG ACA GTG CTG GCC GAG GTG AAC ATT AAC AAC AGT GTA TTC AGA CAG TAC TTT TTT GAG ACC AAG TGC CGA GCC TCC AAT CCT GTT
 740        750        760        770        780        790        800        810        820

                70                              80                              90
glu ser gly cys arg gly ile asp ser lys his trp asn ser tyr cys thr thr thr his thr phe val lys ala leu thr thr asp glu
GAG AGT GGG TGC CGG GGC ATC GAC TCC AAA CAC TGG AAC TCA TAC TGC ACC ACG ACT CAC ACC TTC GTC AAG GCG TTG ACA ACA GAT GAG
 830        840        850        860        870        880        890        900        910

                100                             110
lys gln ala ala trp arg phe ile arg ile asp thr ala cys val cys val leu ser arg lys ala thr arg arg gly OP
AAG CAG GCT GCT TGG AGG TTC ATC CGG ATA GAC ACA GCC TGT GTG TGT GTG CTC AGC AGG AAG GCT ACA AGA AGA GGC TGA CTTGCCTGCAGC
 920        930        940        950        960        970        980        990        1000       1010

CCCCTTCCCCACCTGCCCCCTCCACACTCTCCTGGGCCCCTCCCTACCTCAGCCTGTAAATTATTTTAAATTATAAGGACTGCATGATAATTTATCGTTTATACAATTTTAAAGACATTA
         1020      1030      1040      1050      1060      1070      1080      1090      1100      1110      1120      1130

TTTATTAAATTTTCAAAGCATCCTGTATACCGAA
         1140      1150      1160
```

Figure 4. Mouse β-NGF cDNA sequence. The nucleotide sequence was translated, and the longest open reading frame containing the β-NGF amino acid sequence (large boxed area) is shown. (■) Potential processing sites; (▨) potential N-directed glycosylation sites. Boxed areas indicate possible initiation methionines.

a

```
TACTTGCTTTAGGAAGTTAAGTATTATGTCTTATTCTCCTTTGTGTCCCTAGCACCTAACACTTAAAACAGTGGCCAGCACAGGACCTGCAAGTTTAAGT    100
GTTTAATTAATGAAATAAATGAATCCCAATTTTGGGATGAGAGAAAGCACTACTTAAGCATCTAGTAGCAATGCAGCCTGGAAAACATTCAAAGTCACGG    200
AATCTCAGATGATCAGAGCCAAAGGGGACCTTAGCTGTCATCTGTGCCAGCTTCTTATCCTATAGAGGAGAAAGCTCAAAGATGAAATGAATCTCCTTCT    300
ATACAGGAGAAGCTCAGAGTGAACTGAATCAGAATGCGGGTGTGTGGGTTCCAGCCTGCAACCTTTCAGGTTTAGCCAAACACCCAGATGAAGGGTTTGT    400
GGACTAGACGAAACCATCTTCCCATGAGTAATGGGACCCAGATAATGCCCACCTCTTACCCTGGGACACGCCATTCTCCCTCTCCCATGCTAACTCCAAC    500
CCTGGGAGAGCATGAAAATGTTCTTTGTCACAGAATGTAACCTTTAAAGAGTGTCTGAGTATGCATTTCATCACTAGCCTTCAACCCCAATTGAGTATTG    600
AAAGGTTTTTCTGGTACTTTCTGGAGCAAGAAGACTATTTTGAGCAAGATGGGAAAGGAAGAAGAATGGAGACATCCCAGGGCTTAATTTCATGATTTCT    700
AGTAACTTGAAGATCACTTTAGAGGTCCTTGCTACCTCCCCATTCTCCAACTCCTCTTCGTGGTTGGCGAATTTGGGGAGCGATGGTGGCTTTTCTGACA    800
TTTGCTTTCATAGCACAAGCTGAGAGGGAGTTGGATGAAGATATGTGGTGGGGATCCACGCTGGAAAAAGATATCACAGGGAGAAGATTTTTTTGAAGTT    900
GAAGAGAGAATACGGACAGGAAAGTTAAGATGTCATTCTAGAACTTTATTGGGAGGGCATCTCCACCCTACAACAAATTCTGTGATGGACATAATCATTC  1,000
ATTCATTTATCCGTAAATATCACCCTCTTGTTCAAAGCCCTCCACTGCCTTCCTAATATCCTGAGGATAAAACCATAGCTCCTTGCTGTGTCTCTGTAGA  1,100
CCTGGCTCTTCCTGGCTCTCCAGCTCATTTTCTAGGTCTCGTTACTTCATGCTCAGAACCTTTGTCTTGTTTCTAGCTCAGGGCCTTTGCACTTGTTCTT  1,200
GCTGCCTAGAATGTTCTCTCCCTCATTCCTTCTCATCCTCCAGATCTCAACTTGAAGGCCATCTCCTCAGAGCTCCTCGCTGAGCGTCCTGTCTACAGTG  1,300
GCCCCTCGATACATCCTGCAGTTGCTCTCTATCATCAGACCCTGTAATTGCCTTCATGGCATATAAAGAATCTGGAGTATCTTGCTTATTTACACAACAC  1,400
TGTAAGCTCCATGAGAGCAGAGGCCTTGTTTGTCTTGTTTACTGCTGCTCAGCACCAAAAACAGTGCCTGGCACATAGTCGGTGCCCAGAAAATATTGTG  1,500
AATGAATGAAGTGCCTACATAGATTACATTATAGAAGTGAGAGGAGAATAGAAAACTTCCATTGTTTCTAGAAACTACAGCCTAAAATTGATTTTTTAAA  1,600
ATTGTATCAGCTCCATAGCTTCCAATCCTAAAATCTGCCTTTCAGTGTGGTACTCTGAGATTCCTGTCTGATTCTGTGAGAGCTCCACATTCTCTCTCAA  1,700
ATGGTCAGTCTGTCTTATTTGTCACCATTACTCATCTGCATTTTTATCAAAGCACCAACTTGCTCTGAATTGTCAGGGATTTTGCGTCTGTATAAGGTAT  1,800
TTTAGGCTGGTTCAGAGTTGGATCTGTTATGTCTGCATGTGTAATGTACTGAACAATTTCTATTTTGATGCCAGATTAGGGATCTGCTGGGGCAAGACTT  1,900
TGGCATGTGTCTAGAAACACCTGCACTAGGTGCAAGATCAGCCATGGACTGTGTCCAGGCTGAAACCAAAAGGTATGGCGCAAGAGTGAGAGGCAGGTGC  2,000
CACCACAGGACCATGAGAGGCCAAGCTCCGGTATAATTTTGGTAGACCAAATTCTAGCTCCTTCCTGGGCCTTGATGCTGGTAAAATCCCAGAACTCAAG  2,100

                                                                                                  yVa
GAAATGGAATTTGTCCTATTGGCACATGCCTCCCCACTGTGTAGGGCACAGGGAATGTGGTGAGGTACAGTCTAATGCCAGCTCTCCCCCTCCACAGAGT  2,200
lLeuAlaSerGlyArgAlaValGlnGlyAlaGlyTrpHisAlaGlyProLysLeuSerSerAlaSerGlyProAsnAsnSerPheThrLysGlyAlaAla
TTTGGCCAGTGGTCGTGCAGTCCAAGGGGCTGGATGGCATGCTGGACCCAAGCTCAGCTCAGCGTCCGGACCCAATAACAGTTTTACCAAGGGAGCAGCT  2,300
PheTyrProGlyHisThrGlu
TTCTATCCTGGCCACACTGAGGTAAGTGCCTAGGCGCTTGGCCTTGCCAAGGTCCTCCCTCTGCAGCTGCCAGAAGCAGGAGTCCCAAGTGACAGGACCT  2,400

GAGAGGGCAAGTCAGAACCAACTGCTGAGCAGCAGGGGCCTAGAGAAGCTTTCAGTGGTGCAGCAAATGCAGCTCCCCTAGCCCACTCAGGAAAGGAAAT  2,500
CTTTTGTGCATGCAGGTTGGCCCCATGGCCCCAGGGACATATCAGGAGTCTGAGCTGCATCAAGGGTCTAAAGTTGCCGAAGCACAGGACCTCGAATAG  2,600
ACAGCGTCAGTCAGACCCTCCACACAACTCTCTGGTCCGACTGAGCAGGGAACTTGCAAAATGGAAGGTTCACAGCCACTCTTTGTGCCCTTCACCAGTG  2,700
GAGTAGAGGGGTAGGGAAGAGTTCCAGATCTGCACACAGGAGGTGTGGCTCCGAGTTCCACTTCTGTTTGCTGTGTACCTGTGATGCAATTACTTAATCT  2,800
TACTACATCTCTGGTTTCTCATTGTTAAAATAGGAATAATATGGCCAACCTCCCAGGTTTGTGTAATGGTTAATTAGAAAATTATACCCAATAAATAATA  2,900
ATTACTATTATCTCAGCTCATTCATGTCAAGGTATGTC ----2.8kb---- GGCAGAGCAGTCCCTACTGGCTTACAGACAAGGGACTTGGCACC  3,000
AGCTCCTTTGAGCAGGGTGATATGATCAGAGCAGGCTTTAAGAAGCTAAATGGGGTGGCCAGGGATAGGAGGAATTGGAAGAAGAGAGATGAGGGTCCAG  3,100
GATATCAGGTGGGAGACTGTTCCCATGACCTGAACTAGGGGCCCAGAGATGAGAATGGAAGATAAAATACAGTTTCAAGATAATGCTACCAAACCAGAAC  3,200
CAAAAGGGCTGGATGGAGGTCAGGGAGGAAAAAGCAGAGGGATAAGCAGAGGTATGGAGCTAGTGGATGAATGAATGATGGTTCCAGTAACAGAAAAAGG  3,300
AAACTGTGAAGAAGCGTGTTTAGGGAGATGAGGAGAGAGGATCAAGAGAGATGGTAGCTGTGAAAACTATCATCAAATGGTTTCAAAATGGGGCTGTTGC  3,400
TATTGTATGCAGAGAAGGAAGATGGGGAGAAGATGATCCTAATTATGATTATGAGTATGAAGATTTTTGTTTTGGTAGTCTTGCATTTAAAATGCCAGTA  3,500
TCTATGGAGAATTGCCCTCAGGAAGCAGGAGGGCAGAACTAGCTTCAAAAGACTAGTTGATTAGAGATCCAGGTATGGGAGTCAATTGCACAGAAATTTC  3,600
GATGGGCTCCTTAGAAGAGGAGGAGCCCACCAAAGGAGAGACCAAGAATCGAAGTGGGAACCTCTGGTAATGGGGAATAGGGAGAGGTATGTCCCTTATG  3,700
TCCATCGAAACCCTTTTATTTAGTTAATTTCATTTTCCTCTTGAAATTCTAGCAAGTGTTACTAATTTGGATATAAGGGAATGAATTTCAGTTTTCTTTC  3,800
TTCTCCTTTAATGTATCTGATAAGAGATATTATGTACCGAGAAGAACAAAGAAGTGAAACATGAAAAGGAGCAAAAGTAAAAGGTTTAGATTTGCTGCCA  3,900
ACTGAATATTCAGTCCGCTAAGAGTTGACCGGACATTTATCGCAAGACCATTAACAGATCTGGTAATGGTCAAAACCCCAGATATGCCATTGCCTCCTTT  4,000
GCATGGTCAGCTGGAGAACTGGAGCAAACAGAAACATTCTTCAAAGAAACCAAAGAGAAAAGAATTTGGTGATTTCCTTCTTGGTATCTCACCAAAGAGC  4,100
TGTCTATTTCCAGCTGAATAAAAGCACAGTATGGAGGAGAAGAATGTGTTGCCTTAAAACAGATTAGAAAAGCAGTGTGTAGCCATATGCACAAAGGCCA  4,200
AAAGAAACCACACACTGTGATGGAAAGAAGGCAGATGGTGAGCTTCTCTTGCCACTCTCATTCATTCTTTCTTTCATTCATTCTCCTTAAATTTTTTGAG  4,300
TACCTACTACATGCCAGACACGGCAACTGGCCCTGGAAATATTGAAATGCAAATTGTGCAGCTTGCAAAGCCCTCTCAGGGAGACAGACAAGTTAATGAT  4,400
TATCCACAACCCTAGGAGAGGTGATAGGGTTATGCACAAACTGCTGTGGAATACAAGGGATGGAGTGCCCAGATACACAAGTGGATGAGACACGGCTCCA  4,500
CCCTCGAGTACCTCACAGTCGGGGTGAGACAATAAAAATGGCTTTGAAGGAGGTCTGAGAGAGGTAAAACCGCGTGAAAGCACAAAAGACAAAGTCCTTG  4,600
ACCACAAGGAGACAGAAATAACTCATGATTTACACAAATTTTGTGAGAACCTACTTCATCCTAACAACTACTCTGGGAGATAGTGTTTTTTGCCCACCTC  4,700
CGATCAATGGGAACGCTAAGGCTCTGGGAAATTAAATGGGTTTCTTCTGCTAACAAGTAGTGGAGTAGGAATCCAAACACGATTTTCTAACCCCCAGACC  4,800
CAATGCAGAGTTCTTGTCATTACACTGATTTAGTAGGAGAGAACACACACACTCACACTCACATGAACACACACATATGCACATACAATTACGCTATTCT  4,900
GAAGTGCAGAATGGCAAAAGATCATTTTTGATTGGAGTGATCTGAGAAGGCTTCGAGGACTAGGTCGTATCTGTGTGGAAACGGAGCAGAATGACCCATA  5,000
AGAGTGGGAAGGAGAGATGGGAGAAAGGATAGAATTGGAAGTTTGTAGAGCACTCTTACAGGAGATATTTCTGATCCTACCAGGAAGATCTATGGAAGCT  5,100
TTACAGAGGAGCTGACGTTTGCTACACATCTACAAGTATGCATAGGAGCTCCGCGGAGGCCAGTGAGAGGCCCTCCAGGAGCAGAACTAATTCCACAATT  5,200
ACTTGACCAAGTTGGGGGATTATTTGTGGGGTAACTGCAGTGCAGTATGGAGTCCTCTTGGGGACAGTTAGAGCCATACCATTTGATCTATAGTCACATA  5,300
AGAACAAACAATAAAAAGAAAAGACATGCTTAAGAGTGAAAGAGAAAGGGAGGGAGAAAAGAAGGAAGGGTGGATGGAAGGACACTAGCTTAGTAAGGGG  5,400
TCAACTTTGGATTCTATTTCTGGTTCAGTTTTCATTTGTGACTTCAGTGCTTTAGTGTTAGTTCATTTTCTGTGAATCAGTTTCCTTATATGTGAAATAA  5,500
ATATGATAAATCCTAATTGAACTCACAGACTCATGAGAAGATAGAAGTGAACACATTTTAAAAACATCACACAAAGAGGAACTATTATGTGGTCCACATT  5,600
TATATATGTGGGGTAGCGTCTGAAGAGGTGCCTGGACTAAGATGGTCCCAGAGCCACAAGGTTTTTGCCAAACATGACGCTTTGTGAATTCATAACAAGG  5,700
GCTCCAAGTCACCAGATCTTAGAGCTGACCCAGTGCACTGTCTGAAAGGGGGTACCAGTTCTGAGGCTTCAAGACATGTCCCCAGCAGATCTTCCCCGTG  5,800
CCTTCCCAGAGGATTCAAAACTGTTGAGCAGGACGGCACCATCACATCAAGGCACAAGTGCCAGGGAGAGGTGTTAAACTCTCCCCACCAACCTCCCTGG  5,900
TACACACATGGACACTTGCCACCTCCCTCAGCCGCCTTAAGCTTCAGAGAACTCAAAGGACTCTGTAAGTGATGTCTCCAAGCTCATATCGAACTACTGG  6,000
GCAAAATTTCAGGGGCTCTGTCACTTCCTGGAGAAGCTCGGATGGGGTGACCACACATCCATACTGCCTGAGTCAGCCCCGGGTTACGCCTGTTGTCCCG  6,100
GTATAACCATTGCTAGCACACCCTTTCCCTCTCAGAAGTGCCCCGGTTTGAATGAAACCTCTTCGTGATCCCCTTGGAGGTCAACTCTGAGGGACCCAGA  6,200

                                                          ValHisSerValMetSerMetLeuPheTyrThrLeuIle
AACTGCCTTTTGACTGCATTTAGTACTCCATGAAGTCACCCTCATTTCTTTTTCATTCCAGGTGCATAGCGTAATGTCCATGTTGTTCTACACTCTGATC  6,300
ThrAlaPheLeuIleGlyIleGlnAlaGluProHisSerGluSerAsnValProAlaGlyHisThrIleProGlnValHisTrpThrLysLeuGlnHisS
ACAGCTTTTCTGATCGGCATACAGGCGGAACCACACTCAGAGAGCAATGTCCCTGCAGGACACACCATCCCCCAAGTCCACTGGACTAAACTTCAGCATT  6,400
erLeuAspThrAlaLeuArgArgAlaArgSerAlaProAlaAlaAlaIleAlaAlaArgValAlaGlyGlnThrArgAsnIleThrValAspProArgLe
CCCTTGACACTGCCCTTCGCAGAGCCCGCAGCGCCCCGGCAGCGGCGATAGCTGCACGCGTGGCGGGGCAGACCCGCAACATTACTGTGGACCCCAGGCT  6,500
uPheLysLysArgArgLeuArgSerProArgValLeuPheSerThrGlnProProArgGluAlaAlaAspThrGlnAspLeuAspPheGluValGlyGly
GTTTAAAAAGCGGCGACTCCGTTCACCCCGTGTGCTGTTTAGCACCCAGCCTCCCCGTGAAGCTGCAGACACTCAGGATCTGGACTTCGAGGTCGGTGGT  6,600
AlaAlaProPheAsnArgThrHisArgSerLysArgSerSerSerHisProIlePheHisArgGlyGluPheSerValCysAspSerValSerValTrpV
GCTGCCCCCTTCAACAGGACTCACAGGAGCAAGCGGTCATCATCCCATCCCATCTTCCACAGGGGCGAATTCTCGGTGTGTGACAGTGTCAGCGTGTGGG  6,700
alGlyAspLysThrThrAlaThrAspIleLysGlyLysGluValMetValLeuGlyGluValAsnIleAsnAsnSerValPheLysGlnTyrPhePheGl
TTGGGGATAAGACCACCGCCACAGACATCAAGGGCAAGGAGGTGATGGTGTTGGGAGAGGTGAACATTAACAACAGTGTATTCAAACAGTACTTTTTTGA  6,800
uThrLysCysArgAspProAsnProValAspSerGlyCysArgGlyIleAspSerLysHisTrpAsnSerTyrCysThrThrThrHisThrPheValLys
GACCAAGTGCCGGGACCCAAATCCCGTTGACAGCGGGTGCCGGGGCATTGACTCAAAGCACTGGAACTCATATTGTACCACGACTCACACCTTTGTCAAG  6,900
AlaLeuThrMetAspGlyLysGlnAlaAlaTrpArgPheIleArgIleAspThrAlaCysValCysValLeuSerArgLysAlaValArgArgAlaEnd
GCGCTGACCATGGATGGCAAGCAGGCTGCCTGGCGGTTTATCCGGATAGATACGGCCTGTGTGTGTGTGCTCAGCAGGAAGGCTGTGAGAAGAGCCTGAC  7,000

CTGCCGACACGCTCCTCCCCCTGCCCCTTCTACACTCTCCTGGGCCCCTCCCTACCTCAACCTGTAAATTATTTTAAATTATAAGGACTGCATGGTAAT  7,100
TTATAGTTTATACAGTTTTAAAGAATCATTATTTATTAAATTTTTGGAAGCATCCTGTGTGCTGATGCTGGTTATTTTTTTTGAGTAAAATCATCTGCAA  7,200
GTCTGAGGAAGATGCAGGGGGAATTGTCTGAAGCACCCCCTGGCTCCTTCTAAGGCCCACCTGTAACCCTTACCCCACCCCCAGCCAGTGCTGCAACTTC  7,300
AGGAAAGGCTAACTGGTTTCAGATATTAGCCTAAGCCAGGCATGATTAGCAAAGGAAGCCTGCTGGATTGCAACTTTGTTCTACATTCCAGAGCCAAAGC  7,400
TACCTAATCATTGGATTACCTGCCCACGGCCTGGAAGCATGGGCCCTGGCTCCTCCCTTGGAGAAGTTAGTGGAGCCTCATCAGGAGGCAGATCACCACT  7,500
GGGCATGGGGCTGCCCTGGCCTCAGAGGCACAGTCTCAGTCCTGGGGGCACTCGATAGACAGGAAGAGGCTTGTACCAAAGATCTGTGAGCTGCTCATTT  7,600
GTAGAGGGAGTGTGCTGTCTGGAGAGTTGTAACAGAGTAACTCACGAAGCCTGCCAAAGTAAGATCTAGGGTTTCATGTTCAGCTGGCCTTAGTACTCGT  7,700
TTTATCCCTCAGCAGCCTCCTGCAGAAAGACTTTCATTACCCCAAGAAAACATTAGTGCCTCCGACACTCTTGCAACCCACTAACAGCCACAGTAGCAAC  7,800
TTCCAGGGGCCCCACCATCAGCCCACAGCCAGGGAAGCCAGAGACGGGAGAGAACTTCTGAAAATATTTTTATTCTTAATTGGATGATTTCCATAAGTGA  7,900
CCAAAGGTGACTGGGGCCGTGGACCTGCAGAGAAGAGGGTGGGGCAGAAAGACTAAGTAAATGGAGTAAGAACCAAATAAACCTGGAGATGTGGAGGAC  8,000
AGGTGGGCCGGCAGGAACCACGGCTTTCTGCTTCTCCCAAACAGGGGAAATGAAACAAGGATGACAAAGAGAAGGCCAGAGGTGGATGTGAAGAAGGGGA  8,100
GGGGAGGACAGAGCAGGAGGAGGAGGGAGGTGTGATCTCTCATCTTTACCAGCACTCTCCAGCCTCCAGGAAAGTTAGCAACCTGGGTGGTTCTTGTGGG  8,200
GCTTTTTGGATGTGCTAATAAACACTGACTCCATCCAGAGCATCTTAGGAGTGAAGGTGACAGAGGGACTCCAGGGTTCAGCCTGCAGCAGGGCTCCACC  8,300
CAGCCTCACACGACCCCCTTCCCCTGCAACTTTCATTTTGTATTTTCCTTGAAGCCACAATATTCCCCCAGACAGAGCACTTATATTAAGACAGAAAATT  8,400
TCCCCTCCTATCCATCCTGTCTCCCCCGAGGATGAAATGCCTACACTGGTTTTCAAGAGAACAGGCCCAAATTGTCCATCCGAGAAAAGAGCAAATGGT  8,500
CTTATTCATTCCCTCTGTAATCCTCGCTGTCCGGCCCCACCTCCAACGCCATCCAAATAGGGCAAATTAGAAACGTAGGGAGCTTTGGCATGAGAGCCAT  8,600
TTGATTAGGGTAGAAGTATGATGAAATGTGCCCTGTTTTTCTATCAGAGCCACTGCAATTGGAATAAAGGTTTATGCTCAAGAAACACCAGCTAGAACA  8,700
TTCTGTGCTGAGATTCTAAGTGAGATAA-AGGCAT                                                                    8,735
```

Figure 5. (*Continued on facing page.*)

5b shows the resulting physical map of clone λhβN8. An 8700-bp nucleotide sequence derived from subcloned, overlapping *Eco*RI and *Hin*dIII fragments is shown in Figure 5a.

Regions within phage λhβN8 that are homologous to our mouse β-NGF cDNA sequence were identified by Southern hybridization experiments (Southern 1975), followed by nucleotide sequence analysis (Maxam and Gilbert 1977) and translation of the sequence into amino acid sequence information. Two regions of extensive homology were identified, 124 bp and 738 bp in length, respectively, which are separated by a 6.5-kb intervening sequence. The 738-bp region encodes a protein that is highly homologous (90%) to mature mouse β-NGF (Fig. 6) and apparently represents the thus far undetermined human β-NGF protein sequence.

Interestingly, we were unable to identify a region on phage λhβN8, or any of our other phage isolates, that is equally homologous to the 5′ end of the mouse β-NGF cDNA. Either the rest of the 5′ end of the human β-NGF gene including regulatory sequences is separated from the 124-bp homologous mRNA sequence by a >6-kb intervening sequence and is therefore not present on λhβN8, or human and mouse sequences have diverged more extensively in this region. A Goldberg-Hogness sequence (TATAAA) is present at nucleotide position 1362, with distantly homologous sequences downstream (1500–1560); thus, the 5′-end sequences may be more distantly related. However, use of the 5′ mouse cDNA probe on human genomic Southern blots at low hybridization stringency detected a homologous sequence elsewhere in the human genome, supporting the first interpretation (data not shown). Isolation of the human prepro-β-NGF cDNA will be necessary to clarify this point. In addition, Southern hybridization experiments using human and mouse chromosomal DNAs indicated the presence of a single β-NGF gene per haploid genome (data not shown).

DISCUSSION

Characteristics of Mouse β-NGF cDNA

The 1164-nucleotide cDNA sequence (Fig. 4) contains a long (327 amino acid) open reading frame that includes the known amino acid sequence of mouse β-NGF (amino acids 1–118; Fig. 4) (Angeletti et al. 1973a,b). The predicted β-NGF sequence differs at one position from that determined by amino acid sequencing procedures: We predict an aspartic acid residue at amino acid 30, rather than asparagine (Angeletti et al. 1973a,b). The β-NGF sequence is flanked at its amino and carboxyl termini by polypeptide extensions that presumably represent portions of a prepro-β-NGF precursor. The boundary sequences include basic amino acids (Lys-Arg, Arg-Arg) at both ends, which are found at precursor processing sites in other systems (Nakanishi et al. 1979; Kakidani et al. 1982; Amara et al. 1982; Noda et al. 1982). Processing at the aminoterminal boundary excludes both basic residues from the mature polypeptide, whereas cleavage at the carboxyterminal end appears to occur between the two arginine residues. The resulting mature β-NGF chain possesses a carboxyterminal arginine residue, thought to be essential for the stability of 7S NGF complex (Smith et al. 1968). The entire carboxyterminal propeptide is only 2 amino acids in length, which is unprecedented in other prohormone systems. One might consider the possibility that the two carboxyterminal amino acids are part of the natural β-NGF molecule and nonspecific cleavage occurred at this site during the extraction of the protein from an MSG homogenate, known to be rich in proteases.

Determination of the exact size of the aminoterminal region is difficult, since no aminoterminal polypeptide sequencing information for in vitro-translated prepro-β-NGF or pro-β-NGF is yet available. Evidence has been presented for the existence of a 22,000-dalton biosynthetic mouse pro-β-NGF precursor (Berger and Shooter

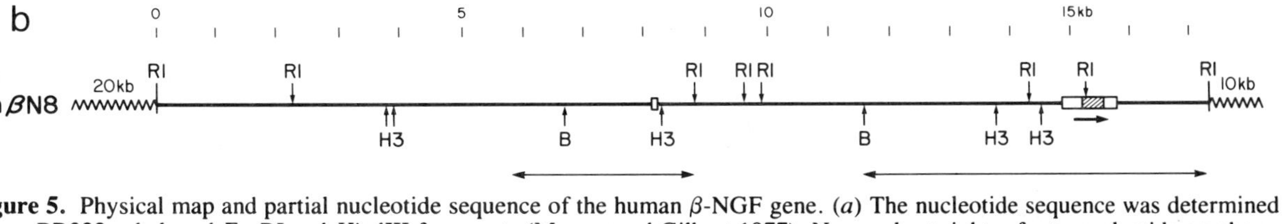

Figure 5. Physical map and partial nucleotide sequence of the human β-NGF gene. (*a*) The nucleotide sequence was determined from pBR322 subcloned *Eco*RI and *Hin*dIII fragments (Maxam and Gilbert 1977). Numerals at right refer to nucleotide numbers. The regions homologous to the mouse β-NGF cDNA clone are translated, and the region corresponding to mature human β-NGF is boxed in. Potential initiation methionines are underlined with dashed lines, at positions analogous to the mouse sequence (see *b*). The position of a transcription initiation sequence, TATAA (double underline), is also shown. Three possible AATAAA polyadenylation signals are underlined. The nucleotide sequence of 2.8 kb of the intervening sequence is not shown due to space considerations, but was completely determined. (*b*) A human gene library (Lawn et al. 1978) was screened (Grunstein and Hogness 1975) using a radiolabeled (Taylor et al. 1976) 470-bp *Rsa*I fragment of pmβN-9G1. λhβN8 was analyzed by restriction enzyme cleavage with *Eco*RI (RI), *Hin*dIII (H3), and *Bam*HI (B); cleavage sites are indicated by vertical arrows. Of the 17.3-kb human DNA insert, 15 kb was sequenced; horizontal arrows below the map represent regions for which the sequence is presented here. (∿∿∿) Charon-4A phage arms; (□) presumptive mature mRNA regions. Hatched area of the larger box codes for mature β-NGF, which would be transcribed from left to right (arrow). The size scale at top is shown in kilobase pairs, from the 5′ end of the phage insert.

1977); however, the nucleotide-sequence-predicted precursor is longer than that previously detected.

Localization of the Methionine Initiation Codon and Signal Sequence

Three methionine residues are candidates for designation as the protein synthesis initiation codon (amino acids −187, −121, and −119); however, several factors strongly implicate amino acid −121 of our sequence as the actual initiation codon employed. The methionine codon closest to the 5′ end of mammalian mRNAs is usually used to initiate protein synthesis; however, there are many exceptions to this general scheme (Kozak 1981). Since β-NGF is a secreted protein, the initiation codon is likely to be followed by a signal sequence for cotranslational transfer of this polypeptide into the lumen of the endoplasmic reticulum (Blobel et al. 1979). The stretch of amino acids following the methionine residue at position −187 (closest to

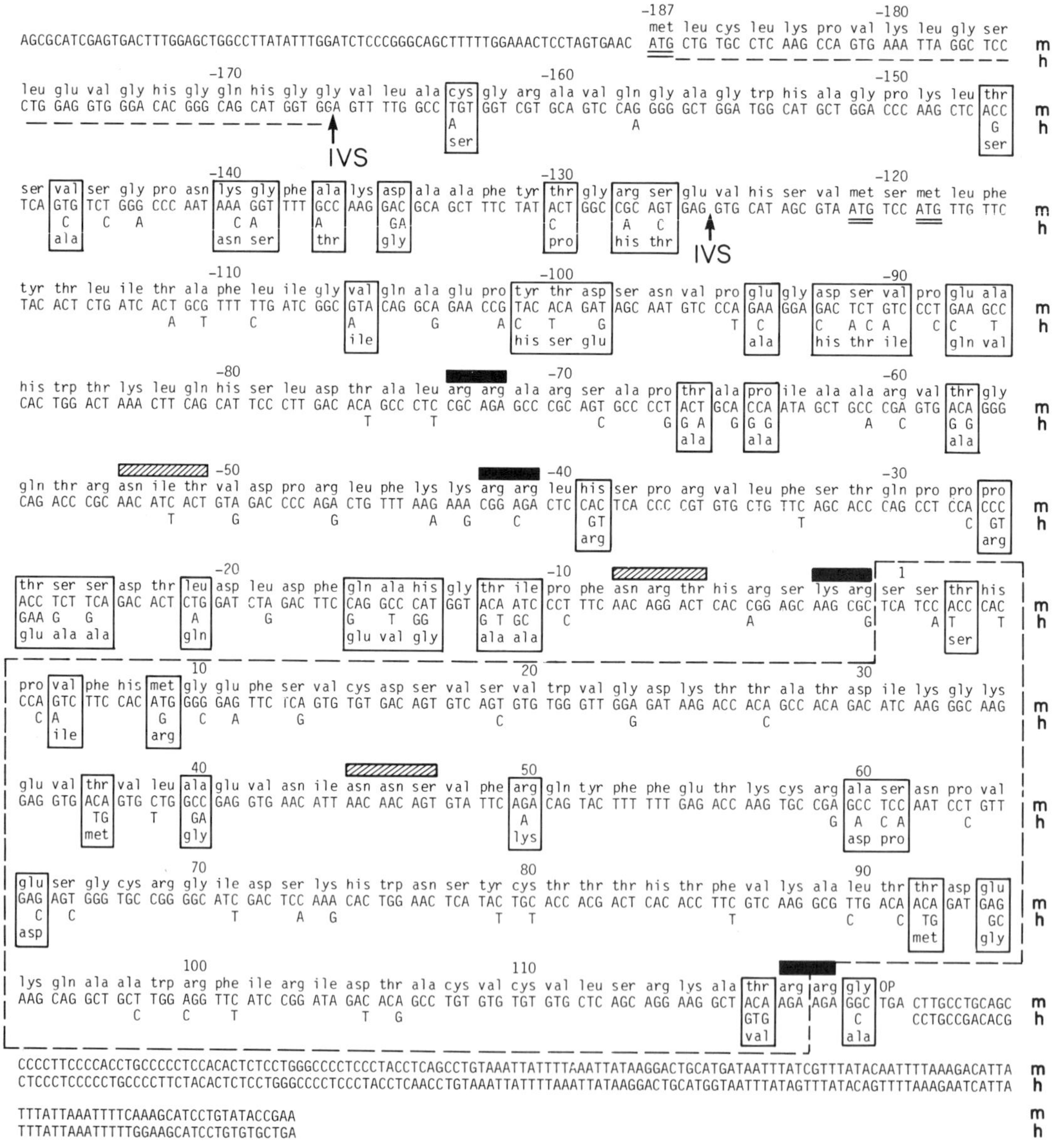

Figure 6. Comparison of cloned mouse pre-pro-β-NGF cDNA with human sequences. The mouse cDNA sequence was derived from overlapping clones (Fig. 3) according to the procedure of Maxam and Gilbert (1977). Only the coding strand is shown; the longest open reading frame starting with a methionine residue is also presented. A termination codon is present upstream at the twelfth nucleotide of the top line. Double underlines indicate the potential initiation codons. Amino acid positions are numbered, beginning with the first aminoterminal residue of the mature β-NGFs, the sequences of which are enclosed by a dashed line. The mouse coding-strand nucleotide and amino acid sequences (m) are shown in their entirety; only differences are shown below for the corresponding human (h) sequences. The positions at which a nucleotide difference causes a change in the coded amino acid are boxed. (▭) Potential N-linked glycosylation sites; (■) potential protease cleavage sites. IVS refers to the position of intervening sequences in the human gene, and OP indicates the opal termination codon, TGA. Sequences upstream from the IVS at position −166 are only known for the mouse cDNA; dashes under amino acids −167 to −186 denote missing human sequences.

the 5′ end) contains a high percentage of polar and charged amino acids and bears no resemblance to any previously described signal sequence (Fig. 7) (Blobel et al. 1979). On the other hand, amino acids between methionine residue −121 or −119 and residue −104 represent an excellent candidate signal sequence. The 16–18 amino acids following residue −121 or −119 include a completely hydrophobic stretch of six amino acids (Ala-Phe-Leu-Ile-Gly-Val). Cleavage by signal peptidase could occur between the small amino acid alanine −104 and the glutamic acid residue at position −103. Signal peptidase cleaves after an identical glutamine-alanine sequence to leave an analogous aminoterminal glutamic acid residue in the case of sheep pre-α lactalbumin (Lingappa et al. 1978). Because methionine −121 follows Kozak's rule in that it possesses a purine nucleotide three residues upstream (Kozak 1981), we favor amino acid −121 over methionine −119 as the actual initiator methionine. The use of methionine −121 for translation initiation of mouse prepro-β-NGF would result in a 27,000-dalton preprohormone and a 25,000-dalton pro-β-NGF if signal peptide processing occurs at residue −104. The possibility, however, that methionine −187 is the actual initiation methionine cannot be entirely excluded. In that case the molecular weight of the β-NGF biosynthetic precursor would be 33,800.

The absence of a classic signal peptide and presence of an internal-signal-type sequence would pose interesting functional questions for such a precursor protein. The availability of the sequence reported here and the corresponding cDNA and chromosomal gene clones will permit further elucidation of the actual size as well as of the specificity and order of intracellular processing events for the β-NGF precursor protein. The particularly interesting question as to whether or not other putative processing products of the β-NGF precursor possess novel biological activities can now be addressed.

Comparison of Human and Mouse Prepro-β-NGF Sequences

Comparison of mouse β-NGF nucleotide and amino acid sequences with corresponding human sequences reveals high levels of homology. Figure 6 shows a comparison of 861 bp composed of the small and large exons of the human gene with the mouse cDNA sequence. The nucleotide sequences are highly homologous overall (86%). Although we have tentatively assigned the amino terminus for mouse prepro-β-NGF to methionine −121, we included the amino acid sequences of the open reading frame that precede amino acid −121 in both mouse cDNA and human gene for comparison. Most of the sequence upstream of amino acid −121 (40 amino acids) corresponds to the small exon in the human gene, and may actually represent 5′-untranslated sequences. The level of homology in this region (positions −126 to −166) is significantly lower than that overall (70%). Highly conserved regions within the putative prepro-β-NGF amino acid sequence are the prepeptide (−121 to −104; 94.5%) and the sequences of mature β-NGF (90%). Within the aminoterminal and carboxyterminal pro-sequences, regions of more extensive divergence separate several areas of high homology between mouse and human. Interestingly, regions that surround two Arg-Arg dipeptides (positions −41/42 and −72/73) and two potential N-linked glycosylation sites at amino acids −8 and −53 are absolutely conserved, which may reflect the functional importance of these sequences. A third Arg-Arg dipeptide is present at the carboxyterminal end of the pro-β-NGF sequence and appears to be a protease cleavage site in both human and mouse sequences for the removal of a carboxyterminal dipeptide. Within the mature β-NGF sequence, the cysteine residues, important for the formation of proper tertiary structure (Angeletti et al. 1973a), are conserved, as are 90% of the other residues; amino acid differences are mostly conservative with respect to the chemical nature of the residues. Further evidence for the structural homology between mouse and human β-NGFs has been obtained by expression of the human β-NGF sequence in *Escherichia coli* and immunological detection of a protein of the correct size using an anti-mouse β-NGF antiserum (A. Ullrich et al., in prep.).

Is β-NGF a Member of the Insulin Gene Family?

Because of low levels of homology in the amino acid sequences of the B chains of NGF and the insulin gene family members insulin, relaxin, and insulinlike growth factors, it has been postulated that β-NGF is a member of the insulin gene family (Bradshaw 1978). We therefore compared the nucleotide sequences of the β-NGF and insulin genes (Lomedico et al. 1979; Ullrich et al. 1980); the level of homology is greater than random in the B-chain region but does not further strengthen the hypothesis that these genes are related. Both genes contain an intervening sequence just upstream of the coding sequences in the 5′-untranslated region; however, this is not a particularly unusual gene feature (Breathnach et al. 1978). The absence of an intervening sequence within the β-NGF-coding region is unlike most of the insulin genes thus far isolated; however, the rat-I insulin gene also lacks an intervening sequence in the coding region (Lomedico et al. 1979), so the significance of this β-NGF gene feature remains uncertain. Isolation and characterization of other genes in the insulin gene family will be necessary to shed additional light on this interesting hypothesis (Bradshaw 1978).

ACKNOWLEDGMENTS

We are grateful to Mark Vasser for carrying out oligonucleotide synthesis and to Dr. Peter Seeburg, Joel Hayflick, and Dr. Ellson Chen for their sequencing assistance. We acknowledge the contribution of Dr. Stacey Sias in the early phases of this work. We also extend thanks to Drs. Xandra O. Breakefield and

Suzanne Pfeffer for their continued interest and useful discussions, to Jeanne Arch for her expert preparation of the manuscript, and to Robert Swanson, President of Genentech, Inc., for his interest and support.

REFERENCES

Amara, S.G., V. Jonmas, M.G. Rosenfeld, E.S. Ong, and R.M. Evans. 1982. Alternative RNA processing in calcitonin gene expression generates mRNAs encoding different polypeptide products. *Nature* **298:** 240.

Angeletti, R.H., M.A. Hermodson, and R.A. Bradshaw. 1973a. Amino acid sequences of mouse 2.5S nerve growth factor. II. Isolation and characterization of the thermolytic and peptic peptides and the complete covalent structure. *Biochemistry* **12:** 100.

Angeletti, R.H., D. Mercanti, and R.A. Bradshaw. 1973b. Amino acid sequences of mouse 2.5S nerve growth factor. I. Isolation and characterization of the soluble tryptic and chymotryptic peptides. *Biochemistry* **12:** 90.

Berger, E.A. and E.M. Shooter. 1977. Evidence for pro-β-nerve growth factor, a biosynthetic precursor to β-nerve growth factor. *Proc. Natl. Acad. Sci.* **74:** 3647.

Blobel, G., P. Walter, C.N. Chang, B.M. Goldman, A.H. Erickson, and V.R. Lingappa. 1979. Translocation of proteins across membranes: The signal hypothesis and beyond. *Soc. Exp. Biol. Semin. Ser.* **33:** 9.

Bradshaw, R.A. 1978. Nerve growth factor. *Annu. Rev. Biochem.* **47:** 191.

Breathnach, R., C. Benoist, K. O'Hare, F. Gannon, and P. Chambon. 1978. Ovalbumin gene: Evidence for a leader sequence in mRNA and DNA sequences at the exon-intron boundaries. *Proc. Natl. Acad. Sci.* **75:** 4853.

Carpenter, G. and S. Cohen. 1979. Epidermal growth factor. *Ann. Rev. Biochem.* **48:** 193.

Crea, R. and T. Horn. 1980. Synthesis of oligonucleotides on cellulose by a phosphotriester method. *Nucleic Acids Res.* **8:** 2331.

Grunstein, M. and D.S. Hogness. 1975. Colony hybridization: A method for the isolation of cloned DNAs that contain a specific gene. *Proc. Natl. Acad. Sci.* **72:** 3961.

Harper, G.P. and H. Thoenen. 1980. The distribution of nerve growth factor in the male sex organs of mammals. *J. Neurochem.* **34:** 893.

Ishii, D.N. and E.M. Shooter. 1975. Regulation of nerve growth factor synthesis in mouse submaxillary glands by testosterone. *J. Neurochem.* **25:** 843.

Kakidani, H., Y. Furutani, H. Takahashi, M. Noda, Y. Morimoto, T. Hirose, M. Asai, S. Inayama, S. Nakanishi, and S. Numa. 1982. Cloning and sequence analysis of cDNA for porcine β-neo-endorphin/dynorphin precursor. *Nature* **298:** 245.

Kozak, M. 1981. Possible role of flanking nucleotides in recognition of the AUG initiator codon by eukaryotic ribosomes. *Nucleic Acids Res.* **9:** 5233.

Lawn, R.M., E.F. Fritsch, R.C. Parker, G.B. Lake, and T. Maniatis. 1978. The isolation and characterization of linked δ- and β globin genes from a cloned library of human DNA. *Cell* **15:** 1157.

Lehrach, H., D. Diamond, J.M. Wozney, and H. Boedtker. 1977. RNA molecular weight determinations by gel electrophoresis under denaturing conditions, a critical reexamination. *Biochemistry* **16:** 4743.

Levi-Montalcini, R. 1982. Developmental neurobiology and the natural history of nerve growth factor. *Annu. Rev. Neurosci.* **5:** 341.

Lingappa, V.R., J.R. Lingappa, R. Prasad, K.B. Ebner, and G. Blobel. 1978. Coupled cell-free synthesis, segregation, and core glycosylation of a secretory protein. *Proc. Natl. Acad. Sci.* **75:** 2338.

Lomedico, P., N. Rosenthal, A. Efstratiadis, W. Gilbert, R. Kolodner, and R. Tizard. 1979. The structure and evolution of the two nonallelic rat preproinsulin genes. *Cell* **18:** 545.

Maniatis, T., E.F. Fritsch, and J. Sambrook, eds. 1982. *Molecular cloning.* Cold Spring Harbor Laboratory, Cold Spring Harbor, New York.

Maxam, A.M. and W. Gilbert. 1977. A new method for sequencing DNA. *Proc. Natl. Acad. Sci.* **74:** 560.

Nakanishi, S., A. Inoue, T. Kita, M. Nakamura, A.C.Y. Chang, S.N. Cohen, and S. Numa. 1979. Nucleotide sequence of cloned cDNA for bovine corticotropin-β-lipoprotein precursor. *Nature* **278:** 423.

Noda, M., Y. Furutani, H. Takahashi, M. Toyosato, T. Hirose, S. Inayama, S. Nakanishi, and S. Numa. 1982. Cloning and sequence analysis of cDNA for bovine adrenal preproenkephalin. *Nature* **295:** 202.

Smith, A.P., S. Varon, and E.M. Shooter. 1968. Multiple forms of the nerve growth factor protein and its subunits. *Biochemistry* **7:** 3259.

Southern, E. 1975. Detection of specific sequences among DNA fragments separated by gel electrophoresis. *J. Mol. Biol.* **98:** 503.

Taylor, J.M., R. Illmensee, and J. Summers. 1976. Efficient transcription of RNA into DNA by avian sarcoma virus polymerase. *Biochim. Biophys. Acta* **4:** 324.

Ullrich, A., T.J. Dull, A. Gray, J. Brosius, and I. Sures. 1980. Genetic variation in the human insulin gene. *Science* **209:** 612.

Ullrich, A., J. Shine, J. Chirgwin, R. Pictet, E. Tischer, W.J. Rutter, and H.M. Goodman. 1977. Rat insulin genes: Construction of plasmids containing the coding sequences. *Science* **196:** 1313.

Wallace, B., M.J. Johnson, T. Hirose, T. Miyake, E.H. Kawashima, and K. Itakura. 1981. The use of synthetic oligonucleotides as hybridization probes. II. Hybridization of oligonucleotides of mixed sequence to rabbit β-globin DNA. *Nucleic Acids Res.* **9:** 879.

Yankner, B.A. and E.M. Shooter. 1982. The biology and mechanism of action of nerve growth factor. *Annu. Rev. Biochem.* **51:** 845.